FULLERENES

FULLERENES
Chemistry, Physics, and Technology

Edited By

KARL M. KADISH
Department of Chemistry
University of Houston

RODNEY S. RUOFF
Department of Mechanical Engineering
Northwestern University

WILEY-INTERSCIENCE

A JOHN WILEY & SONS, INC., PUBLICATION

New York · Chichester · Weinheim · Brisbane · Singapore · Toronto

For ordering and customer service please call 1-800-Call-Wiley.

Library of Congress Cataloging-in-Publication Data:

Fullerenes : chemistry, physics and technology / edited by Karl M. Kadish and Rodney S. Ruoff.
    p. cm.
   Includes index.
   ISBN 0-471-29089-0 (cloth : alk. paper)
   1. Fullerenes. I. Kadish, Karl M., 1945– II. Ruoff, Rodney S.
QD181.C1 F843 2000
546′.681—dc21   00-033402

Printed in the United States of America.

10 9 8 7 6 5 4 3 2 1

CONTENTS

CONTRIBUTORS

TAKESHI AKASAKA, Graduate School of Science and Technology, Niigata University, Niigata, Japan

ALAN L. BALCH, Department of Chemistry, University of California, Davis, California

DIETHARD K. BOHME, Department of Chemistry and Centre for Research in Earth and Space Science, York University, Toronto, Ontario, Canada

OLGA V. BOLTALINA, Department of Chemistry, Moscow State University, Moscow, Russia

VICTOR BUNTAR, Atomic Institute of the Austrian Universities, Vienna, Austria

LONG Y. CHIANG, Center for Condensed Matter Sciences, National Taiwan University, Taipei, Taiwan

DENNIS W. CHOI, Departments of Neurology and Chemistry, Washington University, St. Louis, Missouri

NASREEN G. CHOPRA, Department of Physics, University of California at Berkeley and Materials Sciences Division, Lawrence Berkeley National Laboratory, Berkeley, California

MARVIN L. COHEN, Department of Physics, University of California at Berkeley and Materials Sciences Division, Lawrence Berkeley National Laboratory, Berkeley, California

SARAH CUDDIHY, Departments of Neurology and Chemistry, Washington University, St. Louis, Missouri

PRADEEP DEOTA, The Maharaja Sayajirao University of Baroda, Baroda, India

FRANÇOIS DIEDERICH, Laboratory of Organic Chemistry, ETH-Zentrum, Zürich, Switzerland

LAURA L. DUGAN, Departments of Neurology and Medicine, Washington University School of Medicine, St. Louis, Missouri

LOURDES E. ECHEGOYEN, Department of Chemistry, University of Miami, Coral Gables, Florida

LUIS ECHEGOYEN, Department of Chemistry, University of Miami, Coral Gables, Florida

MALCOLM L. H. GREEN, Wolfson Catalysis Centre, Carbon Nanotechnology Group, Inorganic Chemistry Laboratory, University of Oxford, Oxford, United Kingdom

JEFFREY C. GROSSMAN, Department of Physics, University of California at Berkeley and Materials Sciences Division, Lawrence Berkeley National Laboratory, Berkeley, California

DIRK M. GULDI, Radiation Laboratory, University of Notre Dame, Notre Dame, Indiana

ALEX V. HAMZA, Lawrence Livermore National Laboratory, University of California, Livermore, California

PREBEN HVELPLUND, Institute of Physics and Astronomy, University of Aarhus, Aarhus C, Denmark

PRASHANT V. KAMAT, Radiation Laboratory, University of Notre Dame, Notre Dame, Indiana

KAORU KOBAYASHI, Department of Chemistry, Graduate School of Science, Tokyo Metropolitan University, Hachioji, Tokyo, Japan

MIKHAIL V. KOROBOV, Department of Chemistry, Moscow State University, Moscow, Russia

TIEN-SUNG LIN, Departments of Neurology and Chemistry, Washington University, St. Louis, Missouri

STEVEN G. LOUIE, Department of Physics, University of California at Berkeley and Materials Sciences Division, Lawrence Berkeley National Laboratory, Berkeley, California

EVA LOVETT, Departments of Neurology and Chemistry, Washington University, St. Louis, Missouri

BEI-WEN MA, Departments of Neurology and Chemistry, Washington University, St. Louis, Missouri

MICHELE MAGGINI, Department of Organic Chemistry, University of Padova, Padova, Italy

SERENA MARGADONNA, Fullerene Science Centre, School of Chemistry, Physics and Environmental Science, University of Sussex, Brighton, United Kingdom

MICHAEL E. MCHENRY, Department of Materials Science and Engineering, Carnegie Mellon University, Pittsburgh, Pennsylvania

MARK S. MEIER, Department of Chemistry, University of Kentucky, Lexington, Kentucky

SHIGERU NAGASE, Department of Chemistry, Graduate School of Science, Tokyo Metropolitan University, Hachioji, Tokyo, Japan

BERTHOLD NUBER, Sony Corporation Frontier Science Laboratories, Yokohama, Japan

EIJI ŌSAWA, Laboratories of Computational Chemistry and Fullerene Science, Department of Knowledge-Based Information Engineering, Toyohashi University of Technology, Toyohashi, Aichi, Japan

CHARLES PISKOTI, Department of Physics, University of California at Berkeley and Materials Sciences Division, Lawrence Berkeley National Laboratory, Berkeley, California

KOSMAS PRASSIDES, Fullerene Science Centre, School of Chemistry, Physics and Environmental Science, University of Sussex, Brighton, United Kingdom

MAURIZIO PRATO, Department of Pharmaceutical Science, University of Trieste, Trieste, Italy

DAVID I. SCHUSTER, Department of Chemistry, New York University, New York, New York

HISANORI SHINOHARA, Department of Chemistry, Nagoya University, Nagoya, Japan

ZDENĚK SLANINA, Laboratories of Computational Chemistry and Fullerene Science, Department of Knowledge-Based Information Engineering, Toyohashi University of Technology, Toyohashi, Aichi, Japan

JEREMY SLOAN, Wolfson Catalysis Centre, Carbon Nanotechnology Group, Inorganic Chemistry Laboratory, University of Oxford, Oxford, United Kingdom

ALLAN L. SMITH, Department of Chemistry, Drexel University, Philadelphia, Pennsylvania

SHEKHAR SUBRAMONEY, DuPont Company, Experimental Station, Wilmington, Delaware

BERTIL SUNDQVIST, Department of Experimental Physics, Umea University, Umea, Sweden

ROGER TAYLOR, Fullerene Science Centre, School of Chemistry, Physics and Environmental Science, University of Sussex, Brighton, United Kingdom

TAKATSUGU WAKAHARA, Graduate School of Science and Technology, Niigata University, Niigata, Japan

LEE Y. WANG, National Taiwan University, Center for Condensed Matter Sciences, 1 Roosevelt Rd. Section 4, Taipei, Taiwan

STEPHEN R. WILSON, Department of Chemistry, New York University, New York, New York

ALEX ZETTL, Department of Physics, University of California at Berkeley and Materials Sciences Division, Lawrence Berkeley National Laboratory, Berkeley, California

XIANG ZHAO, Laboratories of Computational Chemistry and Fullerene Science, Department of Knowledge-Based Information Engineering, Toyohashi University of Technology, Toyohashi, Aichi, Japan

CHAPTER 1

ELECTROCHEMISTRY OF FULLERENES

LUIS ECHEGOYEN, FRANÇOIS DIEDERICH, and LOURDES E. ECHEGOYEN

1.1 INTRODUCTION

The high electron affinity of fullerenes was theoretically predicted before their solution electrochemistry was actually recorded. Buckminsterfullerene (C_{60}), for example, was described as having a closed-shell configuration consisting of 30 bonding molecular orbitals (MO) with 60 π-electrons [1]. The completely occupied HOMO (highest occupied molecular orbital) was calculated to be 1.5–2.0 eV lower than the antibonding LUMO (lowest unoccupied molecular orbital) [1,2]. The LUMO was predicted to be triply degenerate and energetically low-lying, making C_{60} a good electron acceptor capable of having six easily accessible negative oxidation states in solution. Similar predictions were made for C_{70} [3,4] and for the most stable isomers of the higher fullerenes, C_{76}, C_{78}, C_{82}, and C_{84} [5]. Electron affinities (EAs) and ionization potentials (IPs) were also calculated for C_{60} through C_{84}, and these are shown in Table 1.1. A quick glance at these values clearly shows that fullerenes should easily accept and donate electrons, and thus they were expected to display very rich electrochemistry. In addition, the combination of electronic properties with the high double-bond character at the junction of two six-membered rings (see Chapter 3 in this volume) provided fullerenes with the extraordinary ability to easily undergo addition reactions. As a consequence, an almost infinite array of mono to multiple adducts could in principle be prepared with controllable three-dimensional properties. For electrochemists, the field was young, open, and very promising.

From the time these calculations were made to the present, over 2000 publications have appeared in the literature on fullerene electrochemistry alone, including several reviews [6–10]. This chapter presents an up-to-date account of

Fullerenes: Chemistry, Physics, and Technology, Edited by Karl M. Kadish and Rodney S. Ruoff. 0-471-29089-0   Copyright © 2000 John Wiley & Sons, Inc.

TABLE 1.1 Estimated Electron Affinities and Ionization Potentials of Selected Fullerenes

	EA (eV)[a]	IP (eV)[a]
C_{60}	2.7	7.8
C_{70}	2.8	7.3
C_{76} (D_2 isomer)	3.2	6.7
C_{78} (C_{2v} isomer)	3.4	6.8
C_{82} (C_2 isomer)	3.5	6.6
C_{84} (D_2 isomer)	3.5	7.0
C_{84} (D_{2d} isomer)	3.3	7.0

[a] Values reported are averages from two levels of HF (MNDO) theory in Ref. 5.

the redox properties of the pristine fullerenes and their large number of derivatives, as revealed by electrochemical studies in solution. A section on electrosynthetic methods is included, with special emphasis on the newly discovered "walk-on-the-sphere" migration of bis adducts of C_{60} [11] or "the shuffle" and, most importantly, the *Retro-Bingel* reaction [12], a reaction with serious potential applications in fullerene derivative synthesis. Other areas such as supramolecular electrochemistry of C_{60} [6] and the electrochemical behavior of fullerene thin films [7–10] have recently been reviewed and, in the interest of brevity, will not be covered here.

1.2 ELECTROCHEMISTRY OF C_{60} AND C_{70}

1.2.1 Reduction of C_{60}

Once the large-scale method for preparation and purification of C_{60} became widely available [13–16], several groups began to explore its electrochemistry in an attempt to confirm the theoretical predictions. However, due to limitations in the solvent potential window, only the first five reduction processes for C_{60} were reported between 1990 and 1991 [17–21]. It was not until 1992 that all six one-electron reductions were electrochemically detected by three independent groups in (chronological order) Miami (Florida) [22], Tokyo (Japan) [23], and Austin (Texas) [24].

The Miami group reported the observation, using cyclic voltammetry (CV) and differential pulse voltammetry (DPV), of six successive, fully reversible, one-electron reductions using the following conditions: 0.1 M electrolyte solution in a mixed solvent system consisting of toluene/acetonitrile in a 5:1 volume ratio, under vacuum, at $-10\,°C$ (see Fig. 1.1) [22]. The potentials measured under these conditions are shown on Table 1.2. As expected on the basis of the triply degenerate LUMO, the potential separation between any two successive reductions is almost constant. This separation is 450 ± 50 mV. On the volta-

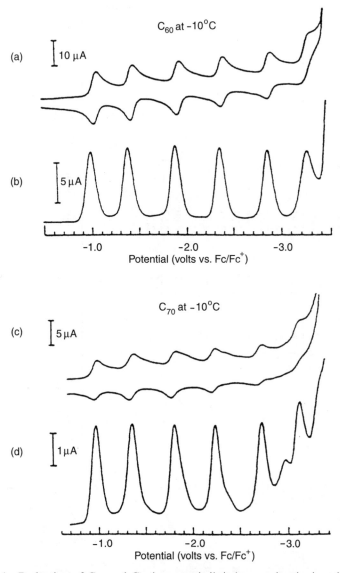

Figure 1.1   Reduction of C_{60} and C_{70} in acetonitrile/toluene using (a,c) cyclic voltammetry at a 100 mV/s scan rate and (b,d) differential pulse voltammetry (50 mV pulse, 50 ms pulse width, 300 ms period, 25 mV/s scan rate). Reprinted with permission from Ref. 22. Copyright © 1992 American Chemical Society.

metric time scale, C_{60}^- through C_{60}^{6-} appear to be chemically stable. However, only C_{60}^- through C_{60}^{4-} are stable when generated by controlled potential coulometry under vacuum, using toluene acetonitrile as solvent.

An exhaustive investigation by Kadish and co-workers [25] of the cathodic

TABLE 1.2 Half-wave Reduction Potentials (in V versus Fc/Fc$^+$) of C$_{60}$ Using Various Solvents, Supporting Electrolytes, and Temperatures

Solvent	Supporting Electrolyte	Temperature (°C)	E_1	E_2	E_3	E_4	E_5	E_6	Reference
PhMe/MeCN[a]	TBAPF$_6$	-10	-0.98	-1.37	-1.87	-2.35	-2.85	-3.26	22
PhMe/DMF	TBAPF$_6$	-60	-0.82	-1.26	-1.82	-2.33	-2.89	-3.34[b]	23
Liquid NH$_3$[c]	KI	-70	-1.04	-1.56	-2.00	-2.37	-2.43	-3.03	24
Benezene	THAClO$_4$	+45	-0.83	-1.29	-1.89	-2.46	-3.07		21
MeCN[d]	TBAClO$_4$	+22		-1.24	-1.73	-2.19	-2.78		25
Pyridine	TBAClO$_4$	+22	-0.86	-1.38	-1.80	-2.33			25
DMF	TBAClO$_4$	+22	-0.77	-1.23	-1.82	-2.36			25
PhCN	TBAClO$_4$	+22	-0.92	-1.34	-1.82				21
THF	TBABF$_4$	-60	-0.96	-1.53	-2.11	-2.60	-3.13	-3.59	26
	TBABF$_4$	+25	-0.93	-1.52	-2.08	-2.57	-3.08		26
	TBAClO$_4$	+22	-0.90	-1.49	-2.06	-2.56			25
DCM	TBAPF$_6$	+22	-1.02	-1.41	-1.87				25
	TBAClO$_4$	+22	-1.00	-1.39	-1.84[e]				25
	TBABF$_4$	+25	-1.01	-1.40					17
TCE	TBAPF$_6$	+25	-1.06						32

PhMe = toluene; MeCN = acetonitrile; DMF = dimethylformamide; PhCN = benzonitrile; THF = tetrahydrofuran; DCM = dichloromethane; TCE = 1,1,2,2-tetrachloroethane; DMA = dimethylamine; ODCB = o-dichlorobenzene; TBA = tetra-n-butyl ammonium; THA = tetra-n-hexyl ammonium.

[a] Glassy carbon electrode. All others use Pt.

[b] Anodic peak potential.

[c] Due to the presence of surface waves upon reoxidation, anodic peak potentials are reported.

[d] Controlled potential electrolysis used to generate C$_{60}$$^-$ due to insolubility of C$_{60}$ in MeCN.

[e] Scan rate > 5 V/s.

4

electrochemistry of C_{60} has demonstrated that the choices of solvent, supporting electrolyte, and temperature have a profound effect on the reduction potentials. The most relevant of these results along with those of other studies are summarized in Table 1.2.

Furthermore, the solubility of the electrogenerated species is also highly dependent on the choice of solvent and supporting electrolyte. For example, $C_{60}{}^0$ is insoluble in DMF and THF, but all of its anions readily dissolve in these solvents. In acetonitrile, $C_{60}{}^0$ is also insoluble, but a soluble dianion can be generated from a suspension in the presence of K^+, Cs^+, or Ca^{2+} cations. A soluble trianion can also be prepared in acetonitrile if TBA^+, K^+, or Ca^+ cations are present. These acetonitrile solutions have been used to study the conductivities of electrodeposited films of C_{60} [27].

1.2.2   Reduction of C_{70}

Like C_{60}, C_{70} was also expected to display six easily accessible one-electron reductions, based on the MO calculations. Although its LUMO was only doubly degenerate, the energy separation between the LUMO and the LUMO + 1 was small enough. In the first report on the electroreduction of C_{70}, only the first and second reduction potentials were observed. That report indicated that these reduction potentials were nearly identical to the corresponding ones of C_{60} in o-dichlorobenzene (ODCB) and in benzonitrile (PhCN) [18]. Later investigations confirmed this observation in acetonitrile/toluene and detected all six redox waves [22]. Figure 1.1c,d shows the CV and DPV of C_{70} where the six reversible one-electron reductions are observed in acetonitrile/toluene. The corresponding potentials are listed in Table 1.3. From the trianion up to and including the hexa-anion, C_{70} becomes increasingly easier to reduce than C_{60}, and unlike C_{60}, all six waves are easily observed even at room temperature [22]. A simple explanation for this phenomenon has been presented by Cox et al. [19], who attribute the difference in reduction potentials after the third electron to a charge separation delocalization model. Since C_{70} is larger than C_{60}, adding the third electron would be easier if the charges are separated by larger distances or if they can be delocalized over a larger number of carbon atoms.

Although no systematic study on the solvent, electrolyte, or temperature dependence of the reduction potentials of C_{70} has been published, a quick glance at Table 1.3 indicates that, like C_{60}, C_{70} displays such dependency.

1.2.3   Oxidation of C_{60} and C_{70}

Calculations by Haddon et al. [1] had indicated that the HOMO of C_{60} would be energetically low-lying and fivefold degenerate. In addition, the IP was estimated to have a high value, 7.8 eV (see Table 1.1). These results suggested that C_{60} should not be readily oxidizable. In fact, initial studies on the anodic electrochemistry of C_{60} indicated irreversible behavior (see Table 1.4) [21,25,28,31]. The first study, performed on a film of C_{60} in acetonitrile, showed a chemically

TABLE 1.3 Half-wave Reduction Potentials (in V versus Fc/Fc$^+$) of C$_{70}$ Using Various Solvents, Supporting Electrolytes, and Temperatures

Solvent[a]	Supporting Electrolyte	Temperature (°C)	E_1	E_2	E_3	E_4	E_5	E_6	Reference
PhMe/MeCN[b]	TBAPF$_6$	−10	−0.97	−1.34	−1.78	−2.21	−2.70	−3.70	22
DMA[c,d]	TBABr	−65	−0.45	−0.90	−1.34	−1.77	−2.33	−2.80	28
ODCB[e]	?	+25	−0.99	−1.44	−1.90				18
THF	?	+25	−0.85	−1.41	−1.95				18
PhCN	?	+25	−0.91	−1.32					18
PhMe/DCM	TBAPF$_6$	RT?	−0.91	−1.28	−1.70				19, 29
DCM	?	+25	−0.97	−1.35	−1.76				18
	TBAPF$_6$	+25	−0.93	−1.31	−1.73	−2.09			30
TCE	TBAPF$_6$	+25	1.02						32

[a] See Table 1.2 for solvent abbreviations.
[b] Glassy carbon electrode. All others use Pt.
[c] Cathodic peak potential.
[d] Versus Ag/AgCl.
[e] Scan rate = 25 mV/s.

6

TABLE 1.4 Half-wave Oxidation Potentials (in V versus Fc/Fc$^+$) of C$_{60}$ and C$_{70}$ Using Various Solvents, Supporting Electrolytes, and Temperatures

	Solvent[a]	Supporting Electrolyte	Temperature (°C)	E_1^{ox}	E_2^{ox}	$\langle E_{ox}^1 - E_{red}^1 \rangle$	Reference
C$_{60}$	MeCN[b]	TBABF$_4$	RT?	+1.60[c]			31
	PhCN	TBAPF$_6$	RT	+1.30[c]			21
	DCE	TBAClO$_4$	22	+1.28[c]			25
	DCM	TBAClO$_4$	22	+1.32[c]			25
	TCE	TBAPF$_6$	RT	+1.26		2.32	32
C$_{60}$ + C$_{70}$	DCM	TBAPF$_6$	RT?	+1.65[d,e]	+1.75[d,e]		28
C$_{70}$	PhCN	TBAPF$_6$	RT	+1.30[c]			21
	TCE	TBAPF$_6$	RT	+1.20	+1.75	2.22	32

[a] See Table 1.2 for solvent abbreviations.
[b] Film of C$_{60}$ in MeCN.
[c] Chemically irreversible.
[d] It was not reported which of these potentials corresponds to C$_{60}$ and which one to C$_{70}$.
[e] Adsorption observed.

irreversible wave at $+1.6$ V versus Fc/Fc^+ [31]. Another report studied the anodic electrochemistry of both C_{60} and C_{70} in benzonitrile/TBAPF$_6$ [21]. Chemically irreversible waves were observed at $+1.30$ V versus Fc/Fc^+ for both fullerenes. Irreversibility remained even after lowering the temperature to $-15\,^{\circ}C$ or increasing the scan rate up to 50 V/s. Peak current intensities indicated that four electrons were transferred during oxidation, a fact that was confirmed by controlled potential coulometry (CPC) at 1.44 V versus Fc/Fc^+. Other studies done with C_{60} in DCM and DCE also showed an irreversible wave at $+1.32$ and 1.28 V, respectively, versus Fc/Fc^+ [25]. Meerholz et al. [28] reported the observation of two oxidation peaks in close proximity at 1.65 and 1.75 V versus Ag/AgCl for a mixture of C_{60} and C_{70}. Although no quantification of anodic versus cathodic peak currents was performed, the authors observed that oxidation currents were approximately double those of the single-electron reduction waves. This fact led them to suggest the existence of two two-electron transfers. However, they also observed adsorption effects upon scan reversal after oxidation.

The first truly reversible oxidation of C_{60} and C_{70} was reported in 1993 by Xie et al. [32], who observed the process by CV and Osteryoung square wave voltammetry (OSWV) (see Fig. 1.2 and Table 1.4). For C_{60}, a one-electron, chemically reversible oxidation wave appeared at $+1.26$ V versus Fc/Fc^+ at room temperature in 1,1,2,2-tetrachloroethane (TCE). Under the same conditions, the first one-electron oxidation of C_{70} occurred at $+1.20$ V, a potential 60 mV more negative (easier to oxidize) than that of C_{60}. Both oxidations were electrochemically quasireversible with $\Delta E_{pp} \cong 80$ mV. In addition, a second oxidation wave was observed for C_{70} close to the limit of the solvent potential window at $+1.75$ mV. However, this wave appeared to be chemically irreversible. Table 1.4 summarizes the oxidation potential results.

The difference between the experimental first reduction and oxidation potentials is a good measure of the HOMO–LUMO energy difference (the HOMO–LUMO gap) in solution. For C_{60} and C_{70}, these gaps are, respectively, 2.32 and 2.22 V at room temperature.

1.3 ELECTROCHEMISTRY OF THE HIGHER FULLERENES, C$_{76}$, C$_{78}$, C$_{82}$, AND C$_{84}$

The values presented in Table 1.1 for the estimated EAs and IPs of fullerenes clearly establish that the higher fullerenes should be reduced and oxidized more readily than C_{60} and C_{70}. Indeed, this has been the case. However, due to the small quantities of these compounds present in fullerite soot and the difficulties with their isolation and purification, reports on their redox properties did not appear until 1992, soon after pure samples in milligram quantities became available [33–35]. Because higher fullerenes may exist as two or more structural isomers, the interpretation of electrochemical studies of these larger cages has been complicated. The presence of two or more isomers with different symme-

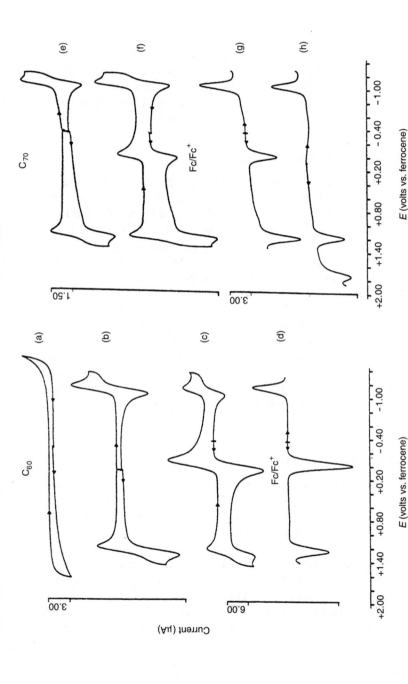

Figure 1.2 CV and OSWV scans showing oxidation and first reduction waves for (a–d) C_{60} and (e–h) C_{70} in TCE at room temperature. CVs were run at a scan rate of 100 mV/s. OSWVs were run at 60 mV/s. (a) CV of background solvent–electrolyte system. (b) CV of C_{60}, no ferrocene added. (c) Same as (b) after addition of ferrocene. (d) OSWV of the solution in (c). (e) CV of C_{70}, no ferrocene added. (f) Same as (e) after addition of ferrocene. (g) OSWV of the solution in (f). (h) OSWV of the solution in (e), without ferrocene added, scanned to more negative potential to show the second oxidation wave for C_{70}. Reprinted with permission from Ref. 32. Copyright © 1993 American Chemical Society.

tries may result in the observation of multiple redox waves, some of which have been difficult or impossible to assign with certainty [36]. Reduction potentials for the higher fullerenes are presented in Table 1.5. Unlike C_{60} and C_{70}, the potential separation between any two successive reductions is not very constant for these larger cages. This can be explained in terms of the loss of the triple or nearly triple degeneracy of the LUMO [5]. Oxidation potentials, along with the corresponding values of the HOMO–LUMO gaps in solution, are presented in Table 1.6. Note that, as in the case of C_{60} and C_{70}, the redox potentials for these higher fullerenes are solvent and/or electrolyte dependent.

1.3.1 Reduction and Oxidation of C_{76}

C_{76} possesses a chiral structure with D_2 symmetry [35]. Its two enantiomers, C_{76} and *ent*-C_{76}, are typically isolated as a 1:1 racemic mixture. Four reversible reduction waves have been observed by CV in PhCN and DCM [37], and up to six waves were detected in a PhMe/MeCN mixture at $-15\,°C$, the sixth reduction being irreversible at the limit of the solvent potential window [38]. Only two reduction waves are observable in TCE by OSWV, but this solvent has been extraordinarily useful for the detection of reversible oxidation waves [39]. The first reduction potentials for C_{76} in the three solvents mentioned above are 100–200 mV more positive than the corresponding ones for C_{60}, not an unexpected result considering the higher electron affinity estimated for C_{76} (see Table 1.1). Up to two oxidation waves are observed in TCE [39,40]. The first one is chemically reversible and electrochemically quasireversible. It occurs at $+0.81$ V, making C_{76} about 450 mV easier to oxidize than C_{60}. The second oxidation in TCE occurs at $+1.30$ V, but it is chemically irreversible. Others have reported chemically irreversible first and second oxidations in various solvents [37,38].

1.3.2 Reduction and Oxidation of C_{78}

Studies on the redox properties of C_{78} have been restricted to the two most abundant constitutional isomers, C_{2v}-C_{78} and D_3-C_{78} (see Tables 1.5 and 1.6). The first study by Selegue et al. [38] on the C_{2v} isomer revealed five reversible plus one irreversible one-electron reductions and an irreversible oxidation step in PhMe/MeCN at $-15\,°C$.

Another study involving OSWV of an isomeric mixture of the C_{2v} and D_3 isomers in TCE showed that reduction waves for the two isomers overlapped at -0.77 and -1.08 V versus Fc/Fc$^+$ [39]. However, four oxidation waves were observed, consisting of two pairs of consecutive small and large waves. Each pair was assigned, based on their current intensity ratio, to the two isomers in the mixture, which exist in a 1:5 ratio of C_{2v}-C_{78}/D_3-C_{78}. The assignment of D_3-C_{78} as the easier to oxidize isomer was also in agreement with the previously calculated HOMO energies for the two isomers [41].

In the most recent study involving C_{78}, a pure sample of the C_{2v} isomer was prepared by the newly discovered *Retro-Bingel* reaction [12], while a pure sample of the D_3 isomer was obtained by high-performance liquid chromatography (HPLC) as described previously [42,43]. Their redox behavior was studied by CV in DCM at room temperature, and it revealed that, indeed, their cathodic electrochemistry is very similar (although not identical) in this solvent [30]. The first two reductions are easier for the D_3 isomer by 60 and 100 mV, respectively, while the third and fourth reductions are nearly identical for the two isomers. Partial confirmation of the anodic results in TCE was also obtained. The D_3 isomer exhibits a reversible one-electron oxidation at +0.74 V. Unfortunately, due to limitations in the solvent (DCM) potential window, neither the second oxidation for the D_3 isomer nor the two oxidations for the C_{2v} isomer were observed.

1.3.3   Reduction and Oxidation of C_{82}

The electrochemistry of C_{82} has also been studied in detail, but with a primary interest to compare its redox properties to those of its fascinating endohedral structures [44–46]. To our knowledge, only one study has been conducted on the electrochemistry of the empty C_{82} cage [47]. These results are summarized in Table 1.5. Up to four reversible reduction waves were observed by OSWV in pyridine, but no oxidation was detected within the limits of the solvent potential window. The reader is referred to Chapter 9 in this volume for details on endohedral C_{82} electrochemistry.

1.3.4   Reduction and Oxidation of C_{84}

Of the 24 isomers of C_{84} that satisfy the "isolated pentagon rule" [33–35], only two have been found in extracts of fullerene soot. They possess D_2 and D_{2d} symmetries. Until 1998, all electrochemical studies performed on C_{84} [36,38–40,48] involved a 2:1 isomeric mixture (D_2:D_{2d}) because their separation by HPLC was not accomplished until very recently [49]. Up to seven redox waves have been observed in PhCN and five in ODCB [36,48]. In the latter solvent, only the fourth wave appears to have two components, indicating a possible reduction potential difference between the two isomers upon addition of a fourth electron [48]. In TCE, only three waves are observed by OSWV in the cathodic scan, but, as mentioned before, this solvent is exceptional to resolve fullerene oxidation waves. The only known reversible oxidation for C_{84} appears at +0.93 V versus Fc/Fc$^+$ using TCE as solvent (see Table 1.6) [39]. The corresponding value for the HOMO–LUMO gap in solution thus gives a value of 1.60 V. Kadish and co-workers were able to resolve ten redox waves in pyridine, the first six of which they were able to assign unequivocally to the first, second, and third one-electron reductions of each of the two isomers. With the aid of electron spin resonance (ESR) studies, the more positive potentials of the three pairs were attributed to the D_{2d} isomer, while the more negative were attributed to the D_2 isomer (see Table 1.5). The last four waves were attributed

TABLE 1.5 Half-wave Reduction Potentials (in V versus Fc/Fc^+) of the Less Abundant Higher Fullerenes C_{76}, C_{78}, C_{82}, and C_{84} Using Various Solvents, Supporting Electrolytes, and Temperatures

	Solvent[a]	Supporting Electrolyte	Temperature (°C)	E_1	E_2	E_3	E_4	E_5	E_6	Reference
C_{76}	PhCN	TBABF$_4$	RT	-0.76[b]	-1.19[b]	-1.72[b]	-2.06[b]			37
		PPN$^+$Cl$^-$	RT?	-0.78	-1.16					40
	DCM	TBABF$_4$	RT	-0.79	-1.12	-1.57				37
		TBAPF$_6$	RT	-0.83	-1.15	-1.64	-2.01			30
	TCE	TBAPF$_6$	RT	-0.83	-1.12					39
		PPN$^+$Cl$^-$	RT?	-0.91						40
	PhMe/MeCN	TBAPF$_6$	-15	-0.83	-1.17[c]	-1.68[c]	-2.10[c]	-2.61[c]	-3.04[c,d]	38
C_{78}	TCE	TBAPF$_6$	RT	-0.77	-1.08					39
		PPN$^+$Cl$^-$	RT?	-0.79	-1.13					40
	PhCN	PPN$^+$Cl$^-$	RT?	-0.64	-1.06					40
C_{2v}-C_{78}	DCM	TBAPF$_6$	RT	-0.70	-1.04	-1.72	-2.07			30
	PhMe/MeCN	TBAPF$_6$	-15	-0.72[c]	-1.08[c]	-1.79[c]	-2.18[c]	-2.45[c]	-2.73[c]	38
D_3-C_{78}	DCM	TBAPF$_6$	RT	-0.64	-0.94	-1.70	-2.05			30
C_{82}	Pyridine	TBAP	RT?	-0.47	-0.80	-1.42	-1.84			47
C_{84}[f]	PhCN[e]	TEAPF$_6$	RT?	-0.63	-0.95	-1.28	-1.58	-1.87		38, 48
	ODCB	TBAPF$_6$	RT?	-1.00[e]	-1.29[e]	-1.60[e]	-1.96[e]	-2.21[e,g]		48
	TCE	TBAPF$_6$	RT	-0.67	-0.96	-1.27				39
	PhMe/MeCN	TBAPF$_6$	+22	-0.67	-1.00	-1.34	-1.72	-1.99	-2.40[h]	36
D_{2d}-C_{84}	Pyridine[i]	TBAP	+22	-0.52	-0.84	-1.20	-1.63[i]	-2.07[i]		36
	Pyridine	TBAP	RT?	-0.46	-0.77	-1.58	-1.98	-2.27[d]		50
D_2-C_{84}	Pyridine[i]	TBAP	+22	-0.61	-0.97	-1.30	-1.52[i]	-1.94[i]		36
	Pyridine	TBAP	RT?	-0.65	-0.98	-1.34	-1.75			50

[a] See Table 1.2 for solvent abbreviations.

[b] Scan rate = 1 V/s.

[c] Scan rate = 2 V/s.

[d] May be irreversible.

[e] Reported value is ± 40 mV due to adsorption and uncompensated resistance complications.

[f] Mixture of two major isomers, D_2-C_{84} and D_{2d}-C_{84}.

[g] Nature of this process is unknown because it was observed at the limit of the solvent potential window.

[h] Tentative assignment to the sixth reduction of C_{84}.

[i] The fourth and fifth reduction potentials are not assigned with certainty because of variation in the heights of the corresponding peaks (see Ref. 36).

TABLE 1.6 Half-wave Oxidation Potentials (in V versus Fc/Fc$^+$) of the Higher Fullerenes, C$_{76}$, C$_{78}$, and C$_{84}$, Using Various Solvents and Temperatures

	Solvent[a]	Supporting Electrolyte	Temperature (°C)	E_1^{ox}	E_2^{ox}	$\langle E_{ox}^1 - E_{red}^1 \rangle$	Reference
C$_{76}$	TCE	TBAPF$_6$	RT	+0.81	+1.30[b]	1.64	39
	DCM	TBAPF$_6$	RT	+0.84		1.67	30
C$_{2v}$-C$_{78}$	TCE	TBAPF$_6$	RT	+0.95	+1.43	1.72	39
	PhMe/MeCN	TBAPF$_6$	−15	+0.90[b]		1.62[c]	38
D$_3$-C$_{78}$	TCE	TBAPF$_6$	RT	+0.70	+1.17	1.47	39
	DCM	TBAPF$_6$	RT	+0.74		1.38	30
C$_{84}$	TCE	TBAPF$_6$	RT	+0.93		1.60	39

[a] See Table 1.2 for solvent abbreviations.

[b] Chemically irreversible.

[c] Calculated from data in Ref. 38.

to either a solubility difference or a chemical process occurring upon addition of a fourth electron to both isomers [36].

Very recently, Anderson et al. [50] were able to separate the D_2 and D_{2d} isomers of C_{84} using HPLC and identified each isomer by the intensity ratio of the corresponding voltametric currents. They conducted OSWV experiments and confirmed that, indeed, as reported by Kadish and co-workers, the D_{2d} isomer is easier to reduce than the D_2 isomer (see Table 1.5). Interestingly, they discovered a dramatic difference in the voltametric behavior of the two isomers that suggests a difference in the electronic (MO) structure between them. There is a relatively constant potential separation of approximately 0.35 V between the four consecutive waves exhibited by the D_2 isomer. On the other hand, the four waves of the D_{2d} isomer are grouped in two sets of two equally spaced waves with a potential gap between the second and third waves of 0.71 V (see Table 1.5 and Fig. 1.3).

1.4 FULLERENE ELECTROCHEMISTRY GENERAL TRENDS: COMPARISON WITH THEORY

In agreement with predictions that EAs (Table 1.1) increase with increasing fullerene size, it has been shown that experimental first one-electron reductions become easier as the number of carbon atoms in the cage increases. This correlation can clearly be seen in Figure 1.4, where a plot of EA versus the first reduction potential is presented. On the other hand, although calculated ionization potentials correlate well with experimental first oxidation potentials, there is no evident trend associated with an increase in fullerene size. The plot of IP versus first oxidation potential (Fig. 1.5) shows that only C_{60} and C_{70}, but not the higher fullerenes, follow the "size trend."

Another good correlation was shown to exist between the calculated and experimental (in solution) HOMO–LUMO energy gaps. The corresponding plot is shown in Figure 1.6. Solvation effects are responsible for the decrease in the experimental values resulting in a slope greater than one.

A final inverse correlation between the experimental HOMO–LUMO gap and the number of carbon atoms has also been obtained. Although it is not a linear relationship, as the number of carbon atoms increases the energy gap decreases, eventually leading to the value for the band gap of graphite. That plot is presented in Figure 1.7.

1.5 ELECTROCHEMISTRY OF C_{60} DERIVATIVES

The ability of C_{60} and other fullerenes to easily undergo addition reactions has proved useful in the production of a myriad of derivatives. A large number of functional groups have been attached covalently to the fullerene moiety in an effort to tune its electronic properties to produce materials such as three-

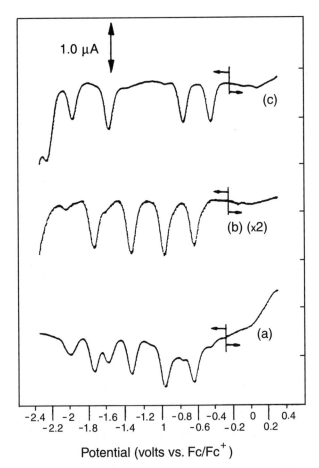

Figure 1.3 Square-wave voltammograms obtained in 0.1 M TBAP/pyridine solution of (a) a mixture of C_{84} isomers, (b) D_2-C_{84}, and (c) D_{2d}-C_{84}. Voltametric data were obtained using a frequency of 15 Hz, a peak amplitude of 25 mV, and a potential step of 2 mV. Limited sample size required the use of a microelectrochemical cell containing only 50 μL of solution. Reprinted with permission from Ref. 50. Copyright © 1998 Elsevier Science S.A.

dimensional electroactive polymers, molecular wires, surface coatings, and electro-optic devices [10]. For the sake of organization, we have decided to classify C_{60} derivatives based on the geometric shape at the point of adduct attachment. Thus, we classify derivatives in three groups. Group one includes those derivatives formed by attachment of two (or more) individual adducts via single bonds to two (or more) carbon atoms in the fullerene, as depicted in Figure 1.8a. These will simply be referred to as *singly bonded functionalized derivatives*. The group includes the dihydrofullerenes, the di- and tetra-alkylfullerenes, and the fluorofullerenes.

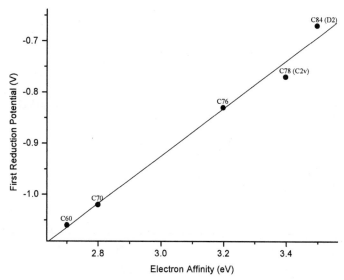

Figure 1.4 Plot of first reduction potential (in TCE) versus estimated electron affinities for the most common fullerenes. A correlation coefficient of 0.99 was found.

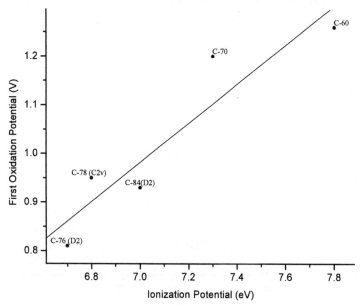

Figure 1.5 Plot of first oxidation potential (in TCE) versus estimated ionization potentials for the most common fullerenes.

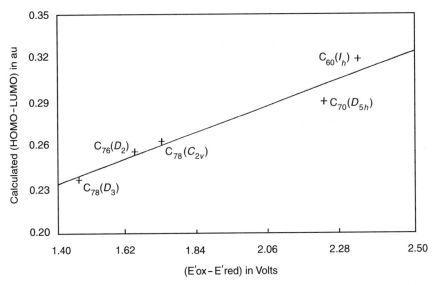

Figure 1.6 Plot of the calculated HOMO–LUMO energy separations versus the experimentally determined potential differences between the first oxidation and the first reduction $(E_{ox}^1 - E_{red}^1)$ for the fullerenes. Reprinted with permission from Ref. 39. Copyright © 1995 American Chemical Society.

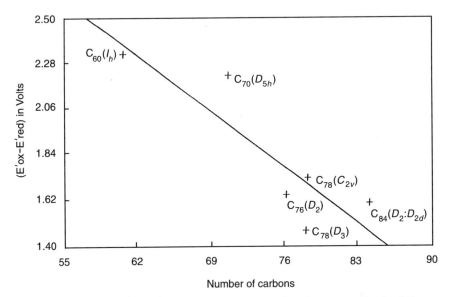

Figure 1.7 Plot of $(E_{ox}^1 - E_{red}^1)$ versus the number of carbon atoms in the fullerene. Reprinted with permission from Ref. 39. Copyright © 1995 American Chemical Society.

(a) Singly-bonded Functionalized Derivatives

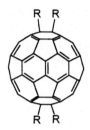

R = H, alkyl group,[a]F[b]   R = alkyl group[a,c]

[a] Alkyl groups may be at a 1,2- or a 1,4-position
[b] Fluorofullerenes are actually multifluorinated
[c] The second set may be at other positions within the fullerene

(b) Cyclopropanated Derivatives

A = C-R$_2$, N-R$_2$, O, transition metal-R$_2$
R = H or any functional group

(c) Cycloaddition Products

A, D, J = C, N, O, Si, or R   R = any functional group

Figure 1.8   Classification of C$_{60}$ derivatives.

Group two consists of all those derivatives in which attachment of the substituent results in the formation of a cyclopropane ring. These will be referred to as *cyclopropanated derivatives*. This group includes methanofullerenes, fulleroids, iminofullerenes, fullerene epoxides, and transition metal derivatives (see Fig. 1.8b). The third group consists of all Diels–Alder addition products, which result in the formation of four-, five-, or six-membered rings and heterocyclic derivatives with rings larger than a cyclopropane ring (see Fig. 1.8c). These will be referred to collectively as *Cycloaddition products*. We hope that organizing C_{60} derivatives in this manner gives the reader a better chance to understand how invaluable electrochemical studies have been in determining how the nature, geometry, structure, and number of addends influence the electrochemical behavior of the C_{60} sphere.

1.5.1   Singly-bonded Functionalized Derivatives of C_{60}

1.5.1.1   Dihydrofullerenes $C_{60}H_2$ exhibits, in general terms, reduction potentials that are cathodically shifted 120–200 mV with respect to the corresponding ones for C_{60} [51–53]. At room temperature, reduction results in the release of H_2 (decomposition) and formation of the parent C_{60} compound. Reduction at $-50\,°C$, however, shows two reversible steps corresponding to the formation of $C_{60}H_2^-$ and $C_{60}H_2^{2-}$, but the trianions and tetra-anions decompose to the corresponding trianions and tetra-anions of C_{60} (see Table 1.7). Analogous decomposition processes have been observed in voltammetric studies of some organometallic derivatives, the epoxyfullerene, $C_{60}O$, and the bis(methanofullerene) (see **19**, Fig. 1.18).

Two chemically irreversible, multielectron, oxidation steps have been observed in PhCN at (peak potential) $+1.13$ and $+1.35$ V versus Fc/Fc^+. Irreversibility persists even at scan rates up to 8 V/s. The first oxidation is consistent with the loss of two protons and two electrons from $C_{60}H_2$ to form C_{60}. The second oxidation has been attributed to the formation of C_{60}^+, since the peak potential for that step is in good agreement with other reported values under similar conditions (see also Table 1.3) [52].

1.5.1.2   Alkylfullerenes   Alkylfullerenes that have been studied by CV or DPV have the general formula R_2C_{60} and R_4C_{60}. The former may exist as either 1,2 or 1,4 isomers, while many more isomers exist for the latter. In benzonitrile, all dialkylfullerenes exhibit at least two reversible reduction waves that are 90–150 mV more negative than the corresponding ones for pure C_{60} [54–56]. Changing the alkyl group produces no significant effect in the reduction potential (see Table 1.7). However, the tetra-alkylfullerenes are approximately 140 mV more difficult to reduce than the dialkyl derivatives and approximately 260 mV more difficult to reduce than the parent C_{60}. Each sequential loss of a double bond seems to induce a cathodic shift, indicating that as the fullerene cage becomes increasingly functionalized, reductions become more difficult [56]. Evidence for the latter has been obtained in studies involving

TABLE 1.7 Half-wave Reduction Potentials (in V versus Fc/Fc$^+$) of the Singly Bonded Di- and Multifunctionalized Fullerene Derivatives Using Various Solvents and Supporting Electrolytes

	Solvent[a]	Supporting Electrolyte	Temperature (°C)	E_1	E_2	E_3	E_4	Reference
C$_{60}$	PhMe/DMF	TBAP	−50	−0.89	−1.34	−1.90	−2.40	51
C$_{60}$H$_2$	PhMe/DMF	TBAP	−50	−1.02	−1.46	−2.07[b]	−2.59[b]	51
	PhCN/MeCN	TBAPF$_6$	RT?	−1.05[b]	−1.44[b]	−1.99[b]	−2.36[b]	52
	PhCN	TBAPF$_6$	RT?	−1.04[b]	−1.43[b]	−2.01[b]		52
	DCM	TBAPF$_6$	RT?	−1.11[b]	−1.57[b]	−2.18[b]		52
	PhMe/DMSO	TBAPF$_6$	+2	−1.14	−1.57	−2.18		53
C$_{60}$	PhCN	TBAP	RT?	−0.88[c]	−1.30[c]	−1.77[c]		56
C$_{60}$(CH$_3$)$_2$	PhCN	TBAP	25	−1.03[c]	−1.45[c]	−2.00[c]		54
C$_{60}$(PhCH$_2$)$_2$	PhCN	TBAP	RT?	−1.00[c]	−1.43[c]	−1.96[c]		55
C$_{60}$(C$_2$H$_5$)$_2$	PhCN	TBAP	RT?	−1.00[c]	−1.43[c]			56
C$_{60}$(C$_4$H$_9$)$_2$	PhCN	TBAP	RT?	−0.99[c]	−1.39[c]			56
C$_{60}$(CH$_3$)$_4$	PhCN	TBAP	RT?	−1.17	−1.57			56
C$_{60}$(n-C$_4$H$_9$)$_4$	PhCN	TBAP	RT?	−1.14	−1.55			56
C$_{60}$F$_{48}$	DCM	TBAPF$_6$	RT?	+0.79[d]	+0.44[d−f]			57
C$_{60}$F$_{47}$	DCM	TBAPF$_6$	RT?		+0.37[d,e,g]	−0.12[d,e,g]		57
C$_{60}$F$_{46}$	DCM	TBAPF$_6$	RT?	+0.51[h]	+0.01[f,h,i]			60

[a] See Table 1.2 for solvent abbreviations.

[b] $E_{1/2}$ recalculated from the original data using the reported value for Fc/Fc$^+$ in 0.1 M TBAP/benzonitrile solution of −0.45 V versus SCE, which appears in Ref. 54.

[c] Product decomposes to the corresponding C$_{60}$ anion.

[d] Versus SCE.

[e] Peak potential.

[f] Irreversible.

[g] C$_{60}$F$_{48}^{2-}$ loses F$^-$ to form C$_{60}$F$_{47}^-$, which can be reduced to C$_{60}$F$_{47}^{2-}$.

[h] Versus Ag/AgCl.

[i] Average of anodic and cathodic peak potentials.

mono through hexakis(methano) adducts of C_{60} [57]. Alkylfullerenes appear to be more stable toward electrochemical reduction than hydrofullerenes. To our knowledge, there have been no reports of decomposition of alkyfullerenes upon electrochemical reduction. Their anions appear to be more stable than those of hydrofullerenes.

1.5.1.3 Fluorofullerenes   Due to the high electronegativity of fluorine, multifluorinated derivatives of C_{60} such as $C_{60}F_{48}$ and $C_{60}F_{46}$ exhibit the highest electron affinities reported for fullerene derivatives. Their reduction potentials occur approximately 1.0 V more positive than pure C_{60} under similar conditions [44–46,60,61]. The actual values are reported in Table 1.7. It is interesting to note that upon addition of a second electron to form $C_{60}F_{48}{}^{2-}$, a loss of F^- occurs, and a new reversible redox wave appears at +0.37 V (versus SCE). This wave corresponds to the step $C_{60}F_{47}{}^-/C_{60}F_{47}{}^{2-}$ [60]. Some of these compounds have been used as cathode materials on solid-state lithium cells [62].

1.5.2 Cyclopropanated Derivatives of C_{60}

1.5.2.1 Epoxyfullerenes   The electrochemistry of $C_{60}O$ is characterized by two factors: (1) irreversibility involving decomposition to the parent C_{60} and (2) polymerization. Three one-electron transfer waves have been observed by CV in ODCB (see Fig. 1.9), DCM, and PhMe/MeCN. Initially, all three reductions were reported as chemically irreversible [63], but later investigations observed the formation of a polymeric film on the surface of the working electrode [64]. Indeed, a more intensive investigation found that the first wave is due to the formation of $C_{60}O^-$, which undergoes decomposition to $C_{60}{}^-$. The second wave corresponds to the irreversible formation of $C_{60}O^{2-}$, which is believed to be the polymer initiator. Wave three represents the formation of $C_{60}O^{3-}$. The process is described in Scheme 1. The polymer film has the structure $(-C_{60}O-C_{60}O-)$, and it may be produced from the reaction of the products of the first and second reductions [65].

1.5.2.2 Transition Metal Derivatives   Direct attachment of electron-rich transition metal complexes to the exterior of the fullerene framework has raised a lot of interest because of the possibility of producing crystalline molecular solids [66,67]. As a consequence, several mono through hexakis adducts of the type shown in Figure 1.10 have been prepared and studied [66–71]. In particular, the electrochemistry of $[(R_3P)_2M]_xC_{60}$ (with M = Ni, Pd, and Pt, $x = 1–4$, and R = Ph or Et) has been studied in detail (see Table 1.8) [68,69]. In the case of the mono-adduct complexes $(Ph_3P)_2PtC_{60}$ and $(Et_3P)_2MC_{60}$, three to four one-electron reduction waves have been observed at potentials more negative than those of C_{60}. Loss of the metal fragment accompanies reduction with dissociation rates being dependent on the phosphine ligand ($Ph_3 > Et_3$), the metal (Ni > Pd > Pt), and the extent of reduction (trianion > dianion > monoanion).

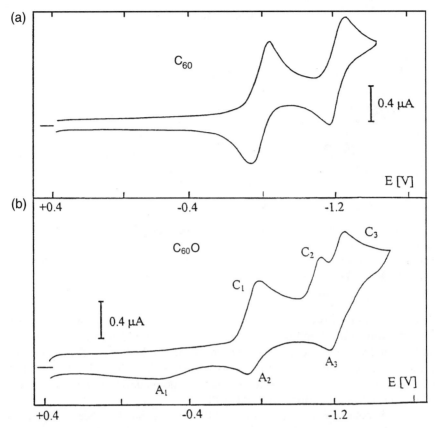

Figure 1.9 CV of (a) C_{60} and (b) $C_{60}O$ in 0.1 M TBAP/4:1 toluene/acetonitrile using a scan rate of 100 mV/s. Reprinted with permission from Ref. 65. Copyright © 1995 American Chemical Society.

As shown in Table 1.8, the reduction potentials for the mono adducts of Pt, Pd, and Ni are essentially the same. Thus, there appears to be negligible extension of d-orbital back-bonding density beyond the carbon–carbon bond where the metal is attached. In general, reductions are C_{60} centered while oxidations are metal centered [68]. Fourier transform Raman studies on some of these complexes have shown evidence of weakening of the C–C bond of the fullerene upon metal-to-fullerene, d-to-π, back-bonding. This symmetry-lowering effect may be responsible for the loss of the metal fragment upon electron reduction [67].

During reduction of the series $[(Et_3P)_2Pt]_xC_{60}$ (with $x = 1$–4), it was found that for each organometallic group added to the fullerene core, a successive cathodic shift of approximately 0.35 V was observed (see Table 1.8), signaling a successive decrease in the electron affinity of C_{60} [66,67]. Even more interesting,

products

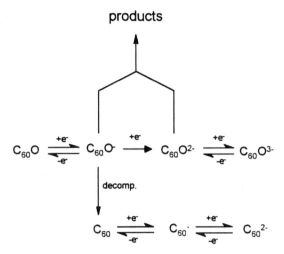

Scheme 1   Decomposition and polymerization process for $C_{60}O$. Reprinted with permission from Ref. 65. Copyright © 1995 American Chemical Society.

voltametric studies of the oxidation of the same series showed a unique stepwise electrochemical disassembly of the molecules. One Pt addend was lost for every additional oxidation step, leading to the final product, free C_{60}, which was produced upon oxidation of the mono adduct at 0.40 V versus Fc/Fc^+ [69]. Changing the metal also produces changes in the oxidation potentials. Ease of oxidation follows the order Ni > Pd > Pt.

The CV behavior of an iridium derivative of C_{60} has also been studied. Although it is really not a cyclopropanated derivative because the metal is attached to C_{60} via a single bond, it is best to describe it here because of its transition metal nature. It has the general formula $(\eta^5 C_9 H_7)Ir(CO)(\eta^2 C_{60})$. In

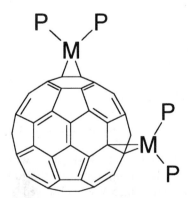

Figure 1.10   A bis-adduct of a C_{60} metal derivative. M = Ni, Pd, or Pt.

TABLE 1.8 Half-wave Reduction Potentials (in V versus Fc/Fc$^+$) of Epoxyfullerene and Some Transition Metal Derivatives of C_{60}

	Solvent[a]	Supporting Electrolyte	Temperature (°C)	E_1	E_2	E_3	E_4	Reference
$C_{60}O$	ODCB	TBAPF$_6$	RT	-1.08^b	-1.48^b	-1.93^b		63
C_{60}	THF	TBAPF$_6$	RT	-0.86	-1.44	-2.00		68
$(Ph_3P)_2Pt–C_{60}$	THF	TBAPF$_6$	RT	-1.21	-1.75	-2.23		68
$[(Et_3P)_2Pt]_n–C_{60}$								
$n = 1$	THF	TBAPF$_{6+}$	RT	-1.20	-1.73	-2.27		68
$n = 2$	THF	TBAPF$_6$	RT	-1.51				68
$n = 3$	THF	TBAPF$_6$	RT	-1.93				68
$n = 4$	THF	TBAPF$_6$	RT	-2.31^b				68
$(Et_3P)_2Pd–C_{60}$	THF	TBAPF$_6$	RT	-1.18	-1.69	-2.23^b		68
$(Et_3P)_2Ni–C_{60}$	THF	TBAPF$_6$	RT	-1.20	-1.74	-2.32		68
$(C_9H_7)Ir(CO)C_{60}$	DCM	TBAPF$_6$	RT	-1.08	-1.43	-1.94^b		70
C_{60}	THF	TBABF$_4$	20	-0.70	-1.30	-1.90	-2.40	71
$C_{60}HRh(CO)(PPh_3)_2$	THF	TBABF$_4$	20	-0.92	-1.30	-2.20	-2.90	71
$C_{60}HIr(CO)(PPh_3)_2$	THF	TBABF$_4$	20	-1.03	-1.55	-2.14	-2.90	71
$C_{60}[HRh(CO)(PPh_3)_2]_2$	THF	TBABF$_4$	20	-0.92	-1.30		-2.90	71
$C_{60}[HIr(CO)(PPh_3)_2]_2$	THF	TBABF$_4$	20		-1.55	-2.14	-2.90	71

^{a} See Table 1.2 for solvent abbreviations.

^{b} Irreversible.

DCM, its CV is characterized by two reversible waves and one irreversible wave (Table 1.8), and, in contrast with the Ni, Pd, and Pt complexes described above, no loss of the Ir fragment was reported. An irreversible oxidation peak is also observed at +0.72 V versus Fc/Fc^+ during the anodic scan. Infrared (IR) and ultraviolet (UV) spectroscopies have confirmed that reductions are C_{60} centered, while the observed oxidation is metal centered. Oxidation is followed by an irreversible loss of CO [70].

Recent comparative studies of another series of mono- and bimetallic complexes of Ir and Rh with C_{60} in which the metal is attached to two adjacent carbons (forming a cyclopropane ring) indicate that, as previously observed for all other transition metal fullerene complexes, reductions are C_{60} centered (see Table 1.8) [66–73]. These complexes have the general formula $C_{60}[HM(CO)(PPh_3)]_x$, where M = Ir or Rh and $x = 1$ or 2. Reductions are negatively shifted by approximately 0.30 V with respect to those of pure C_{60} and are also accompanied by breakage of the C–M bonds. As a consequence, voltammograms also display the corresponding reduction waves of pure C_{60}. Oxidations are metal centered and are also accompanied by the loss of the metal fragment(s). Both the mono and bis adducts of Rh and Ir exhibit a two-electron oxidation peak at +1.13 and +1.22 V versus Fc/Fc^+, respectively. The bis adducts exhibit an additional peak at +0.80 and +1.00 V, respectively, which corresponds to the oxidation of the second addend [69].

1.5.2.3 Methanofullerenes and Fulleroids

Four different structures are possible for homo-cyclopropanated derivatives, depending on whether the two bridge-head carbon atoms on the surface of the fullerene are bonded or not. Of the four, only two have been identified: the [5,6]open or fulleroid and the [6,6]closed or methanofullerene. The fulleroid is the kinetic product [72,73] that rearranges into a methanofullerene (the thermodynamic product) under a variety of reaction conditions (it can be done by light/thermal, pure light, or electrochemically) (see Scheme 2) [74,75]. The reason why these two are the

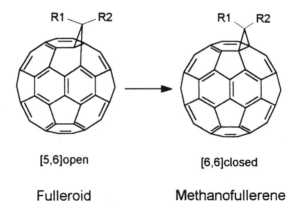

R1   R2   R1   R2

[5,6]open   [6,6]closed

Fulleroid   Methanofullerene

Scheme 2 Fulleroid–methanofullerene isomerization. Reprinted with permission from Ref. 7. Copyright © 1998 American Chemical Society.

only observed isomers has been attributed to the preservation of the [5] radialene substructure within C_{60} [76,77]. Formation of methanofullerenes results in a partial loss of conjugation of the C_{60} sphere, from the 60 π-electron to a 58 π-electron configuration. In contrast, formation of fulleroids ([5,6]open) does not result in a reduction of the parent $C_{60}\pi$ electron configuration. They are the only derivatives that retain the 60 π-electron configuration of C_{60} [78]. Despite initial belief that a loss of conjugation would affect the electronegativity of these derivatives, studies on the electrochemically induced methanofullerene–fulleroid rearrangement have shown that there is no significant difference in the reduction potentials of the two isomers. Depending on the substituent, temperature, and/or solvent, differences might be observed after the third [26,79], fourth, [26] or fifth [80,81] electron reduction.

Table 1.9 contains reduction potentials of several selected fulleroid and methanofullerene derivatives. The table is not exhaustive due to the large number of derivatives that have been prepared and the similarities in their electrochemical properties. In general terms and with few exceptions, however, saturation of a double bond in C_{60} upon formation of a methanofullerene causes a cathodic shift in reduction potentials ranging between 100 and 150 mV with respect to pure C_{60} under the same conditions [72,73,82,89,91]. For example, $C_{61}H_2$ and $C_{61}Ph_2$ exhibit reduction potentials that are shifted approximately 100 mV relative to C_{60}. The reductions are, however, essentially the same relative to each other, indicating no difference in the electron-donating effect between the two groups [72,82].

The shifts for other derivatives vary depending on the electron-donating or electron-withdrawing nature of the substituent, and also on its orientation [63,83–86]. In a study involving $C_{61}RR'$ (where R = CN, NO_2, or CO_2Et, and $R' = CN$, NO_2, or CO_2Et) it became evident that in order to compensate for the effect of saturating a fullerene double bond, both R and R' had to be very strong electron-withdrawing groups such as the cyano group. For the dicyanomethanofullerene, the first reduction potential occurred 156 mV more positive than that of C_{60}, making it a better electron acceptor and the most electronegative methanofullerene prepared so far. Any other combination of R and R' produced reduction potentials more negative than those of C_{60} (see Table 1.9) [83].

Eiermann et al. [84–86] have studied the effect of the orientation of addends on the electrochemistry of C_{60}. When the phenyl groups that contain the electron-withdrawing groups lie parallel to the surface of C_{60}, the redox potentials fall well within the "standard range" for methanofullerenes (30–100 mV more negative than C_{60}, in this case). However, when the phenyl rings are held rigidly perpendicular to the surface of the sphere, reductions become more positive by 40–80 mV than the corresponding ones for pure C_{60}. The authors have attributed this to an orbital through-space interaction phenomenon called "periconjugation" (different from "spiroconjugation") between the p_z orbitals of the phenyl rings of the addends and the p_z orbitals of the C_{60} carbon atoms adjacent to the bridge. Modified neglect of diatomic overlap (MNDO) calculations used to model the orbital interactions support the electrochemical find-

TABLE 1.9 Half-wave Reduction Potentials (in V versus Fc/Fc$^+$) of Some Methanofullerenes, Fulleroids, and Iminofullerenes

	Solvent[a]	Supporting Electrolyte	Temperature (°C)	E_1	E_2	E_3	E_4	Reference
C$_{60}$[b]	THF	TBAPF$_6$	RT?	-0.85	-1.45	-2.04	-2.54	82
C$_{61}$H$_2$[b]	THF	TBAPF$_6$	RT?	-0.97	-1.55	-2.12	-2.60	72
C$_{61}$Ph$_2$[b]	THF	TBAPF$_6$	RT?	-0.96	-1.54	-2.09	-2.55	82
C$_{60}$	PhMe/DMF	TBABF$_4$	-72	-1.12				83
C$_{61}$RR′								
R = R′ = CN	PhMe/DMF	TBABF$_4$	-72	-0.97[c]				83
R = R′ = CO$_2$Et	PhMe/DMF	TBABF$_4$	-72	-1.16[c]				83
R = CN, R′ = CO$_2$Et	PhMe/DMF	TBABF$_4$	-72	-1.11[c]				83
R = NO$_2$, R′ = H	PhMe/DMF	TBABF$_4$	-72	-1.16[c]				83
C$_{60}$	ODCB	TBAPF$_6$	RT	-1.17	-1.53	-1.99	-2.47	99
3	ODCB	TBAPF$_6$	RT	-1.13	-1.52	-1.96	-2.43	99
4	ODCB	TBAPF$_6$	RT	-1.17	-1.53	-1.99	-2.45	99
5	ODCB	TBAPF$_6$	RT	-1.17	-1.53	-2.00	-2.45	99
6	ODCB	TBAPF$_6$	RT	-1.15	-1.52	-1.98	-2.47	99

[a] See Table 1.2 for solvent abbreviations.

[b] Recalculated from the original data to report versus Fc/Fc$^+$.

[c] Irreversible at room temperature.

ings. Similar results were obtained for three different quinone-type methanofullerene derivatives, in which their better electron acceptor ability than pure C$_{60}$ could be attributed to periconjugation [84,85]. Unfortunately, some of these compounds behave electrochemically irreversibly with a subsequent opening of the cyclopropane ring. The latter finding has been supported by ESR measurements [86].

Several studies have been conducted to determine the effect of the number of addends and the addition pattern on the redox properties of C$_{60}$ [57–59,82,87]. Boudon et al. [57–59], for example, reported that as the number of addends increases from one to six, the C$_{60}$-centered electron reductions become increasingly difficult and irreversible. Not surprisingly, oxidations become easier to conduct when going from the mono to the tris adduct, remaining the same for the rest of the adducts [57–59]. Consequently, they concluded that as the fullerene cage becomes increasingly functionalized, a stepwise loss of conjugation occurs, and the LUMO becomes increasingly higher in energy. Similar results were also obtained by Guldi et al. [87] for the series C$_{60}$[C(CO$_2$Et)$_2$]$_x$ (with $x = 1$–3), and others. As for the nature of the addition pattern, different bis-adduct regioisomers first reduce at almost identical potentials (-1.10 to -1.13) [59], indicating a very small influence from regioisomerism. Furthermore, Cardullo et al. reported that an all equatorial multiple-cyclopropane addition pattern (from bis- to tetrakis-adducts) results in only a small effect on the ability of the fullerene to accept electrons. The shifts in the first reduction potential average only 36 mV per addend [58].

Other interesting molecular architectures have been built combining methanofullerenes with other groups such as crown ethers [80,90] and porphyrins [91–93]. Three of these compounds are depicted in Figure 1.11. The electrochemistry of **1a** and **1b**, however, does not deviate in any way from that of other methanofullerenes; that is, reduction potentials are simply cathodically shifted approximately 40–50 mV relative to pure C$_{60}$ [80,90,91–93]. Due to the large distance between the crown ether and the fullerene moieties in **1a**, cation binding by the crown does not measurably perturb the fullerene-centered electrochemistry.

A recent investigation has demonstrated that a significant perturbation of the electronic structure of the fullerene occurs only when the bound cation is closely and tightly positioned on the fullerene surface as in the *trans*-1 fullerenocrown bis adduct **2** [90]. The CV of **2** (Fig. 1.12) clearly shows a significant shift (90 mV) upon K$^+$ complexation by the crown ether, relative to the uncomplexed molecule. The *trans*-2 and *trans*-3 isomers of **2** (not shown) also exhibit shifts of 50 and 40 mV, respectively, upon K$^+$ complexation. The latter decrease in anodic shift can be explained by an increase in the average distance from the bound cation to the fullerene surface, as evidenced by molecular modeling.

1.5.2.4 *Imino[60]fullerenes*   Like their C-bridged analogs, the N-bridged fullerenes exist either as [6,6]closed (epiminofullerenes) or [5,6]open (azahomo-

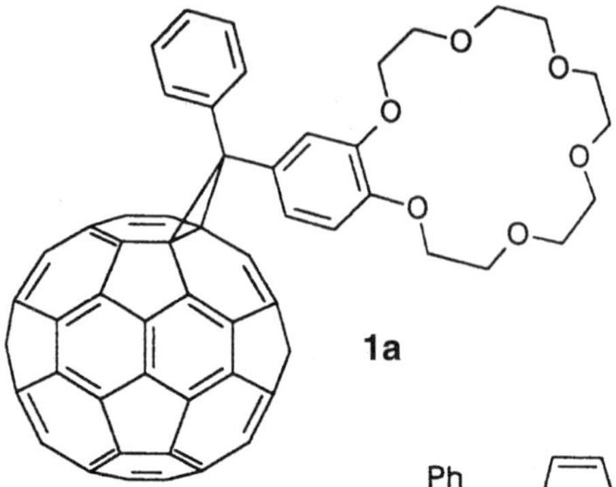

1a

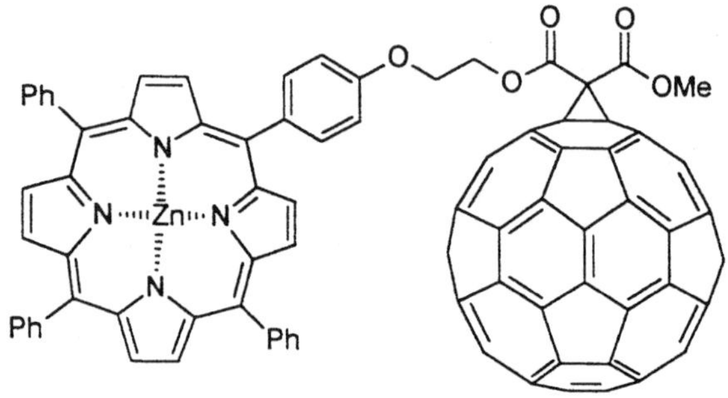

1b

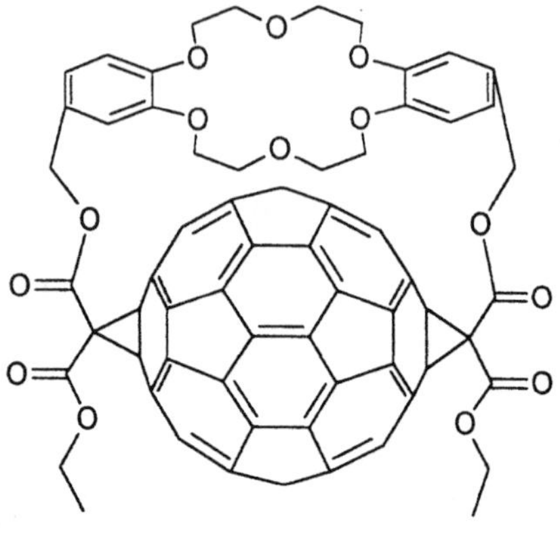

2

Figure 1.11   Selected crown ethers and porphyrin cyclopropano derivatives of C_{60}.

fullerenes). The redox properties of some of these derivatives have been investigated, and, in general, it has been found that they reduce at potentials that are
closer to the reduction potentials of pure C_{60} but less negative than those of
methanofullerenes and fulleroids (see Table 1.9) [94–99]. It has been proposed
that the better electron-withdrawing nature of N makes the carbon cage in
iminofullerenes a better electron acceptor [94–99]. Reduction potentials vary
depending on the addend, but one particular case, compound **3** in Figure 1.13,

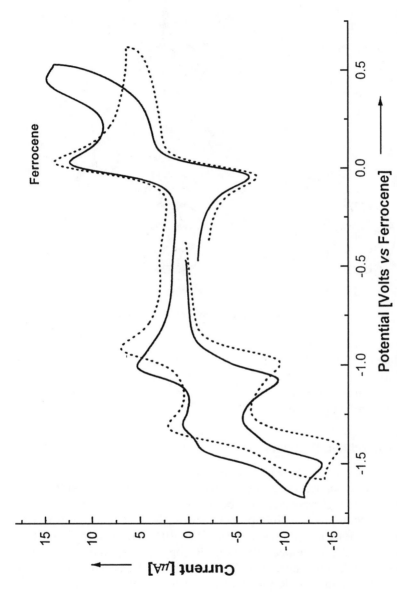

Figure 1.12 CVs of the *trans*-1-fullerene bis adduct **2**. *Solid line*: solution of free **2** with one equivalent of [2.2.2]cryptand present to ensure absence of any K[+] complex; *dotted line*: solution of **2** with 10 equivalents of KPF$_6$. Scan rate = 100 mV. Reprinted with permission from Ref. 90. Copyright © 1998 Wiley–VCH Verlag GmbH.

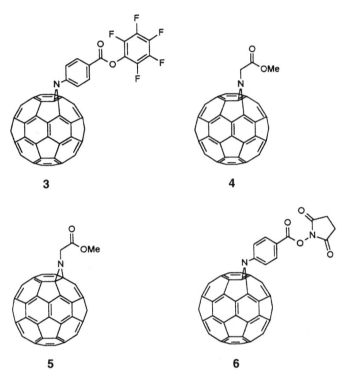

Figure 1.13 Some selected imino[60]fullerenes.

shows reduction potentials that are approximately 40 mV less negative than those of pure C_{60} [99]. In addition, there appears to be no significant difference between the redox potentials of epiminofullerenes and azahomofullerenes (structures **4** and **5** in Fig. 1.13).

1.5.3 Cycloaddition Products

The electrochemistry of a large number of derivatives bearing a four-, five-, or six-membered ring connection to the surface of C_{60} has been studied [63,101–110]. Many of these compounds have been designed to combine the fullerene electronic and photophysical [100] properties with those of other electroactive centers and chromophores such as ferrocene [101–103], tetrathiafulvalene [102], porphyrins [104–107], quinones [108–110], [Ru(bpy)$_3$] [111], and others [111–114] to form conducting salts, photo-induced charge transfer complexes, or nonlinear optical materials. Representative structures are depicted in Figure 1.14, and their corresponding redox potentials are shown in Table 1.10. With the few exceptions that will be discussed below, most of these derivatives reduce at potentials within the range observed for all other derivatives of C_{60}. However, one important fact is that the nature of the fused addend has a significant

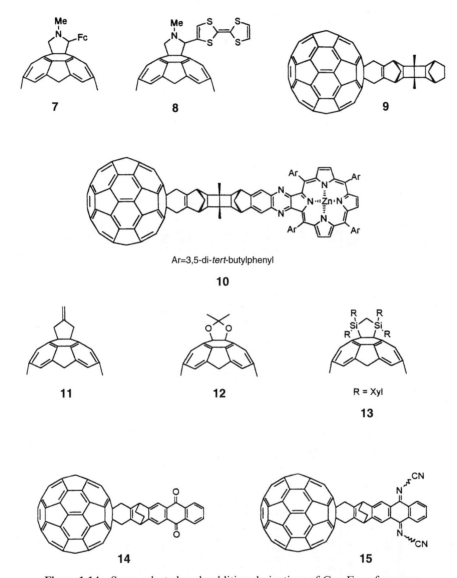

Figure 1.14 Some selected cycloaddition derivatives of C$_{60}$. Fc = ferrocene

influence on the first reduction potential. Cardullo et al. determined that for bis- and tris-adducts a fused cyclohexene ring shifts the fullerene-centered first reduction potential more cathodically than a fused cyclopropane ring [58].

One study investigated the electron-withdrawing effect of the heteroatoms oxygen and silicon (compounds **12** and **13** in Fig. 1.14) on the electrochemistry of fullerene [63]. Reduction potentials are shifted the most for silicon and the least for oxygen, demonstrating that the electron-withdrawing effect follows the order O > C > Si. These results are also presented in Table 1.10.

TABLE 1.10 Half-wave Reduction Potentials (in V versus Fc/Fc$^+$) of Selected Cycloaddition Product Fullerene Derivatives Using Various Solvents and Supporting Electrolytes

	Solvent[a]	Supporting Electrolyte	Temperature (°C)	E_1	E_2	E_3	E_4	E_5	Reference
C$_{60}$	PhMe/MeCN	TBAP	-45	-0.94	-1.33	-1.83	-2.28	-2.74	102
7	PhMe/MeCN	TBAP	-45	-1.08	-1.47	-2.03	-2.44	-3.14	102
8	PhMe/MeCN	TBAP	-45	-1.03	-1.43	-1.99	-2.41	-3.09	102
9	PhMe/MeCN	TBAPF$_6$	RT?	-0.58[b]	-1.01[b]	-1.40[b]			107
10	PhMe/MeCN	TBAPF$_6$	RT?	-0.58[b]	-1.03[b]	-1.47[b]			107
C$_{60}$	ODCB	TBAPF$_6$	RT	-1.13	-1.50	-1.95			63
11	ODCB	TBAPF$_6$	RT	-1.23	-1.58	-2.11			63
12	ODCB	TBAPF$_6$	RT	-1.13	-1.50	-1.95			63
13	ODCB	TBAPF$_6$	RT	-1.28	-1.66	-2.16			63
C$_{60}$	PhMe/MeCN	TBAPF$_6$	RT	-0.99	-1.39	-1.90	-2.37	-2.87	108
14	PhMe/MeCN	TBAPF$_6$	RT	-1.12[c]	-1.48[c-e]	-2.06[c-e]	-2.56[c]		108
15	PhMe/MeCN	TBAPF$_6$	RT	-0.78[d]	-1.13[c-e]	-1.52[c]	-2.09[c]	-2.48[c]	108

[a] See Table 1.2 for solvent abbreviations.
[b] Versus Ag/AgCl.
[c] Fullerene-based reduction.
[d] Addend-based reduction.
[e] Two-electron process.

Another recent study by Echegoyen and co-workers [108] has provided the first complete electrochemical characterization of C_{60} acceptor dyads (compounds 14 and 15 in Fig. 1.14), where an unambiguous assignment of the location of each reduction step at one of the electrophores has been made. Results are presented in Table 1.10. In the case of 14, where the addend is an anthraquinone, the first reduction is fullerene based. For 15, where the anthracene addend bears the more electron-withdrawing cyanoimino quinone groups, the first reduction is addend based. All other redox processes follow the expected order, with reductions of both the fullerene and the addend occurring almost simultaneously. ESR was used to determine the exact location of the electron at each reduction step.

Interestingly, the penta-anion of 14 is expected to have three electrons localized on the C_{60} moiety and two on the anthraquinone based on the reduction potentials obtained by CV. On the ESR spectrum, however, the penta-anion shows a single resonance based on the anthraquinone group, thus contradicting the voltametric results and suggesting that there are four electrons on the fullerene core. Calculations are underway to try to explain these findings [108].

1.5.4 C_{60} Dimers

Connecting two fullerenes in close proximity might introduce interesting electronic properties, since the presence of an electron-deficient sphere may influence the electrochemistry of the other. Determining this "communication" between fullerene cores has been the driving force behind the preparation of several C_{60} dimers, using different molecular bridges such as acetylene [57,115] and phenylene [89] spacers. Other dimers, including C_{120} [116], $(C_{59}N)_2$ [117], and $C_{120}O$ [118] (structures 16, 17, and 18, respectively, in Fig. 1.15) have also been prepared. With two exceptions, compounds 17 and 18, the electrochemistry of these dimers reveals simultaneous, superimposable redox processes for the individual cores, as demonstrated by peak currents being twice those of the corresponding monomers. This indicates that the spheres do not interact and thus behave totally independent of each other.

The dimer 16 has been reported to undergo dissociation into two C_{60} molecules upon one-electron voltametric reduction.

Compound 17 was studied by CV and chronoamperometry. Three overlapping (closely spaced) pairs of reversible two-electron reductions were observed, suggesting weakly interacting electrophores [117].

The CV and DPV of 18 show two reversible asymmetric peaks and a set of two reversible peaks. Deconvolution of the asymmetric peaks indicate that each one corresponds to two closely spaced waves, which are the result of superposition of two independent processes occurring at very close reduction potentials. The fact that these processes can be resolved indicates that the two fullerene cores are communicating. Observation of a triplet feature by ESR supports this conclusion [118].

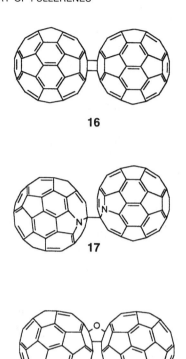

16

17

18

Figure 1.15 Some selected [60]fullerene dimers.

1.6 ELECTROCHEMISTRY OF C_{70} DERIVATIVES

Despite the fact that only a few studies have been carried out so far on the redox properties of C_{70} derivatives, it has become evident that the electrochemistry of C_{70} mono-adduct derivatives is very similar to that of C_{60} mono adducts. In general, shifts of 70–130 mV have been observed for two isomers of $C_{70}H_2$ [51], three isomers of isoxalene fused C_{70} [119], and the two methano[70]fullerenes, $C_{71}R_2$ with $R = CO_2(C_2H_5O)_2H$ or $CO_2CH_2CO_2Et$ (see Table 1.11) [120].

Very recently, a study was conducted on the changes in the redox properties of C_{70} as a function of the number of addends [30]. As a general trend, it was found that the first reduction becomes increasingly more difficult with increasing number of addends, whereas the first oxidation becomes facilitated. The average increase in reduction potential is approximately 120 mV from C_{70} to mono adduct, approximately 50 mV from mono to bis adduct, approximately 30 mV from bis to tris adduct, and approximately 80 mV from tris to tetrakis adduct. One tetrakis adduct is 300 mV more difficult to reduce and 270 mV easier to oxidize than pristine C_{70} [30].

TABLE 1.11 Half-wave Reduction Potentials (in V versus Fc/Fc$^+$) of C_{70} Derivatives

	Solvent[a]	Supporting Electrolyte	Temperature (°C)	E_1	E_2	E_3	E_4	Reference
C_{70}	PhMe/DMF	TBAP	-50	-0.91	-1.34	-1.83		51
7,8-$C_{70}H_2$	PhMe/DMF	TBAP	-50	-1.04	-1.48	-1.96		51
1,9-$C_{70}H_2$	PhMe/DMF	TBAP	-50	-1.03	-1.52	-1.93		51
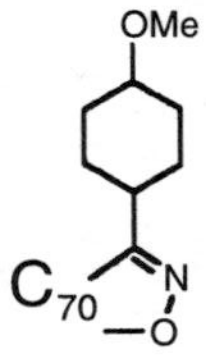	PhMe/MeCN	?	RT?	-0.98	-1.37	-1.79	-2.24	119
C_{70}	DCM	TBAPF$_6$	+25	-0.93	-1.31	-1.73	-2.09	120
$C_{71}R_2$ R = $CO_2(C_2H_4O)_2H$	DCM	TBAPF$_6$	+25	-1.02	-1.39			120
$C_{71}R_2$	DCM	TBAPF$_6$	+25	-1.05	-1.44[b]	-1.84	-2.02	30
$C_{72}R_4$	DCM	TBAPF$_6$	+25	-1.11	-1.47[b]	-1.88	-2.06	30
$C_{73}R_6$	DCM	TBAPF$_6$	+25	-1.13	-1.52	-2.15		30
$C_{74}R_8$	DCM	TBAPF$_6$	+25	-1.23	-1.45	-1.65		30
R = $CO_2CH_2CO_2Et$								

[a] See Table 1.2 for solvent abbreviations.
[b] Irreversible.

Note in Table 1.11 that the second reduction steps of both $C_{71}R_2$ and $C_{72}R_4$ for $R = CO_2CH_2CO_2Et$ are irreversible. An EC mechanism was proposed for this reduction step, in which an electron transfer step (E) is followed by a chemical reaction (C). In fact, addition of two electron equivalents to $C_{71}R_2$ by controlled potential electrolysis (CPE) results in the loss of the addend and conversion (in 70% yield) into pure C_{70} (see Section 1.8.3) [12].

1.7   ELECTROCHEMISTRY OF C_{76} AND C_{78} DERIVATIVES

Only one study has been published on the electrochemistry of C_{76} and C_{78} derivatives [30]. In that report, three constitutionally isomeric pairs of diastereomeric mono adducts of C_{76} bearing bis(alcoxycarbonyl)methano addends were investigated by steady-state voltammetry (SSV). Two of the pairs of mono adducts were 40 mV easier to reduce than the parent C_{76} fullerene, in contrast to most other methanofullerene derivatives of C_{60} and C_{70}. Furthermore, the four reduction steps observed for these isomers are reversible at all scan rates, demonstrating a higher chemical stability than their C_{60} and C_{70} counterparts.

As in the case of C_{60} and C_{70}, the reduction of functionalized C_{78} is more difficult than the pristine fullerene, whereas the oxidation is facilitated. The redox properties of three bis adducts and three tris adducts, consisting of bis(malonate)methano addends connected to C_{78}, were investigated. All of these derivatives exhibited a one-electron reversible (or quasireversible) first reduction step, while the second reduction was irreversible in half of the compounds. After the second reduction, the latter underwent a chemical transformation resulting, eventually, in the generation of the parent C_{2v}-C_{78}. The use of controlled potential electrolysis to generate the pure fullerenes (C_{2v}-C_{78} and D_3-C_{78} in this case) will be discussed in Section 1.8.3. The CV of one of the bis adducts is presented in Figure 1.16, along with that of the parent C_{2v}-C_{78} for comparison. For this derivative, all four reduction steps, as well as the one-electron oxidation, are reversible or quasireversible. Coulometric experiments showed that the dianion is stable but the trianion is not, with decomposition leading to C_{2v}-C_{78} if exhaustively electrolyzed.

1.8   ELECTROLYSIS OF FULLERENES AS A SYNTHETIC TOOL

1.8.1   Electrosynthesis of C_{60} Derivatives

The high stability of coulometrically generated mono- to tetra-anions of fullerenes and the ease with which they can be produced under controlled conditions have offered a variety of possibilities for the preparation of several derivatives. In fact, studies on the protonation of C_{60}^- and C_{60}^{2-} showed that although C_{60}^- is a weak base, C_{60}^{2-} behaves as a relatively strong base, with pK_{a2} values of 16 in DMSO and 10 in ODCB [121–123]. Thus, once reduced beyond the second reduction step, C_{60} becomes a strong nucleophile. This finding has been exploited to electrosynthesize several derivatives of C_{60} such as

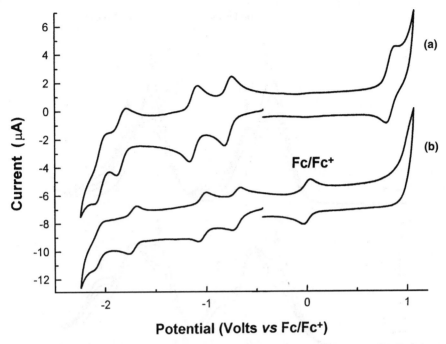

Figure 1.16 CV in DCM of (a) bis adduct of C_{2v}-C_{78} and (b) pure C_{2v}-C_{78}. Scan rate = 100 mV/s. Reprinted with permission from Ref. 30. Copyright © 1998 American Chemical Society.

$C_{60}R_2$ (R = CH$_3$, C_2H_5, or n-C_4H_9) [54,56,124], $C_{60}R_4$ (R = CH$_3$ or n-C_4H_9) [56,124], and $C_{61}RR'$ (methanofullerenes) (R = H or CO_2Et and R' = H, CO_2Et, t-butyl, or CN) [125].

The electrosynthetic method to produce the above consists of dissolving a sample of C_{60} in the chosen solvent/supporting electrolyte combination and exhaustively reducing it, via CPE, to produce $C_{60}{}^{2-}$. The reduction potential depends on the solvent and supporting electrolyte, and it should be set a few millivolts beyond the corresponding $E_{1/2}$ determined by CV. The next step, in the absence of any applied potential, is the addition of an alkyl halide for preparing $C_{60}R_2$ and $C_{60}R_4$ or a *gem*-dihalo compound for preparing methanofullerenes. While the reaction proceeds, it can be monitored in situ by observing the growth of the voltametric (CV or DPV) peaks of the products at the expense of those of the reactant (see Fig. 1.17). Once the reaction is complete, the solvent is evaporated, and the resulting solid is washed with MeCN to remove the supporting electrolyte and dried under vacuum. Column chromatography is used to separate any unreacted C_{60} [55,126].

The formation of $C_{60}R_2$ and $C_{60}RR'$ from $C_{60}{}^{2-}$ and alkyl halides (RX and RX') occurs in two steps. The first step results in the formation of $C_{60}R^-$ via an electron transfer mechanism from $C_{60}{}^{2-}$ to RX. In the second step, $C_{60}R^-$ undergoes an S_N2 reaction with R'X to yield $C_{60}RR'$ [127].

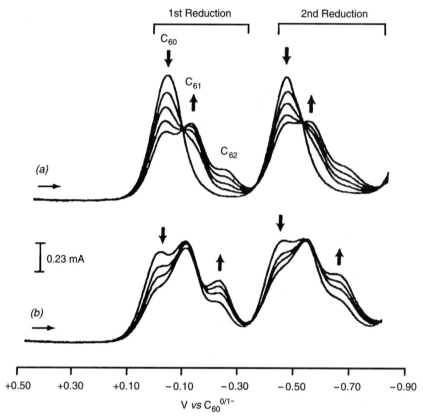

Figure 1.17 OSWVs of C_{60} in 0.1 M TBAPF$_6$/acetonitrile:toluene (1:2) containing I$_2$CHCMe$_3$. C_{61} and C_{62} represent C_{61}HCMe$_3$ and C_{62}H$_2$(CMe$_3$)$_2$, respectively. Voltammograms recorded (a) during the first 80 min and (b) during the last 60 min of the course of the reaction. Sweep width = 25 mV; frequency = 15 Hz; potential step = 4 mV. Reprinted with permission from Ref. 125. Copyright © 1996 The Royal Society of London.

1.8.2 Electrochemically Induced "Shuffle"

During the course of an investigation to determine the electrochemical properties of a series of bis(methanofullerenes), an observation was made concerning the reported electrochemistry of the cis-2 bis(methanofullerene) isomer, **19**, shown in Figure 1.18 (the nomenclature used to describe bis adducts of C_{60}, introduced by Hirsch, is also shown in Fig. 1.18) [128]. Of all the bis(methanofullerene) isomers studied in that report, the cis-2 was the only one that exhibited a chemically irreversible second one-electron reduction, as determined by using CV and SSV on a rotating disk electrode. That observation prompted a study

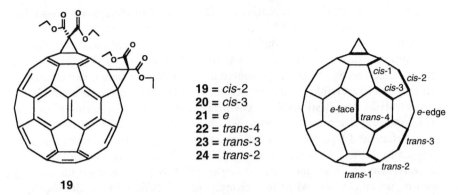

19 = *cis*-2
20 = *cis*-3
21 = *e*
22 = *trans*-4
23 = *trans*-3
24 = *trans*-2

19

Figure 1.18 Representation of the C_{60} bis(malonate)adducts **19–24**, which were subjected to CPE at the first and second reduction potentials. Reprinted with permission from Ref. 11. Copyright © 1998 American Chemical Society.

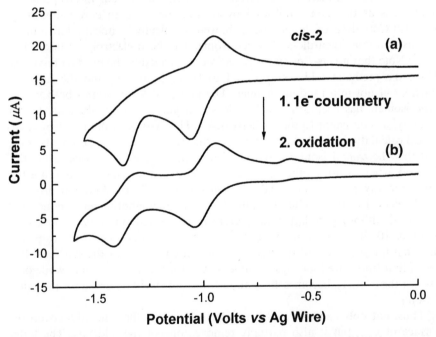

Figure 1.19 (a) CV of *cis*-2 bis adduct **19** in 0.1 M TBAPF$_6$/DCM. (b) Voltammogram recorded after a one electron/molecule CPE followed by reoxidation, clearly showing that the second reduction has become more chemically reversible. Scan rate = 100 mV/s. Reprinted with permission from Ref. 11. Copyright © 1998 American Chemical Society.

that has been reported very recently [11]. During that investigation, it was discovered that upon CPE of **19** with one electron per molecule (to form the monoanion) followed by reoxidation, the *cis*-2 isomer disappeared completely, and a mixture of other bis isomers was formed. The CVs of the "before and after 1e$^-$-electrolysis" of **19** are presented in Figure 1.19, where it can clearly be seen that the irreversible second reduction wave becomes more reversible after the electrolysis. Unquestionably, a chemical reaction must be taking place following the one-electron coulometry, which "shuffles" the adducts on the fullerene surface and leads to the other bis-adduct isomers. The product mixture, analyzed by HPLC, consisted of the *e*-isomer (57%), the *trans*-3 isomer (31%), the *trans*-4 isomer (8%), and the *cis*-3 isomer (4%). No *cis*-2 isomer, the starting material, was recovered. Most interestingly, one-electron CPE of all other synthetically available bis adducts **20–24** did not produce any isomerization, and only the starting materials were recovered. Therefore, isomeric conversion upon one-electron CPE occurs only for the *cis*-2 bis(methanofullerene) **19**.

The most remarkable observations took place during a two-electron CPE of each of the different isomers. In most cases there remained a substantial amount of current after two electrons had been transferred. Interrupting the electrolysis at this point, followed by reoxidation, column chromatography, and HPLC analysis of the products, showed practically identical relative product distributions, regardless of which isomer had been electrolyzed (see Table 1.12). Note that no *cis*-2 or *cis*-3 isomers were detected in the product mixtures. Thus, the methano adducts appeared to be migrating around the fullerene sphere, but not in a random manner. The preferred formation of both *e*- and *trans*-isomers has been attributed, at least partially, to their inherently higher stability, as compared to the *cis*-derivatives. This had been predicted by theoretical calculations [129]. Another noteworthy result of these experiments is the fact that the *trans*-1 isomer is produced in approximately 10% yield using this method, whereas using other methods it is produced in quantities that are too small for any practical use (0.8–2%) [90,130–132]. Two additional products were detected in the product mixture, the mono-methanofullerene adduct and pure C_{60}. Although the latter was present in very small amounts, a substantial amount (10–30%) of the mono adduct was formed. It is noteworthy to recognize that the 10% yield of the *trans*-1 isomer, if corrected for the statistical factor of four (there are four sites possible on C_{60} for the *trans*-2 isomer while only one for the *trans*-1), becomes 40%, almost the same yield observed for the *trans*-2 isomer.

Thus, not only can a two-electron electrolysis "shuffle" the adducts on the surface of C_{60}, but it also partially removes one of the addends. The latter process is the reversal of the Bingel reaction [130] and the subject of the next section. Although no mechanistic information was provided in that study, a detailed analysis of products obtained after submitting mixtures of selected isomers to the same electrolysis protocol described above proved that the "shuffling" occurs as an intramolecular process; that is, crossover experiments were negative [11].

TABLE 1.12 Relative Distribution of C_{60} Bis(malonate) Bis adducts After the Two-Electron Controlled Potential Electrolysis

Starting Material	cis-(1–3)	e-	trans-4	trans-3	trans-2	trans-1	Mono-	Yield
cis-2 (19)	—	23	5	8	52	12	29	49
cis-3 (20)	—	22	6	10	51	11	28	53
e- (21)	—	26	9	12	44	9	10	44
trans-4 (22)	—	23	10	12	47	8	14	63
trans-3 (23)	—	23	9	12	46	10	23	43
trans-2 (24)	—	22	6	9	53	10	19	39

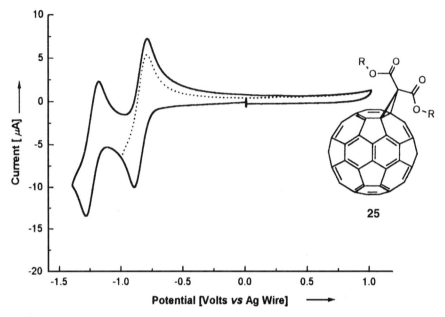

Figure 1.20 CV of **25** in 0.1 M TBAPF$_6$/DCM. The dotted and solid lines correspond to scans with switching potentials of -1.0 and 1.4 V, respectively. Reprinted with permission from Ref. 12. Copyright © 1998 Wiley–VCH Verlag GmbH.

1.8.3 Electrochemical Removal of Methano Adducts: The *Retro-Bingel* Reaction

As mentioned in the previous section, reductive electrolysis of ester-containing methanofullerenes can also result in the removal of the addends. The simple mono adduct **25**, when analyzed by CV, exhibits two reversible one-electron reduction waves as shown in Figure 1.20. When one equivalent of electrons per molecule are added by CPE at approximately the reduction potential for the first reduction, the entire amount of starting material is recovered after reoxidation; that is, no removal of the addend is observed. However, when CPE is conducted at approximately the reduction potential for the formation of the dianion, the current does not decrease to background levels after two equivalents of electrons have been added. As in the case of the bis adducts, a substantial amount of current remains. If allowed to proceed to completion so that the current decreases to background level, the total electron count is four electrons per molecule. Upon reoxidation, however, only 1.5–2 electrons are recovered. Analysis of the product shows that the material is 82% C$_{60}$. This process has been called the *Retro-Bingel* reaction [12].

The *e-* and *trans*-3 bis adducts, **21** and **23**, which had been used for the "shuffle" experiments above, were also converted cleanly and in 75% yield into pure C$_{60}$ when, instead of interrupting the CPE after two electron equivalents

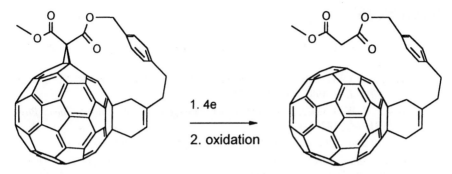

Scheme 3   A selective Retro-Bingel process.

had been added, it was allowed to continue until background level current was reached. The total charge transferred for the bis adducts in this case was approximately six electrons per molecule.

Testing the reaction with other derivatives has shown that the reaction appears to be very general. As mentioned above, a derivative of C_{70}, $C_{71}(CO_2CH_2CO_2Et)_2$, was converted in 70% yield to C_{70} upon a two-electron CPE. Other compounds, which are either not available by other routes or very difficult to prepare or isolate, have also been obtained with this procedure. These include two enantiomers of C_{76}, (^{f}A)-C_{76} and (^{f}C)-C_{76} [12], a C_2-symmetric bis(methano) adduct of C_{2v}-C_{78} and pure C_{2v}-C_{78} [30].

In view of the difficulties encountered when using traditional methods to separate and purify the higher fullerenes, the *Retro-Bingel* reaction seems to offer an excellent, easy to conduct alternative. The method consists of derivatizing the higher fullerene isomer mixture via the *Bingel* reaction [126] using chiral malonates. This is followed by chromatographic separation of the different isomers and *Retro-Bingel* to the pure, pristine higher fullerene materials.

The use of the *Retro-Bingel* reaction as a protection–deprotection tool in fullerene derivative synthesis has also been preliminarily studied for the compound shown in Scheme 3.

1.9 CONCLUSIONS AND OUTLOOK

Electrochemistry has undoubtedly represented a remarkable tool in the study of the electronic properties of fullerenes and their derivatives. Voltametric studies have been used not only to provide the experimental demonstration of predicted electronic structures, but to assess the properties and stability of many new fullerene-based materials. Furthermore, with the discovery of the "shuffle" and the *Retro-Bingel* reaction, electrochemistry has begun to play a fundamental role in the establishment of fundamental mechanistic issues. In the area of fullerene electrochemistry, the future continues to be, as in the beginning, open and very promising, with progress being limited only by human imagination.

REFERENCES

1. R. C. Haddon, L. E. Brus, and K. Raghavachari, *Chem. Phys. Lett.* **1986**, *125*, 459–464.

2. S. H. Yang, C. L. Pettiette, J. Conceicao, O. Cheshnovsky, and R. E. Smalley, *Chem. Phys. Lett.* **1987**, *139*, 233–238.

3. J. Baker, P. W. Fowler, P. Lazzeretti, M. Malagoli, and R. Zanasi, *Chem. Phys. Lett.* **1991**, *184*, 182–186.

4. J. R. Colt and G. E. Scuseria, *J. Phys. Chem.* **1992**, *96*, 10265–10268.

5. J. Cioslowski, *in Electronic Structure Calculations on Fullerenes and Their Derivatives.* Oxford University Press, New York, 1995, Chaps. 3–5 and references therein.

6. P. L. Boulas, M. Gomez-Kaifer, and L. Echegoyen, *Angew. Chem. Int. Ed.* **1998**, *37*, 216–247.

7. L. Echegoyen and L. E. Echegoyen, *Acc. Chem. Res.* **1998**, *31*, 593–601.

8. J. Chlistunoff, D. Cliffel, and A. J. Bard, *Thin Solid Films* **1995**, *257*, 166–184.

9. D. Chlistunoff, D. Cliffel, and A. J. Bard, in *Handbook of Organic Conductive Molecules and Polymers*, Vol. 1, H. S. Nalwa (ed.). Wiley, New York, 1997, pp. 333–412.

10. M. Prato, *J. Mater. Chem.* **1997**, *7*, 1097–1109 and references therein.

11. R. Kessinger, M. Gómez-López, C. Boudon, J.-P. Gisselbrecht, M. Gross, L. Echegoyen, and F. Diederich, *J. Am. Chem. Soc.* **1998**, *120*, 8545–8546.

12. R. Kessinger, J. Crassous, A. Herrmann, M. Rüttimann, L. Echegoyen, and F. Diederich, *Angew. Chem. Int. Ed.* **1998**, *37*, 1919–1922.

13. W. Krätschmer, L. D. Lamb, K. Fostiropoulus, and D. R. Huffman, *Nature* **1990**, *347*, 354–358.

14. H. Ajie, M. M. Alvarez, S. J. Anz, R. D. Beck, F. Diederich, K. Fostiropoulus, D. R. Huffman, W. Krätschmer, Y. Rubin, K. E. Schriver, D. Sensharma, and R. L. Whetten, *J. Phys. Chem.* **1990**, *94*, 8630–8633.

15. R. Taylor, J. P. Hare, A. K. Abdul-Sada, and H. W. Kroto, *J. Chem. Soc. Chem. Commun.* **1990**, 1423–1425.

16. R. D. Johnson, G. Meijer, and D. S. Bethune, *J. Am. Chem. Soc.* **1990**, *112*, 8983–8984.

17. R. E. Haufler, J. Conceicao, L. P. F. Chibante, Y. Chai, N. E. Byrne, S. Flanagan, M. M. Haley, S. C. O'Brien, C. Pan, Z. Xiao, W. E. Billups, M. A. Ciufolini, R. H. Hauge, J. L. Margrave, L. J. Wilson, R. F. Curl, and R. E. Smalley, *J. Phys. Chem.* **1990**, *94*, 8634–8636.

18. P.-M. Allemand, A. Koch, F. Wudl, Y. Rubin, F. Diederich, M. M. Alvarez, S. J. Anz, and R. L. Whetten, *J. Am. Chem. Soc.* **1991**, *113*, 1050–1051.

19. D. M. Cox, S. Bethal, M. Disko, S. M. Gorun, M. Greaney, C. S. Hsu, E. B. Kollin, J. Millar, J. Robbins, W. Robbins, R. D. Sherwood, and P. Tindall, *J. Am. Chem. Soc.* **1991**, *113*, 2940–2944.

20. D. Dubois, K. M. Kadish, S. Flanagan, R. E. Haufler, L. P. F. Chibante, and L. J. Wilson, *J. Am. Chem. Soc.* **1991**, *113*, 4364–4366.

21. D. Dubois, K. M. Kadish, S. Flanagan, and L. J. Wilson, *J. Am. Chem. Soc.* **1991**, *113*, 7773–7774.

22. Q. Xie, E. Perez-Cordero, and L. Echegoyen, *J. Am. Chem. Soc.* **1992**, *114*, 3978–3980.

23. Y. Ohsawa and T. Saji, *J. Chem. Soc. Chem. Commun.* **1992**, 781–782.

24. F. Zhou, C. Jeboulet, and A. J. Bard, *J. Am. Chem. Soc.* **1992**, *114*, 11004–11006.

25. D. Dubois, G. Moninot, W. Kutner, M. T. Jones, and K. M. Kadish, *J. Phys. Chem.* **1992**, *96*, 7137–7145.

26. F. Paolucci, M. Marcaccio, S. Roffia, G. Orlandi, F. Zerbetto, M. Prato, M. Maggini, and G. Scorrano, *J. Am. Chem. Soc.* **1995**, *117*, 6572–6580 (1995).

27. W. Koh, D. Dubois, W. Kutner, M. T. Jones, and K. M. Kadish, *J. Phys. Chem.* **1993**, *97*, 6871–6879.

28. K. Meerholz, P. Tschuncky, and J. Heinze, *J. Electroanal. Chem.* **1993**, *347*, 425–433.

29. S. M. Gorum, M. A. Greaney, D. M. Cox, R. Sherwood, C. Day, V. Day, and R. Upton, *Mater. Res. Soc. Symp. Proc.* **1991**, *206*, 659.

30. C. Boudon, J.-P. Gisselbrecht, M. Gross, A. Herrmann, M. Rüttimann, J. Crassous, F. Cardullo, L. Echegoyen, and F. Diederich, *J. Am. Chem. Soc.* **1998**, *120*, 7860–7868.

31. C. Jeboulet, A. Bard, and F. Wudl, *J. Am. Chem. Soc.* **1991**, *113*, 5456–5457.

32. Q. Xie, F. Arias, and L. Echegoyen, *J. Am. Chem. Soc.* **1993**, *115*, 9818–9819.

33. F. Diederich and R. L. Whetten, *Acc. Chem. Res.* **1992**, *25*, 119–126.

34. C. Thilgen, A. Herrmann, and F. Diederich, *Angew. Chem. Int. Ed. Engl.* **1997**, *36*, 2268–2280.

35. R. Ettl, I. Chao, F. Diederich, and R. L. Whetten, *Nature* **1991**, *353*, 149–153.

36. P. L. Boulas, M. T. Jones, R. S. Ruoff, D. C. Lorents, R. Malhotra, D. S. Tse, and K. M. Kadish, *J. Phys. Chem.* **1996**, *100*, 7573–7579 (1996).

37. Q. Li, F. Wudl, C. Thilgen, R. L. Whetten, and F. Diederich, *J. Am. Chem. Soc.* **1992**, *114*, 3994–3996.

38. J. P. Selegue, J. P. Shaw, T. F. Guarr, and M. S. Meier, Purification and Characterization of the Larger Fullerenes: New aspects of C_{76}, C_{78}, and C_{84} in *Recent Advances in the Chemistry and Physics of Fullerenes and Related Materials*, Vol. 94-24, K. M. Kadish and R. S. Ruoff (eds.). The Electrochemical Society, Pennington, NJ, 1994, pp. 1274–1291.

39. Y. Yang, F. Arias, L. Echegoyen, L. P. F. Chibante, S. Flanagan, A. Robertson, and L. Wilson, *J. Am. Chem. Soc.* **1995**, *117*, 7801–7804.

40. J. A. Azamar-Barrios, P. E. Muñoz, and A. Pénicaud, *J. Chem. Soc. Faraday Trans.* **1997**, *93*, 3119–3123.

41. J. R. Colt and G. E. Scuseria, *Chem. Phys. Lett.* **1992**, *199*, 505–512.

42. A. Herrmann and F. Diederich, *J. Chem. Soc. Perkin Trans. 2* **1997**, 1679–1684.

43. F. Diederich, R. L. Whetten, C. Thilgen, R. Ettl, I. Chao, and M. M. Alvarez, *Science* **1991**, *254*, 1768–1770.

44. T. Suzuki, K. Kikuchi, K. Oguri, Y. Nakao, S. Suzuki, Y. Achiba, K. Yamamoto, H. Funasaka, and T. Takahashi, *Tetrahedron* **1996**, *52*, 4973–4982.

45. M. R. Anderson, H. C. Dorn, S. Stevenson, P. M. Burbank, and J. R. Gibson, *J. Am. Chem. Soc.* **1997**, *119*, 437–438.

46. E. Yamamoto, M. Tansho, T. Tomiyama, H. Shinohara, Y. Kawahara, and Y. Kobayashi, *J. Am. Chem. Soc.* **1996**, *118*, 2293–2294.

47. P. B. Burbank, J. R. Gibson, H. C. Dorn, and M. R. Anderson, *J. Electroanal. Chem.* **1996**, *417*, 1–4.

48. M. S. Meier, T. F. Guarr, J. P. Selegue, and V. K. Vance, *J. Chem. Soc. Chem. Commun.* **1993**, 63–65.

49. T. J. S. Dennis, T. Kai, T. Tomiyama, and H. Shinohara, *Chem. Commun.* **1998**, 619–620.

50. M. R. Anderson, H. C. Dorn, S. A. Stevenson, and S. M. Dana, *J. Electroanal. Chem.* **1998**, *444*, 151–154.

51. P. Boulas, F. D'Souza, C. C. Henderson, P. A. Cahill, M. T. Jones, and K. M. Kadish, *J. Phys. Chem.* **1993**, *97*, 13435–13437.

52. T. F. Guarr, M. S. Meier, V. K. Vance, and M. Clayton, *J. Am. Chem. Soc.* **1993**, *115*, 9862–9863.

53. M. E. Niyazimbetov, D. H. Evans, S. A. Lerke, P. A. Cahill, and C. C. Henderson, *J. Phys. Chem.* **1994**, *98*, 13093–13098.

54. C. Caron, R. Subramanian, F. D'Souza, J. Kim, W. Kutner, M. T. Jones, and K. M. Kadish, *J. Am. Chem. Soc.* **1993**, *115*, 8505–8506.

55. R. Subramanian, K. M. Kadish, M. N. Vijayashree, X. Gao, M. T. Jones, M. D. Miller, K. L. Krause, T. Suenobu, and S. Fukuzumi, *J. Phys. Chem.* **1996**, *100*, 16327–16335.

56. F. D'Souza, C. Caron, R. Subramanian, W. Kutner, M. T. Jones, and K. M. Kadish, in *Recent Advances in the Chemistry and Physics of Fullerenes and Related Materials*, Vol. 94–24, K. M. Kadish and R. S. Ruoff (eds.). The Electrochemical Society, Pennington, NJ, 1994, pp. 768–778.

57. C. Boudon, J.-P. Gisselbrecht, M. Gross, L. Isaacs, H. L. Anderson, R. Faust, and F. Diederich, *Helv. Chim. Acta* **1995**, *78*, 1334–1344.

58. F. Cardullo, P. Seiler, L. Isaacs, J.-F. Nierengarten, R. F. Haldimann, F. Diederich, T. Mordasini-Denti, W. Thiel, C. Boudon, J.-P. Gisselbrecht, and M. Gross, *Helv. Chim. Acta* **1997**, *80*, 343–371.

59. J.-F. Nierengarten, T. Habicher, R. Kessinger, F. Cardullo, V. Gramlich, J.-P. Gisselbrecht, C. Boudon, and M. Gross, *Helv. Chim. Acta* **1997**, *80*, 2238–2276.

60. F. Zhou, G. J. Van Berkel, and B. T. Donovan, *J. Am. Chem. Soc.* **1994**, *116*, 5485–5486.

61. N. Liu, H. Touhara, Y. Morio, D. Komichi, F. Okino, and S. Kawasaki, *J. Electrochem. Soc.* **1996**, *143*, L214–L217.

62. N. Liu, H. Touhara, F. Okino, and S. Kawasaki, *J. Electrochem. Soc.* **1996**, *143*, 2267–2272.

63. T. Suzuki, Y. Maruyama, T. Akasaka, W. Ando, K. Kobayashi, and S. Nagase, *J. Am. Chem. Soc.* **1994**, *116*, 1359–1363.

64. M. Ferduco, D. A. Costa, A. L. Balch, and W. R. Fawcett, *Angew. Chem. Int. Ed. Engl.* **1995**, *34*, 194–196.

65. K. Winkler, D. A. Costa, A. L. Balch, and W. R. Fawcett, *J. Phys. Chem.* **1995**, *99*, 17431–17436.

66. P. J. Fagan, J. C. Calabrese, and B. Malone, *Acc. Chem. Res.* **1992**, *25*, 134–142.

67. A. L. Balch and M. Olmstead, *Chem. Rev.* **1998**, *98*, 2123–2165.

68. S. A. Lerke, B. A. Parkinson, D. H. Evans, and P. J. Fagan, *J. Am. Chem. Soc.* **1992**, *114*, 7807–7813.

69. S. A. Lerke, D. H. Evans, and P. J. Fagan, *J. Electroanal. Chem.* **1995**, *383*, 127–132.

70. R. S. Koefod, C. Xu, W. Lu, J. R. Shapley, M. Hill, and K. R. Mann, *J. Phys. Chem.* **1992**, *96*, 2928–2930.

71. L. I. Denisovich, S. M. Peregudova, A. V. Usatov, A. L. Sigan, and Y. N. Novikov, *Russ. Chem. Bull.* **1997**, *46*, 1251–1257.

72. T. Suzuki, Q. Li, K. C. Khemani, and F. Wudl, *J. Am. Chem. Soc.* **1992**, *114*, 7301–7302.

73. L. Isaacs, A. Wehrsig, and F. Diederich, *Helv. Chim. Acta* **1993**, *76*, 1231–1250.

74. J. Osterodt, M. Nieger, P.-M. Windschief, and F. Vögtle, *Chem. Ber.* **1993**, *126*, 2331–2336.

75. A. B. Smith III, R. M. Strongin, L. Brard, G. T. Furst, W. J. Romanow, K. G. Owens, and R. C. King, *J. Am. Chem. Soc.* **1993**, *115*, 5829–5830.

76. L. Isaacs and F. Diederich, *Helv. Chim. Acta* **1993**, *76*, 2454–2464.

77. R. J. Taylor, *J. Chem. Soc. Perkin Trans.* **1992**, 3–6.

78. F. Wudl, *Acc. Chem. Res.* **1992**, *25*, 157–161.

79. M. Eiermann, F. Wudl, M. Prato, and M. Maggini, *J. Am. Chem. Soc.* **1994**, *116*, 8364–8365.

80. F. Arias, L. Echegoyen, S. R. Wilson, Q. Lu, and Q. Lu, *J. Am. Chem. Soc.* **1995**, *117*, 1422–1427.

81. F. Arias, Q. Xie, Y. Wu, Q. Lu, S. R. Wilson, and L. Echegoyen, *J. Am. Chem. Soc.* **1994**, *116*, 6388–6394.

82. T. Suzuki, Q. Li, K. C. Khemani, F. Wudl, and Ö. Almarsson, *Science* **1991**, *254*, 1186–1188.

83. K. M. Keshavarz, B. Knight, R. C. Haddon, and F. Wudl, *Tetrahedron* **1996**, *52*, 5149–5159.

84. M. Eiermann, R. C. Haddon, B. Knight, Q. C. Li, M. Maggini, N. Martin, T. Ohno, M. Prato, T. Suzuki, and F. Wudl, *Angew. Chem. Int. Ed. Engl.* **1995**, *34*, 1591–1594.

85. T. Ohno, N. Martin, B. Knight, F. Wudl, T. Suzuki, and H. Yu, *J. Org. Chem.* **1996**, *61*, 1306–1309.

86. B. Knight, N. Martin, T. Ohno, E. Orti, C. Rovira, J. Veciana, J. Vidal-Gancedo, P. Viruela, R. Viruela, and F. Wudl, *J. Am. Chem. Soc.* **1997**, *119*, 9871–9882.

87. D. M. Guldi, H. Hungerbühler, and K.-D. Asmus, *J. Phys. Chem.* **1995**, *99*, 9380–9385.

88. H. L. Anderson, C. Boudon, F. Diederich, J.-P. Gisselbrecht, M. Gross, and P. Seiler, *Angew Chem. Int. Ed. Engl.* **1994**, *33*, 1628–1631.

89. T. Suzuki, Q. Li, K. C. Khemani, F. Wudl, and Ö. Almarsson, *J. Am. Chem. Soc.* **1992**, *114*, 7300–7301.

90. J.-P. Bourgeois, L. Echegoyen, M. Fibbioli, E. Pretsch, and F. Diederich, *Angew. Chem. Int. Ed.* **1998**, *37*, 2118–2121.

91. E. Dietel, A. Hirsch, J. Zhou, and A. Rieker, *J. Chem. Soc. Perkin Trans. 2* **1998**, 1357–1364.

92. E. Dietel, A. Hirsch, E. Eichhorn, A. Rieker, S. Hackbarth, and B. Röder, *Chem. Commun.* **1998**, 1981–1982.

93. J.-P. Bourgeois, F. Diederich, L. Echegoyen, and J.-F. Nierengarten, *Helv. Chim. Acta* **1998**, *81*, 1835–1844.

94. M. Prato, Q. C. Li, F. Wudl, and V. Lucchini, *J. Am. Chem. Soc.* **1993**, *115*, 1148–1150.

95. C. J. Hawker, P. M. Seville, and J. W. White, *J. Org. Chem.* **1994**, *59*, 3503–3505.

96. T. Ishida, K. Tanaka, and T. Nogami, *Chem. Lett.* **1994**, 561–562.

97. C. J. Hawker, K. L. Wooley, and J. M. J. Fréchet, *J. Chem. Soc. Chem. Commun.* **1994**, 925–926.

98. S. Kuwashima, M. Kubota, K. Kushida, T. Ishida, M. Ohashi, and T. Nogami, *Tetrahedron Lett.* **1994**, *35*, 4371–4374.

99. J. Zhou, A. Rieker, T. Grösser, A. Skiebe, and A. Hirsch, *J. Chem. Soc. Perkin Trans. 2*, **1997**, 1–5.

100. D. M. Guldi, M. Maggini, G. Scorrano, and M. Prato, *J. Am. Chem. Soc.* **1997**, *119*, 974–980, and references therein.

101. M. Maggini, A. Karlsson, G. Scorrano, G. Sandonà, G. Farnia, and M. Prato, *J. Chem. Soc. Chem. Commun.* **1994**, 589–590.

102. M. Prato, M. Maggini, C. Giacometti, G. Scorrano, G. Sandonà, and G. Farnia, *Tetrahedron* **1996**, *52*, 5221–5234.

103. M. Prato and M. Maggini, *Acc. Chem. Res.* **1998**, *31*, 519–525.

104. H. Imahori and Y. Sakata, *Adv. Mater. Res.* **1997**, *9*, 537–546.

105. K. Dürr, S. Fiedler, T. Linssen, A. Hirsch, and M. Hanack, *Chem. Ber.* **1997**, *130*, 1375–1378.

106. T. Linssen, K. Dürr, M. Hanack, and A. Hirsch, *J. Chem. Soc. Chem. Commun.* **1995**, 103–104.

107. M. G. Ranasinghe, A. M. Oliver, D. F. Rothenfluh, A. Salek, and M. N. Paddon-Row, *Tetrahedron Lett.* **1996**, *37*, 4797–4800.

108. M. Diekers, A. Hirsch, S. Pyo, J. Rivera, and L. Echegoyen, *Eur. J. Org. Chem.*, **1998**, 1111–1121.

109. B. Illescas, N. Martín, and C. Seoane, *Tetrahedron Lett.* **1997**, *38*, 2015–2018.

110. M. Iyoda, F. Sultana, S. Sasaki, and M. Yoshida, *J. Chem. Soc. Chem. Commun.* **1994**, 1929–1930.

111. M. Maggini, D. M. Guldi, S. Mondini, G. Scorrano, F. Paolucci, P. Ceroni, and S. Roffia, *Chem. Eur. J.* **1998**, *4*, 1992–2000.

112. B. Illescas, N. Martín, and C. Seoane, *Tetrahedron Lett.* **1995**, *36*, 8307–8310.

113. D. H. Evans and S. Lerke, Reversible Reduction Potentials of Some New Organo Fullerenes in *Recent Advances in the Chemistry and Physics of Fullerenes and Related Materials*, Vol. 94-24, K. M. Kadish and R. S. Ruoff (eds.). The Electrochemical Society, Pennington, NJ, 1994, pp. 1087–1097.

114. N. S. Sariciftci, F. Wudl, A. J. Heeger, M. Maggini, G. Scorrano, M. Prato, J. Bourassa, and P. C. Ford, *Chem. Phys. Lett.* **1995**, *247*, 510–514.

115. P. Timmerman, L. E. Witschel, F. Diederich, C. Boudon, J.-P. Gisselbrecht, and M. Gross, *Helv. Chim. Acta* **1996**, *79*, 6–20.

116. G.-W. Wang, K. Komatsu, Y. Murata, and M. Shiro, *Nature* **1997**, *387*, 583–586.

117. J. C. Hummelen, B. Knight, J. Pavlovich, R. Gonzalez, and F. Wudl, *Science* **1995**, *269*, 1554–1556.

118. A. L. Balch, D. A. Costa, W. R. Fawcett, and K. Winkler, *J. Phys. Chem.* **1996**, *100*, 4823–4827.

119. M. S. Meier, M. Poplawska, A. L. Compton, J. P. Shaw, J. P. Selegue, and T. F. Guarr, *J. Am. Chem. Soc.* **1994**, *116*, 7044–7048.

120. J.-F. Nierengarten, A. Herrmann, R. R. Tykwinski, M. Rüttimann, F. Diederich, C. Boudon, J.-P. Gisselbrecht, and M. Gross, *Helv. Chim. Acta* **1997**, *80*, 293–316.

121. D. E. Cliffel and A. J. Bard, *J. Phys. Chem.* **1994**, *98*, 8140–8143.

122. M. E. Niyazymbetov, D. E. Evans, S. A. Lerke, P. A. Cahill, and C. C. Henderson, *J. Phys. Chem.* **1994**, *98*, 13093–13098.

123. M. E. Niyazymbetov and D. H. Evans, *J. Electrochem. Soc.* **1995**, *142*, 2655–2658.

124. K.-M. Mangold, W. Kutner, L. Dunsch, and J. Fröhner, *Synth. Methods* **1996**, *77*, 73–76.

125. P. L. Boulas, Y. Zuo, and L. Echegoyen, *Chem. Commun.* **1996**, 1547–1548.

126. F. Arias, P. Boulas, Y. Zuo, O. Dominguez, M. Gomez-Kaifer, and L. Echegoyen, Synthesis and Electrosynthesis of Methano [60] Fullerenes, Bis-Aza-Fulleroid Crown Ethers, and Phenanthrolyl [60] Fullerene in *Recent Advances in the Chemistry and Physics of Fullerenes and Related Materials*, Vol. 3, K. M. Kadish and R. S. Ruoff (eds.), The Electrochemical Society, Pennington, NJ, 1996, pp. 165–176.

127. S. Fukuzumi, T. Suenobu, T. Hirasaka, R. Arakawa, and K. M. Kadish, *J. Am. Chem. Soc.* **1998**, *120*, 9220–9227.

128. F. Cardullo, P. Seiler, L. Isaacs, J.-F. Nierengarten, R. F. Haldimann, F. Diederich, T. Mordasini-Denti, W. Thiel, C. Boudon, J.-P. Gisselbrecht, and M. Gross, *Helv. Chim. Acta* **1997**, *80*, 343–371.

129. A. Hirsch, I. Lamparth, and H. R. Karfunkel, *Angew. Chem. Int. Ed. Engl.* **1994**, *33*, 437–438.

130. C. Bingel, *Chem. Ber.* **1993**, *126*, 1957–1959.

131. B. Kräutler, T. Müller, J. Maynollo, K. Gruber, C. Krattky, P. Ochsenbein, D. Schwarzenbach, and H.-B. Bürgi, *Angew. Chem. Int. Ed. Engl.* **1996**, *35*, 1204–1206.

132. Y. Rubin, *Chem. Eur. J.* **1997**, *3*, 1009–1016.

CHAPTER 2

SOLUBILITY OF THE FULLERENES

MIKHAIL V. KOROBOV and ALLAN L. SMITH

2.1 INTRODUCTION

Although C_{60}, buckminsterfullerene, was discovered in 1985 [1], it was not until Krätschmer, Lamb, Fostiropoulos, and Huffman [2] discovered in 1990 that C_{60} was soluble in benzene that the scientific literature exploded with experimental studies of these remarkable molecules. (Fullerenes must be efficiently dissolved in order to extract them from soot and functionalize them by organic reactions.) The dependence of fullerene solubility on both temperature and organic structure of the solvent must be understood in order to separate different members of the fullerene family from each other and from their precursors or derivatives.

The practical problem of fullerene solubility has partially been solved just by collecting experimental data. The solubility of C_{60} is known in almost 150 solvents, probably the highest number among all chemical substances. Effective, relatively inexpensive, and readily available solvents were soon found: aromatic solvents (toluene, 1,2-dimethylbenzene, and 1,2-dichlorobenzene) and carbon disulfide. However, problems familiar to physical chemists studying solubility phenomena for years have remained to confront fullerene researchers. There is still no good theory to explain or to predict absolute values of fullerene solubility and changes in solubility when changing the solvent or the fullerene itself. This serious restriction is one of the reasons why separating fullerenes from each other on a large scale without chromatography is still difficult.

In this chapter, we present quantitative solubility data and studies of solubility phenomena by methods of physical chemistry. We start with a summary of all data on room temperature solubility of C_{60} and C_{70} published through early 1998. As far as we can tell, published quantitative solubility data on other fullerenes or on fullerene derivatives do not exist, except for the water solubilities of some fullerene derivatives. Next, we discuss trends in fullerene solu-

Fullerenes: Chemistry, Physics, and Technology, Edited by Karl M. Kadish and Rodney S. Ruoff.
0-471-29089-0 Copyright © 2000 John Wiley & Sons, Inc.

bilities and different correlations between solubilities and physico chemical properties of solvents. In assessing these trends, we have paid special attention to thermodynamic considerations and models. Separate sections are devoted to the solubility of C_{60} in water, the formation of charge-transfer complexes, the use of reversible functionalization or electrochemical reduction of fullerenes to promote solubilization, and the role of aggregation of fullerenes in liquid solutions.

We then describe the peculiarities in the temperature dependence of the solubility of fullerenes. Both the abnormal decrease of solubility with temperature and the occurrence of temperatures of maximum solubility, a puzzle for some years, have been explained from basic thermodynamic equations. This simple theory is thus presented. Since solid solvates of fullerenes with the solvents played an important part in the explanations given, their thermodynamic and structural properties are also reviewed. Indeed, the first published photograph of "solid C_{60}," on the cover of the 1990 issue of *Nature* in which the article by Krätschmer et al. appeared, is not really of pure C_{60} crystals but of a benzene solvate of composition $C_{60} \cdot 4C_6H_6$.

2.2   SOLUBILITY OF C_{60} AND C_{70}

Solubilities of C_{60} at 298 K are presented in Table 2.1. In the preparation of this table, the corresponding table from the review paper of Beck and Mandi [3] was used, but more recent data and comments have been added. Results from different laboratories are in reasonably good agreement, although measurements of the solubility of C_{60} have been described by several authors as difficult and sometimes irreproducible. Typical methods of determining the solubility include stirring of the samples for a period of time from 5 to 48 hours and determining the saturated concentration by ultraviolet/visible spectrophotometry [4,5], use of high-performance liquid chromatography (HPLC) [6], or by simple weighing [7,8]. Because molar absorptivities in the ultraviolet (UV) are much greater than those in the visible, the UV spectrum has been used when solubilities are low and the visible spectrum when solubilities are larger. The Lambert–Beer law was satisfied for C_{60} concentrations up to the saturation limit [9]. More careful measurements [10] demonstrated the necessity of longer stirring in some cases (e.g., up to 20 days for C_{60} in CS_2) to reach a time-independent saturated concentration.

The solvents in Table 2.1 do not form covalently bonded chemical compounds with C_{60} at room temperature. Only slight shifts in infrared (IR), UV, or nuclear magnetic resonance (NMR) spectra due to interaction of C_{60} with the solvent were found. For some fullerene–solvent systems, dissolution of the solid is accompanied by charge transfer or even chemical reaction of the fullerene. These systems are discussed later.

Column 4 of Table 2.1 shows the solid phase in equilibrium with the saturated liquid solution at $T = 298$ K. As is shown later, for many fullerene–

TABLE 2.1   Solubility of C_{60} in Organic Solvents at $T = 298\,K$

| Solvent | Solubility of C_{60} | | Equilibrium Solid Phase mol C_{60}:mol Solvent | Reference |
	$10^4 \times$ Mole Fraction	$10^3 \times$ Molarity		
AROMATIC HYDROCARBONS				
Benzene	2.11	2.36	1:4	6
	1.86	2.08	1:4	7
	1.79	2.00	1:4	4[a]
	1.09	1.22	1:4	3
	1.73	1.94	1:4	11[a]
	2.30	2.58	1:4	12
Toluene	4.14	3.89	C_{60}	6
	4.29	4.03	C_{60}	7
	3.38	3.18	C_{60}	10
	3.18	2.99	C_{60}	4
	3.34	3.15	C_{60}	13
	4.29	4.03	C_{60}	5
	3.54	3.33	C_{60}	11[a]
	4.71	4.44	C_{60}	12
1,2-Dimethylbenzene	14.7	12.1	1:2	7
	12.4	10.2	1:2	14
	15.7	12.9	1:2	5
1,3-Dimethylbenzene	2.38	1.94	1:2(?)	7
	4.81	3.93	1:2(?)	14
1,4-Dimethylbenzene	10.0	8.19		7
	5.34	4.36		14
1,2,3-Trimethylbenzene	8.77	6.53		7
1,2,4-Trimethylbenzene	33.5	24.9		7
1,3,5-Trimethylbenzene	2.90	2.08	1:0.5(?)	6
	1.93	1.38		4
	3.28	2.36		7
1,2,3,4-Tetramethylbenzene	12.0	8.06		7
1,2,3,5-Tetramethylbenzene	43.3	29.0		7
Tetralin	30.1	22.2		6
	27.9	20.3		15
	29.7	21.8		11[a]
Ethylbenzene	4.42	3.61		7
	3.67	3.00		14
n-Propylbenzene	2.9	2.08		7
Isopropylbenzene	2.32	1.67		7
n-Butylbenzene	4.12	2.64		7
sec-Butylbenzene	2.38	1.53		7
tert-Butylbenzene	1.93			7
Fluorobenzene	0.77	0.82		6
	1.56	1.67		7

TABLE 2.1 (continued)

Solvent	Solubility of C_{60}		Equilibrium Solid Phase mol C_{60}:mol Solvent	Reference
	$10^4 \times$ Mole Fraction	$10^3 \times$ Molarity		
Chlorobenzene	9.88	9.72		6
	8.04	7.92		7
Bromobenzene	4.82	4.58	1:2	6
	4.09	3.89	1:2	7
Iodobenzene	3.26	2.92		7
1,2-Dichlorobenzene	42.0	37.5	1:2	6
	38.3	34.2	1:2	7
	35.1	31.8	1:2	15
	36.6	32.5	1:2	16
1,2-Dibromobenzene	23.1	19.2		7
1,3-Dichlorobenzene	3.80	3.33	1:2	7
	7.98	6.99	1:2	16
1,3-Dibromobenzene	23.1	19.2		7
1,2,4-Trichlorobenzene	14.7	11.8		6
	18.0	14.4		7
	8.4	6.73		8
	36.7	29.6		12
Styrene	5.97	5.21		17
o-Cresol	0.02	0.02		6
Benzonitrile	0.58	0.57		6
Nitrobenzene	1.14	1.11		6
Anisole	8.45	7.78		6
	10.1	9.3		11[a]
p-Bromoanisole	29.9	23.3		11[a]
m-Bromoanisole	28.4	22.5		11
Benzaldehyde	0.59	0.54		14
Phenyl isocyanate	3.68	3.39		14
Thiophenol	9.84	9.60		17
1-Methyl 2-nitrobenzene	3.98	3.38		14
1-Methyl,3-nitrobenzene	3.88	3.28		14
Benzyl chloride	3.83	3.33		14
Benzyl bromide	8.15	6.86		14
1,1,1-Trichloromethylbenzene	9.43	6.67		14
1-Methylnaphthalene	64.7	45.8		6
	65.1	46.1		7
1-Phenylnaphthalene	129	69.4		6
1-Chloronaphthalene	95.6	70.8		6
1-Bromo,2-methylnaphthalene	74.8	48.3		7

TABLE 2.1 (continued)

Solvent	Solubility of C_{60}		Equilibrium Solid Phase mol C_{60}:mol Solvent	Reference
	$10^4 \times$ Mole Fraction	$10^3 \times$ Molarity		
ALKANES				
n-Pentane	0.008	0.007	1:1	6
	0.006	0.006	1:1	4
	0.005	0.004	1:1	18
	.011	.010	1:1	12
n-Hexane	0.078	0.060	1:1	6
	0.073	0.056	1:1	4
	0.066	0.051	1:1	18
	0.095	0.072	1:1	13
	0.084	0.064	1:1	12
2-Methylpentane	0.032	0.026		18
3-Methylpentane	0.045	0.035		18
n-Heptane	0.098	0.067	1:1	18
	0.609	0.42	1:1	11[a]
n-Octane	0.056	0.035	1:1	4
	0.045	0.028	1:1	18
	0.68	0.42	1:1	11[a]
	0.057	0.035	1:1	12
Isooctane	0.059	0.036		4
	0.064	0.039		12
n-Nonane	0.154	0.086		18
	0.084	0.047		12
n-Decane	0.192	0.099		6
	0.189	0.097		4
	0.195	0.100		12
Dodecane	0.287	0.126		4
	0.325	0.143		12
Tetradecane	0.455	0.175		4
	0.607	0.233		12
CYCLIC ALKANES				
Cyclopentane	0.003	0.003		6
Cyclohexane	0.054	0.050	1:13	6
	0.077	0.071	1:13	4
	0.054	0.050	1:13	14
	0.054	0.050	1:13	13
	0.052	0.048	1:13	19
	0.081	0.075	1:13	12
Cyclohexene	1.66	1.68		17
1-Methyl,1-cyclohexene	1.69	1.43		17
Methylcyclohexane	0.31	0.24		17
1,2-Dimethylcyclohexane, mixture of *cis* and *trans*	0.26	0.18		17

TABLE 2.1 (continued)

Solvent	Solubility of C_{60}		Equilibrium Solid Phase mol C_{60}:mol Solvent	Reference
	$10^4 \times$ Mole Fraction	$10^3 \times$ Molarity		
Ethylcyclohexane	0.49	0.35		17
3:7 Mixture of *cis* and *trans* decalins	9.8	6.39		6
	4.2	2.6		20
cis-Decalin	4.7	3.1		6
trans-Decalin	2.9	1.8		6
HALOALKANES				
Dichloromethane	0.23	0.36		6
	0.23	0.35		4
	0.20	0.32		14
Trichloromethane	0.17	0.22		6
	0.19	0.24		17
	0.57	0.71		12
Tetrachloromethane	0.43	0.44	1:13	6
	0.60	0.62	1:13	4
	0.14	0.14	1:13	21[b]
Dibromomethane	0.35	0.50		14
Tribromomethane	6.83	7.83		14
Iodomethane	0.67	1.07		14
Diiodomethane	0.18	0.17		14
Bromochloromethane	0.68	1.04		17
Bromoethane	0.07	0.10		17
Iodoethane	0.31	0.39		17
Trichloroethylene	1.74	1.94	1:1	6
Tetrachloroethylene	1.70	1.67		6
Dichlorodifluoroethane	0.04	0.03		6
1,1,2-Trichlorotrifluoroethane	0.02	0.02		6
1,1,2,2-Tetrachloroethane	7.78	7.36		6
1,2-Dibromoethylene	2.16	2.61		14
1,2-Dichloroethane	0.09	0.12		14
1,2-Dibromoethane	0.60	0.69		6
	0.65	0.75		14
1,1,1-Trichloroethane	0.21	0.21		14
1-Chloropropane	0.03	0.03		17
1-Bromopropane	0.07	0.08		14
1-Iodopropane	0.24	0.25		14
2-Iodopropane	0.17	0.17		14
1,2-Dichloropropane	0.14	0.14		14
1,3-Dichloropropane	0.15	0.15		14
($\pm$)-1,2-Dibromopropane	0.52	0.50		14
1,3-Dibromopropane	0.57	0.56		17

TABLE 2.1 (continued)

Solvent	Solubility of C_{60}		Equilibrium Solid Phase mol C_{60}:mol Solvent	Reference
	$10^4 \times$ Mole Fraction	$10^3 \times$ Molarity		
1,3-Diiodopropane	4.41	3.84		14
1,2,3-Trichloropropane	1.15	1.08		14
1,2,3-Tribromopropane	11.4	9.74		14
1-Chloro-2-methylpropane	0.04	0.04		14
1-Bromo-2-methylpropane	0.13	0.12		17
1-Iodo-2-methylpropane	0.54	0.47		14
2-Chloro-2-methylpropane	0.02	0.014		17
2-Bromo-2-methylpropane	0.09	0.083		17
2-Iodo-2-methylpropane	0.39	0.32		17
Bromobutane	1.79	1.67		11[a]
Cyclopentyl bromide	0.62	0.57		14
Cyclohexyl chloride	0.87	0.74		14
Cyclohexyl bromide	3.77	3.06		14
Cyclohexyl iodide	14.5	11.2		14
Bromoheptane	5.01	3.19		11[a]
Bromooctane	8.22	4.72		11[a]
1-Bromotetradecane	23.4	8.6		11[a]
1-Bromooctadecane	27.6	8.6		11[a]
ALCOHOLS				
Methanol	0.00002	0.000046		18
Ethanol	0.001	0.0014		6
	0.00082	0.0011		18
1-Propanol	0.004	0.0057		18
1-Butanol	0.012	0.013		18
1-Pentanol	0.045	0.042		18
1-Hexanol	0.073	0.058		18
1-Octanol	0.103	0.065		18
2-Propanol	0.0023	0.0029		18
2-Butanol	0.0046	0.0050		18
2-Pentanol	0.027	0.025		18
3-Pentanol	0.043	0.040		18
1,3-Propandiol	0.00094	0.0013		18
1,4-Butandiol	0.0024	0.0030		18
1,5-Pentandiol	0.0064	0.0061		18
OTHER POLAR SOLVENTS				
Nitromethane	0.000	0.000		6
	0.15	0.3		11[a]
Nitroethane	0.002	0.003		6
Acetone	0.001	0.001		6
Acetonitrile	0.000	0.000		6
Acrylonitrile	0.0036	0.0055		17

TABLE 2.1 (continued)

Solvent	Solubility of C_{60}		Equilibrium Solid Phase mol C_{60}:mol Solvent	Reference
	$10^4 \times$ Mole Fraction	$10^3 \times$ Molarity		
n-Butylamine	5.06	5.12		22
2-Methyloxyethyl ether	0.064	0.044		14
N,N-dimethylformamide	0.029	0.038		14
Dioxane	0.049	0.057		4
Water	3.2×10^{-22}	1.8×10^{-21}		18
MISCELLANEOUS				
Carbon disulfide	6.60	11.0		6
	4.31	7.17		4
	6.43	10.7		5
	6.26	10.4		23
	9.87	16.4		12
Thiophene	0.44	0.56		7
	0.27	0.33		22
Tetrahydrofuran	0.07	0.08		6
	0.6	0.8		11[a]
	0.04	0.05		12
2-Methylthiophene	9.13	9.44		6
Pyridine	1.00	1.24		6
	0.34	0.42		7
	0.33	0.42		11[a]
Piperidine	72	74		11[a]
2,4,6-Trimethylpyridine	15.9	12.1		11[a]
Pyrrolidine	54	66		11[a]
N-Methyl-2-pyrrolidone	1.19	1.24		6
Tetrahydrothiophene	0.04	0.042		6
	0.14	0.158		22
Quinoline	11.8	10.0		7

[a] All measurements taken at 303 K.
[b] Measurement taken at 291 K.

solvent systems, this phase is a solid solvate rather than pristine C_{60}. For other cases, this equilibrium solid phase has never been identified.

The solubilities of C_{70} at 298 K are presented in Table 2.2.

2.3 SOLUBILITY OF C_{60} AND SOME DERIVATIVES IN WATER

C_{60} is an extremely hydrophobic molecule. Getting fullerenes to dissolve in water and other polar solvents is challenging because of this hydrophobicity. By measuring C_{60} solubility in a series of small straight-chain alcohols and

TABLE 2.2 Solubility of C_{70} at $T = 298$ K

| Solvent | Solubility of C_{70} | | Equilibrium Solid Phase | Reference |
	mole fraction $\times 10^4$	Mole/L $\times 10^3$		
Toluene	1.35	1.27	1:1	10
	1.80	1.70	1:1	5
	1.77	1.67	1:1	24
Benzene	1.38	1.55		24
1,2-Dimethylbenzene	22.4	18.6	1:3(?)	5
1,4-Dimethylbenzene	5.83	4.74		24
1,3,5-Trimethylbenzene	2.43	1.75		24
1,2-Dichlorobenzene	34.8	31.0	1:3(?)	15
	48.3	43.1	1:3(?)	24
	40.2	35.7		16
1,3-Dichlorobenzene	25.4	22.3		16
n-Pentane	0.0028	0.0024		24
n-Hexane	0.020	0.015		24
n-Heptane	0.082	0.056		24
n-Octane	0.082	0.050		24
n-Decane	0.12	0.063		24
Dodecane	0.27	0.12		24
Cyclohexane	0.10	0.095		24
Acetone	0.00098	0.0023		24
Isopropanol	0.0019	0.0025		24
Carbon tetrachloride	0.14	0.14		24
Carbon disulfide	7.07	11.8		24
	10.9	18.2		15
Tetralin	20.1	14.6		24
Dichloromethane	0.061	0.095		24
Water	2.8×10^{-21}	1.6×10^{-10}		18

extrapolating to zero carbon content, Heymann [18] has estimated the mole fraction of C_{60} in a saturated aqueous solution to be approximately 3×10^{-22} and an order of magnitude larger for C_{70}.

The interest in water solubility increased dramatically after the bioactivity of C_{60} was reported [25–27], including the assumption of complementarity of the C_{60} with the active site of the human immunodeficiency protein (HIV) [26] and preparation of a diamido diacid diphenyl fulleroid derivative, designed specifically to inhibit an HIV enzyme [27].

Several methods of solubilization were developed. Andrievsky et al. [28] have reported the formation of stable, finely dispersed aqueous colloidal solutions with particle sizes of approximately 0.22 μm and concentrations of 7×10^{-6} mol/L of a mixture with the C_{60}/C_{70} ratio 10:7. The colloidal solution was prepared from the two-phase system C_{60}–toluene and water by ultrasonic treatment for several hours until the toluene was completely removed. The solution was unaffected by boiling and changing of pH. The reverse extraction of fullerenes by toluene was not possible. The preparation of a water suspension of C_{60} has also been described by Scrivens et al. [29].

Another approach to solubilization employs the formation of host–guest type 1:1 and 1:2 complexes of fullerene C_{60} with γ-cyclodextrin [30–34], which are soluble in water. A true magenta-colored solution could be prepared by different methods, the simplest one consisting of a ball-milled mixture of C_{60} and γ-cyclodextrin [32] in water. Andersson et al. [33] estimated the solubility of C_{60} to be about 10^{-4} mol/L. They also reported some selectivity of inclusion: the host–guest complex was readily formed with C_{60}, but not with C_{70}. The formation of host–guest complexes of C_{60} with calixarenes has been used by Atwood et al. [35] to purify fullerene mixtures.

It is also possible to design water-soluble derivatives or copolymers of fullerenes [27,36–41]. The idea is to introduce functional groups with the appropriate structure in order to make the resulting compound hydrophilic. Attachment of only a single hydrophilic addend promotes water solubility, but strong attractive forces remain between the fullerene units resulting in the formation of colloidal fullerene clusters. Formation of these clusters can be prevented by increasing the number of addends at the fullerene core [40].

Tomioka et al. [36] synthesized fullerene derivatives bearing carboxy groups (e.g., $1',1'$-bis(4-carboxymethylphenyl)-1,2-methano-fullerene C_{60}). The room temperature solubility of the named compound in a tetrahydrofuran/water mixture (4:1, by volume) was $(1.4–1.8) \times 10^{-2}$ mol(C_{60})/L. Even small structural changes in the derivative result in an unexpectedly large negative effect on solubility. The solubility also decreased sharply as the ratio of THF/water was decreased. Comparable water solubilities can be reached with certain copolymers [41].

Surfactants have been employed successfully to solubilize both C_{60} and C_{70} in aqueous media [42–44]. With polyvinylpyrrolidone, the achieved room temperature solubilities were 5.6×10^{-4} and 2.4×10^{-4} mol/L for C_{60} and C_{70} respectively [42]. These numbers cannot be considered as true solubilities since micelle solutions were formed.

2.4   FORMATION OF FULLERENE–SOLVENT CHARGE-TRANSFER COMPLEXES IN SOLUTION

Because C_{60} is a good electron pair acceptor (Lewis acid), it can form charge-transfer complexes with electron pair donors (Lewis bases) [45–47]. Dissolution of fullerenes in tertiary amines and substituted anilines is accompanied by the formation of such donor–acceptor complexes in the liquid state as evidenced by the appearance of new visible and near-infrared absorption bands. Stability constants have been determined for these complexes using the Benesi–Hildebrand plot [48–53]. The room temperature solubilities of C_{60} in aniline, N-

methylaniline, and N,N'-dimethylaniline were found to be 1.05, 1.16, 3.89 mg/mL, respectively. These solubilities definitely correspond to saturation relative to the solid donor–acceptor complexes rather than the pure fullerene. Complexation with aniline was applied to separate C_{60} from different endohedral complexes [54].

2.5 SOLUBILIZING FULLERENES THROUGH CHEMICAL REACTIONS

The solubility of fullerenes is strongly changed by functionalizing them. One strategy for extracting C_{60} and other fullerenes from graphitic soot has been to form derivatives through chemical reactions, which can subsequently be reversed to recover the pure fullerenes. The charge-transfer complexes of the previous section are an example of this method, but there are other examples. In the presence of a catalyst (e.g., $AlCl_3$), the Friedel–Crafts reaction takes place [55]. Nie et al. [56] realized that the kinetic instability of Diels–Alder adducts of C_{60} [57] can be exploited to develop a method of purifying fullerenes without chromatography. They found [58] that cyclopentadiene-functionalized silica gel reacted quickly with C_{60} at room temperature, but the reverse reaction liberating C_{60} occurs readily at $100\,^{\circ}C$. Yi-Zhong et al. [59] have proposed a strategy involving a readily removable, polar solubilizing functional group to carry out syntheses of fullerene derivatives with intrinsically low solubilities.

Diener and Alford [60] recognized that the computed HOMO–LUMO gaps of fullerenes provide a way to classify and predict solubility. Those fullerenes that have relatively large HOMO–LUMO gaps, such as C_{60} and C_{70}, are soluble in many organic solvents. Those that have small HOMO–LUMO gaps, such as C_{74}, are kinetically unstable and react to form insoluble, polymerized solids. Graphitic soot contains higher fullerenes as polymerized solids, which can be solubilized by electrochemical reduction under anaerobic conditions to produce closed-shell anions. Using this method, Diener and Alford have isolated and characterized normally insoluble fullerenes such as C_{74}.

Chen et al. [61] have made naked metallic and semiconducting single-walled carbon nanotubes (SWNTS) soluble in organic solvents by derivatization with thionyl chloride and octadecylamine. The long-chain amine groups add to the open ends of the shortened SWNTs. The solubilities of the derivatized SWNTs exceed 1 mg/mL in 1,2-dichlorobenzene and CS_2. Both ionic (charge-transfer) and covalent solution phase chemistry of the SWNTs were demonstrated. Reaction of these soluble species led to functionalization of the nanotube walls.

2.6 FULLERENE CLUSTERS IN SOLUTION

Experimental studies of the aggregation of fullerenes in solution are limited. In an early study, Honeychuck et al. [62] determined the molecular weight of C_{60}

in toluene (710 ± 10 g/mol) and in chlorobenzene (930 ± 5 g/mol) by vapor osmometry. The latter result was interpreted as a partial aggregation of C_{60} in chlorobenzene. Several studies [9,63] have found that in UV/visible spectrophotometry the Lambert–Beer law is strictly valid for C_{60} in a number of solvents up to the saturation limit—strong evidence that the solute molecules are not aggregating in the solvent. An indirect proof of the lack of aggregation came from the studies of Rubtsov et al. [64] of the rotational dynamics of C_{60} in various solvents. These authors did not need to introduce any fullerene species other than monomers in order to explain their experimental data.

Fullerene dimers and polymers have been observed in the solid state after special temperature/pressure, photochemical, or chemical treatment [65–67]. However, these species are held together by covalent bonding, as was proved by IR or Raman spectra; they are not aggregates because they are not held together by weak intermolecular forces. No spectral features attributable to fullerene dimers have been observed in saturated liquid solutions of fullerenes.

On the other hand, the formation of fullerene clusters in solutions has been suggested in different systems and appears to be a convenient way to discuss properties of the dissolved fullerene material [10,68–73]. These different cases must be analyzed separately, since experimental support and theoretical reasoning presented by the authors vary greatly from case to case.

Bezmelnitsin et al. [68] originally proposed the formation of huge fullerene clusters (5–40 C_{60} units) in the saturated solutions of C_{60} with toluene, carbon disulfide, and hexane as a reason for the unusual temperature dependence of solubility of C_{60} [74]. However, this theory was not able to explain the experimentally observed differences in the temperature of maximum solubility and the changes of the slope of the solubility curves in different solvents (vide infra). In their review of C_{60} solubility, Beck and Mandi [3] state that no experimental evidence of cluster formation was present in the systems observed.

Sun and co-workers [69,70] and Ghosh et al. [71] observed the formation of fullerene C_{70} and C_{60} clusters in room temperature solvent mixtures. Each of the mixtures consists of two solvents in which fullerenes have very different solubilities. A good example is a mixture of toluene and acetonitrile, in which fullerenes are soluble and practically insoluble, respectively. The clusters were generated by "fast" or "slow" addition of acetonitrile to toluene solution of fullerenes [70]. Dramatic changes in the UV absorption spectra were attributed to reversible formation of clusters. Light scattering experiments confirmed the presence of the clusters and showed that the particle size varied from 100 to 1000 nm, depending on the concentration of fullerenes and the solvent composition. The cluster solutions were stable, showing no precipitation over time. Based on the solubility measurements performed by Mandi et al. [51] for C_{60} in the toluene/acetonitrile mixture, it can be stated that the phenomenon of aggregation took place at concentrations close to the saturated concentration of fullerenes in the particular mixed solvent. By comparing data from these references it can also be shown that the relative threshold of the cluster formation depends on the composition of the solvent; for example, with the increase of acetonitrile concentration, the formation started at lower concentrations of fullerene C_{60} relative to saturation. One possible explanation is the formation of a colloidal solution of fullerenes with micelles containing both fullerene molecules and molecules of different solvents. The formation of fullerene aggregates with no solvent inclusions has also been suggested [69–71].

Using light scattering, Ying et al. [72] reported slow aggregation (22 days) of C_{60} in a room temperature benzene solution. In this case, the aggregation was reversible and the fullerene clusters were rather unstable: they could be dispersed simply by shaking the solution by hand. It should be noted, that contrary to the authors' claim, the phenomenon was observed in saturated or even supersaturated solutions rather than in "fairly diluted solutions." The concentrations of C_{60} were from 1.93 to 2.93×10^{-3} M (see Table 2.1 for comparison), while Ying et al. used the data from the earlier paper of Ajie et al. [73] where the saturated concentration of C_{60} in benzene was overestimated.

Ahn et al. [75] have used photoluminescence to study the aggregation of C_{60} in toluene, benzene, and carbon disulfide solutions at temperatures below 220 K. The aggregates did not survive heating. Although these authors prepared unsaturated solutions at room temperature, it is worth noting that the solubility of C_{60} in all three solvents decreases with decreasing temperature below 298 K, and thus at 220 K the solutions could have been supersaturated.

Tomiyama et al. [10] have reported minor deviations from the Lambert–Beer law for C_{60} in carbon disulfide, and they explain this by "the tendency for the formation of small aggregates in the solution."

In conclusion, it can be stated that no evidence for the formation of thermodynamically stable, chemically bonded fullerene oligomers in liquid solutions has been presented in the literature. However, formation of van der Waals aggregates or colloidal particles has been observed at concentrations close to that of saturation in a number of systems, although the nature of these aggregates is not yet clear.

2.7 TRENDS AND CORRELATIONS IN FULLERENE SOLUBILITY

The first attempt to explain trends in C_{60} solubility was done by Sivaraman et al. [4], who plotted the solubility versus the Hildebrand solubility parameter of a series of solvents. In a study of C_{60} solubility in 47 solvents, Ruoff et al. [6] examined the dependence of solubility on the polarizability, polarity, molecular size, and cohesive energy density of the solvents. They found no distinctive parameter that universally explains the solubility of C_{60}. However, by comparing the value of these solvent parameters with those of solid C_{60} itself, they conclude that "all other things being equal, a solvent with a solvent parameter whose value is close to that of C_{60} will win over a solvent whose solvent parameter differs significantly."

With no single parameter to correlate C_{60} solubilities, it makes sense to use multivariate statistical methods to achieve a better correlation. Murray et al.

[76] used computed molecular surface areas and electrostatic potentials calculated on the solvent molecular surface to achieve such a correlation. The resulting equation predicted the C_{60} solubility for 22 solvents with a linear correlation coefficient of 0.954. Their results show that C_{60} solubility is enhanced by the solvent molecules having large surface areas. Smith et al. [77] employed the theoretical linear solvation energy approach of Famini et al. [78] with the C_{60} solubility dataset of Beck and Mandi [3]. The solvent parameters were computed using the MOPAC semiempirical quantum chemistry software: (1) volume per molecule; (2) polarizability index, the ratio of the molecular polarizability to the molecular volume; (3) covalent acidity and basicity, computed from HOMO and LUMO energies of the solvent; and (4) electrostatic acidity and basicity, computed from the maximum positive and negative charge on a solvent atom. Parameters for C_{60} were also computed, and a comparison shows that (1) the molecular volume of C_{60} is more than four times as large as the average solvent and 80% larger than the largest solvent, (2) C_{60} has a higher polarizability index than any solvent, (3) C_{60} is a stronger Lewis acid than any solvent, and (4) C_{60} is a stronger Lewis base than most solvents. The multilinear regression fit for 101 solvents gives a correlation coefficient $R^2 = 0.67$, with no attempt to remove outliers. Statistically significant correlation coefficients are positive for molar volume and both Lewis acidity and Lewis basicity, and negative for electrostatic basicity. Thus, a good solvent for C_{60} should have a large molar volume (and, by implication, a large polarizability) and should also be both a good Lewis acid and a good Lewis base, but should have minimal polarity or regions of negative charge.

Marcus [79] submitted the solubilities of C_{60} at 298 and 303 K to stepwise linear regression with respect to solvent properties. He also carried out a similar analysis for SF_6, an approximately spherical molecule that also forms a van der Waals solid as does C_{60}. For C_{60}, the solvent's polarizability helps its solubility, whereas the solvent's polarity counteracts its ability to dissolve C_{60} (at 298 K, 61 solvents, $R^2 = 0.86$, with outliers removed). On the contrary, the solubility of SF_6 is mainly affected (adversely) by the surface tension of the solvent, and somewhat assisted by its polarizability and hampered by its polarity (at 298 K, 38 solvents, $R^2 = 0.95$).

A basic problem in applying linear free-energy relationships to the correlation of C_{60} solubilities is that this analysis is based on the assumption that the solid phase in equilibrium with the saturated solution is pure solute. For many solvents, this is not the case for C_{60} because the equilibrium solid phase at room temperature is solvated C_{60} (vide infra). If incongruent melting temperatures and decomposition enthalpies are known for the solvate, hypothetical solubilities of pure C_{60} may be computed. It is these hypothetical solubilities that should be included in a multivariate analysis.

Talukdar et al. [11] present what they term "a rational approach" to C_{60} solubility, based on the hypothesis that "solubility is largely controlled by efficient charge-transfer interactions between electron-deficient C_{60} with suitable n and π donors." Without using multivariate analysis, they identify trends in solubility (in mg/mL, not mole fraction) for series of organic solvents with differing electron donor capacity. C_{60} solubilities in 20 solvents are presented, and chemical shift measurements are presented to support their arguments. While this approach holds some conceptual merit, it cannot be argued that the ability to form charge-transfer complexes is the only (or even the most important) factor in making a good C_{60} solvent. Both studies quoted above show that molecular size and polarizability of the solvent are the dominant factors. In the multivariate analysis it may be useful to include as a variable some accepted measure of a solvent's ability to form charge-transfer complexes, such as Gutmann's solvent donicity number [45]. The claims of Talukdar et al. that piperidine and pyrrolidine, both secondary amines, are excellent solvents for C_{60} is not true, since secondary amines have been shown to undergo nucleophilic additions with the electron-deficient C_{60} [80].

2.8 TEMPERATURE DEPENDENCE OF SOLUBILITY

The abnormal temperature dependence of solubility of fullerene C_{60} in a number of solvents, first reported by Ruoff et al. [74], attracted attention to this phenomenon and provoked experimental studies in different laboratories. This "abnormality" included a decrease of solubility with increasing temperature, very unusual for nonelectrolytes, and the appearance of a temperature of maximum solubility. So far, temperatures of maximum solubility have been observed in the systems C_{60} with toluene, CS_2 [5,23,74], 1,2-dimethylbenzene [5], 1,2-dichlorobenzene, tetralin, 1,2-diphenylacetone [15], and for C_{70} with 1,2-dichlorobenzene, tetralin, and 1,2-diphenylacetone [15].

The nature of solubility maxima, first attributed directly [74] or indirectly [68] to the orientational phase transition in pure C_{60}, was later explained on routine thermodynamic grounds [81–84]. Simple thermodynamic equations combined slopes of the temperature dependencies of solubility, enthalpies of dissolution, and enthalpies of the incongruent decomposition (incongruent melting) of the solid solvates of fullerene with the solvents. The latter effect proved to be the reason for the appearance of temperatures of maximum solubility [83,85].

2.8.1 Thermodynamic Model

A maximum of solubility, for example, in the system C_{60}–B, is due to the formation/decomposition of a solid solvate (I), which melts incongruently to yield another solvate (II) and a liquid saturated solution of C_{60} in a solvent B. At temperatures below the incongruent melting point a saturated solution is in equilibrium with the solvate (I) (n moles of B per mole of C_{60}, where n is an integer or a rational fraction). Above the incongruent melting point the saturated solution is in equilibrium with the solvate (II). In order to observe a temperature of maximum solubility, the incongruent melting must occur below the boiling point of the solvent. In the simplest case the solvate (II) is pure C_{60}.

Below the incongruent melting point, where C_{60} forms the solvated crystal (I) with the solvent B, the basic equation of phase equilibrium would be

$$\mu(C_{60} \cdot nB) = \mu(C_{60}, \text{ in sat'd. liq. soln.}) + n\mu(B, \text{ in sat'd. liq. soln.}) \quad (2.1)$$

where μ is a chemical potential. After temperature differentiation and rearrangement, Eq. (2.1) leads to Eq. (2.2):

$$\frac{dx}{dT} = (H(C_{60}, \text{ liquid}) - H(C_{60}, \text{ solid}))$$

$$+ n(H(B, \text{ liquid}) - H(B, \text{ solid})) \Big/ n\frac{\partial\mu}{\partial x}[1/n - x/(n - x)]T \quad (2.2)$$

which relates x, the mole fraction of C_{60} in the saturated liquid solution, to the temperature T. Here $H(C_{60}, \text{ liquid})$, $H(C_{60}, \text{ solid})$, $H(B, \text{ liquid})$, and $H(B, \text{ solid})$ are partial molar enthalpies of C_{60} and of the solvent B in the equilibrium liquid and solid phases, respectively, and μ is a chemical potential of C_{60} in the saturated liquid solution. The solid compound of C_{60} with the solvent in Eq. (2.2) is $C_{60} \cdot nB$. The numerator in Eq. (2.2) is evidently the integral enthalpy of dissolution of a solvated crystal $C_{60} \cdot nB$ into the saturated solution.

We now subtract and add the term $\{H(C_{60}, \text{ solid, pure}) + nH(B, \text{ liquid, pure})\}$ from the numerator of Eq. (2.2), where $H(B, \text{ liquid, pure})$ and $H(C_{60}, \text{ solid, pure})$ are molar enthalpies of pure liquid B and pure solid C_{60}, respectively. We rewrite the result as

$$\frac{dx}{dT} = \{\Delta H(C_{60}, \text{ dissolution}, x) - \Delta_f H(C_{60} \cdot nB, \text{ solid})$$

$$+ n(\Delta H(B, \text{ dissolution}, x)\} \Big/ n\frac{\partial\mu}{\partial x}[1/n - x/(n - x)]T \quad (2.3)$$

Here $\Delta H(C_{60}, \text{ dissolution}, x)$ and $\Delta H(B, \text{ dissolution}, x)$ are partial molar enthalpies of dissolution of pure solid C_{60} and pure liquid solvent B into the saturated liquid solutions with composition x at temperature T. The term $\Delta_f H(C_{60} \cdot nB)$ is the enthalpy of formation of $C_{60} \cdot nB$ from the pure solid C_{60} and pure liquid B, the formation reaction being

$$C_{60}(\text{solid, pure}) + nB(\text{liquid, pure}) = C_{60} \cdot nB(\text{solid}) \quad (2.4)$$

Saturated solutions of fullerenes are rather dilute (see Table 2.1). Hence, it may be assumed that $1/n$ in the denominator of Eq. (2.2) is much larger than $x/(n - x)$ and $\Delta H(B, \text{ dissolution}, x)$ is zero. Further simplifications are possible if the concentration dependence of the activity coefficient, $\gamma(C_{60})$, can be neglected in the saturated and nearly saturated solutions. This corresponds to the so-called Henry's law behavior typical for the solute in the diluted solutions. Hence,

$$\gamma(C_{60}) = \text{constant}$$
$$d \ln \gamma(C_{60})/dx = 0, \quad d\mu/dx = RT/x$$

(2.5)

The strictly thermodynamic equation, Eq. (2.3), along with the simplifications of Eq. (2.5) describe the temperature dependence of solubility of a fullerene when the equilibrium solid phase is a solvated crystal. Taking into consideration Eq. (2.5), Eq. (2.3) may be rewritten as

$$d \ln x(C_{60})/dT = \{\Delta H(C_{60}, \text{dissolution}, x) - \Delta_f H(C_{60} \cdot nB)\}/RT^2 \quad (2.6)$$

In Eq. (2.6) both terms in the numerator are negative for fullerenes. If pristine C_{60} is the equilibrium phase, Eq. (2.3) may be rewritten to obtain the normal form of the van't Hoff equation:

$$d \ln x(C_{60})/dT = \Delta H(C_{60}, \text{dissolution}, x)/RT^2 \quad (2.7)$$

Equations (2.6) and (2.7), along with the thermodynamic properties of fullerenes, explain the abnormal temperature dependence of fullerene solubility.

As shown by calorimetric measurements (see Table 2.7), the enthalpies of dissolution of C_{60} and C_{70} are negative, at least in aromatic solvents. Hence, at high temperatures where pristine C_{60} is the equilibrium phase, Eq. (2.7) predicts a decrease of solubility with increasing temperature. In contrast, for lower temperatures when the solid phase is solvated C_{60}, the sign of the difference $\{\Delta H(C_{60}, \text{dissolution}, x) - \Delta H(C_{60} \cdot nB, \text{sol})\}$ in Eq. (2.6) determines the temperature dependence of solubility. There is an explanation of how to achieve a temperature of maximum solubility. If the enthalpy of formation of the solvated phase (reaction in Eq. (2.4)) is more exothermic than the enthalpy of dissolution of pure C_{60}, the numerator of Eq. (2.6) is positive and the solubility increases with increasing temperature. Approaching the incongruent melting point from below, the solvated crystal decomposes, giving pristine C_{60} as the high-temperature equilibrium solid phase. This decomposition at the incongruent melting point has been characterized using optical microscopy and X-ray diffraction [95]. The change from positive to negative temperature dependence makes the incongruent melting point a temperature of maximum solubility. The situation described is typical, but not the only one possible at the incongruent melting point (e.g., the equilibrium solid phases at higher temperatures may be another solvated crystal rather than a pristine fullerene). Different types of solubility curves are discussed by Avramenko et al. [86].

It should be emphasized that it is the unusual thermodynamic properties of fullerenes—namely, their negative enthalpies of dissolution—and the unusual stability of their solvated crystals, rather than the nature of the universal thermodynamic equations (2.1) to (2.6), which causes their abnormal solubility behavior. Further details and discussion can be found in the original papers.

2.8.2 Comparison with the Experimental Data

Such comparison was possible in five systems (Table 2.3). Subjects of comparison were as follows:

- Temperature of maximum solubility (column 2) and incongruent melting temperatures (column 3).
- Changes of enthalpy at the temperature of maximum solubility, calculated from solubility data (column 4) and from differential scanning calorimetry data (column 5).
- Enthalpies of solution of the pure fullerenes, calculated from solubility (column 6) and solution calorimetry (column 7) data.

It can be stated that qualitative agreement exists between the results from different methods. In all the systems where temperatures of maximum solubility were found, an incongruent melting point of a certain solvate existed at temperatures 5–15 degrees above this temperature. In the system C_{70}–toluene, where no solvates were identified at temperatures below the boiling point of toluene, no maximum was found [5]. This confirms that formation and incongruent melting of solid solvates is the cause of abnormal temperature dependencies of solubility.

Enthalpies of solution of pure fullerenes, measured by different methods (columns 6 and 7, Table 2.3) seem to be in reasonable agreement in all the systems studied. Changes of enthalpy at the incongruent melting point, measured by differential scanning calorimetry (DSC), fall in line with the solubility data in the systems of C_{60} with toluene, 1,2-dimethylbenzene and 1,2-dichlorobenzene, and bromobenzene. A metastable solvate of C_{60} with 1,2-dichlorobenzene with a lower incongruent melting point was definitely formed in the course of solubility measurements [15]. Formation of a stable solvate in this system needed at least a few weeks, that is, much more time than the duration of the solubility experiment. The solubility curve for the system C_{70}–1,2-dimethylbenzene, obtained by Zhou et al. [5], showed evidence of an incongruent decomposition at $T = 283$ K. The decomposition temperature of the second solvate ($T = 369$ K) was above the temperature interval of solubility measurements. The negative enthalpy of dissolution of C_{70} in 1,2-dimethylbenzene (Table 2.3), combined with the enthalpy of incongruent melting at $T = 369$ K should give a positive slope of the temperature dependence of solubility ($\ln x$ versus $1/T$), with the corresponding ΔH equal to 4.2 kJ/mol at temperatures between 283 and 369 K, and a positive slope ($\Delta H = 20.2$ kJ/mol) below $T = 283$ K. The experimental slope below $T = 283$ K is smaller, but close to the experimental value. Change of the slope of the solubility curve at $T = 283$ K corresponded to the enthalpy

TABLE 2.3 Temperatures of Maximum Solubility, Slopes, and Changes of the Slope of ln x Versus $1/T$ Solubility Curves, Compared to DSC and SC Data

System	T_{tms} (K)[a]	T_{imp} (K)[b]	ΔH_{tms} (kJ/mol)[c]	ΔH_{dsc} (kJ/mol)[d]	ΔH_{soln} (kJ/mol)[e]	ΔH_{soln} (kJ/mol)[f]
C_{60}–toluene	278 [74]	285	35 [74]	30	-12 ± 2 [74]	-9.2 ± 0.2
C_{60}–1,2-dimethylbenzene	303 [5]	322	34 [5]	31		-14.28 ± 0.44
C_{60}–1,2-dichlorobenzene	310 [15]	322	24 [15]	19	-13.3 [23]	-14.45 ± 0.33
C_{70}–1,2-dimethylbenzene	~280 [15]	282	11 [5]	18		-18.84 ± 0.62
C_{70}–1,2-dichlorobenzene	328 [15]	335	23 [15]	8	-13.6 [23]	-18.67 ± 0.34

[a] Temperature of maximum solubility, from solubility measurements.
[b] Incongruent melting point, from DSC.
[c] Enthalpy change for incongruent melting, from slopes of temperature–solubility curves.
[d] Enthalpy change for incongruent melting, estimated from DSC.
[e] Enthalpy of solution above the incongruent melting point, from slope of temperature–solubility curve.
[f] Enthalpy of solution above the incongruent melting point, from solution calorimetry. All values taken from Korobov et al. [87].

of 11 kJ/mol, while the enthalpy of incongruent melting was 18 kJ/mol (Table 2.3). More work is needed in order to understand the cause of disagreement in the system $C_{70}-o$-dichlorobenzene.

2.9 THERMODYNAMICS OF SOLID SOLVATES

As shown earlier, solid solvates of fullerenes C_{60} and C_{70} play a significant role in the solubility phenomena. The formation of solvates has been reported by a number of groups. Measured physical properties of these solvates include:

- DSC data on thermodynamic stability and composition of solvates [19,81,83,84,87–93].
- Heat capacity measurements [19,91–93].
- Determination of solution enthalpies [84,87,91].
- X-ray powder [88–90,94–97] and sometimes X-ray single crystal data [84,91,98,99].
- Solid-state NMR results investigating possible rotation of the fullerene or solvent units inside the solvate [100,101].

The solvates are held together with weak van der Waals interactions. No charge transfer, detected by changes in infrared spectra, was reported. Data on the stability of the solvated crystals of C_{60}, both measured by DSC and calculated through standard thermodynamic equations, are summarized in Table 2.4.

DSC measurements on solvates can give the temperature and the enthalpy of one of the two isothermal transitions, namely, of incongruent melting

$$C_{60} \cdot nB(s) = nB(liq.) + C_{60}(s) \qquad (2.8)$$

or of desolvation

$$C_{60} \cdot nB(s) = nB(gas) + C_{60}(s) \qquad (2.9)$$

depending on whether the incongruent melting occurs below or above the normal boiling point of the solvent, B. The recovered $C_{60}(s)$ has face-centered cubic (fcc) structure, which is the equilibrium phase of pristine C_{60} above 259 K.

The enthalpy of incongruent melting measured for reaction (2.8) corresponds to a process that forms solid C_{60} and a saturated solution of C_{60} in B, rather than pure B. However, since the resulting saturated solutions are rather dilute (see Table 2.1), there is good reason to replace the saturated solution with the pure liquid B in Eq. (2.8).

The determination of composition of solvates (column 2 in Table 2.4) presented a significant problem in a number of cases, mainly due to the low stability of the solvates and possible uncontrolled loss of the solvent from the dried samples. In some cases (e.g., C_{60} with 1,3,5-trimethylbenzene), it is still not clear whether solid solvate with a fixed composition or solid solutions with variable composition are formed [87]. The other example is a system of C_{60} with CCl_4 and C_6H_{12}, where according to different authors, the solid phase with the composition 1:13.7, 1:13, 1:12, or 1:10 was formed [19,91,92,96,97,102].

Temperatures of the incongruent melting or desolvation, measured by DSC (column 3) may be overestimated by 5–15 degrees due to the nonzero heating rate characteristic for this method [92]. Such correction may lead to better agreement between the incongruent melting temperatures and temperatures of maximum solubility.

In some systems (e.g., C_{60} with CCl_4), more than one solvate is formed. With a stepwise increase of temperature, incongruent melting (desolvation) is observed. In this case column 3 contains all the transition temperatures found. They do not necessarily correspond to the reactions (2.8) or (2.9). For example, the upper line for the system of C_{60} with CCl_4 corresponds to the incongruent melting of the solvate with the composition 1:13 and the formation of the another solvate with the composition 1:2. The next line presents the temperature of desolvation (reaction (2.9)) of the 1:2 solvate.

Column 4 contains enthalpies of incongruent melting (Eq. (2.8)) for all of the solvates. If this incongruent melting was observed in an experiment, the experimental value is given. If step by step decomposition of the solvate takes place, enthalpies of all the steps were summed in order to get the value presented. If the enthalpies of desolvation were measured in the experiment, the enthalpies of evaporation of the solvent at normal boiling point [103] were subtracted in order to get the desired value. In the last two cases one has to address the original papers in order to find the enthalpies measured. The numbers in parentheses are the enthalpies of reaction (2.8) per mole of a solvate, that is, $\Delta H / n$.

Table 2.4 also includes entropies of the reaction (2.8) (column 5) calculated using the relationship

$$\Delta S(2.8) = \Delta H(2.8) / T \tag{2.10}$$

valid at the transition temperature. The comments in the preceding paragraph relating to enthalpies are equally valid in the case of entropies.

Column 6 contains ΔG (298 K) of the reaction (2.8), calculated under the assumption that both $\Delta H(2.8)$ and $\Delta S(2.8)$ are temperature independent from room temperature up to the decomposition temperature of the solvate. With the knowledge of $\Delta G(298\ \text{K})$ it was possible to calculate the decomposition vapor pressure of a solvent over a solvate at room temperature (column 7), which characterizes the solvate's ability to survive outside solution.

Three groups of the solvates can be identified. The first group includes solvates of C_{60} with normal alkanes. A number of such compounds are known [88], though only in two cases, with n-heptane and n-octane, are the thermodynamic data available. Solvates with chloroalkanes and chloroalkenes also

TABLE 2.4 Thermodynamics of Solid Solvates of C_{60}

Solvent	C_n:S^a	T (K)	$\Delta H(2.8)$ (kJ/mol)[b]	$\Delta S(2.8)$ (J/(mol·K))[b]	$\Delta G°$ (kJ/mol)	P (kPa)[c]
Benzene [87]	1:3.8 ± 0.2	322 ± 1	41 ± 1	127	3.0	9.3
	(1:4)		(10)	(32)		
Toluene [87]	1:1.8 ± 0.2	285 ± 1	30 ± 2	105		
	(1:2)		(15)	(50)		
1,2-Dimethylbenzene [87]	1:2.11 ± 0.18	322.0 ± 2.6	31.3 ± 1.4	97	2.3	0.5
	(1:2)		(15.5)	(43)		
1,3-Dimethylbenzene [87]	1:2.3–3.6	294 ± 2.3	40	124		
	(1:2)		(20)	(62)		
	(1:0.5)	370 ± 4	11.9 ± 0.7	32	2.3	0.2
			(24)	(64)		
1,2-Dichlorobenzene [87]	1:2	322	18.5 ± 2.3	57	1.4	0.14
	1:2	342	18.9	55	2.4	0.11
			(9)	(28)		
1,3-Dichlorobenzene [87]	(1:2)	308	40.7	122	4.2	0.11
			(20)	(61)		
	(1:0.5)	408	12.2	30	3.2	0.02
			(24)	(60)		
1,3,5-Trimethylbenzene [87]	?	292–300	25–44	146	6.4	0.0008
	1:0.5	460.8 ± 1.5	16.3 ± 0.5	33		
			(33)	(66)		
1,2,4-Trimethylbenzene [87]	1:1.7–2.4	322	38.2 ± 0.4	119	2.8	0.2
	(1:2)		(19)	(59)		
Bromobenzene [84]	1:2	350 ± 1	42 ± 2	121	6.2	0.2
			(21)	(60)		

Octane [110]	1:1	398	15	39	3.9	0.4
Heptane [89]	1:1	383	3.5	10	0.6	4.8
1,1-Dichloroethane [88]	1:1	343	12	35	1.6	5.6
1,1,2-Dichloroethane [88]	1:1	430	15	25	7.2	0.2
Trichloroethylene [88]	1:1	416	12	18	7.0	0.6
Carbon tetrachloride [92, 102]	1:13	319	59.2	180	5.4	12.8
			(4.6)	(13.8)		
	1:2	402	9.2	23	2.2	0.7
			(4.6)	(11.5)		
Cyclohexane [19]	1:13	350	73	208		
	1:2					

[a] Solvate composition, moles C_{60} per mole solvent; inferred stoichiometric composition is given in parentheses.

[b] Numbers in parentheses are ΔH or ΔS per mole of solvent.

[c] Decomposition pressure of solvate.

belong to the same group [88,104]. All of them have the same simple mole ratio (1:1, $n = 1$), and possibly the same type of crystal structure [105]. The second group includes solid solvates with aromatic solvents [81,83,84,87]. Here, at least three compositions were identified with the approximate molar ratios 1:4, 1:2, and 1:0.5. X-ray single crystal data are very limited [84,98,99], although it was emphasized that the $C_{60} \cdot 2C_6H_5Br$ structure is similar to those observed for $C_{60} \cdot 2D$, where D = ferrocene and P_4 [84]. Both groups of solvates can be thought of as host (C_{60})–guest (solvent) compositions, with the C_{60} molecules forming close-packed layers.

So-called massively solvated crystals of C_{60} with cyclohexane and CCl_4 form the third group [19,91,92,96,97,104]. In these solvates, C_{60} units are separated from each other by the solvent molecules, which are massively enclosed in the crystal. The distance of closest contact between C_{60} molecules is larger than the van der Waals diameter of C_{60}. The ideal composition of the solvates was claimed to be (1:13, $n = 13$) [91], in which 24 cyclohexane molecules surround the C_{60}. The crystal is a result of aggregation of the massively solvated C_{60} molecules, rather than inclusion of C_{60} in a CCl_4 or C_6H_{12} host structure. That is probably an analog of AB_{13} compound formed by round particles with the typical van der Waals diameter ratio (≈ 0.6) and known in several alloys and binary hard-sphere mixtures [91]. With both CCl_4 and C_6H_{12}, the massively solvated crystal melts incongruently with the formation of the other solvate, $C_{60} \cdot 2CCl_4$ or $C_{60} \cdot 2C_6H_{12}$, which belongs to the first group.

From the thermodynamic point of view, all three groups of solvates are characterized by relatively low enthalpies of formation. Enthalpies of reaction (2.8) are about 10–15 kJ per mole of solvent for the first two groups and about 5 kJ per mole for the massively solvated crystals. However, entropies of reactions are significantly different in different groups. Formation of solvates of C_{60} with aromatics causes "ordering" and significant loss of entropy. The typical entropy of reaction (2.8) per mole of solvent is about 50 J/(mol · K), which is above or comparable with the entropy of melting of a pure solvent. The only strange exception from this rule is the $C_{60} \cdot 2(1,2\text{-}C_6H_4Cl_2)$ solvate. Due to the significant entropy of the reaction (2.8), solvates with aromatics undergo incongruent melting. Solvates with different fullerene to solvate ratios are formed, dependent on the structure of the solvent molecule; for example, ortho-derivatives and meta-derivatives of the same molecule form different solvates [87]. No orientational phase transitions have been observed in these solvates.

In contrast, solvates of the first group are relatively less ordered, "clathrate-type" compounds [105]. The entropy of reaction (2.8) is smaller and significantly less than the entropy of melting; for example, in case of heptane the difference is 68 J/(mol · K). Simple calculations show that at temperatures below 90 K the 1:1 solvate of C_{60} with octane must decompose into solid C_{60} and solid octane. For heptane, this temperature is 157 K. For the first group, the influence of the chemical identity of the solvents on solvate composition is less pronounced: the same composition is formed in all cases. Most of the solvates undergo orientational phase transitions at temperatures lower than that for pristine C_{60}.

Ceolin et al. [88] claimed that solvates may be classified by their molecular volumes. According to these authors, solvates with aromatics and massively solvated crystals have positive excess molar volume (molar volume minus sum of molar volumes of components), while 1:1 solvates with alkanes have negative excess molar volumes. It is worth noting, however, that the data on volumes in the former group are controversial (e.g., if one assumes that the massively solvated crystal with cyclohexane in the work of Jansen and Waidmann [96] includes not 12 but 13 cyclohexane units [91], the excess volume will change the sign and will become negative). It is yet to be proved if solvates with positive excess molar volumes really exist.

For their classification of solvate structures, Slovokhotov et al. [106] have used the dilution parameter, ρ, where

$$\rho = [V/Z - V(C_{60})]/V(C_{60}) \qquad (2.11)$$

where V/Z is the unit cell volume per one C_{60} moiety in the solvate, and $V(C_{60}) = 524 \text{ Å}^3$ is the van der Waals volume of C_{60}. Both solvates of the first group and aromatic solvates belong to the group with $\rho = 0.6$–1.0. This condition was described as "structures with two-dimensional layers of closely packed C_{60} spheres with relatively small quasi-spherical molecular guests [106] between the layers." For massively solvated crystals, $\rho > 4$.

All of the solvates in Table 2.4 (except that with toluene) have thermodynamically stable phases at room temperature and normal pressure relative to both reactions (2.8) and (2.9). Two solvates with the highest numbers in column 7, 1:4 with benzene and 1:13 with CCl_4, were claimed in the literature to be "unstable under air without mother liquor" [88], which in this case means rapid loss of the solvent due to its significant (though less than atmospheric) pressure under the solvate. Massively solvated crystal with cyclohexane probably belongs to the same class.

No thermodynamic information is available for some of the C_{60} solvates identified. X-ray powder structures of 1:2 solvates of C_{60} with $CHBr_3$, CH_2Cl_2, and CH_2Cl_2 have been reported [96]. A 2:1 solvate of C_{60} with diethyl ether was obtained by Ozlany et al. [107].

Data on the stability of the fullerene solvates with C_{70} are available only for some aromatic solvents (Table 2.5). The same trends are observed here, though it is worth noting that the toluene solvate belongs to the "clathrate group" of solvates and represents a certain extreme case with a very low enthalpy and *negative* entropy of the reaction (2.1). Using heat capacity, DSC, and NMR methods, Cheng et al. [93] demonstrated that this solvate "is a physical 1:1 mixture of the molecules" and "the toluene molecules seem to gain liquid-like mobility" inside the solvate. A massively solvated crystal of C_{70} with cyclohexane (1:12) was obtained and its X-ray powder structure was described by Jansen and Waidmann [96].

TABLE 2.5 Thermodynamics of Solid Solvates of C_{70}

Solvent	$(C_n{:}S)^a$	T (K)	$\Delta H(2.8)$ (kJ/mol)[b]	$\Delta S(2.8)$ (J/(mol·K))	ΔG° (kJ/mol)	P (kPa)
Toluene [93]	1:1	452	2.9	−7.4	5.1	0.48
1,2-Dimethylbenzene [87]	1:(3–4)	283.0 ± 1.2	41 ± 2	128	3.4	
	1:2.0 ± 0.5	368.7 ± 0.8	23.0 ± 1.0	63	4.4	0.36
			(11.5)	(31.5)		
1,2-Dichlorobenzene [87]		327.9 ± 1.1	30 ± 1	80	6.1	
		397.5 ± 1.0	21.4 ± 1.1	53	5.4	
Bromobenzene [87]	1:2	290 ± 1	16.2 ± 1.2	56	−0.4	
			(8)	(28)		

^{a} Solvate composition, moles C_{60} per mole solvent; inferred stoichiometric composition is given in parentheses.

^{b} Numbers in parentheses are ΔH or ΔS per mole of solvent.

^{c} Decomposition pressure of solvate.

The first data on ternary solvated crystals of C_{60}–C_{70} mixtures were obtained with aromatic solvents [108] and cyclohexane [109]. In both papers, ternary solvates were considered as nearly ideal mixtures of the binary solid solvates.

2.10 HYPOTHETICAL SOLUBILITIES

The formation of solvates determines the unusual temperature dependence of fullerene solubility and influences the absolute values of solubility at constant temperature. In order to compare solubilities in different solvents, we must be sure that they are saturated concentrations relative to the same solid phase, namely, to pure solid C_{60} or C_{70}. Formation of solid solvates lowers the solubility relative to the hypothetical case of pure C_{60} or C_{70} in equilibrium with the saturated solution. Simple thermodynamic considerations lead to the equation

$$\ln(x'/x) = \Delta G^{\circ}/RT \qquad (2.12)$$

Here x' is the hypothetical solubility relative to pure C_{60} or C_{70} and x is the experimentally measured solubility, respectively. ΔG° is the Gibbs energy of decomposition of solid phase in equilibrium with the saturated solution at $T = 298$ K, according to reaction (2.8) (see Table 2.6). It is worth noting that one need not know the composition of solvate in order to use this equation. Though nonequilibrium solubilities x' relative to the pure fullerene cannot be

TABLE 2.6   Measured and Hypothetical Solubilities of C_{60} and C_{70}

Solvent (Fullerene)	Equilibrium Solid Phase at $T = 298$ K	ΔG° (kJ/mol)	$10^4 \, x$	$10^4 x'$
Benzene (C_{60})	1:4	3.0	2.0	6.9
Toluene (C_{60})	C_{60}		4.2	4.2
1,2-Dimethylbenzene (C_{60})	1:2	2.3	14.2	37.7
1,3-Dimethylbenzene (C_{60})	1:2	2.3	3.6	10.4
1,3,5-Trimethylbenzene (C_{60})	1:0.5	6.4	3.3	43
1,2,4-Trimethylbenzene (C_{60})	1:2	2.8	33.5	105.7
Bromobenzene (C_{60})	1:2	6.2	4.4	54.6
1,2-Dichlorobenzene (C_{60})	1:2	1.4	38.3	68.3
1,3-Dichlorobenzene (C_{60})	1:2	4.2	3.8	19.7
CCl_4 (C_{60})	1:13	5.4	0.43	3.8
Heptane (C_{60})	1:1	0.6	0.098	0.12
Octane (C_{60})	1:1	3.9	0.045	0.22
Toluene (C_{70})	1:1	5.1	1.6	7.2
1,2-Dimethylbenzene (C_{70})	1:2	4.4	22	130

observed in the experiment, it is these solubilities that should be used in establishing correlations or trends in solubility while going from solvent to solvent. The determination of $\Delta G°$ is discussed in a previous section. Solubilities x and x' for the solvents for which $\Delta G°$ is available are presented in Table 2.6.

Recalculations of solubility bring some order to the solubility data. If trends in x', and not x, are examined it can be stated that:

- Solubility increases along the series of alkanes (octane was the only exception with x lower than for heptane, as was extensively discussed by Heymann [18].
- Solubility increases along the series of aromatic solvents from toluene to trimethylbenzene with increasing number of methyl substituent (1,3,5-trimethylbenzene was the exception, when x were compared).
- C_{70} is more soluble than C_{60}, at least in aromatic solvents (toluene broke this rule before).

However, there is still no explanation for the large difference between the solubility of C_{60} in positional isomers, 1,2- and 1,3-dimethylbenzene, 1,2- and 1,3-dichlorobenzene, and 1,3,5- and 1,2,4-trimethylbenzene. The chemical properties in the liquid phase within the two pairs of solvents are closely similar and can hardly explain this difference. It was hoped that small differences in geometry caused by the positions of the substituents on the benzene ring would bring different types of solvates with different stability into equilibrium with the saturated solution at room temperature. The latter is true, yet the difference in x' still exists, although it is smaller than the difference in x. The corresponding data in Table 2.6 show that both the enthalpy and the entropy of solution make a contribution to this difference. Surprisingly, there is no difference between the solubility of C_{60} in 1,2- and in 1,3-dibromobenzenes (see Table 2.1).

2.11 ENTHALPIES OF SOLUTION

The enthalpies of solution of C_{60}, C_{70}, and their solvates in a number of solvents were measured by direct calorimetry. The data are presented in Table 2.7. Most of the measurements were performed for the aromatic solvents at $T = 298$ K. The results for the C_{60}–1,2-dimethylbenzene system, where the enthalpy was measured at two temperatures, give an idea about possible temperature dependence of this value. The important result is that all the enthalpies of solution are negative for the pristine C_{60} and C_{70}. According to Korobov et al. [87], the dissolution of C_{70} is more exothermic than C_{60}. Methyl groups as substituents increase the interaction with C_{60} as indicated by an increasingly more exothermic solution enthalpy in the series toluene–1,2-dimethylbenzene–1,2,4-trimethylbenzene. The solution enthalpy of both C_{60} and C_{70} in 1,3-dichlorobenzene is less exothermic than for 1,2-dichlorobenzene by 1.5 kJ/mol [87]. The striking disagreement for the C_{70} systems between the results of

TABLE 2.7   Enthalpies and Entropies of Solution of C_{60}, C_{70}, and Their Solvates

Solute/Solvent	ΔH (kJ/mol)[a] (Cal)	ΔH (kJ/mol)[b] (TDS)	ΔS (J/(mol · K))[a] (Cal)
C_{60}/toluene	−8.6 ± 0.7 [83]	−12 ± 2 [5,74]	−95.5
	−8.3 ± 0.7 [82]		
	−7.54 ± 0.02 [111]	−9.8 [13]	
C_{60}/1,2-dimethylbenzene	−14.53 ± 0.30	−11.6 [5]	−87.7
	−12.08 ± 0.16 (323 K) [87]		
C_{60}/1,2,4-trimethylbenzene	−17.72 ± 0.40 [87]		−96.6
C_{60}/1,2-dichlorobenzene	−14.00 ± 0.15 [87]	−13.3 [15]	−86.6
C_{60}/1,3-dichlorobenzene	−12.49 ± 0.18 [87]		−92.0
C_{60}/bromobenzene	−11.5 ± 2.0 [84]		−81.9
C_{60} · $2C_6H_5Br$/bromobenzene	28 ± 1 [84]		
C_{60}/CS_2	−20 ± 1 [82]		
C_{60}/hexane		−6.0 [13]	
C_{70}/CS_2	−9 ± 1 [82]		
C_{70}/toluene	−15.21 ± 0.70 [87]		−108.9
	−1.38 ± 0.34 [111]		
C_{70}/1,2-dimethylbenzene	−18.84 ± 0.62 [87]		−99.3
	−1.93 ± 0.03 [111]		
C_{70}/1,2-dichlorobenzene	−18.67 ± 0.17 [87]	−13.6 [15]	−88.5
C_{70}/1,3-dichlorobenzene	−17.60 ± 0.36 [87]		
C_{60} · $13(C_6H_{12})$/cyclohexane	16 [91]	55 [91]	
		48.2 [13]	

[a] Solution enthalpies and entropies determined from calorimetry.

[b] Solution enthalpies determined from temperature dependencies of solubility.

Korobov et al. [87] and Yin et al. [111] was explained by Korobov et al. as the accidental formation of solvates prior to solution of samples by exposure to the vapor in the experiments of Yin et al. Having positive enthalpies of solution, these solvates could decrease the effective enthalpies measured.

Solution enthalpies can also be derived from the temperature dependencies of solubility using the integrated form of the van't Hoff equation. These values are usually much less accurate, though the agreement with the calorimetric results is reasonable, except the case of a solvate of cyclohexane [112].

Table 2.7 also includes entropies of solution, calculated by the equation

$$\Delta S_{sol} = R \ln x' + \Delta H_{sol}/T \qquad (2.13)$$

where x' are hypothetical solubilities, at saturated concentrations relative to pure fullerene (see Table 2.6).

2.12   THERMODYNAMICS OF SOLUBILITY

Attempts to explain the trends in room temperature solubility of C_{60} have already been reviewed. The thermodynamic approach was represented by the Hildebrand regular solution model, which was tried by several authors [4,6,18]. Heymann [18] used the Hildebrand–Scatchard equation to correlate experimental solubilities of alkanes and alcohols with δ, the Hildebrand solubility parameter. The qualitative correlation was reasonable, but one needs the free energy of melting of C_{60} to be 108 kJ/mol in order to account for the measured solubilities. Such a high value looks unreal, compared to 8 kJ/mol, calculated with the parameters of melting from Smith et al. [113].

It is worth noting that solubility data for fullerenes do not fall in line with the basic concepts of Hildebrand's regular solution model. This model is based on the assumption that the entropy of mixing in the solution is equal to the ideal entropy of mixing with the excess entropy of mixing equal to zero. The excess enthalpy of mixing should be positive, if the Hildebrand–Scatchard equation is valid. This makes both the enthalpy and the entropy of solution positive in the regular model:

$$\Delta H_{sol} = \Delta H_{melt} + \Delta H_{ex} > 0$$

$$\Delta S_{sol} = \Delta S_{melt} + \Delta S_{ex} > 0$$

where subscripts "sol", "melt," and "ex" mean "solution," "melting," and "excess," respectively. In contrast, the remarkable feature of fullerene solubility is that both the solution enthalpies and the solution entropies are significantly negative [5,15,74]:

$$\Delta H_{sol} < 0 \quad \text{and} \quad \Delta S_{sol} < 0$$

This was proved experimentally for aromatic solvates, cyclohexane, and CS_2 (Table 2.7). If one assumes large, nearly constant negative entropy of solution for C_{60} in the whole family of alkanes and alcohols, this could explain the strange free energy of melting calculated by Heymann [18]. What was supposed to be "the free energy of melting" could actually be the sum $\{\Delta G_{melt} + (-T \, \Delta S_{ex})\}$. One must keep in mind that a liquid state has never been observed for fullerenes. According to Hagen et al. [114], the liquid state may be metastable due to the short range of the C_{60}–C_{60} interaction. This introduces uncertainty into the discussion of ΔG_{melt}.

Within experimental error, plots of $\ln(x)$ versus $1/T$ for solubility are linear [5,15,74], implying that both ΔH_{sol} and ΔS_{sol} are independent of both concentration and temperature, to within the precision of the data. This is typical Henry's law behavior, characteristic of dilute solutions. Ruelle et al. [115,116] have demonstrated that the partial molar volume of C_{60} is independent of concentration over the range from zero to the saturated solution for a number of solvents. The values of these "partial molar volumes at infinite dilution" were surprisingly low, lower than the molar volume of solid fcc C_{60}. A positive correlation was found between the partial molar volume of C_{60} in a given solvent and the molar volume of that pure solvent. At least in aromatic solvents, both C_{60} and C_{70} have an activity coefficient that is much higher than unity and nearly constant at constant temperature for the whole concentration range below saturation [81].

Entropy loss due to solvation dominates the low solubility of fullerenes [81,91]. Nagano et al. [91] compared such behavior to the similar properties of hydrophobic solutes in aqueous solution. The scaled particle theory [117–119] emphasizes that the large entropy loss of hydrophobic hydration is due to the smaller size of water compared to the size of the solute molecule.

Using the results of their small-angle neutron scattering experiment in C_{60}/CS_2 solutions, Gripon et al. [23] tried to establish a correlation between solubility and C_{60}–C_{60} interaction in the solution. The latter was represented by the measured second virial coefficient and the related value of the stickiness parameter. Smorenburg et al. [120] have computed the structure factors C_{60}–C_{60}, C_{60}–CS_2, CS_2–CS_2 of a C_{60} solution in CS_2 using a hard-sphere potential.

2.13 CONCLUSION

In closing, we would like to identify the challenging problems still remaining in this field. The first one is the separation of fullerenes from each other by preferential dissolution in a certain solvent. One such promising attempt is already known [121]. It is doubtful that really powerful nonchromatographic procedures for separating fullerenes can be developed. However, all opportunities to enrich the solution or solid phase with, say, C_{70} relative to C_{60}, by choosing a proper solvent and temperature, should be investigated. From the point of view of thermodynamics, this means that multicomponent (at least three component) phase diagrams of the systems of fullerenes with the solvent have to be studied and modeled. Powerful computer modeling will be possible if the data on the interaction of fullerenes in the liquid solutions, solid phase, and the formation and interaction of the solid solvates in the ternary systems become available. It is clear from the existing data that C_{70} is more soluble than C_{60} in 1,2-dimethylbenzene and that the ratio of solubilities increases with increasing temperature. The interactions in the ternary system, however, could spoil this clear picture.

The second and more exciting problem is the development of a theory capable of predicting solubilities of fullerenes. In the same way that the very simple and useful Hildebrand model was based mainly on the data on the solubility of iodine, a theory of fullerene solubility should be developed, using C_{60} solubility data at least for certain classes of solvents. Aromatic solvents, for which much thermodynamic information is available, look like appropriate candidates. The theory first has to give an explanation for the loss of entropy in the solution process. Prediction of trends in solubility behavior and absolute values of solubility are also in demand.

More theoretical and experimental work is needed in order to follow the trends in stability and structure of solid solvates of fullerenes. Certain parameters of the solvated crystals will probably help to understand solute–solvent interaction in the liquid solutions. Massively solvated crystals, where fullerene molecules are already completely surrounded by solvent molecules, are of special interest in this case. Another interesting issue is whether orientational phase transitions occur in solid solvates. As has already been stated, the temperature of the transition was shifted in some solvates relative to fcc C_{60}, while in aromatic solvates, this transition was not at all observed. Detailed study of the orientational phase transition both in pristine fullerenes and in the solvates could clear up the transition's mechanism on the molecular level. Solid-state NMR measurements at different temperatures would be useful in this case.

New reports on the formation of the aggregates in the solutions will probably appear. It is important to discover the nature of different processes that lead to the formation of clusters. Large kinetically stable clusters, observed in mixed solvents, are first candidates for such investigations.

ACKNOWLEDGMENTS

M. V. K. thanks RFBR, Grant 970332492, and the Russian State Program "Fullerenes and Atomic Clusters", Grant 96117, for financial support.

REFERENCES

1. H. W. Kroto, J. P. Heath, S. C. O'Brien, R. F. Curl, and R. E. Smalley, *Nature* **1985**, *318*, 162.
2. W. Krätschmer, L. D. Lamb, K. Fostiropoulos, and D. R. Huffman, *Nature* **1990**, *347*, 354–358.
3. M. T. Beck and G. Mandi, *Fullerene Sci. Technol.* **1997**, *5*, 291–310.
4. N. Sivaraman, B. Dhamodaran, I. Kaliappan, T. G. Srinivasan, P. R. Vasudeva Rao, and C. K. Mathews, *J. Org. Chem.* **1992**, *57*, 6077–6079.
5. X. Zhou, J. Liu, Z. Jin, Z. Gu, Y. Wu, and Y. Sun, *Fullerene Sci. Technol.* **1997**, *5*, 285–290.
6. R. S. Ruoff, D. S. Tse, R. Malhotra, and D. C. Lorents, *J. Phys. Chem.* **1993**, *97*, 3379–3383.
7. W. A. Scrivens and J. M. Tour, *J. Chem. Soc. Chem. Commun.* **1993**, *15*, 1207–1209.

8. T. M. Letcher, P. B. Crosby, U. Domanska, P. W. Fowler, and A. C. Legon, *S. Afr. J. Chem.* **1993**, *46*, 41.

9. J. Catalan, J. L. Saiz, J. L. Laynez, N. Jagerovic, and J. Elguero, *Angew. Chem. Int. Ed. Engl.* **1995**, *34*, 105–107.

10. T. Tomiyama, S. Uchiyama, and H. Shinohara, *Chem. Phys. Lett.* **1997**, *264*, 143.

11. S. Talukdar, P. Pradhan, and A. Banerji, *Fullerene Sci. Technol.* **1997**, *5*, 547–557.

12. K. Kimata, T. Hirose, K. Moriuchi, K. Hosoya, T. Araki, and N. Tanaka, *Anal. Chem.* **1995**, *67*, 2556.

13. W. Chen and Z. Xu, *Fullerene Sci. Technol.* **1998**, *6*, 695.

14. M. T. Beck, G. Mandi, and S. Keki, in *Recent Advances in the Chemistry and Physics of Fullerenes and Related Materials*, R. S. Ruoff and K. M. Kadish, eds., Electrochemical Society, Pennington, NJ, **1995**, 1510.

15. R. J. Doome, S. Dermaut, A. Fonseca, M. Hammida, and J. B. Nagy, *Fullerene Sci. Technol.* **1997**, *5*, 1593–1606.

16. A. L. Smith, J. Tian, and M. V. Korobov, in preparation.

17. M. T. Beck and G. Mandi, in *Recent Advances in the Chemistry and Physics of Fullerenes and Related Materials*, R. S. Ruoff and K. M. Kadish, eds., Electrochemical Society, Pennington, NJ, **1996**, 32.

18. D. Heymann, *Carbon* **1996**, *34*, 627–631.

19. Y. Nagano, T. Tamura, and T. Kiyobayashi, *Chem. Phys. Lett.* **1994**, *228*, 125–130.

20. K. Lozano, L. P. F. Chibante, X. Y. Sheng, A. Gaspar-Rosas, and E. V. Barrera, "Physical Examination and Handling of Wet and Dry C60," in *Processing and Handling of Powders and Dusts*, T. P. Battle and H. Henein (eds.). Minerals, Metals, and Materials Society, Warrendate PA, 1997.

21. Y. Nagano and T. Nakamura, *Chem. Phys. Lett.* **1997**, *265*, 358–360.

22. T. M. Letcher, P. B. Crosby, U. Domanska, P. W. Fowler, and A. C. Legon, *S. Afr. J. Chem.* **1993**, *46*, 41.

23. C. Gripon, L. Legrand, I. Rosenman, F. Boue, and C. J. Regnaut, *J. Crystal Growth* **1998**, *183*, 258.

24. N. Sivaraman, R. Dhamodaran, I. Kaliappan, T. G. Srinavasan, P. R. Vasudeva Rao, and C. K. Mathews, "Solubility of C60 and C70 in Organic Solvents", in *Recent Advances in the Chemistry and Physics of Fullerenes and Related Materials*, K. M. Kadish and R. S. Ruoff, eds., The Electrochemical Society, Pennington, NJ, **1994**, 156–165.

25. F. Diederich and C. Thilgen, *Science* **1996**, *271*, 317–323.

26. S. Friedman, D. DeCamp, R. Sijbesma, G. Srdanov, F. Wudl, and G. Kenyon, *J. Am. Chem. Soc.* **1993**, *115*, 6506.

27. R. Sijbesma, G. Srdanov, F. Wudl, J. Castro, C. Wilkins, S. Fredman, and G. Kenyon, *J. Am. Chem. Soc.* **1993**, *115*, 6510.

28. G. Andrievsky, M. Kosevich, O. Vivk, V. Shelkovsky, and L. Vashchenko, *J. Chem. Soc. Chem. Commun.* **1995**, 1281.

29. W. Scrivens, J. Tour, and K. Creek, *J. Am. Chem. Soc.* **1994**, *116*, 4517.

30. T. Anderson, K. Nillson, M. Sundahl, G. Westman, and O. Wennerstrom, *J. Chem. Soc. Chem. Commun.* **1992**, 604.

31. P. Boulas, W. Kutner, T. Jones, and K. Kadish, *J. Phys. Chem.* **1994**, *98*, 1282.

32. A. Buvari-Barcza, L. Barcza, T. Braun, I. Konkoly-Thege, K. Ludanyi, and K. Vekey, *Fullerene Sci. Technol.* **1997**, *5*, 311.

33. T. Andersson, G. Westman, O. Wennerstrom, and M. Sundahl, *J. Chem. Soc. Perkin Trans. 2* **1994**, 1097.

34. R. M. Williams and J. W. Verhoeven, *Rec. Trav. Chim. Pays-Bas* **1992**, *111*, 531–532.

35. J. L. Atwood, G. A. Koutsantonis, and C. L. Raston, *Nature* **1994**, *368*, 229–231.

36. H. Tomioka and K. Yamamoto, *J. Chem. Soc. Perkin Trans.* **1995**, *1*, 63.

37. I. Lamparth and A. Hirsch, *J. Chem. Soc. Chem. Commun.* **1994**, 1727.

38. H. Tokuyama, S. Yamado, E. Nakamura, T. Shiraki, and Y. Sugiura, *J. Am. Chem. Soc.* **1993**, *115*, 7918.

39. D. M. Guldi, H. Hungerbuhler, and K.-D. Asmus, *J. Phys. Chem.* **1995**, *99*, 13487–13493.

40. D. Guldi and K.-D. Asmus, in *Recent Advances in the Chemistry and Physics of Fullerenes and Related Materials*, R. S. Ruoff and K. M. Kadish, eds., Electrochemical Society, Pennington, NJ, **1997**, 82.

41. Y. P. Sun, G. Lawson, N. Wang, B. Liu, and D. Moton, in *Recent Advances in the Chemistry and Physics of Fullerenes and Related Materials*, K. Kadish and R. S. Ruoff, eds., Electrochemical Society, Pennington, NJ, **1997**, 645.

42. A. Beeby, J. Eastoe, and R. K. Heenan, *J. Chem. Soc. Chem. Commun.* **1994**, 173.

43. T. Yagami, K. Fukuhara, S. Sueyoshi, and N. Miyata, *J. Chem. Soc. Chem. Commun.* **1994**, 517.

44. K. Hungerbuhler, D. Guldi, and K.-D. Asmus, *J. Am. Chem. Soc.* **1993**, *115*, 3386.

45. P. M. Allemand, K. C. Khemani, A. Koch, F. Wudl, K. Holczer, S. Donovan, G. Gruner, and J. D. Thompson, *J. Am. Chem. Soc.* **1991**, *113*, 1050.

46. R. J. Sension, A. Z. Szarka, G. R. Smith, and R. M. Hochstrasser, *Chem. Phys. Lett.* **1991**, *185*, 179.

47. Y. Wang, *J. Phys. Chem.* **1992**, *96*, 764–767.

48. Y. Wang and L. T. Cheng, *J. Phys. Chem.* **1992**, *96*, 1530–1532.

49. R. Seahadri, F. D'Souza, V. Krishnan, and C. Rao, *Chem. Lett.* **1993**, 217.

50. Y.-P. Sun, C. E. Bunker, and B. Ma, *J. Am. Chem. Soc.* **1994**, *116*, 9692–9699.

51. G. Mandi and M. Beck, in *Recent Advances in the Chemistry and Physics of Fullerenes and Related Materials*, K. M. Kadish and R. S. Ruoff, eds., Electrochemical Society, Pennington, NJ, **1997**, 382.

52. H. Klos, I. Rystan, W. Schutz, B. Gotschby, A. Skiebe, and A. Hirsch, *Chem. Phys. Lett.* **1994**, *224*, 333.

53. H. N. Ghosh, D. K. Palit, A. V. Sapre, and J. P. Mittal, *Chem. Phys. Lett.* **1997**, *265*, 365–373.

54. Y. Kubozono, H. Maceda, Y. Takebayashi, K. Hiraoka, T. Nakai, S. Kashino, S. Ukita, and T. Sogabe, *J. Am. Chem. Soc.* **1996**, *118*, 6998.

55. G. M. Olah, I. Busci, D. Ha, R. Anizfeld, C. S. Lee, and G. K. Surya Prahasb, *Fullerene Sci. Technol.* **1997**, *5*, 389.

56. B. Nie, K. Hasan, M. D. Greaves, and V. M. Rotello, *Tetrahedron Lett.* **1995**, *36*, 3617–3618.

57. L. M. Giovane, J. W. Barco, T. Yadav, A. L. Lafleur, J. A. Marr, J. B. Howard, V. M. Rotello, *J. Phys. Chem.* **1993**, *97*, 8560–8561.

58. B. Nie and V. M. Rotello, *J. Org. Chem.* **1996**, *61*, 1870–1871.

59. A. Yi-Zhong, G. A. Ellis, A. L. Viado, and Y. Rubin, *J. Org. Chem.* **1995**, *60*, 6353–6361.

60. M. D. Diener and J. M. Alford, *Nature* **1998**, *393*, 668–671.

61. J. Chen, M. A. Hamon, H. Hu, Y. S. Chen, A. M. Rao, P. C. Eklund, and R. Haddon, *Science* **1998**, *282*, 95–98.

62. R. V. Honeychuck, T. W. Cruger, and J. Milliken, *J. Am. Chem. Soc.* **1993**, *115*, 3034–3035.

63. R. N. Thomas, *Anal. Chim. Acta* **1994**, *289*, 57–67.

64. I. Rubtsov, D. Khudiakov, V. Nadtochenko, A. Lobach, and A. Moravskii, *Chem. Phys. Lett.* **1994**, *229*, 517.

65. A. M. Rao, P. Zhou, K.-A. Wang, G. T. Hager, J. M. Holden, Y. Wang, W.-T. Lee, X.-X. Bi, P. C. Eklund, D. S. Cornett, M. A. Duncan, and I. J. Amster, *Science* **1993**, *259*, 955–957.

66. A. Rao, P. Eklund, J. Hodeau, L. Marques, and M. Nunez-Regueiro, *Phys. Rev. B* **1997**, *57*, 4766.

67. G. Wang, K. Komatsu, Y. Murata, and M. Shiro, *Nature* **1997**, *387*, 583.

68. V. N. Bezmelnitsin, A. V. Eletskii, and E. V. Stepanov, *J. Phys. Chem.* **1994**, *98*, 6665–6667.

69. Y. Sun and C. Bunker, *Nature* **1993**, *365*, 398.

70. Y. Sun, B. Ma, C. Bunker, and B. Liu, *J. Am. Chem. Soc.* **1995**, *117*, 12705.

71. H. N. Ghosh, A. V. Sapre, and J. P. Mittal, *J. Phys. Chem.* **1996**, *100*, 9439–9443.

72. Q. Ying, J. Maracek, and B. Chu, *Chem. Phys. Lett.* **1994**, *219*, 214.

73. H. Ajie, M. M. Alvarez, S. J. Anz, R. D. Beck, F. Diederich, K. Fostiropoulos, D. R. Huffman, W. Krätschmer, Y. Rubin, K. E. Schriver, K. Sensharma, and R. L. Whetten, *J. Phys. Chem.* **1990**, *94*, 8630–8633.

74. R. S. Ruoff, R. Malhotra, D. L. Huestis, D. S. Tse, and D. C. Lorents, *Nature* **1993**, *362*, 140–141.

75. J. S. Ahn, K. Suzuki, Y. Iwasa, N. Otsuka, and T. Mitani, Proc. SPIE Fullerenes and Photonics IV. SPIE, Bellingham, WA **1997**, 3142.

76. J. S. Murray, S. G. Gagarin, and P. Politzer, *J. Phys. Chem.* **1995**, *99*, 12081–12083.

77. A. L. Smith, L. Y. Wilson, and G. R. Famini, "A Quantitative Structure-Property Relationship Study of C60 Solubility", in *Recent Advances in the Chemistry and Physics of Fullerenes and Related Materials*, R. S. Ruoff and K. D. Kadish, eds., Electrochemical Society, Pennington, NJ, **1996**, 53–62.

78. G. R. Famini, B. C. Marquez, and L. Y. Wilson, *J. Chem. Soc. Perkin Trans. 2* **1993**, 773–782.

79. Y. Marcus, *J. Phys. Chem.* **1997**, *101*, 8617–8623.

80. A. Hirsch, *The Chemistry of the Fullerenes*. Thieme, New York, 1994.

81. M. V. Korobov, A. L. Mirakian, N. V. Avramenko, and R. S. Ruoff, *Dokl. Akad. Nauk.* **1996**, *349*, 346.

82. A. L. Smith, E. Walter, M. V. Korobov, and O. Gurvich, *J. Phys. Chem.* **1996**, *100*, 6775.

83. G. Olofsson, I. Wadso, and R. S. Ruoff, "Incongruent Melting Transition for C60 Cosolvates of Some Aromatic Solvents", in *Recent Advances in the Chemistry and Physics of Fullerenes and Related Materials*, K. M. Kadish and R. S. Ruoff, eds., Electrochemical Society, Pennington, NJ, **1996**, 17–31.

84. M. V. Korobov, et al., *J. Phys. Chem.* **1998**, *102*, 3712–3717.

85. M. V. Korobov, A. L. Mirakian, N. V. Avramenko, I. L. Odinec, and R. S. Ruoff, "Formation of Cosolvates explains the solubility behavior of C60: does the model work?", in *Recent Advances in the Chemistry and Physics of Fullerenes and Related Materials*, R. S. Ruoff and K. M. Kadish, eds., Electrochemical Society, Pennington, NJ, **1996**, 5–16.

86. N. Avramenko, A. Mirakian, and M. Korobov, *High Pressures* **1998**, *30*, 371.

87. M. V. Korobov, A. L. Mirakian, N. V. Avramenko, G. Olofsson, A. L. Smith, and R. S. Ruoff, *J. Phys. Chem.* **1999**, *103B*, 1339–1346.

88. R. Ceolin, F. Michaud, V. Agafonov, J. L. Tamarit, A. Dworkin, and H. Swarc, in *Fullerenes Vol 5: Recent Advances in the Chemistry and Physics of Fullerenes and Related Materials*, K. M. Kadish and R. S. Ruoff, eds., Electrochemical Society, Pennington, NJ, **1997**, 373.

89. R. Ceolin, V. Agafonov, B. Bachet, A. Gonthier-Vassal, H. Szwarc, S. Toscani, G. Keller, C. Farbe, and A. Rassat, *Chem. Phys. Lett.* **1995**, *244*, 100–104.

90. R. Ceolin, V. Agafonov, B. Bachet, A. Gonthier-Vassal, H. Szwarc, G. Keller, C. Fabre, and A. Rassat, *Fullerene Sci. Technol.* **1997**, *5*, 559.

91. Y. Nagano, Y. Miyazaki, T. Tamura, T. Nakamura, and T. Kimura, "Thermodynamic Properties of Massively Solvated C_{60} Crystal: $C_{60}(CCl_4)_{13}$", in *Fullerenes Vol 5: Recent Advances in the Chemistry and Physics of Fullerenes and Related Materials*, K. M. Kadish and R. S. Ruoff, eds., Electrochemical Society, Pennington, NJ, **1997**, 364.

92. Y. Nagano and T. Tamura, *Chem. Phys. Lett.* **1996**, *252*, 362–366.

93. J. Cheng, Y. Jin, A. Xenopoulos, W. Chen, and B. Wunderlich, *Mol. Cryst. Liq. Cryst.* **1994**, *250*, 359.

94. S. Pekker, G. Faigel, K. Fodor-Csorba, L. Granasy, E. Jakab, and M. Tegze, *Solid State Commun.* **1992**, *83*, 423–426.

95. A. Talyzin, *J. Phys. Chem. B* **1997**, *101*, 9679.

96. M. Jansen and G. Waidmann, *Z. Anorg. Allg. Chem.* **1995**, *621*, 14–18.

97. A. Gangopadhyay, J. Schilling, M. Leo, W. Buhro, K. Robinson, and T. Kowalewski, *Solid State Commun.* **1995**, 597.

98. A. Balch, J. W. Lee, B. C. Noll, and M. M. Olmstead, *J. Chem. Soc. Chem. Commun.* **1993**, 56–58.

99. M. F. Meidine, P. B. Hitchcock, H. W. Kroto, R. Taylor, and D. R. M. Walton, *J. Chem. Soc. Chem. Commun.* **1992**, 1534–1537.

100. J. Cheng, Y. Jin, A. Xenopoulos, W. Chen, and B. Wunderlich, *Mol. Cryst. Liq. Cryst.* **1994**, *250*, 359–371.

101. H. He, J. Barras, J. Foulkes, and J. Klinowski, *J. Phys. Chem.* **1997**, *101*, 117–122.

102. M. Barrio, D. O. Lopez, J. L. Tamarit, H. Szware, S. Toscani, and R. Ceolin, *Chem. Phys. Lett.* **1996**, *260*, 78–81.

103. *CRC Handbook of Chemistry and Physics*, 77th ed., D. R. Lide (ed.). CRC Press, Boca Raton, FL, 1996.

104. V. Gritsenko, O. Dyachenko, N. Kuschc, N. Spitsina, E. B. Yagubskii, N. V. Avramenko, and M. Phrolova, *Russian Chem. Bull.* **1994**, *43*, 1183.

105. S. Pekker, G. Faigel, G. Ogzlanyi, M. Tegze, T. Kemeny, and E. Jakab, *Synth. Metals* **1993**, *55–57*, 3014–3020.

106. Y. L. Slovokhotov, I. V. Moskaleva, V. I. Shilnikov, E. F. Valeev, Y. N. Novikov, A. I. Yanovsky, and Y. T. Struchkov, *Mol. Mat.* **1996**, *8*, 117.

107. G. Ozlany, G. Bortel, G. Faigel, S. Pekker, and M. Tegze, *J. Phys. Condensed Matter* **1993**, *5*, 165.

108. M. Korobov, N. Avramenko, E. Stukalin, and R. Ruoff, "Ternary Systems of C60 and C70 with Aromatic Solvents", in *Recent Advances in the Chemistry and Physics of Fullerenes and Related Materials*, K. M. Kadish, Ed. Electrochemical Society, Pennington, NJ, **1998**, 559.

109. M. Jansen, K. Kniep, and G. Waidmann, *Fullerene Sci. Technol.* **1997**, *5*, 699.

110. R. Ceolin, *Fullerene Sci. Technol.* **1997**, *5*, 559.

111. J. Yin, B.-H. Wang, Z. Li, Y.-M. Zhang, X.-H. Zhou, and Z.-N. Gu, *J. Chem. Thermodyn.* **1996**, *28*, 1145.

112. Y. Jin, Y. Cheng, M. Varma-Nair, G. Liang, Y. Fu, B. Wunderlich, X.-D. Xiang, R. Mostovoy, and A. Zettl, *J. Phys. Chem.* **1992**, *96*, 5151.

113. A. L. Smith, D. Li, B. King, and G. Zimmerman, "Solubility, Diffuse Reflectance Infrared Spectroscopy, and High-Temperature Gas Phase Studies of C60 and C70", in *Fullerenes: Chemistry, Physics and New Directions*, K. M. Kadish and R. S. Ruoff, eds., Electrochemical Society, Pennington, NJ, **1994**, 443–458.

114. M. Hagen, E. Meijer, G. Mooij, D. Frenkel, and H. Lekkerkerker, *Nature* **1993**, *365*, 425.

115. P. Ruelle, A. Farina-Cuendet, and U. Kesselring, *J. Chem. Soc. Chem. Commun.* **1995**, 1161.

116. P. Ruelle, A. Farina-Cuendet, and V. Kesselring, *J. Am. Chem. Soc.* **1996**, *118*, 1777–1784.

117. R. Perotti, *J. Phys. Chem.* **1963**, *67*, 1840.

118. R. Perotti, *J. Phys. Chem.* **1965**, *69*, 281.

119. M. Lucas, *J. Phys. Chem.* **1976**, *80*, 359.

120. H. Smorenburg, R. Crevecouer, and L. de Schepper, and L. de Graaf, *Phys. Rev. E* **1995**, *52*, 2742.

121. R. J. Doome, A. Fonseca, H. Richter, J. B. Nagy, P. A. Thiry, and A. A. Lucas, *J. Phys. Chem. Solids* **1997**, *58*, 1839–1843.

CHAPTER 3

ORGANIC CHEMISTRY OF FULLERENES

STEPHEN R. WILSON, DAVID I. SCHUSTER, BERTHOLD NUBER,
MARK S. MEIER, MICHELE MAGGINI, MAURIZIO PRATO, and
ROGER TAYLOR

3.1 INTRODUCTION

The unique structural properties of C_{60}, the most abundant, the least expensive, and therefore the most thoroughly investigated of the fullerenes, govern its chemical behavior. The six-membered rings are not classically aromatic in that they contain alternating single and double bonds. The C–C double bonds are located exclusively between two six-membered rings (so-called 6,6-bonds, bond length 1.38 Å) and are exocyclic with respect to the five-membered rings. Thus, the C–C bonds between five- and six-membered rings, the 5,6-bonds, are essentially single bonds with length 1.45 Å. The overall fullerene structures can therefore be viewed as fused 1,3,5-cyclohexatrienes and [5] -radialenes. The structures of higher fullerenes are built using increasingly larger chains of six-membered rings, while the number of five-membered rings remains constant at twelve.

Second, the bonds at each of the carbons on the fullerene surface deviate substantially from the ideal for sp^2-hybridized carbons. The pyramidalization angle, defined by Haddon and Raghavachari [1] as $(\theta_{\sigma\pi} - 90)^\circ$, is 11.6° for C_{60}, the largest value of any of the fullerenes ($\theta_{\sigma\pi} = 0°$ for graphite). This pyramidalization angle induces strain of 8.5 kcal/mol/per carbon atom, the single largest contribution to the total heat of formation of C_{60} (10.16 kcal/mol/per carbon atom) [2]. Exohedral additions to the sphere are therefore driven by relief of strain, while endohedral additions cause an increase in strain and are consequently strongly inhibited. For example, even when isolated nitrogen atoms are inserted into C_{60} using a beam technique, they do not bond to carbon

Fullerenes: Chemistry, Physics, and Technology, Edited by Karl M. Kadish and Rodney S. Ruoff.
0-471-29089-0 Copyright © 2000 John Wiley & Sons, Inc.

and remain spectroscopically indistinguishable from nitrogen atoms *in vacuo* [3]. Similarly, fluorine atoms and methyl radicals within the C_{60} sphere are predicted to be chemically inert [4].

Third, theory predicts three low-lying degenerate lowest unoccupied molecular orbitals (LUMOs) in C_{60}, which accounts for its exceptionally low reduction potential (-0.99 eV versus SCE) and its ability to reversibly accept up to six electrons electrochemically, up to the hexa-anion [5,6]. In contrast, oxidation of C_{60} is relatively difficult (IP $+1.76$ eV) [5], although C_{60} radical cations can be generated by photoinduced electron transfer and by other methods [7].

Certain general patterns of reactivity emerge from these basic properties of C_{60}, and to a lesser extent apply to C_{70} and higher fullerenes:

1. C_{60} behaves essentially as an electron-deficient alkene, not as an aromatic hydrocarbon. Its characteristic reactions involve cycloadditions as well as additions of nucleophiles and free radicals across the bonds between two six-membered rings, that is, 6,6-bonds. Hydroborations, hydrometallations, hydrogenations, halogenations, and metal complex formation are also observed [8].

2. The main driving force for additions to fullerenes is the relief of strain, leading to change in hybridization at the reacting carbons on the surface of the sphere from sp^2 to sp^3, which is readily detected by ^{13}C NMR spectrometry. X-ray structures of monoadducts of C_{60} and C_{70} indicate distortions to relieve strain on the fullerene surface, which have strong implications with respect to subsequent additions [9]. Since fullerene additions are generally exothermic and exergonic, due to relief of strain, it is often difficult to stop at the monoadduct stage. Multiadducts are well known and have very interesting structures and properties, as will be seen below. At an advanced stage of the addition sequence, the remaining unreacted six-membered rings become essentially isolated aromatic rings, and consequently addition comes to a stop.

3. The regioselectivity of addition reactions is governed by avoidance of products with a 5,6 double bond, since each such bond carries a price tag of 8.5 kcal/mol. Thus, 1,2-addition occurs to give ring-closed 6,6-adducts, with two sp^3 carbon atoms on the fullerene, and in some cases ring-opened 6,5-adducts known as fulleroids, analogous to bridged annulenes, where all fullerene atoms are still sp^2 hybridized. Generally, 6,5-open adducts are thermodynamically more stable than corresponding 6,5-closed structures (which necessarily contain 5,6 double bonds) but are less stable than the corresponding 6,6-closed adducts [10]. Isomerization of 6,5-open to 6,6-closed adducts in some cases can be brought about by light or heat [11].

4. Bis-addition to C_{60}, even when restricted to 6,6-bonds, is a complicated process since there are eight different sites for reaction. The selectivity of the second addition reaction is governed by differences in bond order and LUMO coefficients at each site [12]. Although at least two of us (DIS and SRW) believe that the term "regioselectivity" is not the proper term to describe discrimina-

tion between the *chemically distinct* 6,6-bonds on the surface of a C_{60} mono-adduct, this term is so firmly entrenched in the fullerene literature that it seems futile to suggest abandoning it at this late stage. Problems in the preparation of bis and higher adducts of C_{60} will be discussed later in this chapter. The isomer problem is clearly much more complex in the case of unsymmetrical addends and in additions to less symmetric higher fullerenes. One significant factor that simplifies reactions of higher fullerenes is that addition to the curved C–C double bonds at the poles is favored over addition to the more planar graphite-like bonds at the equator, because of the relief of strain.

The following sections will summarize recent progress in several areas involving synthesis of functionalized derivatives of fullerenes, with emphasis on reactions involving C_{60}. Because of space limitations, we have necessarily concentrated on processes of widest general applicability and interest.

3.2 CYCLOADDITIONS TO FULLERENES   (David I. Schuster)

Cycloadditions along with nucleophilic additions represent the most powerful methods for synthesizing a large variety of functional derivatives of fullerenes. The literature through 1994 on thermal and photochemical cycloadditions to C_{60} has been thoroughly reviewed by Hirsch [8] and therefore this section will concentrate on papers published between 1995 and 1998. As will be seen, a large number of interesting and unusual fullerene cycloadducts have been prepared in the last few years, many of which were designed for possible biological applications.

3.2.1 Diels–Alder [4+2] Cycloadditions

3.2.1.1 Additions to C_{60} Tables 3.1, 3.2, and 3.3 summarize much of the work reported in the last four years. In the reactions reported in Table 3.1, the dienes shown were added directly to C_{60} under the conditions indicated, while in Table 3.2 the dienes were generated *in situ* from the reactants shown. Use of this approach to prepare a variety of novel donor–acceptor dyads is summarized in Table 3.3. Yields of the cycloadducts are based on consumed C_{60} unless otherwise noted.

The first entry in Table 3.1 involves the addition of five analogs of Danishefsky's diene **1**, to give fullerene-fused cyclohexanones **2** [13]. These compounds are extremely useful for further chemical transformations [14]. The next entries, involving acyclic dienes **3** and a cyclic diene **5** with electron-withdrawing substituents [13], demonstrate that both electron-deficient and electron-rich dienes can successfully be added to C_{60}. This is readily rationalized in terms of interaction of the highest occupied molecular orbital (HOMO) with the very low-lying lowest unoccupied molecular orbital (LUMO) of C_{60} [13]. While the bicyclic diene **7** failed to add to C_{60}, for reasons that are not obvious, the tri-

TABLE 3.1 Diels–Alder Addition to C$_{60}$

Diene	Conditions	Product (Yield/*Reference Number*)
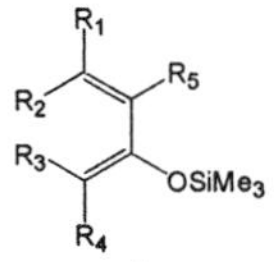 **1** R$_1$-R$_5$=H R$_1$=OMe; R$_2$-R$_5$=H R$_1$=Ph; R$_2$-R$_5$=H R$_1$,R$_2$=Me; R$_3$-R$_5$=H R$_1$=H; R$_2$-R$_3$=-CH$_2$CH$_2$-; R$_4$=R$_5$=H	Toluene, reflux, then 1 N HCl/THF	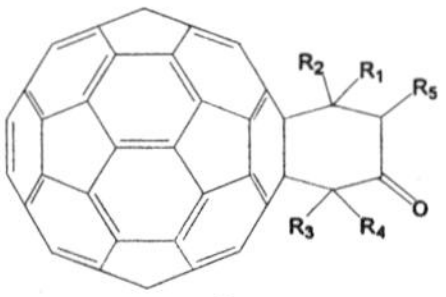 **2** (28-48% / *13*)
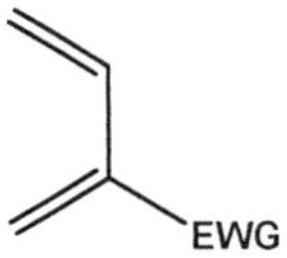 **3** EWG=CO$_3$Et, COMe, CN, SO$_2$Ph, NO$_2$	Toluene, reflux	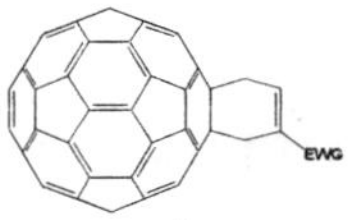 **4** (39–79% / *13*)
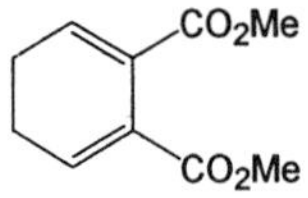 **5**	*o*-DCB, reflux	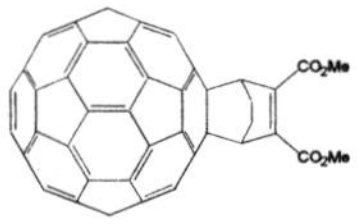 **6** (27% / *13*)
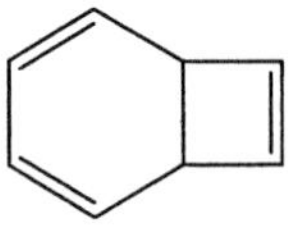 **7**		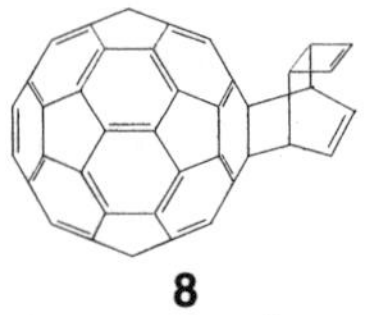 **8** (no reaction / *15*)
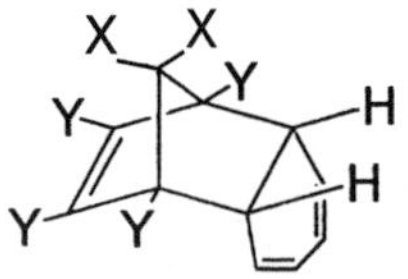 **9** X=OCH$_3$; Y=Cl	*o*-Xylene, 50 °C 50 h	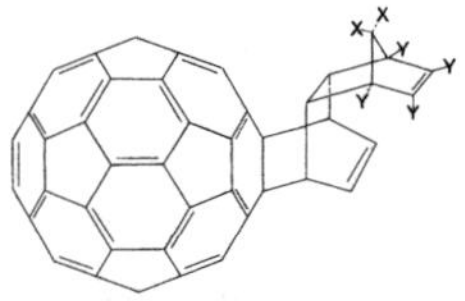 **10** (25% / *15*)
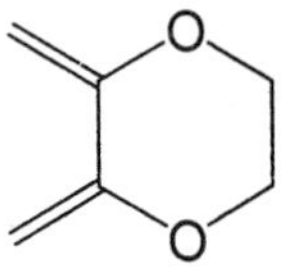 **11**	Toluene, 25 °C 12 h (dark)	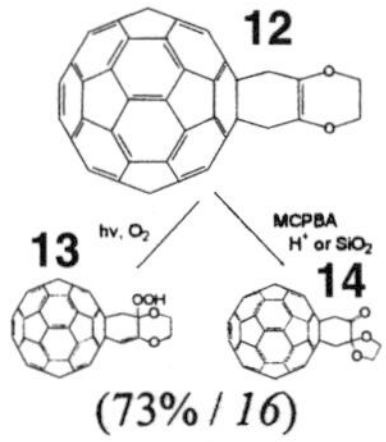 **12 13 14** (73% / *16*)

TABLE 3.1 (continued)

Diene	Conditions	Product (Yield/*Reference Number*)
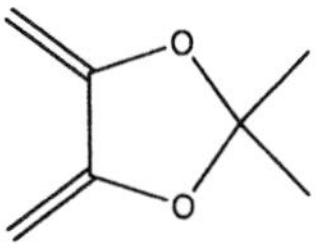 **15**	*o*-DCB, 300 MPa, 24 h; then TFA	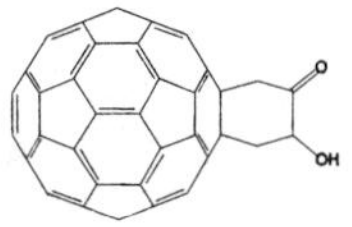 **16** (33% / *16*)
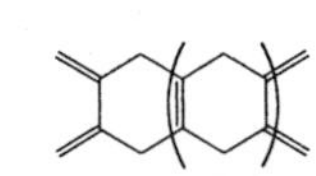 **17** 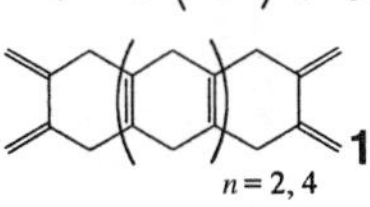 **18** $n = 2, 4$	Toluene, reflux, 8 h	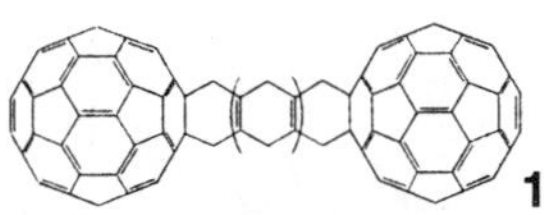 **19** (40-72% / *17*) n = 2,4,6
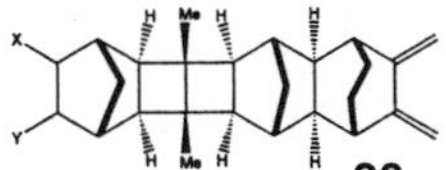 **20** X,Y = H 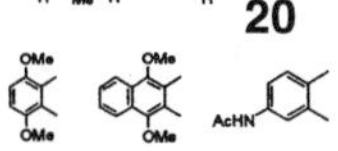 A B C D 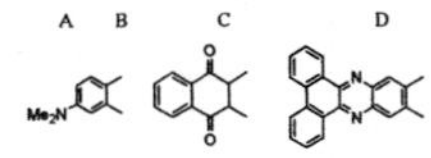 E F G	Toluene, reflux	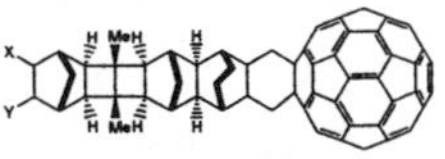 **21** *Ref. 18* X,Y = H 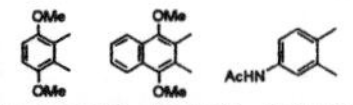 A(86%) B(100%) C(59%) D (41%) 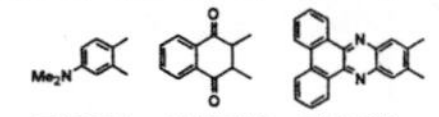 E(60%) F (58%) G(4.2%)
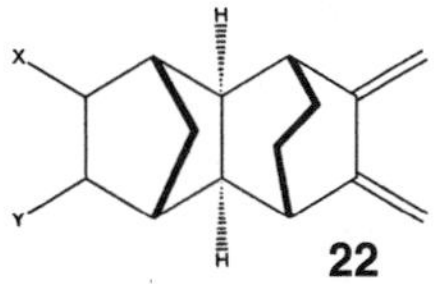 **22** X,Y 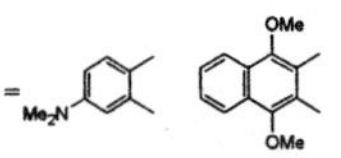 = A B 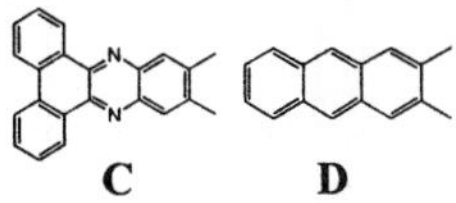 C D	Toluene, reflux	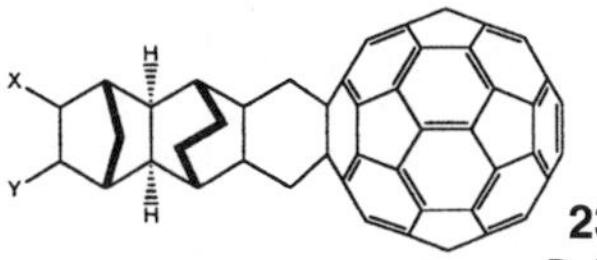 **23** *Ref. 18* X,Y 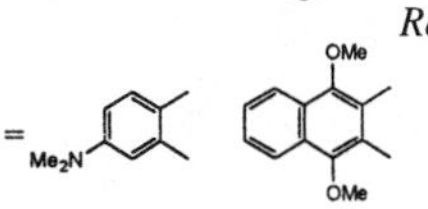 = A(61%) B(36%) 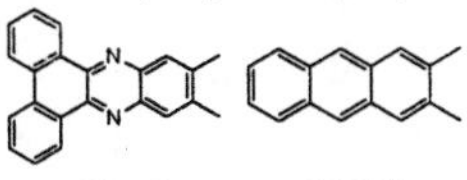 C(34%) D(52%)
 **24**	Toluene, reflux	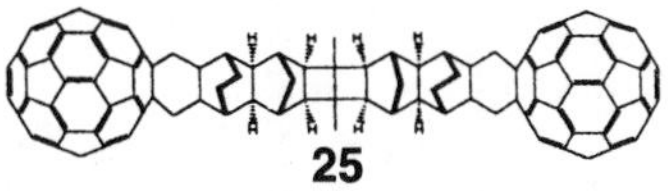 **25** (81% / *18*)

TABLE 3.2 In Situ Generation of Dienes: [4+2] Additions to C$_{60}$

Diene	Precursor	Conditions	Product	Yield	Reference
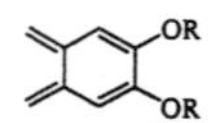 **26** (R-R = -CH$_2$CH$_2$(OCH$_2$CH$_2$O)-CH$_2$CH$_2$-)	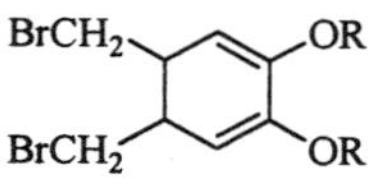 **27**		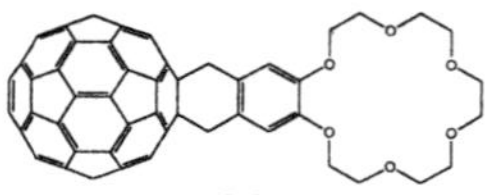 **28**		20
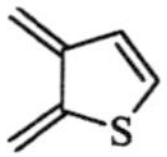 **29**	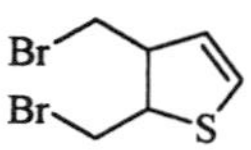 **30**	KI, 18-crown-6 Toluene, reflux	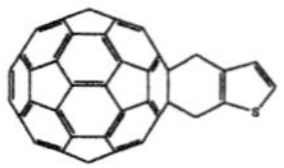 **31**	36%	21
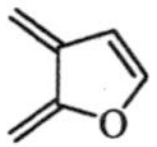 **32**	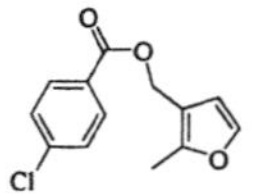 **33**	FVP, then PhCl, −20°C to RT	 **34**	35%	21
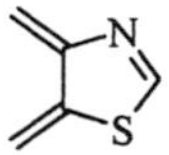 **35**	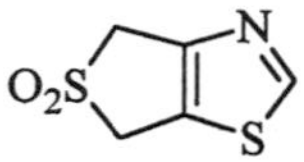 **36**	o-DCB Reflux, 2 h	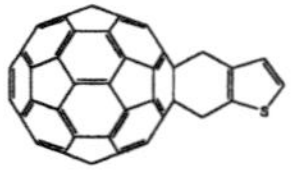 **37**	43%	21
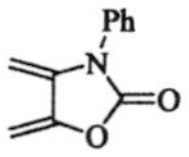 **38**			No Reaction		

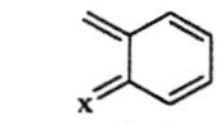

Reactants	Conditions	Product	Yield	Ref
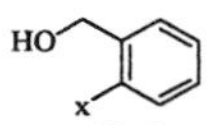 **39** (a, x=O; b, x=S) / **40** (a, x=O; b, x=S)	o-DCB 180 °C, 0.5 h	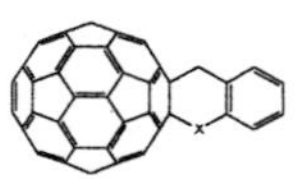 **41**	A = 50% B = 54%	21
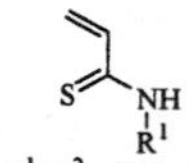 **42** (R^1, R^2 = CH$_3$, Ph) / **43** (R^1, R^2 = CH$_3$, Ph)	1. R^2COCl, pyr. 2. 65 °C, 0.5 h	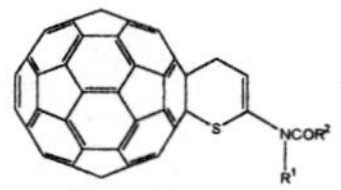 **44**	43–69%	21
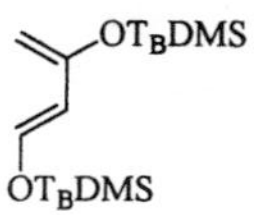 **45** (OT$_B$DMS, OT$_B$DMS)	o-DCB RT, overnight; then 10% HCl	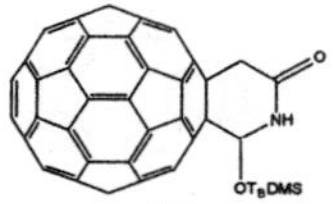 **46**	64%	21
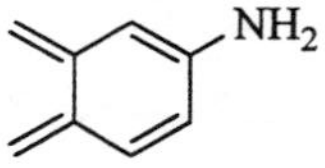 **47** (NH$_2$) / **48** (NH$_2$, O$_2$S)	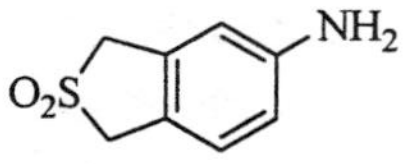 1,2,4-C$_6$H$_3$Cl$_3$ 214 °C, 20 h	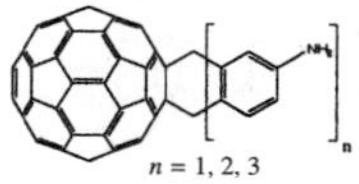 **49** (n = 1, 2, 3)	49%, n = 1 23%, n = 2	22
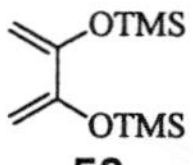 **50** (OTMS, OTMS) / **51** (OTMS, OTMS)	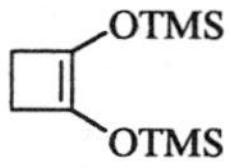 o-DCB, 180 °C, then HCl/ MeOH	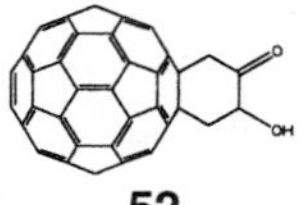 **52**	89%	15, 23
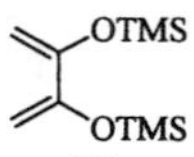 **50** (OTMS, OTMS) / **51** (OTMS, OTMS)	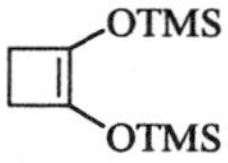 o-DCB, 180 °C, followed by Br$_2$, −77 °C to RT, Et$_3$N/ THF	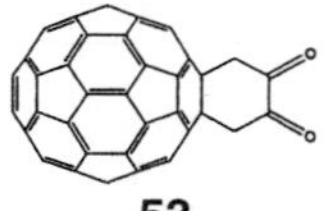 **53**		15, 23

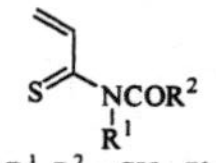

97

TABLE 3.2 (continued)

Diene	Precursor	Conditions	Product	Yield	Reference
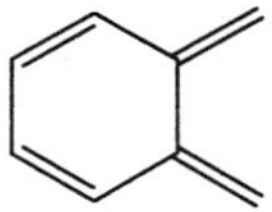 54	55 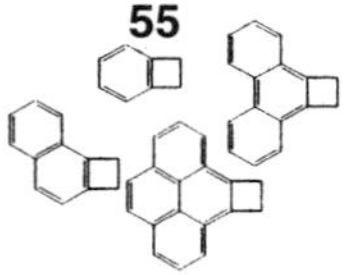	Tetralin, reflux 1 h	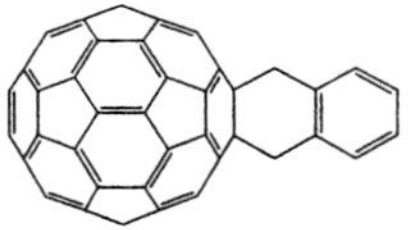 56 and analogous products	20–30%	24
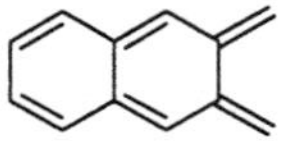 57	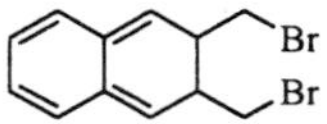 58 (Br, Br)	KI, 18-crown –6 Toluene, reflux	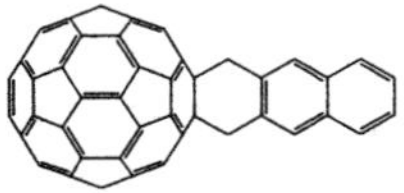 59	20%	24
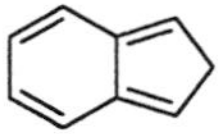 60	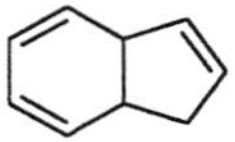 60	o-DCB, reflux		30–35%	25
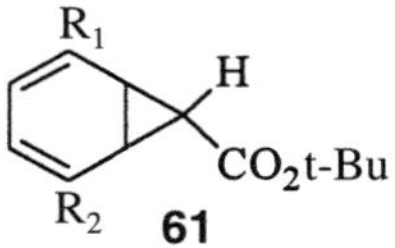 61 (R_1, H, CO_2t-Bu, R_2)	N_2CHCO_2-t-Bu + PhR	$Rh_2(OAc)_4$	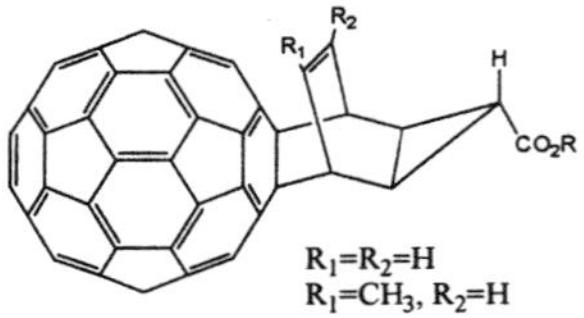 $R_1=R_2=H$; $R_1=CH_3$, $R_2=H$	A = 24% B = 43%	26

98

TABLE 3.3 Donor–Acceptor Dyads

Reactant	Conditions	Product	Reference
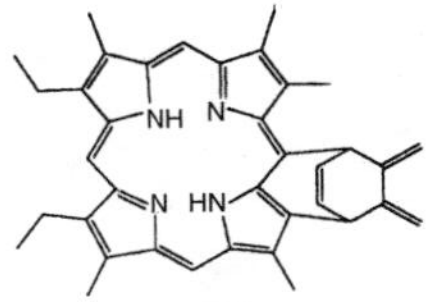 **62**	Toluene, reflux 3 h	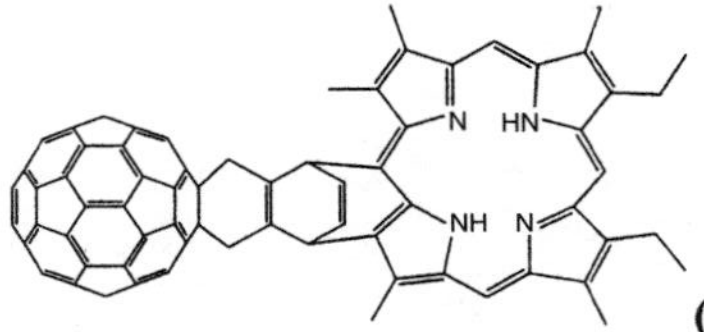 (34%) **63**	27
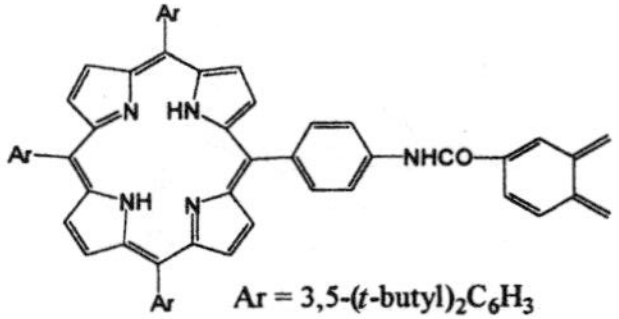 Ar = 3,5-(t-butyl)$_2$C$_6$H$_3$ **64**	Toluene	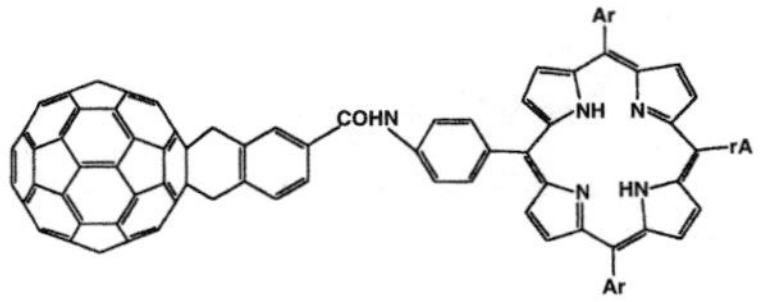 (50%) **65**	28
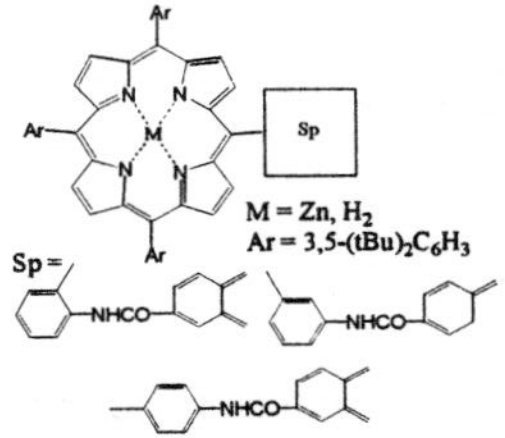 M = Zn, H$_2$ Ar = 3,5-(tBu)$_2$C$_6$H$_3$ Sp =		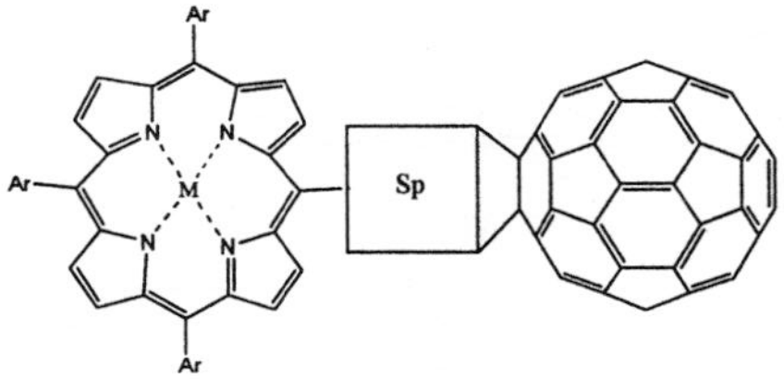 Sp **66**	29

99

TABLE 3.3 (continued)

Reactant	Conditions	Product	Reference
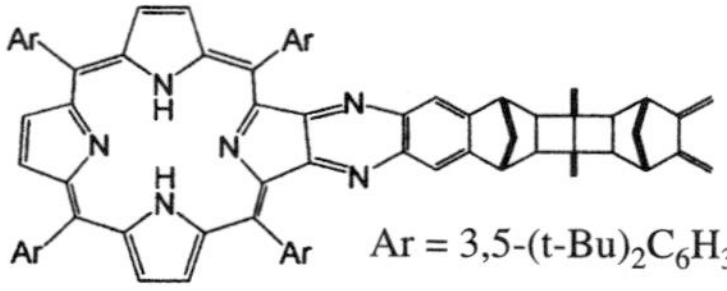 Ar = 3,5-(t-Bu)$_2$C$_6$H$_3$ **68** **70**	o-DCB 110–115 °C Toluene, reflux 3 h Toluene, reflux 3 h	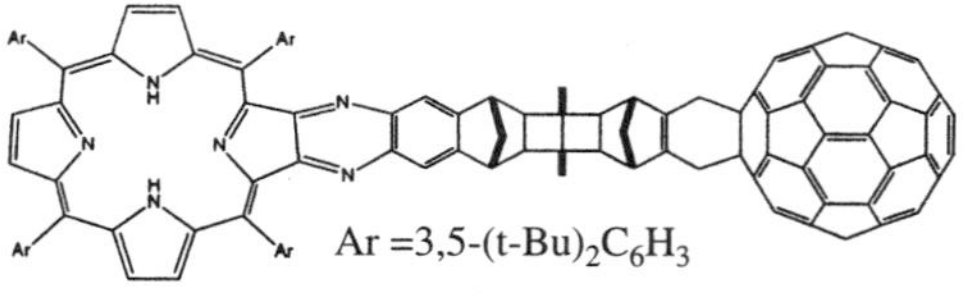 Ar = 3,5-(t-Bu)$_2$C$_6$H$_3$ **67 (84%)** **69 (49%)** **71 (42%)**	30 31 31

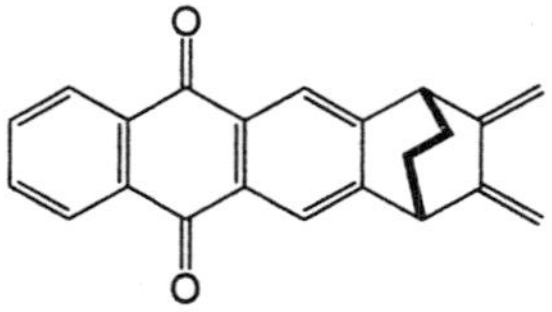

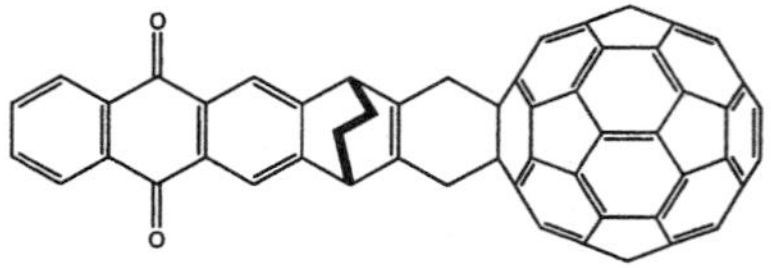

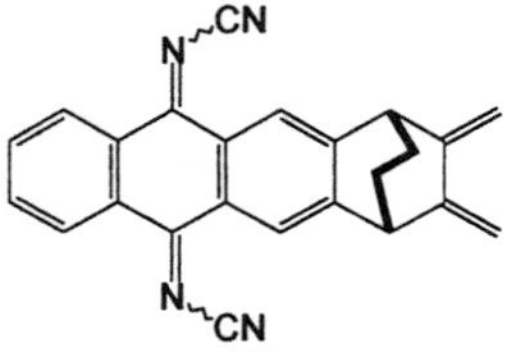

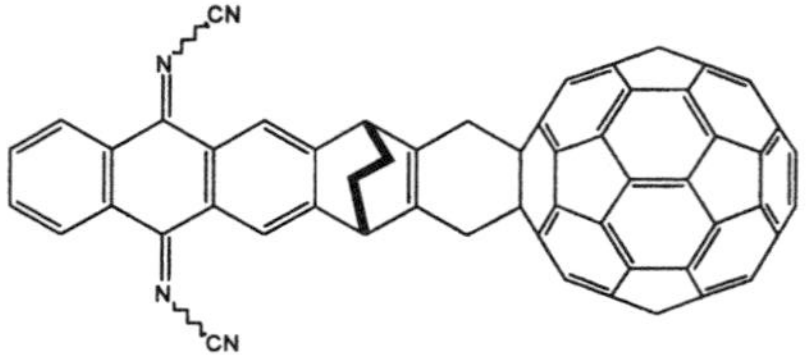

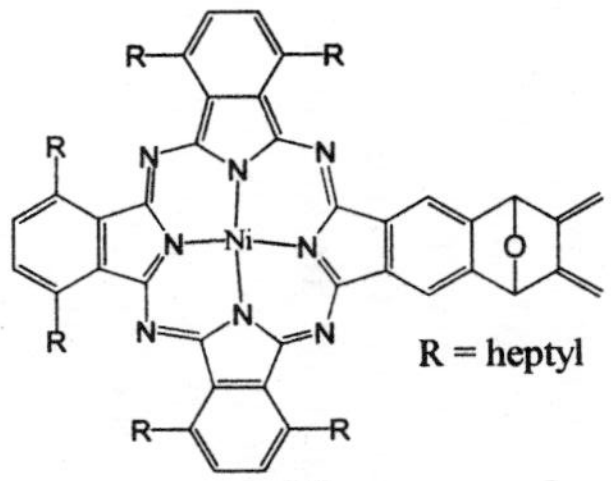

32

Toluene, reflux

72   R = heptyl

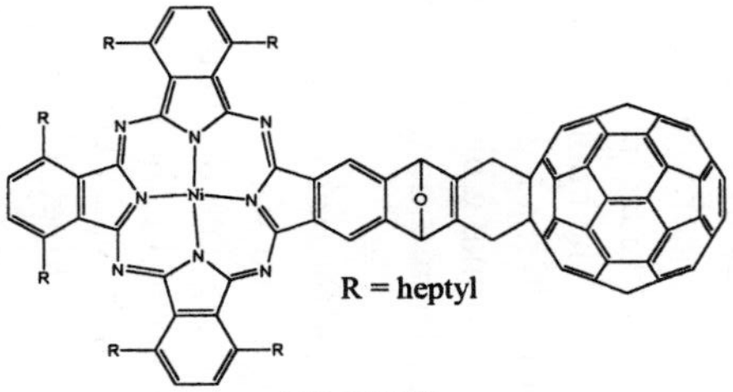

73 (75%)   R = heptyl

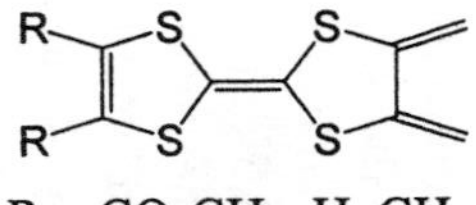

33

PhCN, reflux 1 h

R = CO$_2$CH$_3$, H, CH$_3$ **74**

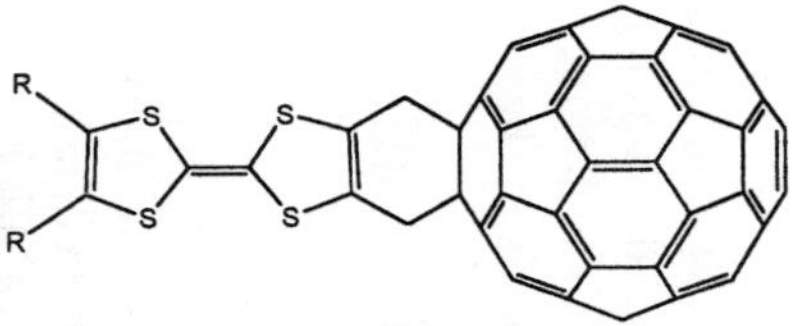

75 (19–58%)

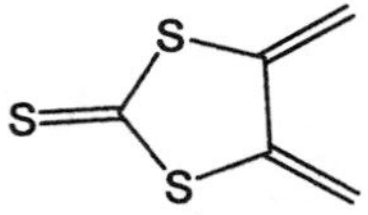

33

PhCN, reflux 2 h

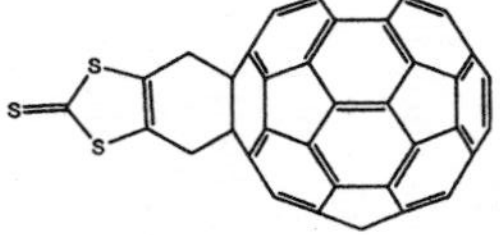

76

77 (41%)

cyclic diene **9** did give a [4+2] adduct **10** in moderate yield [15]. The ability to introduce useful functional groups is illustrated with dienes **11** and **15**, from which the dioxolene derivative **12** and acyloin **16** were prepared; **12** could be further elaborated functionally as shown [16]. Construction of bis-adducts **19** in which two C_{60} units are separated by a semirigid spacer was accomplished using dienes **17** and **18** [17]. The next series of entries in Table 3.1 involves synthesis of "ball-and-chain" adducts **21** and **23** containing a series of fused bicyclic linkers from addition to C_{60} of dienes **20** and **22** by Paddon-Row and co-workers [18]. It is possible using this technique to incorporate a variety of terminal aromatic groups, such as 1,4-dimethoxybenzene, naphthoquinone, N,N-dimethylaniline, anthracene, and phenanthroline. The ability of such compounds to undergo photoinduced electron transfer reactions has been examined [19]. Using the bis-diene **24**, this technique was used to make bis-adduct **25** [18]. It should be noted that the yields in these Diels–Alder reactions are acceptable to very good.

The dienes shown in Table 3.2 were all generated *in situ* from appropriate precursors, usually in the presence of C_{60}. This approach was used by the Diederich group [20] to prepare the fused furan and thiophene derivatives **31**, **34**, and **37**. Possible limitations of this approach are shown by the failure of diene **38** to add to C_{60}, probably due to the relatively large separation of the diene termini and/or the strain of the corresponding [4+2] adduct. Cycloadducts **41**, **44**, and **46**, in which a heterocyclic system is directly fused to C_{60}, can be obtained in good yields from heterodienes **39**, **42**, and **45**, extending the applicability of this approach [21]. Other examples of reactive quinodimethanes are **47**, **50**, **54**, and **57** prepared, respectively, by thermolysis of cyclic sulfones **48**, substituted cyclobutenes **51**, substituted benzocyclobutenes **55**, and 1,4-elimination of a dibromide **58**. These allow entry, respectively, into anilino adducts **49** [22], fullerenoacyloins and fullerenodiones **52** and **53** [15,23], and a variety of aromatic ring-fused adducts **56** and **59** [24]. Finally, addition to C_{60} of isoindene **60** and a norcaradiene derivative **61** have been reported [25,26].

Diels–Alder reactions have recently been used by several groups to prepare porphyrin–C_{60} dyads, which are of great interest because of their potential activity as photosynthetic model systems due to anticipated intramolecular electron-transfer capability, and as potential antitumor agents due to their potential activity as photosensitizers for generation of singlet oxygen [26a]. The first two such Diels–Alder-derived porphyrin–C_{60} dyads reported were **63** and **65**, prepared by the Gust and Imahori groups from **62** and **64**, respectively [27,28]. By changing the bond connectivity in their spacer, Imahori and co-workers systematically changed the spacing and relative orientation of the porphyrin and fullerene moieites in dyads **66**, which are analogs of **65** [29]. In these and related compounds, a 3,5-di-*tert*-butylphenyl group is used instead of a simple phenyl substituent on the porphyrin to improve solubility characteristics. Using their "ball-and-chain" methodology, the Paddon-Row group has prepared dyad **67** [30]. The ability of these dyads, as well as related porphyrin-substituted fulleropyrrolidines prepared by Prato reactions (see Section 3.2.5), to undergo photoinduced intramolecular electron transfer has been examined, as a function of the spatial disposition of the two chromophores. It is clear that such dyads prefer to adopt a conformation in which the porphyrin and fullerene moieties are as close together in space as allowed by the structural constraints of the spacer [27–29]. Hirsch and co-workers have reported the preparation of related quinoid-type donor–acceptor dyads **69** and **71** from **68** and **70**, respectively, as well as the phthalocyanine–C_{60} dyad **73** from **72** in 75% yield [31,32]. Recent examples of donor–acceptor dyads are the tetrathiafulvalene–C_{60} dyads **75** and **77**, prepared from dienes **74** and **76**, respectively, which in turn were generated from the corresponding cyclic sulfones [33].

These examples amply illustrate the huge variety of structurally fascinating Diels–Alder adducts of C_{60} that have been prepared only between 1995 and 1998. We can anticipate the preparation of many more interesting adducts in the years to come. It must be emphasized that the adducts shown in Tables 3.1–3.3, with the exception of **49**, represent addition of a single diene moiety to the fullerene 6,6-bond. As with all fullerene additions, multiadditions can occur. It is possible to minimize the formation of bis-, and tris-, and higher adducts by using an excess of C_{60}, while yields of higher adducts can be improved using a large excess of the diene component. This has been accomplished by Kräutler, who has prepared the hexa-adduct **78** of 2,3-dimethyl-1,3-butadiene in good yield [34]. It is not yet known if the remarkable regiocontrol observed in this reaction, in which addition occurs across the six symmetry-equivalent 6,6-bonds of C_{60}, is due to thermodynamic or kinetic factors. The fullerene portion of this symmetrical hexa-adduct consists of 12 sp^3 carbons, which span a multiply bridged cyclophane comprising eight symmmetrically arranged benzenoid rings; the 2-buteno bridges span the network of an octahedron while the benzenioid rings surround a cube, a previously unknown superposition of two basic platonic bodies (see structures **79** and **80**, respectively).

While all the above reactions occur thermally, as expected for [4+2] cycloadditions, one example of a photochemical [4+2] cycloaddition to C_{60} has been reported [35]. When a mixture of C_{60} and anthracene in benzene was excited with light >500 nm, where only C_{60} absorbs, an adduct was obtained that appears to be the same as that obtained thermally, namely, the [4+2] adduct. It was shown that this reaction occurs between triplet excited states of C_{60} and anthracene in its ground state.

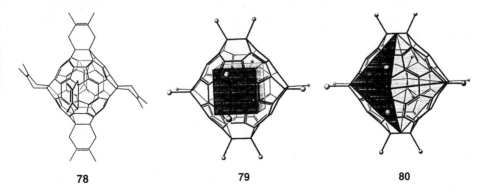

78   **79**   **80**

3.2.1.2 Additions to C$_{70}$ Because of the lower symmetry of C$_{70}$ (D_{5h}) compared to C$_{60}$ (I_h), [4+2] addition of dienes to C$_{70}$ across 6,6-bonds can lead to four isomeric monoadducts. Diederich and co-workers [36a] first studied [4+2] cycloadditions to C$_{70}$ using a reactive *ortho*-quinodimethane. They showed that three monoadducts were formed, namely, **81**, **82**, and **83**, from addition across C(1)–C(2), C(5)–C(6), and C(7)–C(21), respectively (see Eq. (3.1)).

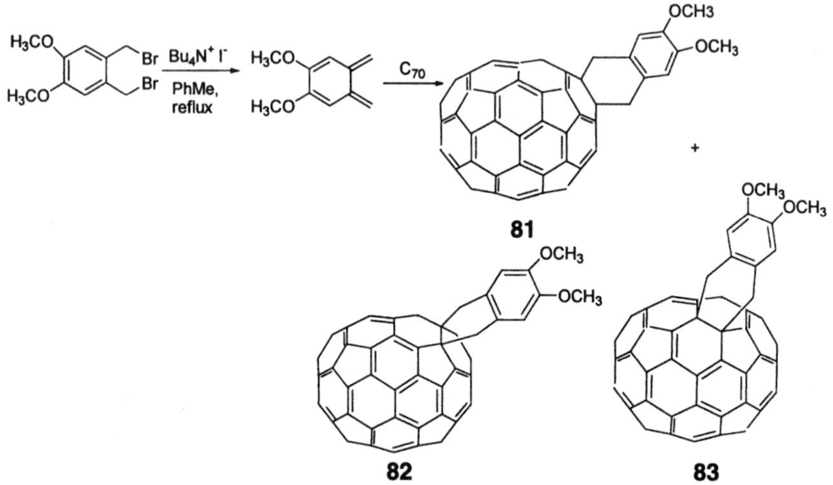

$$(3.1)$$

Given that the most reactive bonds of C$_{70}$ are those near the poles, which possess highest local curvature, it is not surprising that the yields of **81–83** were 24%, 10%, and 1–2%, respectively. Low temperature X-ray crystal structures have been obtained for **81** and **82** [36b]. On the other hand, Taylor and co-workers [37] have found that [4+2] addition of pentamethylcyclopentadiene to C$_{70}$ gives a single isolated adduct, which was shown by nuclear magnetic resonance (NMR), including nuclear overhauser effect (NOE) experiments, to be the 1,2-adduct with the methyl group pointing away from the fullerene (exo) (see Eq. (3.2)). ^{1}H NMR spectra indicated traces of a second product, possibly a 5,6-adduct.

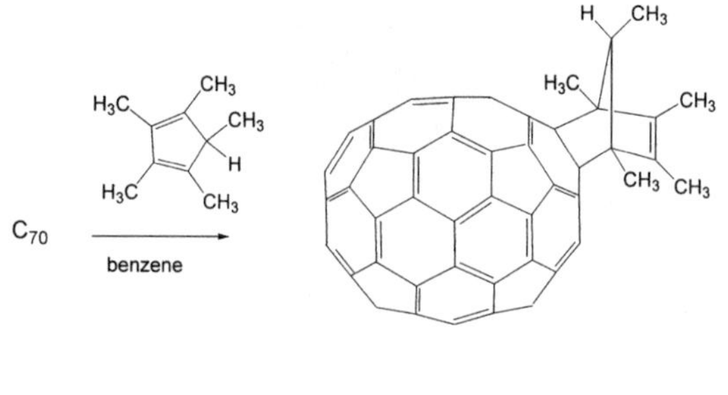

$$(3.2)$$

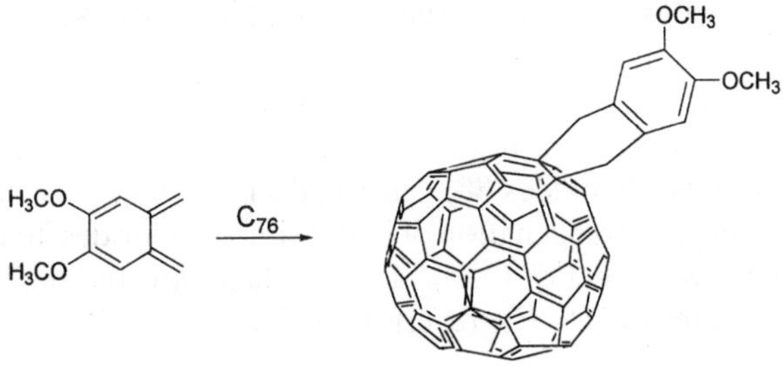

84

A single example of [4+2] cycloaddition to D_2-C_{76} has been reported by the Diederich group [36a] using the same quinodimethane used for reaction with D_{5h}-C_{70} (see Eq. (3.1)). At least six isomeric monoadducts are formed with C_{76}, of which one (**84**) has been fully characterized. ^1H NMR spectra of various chromatography fractions indicate one of the adducts has C_2 symmetry while the other four have C_1 symmetry. Once again, it appears that addition occurs at the 6,6-bonds with greatest curvature on the fullerene surface.

3.2.2 [3+2] Cycloadditions

3.2.2.1 Additions to C_{60} Hirsch [8] has thoroughly reviewed the early literature through 1994 on 1,3-dipolar cycloadditions to C_{60} using diazo compounds, alkyl azides, nitrile oxides, and azomethine ylides generated from a variety of starting materials. C_{60} behaves as a very reactive dipolarophile. Eguchi et al. [21] have published a more up-to-date review. By far, the greatest activity in this area in the past five years involves addition of nitrile ylides to C_{60}, to which a separate section of this chapter is devoted (see Section 3.2.6).

In the case of diazo compounds, it has been shown by many people that addition occurs across a 6,6-bond to give initially a pyrazoline **85**, which then undergoes either thermal or photochemical loss of nitrogen to give a mixture of 1,2-bridged methanofullerenes **87** and 1,6-open methanofulleroids **86** (see Eq. (3.3)) [38,39]. In these products, no double bond is present at a 5,6-ring fusion. It has also been shown that 1,6-methanofulleroids **86** rearrange either thermally, electrochemically, or photochemically to **87** by a 1,5-sigmatropic shift [11,40]. The latter is an example of the well-known Zimmerman di-pi-methane rearrangement [41].

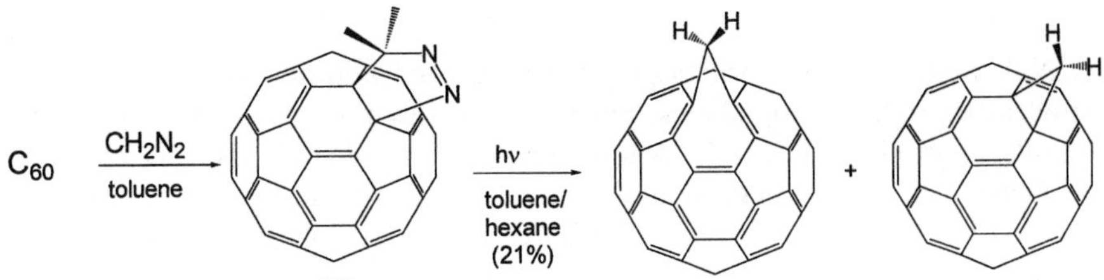

(3.3)

Thermal addition of azides to C_{60} follows a similar course. Since this subject is closely connected to the preparation of azafullerenes, azide additions to both C_{60} and C_{70} will be discussed more fully in Section 3.2.5.2.

Meier and Poplawska [42] have provided more details about the thermal addition to C_{60} of nitrile oxides **88**, generated in situ by either dehydration of nitroalkanes or dehydrohalogenation of hydroximoyl chlorides, to give fullerene-fused isoxazolines **89** (see Eq. (3.4)). The products are thermally unstable at elevated temperatures ($>280\,°C$), reverting to C_{60}.

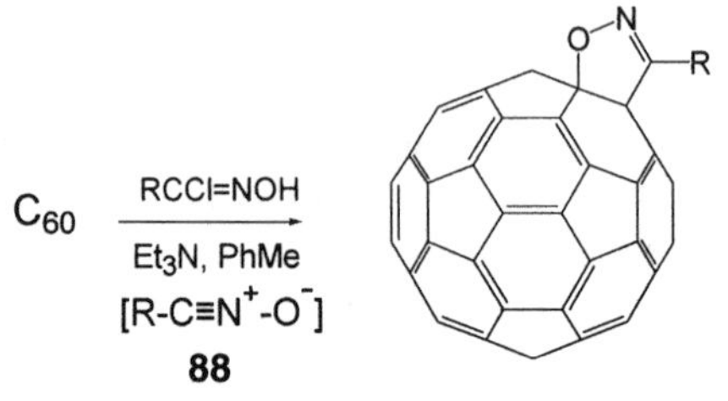

$$C_{60} \xrightarrow[\substack{Et_3N,\ PhMe \\ [R-C\equiv N^+-O^-] \\ \textbf{88}}]{RCCl=NOH} \tag{3.4}$$

89

A new type of 1,3-dipolar [3+2] addition to C_{60} was reported by Matsubara et al. [43], involving formation of 3-substituted-1–phenyl-2-pyrazolines **91** from 3-substituted-1-phenyl nitrilimines **90**, as shown in Eq. (3.5). This cycloaddition occurs as usual across a 6,6-bond.

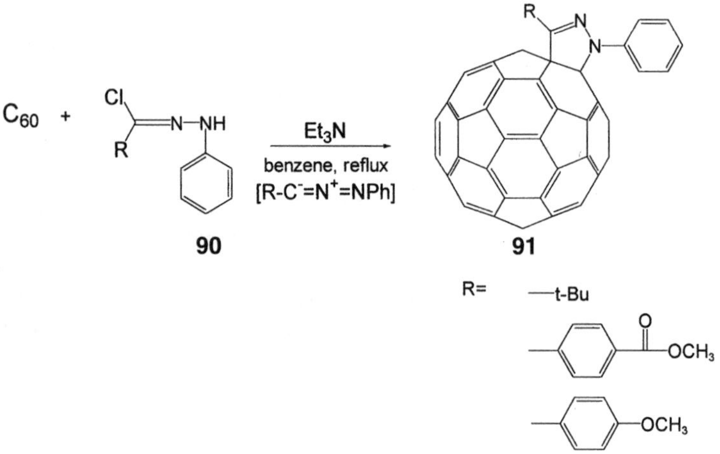

$$C_{60} + \underset{\textbf{90}}{\overset{\text{Cl}}{\underset{R}{\bigvee}}=N-NH} \xrightarrow[\substack{\text{benzene, reflux} \\ [R-C^-=N^+=NPh]}]{Et_3N} \tag{3.5}$$

91

R= —t-Bu

—⟨⟩—C(O)—OCH₃

—⟨⟩—OCH₃

Nakamura and co-workers [44] have reported addition to C_{60} of dipolar vinylcarbene-like intermediates generated by thermolysis of cyclopropenone acetals, as shown in Eq. (3.6). A mixture of [3+2] and [1+2] adducts **92** and **93** is obtained in a ratio dependent on the temperature. At $140\,°C$ the former is favored by 93:7, while at $80\,°C$ the latter is favored by 10:1. In more recent work in which two cyclopropenone acetal moieties were connected by a poly-

methylene tether of varying length, bis-[3+2] cycloaddition occurred in a regioselective and stereoselective manner, which correlated with the computed
conformational strain of the tether (see section 3.6) [45].

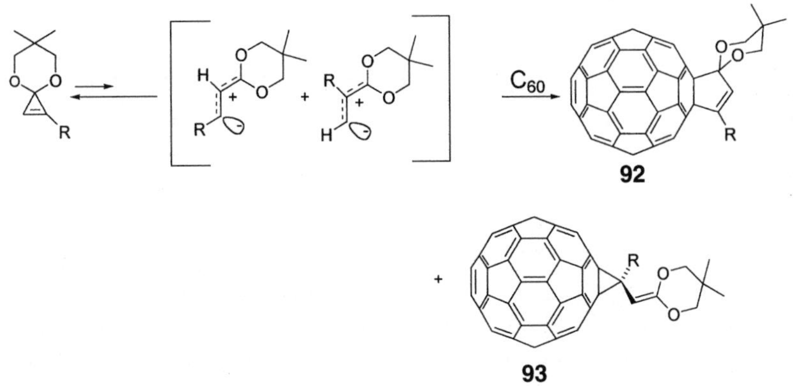

92

(3.6)

93

3.2.2.2 Additions to C$_{70}$ As far as 1,3-dipolar cycloadditions to C$_{70}$ are
concerned, the situation is more complex due to the presence of four types of
6,6-bonds, as previously discussed. Furthermore, addition of unsymmetrical
dipolar reagents across the most reactive 6,6-bond in C$_{70}$, namely, C(1)–C(2),
can give regioisomeric adducts. Thus, complex mixtures of C$_{70}$ cycloadducts
are expected and indeed are observed. In the case of addition of diazomethane
to C$_{70}$, Smith et al. [39] found that three isomeric pyrazolines were formed in a
ratio of 12:1:2. These were shown by ^{1}H NMR to be the two regioisomeric 1,2-
adducts **94** and **95** and the symmetrical 5,6-adduct **96** (Eq. (3.7)).

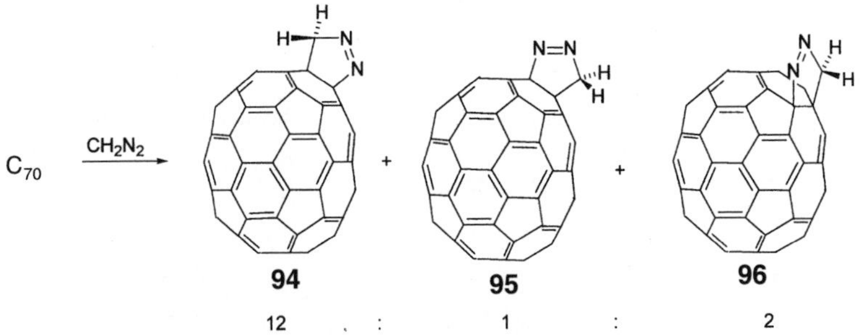

C$_{70}$ $\xrightarrow{CH_2N_2}$

94   **95**   **96**   (3.7)

12   :   1   :   2

The regioselectivity in formation of the 1,2-adducts can be rationalized in terms
of polarization of the C(1)–C(2) double bond of C$_{70}$ due to surface curvature,
such that the carbon atom nearer the pole (C1) is positive with respect to the
carbon atom (C2) nearer the equator (see the discussion in Section 3.2.4 in
connection with regioselective azide additions to C$_{70}$).

In contrast to the additions of diazomethane and azides, little selectivity was
observed in additions of nitrile oxides to C$_{70}$ (Eq. (3.8)), in which the two
regioisomeric 1,2-adducts and a 5,6-adduct were produced in relative yields of

nearly 1:1:1 [46].

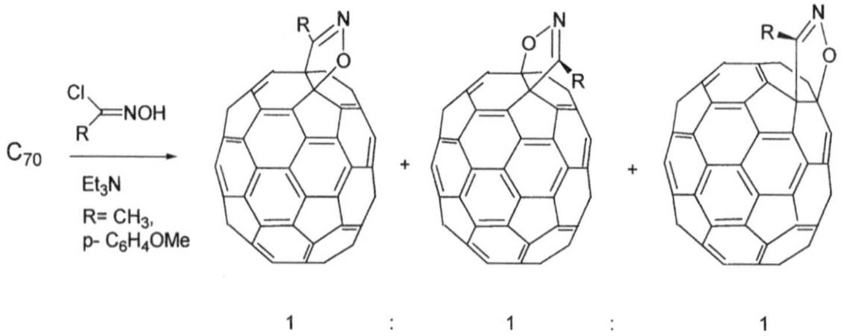

C_{70}
Cl
=NOH
R
Et_3N
R= CH_3,
p- C_6H_4OMe
R
N
O
+
N
O
R
+
N
R
O
1 : 1 : 1

(3.8)

In this case, little selectivity is observed between attachment of the C and O atoms of the dipole to C1 on the C_{70} surface. In competitive experiments, it was shown that the rate of addition of nitrile oxides to C_{70} is about half of that to C_{60}, essentially the same finding as that of Wilson and Lu [47] using *N*-methylazomethine ylide. In the latter reaction, three isomeric monoadducts were obtained in a 46:41:13 ratio. Their structures were assigned on the basis of ^{1}H NMR as the 1,2-, 5,6-, and 7,21-fullero[70]pyrrolidines **97–99**, as shown in Eq. (3.9). This represents the first example of 1,3-dipolar cycloaddition across the relatively unreactive C(7)–C(21) bond of C_{70}.

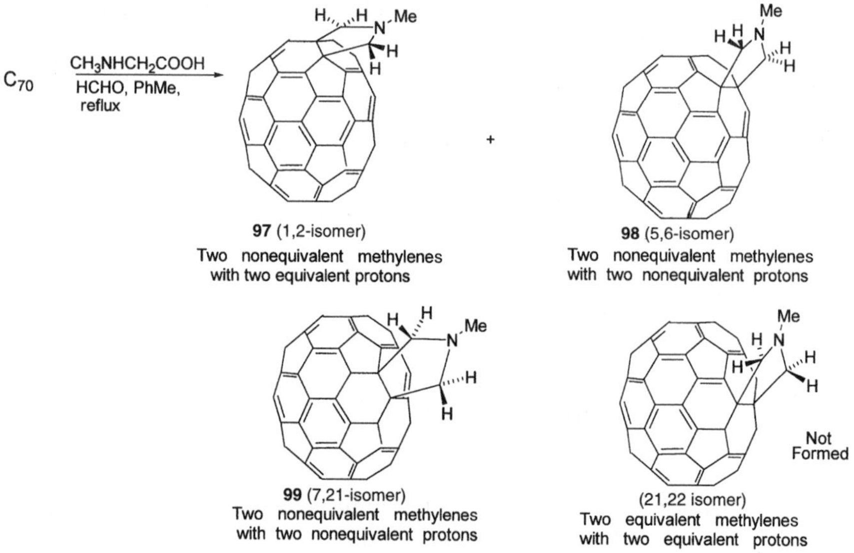

C_{70}
CH_3NHCH_2COOH
HCHO, PhMe,
reflux
+
97 (1,2-isomer)
Two nonequivalent methylenes
with two equivalent protons
98 (5,6-isomer)
Two nonequivalent methylenes
with two nonequivalent protons
99 (7,21-isomer)
Two nonequivalent methylenes
with two nonequivalent protons
Not
Formed
(21,22 isomer)
Two equivalent methylenes
with two equivalent protons

(3.9)

3.2.3 Photochemical [2+2] Cycloadditions to Fullerenes

It was established some time ago that cyclic α, β-unsaturated ketones (enones) undergo [2+2] photocycloaddition to C_{60} at wavelengths (ca. 300 nm), where the enones are the principal light-absorbing species [48]. Under conditions where monoaddition predominates, both *cis*- and *trans*-fused monoadducts are obtained (see Eq. (3.10)). In a recent full report, Schuster et al. [49] showed that this reaction is entirely general, as a variety of monocyclic, bicyclic, and tetra-cyclic (steroid) enones undergo photoaddition to C_{60}. ^{13}C NMR spectra demonstrate that the products are indeed closed 6,6-adducts (see above). The fact that two diastereomeric C_{60} cycloadducts are obtained even with 2-cycloheptenone, which does not photoadd to simple alkenes, is consistent with the general conclusion that C_{60} behaves as an electron-deficient alkene. The observation of enhanced CD bands at long wavelengths in *trans*- but not *cis*-fused enone–C_{60} adducts suggests the presence in the strained *trans*-fused adducts of a new chromophore, which has been suggested [48,49] to be a twisted fullerene with local C_2 symmetry. In addition, distinct ^{3}He NMR signals are seen for the stereoisomeric products from addition of 3-methylcyclohex-2-enone and 2-cycloheptenone to ^{3}He@C_{60} [49].

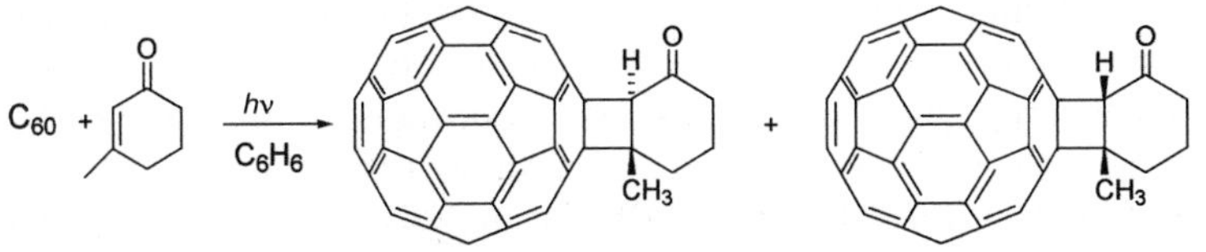

$$ \text{(3.10)} $$

Analogous photocycloaddition of enolic 1,3-diones to C_{60} would provide a route to C_{60}-fused cyclooctanediones by a de Mayo-type photoaddition-ring opening sequence (see Eq. (3.11)) [50].

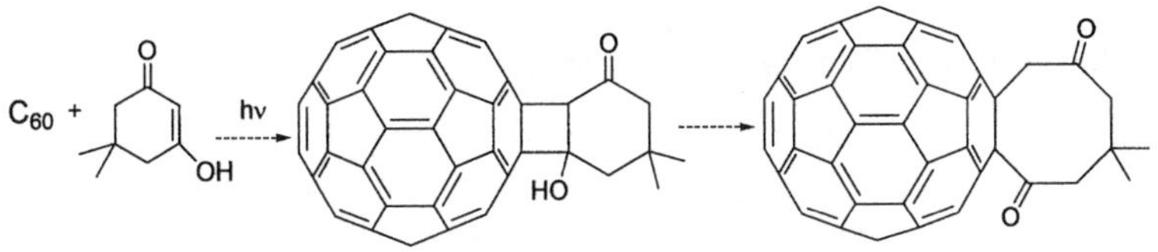

$$ \text{(3.11)} $$

However, photoaddition of cyclohexane-1,3-dione and dimedone took an entirely different and unexpected course, as shown in Eq. (3.12), producing the furanyl-fused fullerenes **100** and **101** [51].

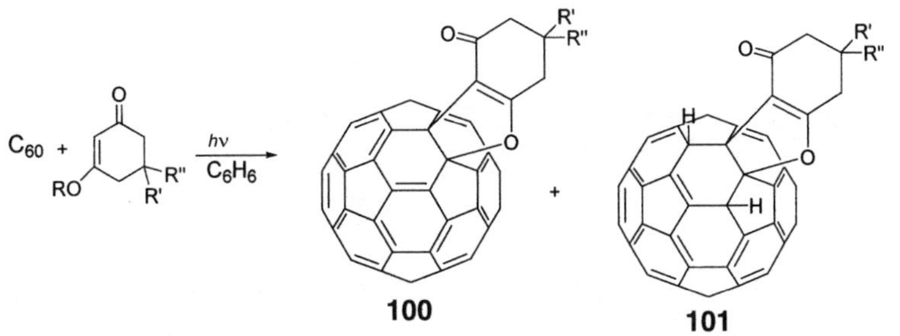

$$ \text{(3.12)} $$

100   **101**

The dehydro adducts **100** can also be obtained by thermal (dark) [3+2] addition of these same diones to C_{60} in the presence of piperidine according to conditions previously described by Eguchi (see below) [21,52]. The yield of reduced adducts **101** (which afford **100** by treatment with DDQ) increases with the efficiency of deoxygenation of the reaction mixture prior to irradiation. Thus, oxygen plays a critical role in this novel cycloaddition reaction. Jensen et al. [51] propose a plausible but as yet unproven mechanism for this reaction.

Earlier examples of both thermal and photochemical [2+2] photoadditions to C_{60} have been reviewed by Hirsch [8], involving benzyne, quadricyclane, and electron-rich alkynes. Arylalkenes have been shown to undergo [2+2] photocycloaddition to C_{60} via a stepwise mechanism [53a], and there is evidence that this reaction may occur with simple alkenes although the products have not been completely characterized [53b].

Preliminary experiments on photocycloaddition of enones to C_{70} using 3-methylcyclohex-2-en-1-one indicate not surprisingly that the reaction is much more complex than the reaction with C_{60} [54]. Since the addend is unsymmetrical, photoaddition across the most reactive 6,6-bond of C_{70}, $C(1)–C(2)$, would give regioisomeric adducts, which furthermore can be *cis*- or *trans*-fused. The most surprising finding is that only four monoadducts of 3-methyl-2-cyclohexenone to C_{70} are formed, according to Buckyclutcher high-performance liquid chromatography (HPLC) analysis. These products have yet to be isolated in sufficient quantities for structural identification. The chemical selectivity as well as regioselectivity on the C_{70} surface in a photochemical cycloaddition reaction is expected to be quite different from that in thermal addition reactions.

3.2.4   Oxidative [3+2] Cycloadditions to C_{60}

As shown in Eq. (3.13), β-dicarbonyl compounds undergo oxidative addition to C_{60} in the presence of piperidine at ambient temperatures [21,52]. The base must be present in excess in order to obtain good yields of products, which are dihydrofuran-fused C_{60} adducts. No adducts are formed using triethylamine instead of piperidine. The mechanism of this novel oxidative cycloaddition reaction is still not clear. As already noted, the same type of product is obtained from β-diketones and C_{60} by ultraviolet (UV) irradiation in the presence of oxygen [51].

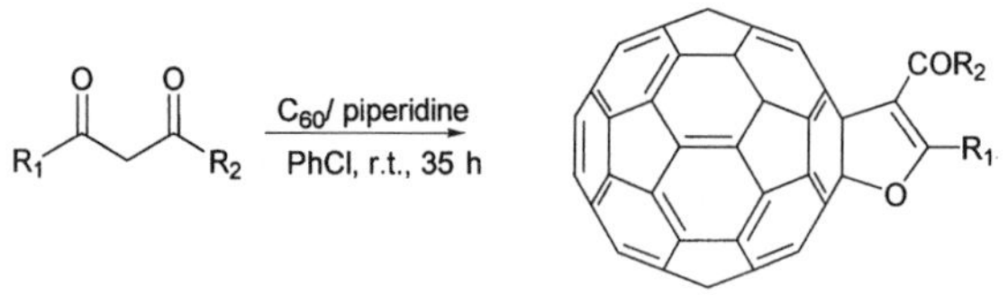

$$(3.13)$$

R_1	R_2	Yield %
Me	O-t-Bu	59
Me	OEt	60
o-C6H4-		39
CH2CO2Me	OMe	27
Me	Me	48
-(CH2)3-		49

Eguchi et al. [21] have shown that this reaction may be broadly applicable for functionalization of fullerenes. Thus, fused heterocycles can be generated in an analogous manner from appropriate precursors, as shown in Eqs. (3.14) and (3.15).

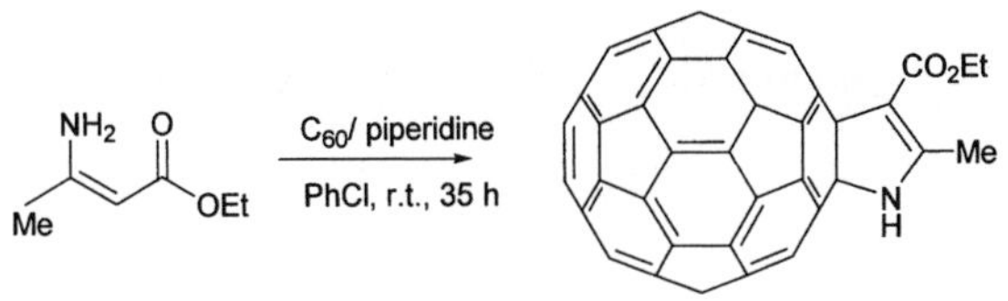

NH2 O
Me
OEt
C60/ piperidine
PhCl, r.t., 35 h
CO2Et
Me
N
H

(3.14)

S
Me
NH2
C60, PhCl
Δ, 8 h
N
Me
S

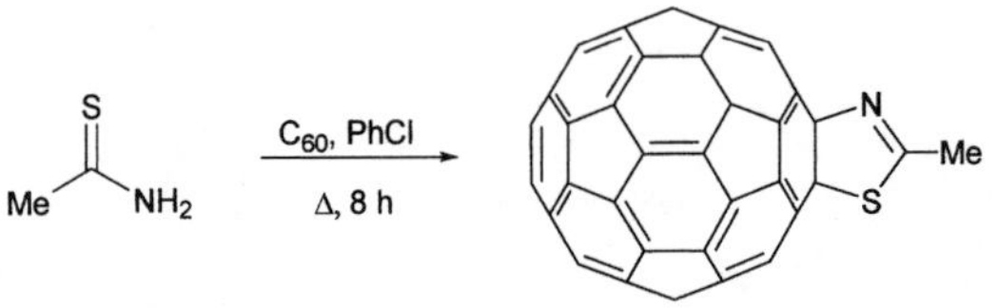

(3.15)

A formal oxidative [3+2] photocycloaddition to C_{60} of amino acid esters has been reported by Gan et al. [55]. As examples, see Eqs. (3.16)–(3.18) for photoaddition to C_{60} of glycine esters (methyl, ethyl, or benzyl), sarcosine, and α-methyl-iminodiacetic methyl ester, respectively, in the presence of oxygen. The products were spectroscopically characterized as fulleropyrrolidines. In the last two reactions, the net result is the loss of two hydrogen atoms, one each from the CH_2 and CH_3 moieties.

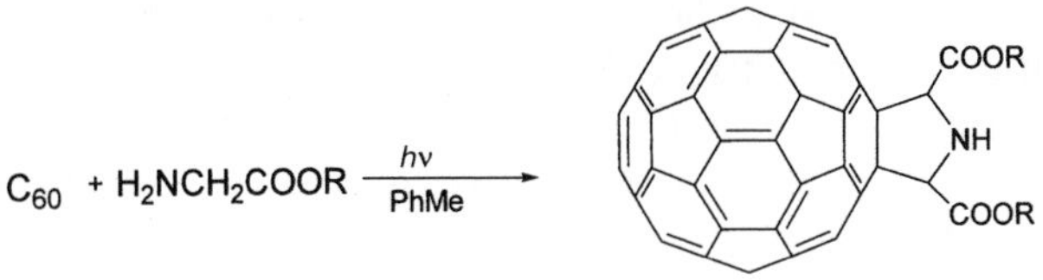

C60 + H2NCH2COOR
hν
PhMe
COOR
NH
COOR

(3.16)

C60 + Me2NCH2COOMe
C60, hν
N–Me
COOMe

(3.17)

C60 + MeN(CH2COOMe)2
C60, hν
N–CH2COOMe
COOMe

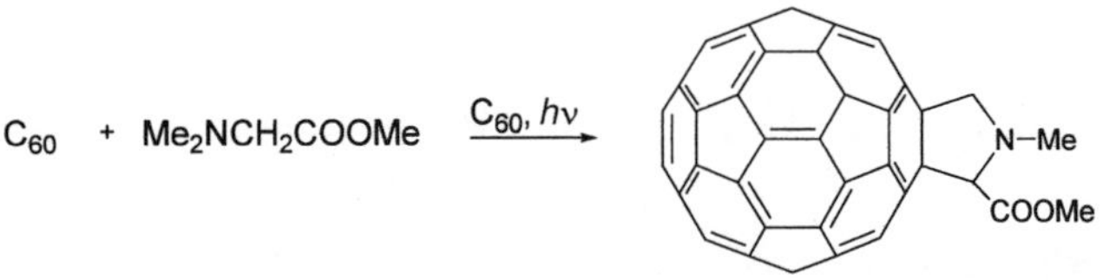

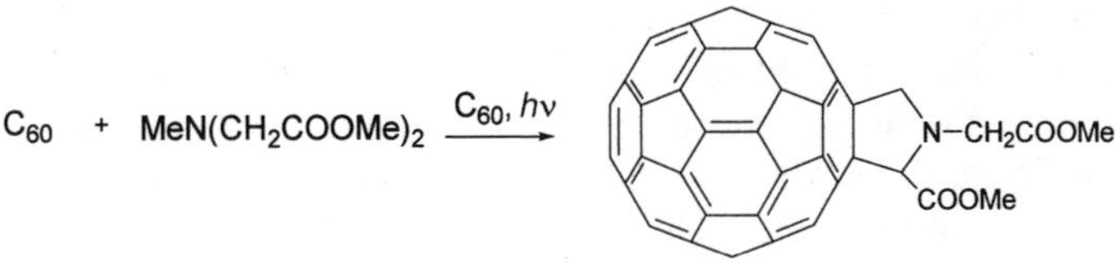

(3.18)

The authors invoke the mechanism in Scheme 1 in which the key intermediate is the α-carbon-centered radical. Two plausible routes leading to this radical involve (1) photoinduced electron transfer followed by H-atom transfer, or (2)

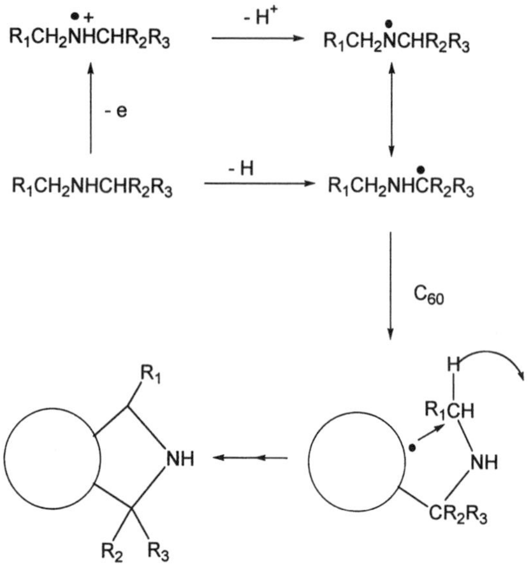

Scheme 1

photosensitized formation of singlet oxygen followed by hydrogen abstraction. Addition of the carbon-centered radical to C_{60} followed by a second H-abstraction would afford the fulleropyrrolidine. Similar pathways afford the products seen in the other reactions. Since β-alanine ethyl ester does not react with C_{60}, formation of a stabilized radical at the α-carbon of the amino acid seems to be required. A complex mixture of adducts was obtained starting with esters of ethylenediaminetetraacetic acid and analogous compounds. If unesterified carboxylic acids are used, decarboxylation occurs (further implicating radical intermediates) and dihydrofullerenes are formed.

3.2.5 Azide Additions and Formation of Heterofullerenes (Berthold Nuber)

3.2.5.1 Introduction In heterofullerenes, one or more carbons of the fullerene core are substituted by a non-carbon atom. Apart from exohedral and endohedral fullerene derivatives, heterofullerenes represent the third fundamental group of functionalized fullerenes, and have been the least explored. Before 1995, only the mass spectrometric detection of borafullerenes [56,57] and niobiumfullerenes [58] was reported. In 1995 two groups independently published the generation of azafullerenes [59,60] under mass spectrometric conditions and soon thereafter two ways of synthesizing macroscopic amounts of monoazafullerenes were found [61,62]. Since all known azafullerene preparations start from exohedral azide adducts, this section describes the addition

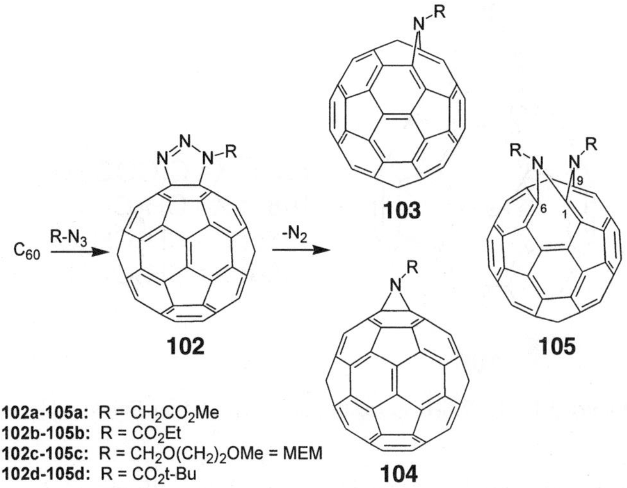

Scheme 2 Reaction of azides R-N₃ with C₆₀.

pathways of azides to C_{60} and C_{70} as well as the preparation and chemistry of aza[60]fullerenes and aza[70]fullerenes.

3.2.5.2 Azide Additions to C_{60} and C_{70}

Azido compounds react with C_{60} in a [3+2]-addition to form products of the kind shown in Scheme 2 [63,64]. The first step is the addition of the 1,3-dipole to a 6,6-bond of C_{60} to afford the triazolines **102**, which are the major products when the reaction is run under moderate conditions (i.e., 60 °C). At temperatures higher than 80 °C the loss of nitrogen occurs [65], leading to the 5,6-fulleroid **103** and the 6,6-aziridino compound **104**. However, the triazolines **102** are stable enough to be isolated and in the case of **102c** it was possible to obtain single crystals suitable for X-ray analysis [9]. Carrying out the azide addition to C_{60} at higher temperatures (e.g., 120 °C) leads (apart from monoadducts) in a remarkable regioselective reaction to the 1,6;1,9-bis(azafulleroid) **105**, which is the only bis-adduct obtained, apart from traces of a by-product [64].

Bis(fulleroids) with a 1,6;7,8-addition pattern (**106a–d**, Fig. 3.1) can be obtained using tether-linked bis(azido) compounds [65a,66–69].

Azido compounds do not react exclusively as 1,3-dipoles. For example, ethyl azidoformate in boiling tetrachloroethane (147 °C) forms the resonance stabilized nitrene :N-COOEt, which reacts with C_{60} by [2+1]-addition to give the monoadducts **103b** and **104b**. The **103b:104b** ratio is about 1:12, which overturns the 5,6-:6,6-ratio of 7:1 from the triazoline route. The nature of the bis-adducts formed by multiple nitrene additions to 6,6-bonds differs fundamentally from the situation described above. These compounds are analogous to the products obtained by twofold Bingel–Hirsch reaction [70,71] and occur in similar relative yields. In addition to the seven isomers formed in this reaction

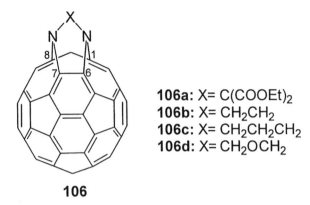

106a: X= C(COOEt)$_2$
106b: X= CH$_2$CH$_2$
106c: X= CH$_2$CH$_2$CH$_2$
106d: X= CH$_2$OCH$_2$

106

Figure 3.1 1,6;7,8-Bis(azafulleroids) with tether-linked addends.

(*cis*-2, *cis*-3, *e*, and *trans*-1 to *trans*-4 [72], Fig. 3.2a), nitrene addition leads to all eight conceivable 6,6-bis-adducts [73] including the *cis*-1 adduct **107** pictured in Figure 3.2b. Compound **107** is one of the few examples [74] of a fullerene compound with open 6,6-bonds. Interestingly, cleaving the *tert*-Bu ester groups of **107b** in the presence of trifluoroacetic acid results in the formation of the analogous [6,6]-closed structure [73].

Since C$_{70}$ has a lower symmetry (D_{5h}) compared to I_h-symmetric C$_{60}$, one can distinguish between five different sets of carbon atoms (A–E, Scheme 3) and four types of 5,6- and 6,6-bonds (1–4, Scheme 3). Since the driving force

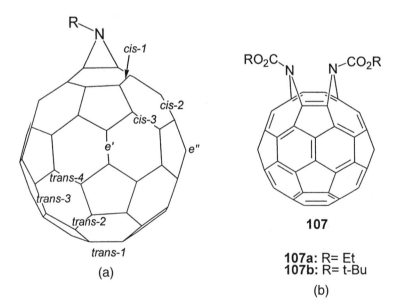

107

107a: R= Et
107b: R= t-Bu

(b)

Figure 3.2 (a) Relative positions of 6,6-double bonds in C$_{60}$; (b) *cis*-1 bis-adduct containing open 6,6-bonds.

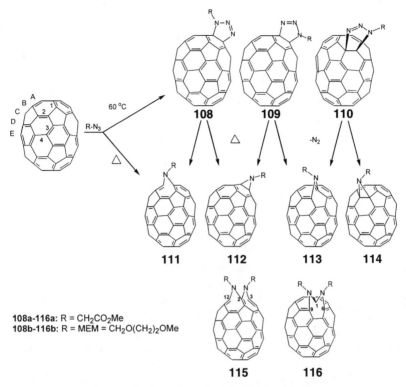

108a-116a: R = CH₂CO₂Me
108b-116b: R = MEM = CH₂O(CH₂)₂OMe

Scheme 3 Reaction of azides R-N₃ with C₇₀.

for additions to 6,6-bonds is the release of strain in the fullerene framework, addends preferably attack the more strained polar regions of C_{70} (i.e., bonds 1 and 2 in Scheme 3) [75–78].

The regiochemistry of azide additions to C_{70} shows the expected behavior [79,80]. Treatment of C_{70} with methyl azidoacetate at moderate temperatures leads to the three isomeric triazolines **108a–110a** (Scheme 3) in a ratio of about 12:3:1 [79]. Carrying out the azide addition at 120 °C [79] or thermal treatment of **108–110** [80] results in the formation of the 5,6- and 6,6-compounds **111–114**. Starting from the isolated triazolines **108–110**, nitrogen extrusion leads regioselectively to the products as indicated in Scheme 3.

As in the C_{60} case, higher adducts are formed along with the monoadducts [62,69]. When the azide addition is carried out at 120 °C, the reaction gives rise to two C_S-symmetric isomers **115** and **116**, showing 2,3;2,12- and 1,6;1,9-addition patterns, respectively, in a ratio of about 4:1 (Scheme 3).

3.2.5.3 Synthesis of the Heterofullerene Dimers (C₅₉N)₂ and (C₆₉N)₂

Aza[60]fullerene Hummelen and Wudl [61] published the first synthesis of $C_{59}N$ in the dimeric form **117** in macroscopic amounts. Heating the cage-opened ketolactam derivative **118** in the presence of *p*-TsOH gives rise to the

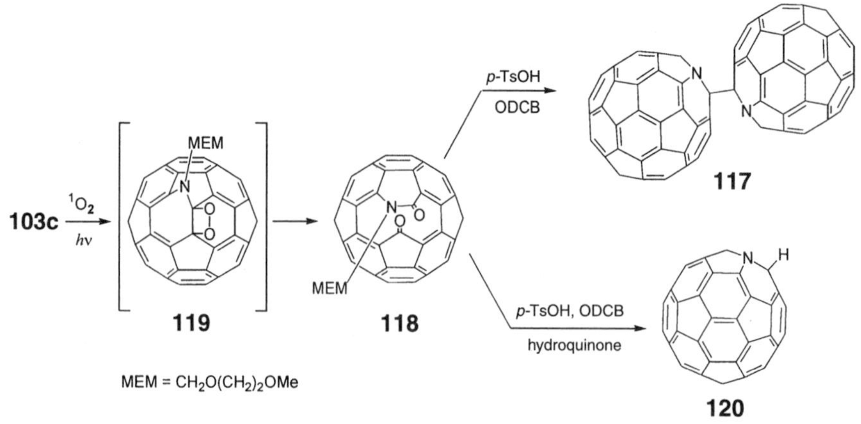

Scheme 4

formation of a green-brown solution of $(C_{59}N)_2$ with yields above 80% [81] (Scheme 4). Compound **118** itself is obtained by photosensitized addition of singlet oxygen to the azafulleroid **103c** [82]. The reaction probably proceeds in a [2+2]-addition mode via the dioxetane derivative **119**, which could not be isolated [83]. ^{13}C NMR measurements unambiguously proved the C_S symmetry of **117** showing the characteristic resonance for the sp^3-carbon on the fullerene at about 90 ppm [84,85].

Heating **118** with p-TsOH in the presence of excess hydroquinone as reduction reagent gives rise to the hydroazafullerene $HC_{59}N$ **120** [86] (Scheme 4).

Based on mass spectrometric investigations, [59] Nuber and Hirsch developed a synthesis of azafullerenes from the bis(azafulleroid) precursor **105c** (Scheme 5) [62]. Treatment of **105c** (see Scheme 2) with butylamine and DBU in toluene and heating the resulting amine adduct **121** in the presence of excess p-TsOH in ODCB led to $(C_{59}N)_2$ **117** [87] in an optimized yield of 26% [69]. The

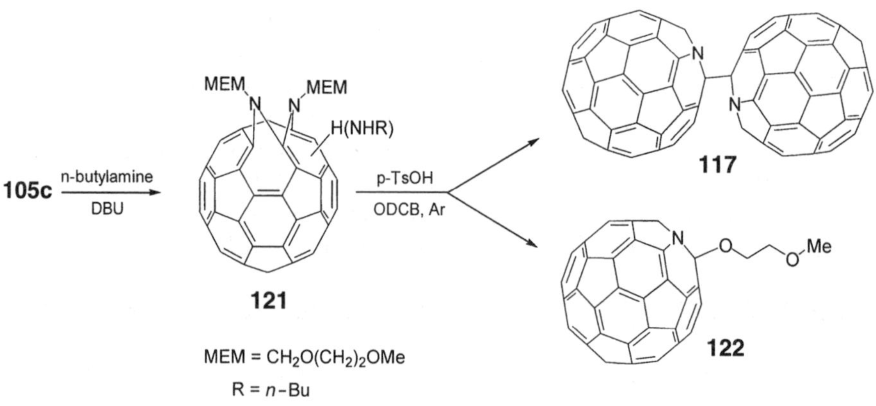

Scheme 5

alkoxy-substituted compound **122** was the first fully characterized monomeric $C_{59}N$ derivative and is formed along with **117** (Scheme 5).

Aza[70]fullerene   The methodology described above was applied to the first synthesis of aza[70]fullerenes $C_{69}N$ [62]. Considering the structures of **115** and **116** (Scheme 3), one expects different $C_{69}N$ isomers depending on which bis(azafulleroid) is used as the precursor [87]. Compound **115** should lead to an isomer in which a carbon of set A (Scheme 3) is replaced by a nitrogen, whereas starting with **116** the nitrogen should occupy a B-position. Starting from a 4:1 mixture of **115b** and **116b**, a mixture of all three possible dimers **123–125** with AA′, AB′, and BB′ substitution patterns (Scheme 6) was obtained, according to HPLC, in a ratio of about 16:3:1 [62,69].

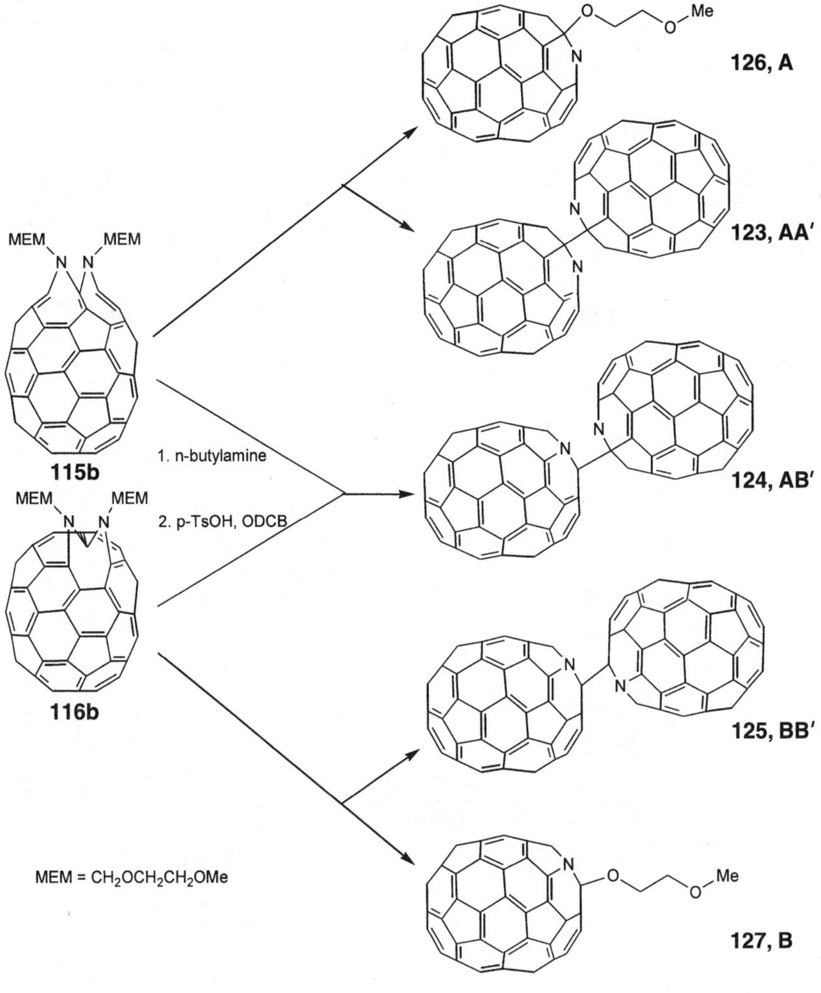

Scheme 6

Analogous to the [60]fullerene case, alkoxy-substituted products are also formed (Scheme 6). The two expected isomers **126** and **127** (A- and B-substituted) are found according to ^{1}H NMR in a ratio of about 7:1 [62,69].

Wudl and co-workers [88,89] also applied their method to C_{70}. Depending on the regiochemistry of the ketolactam precursors **128**, **129**, and **130**, $C_{69}N$ dimers with AA'-, BB'-, and CC'-substitution patterns are obtained (Scheme 7).

Heating equimolar amounts of $(C_{59}N)_2$ and $(C_{69}N)_2$, as well as starting from their respective precursors [89], results in the formation of mixed aza-fullerene dimers $C_{69}N-C_{59}N$.

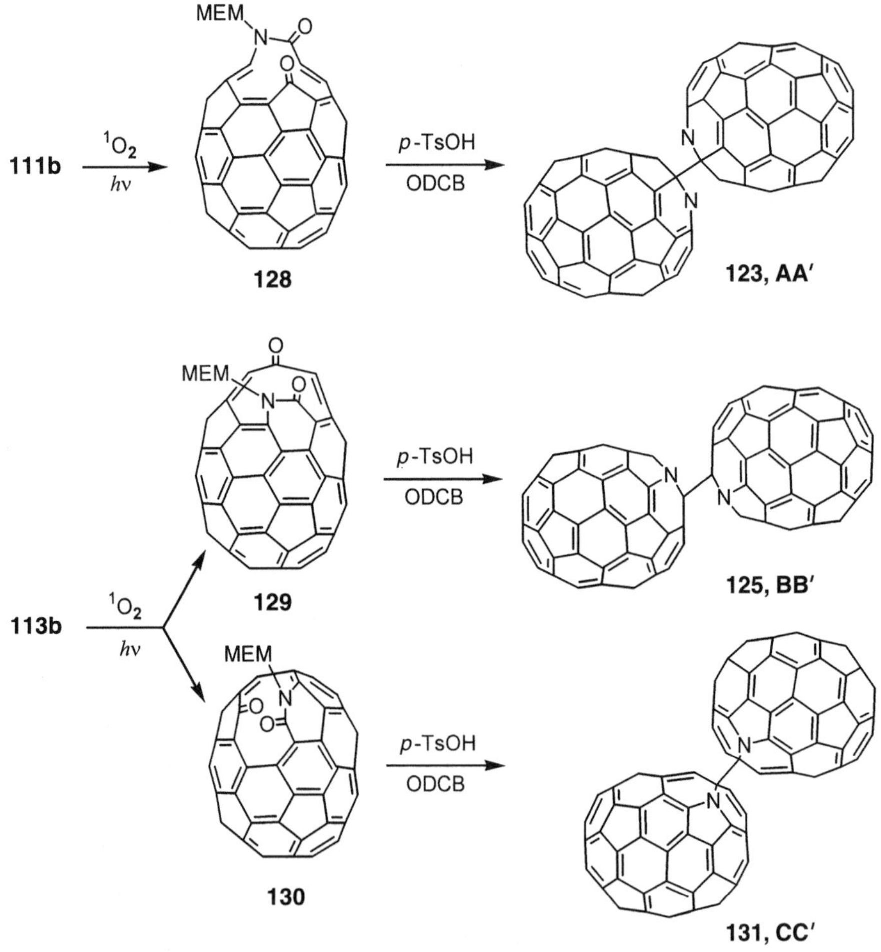

Scheme 7

3.2.5.4 Chemistry of Azafullerenes The length of the C—C′ dimer bond in $(C_{59}N)_2$ was calculated to be 160.9 pm (average C—C bond 154 pm) with a bond energy of only 18 kcal/mol [90,91]. Thus, heat or light should easily cleave this bond to give monomeric radicals, which were indeed detected by ESR spectroscopy in the case of $(C_{59}N)_2$ [92], as well as $(C_{69}N)_2$ [88]. Wudl and co-workers report that heating $(C_{59}N)_2$ in refluxing ODCB in the presence of excess diphenylmethane leads to the formation of **132** (Scheme 8) in a reaction for which a free radical chain mechanism was proposed [93].

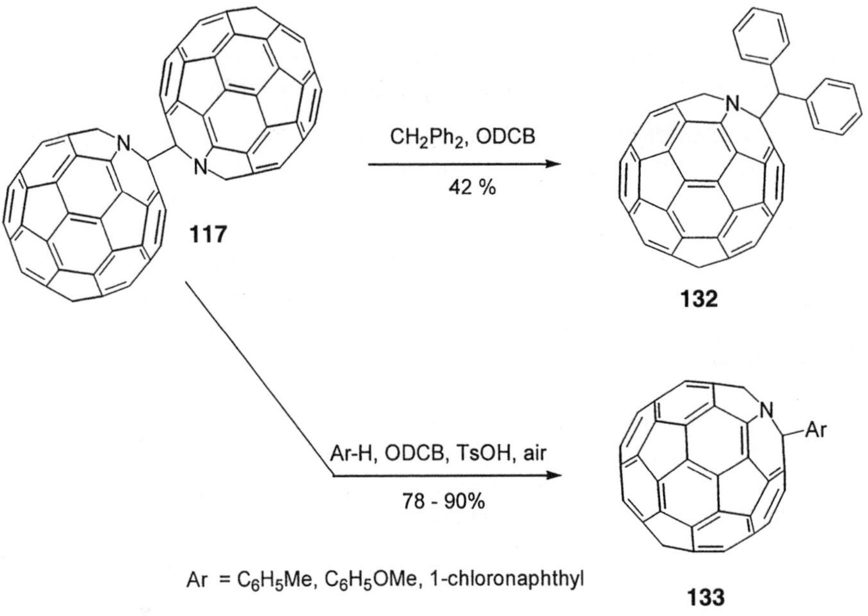

Ar = C_6H_5Me, C_6H_5OMe, 1-chloronaphthyl

Scheme 8

Heating $(C_{59}N)_2$ in the presence of aromatic compounds (e.g., anisole, toluene, and 1-chloronaphthalene), air, and excess p-TsOH gives the arylated monomers **133** in high yields, up to 90% (Scheme 8) [94]. The proposed mechanism of this reaction involves an electrophilic aromatic substitution by $C_{59}N^+$, which is presumed to be formed via thermal homolysis of the dimer followed by oxidation of the $C_{59}N$ radical.

The arylated compounds **133** represent ideal starting materials to explore the special features of the $C_{59}N$ core. Reuther and Hirsch [95] reported recently on the reaction of various compounds $ArC_{59}N$ with ICl at room temperature, which leads in 50–60% yield to the tetrachlorinated species **134** (Eq. (3.19)). This system possesses very interesting structural features, since it contains an intact pyrrole moiety decoupled from the conjugated π-system of the fullerene cage (Eq. (3.19)).

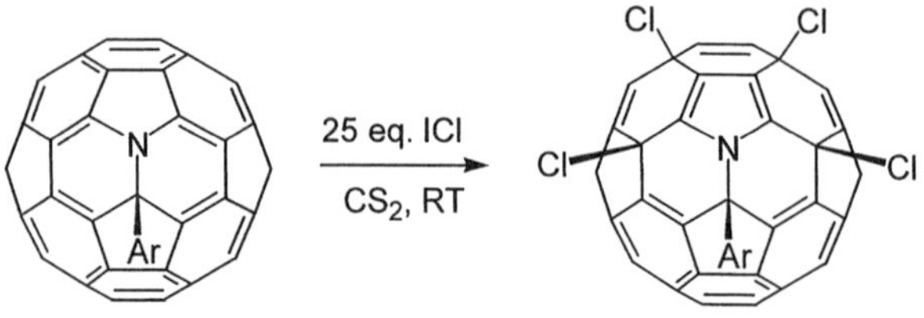

$$(3.19)$$

Ar = C_6H_5Me, C_6H_5OMe, $C_6H_5O(CH_2)_2CHMe_2$

3.2.6 Addition of Azomethine Ylides: Fulleropyrrolidines and Fulleroprolines (Michele Maggini and Maurizio Prato)

3.2.6.1 Introduction Fulleropyrrolidines are a class of organofullerenes in which a pyrrolidine ring is fused to a ring junction of either [60]fullerene or [70]fullerene [96–98]. Whereas monofunctionalization of the highly symmetrical C_{60} affords a single fulleropyrrolidine, the same addition to C_{70} gives rise to mixtures of isomers [99]. While fulleropyrrolidines retain the basic properties of the parent fullerene, they are much easier to handle in terms of solubility in polar solvents and ability to form mono- or multilayers on surfaces or assemblies in supramolecular arrays.

3.2.6.2 Cycloaddition of Azomethine Ylides to C_{60} Fulleropyrrolidines are smoothly prepared via 1,3-dipolar cycloaddition of azomethine ylides to C_{60} [100–102]. These reactive intermediates can be generated in many ways from readily accessible starting materials following well-established protocols [103]. One of the easiest approaches to produce azomethine ylides involves decarboxylation of immonium salts derived from condensation of α-amino acids with aldehydes (ketones can also be used) [104–106]. The C_{2v} symmetrical *N*-methylfulleropyrrolidine **134** [100], for instance, can be prepared in a one-pot synthesis and 41% yield by heating paraformaldehyde, *N*-methylglycine, and C_{60} in refluxing toluene (Scheme 9).

The reaction is simple and convenient with specific advantages being several-

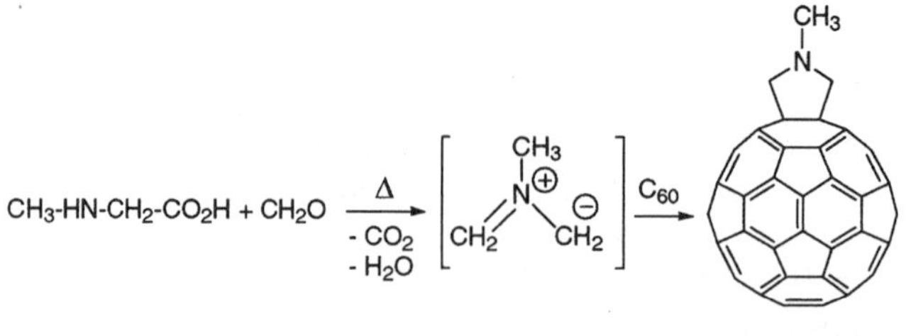

Scheme 9

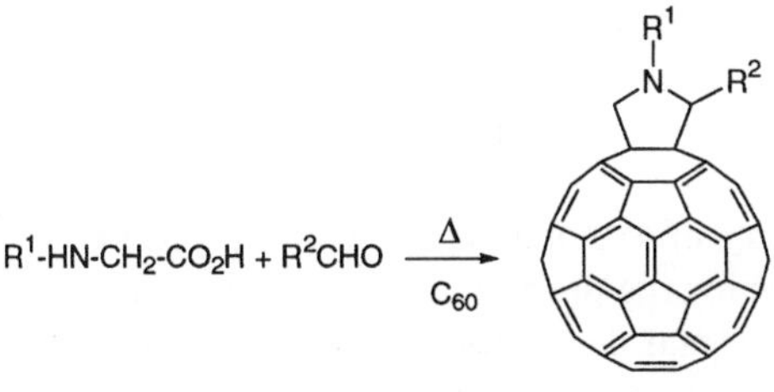

R^1-HN-CH$_2$-CO$_2$H + R^{2}CHO $\xrightarrow[\text{C}_{60}]{\Delta}$

135

Scheme 10

fold: (1) Monofunctionalization affords a single isomer, namely, the product of cycloaddition across a 6,6-ring junction of the fullerene [100,107]. (2) The use of different aldehydes leads to a variety of 2-substituted fulleropyrrolidines, whereas reaction of N-functionalized glycines with aldehydes gives N-substituted fulleropyrrolidines **135** (Scheme 10). Clearly, the use of both of the above reactants allows the concomitant introduction of two functional chains in one step. (3) The reaction can be performed in the presence of many other functional groups.

Mention must be made of a minor limitation of the method. When $R^2 \neq H$ a new stereocenter at position 2 of the pyrrolidine ring is produced. If R^1 or R^2 is enantiopure, a mixture of diastereoisomers is therefore expected [108–111].

Usually, a careful control of the stoichiometry of the reagents and of the reaction conditions allows isolation of the monoaddition product with multiple adducts present only in traces. On the other hand, when the reaction is carried out using an excess of ylide precursors, diadducts as well as higher adducts can be isolated. Mass spectrometric analysis showed that up to nine pyrrolidine rings can be introduced on the C$_{60}$ frame when a large excess of reagents is employed. Purification and correct structure determination, as well as the study of the preferred patterns for isomer formation, are currently challenging aspects of multiple azomethine ylide additions to the fullerene core [112–114].

Another method of choice for the synthesis of fulleropyrrolidines is based on the thermal ring opening of aziridines (Scheme 11) [100]. Aziridines open quite

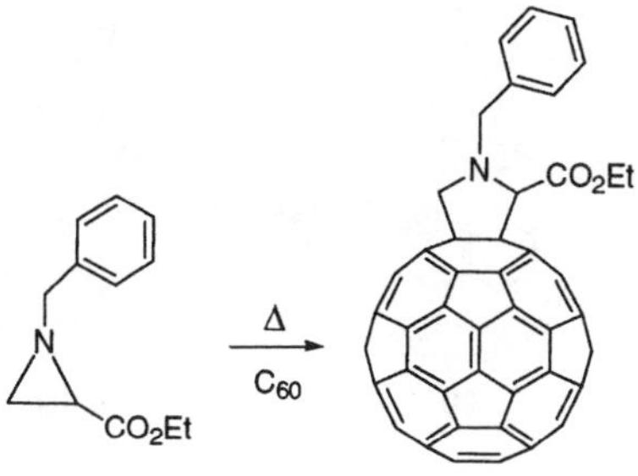

136

Scheme 11

readily if the carbon atoms are functionalized with electron-withdrawing substituents. Thus, when N-benzyl-2-carboethoxy aziridine is heated in chlorobenzene at reflux, N-benzylfulleropyrrolidine **136** is isolated in 40% yield after chromatography.

The rich chemistry of azomethine ylides has promoted the use of a variety of alternative methods for the preparation of substituted fulleropyrrolidines. Acid-catalysis [101] or thermal desilylation [115] of trimethylsilyl amino derivatives, tautomerization of α-amino esters, immonium salts [116], and imines [117,118], or reaction with aldehydes in the presence of aqueous ammonia, [119] leads to fulleropyrrolidines. Photochemical treatment of tertiary amines, [120–122] amino acid esters, or aminopolycarboxylic esters [55,123] with C_{60} has also been employed to synthesize pyrrolidine derivatives.

The useful derivative **139**, precursor to N-H fulleropyrrolidine **140**, can be obtained upon treatment of N-trityloxazolidinone **137** with C_{60} in refluxing toluene [100,124], followed by acid-catalyzed removal of the trityl protecting group from the resulting fulleropyrrolidine **138** (Scheme 12). The highly insoluble ammonium salt **139** can be collected by filtration and treated with excess pyridine (or triethylamine) when pyrrolidine **140** is needed.

An alternative route to **140** involves the direct condensation of glycine with paraformaldehyde in the presence of C_{60} (Scheme 12). Fulleropyrrolidine **140** is

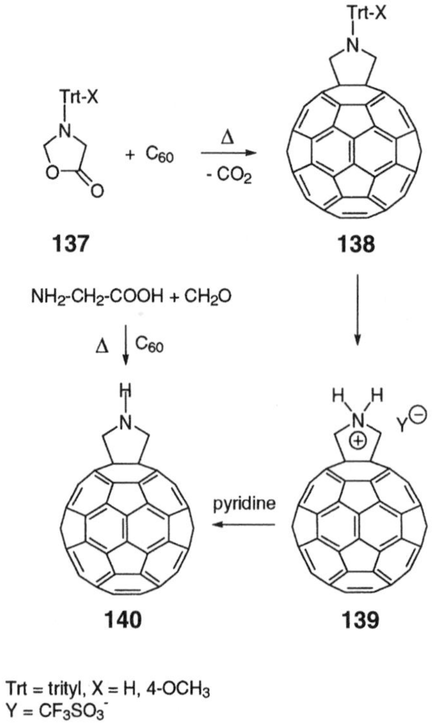

Trt = trityl, X = H, 4-OCH$_3$
Y = CF$_3$SO$_3^-$

Scheme 12

stable in solution at room temperature but concentration or prolonged heating affords intractable material. This is in line with the known reactivity of the fullerene core toward secondary amines [8] although substituted N-H fullero-pyrrolidines can be isolated as solids when an alkyl or aryl group is present at either position 2 or 5 of the pyrrolidine ring [102,115,117,123,125–131]. Pyrrolidine **140** has been efficiently acylated to give amides that have been used for preparation and deposition of Langmuir–Blodgett films [124,132]. Recently, a maleimidopropionyl amide, from **140** and maleimidopropionyl chloride, was used to prepare a fullerene-modified protein [133].

3.2.6.3 Functionalized Fulleropyrrolidines

Soluble Fulleropyrrolidines The solubility of fullerenes increases upon functionalization. Generally, bulky or long-chain substituents prevent close contact among fullerene spheres, thus improving solubility. Several fulleropyrrolidines bearing solubilizing pendant groups have been produced (a few selected examples are shown in Fig. 3.3). These groups not only improve solubility but serve,

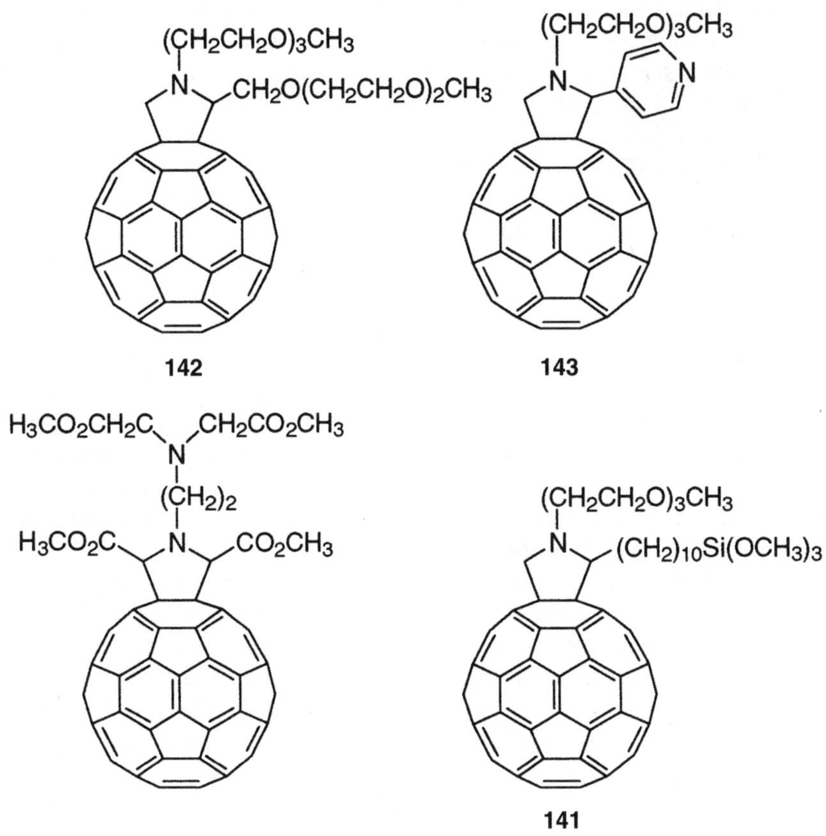

Figure 3.3

as in the case of derivative **141** [134,135], to anchor the fullerene to inert matrices or to surfaces for further studies and production of devices for various applications.

The synthesis of water-soluble fullerene derivatives has been an essential prerequisite for assessing the participation of the C_{60} core in biologically relevant processes, such as singlet oxygen generation, radical scavenging properties, and HIV-1 protease inhibition [136]. A few water-soluble fulleropyrrolidines, such as **142** and **143**, have been prepared and tested in preliminary biological assays against a variety of microorganisms with promising results [137].

Spin-Labeled Fulleropyrrolidines A TEMPO group (2,2,6,6-tetramethyl-piperidine-1-oxyl) has been covalently linked to C_{60} through cycloaddition of azomethine ylides [105,125–128,138,139]. Figure 3.4 shows derivatives **144–149** in which the nitroxide moiety is attached to different positions of the pyrrolidine ring.

Upon laser illumination, the excited quartet state, derived from interaction of triplet (fulleropyrrolidine) with doublet (nitroxide), was observed by electron paramagnetic resonance (EPR) in liquid solutions [125], in matrices where the molecular motion is not completely frozen [127] as well as in a rigid matrix [138]. The dependence of the interaction between excited triplet C_{60} and the TEMPO radical has also been investigated as a function of the relative distance and orientation of the partners. The nitroxide unpaired spin has been a useful paramagnetic probe for studying C_{60} anions [128].

The nitroxide radical also proved to be a suitable probe for the study of photoinduced intramolecular electron-transfer processes from ferrocene to fullerene in derivatives **147** and **148** [140]. The interesting C_{60}–nitroxide **149** was prepared by chloroperbenzoic acid oxidation of the corresponding NH-pyrrolidine derivative, which, in turn, was synthesized by the decarboxylative method using α-aminoisobutyric acid, methyl nonyl ketone, and C_{60} [105,141].

Covalently and Noncovalently Linked Donor–Acceptor Model Systems Intramolecular energy- or electron-transfer processes within donor–acceptor (D-A) model systems have been the subject of extensive theoretical and experimental studies [142]. Factors such as structure and distance between D and A, solvent effects, and D-A energetics, have systematically been exploited in order to control charge-separation dynamics and efficiencies.

A very active front in fullerene research relates to the combination of the absorption properties and the electron-accepting properties of C_{60} with those of other covalently linked electro- and/or photoactive addends [29,97]. The long-term objective of these studies is to make use of the long-lived charge-separated state for artificial photosynthesis or for useful functions such as conversion of solar light into electric current. A wide range of donor-linked C_{60} derivatives have been prepared via azomethine ylide cycloaddition. Donors include ferrocene [116,140,143–145], electron-rich aromatics [146,147], ruthenium(II) trisbipyridine complexes [106,148,149], tetrathiafulvalene [116,150,151], benzo-

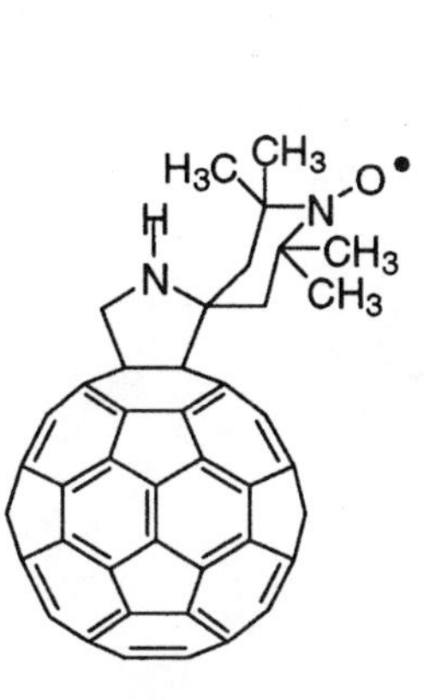

Figure 3.4

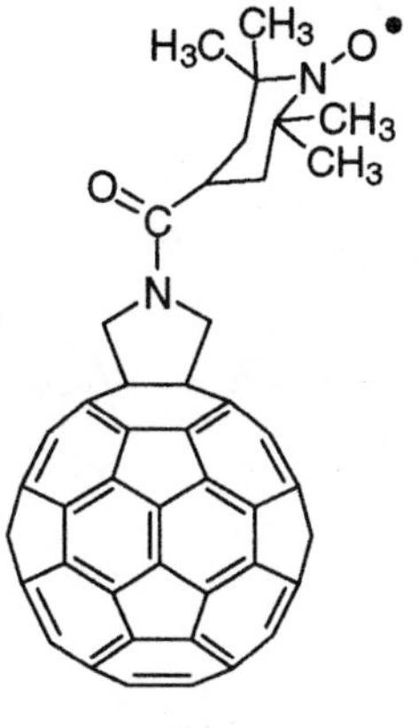

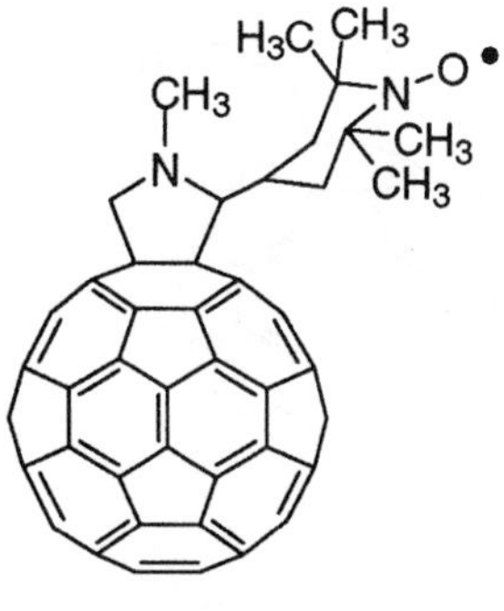

quinone [130], and porphyrins with one [107,152–157] or two [158] C_{60} units (Fig. 3.5).

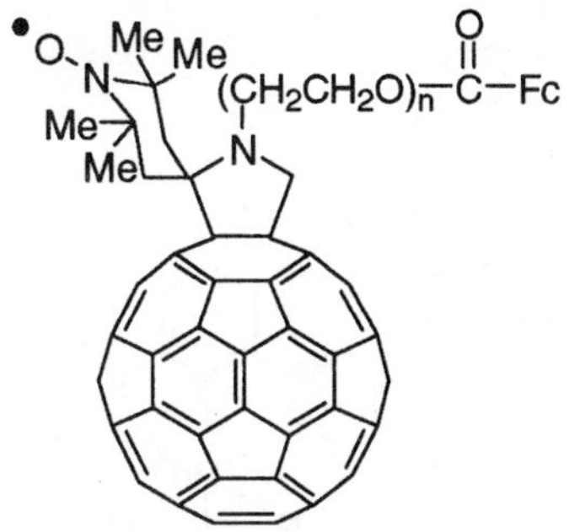

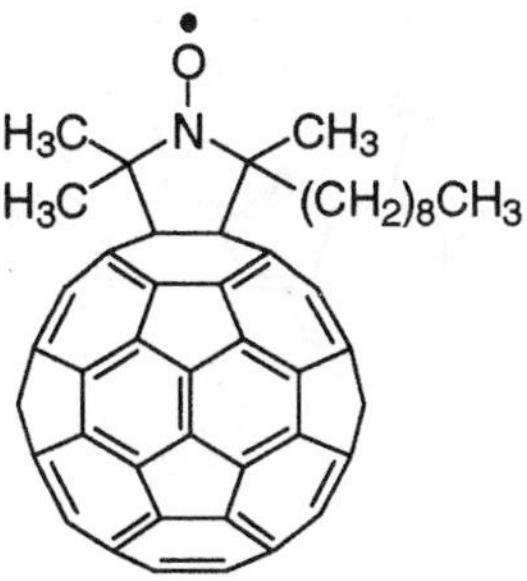

A number of these D-A systems have been investigated by means of laser flash photolysis and pulse radiolysis to probe their photophysical properties, and for assessing whether or not functionalized fullerenes are suitable electron-accepting chromophores. In an early contribution, photoinduced intramolecular electron transfer in a fullereropyrrolidine–aniline dyad was examined [146], and it was concluded that C_{60} adducts may not be suitable electron acceptors for long-range electron transfer in bichromophoric systems. However, later results showed that C_{60} can accelerate photoinduced charge separation (CS) and decelerate charge recombination (CR) [29,159]. The extent and rate of CR could be decreased by modulating the separation (or the spatial orientation) of the D and A by means of flexible or rigid hydrocarbon and even steroid bridges [156], while maintaining a high rate of charge separation [29,98,143,145,147].

Currently, the most promising research in the field of C_{60}-based D-A systems

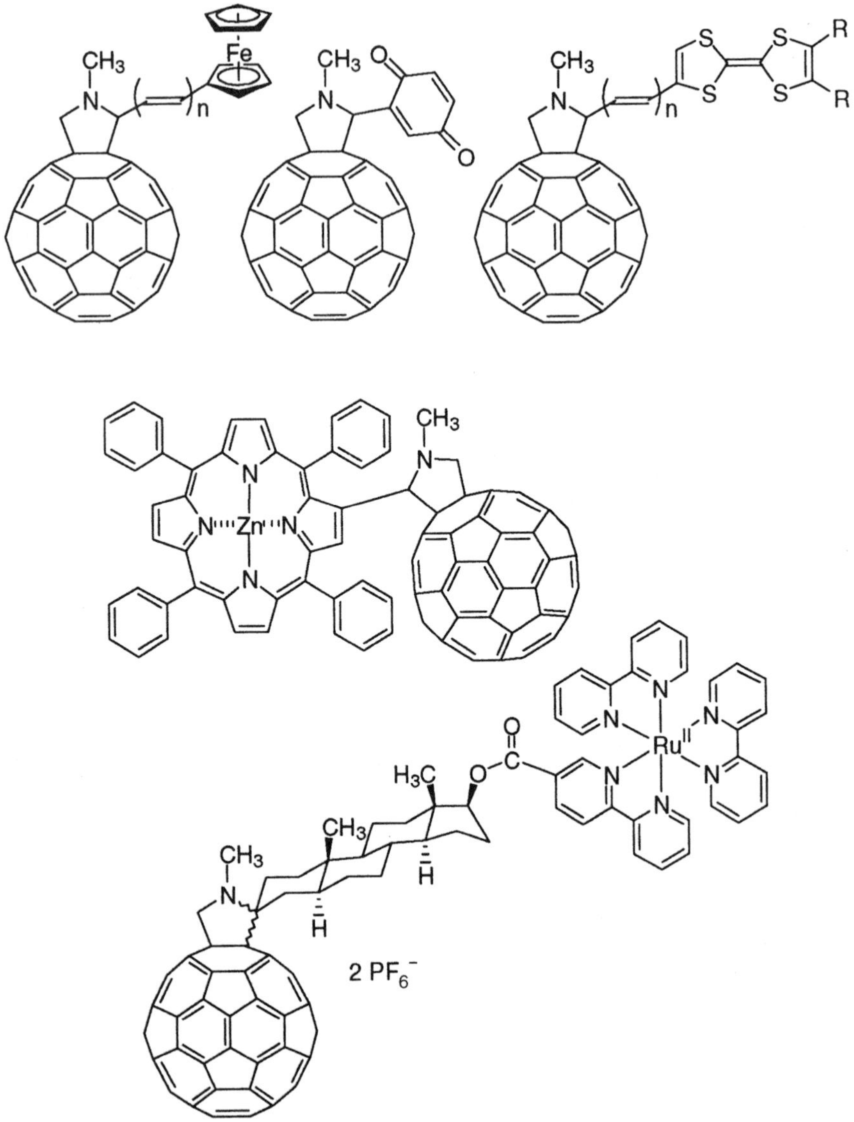

Figure 3.5

relies on (1) the arrangement of several redox-active building blocks in which multistep electron transfer can occur; (2) the design of supramolecular systems of noncovalently attached chromophores; and (3) the enhancement of the electron-accepting properties of the functionalized fullerene. These objectives are illustrated by two recent reports on the porphyrin–pyromellitimide–fulleropyrrolidine triad **150** [147] and the fulleropyrrolidine–porphyrin–carotene triad **151** [157] (Fig. 3.6).

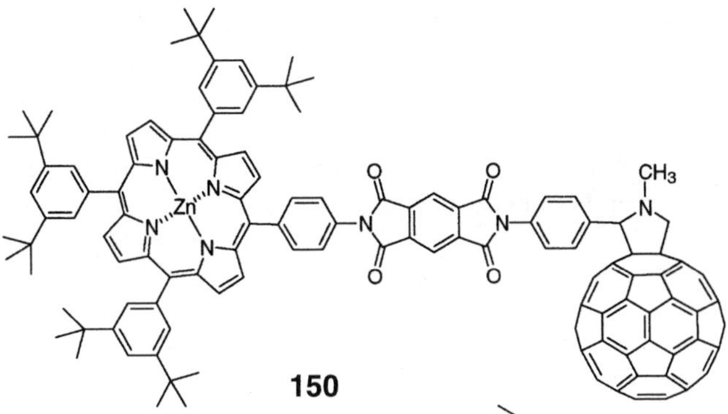

150

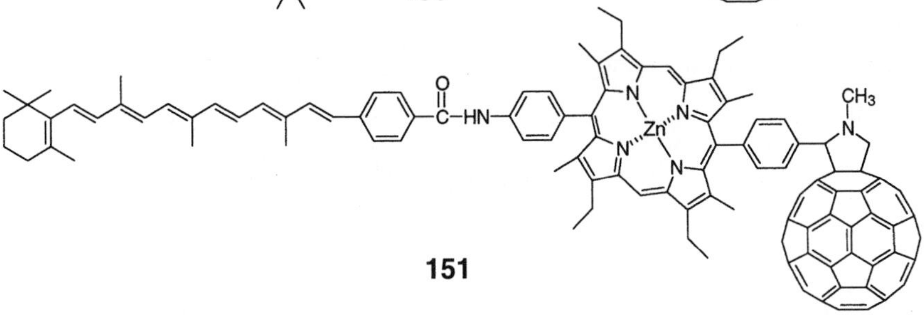

151

Figure 3.6

The system shown in Figure 3.7, in which a pyridine-functionalized fullero-pyrrolidine coordinates the zinc metal of a tetraphenyl porphyrin with high affinity, represents a case where ligation is used to associate a donor and an acceptor noncovalently [160].

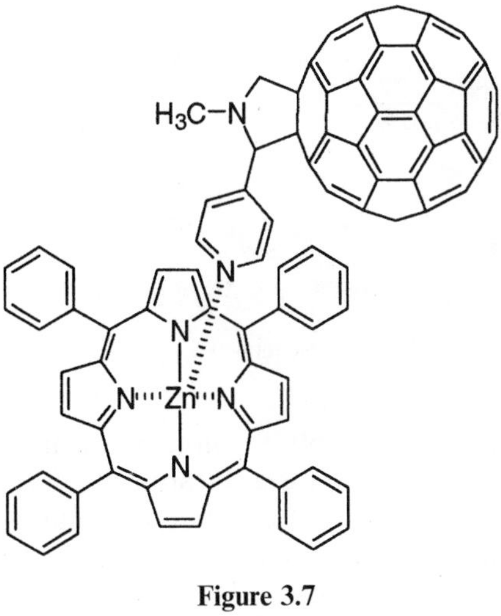

Figure 3.7

In such systems intramolecular photoinduced electron transfer from the porphyrin to C_{60} would occur very fast. Subsequently, the corresponding CS pair should separate, so that charge recombination becomes less probable. Time-resolved photolytic and EPR experiments showed that in CH_2Cl_2 and polar solvents a long-lived CS state is attained with lifetimes of several hundreds of microseconds in benzonitrile [160].

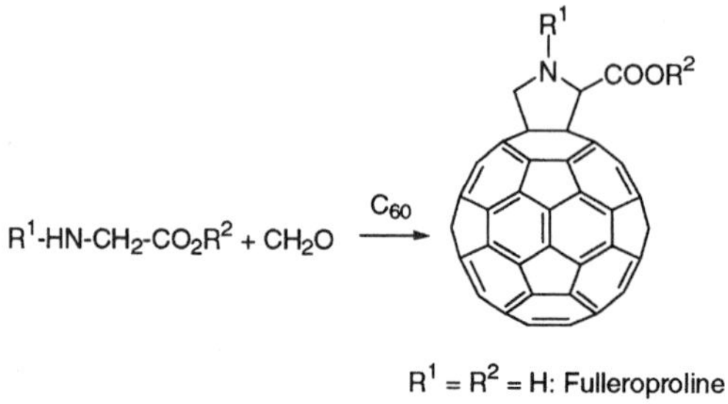

$R^1 = R^2 = H$: Fulleroproline

Scheme 13

3.2.6.4 Fulleroprolines

Azomethine ylides derived from amino acid esters and paraformaldehyde add thermally to C_{60} to yield derivatives of fulleroproline, a new unnatural amino acid in which a proline ring is fused directly to C_{60} (Scheme 13) [108,161]. Fulleroproline derivatives also can be prepared via thermal ring opening of 2-substituted carboalkoxy aziridines [108] or photolysis of N,N-dialkylamino acid esters [55] in the presence of C_{60}.

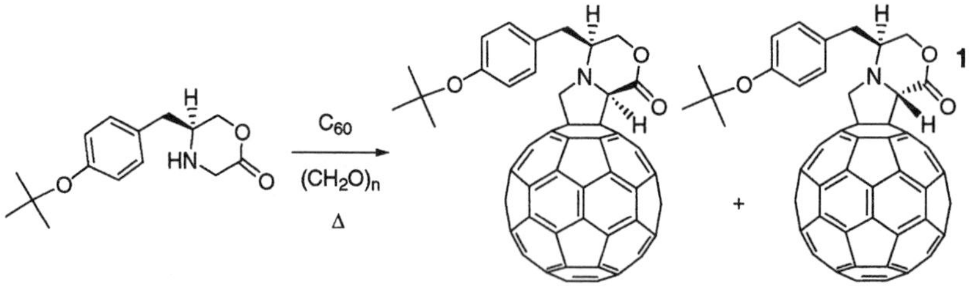

152a (minor)   **152b** (major)

Scheme 14

The use of fulleroproline in peptide synthesis requires that this new amino acid be available in optically pure form. In principle, this can be accomplished by asymmetric azomethine ylide cycloaddition or enantioselective HPLC separations of the fulleroproline enantiomers. While both approaches have been pursued with promising results [108,161], chromatographic separation of diastereoisomers has been employed for the synthesis of optically pure oligopeptides incorporating fulleroprolines [108,110]. Scheme 14 shows the preparation of optically pure fulleroproline derivatives **152**, and Scheme 15 shows the preparation of dipeptide **154** from the *N*-protected fulleroproline **153**.

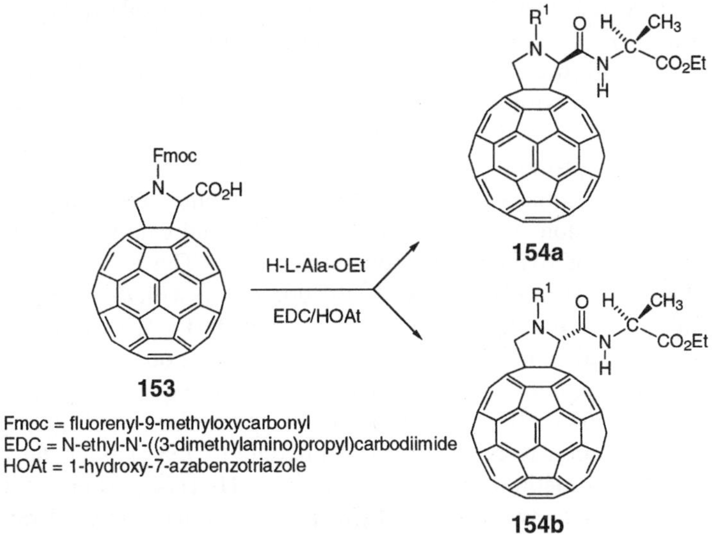
Fmoc
N
CO2H
153
H-L-Ala-OEt
EDC/HOAt
R1
O
H
CH3
N
C
CO2Et
H
154a
154b
Fmoc = fluorenyl-9-methyloxycarbonyl
EDC = N-ethyl-N'-((3-dimethylamino)propyl)carbodiimide
HOAt = 1-hydroxy-7-azabenzotriazole

Scheme 15

The absolute configuration of **152a–b** and **154a–b** was determined by means of NOE analysis (**152a–b**) and CD spectroscopy (**152a–b**, **154a–b**).

3.3 HYDROGENATED FULLERENES (Mark S. Meier)

Hydrogenated fullerenes are compounds in which one or more of the C–C double bonds have been hydrogenated. A detailed review article dedicated to hydrogenated fullerenes was published in 1997 [162]. While a wide range of conditions have been used to make hydrogenated fullerenes [162], this section concentrates on the preparatively useful methods.

3.3.1 $C_{60}H_n$ Compounds

Although hydrogenated fullerenes have been reported for C_{60}, C_{70} [162], C_{76}, C_{78}, and C_{84} [163], this discussion is limited to the best characterized hydrogenated compounds, namely, those derived from C_{60} and C_{70}. The simplest hydrogenated derivative of a fullerene is $C_{60}H_2$, first reported in 1993 by Henderson and Cahill [164]. Its preparation involved treatment of C_{60} with borane·THF, followed by protonolysis of the organoborane adduct. $C_{60}H_2$ (**155**) was isolated in 10–30% yield following HPLC chromatography.

Even for this simple compound, 37 isomers are theoretically possible [165]. Only the 1,2-isomer was formed in the hydroboration reaction, which was proved by preparation of $C_{60}HD$ and measurement of an H-D NMR coupling constant (2.4 Hz) consistent with a 1,2-isomer.

Since the intial preparation of **155**, a number of other synthetic methods have been reported, including electrochemical reduction of C_{60} [166], hydrozirconation [167], diimide [167], and chromous ion reduction [168], photoinduced electron transfer [169], heterogeneous catalytic hydrogenation [168,170], reduction with a stoichiometric amount of hydrogen on palladium [171], and dissolving metal reduction with Zn metal [172].

Hydroboration/protonolysis of 1,2-$C_{60}H_2$ (**155**) led to the formation of a mixture of $C_{60}H_4$ isomers [173], with the 1,2,3,4-isomer **156** being the major product. Isomerization of the mixture over a Pt metal catalyst indicated that **156** is also the thermodynamically most stable isomer. This is in agreement with calculations at the HF/6-31G* level. Reduction of C_{60} with diimide produces a mixture of isomers, but **156** is the major product. In the Zn(Cu) reduction of C_{60}, the *trans*-3 (**157**) and *e* (**158**) isomers of $C_{60}H_4$ were obtained in a 1:1 ratio, although smaller amounts of other isomers were detected as well [168].

Zn(Cu) reductions of C_{60} have provided an interesting set of hydrogenated fullerenes. In addition to being an excellent route to **155** (1 hour, 66% yield), **157**, and **158**, this method can also produce $C_{60}H_6$ [174]. After a total of 4 hours of reaction time, two isomers of $C_{60}H_6$ were produced in a 6:1 ratio (35% yield). The major $C_{60}H_6$ species was identified as the D_3-symmetric, *trans*-3, *trans*-3 isomer **159**, and the minor isomer was identified as the C_3 isomer **160**

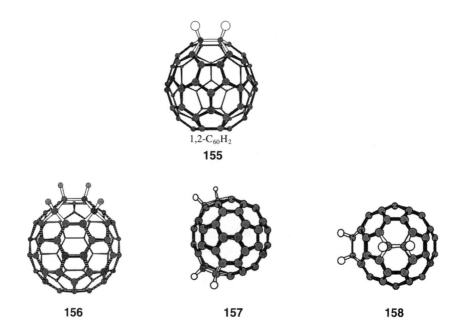

1,2-$C_{60}H_2$
155

156   **157**   **158**

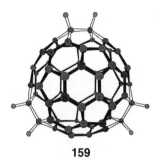

159   **160**

[175]. Calculations suggest that **159** and **160** are not the thermodynamically most stable $C_{60}H_6$ species [176], but the predicted thermodynamic isomers of $C_{60}H_6$ cannot be produced from reduction of **157** and **158**, respectively, without significant rearrangement.

$C_{60}H_{18}$ has also been made by hydrogen transfer from 9,10-dihydroanthracene [177,178], and has been detected in a variety of other reductions of C_{60}. High-pressure/high-temperature hydrogenation of C_{60} produced sufficient amounts of $C_{60}H_{18}$ for characterization [179]. Analysis of the ^{1}H NMR spectrum suggested a structure with C_3 symmetry, which led to the assignment of the dome-shaped structure **161**. Selective formation of this isomer has been

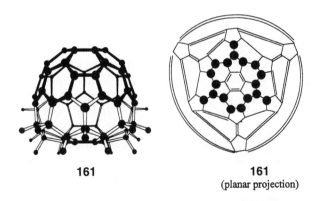

161   **161**
                         (planar projection)

rationalized as resulting from a process in which the site of addition of each new pair of hydrogens is guided by a combination of strain and bond order considerations.

3.3.2 $C_{70}H_n$ Compounds

C_{70} chemistry has always lagged behind that of C_{60}, due to the higher cost of C_{70} and its lower symmetry (D_{5h}) compared to C_{60} (I_h). Because of the latter characteristic, the number of possible isomeric species is much higher for $C_{70}H_n$

than for $C_{60}H_n$ compounds. The low symmetry also leads to difficulty in assigning structures to isomeric $C_{70}H_n$ compounds.

Preparation of the simplest C_{70} hydride, $C_{70}H_2$, was reported in 1994 by the groups of Taylor [168] and Cahill [180], using diimide reduction and hydroboration/protolysis, respectively. The dimide reaction gives a mixture of eight products, while the 1,2-isomer **162** is formed preferentially by the latter procedure. The yield of the 5,6-isomer **163** is about 5% in diimide [168] as well as Zn(Cu) [174,181] reductions, but is 33% in the hydroboration/protonolysis route [180]. The preferential formation of the 1,2-isomer **162** over other isomers seems reasonable, since the carbon atoms in the C1–C2 bond are the most pyramidal (and therefore the most reactive) in the molecule, followed by the atoms in the C5–C6 bond [182].

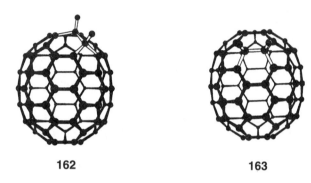

162   **163**

Reduction of C_{70} with Zn(Cu) can also be used to produce $C_{70}H_4$, $C_{70}H_8$, and $C_{70}H_{10}$, with smaller amounts of more highly hydrogenated material [174,181]. One major isomer of $C_{70}H_4$ (**164**) and of $C_{70}H_8$ (**165**) are produced.

It is clear that $C_{70}H_8$ (**165**) does not result from reduction of **162**, **163**, or **164**. This was confirmed by purifying a sample of **162** (containing a small amount of **163**) and resubjecting it to the Zn(Cu) reduction conditions. No **165**

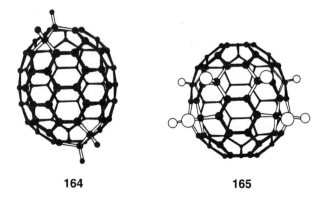

164   **165**

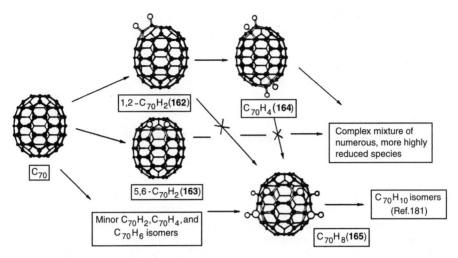

Figure 3.8 Different manifolds in the reduction of C_{70} with Zn(Cu).

is formed in these experiments. It appears that there are several reduction manifolds in operation at the same time. One manifold includes **162**, and products derived from **162** (including **164**). A second manifold includes **163** (the 5,6-isomer) and products derived therefrom. A third manifold does not produce a $C_{60}H_2$ species that is detectable under the reaction conditions. It appears that the $C_{70}H_2$, $C_{70}H_4$, and $C_{70}H_6$ species on this manifold undergo reduction at rates comparable to their rates of formation, and the only compound on this manifold that accumulates to a significant concentration is **165**. Essentially, **165** appears with no accompanying precursors. These reduction manifolds are summarized in Figure 3.8.

3.3.3 Highly Hydrogenated Fullerenes

The first preparation of a highly hydrogenated fullerene via the Birch reduction was reported in 1990 soon after the intial isolation of macroscopic quantities of C_{60} [183]. The highly air-sensitive nature of $C_{60}H_{36}$ has made it quite difficult to characterize, but the broad lines observed in the ^1H and ^{13}C NMR spectra suggest that this material is a complex mixture of isomers. An impressive number of isomers ($>10^{14}$) are possible [165].

A few reports exist of species with molecular formulas as high as $C_{60}H_{38}$ to $C_{60}H_{46}$ [184–186], but these compounds have not yet been characterized in detail.

3.3.4 Properties

Fullerene C–H bonds are notably acidic. The first observation of this characteristic came with the preparation of *tert*-BuC$_{60}$H, where the pK_a was determined to be 5.6. The pK_a of $C_{60}H_2$ itself has been determined to be 4.7 [187], making this hydrocarbon as acidic as acetic acid. The second ionization constant for $C_{60}H_2$ is ~16 [187]. The pK_a of the $C_{60}H$ radical has been calculated from electrochemical data to be ~9 [187]. The pK_a of the $C_{60}H$ radical (as an inclusion complex in γ-cyclodextrin) has been determined to be 4.5 [188].

3.4 HALOGENATION AND ARYLATION OF FULLERENES (Roger Taylor)

3.4.1 Halogenation of Fullerenes

Halogenation of fullerenes is a radical reaction and can be achieved for fluorination, chlorination, and bromination, but not iodination. The latter failure may be attributed to a combination of the weak C–I bond and the need for the bulky addends to be well separated, which would result in substantial bond reorganization with double bonds being placed in pentagonal rings, which is unfavorable [189]. The halogenofullerenes eliminate halogen on heating to approximately 150 °C, the stability order being fluoro- > chloro- > bromofullerenes; in the case of the latter the decomposition is more rapid, the fewer bromines that are present. The solubility order is fluoro- > chloro- > bromofullerenes, and indeed the bromofullerenes are difficult to use in syntheses because of their low solubility. The halogenofullerenes are much more reactive than the corresponding halogenoalkanes and are susceptible toward nucleophilic substitution by atmospheric water, especially in the presence of a cosolvent.

3.4.1.1 Fluorination Fluorination can be achieved with either fluorine gas [190–200], XeF_2 [191], or metal fluorides [201–204]; KrF_2, BrF_5, and IF_5 have been used in limited experiments [205,206]. Each process has to be carried out under heterogeneous conditions, because fullerenes are insoluble in fluorine-resistant solvents. Control of fluorination is impossible using fluorine gas or the volatile fluorides, because fluorine cannot penetrate the tightly packed fullerene lattice. Fluorination therefore takes place stepwise with the outer fullerene molecules becoming substantially fluorinated and they then expand away from the bulk. Hence, the product is a mixture of highly fluorinated and unfluorinated molecules [199]. Since the packing in [70]fullerene is not as good as in [60]fullerene, it fluorinates faster than the latter [195,207], this being the reverse reactivity order to that observed in most reactions. For the same reason fluorination is slower the purer the fullerene. Only $C_{60}F_{48}$ has been obtained as a pure compound from fluorine gas fluorination (at high temperature) [198,200]. Under more drastic conditions (F_2/hv or KrF_2/HF), fluorination continues to the point where cage-opening occurs, giving hyperfluorinated derivatives of composition up to $C_{60}F_{78}$ [194,205].

Fluorination by heated metal fluorides under vacuum has the advantage that

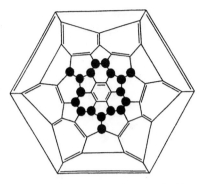

Figure 3.9   Schlegel diagram for C_{3v}-$C_{60}F_{18}$ ($\bullet$ = F).

the fluorofullerenes are more volatile than the parent fullerene and are pumped away from the reaction zone as fluorination proceeds. Thus, by using MnF_3, CoF_3, K_2PtF_6, and so on, it has been possible to prepare, isolate, and characterize $C_{60}F_{18}$ [203], $C_{60}F_{18}O$ [204], $C_{60}F_{36}$ (both T and C_3 isomers in approximately 1:3 ratio) [201,202], $C_{70}F_n$ ($n = 34, 36, 38, 40, 42$)[208], and $C_{84}F_{40}$ [208]; many if not all of these derivatives are isostructural with their hydrogenated counterparts. This is further supported by the similarities of the ^{3}He NMR spectra for the species $C_{60}X_{18}$ (X = F, H) and likewise for $C_{60}X_{36}$ [209]. There are a number of isomers of a given $C_{70}F_n$, each having significantly different polarities. The color intensity decreases with fluorination level: $C_{60}F_{18}$ is green-yellow, $C_{60}F_{36}$ is lemon-white, and $C_{60}F_{48}$ is pure white. Because of the high polarity of fluorine, derivatives of very similar fluorine content can readily be separated by HPLC, and, for example, the derivatives of C_{70} consist of many isomers which themselves have been isolated and characterized [208]. Figures 3.9–3.12 show the Schlegel diagrams for $C_{60}F_{18}$, $C_{60}F_{36}$ (T and C_3), and $C_{60}F_{48}$.

Due to their high polarity, the fluorofullerenes readily form charge-transfer complexes; for example, $C_{60}F_{48}$ forms a bright red solution in toluene [200].

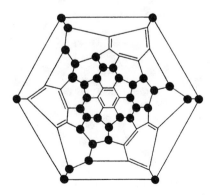

Figure 3.10   Schlegel diagram for T-$C_{60}F_{36}$ ($\bullet$ = F).

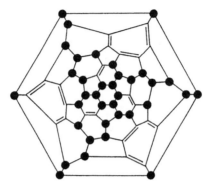

Figure 3.11   Schlegel diagram for the C_3-isomer of $C_{60}F_{36}$ ($\bullet = F$).

$C_{60}F_{18}$ is one of the most polar compounds known and is calculated to have a dipole moment of approximately 6 Debye units [210].

Just as hydrogenation is accompanied by the formation of CH_2 adducts, so too CF_2 adducts are produced during fluorination [204]. These arise probably from reaction with difluorocarbene produced by fragmentation of some hyperfluorinated cages. The radical nature of fluorination is confirmed by the addition of both F and Cl to the cage when fluorination is carried out by IF_5 in the presence of CCl_4, CCl_3F being also detected as a by-product [206].

3.4.1.2 Chlorination   Chlorination of [60]fullerene with chlorine gas at 250 °C gives orange $C_{60}Cl_{24}$ [211], the structure of which has not been determined, but is probably isostructural with $C_{60}Br_{24}$ (see below). A fast atom bombardment (FAB) mass spectrum of $C_{60}Cl_{24}$ has been obtained, the only known example of a mass spectrum of a polychlorinated fullerene [206]. Reac-

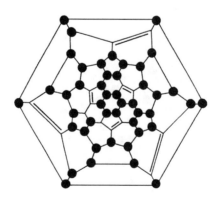

Figure 3.12   Schlegel diagram for one D_3-isomer of $C_{60}F_{48}$ ($\bullet = F$).

tion of [60]fullerene with liquid chlorine at $-35\,^{\circ}C$ produces a brown solid of approximate composition $C_{60}Cl_{12}$ and this dechlorinates on heating at 200–350 $^{\circ}C$, that is, at a lower temperature than $C_{60}Cl_{24}$ [211]. This follows the general pattern of fullerene derivative stability, namely, higher stability for compounds having a larger number of addends.

Reaction of [60]fullerene with ICl does not give the anticipated iodination (for reasons noted above) but instead produces $C_{60}Cl_6$ (with $C_{60}Cl_{12}$ as a by-product) [212]. Addition in a given hexagon occurs in a 1,4- rather than 1,2-manner due to steric considerations, producing ultimately $C_{60}Cl_5^{\cdot}$, an unstable radical that adds a further chlorine to give the product. This necessitates the final placement of two chlorines adjacent to each other, the product being isostructural with $C_{60}Br_6$ (see below). Similar behavior could be anticipated for chlorination of [70]fullerene, but here the unfavorable π-bond location in the equatorial region (which arises due to the need to have double bonds exocyclic to the pentagons in the end caps) causes the 1,4-addition to take place here instead to give $C_{70}Cl_{10}$ [213]. This also results in two chlorines ending up adjacent to each other. Thus, in both $C_{60}Cl_6$ and $C_{70}Cl_{10}$ a region of instability and consequently higher reactivity is produced relative to the rest of the molecule.

The chlorination reaction is slower if carried out in toluene, confirming it to be a radical reaction, since toluene is a good radical scavenger.

3.4.1.3 *Bromination*
Bromination of fullerenes is achieved by reaction with bromine, either neat (giving $C_{60}Br_{24}$ [214]), in either CS_2 or $CHCl_3$ (giving $C_{60}Br_8$ [216]), or in either benzene or CCl_4 (giving $C_{60}Br_6$ [215]). These compounds have been very useful for providing fundamental information concerning the multiple addition patterns to fullerenes, but their low solubilities are such that no synthetic work has been undertaken with them. The stability order is $C_{60}Br_6 < C_{60}Br_8 < C_{60}Br_{24}$; on heating $C_{60}Br_6$ degrades into [60]fullerene after first rearranging into $C_{60}Br_8$ [215]. The eclipsing interaction between the *cis*-bromines in $C_{60}Br_6$ causes the C(2)–Br bond length to be greater (2.032 Å) than the average (1.963 Å) for other types of C–Br bonds. The single crystal X-ray structures for $C_{60}Br_{24}$, $C_{60}Br_8$, and $C_{60}Br_6$ are shown in Figures 3.13, 3.14, and 3.15, respectively.

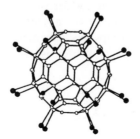

Figure 3.13 X-ray structure for $C_{60}Br_{24}$ ($\bullet$ = Br).

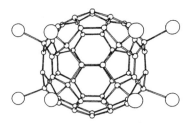

Figure 3.14 X-ray structure for $C_{60}Br_8$ ($\circ$ = Br).

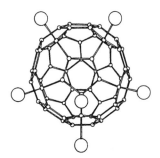

Figure 3.15 X-ray structure for $C_{60}Br_6$ ($\circ$ = Br).

3.4.2 Nucleophilic Substitution of Halogenofullerenes

Halogenofullerenes, especially fluorofullerenes, react readily with a range of nucleophiles [216], due to the high electrophilicity of the cage, further enhanced by the electron-withdrawing halogens. The reactivity order is C–F > C–Cl > C–Br. Most work has concentrated on the reaction with water and with aromatics.

3.4.2.1 Substitution by Water The reaction with water is particularly facile and is greatly accelerated by the presence of cosolvents since halogeno-fullerenes (especially the fluorofullerenes) are hydrophobic. The most probable substitution mechanism involves addition–elimination (Fig. 3.16) [216]. The reaction products are either hydroxy derivatives or epoxides, believed to be formed by subsequent loss of HX from adjacent X and OH groups [216]; the presence of up to 18 epoxy groups has been detected [199]. A number of these hydroxides and epoxides have now been isolated and characterized, such as

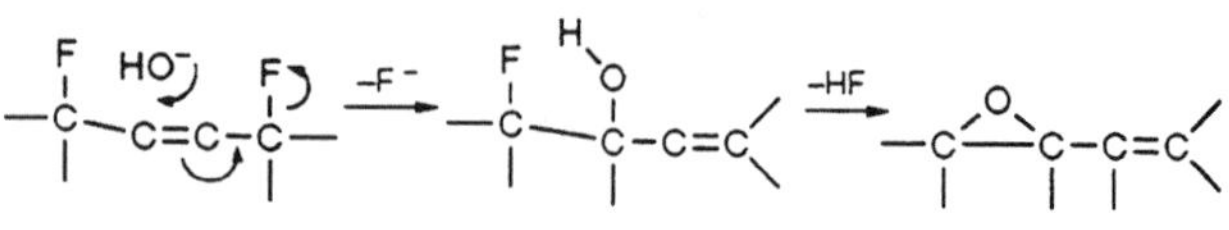

$$-\overset{\text{F}}{\underset{}{\text{C}}}-\text{C}=\text{C}-\overset{\text{F}}{\underset{}{\text{C}}}- \xrightarrow{-\text{F}^-} -\overset{\text{F}}{\underset{}{\text{C}}}-\overset{\text{H}}{\underset{}{\text{C}}}\overset{\text{O}}{-}\text{C}=\text{C} \xrightarrow{-\text{HF}} -\text{C}\overset{\text{O}}{-}\text{C}-\text{C}=\text{C}$$

Figure 3.16 Proposed addition–elimination mechanism producing epoxides from fluoro-fullerenes.

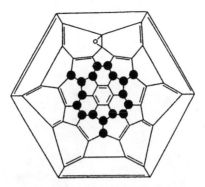

Figure 3.17 Structure of $C_{60}F_{18}O$.

$C_{60}F_{35}OH$, $C_{60}F_{34}O$, $C_{60}F_{33}O \cdot OH$, $C_{60}F_{32}O_2$, and $C_{70}F_{37}OH$ [208,217]. As an example, the structure of $C_{60}F_{18}O$ is shown as a Schlegel diagram in Figure 3.17.

A nucleophilic substitution may be responsible for the sharp singlet at approximately -152 ppm obtained in the ^{19}F NMR spectrum of fluorofullerenes dissolved in moisture-containing solvents such as diethyl ether and THF, since a strong peak at approximately -190 ppm due to HF is also obtained. The singlet is due evidently to a symmetrical species, as yet unidentified ($C_{60}F_{60}$ was considered to be an attractive suspect [191], but subsequent mass spectrometry indicates that this is unlikely).

3.4.3 Substitution by Aryl Groups

This constitutes a major area of fullerene chemistry since the number of potential derivatives is very large. The reaction is facilitated by Lewis acid catalysts (the effectiveness of catalysts is $AlCl_3 > FeCl_3 > TiCl_4$), occurs more readily with aromatics containing electron-supplying groups, and is subject to steric hindrance. Thus, mesitylene, which lacks unhindered sites, will not react. Reaction takes place by a novel S_N1 substitution involving *frontside* attack only. The formation of a carbocation (free or partial) is unexpected because of the electron-withdrawing nature of the cage. However, this is mitigated by the less electron-withdrawing sp^3 carbons, and the catalyst-aided polarization of the carbon–halogen bonds. Direct replacement is supported both by the symmetry of $C_{60}F_{15}Ph_3$ [218] (see below) and by the isolation of the intermediate carbocation from $C_{60}Ph_5Cl$ [219].

3.4.3.1 Substitution in Fluorofullerenes
Only one example of this process is known, involving the reaction of $C_{60}F_{18}$ with benzene/$FeCl_3$ to give the uniquely shaped *triumphene* $C_{60}F_{15}Ph_3$ [218], Figure 3.18. The C_{3v} symmetry can only arise from direct substitution, and the most accessible fluorines are the

Figure 3.18 Space-filling model of $C_{60}F_{15}Ph_3$.

ones replaced. The reaction is also accompanied by the remarkable formation of $C_{60}Ph_{18}$ in which the phenyl groups occupy all of the previously fluorinated sites. $C_{60}Ph_{18}$ is also remarkable since it is believed to be the most phenylated organic compound known.

Thus far, phenyl substitution in either $C_{60}F_{36}$ or $C_{60}F_{38}$ has been unsuccessful, presumably due to steric hindrance.

3.4.3.2 Substitution in Chlorofullerenes Reaction of $C_{60}Cl_6$ with aromatics/FeCl$_3$ gives $C_{60}Ar_5Cl$ (see Fig. 3.19). Further reaction of this deriv-

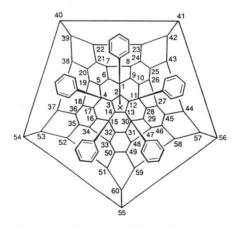

Figure 3.19 Schlegel diagram of the product (X = Cl) of reaction of $C_{60}Cl_6$ with benzene/FeCl$_3$.

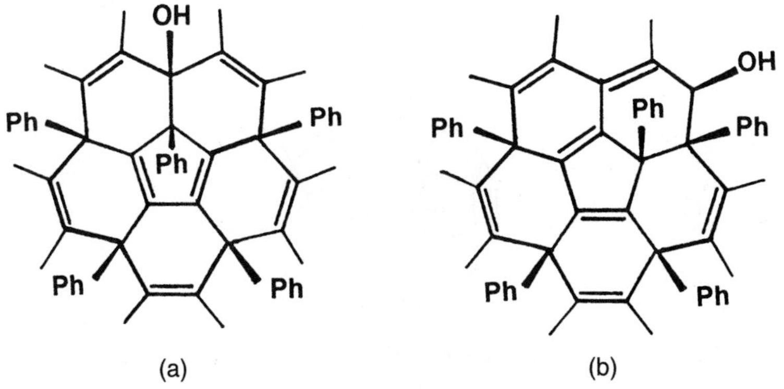

Figure 3.20   Fullerenols produced from reaction of $C_{60}Ph_5Cl$ with $AlCl_3$/water.

ative with PPh_3 results in replacement of the remaining chlorine by hydrogen, derived from traces of water in the solvent [220]. The failure to substitute the sixth chlorine suggests steric hindrance, because reaction of $C_{60}Cl_6$ with allyl bromide and ferric chloride results in replacement of all six chlorines by the smaller allyl groups [221].

The effect of steric hindrance in these reactions is also shown by the reaction with trimethylsilylbenzene, which would normally involve C–Si cleavage rather than C–H cleavage [222]. However, in the reaction with the fullerene the latter occurs only, due to the difficulty of the fullerene cage approaching the aromatic *ipso* ring carbon adjacent to the bulky trimethylsilyl group.

3.4.3.3 Other Substitution Reactions

Various other products are obtained in the reaction with benzene. These include symmetrical $C_{60}Ph_2$ and unsymmetrical $C_{60}Ph_4$ [223]. Reaction of $AlCl_3$ with $C_{60}Ph_5Cl$, followed by a trace of water produces two fullerenols (Fig. 3.20), resulting from formation and subsequent rearrangement of a carbocation having an antiaromatic pentagonal ring [224].

On standing in air and light, $C_{60}Ph_5H$ forms two oxidized products. The minor component is unsymmetrical $C_{60}Ph_4O_2$, which eliminates two molecules of CO during electron ionization (EI) mass spectrometry to give phenylated C_{58} derivatives [225]. The ability to readily eliminate C_2 depends on there being a 6,5-*double* bond in the precursor molecules [226], as is the case here. The major component is a benzo[*b*]furano[60]fullerene (Fig. 3.21), which is obtained also by similar treatment of the $C_{60}Ph_5Cl$ precursor. The mechanism of its formation is not known, but the hydride precursor may form a hydroperoxide which then eliminates water [227].

Another surprising product, phenylated isoquinolono-3,4,1,2-[60]fullerene (Fig. 3.22) is obtained from the reaction between $C_{60}Ph_5Cl$ and $BrCN/FeCl_3$

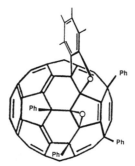

Figure 3.21   A benzo[*b*]furano[60]fullerene produced by standing $C_{60}Ph_5H$ in air.

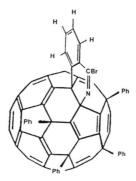

Figure 3.22   An isoquinolono[60]fullerene precursor produced by reaction of $C_{60}Ph_5Cl$ with $BrCN/FeCl_3$.

and arises from nucleophilic substitution of Cl by BrCN, followed by electrophilic substitution of the cyano carbon into the *ortho* position of the adjacent phenyl ring, and hydrolysis of the C–Br bond [228].

Reaction of $C_{70}Cl_{10}$ with benzene/$FeCl_3$ produces $C_{70}Ph_{10}$ (Fig. 3.23) and $C_{70}Ph_8$ (Fig. 3.24) [229], together with smaller amounts of $C_{70}Ph_{2/4/6}$; rotation of the adjacent phenyl groups in $C_{70}Ph_{10}$ is sterically restricted. In $C_{70}Ph_8$ there is a double bond in a pentagonal ring, making this a high-reactivity region. Thus, for example, [4+2] cycloaddition with anthracene occurs readily across this bond [230].

A by-product of the phenylation reaction is $C_{70}Ph_9OH$ (Fig. 3.25), the first monohydroxyfullerene to be prepared; the adjacency of OH and phenyl groups results in restricted rotation of the latter [231]. This compound may arise from atmospheric oxidation of $C_{70}Ph_9H$, which can also be isolated as a by-product. Two isomeric and symmetrical diols $C_{70}Ph_8(OH)_2$, one of which is shown in Figure 3.26, are also obtained from reacting a dichloromethane solution of $C_{70}Ph_8$ with 18-crown-6 and $KMnO_4$ [224]. The isomer shown possesses typical fullerenol properties in being very insoluble in many organic solvents.

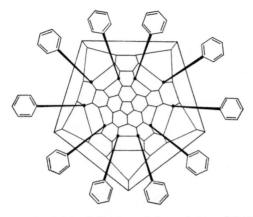

Figure 3.23   Schlegel diagram of the structure of $C_{70}Ph_{10}$.

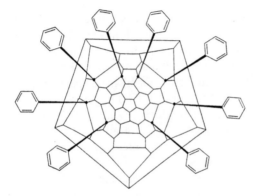

Figure 3.24   Schlegel diagram of the structure of $C_{70}Ph_8$.

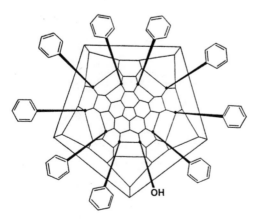

Figure 3.25   Schlegel diagram of the structure of $C_{70}Ph_9OH$.

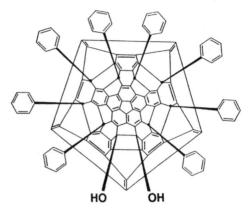

Figure 3.26   Schlegel diagram of one isomer of $C_{70}Ph_8(OH)_2$.

On standing in air, $C_{70}Ph_8$ undergoes a ring-opening oxidation to give the bis(lactone) $C_{70}Ph_8O_4$ (Fig. 3.27), which has an 11-membered hole in the cage [232]. The mechanism is believed to involve oxygen insertion into 6:5-bonds followed by oxidation of the intermediate vinyl ether, a general process known to be rapid [233]. Under conditions of EI mass spectrometry, $C_{70}Ph_8O_4$ readily loses two molecules of CO_2 to give $C_{68}Ph_8$ (cf. the formation of $C_{58}Ph_2$ described above), the structure of which is believed to be as shown in Figure 3.28. Although this structure has two pairs of adjacent pentagons, strain is relieved through the adjacency of the seven-membered ring, and the sp^3-hybridized carbons. It is the unique presence of a 6:5-double bond in $C_{70}Ph_8$ that facilitates ring contraction [226]. In general, the presence of addends on the cages is likely to improve the prospects of isolating irregular fullerenes; the possibility of fullerenes having seven-membered rings was predicted somewhat earlier [234] and has been supported recently by calculations [235].

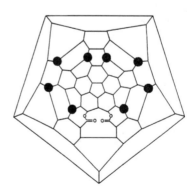

Figure 3.27   Schlegel diagram of the structure of the bis(lactone) $C_{70}Ph_8O_4$ ($\bullet$ = Ph).

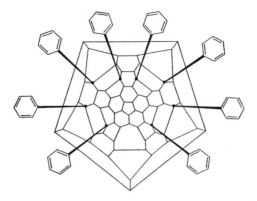

Figure 3.28   Schlegel diagram of the proposed structure of $C_{68}Ph_8$.

Arylfullerenes cannot be obtained directly from bromofullerenes because these are too insoluble, but they can be obtained indirectly by generating the bromofullerenes in the presence of aromatics and $FeCl_3$. The poor lack of control of the reaction leads to a large range of products, but these can be separated by HPLC if required. Products that have been detected, and in some cases isolated from this procedure (using benzene as the aromatic), include $C_{60}Ph_n$ ($n = 4, 6, 8, 10, 12$), $C_{60}Ph_nO_2$ ($n = 4, 6, 8, 10, 12$), $C_{60}Ph_nOH$ ($n = 7, 9, 11$), $C_{60}Ph_nH_2$ ($n = 4, 10$), $C_{60}Ph_4H_4$, $C_{60}Ph_5H_3$, $C_{60}Ph_nO_2H$ ($n = 5, 9$), $C_{60}Ph_4C_6H_4O_2$, $C_{60}Ph_9OH_3$, and $C_{60}Ph_{11}O_3H_2$ [235]. If either toluene or chlorobenzene is used instead of benzene, the main products are $C_{60}(MeC_6H_4)_4$ and $C_{60}(ClC_6H_4)_5H$, respectively.

3.5 NUCLEOPHILIC ADDITIONS   (Stephen R. Wilson)

3.5.1 Michael Additions

The electron-deficient double bonds of C_{60} are quite susceptible to nucleophilic attack. Indeed, some of the first reported reactions of fullerenes were of this type [236,237] (Eq. (3.20)):

$$C_{60} + PhMgX \rightarrow C_{60}Ph_n$$
$$C_{60} + RNH_2 \rightarrow C_{60}(NHR)_n$$

(3.20)

The additions of primary amines and MeO^- to C_{60} were studied in detail by electrospray mass spectrometry [238,239]. In most cases, the reaction is so facile that multiple additions occur (Eq. (3.21)) [240].

$$C_{60} + HO^- \rightarrow C_{60}(OH)_n$$

(3.21)

In the case of ethylenediamines, oxidative addition/cyclization occurs to give fullerene-fused heterocycles (Eq. (3.22)) [241]:

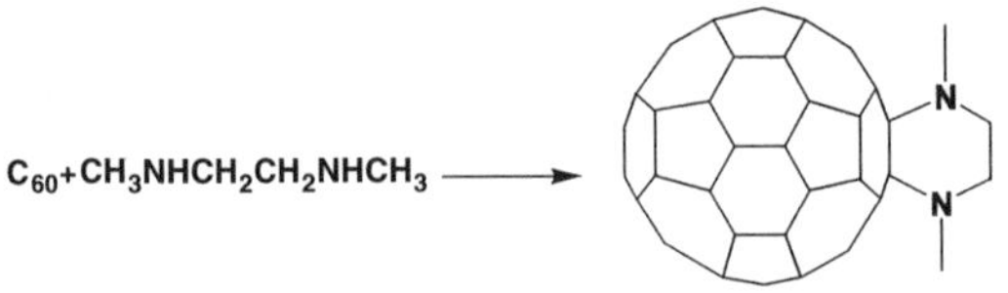

$$C_{60} + CH_3NHCH_2CH_2NHCH_3 \longrightarrow$$

(3.22)

Direct mono-addition of Grignard and alkyllithium reagents to C_{60} has been adapted as a preparatively useful process (Eq. (3.23)) [242,243].

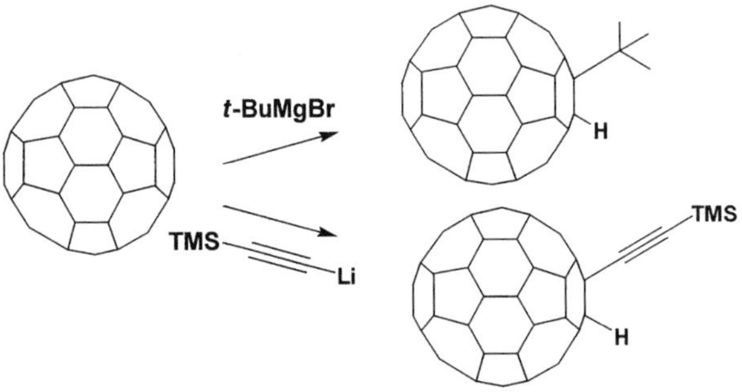

(3.23)

Recently, the addition of excess PhMgBr/CuBr to C_{60} was reported to give 94% of a penta-addition product [244] (Eq. (3.24)).

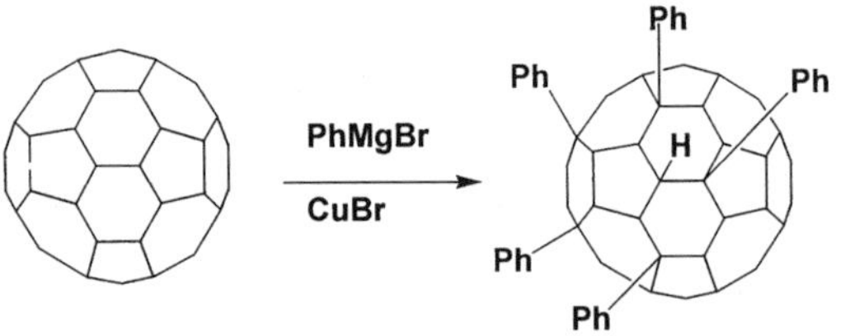

(3.24)

In a similar way, addition of the more hindered fluorenyl anion gave an analogous tetra-adduct, with the fifth group being added in a reversible manner (Eq. (3.25)) [245].

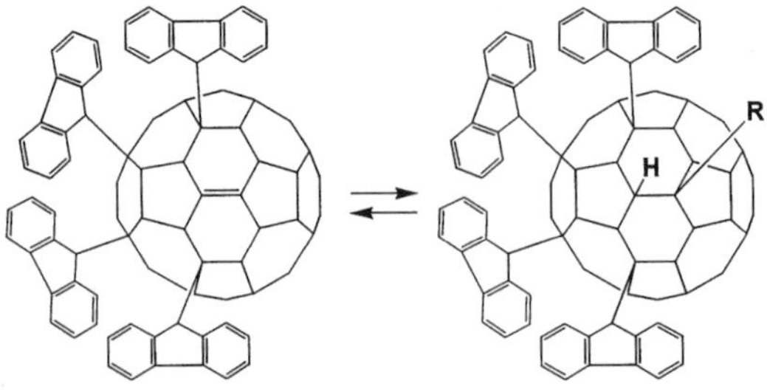

(3.25)

Addition of other types of nucleophiles such as phosphine anions also occurs

(Eq. (3.26)) [246].

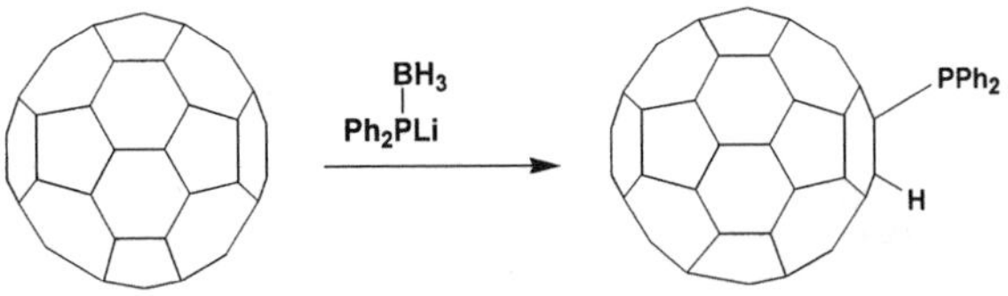

$$(3.26)$$

The addition of enolates and other "soft" anions to C_{60} often results in the occurrence of subsequent reactions (Eq. (3.27)), including oxidative cyclization of the initial adducts [247], oxidation of the nucleophile, as in the case of thiols such as $PhCH_2SH$ [248], or tandem-Michael addition reactions [249]. The redox potential of the fullerene anion is near enough to that of many synthetically useful carbanions to cause a switch to a single electron transfer mechanism which can complicate the reaction.

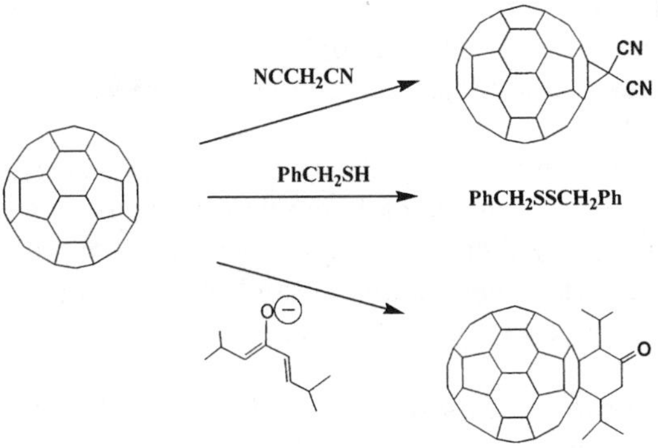

$$(3.27)$$

3.5.2 Bingel–Hirsch Reaction

This section is not intended to be a comprehensive survey of the widely used Bingel–Hirsch reaction, as several extensive reviews of this reaction have already been published [8]. Rather, an overview and introduction to this reaction are presented, along with some of its many variations and strategies, which should serve as a useful guide to key literature in this field. A major reason the Bingel–Hirsch reaction (Eq. (3.28)) has been found to be so useful is that nucleophilic attack on fullerenes by an ionic pathway occurs under mild reaction conditions and produces exclusively methanofullerene products from addition across 6,6-bonds [70,77].

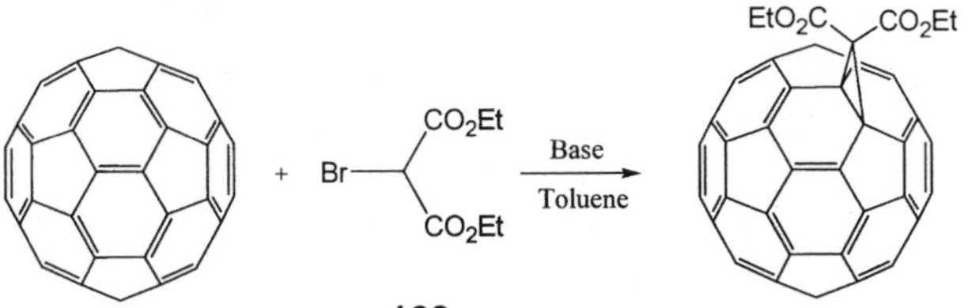

$$(3.28)$$

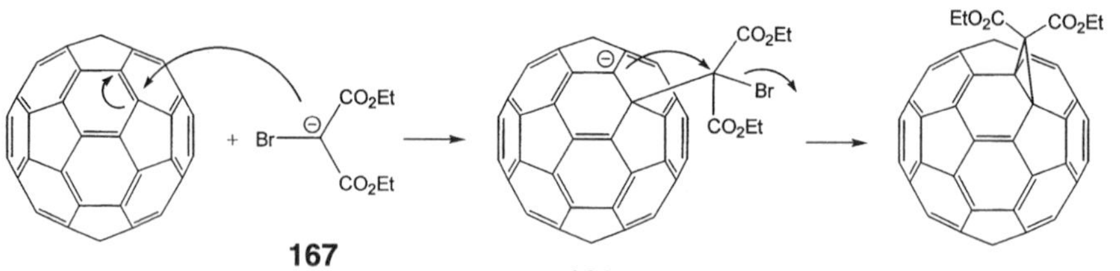

Figure 3.29 The mechanism of the Bingel–Hirsch reaction.

The "classic" reaction conditions for the Bingel–Hirsch reaction involve treatment of diethylbromomalonate **166** with sodium hydride in the presence of C_{60} [255,256]. Deprotonation of **166** leads to formation of the nucleophilic anion **167**, which attacks C_{60}, giving intermediate **168**. Elimination of Br^- by cyclization gives methanofullerene **169** (see Fig. 3.29). NaH was selected as the base in this reaction because it is non-nucleophilic, not because of its base strength. Primary and secondary amines can not be used as bases since they are known to add to C_{60} [236,250].

This section focuses on recent applications of the Bingel–Hirsch reaction, in particular, (1) variations that use carbanionic precursors to methanofullerenes other than malonates, (2) alternative ways of accomplishing the same reactions other than starting with bromomalonates, and (3) use of the Bingel–Hirsch reaction as a route to a variety of new and interesting fullerene materials.

3.5.2.1 Carbanionic Precursors to Methanofullerenes Other than Malonates
In Bingel's original work [77], he explored not only bromomalonate as a precursor to methanofullerenes but also bromoketoesters and bromoketones. For example, the treatment of **170a–c** with NaH in the presence of C_{60} leads to the formation of a methanofullerene products **171a–c** (Eq. (3.29)). Another example involved the use of malonic acid half esters **172**, wherein the intermediate decarboxylated in situ [251] (Eq. 3.30).

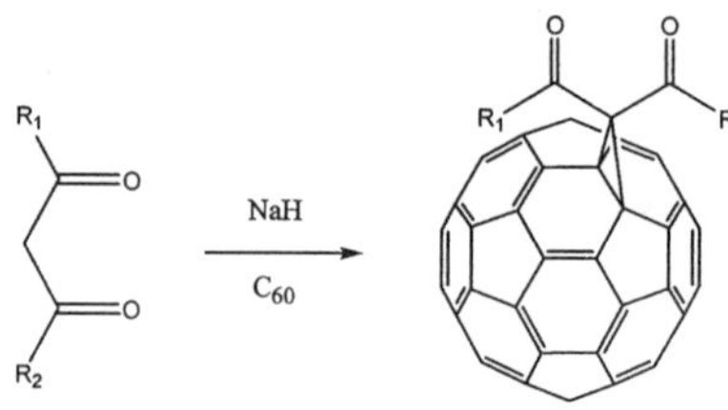

$$(3.29)$$

170a,b,c   **171a,b,c**

a. R_1 = Me, R_2 = OMe
b. R_1 = Ph, R_2 = H
c. R_1 = R_2 = Ph

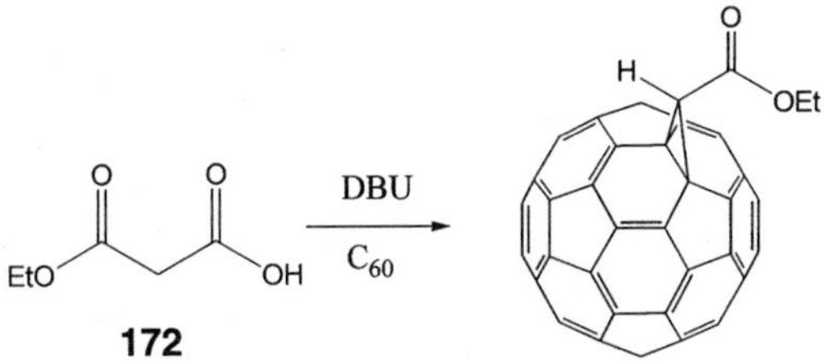

$$(3.30)$$

172

The reaction of phosphonium ylides with C_{60} also causes smooth transformation to methanofullerenes via an analogous addition/elimination process (Eq. (3.31)) [252].

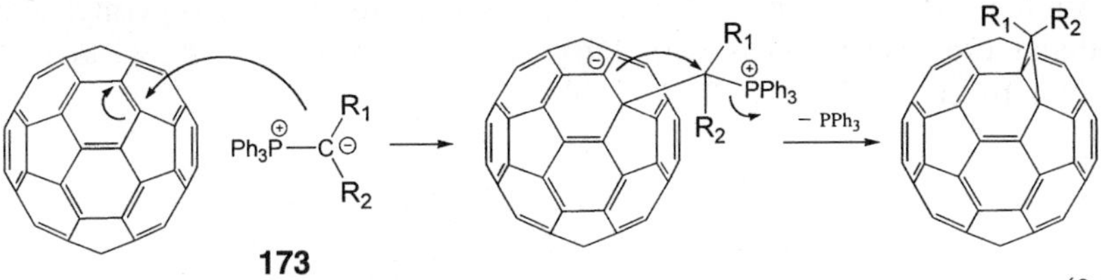

173

$$(3.31)$$

A series of ylides $RCH=PPh_3$ **173**, where R is alkyl, aryl, or thiophenyl, react very fast with C_{60}, in some cases giving only multiple adducts. Yields in this reaction are 15–25%. A similar process using sulfur ylides **174a–e** is more general and gives even higher yields of methanofullerenes [253]. For example, the reactions in Eq. (3.32) proceed in 37–87% yield.

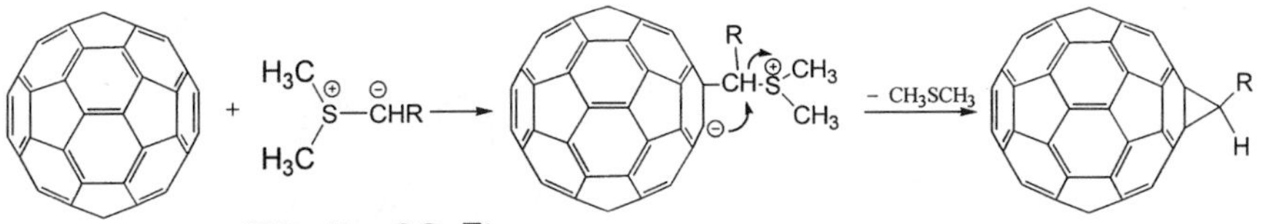

174a. R = CO_2Et
174b. R = CO_2-t-Bu
174a. R = CH=CHCOOEt
174d. R = $CON(Et)_2$
174e. R = COPh

$$(3.32)$$

An important example is compound **174b** (R = *tert*-butylcarboxy), which serves as a precursor to methanofullerenecarboxylic acid, which is a very useful synthon for synthesis of a range of fullerene derivatives. In another example, the 2-bromo heterocyclic compound **175** has been used as a precursor for chiral methanofullerenes (Eq. (3.33)) [254].

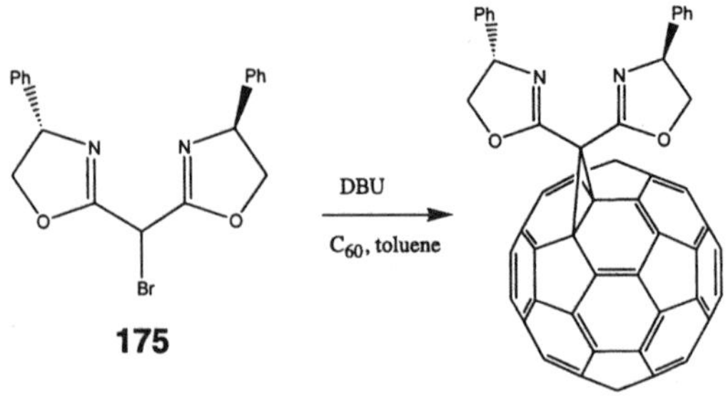

(3.33)

175

Since both enantiomeric forms of chiral compound **175** are available, enantiopure C_{60} adducts can be generated in about 50% yield. Heterocycle **175** is a
"masked" ester available from the corresponding malonate [255].

Another precursor for a Bingel–Hirsch type reaction is dipyridylchloro-
methane **176**, which reacts with DBU and C_{60} to give a 32% yield of the
methanofullerene **177** (Eq. (3.34)) [256].

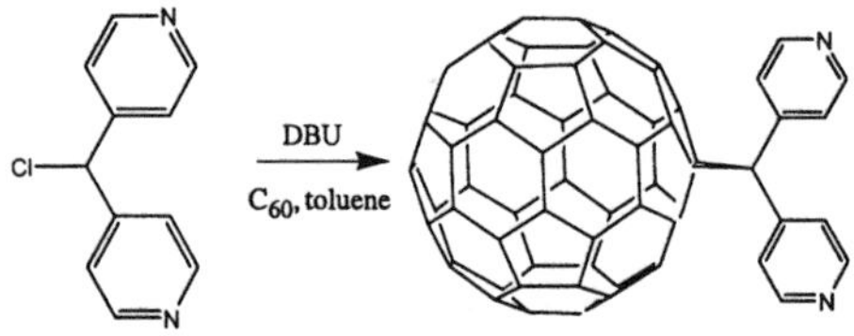

(3.34)

176   **177**

Because of the insolubility of **177**, the reaction was considerably improved by
using the more soluble pentamethanofullerene precurser **178** to give **179** (Eq.
(3.35)). The Pyr–C–Pyr methanofullerene adduct **179** was used to form dimeric
Pt complexes [256].

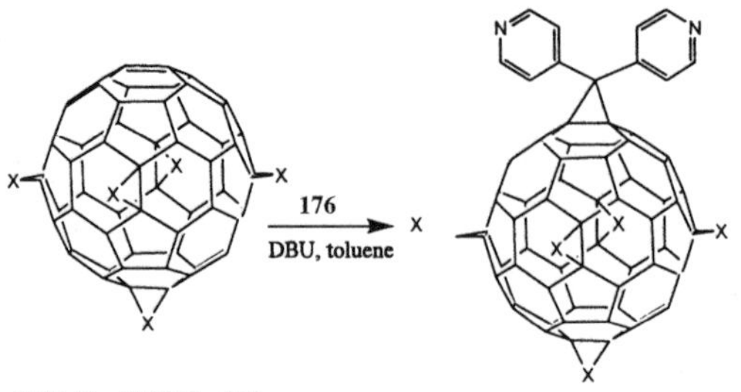

(3.35)

178 X=C(CO$_2$Et)$_2$

179 X=C(CO$_2$Et)$_2$

3.5.2.2 Alternative Bingel–Hirsch Conditions

In cases where complex
malonates are used in the Bingel–Hirsch reaction, the synthesis of bromomalonates is often either not convenient or else results in low yields because of difficulty in separation of unsubstituted, monobromo, and dibromomalonates. In

such situations, approaches that generate the reactive monohalomalonate intermediate in situ have been developed. The fact that a mixture of unsubstituted, monobromo, and dibromomalonates gives only the desired methanofullerene adduct on treatment with DBU in the presence of C_{60} suggested the prospect of generating the desired intermediate from the malonate precursor in situ [257]. When diethyl malonate was treated with DBU in the presence of C_{60} and the bromine source CBr_4, the reaction proceeded very smoothly and in good yield to give the methanofullerenes **180a–d** (Eq. (3.36)) in good yields.

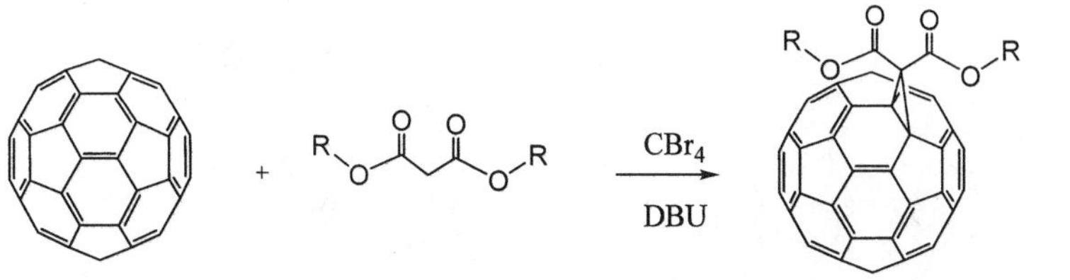

$$\text{(3.36)}$$

180a. R = Et
180b. R = octadecyl
180c. R = Me
180d. R = dendrimer

This modification is mild and very general and has been widely employed in tether directed cyclizations (see Section 3.6). A related process was developed by Nierengarten and Nicound [251], who employed DBU and I_2 as a reagent for formation of the mono-iodomalonate **182** from **181** in situ in good yields (Eq. (3.37)) [251].

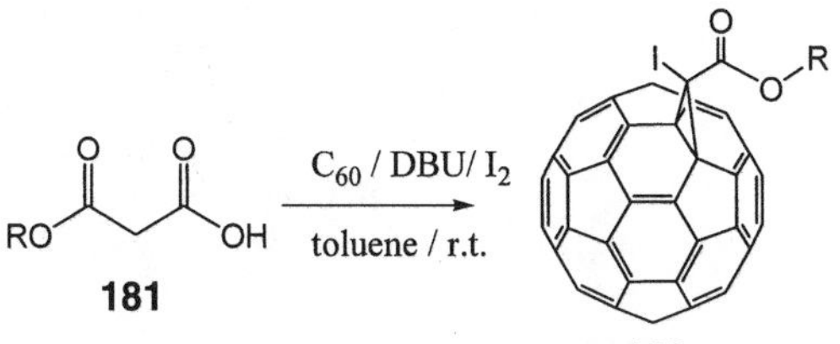

$$\text{(3.37)}$$

3.5.2.3 The Bingel–Hirsch Reaction as a Route to New Fullerene Materials

Products of the Bingel–Hirsch reaction can provide an anchor point on the fullerene core for attachment of diverse groups and thereby provide a route to a wide variety of new fullerene derivatives with useful properties. For example, the hydroysis of methanofullerene **169** to the corresponding malonic acid **183** or the cleavage of methanofullerene *tert*-butyl ester **184** to carboxylic acid **185** provides two types of functionalized fullerenes for linking to many types of side chains (Fig. 3.30). Fullerene diacid **183** can be activated with DCC/N-hydroxysuccinimide to produce compound **186** as a result of decarboxylation and active ester formation [258].

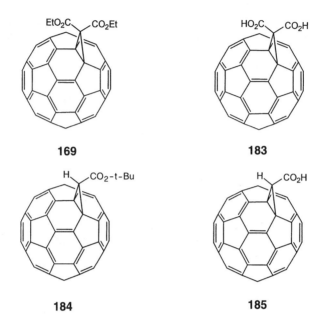

Figure 3.30

Compound **186** can be coupled with *n*-butylamine, benzylamine, or morpholine to give amides **187a,b,c** (in 5%, 15%, and 12% yields, respectively, Eq. (3.38)). Wilson and co-workers [253] reported another approach using acid **185**, which could be converted to acid chloride **188** with SOCl$_2$ and then reacted with benzyl amine to produce amide **187b** (Eq. (3.39)).

Acid **185** was also coupled using DCC to a series of monoprotected glycols to give C_{60} derivatives, which were then converted to a series of fullerene–porphyrin dyads such as **189** (Eq. (3.40)) [259]. Compound **185** was also used as a synthon for the preparation of a fullerene/peptide T analog **190** (Eq. (3.41)) [260].

(3.40)

189 (porphyrin-C_{60} dyads)

185

190

(3.41)

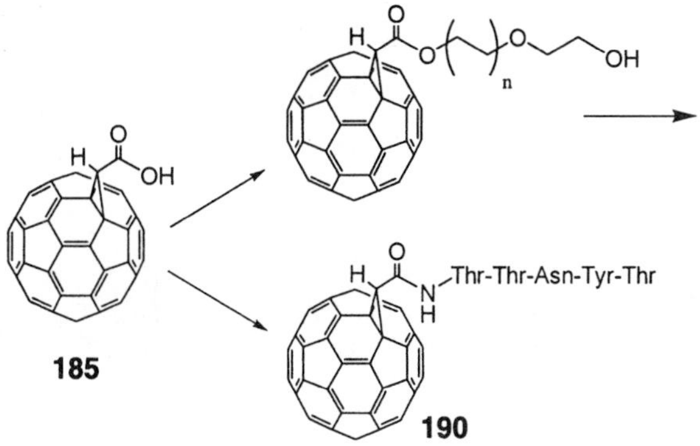

In all of these examples, the activated methanofullerene acyl group (or its corresponding tetrahedral intermediate) is unavoidably hindered by the nearby fullerene sphere during attack by the nucleophile. In order to prepare a more reactive, less sterically hindered synthon, Diederich and co-workers used the Bingel–Hirsch reaction to prepare carboxylic acid **191** (Eq. (3.42)) [261], in which the acyl group is further distanced from the sphere. Coupling of this acid with morpholine via the DCC-mediated method gives a 72% yield of amide **192**. Many other applications of linker **191** have been reported by the Diederich group [261].

191 **192**

(3.42)

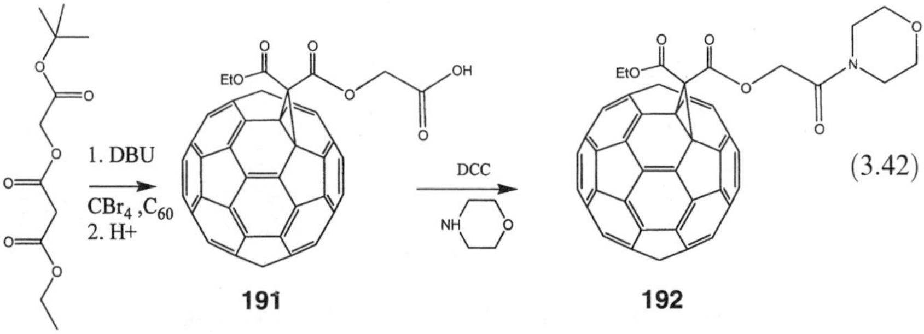

Hirsch recently reported the use of a new general "adapter" molecule for the Bingel–Hirsch reaction. The bis-(malonylcarboxylic acid) **193** was first linked to various dendrimeric polyamides through the free amino group, and then reacted with C_{60} to give a methanofullerene **194** (see structure A, Fig. 3.31) in 26% yield (Eq. (3.43)) [262].

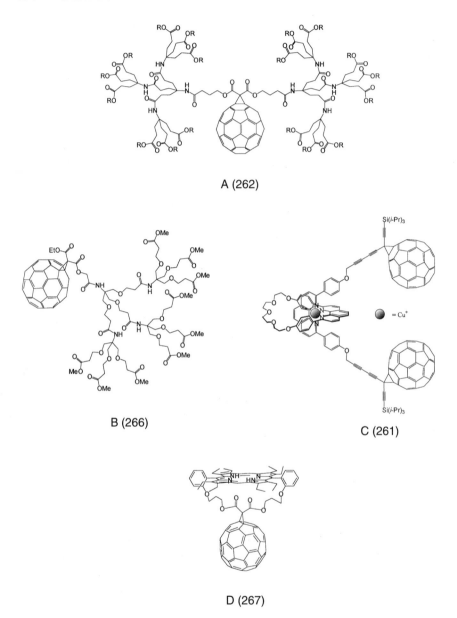

Figure 3.31 Aesthetically pleasing fullerene compounds.

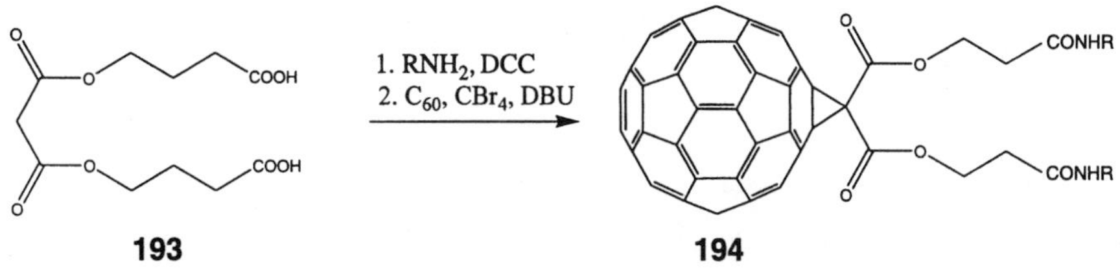

$$(3.43)$$

Since the Bingel–Hirsch reaction is so reliable, it has been employed in a number of tether-directed reactions, in which reactive groups placed at both ends of the tether can be induced to react selectively at specific sites on the sphere. This type of process is discussed in detail in Section 3.6, so only one prototypical example from Diederich's group will be described here, beginning with isomeric *o*-, *m*-, and *p*-hydroxymethyl benzyl alcohols **195a,b,c** (Eq. (3.44)) [263].

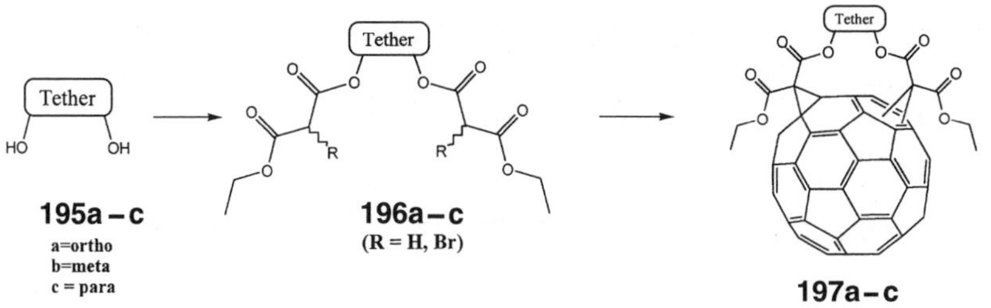

$$(3.44)$$

Diols **195a–c** were converted to bis(2-bromomalonates) **196a–c** by treatment with ethyl malonyl chloride followed by DBU/CBr$_4$ (Eq. (3.44)). The bromo-malonates were then reacted with C$_{60}$ and DBU to produce cyclized adducts **197a–c**. Each tethered reagent showed selectivity in adduct formation. The *ortho*-isomer **196a** gave 33% of the *cis*-2 isomer **197a**, whose structure was confirmed by X-ray crystallography (Fig. 3.32). The analogous *meta*-bis(2-bromomalo-nate) **196b** under the same conditions gave 32% of the same fullerene linkage, namely, the *cis*-2 isomer **197b**. On the other hand, the *para*-bis(2-bromomalo-nate) compound **196c** gave two products, *trans*-4-**197c** in 33% yield and *e*-**197c** in 8% yield. (See Section 3.2.5.2 for definitions and further examples.)

The electrochemical "reverse Bingel–Hirsch" reaction was first reported by Wilson and co-workers [264], who noticed that reduction of Bingel–Hirsch ad-duct **180a** with two electrons caused rapid methanofullerene ring opening and reverse Michael addition to give C$_{60}$ as the product. This allows the Bingel–Hirsch addition product to serve as a blocking or protecting group for further fullerene functionalization. An application of this reaction to the preparation of pure enantiomers of C$_{76}$ was recently reported [265].

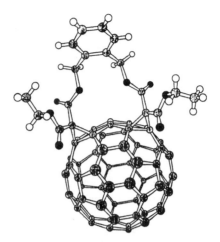

Figure 3.32   X-ray structure of *cis*-2-linked methanofullerene bis-adduct [26].

Finally, the widely adaptable Bingel–Hirsch reaction has been used to prepare a variety of novel new fullerene materials. A few of these "aesthetically pleasing" structures are shown in Figure 3.31 [261,262,266,267].

3.6   TETHER-DIRECTED BIS-ADDITIONS TO FULLERENES

The addition of two or more groups to the C_{60} surface can yield a mixture of eight isomeric bis-adducts from additions of symmetrical addends, considering only 6,6-closed structures (Fig. 3.33). (See earlier discussion in Section 3.2.5.2). Hirsch et al. [70] have reported that there is molecular orbital control that favors *equatorial* and *trans*-3 isomers as major bis-addition products in the Bingel–Hirsch reaction, although all the possible bis-adducts are formed with the exception of *cis*-1 (for steric reasons). Several strategies have been attempted to obviate this problem. The most effective of these to date is tether-directed bis-addition, an approach in which identical reactive groups are attached to the

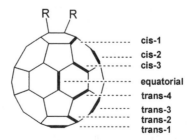

Figure 3.33   Positions for bis-addition on C_{60}.

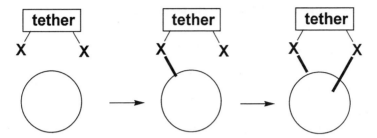

Figure 3.34   Tether-directed bis-addition to C_{60}.

end of a tether or linker, which then directs attachment of the ligand selectively to two specific positions on the fullerene sphere (Fig. 3.34).

Several examples of this approach have been reported. Depending on the constitution of the tether group, one or at worst a small number of isomers are formed in the reaction. One of the earliest examples of tether-directed bis-addition was described in Section 3.5.2.1 [263a].

The following section summarizes the published work relating to (1) the choice of tether and consequent selectivity in the attachment to the fullerene, and (2) the particular reactions that are appropriate. A recent review on this subject by Diederich and Kessinger has appeared [263b].

First, reaction at a specific position on C_{60} must relate to the trajectory of attack in the reaction under consideration, and the distance between the reactive ends of the tether (Fig. 3.34). The design of rigid or semirigid scaffolding for linking molecules has been the subject of computational work (see the program DESIGN for one approach [268]). Figure 3.35 illustrates the three major classes of fullerene reactions that have been used in tether-directed chemistry, namely, Diels–Alder reaction, 1,3-dipolar cycloaddition, and the Bingel–Hirsch reaction, together with the respective distances between the closest atoms of the reacting moiety. The *exact* spatial presentation of the two

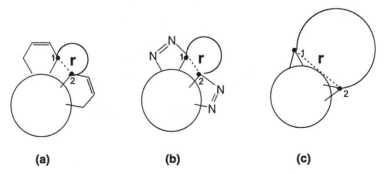

(a)   **(b)**   **(c)**

Figure 3.35   End–end distances for three types of fullerene reactions: (a) Diels–Alder, (b) 1,3-dipolar addition, and (c) Bingel–Hirsch.

TABLE 3.4   Interatomic 1–2 Distances (R) for Bis-Addition Reactions (See Figure 3.35)

Isomer (see figure 3.33)	Distance (Å) Diels–Alder or 1,2-Dipolar Addition	Distance (Å) Bingel–Hirsch Reaction
cis-1	2.53	1.95
cis-2	3.10	4.47–5.62
cis-3	5.5	5.21
Equatorial	6.52	7.56
trans-4	7.51	7.96
trans-3	8.69	8.13
trans-2	8.68	9.60
trans-1	9.13	10.04

reactive groups by the tether molecule to the fullerene must involve consideration not only of the three-dimensional (3D) projections of the groups, but the trajectory of attack in the transition state of the bond-forming reaction. On the other hand, a simple analysis of the interatomic distances (Fig. 3.35) for reaction at the eight possible 6,6-bond positions is highly informative. This information is collected in Table 3.4. An estimate of the possible isomer distribution for the major fullerene reactions using tethers of different lengths can then be deduced from the distances collected in Table 3.4.

A recent report described a more detailed theoretical analysis [269a]. Using a computer model of a published [269b] tether-directed Diels–Alder addition, Friedman and Kenyon computed the distances between the diene termini and target bonds (Fig. 3.36, distances a and b) using all possible conformations achievable by rotation about the seven flexible bonds in the starting structure. The results indicate that by careful analysis of the distances a and b between the tether ends and the isomeric Diels–Alder target reaction positions in Figure 3.36, only one best solution—namely, equatorial attack—is predicted in this

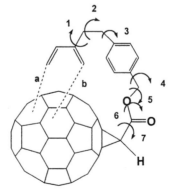

Figure 3.36   Computational model for tethers. Distance a/b shows target distances for Diels–Alder reaction.

TABLE 3.5   Tether-Directed Additions to C_{60} (Linker Structures in the adducts are shown in Figure 3.37a and 3.37b)

Linker	Isomer (% Yield)	Reference
198a	Equatorial (23)	273
198b	Equatorial (50)	273
199a	*cis*-2 (10), *cis*-3 (8)	278
199b	*cis*-2 (20), *cis*-3 (9)	278
199c	Many products	278
199d	Equatorial (30)	278
200a	*cis*-1 (61)	274
200b	*cis*-1 (45)	274
200c	*cis*-1 (60)	274
200d	*cis*-1 (24)	274
200e	*cis*-1 (85)	274
201	*trans*-1 (30)	275
202	*trans*-3 (20)	275
203	*trans*-3 (20)	277
204	*trans*-4 (22)	276
205a	*cis*-2 (16)	261
205b	*cis*-2 (16)	261
205c	*cis*-2 (10)	261
206a	*cis*-2 (—)	264
206b	*cis*-2 (—)	264
206c	*cis*-2 (—)	264
207	*cis*-2 (20), *cis*-3 (13)	264
208a	*cis*-1 (85)	271
208b	*cis*-1 (86)	271
209a	*cis*-1 (14), *cis*-2 (23)	270
209b	*cis*-1 (17),	270
209c	0	270
209d	*cis*-3 (41)	270
210a	*cis*-1 (—)	272
210b	*cis*-1 (—)	272
211	Equatorial (44)	261b
212a	Equatorial (9), *trans*-4 (9)	261b
212b	*trans*-3 (~25%, in/out mix)	261b
213	—	281

case. The equatorial reaction product was in fact the only experimentally observed product [269b]. A summary of the published examples of tether-directed bis-additions to C_{60} is given in Table 3.5 and figure 3.37.

A general solution to the bis-addition isomer problem will be at hand when a generic linker is found for attack at each position on the sphere (cf. Figures 3.33 and 3.34), and methods are available for easy removal of the tether group. No real general conclusion may be drawn from the findings to date other than that

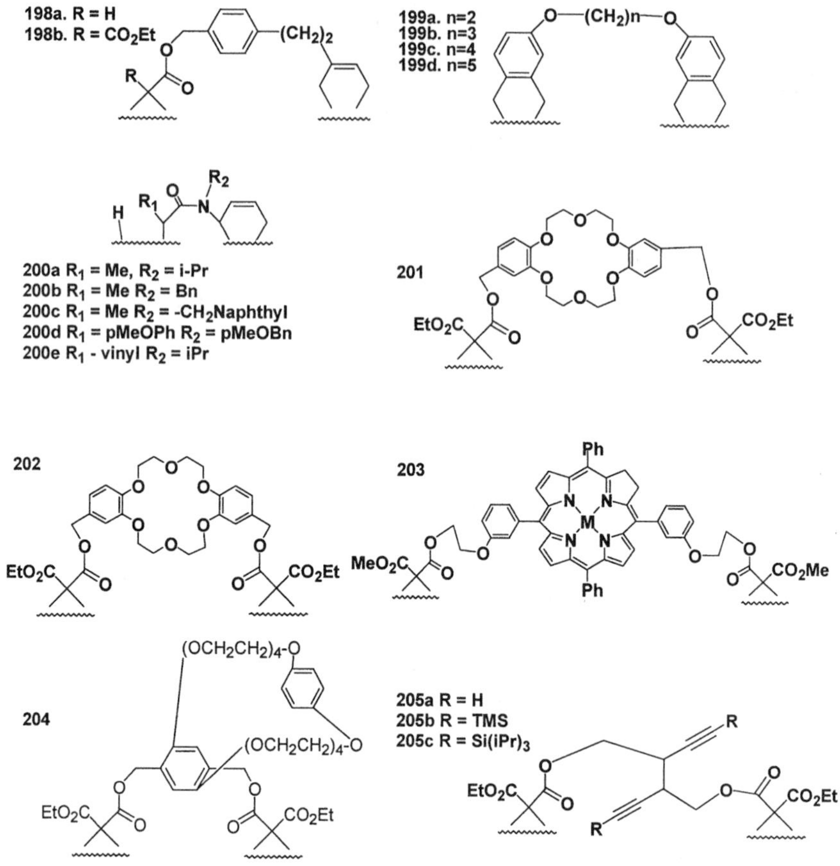

(a)

Figure 3.37 Tether (linker) structures for tether-directed additions in Table 3.5 (continued on page 161)

a range of reactions seem to work well, and that much more work needs to be done in this area.

Finally, some examples of tether-directed tris-addition reactions have also appeared [279,280].

3.7 OPENING HOLES IN FULLERENES

From the beginning of fullerene chemistry, imaginative scientists dreamed of cracking open the fullerene to see what is inside. While it is commonly assumed that "nature abhors a vacuum," what else could be there? Early experiments by Smalley and his group [282] described the formation of endohedral

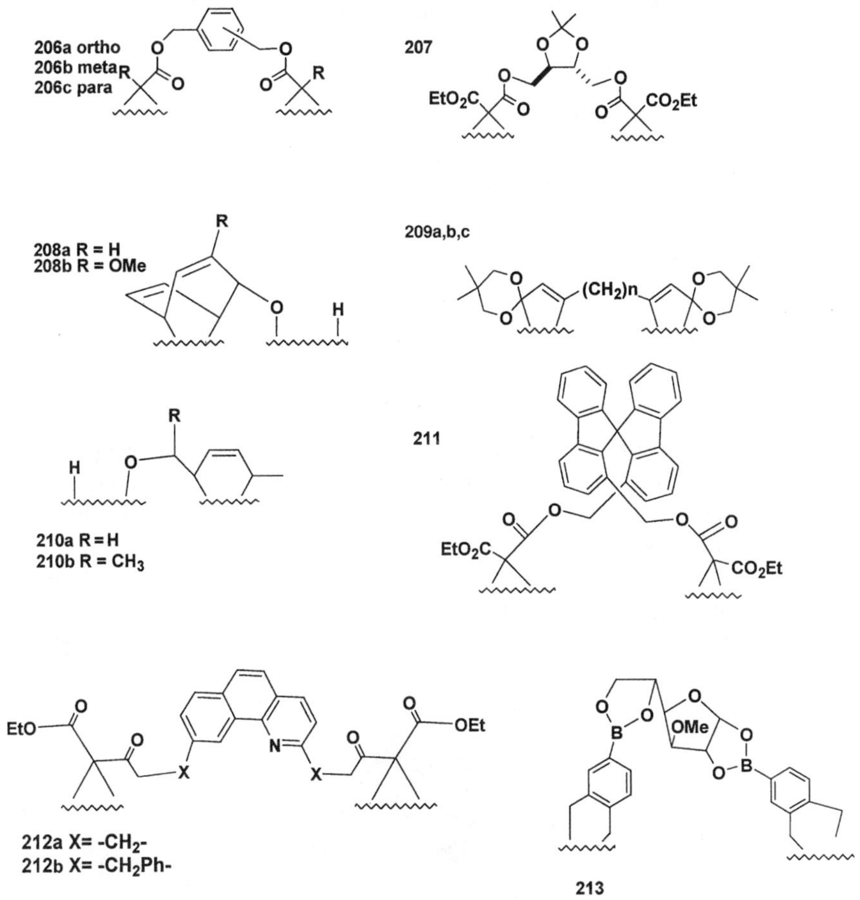

Figure 3.37 (continued)

metallofullerenes. Using a mass spectrometric technique called "laser shrink-wrapping," successive loss of two carbon atoms at a time around the encapsulated metal atom of endohedral metallofullerenes showed that, as one might expect, the metal atom trapped inside was eventually released when enough of the fullerene cage was stripped away.

In a similar type of experiment, Saunders et al. [283a] showed that C_{60} produced by the Huffman–Krätschmer process (in about 400 torr He gas) had He atoms trapped inside the cage. Encapsulated helium could be detected by mass spectrometry after release from the sample. Heating the C_{60} sample at high enough temperatures evidently formed cracks in the cage, allowing release of helium atoms as shown in Table 3.6. As one can see from the amount of helium-4 detected, cracks in the cage begin to appear at $\sim 600\,°C$.

TABLE 3.6 Release of Helium-4 from C$_{60}$ upon Heating

Heating Time (min)	Temperature (°C)	^{4}He ($\times 10^{-9}$ cm^{-1})
55	110	0.2
55	230	1
55	350	4.7
115	510	9.6
50	650	26
50	780	310
90	850	950
60	940	90

Theoretical calculations have predicted possible pathways for ring opening of C$_{60}$ for release of noble gases. Murry and Scusaria [284] calculated the energetics of cracking holes in a fullerene and proposed a "window mechanism" for opening fullerene holes as shown in Figure 3.38. Thiel and co-workers have disagreed with this hypothesis [285], but in any case, the details of how the fullerene cage opens at high temperature are still not known. Reversible cage opening at high temperatures has been used to insert noble gas atoms inside the cage [283b].

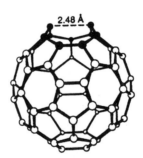

Figure 3.38 The window in the fullerene cage suggested by Murry and Scusaria [284].

It was not until 1995 that Wudl and co-workers succeeded in preparing and isolating compound **214**, a molecule with a permanent 11-membered hole in the cage [61,286]. Compound **214** was prepared by a two-step process by addition of MEM-N$_3$ to C$_{60}$, followed by reaction with singlet oxygen (see Section 3.2.5).

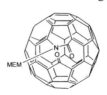

214

Computer models indicate that the hole in the fullerene compound **214** is only about 0.3×0.5 Å, which is not large enough to allow any atoms or molecules to enter or leave the cage.

A second report in 1995 described a C_{70}–fullerene compound containing a hole. During experiments with the preparation and characterization of polyphenylated fullerenes, Taylor and co-workers [287] observed that when a CCl_4 solution of compound $C_{70}Ph_8$ (see Fig. 3.24, Section 3.4.3.3) was allowed to stand in air for two weeks, a new compound, $C_{70}Ph_8O_4$ (see Fig. 3.27, Section 3.4.3.3), was formed in quantitative yield. Spectroscopic analysis indicated that the new compound has an 11-membered hole in the C_{70} cage.

An elegant example of a directed synthesis of a fullerene compound containing a hole was reported by Rubin and co-workers [288]. Photolysis of fullerene diene **215** leads to [4+4] cyclization followed by [2+2+2] retrocyclization which gave as the product the novel diene **216** (Eq. (3.45)).

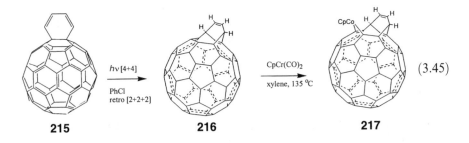

(3.45)

215   **216**   **217**

Formation of the cobalt complex from **216** leads to C–C bond insertion and ring opening to produce metal complex **217**. The X-ray crystal structure of **217** reveals a sizable hole in the fullerene wall. Extension of this process by combining two such ring opening/metal insertions could provide a method to "open a seam" down the entire side of the fullerene. Such prospects and speculations have been reviewed [289].

There is still plenty of opportunity for imaginative approaches to cracking open fullerene cages. While slicing C_{60} open has not yet been achieved, the piece that would result from cutting a fullerene sphere in half ($C_{30}H_{12}$) has been prepared by an independent route (Fig. 3.39) [290].

Figure 3.39 $C_{30}H_{12}$, a semibuckminsterfullerene molecule.

3.8 PERSPECTIVES

We have covered a wide range of fullerene organic chemistry in this chapter, but some important aspects have been omitted for lack of space. These topics include (1) oxidation chemistry, i.e., formation of fullerene polyols and epoxides; (2) chiral fullerene compounds; (3) organometallic aspects, including metals both inside and outside the ball; (4) higher fullerene chemistry, particularly selective reactions of fullerenes larger than C_{70}; and (5) directed synthesis of fullerenes from precursors other than graphite.

In addition, we have not mentioned the newest frontier of fullerene organic chemistry that is waiting to be exploited, namely, the chemistry of carbon nanotubes. The first examples of nanotube chemistry are just emerging and we expect that this will be an active area over the next few years.

The title of this chapter is almost redundant. Organic Chemistry is by definition the chemistry of carbon. The remarkable discovery of these novel molecular forms of carbon—the fullerene family—has led to the emergence of this new and important field of organic chemical exploration. We are certain that this field will continue to expand for the indefinite future.

ACKNOWLEDGMENTS

SRW and DIS are grateful to the National Science Foundation and the Petroleum Research Fund, American Chemical Society, for support of research on fullerenes at New York University. They are also grateful to the following student members of the NYU Fullerene Group for assistance in the preparation of this chapter: Xuefei Tan, Peng Cheng, Shaun MacMahon, Glenn Mann, Hai Mi, and Bradley Pua.

REFERENCES

1. R. C. Haddon and K. Raghavachari, 'Electronic Structure of the Fullerenes: Carbon Allotropes of Intermediate Hybridization' in *Buckminsterfullerenes*, W. E. Billups and M. A. Ciufolini (eds.). VCH Publishers, New York, 1993, Chap. 7.
2. H. D. Beckhaus, C. Rüchardt, M. Kao, F. Diederich, and C. S. Foote, *Angew. Chem. Int. Ed. Engl.* **1992**, *31*, 63.
3. B. Pietzak, M. Waiblinger, T. Almeida Murphy, A. Weidinger, M. Hohne, E. Dietel, and A. Hirsch, *Chem. Phys. Lett.* **1997**, *279*, 259.
4. H. Mauser, A. Hirsch, N. J. R. van Eikema Hommes, and T. Clark, *J. Mol. Model.* **1992**, *3*, 415.
5. L. J. Wilson, S. Flanagan, L. P. F. Chibante, and J. M. Alford, 'Fullerene Electrochemistry: Detection, Generation, and Study of Fulleronium and Fulleride Ions in Solution' in *Buckminsterfullerenes*, W. E. Billups and M. A. Ciufolini (eds.). VCH Publishers, New York, 1993, Chap. 11.

6. F. Arias, L. Echegoyen, S. R. Wilson, Q. Lu, and Q. Lu, *J. Am. Chem. Soc.* **1995**, *117*, 1422.

7. S. Nonell, J. W. Arbogast, and C. S. Foote, *J. Phys. Chem.* **1992**, *96*, 4169. See also G. Lem, D. I. Schuster, S. H. Courtney, Q. Lu, and S. R. Wilson, *J. Am. Chem. Soc.* **1995**, *117*, 554.

8. For reviews of the earlier literature, see (a) A. Hirsch, *The Chemistry of the Fullerenes.* G. Thieme Verlag, Stuttgart and New York, 1994. (b) A. Hirsch, *Synthesis*, 895 (1995). See also F. Diederich and C. Thilgen, *Science* **1996**, *271*, 317.

9. B. Nuber, F. Hampel, and A. Hirsch, *Chem. Commun.* **1996**, 1799.

10. M. Prato, V. Lucchini, M. Maggini, E. Stimpfli, G. Scorrano, M. Eiermann, T. Suzuki, and F. Wudl, *J. Am. Chem. Soc.* **1993**, *115*, 8479.

11. M. Eiermann and F. Wudl, *J. Am. Chem. Soc.* **1994**, *116*, 8364. See also R. A. J. Janssen, J. C. Hummelen, and F. Wudl, *J. Am. Chem. Soc.* **1995**, *117*, 544.

12. A. Hirsch, I. Lamparth, and G. Schick, *Liebigs Ann.* **1996**, 1725.

13. M. Ohno, T. Azuma, S. Kojima, Y. Shirakawa, and S. Eguchi, *Tetrahedron* **1996**, *52*, 4983.

14. Y. Rubin, 'Recent Aspects of the Functionalization Chemistry of Buckminsterfullerene (C_{60}): Preparation of New Materials and Compounds of Biological Interest' in *The Chemical Physics of Fullerenes 10 (and 5) Years Later*, W. Andreoni (ed.). Kluwer Academic Publishers, The Netherlands, 1996, p. 295.

15. G. Mehta and M. B. Viswanath, *Synlett*, **1995**, 679.

16. G. Torres-Garcia and J. Mattay, *Tetrahedron* **1996**, *52*, 5421. See also J. Averdung, G. Torres-Garcia, H. Luftmann, I. Schlachter, and J. Mattay, *Fullerene Sci. Technol.* **1996**, *4*, 633.

17. L. A. Paquette and R. J. Graham, *J. Org. Chem.* **1995**, *60*, 2598.

18. J. M. Lawson, A. M. Oliver, D. F. Rothenfluh, Y.-Z. An, G. A. Ellis, M. G. Ranasinghe, S. I. Khan, A. G. Franz, P. S. Ganapathi, M. J. Shephard, M. N. Paddon-Row, and Y. Rubin, *J. Org. Chem.* **1996**, *61*, 5032.

19. R. M. Williams, M. Koeberg, J. M. Lawson, Y.-Z. An, Y. Rubin, M. N. Paddon-Row, and J. W. Verhoeven, *J. Org. Chem.* **1996**, *61*, 5055.

20. F. Diederich, U. Jonas, V. Gramlich, A. Herrmann, H. Ringsdorf, and C. Thilgen, *Helv. Chim. Acta* **1993**, *76*, 2445. See also U. Jonas, F. Cardullo, P. Belik, F. Diederich, A. Gügel, E. Harth, A. Herrmann, L. Isaacs, K. Müllen, H. Ringsdorf, C. Thilgen, P. Uhlmann, A. Vasella, C. A. A. Waldraff, and M. Walter, *Chem. Eur. J.* **1995**, *1*, 243.

21. S. Eguchi, M. Ohno, S. Kojima, N. Koide, A. Yashiro, Y. Shirakawa, and H. Ishida, *Fullerene Sci. Technol.* **1996**, *4*, 303. See also M. Ohno, S. Kojima, Y. Shirakawa, and S. Eguchi, *Tetrahedron Lett.* **1995**, *36*, 6899 and M. Ohno, N. Koide, H. Sato, and S. Eguchi, *Tetrahedron* **1997**, *27*, 9075.

22. M. Walter, A. Gügel, J. Spickermann, P. Belik, A. Kraus, and K. Müllen, *Fullerene Sci. Technol.* **1996**, *4*, 101.

23. J. Mattay, G. Torres-Garcia, J. Averdung, C. Wolff, I. Schlachter, H. Luftmann, C. Siedschlag, P. Luger, and M. Ramm, *J. Phys. Chem. Solids* **1997**, *58*, 1929.

24. Y. Nakamura, T. Minowa, S. Tobita, H. Shizuka, and J. Nishimura, *J. Chem. Soc. Perkin Trans. 2* **1995**, 2351.

25. A. Puplovskis, J. Kacens, and O. Neilands, *Tetrahedron Lett.* **1997**, *38*, 285.

26. W. Duczek, W. Radeck, H.-J. Niclas, M. Ramm, and B. Costisella, *Tetrahedron Lett.* **1997**, *38*, 6651.

26a. For a recent review of this field, see N. Martin, L. Sánchez, B. Illescâs, and I. Pérez, *Chem. Rev.* **1998**, *98*, 2527.

27. P. A. Liddell, J. P. Sumida, A. N. MacPherson, L. Noss, G. R. Seely, K. N. Clark, A. L. Moore, T. A. Moore, and D. Gust, *Photochem. Photobiol.* **1994**, *60*, 537.

28. H. Imahori, K. Hagiwara, T. Akiyama, S. Taniguchi, T. Okada, and Y. Sakata, *Chem. Lett.* **1995**, 265.

29. H. Imahori and Y. Sakata, *Adv. Mat.* **1997**, *9*, 537. See also H. Imahori, K. Hagiwara, M. Aoki, T. Akiyama, S. Taniguchi, T. Okada, M. Shirakawa, and Y. Sakata, *J. Am. Chem. Soc.* **1996**, *118*, 11771.

30. M. G. Ranasinghe, A. M. Oliver, D. F. Rothenfluh, A. Salek, and M. N. Paddon-Row, *Tetrahedron Lett.* **1996**, *37*, 4797.

31. M. Diekers, A. Hirsch, S. Pyo, J. Rivera, and L. Echegoyen, *Eur. J. Org. Chem.* **1998**, 1111.

32. T. G. Linssen, K. Dürr, M. Hanack, and A. Hirsch, *J. Chem. Soc. Chem. Commun.* **1995**, 103.

33. J. Llacay, J. Veciana, J. Vidal-Gancedo, J. L. Bourdelande, R. Gonzalez-Moreno, and C. Rovira, *J. Org. Chem.* **1998**, *63*, 5201.

34. B. Kräutler and J. Maynollo, *Angew. Chem. Int. Ed. Engl.* **1995**, *34*, 87. See also B. Kräutler and J. Maynollo, *Tetrahedron* **1996**, *52*, 5033.

35. N. F. Goldshleger, N. N. Denisov, A. S. Lobach, V. A. Nadtochenko, Y. M. Shulga, and V. N. Vasilets, *Mol. Mat.* **1996**, *8*, 73.

36. (a) A. Herrmann, F. Diederich, C. Thilgen, H.-U. ter Meer, and W. H. Müller, *Helv. Chim. Acta* **1994**, *77*, 1689. (b) P. Seiler, A. Herrmann, and F. Diederich, *Helv. Chim. Acta* **1995**, *78*, 344.

37. M. F. Meidine, A. G. Avent, A. D. Darwish, H. W. Kroto, O. Ohashi, R. Taylor, and D. R. M. Walton, *J. Chem. Soc. Perkin Trans. 2* **1994**, 1189.

38. For some early contributions in this area, see: T. Suzuki, L. Qi, K. Khemani, F. Wudl, and O. Almarsson, *Science* **1991**, *254*, 1186; T. Suzuki, L. Qi, K. C. Khemani, and F. Wudl, *J. Am. Chem. Soc.* **1992**, *114*, 7301; F. Wudl, *Acc. Chem. Res.* **1992**, *25*, 157.

39. A. B. Smith III, R. M. Strongin, L. Brard, G. T. Furst, W. J. Romanow, K. G. Owens, and R. C. King, *J. Am. Chem. Soc.* **1993**, *115*, 5829.

40. Q. Lu, Ph.D. Dissertation, New York University, 1995. D. I. Schuster, S. R. Wilson, and Q. Lu, unpublished results.

41. H. E. Zimmerman, 'The Di-π-Methane (Zimmerman) Rearrangement' in *Rearrangements in Ground and Excited States*, Vol. 3, P. de Mayo (ed.). Academic Press, New York, 1980.

42. M. S. Meier and M. Poplawska, *Tetrahedron* **1996**, *52*, 5043. See also H. Irngartinger, C.-M. Köhler, U. Huber-Patz, and W. Krätschmer, *Chem. Ber.* **1994**, *127*, 581.

43. Y. Matsubara, H. Muraoka, H. Tada, and Z. Yoshida, *Chem. Lett.* **1996**, 373.

44. H. Tokuyama, H. Isobe, and E. Nakamura, *Bull. Chem. Soc. Jpn.* **1995**, *68*, 935.

45. H. Isobe, H. Tokuyama, M. Sawamura, and E. Nakamura, *J. Org. Chem.* **1997**, *62*, 5034.

46. M. S. Meier, M. Poplawska, A. L. Compton, J. P. Shaw, J. P. Selegue, and T. F. Guarr, *J. Am. Chem. Soc.* **1994**, *116*, 7044.

47. S. R. Wilson and Q. Lu, *J. Org. Chem.* **1995**, *60*, 6496.

48. S. R. Wilson, N. Kaprinidis, Y. Wu, and D. I. Schuster, *J. Am. Chem. Soc.* **1993**, *115*, 8495. See also S. R. Wilson, Y. Wu, N. Kaprinidis, D. I. Schuster, and C. J. Welch, *J. Org. Chem.* **1993**, *58*, 6548.

49. D. I. Schuster, J. Cao, N. Kaprinidis, Y. Wu, A. W. Jensen, Q. Lu, H. Wang, and S. R. Wilson, *J. Am. Chem. Soc.* **1996**, *118*, 5639.

50. A. C. Weedon, '[2+2]-Photocycloaddition Reactions of Enolized 1,3-Diketones and 1,2-Diketones with Alkenes: The de Mayo Reaction' in *CRC Handbook of Organic Photochemistry and Photobiology*, W. M. Horspool and P.-S. Song (eds.). CRC Press, Boca Raton, FL, 1995, Chap. 54.

51. A. W. Jensen, A. Khong, M. Saunders, S. R. Wilson, and D. I. Schuster, *J. Am. Chem. Soc.* **1997**, *119*, 7303.

52. M. Ohno, A. Yashiro, and S. Eguchi, *Chem. Commun.* **1996**, 291.

53. (a) G. Vassilikogiannakis and M. Orfanopoulos, *J. Am. Chem. Soc.* **1997**, *119*, 7394. (b) G. Lem and D. I. Schuster, unpublished results.

54. Y. Wang, S. R. Wilson, and D. I. Schuster, unpublished results.

55. L. Gan, D. Zhou, C. Luo, H. Tan, C. Huang, M. Lu, J. Pan, and L. Wu, *J. Org. Chem.* **1996**, *61*, 1954. See also L. Gan, J. Jiang, W. Zhang, Y. Su, Y. Shi, C. Huang, J. Pan, M. Lu, and Y. Wu, *J. Org. Chem.* **1998**, *63*, 4240.

56. T. Guo, C. Jin, and R. E. Smalley, *J. Phys. Chem.* **1991**, *95*, 4948.

57. V. Chai, T. Guo, C. Jin, R. E. Haufler, L. P. F. Chibante, I. Fure, L. Wang, J. M. Alford, and R. E. Smalley, *J. Phys. Chem.* **1991**, *95*, 7564.

58. D. E. Clemmer, J. M. Hunter, K. B. Shellmov, and M. F. Jarrold, *Nature* **1994**, *372*, 248.

59. I. Lamparth, B. Nuber, G. Schick, A. Skiebe, T. Grösser, and A. Hirsch, *Angew. Chem.* 107, 2473 (1995); *Angew. Chem. Int. Ed. Engl.* **1995**, *35*, 2257.

60. J. Averdung, H. Luftmann, I. Schlachter, and J. Mattay, *Tetrahedron* **1995**, *51*, 6977.

61. J. C. Hummelen and F. Wudl, *Science* **1995**, *269*, 1554.

62. B. Nuber and A. Hirsch, *Chem. Commun.* **1996**, 1421.

63. M. Prato, Q. Chan, F. Wudl, and V. Lucchini, *J. Am. Chem. Soc.* **1993**, *115*, 1148.

64. T. Grösser, M. Prato, V. Lucchini, A. Hirsch, and F. Wudl, *Angew. Chem. Int. Ed. Engl.* **1995**, *34*, 1343.

65. For discussions of this and related mechanisms see: (a) C. K.-F. Shen, H.-H. Yu, G.-C. Juo, K.-M. Chien, G. R. Her, and T.-Y. Luh, *Chem. Eur. J.* **1996**, *3*, 744; (b) R. F. Haldimann, F.-G. Klärner, and F. Diederich, *Chem. Commun.* **1997**, 237; (c) G. Schick, Ph.D. Thesis, Friedrich Alexander University, Erlangen, Germany, 1997.

66. G. X. Dong, J. S. Li, and T. H. Chan, *J. Chem. Soc. Chem. Commun.* **1995**, 1725.

67. L.-L. Shiu, K. M. Chien, T. S. Liu, T.-I. Lin, G. R. Her, and T.-Y. Luh, *J. Chem. Soc. Chem. Commun.* **1995**, 1195.

68. C. K.-F. Shen, K.-M. Chien, C.-G. Juo, G.-R. Her, and T.-Y. Luh, *J. Org. Chem.* **1996**, *61*, 9242.

69. B. Nuber, Ph.D. Thesis, Friedrich Alexander University, Erlangen, Germany, 1998.

70. A. Hirsch, I. Lamparth, and H. R. Karfunkel, *Angew. Chem. Int. Ed. Engl.* **1994**, *33*, 437.

71. A. Hirsch, I. Lamparth, T. Grösser, and H. R. Karfunkel, *J. Am. Chem. Soc.* **1994**, *116*, 9385.

72. For nomenclature of multiple fullerene adducts see Ref. 12.

73. G. Schick, A. Hirsch, H. Mauser, and T. Clark, *Chem. Eur. J.* **1996**, *2*, 935.

74. For another example see: M. R. Banks, J. I. Cadogan, I. Gosnay, A. J. Henderson, P. K. G. Hodgson, W. G. Kerr, A. Kerth, P. R. R. Langridge-Smith, J. R. A. Millar, A. R. Mount, J. A. Parkinson, A. T. Taylor, and P. Thornburn, *Chem. Commun.* **1996**, 507.

75. A. L. Balch, V. J. Catalano, J. W. Lee, M. M. Olmstead, and S. R. Parkin, *J. Am. Chem. Soc.* **1991**, *113*, 8953.

76. A. Hirsch, T. Grösser, A. Skiebe, and A. Soi, *Chem. Ber.* **1993**, *126*, 1061.

77. C. Bingel, *Chem. Ber.* **1993**, *126*, 1957.

78. A. B. Smith III, R. M. Strongin, L. Brard, G. T. Furst, W. J. Romanow, K. G. Owens, and R. C. King, *J. Am. Chem. Soc.* **1994**, *116*, 2187.

79. B. Nuber and A. Hirsch, *Fullerene Sci. Technol.* **1996**, *4*, 715.

80. C. Bellavia-Lund and F. Wudl, *J. Am. Chem. Soc.* **1997**, *119*, 943.

81. For a discussion of the mechanism see Ref. 61.

82. J. C. Hummelen, M. Prato, and F. Wudl, *J. Am. Chem. Soc.* **1995**, *117*, 7003.

83. For some stable fullerene dioxetane derivatives see: (a) I. Lamparth, A. Herzog, and A. Hirsch, *Tetrahedron* **1996**, *52*, 5065; (b) Ref. 65c; (c) See also Ref. 69.

84. C. Bellavia-Lund, M. Keshavarz-K, T. Collins, and F. Wudl, *J. Am. Chem. Soc.* **1997**, *119*, 8101.

85. F. Rachdi, L. Hajji, H. Dolt, M. Ribet, T. Yildirim, J. E. Fischer, C. Goze, M. Mehring, A. Hirsch, and B. Nuber, *Carbon* **1998**, *36*, 607.

86. M. Keshavarz-K, R. Gonzalez, R. G. Hicks, G. Srdanov, V. I. Srdanov, T. G. Collins, J. C. Hummelen, C. Bellavia-Lund, J. Pavlovich, and F. Wudl, *Nature* **1996**, *383*, 147.

87. For a discussion of the mechanism see A. Hirsch and B. Nuber *Acc. Chem. Res.* **1999**, *32*, 795.

88. K. Hasharoni, C. Bellavia-Lund, M. Keshavarz-K, G. Srdanov, and F. Wudl, *J. Am. Chem. Soc.* **1997**, *119*, 11128.

89. J. C. Hummelen, C. Bellavia-Lund, and F. Wudl, "Heterofullerenes", in *Fullerenes and Related Structures: Topics in Current Chemistry 199*, Springer, New York, 1999, 93.

90. A. Cuioni and W. Andreoni, "Recent Progress in the Theoretical Approach to Azafullerenes", in *Mol. Nanostructures*. H. Kuzmany, V. Finle, M. Mehring, and R. Siegmar, Eds., World Scientific Publishing Co., Singapore, 1998, 81.

91. W. Andreoni, A. Curioni, K. Holczer, K. Prassides, M. Keshavarz-K, J. C. Hummelen, and F. Wudl, *J. Am. Chem. Soc.* **1996**, *118*, 11335.

92. A. Gruss, K.-P. Dinse, A. Hirsch, B. Nuber, and U. Reuther, *J. Am. Chem. Soc.* **1997**, *119*, 8728.

93. C. Bellavia-Lund, R. Gonzalez, J. C. Hummelen, R. G. Hicks, A. Sastre, and F. Wudl, *J. Am. Chem. Soc.* **1997**, *119*, 2946.

94. B. Nuber and A. Hirsch, *Chem. Commun.* **1998**, 405.

95. U. Reuther and A. Hirsch, *Chem. Commun.* **1998**, 1401.

96. M. Prato and M. Maggini, *Acc. Chem. Res.* **1998**, *31*, 519.

97. For reviews, see M. Prato, *J. Mater. Chem.* **1997**, *7*, 1097; and N. Martin, L. Sánchez, B. Illescas, and I. Pérez, *Chem. Rev.* **1998**, *98*, 2527.

98. D. M. Guldi and M. Maggini, *Gazz. Chim. It.* **1997**, *127*, 779.

99. S. R. Wilson and Q. Lu, *J. Org. Chem.* **1995**, *60*, 6496.

100. M. Maggini, G. Scorrano, and M. Prato, *J. Am. Chem. Soc.* **1993**, *115*, 9798.

101. X. Zhang, M. Willems, and C. S. Foote, *Tetrahedron Lett.* **1993**, *34*, 8187.

102. S. R. Wilson, Y. Wang, J. Cao, and X. Tan, *Tetrahedron Lett.* **1996**, *37*, 775.

103. O. Tsuge and S. Kanemasa, *Adv. Heterocyclic Chem.* **1989**, *45*, 231.

104. D.-G. Zheng, C.-W. Li, and Y.-L. Li, *Synth. Commun.* **1998**, *28*, 2007.

105. Y. Li, Z. Mao, J. Xu, J. Yang, Z. Guo, D. Zhu, J. Li, and B. Yin, *Chem. Phys. Lett.* **1997**, *265*, 361.

106. M. Maggini, D. M. Guldi, S. Mondini, G. Scorrano, F. Paolucci, P. Ceroni, and S. Roffia, *Chem. Eur. J.*, **1998**, *4*, 1992.

107. Y. Sun, T. Drovetskaja, R. D. Bolskar, R. Bau, P. D. W. Boyd, and C. A. Reed, *J. Org. Chem.* **1997**, *62*, 3642.

108. A. Bianco, M. Maggini, G. Scorrano, C. Toniolo, G. Marconi, C. Villani, and M. Prato, *J. Am. Chem. Soc.* **1996**, *118*, 4072.

109. F. Novello, M. Prato, T. Da Ros, M. De Amici, A. Bianco, C. Toniolo, and M. Maggini, *Chem. Commun.* **1996**, 903.

110. A. Bianco, T. Bertolini, M. Crisma, G. Valle, C. Toniolo, M. Maggini, G. Scorrano, and M. Prato, *J. Pept. Res.* **1997**, *50*, 159.

111. X. Tan, D. I. Schuster, and S. R. Wilson, *Tetrahedron Lett.* **1998**, *39*, 4187.

112. Q. Lu, D. I. Schuster, and S. R. Wilson, *J. Org. Chem.* **1996**, *61*, 4764.

113. L. Pasimeni, A. Hirsch, I. Lamparth, A. Herzog, M. Maggini, M. Prato, C. Corvaja, and G. Scorrano, *J. Am. Chem. Soc.* **1997**, *119*, 12896.

114. J. R. Cross, A. Jimenez-Vasquez, Q. Lu, M. Saunders, D. I. Schuster, S. R. Wilson, and H. Zhao, *J. Am. Chem. Soc.* **1996**, *118*, 11454.

115. M. Iyoda, F. Sultana, and M. Komatsu, *Chem. Lett.* **1995**, 1133.

116. M. Prato, M. Maggini, C. Giacometti, G. Scorrano, G. Sandonà, and G. Farnia, *Tetrahedron* **1996**, *52*, 5221.

117. L.-H. Shu, G.-W. Wang, S.-H. Wu, H.-M. Wu, and X.-F. Lao, *Tetrahedron Lett.* **1995**, *36*, 3871.

118. S.-H. Wu, W.-Q. Sun, D.-W. Zhang, L.-H. Shu, H.-M. Wu, J.-F. Xu, and X.-F. Lao, *J. Chem. Soc. Perkin Trans. 1* **1998**, 1733.

119. A. Komori, M. Kubota, T. Ishida, H. Niwa, and T. Nogami, *Tetrahedron Lett.* **1996**, *37*, 4031.

120. G. E. Lawson, A. Kitaygorodskiy, B. Ma, C. E. Bunker, and Y.-P. Sun, *J. Chem. Soc. Chem. Commun.* **1995**, 2225.

121. K.-F. Liou and C.-H. Cheng, *Chem. Commun.* **1996**, 1423.

122. S.-H. Wu, D.-W. Zhang, G.-W. Wang, L.-H. Shu, H.-M. Wu, J.-F. Xu, and X.-F. Lao, *Synth. Commun.* **1997**, *27*, 2289.

123. D. Zhou, H. Tan, C. Luo, L. Gan, C. Huang, J. Pan, M. Lü, and Y. Wu, *Tetrahedron Lett.* **1995**, *36*, 9169.

124. M. Maggini, A. Karlsson, L. Pasimeni, G. Scorrano, M. Prato, and L. Valli, *Tetrahedron Lett.* **1994**, *35*, 2985.

125. C. Corvaja, M. Maggini, M. Prato, G. Scorrano, and M. Venzin, *J. Am. Chem. Soc.* **1995**, *117*, 8857.

126. C. Corvaja, M. Maggini, M. Ruzzi, and G. Scorrano, *Appl. Magn. Res.* **1997**, *13*, 337.

127. C. Corvaja, M. Maggini, M. Ruzzi, G. Scorrano, and A. Toffoletti, *Appl. Magn. Res.* **1997**, *12*, 477.

128. F. Arena, F. Bullo, F. Conti, C. Corvaja, M. Maggini, M. Prato, and G. Scorrano, *J. Am. Chem. Soc.* **1997**, *119*, 789.

129. P. de la Cruz, A. de la Hoz, F. Langa, B. Illescas, and N. Martin, *Tetrahedron* **1997**, *53*, 2599.

130. M. Iyoda, F. Sultana, A. Kato, M. Yoshida, Y. Kuwatani, M. Komatsu, and S. Nagase, *Chem. Lett.* **1998**, 63.

131. Z. Guo, Y. Li, J. Xu, Z. Mao, Y. Wu, and D. Zhu, *Synth. Commun.* **1998**, *28*, 1957.

132. M. Maggini, L. Pasimeni, M. Prato, G. Scorrano, and L. Valli, *Langmuir* **1994**, *10*, 4164.

133. A. Kurz, C. M. Halliwell, J. J. Davis, A. H. O. Hill, and G. W. Canters, *J. Chem. Soc. Chem. Commun.* **1998**, 433.

134. R. Signorini, M. Zerbetto, M. Meneghetti, R. Bozio, M. Maggini, C. D. Faveri, M. Prato, and G. Scorrano, *Chem. Commun.* **1996**, 1891.

135. M. Maggini, C. De Faveri, G. Scorrano, M. Prato, G. Brusatin, M. Guglielmi, M. Meneghetti, R. Signorini, and R. Bozio, *Chem. Eur. J.* **1999**, *5*, 2501.

136. A. W. Jensen, S. R. Wilson, and D. I. Schuster, *Bioorg. Med. Chem.* **1996**, *4*, 767.

137. T. Da Ros, M. Prato, F. Novello, M. Maggini, and E. Banfi, *J. Org. Chem.* **1996**, *61*, 9070.

138. N. Mizuochi, Y. Ohba, and S. Yamauchi, *J. Phys. Chem. A* **1997**, *101*, 5966.

139. Y. L. Li, J.-H. Xu, D.-G. Zheng, J.-K. Yang, C.-Y. Pan, and D.-B. Zhu, *Solid State Commun.* **1997**, *101*, 123.

140. F. Conti, C. Corvaja, C. Gattazzo, A. Toffoletti, P. Bergo, M. Maggini, G. Scorrano, and M. Prato, unpublished work.

141. D.-G. Zheng, Y.-L. Li, Z. Mao, and D.-B. Zhu, *Synth. Commun.* **1998**, *28*, 879.

142. J. Deisenhofer and J. R. Norris, *The Photosynthetic Reaction Center.* Academic Press, San Diego, 1993.

143. D. M. Guldi, M. Maggini, G. Scorrano, and M. Prato, *Res. Chem. Intermed.* **1997**, *23*, 561.

144. M. Maggini, A. Karlsson, G. Scorrano, G. Sandonà, G. Farnia, and M. Prato, *J. Chem. Soc. Chem. Commun.* **1994**, 589.

145. D. Guldi, M. Maggini, G. Scorrano, and M. Prato, *J. Am. Chem. Soc.* **1997**, *119*, 974.

146. R. M. Williams, J. M. Zwier, and J. W. Verhoeven, *J. Am. Chem. Soc.* **1995**, *117*, 4093.

147. R. M. Williams, M. Koeberg, J. M. Lawson, Y.-Z. An, Y. Rubin, M. N. Paddon-Row, and J. W. Verhoeven, *J. Org. Chem.* **1996**, *61*, 5055.

148. M. Maggini, A. Donò, G. Scorrano, and M. Prato, *J. Chem. Soc. Chem. Commun.* **1995**, 845.

149. N. S. Sariciftci, F. Wudl, A. J. Heeger, M. Maggini, G. Scorrano, M. Prato, J. Bourassa, and P. C. Ford, *Chem. Phys. Lett.* **1995**, *247*, 210.

150. N. Martin, L. Sánchez, C. Seoane, R. Andreu, J. Garín, and J. Orduna, *Tetrahedron Lett.* **1996**, *37*, 5979.

151. N. Martin, I. Perez, L. Sánchez, and C. Seoane, *J. Org. Chem.* **1997**, *62*, 5690.

152. C. A. Reed and P. D. W. Boyd, *J. Phys. Chem.* **1996**, *100*, 15926.

153. T. Drovetskaya, C. A. Reed, and P. Boyd, *Tetrahedron Lett.* **1995**, *36*, 7971.

154. H. Imahori and Y. Sakata, *Chem. Lett.* **1996**, 199.

155. T. Akiyama, H. Imahori, A. Ajawakom, and Y. Sakata, *Chem. Lett.* **1997**, 907; H. Imahori, K. Yamada, M. Hasegawa, S. Taniguchi, T. Okada, and Y. Sakata, *Angew. Chem. Int. Ed. Engl.* **1997**, *36*, 2626.

156. R. Fong II, S. R. Wilson, and D. I. Schuster, *Org. Lett.* **1999**, *1*, 729.

157. P. A. Liddell, D. Kuciauskas, J. P. Sumida, B. Nash, D. Nguyen, A. L. Moore, T. A. Moore, and D. Gust, *J. Am. Chem. Soc.* **1997**, *119*, 1400.

158. S. Higashida, H. Imahori, T. Kaneda, and Y. Sakata, *Chem. Lett.* **1998**, 605.

159. H. Imahori, K. Hagiwara, T. Akiyama, M. Aoki, S. Taniguchi, T. Okada, M. Shirakawa, and Y. Sakata, *Chem. Phys. Lett.* **1996**, *263*, 545.

160. T. Da Ros, M. Prato, D. M. Guldi, E. Alessio, M. Ruzzi, and L. Pasimeni, *Chem. Commun.* **1999**, 635.

161. M. Maggini, G. Scorrano, A. Bianco, C. Toniolo, R. P. Sijbesma, F. Wudl, and M. Prato, *J. Chem. Soc. Chem. Commun.* **1994**, 305.

162. N. F. Gol'dshleger and A. P. Moravshii, *Russ. Chem. Rev.* **1997**, *66*, 323.

163. A. D. Darwish, H. W. Kroto, R. Taylor, and D. R. M. Walton, *J. Chem. Soc. Perkin Trans. 2* **1996**, 1415.

164. C. C. Henderson and P. A. Cahill, *Science* **1993**, *259*, 1885.

165. K. Balasubramanian, *Chem. Phys. Lett.* **1992**, *182*, 257.

166. D. E. Cliffel and A. J. Bard, *J. Phys. Chem.* **1994**, *98*, 8140.

167. S. Ballenweg, R. Gleiter, and W. Krätschmer, *Tetrahedron Lett.* **1993**, *34*, 3737.

168. A. G. Avent, A. D. Darwish, D. K. Heimbach, H. W. Kroto, M. F. Meidine, J. P. Parsons, C. Remars, R. Roers, O. Ohashi, R. Taylor, and D. R. M. Walton, *J. Chem. Soc. Perkin Trans. 2* **1994**, 15.

169. S. Fukuzumi, T. Suenobu, S. Kawamura, A. Shida, and K. Mikami, *J. Chem. Soc. Chem. Commun.* **1997**, 291.

170. L. Becker, T. P. Evans, and J. L. Bada, *J. Org. Chem.* **1993**, *58*, 7630.

171. B. Morosin, C. Henderson, and J. E. Schirber, *Appl. Phys. A* **1994**, *59*, 178.

172. M. S. Meier, V. K. Vance, P. K. Corbin, M. Clayton, M. Mollman, and M. Poplawska, *Tetrahedron Lett.* **1994**, *35*, 5789.

173. C. C. Henderson, C. M. Rohlfing, R. A. Assink, and P. A. Cahill, *Angew. Chem. Int. Ed. Engl.* **1994**, *33*, 786.

174. R. G. Bergosh, J. A. Laske Cooke, M. S. Meier, H. P. Spielmann, and B. R. Weedon, *J. Org. Chem.* **1997**, *62*, 7667.

175. M. S. Meier, B. R. Weedon, and H. P. Spielmann, *J. Am. Chem. Soc.* **1996**, *118*, 11682.

176. P. A. Cahill, *Chem. Phys. Lett.* **1996**, *254*, 257.

177. C. Rüchardt, M. Gerst, J. Ebenhoch, H.-D. Beckhaus, E. E. B. Campbell, R. Tellgmann, H. Schwarz, T. Weiske, and S. Pitter, *Angew. Chem. Int. Ed. Engl.* **1993**, *32*, 584.

178. M. Gerst, H.-D. Beckhaus, C. Rüchardt, E. E. B. Campbell, and R. Tellmann, *Tetrahedron Lett.* **1993**, *34*, 7729.

179. A. D. Darwish, A. G. Avent, R. Taylor, and D. R. M. Walton, *J. Chem. Soc. Perkin Trans. 2* **1996**, 2051.

180. C. C. Henderson, C. M. Rohlfing, K. T. Gillen, and P. A. Cahill, *Science* **1994**, *264*, 397.

181. H. P. Spielmann, G.-W. Wang, M. S. Meier, and B. R. Weedon, *J. Org. Chem.* **1998**, *63*, 3865.

182. R. C. Haddon, *Science* **1993**, *261*, 1545.

183. R. E. Haufler, J. Conceicao, L. P. F. Chibante, Y. Chai, N. E. Byrne, S. Flanagan, M. M. Haley, S. C. O'Brien, C. Pan, Z. Xiao, W. E. Billups, M. A. Ciufolini, R. H. Hauge, J. L. Margrave, L. J. Wilson, R. F. Curl, and R. E. Smalley, *J. Phys. Chem.* **1990**, *94*, 8634.

184. K. Shigematsu, K. Abe, M. Mitani, and K. Tanaka, *Chem. Express* **1992**, *7*, 37.

185. K. Shigematsu, K. Abe, M. Mitani, and K. Tanaka, *Chem. Express* **1992**, *7*, 957.

186. A. D. Darwish, A. K. Abdul-Sada, J. Langley, H. W. Kroto, R. Taylor, and D. R. M. Walton, *J. Chem. Soc. Perkin Trans. 2* **1995**, 2359.

187. M. E. Niyazymbetov, D. H. Evans, S. A. Lerke, P. A. Cahill, and C. A. Henderson, *J. Phys. Chem.* **1994**, *98*, 13093.

188. V. Ohlendorf, A. Willnow, H. Hungerbühler, D. M. Guldi, and K.-D. Asmus, *J. Chem. Soc. Chem. Commun.* **1995**, 759.

189. R. Taylor, *Tetrahedron Lett.* **1991**, 3731.

190. H. Selig, C. Lifshitz, T. Peres, J. E. Fischer, A. R. McGhie, W. J. Romanov, J. P. McCauley, and A. B. Smith, *J. Am. Chem. Soc.* **1991**, *113*, 5475.

191. J. H. Holloway, E. G. Hope, R. Taylor, G. J. Langley, A. G. Avent, T. J. Dennis, J. P. Hare, H. W. Kroto, and D. R. M. Walton, *J. Chem. Soc. Chem. Commun.* **1991**, 966.

192. A. A. Tuinman, P. Mukherjee, J. L. Adcock, R. L. Hettich, and R. N. Compton, *J. Phys. Chem.* **1992**, *96*, 7584.

193. S. K. Chowdhury, S. Cameron, D. M. Cox, K. Kniaz, R. A. Strongin, M. A. Cichy, J. E. Fischer, and A. B. Smith, *Org. Mass. Spectrom.* **1993**, *28*, 860.

194. A. A. Tuinman, A. A. Gakh, J. L. Adcock, and R. N. Compton, *J. Am. Chem. Soc.* **1993**, *115*, 5885.

195. H. Hamwi, C. Fabre, P. Chaurand, S. Della-Negra, C. Ciot, D. Djurado, J. Dupuisa, and A. Rassat, *Fullerene Sci. Technol.* **1993**, *1*, 499.

196. R. Taylor, G. J. Langley, J. H. Holloway, E. G. Hope, H. W. Kroto, and D. M. Walton, *J. Chem. Soc. Chem. Commun.* **1993**, 875.

197. K. Kniaž, J. E. Fischer, H. Selig, G. B. M. Vaughan, W. J. Romanov, D. M. Cox, S. K. Chowdhury, J. P. McCauley, R. M. Strongin, and A. B. Smith III, *J. Am. Chem. Soc.* **1993**, *115*, 6060.

198. A. A. Gakh, A. A. Tuinman, J. L. Adcock, R. A. Sachleben, and R. A. Compton, *J. Am. Chem. Soc.* **1994**, *116*, 819.

199. R. Taylor, G. J. Langley, J. H. Holloway, E. G. Hope, A. K. Brisdon, H. W. Kroto, and D. R. M. Walton, *J. Chem. Soc. Perkin Trans. 2* **1995**, 181.

200. O. V. Boltalina, V. F. Bagryantsev, V. A. Seredenko, L. V. Sidorov, A. S. Zapolskii, and R. Taylor, *J. Chem. Soc. Perkin Trans. 2* **1996**, 2275.

201. O. V. Boltalina, A. Ya. Borschevskii, L. N. Sidorov, J. M. Street, and R. Taylor, *Chem. Commun.* **1996**, 529.

202. O. V. Boltalina, J. M. Street, and R. Taylor, *J. Chem. Soc. Perkin Trans. 2* **1998**, 649.

203. O. V. Boltalina, V. Yu. Markov, R. Taylor, and M. P. Waugh, *Chem. Commun.* **1996**, 2549.

204. A. G. Avent, O. V. Boltalina, P. W. Fowler, A. Yu. Lukonin, V. K. Pavlovich, J. P. B. Sandall, J. M. Street, and R. Taylor, *J. Chem. Soc. Perkin Trans. 2* **1998**, 1319.

205. O. V. Boltalina, A. K. Abdul-Sada, and R. Taylor, *J. Chem. Soc. Perkin Trans. 2* **1995**, 981.

206. A. J. Adamson, J. H. Holloway, E. G. Hope, and R. Taylor, *Fullerene Sci. Technol.* **1997**, *5*, 629.

207. R. Taylor and D. R. M. Walton, *Nature* **1993**, *363*, 685.

208. R. Taylor, J. M. Street, and O. V. Boltalina, *J. Chem. Soc. Perkin Trans. 2* **1999**, 1475.

209. O. V. Boltalina, M. Buehl, A. Khong, M. Saunders, and R. Taylor, *J. Chem. Soc. Perkin Trans. 2* **1999**, 1475.

210. J. P. B. Sandall and P. W. Fowler, private communication to R. Taylor.

211. G. A. Olah, I. Bucsi, C. Lambert, R. Anisfeld, N. P. Trivedi, D. K. Sensharma, and G. K. S. Prakash, *J. Am. Chem. Soc.* **1991**, *113*, 9385.

212. P. R. Birkett, A. G. Avent, A. D. Darwish, H. W. Kroto, R. Taylor, and D. R. M. Walton, *J. Chem. Soc. Chem. Commun.* **1993**, 1230.

213. P. R. Birkett, A. G. Avent, A. D. Darwish, H. W. Kroto, R. Taylor, and D. R. M. Walton, *J. Chem. Soc. Chem. Commun.* **1995**, 683.

214. F. N. Tebbe, R. L. Harlow, D. B. Chase, L. E. Firment, E. R. Holler, B. Malone, P. J. Skrusic, and E. Wasserman, *J. Am. Chem. Soc.* **1991**, *113*, 9900. See also F. N. Tebbe, R. L. Harlow, D. B. Chase, D. L. Thorn, G. C. Campbell, J. C. Calabrese, N. Herron, R. J. Young, and E. Wasserman, *Science* **1992**, *256*, 822.

215. P. R. Birkett, P. B. Hitchcock, H. W. Kroto, R. Taylor, and D. R. M. Walton, *Nature* **1992**, *357*, 479.

216. R. Taylor, J. H. Holloway, E. G. Hope, G. J. Langley, A. G. Avent, T. J. Dennis, J. P. Hare, H. W. Kroto, and D. R. M. Walton, *Nature* **1992**, *355*, 27; *J. Chem. Soc. Chem. Commun.* **1992**, 665.

217. O. V. Boltalina, J. H. Holloway, E. G. Hope, J. M. Street, and R. Taylor, *J. Chem. Soc. Perkin Trans. 2* **1998**, 1845.

218. O. V. Boltalina, J. M. Street, and R. Taylor, *Chem. Commun.* **1998**, 1827.

219. A. G. Avent, P. R. Birkett, H. W. Kroto, R. Taylor, and D. R. M. Walton, *Chem. Commun.* **1998**, 2153.

220. A. G. Avent, P. R. Birkett, J. D. Crane, A. D. Darwish, G. J. Langley, H. W. Kroto, R. Taylor, and D. R. M. Walton, *J. Chem. Soc. Chem. Commun.* **1994**, 1463. See also P. R. Birkett, A. G. Avent, A. D. Darwish, I. Hahn, J. O'Loughlin, H. W. Kroto, G. J. Langley, R. Taylor, and D. R. M. Walton, *J. Chem. Soc. Perkin Trans 2* **1997**, 1121.

221. A. K. Abdul-Sada, A. G. Avent, P. R. Birkett, H. W. Kroto, R. Taylor, and D. R. M. Walton, *Chem. Commun.* **1998**, 393.

222. R. Taylor, *Electrophilic Aromatic Substitution.* Wiley, Chichester, 1989, Chap. 11.

223. P. R. Birkett, A. G. Avent, A. D. Darwish, H. W. Kroto, R. Taylor, and D. R. M. Walton, *J. Chem. Soc. Perkin Trans. 2* **1997**, 457.

224. A. G. Avent, P. R. Birkett, A. D. Darwish, and R. Taylor, unpublished work.

225. A. D. Darwish, P. R. Birkett, G. J. Langley, H. W. Kroto, R. Taylor, and D. R. M. Walton, *Fullerene Sci. Technol.* **1997**, *5*, 705.

226. R. Taylor, *Lectures Notes on Fullerene Chemistry: A Handbook for Chemists.* Imperial College Press, London, 1998.

227. A. D. Darwish, A. G. Avent, H. W. Kroto, R. Taylor, and D. R. M. Walton, *Chem. Commun.* **1997**, 1579.

228. A. K. Abdul-Sada, A. G. Avent, P. R. Birkett, H. W. Kroto, R. Taylor, and D. R. M. Walton, *Chem. Commun.* **1998**, 307.

229. P. R. Birkett, A. G. Avent, A. D. Darwish, H. W. Kroto, R. Taylor, and D. R. M. Walton, *J. Chem. Soc. Perkin Trans. 2* **1996**, 5235.

230. A. G. Avent, P. R. Birkett, A. D. Darwish, H. W. Kroto, R. Taylor, and D. R. M. Walton, *Fullerene Sci. Technol.* **1997**, *5*, 643.

231. P. R. Birkett, A. G. Avent, A. D. Darwish, H. W. Kroto, R. Taylor, and D. R. M. Walton, *Chem. Commun.* **1996**, 1231.

232. P. R. Birkett, A. G. Avent, A. D. Darwish, H. W. Kroto, R. Taylor, and D. R. M. Walton, *Chem. Commun.* **1996**, 1869.

233. R. Taylor, *J. Chem. Res. (S)* **1987**, 178.

234. R. Taylor, *Interdisciplinary Sci. Rev.* **1992**, *17*, 171.

235. A. Ayuela, P. W. Fowler, D. Mitchell, R. Schmidt, G. Seifert, and F. Zerbetto, *J. Phys. Chem.* **1996**, *100*, 15634.

236. F. Wudl, A. Hirsch, K. C. Khemani, T. Suzuki, P.-M. Allemand, A. Koch, H. Eckert, G. Srdanov, and H. M. Webb, "Survey of Chemical Reactivites of C_{60}, Electrophile and Dieno-polarophile Par Excellence" in *Fullerenes: Synthesis, Properties, and Chemstry of Large Carbon Clusters*, G. S. Hammond and V. S. Kuck (eds.), ACS Symposium Series No. 481. American, Chemical Society, Washington DC, 1992, p. 161.

237. A. Hirsch, Q. Li, and F. Wudl, *Angew. Chem. Int. Ed. Engl.* **1991**, *30*, 1309.

238. S. R. Wilson and Y. Wu, *Org. Mass Spectrom.* **1994**, *29*, 186.

239. S. R. Wilson and Y. Wu, *J. Am. Chem. Soc.* **1993**, *115*, 10334.

240. J. Li, A. Takeuchi, M. Ozawa, S. Li, K. Saigo, and K. Kitazawa, *Chem. Commun.* **1993**, 1784.

241. K.-D. Kample, N. Egger, and M. Vogel, *Angew. Chem. Int. Ed. Engl.* **1993**, *32*, 1174.

242. A. Hirsch, A. Soi, and H. R. Karfunkel, *Angew. Chem. Int. Ed. Engl.* **1992**, *31*, 766.

243. K. Kamatsu, Y. Murata, N. Takimoto, S. Mori, N. Sugita, and T. S. M. Wan, *J. Org. Chem.* **1994**, *59*, 6101.

244. M. Sawamura, H. Iikura, and E. Nakamura, *J. Am. Chem. Soc.* **1996**, *118*, 12850.

245. Y. Murat, M. Shiro, and K. Komatsu, *J. Am. Chem. Soc.* **1997**, *119*, 8117.

246. S. Yamago, M. Yanagawa, and E. Nakamura, *J. Chem. Soc. Chem. Commun.* **1994**, 2093.

247. G. A. Burley, P. A. Keller, S. G. Pyne, and G. E. Ball, *Chem. Commun.* **1998**, 2539.

248. R. Subramanian, P. Boulas, M. N. Vijayashree, F. D'Souza, M. T. Jones, and K. M. Kadish, *Proc. Electrochem. Soc.* **1994**, *94-24*, 779.

249. P. S. Ganapathi, S. H. Friedman, G. L. Kenyon, and Y. Rubin, *J. Org. Chem.* **1995**, *60*, 2954.

250. A. Hirsch, Q. Li, and F. Wudl, *Angew. Chem. Int Ed. Engl.* **1991**, *30*, 1309.

251. J.-F. Nierengarten and J.-F. Nicound, *Tetrahedron Lett.* **1997**, *38*, 7737.

252. H. J. Bestmann, D. Hadawi, T. Roder, and C. Moli, *Tetrahedron Lett.* **1994**, *35*, 9017.

253. Y. Wang, J. Cao, D. I. Schuster, and S. R. Wilson, *Tetrahedron Lett.* **1995**, *38*, 6843.

254. F. Djojo and A. Hirsch, *Chem. Eur. J.* **1998**, *4*, 344.

255. T. G. Grant and A. I. Meyers, *Tetrahedron* **1994**, *50*, 2297.

256. T. Habicher, J.-F. Nierengarten, V. Gramlich, and F. Diederich, *Angew. Chem. Int. Ed. Engl.* **1998**, *37*, 1916.

257. X. Camps and A. Hirsch, *J. Chem Soc. Perkin Trans. 1* **1997**, 1595.

258. I. Lamparth, G. Schick, and A. Hirsch, *Liebigs Ann.* **1997**, 253.

259. P. S. Baran, R. R. Monaco, A. U. Khan, D. I. Schuster, and S. R. Wilson, *J. Am. Chem. Soc.* **1997**, *119*, 8363.

260. C. Toniolo, A. Bianco, M. Maggini, G. Scorrano, M. Prato, M. Marastoni, R. Tomatis, S. Spisani, G. Palu, and E. Blair, *J. Med. Chem.* **1994**, *45*, 584.

261. (a) J.-F. Nierengarten, A. Herrmann, R. R. Tykwinski, M. Rüttimann, F. Diederich, C. Boudon, J.-P. Gisselbrecht, and M. Gross, *Helv. Chim. Acta* **1997**, *80*, 293. (b) J.-F. Nierengarten, T. Habicher, R. Kessinger, F. Cardullo, F. Diederich, V. Gramlich, J.-P. Gisselbrecht, C. Boudon, and M. Gross, *Helv. Chim. Acta* **1997**, *80*, 2238.

262. M. Brettreich and A. Hirsch, *Tetrahedron Lett.* **1998**, 2731.

263. (a) J.-F. Nierengarten, V. Gramlich, F. Cardullo, and F. Diederich, *Angew. Chem. Int. Ed. Engl.* **1996**, *35*, 2101. (b) F. Diederich and R. Kessinger, *Acc. Chem. Res.* **1999**, *32*, 537.

264. F. Arias, Y. Yang, L. Echegoyen, Q. Lu, and S. R. Wilson, *Proc. Electrochem. Soc.* **1996**, *95-10*, 200.

265. R. Kessinger, J. Crassous, A. Herrmann, M. Rüttimann, L. Echegoyen, and F. Diederich, *Angew. Chem. Int. Ed. Engl.* **1998**, *37*, 1919.

266. F. Diederich, C. Dietrich-Buchecker, J.-F. Niergartent, and J.-P. Sauvage, *Chem Commun.* **1995**, 781.

267. P. Cheng, S. R. Wilson, and D. I. Schuster, *Chem Commun.* **1999**, 89.

268. S. R. Wilson, W. Tam, M. J. DiGrandi, and W. Cui, *Tetrahedron* **1993**, *49*, 3655.

269. (a) S. H. Friedman and G. L. Kenyon, *J. Am. Chem. Soc.* **1997**, *119*, 447. (b) L. Isaacs, R. F. Haldemann, and F. Diederich, *Angew. Chem. Int. Ed. Engl.* **1994**, *33*, 2339.

270. (a) E. Nakamura, H. Isobe, H. Tokuyama, and M. Sawamura, *J. Chem. Soc., Chem. Commun.* **1996**, 1747. (b) H. Isobe, H. Tokuyama, M. Sawamura, and E. Nakamura, *J. Org. Chem.* **1997**, *62*, 5034.

271. H. Takeshita, J.-F. Liu, N. Kato, and A. Mori, *Tetrahedron Lett.* **1994**, 6305.

272. Q. Lu, Ph.D. Thesis, New York University, 1994.

273. L. Isaacs, F. Diederich, and R. F. Haldimann, *Helv. Chim. Acta* **1997**, *80*, 317.

274. A. Franz, Y.-Z. An, P. D. Ganapathi, R. Neier, and Y. Rubin, *Proc. Electrochem. Soc.* **1996**, *96-10*, 1326.

275. J.-P. Bourgeoise, L. Echegoyen, M. Fibbioli, E. Pretsch, and F. Diederich, *Angew. Chem. Int. Ed. Engl.* **1998**, *37*, 2118.

276. P. R. Ashton, F. Diederich, M. Gomez-Lopez, J.-F. Nierengarten, J. A. Preece, F. M. Raymo, and J. F. Stoddart, *Angew. Chem. Int. Ed. Engl.* **1997**, *36*, 1448.

277. E. Dietel, A. Hirsch, E. Eichhorn, A. Rieker, S. Hackbarth, and B. Roder, *Chem Commun.* **1998**, 1981.

278. M. Taki, S. Sugita, Y. Nakamura, E. Kasashima, E. Yashima, Y. Okamoto, and J. Nishimura, *J. Am. Chem. Soc.* **1997**, *119*, 926.

279. F. Cardullo, P. Seiler, L. Isaacs, J.-F. Nierengarten, R. F. Haldimann, F. Diederich, T. Mordasini-Denti, W. Thiel, C. Boudon, J.-P. Gisselbrecht, and M. Gross, *Helv. Chim. Acta* **1997**, *80*, 343.

280. F. Cardullo, L. Isaacs, F. Diederich, J.-P. Gisselbrecht, C. Boudon, and M. Gross, *Chem. Commun.* **1996**, 797.

281. T. Ishi-I, K. Nakashima, and S. Shinkai, *Chem. Commun.* **1998**, 1047.

282. Y. Chao, T. Guo, C. Jin, R. E. Haufler, L. P. F. Chibante, J. Fure, L. Wang, J. M. Alford, and R. E. Smalley, *J. Phys. Chem.* **1991**, *95*, 7564.

283. (a) M. Saunders, H. A. Jiménez-Vázquez, R. J. H. Cross, and R. J. Poreda, *Science* **1993**, *259*, 1428. (b) M. Saunders, R. J. Cross, H. A. Jiménez-Vázquez, R. Shimshi, and A. Khong, *Science* **1996**, *271*, 1693.

284. R. L. Murry and G. E. Scuseria, *Science* **1994**, *263*, 791.

285. S. Patchkovskii and W. Thiel, *J. Am. Chem. Soc.* **1998**, *120*, 556.

286. J. C. Hummelen, M. Prato, and F. Wudl, *J. Am. Chem. Soc.* **1995**, *117*, 7003.

287. P. R. Birkett, A. G. Avent, A. D. Darwish, H. W. Kroto, R. Taylor, and D. R. M. Walton, *J. Chem. Soc. Chem. Commun.* **1995**, 1869.

288. M. J. Arce, A. L. Viado, Y. S. An, S. I. Khan, and Y. Rubin, *J. Am. Chem. Soc.* **1996**, *118*, 3775.

289. Y. Rubin, *Chem. Eur. J.* **1997**, *3*, 1009.

290. P. W. Rabideau, A. H. Abdourazak, H. E. Folson, Z. Marcinow, A. Sygula, and R. Sygula, *J. Am. Chem. Soc.* **1994**, *116*, 7891.

CHAPTER 4

STRUCTURAL INORGANIC CHEMISTRY OF FULLERENES AND FULLERENE-LIKE COMPOUNDS

ALAN L. BALCH

4.1 INTRODUCTION

With the availability of synthetically useful amounts of C_{60} (as well as the higher fullerenes C_{70} and, to a much more limited extent, C_{76}, C_{78}, and C_{84}), the reactivities of these all-carbon molecules with an array of inorganic compounds have been examined. Metals have been bound to the outside of fullerenes, and metal and nonmetal atoms have been encapsulated within fullerenes. Some inorganic molecules with fullerene-like features have been prepared. This chapter focuses on some aspects of the structural chemistry that involves inorganic fullerene chemistry.

A number of reviews are available that cover related chemistry [1]. A comprehensive review of transition metal fullerene chemistry with extensive coverage in an element-by-element fashion has recently be published [2]. Aspects of the organometallic reactivity of fullerenes have been covered in two reviews [3,4]. Another article deals with both endo- and exohedral metal fullerene interactions [5]. The intercalation chemistry of the fullerenes, which has led to the formation of superconducting A_3C_{60} (A = alkali metal ion), has also been reviewed [6]. The formation and reactivity of endohedral species in which an atom, generally a nontransition metal atom, is encapsulated within a fullerene are covered here briefly, since several reviews are available in this volume and elsewhere [7–9].

The structures of C_{60}, C_{70}, and C_{84} are shown in Figure 4.1. The structure of C_{60} consists of 60 identical carbon atoms, but there are two distinct types of C–C bonds. Those bonds at the 6:6 ring junctions are shorter (bond length 1.38 Å) than the bonds at the 6:5 ring junctions (bond length 1.45 Å). The

Fullerenes: Chemistry, Physics, and Technology, Edited by Karl M. Kadish and Rodney S. Ruoff.
0-471-29089-0 Copyright © 2000 John Wiley & Sons, Inc.

(a) C60   (b) C70

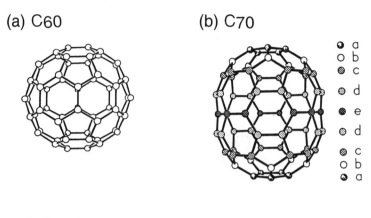

○ a
○ b
◐ c
◎ d
◉ e
◉ d
◐ c
○ b
◑ a

(c) C84 (2 of 24 isomers)

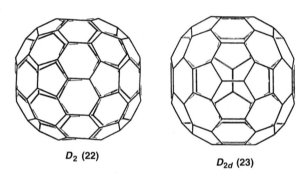

D₂ (22)

D₂d (23)

Figure 4.1 Ideal structures for (a) C_{60}, (b) C_{70} with the five types of carbon atoms labeled a–e, and (c) the two most prevalent isomers (of the 24 structures that satisfy the isolated pentagon rule) of C_{84}.

bonds at the 6:6 ring junction behave as olefinic units, and metal ions commonly coordinate to them in an η^2-fashion. C_{70} has D_{5h} symmetry and consists of two C_{60}-like hemispheres that are bridged by a set of an additional 10 carbon atoms (those labeled e in Fig. 4.1). There are five types of carbon atoms (labeled a–e) that form nine layers. There are eight types of C–C bonds: four occur between 6:6 ring junctions, and four involve 6:5 ring junctions. The C–C bond lengths at the 6:6 ring junctions are shorter than those at the 6:5 ring junctions, and the shortest C–C bonds are found at the curved poles of the C_{70} molecule. The C_a–C_b and C_c–C_c bonds at the poles of the molecule have the highest π-bond orders and are expected to be the most reactive by that criterion [10]. There is only a single structure that obeys the isolated pentagon rule for C_{60} and C_{70}, but for C_{84}, there are 24 isomeric structures that obey that rule [11]. At least five isomers are present in the arc-generated material that is usually utilized by synthetic chemists, but within this mixture, the $D_2(22)$ and $D_{2d}(23)$ isomers shown in Figure 4.1 are the most abundant [12,13].

The reactions of inorganic and organometallic compounds with fullerenes result in the formation of an extensive group of new compounds. Here we consider the structures of the products in the following groupings: covalent additions including σ-additions, π-additions, and other forms of coordination; the formation of fulleride salts; and co-crystallization of fullerenes with inorganic compounds. A brief section discusses some aspects of the chemistry of endohedral fullerenes. Another section focuses on inorganic compounds with large, cage-like structures that resemble the fullerenes, and the final section concerns inorganic nanoscale materials that are related to the fullerenes.

4.2 COVALENT ADDITIONS OF INORGANIC COMPOUNDS

4.2.1 General

The reactivity of C_{60} is complicated by the frequent occurrence of multiple additions to the 30 identical 6:6 ring junctions and by the regiochemistry of these addition reactions. Frequently, the multiplicity of additions can be controlled by manipulation of the reaction stoichiometry, but the regiochemistry is generally subject to less external control. Scheme 1 shows the eight possible regioisomers that can form when double addition is restricted to the olefinic bonds at the 6:6 ring junctions. The nomenclature of Hirsh et al. [14] is useful in identifying these. There are three *cis* isomers with the addenda on the same hemisphere, the unique equatorial isomer, and four *trans* isomers with the addenda on the opposite hemisphere.

4.2.2 σ-Additions

The osmylation of C_{60} demonstrates many of the issues that occur when reactive molecules add to C_{60} under kinetic control [15]. Treatment of C_{60} with

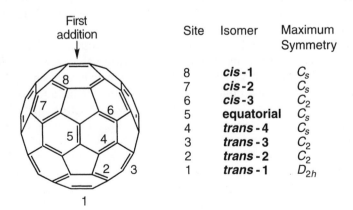

	Site	Isomer	Maximum Symmetry
	8	*cis*-1	C_s
	7	*cis*-2	C_s
	6	*cis*-3	C_2
	5	equatorial	C_s
	4	*trans*-4	C_s
	3	*trans*-3	C_2
	2	*trans*-2	C_2
	1	*trans*-1	D_{2h}

Scheme 1 Regiochemistry of double addition to C_{60}.

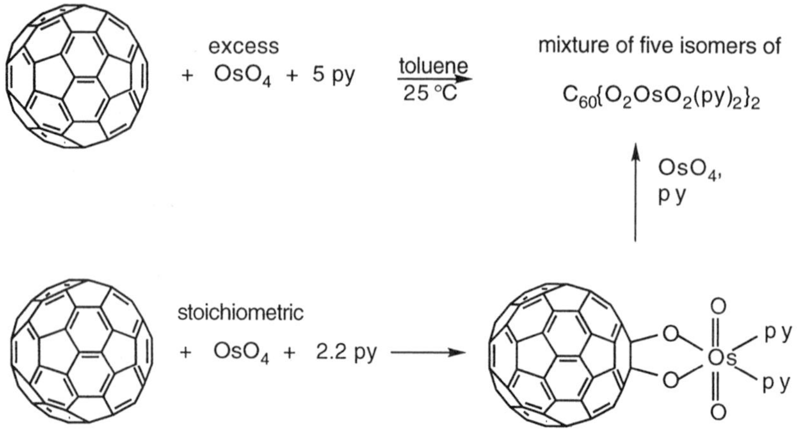

Scheme 2 Osmylation of C_{60}.

excess osmium tetroxide in the presence of pyridine (py) yields a mixture of products as shown in Scheme 2. The composition of the mixture can be controlled by attention to the stoichiometry during the addition process. Additionally, since the single addition product has a higher solubility in toluene, it can be separated from the double addition products. Exchange of 4-*t*-butylpyridine for the pyridine ligands gives an even more soluble derivative that can be crystallized. The structure of this adduct, $C_{60}O_2OsO_2(4$-t-butylpyridine$)_2 \cdot 2.5$ toluene, as determined by a single-crystal X-ray diffraction study, is shown in Figure 4.2 [16]. Two of the oxygen atoms of the osmyl group have been added to a C–C bond at a 6:6 ring fusion in the fullerene. The addition of this osmyl group to the fullerene produces a structural perturbation of the soccer ball structure that is restricted to the vicinity of the site of addition. The unreacted atoms of the fullerene lie in a shell with a radius of 3.46–3.56 Å about the center with an average center-to-carbon distance of 3.512 Å. However, the two carbon atoms that are connected to the osmyl function are pulled away from the core of the fullerene so that they are 3.80(2) and 3.81(2) Å from the center.

Double addition of osmium tetroxide to C_{60} produces five of the eight possible regioisomers (see Scheme 1) that can form through exclusive addition to the 6:6 ring junctions. The $C_{60}\{O_2OsO_2(py)_2\}_2$ isomers have been separated by high-pressure liquid chromatography, and four of the double addition products have been examined by ^{13}C NMR and ^{1}H NMR spectroscopy [17]. These spectroscopic data indicate that two of the isolated isomers have C_s symmetry and that the other two have C_2 symmetry. 2D NMR studies on samples that were ^{13}C-enriched in the fullerene portion led to the identification of one of the C_s isomers as the equatorial isomer and one of the C_2 isomers as the *trans*-3 isomer (see Fig. 4.2). With the assumption that steric effects preclude the formation of the three *cis* isomers, it was concluded that the other C_2 isomer was

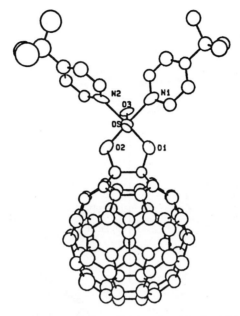

Figure 4.2 The structure of $C_{60}O_2OsO_2$(4-t-butylpyridine)$_2$. This was the first compound for which the soccer ball structure of C_{60} was determined by X-ray crystallography. Reprinted with permission from Ref. 16. Copyright © 1991 Science.

the *trans*-2 isomer and that the remaining C_s isomer was the *trans*-4 isomer. The most rapidly eluting isomer was not sufficiently soluble for NMR analysis, and consequently this isomer was believed to be the nonpolar, *trans*-1 isomer.

The *trans*-2 and *trans*-3 double addition products, $C_{60}O_2OsO_2(py)_2$, are chiral molecules with C_2 symmetry, and it has been possible to resolve the two enantiomers for each through the use of chiral ligands during the addition process [18]. The enantioselectivity in the bis-osmylation process has been attributed to attractive electronic interactions between the ligands coordinated to the osmium reagent and the fullerenes rather than repulsive, steric effects. Moreover, there is ample crystallographic evidence for attractive interactions between aromatic groups and fullerene surfaces.

Additional examples of σ-additions of inorganic and organometallic compounds to C_{60} are given in Scheme 3 [19–23]. The occurrence of these reactions generally give 1,2-additions at 6:6 ring junctions. Another of the products of 1,2-additions has been characterized crystallographically. The structure of that compound, $C_{60}S_2Fe_2(CO)_6$, which forms by opening of the S–S bond in the precursor and the addition of two sulfur atoms to the fullerene, is shown in Figure 4.3. With bulky substituents 1,2-addition becomes disfavored. Thus, in the case of $C_{60}\{Si(SiMe_3)_3\}_2$ addition has proceeded to give a product with the 1,16-addition pattern, as seen in the crystallographically determined structure shown in Figure 4.4 [23].

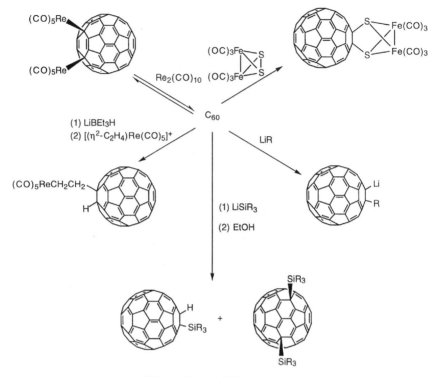

Scheme 3   σ-Additions to C_{60}.

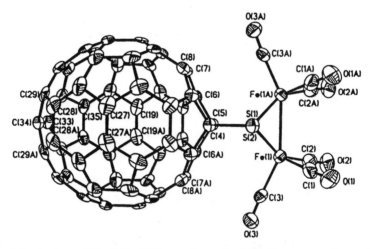

Figure 4.3   A view of $C_{60}S_2Fe_2(CO)_6$ as determined by X-ray crystallography. Reprinted with permission from Ref. 20. Copyright © 1996 American Chemical Society.

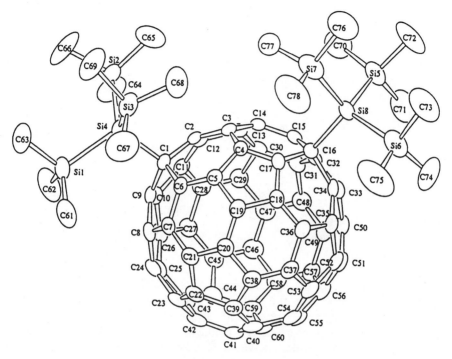

Figure 4.4 The structure of $C_{60}\{Si(SiMe_3)_3\}_2$ as determined by X-ray crystallography. Reprinted with permission from Ref. 23. Copyright © 1998 Elsevier.

4.2.3 η^2-Additions

The formation of complexes in which the fullerene is bound to a transition metal complex in η^2 fashion through a 6:6 ring junction is an ubiquitous occurrence in the reactions of transition metal complexes with fullerenes. The reactions that lead to the formation of four such adducts are shown in Scheme 4, and the structures of the products, as determined by X-ray crystallography, are shown in Figure 4.5. The crystallographic work on $(\eta^2\text{-}C_{60})Pt(PPh_3)_2$ was the initial demonstration that metal complexes are bound to the outer surface of C_{60} through η^2-coordination to a 6:6 ring junction [24]. The stability of these adducts ranges from very stable for $(\eta^2\text{-}C_{60})Mo(CO)_3(dppe)$ (dppe is bis(diphenylphosphino)ethane), which requires heating to 80 °C for 8 h to affect substitution of the fullerene, to very unstable for $(\eta^2\text{-}C_{60})Ir(CO)Cl(PPh_3)_2$, which is in equilibrium with its constituents in solution [25]. Table 4.1 presents structural data for the entire group of complexes that have been characterized crystallographically and shown to contain a single $(\eta^2\text{-}C_{60})M$ unit. Binding of metal centers to the outside of C_{60} results in elongation of the coordinated C–C bond and further pyramidalization of the fullerene carbon atoms (e.g., from 11.6° in C_{60} itself to 15.4° in $(\eta^2\text{-}C_{60})Pt(PPh_3)_2$ [26].

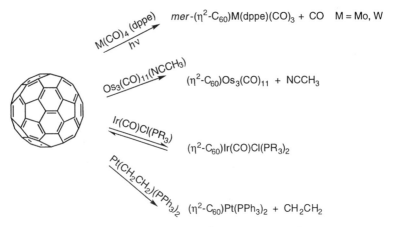

Scheme 4   Formation of η^2-C_{60} metal complexes.

The phenyl rings in the ligands of some of these fullerene complexes make face-to-face contact with the fullerene. Such is the case with both (η^2-C_{60})Ir(CO)Cl(PPh$_3$)$_2$ and (η^2-C_{60})Mo(CO)$_3$(dppe) in Figure 4.5. The presence of such face-to-face arene/fullerene contacts suggested that new phosphine ligands could be designed that would facilitate the formation of these π–π contacts. Thus, the new ligand Ph$_2$PCH$_2$C$_6$H$_4$OCH$_2$C$_6$H$_5$ (Ph$_2$Pbob) was synthesized with arms that had flexible connectors that allow the flat aromatic portions to surround the curved exterior of a fullerene [27]. The new ligand was used to form Ir(CO)Cl(Ph$_2$Pbob)$_2$, which was then added to C_{60} to yield the complex shown in Figure 4.6. In this solid, the fullerene is bound to iridium in the expected η^2 fashion at a 6:6 ring junction and the benzyloxybenzyl arms reach out to embrace the fullerene portion of the adjacent molecule. A drawing of an array of three of these molecules, which form an extended chain in the solid, is shown in Figure 4.7. Two of the phenyl rings of each benzyloxybenzyl group make face-to-face π contact with an adjacent fullerene, while another phenyl ring makes face-to-face contact with the fullerene within the molecule. A second generation of ligands of this type has also been made [28].

Multiple addition of metal complexes of C_{60} produces another set of complexes with (η^2-C_{60})M coordination. The regiochemistry of these additions depends on the nature of the metal complex involved. For Ir(CO)Cl(PPh$_3$)$_2$, where the ligands are fairly bulky and the addition occurs under thermodynamic control, the *trans*-1 isomer is preferentially crystallized from the complex mixture of adducts that exists in solution [29,30]. The structures of two examples of such *trans*-1 complexes are shown in Figure 4.8. On the other hand, when polynuclear complexes with bridging ligands or metal–metal bonding are involved, multiple additions occur at 6:6 ring junctions in close proximity (i.e., on a common hexagonal face of the fullerene). Thus, Ir$_2$(μ-OMe)(μ-OPh)(η^4-C_8H_{12})$_2$ and Ir$_2$(Cl)$_2$(η^4-C_8H_{12})$_2$ add to C_{60} to produce the compounds shown in

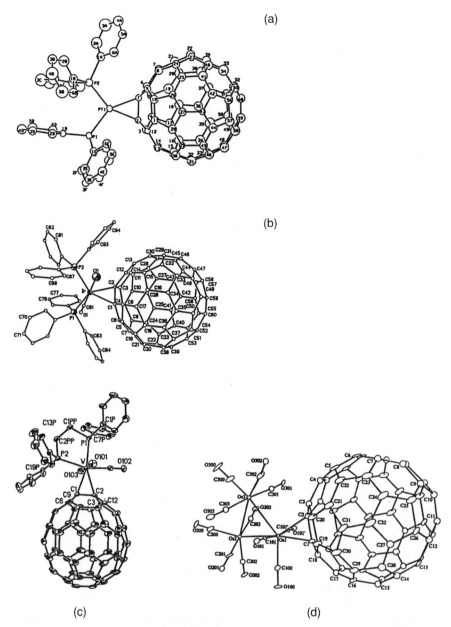

Figure 4.5 Drawings of the crystallographically determined structures of four adducts of C_{60} with simple η^2-coordination. (a) $(\eta^2\text{-}C_{60})Pt(PPh_3)_2$. Reprinted with permission from Ref. 24. Copyright © 1991 Science. (b) $(\eta^2\text{-}C_{60})Ir(CO)Cl(PPh_3)_2$. Reprinted with permission from Ref. 25. Copyright © 1991 American Chemical Society. (c) $(\eta^2\text{-}C_{60})Os_3(CO)_{11}$. Reprinted with permission from J. T. Park, H. Song, J.-J. Cho, M.-K. Chung, J.-H. Lee, and I. H. Shu, *Organometallics* **1998**, *17*, 227. Copyright © 1998 American Chemical Society. (d) $(\eta^2\text{-}C_{60})Mo(CO)_3(dppe)$. Reprinted with permission from H.-F. Hsu, Y. Du, T. E. Albrecht-Schmitt, S. R. Wilson, and J. R. Shapley, Organometallics **1998**, *17*, 1756. Copyright © 1998 American Chemical Society.

185

TABLE 4.1 X-Ray Crystal Structure Data for Complexes Containing a Single $(\eta^2\text{-}C_{60})M$ Unit

Compound	M–C	C–C (Fullerene)	Reference
mer-$(\eta^2\text{-}C_{60})$	2.309(7)	1.483(10)	*a*
$\text{Mo(CO)}_3(\text{Ph}_2\text{P}(\text{CH}_2)_2\text{PPh}_2)$	2.306(7)		
$(\eta^2\text{-}C_{60})\text{W(CO)}_3$	2.291(6)	1.497(8)	*a*
$(\text{Ph}_2\text{P}(\text{CH}_2)_2\text{PPh}_2)$	2.296(5)		
$(\eta^2\text{-}C_{60})\text{W(CO)}_2$	2.304(26)	1.429(30)	*b*
(1,10-phenanthroline)(dibutylmaleate)	2.300(19)		
$(\eta^2\text{-}C_{60})\text{Ru(NO)Cl(PPh}_3)_2$	2.216(3)	1.489(7)	*c*
	2.219(3)		
$(\eta^2\text{-}C_{60})\text{Os}_3\text{(CO)}_{11}$	2.21(2)	1.42(3)	*d*
	2.26(2)		
$(\eta^2\text{-}C_{60})\text{Ir(CO)Cl(PPh}_3)_2$	2.19(2)	1.53(3)	*e*
	2.19(2)		
$(\eta^2\text{-}C_{60})\text{Ir(CO)Cl(bobPPh}_3)_2$	2.194(10)	1.48(2)	*f*
$(\eta^2\text{-}C_{60})\text{RhH(CO)(PPh}_3)_2$	2.165(8)	1.479(11)	*g*
	2.151(8)		
$(\eta^2\text{-}C_{60})\text{Rh(acac)(3,5-diMepy)}_2$	2.08(11)	1.50(3)	*h*
$(\eta^2\text{-}C_{60})\text{Pd(PPh}_3)_2$	2.123(14)	1.447(25)	*i*
	2.086(16)		
$(\eta^2\text{-}C_{60})\text{Pt(PPh}_3)_2$	2.145(21)	1.502(30)	*j*
	2.115(23)		
$(\eta^2\text{-}C_{60})\text{Pt(+DIOP)}$	2.09(2)	1.51(2)	*k*
	2.12(2)		

a H.-F. Hsu, Y. Du, T. E. Albrecht-Schmitt, S. R. Wilson, and J. R. Shapley, *Organometallics* **1998**, *17*, 1756.

b K. Tang, S. Zheng, X. Jin, H. Zeng, Z. Gu, X. Zhou, and Y. Tang, *J. Chem. Soc. Dalton Trans.* **1997**, 3585.

c A. N. Chernega, M. L. H. Green, J. Haggitt, and A. H. H. Stephens, *J. Chem. Soc. Dalton Trans.* **1998**, 755.

d J. T. Park, H. Song, J.-J. Cho, M.-K. Chung, J.-H. Lee, and I. H. Shu, *Organometallics* **1998**, *17*, 227.

e A. L. Balch, V. J. Catalano, and J. W. Lee, *Inorg. Chem.* **1991**, *30*, 3980.

f A. L. Balch, V. J. Catalano, J. W. Lee, and M. M. Olmstead, *J. Am. Chem. Soc.* **1992**, *114*, 5455.

g A. L. Balch, J. W. Lee, B. C. Noll, and M. M. Olmstead, *Inorg. Chem.* **1993**, *32*, 55.

h Y. Ishii, H. Hoshi, Y. Hamada, and M. Hidai, *Chem. Lett.* **1994**, 801.

i V. V. Bashilov, P. V. Petrovskii, V. I. Sokolov, S. V. Lindeman, I. A. Guzey, and Y. T. Struchkov, *Organometallics* **1993**, *12*, 991.

j P. J. Fagan, J. C. Calabrese, and B. Malone, *Science* **1991**, *252*, 1160.

k V. V. Bashilov, P. V. Petrovskii, V. I. Sokolov, F. M. Dolgushin, A. I. Yanovsky, and Y. T. Struchkov, *Russ. Chem. Bull.* **1996**, *45* 1207.

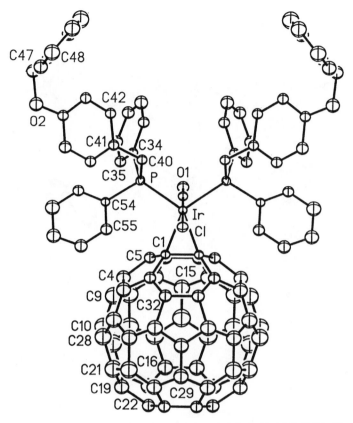

Figure 4.6 A view of an individual molecule of $(\eta^2\text{-}C_{60})\text{Ir(CO)Cl(Ph}_2\text{Pbob)}_2$. The molecule possesses crystallographic mirror symmetry. Reprinted with permission from Ref. 27. Copyright © 1992 American Chemical Society.

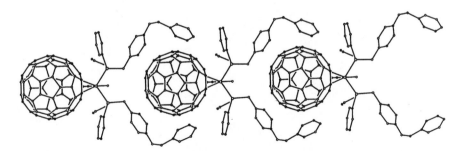

Figure 4.7 A diagram of $(\eta^2\text{-}C_{60})\text{Ir(CO)Cl(Ph}_2\text{Pbob)}_2$, which shows three individual molecules. The phenyl rings of the Ph_2Pbob ligand encircle the fullerene on an adjacent molecule through face-to-face, π–π contact. This motif creates infinite chains of molecules in the solid. Reprinted with permission from Ref. 27. Copyright © 1992 American Chemical Society.

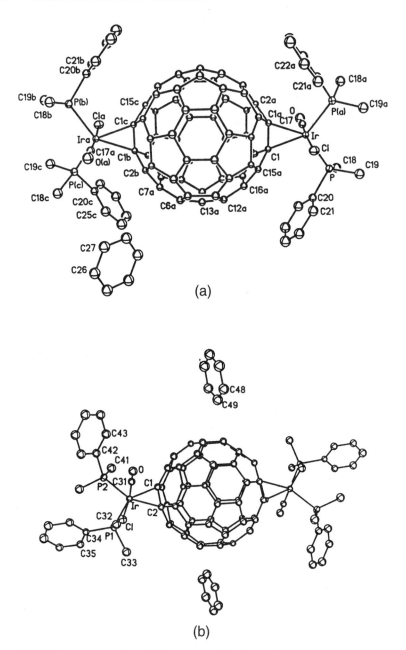

(a)

(b)

Figure 4.8   The structures of two different crystalline forms of $C_{60}\{Ir(CO)Cl(PPhMe_2)_2\}_2$: (a) $C_{60}\{Ir(CO)Cl(PPhMe_2)_2\}_2 \cdot C_6H_6$ and (b) $C_{60}\{Ir(CO)Cl(PPhMe_2)_2\}_2 \cdot 2C_6H_6$. Reprinted with permission from Ref. 29. Copyright © 1992 American Chemical Society.

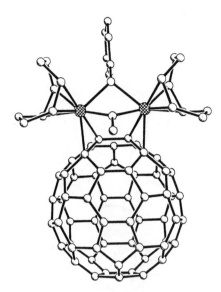

Figure 4.9 The crystallographically determined structure of $C_{60}\{Ir_2(\mu\text{-OMe})(\mu\text{-OPh})\cdot(\eta^4\text{-}C_8H_{12})_2\}$. The iridium atoms are shown as hatched circles, the oxygen atoms as open circles, and the carbon atoms as partially shaded circles. Reprinted with permission from Ref. 31. Copyright © 1998 Pergamon Press.

Figures 4.9 and 4.10 [31,32]. In $(C_{60})Ir_2(\mu\text{-OMe})(\mu\text{-OPh})(\eta^4\text{-}C_8H_{12})_2$ the two iridium atoms are each bound to the fullerene in η^2 fashion on a single hexagonal face. In $(C_{60})\{Ir_2(Cl)_2(\eta^4\text{-}C_8H_{12})_2\}_2$ there is a similar attachment of each binuclear unit to the fullerene, but the two bulky addenda are positioned at opposite ends of the soccer bowl. Figures 4.11 and 4.12 show the structures of other C_{60} adducts in which two metal centers are attached to the fullerene through coordination on a single hexagon [33].

C_{60} reacts with $Ru_3(CO)_{12}$ and related carbonyl clusters to form adducts with three ruthenium atoms bound to a common hexagon of the fullerene [34]. The structures of three such adducts are shown in Figures 4.13 and 4.14.

Treatment of C_{60} with an excess of $Pt(PEt_3)_4$ results in the formation of $C_{60}\{Pt(PEt_3)_2\}_6$ in which the six platinum atoms are coordinated to C_{60} in η^2 fashion in an octahedral array [35]. The structure of the adduct is shown in Figure 4.15. This arrangement places the $(PEt_3)_2$ moieties as far apart as possible and leaves localized benzenoid units within the fullerene [36].

Complexes with η^2-coordination of a metal center to the fullerene are also formed with C_{70} [37]. Figure 4.16 shows the structures of three such adducts, while Table 4.2 gives some bond distances. In each case, the metal atom is bound to a C_a–C_b bond at one pole of the oblong fullerene. Such bonds are the most pyramidalized bonds in this fullerene, and the ones that also have highly localized, olefinic character.

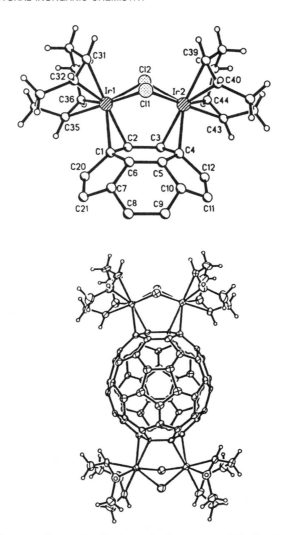

Figure 4.10 The crystallographically determined structure of $C_{60}\{Ir_2Cl_2(\eta^4\text{-}C_8H_{12})_2\}_2 \cdot 2C_6H_6$ (from data in Ref. 32). The top view shows details of the iridium/fullerene coordination, while the lower view shows the entire molecule. Reprinted with permission from Ref. 32. Copyright © 1993 American Chemical Society.

Multiple additions are also observed for additions to C_{70}. So far, additions of relatively large transition metal complexes occur at opposite poles of the molecule. For double addition, there are three possible isomers, if reactivity is confined to the C_a–C_b bonds at opposite ends of the molecule. For $Ir(CO)Cl(PPhMe_2)_2$ addition occurs to give a single isomer, as seen in Figure 4.17 [38].

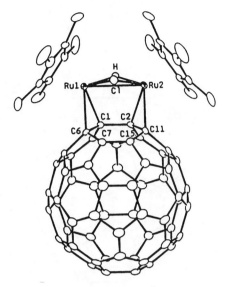

Figure 4.11 A drawing of $C_{60}Ru_2(\mu\text{-Cl})(\mu\text{-H})(\eta^5\text{-}C_5Me_5)_2$ as determined by X-ray crystallography. Reprinted with permission from Ref. 33. Copyright © 1995 American Chemical Society.

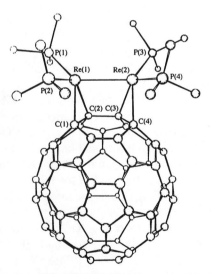

Figure 4.12 A drawing of $C_{60}Re_2H_8(PMe_3)_4$ as determined by X-ray crystallography. The rhenium bound hydrogen atoms were not located. The Re–Re distance, 2.8945(5) Å, is compatible with the presence of a Re–Re single bond. Reprinted with permission from A. N. Chernega, M. L. H. Green, J. Haggitt, and A. H. H. Stephens, *J. Chem. Soc. Dalton Trans.* **1998**, 755. Copyright © 1998 Royal Society of Chemistry.

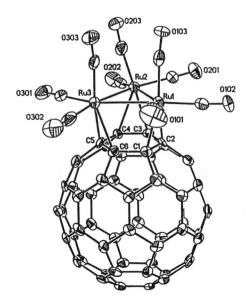

Figure 4.13   The structure of $Ru_3(CO)_9(\mu_3\text{-}\eta^2, \eta^2, \eta^2 C_{60})$ as determined by X-ray crystallography. Reprinted with permission from Ref. 34. Copyright © 1996 American Chemical Society.

Addition of four platinum centers to C_{70} occurs in a stepwise fashion to yield the four complexes $C_{70}\{Pt(PPh_3)_2\}_n$, where n is 1, 2, 3, and 4. The first two additions occur at $C_a\text{-}C_b$ bonds while the second two occur at the $C_c\text{-}C_c$ bonds. The final product, $C_{70}\{Pt(PPh_3)_2\}_4$, has been crystallized, and its structure is shown in Figure 4.18 [39].

The presence of metal–metal bonding can lead to the production of complexes in which metal atoms are bound to adjacent olefinic sites within C_{70}. Thus, reaction of C_{70} with $Ru_3(CO)_{12}$ yields $(C_{70})Ru_3(CO)_9$ in which a triangle of ruthenium atoms is bound to a hexagonal face of the fullerene at one pole, as seen in Figure 4.19 [40]. Two ruthenium atoms are bound to $C_a\text{-}C_b$ bonds while the third is bound to a $C_c\text{-}C_c$ bond of the fullerene. Addition of two Ru_3 units to C_{70} produces $(C_{70})\{Ru_3(CO)_9\}_2$, whose structure is shown in Figure 4.20. The two ruthenium clusters reside at opposite poles of the fullerene and occupy hexagonal faces as they do in $(C_{60})Ru_3(CO)_9$ and in $(C_{70})Ru_3(CO)_9$.

Only one crystallographic study has been made of a fullerene larger than C_{70}. Black crystals of $(\eta^2\text{-}C_{84})Ir(CO)Cl(PPh_3)_2 \cdot 4C_6H_6$ are formed when $Ir(CO)Cl(PPh_3)_2$ is added to a solution of a mixture of isomers of C_{84} [41]. This mixture consists largely of the $D_2(22)$ and $D_{2d}(23)$ isomers shown in Figure 4.1. The result of a single-crystal X-ray diffraction study of $(\eta^2\text{-}C_{84})Ir(CO)Cl(PPh_3)_2 \cdot 4C_6H_6$ is shown in Figure 4.21. Upon crystallization, the $D_{2d}(23)$ isomer of C_{84} separates from the mixture of five or more isomers. The iridium ion is coordinated to the $C(32)\text{-}C(53)$ bond, which is one of the 19 different types of C–C

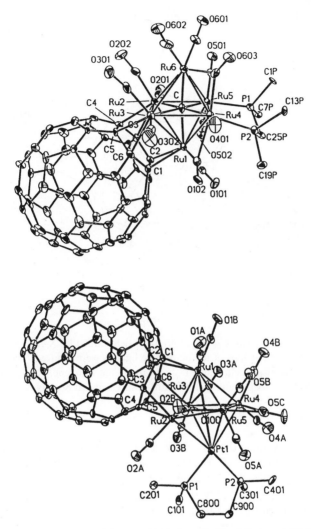

Figure 4.14 The structures of (a) $Ru_6C(CO)_9(\mu_3\text{-}\eta^2, \eta^2, \eta^2C_{60})$ and (b) $Ru_5PtC(CO)_9 \cdot$ (dppe)$(\mu_3\text{-}\eta^2, \eta^2, \eta^2C_{60})$ as determined by X-ray crystallography. Reprinted with permission from K. Lee, H.-F. Hsa, and J. R. Shapley, *Organometallics* **1997**, *16*, 3876; and from K. Lee and J. R. Shapley, *Organometallics* **1998**, *17*, 3020. Copyright © 1997, 1998 American Chemical Society.

bonds in C_{84}. Hückel calculations indicate that this is the bond with the highest π-bond order. Consequently, it should be the most reactive bond in the fullerene [18]. As a consequence of coordination, the C(32)–C(53) bond (bond length 1.455(6) Å) is elongated relative to its unreacted counterpart, the C(42)–C(43) bond (bond length, 1.332(11) Å) at the opposite side of the fullerene. The

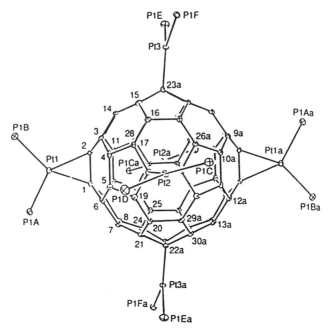

Figure 4.15 The structure of $C_{60}\{Pt(PEt_3)_2\}_6$ as determined by X-ray crystallography. For simplicity the ethyl groups of the phosphine ligands were omitted. The six platinum atoms are arranged in an octahedral fashion about the fullerene. Reprinted with permission from Ref. 35. Copyright © 1991 American Chemical Society.

separation of C_{84} isomers that is achieved through crystallization of the adduct, $(\eta^2\text{-}C_{84})Ir(CO)Cl(PPh_3)_2 \cdot 4C_6H_6$, is not complete. Examination of residual electron density within the fullerene portion of the adduct indicates that another isomer of C_{84} is probably present. Recently, a chromatographic separation of the isomers of C_{84} has been reported [12].

4.2.4 Coordination Modes for Chemically Modified Fullerenes

Several chemically modified fullerenes have also been found to form complexes with transition metals. $Ir(CO)Cl(PPh_3)_2$ binds to fullerene oxides, and the resulting adducts give structural information about the location of the oxygen atoms. Figure 4.22 shows the structure of $(\eta^2\text{-}C_{60}O)Ir(CO)Cl(PPh_3)_2$ as determined by X-ray crystallography [42], while Figure 4.23 shows the structure of the predominant form of $(\eta^2\text{-}C_{60}O_2)Ir(CO)Cl(PPh_3)_2$ [43]. In both of these compounds there is some degree of disorder in the location of the oxygen atoms, but it is clear that these atoms are present as epoxide groups, which are more localized in these adducts than in crystalline $C_{60}O$ itself, where the oxygen atom is disordered over the entire surface of the fullerene [44]. $C_{70}O$ also reacts

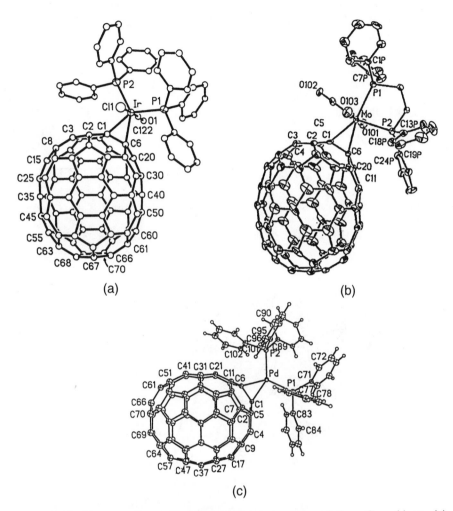

Figure 4.16 The crystallographically determined structures of three C_{70} adducts. (a) $(\eta^2\text{-}C_{70})Ir(CO)Cl(PPh_3)_2 \cdot 2.5C_6H_6$ as determined by X-ray crystallography. This was the first crystal structure determination for a higher fullerene. Reprinted with permission from Ref. 37. Copyright © 1991 American Chemical Society. (b) $(\eta^2\text{-}C_{70})Mo(CO)_3 \cdot$ (dppe). Reprinted with permission from H.-F. Hsu, Y. Du, T. E. Albrecht-Schmitt, S. R. Wilson, and J. R. Shapley, *Organometallics* **1998**, *17*, 1756. Copyright © 1998 American Chemical Society. (c) $(\eta^2\text{-}C_{70})Pd(PPh_3)_2 \cdot CHCl_3$. Reprinted with permission from Ref. 83. Copyright © 1998 Elsevier.

with $Ir(CO)Cl(PPh_3)_2$ to form crystals of $(\eta^2\text{-}C_{70}O)Ir(CO)Cl(PPh_3)_2 \cdot 5C_6H_6$, which have also been examined by X-ray crystallography. The structure of this adduct reveals that $C_{70}O$ consists of two isomers with the epoxide oxygen atom sitting over either a C_a–C_b or a C_c–C_c bond, as shown in Figure 4.24. The

TABLE 4.2 X-Ray Crystallographic Data for Complexes Containing a Single $(\eta^2\text{-}C_{70})M$ Unit

Compound	M–C	C–C	Reference
$(\eta^2\text{-}C_{70})Mo(CO)_3(PhPCH_2CH_2PPh_2)$	2.290(7) 2.278(6)	1.502(10)	a
$(\eta^2\text{-}C_{70})Ir(CO)Cl(PPh_3)_2$	2.19(2) 2.18(2)	1.46(3)	b
$(\eta^2\text{-}C_{70})Pd(PPh_3)_2$	2.106(11) 2.123(11)	1.48(2)	c

[a] H.-F. Hsu, Y. Du, T. E. Albrecht-Schmitt, S. R. Wilson, and J. R. Shapley, *Organometallics* **1998**, *17*, 1756.

[b] A. L. Balch, V. J. Catalano, J. W. Lee, M. M. Olmstead, and S. R. Parkin, *J. Am. Chem. Soc.* **1991**, *113*, 8954.

[c] M. M. Olmstead, L. Hao, and A. L. Balch, *J. Organometal. Chem.*, **1999**, *578*, 85.

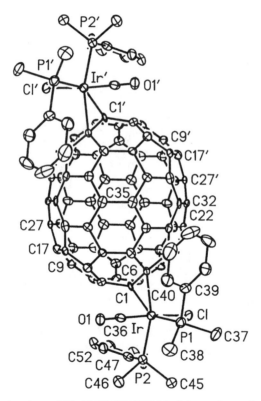

Figure 4.17 The structure of $C_{70}\{Ir(CO)Cl(PPhMe_2)_2\}_2$ as determined by X-ray crystallography. Reprinted with permission from Ref. 38. Copyright © 1992 VCH Verlagsgesellschaft mbH.

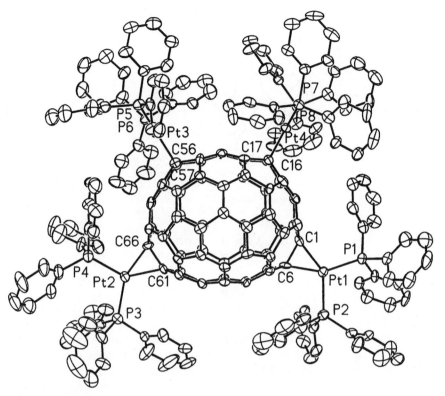

Figure 4.18 The crystallographically determined structure of $(C_{70})\{Pt(PPh_3)_2\}_4$. Reprinted with permission from Ref. 39. Copyright © 1996 VCH Verlagsgesellschaft mbH.

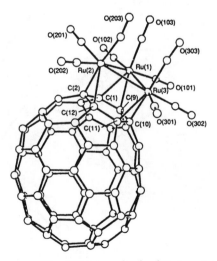

Figure 4.19 The structure of $Ru_3(CO)_9(\mu_3\text{-}\eta^2, \eta^2, \eta^2 C_{70})$ as determined by X-ray crystallographic study. Reprinted with permission from Ref. 40. Copyright © 1997 Royal Society of Chemistry.

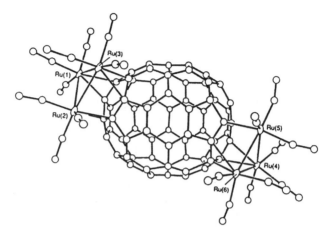

Figure 4.20   The structure of $\{Ru_3(CO)_9\}_2C_{70}$ as determined by X-ray crystallographic study. Reprinted with permission from Ref. 40. Copyright © 1997 Royal Society of Chemistry.

iridium complex binds to the fullerene portion in the usual η^2 fashion at a C_a–C_b bond.

Chemically modified fullerenes also present novel bonding situations that allow coordination in fashions other than the usual η^2 fashion. The reaction of $[Ph_2Cu]^-$ with C_{60} followed by hydrolysis produces $C_{60}Ph_5H$, which has a

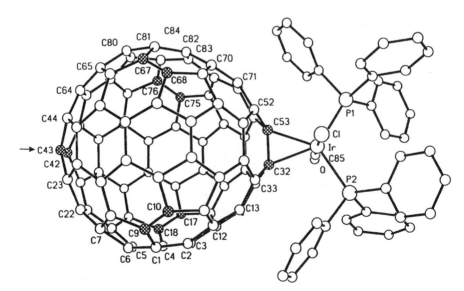

Figure 4.21   The structure of $(\eta^2\text{-}C_{84})Ir(CO)Cl(PPh_3)_2 \cdot 4C_6H_6$ as determined by X-ray crystallography. The arrow points to the C(42)–C(43) bond, which is directly opposite the C–C bond to which the iridium atom is coordinated. Reprinted with permission from Ref. 41. Copyright © 1994 American Chemical Society.

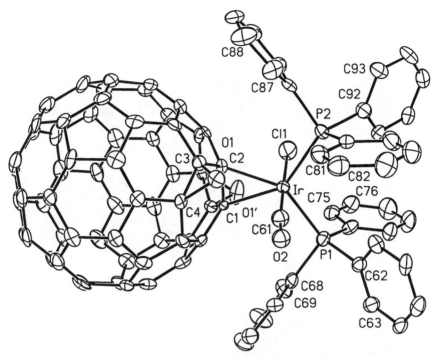

Figure 4.22 The structure of $(\eta^2\text{-}C_{60}O)Ir(CO)Cl(PPh_3)_2$ as determined by X-ray crystallography. The epoxide oxygen atom is disordered over two similar but not identical positions with site occupancies of 0.644 for O(1) and 0.025 for O(1′). Reprinted with permission from Ref. 42. Copyright © 1994 American Chemical Society.

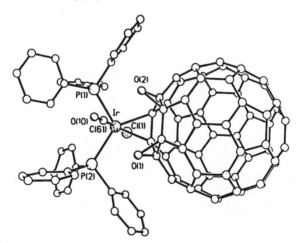

Figure 4.23 A view of the predominant form of $(\eta^2\text{-}C_{60}O_2)Ir(CO)Cl(PPh_3)_2$ that is present in the crystalline adduct as determined by X-ray crystallography. The disorder in the seven oxygen sites in the adduct has been analyzed quantitatively. The form shown here represents 60% of the contents of the crystal. Reprinted with permission from Ref. 43. Copyright © 1995 American Chemical Society.

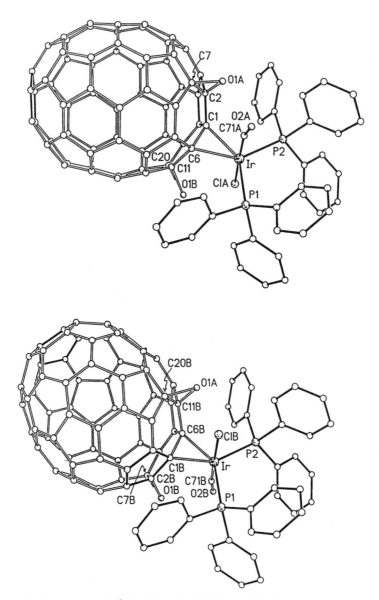

Figure 4.24 Drawings of $(\eta^2\text{-}C_{70}O)Ir(CO)Cl(PPh_3)_2$ that share a common site in crystalline $(\eta^2\text{-}C_{70}O)Ir(CO)Cl(PPh_3)_2 \cdot 5C_6H_6$. Reprinted with permission from A. L. Balch, D. A. Costa, and M. M. Olmstead, *Chem. Commun.* **1996**, 2449. Copyright © 1996 Royal Society of Chemistry.

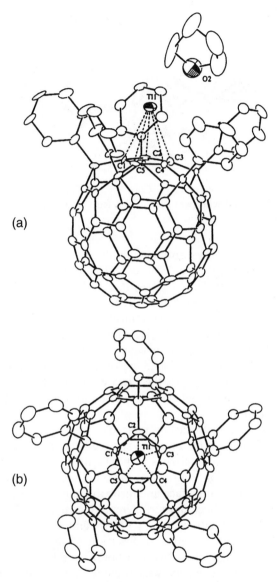

Figure 4.25 Two drawings of the structure of $(\eta^5\text{-}Ph_5C_{60})Tl$ as determined by X-ray crystallography. Reprinted with permission from Ref. 45. Copyright © 1996 American Chemical Society.

unique pentagonal face that is surrounded by five phenyl groups [45]. Deprotonation of $C_{60}Ph_5H$ forms $[C_{60}Ph_5]^-$, which is able to function as an η^5-ligand and forms adducts with Et_3PCu^+ and Tl^+. The η^5-bonding mode has been confirmed by X-ray crystallography for the thallium complex, whose structure is shown in Figure 4.25. The thallium atom is situated 2.60 Å above the center

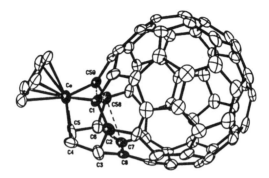

Figure 4.26   The structure of $(\eta^5\text{-}C_5H_5)Co(\eta^1,\eta^1,\eta^2\text{-}C_{64}H_4)$ as determined by X-ray crystallography. Reprinted with permission from Ref. 46. Copyright © 1996 American Chemical Society.

of the pentagonal face that is surrounded by the phenyl groups. The Tl–C bond lengths fall in a narrow range, 2.85(1) to 2.90(1) Å, and are comparable to the corresponding distances in $Tl(\eta^5\text{-}C_5Bz_5)$. It is likely that other transition metals will also coordinate to this modified fullerene in η^5 fashion.

Figure 4.26 shows the crystallographically determined structure of a remarkable molecule, $(\eta^5\text{-}C_5H_5)CoC_{64}H_4$, that contains a cobalt atom that is incorporated into part of a carbon cage structure [46]. This compound is formed by initial photolysis of 1,2-(3,5-cyclohexadieno) C_{60} followed by treatment with $(\eta^5\text{-}C_5H_5)Co(CO)_2$. In the process one of the C–C bonds in the original fullerene is broken, but the two carbon atoms are subsequently coordinated to the cobalt atom.

A number of curved hydrocarbon molecules, such as corannulene, $C_{20}H_{10}$, that represent fragments of the fullerene surface have been prepared and compared to their fullerene counterparts [47]. The availability of such molecules allows chemical exploration not only of the outer surface of curved hydrocarbons but also of the interior surface and, probably most importantly, the edges. The reaction of the semibuckminsterfullerene, $C_{30}H_{12}$, with $Pt(C_2H_4)(PPh_3)_2$ does not result in coordination of a $Pt(PPh_3)_2$ unit to a 6:6 ring junction as happens with fullerenes, but rather to insertion of the platinum into one of the C–C bonds on the edge of the hydrocarbon, as shown in Scheme 5 [48]. The structure of the platinum complex, which retains considerable curvature in the hydrocarbon portion, is shown in Figure 4.27. The unusual C–C bond breaking seen in Scheme 5 has been attributed to relief of the strain present in the five-membered rings at the edge of the parent hydrocarbon. This reaction serves as a model for one of the proposed mechanisms of carbon nanotube formation [49]. In that mechanism, a metal atom binds to the edge of a growing nanotube and migrates about that edge as growth occurs. Related reactivity at the edge of another fullerene fragment, $C_{36}H_{14}$, has been observed to result in facile oxidation, with cleavage of a C–C bond within a six-membered ring, to produce the nearly planar dione $C_{36}H_{14}O_2$ [50].

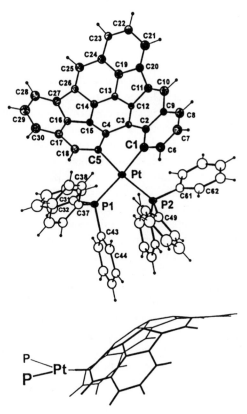

Figure 4.27 The structure of $(C_{36}H_{12})Pt(PPh_3)_2$, which shows that one of the C–C bonds at the edge of the fullerene fragment has been oxidatively added to the platinum atom. Reprinted with permission from Ref. 48. Copyright © 1998 American Chemical Society.

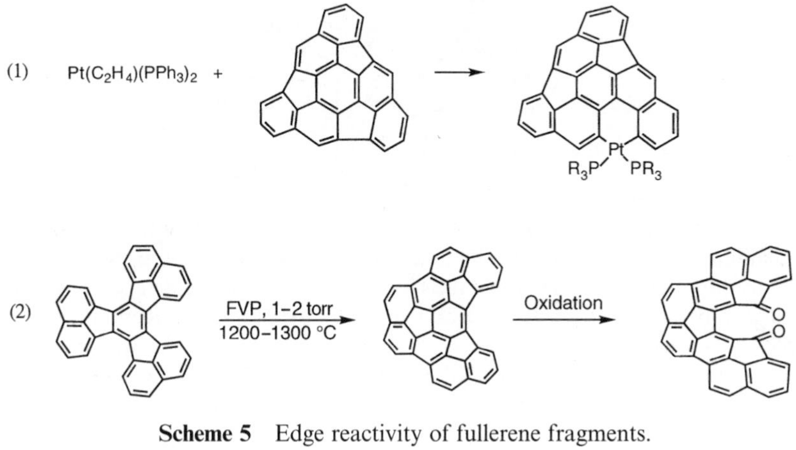

(1) $Pt(C_2H_4)(PPh_3)_2$ +

(2)

FVP, 1–2 torr
1200–1300 °C

Oxidation

Scheme 5   Edge reactivity of fullerene fragments.

4.3 FULLERIDE SALTS

The fullerenes undergo reduction to form a family of fulleride ions. For C_{60} the lowest unoccupied molecular orbitals (LUMOs, the triply degenerate t_{1u} orbitals) are relatively low in energy, and C_{60} is readily reduced. As predicted by the nature of the LUMO, six-reversible one-electron reductions can be observed for C_{60} [51,52]. The reduction potentials are uniformly spaced with approximately a 0.5 V difference between successive reduction waves as observed by cyclic voltammetry. The reduction potentials show only modest variations when the solvent and/or supporting electrolyte are changed. In toluene/acetonitrile (4:1, v/v) with tetra(n-butyl)ammonium hexafluorophosphate as supporting electrolyte, the reduction potentials for C_{60} are -0.98, -1.37, -1.87, -2.35, -2.85, and -3.26 V versus a ferrocene/ferrocinium electrode. Salts containing the fulleride ions have been prepared, and this section gives an overview of the structural chemistry of these salts, particularly with inorganic ions.

Solids containing fulleride ions have been obtained principally through two different routes: intercalation of metal atoms into the solid fullerene via vapor transport or by solution techniques. The intercalation route has been heavily investigated as a route to salts of the type A_3C_{60}, where A is an alkali metal ion, which exhibit superconductivity at temperatures up to approximately 40 K. A number of excellent reviews cover the physics and chemistry of this type of intercalation product as well as the nonsuperconducting phases, A_4C_{60} and A_6C_{60} [6,53–56]. Figure 4.28 shows the face-centered cubic (fcc) structure for

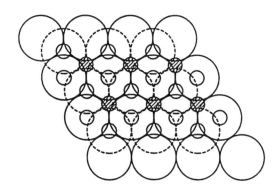

Figure 4.28 A diagram that shows the face-centered cubic structure for C_{60} and site of intercalation. The large circles are the fullerene molecules; the small open circles show the tetrahedral interstitial sites, while the hatched circles show the octahedral sites. Reprinted with permission from R. C. Haddon, A. F. Hebard, M. J. Rosseinsky, D. W. Murphy, S. H. Glarum, T. T. Palstra, A. P. Ramirez, S. J. Duclos, R. M. Flemming, T. Siegrist, and R. Tycko, in *Fullerenes*, G. S. Hammond and V. J. Kuck (eds.), American Chemical Society Symposium Series 481, p. 71. 1992. Copyright © 1992 American Chemical Society.

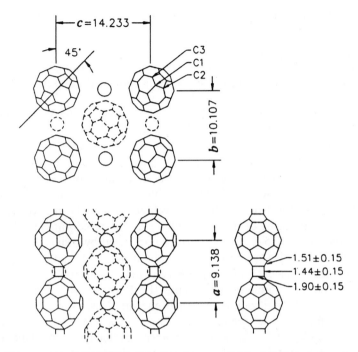

Figure 4.29 Structure of $\{Rb[C_{60}]\}_n$, showing covalently bonded chains along the a direction. Solid (broken) lines show fullerenes and cations centered at $x = 0(x = \frac{1}{2})$ in the b–c plane at the top, and $y = 0(y = \frac{1}{2})$ in the a–c plane in the middle. Solid line through the upper-left fullerene denotes the molecular mirror plane, and the fragment to the right is a projection of one chain perpendicular to that plane. (All distances given in Å.) Reprinted with permission from Ref. 58. Copyright © 1994 Nature.

C_{60} and the interstitial sites where metal ions can intercalate. Since a number of reviews are available on these intercalated species, this section concentrates on structural studies of salts of the fulleride anion and dianion, where comparisons can be made between materials prepared by solution phase methods and materials made by intercalation.

4.3.1 Monoanions

Intercalation of alkali metal atoms into C_{60} yields an array of different phases with the composition AC_{60} [57]. Above 400 °C, a face-centered cubic structure with freely rotating fulleride ions is stable, while below 400 °C, an orthorhombic phase, $\{A[C_{60}]\}_n$, forms with a polymeric chain of fulleride ions that are connected through C–C bonds [58–60]. Figure 4.29 shows a drawing of the structure of the air stable, polymeric phase, $\{Rb[C_{60}]\}_n$. In this solid a [2+2] cycloaddition between adjacent fullerides forms the four-membered rings that covalently link the fulleride ions [58]. Metallic conductivity has been observed for $\{A[C_{60}]\}_n$ (A = K, Rb, and Cs) [61]. The analysis of the X-ray powder data for $\{Rb[C_{60}]\}_n$ suggests a remarkably long fullerene C–C bond (estimated length 1.65–2.15 Å) at the four-membered ring junction [58,62]. Other metastable phases of AC_{60} are known, including a form with discrete, dimeric fullerene units [63,64].

Solution phase methods have been used to prepare the solid salts, $[(\eta^2\text{-}C_5H_5)_2Co^+][C_{60}{}^-]$, $[Na(\text{dibenzo-18-crown-6})^+][C_{60}{}^-]$, and $[\text{bis}(N\text{-methylimidazole})\text{-(tetraphenylporphyrinato)tin(IV)}^{+2}][C_{60}{}^-]_2$, which are believed to contain isolated $[C_{60}]^-$ units [65,66]. These have been characterized extensively by electron paramagnetic resonance (EPR) and ultraviolet/visible (UV/VIS) spectroscopy, and $[(\eta^2\text{-}C_5H_5)_2Co^+][C_{60}{}^-]$ has been examined by single-crystal X-ray diffraction. Electrocrystallization techniques have been effective in producing crystalline forms of $[C_{60}]^-$ [67,68]. The air stable salt $[Ph_4P]_2[C_{60}][I]_n$, with n in the range 0.15–0.35, has been characterized by a single-crystal X-ray diffraction study [67]. Despite the nonstoichiometric formulation, the spectroscopic data are consistent with the presence of only $[C_{60}{}^-]$ (no C_{60} or $[C_{60}{}^{2-}]$) in the solid. The stoichiometric compound $[Ph_4P^+]_2[C_{60}{}^-][Cl^-]$ has also been formed by electrocrystallization and has been characterized by single-crystal X-ray diffraction [69]. Other related salts, $[Ph_4P^+]_2[C_{60}{}^-][X^-]$, with X = Br or I, and $[Ph_4As^+]_2[C_{60}{}^-][Cl^-]$, have similarly been characterized [68]. Due to issues of disorder and crystal quality, these structure determinations have not been able to examine the fulleride geometry in sufficient precision to detect any possible distortion that results from addition of an electron to C_{60}. However, in each of the crystallographically characterized salts, the fulleride ions are well separated and direct covalent C–C bonding like that found in $\{Rb[C_{60}]\}_n$ is not present.

Dark red crystals of $[(\eta^5\text{-}C_5Me_5)_2Ni^+][C_{60}{}^-] \cdot CS_2$ form from carbon disulfide solutions of $(\eta^5\text{-}C_5Me_5)_2Ni$ and C_{60} [70]. An X-ray crystallographic study shows that the fulleride has undergone axial compression so that there is a short axis dimension, 6.878(6) Å (from the midpoint of a C–C bond at a 6:6 ring junction to the opposite midpoint), and two corresponding longer orthogonal distances, 6.965(5) and 6.976(5) Å. This structural change may result from a Jahn Teller distortion, but the framework of $[(\eta^5\text{-}C_5Me_5)_2Ni]^+$ cations may also reshape the fulleride ion. A stereoscopic drawing of the structure is shown in Figure 4.30.

4.3.2 Dianions

Na_2C_{60}, which is not superconducting, is formed via intercalation and possesses a fluorite structure with all tetrahedral holes filled in the fcc close-packed structure [71]. Treatment of Na_2C_{60} with ammonia at room temperature produces $[Na(NH_3)_4{}^+]_2[C_{60}{}^{2-}]$ in which the close-packed structure is expanded so that the intralayer $[C_{60}{}^{2-}] \cdots [C_{60}{}^{2-}]$ separation is 12.22 Å while the corresponding interlayer separation is 10.24 Å [72].

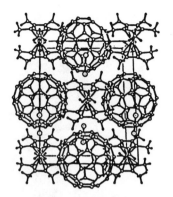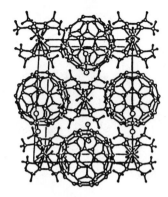

Figure 4.30 A stereoscopic drawing of the solid-state structure of $[(\eta^5\text{-}C_5Me_5)_2Ni^+]\cdot$ $[C_{60}^-]\cdot CS_2$. Reprinted with permission from Ref. 70. Copyright © 1995 American Chemical Society.

Crystals of $[(Ph_3P)_2N^+]_2[C_{60}^{2-}]$ have been obtained by mixing $[(Ph_3P)_2N]Cl$ with $[Na(crown)^+]_2[C_{60}^{2-}]$ in acetonitrile [73, 74]. Within the solid, the fulleride ions are ordered and the data are of sufficient precision to reveal a significant distortion in the fulleride, which has been described as an "axial elongation with an apparent rhombic squash" with a long diameter of 7.126(5) Å and a short diameter of 7.040(5) Å [73,74]. A bar graph that shows the distribution of distances within this fulleride is contained in Figure 4.31. The salt $[Ba(NH_3)_7^{2+}][C_{60}^{-2}]\cdot NH_3$ has been prepared by the reaction of C_{60} and

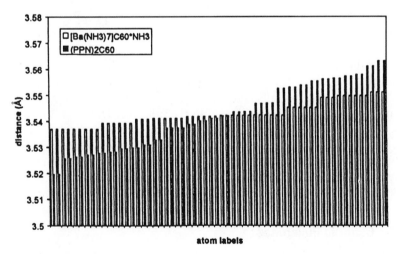

Figure 4.31 A comparison of the distances from the center of the fulleride ion to the carbon atoms in $[Ba(NH_3)_7^{2+}][C_{60}^{2-}]\cdot NH_3$ (*hollow bars*) and $[(Ph_3P)_2N^+]_2[C_{60}^{2-}]$ (*solid bars*). Reprinted with permission from Ref. 75. Copyright © 1998 American Chemical Society.

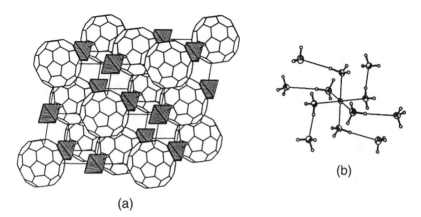

(a)

(b)

Figure 4.32 A view of the structure of $[Ni(NH_3)_6{}^{2+}][C_{60}{}^{2-}] \cdot 6NH_3$ as determined by X-ray crystallography. (a) The packing within the solid. (b) The hydrogen bonding that occurs between the cation and the adjacent ammonia molecules. Reprinted with permission from K. Himmel and M. Jansen, *Chem. Commun.* **1998**, 1205. Copyright © 1998 Royal Society of Chemistry.

barium metal in liquid ammonia and has also been characterized by X-ray crystallography [75]. In this salt the fulleride ion is also ordered, but the structural analysis shows a much less pronounced distortion than that seen in $[(Ph_3P)_2N^+]_2[C_{60}{}^{2-}]$. Relevant data for comparison are shown in Figure 4.31. The structure of the related salt, $[N \cdot (NH_3)_6]C_{60} \cdot 6NH_3$ is shown in Figure 4.32.

Another $[C_{60}{}^{2-}]$ containing salt, $[K([2.2.2.]crypt)^+]_2[C_{60}{}^{2-}] \cdot 4(C_6H_5CH_3)$, has been formed by reaction of C_{60} with potassium metal in the presence of [2.2.2.]crypt in dimethylformamide [76]. This compound crystallizes in a layered structure that consists of fulleride layers with distorted hexagonal close packing separated by layers of the cationic complexes.

It is notable that the salts containing $[C_{60}{}^-]$ and $[C_{60}{}^{2-}]$ so far do not have metal centers that directly coordinate to the fullerene. However, the salts that have been structurally characterized do not involve transition metal compounds with vacant sites available for coordination. Recently, however, electrochemical reduction of fullerenes in the presence of transition metal complexes has been shown to result in the deposition of redox active films that adhere to the electrode surface and appear to involve covalent metal fulleride coordination [77].

4.3.3 Cationic Fullerenes

Although reductions of fullerenes are relatively readily achieved, oxidations are more difficult. Cyclic voltammetry studies show that C_{60}, C_{70}, and C_{76} undergo a single, one-electron oxidation at 1.26, 1.20, and 0.81 V versus a ferrocene/ferrocinium electrode in tetrachloroethylene solution with tetra(*n*-butyl)ammo-

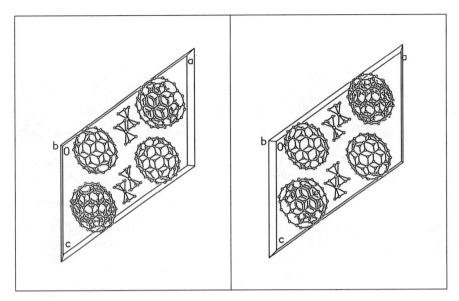

Figure 4.33 A diagram that shows the packing of fullerene and ferrocene molecules in $C_{70} \cdot 2\{(\eta^5\text{-}C_5H_5)_2Fe\}$. Reprinted with permission from Ref. 83. Copyright © 1998 Elsevier.

nium hexafluorophosphate as supporting electrolyte. Only one salt that contains a fullerene cation, $[CB_{11}H_6Br_6^-]\{C_{76}^+]$, has been isolated [78]. C_{76} was utilized in this work because of the greater ease of oxidation of this particular fullerene.

4.4 COCRYSTALLIZATION OF FULLERENES WITH INORGANIC COMPOUNDS

Fullerenes have a marked propensity to form crystals that incorporate additional molecules into the lattice. Frequently, when C_{60} crystallizes from solution, molecules of the solvent are incorporated into the solid, as is the case with $C_{60} \cdot 4(C_6H_6)$ [79–81]. While the incorporation of solvate molecules into the crystal lattices is fairly common, C_{60} (and C_{70}) has a remarkable ability to form inclusion compounds with an interesting array of other compounds. In a number of cases the fullerene orientationally ordered within these solids.

Ferrocene is not a sufficiently good reducing agent to reduce C_{60} to $[C_{60}]^-$ (or C_{70} to $[C_{70}]^-$). However, solutions of C_{60} (or C_{70}) and ferrocene deposit black crystals of $C_{60} \cdot 2\{(\eta^5\text{-}C_5H_5)_2Fe\}$ [82] (or $C_{70} \cdot 2\{(\eta^5\text{-}C_5H_5)_2Fe\}$ [83]). A drawing of the structure of $C_{70} \cdot 2\{(\eta^5\text{-}C_5H_5)_2Fe\}$ is shown in Figure 4.33. The structure of $C_{60} \cdot 2\{(\eta^5\text{-}C_5H_5)_2Fe\}$ is similar. The C_{70} molecule makes close face-to-face contact with the ferrocene molecules through the five-membered

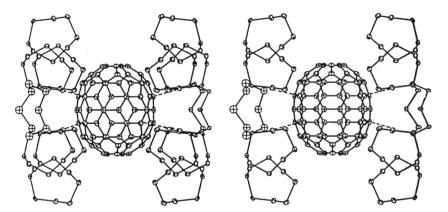

Figure 4.34 A stereoscopic drawing that shows the packing of molecules of C_{70} and S_8 in $C_{70} \cdot 6(S_8)$ as determined by X-ray crystallography. Fourteen individual molecules of S_8 surround each C_{70} molecule. Reprinted with permission from Ref. 85a. Copyright © 1998 Neue Schweizerische Chemische Gesellschaft.

rings along the five fold axis of the fullerene. Similarly, benzene solutions of $Fe_4(CO)_4(\eta^5\text{-}C_5H_5)_4$ and C_{60} produce crystals of $C_{60} \cdot \{Fe_4(CO)_4(\eta^5\text{-}C_5H_5)_4\} \cdot 3C_6H_6$ in which the fullerene is also orientationally ordered [84]. The infrared spectrum of the product reveals that there is a small degree of charge transfer from the fullerene to the organometallic component.

The fullerenes C_{60}, C_{70}, and C_{78} have been found to form cocrystallized compounds with cyclo-octasulfur (S_8). Thus, slow evaporation of carbon disulfide solutions of C_{70} and S_8 yield crystals of $C_{70} \cdot 6(S_8)$ [85]. The fullerene component, which is surrounded by 14 molecules of S_8, is fully ordered, while there is disorder in the positions of some of the S_8 molecules. A stereoscopic view of the arrangement of S_8 molecules about a C_{70} molecule is shown in Figure 4.34. Similar structures have been found crystallographically for $C_{60} \cdot (S_8)^{86}$ and for $C_{76} \cdot 6(S_8)$ [87]. The latter structure possesses an array of sulfur molecules that is similar to that shown in Figure 4.34. The C_{76} molecules are disordered, but the data have been analyzed to show an "overriding dominance of the D_2 isomer" rather than the T_d isomer of C_{76}.

Mixing a toluene solution of C_{60} with a carbon disulfide solution of white phosphorus (P_4) produces air-stable, blue-black crystals of $C_{60} \cdot 2P_4$ [88]. The structure derived from an Rietveld analysis of the X-ray powder diffraction data is shown in Figure 4.35.

Interestingly, mass spectroscopic experiments have also given information about the clustering of P_4 and S_8 molecules about a single molecule of C_{60} [89]. Clusters are formed in a condensation cell from vapors of C_{60} and red phosphorus (or sulfur). The spectrum obtained for the C_{60}/P_4 system shows the existence of an array of clusters with a prominent maximum for a cluster with the composition $C_{60} \cdot 32(P_4)$. Structures consisting of 32 P_4 tetrahedra positioned above the 32 pentagonal and hexagonal faces of the fullerene have been con-

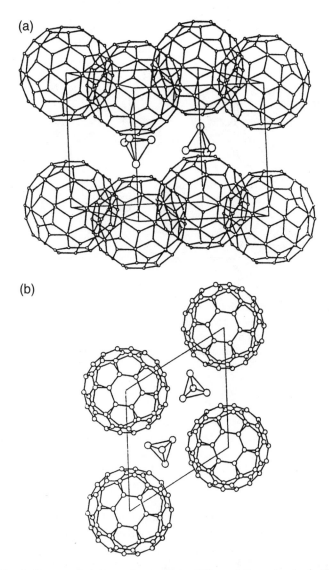

(a)

(b)

Figure 4.35 A diagram that shows the packing of C_{60} and P_4 molecules in $C_{60} \cdot 2(P_4)$ as determined by X-ray crystallography. Only one orientation of the P_4 molecules is shown here. Reprinted with permission from Ref. 88. Copyright © 1994 Royal Society of Chemistry.

sidered for this magic number cluster. For the C_{60}/S_8 system the maximum in spectral intensity occurs for a cluster with the composition $C_{60} \cdot 12(S_8)$.

Reaction of C_{60} with $(PhCN)_2PdCl_2$, a labile complex that easily forms complexes with olefins, in benzene results in the formation of black crystals of $C_{60} \cdot 2(Pd_6Cl_{12}) \cdot 2.5C_6H_6$ [90]. The product consists of individual molecules of C_{60}, benzene, and Pd_6Cl_{12}. The last of these components has a cubic structure

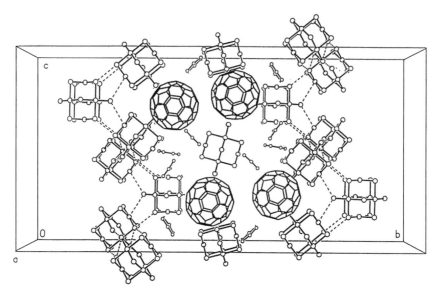

Figure 4.36 A section through the structure of $C_{60} \cdot 2(Pd_6Cl_{12}) \cdot 2.5C_6H_6$, which shows the relative locations of the three molecular components. The C_{60} molecules are orientationally disordered and are shown with idealized drawings of the molecules at their experimentally determined locations. Reprinted with permission from Ref. 90. Copyright © 1996 American Chemical Society.

with the six palladium atoms at the faces of the cube and the chloride ions at the midpoints of the edges. The cluster is of comparable size to that of the fullerene with an average *trans* $Cl \cdots Cl$ distance of 6.54 Å and an average *trans* $Pd \cdots Pd$ distance of 4.67 Å. A view of the solid is shown in Figure 4.36. There is significant attraction between Pd_6Cl_{12} and aromatic molecules, since a variety of binary and ternary cocrystals of Pd_6Cl_{12} with aromatic molecules (benzene, mesitylene, durene, and hexamethylbenzene) have been prepared.

Toluene solutions of C_{60} and Ph_3PAuCl yield crystalline $C_{60} \cdot 4\{(Ph_3P)AuCl\} \cdot 0.1C_6H_5CH_3$ [91]. The X-ray crystal structure of the solid shows that it consists of isolated C_{60} and Ph_3PAuCl molecules without any covalent interaction between them.

Solutions of C_{60}, as well as of C_{70} and $C_{60}O$, and cobalt(II) octaethylporphyrin ($Co^{II}(OEP)$) produce crystalline precipitates: $C_{60} \cdot 2\{Co^{II}(OEP)\} \cdot CHCl_3$, $C_{70} \cdot Co^{II}(OEP) \cdot (C_6H_6)$, and $C_{60}O \cdot 2\{Co^{II}(OEP)\} \cdot CHCl_3$ [92]. X-ray crystallographic studies reveal that there is no direct covalent link between the cobalt porphyrin and the fullerene in these compounds. The fullerenes do closely approach the porphyrin planes as shown in Figure 4.37 for $C_{60} \cdot 2\{Co^{II}(OEP)_2\} \cdot CHCl_3$. The distances between the cobalt atoms and adjacent carbon atoms are in the range 2.7–2.9 Å, which is short for van der Waals contacts but too long to represent any chemical bonding between the units.

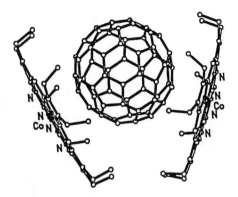

Figure 4.37 A drawing of the interaction of the two porphyrins with C_{60} in $C_{60} \cdot \{Co^{II}(OEP)\} \cdot CHCl_3$. Reprinted from M. M. Olmstead, D. A. Costa, K. Maitra, B. C. Noll, S. L. Phillips, P. M. van Calcar, A. L. Balch, *J. Am. Chem. Soc.* **1999**, *121*, 7090.

Such interactions may also be involved in the use of porphyrin-modified materials for the chromatographic separation of fullerenes [93]. In addition to the close but asymmetrical fullerene/porphyrin contact, these compounds also show back-to-back $Co^{II}(OEP)/Co^{II}(OEP)$ interactions. The fullerene/$Co^{II}(OEP)$ and $Co^{II}(OEP)/Co^{II}(OEP)$ interactions are arranged to produce continuous chains of molecules in the solid state.

Some cocrystallized compounds are formed by gas/solid reactions. $C_{60} \cdot 2I_2$ is formed by vapor phase intercalation of molecular iodine into crystalline C_{60} [94]. Rietveld analysis of the powder diffraction pattern results in a structure that consists of a simple hexagonal lattice of C_{60} molecules with layers of iodine molecules interspersed between the layers of C_{60} molecules.

It is also possible to have cocrystallization occur from solution in which *an ionic compound* and C_{60} form a new solid. Thus, solutions of C_{60} and silver nitrate form black $C_{60} \cdot 5\{Ag(NO_3)\}$, whose structure is shown in Figure 4.38 [95]. The solid contains a continuous ionic network of silver nitrate, which encapsulates the fullerene to form a zeolite-like unit. The coordination of the fullerene by silver involves both η^1-and η^2-bonding.

4.5 ENDOHEDRAL MOLECULES

Since two chapters in this volume discuss the formation of metallofullerenes, endohedral fullerenes with metals on the inside rather than the outside, only a few comments about the endohedral species are made here. Theoretical and experimental work indicates that the insides of the fullerenes are much less reactive than the outsides. This differential reactivity comes about because reactions on the exterior result in further pyramidalization of the carbon atoms and a consequent release in strain when a reaction occurs on the outside of the

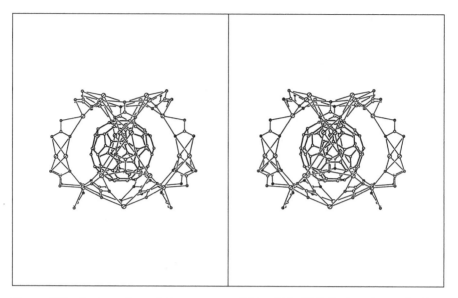

Figure 4.38 A stereoview of the structure of $C_{60} \cdot 5\{Ag(NO_3)\}$ that shows the four straps of the silver nitrate network that surround the fullerene. Only one of the four orientations of the fullerene is shown. Reprinted with permission from Ref. 95. Copyright © 1999 VCH Verlagsgesellschaft mbH.

fullerene [96]. In contrast, creation of a new covalent bond on the inside of the fullerene results in additional strain with the carbon cage. Basically, the π-type orbitals of a fullerene are more localized on the outer surface than on the inner surface. Consequently, it is possible to entrap some unusual species on the insides of fullerenes.

Metallofullerenes involving readily ionized metals—the alkali metals, alkaline earth metals, and the lanthanides—have been given considerable study. In some cases the metallofullerenes have been isolated in pure form but in small quantities. However, in many cases the endohedral species can be examined as a minor component mixed with the parent fullerene because the endohedral has a distinctive spectroscopic signature.

Fullerene preparation generally occurs in a helium atmosphere and helium atoms are found to be incorporated into a small fraction of the fullerenes that are formed. Considerable study of these have been made using both mass spectrometry and ^{2}He NMR spectroscopy [97]. Despite problems of low abundance of these He@fullerenes, ^{2}He NMR spectroscopy is an effective probe of such species [98]. The chemical shift is sensitive to many factors, including the number of carbon atoms in the fullerene, the presence of additional groups, and the redox state of the fullerene [99].

Heating C_{70} with ^{22}Ne at high pressures produces evidence for the formation of both Ne@C_{70} and the Ne$_2$@C_{70}, with the latter containing the "artificial molecule" Ne_2, which is stabilized by confinement within the fullerene [100]. The abundance of the later was higher than expected if incorporation of each atom of neon were merely a statistical event.

Additionally, nitrogen atoms have been found trapped with C_{60}, and spectroscopic evidence shows that these atoms are present in an unperturbed, $S = \frac{3}{2}$, spin state [101,102]. Fullerenes with nitrogen atoms inside are readily detected through EPR spectroscopy, and the ability to observe an EPR spectrum that shows little perturbation from that of a nitrogen atom in a vacuum presents one of the most convincing experimental demonstrations of the unreactive nature of the inside of these carbon cages.

4.6 INORGANIC FULLERENE-LIKE MOLECULES

Attention has been given to the possible preparation of several fullerene-like molecules that would involve elements other than carbon. The possibilities of constructing the silicon [103] and boron–nitrogen [104] analogs of C_{60} have been examined theoretically. For the hypothetical species Si_{60}, a closed, layered structure as well as an open, cage structure, analogous to that of C_{60}, have been considered.

Experimentally, Zintl phases have been found to contain fullerene-like structures [105]. The hexagonal phases, $Na_{98}In_{97}Z_2$ with Z = Ni, Pd, or Pt, consist of a set of nested arrays that can be formulated as $Ni@In_{10}@Na_{39}@In_{74}$. A view of the In_{74} fullerene (with D_{3h} symmetry) and a drawing that shows its relationship to the overall structure of the solid are shown in Figure 4.39. Another phase, orthorhombic $Na_{172}In_{197}Z_2$, also contains a C_{60}-like polyhedron with the composition $In_{48}Na_{12}$.

Molybdenum oxide chemistry has yielded an inorganic super fullerene, $[Mo^{VI}_{72}Mo^V_{60})_{372}(CH_3CO_2)_{30}(H_2O)_{72}]^{42+}$ [106]. This ion has been assembled by solution phase techniques from ammonium molybdate and crystallized as the ammonium salt. The structure consists of 12 Mo_{11} units positioned about fivefold symmetry axes, as shown in Figure 4.40.

A family of metal–carbon clusters, "metcars," which contain polyhedral arrays of carbon and metal atoms have been discovered by Castleman and co-workers [107]. For example, $[Ti_8C_{12}]^+$ has been prepared by reaction of titanium atoms with hydrocarbon (CH_4, C_2H_2, C_6H_6, etc.) vapors and identified by mass spectrometry. In the gas phase ammonia reacts with $[Ti_8C_{12}]^+$ to form a series of adducts, $[Ti_8(NH_3)_nC_{12}]^+$ with $n = 0$ to 8, in which the ammonia molecules presumably are coordinated to the titanium atoms. Hence, it appears that the titanium atoms are in exposed sites on the surface of the cluster. Figure 4.41 shows the T_h and T_d structures proposed for $[Ti_8C_{12}]^+$ [108].

Fullerenes with metal atoms incorporated into the carbon framework itself have also been detected. Pulsed laser vaporization of niobium carbide/graphite rods yields $[NbC_n]^+$ clusters that have been studied by injection-ion drift-tube mass spectrometry [109]. Fullerene, bicyclic ring, and monocyclic ring struc-

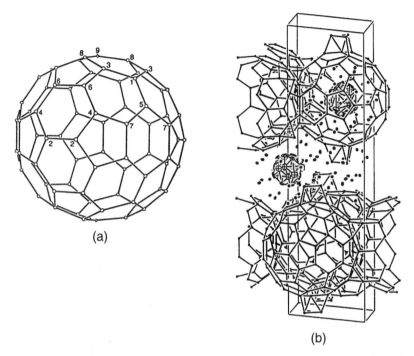

(a)

(b)

Figure 4.39 Two views of the structure of $Na_{96}In_{97}Z_2$ ($Z = Ni$, Pd, or Pt) as determined by X-ray crystallography. (a) The packing within the solid. (b) The fullerene-like cage of the In_{74} portion. The In_{74} fullerene is not hollow but contains an array described as $Z@In_{10}@Na_{39}@In_{74}$. Reprinted with permission from Ref. 105a. Copyright © 1993 Science.

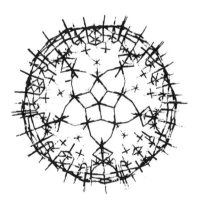

Figure 4.40 A diagram of the structure of $[Mo^{VI}_{72}Mo^{V}_{60})_{372}(CH_3CO_2)_{30}(H_2O)_{72}]^{42+}$ viewed down a fivefold rotation axis. Reprinted with permission from Ref. 106. Copyright © 1998 VCH Verlagsgesellschaft mbH.

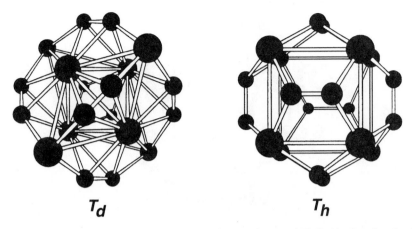

T_d T_h

Figure 4.41 Proposed T_h and T_d structures for the metcar, $[Ti_8C_{12}]^+$. Reprinted with permission from Ref. 108b. Copyright © 1996 American Chemical Society.

tures have been proposed for the variety of species that form in this process. For the species $[NbC_n]^+$ with an odd number of carbon atoms (n) a fullerene-type structure with the niobium atoms as an integral part of the fullerene cage has been proposed; while for $[NbC_n]^+$ with an even number of carbon atoms, an endohedral structure is believed to be present. Similar structures have been proposed for $[La_2C_n]^+$ ($n = 28$–100) clusters [110].

Cage substitution of metal atoms for carbon in fullerenes has also been achieved by photofragmentation of gas phase metal–fullerene clusters, $[C_{60}M_m]^+$, where M = Fe, Co, Ni, and Rh and m ranges from 0 to 30 [111]. The parent clusters were prepared by mixing the appropriate metal vapor with fullerene vapor and analyzed by time of flight mass spectrometry. Subsequent photolysis of these clusters produces new clusters, $[C_{59-2n}M]^+$, in which one of the carbon atoms of the fullerene cage is replaced by a metal atom.

Related clusters, $[(C_{59}Pt]^+$ and $[C_{59}Ir]^+$, have been obtained in the gas phase by laser ablation of electrochemically prepared films that are generated by simultaneous reduction of fullerenes and transition metal complexes such as (pyridine)$_2$PtCl$_2$ and Ir(CO)$_2$Cl(p-toluidine) [112,113]. Density functional calculations on the corresponding neutral clusters suggest that $[(C_{59}Pt]$ and $[C_{59}Ir]$ have the geometry shown in Figure 4.42 with a three-coordinate metal atom that protrudes away from the fullerene core [113].

4.7 SUMMARY

This chapter has concentrated on the structural aspects of molecules and molecular arrays that are formed between the fullerenes, principally C$_{60}$, and inorganic compounds. The reactivity of the outer surface of the fullerenes has

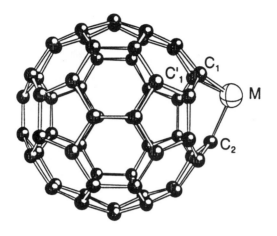

$$C_{59}Ir, \; C_{59}Pt$$

Figure 4.42 The structure of $[C_{59}Pt]$ and $[C_{59}Ir]$ as obtained from density functional calculations. Selected bond lengths (Å) and angles (°): C_1–Pt = 1.992, C_2–Pt = 1.934, C_1–Pt–C_1' = 78.5, C_1–Pt–C_2 = 96.7; C_1–Ir = 1.954, C_2–Ir = 1.905, C_1–Ir–C_1' = 81.7, C_1–Ir–C_2 = 92.3. Reprinted with permission from Ref. 112. Copyright © 1999 Royal Society of Chemistry.

led to a variety of covalently bonded adducts. Coordination of metal centers directly to C_{60} largely occurs in an η^2 fashion, but recent work has begun to uncover different modes of coordination. A number of fulleride salts have been structurally characterized and shown to have varying degrees of separation of the fulleride ions. Fullerenes have a notable propensity to cocrystallize with inorganic and organometallic species as well as with organic molecules. Continued development in this area is expected to yield fullerene-based compounds and materials that will have not only intriguing structures but utility where the remarkable physical properties of these newly created substances can be exploited.

ACKNOWLEDGMENTS

The author is pleased to express appreciation for all of the work of his collaborators in the fullerene field. Their names are given in the citations, but special recognition is given to Dr. Marilyn M. Olmstead for her numerous contributions. Research at the University of California–Davis on fullerenes has generously been supported by the National Science Foundation (currently grant CHE 9610507).

REFERENCES

1. A. L. Balch, Reactions of Fullerenes with Inorganic and Organometallic Compounds in *The Chemistry of Fullerenes*, R. Taylor (ed.), Advanced Series in Fullerenes, Vol. 4. World Scientific Publishing Co., Singapore, 1995, p. 220.

2. A. L. Balch and M. M. Olmstead, *Chem. Rev.* **1998**, *108*, 2123.

3. J. R. Bowser, *Adv. Organometal. Chem.* **1994**, *36*, 57.

4. A. H. H. Stephens and M. L. H. Green, *Adv. Inorg. Chem.* **1997**, *44*, 1.

5. W. Sliwa, *Trans. Met. Chem.* **1996**, *21*, 583.

6. M. J. Rosseinsky, *J. Mater. Chem.* **1995**, *5*, 1497.

7. T. Akasaka, in *Fullerenes:* Chemistry, Physics, and Technology, K. M. Kadish and R. Ruoff, (eds.). Wiley, New York, 1999, p. 395.

8. H. Shinohara, in *Fullerenes:* Chemistry, Physics, and Technology, K. M. Kadish and R. Ruoff, (eds.). Wiley, New York, 1999, p. 357.

9. F. T. Edelmann, *Angew. Chem. Int. Ed. Engl.* **1995**, *34*, 981. S. Nagase, K. Kobayashi, and T. Akasaka, *Bull. Chem. Soc. Jpn.* **1996**, *69*, 2131. T. Braun, *ACH—Models Chem.* **1995**, *132*, 245.

10. R. Taylor, *J. Chem. Soc. Perkins Trans. 2* **1993**, 813.

11. P. W. Fowler and D. E. Manolopoulos, *An Atlas of Fullerenes*. Clarendon Press, Oxford, 1995.

12. T. J. S. Dennis, T. Kai, T. Tomiyama, and H. Shinohara, *Chem. Commun.* **1998**, 619.

13. M. Sanders, H. A. Jiménez-Vázquez, R. J. Cross, W. E. Billups, C. Gesenberg, A. Gonzalez, W. Luo, R. C. Haddon, F. Diederich, and A. Herrmann, *J. Am. Chem. Soc.* **1995**, *117*, 9305.

14. A. Hirsch, I. Lamparth, and H. R. Karfunkel, *Angew. Chem. Int. Ed. Engl.* **1994**, *33*, 437.

15. J. M. Hawkins, *Acc. Chem. Res.* **1992**, *25*, 150.

16. J. M. Hawkins, A. Meyer, T. A. Lewis, S. Loren, and F. J. Hollander, *Science* **1991**, *252*, 312.

17. J. M. Hawkins, S. Loren, A. Meyer, and R. Nunlist, *J. Am. Chem. Soc.* **1991**, *113*, 7770.

18. J. M. Hawkins, A. Meyer, and M. A. Solow, *J. Am. Chem. Soc.* **1993**, *115*, 7499.

19. S. Zhang, T. L. Brown, Y. Du, and J. R. Shapley, *J. Am. Chem. Soc.* **1993**, *115*, 6705.

20. M. D. Westmeyer, T. B. Rauchfuss, and A. K. Verma, *Inorg. Chem.* **1996**, *35*, 7140.

21. W. Beck, H. Bentele, and S. Hüffer, *Chem. Ber.* **1995**, *128*, 1059.

22. P. J. Fagan, P. J. Krusic, D. H. Evans, S. A. Lerke, and E. Johnson, *J. Am. Chem. Soc.* **1992**, *114*, 9697.

23. T. Kusukawa and W. Ando, *J. Organometal. Chem.* **1998**, *561*, 109.

24. P. J. Fagan, J. C. Calabrese, and B. Malone, *Science* **1991**, *252*, 1160.

25. A. L. Balch, V. J. Catalano, and J. W. Lee, *Inorg. Chem.* **1991**, *30*, 3980.

26. R. C. Haddon, *Science* **1993**, *261*, 1545.

27. A. L. Balch, V. J. Catalano, J. W. Lee, and M. M. Olmstead, *J. Am. Chem. Soc.* **1992**, *114*, 5455.

28. V. J. Catalano and N. Parodi, *Inorg. Chem.* **1997**, *36*, 537.

29. A. L. Balch, J. W. Lee, B. C. Noll, and M. M. Olmstead, *J. Am. Chem. Soc.* **1992**, *114*, 10984.

30. A. L. Balch, J. W. Lee, B. C. Noll, and M. M. Olmstead, *Inorg. Chem.* **1994**, *33*, 5238.

31. M. Soimasuo, T. T. Pakkanen, M. Ahlgrén, and T. A. Pakkanen, *Polyhedron* **1998**, *17*, 2073.

32. M. Rasinkangas, T. T. Pakkanen, T. A. Pakkanen, M. Ahlgrén, and J. Rouvinen, *J. Am. Chem. Soc.* **1993**, *115*, 4901.

33. I. J. Mavunkal, Y. Chi, S.-M. Peng, and G.-H. Lee, *Organometallics* **1995**, *14*, 4454.

34. H.-F. Hsu and J. R. Shapley, *J. Am. Chem. Soc.* **1996**, *118*, 9192.

35. P. J. Fagan, J. C. Calabrese, and B. Malone, *J. Am. Chem. Soc.* **1991**, *113*, 9408.

36. F. Djojo, A. Herzog, I. Lampath, F. Hampel, and A. Hirsch, *Chem. Eur. J.* **1996**, *2*, 1537.

37. A. L. Balch, V. J. Catalano, J. W. Lee, M. M. Olmstead, and S. R. Parkin, *J. Am. Chem. Soc.* **1991**, *113*, 8953.

38. A. L. Balch, J. W. Lee, and M. M. Olmstead, *Angew. Chem. Int. Ed. Engl.* **1992**, *31*, 1356.

39. A. L. Balch, L. Hao, and M. M. Olmstead, *Angew. Chem. Int. Ed. Engl.* **1996**, *35*, 188.

40. H.-F. Hsu and J. R. Shapley, *J. Chem. Soc. Chem. Commun.* **1997**, 1125.

41. A. L. Balch, A. S. Ginwalla, B. C. Noll, and M. M. Olmstead, *J. Am. Chem. Soc.* **1994**, *116*, 2227.

42. A. L. Balch, D. A. Costa, J. W. Lee, B. C. Noll, and M. M. Olmstead, *Inorg. Chem.* **1994**, *33*, 2071.

43. A. L. Balch, D. A. Costa, B. C. Noll, and M. M. Olmstead, *J. Am. Chem. Soc.* **1995**, *117*, 8926.

44. G. B. M. Vaughan, P. A. Heiney, D. E. Cox, A. R. McGhie, D. R. Jones, R. M. Strongin, M. A. Cichy, and A. B. Smith III, *J. Chem. Phys.* **1992**, *168*, 185.

45. M. Sawamura, H. Iikura, and E. Nakamura, *J. Am. Chem. Soc.* **1996**, *118*, 12850.

46. M.-J. Arce, A. L. Viado, Y.-Z. An, S. I. Kahn, and Y. Rubin, *J. Am. Chem. Soc.* **1996**, *118*, 3775.

47. L. T. Scott, *Pure Appl. Chem.* **68**, 291 (1996). P. W. Rabideau and A. Sygula, *Acc. Chem. Res.* **1996**, *29*, 235. D. M. Forkey, S. Attar, B. C. Noll, R. Koerner, M. M. Olmstead, and A. L. Balch, *J. Am. Chem. Soc.* **1997**, *119*, 5766.

48. R. M. Shaltout, R. Sygula, A. Sygula, F. R. Fronczek, G. G. Stanley, and P. W. Rabideau, *J. Am. Chem. Soc.* **1998**, *120*, 835.

49. A. Thess, R. Lee, P. Nikolaev, H. Dai, P. Petit, J. Robert, C. Xu, Y. H. Lee, S. G. Kim, A. G. Rinzler, D. T. Colbert, G. E. Scuseria, D. Tománek, J. E. Fischer, and R. E. Smalley, *Science* **1996**, *273*, 483.

50. S. Attar, D. M. Forkey, M. M. Olmstead, and A. L. Balch, *Chem. Commun.* **1998**, 1255.

51. J. Chlistunoff, D. Cliffel, and A. J. Bard, Electrochemistry of Fullerenes in *Handbook of Organic Conductive Molecules and Polymers*. Vol. 1, *Charge-Transfer Salts, Fullerenes and Photoconductors*, H. S. Nalwa (ed.). Wiley, New York, 1997, p. 382.

52. L. Echegoyen, in *Fullerenes: Chemistry, Physics, and Technology*, K. M. Kadish and R. Ruoff, (eds.) Wiley, New York, 1999, p. 1.

53. R. C. Haddon, *Acc. Chem. Res.* **1992**, *25*, 127.

54. O. Gunnarsson, *Rev. Mod. Phys.* **1997**, *69*, 575.

55. V. Buntar and H. W. Weber, *Supercond. Sci. Technol.* **1996**, *9*, 599.

56. K. Tanigaki and K. Prassides, *J. Mater. Chem.* **1995**, *5*, 1515.

57. A. V. Nikolaev, K. Prassides, and K. H. Michel, *J. Chem. Phys.* **1998**, *108*, 4912.

58. P. W. Stephens, G. Bortel, G. Faigel, M. Tegze, A. Jánossy, S. Pekker, G. Oszianyl, and L. Forró, *Nature* **1994**, *370*, 636.

59. S. Pekker, A. Jánossy, L. Mihaly, O. Chauvet, M. Carrard, and L. Forró, *Science* **1994**, *265*, 1077.

60. D. Koller, M. C. Martin, P. W. Stevens, L. Mihaly, S. Pekker, A. Janossy, O. Chauvet, and L. Forro, *Appl. Phys. Lett.* **1995**, *66*, 1015.

61. F. Bommeli, L. Degiorgi, P. Wachter, Ö. Legeza, A. Jánossy, G. Oszlanyi, O. Chauvet, and L. Forro, *Phys. Rev. B* **1995**, *51*, 14794.

62. C. H. Choi and M. Kertesz, *Chem. Phys. Lett.* **1998**, *282*, 318.

63. G. Oszlányi, G. Bortel, G. Faigel, M. Tegze, L Gránásy, S. Pekker, P. W. Stephens, G. Bendele, R. Dinnebier, G. Mihály, A. Jánossy, O. Chauvet, and L. Forró, *Phys. Rev. B* **1995**, *51*, 12228.

64. Q. Zhu, D. E. Cox, and J. E. Fischer, *Phys. Rev. B* **1995**, *51*, 3966.

65. A. L. Balch, J. W. Lee, B. C. Noll, and M. M. Olmstead, in *Recent Advances in the Chemistry and Physics of Fullerenes and Related Materials*, R. S. Ruoff and K. M. Kadish (eds.) Electrochemical Society Proceedings Volume 94-24. The Electrochemical Society, Pennington, NJ, 1994, p. 1231.

66. J. Stinchcombe, A. Penicaud, P. Bhyrappa, P. D. W. Boyd, and C. A. Reed, *J. Am. Chem. Soc.* **1993**, *115*, 5212.

67. A. Pénicaud, A. Peréz-Benítez, R. Gleason, E. P. Muñoz, and R. Escudero, *J. Am. Chem. Soc.* **1993**, *115*, 10392.

68. V. V. Gritenko, O. A. Dyachenko, G. V. Shilov, N. G. Spitsyna, and E. B. Yagubskii, *Russ. Chem. Bull.* **1997**, *46*, 1878.

69. U. Bilow and M. Jansen, *J. Chem. Soc. Chem. Commun.* **1994**, 403.

70. W. C. Wan, X. Liu, G. M. Sweeney, and W. E. Broderick, *J. Am. Chem. Soc.* **1995**, *117*, 9580.

71. R. Tycko, G. Dabbagh, M. J. Rosseinsky, D. W. Murphy, R. M. Flemming, A. P. Ramirez, and J. C. Tully, *Science* **1991**, *253*, 884.

72. A. J. Fowkes, J. M. Fox, P. F. Henry, S. J. Heyes, and M. J. Rosseinsky, *J. Am. Chem. Soc.* **1997**, *119*, 10413.

73. P. Paul, Z. Xie, R. Bau, P. D. W. Boyd, and C. A. Reed, *J. Am. Chem. Soc.* **1994**, *116*, 4145.

74. P. D. W. Boyd, P. Bhyrappa, P. Paul, J. Stinchcombe, R. D. Bolskar, Y. Sun, and C. A. Reed, *J. Am. Chem. Soc.* **1995**, *117*, 2907.

75. K. Himmel and M. Jansen, *Inorg. Chem.* **1998**, *37*, 3437.

76. T. F. Fässler, A. Spiekermann, M. E. Spahr, and R. Nesper, *Angew. Chem. Int. Ed. Engl.* **1997**, *36*, 486.

77. A. L. Balch, D. A. Costa, and K. Winkler, *J. Am. Chem. Soc.* **1998**, *120*, 9614.

78. R. D. Bolskar, R. S. Mathur, and C. A. Reed, *J. Am. Chem. Soc.* **1996**, *118*, 13093.

79. M. F. Meidine, P. B. Hitchock, H. W. Kroto, R. Taylor, and D. M. R. Walton, *J. Chem. Soc. Chem. Commun.* **1992**, 1534.

80. A. L. Balch, J. W. Lee, B. C. Noll, and M. M. Olmstead, *J. Chem. Soc. Chem. Commun.* **1993**, 56.

81. H. B. Bürgi, R. Restori, D. Schwarzenbach, A. L. Balch, J. W. Lee, B. C. Noll, and M. M. Olmstead, *Chem. Mater.* **1994**, *6*, 1325.

82. J. D. Crane, P. B. Hitchcock, H. W. Kroto, R. Taylor, and D. R. M. Walton, *J. Chem. Soc. Chem. Commun.* **1992**, 1764.

83. M. M. Olmstead, L. Hao, and A. L. Balch, *J. Organometal. Chem.* **1999**, *578*, 85.

84. J. D. Crane and P. B. Hitchcock, *J. Chem. Soc. Dalton Trans.* **1993**, 2537.

85. (a) H. B. Bürgi, P. Venugopalan, D. Schwarzenbach, F. Diederich, and C. Thilgen, *Helv. Chim. Acta* **1993**, *76*, 2155. (b) G. Roth and P. Adelmann, *J. Phys. I Fr.* **1992**, *2*, 1541.

86. G. Roth and P. Adelmann, *Appl. Phys. A* **1993**, *56*, 169. G. Roth and P. Adelmann, *Mater. Lett.* **1993**, *16*, 357.

87. R. H. Michel, M. M. Kappes, P. Adelmann, and G. Roth, *Angew. Chem. Int. Ed. Engl.* **1994**, *33*, 1651.

88. R. E. Douthwaite, M. L. H. Green, S. J. Heyes, M. J. Rosseinsky, and J. F. C. Turner, *J. Chem. Soc. Chem. Commun.* **1994**, 1367.

89. F. Tast, N. Malinowski, M. Heinebrodt, I. M. L. Billas, and T. P. Martin, *J. Chem. Phys.* **1997**, *106*, 9373.

90. M. M. Olmstead, A. S. Ginwalla, B. C. Noll, D. S. Tinti, and A. L. Balch, *J. Am. Chem. Soc.* **1996**, *118*, 7737.

91. R. Fabiański and B. Kuchta, *Adv. Mater. Optics Electron.* **1996**, *6*, 297. S. Król, A. Łapiński, and A. Graja, *Adv. Mater. Optics Electron.* **1996**, *6*, 255. A. L. Balch, M. M. Olmstead, *Coord. Chem. Rev.* **1999**, *185–186*, 601.

92. M. M. Olmstead, D. A. Costa, K. Maitra, B. C. Noll, S. L. Phillips, P. M. van Calcar, A. L. Balch, *J. Am. Chem. Soc.* **1999**, *121*, 7090.

93. J. Xiao, M. R. Savina, G. B. Martin, A. H. Francis, and M. E. Meyerhoff, *J. Am. Chem. Soc.* **1994**, *116*, 9341.

94. Q. Zhu, D. E. Cox, J. E. Fischer, K. Knaiz, A. R. McGhie, and O. Zhou, *Nature* **1992**, *355*, 712.

95. M. M. Olmstead, K. Maitra, and A. L. Balch, *Angew. Chem. Int. Ed. Engl.* **1999**, *39*, 231.

96. H. Mauser, A. Hirsch, N. J. R. V. Hommes, and T. Clark, *J. Mol. Modeling* **1997**, *3*, 415.

97. M. Saunders, H. A. Jiménez-Vázquez, R. J. Cross, and R. J. Poreda, *Nature* **1994**, *367*, 256.

98. A. Khong, H. A. Jiménez-Vázquez, M. Saunders, R. J. Cross, J. Laskin, T. Peres, C. Lifshitz, R. Strongin, and A. B. Smith III, *J. Am. Chem. Soc.* **1998**, *120*, 6380.

99. E. Shabtai, A. Weitz, R. C. Haddon, R. E. Hoffman, M. Rabinovitz, A. Khong, R. J. Cross, M. Saunders, P.-C. Cheng, and L. T. Scott, *J. Am. Chem. Soc.* **1998**, *120*, 6389.

100. J. Laskin, T. Peres, C. Lifshitz, M. Saunders, R. J. Cross, and A. Khong, *Chem. Phys. Lett.* **1998**, *285*, 7.

101. B. Pietzak, M. Waiblinger, T. A. Murphy, A. Weidinger, M. Höhne, E. Dietel, and A. Hirsch, *Chem. Phys. Lett.* **1997**, *279*, 259.

102. H. Mauser, N. J. R. V. Hommes, T. Clark, A. Hirsch, B. Pietzak, A. Weidinger, and L. Dunsch, *Angew. Chem. Int. Ed. Engl.* **1997**, *36*, 2835.

103. C. Zybill, *Angew. Chem. Int. Ed. Engl.* **1972**, *31*, 173. T. Lange and T. P. Martin *Angew. Chem. Int. Ed. Engl.* **1972**, *31*, 172. M. C. Piqueras, R. Crespo, E. Orti, and F. Tomás, *Chem. Phys. Lett.* **1993**, *213*, 509.

104. I. Silaghi-Dumitrescu, I. Haiduc, and D. B. Sowerby, *Inorg. Chem.* **1993**, *32*, 3755.

105. (a) S. C. Sevov and J. D. Corbett, *Science* **1993**, *252*, 880. (b) S. C. Sevov and J. D. Corbett, *J. Solid State Chem.* **1996**, *123*, 344.

106. A. Müller, E. Krickemeyer, H. Bögge, M. Schmidtmann, and F. Peters, *Angew. Chem. Int. Ed. Engl.* **1998**, *37*, 3360.

107. B. C. Guo, K. P. Kerns, and A. W. Castleman, Jr., *Science* **1992**, *255*, 1411. B. C. Guo, S. Wei, J. Purnell, S. Buzza, and A. W. Castleman, Jr., *Science* **1992**, *256*, 515. S. Wei, B. C. Guo, J. Purnell, S. Buzza, and A. W. Castleman, Jr., *Science* **1992**, *256*, 818. B. C. Guo, K. P. Kerns, and A. W. Castleman, Jr., *J. Am. Chem. Soc.* **1993**, *115*, 7415. K. P. Kerns, B. C. Guo, H. T. Deng, and A. W. Castleman Jr., *J. Am. Chem. Soc.* **1995**, *117*, 4026. H. T. Deng, K. P. Kerns, and A. W. Castleman, Jr., *J. Am. Chem. Soc.* **1996**, *118*, 446.

108. (a) I. Dance, *J. Chem. Soc. Chem. Commun.* **1992**, 1779. (b) I. Dance, *J. Am. Chem. Soc.* **1996**, *118*, 6309.

109. D. E. Clemmer, J. M. Hunter, K. B. Shellmov, and M. F. Jarrold, *Nature* **1994**, *372*, 248.

110. K. B. Shelimov and M. F. Jarrold, *J. Am. Chem. Soc.* **1995**, *117*, 6404.

111. W. Branz, I. M. L. Billas, N. Malinowski, F. Tast, M. Heinebrodt, and T. P. Martin, *J. Chem. Phys.* **1998**, *109*, 3425.

112. J. M. Poblet, J. Muñoz, K. Winkler, M. Cancilla, A. Hayashi, C. B. Lebrilla, and A. L. Balch, *Chem. Commun.* **1999**, 493.

113. A. L. Balch, D. A. Costa, and K. Winkler, *J. Am. Chem. Soc.* **1998**, *120*, 9614.

CHAPTER 5

PHOTOPHYSICAL PROPERTIES OF PRISTINE FULLERENES, FUNCTIONALIZED FULLERENES, AND FULLERENE-CONTAINING DONOR-BRIDGE ACCEPTOR SYSTEMS

DIRK M. GULDI and PRASHANT V. KAMAT

5.1 INTRODUCTION

Since the initial discovery of fullerenes [1,2] chemists and physicists worldwide have studied solid state properties ranging from superconductivity and nanostructured devices to endohedral fullerene chemistry [3–5]. The three-dimensionality of C_{60}, arranging 60 individual carbon atoms in alternating pentagon and hexagon architectures, evoked a lively interest to relate their properties to conventional two-dimensional π-systems.

The three-dimensional structure leads to a substantial curvature of the carbon network. This leads to a strong pyramidalization of the individual carbon atoms [6]. In particular, an average σ-bond hybridization of $sp^{2.278}$ and a fractional s-character of the p-orbitals of 0.085 are reported (Fig. 5.1) [7]. The combination of these effects influences the fullerene's reactivity relative to planar aromatic hydrocarbons. Accordingly, a very reactive exterior surface but nearly inert interior part are found. While the exterior reacts with a large number of radicals with nearly diffusion-controlled kinetics (radical sponge) [8], the interior has recently been shown to stabilize atomic nitrogen and, therefore, has been compared with a Faraday cage [9].

The electronic configuration of fullerenes, namely, a fivefold degenerate highest occupied molecular orbital (HOMO; h_u) and a triply degenerate lowest unoccupied molecular orbital (LUMO; t_{1u}), are separated by a small energy gap of approximately 1.8 eV [10]. Redox chemistry studies have revealed the ability

Fullerenes: Chemistry, Physics, and Technology, Edited by Karl M. Kadish and Rodney S. Ruoff.
0471-290889-0 Copyright © 2000 John Wiley & Sons, Inc.

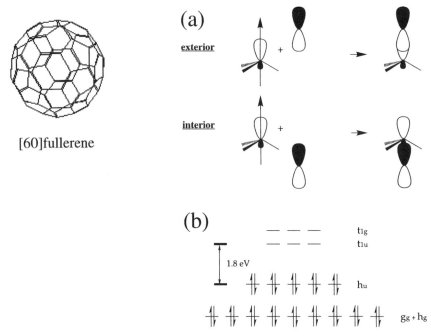

[60]fullerene

Figure 5.1 Illustration of the (a) pyramidalization of the individual carbon atoms and (b) p-orbital energy levels in C_{60}.

of fullerenes to accommodate up to six electrons in subsequent reversible reduction steps leading to the generation of the corresponding mono-, di-, tri-, tetra-, penta-, and hexa-anion [11–13]. Similarly, n-type doping of fullerenes with alkali metals, alkali-earth metals, and various organic electron donors leads also to the occupation of the (t_{1u}) LUMO [14].

Moreover, an exceptionally low reorganization energy of the fullerene core in electron transfer reactions is directly associated with a high degree of charge and energy delocalization within the carbon framework [15]. Accordingly, charge separation in the sense of a forward electron transfer is strongly favored. At the same time, charge recombination (i.e., back electron transfer) experiences a noticeable slowdown relative to conventionally used electron acceptors, such as two-dimensional planar molecules (quinones, etc.).

Under anaerobic conditions the singly reduced π-radical anion, $C_{60}^{\bullet-}$, is stable, but upon reaction with molecular oxygen the regeneration of the singlet ground state and formation of $O_2^{\bullet-}$ are noted [16]. The optical absorption spectrum of the π-radical anion, showing a narrow band in the near infrared (IR) at 1080 nm (Fig. 5.2), was first observed in low-temperature matrix using γ-irradiation [17]. This band is a diagnostic probe that helped in assisting the identification of $C_{60}^{\bullet-}$ in more complex inter- and intramolecular transfer reactions. The ultraviolet/visible (UV/VIS) region can similarly be employed as a probe for electron transfer (ET) reactions. For example, bleaching of the sin-

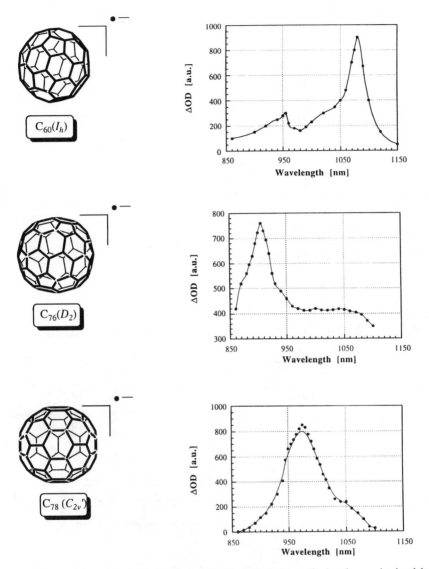

Figure 5.2 NIR spectrum of the $C_{60}{}^{\bullet-}$, $C_{76}{}^{\bullet-}$, and $C_{78}{}^{\bullet-}$ radical anions, obtained by pulse radiolytic reduction in a toluene/2-propanol/acetone (8:1:1) solvent mixture.

glet ground state absorption around 265 nm and 330 nm point to the π-radical anion.

Oxidation, on the other hand, has been found to be limited to an irreversible one-electron process resulting in the formation of the highly unstable $C_{60}{}^{\bullet+}$ π-radical cation. The oxidation potential of C_{60} is 2.15 V vs. NHE, as, for example, measured in benzonitrile, and relates well to the fully occupied and, in turn, resonance stabilized HOMO [11]. Similar to the one-electron reduced form, the $C_{60}^{\cdot+}$ π-radical cation displays a characteristic band in the near IR around 980 nm [17]. In summary, $C_{60}^{\cdot-}$ and $C_{60}^{\cdot+}$ electronic absorption features, substantially different from those of neutral C_{60}, allow an accurate analysis of inter- and intramolecular ET dynamics in C_{60}-containing systems (see below).

This chapter summarizes the intriguing properties of fullerenes as a photosensitizer and a multifunctional electron storage moiety. The chemical modification of fullerenes gives the opportunity to combine the properties of the fullerenes with those of other interesting materials such as electro- or photoactive species, which are discussed in the current work. Nevertheless, the reader is directed to a series of excellent review articles, which have appeared in recent years [18–21].

5.2 PRISTINE [60]FULLERENE

5.2.1 Singlet and Triplet Excited State Properties

Laser flash photolysis is a convenient method to characterize the singlet and triplet excited state behavior of fullerenes (Fig. 5.3) [21,22]. Various spectro-

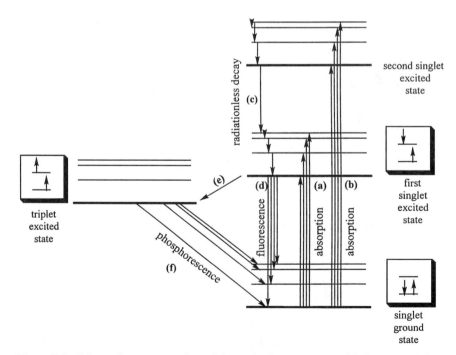

Figure 5.3 Schematic representation of the excitation processes of fullerenes and functionalized fullerene derivatives.

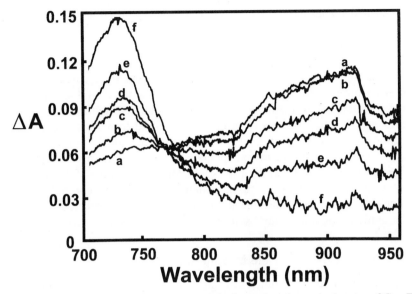

Figure 5.4 Difference absorption spectra of singlet and triplet excited states of C_{60}. The spectra were recorded following the 355 nm laser pulse (pulse width 18 ps) excitation of C_{60} at time intervals: (a) 0, (b) 100, (c) 300, (d) 500, (e) 900, and (f) 1900 ps. Reprinted with permission from Ref. 74. Copyright © 1992 American Chemical Society.

scopic techniques such as picosecond and nanosecond laser flash photolysis, Raman spectroscopy, EPR, and optically detected magnetic resonance (ODMR) have been employed extensively to characterize the excited states and the photochemical reactivity of C_{60} [23–45] and larger fullerenes [46,47]. Figure 5.4 displays the absorption changes recorded following picosecond laser pulse excitation of C_{60} in deaerated toluene solution at different times. The transient absorption spectrum recorded immediately after the laser pulse corresponds to the singlet excited state. The singlet excited C_{60} is characterized by a broad absorption maximum around 920 nm. On the other hand, the singlet excited states of larger fullerenes absorb rather weakly in the 450 to 950 nm region (see Table 5.1). In general, the decay of the singlet excited fullerenes follows clean first-order kinetics ($\tau = 1.25$ ns). A growth of absorption in the 740 nm region parallels the decay of the singlet excited state. This is attributed to an inter-system crossing leading to the formation of the energetically lower-lying triplet excited state:

$$C_{60} \xrightarrow{h\nu} (^{1}C_{60}{}^{*}) \xrightarrow{ISC} (^{3}C_{60}{}^{*}) \tag{5.1}$$

The intersystem crossing is expected to be the major deactivation step for the singlet excited state and to exhibit significant amounts of triplet formation. The rate constant for the intersystem crossing process, as measured from the decay of the excited singlet state of C_{60}, is 8.0×10^8 s^{-1}. Larger fullerenes also exhibit similar intersystem crossing behavior as that of pristine C_{60}. With the exception of C_{70}, all larger fullerenes investigated so far exhibit very low quantum yields for the triplet excited state generation.

Nanosecond laser excitation of deaerated solutions of C_{76} resulted in transient absorption changes that display broad maxima at 555 and 650 nm [46,47]. The absorption band in general resembled the one detected in the picosecond experiments. Kinetic analysis of both bleaching and formation traces of time-absorption profiles (C_{76} and C_{78}) revealed a recovery of the transient absorption to the original baseline in an exponential manner. The triplet excited states of C_{76} and C_{78} decayed with half-lives typically in the range of 3.2 and 3.8 µs, respectively. These values are smaller than the previously reported values for pristine C_{60} in methylcyclohexane and toluene and are influenced by the triplet–triplet annihilation and ground state quenching processes.

5.2.2 Triplet Quantum Yield Measurements

Triplet quantum yields of fullerenes can be determined from the triplet–triplet energy transfer method. By employing the triplet excited state fullerene as the donor species and a known reference compound, such as a crown ether functionalized squaraine dye as the acceptor species, one can determine the triplet quantum yield of various fullerenes [48]. For C_{60} and C_{70} one obtains relatively high triplet quantum yield ($\Phi_T > 0.9$) [27] as compared to the larger fullerenes, for example, Φ_T of C_{76} and C_{78} were 0.05 and 0.12, respectively [47,49,50]. These results suggest that nonradiative decay processes are dominant in the deactivation of the singlet excited states of larger fullerenes.

5.2.3 Pulse Radiolytic Generation of Excited Triplet States

Pulse radiolysis is also a convenient technique to generate and characterize triplet excited states of fullerenes, in nonaqueous media such as toluene or benzene [39,51–54]. Radiolysis of a nitrogen-saturated toluene solution containing biphenyl yields long-lived excited triplets of biphenyl (^{3}BP*). In the presence of fullerene it is possible to induce intermolecular triplet–triplet energy transfer from the ^{3}BP* ($E_T = 64.5$ kcal/mol) to the lower-lying triplets ($E_T = 35.34$ kcal/mol) of fullerenes. This route provides an elegant method for forming triplet excited states of the fullerenes bypassing the initial singlet excited state (e.g., see Refs. 39 and 47). The quenching rate constants of the energy transfer can be determined from the decay kinetics of the transient absorption decay of ^{3}BP* at 360 nm on the growth in fullerene triplet absorption, following pulse irradiation. The triplet–triplet energy transfer rate constants summarized in Table 5.1a indicate that the energy transfer occurs with nearly diffusion-controlled rate. The excited triplet of fullerenes decayed via a dose-dependent first-order kinetics, on a time scale of 100 µs, to regenerate the ground state.

TABLE 5.1a   Photophysical Properties of C_{60}, C_{70}, and a Monofunctionalized Fullerene Derivative, $C_{60}C(COOEt)_2$ [21,22,38,93,94,158]

Property	C_{60} 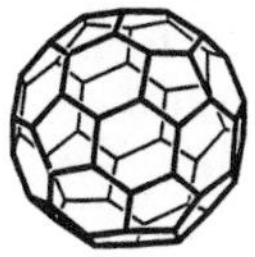	C_{70} 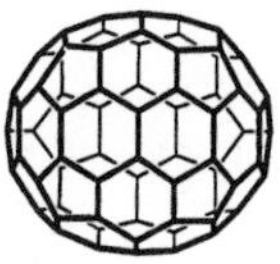	$C_{60}C(COOEt)_2$ 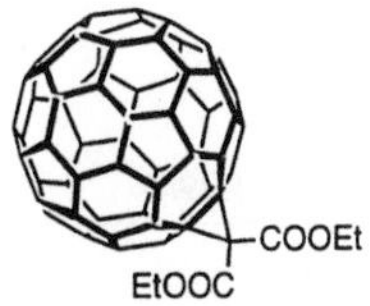
E (singlet)	1.99 eV	1.90 eV	1.79 eV
E (triplet)	1.57 eV	1.60 eV	1.502 eV
λ_{max} (singlet)	920 nm	660 nm	920 nm
λ_{max} (triplet)	747 nm	400 nm	720 nm
ε (triplet)	20,000 $M^{-1}cm^{-1}$	14,000 $M^{-1}cm^{-1}$	14,000 $M^{-1}cm^{-1}$
τ (singlet)	1.3 ns	0.7 ns	2.23 ns
τ (triplet)	135 μs	11.8 ms	40 μs
$E_{1/2}$ ($^{1*}C_{60}/C_{60}{}^{\bullet-}$)	1.44 V vs SCE	1.39 vs SCE	1.16 vs SCE
$E_{1/2}$ ($^{3*}C_{60}/C_{60}{}^{\bullet-}$)	1.01 vs SCE	1.09 vs SCE	0.86 vs SCE
Φ (fluorescence)	1.0×10^{-4}	3.7×10^{-4}	6.0×10^{-4}
Φ (triplet)	0.96	0.9	0.96
k_q (oxygen)	$1.6 \times 10^9 \ M^{-1}s^{-1}$	$1.9 \times 10^9 \ M^{-1}s^{-1}$	$1.8 \times 10^9 \ M^{-1}s^{-1}$
k_q (DABCO)	$2.5 \times 10^9 \ M^{-1}s^{-1}$		$7.7 \times 10^7 \ M^{-1}s^{-1}$
k_{ET} (biphenyl)	$1.7 \times 10^{10} \ M^{-1}s^{-1}$	$2.0 \times 10^{10} \ M^{-1}s^{-1}$	$1.8 \times 10^{10} \ M^{-1}s^{-1}$

TABLE 5.1b   Deactivation of Triplet Excited C_{60} in the Presence of Quenchers

Quencher	Bimolecular Quenching Rate Constant $10^9 \ M^{-1}s^{-1}$	Reference
$^3C_{60}{}^*$	5.4	35
C_{60}	0.015	35,39
C_{70}	2.4	35
C_{84}	4.7	49
O_2	1.6	95

5.2.4 Deactivation of Excited States

The deactivation processes that commonly determine the fate of excited fullerenes are summarized in Figure 5.5. While intersystem crossing is a major pathway for the deactivation of the singlet excited state, the triplet excited states of fullerenes are susceptible to a variety of deactivation processes. These include ground state quenching [39], triplet–triplet annihilation [23], and quenching by molecular oxygen [55].

These deactivation processes directly compete with the intrinsic decay of the triplet excited state to the singlet ground state via a spin-forbidden intersystem

TABLE 5.1c Comparison of Kinetic and Thermodynamic Properties of Fullerenes [39,46,49]

Fullerene	E°_{red} (V vs Fc/Fc$^+$)	Rate Constant (k) for T-T Energy Transfer from ^{3}(BP)* (10^{10} M^{-1}s^{-1})	Rate Constant (k) for Reduction with (CH$_3$)$_2^{\cdot}$COH (10^{10} M^{-1}s^{-1})
C$_{60}$	-1.06	1.7	0.85
C$_{70}$	-1.02	2.0	0.8
C$_{76}$(D_2)	-0.83	1.1	1.3
C$_{78}$($C_{2v'}$)	-0.77	1.3	1.6
C$_{84}$	-0.67	0.4	

crossing and, thus, control the lifetime of the excited state. On the basis of gas phase studies it has been possible to establish a simple exponential dependence of the ^{3}C$_{60}^*$ lifetime on the internal vibrational energy content [44]. The solvent-dependent spectral sensitivity of triplet excited C$_{60}$ and C$_{70}$ arises from environmentally induced quenching of the electronic angular momentum. Resonance enhancement of both totally and non-totally symmetric modes indicates that extensive intersystem mixing is an important feature of the spectroscopy of fullerenes.

A number of researchers have obtained bimolecular rate constants for the self-quenching and O$_2$ quenching of triplet excited fullerenes (see Tables 5.1a–5.1c). Rapid intermolecular energy transfers between isoenergetic triplet states dominate at high excited state concentrations while ground state quenching and

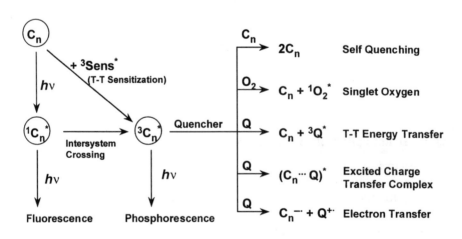

Excited State Processes of Fullerenes (n = 60, 70, 76, 78, 84 etc.)

Figure 5.5 Excited state processes of fullerenes.

oxygen quenching dominate at higher ground state concentrations. Interestingly, efficient quenching of triplet excited C_{60} is also seen with other fullerenes. The bimolecular rate constant for the quenching of $^3C_{60}^*$ by a ground state fullerene molecule increases with increasing number of carbon atoms in the unexcited fullerene molecule. This increase in the quenching rate constant reflects the increasing energy gap between the donor and acceptor triplets; that is, larger fullerenes have lower triplet energies. A decreasing pattern in the energy of lowest unoccupied molecular orbitals of the fullerenes has also been observed in the electrochemical reduction of larger fullerenes.

5.2.5 Excited State Behavior in Polymer Films

Photoactive polymers have important applications in photoresists, xerography, photocuring of paints and resins, and solar energy conversion systems [56–60]. One convenient method of developing such photoactive polymers is to incorporate photoactive guest molecules in a polymer film via electrostatic or hydrophobic interactions. The type and degree of interaction with the polymer influences the excited state properties of the photoactive guest molecule. The polymeric environment is also useful for effecting and controlling photochemical processes more efficiently than can be accomplished in homogeneous solutions.

Photochemical studies of fullerenes in polymer films have been investigated by a number of researchers [61–70]. Enhanced photoconductivity in C_{60}-doped polymers has been attributed to the formation of excited charge-transfer products [58]. The photoinduced charge transfer between excited C_{60} and a conducting polymer such as poly(3-octylthiophene) or poly(vinyl carbazole) has been investigated by (UV) laser flash photolysis [66,69,71–73]. Electron-transfer products have also been identified in UV-excited polyvinyl carbazole solutions containing C_{60}. The fast recombination process in the photoexcited polymers is a limiting factor in extending the lifetime of charge-transfer products. On the other hand, nonreactive polymers such as polymethylmethacrylate [59,61,74] or polystyrene [60,62] have been shown to stabilize the triplet excited states and facilitate photochemical electron transfer between C_{60} (or C_{70}) and an electron donor. Efforts have also been directed toward the development of photoactive fullerene polymers by copolymerizing with C_{60} [75,76]. Interaction of C_{60} with the functional groups of a polymer can be a crucial factor in controlling the excited state charge transfer.

Comparison of the excited state behavior of C_{60} in a neutral polystyrene (PS) and reactive poly(9-vinyl carbazole) (PVCz) polymeric environment provides a convenient way to study how a polymer environment influences the excited state properties of doped fullerenes. While PS exhibits little absorption in the 300–350 nm region, PVCz shows significant absorption below 360 nm. C_{60} incorporated in the PS film is found to retain all the characteristic absorption bands of C_{60} with a sharp absorption peak at 336 nm and a broad absorption in the 500–600 nm region. In the case of PVCz films the absorption band in the

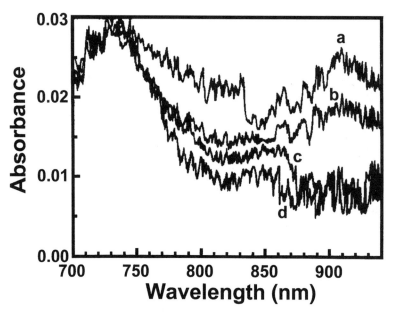

Figure 5.6 Time-resolved difference absorption spectra of singlet and triplet excited states of C_{60}-doped PVCz film ($C_{60} = 100$ nmol; C_{60} polymer ratio was 1:1.6). The spectra were recorded following the 355 nm laser pulse excitation at time intervals: (a) 0, (b) 400, (c) 1150, and (d) 4900 ps. Reprinted with permission from Ref. 68. Copyright © 1995 American Chemical Society.

visible is rather broad (400–650 nm). The absorption in the visible is significantly increased in C_{60}-doped PVCz films. These changes in the absorption characteristics indicate charge-transfer interaction between C_{60} and PVCz. Similar charge-transfer complex bands can also be observed upon doping a nonreactive polymer host, for example, polymethylmethacrylate, with electron donors such as dimethylaniline together with C_{60}.

The absorption characteristics of $^1C_{60}{}^*$ and $^3C_{60}{}^*$ in polystyrene or polymethylmethacrylate films are similar to those observed in nonpolar solvents. The isosbestic point observed at 790 nm suggests that there are only two transient states, which contribute to the absorption in the 700–960 nm region. Furthermore, the decay of $^1C_{60}{}^*$ at 910 nm closely matches the growth of $^3C_{60}{}^*$ at 740 nm with a $^1C_{60}{}^*$ lifetime of 1.2 ns. On the other hand, a triplet excited charge-transfer complex between C_{60} and carbazole and/or the electron-transfer products formed following the laser excitation contribute to the absorption in the 700–750 nm region (Fig. 5.6). The yields of both singlet and triplet excited states in PVCz films were also lower than in polystyrene film. The possible photochemical reaction, which leads to the charge separation in the C_{60}-doped PVCz films, are summarized in Scheme 1.

$$C_{60} + PVCz \Leftrightarrow [C_{60} : PVCz]$$

$$\downarrow hv \qquad\qquad \downarrow hv$$

$${}^3C_{60}{}^* + PVCz \rightarrow {}^1C_{60}{}^* + PVCz \qquad {}^1[C_{60} : PVCz]^* \rightarrow {}^3[C_{60} : PVCz]^*$$

$$[C_{60}{}^{\cdot-} : PVCz^+]$$

$$\downarrow$$

$$[C_{60} + PVCz]$$

Scheme 1

Both the complexed and uncomplexed forms of fullerenes are capable of undergoing photoinduced electron-transfer processes in the PVCz film. Because of the fast recombination within the complex, it was not possible to observe long-lived products in the C_{60}–PVCz system. Similar fast recombination of charge-transfer products has also been noted in C_{60}–amine complexes [24,77,78] and C_{60}-doped poly(3-octylthiophene). However, in the quenching experiment of ${}^3C_{60}{}^*$ with dimethylaniline in a polymethylmethacrylate film, it was possible to characterize electron-transfer products. It was concluded in this study that uncomplexed C_{60} was responsible for the net electron transfer and that the polymer matrix stabilized the electron-transfer products. Formation of fullerene-based macromolecules has also been observed in the photochemical reaction with amines [79].

5.3 INTERMOLECULAR ENERGY AND ELECTRON-TRANSFER REACTIONS

5.3.1 Energy Transfer from Donor Moieties

Excited state energy transfer from three different pyrazine derivatives was probed by emission and transient absorption spectroscopy (Fig. 5.7). Once photo-excited, these triplet excited pyrazines undergo rapid intermolecular energy transfer to a monofunctionalized pyrrolidinofullerene with bimolecular rate constants ranging from 1.1×10^{10} to 3.64×10^9 $M^{-1}s^{-1}$ [80]. The fullerene triplet excited state is the final product of these bimolecular energy-transfer reactions. Only the latter value is much lower than those observed for energy-transfer reactions from photoexcited biphenyl [39,51–53]. This difference is related to the driving forces $(-\Delta G)$ of the respective energy-transfer reactions. In contrast, the rate constants for the earlier two pyrazine models differ only slightly from that reported for biphenyl. Considering a diffusion-controlled rate of 8.8×10^9 $M^{-1}s^{-1}$ in methylcyclohexane, it is suggested that intermolecular energy transfer between the fullerene model and these two pyrazine derivatives takes place

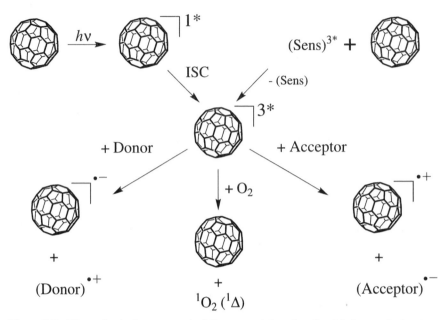

Figure 5.7 Photophysical processes in fullerene and functionalized fullerene derivatives.

with dynamics that are clearly governed by diffusion. More importantly, the driving forces for these reactions are highly exothermic and thus support bimolecular kinetics that are either in an "inverted" region or within diffusion-controlled limits.

5.3.2 Energy Transfer to Acceptor Moieties

Another remarkable property of fullerenes is their ability to generate singlet oxygen, $^1O_2(^1\Delta)$, via bimolecular energy transfer between triplet excited fullerenes and molecular oxygen, $^3O_2(^3\Sigma)$ [21,81]. The reactivities of the triplet excited states of fullerenes with molecular oxygen are generally probed by monitoring the fate of the triplet–triplet absorption as a function of oxygen concentration or emission from the $^1O_2(^1\Delta)$ state with a peak at 1272 nm.

It should be noted that, in thin films, singlet oxygen is potent since it can also react with the fullerene core [82,83]. During the photochemical reaction with O_2, C_{60} acts first as a generator of molecular oxygen in the $^1\Delta$ singlet electronic state, which subsequently reacts with C_{60} itself via addition to the reactive surface [82].

5.3.3 Electron Transfer at a Semiconductor Interface

Semiconductor clusters such as ZnO or TiO_2 are excellent choices to carry out reduction in a controlled way since the energy of their conduction bands

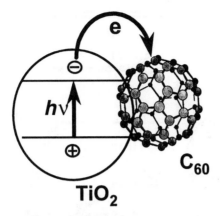

Figure 5.8 Schematic diagram illustrating the reduction of C_{60} using UV-irradiated TiO_2 nanoparticles.

($E_{CB} = -0.5V$ versus NHE) thermodynamically favors only one-electron reduction ($E_{1/2}(C_{60}/C_{60}^{\bullet-}) = -0.2$ V versus NHE) and not higher reductions (Fig. 5.8).

The reduction of C_{60} was first demonstrated by carrying out laser flash photolysis experiments in which the ZnO colloids were excited with a nanosecond nitrogen laser [84]. Mechanistic studies have been carried out to further elucidate photoelectrochemical reduction of C_{60} [85,86].

Along with the $^3C_{60}{}^*$ formation a long-lived transient is also formed in the presence of TiO_2. This long-lived transient, which is formed within the laser pulse duration, is attributed to $C_{60}^{\bullet-}$ (reaction 5.2).

$$TiO_2(e_{CB}) + C_{60} \rightarrow TiO_2 + C_{60}^{\bullet-} \tag{5.2}$$

The time-resolved transient absorption spectra recorded following the 337 nm laser pulse (pulse width 6 ns) excitation of TiO_2 colloids in the presence of C_{60} are shown in Figure 5.9. The transient absorption spectrum recorded immediately after laser pulse excitation shows maxima at 360, 400, and 740 nm. These spectral features are characteristic of $^3C_{60}{}^*$ and arise as a result of direct excitation of C_{60}. On the other hand, the transient absorption spectrum recorded 100 μs after laser pulse excitation (spectrum (b) in Fig. 5.9), with absorption maxima at 350, 410, and ~500 nm, is distinctively different from spectrum (a), but closely matches the difference absorption spectrum of the C_{60} π-radical anion. A maximum quantum yield of 0.24 for the reduction of C_{60} was reported in these studies. Steady-state UV photolysis of TiO_2 suspension can also be employed to generate stable monoanions of fullerenes. By using a similar approach, Stasko et al. [86] have investigated the EPR properties of the C_{60} anion.

Semiconductor supports, such as TiO_2, have been shown to participate in the surface photochemical processes, resulting in the oxidation of adsorbed sub-

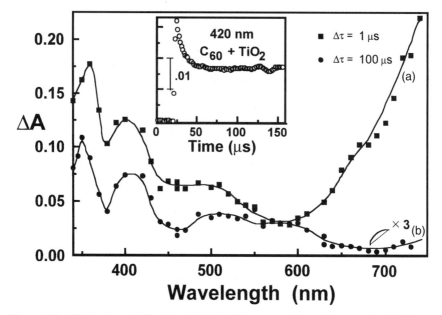

Figure 5.9 Reduction of C_{60} in colloidal TiO_2 suspension. Time-resolved difference absorption spectra recorded (a) immediately (1 μs) and (b) 100 μs following the 337 nm laser pulse (pulse width 6 ns ~2 mJ) excitation of a degassed sample of 0.05 mM C_{60} and 0.04 M TiO_2 in 50:50 ethanol/benzene. Reprinted with permission from Ref. 85. Copyright © 1994 American Chemical Society.

strates (Fig. 5.10) [87]. Anatase TiO_2 is a large bandgap semiconductor ($E_g = 3.2$ eV) that requires UV light to initiate bandgap excitation. By exciting the adsorbed substrate with visible light we can therefore avoid direct excitation of the semiconductor support and selectively probe the excited behavior of molecules of interest. This phenomenon, commonly referred to as *dye sensitization*, is beneficially used to develop photochemical solar cells [87].

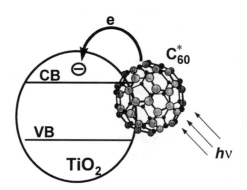

Figure 5.10 Photoinduced charge transfer between excited C_{60} and TiO_2 nanoparticle.

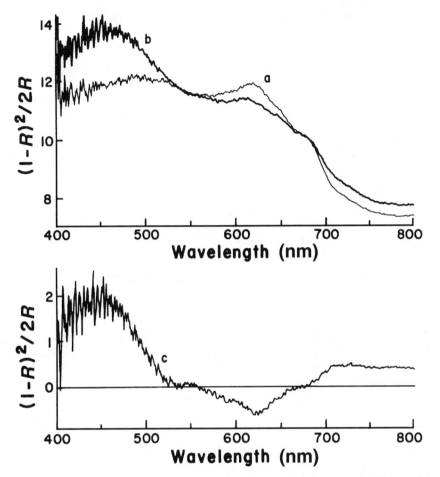

Figure 5.11 The diffuse reflectance absorption spectra of C_{60}-coated TiO_2 particles (0.09 mmol/g support) recorded (a) before and (b) after the photolysis with 532 nm laser for 3 min. Spectrum (c) is the difference between spectra (a) and (b). Reprinted with permission from Kamat et al., *J. Phys. Chem. B* **101**, 4422 (1997). Copyright © 1997 American Chemical Society.

The absorption changes observed following the 532 nm laser excitation of a C_{60}-coated TiO_2 sample is shown in Figure 5.11. The decreased absorption in the 620 nm region and an increased absorption in the 400 nm region indicate irreversible chemical changes occurring with C_{60} on the TiO_2 surface. Diffuse reflectance laser flash photolysis of C_{60}-coated TiO_2 samples indicates formation of a transient that is insensitive to oxygen. At low surface coverage, the transient absorption spectra of both degassed and air equilibrated samples of C_{60}/TiO_2 are similar. Some irreversible changes were also seen on an Al_2O_3 surface, but this fraction was small compared to the C_{60}/TiO_2 sample. The higher yields of the oxidation product observed on the TiO_2 surface are attributed to the greater reactivity of the semiconducting support material. The linear dependence of the transient formed on the TiO_2 surface on the square of the laser dose shows that the formation of this oxidation product is a biphotonic process.

The photoejection of electrons from excited C_{60} into TiO_2 particles is considered to be the primary step responsible for photooxidation.

$$C_{60} + h\nu \rightarrow {}^1C_{60}{}^* \rightarrow {}^3C_{60}{}^* \qquad (5.3)$$

$$(C_{60}/TiO_2) + nh\nu \rightarrow C_{60}{}^{\cdot +} + TiO_2(e) \qquad (5.4)$$

In the presence of oxygen the back electron transfer between the injected electron and $C_{60}{}^{\cdot +}$ is suppressed as the oxygen scavenges electrons.

$$O_2 + TiO_2(e) \rightarrow O_2{}^{\cdot -} + TiO_2 \qquad (5.5)$$

Since the oxidation potential of ${}^1C_{60}{}^*$ is 0.1 V versus NHE ($E_s = 1.5$ eV) we would not expect C_{60} to directly inject electrons from its singlet excited state into TiO_2 particles. Therefore, more than a single photon is necessary to induce the photoinjection process. The photoejection of electrons into TiO_2 particles is likely to be followed by the interaction of $C_{60}{}^{\cdot +}$ π-radical cation with either the lattice oxygen (of TiO_2) or chemisorbed oxygen that results in the formation of an epoxide type species (reaction 5.6):

$$C_{60}{}^{\cdot +} + TiO_2 \rightarrow (C_{60}O \cdots TiO) \qquad (5.6)$$

The difficulty in extracting the oxidation product by conventional methods suggests a strong binding of the epoxide to the TiO_2 surface. These observations are similar to the oxidation of fullerenes to their epoxides with UV irradiation in solutions [88–90]. For example, photooxidation of C_{60} in benzene solution has been carried out by irradiating oxygenated solutions with UV light and the photoproduct, fullerene epoxide ($C_{60}O$), has been isolated. Electron paramagnetic resonance (EPR) signals corresponding to the $C_{60}-O_n$ system have also been observed by Pace and co-workers [91]. The formation of EPR active species on oxides such as TiO_2 was recently observed by Anpo and co-workers [92]. These researchers found C-centered radicals to be the likely precursor intermediates for the epoxide formation. The distinct stability of these radicals is attributed to the large π-system of fullerenes. The carbon-centered radicals generated via a biphotonic process act as precursors for the epoxide formation on the TiO_2 surface.

5.3.4 Electron Transfer from Sacrificial Electron Donors

As pointed out earlier, a bimolecular transfer reaction is limited to the triplet excited state and is thought to lead to the formation of an ion radical pair (one-electron reduced $(C_{60}^{\cdot-})$ and the oxidized form of the applied quencher $(Q^{\cdot+})$. The diagnostic peak of the fullerene π-radical anion in the near IR is a sensitive marker whose detection points to the generation of a charge-separated radical pair, for example $\{C_{60}^{\cdot-}/Q^{\cdot+}\}$. A possible mechanism involves the sequence of a photoinduced electron transfer and diffusional dissociation of the contact pair. The dissociation depends on the ability of the environment to retard the exergonic back electron transfer. Stabilization of the radical pair can be achieved, for example, via strong dipole–dipole interactions and fine tuning the redox potential of the reduced electron acceptor $(C_{60}^{\cdot-})$ and oxidized electron donor $(Q^{\cdot+})$.

The low reduction potential of $^3C_{60}^*$ [93] and $^3C_{70}^*$ [94] enables reductive quenching reactions with a number of suitable electron donors (D) (Tables 5.2 and 5.3). Since primary and secondary amines are known to undergo addition reaction with the highly reactive fullerene core, bimolecular quenching is pri-

TABLE 5.2   Rate Constant for the Reductive Quenching of C_{60}
[21,42,78,79,95,97,99–110]

Quenching Donor	Toluene, $k_q \times 10^9$ M^{-1}s^{-1}	Benzonitrile, $k_q \times 10^9$ M^{-1}s^{-1}
Tetramethyphenyldiamine		5.2
N,N-diethylaniline		5.1
N,N-dimethylaniline		2.6
Diphenylamine		0.87
Triphenylamine		0.89
Tripropylamine		0.94
Triethylamine	2.9	0.96
Aniline		0.22
Dibutylamine		0.0072
Dipropylamine		0.0065
Diazabicyclooctane	2.5	0.18[a]
Polysilane		0.21[b]
Ferrocene	3.3	
Tetrathiafulvalene	9.9	5.0
BEDT-tetrathiafulvalene	5.9	3.8
β-Carotene	7.8	5.3
Tetramethylbenzidine	6.7	1.3
Zn phthalocyanine	2.1	1.6
H$_2$ phthalocyanine	2.8	3.0
Quaterthiophene	0.31	
TEMPO	3.3[c]	
HTEMPO	1.9[c]	
DTBN	3.0[c]	

[a] n-Propanol.
[b] CH$_3$CN, BZCN.
[c] CH$_2$Cl$_2$.

TABLE 5.3 Rate Constant for the Reductive Quenching of C_{70} [94]

Quenching Donor	Oxidation Potential (V vs SCE)	Benzonitrile, $k_q \times 10^9$ M^{-1}s^{-1}
Tetramethyphenyldiamine	0.14	5.1
N,N-diethylaniline	0.64	7.1
N,N-dimethylaniline	0.71	4.8
Diphenylamine	0.84	3.5
Triphenylamine	0.85	1.4
Triethylamine	0.96	0.95
Aniline	0.98	0.15
Dibutylamine	1.17	0.10
Dipropylamine	1.23	0.089
Diethylamine	1.31	0.11
Dibenzylamine	1.36	0.019
Pyrene	1.36	0.00012

marily investigated with tertiary amines, including DABCO [95,96]. Neither this bicyclic amine nor its oxidized from, DABCO$^{+\cdot}$, adds to the ground state C_{60} and thus they are considered to be an excellent choices for inducing intermolecular electron-transfer reactions. Also, ferrocene (FC) [97,98], tetrathiafulvalene (TTF) [99], bis(ethylenedithio)tetrathiafulvalene (BEDT-TTF) [99], 3,3',5,5'-tetramethybenzidine (NTMB) [100,101], stable nitroxide radicals [102], polysilane [103], oligothiophene [104,105], and quaterthiophene moieties [104,105] are all excellent electron donors. In general, in polar solutions, including benzonitrile, efficient charge separation takes place, which can conveniently be confirmed through formation of the characteristic $(C_{60}^{\cdot-})$ radical anion absorption at 1080 nm. Experiments in the latter give rise to a bimolecular rate constant for the charge recombination on the order of 10^{10} M^{-1}s^{-1}. On the other hand, the insufficient solvent polarity of benzene, and similarly toluene, retards the stabilization. Accordingly, a simple collisional deactivation route is proposed.

Since fullerenes are mildly electrophilic molecules, they were also found to form charge-transfer complexes with aromatic amines such as N,N-dimethylaniline (DMA) [42,79,106], N,N-diethylaniline (DEA) [107], diphenylamines (DPA) [78], triphenylamine (TPA) [78], triethylamine (TEA) [79,106], N,N-dimethyl-1-naphthylamine [108], NADH (1-benzyl-1,4-dihydronictiamide), and NAD analog [109] in toluene solutions. These complexes are generally very weak, with formation constants on the order of 0.1 or less. Therefore, high amine concentrations are required to study the characteristics and properties of these fullerene-containing charge-transfer complexes. At low amine concentrations the fullerene excited states are effectively quenched through electron-transfer interactions. In aliphatic hydrocarbons, the transient formation of an exciplexes governs the quenching of the monomer excited states. In this context, it should be mentioned that the exhaustive illumination of a fullerene solution containing TEA, trimethylamine (TMA), and DMA triggers the photoreduction of the fullerene core and gives rise to the formation of pyrrolidine adducts [110]. A proposed mechanism therefore suggests a sequential intermolecular electron transfer, followed by a deprotonation yielding an α-aminoalkyl and HC_{60} radical pair whose formation is, finally, succeeded by an intramolecular combination of the radical pair.

Besides the photoinduced $[2 + 2]$ cycloaddition to the fullerene cage, the addition of ketene silyl acetals is also reported to form a stable and isolable monofunctionalized fullerene derivative [111]. Again, an intermolecular electron transfer to the fullerene triplet has been proposed to be the initial key step. A reasonable product yield underlines the importance of this reaction as a versatile and valuable alternative to thermally induced fullerene functionalization.

Intermolecular electron transfer was also studied with a zinc phthalocyanine (ZnPc) [112]. These studies unravel, however, an interesting twist. Excitation of pristine C_{60} in benzonitrile gives rise to the rapid formation of the fullerene diagnostic π-radical anion band, in line with the oxidation potential of the $ZnPC/ZnPC^{\bullet+}$ couple. As a direct consequence of raising the energy levels for the charge-separated radical pair in nonpolar solvents, the electron transfer is shut off and, instead, the intriguing energy transfer route is activated.

Similarly, the character of an intermolecular reaction between β-carotene and the triplet excited fullerene depends strongly on the solvent polarity [113]. For example, while in polar benzonitrile radical pair formation dominates, the unfavorable thermodynamic forces in benzene activate the energy transfer route.

C_{60} forms ground-state contact complexes with substituted naphthalenes, as evidenced by a new absorption band, significantly different from those of the individual fullerene and naphthalene precursor absorption [114]. In characterizing this charge-transfer effect, the corresponding absorption maximum was analyzed upon tuning the ionization potential of the naphthalene. This was done by introducing appropriate functionalizing groups. A successive red shift upon decreasing the ionization potential of the resulting naphthalene derivatives substantiates the postulated charge-transfer interactions. Absorption of light leads to the population of an excited state with a true charge-transfer character. Interestingly, the photoexcited C_{60}–naphthalene complexes decay via a locally excited fluorescent state localized at C_{60}, which, in turn, is the precursor state to the fullerene triplet excited state.

In thin films, C_{60} gives rise to weak interactions with organic photoconductors, such as poly(3-octylthiophene) and copper phthalocyanine (CuPc) [115]. At room temperature, an interdiffusion between C_{60} and a polymer film of poly(3-octylthiophene) impacts the film stability. CuPc, on the other hand, builds a stable interface with the fullerene. The electronic levels are close to the optical gap and, in turn, favor an electron transfer to C_{60} via some excitonic states of the photoconductor. In conclusion, a heterojunction between CuPC and C_{60} appears to be a suitable candidate for the construction of a cell that may be employable in the future in photovoltaic devices.

5.3.5 Electron Transfer to an Acceptor Molecules

Oxidative quenching of triplet excited C_{60} has been studied with electron acceptors ranging from tetracyanoquinondimethane (TCNQ) [31,116–118] and tetracyanoethylene (TCNE) [28,119] to chloraniline (ClA) [120–122]. Although very fast quenching occurs, this process, surprisingly, is not accompanied by the recombination of the free radical pairs. This behavior stems from the triplet exciplex formation, which prevents escaping from the cage. Only in the case of TCNE the exciplex has been considered as an ion–radical pair. In a complementary EPR investigation, it was shown that the stable radical anion $TCNE^{\cdot-}$ interacts with $^3C_{60}^*$. Chemically induced dynamic electron polarization of the reduced acceptor, $TCNE^-$, has been demonstrated in these studies [123].

An indirect approach, in light of oxidizing C_{60}, has been demonstrated via cosensitization. In particular, the photosensitized electron transfer takes place between the singlet excited state of N-methylacridinium heaxfluorophosphate (MA) and biphenyl (BP) [124]. In this process, the primary MA acceptor is excited and abstracts an electron from the BP donor. The lifetime of the generated π-radical cation ($BP^{\cdot+}$) enables a secondary electron transfer from C_{60}.

A similar concept has also been forwarded employing pulse radiolysis, rather than photolysis to generate $BP^{\cdot+}$ and various polycyclic arene π-radical cations [125,126]. These display fine-tuned oxidation potentials that all enable the subsequent generation of $C_{60}^{\cdot+}$.

5.4 WATER-SOLUBLE SUPERSTRUCTURES (CYCLODEXTRINE, SURFACTANTS, VESICLES)

The desire to study the photochemical reactivity of fullerenes in polar solvents and particularly in aqueous solution emerges from the quest to design novel materials that are promising in drug delivery or in preventing/repairing DNA damage [127–130]. In the case of fullerenes, their participation in the generation of singlet oxygen, with a quantum yield near unity, highlights their potential application, for example, in antitumor activity. Successful attempts to overcome the difficulty of the strong hydrophobicity of this carbon allotrope have been twofold. One approach is to incorporate C_{60} into water-soluble superstructures such as γ-cyclodextrin [131] and calixarines [132] or amphiphilic surfactant structures (Table 5.4) [133]. The other approach demands functionalization of the fullerene core with hydrophilic ligands, which will be discussed later in this chapter.

The preparation of a water-soluble C_{60} complex has been achieved by generating a C_{60}/γ-cyclodextrin (γ-CD) host–guest complex where the fullerene core is embedded between the cavities of two γ-CD molecules [131,134]. But only γ-cyclodextrin is capable of accommodating a single fullerene molecule and, thus, to provide water solubility to this host–guest complex, while the smaller cavities of the α and β analogs fail to incorporate the hydrophobic

TABLE 5.4 Water-Soluble Host–Guest Structures for the Solubilization of Fullerenes [39,46,49]

Compound	Host Structure	Structure	Guest Molecule
γ-Cyclodextrin	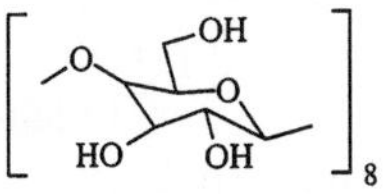	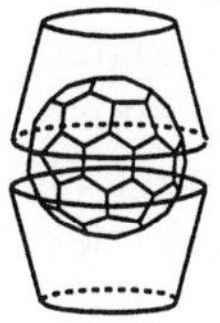	C_{60} Monofunctionalized C_{60} derivatives
Calix[8]arene	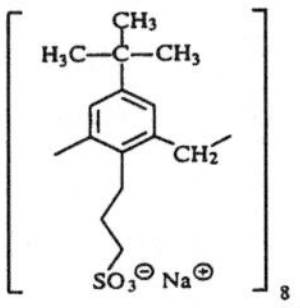		C_{60}, C_{70} Monofunctionalized C_{60} derivatives
Triton X-100	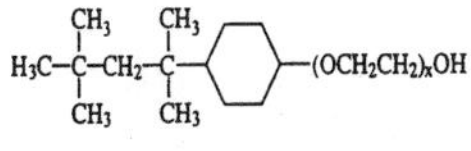	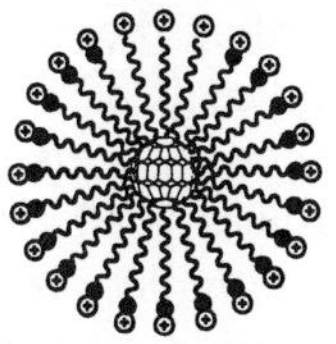	C_{60}, C_{70}, C_{76}, C_{78} Monofunctionalized C_{60} derivatives Bisfunctionalized C_{60} derivatives
Cetrimonium	$CH_3(CH_2)_{15}N^+(CH_3)_3$		C_{60}, C_{70}, C_{76}, C_{78} Monofunctionalized C_{60} derivatives
DODAB	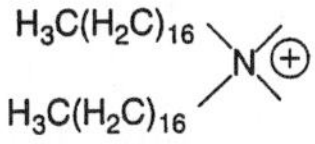		C_{60}, C_{70} Monofunctionalized C_{60} derivatives Bisfunctionalized C_{60} derivatives
Lecithin	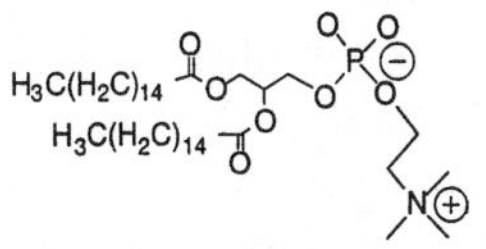	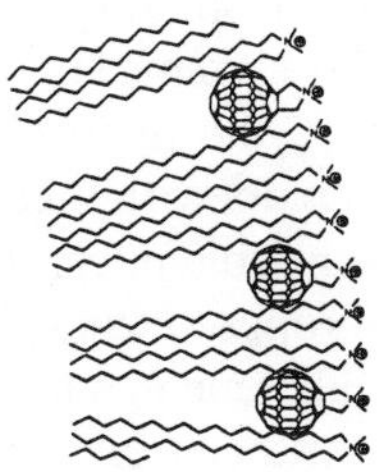	C_{60}, C_{70} Monofunctionalized C_{60} derivatives Bisfunctionalized C_{60} derivatives Monofunctionalized C_{70} derivatives
DHP	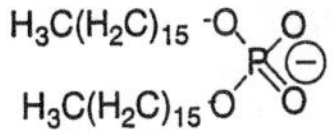		C_{60}, C_{70} Monofunctionalized C_{60} derivatives Bisfunctionalized C_{60} derivatives

fullerene core. Similarly, attempts to incorporate any of the larger fullerenes, even the slightly larger C_{70} moiety, were without success. Although the reactivity of the fullerene in this complex is noticeably lower than in the unprotected form, C_{60}/γ-CD is still susceptible for easy reduction by radicals generated in the aqueous phase, or to excitation and quenching processes [16,135].

Excitation of a nitrogen-purged solution of C_{60}/γ-CD leads to the formation of the triplet state, exhibiting a lifetime of approximately 83 µs. The long triplet lifetime, even at high complex concentration, is rationalized in terms of a quantitative separation of the fullerene cores, excluding the triplet–triplet or ground state triplet annihilation routes [131,136]. Addition of a quencher led to the same observation, as described above for the homogeneous solution of C_{60} [96,137]. Along with the quenching of the triplet state at 750 nm, the grow-in of a new absorption attributable to $C_{60}{}^{\bullet-}$ takes place around 1080 nm. The lifetime of $C_{60}{}^{\bullet-}$ revealed a strong dependence on the nature of the quencher. Successful charge separation results were reported for hexacyanoferrate(II), DABCO, and I_2. Regarding the redox potentials of these oxidized quenchers, the iodine atoms reoxidize $(C_{60}{}^{\bullet-})$ very rapidly (0.4 µs), whereas the corresponding back reaction with $[Fe(CN)_6]^{3-}$ and $DABCO^{\bullet+}$ are slowed down considerably, with half-lives of 62 and 59 µs, respectively.

Although the fullerene moiety is embedded between two host molecules of γ-cyclodextrin, the formation of donor–acceptor complexes with electron-rich amines, such as triethylamine, pyridine, hexamethylenetetramine, acrylamide, and sulfur compounds (thiodiglycolic acid and 2,2′-thiodiethanol) can still be observed [138]. However, corresponding picosecond studies showed that the triplet excited yields of the resulting 1:1 complexes are remarkably decreased without the observation of a solvent-separated ion pair.

Stabilization of fullerene monomers in aqueous media is also pursued by incorporation into aqueous micelles of Triton X-100 [133,139–141], poly(vinylpyrrolidone) [142], and cetyltrimethylammonium chloride [133,139–141]. Photoinduced excitation studies show that $C_{60}{}^{\bullet-}$ can be formed efficiently in the presence of electron donors via the triplet route in a micellar medium. With ascorbate as an electron donor a maximum lifetime for $C_{60}{}^{\bullet-}$ of $\tau = 400$ s was reported in micellar systems [143]. In light of the heterogeneous nature of the core shell structures, the hydrophobic core acts as a protective shield to yield long-lived radical pairs with lifetimes much longer than that of the micelle ($\tau = 5$ s). Strong hydrophobic interactions between the surfactant and the fullerene core have been proposed to explain the strongly enhanced lifetime of the one-electron reduced fullerene in surfactant systems.

Incorporation into artificial bilayer membranes, despite being successful in principle, nevertheless, disclosed a number of unexpected complications [133,144–146]. The most dominant parameter, in this view, is the strong aggregation forces among the fullerene cores, inducing a profound effect on the efficiency of fullerene reduction. The lack of appropriately structured subunits within the vesicular hosts assisting in keeping the fullerene units apart was found to be the reason for instantaneous cluster formation. At the conclusion of these studies, the systematic variation of the functionalizing addends covalently linked to the fullerene core was deemed to be crucial, to ensure a facilitated incorporation into different vesicular matrices, to obstruct cluster formation, and to probe these systems in electron-transfer studies.

Furthermore, it has been shown that C_{70} plays a key function in mediating electron transfer through an artificial lipid bilayer membrane in which the fullerene is embedded in the form of aggregates within the membrane interior [144]. On one side of the lipid bilayer an aqueous solution of a strong electron donor was placed, while the three-component cell was completed with an aqueous solution containing anthraquinone 2-sulfonate, an adequate electron acceptor, on the other side of the membrane. Upon removal of oxygen and illumination, a photocurrent was observed between the two aqueous phases. In principle, the two aqueous layers constitute a redox gradient allowing the electron transfer to take place by donating sequentially an electron from the ascorbate to the fullerene and, finally, to the anthraquinone 2-sulfonate. To confirm a possible electron-hopping mechanism or, alternatively, a diffusion of a one-electron reduced fullerene moiety, the same studies were carried out in blank experiments without a bilayer-incorporated fullerene moiety and without removal of oxygen. While the earlier setup is meant to interrupt the redox gradient across the cell, oxygen quenches the excited and reduced states. Indeed, no photocurrent was observed in the two blanks, pointing to the key character of the fullerene. A similar across-membrane electron transport was reported, focusing on an intermolecular electron transport between a porphyrin, attached to the bilayer–water interface, and fullerenes, embedded in the lipid bilayer membrane.

Also, several larger fullerenes, namely, $C_{70}(D_{5h})$, $C_{76}(D_2)$, and $C_{78}(C_{2v'})$, have been solubilized, besides $C_{60}(I_h)$, via capping with suitable surfactants [139]. Excited states and reduced states of these surfactant systems were probed by photolytic and radiolytic techniques. Flash photolytic irradiation of surfactant solutions of C_{60}, C_{70}, C_{76}, and C_{78} with a 355 nm laser source gives rise to the rapid grow-in of transient singlet–singlet absorption with λ_{max} at 920, 660, 650, and 890 nm, respectively. The associated intersystem crossing rates to the energetically lower-lying triplet excited states are typically of the order of 9.0×10^8 s^{-1} and, therefore, are similar to homogeneous organic solutions. Reductive quenching of the triplet excited states by DABCO occurs with $(0.8–7.4) \times 10^7$ M^{-1}s^{-1} and yields the fullerene π-radical anions. Direct reduction of surfactant-capped fullerene monomers has been studied by means of time-resolved pulse radiolysis with measurements being conducted in the near-IR region. Radiolytic reduction of $(C_{76})_{surfactant}$ and $(C_{78})_{surfactant}$ in aqueous media assisted in confirming the flash photolytic data and resulted in the formation of characteristic π-radical anion bands positioned at 920 and 980 nm, respectively.

5.5 FUNCTIONALIZED FULLERENE DERIVATIVES

The various types of functionalized fullerene derivatives so far synthesized include (1) cyclic adducts formed via cycloaddition, (2) C_{60}–R_X type derivatives

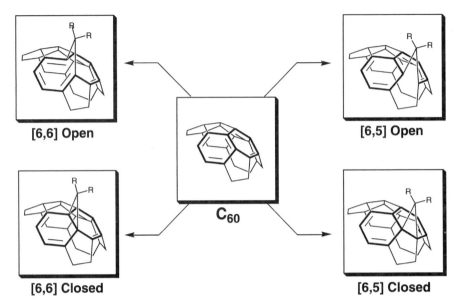

Figure 5.12   The [6,6]-open, [6,6]-closed, [6,5]-open, and [6,5]-closed isomers resulting from the monofunctionalization of C_{60}.

resulting from radical addition or by reaction of C_{60} anions with various alkyl halides [147], (3) adducts involving triangular bridging of a C–C bond [148–150], and (4) organometallic derivatives [151]. The driving force for these additions is related to a relief of strain energy associated with the fullerene surface curvature and the deviation from planarity of the C_{60} double bonds. Among these reactions, cycloadditions provide a valuable tool for functionalization of the fullerene core and have subsequently been used to produce a large array of modified fullerenes with practical applications [19,20,152,153]. In principle, more than a single addition is possible to the C_{60} core, which has 30 double bonds all sharing the same reactivity. However, careful control of the stoichiometry of the reagents and of the reaction conditions allows the isolation of monoaddition products.

Covalent attachment of a single functionalizing addend yields up to four different isomeric products (Fig. 5.12) [150,154,155]. Among these four, the [6,6]-closed and the [6,5]-open are the most important isomers, in which the addend is placed at the junctions of two hexagons or across the single bonds at the hexagon–pentagon junction, respectively. The structure of the [6,5]-open isomer is very appealing, since the fullerene core preserves the π-system of pristine C_{60} but still enables attachment of virtually any functionalizing group. Typically, the [6,5]-open derivatives are very unstable and undergo conversion into the thermodynamically more stable [6,6]-closed isomers [156].

Recently, the isolation and characterization of C_{60} polyadducts has become another research field of high interest [150,154,155]. In particular, the potential use of the C_{60} framework as a template for synthesizing novel and three-

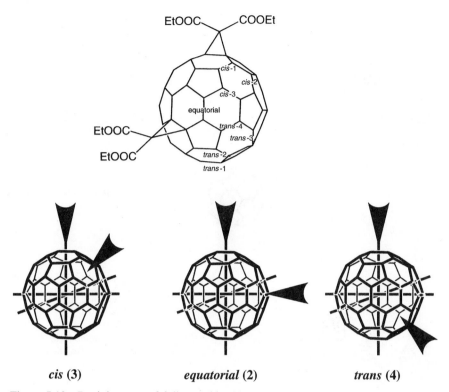

Figure 5.13 Regioisomers of fullerene bisadducts and corresponding configuration of *cis*-, equatorial-, and *trans*-isomers.

dimensionally structured materials appears to be very appealing. Functionalization of a C=C double bond, located at the junction of two hexagons, basically sustains the fullerene structure. The reduction of the symmetry in monoadducts ($C_{2v'}$) results, however, in a regioselectivity of subsequent functionalization reactions at specific positions. This reveals that some bonds in the functionalized fullerene become preactivated, leading to a preferential substitution at certain C=C double bonds. Hence, the first step in the synthesis is the formation of a monoadduct, which upon further addition is successively transformed into well-defined bis- and tris-regioisomers. For example, when two identical addends are placed on the C_{60} core, up to eight isomeric bisadducts can be generated (Fig. 5.13). Their nomenclature is based on the following rules and refers to the three-dimensional geometry of the fullerene: The second addition can take place within either the same hemisphere (*cis*-isomer), at the equator (equatorial-isomer), or within the opposite hemisphere (*trans*-isomer).

5.5.1 Electroinactive, Photoinactive Moieties

The simplest case describes a covalently attached addend that lacks any VIS absorption and is redoxinactive (Fig. 5.14), in the sense of accepting or donat-

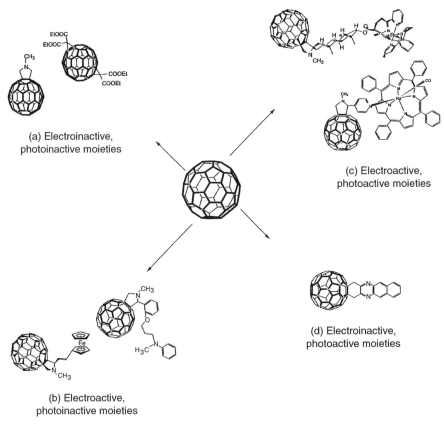

(a) Electroinactive,
photoinactive moieties

(c) Electroactive,
photoactive moieties

(d) Electroinactive,
photoactive moieties

(b) Electroactive,
photoinactive moieties

Figure 5.14 Overview of functionalized fullerene derivatives: (a) electroinactive and photoinactive moieties, (b) electroactive and photoinactive moieties, (c) electroactive and photoactive moieties, and (d) electroinactive and photoactive moieties.

ing an electron to the synthetically bound fullerene moiety. It should be emphasized that these derivatives evolve as important reference compounds for understanding the intramolecular dynamics within the more complex donor–acceptor dyads.

5.5.1.1 Methanofullerene Functionalization of C_{60}, by means of converting the fullerene core into a 58 π-electron system ([6,6]-closed isomer), leads to a significant perturbation of the fullerene π-systems [157]. This sensitively shows up in many electronically related parameters, such as optical absorption of triplet excited states and π-radical anions, and in all the redox properties. In general, the degree of perturbation relative to pristine C_{60} increases with the number of functionalizing addends (Table 5.5) [54,158,159].

The changes observed for the triplet excited states, the reductive quenching by DABCO, the transient absorption spectra, and the electrochemical

TABLE 5.5 Photophysical Properties of *cis-3*-$C_{60}[C(COOEt)_2]_2$, *e*-$C_{60}[C(COOEt)_2]_2$, *trans-3*-$C_{60}[C(COOEt)_2]_2$, *trans-2*-$C_{60}[C(COOEt)_2]_2$, **and** *e, e, e*-$C_{60}[C(COOEt)_2]_3$ [158]

Property	*cis-3* $C_{60}[C(COOEt)_2]_2$ 	*e*-$C_{60}[C(COOEt)_2]_2$ 	*trans-3*- $C_{60}[C(COOEt)_2]_2$ 	*trans-2*- $C_{60}[C(COOEt)_2]_2$ 	*e,e,e*- $C_{60}[C(COOEt)_2]_3$ 
E (singlet)		1.775 eV	1.765 eV	1.770 eV	1.765 eV
E (triplet)		1.50 eV	1.502 eV	1.504 eV	1.498 eV
λ_{max} (singlet)		885 nm	895 nm	875 nm	868 nm
λ_{max} (triplet)	690 nm	710 nm	705 nm	690 nm	650 nm
ε (triplet)	8400 M^{-1}cm^{-1}	9700 M^{-1}cm^{-1}	8000 M^{-1}cm^{-1}	8200 M^{-1}cm^{-1}	7500 M^{-1}cm^{-1}
τ (singlet)		4.2 ns	2.8 ns	2.4 ns	4.52 ns
$E_{1/2}$ ($^{1*}C_{60}/C_{60}{}^{\bullet-}$)		1.02 V vs SCE	0.97 V vs SCE	1.01 vs SCE	0.91 vs SCE
$E_{1/2}$ ($^{3*}C_{60}/C_{60}{}^{\bullet-}$)		0.74 vs SCE	0.70 vs SCE	0.74 vs SCE	0.64 vs SCE
k_q (DABCO)		6.4×10^{6} M^{-1}s^{-1}	1.4×10^{6} M^{-1}s^{-1}	5.8×10^{6} M^{-1}s^{-1}	1.3×10^{6} M^{-1}s^{-1}

reduction sensitively reflect the site and degree of functionalization of C_{60}. For example, a strong blue shift for the triplet excited state of almost 100 nm, relative to pristine C_{60}, is noted. This also coincides with a slowdown of the rate for reductive quenching of the triplet excited states by DABCO. The latter amounts to nearly three orders of magnitude for $e, e, e\text{-}(^3C_{60}{}^*)[C(COOEt)_2]_3$ as compared to $^3C_{60}{}^*$. Similarly, an almost linear relationship is observed between the energy of the characteristic π-radical anion absorption band in the near IR and the number of bis(ethoxy-carbonyl)methylene groups attached to the fullerene. These absorptions range from 1080 to 1015 nm for $C_{60}{}^{\cdot-}$ and $e, e, e\text{-}(C_{60}{}^{\cdot-})[C(COOEt)_2]_3$, respectively. A correlation, linking the reductive quenching rates of triplet excited fullerenes with the electrochemically determined redox potentials and optical parameters, shows that functionalization of pristine C_{60} obstructs the ease of reduction in both the ground and the excited state. The optical and redox properties were most significantly affected upon lifting the first C=C double bond. Higher degrees of functionalization exhibit, however, lesser effects, which parallels the results regarding the loss in the heat of formations relative to pristine C_{60} by introduction of a sp^3-hybridized carbon, locally reducing the fullerene curvature. Subsequent bis- and trisfunctionalization exert only a minor impact on the perturbation of the fullerene π-system. It is interesting to note that among the various bisadducts the e-$C_{60}[C(COOEt)_2]_2$ isomer exhibits the least distortion, compared to pristine C_{60}.

Comparing the HOMOs and LUMOs of a monoadduct with those of pristine C_{60} and of various bisadducts is helpful in rationalizing potential resonance stabilization or even some destabilization effects within the fullerene cores of the various derivatives [157]. The molecular orbitals of the monoadduct $C_{60}C(COOEt)_2$ disclose in the singlet ground state a significant electron deficit, especially in the equatorial area. Reduction, in light of a one-electron addition into an accepting LUMO, has been proposed to lead to an electron distribution that shows, on the other hand, a notable localization in the equatorial area [160]. This may serve as supportive evidence to explain the substantial optical differences between ground state, reduced state, and excited state spectra of pristine C_{60} and monofunctionalized fullerene derivatives. On the other hand, the selective induction of a second functionalizing addend, especially into the equatorial area, has two major consequences. It reintensifies the electron distribution of the singlet ground state in the close proximity of the equatorial position, and it homogenizes the corresponding LUMO level. Therefore, an outset is created, which, in essence, resembles the one known for pristine C_{60}. It is interesting to note that this effect reaches its maximum for the e-$C_{60}[C(COOEt)_2]_2$ adduct. The latter argument emerges from the analysis of the corresponding electron distribution in the LUMOs of the *trans*-3 and *trans*-4-adducts. Unmistakably, these LUMO levels imply, once again, a strong perturbation of the π-electron density, comparable to that shown for the monoadduct.

Three different monofunctionalized methanofullerenes, carrying a bis(4-*tert*-butylbenzoate), a bis(4-*tert*-butyldiphenylsilyoxymethy)phenyl, and an (ethoxy-carbonyl)methylcarboxylate functionality rather than the bis(ethoxy-carbonyl)-methylene groups ($C_{60}C(COOEt)_2$), were characterized in light of their singlet ground state and excited state behaviors [161]. In summary, the photophysical data are virtually identical to those described above for the bis(ethoxy-carbonyl)methylene derivative, $C_{60}C(COOEt)_2$.

5.5.1.2 Pyrrolidinofullerene

These derivatives, prepared via 1,3-dipolar cycloaddition of azomethine ylides to C_{60}, colloquially termed pyrrolidino-fullerenes, retain the basic electronic and electrochemical properties of the methanofullerenes along with the features of the added functional groups [154]. The latter effect impacts, for example, the fluorescence and excited state spectra, relative to the methanofullerenes [159,162–164]. Specifically, red-shifted emission and blue-shifted triplet excited state spectra are found. This alteration is attributed to the weaker electron-donating ability of the adjacent pyrrolidine ring, relative to the cyclopropyl ring. To support this assignment, the excited state characteristics of the pyrrolidinofullerene are compared with those of two other [6,6]-closed isomers, namely, an aziridinofullerene and a dihydrofullerene. The emission and excited state absorption spectra of these derivatives are, indeed, governed by the electron-donating and electron-withdrawing nature of the corresponding functionalizing addends, an aziridine and a cyclohexanone group, respectively.

5.5.1.3 Dihydrofullerene

The singlet ground state and lowest triplet state properties of three dihydrofullerenes are compared to those of [6,6]-closed methanofullerenes and [6,5]-open methanofullerides. Quantitatively, they are very similar [165,166]. A closer analysis reveals, however, a series of striking differences for the $C_{60}H_4$ derivative. Specifically, the ground state absorptions lack the C_{2v} characteristics of other monofunctionalized derivatives. Furthermore, the extinction coefficients of the triplet–triplet absorption around 700 nm are remarkably reduced with $\varepsilon = 4800$ $M^{-1}cm^{-1}$ compared to methanofullerenes and pyrrolidinofullerenes, which range typically between $\varepsilon = 11,000$ $M^{-1}cm^{-1}$ and $\varepsilon = 16,000$ $M^{-1}cm^{-1}$.

5.5.1.4 Aziridinofullerene

Aziridinofullerenes have recently gained considerable attention, since they play an important role as starting material for the synthesis of heterofullerenes, such as $C_{59}N^{\bullet}$ and $(C_{59}N)_2$ [167]. The [6,5]-open structure gains some degree of stabilization relative to the methanofullerene series, although the [6,6]-closed configuration is still the thermodynamically more stable isomer. In contrast to the methanofullerenes, the [6,5]-open aziridinofullerene isomer does not isomerize, upon illumination, but rather exhibits conversion to a ring-opened ketolactam [168,169].

The spectroscopic properties and kinetic data lead to the conclusion that the electron pair on the nitrogen may perturb the delocalized π-electrons [170]. A number of important consequences, regarding the electronic structure of the various fullerene derivatives, derive from this assumption. The [6,5]-open methanofullerene (see above) appears to be truly isoelectronic with the 60 π-electron core of pristine C_{60}. In sharp contrast, the [6,6]-closed isomer can best be represented by a 58 π-electron system. Based on the presented experimental data, the scenario for the corresponding aziridinofullerene isomers is quite different. Considering the free electron pair located at the nitrogen atom, it can be postulated that the [6,6]-closed isomer contains 60 π-electrons, two of them deriving from the nitrogen and the remaining 58 from the functionalized fullerene core. Finally, the [6,5]-open isomer actually carries 62 π-electrons. It is interesting to note that, the nitrogen electron pair is in conjugation with the fullerene core and thus extends the π-system in this derivative.

Although the [6,5]-open methanofullerene reveals the best electron-accepting properties, its photoinstability makes it a rather ineffective target molecule for photoinduced electron-transfer processes. In contrast, the [6,6]-closed aziridinofullerene, similar to [6,6]-closed methanofullerene, is characterized by a high photostability and appears to be the most promising monofunctionalized fullerene derivative. Furthermore, it enables attachment of virtually any functionalizing group.

5.5.2   Fullerenes as Photosensitizers and Electron Acceptors

Of increasing complexity are donor–acceptor superstructures with electroactive addends (Fig. 5.15). For example, in dimethylaniline (DMA) [171–173], ferrocene (FC) [98], and tetrathiofulvalene (TTF) [174] based multicomponent supermolecules, fullerenes are implemented as photosensitizers that are thought to undergo intramolecular electron-transfer processes from the adjacent electroactive moiety (Table 5.6). Accordingly, these are classified as electroactive but, due to their moderate VIS absorption characteristics, photoinactive functionalities.

5.5.2.1 Dimethylaniline   In the first generation of a fullerene-bridge–acceptor dyad, the electron-donating aniline group was covalently linked to the fullerene core via a saturated heterocyclic bridge, for example, a pyrrolidinofullerene that bears an aniline functionality at an α-carbon atom [171]. The fullerene fluorescence quantum yield is reduced only in polar solvents, relative to a fullerene model. In parallel to this quenching a strong decrease of the fullerene triplet population was noticed. The exclusive generation of the charge-separated radical pair in polar solvents is rationalized as a direct consequence of a more exergonic reaction, in which the singlet excited state accepts an electron from the aniline donor. In contrast, in nonpolar solvents the population of the singlet excited state is followed by a quantitative intersystem crossing to the energetically lower-lying triplet excited state.

The concept of dimethylaniline–fullerene donor–bridge–acceptor dyads was further developed by extending the separation between the two redoxactive moieties to retard the rapidly occurring back electron transfer. In this light, two new systems were designed implying a 3-σ-bond (Table 5.6, dyad **1**) and an 11-

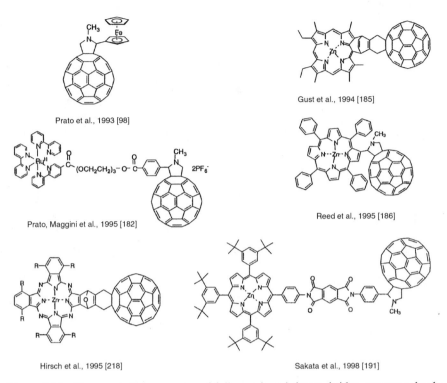

Figure 5.15 Representative examples of fullerene-based donor–bridge–acceptor dyads and triads.

σ-bond (dyad **2**) separation [172]. The earlier configuration is based on replacing the methyl group of the pyrrolidine ring with an N,N,N',N'-tetramethyl-p-phenlenediamine (TMPD), while the latter system implemented the separation of the DMA moiety via a "norbornylogous" bridge. The stronger TMPD donor gives rise to a fast intramolecular charge separation, irrespective of the solvent polarity. In contrast, the 11-σ-bond system followed the first generation of DMA hybrids, because a successful electron transfer only occurs, although much slower, in polar solvents.

Remarkably, large effects of orientation on electron-transfer processes have been reported in a series of fullerene–aniline systems, controlling the rate and efficiency for photoinduced electron transfer in these D-B-A systems. The difference in orientation is achieved by attaching an anilinic donor to the *ortho*- as well *para*-positions of the phenyl groups in the first generation of DMA hybrids, linked by methylene chains of various length. Computational studies suggest that the most stable conformations in the case of *para*-substituted dyads are an extended alignment, while both the *ortho*-substituted dyads adapt a folded conformation. Large values for the rate of charge separation were noticed for the *ortho*-substituted compounds, which has been attributed to the

TABLE 5.6 Rate Constants of Charge Separation and Charge Recombination in Fullerene-Containing Dyads and Triads

Dyad	Dyad Number	Solvent	Charge Separation	Charge Re-combination
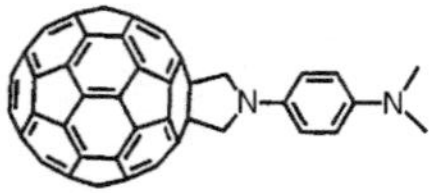	1	Toluene [172]	1.9×10^{10} s^{-1}	
		Benzonitrile	1.0×10^{10} s^{-1}	
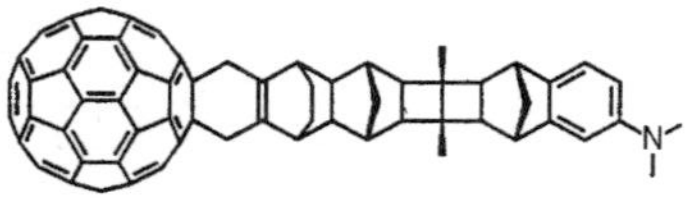	2	Benzonitrile [172]	5.5×10^{9} s^{-1}	4.0×10^{6} s^{-1}
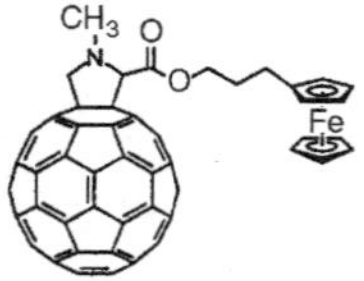	3	Toluene [98]	6.9×10^{9} s^{-1}	1.1×10^{6} s^{-1}
		Benzonitrile	2.6×10^{10} s^{-1}	3.8×10^{5} s^{-1}
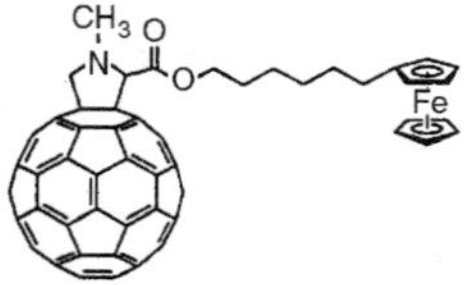	4	Toluene [98]	3.4×10^{9} s^{-1}	6.5×10^{5} s^{-1}
		Benzonitrile	6.9×10^{10} s^{-1}	2.7×10^{5} s^{-1}
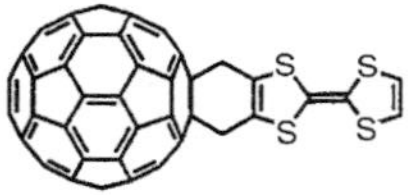	5	Benzonitrile [174]	3.4×10^{5} s^{-1}	9.2×10^{3} s^{-1}

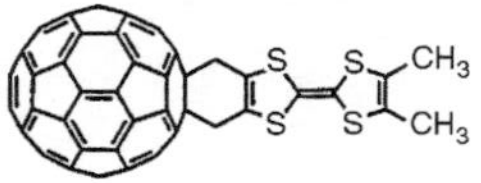	**6**	Benzonitrile [174]	$4.0 \times 10^{6}\ s^{-1}$	$8.7 \times 10^{3}\ s^{-1}$
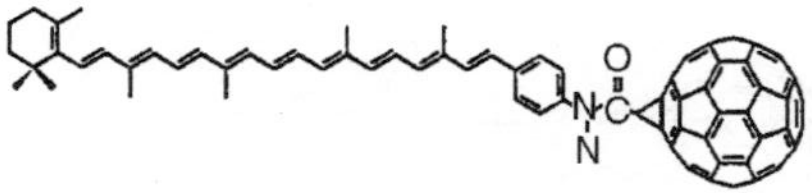	**7**	CS_2 [186]	$1.3 \times 10^{12}\ s^{-1}$	$1.9 \times 10^{9}\ s^{-1}$
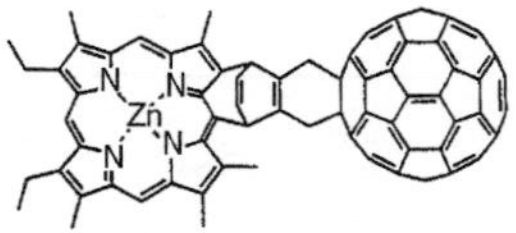	**8**	Benzonitrile [185]	$2.0 \times 10^{11}\ s^{-1}$	
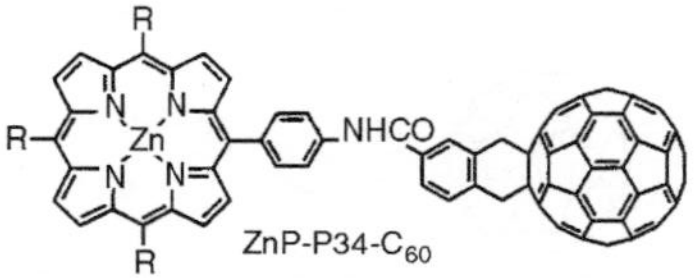	**9**	Benzene [178]	$1.2 \times 10^{10}\ s^{-1}$	$< 2.0 \times 10^{9}\ s^{-1}$
		THF	$9.0 \times 10^{10}\ s^{-1}$	$2.0 \times 10^{9}\ s^{-1}$
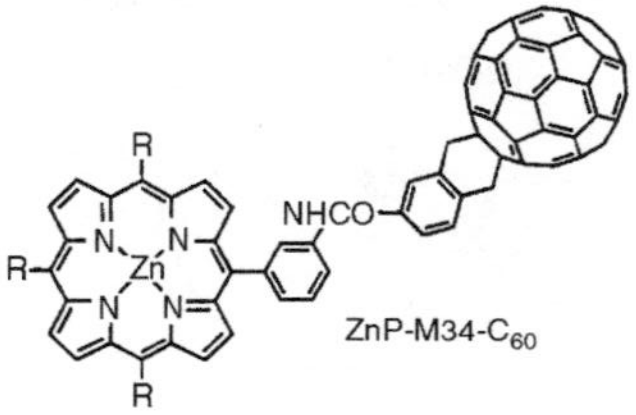	**10**	Benzene [178]	$4.5 \times 10^{9}\ s^{-1}$	$< 1.3 \times 10^{9}\ s^{-1}$
		THF	$4.5 \times 10^{9}\ s^{-1}$	$2.2 \times 10^{8}\ s^{-1}$

TABLE 5.6 (continued)

Dyad	Dyad Number	Solvent	Charge Separation	Charge Re-combination
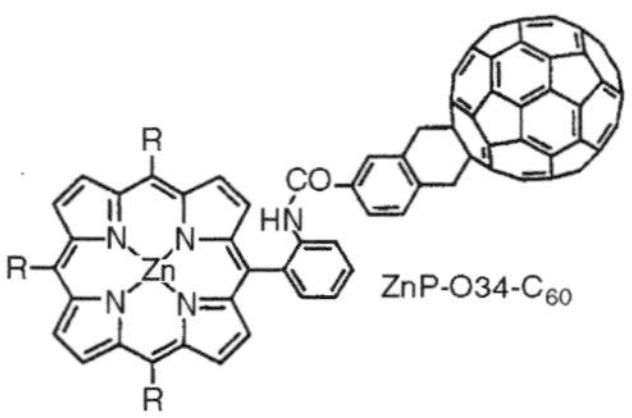 ZnP-O34-C$_{60}$	11	Benzene [178]	1.0×10^{10} s^{-1}	2.9×10^{9} s^{-1}
		THF	2.0×10^{10} s^{-1}	3.3×10^{9} s^{-1}
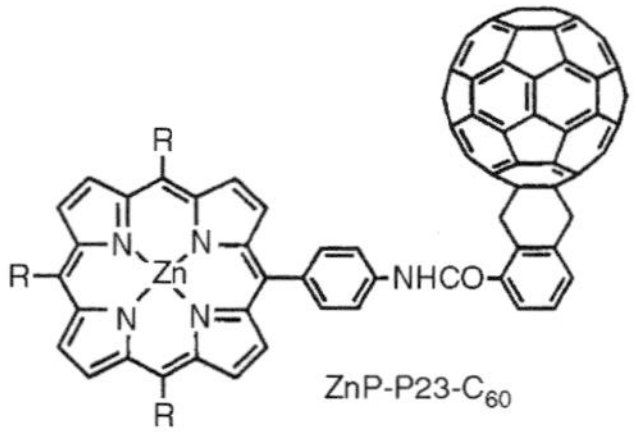 ZnP-P23-C$_{60}$	12	Benzene [178]	1.7×10^{10} s^{-1}	3.3×10^{8} s^{-1}
		THF	2.2×10^{10} s^{-1}	1.4×10^{9} s^{-1}
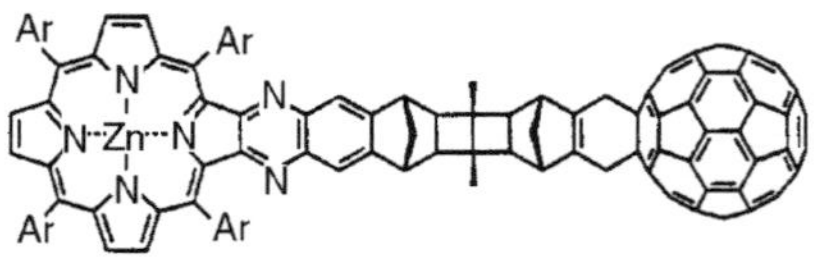	13	Toluene [192]	4.4×10^{9} s^{-1}	
		Benzonitrile	1.0×10^{10} s^{-1}	2.4×10^{6} s^{-1}
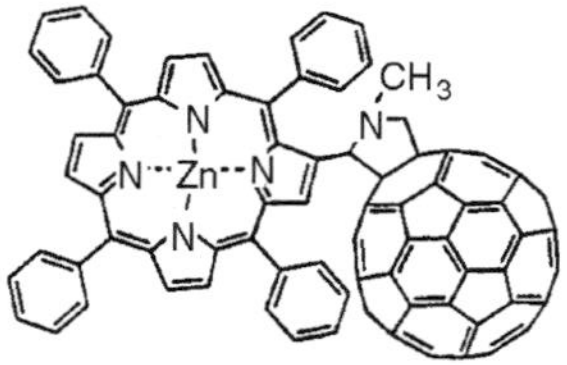	14	Benzonitrile [186]	4.8×10^{11} s^{-1}	2.2×10^{10} s^{-1}

	Structure	Solvent		
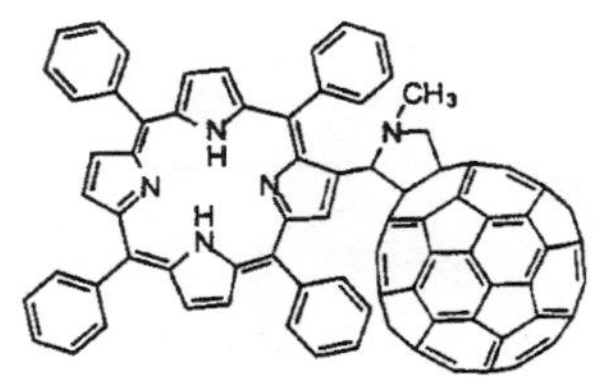 **15**		Benzonitrile [186]	1.8×10^{11} s^{-1}	3.4×10^9 s^{-1}
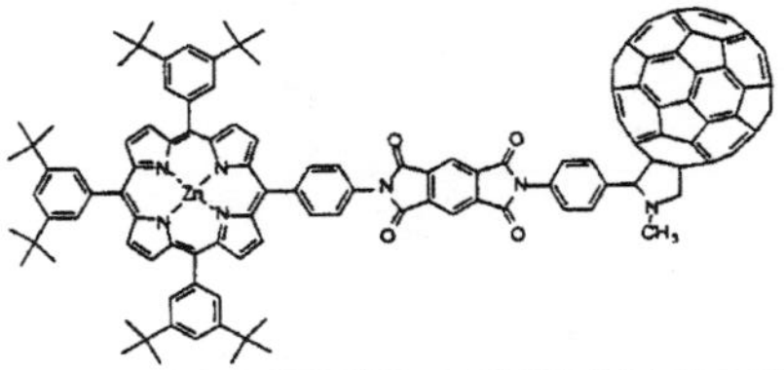 **16**		Dioxane [191]	1.6×10^{10} s^{-1}	7.7×10^8 s^{-1}
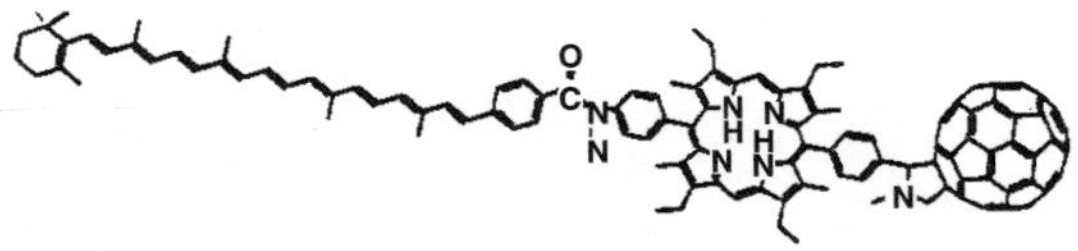 **17**		MTHF [179]	1.0×10^{11} s^{-1}	5.8×10^6 s^{-1}
		Benzonitrile		1.3×10^6 s^{-1}
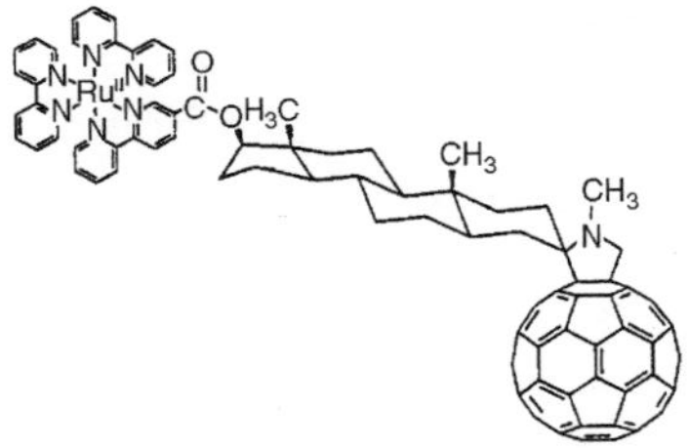 **18**		CH$_2$Cl$_2$ [183]	2.1×10^9 s^{-1}	3.2×10^6 s^{-1}
		Acetonitrile	5.1×10^9 s^{-1}	6.9×10^6 s^{-1}

TABLE 5.6 (continued)

Dyad	Dyad Number	Solvent	Charge Separation	Charge Re-combination
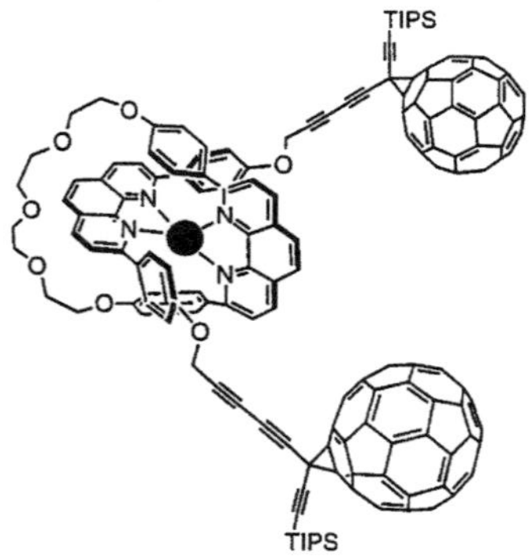	**19**	CH$_2$Cl$_2$ [201]	5.8×10^8 s^{-1}	5.8×10^5 s^{-1}
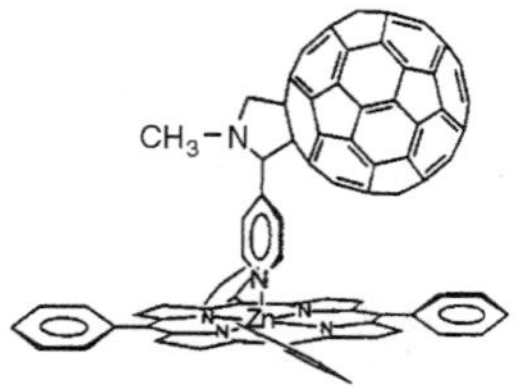	**20**	Toluene [205]	5.5×10^9 s^{-1}	
		CH$_2$Cl$_2$	7.1×10^9 s^{-1}	
		Benzonitrile		1.5×10^9 M^{-1}s^{-1}

configurational flexibility of the aniline moiety. In such a folded configuration it can be thought that the distance between the donor and acceptor is shortened and thereby facilitates the interaction between the two redoxactive moieties.

In a different study, it was shown that the same *para*-substituted fullerene–aniline dyad and a reference model compound (pyrrolidinofullerene) form stable, optically transparent clusters in 75–95% (v/v) acetonitrile–toluene mixtures [175]. Clustering of both the dyad and the reference results in a red-shifted emission maximum, giving rise, however, to a strongly decreased quantum yield for the fluorescence emission of the earlier. The large quenching of fluorescence, noticed for the dyad cluster, is attributed to an electron-transfer process from the appended aniline functional moiety, which is known to be a good electron donor, to the electron-accepting fullerene cluster. In acetonitrile, the growth of a broad absorption was observed following the excitation of the fullerene–aniline dyad cluster. Charge-transfer interaction between the photoexcited fullerene and aniline moieties is responsible for the production of fullerene π-radical anions. This was independently confirmed by recording the diagnostic absorption band of the fullerene π-radical anion in the near-IR region (1010 nm).

5.5.2.2 Tetrathiafulvalene Tetrathiafulvalene gives rise to a remarkable electron-donating phenomenon [174]. In particular, the one-electron oxidized form, namely, the 1,3-dithiolium cation, displays, in contrast to the ground state, a $(4n + 2)$ aromatic character. In light of the electron-transfer processes, this is an important requisite assisting in the stabilization of charge-separated radical pairs. Accordingly, a new family of donor–acceptor fused dyads encompasses the linkage of this organic electron donor, with the fullerene moiety. Dyads **5** and **6** give rise to EPR values and ^{32}S hyperfine coupling constants that unravel a spin density distribution located on the TTF and [60]fullerene moieties, for the cationic and anionic species, respectively. In this line, the lifetime of the triplet excited state relates to the donor strength of the TTF derivative, thus controlling the rate of electron transfer. In nanosecond-resolved flash photolysis, a rapid quenching of the triplet excited states generates transient charge-separated open-shell species with lifetimes typically around 75 μs.

5.5.2.3 Ferrocene Another family of electron acceptors displaying tunable donor strength consists of the ferrocenes. The fluorescence of a series of fullerene–ferrocene dyads (**3** and **4**), with variable spacing blocks, were substantially quenched in methylcyclohexane and in polar solvents [97,98,176]. In transient absorption studies, excitation of the pyrrolidinofullerene reference revealed the immediate formation of the singlet excited state, with λ_{max} around 886 nm, followed by a quantitative intersystem crossing to the triplet excited state with λ_{max} around 705 nm. Picosecond-resolved photolysis of the investigated fullerene–ferrocene dyads in toluene showed light-induced formation of the singlet excited state, which, however, undergoes rapid intramolecular quenching, with rate constants ranging between 2.3×10^9 s^{-1} and 28×10^9 s^{-1}. Evidence for a long-lived charge separated state in degassed benzonitrile

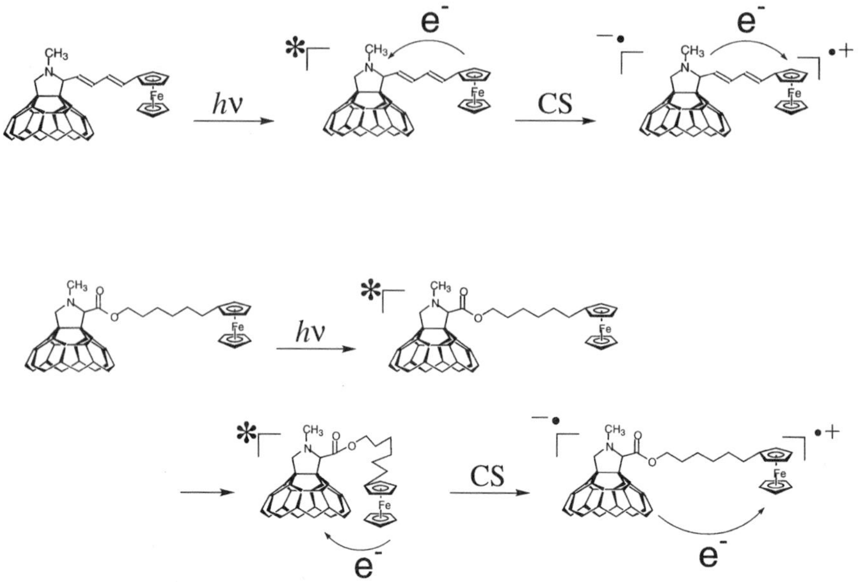

Figure 5.16 Quenching mechanism for photoexcited fullerene–ferrocene dyads.

$(\tau_{1/2} = 2.5$ μs for dyad **4**) stems from complementary nanosecond-resolved photolysis for the flexible spaced ferrocene dyads. In contrast, rigidly spaced dyads fail to disclose any measurable stability for the presumably formed radical pair. The nature of the spacer between C_{60} and ferrocene, weak electronic ground state interactions, steady-state fluorescence, and picosecond-resolved photolysis suggest two different quenching mechanism: through bond electron transfer for rigid spaced dyads and formation of a transient intramolecular exciplex for flexible dyads (Fig. 5.16). To summarize, in polar solvents a flexible spacer facilitates intramolecular electron transfer and, in parallel, stabilizes the radical pair relative to a rigid spacer, which is short or conjugated.

5.5.2.4 Anthraquinone Anthraquinone is a weaker electron acceptor than C_{60}. The placement of two cyanoimino groups at the 9,10-positions of the anthracene lowers the oxidation potential and, therefore, improves the acceptor properties of the resulting anthraquinone moiety [177]. The case of a C_{60}–donor system is unprecedented in the literature and the envisaged oxidative pathway, relative to the fullerene, is so far based on a direct extrapolation of the electrochemical data. It will be seen in future photophysical studies whether the moderate donating properties, established in the intermolecular reaction route, will also apply for the intramolecular route in these newly synthesized donor–acceptor systems. In a direct extension of this concept the discussed dyad has been proposed as a precursor for a triad, utilizing a bisfunctionalized fullerene derivative as a multifunctional spacer.

5.5.3 Fullerenes as Electron Acceptor Moieties

The redox properties of pristine fullerenes and monofunctionalized fullerene derivatives in their ground and excited states have drawn much attention to design devices that are able to perform as molecular switches, receptors, photoconductors, and photoactive dyads. Typically, applications are driven by the excellent electron-accepting properties of fullerenes. This property, together with their low reorganization energy, makes C_{60} and its derivatives good candidates as building block systems for the conversion of light. In a conceptional extension of the above highlighted results, the linkage of a fullerene to a number of interesting photo- and electroactive species is carried out to improve the light-harvesting efficiency of the fullerene dyad. Their subsequent study of intramolecular transfer dynamics between the two moieties will be elucidated in this chapter as a function of excited state energy of the antenna molecule, donor–acceptor distance, and solvent polarity.

In light of the moderate absorption features of fullerenes in the VIS, functionalization of pristine C_{60} with chromophoric addends, such as metalloporphyrins [178–181] or metal-to-ligand-charge-transfer (MLCT) transition metal complexes [182–184], developed as an important objective. This is meant to promote primarily the VIS absorption characteristics of the resulting dyads and, importantly, to improve the light-harvesting efficiency of the fullerene core. Specifically, suitable photosensitizers assist to extend the absorption further into the red, yielding, upon excitation, energetically low-lying excited states. As a direct consequence, the role of the fullerene is significantly changed. Under these circumstances, that is, bearing a photoactive and electroactive moiety, fullerenes operate exclusively as electron and energy acceptor moieties.

5.5.3.1 Porphyrin The initial report of the synthesis and photophysical study of a fullerene linked to a porphyrin pigment (dyad **8**) [185,186] was followed by the covalently linked carotenoid–fullerene dyad **7**. The intramolecular dynamics were compared with another porphyrin–fullerene dyad [187]. In general, the porphyrin or carotene singlet excited states decay via rapid singlet energy transfer to the fullerene or alternatively via electron transfer to the fullerene generating a charge-separated state. Under circumstances that slow down the electron-transfer route, the fullerene deactivates internally to the energetically lower-lying triplet state. A triplet–triplet energy transfer to the attached pigment concludes this sequence. Based on these dyad principles, molecular triad **17** consisting of a diarylporphyrin (P), a carotenoid polyene (C), and a fullerene (C_{60}) was developed [179,188]. The C–P–C_{60} triad (**17**) undergoes in 2-methyltetrahydrofuran a relay of two photoinduced electron-transfer events to yield initially the C–($P^{\bullet+}$)–($C_{60}^{\bullet-}$) pair, which then transforms into ($C^{\bullet+}$)–P–($C_{60}^{\bullet-}$). The final charge-separated state, formed with an overall quantum yield of 0.14, decays with a time constant of 170 ns (Table 5.6). The associated charge recombination process surprisingly yields the carotenoid triplet state, rather than the ground state. Further evidence for this reaction mechanism stems from time-resolved EPR experiments. Specifically, detection of the spin-polarized radical pair and the carotenoid triplet states has verified that the carotenoid triplet indeed is formed from $(C^{\bullet+})-P-(C_{60}^{\bullet-})$ by radical pair recombination.

Separation of the two chromophores, namely, a fullerene and a zinc porphyrin, by a phenyl spacer enabled the systematic variation of the linkage, for example, positioning at the *ortho-*, *meta-*, or *para-*position of the phenyl ring (dyads **9–12**) [178,189]. Despite the importance of the linkage position, in any of the three systems a rapidly occurring charge separation is always succeeded by a complete recovery of the ground state. Supported by a sufficiently large driving force, in polar THF the $(ZnTPP^{\bullet+})-(C_{60}^{\bullet-})$ pair evolves from the singlet excited state of both moieties, fullerene and porphyrin. In benzene, an efficient mixing of the fullerene and porphyrin singlet excited states and a lower driving force leads to a domination of the energy-transfer route over the electron transfer. In light of stabilizing the charge-separated radical pair, it is important to note that, particularly, the *meta*-linkage (dyad **10**) provides an onset that slows down both processes, charge separation and charge recombination, relative to the corresponding *para-* and *ortho*-isomers (dyads **11** and **12**). In addition, it was shown that linking two fullerene acceptors to the porphyrin chromophore enhanced the excited state interaction, by means of a stronger quenching of the porphyrins singlet excited state emission [190].

In contrast to the above C–P–C$_{60}$ triad (**17**), another sequential electron-transfer relay system (triad **16**) has been reported, which relies on separating the porphyrin donor from the fullerene acceptor pair with a pyromellitimide functionality [191]. Upon exciting the porphyrin antenna, the adjacent imide moiety accepts an electron, exhibiting a characteristic absorption around 720 nm, concomitant to the oxidation of the ZnTPP chromophore. This transient state transfers rapidly the charge to the end terminus, namely, the fullerene moiety, and creates a spatially separated $(ZnTPP^{\bullet+})-In-(C_{60}^{\bullet-})$ radical pair, with an overall quantum yield of only 0.46. The resulting charge-separated state undergoes a remarkably slow back electron transfer with a rate of 7.7×10^8 s^{-1}.

Long-ranged photoinduced energy and electron transfers have also been investigated in a dyad containing a zinc tetraphenylporphyrin donor (ZnTPP) and a fullerene acceptor (C$_{60}$), separated by a saturated norbornylogous bridge nine sigma bonds in length (dyad **13**) [192]. While in nonpolar solvents singlet–singlet energy transfer prevails, in polar benzonitrile efficient charge separation occurs with a rate constant of 1×10^{10} s^{-1} to yield $(ZnTPP^{\bullet+})-(C_{60}^{\bullet-})$. The charge-separated radical pair exhibits a remarkably long lifetime of 420 ns.

Polyether chains as linkage blocks between the porphyrin antenna and the electron-accepting fullerene moiety have been used for the convergent synthesis of several different fullerene hybrids [193]. These linking blocks provide tunable molecular topologies, an effect, which is based on metal cation complexation, to bring the two moieties closer together through space. Electrochemical, photophysical, computational, and NMR studies provide evidence for the complexation of potassium cations, which, subsequently, alters the fullerene–

porphyrin distance and the intramolecular dynamics, by means of a substantial acceleration.

Recently, a report on a substantially different strategy focusing on fullerene–porphyrin dyads has been published [180]. This work implies the utilization of an hexakisadduct as an electron acceptor rather than attaching the porphyrin chromophore to a monofunctionalized fullerene core. However, no photophysical measurements were performed so far with this modified fullerene core.

5.5.3.2 *Ruthenium(II) complex*

The strong emission of the ruthenium metal-to-ligand charge-transfer states $(E = 1.96$ eV$)$, generated with high quantum yields, facilitates photoinduced intramolecular electron transfer to electron acceptors, such as the discussed fullerenes. The first study, following the linkage of a fullerene to a ruthenium complex, reported evidence for photoinduced electron transfer from the MLCT excited state of a covalently, but flexible, attached ruthenium(II) complex to a fullerene moiety [182,194]. Photoinduced optical absorption studies showed an excitation spectrum with a long-lived photoinduced absorption (PIA) band at 668 nm, which has been assigned to the oxidized ruthenium chromophore. As a thin solid film, a photoinduced electron spin resonance signal was noted, which exhibits signatures of both the oxidized ruthenium(III) complex and the reduced C_{60} π-radical anion with a lifetime of the order of milliseconds at 80 K. In solution, spectral characteristics for a radical pair are only noted in polar solvents. The flexible spacer has a key character to retard fast charge recombination. Therein, the configurational freedom, provided by the polyglycol spacer, is beneficial to ensure separating the pair.

In an extension of the former work, a pyrrolidinofullerene is rigidly linked via an androstane spacer to a ruthenium(II) trisbipyridine chromophore (dyad **18**, Fig. 5.17) [183]. The emission of the ruthenium MLCT state in CH_2Cl_2 and CH_3CN was substantially quenched, relative to a ruthenium(II) trisbipyridine model. Picosecond-resolved photolysis of the dyad shows that the light-induced formation of the photoexcited ruthenium center is followed by a rapid intramolecular electron transfer to the fullerene core, yielding the respective radical anion. Nanosecond-resolved photolysis confirms formation of the characteristic fullerene π-radical anion band at $\lambda_{max} = 1040$ nm and reveals lifetimes of the charge-separated state of 304 ns and 145 ns in CH_2Cl_2 and CH_3CN, respectively.

With the scope of extending the ground state absorption to the red region of the visible spectrum and, in turn, improving the light sensitization, the excited states properties of a dinuclear [ruthenium(II)]$_2$–androstane–C_{60} donor–bridge–acceptor dyad were studied under various conditions and compared with those of the mononuclear ruthenium(II)–androstane–C_{60} dyad [184]. The design of a dinuclear ruthenium chromophore proves to be a conceptional success. However, the steady-state luminescence quantum yield of the dinuclear dyad indicates a very inefficient intramolecular quenching reaction, while the respective mononuclear dyad is, in general, substantially quenched relative to a

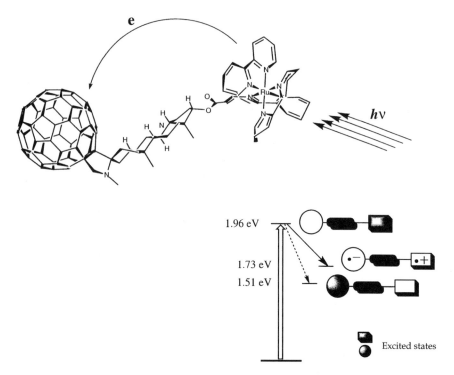

Figure 5.17 Schematic representation of photoinduced electron-transfer events from a ruthenium excited MLCT state to the ground state fullerene.

ruthenium model complex. Picosecond- and nanosecond-resolved photolysis sheds light on the nature of these slow intramolecular dynamics and shows that the photoexcited MLCT states transform, in the case of the dinuclear dyad, into the fullerene triplet excited state. The difference in reactivity relative to the mononuclear dyad has been related to substantially different MLCT excited state energies of the corresponding antenna molecule in the mono- and dinuclear-ruthenium(II) dyad with 1.96 eV and 1.68 eV, respectively. Furthermore, the energy of the radical pair for the dinuclear dyad (in acetonitrile, 1.98 eV) is unfavorably impacted and, thus, its formation would be highly endoergic.

Solvent-dependent conformational changes of a peptide bridge, separating a ruthenium(II) trisbipyridine donor unit from a fulleropyrrolidine acceptor unit, is shown to influence the electron-transfer process that occurs upon photo-excitation of the ruthenium chromophore [195]. A strong protic solvent disrupts the helical secondary structure of the peptide spacer that locates the Ru(II) and C_{60} moieties into a distance that favors their mutual electronic interaction. The reported results strongly support the view that, upon disruption of the 3_{10}-helical structure, the separation between the two components, ruthenium antenna and C_{60}, tends to increase to a point that disfavors their mutual electronic interactions. Therefore, it is reasonable to assume that an unfolding of this type of helix would result in a statistically unordered conformation and consequently a greater average distance between the two termini.

Recently, a new class of fullerene-substituted oligopyridine ligands was synthesized, which successfully coordinate to a ruthenium(II) cation to yield interesting and appealing D–B–A dyads and triads [196,197]. In order to retard undesired back electron transfer reactions, a terpyridine (tpy) metal binding ligand was selected.

5.5.4 Fullerenes as Energy Acceptors

5.5.4.1 Anthracene and Pyrene Fluorophore–fullerene adducts have been prepared employing electron-rich silylbutadienes and fullerenes as precursors via a Diels–Alder addition reaction [198]. A series of different anthracene and pyrene derivatives, which are covalently linked to the fullerene core, disclose in their singlet ground states superimposable features of the corresponding precursors states and, thus, rule out any significant electronic ground state interaction. Once photoexcited, the emission of the anthracene and pyrene moieties decreased, remarkably, by a factor of 10^3–10^4 relative to the reference compound. An energy-transfer route has been proposed to be the major deactivation pathway of these bichromophoric systems, generating the long-lived fullerene triplet.

In an alternative approach to designing arene–fullerene systems, a thermal reaction of C_{60} with dihydrocyclobutaarenes is pursued, which affords dyads that contain arenes ranging from benzene to pyrene [199,200]. Similar to the above Dields–Alder products, the fluorescence of the attached aromatic arenes are quantitatively quenched. This has been ascribed to a very rapid intramolecular singlet–singlet energy transfer that transforms the short-lived and moderately redox-active excited states of the investigated arenes into the highly reactive fullerene triplet excited state. In contrast, the singlet ground state and excited state properties of the fullerene moiety are insignificantly affected by the presence of the attached arenes and, more importantly, virtually identical to a fullerene reference compound that lacks any of the arene functionalities. This, again, is indicative of little intramolecular electronic interaction between the spatially separated photoactive fullerene and arene moieties, although carrying a different spacer unit.

5.5.4.2 Pyrazine Functionalization of pristine C_{60} with three different pyrazine derivatives, carrying substituents with different electron-inducing abilities, is another strategy that assists in promoting the UV/VIS absorption characteristics. This also improves the light-harvesting efficiency of the resulting dyads relative to pristine C_{60} [80]. Photoexcitation of the pyrazine moieties in these dyads leads to the formation of their singlet excited states. In contrast to the pyrazine models, but in analogy to the fullerene–arene dyads, photoexcitation of all the fullerene pyrazine dyads is followed by rapid intramolecular deactivation processes of the latter via energy transfer to the fullerene ground state with half-lives ranging typically between 37 and 100 ps.

5.6 SUPER- AND SUPRASTRUCTURES

Relatively little work has been directed to exploring the photochemical and photophysical properties of supramolecular assemblies containing fullerenes, as electron acceptor units in self-assembled arrays. A few approaches, outlined below, demonstrate the successful utilization of this three-dimensional electron acceptor [201,202]. Specifically, self-assembling motifs, such as threading of a rotaxane or a catenane and coordination of a fullerene ligand to a transition metal center, attracted considerable attention, not only for their fascinating topological features but also for the efficiency of photodriven processes.

5.6.1 Rotaxane

In this novel, self-assembling array a three-component precursor, consisting of a macrocycle coordinating ring, a redoxactive copper(I) center and bisfunc-tionalized fragment are threaded inside the ring [201,202]. The resulting complex reacts with a monofunctionalized fullerene derivative to afford a soluble rotaxane with two fullerenes as end-terminating stoppers (Table 5.6, dyad **19**). Both excited states, the MLCT state of the macrocyclic copper(I) complex and the fullerene moiety, are substantially quenched. Compared to the excited state energies, the outcome of this rapid intramolecular deactivation routes is, not surprisingly, very different in character and product formation. For example, deactivation of the fullerene excited state follows an energy-transfer mechanism to the adjacent Cu(I) complex. The later excited state, in contrast, is mainly quenched by electron transfer to form the charge-separated radical pair comprising an oxidized metal center, namely, Cu(II), and a one-electron reduced fullerene π-radical anion, $C_{60}{}^{\bullet-}$.

5.6.2 Catenane

This exotic molecule, which exhibits an unusual topology and an aesthetic appeal, was obtained via the template-directed formation of cyclobis(paraquat-p-phenylene) around the C_{60} appended macrocyclic polyether [203]. ^{1}H NMR investigation gave rise to the formation of intramolecular stacks of the following type: D–A–D–A. In essence, they imply the interaction between the π-electron-rich fullerene moiety and π-electron-deficient bipyridinium units. Photophysical measurements of this truly unique and exciting array are currently under investigation.

5.6.3 Coordinative Bonds Between a Fullerene Ligand and ZnTPP

A photoactive supramolecular system, in which donor and acceptor moieties are linked via coordinative association, bears a number of particularly appealing features. These emerge primarily from their relative ease of synthesis and their potential key function as new and intelligent material. In such a multicomponent array, photoillumination triggers a sequential rapid electron transfer and a diffusional splitting of the charge-separated radical pair, thus mimicking a key step in natural photosynthesis. With this scope an assembly of a rigidly connected dyad is reported, in which a zinc tetraphenyl porphyrin (ZnTPP) is noncovalently linked to a C_{60} fullerene ligand via axial pyridine coordination to the metal (dyad **20**) [204,205].

In benzonitrile, differential absorption changes in the NIR reveal the participation of two components in the formation of the $C_{60}{}^{\cdot-}$. The faster process points to a participation of the $^1(\pi\text{-}\pi^*)$ ZnTPP in an intramolecular process, while the slower one involves an intermolecular quenching of the lower-lying $^3(\pi\text{-}\pi^*)$ ZnTPP state. For the bimolecular quenching reaction a rate constant of 1.7×10^{10} $M^{-1}s^{-1}$ was resolved. Kinetic analysis of the 1010 nm transient absorption ($C_{60}{}^{\cdot-}$) gives rise to a remarkable lifetime of the separated radical pair in deoxygenated benzonitrile of several hundred microseconds. Charge recombination obeys second-order kinetics with a rate constant of 1.5×10^9 $M^{-1}s^{-1}$.

According to the described results, two different pathways (Fig. 5.18) can be envisioned for the electron-transfer processes. In one scenario, illumination of the porphyrin antenna is followed by a fast intramolecular ET inside the com-

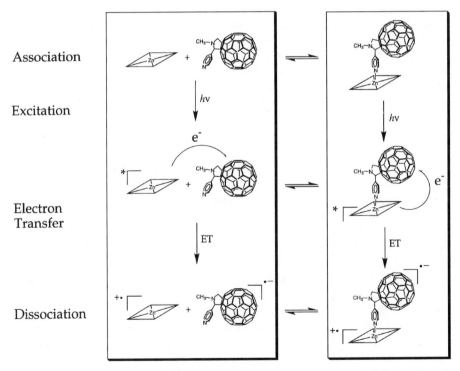

Figure 5.18 Schematic illustration of complex association between a fullerene ligand and ZnTPP, photoexcitation of the ZnTPP chromophore, intra- and intermolecular electron transfer, and complex dissociation of the radical pair.

plex. Alternatively, the free ZnTPP porphyrin is excited and is subject to an intermolecular ET when the acceptor molecules approach closely enough during the molecular diffusion. In coordinating media, the solvent can displace the fullerene ligand from the Zn metal, thus improving the efficiency of the intermolecular route.

5.7 HYDROPHILIC MOIETIES

The remarkable redox properties and energy storage capacities of pristine fullerenes evoked a lively interest in employing these compounds in bimolecular electron- and energy-transfer reactions. The desire to utilize this unique reactivity also in a polar (especially aqueous) environment (where C_{60} is practically insoluble) has triggered numerous successful attempts to modify the hydrophobic fullerene core by functionalization. Covalent attachment of hydrophilic addends across the double bonds located at the junctions of two hexagons afforded novel and innovative materials with appealing characteristics ranging from excellent water solubility to drug delivery and involvement in biologically and even medically relevant processes [153,154].

Enhancement of the fullerene water solubility has been achieved, for example, by monofunctionalization with negatively charged carboxylic, positively charged quaternary ammonium, and noncharged ethyleneglycol groups [206,207]. However, strong adsorption forces between the fullerene cores still exist, which leads to an instantaneous and irreversible formation of micellar clusters of indefinite sizes in aqueous solution (Fig. 5.19). The far most important consequence of this aggregation is that the lifetime of the triplet excited state is reduced by orders of magnitude relative to a monomer solution. Another characteristic of, for example, negatively charged fullerene clusters, $\{C_{60}C(COO^-)_2\}_n$, is their effective shielding against reduction by electron transfer. One powerful possibility to overcome these cluster-specific problems is encapsulation of the respective monomer species in suprastructures such as γ-cyclodextrine or surfactants [208]. Comparison of capped fullerene with γ-CD-incorporated complexes provided unambiguous evidence for the presence of fullerene monomers, probably, in a core(fullerene)–shell(surfactant) type structure.

Substitution of the negative carboxylic acid functionalities, with the above cited positively charged pyrrolidinium group or a nonionic ethyleneglycol chain, helps in principle to prevent a potential charge repulsion with, for example, reducing agents such as hydrated electrons [207]. This proves, however, to be an insufficient attempt to exclude fullerene clustering and points again to their dominant association forces governed by π–π interaction in aqueous media.

In this context, an amino acid derivative of C_{60}, which was synthesized by direct additions of the corresponding amino acid to the fullerene core, evoked great interdisciplinary interest [209]. Specifically, its biological activity upon

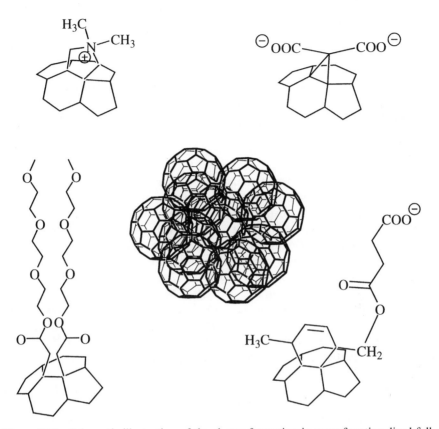

Figure 5.19 Schematic illustration of the cluster formation in monofunctionalized fullerene derivatives.

low-energy visible light illumination, leading to a remarkable site specificity of DNA cleavage at guanine bases, is unprecedented [210]. Although this speaks for a significant photodynamic action, the lack of photocytotoxic effects, which were even found at concentrations of up to 50 μM, curtails their potential application. Nevertheless, it is mandatory to assume that the reported reactivity is based on a cluster-specific behavior and, therefore, it remains to be tested in which way a truly monomeric fullerene participates.

Introduction of a second and subsequently a third hydrophilic ligand at the C_{60} core leads to three-dimensional architectures that display an extremely satisfying water solubility of up to ~1 mM [211,212]. In essence, these derivatives experience an enhanced surface coverage of the hydrophobic fullerene surface. Therefore, fullerene aggregation, due to intermolecular π–π association, is impacted to a noticeable extent. By probing a series of $C_{60}[C(COO^-)_2]_2$ and e,e,e-$C_{60}[C(COO^-)_2]_3$ in light- and radical-induced studies, one can demonstrate that micellar aggregation plays a minor role in the reactivity of these functionalized fullerenes.

Steady-state and time-resolved measurements with a series of bisfunctional-ized fullerene derivatives and a single trisfunctionalized adduct give rise to their monomeric appearance in aqueous media. An earlier observation regarding their water-insoluble precursors, namely, malonic acid diethyl ester derivatives, demonstrated that functionalization of C_{60} leads to a perturbation of the full-erenes π-systems. Its degree, relative to pristine C_{60}, intensifies with increasing number of functionalized addends. The overall impact, which is mainly gov-erned by electronic factors, is valid as well for the hydrolyzed and water-soluble derivatives. This can be appreciated from a deacceleration of the reduction processes in the singlet ground and triplet excited state and comparable blue shifts of the respective absorption bands in the near IR. Besides the perturbation of the π-system, polyfunctionalization reduces the number of reactive sites on the fullerene surface due to sp^3-carbon hybridization, and it is safe to assume that this leads eventually to fullerene derivatives that show less and less resem-blance to the original C_{60} with increasing number of addends.

Polyfunctionalization is a powerful mediator to further improve the water solubility of the fullerene core. For example, $C_{60}(OH)_{18}$ is impressively soluble in neutral aqueous solution up to 3.85×10^{-2} M, but dynamic light scattering experiments disclose evidence of the formation of fullerene aggregates at high solute concentration [213]. This hydrophilic fullerene derivative emits a very weak fluorescence. Photolysis of $C_{60}(OH)_{18}$ in aqueous solution yields the immediate formation of a transient singlet excited state with a broad singlet–singlet absorption in the 550–800 nm region. The energetically higher-lying singlet excited state transforms via intersystem crossing to the broadly absorb-ing (550–800 nm) triplet excited state. In contrast, at low solute concentration, the features of the triplet–triplet absorption differ significantly and exhibit an absorption maximum at 650 nm concomitant to a shoulder at 570 nm. The π-radical anion, $(C_{60}{}^{\bullet -})(OH)_{18}$, generated by electron transfer from hydrated electrons and $(CH_3)_2C^{\bullet}(OH)$ radicals, absorbs with λ_{max} at 870, 980, and 1050 nm. Based on electron-transfer studies with suitable electron donor/acceptor substrates, the reduction potential of the $C_{60}(OH)_{18}/[(C_{60}{}^{\bullet -})(OH)_{18}]$ couple was estimated to be in the range between -0.358 and -0.465 V versus NHE. Overall, the photophysical and electrochemical properties of $C_{60}(OH)_{18}$ fall well in line with the above trend, that a highly functionalized fullerene core has very little resemblance to pristine C_{60}. This assumption is further corroborated by some electronic and kinetic features stemming from time-resolved studies on $C_{60}H_{18}$ [214], $C_{60}H_{36}$ [214], and $C_{60}(C_6H_5)_5Cl$ [215].

5.8 HETEROFULLERENE

Incorporation of a heteroatom into the fullerene carbon network is expected to alter the electronic and geometric properties of the molecule [167]. In particu-lar, due to the EPR active nitrogen probe, the respective $C_{59}N^{\bullet}$ radical form is quite unstable and thus dimerizes [216,217]. With the help of two independent studies, it was demonstrated that photolysis of the parent dimer, $(C_{59}N)_2$, leads to a cleavage of the respective inter-fullerene C–C single bond. The later relaxes and recombines efficiently to the starting dimer.

ACKNOWLEDGMENT

This work was supported by the Office of Basic Energy Sciences of the U.S. Department of Energy (contribution No. NDRL-4125 from the Notre Dame Radiation Laboratory).

REFERENCES

1. H. W. Kroto, J. R. Heath, S. C. O'Brien, R. F. Curl, and R. E. Smalley, *Nature* **1985**, *318*, 162–164.
2. W. Krätschmer, L. D. Lamb, K. Fostiropoulos, and D. R. Huffman, *Nature* **1990**, *347*, 354–358.
3. R. C. Haddon, A. S. Perel, R. C. Morris, T. T. M. Plastra, A. F. Hebrad, and R. M. Fleming, *Appl. Phys. Lett.* **1995**, *67*, 121–123.
4. R. C. Haddon, T. Siegrist, R. M. Fleming, P. M. Bridenbaugh, and R. A. Laudise, *J. Mater. Chem.* **1995**, *5*, 1719–1724.
5. R. C. Haddon, A. F. Hebard, M. J. Rosseinsky, D. W. Murphy, S. J. Duclos, K. B. Lyons, B. Miller, J. M. Rosamilia, R. M. Fleming, A. R. Kortan, S. H. Glarum, A. V. Makhija, A. J. Muller, R. H. Eick, S. M. Zahurak, R. Tycko, G. Dabbagh, and F. A. Thiel, *Nature* **1991**, *350*, 320–322.
6. R. C. Haddon, *Science* **1993**, *261*, 1545–1550.
7. R. C. Haddon, *Acc. Chem. Res.* **1988**, *21*, 243–249.
8. J. R. Morton, F. Negri, and K. F. Preston, *Acc. Chem. Res.* **1998**, *31*, 63–69.
9. H. Mauser, N. J. R. van Eikema Hommes, T. Clark, A. Hirsch, B. Pietzak, A. Weidinger, and L. Dunsch, *Angew. Chem. Int. Ed. Engl.* **1997**, *36*, 2835–2838.
10. R. C. Haddon, L. E. Brus, and K. Raghavachari, *Chem. Phys. Lett.* **1986**, *131*, 165.
11. L. Echegoyen and L. E. Echegoyen, *Acc. Chem. Res.* **1998**, *31*, 593–601.
12. D. Dubois, K. M. Kadish, S. Flanagan, and L. J. Wilson, *J. Am. Chem. Soc.* **1991**, *113*, 7773–7774.
13. D. Dubois, K. M. Kadish, S. Flanagan, R. E. Haufler, L. P. F. Chibante, and L. J. Wilson, *J. Am. Chem. Soc.* **1991**, *113*, 4364–4366.
14. R. C. Haddon, *Acc. Chem. Res.* **1992**, *25*, 127–133.
15. H. Imahori, K. Hagiwara, T. Akiyama, M. Aoki, S. Taniguchi, T. Okada, M. Shirakawa, and Y. Sakata, *Chem. Phys. Lett.* **1996**, *263*, 545–550.
16. V. Ohlendorf, A. Willnow, H. Hungerbühler, D. M. Guldi, and K.-D. Asmus, *J. Chem. Soc. Chem. Commun.* **1995**, 759–760.
17. T. Kato, T. Kodama, T. Shida, T. Nakagawa, Y. Matsui, S. Suzuki, H. Shiromaru, K. Yamauchi, and Y. Achiba, *Chem. Phys. Lett.* **1991**, *180*, 446–450.

18. Y.-P. Sun, in *Molecular and Supramolecular Photochemistry, Vol. 1*, V. Ramamurthy (ed.). Marcel Dekker, New York, **1997**. 325–390.

19. M. Prato, *J. Mater. Chem.* **1997**, *7*, 1097–1109.

20. H. Imahori and Y. Sakata, *Adv. Mater.* **1997**, *9*, 537–546.

21. C. S. Foote, *Top. Curr. Chem.* **1994**, *169*, 347–363.

22. P. V. Kamat and K.-D. Asmus, *Interface* **1996**, *5*, 22–25.

23. T. W. Ebbesen, K. Tanigaki, and S. Kuroshima, *Chem. Phys. Lett.* **1991**, *181*, 501–504.

24. R. J. Sension, C. M. Phillips, A. Z. Szarka, W. J. Romanow, A. R. McGhie, J. P. J. McCauley, A. B. Smith III, and R. M. Hochstrasser, *J. Phys. Chem.* **1991**, *95*, 6075–6078.

25. D. K. Palit, A. V. Sapre, J. P. Mittal, and C. N. R. Rao, *Chem. Phys. Lett.* **1992**, *195*, 1–6.

26. P. A. Lane, L. S. Swanson, Q. X. Ni, J. Shinar, J. Engel, T. J. Barton, and L. Jones, *Mater. Phys. Rev. Lett.* **1992**, *68*, 887–890.

27. L. Biczok, H. Linschitz, and R. I. Walter, *Chem. Phys. Lett.* **1992**, *195*, 339–346.

28. C. A. Steren, P. R. Levstein, H. van Willigen, H. Linschitz, and L. Biczok, *Chem. Phys. Lett.* **1993**, *204*, 23–28.

29. Y. Zeng, L. Biczok, and H. Linschitz, *J. Phys. Chem.* **1992**, *96*, 5237–5239.

30. M. Terazima, K. Sakurada, N. Hirota, H. Shinohara, and Y. Saito, *J. Phys. Chem.* **1993**, *97*, 5447–5450.

31. V. A. Nadtochenko, I. V. Vasil'ev, N. N. Denisov, I. V. Rubtsov, A. S. Lobach, A. P. Moravskii, and A. F. Shestakov, *J. Photochem. Photobiol. A* **1993**, *70*, 153–156.

32. V. Brezova, A. Stasko, P. Rapta, D. M. Guldi, K.-D. Asmus, and K.-P. Dinse, *Magn. Reson. Chem.* **1997**, *35*, 795–801.

33. V. Brezova, A. Stasko, P. Rapta, G. Domschke, A. Bartl, and L. Dunsch, *J. Phys. Chem.* **1995**, *99*, 16234–16241.

34. H. T. Etheridge and R. B. Weisman, *J. Phys. Chem.* **1995**, *99*, 2782–2787.

35. M. R. Fraelich and R. B. Weisman, *J. Phys. Chem.* **1993**, *97*, 11145–11147.

36. J. V. Caspar and L. S. Wang, *Chem. Phys. Lett.* **1994**, *218*, 221–228.

37. K. Tanigaki, T. W. Ebbesen, and S. Kuroshima, *Chem. Phys. Lett.* **1991**, *185*, 189–192.

38. J. S. Arbogast and C. S. Foote, *J. Am. Chem. Soc.* **1991**, *113*, 8886–8889.

39. N. M. Dimitrijevic and P. V. Kamat, *J. Phys. Chem.* **1992**, *96*, 4811–4814.

40. S. M. Argentine, K. T. Kotz, T. Rudalevige, D. Zaziski, A. H. Francis, R. Zand, and J. A. Schleuter, *Res. Chem. Intermed.* **1997**, *23*, 601–620.

41. H. Levanon, V. Meiklyar, S. Michaeli, and D. Gamliel, *J. Am. Chem. Soc.* **1993**, *115*, 8722–8727.

42. R. Seshadri, C. N. R. Rao, H. Pal, T. Mukherjee, and J. P. Mittal, *Chem. Phys. Lett.* **1993**, *205*, 395–398.

43. M. Terazima, N. Hirota, H. Shinohara, and Y. Saito, *Chem. Phys. Lett.* **1992**, *195*, 333–338.

44. H. T. Etheridge, A. B. Smith III, R. D. Averitt, N. J. Halas, and R. B. Weisman, *J. Phys. Chem.* **1995**, *99*, 11306–11308.

45. G. Sauve and P. V. Kamat, in *Recent Advances in the Chemistry and Physics of Fullerenes and Related Materials.*, *Vol. 95-10*, K. M. Kadish and R. S. Ruoff (eds.). The Electrochemical Society, Pennington, NJ, 1995, pp. 399–405.

46. D. M. Guldi, D. Liu, and P. V. Kamat, *J. Phys. Chem. A* **1997**, *101*, 6195–6201.

47. R. V. Bensasson, E. Bienvenue, J.-M. Janot, E. J. Land, S. Leach, and P. Seta, *Chem. Phys. Lett.* **1998**, *283*, 221–226.

48. G. Sauve, P. V. Kamat, K. G. Thomas, K. J. Thomas, S. Das, and M. V. George, *J. Phys. Chem.* **1996**, *100*, 2117–2124.

49. G. Sauve, P. V. Kamat, and R. S. Ruoff, *J. Phys. Chem.* **1995**, *99*, 2162–2165.

50. M. Terazima, N. Hirota, H. Shinohara, and K. Asato, in *Fullerenes, Vol. 95-10*, R. Ruoff and K. Kadish (eds.). The Electrochemical Society, Pennington, NJ, 1995, pp. 267–273.

51. R. V. Bensasson, T. J. Hill, C. Lambert, E. J. Land, S. Leach, and T. G. Truscott, *Chem. Phys. Lett.* **1993**, *201*, 326–335.

52. R. V. Bensasson, T. J. Hill, C. Lambert, E. J. Land, S. Leach, and T. G. Truscott, *Chem. Phys. Lett.* **1993**, *206*, 197–202.

53. K. I. Priyadarsini, H. Mohan, P. R. Birkett, and J. P. Mittal, *J. Phys. Chem.* **1996**, *100*, 501–506.

54. D. M. Guldi, H. Hungerbühler, and K. D. Asmus, *J. Phys. Chem.* **1995**, *99*, 9380–9385.

55. A. A. J. Krasnovsky and C. S. Foote, *J. Am. Chem. Soc.* **1993**, *115*, 6013–6016.

56. K. Yoshino, X. H. Yin, K. Muro, S. Kiyomatsu, S. Morita, A. A. Zakhidov, T. Noguchi, and T. Ohnishi, *Springer Ser. Solid State Sci.* **1993**, *91*, 286–291.

57. G. Yu and A. J. Heeger, *J. Appl. Phys.* **1995**, *78*, 4510–4515.

58. Y. Wang, R. West, and C. H. Yuan, *J. Am. Chem. Soc.* **1993**, *115*, 3844–3845.

59. S. M. Silence, C. A. Walsh, J. C. Scott, and W. E. Moerner, *Appl. Phys. Lett.* **1992**, *61*, 2967–2969.

60. C. E. Bunker, G. E. Lawson, and Y. P. Sun, *Macromolecules* **1995**, *28*, 3744–3746.

61. G. Agostini, C. Corvaja, M. Maggini, L. Pasimeni, and M. Prato, *J. Phys. Chem.* **1996**, *100*, 13416–13420.

62. J. L. Bourdelande, J. Font, and R. Gonzalez-Moreno, *J. Photochem. Photobiol. A* **1995**, *90*, 65–67.

63. S. V. Frolov, P. A. Lane, M. Ozaki, K. Yoshino, and Z. V. Vardeny, *Chem. Phys. Lett* **1998**, *286*, 21–27.

64. A. Ito, T. Morikawa, and T. Takahashi, *Chem. Phys. Lett.* **1993**, *211*, 333–336.

65. R. A. J. Janssen, M. P. T. Christiaans, C. Hare, N. Martin, N. S. Sariciftci, A. J. Heeger, and F. Wudl, *J. Chem. Phys.* **1995**, *103*, 8840–8845.

66. N. S. Sariciftci, L. Smilowitz, A. J. Heeger, and F. Wudl, *Science* **1992**, *258*, 1474–1476.

67. N. S. Sariciftci, L. Smilowitz, D. Braun, G. Srdanov, F. Wudl, and A. J. Heeger, *Synth. Met.* **1993**, *56*, 3125–3130.

68. G. Sauve, N. M. Dimitrijevic, and P. V. Kamat, *J. Phys. Chem.* **1995**, *99*, 1199–1203.

69. A. Watanabe and O. Ito, *J. Chem. Soc. Chem. Commun.* **1994**, 1285–1286.

70. Y. Wang and A. Suna, *J. Phys. Chem. B* **1997**, *101*, 5627–5638.

71. B. Kraabel, C. H. Lee, D. McBranch, D. Moses, N. S. Sariciftci, and A. J. Heeger, *Chem. Phys. Lett.* **1993**, *213*, 389–394.

72. R. A. J. Janssen, J. C. Hummelen, K. Lee, K. Pakbaz, N. S. Sariciftci, A. J. Heeger, and F. Wudl, *J. Chem. Phys.* **1995**, *103*, 788–793.

73. A. Itaya, I. Suzuki, Y. Tsuboi, and H. Miyasaka, *J. Phys. Chem.* **1997**, *101*, 5118–5123.

74. M. Gevaert and P. V. Kamat, *J. Phys. Chem.* **1992**, *96*, 9883–9888.

75. C. J. Hawker, *Macromolecules* **1994**, *27*, 4836–4837.

76. L. Y. Chiang, L. Y. Wang, S.-M. Tseng, J.-S. Wu, and K.-H. Heieh, *J. Chem. Soc. Chem. Commun.* **1994**, 2675–2676.

77. R. J. Sension, A. Z. Szarka, G. R. Smith, and R. M. Hochstrasser, *Chem. Phys. Lett.* **1991**, *185*, 179–183.

78. H. N. Ghosh, H. Pal, A. V. Sapre, and J. P. Mittal, *J. Am. Chem. Soc.* **1993**, *115*, 11722–11727.

79. B. Ma, G. E. Lawson, C. E. Bunker, A. Kitaygorodskiy, and Y.-P. Sun, *Chem. Phys. Lett.* **1995**, *247*, 51–56.

80. D. M. Guldi, G. Torres-Garcia, and J. Mattay, *J. Phys. Chem. A* **1998**, *102*, 9679–9685.

81. G. Black, E. Dunkle, E. A. Dorko, and L. A. Schlie, *J. Photochem. Photobiol. A* **1993**, *70*, 147–151.

82. C. Taliani, G. Ruani, R. Zamboni, R. Danieli, S. Rossini, V. N. Denisov, V. M. Burlakov, F. Negri, G. Orlandi, and F. Zerbetto, *J. Chem. Soc. Chem. Commun.* **1993**, 220–222.

83. D. I. Schuster, P. S. Baran, R. K. Hatch, A. Khan, and S. R. Wilson, *Chem. Commun.* **1998**, 2493–2494.

84. P. V. Kamat, *J. Am. Chem. Soc.* **1991**, *113*, 9705–9707.

85. P. V. Kamat, I. Bedja, and S. Hotchandani, *J. Phys. Chem.* **1994**, *98*, 9137–9142.

86. A. Stasko, V. Brezova, S. Biskupic, K.-P. Dinse, P. Schweitzer, and M. Baumgarten, *J. Phys. Chem.* **1995**, *99*, 8782–8789.

87. P. V. Kamat, *Chem. Rev.* **1993**, *93*, 267–300.

88. J. M. Wood, B. Kahr, S. H. Hoke II, L. Dejarme, R. G. Cooks, and D. Ben-Amotz, *J. Am. Chem. Soc.* **1991**, *113*, 5907.

89. K. M. Creegan, J. L. Robbins, W. K. Robbins, J. M. Millar, R. D. Sherwood, P. J. Tindall, D. M. Cox, A. B. Smith III, J. McCauley, D. R. Jones, and R. T. Gallagher, *J. Am. Chem. Soc.* **1992**, *114*, 1103–1105.

90. D. Heyman and L. P. F. Chibante, *Chem. Phys. Lett.* **1993**, *207*, 339–342.

91. M. D. Pace, T. C. Chestidis, J. J. Yin, and J. Milliken, *J. Phys. Chem.* **1992**, *96*, 6855.

92. M. Anpo, S. G. Zhang, S. Okamoto, H. Yamashita, and Z. Gu, *Res. Chem. Intermed.* **1995**, *21*, 631–642.

93. J. W. Arbogast, A. P. Darmanyan, C. S. Foote, Y. Rubin, F. N. Diederich, M. M. Alvarez, S. J. Anz, and R. L. Whetten, *J. Phys. Chem.* **1991**, *95*, 11–12.

94. T. Osaki, Y. Tai, M. Tazawa, S. Tanemura, K. Inukai, K. Ishiguro, Y. Sawaki, Y. Saito, H. Shinohara, and H. Nagashima, *Chem. Lett.* **1993**, 789–792.

95. J. W. Arbogast, C. S. Foote, and M. Kao, *J. Am. Chem. Soc.* **1992**, *114*, 2277–2279.

96. D. M. Guldi, R. E. Huie, P. Neta, H. Hungerbühler, and K.-D. Asmus, *Chem. Phys. Lett.* **1994**, *223*, 511–516.

97. D. M. Guldi, M. Maggini, G. Scorrano, and M. Prato, *Res. Chem. Intermed.* **1997**, *23*, 561–573.

98. D. M. Guldi, M. Maggini, G. Scorrano, and M. Prato, *J. Am. Chem. Soc.* **1997**, *119*, 974–980.

99. M. M. Alam, A. Watanabe, and O. Ito, *J. Photochem. Photobiol. A* **1997**, *104*, 59–64.

100. Y. Sasaki, Y. Yoshikawa, A. Watanabe, and O. Ito, *J. Chem. Soc. Faraday Trans.* **1995**, *91*, 2287–2290.

101. O. Ito, Y. Sasaki, Y. Yoshikawa, and A. Watanabe, *J. Phys. Chem.* **1995**, *99*, 9838–9842.

102. A. Samanta and P. V. Kamat, *Chem. Phys. Lett.* **1992**, *199*, 635–639.

103. A. Watanabe and O, Ito, *J. Phys. Chem.* **1994**, *98*, 7736–7740.

104. M. Bennati, A. Grupp, P. Bäuerle, and M. Mehring, *Chem. Phys.* **1994**, *185*, 221–227.

105. M. Bennati, A. Grupp, P. Bäuerle, and M. Mehring, *Mol. Cryst. Liq. Cryst.* **1994**, *256*, 751–756.

106. Y.-P. Sun, B. Ma, and G. E. Lawson, *Chem. Phys. Lett.* **1995**, *233*, 57–62.

107. Y. Wang, *J. Phys. Chem.* **1992**, *96*, 764–767.

108. E. Schaffner and H. Fischer, *J. Phys. Chem* **1993**, *97*, 13149–13151.

109. S. Fukuzumi, T. Suenobu, M. Patz, T. Hirasaka, S. Itoh, M. Fujitsuka, and O. Ito, *J. Am. Chem. Soc.* **1998**, *120*, 8060–8068.

110. G. E. Lawson, A. Kitaygorodskiy, B. Ma, C. E. Bunker, and Y.-P. Sun, *J. Chem. Soc. Chem. Commun.* **1995**, 2225–2226.

111. K. Mikami, S. Matsumoto, A. Ishida, S. Takamuku, T. Suenobu, and S. Fukuzumi, *J. Am. Chem. Soc.* **1995**, *117*, 11134–11141.

112. T. Nojiri, M. M. Alam, H. Konami, A. Watanabe, and O. Ito, *J. Phys. Chem. A* **1997**, *101*, 7943–7947.

113. Y. Sasaki, M. Fujitsuka, A. Watanabe, and O. Ito, *J. Chem. Soc. Faraday Trans.* **1997**, *93*, 4275–4279.

114. R. D. Scurlock and P. R. Ogilby, *J. Photochem. Photobiol. A* **1995**, *91*, 21–25.

115. C. Schlebusch, B. Kessler, S. Cramm, and W. Eberhardt, *Synth. Met.* **1996**, *77*, 151–154.

116. V. A. Nadtochenko, N. N. Denisov, I. V. Rubtsov, A. S. Lobach, and A. P. Moravskii, *Chem. Phys. Lett.* **1993**, *208*, 431–435.

117. V. A. Nadtochenko, N. N. Denisov, I. V. Rubtsov, A. S. Lobach, and A. P. Moravsky, *Russ. Chem. Bull.* **1993**, *42*, 1171–1173.

118. V. A. Nadtochenko, N. N. Denisov, A. S. Lobach, and A. P. Moravskii, *Zh. Fiz. Khim.* **1994**, *68*, 228–231.

119. C. A. Steren and H. van Willigen, *Proc. Indian Acad. Sci., Chem. Sci.* **1994**, *106*, 1671–1679.

120. C. A. Steren, H. van Willigen, L. Biczók, N. Gupta, and H. Linschitz, *J. Phys. Chem.* **1996**, *100*, 8920–8926.

121. S. Michaeli, V. Meiklyar, M. Schulz, K. Möbius, and H. Levanon, *J. Phys. Chem.* **1994**, *98*, 7444–7447.

122. S. Michaeli, V. Meiklyar, B. Endeward, K. Möbius, and H. Levanon, *Res. Chem. Intermed.* **1997**, *23*, 505–517.

123. C. A. Steren, H. van Willigen, and M. Fanciulli, *Chem. Phys. Lett.* **1995**, *245*, 244–248.

124. S. Nonell, J. W. Arbogast, and C. S. Foote, *J. Phys. Chem.* **1992**, *96*, 4169–4170.

125. D. M. Guldi, P. Neta, and K.-D. Asmus, *J. Phys. Chem.* **1994**, *98*, 4617–4621.

126. D. M. Guldi and K.-D. Asmus, *J. Am. Chem. Soc.* **1997**, *119*, 5744–5745.

127. A. S. Boutorine, H. Tokuyama, M. Takasugi, H. Isobe, E. Nakamura, and C. Helene, *Angew. Chem.* **1994**, *106*, 2526–2529.

128. R. Sijbesma, G. Srdanov, F. Wudl, J. A. Castoro, C. Wilkins, S. H. Friedman, D. L. DeCamp, and G. L. Kenyon, *J. Am. Chem. Soc.* **1993**, *115*, 6510–6512.

129. R. Bernstein, F. Prat, and C. S. Foote, *J. Am. Chem. Soc.* **1999**, *121*, 464–465.

130. Y. N. Yamakoshi, T. Yagami, S. Sueyoshi, and N. Miyata, *J. Org. Chem.* **1996**, *61*, 7236–7237.

131. T. Andersson, K. Nilsson, M. Sundahl, G. Westman, and O. Wennerström, *J. Chem. Soc. Chem. Commun.* **1992**, 604–606.

132. R. M. Williams and J. W. Verhoeven, *Recl. Trav. Chim. Pays-Bas* **1992**, *111*, 531–532.

133. H. Hungerbühler, D. M. Guldi, and K.-D. Asmus, *J. Am. Chem. Soc.* **1993**, *115*, 3386–3387.

134. M. Sundahl, T. Andersson, K. Nilsson, O. Wennerström, and G. Westman, *Synth. Met.* **1993**, *55*, 3252–3257.

135. K. I. Priyadarsini, H. Mohan, J. P. Mittal, D. M. Guldi, H. Hungerbühler, and K.-D. Asmus, *J. Phys. Chem.* **1994**, *98*, 9565–9569.

136. K. I. Priyadarsini, H. Mohan, A. K. Tyadi, and J. P. Mittal, *J. Phys. Chem.* **1994**, *98*, 4756–4759.

137. N. M. Dimitrijevic and P. V. Kamat, *J. Phys. Chem.* **1993**, *97*, 7623–7626.

138. K. I. Priyadarsini and H. Mohan, *J. Photochem. Photobiol. A* **1995**, 63–67.

139. D. M. Guldi, *J. Phys. Chem. B* **1997**, *101*, 9600–9605.

140. J. Eastoe, E. R. Crooks, A. Beeby, and R. K. Heenan, *Chem. Phys. Lett* **1995**, *245*, 571–577.

141. E. R. Crooks, J. Eastoe, and A. Beeby, *J. Chem. Soc. Faraday Trans.* **1997**, *93*, 4131–4136.

142. Y. N. Yamakoshi, T. Yagami, K. Fukuhara, S. Sueyoshi, and N. Miyata, *J. Chem. Soc. Chem. Commun.* **1994**, 517–518.

143. A. Beeby, J. Eastoe, and E. R. Crooks, *Chem. Comun.* **1996**, 901–902.

144. S. Niu and D. Mauzerall, *J. Am. Chem. Soc.* **1996**, *118*, 5791–5795.

145. R. V. Bensasson, E. Bienvenue, M. Dellinger, S. Leach, and P. Seta, *J. Phys. Chem.* **1994**, *98*, 3492–3500.

146. K. C. Hwang and D. C. Mauzerall, *Nature* **1993**, *361*, 138–140.

147. R. Taylor and D. R. M. Walton, *Nature* **1993**, *363*, 685–693.

148. A. Hirsch, *Synthesis* **1995**, 895–913.

149. F. Diederich and C. Thilgen, *Science* **1996**, *271*, 317–323.

150. F. Diederich, *Pure Appl. Chem.* **1997**, *69*, 395–400.

151. A. L. Balch and M. M. Olmstadt, *Chem. Rev.* **1998**, *98*, 2123–2165.

152. N. Martín, L. Sánchez, B. Illescas, and I. Pérez, *Chem. Rev.* **1998**, *98*, 2527.

153. A. W. Jensen, S. R. Wilson, and D. I. Schuster, *Bioorganic Medicinal Chem.* **1996**, *4*, 767–779.

154. M. Prato and M. Maggini, *Acc. Chem. Res.* **1998**, *31*, 519–526.

155. A. Hirsch, *The Chemistry of the Fullerenes.* Georg Thieme Verlag, Stuttgart, 1994.

156. F. Wudl, *Acc. Chem. Res.* **1992**, *25*, 157–161.

157. A. Hirsch, I. Lamparth, and G. Schick, *Liebigs Ann.* **1996**, 1725–1734.

158. D. M. Guldi and K.-D. Asmus, *J. Phys. Chem. A* **1997**, *101*, 1472–1481.

159. B. Ma, C. E. Bunker, R. Guduru, X.-F. Zhang, and Y.-P. Sun, *J. Phys. Chem. A* **1997**, *101*, 5626–5632.

160. M. Brustolon, A. Zoleo, G. Agostini, and M. Maggini, *J. Phys. Chem. A* **1998**, *102*, 6331–6339.

161. R. V. Bensasson, E. Bienvenue, C. Fabre, J.-M. Janot, E. J. Land, S. Leach, V. Leboulaire, A. Rassat, S. Roux, and P. Seta, *Chem. Eur. J.* **1998**, *4*, 270–278.

162. D. Zhou, L. Gan, H. Tan, C. Luo, C. Huang, G. Yao, and B. Zhang, *J. Photochem. Photobiol. A* **1996**, *99*, 37–43.

163. C. Luo, M. Fujitsuka, A. Watanabe, O. Ito, L. Gan, Y. Huang, and C.-H. Huang, *J. Chem. Soc. Faraday Trans.* **1998**, *94*, 527–532.

164. D. M. Guldi and M. Maggini, *Gazz. Chim. Ital.* **1997**, *127*, 779–785.

165. J. L. Anderson, Y.-Z. An, Y. Rubin, and C. S. Foote, *J. Am. Chem. Soc.* **1994**, *116*, 9763–9764.

166. R. V. Bensasson, E. Bienvenue, J.-M. Janot, S. Leach, P. Seta, D. I. Schuster, S. R. Wilson, and H. Zhao, *Chem. Phys. Lett.* **1995**, *245*, 566–570.

167. J. C. Hummelen, B. Knight, J. Pavlovich, R. Gonzalez, and F. Wudl, *Science* **1995**, *269*, 1554–1556.

168. M. Prato, Q. C. Li, F. Wudl, and V. Lucchini, *J. Am. Chem. Soc.* **1993**, *115*, 1148–1150.

169. M. Eiermann, F. Wudl, M. Prato, and M. Maggini, *J. Am. Chem. Soc.* **1994**, *116*, 8364–8365.

170. D. M. Guldi, H. Hungerbühler, K.-D. Asmus, I. Carmichael, and M. Maggini, *J. Phys. Chem.*, **2000**, in press.

171. R. M. Williams, J. M. Zwier, and J. W. Verhoeven, *J. Am. Chem. Soc.* **1995**, *117*, 4093–4099.

172. R. M. Williams, M. Koeberg, J. M. Lawson, Y.-Z. An, Y. Rubin, M. N. Paddon-Row, and J. W. Verhoeven, *J. Org. Chem.* **1996**, *61*, 5055–5062.

173. M. J. Shephard and M. N. Paddon-Row, *Aust. J. Chem.* **1996**, *49*, 395–403.

174. J. Llacay, J. Veciana, J. Vidal-Gancedo, J. L. Bourdelande, R. Gonzalez-Moreno, and C. Rovira, *J. Org. Chem.* **1998**, *63*, 5201–5210.

175. K. G. Thomas, V. Biju, M. V. George, D. M. Guldi, and P. V. Kamat, *J. Phys. Chem. B* **1999**, *102*, 8864–8869.

176. R. Deschenaux, M. Even, and D. Guillon, *Chem. Commun.* **1998**, 537–538.

177. M. Diekers, A. Hirsch, S. Pyo, J. Rivera, and L. Echegoyen, *Eur. J. Org. Chem.* **1998**, 1111–1121.

178. H. Imahori, K. Hagiwara, M. Aoki, T. Akiyama, S. Taniguchi, T. Okada, M. Shirakawa, and Y. Sakata, *J. Am. Chem. Soc.* **1996**, *118*, 11771–11782.

179. P. A. Liddell, D. Kuciauskas, J. P. Sumida, B. Nash, D. Nguyen, A. L. Moore, T. A. Moore, and D. Gust, *J. Am. Chem. Soc.* **1997**, *119*, 1400–1405.

180. E. Dietel, A. Hirsch, J. Zhou, and A. Rieker, *J. Chem. Soc. Perkin Trans. 2* **1998**, 1357–1364.

181. P. Cheng, S. R. Wilson, and D. I. Schuster, *Chem. Commun.* **1999**, 89–90.

182. M. Maggini, A. Dono, G. Scorrano, and M. Prato, *J. Chem. Soc. Chem. Commun.* **1995**, 845–846.

183. M. Maggini, D. M. Guldi, S. Mondini, G. Scorrano, F. Paolucci, P. Ceroni, and S. Roffia, *Chem. Eur. J.* **1998**, *4*, 1992–2000.

184. D. M. Guldi, M. Maggini, E. Menna, G. Scorrano, P. Ceroni, M. Marcaccio, F. Paolucci, and S. Roffia, *J. Phys. Chem.* **2000**, in press.

185. P. A. Liddell, J. P. Sumida, A. N. Macpherson, L. Noss, G. R. Seely, K. N. Clark, A. L. Moore, T. A. Moore, and D. Gust, *Photochem. Photobiol.* **1994**, *60*, 537–541.

186. D. Gust, T. A. Moore, and A. L. Moore, *Res. Chem. Intermed.* **1997**, *23*, 621–651.

187. D. Kuciauskas, S. Lin, G. R. Seely, A. L. Moore, T. A. Moore, D. Gust, T. Drovetskaya, C. A. Reed, and P. D. W. Boyd, *J. Phys. Chem.* **1996**, *100*, 15926–15932.

188. D. Carbonera, M. Di Valentin, C. Corvaja, G. Agostini, G. Giacometti, P. A. Liddell, D. Kuciauskas, A. L. Moore, T. A. Moore, and D. J. Gust, *J. Am. Chem. Soc.* **1998**, *120*, 4398–4405.

189. H. Imahori, K. Hagiwara, T. Akiyama, S. Taniguchi, T. Okada, and Y. Sakata, *Chem. Lett.* **1995**, 265–266.

190. S. Higashida, H. Imahori, T. Kaneda, and Y. Sakata, *Chem. Lett.* **1998**, 605–606.

191. H. Imahori, K. Yamada, M. Hasegawa, S. Taniguchi, T. Okada, and Y. Sakata, *Angew. Chem. Int. Ed. Engl.* **1997**, *36*, 2626–2629.

192. T. D. M. Bell, T. A. Smith, K. P. Ghiggino, M. G. Ranasinghe, M. J. Shephard, and M. N. Paddon-Row, *Chem. Phys. Lett.* **1997**, *268*, 223–228.

193. P. S. Baran, R. R. Monaco, A. U. Khan, D. I. Schuster, and S. R. Wilson, *J. Am. Chem. Soc.* **1997**, *119*, 8363–8364.

194. N. S. Sariciftci, F. Wudl, A. J. Heeger, M. Maggini, G. Scorrano, M. Prato, J. Bourassa, and P. C. Ford, *Chem. Phys. Lett.* **1995**, *247*, 510–514.

195. A. Polese, S. Mondini, A. Bianco, C. Toniolo, G. Scorrano, D. M. Guldi, and M. Maggini, *J. Am. Chem. Soc.* **1999**, *121*, 3446–3452.

196. D. Armspach, E. C. Constable, F. Diederich, C. E. Housecraft, and J.-F. Nierengarten, *J. Chem. Soc. Chem. Commun.* **1996**, 2009–2010.

197. D. Armspach, E. C. Constable, F. Diederich, C. E. Housecroft, and J.-F. Nierengarten, *Chem. Eur. J.* **1998**, *4*, 723–733.

198. T. Gareis, O. Köthe, and J. Daub, *Eur. J. Org. Chem.* **1998**, 1549–1557.

199. Y. Nakamura, T. Minowa, S. Tobita, H. Shizuka, and J. Nishimura, *J. Chem. Soc. Perkin Trans. 2* **1995**, 2351–2357.

200. Y. Nakamura, T. Minowa, Y. Hayashida, S. Tobita, H. Shizuka, and J. Nishimura, *J. Chem. Soc. Faraday Trans.* **1996**, *92*, 377–382.

201. N. Armaroli, F. Diederich, C. O. Dietrich-Buchecker, L. Flamigni, G. Marconi, J.-F. Nierengarten, and J.-P. Sauvage, *Chem. Eur. J.* **1998**, *4*, 406–416.

202. F. Diederich, J.-F. Nierengarten, and J.-P. Sauvage, *J. Chem. Soc. Chem. Commun.* **1995**, 781–782.

203. P. R. Ashton, F. Diederich, M. Gomez-Lopez, J.-F. Nierengarten, J. A. Preece, F. M. Raymo, and J. F. Stoddart, *Angew. Chem. Int. Ed. Engl.* **1997**, *36*, 1448–1451.

204. N. Armaroli, F. Diederich, L. Echegoyen, T. Habicher, L. Flamigni, G. Marconi, and J.-F. Nierengarten, *New. J. Chem.* **1999**, 77–83.

205. T. Da Ros, M. Prato, D. M. Guldi, E. Alessio, M. Ruzzi, and L. Pasimeni, *Chem. Commun.* **1999**, 635–636.

206. D. M. Guldi, H. Hungerbühler, and K.-D. Asmus, *J. Phys. Chem.* **1995**, *99*, 13487–13493.

207. D. M. Guldi, H. Hungerbühler, and K.-D. Asmus, *J. Phys. Chem. A* **1997**, *101*, 1783–1786.

208. D. M. Guldi, *J. Phys. Chem. A* **1997**, *101*, 3895–3900.

209. H. Tokuyama, S. Yamago, E. Nakamura, T. Shiraki, and Y. Sugiura, *J. Am. Chem. Soc.* **1993**, *115*, 7918–7919.

210. A. Skiebe and A. Hirsch, *J. Chem. Soc. Chem. Commun.* **1994**, 335–336.

211. D. M. Guldi, *Res. Chem. Intermed.* **1997**, *23*, 653–673.

212. D. M. Guldi, H. Hungerbühler, and K.-D. Asmus, *J. Phys. Chem. B* **1999**, *103*, 1444–1453.

213. H. Mohan, D. K. Palit, J. P. Mittal, L. Y. Chiang, K.-D. Asmus, and D. M. Guldi, *J. Chem. Soc. Faraday Trans.* **1998**, *94*, 359–363.

214. D. K. Palit, H. Mohan, and J. P. Mittal, *J. Phys. Chem. A* **1998**, *102*, 4456–4461.

215. D. K. Palit, H. Mohan, P. R. Birkett, and J. P. Mittal, *J. Phys. Chem. A* **1997**, *101*, 5418–5422.

216. A. Gruss, K.-P. Dinse, A. Hirsch, B. Nuber, and U. Reuther, *J. Am. Chem. Soc.* **1997**, *119*, 8728–8729.

217. K. Hasharoni, C. Bellavia-Lund, M. Keshavarz-K., G. Srdanov, and F. Wudl, *J. Am. Chem. Soc.* **1997**, *119*, 11128–11129.

218. J. G. Linssen, K. Dürr, M. Hanack, and A. Hirsch, *J. Chem. Soc., Chem. Commun.* **1995**, 103–104.

CHAPTER 6

CALCULATIONS OF HIGHER FULLERENES AND QUASI-FULLERENES

ZDENĚK SLANINA, XIANG ZHAO, PRADEEP DEOTA, and EIJI ŌSAWA

6.1 INTRODUCTION

Higher fullerenes are of considerable interest for both experiment [1] and theory [2]. Experimental studies have been based on sophisticated liquid chromatography [3], ^{13}C nuclear magnetic resonance (NMR) [4], and, more recently, on ^{3}He NMR spectroscopy [5]. This chapter concentrates on the computational approaches to higher fullerenes. For a wider perspective, we also cover *quasi-fullerenes* [6], that is, the cages containing not only traditional five- and six-membered rings but also other cycles.

The very early history [7–10] of carbon clusters (i.e., before fullerenes) goes back to Hahn and co-workers [11] in the 1940s (though their upper observation limit was only C_{15}). In the 1950s and 1960s experiments [12–14] had already reached to C_{33}. At about that time simple computations had been applied [15,16] to the problem, in particular, by Pitzer and Clementi [15], and later by Hoffmann [17]. The highest computational level then applied was represented [18] by the semiempirical MINDO/2 method. Nevertheless, the computed relative energies for small carbon clusters were essentially correct. On the other hand, much simpler Hückel-type calculations of aromatic systems missed suggestions [19,20] of three-dimensional carbonaceous objects and especially the first visualization [20] of C_{60}. Then, after a few years, the first Hückel calculations of C_{60} were performed [21,22]; however, they also did not attract any experimental interest.

The picture is very different now. Computational support of experiments in fullerene research is common and widespread; it is used as a complementary tool for rational and efficient understanding of the unique species. The fullerene

Fullerenes: Chemistry, Physics, and Technology, Edited by Karl M. Kadish and Rodney S. Ruoff.
0-471-29089-0 Copyright © 2000 John Wiley & Sons, Inc.

family has been expanding along three lines—derivatization or functionalization of the C_{60} and C_{70} cages, their inflation toward higher fullerenes and nanotubes, and their compression toward smaller cages like C_{36} [23]. In recent years knowledge of higher fullerenes has developed considerably [4,24,25]. Several experimental groups have constantly been working on isolation and characterization of higher fullerenes C_n with n reaching well over 90. The ^{13}C NMR analysis is considerably complicated. One observed spectrum can sometimes be consistent with several different structures. In spite of all the difficulties more than 20 stable fullerenes C_n have presently been characterized (n from 60 to 96). Clearly, the number of topologically possible isomeric cages is increasingly high for n around 90—for example, there are 99,918 isomeric closed cages [26] of C_{90} built purely from pentagons and hexagons (this count increases to 199,128 when enantiomers are regarded as distinct). A need for some simplifying concepts is self-evident: Neither in experiment nor in computations can such huge counts be considered. There is a powerful though empirical way in which to scale down the isomeric counts, namely, the so-called isolated pentagon rule (IPR) approach. Within the IPR concept, the structures in which all the pentagons are isolated (i.e., surrounded by hexagons only) are considered as particularly stable. An early heuristic discussion of the IPR concept was presented by Kroto in 1987 [27].

Quantum-chemical computations of higher fullerenes at both semiempirical and *ab initio* levels are typically carried out within the IPR concept. Very recently specific non-IPR structures [28,29] have been considered too. However, even if the IPR concept is strictly applied, the computations almost always end with several isomeric structures of relatively low energy. Moreover, the potential energy itself cannot decide the stability question, as entropy can overcompensate the enthalpy part of the Gibbs function at high temperatures, and in fact an isomeric mixture is produced [30]. Several such mixtures of fullerene isomers have been studied extensively—C_{78}, C_{80}, C_{82}, C_{84}, C_{86}, C_{88}, C_{90}, and so on. The computations have clearly demonstrated that the higher isomeric systems can hardly be understood without considering temperature effects, or in other words, the entropy contributions. This requirement naturally comes from the very fact that fullerenes are routinely synthesized at very high temperatures. This unique feature even allows for a very peculiar physicochemical event [31]—that a structure other than the ground-state structure dominates the high-temperature mixture.

There has been a general, established computational scheme for the fullerene computations. The topologically possible structures are first generated by a combinatorial treatment. At this stage, only topology–bond connectivity-matters, while any energetics is ignored. Then, as the second step, the geometry optimizations can start, possibly at a molecular-mechanical level initially. In any case, the structures must be reoptimized at least with semiempirical methods like AM1, PM3, or SAM1. This requirement is essential as true electronic effects, and Jahn–Teller ones in particular, are never reflected in molecular-mechanical schemes. The geometry optimizations are followed by semiempirical harmonic vibrational analysis (vibrational analysis of fullerenes at a sophisticated *ab initio* level is still too demanding on computational resources). However, the interisomeric energetics is to be checked further with affordable *ab initio* approximations [32]. Depending on the system, those higher treatments can be Hartree–Fock (HF) self-consistent field (SCF) approaches like HF/4-31G or HF/6-31G*, or better a correlated procedure like B3LYP/6-31G* or even the second-order Møller–Plesset perturbation treatment, though with the frozen-core option (MP2 = FC/6-31G*).

The entropy contributions are also included in order to obtain realistic values of the relative concentrations (mole fractions) of the fullerene isomers. The entropy or temperature effects are included by means of the isomeric partition functions (the rigid-rotor and harmonic-oscillator approximation). Among other important contributions, chirality factors emerge as important and sometimes even decisive terms. Special attention has to be given to the symmetries of the optimized structures. Other potentially important aspects of the thermodynamic treatment are still under development, like anharmonicities of the vibrations, nonrigidity of the rotations, or electronic excitations. This chapter surveys the applications of the combined quantum-chemical and statistical-mechanical treatment to several groups of isomers of higher fullerenes and also relates the computations to the available observations. In particular, it concentrates on the IPR isomeric mixtures of C_{76} [33–42], C_{78} [43–46], C_{80} [47–50], C_{82} [51–56], C_{84} [57–63], C_{86} [64–67], C_{88} [65–67], and C_{90} [68–70]. However, some non-IPR and *quasi*-fullerene species are covered too, including the widely discussed C_{36}.

6.2 ENERGETICS OF FULLERENES

The very first MINDO/2 computations of one-, two-, and three-dimensional small carbon clusters C_n pointed out [18,71] a simple, smooth dependency of the relative heats of formation $\Delta H^{\circ}_{f,298}/n$ on the number of carbons n (Fig. 6.1). Of course, at that time virtually nothing was known for n over 8 (Fig. 6.2), the computational upper limit of semiempirical quantum chemistry of those days. After two decades, the curve was extended into the fullerene domain, for example, with *ab initio* HF/STO-3G computations [72,73], or with simple semiempirical tight-binding computations [74], or with the MNDO method [75,76]. The qualitative picture is always the same. Although a shallow minimum could possibly be seen in some of the curves for C_{60}, basically, they are still smooth decreasing dependencies.

There is a simple way in which to rationalize the finding. Let us limit our reasoning to the IPR fullerenes. We deal with two types of bonds, frequently called the 5,6-bonds (between pentagons and hexagons) and 6,6-bonds (shared by two hexagons). Let us suppose, moreover, that those two types of bonds can be represented [77] by some uniform dissociation energies, $H_{5,6}$ and $H_{6,6}$. In a general IPR fullerene C_n, we always have 60 5,6-bonds, while the number of

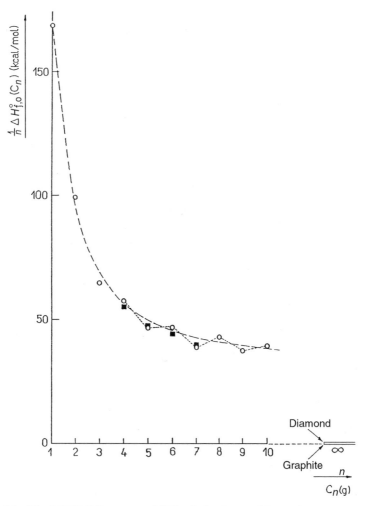

Figure 6.1 The MINDO/2 computed [71] relative heats of formation at room temperature for a few first gas-phase carbon clusters (linear, cyclic, polyhedral).

6,6-bonds is variable, $(3n/2) - 60$. Now, we can readily write for the atomization heat of the considered C_n

$$\Delta H_{at} = 60H_{5,6} + \left(\frac{3n}{2} - 60\right)H_{6,6} \qquad (6.1)$$

The atomization and formation heats for carbon aggregates are linked by the heat of vaporization of carbon, ΔH_{vap}°:

$$\frac{\Delta H_f^\circ}{n} = -\frac{\Delta H_{at}}{n} + \Delta H_{vap}^\circ \qquad (6.2)$$

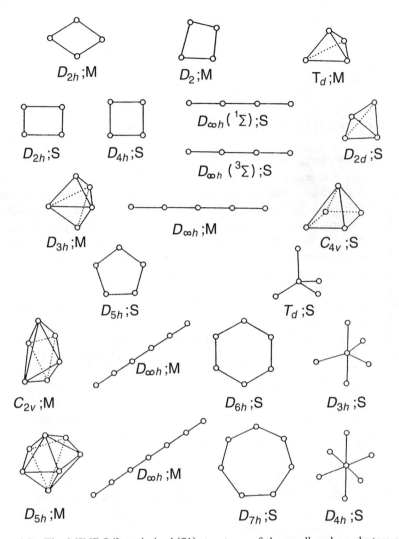

Figure 6.2 The MINDO/2 optimized [71] structures of the small carbon clusters considered in Figure 6.1.

or simply in terms of the bond energies:

$$\frac{\Delta H_f^\circ}{n} = -\frac{3}{2} H_{6,6} + \Delta H_{vap}^\circ + \frac{60}{n}(H_{6,6} - H_{5,6}) \qquad (6.3)$$

This functional dependency can formally be expressed by

$$\frac{\Delta H_f^\circ}{n} = A + \frac{B}{n} \qquad (6.4)$$

where B is a positive constant and A is a relatively small number. Hence, Eq. (6.4) is indeed a smoothly decreasing curve of the type shown in Figure 6.1. From such a curve, one cannot read a particular stability of C_{60} or C_{70}. We have to consider entropy and/or kinetic effects instead, this being done only rarely at present.

6.3 ALGEBRAIC AND TOPOLOGICAL CONSIDERATIONS

Although the computational studies of fullerenes are based primarily on numerical quantum chemistry, they still use the results of topological techniques. In particular, outputs from enumeration studies and exhaustive generations of fullerene cage topologies are applied. Although there are various topological algorithms, producing equivalent results, the most commonly used data are from Fowler and co-workers [26,78–98], summarized in *An Atlas of Fullerenes* by Fowler and Manolopoulos [26]. Moreover, several other groups have developed the topological approach to fullerenes, so that a rich knowledge has already been established [98–134].

The numbers and connectivity patterns of the IPR fullerene cages have represented the most frequently required and applied topological results. As already mentioned, the special cages in which each pentagon is surrounded by hexagons only should exhibit particularly low energies, although exceptions from the empirical IPR rule are not excluded. Generally, if we do not require the IPR structures and consider general fullerenes built from pentagons and hexagons, the cage isomerism first appears for C_{28} and therefrom is always present. However, the IPR structures start only with C_{60}, as the Euler theorem implies just 12 pentagons for any fullerene cage.

The topological and combinatorial approaches can offer various interesting results on the topological symmetries of fullerenes. In particular, it is possible to clarify conditions for the icosahedral symmetry. For that purpose, the concept of Goldberg polyhedra [80,135,136], developed earlier in pure mathematics, is especially convenient. The Goldberg polyhedra are built from $20(b^2 + bc + c^2)$ vertices, where b and c are nonnegative integers. If $b = c$ or $bc = 0$, the point group of symmetry is I_h, otherwise the symmetry is reduced to its rotational subgroup I (though it is customary to call both cases icosahedral symmetry). For example, C_{60} comes from the choice $b = c = 1$. Other I_h symmetry cases are then, for example, $n = 20, 80, 180, 240, 320, 500, 540$. Of course, those are topological symmetries; they can in reality be lowered by Jahn-Teller effects or because of general energy reasons. However, the algebraic treatment can actually point out which of the Goldberg structures has a closed- or open-shell electronic configuration at the HMO level. Thus, the Jahn-Teller effect can be predicted. The governing parameter is $(b - c)$: if it is divisible by 3, the cluster has a multiple of 60 atoms and it is a closed shell; otherwise the species has $(60n + 20)$ atoms and it is an electronic open shell (so that a Jahn-Teller distortion has to appear, resulting in a symmetry reduction, and also a lower stability). Hence, the topological treatment predicts closed-shell configurations among the Goldberg polyhedra for $n = 60, 180, 240, 420, 540$, and so on. Recently, some of the Goldberg-type polyhedra were optimized [137] at the MNDO level. The computations predict strong curvature at the pentagons and show that an evenly distributed curvature (spherical shape) is strongly disfavored. The MNDO findings are supported by *ab initio* SCF calculations [137]; however, they do not agree with the conclusions from a density functional treatment [138,139].

The Stone–Wales transformation [99] or rearrangement represents another useful concept, important for topological and also for mechanistic reasoning. It is a rather hypothetical cage isomerization process for rearrangements of the rings in fullerenes. Such a pairwise interchange of two pentagons and two hexagons is also called the pyracylene transformation [90]. The scheme can be further generalized if other types of rings are taken into consideration [63,131,132]. It can be viewed as a movement of two atoms during which two bonds are broken. It is not necessarily a convenient, feasible kinetic process—it is thermally (but not photochemically) forbidden.

The kinetic barriers for the Stone–Wales rearrangement, however, can be reduced by some convenient catalytic agent [140,141], even through an autocatalysis by free carbon atoms (or small carbon aggregates) present in the reaction mixture at very high temperatures. Hence, the model calculations [140,141] suggest that catalysis could be of crucial importance not only for production of nanotubes but also of fullerenes themselves. The mechanism of fullerene synthesis is virtually unknown, but various catalytic activities should always be considered when the mechanism is designed.

There are of course tremendous experimental difficulties in elucidation of the mechanism, and therefore computational support and insight are essential and productive. There are some generally supposed steps in the mechanism, like production of small linear, cyclic and polycyclic species, and their combination into cages. At some later stages, multiple isomerizations should transform general, non-IPR or even *quasi*-fullerenic cages into a few very stable IPR structures. Although kinetics in electronically excited states could bring some reduction of the barriers (as well as charging), a catalytic action is most likely to be the decisive step. The catalytic action should happen through participation of either elemental atoms or their small clusters. In particular, carbon, nitrogen, oxygen, and hydrogen atoms are relevant because they are always present in the reaction mixture at least in small amounts. Graphite always contains small amounts of many chemical elements [142], and N_2 and O_2 also exist as impurities in the inert gas used as the medium for the fullerene synthesis. Moreover, some metals may originate from the electrical wires in the apparatus. During the process of graphite evaporation in an electric arc, all the components are atomized.

The free atoms can create intermediate complexes with the cages and influence the structure and energetics of the activated complexes. Even carbon atoms themselves can act this way [143,144]. Our preliminary computations, in particular, indicate a catalytic activity for the complexes enhanced by oxygen and nitrogen atoms [140,141]. Also in this case, one should consider not only the ground state but also various electronically excited potential hypersurfaces. In addition to neutral species, positively and/or negatively charged intermediate complexes should be computed. This systematic computational treatment should point out the particularly potent catalytic agents—first for the fullerene synthesis and later for the catalytic control of nanotube productions. Nothing is yet known about the molecular form of the metal catalysts acting in the nanotube synthesis. It actually might be a fullerene cage covered by metal atoms. These metal-covered fullerenes have been known from the experiments of Martin and co-workers [145–148]. Such metal-covered fullerenes can fit the diameters of nanotubes. Overall, understanding the catalytic action represents a key step in mastering fullerene synthesis, although progress is expected from numerical quantum chemistry rather than from topological approaches. The computations are indeed more suitable for the solution, although the complexity will require a full-scale application of all the sophisticated tools of computational chemistry.

The Stone–Wales transformation can be used for a stepwise, and even exhaustive, generation of various isomeric cages. Its application disregards all kinetic aspects; the transformation simply serves as a systematic, though formal structure generator [63,131,132] while the related potential-energy barriers are ignored. This application offers yet another method of exhaustive or nearexhaustive generation of topologically possible cages. It was recently applied to cage generations [132,149,150] for C_{32} and C_{36}. In this way, altogether 199 C_{32} were generated [132]: It was a generalized search as four- and seven-membered rings were also considered. The generalized Stone–Wales rearrangement improves the chances for search completeness. This point is quite pertinent as demonstrated recently [90] with the conventional transformation. The conventional Stone–Wales transformations for the C_{84} IPR structures can actually be decomposed [90] into two disjoint families. This type of decomposition can supply some arguments [63] for kinetic control of production of some of the isomers.

The leapfrog transformation represents [80] still another useful topological concept, serving in the generation and classification of fullerene cages. It is based on a topological concept dubbed dual. Two polyhedra are defined as duals if the vertices in one polyhedron correspond to the face centers of the other polyhedral body (e.g., cube and octahedron are duals). The duals of fullerene cages will thus have triangular faces, and on the contrary, five- and sixcoordinated vertices. The concept of duals also allows for a rational discussion of the topological links between fullerenes and boron hydrides as given in the study of Lipscomb and Massa [101]. Clearly the leapfrog can be used for production of another, higher fullerene, if performed in two subsequent steps. First, we cap each face in a fullerene by a central atom (thus creating a pentagonal or hexagonal pyramid). Second, we create the dual of the triangular polyhedron produced by the first step. The result is obviously again a fullerene cage, however, with three times more atoms. Moreover, both the starting and final fullerene cages belong to the same symmetry group. The leapfrog transformation was also used to rationalize [80] a $60 + 6k$ stability rule (k is zero or any integer greater than 1). According to this rule, C_{78}, C_{84}, or C_{90} should be particularly important higher fullerenes, which is true.

Among all the productive topological applications, the exhaustive cage enumerations, and especially the IPR cage enumerations, represent a particularly important result. Manolopoulos et al. [103] suggested the concept of two-dimensional fullerene representation by their ring spiral. The concept has been developed into a reliable enumeration tool, though not strictly rigorous. The authors realized that fullerenes, at least below some dimensional threshold, could be peeled like an orange—each face, after the first, borders its immediate predecessor, so that the rings come off the cage in one continuous spiral. It is not immediately obvious if this is true for any cage regardless of its dimension (i.e., the existence of just one continuous spiral). It is already known that this presumption is actually not always satisfied. A counterexample is already known [97]—a fullerene with 380 atoms. It is the smallest tetrahedral fullerene without a spiral or unspirable fullerene (T symmetry); that is, the structure would not be counted by the spiral algorithm. It is not clear, however, whether this is the smallest possible unspirable fullerene of any symmetry. The C_{380} structure looks more like a giant tetrahedron than a sphere (Fig. 6.3), and this special shape is apparently important for its unspiralability. In particular, note the four symmetry-equivalent sets of three fused pentagons at the vertices of the master tetrahedron. Clearly, the fullerene-cage enumeration algorithm by Fowler and Manolopoulos [26,95], based on the ring-spiral concept, is a rather practical, heuristic tool. Although not exact in a rigorous mathematical sense, it

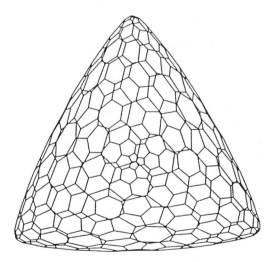

Figure 6.3 The smallest tetrahedral fullerene without a spiral [97]—a T symmetry C_{380}.

is still extremely powerful from the application point of view. It may fail only for the n values that are not really relevant at present.

The connectivity topology in the spiral algorithm can be reconstructed from the sequence of rings in the spiral. This suggests that all possible C_n fullerene cages can be produced if we consider all the ways in which 12 pentagons and $(n/2 - 10)$ hexagons can be combined into a spiral. The related combinatorial term then reads

$$S_n = \frac{(n/2 + 2)!}{12! \times (n/2 - 10)!} \tag{6.5}$$

However, the number of spirals, S_n, is actually higher than the resulting number of related fullerenes. The correspondence between the ring spirals and the distinct fullerenes is not one to one. The duplicities must be removed by an elimination procedure. A spiral can be started in a number of possible ways, but this problem can also be eliminated through a uniqueness test. The uniqueness test can be based on some graph invariant for fullerenes [116], for example, on the eigenvalues of the adjacency or Hückel matrix (the moments of inertia could serve too). Thus, we are dealing with spectrally distinct fullerene isomers. Numerical imprecisions could be a factor in such tests, though not for smaller fullerenes. Hence, the problem of isospectral or *quasi*-isospectral structures could be of interest as a part of the problem of the discrimination power of various graph invariants. Let us also mention that not every spiral combinatorially generated can be really closed.

As can be expected, the total number of the isomeric fullerene cages C_n increases rapidly with the carbon content n. In particular, it has been firmly established that the number of C_{60} isomers is 1812 (or 3532, if enantiomers are distinguished). The important term had been reached in several iterations. First, Manolopoulos et al. [103] concluded there were only 1760 isomers. Then Liu et al. [104,105], using an independent algorithm, asserted there were 1790 isomers (the discrepancy could be related [106] to the fact that the spiral algorithm produced a lower bound rather than the exact value). Finally, however, both groups concluded [107,108] the same value of 1812. For example, for $n = 100, 50, 40, 30,$ and 20, the computed [26] number of all isomeric fullerenes (ignoring their possible enantiomerism) is 285,913, 271, 40, 3, and 1, respectively. Let us stress that these counts consider only pentagon/hexagon rings, with no other cycles allowed, that is, conventional or classical fullerenes only, no *quasi*-fullerenes. It may be of interest to note that, from the quantum-chemical point of view, a one-to-one correspondence is unnecessary between topologically generated structures and local energy minima on a potential hypersurface. The structure after the geometry optimization may occur as a transition state, the particular stationary point may not exist at all, or there may be more than one optimized structure with the required connectivity. A higher saddle point with a nonzero number of imaginary vibrational frequencies (instead of an expected local energy minimum), for example, was reported

[69] in the C_{94} IPR set. After all, there must be numerous transition states mediating interconversions of different fullerene isomers. It is surprising, however, if only a higher saddle point with the given topology is found.

Odd-numbered carbon cages have been rare and thus not usually considered in the enumerations. There are some experimental findings [151,152] of such odd-numbered species and also their computations [153,154]. Once we move to the odd-numbered cages, the picture becomes more diversified as we have to allow for other types of coordination in addition to the three-coordinated carbon atoms. For example, if we allow for just one two-coordinated carbon in the odd-numbered cages (and keep the condition of pentagons/hexagons only), the number of pentagons is reduced to 10 [109,154]. If we go on and allow for two tetra-coordinated carbons and one two-coordinated carbon atom, the number of pentagons increases to 14. Hence, an extension of the IPR concept and generalized enumerations could be developed along the line as well. There are still more derivatized cages, dubbed [110] decorated fullerenes (cages built from water molecules and held together by hydrogen bonds have attracted attention from time to time [111,155]). It is a related topological problem and one can derive from any fullerene a related hydrogen-bonded cage. If we take n water molecules, we deal with $3n/2$ bonds forming the cage, and $n/2$ H atoms must be directed outward (or possibly inside the cage) as they cannot be involved in the hydrogen bond network. Such structures can play a role in atmospheric processes [156].

Given the available computational resources, numerical quantum chemistry has to concentrate primarily on the IPR structures. Especially for fullerenes with n above 100 it is only a tractable option. The requirement of isolated pentagons can readily be built into the spiral algorithm [26]. At first, from all possible spirals we can immediately eliminate those with at least one pentagon–pentagon junction present. This is of course not sufficient because we still have to eliminate spirals with the secondarily generated pentagon–pentagon junctions. As already mentioned, there cannot be any IPR structure for $n < 60$ and indeed, the smallest possible IPR fullerene is buckminsterfullerene C_{60}. The second smallest IPR fullerene comes after an interval of 10 carbons, that is, for $n = 70$, and it is still a unique structure (no IPR cage isomerism yet) like those for $n = 72$ and 74. Beyond this threshold, however, the picture becomes increasingly more diversified: For example, for $n = 76, 78, 80, 82$, and 84, there are 2, 5, 7, 9, and 24 IPR structures, respectively. It is not necessarily a monotonous series—for $n = 86$ the isomeric count drops to 19, and then again increases: 35 and 46 IPR isomers for $n = 88$ and 90, respectively. Soon however, even the IPR isomers become too numerous to be handled by numerical quantum chemistry without preselection. For example, for $n = 140$ we already face [26] 121,354 IPR cages (242,126 if enantiomers are regarded as two distinct species).

There are numerous other enumeration algorithms available, for example, the net-drawing method [128–130]. Liu et al. [104,105] generate the cages ring by ring in all possible ways starting from a seed, for example, a single pentagon.

In each step, the program searches for unsaturated vertices of degree 2 and then adds a segment to the two unsaturated vertices to create a ring. Duplicities are to be eliminated using a graph-isomorphism testing program. Dias [115] presented a generating algorithm based on successive circumscribing and final capping. Pólya's theorem [157,158] has closely been related [30,159] to enumeration studies in chemistry. This elegant mathematical theorem can also be applied to some enumerations of fullerene species. In particular, Balasubramanian [117,120,126,127] has systematically followed this line and applied Pólya's theorem to various functionalized fullerenes. For example, for the derivatives $C_{60}X_n$ the numbers of possible isomers [117] are equal to 37, 577, and 1,971,076,398,255,692 for $n = 2$, 3, and 30, respectively. Optical isomers can be distinguished: For example, there are 14 d,l pairs among the $C_{60}X_2$ isomers.

6.4 THE COMBINED QUANTUM-CHEMICAL AND STATISTICAL-MECHANICAL TREATMENT

After the topological screening and generation of structural candidates, the quantum-chemical calculations can be performed, representing the most demanding step on computational resources. It should be stressed that computational thermochemistry of fullerenes is still not well developed [160,161], among other things, because altogether we have just two solitary experimental figures [162–167]—the heats of formation of C_{60} and C_{70}. The thermochemistry of no other higher fullerene has ever been measured and, thus, there is a shortage of data on which we can test quantum-chemical computations. Hence, our computed interisomeric energetics are only predictions. Once the quantum-chemical part is completed, its outputs actually serve as inputs for the subsequent statistical-mechanical procedures.

The quantum-chemical geometry optimizations can be facilitated by some preliminary molecular-mechanical treatment [168,169], although this technique cannot always be a reliable source of energetics. In spite of a remarkable success in computing energetics of C_{60} and C_{70} [162–167], molecular mechanics is likely to be misleading in any Jahn-Teller situation. In general, semiempirical quantum-chemical geometry optimizations [170–172] are the most typical computational step, though geometry optimizations at the *ab initio* Hartree–Fock SCF level [2,32] can also be performed at present. There has been a standard triad [173–175] of semiempirical quantum-chemical methods: MNDO, AM1, and PM3. However, the triad has recently been complemented [176] with a fourth method dubbed SAM1 (semi-*ab-initio* model 1), expanding beyond the original semiempirical scheme. The SAM1 method, together with the original triad, is available in the AMPAC program package [177]; the three original semiempirical methods are also in the MOPAC package [178]. Both semiempirical and *ab initio* procedures are extensively implemented in the Gaussian series, for example, in G94 [179] and, in a reduced, simplified form, in the Spartan package [180].

The geometry optimizations are now routinely carried out with the analytical energy gradient. They tend to be performed in Cartesian coordinates, that is, without any symmetry constraints. The symmetry constraints can be imposed, if internal coordinates are used, through the so-called Z-matrix formalism; however, this option is less common at present. The computed energetics should further be checked at still more sophisticated levels of theory, although with the fixed molecular geometries optimized previously by a simpler procedure. Such higher treatments can be done either at *ab initio* Hartree–Fock SCF levels (like HF/3-21G, HF/4-31G, and HF/6-31G*) or even with density functional theory (e.g., with Becke's three-parameter functional with the non-local Lee–Yang–Parr correlation functional in the standard 6-31G* basis set, B3LYP/6-31G*). Very rarely, even higher approximations like the second-order Møller–Plesset perturbation treatment [181] are applied to fullerenes. In fact, it frequently happens that the separation energetics for fullerene isomers from different approximations agree quite well [68]; see Figure 6.4.

The geometry optimization procedures should primarily lead to local energy minima, although it may sometimes locate a higher saddle point. Hence, the nature of the located stationary points must be checked accordingly, as only the local energy minima are relevant for interpretation of the NMR data. This check is based on harmonic vibrational analysis, mostly carried out with a numerically constructed force-constant matrix, that is, the second-derivative matrix (numerical differentiation of the analytical gradient). After elimination of the six modes representing overall translations and rotations, the check itself consists of a search for any imaginary vibrational frequency. Saddle points always exhibit one or more imaginary vibrational frequencies. However, if all

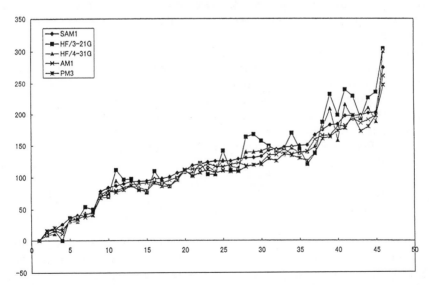

Figure 6.4 The relative energetics (kJ/mol) of C_{90} computed [68] by several selected methods.

the computed vibrational frequencies are real, we indeed deal with a local energy minimum, and the structure is significant for subsequent stability reasoning. The computed vibrational frequencies can also be used in simulation of the vibrational spectra, and in construction of the vibrational partition functions. For the spectra simulation the vibrational frequencies are to be combined with the computed infrared (IR) intensities or Raman activities. The computed vibrational frequencies should, however, be scaled down for the spectral simulations [182–185]. On the other hand, for thermodynamic purposes at higher temperatures the scaling is not important [186] (and the scaling factors for fullerenes are not yet well established).

Although the Cartesian coordinates have at present been routinely used for the geometry optimizations as they streamline the inputs, they also induce a difficulty in recognition of symmetries of the optimized structures. In order to handle the symmetry recognition problem, we have been using a new procedure [49] that actually considers precision of the computed coordinates as a variable parameter, ε (the value of which need not be exactly specified). The individual symmetry elements are recognized within this imprecision ε and are then combined into a point group of symmetry. The origin of the coordinate system is placed in the center of charge; the point is only a candidate for possible center of symmetry. Candidates for the C_2 axes are either lines connecting any two nuclei or perpendicular bisectors of the distance between any two nuclei of the same kind. Next, C_n axes with $n > 2$, S_n and S_{2n} axes, and planes of symmetry are investigated. The symmetry operations extracted this way have to create one of the known symmetry groups. For each symmetry operation considered, coordinates of the interrelated atoms before and after the symmetry operation are checked within the ε margin. If the coordinates are identical within the accuracy ε (i.e., the largest difference is smaller than ε), a symmetry element has been found at the accuracy level ε. This approach is more reliable than the standard symmetry search without the flexible accuracy measure.

The previous computational procedures supply a set of, say, m isomeric fullerene structures (i.e., with the same carbon content n), properly described by the computations as far as their structures, vibrations, and energetics are concerned. Their relative concentrations can be expressed as the mole fractions, w_i, using the isomeric (molecular) partition functions q_i. In terms of q_i and the ground-state energy changes $\Delta H^{\circ}_{0,i}$ the mole fractions [30,187,188] are given by

$$w_i = \frac{q_i \exp(-\Delta H^{\circ}_{0,i}/RT)}{\sum_{j=1}^{m} q_j \exp(-\Delta H^{\circ}_{0,j}/RT)} \tag{6.6}$$

where R is the gas constant and T the absolute temperature. The partition functions are to be constructed within the rigid-rotor and harmonic-oscillator (RRHO) approximation, owing to the present computational limitations. Otherwise, there is only one presumption behind Eq. (6.6)—the presumption of noninteracting particles (or, more specifically, the ideal gas behavior) and the condition of the interisomeric thermodynamic equilibrium. With these presumptions, Eq. (6.6) can be derived from first principles, that is, from equilibrium thermodynamics.

Equation (6.6) refers to the relative ground-state energies $\Delta H^{\circ}_{0,i}$ (i.e., the enthalpies at the absolute zero temperature). However, all semiempirical methods are parameterized for room temperature; that is, they produce the conventional heats of formation at room temperature $\Delta H^{\circ}_{f,298}$ (or the related separation or relative terms $\Delta H^{\circ}_{f,298,r}$). It is just a convention because the terms are more common in practice. For our purpose, however, we have to convert them to the heats of formation at the absolute zero temperature, in order to get the $\Delta H^{\circ}_{f,0}$ terms acting in Eq. (6.6). We can eventually also extract the vibrational zero-point energies and end with the relative potential energies $\Delta E^{\circ}_{r,i}$. The last mentioned $\Delta E^{\circ}_{r,i}$ terms actually correspond exactly to the energy terms commonly produced by *ab initio* computations. There is, however, still another reason to mention the $\Delta E^{\circ}_{r,i}$ terms: They appear in the simple Boltzmann factors,

$$w'_i = \frac{\exp(-\Delta E^{\circ}_{r,i}/RT)}{\sum_{j=1}^{m} \exp(-\Delta E^{\circ}_{r,j}/RT)} \tag{6.7}$$

Clearly, we can obtain Eq. (6.7) if we neglect any rotational–vibrational motion in Eq. (6.6), especially if we put $q_i = 1$ throughout. It is important to realize that the simple Boltzmann factors can never cross. Thus, they cannot represent a complicated system because they do not allow for relative stability interchanges (for an illustration [132], see Fig. 6.5).

It has already been indicated that fullerene cages are frequently chiral. Their chirality must be respected accordingly in the relative stability evaluations. There is, of course, no asymmetric carbon atom in the conventional sense in the fullerene cages (built from three-coordinated carbon atoms). Nevertheless, some of the structures are still chiral as a whole; that is, they are not superimposable on their mirror image. This structural asymmetry can readily be recognized from the point group of symmetry. It is enough to check for the presence of no reflection symmetry, that is, absence of rotation–reflection axes S_n. In fact, only the C_n, D_n, T, O, and I groups obey the requirement. For an enantiomeric pair its partition function q_i in Eq. (6.6) has to be doubled. It can formally be done by means of the chirality partition function [189]. In this approach we tacitly assume the equal presence of both optical isomers under the fullerene synthesis conditions.

6.5 C₃₆ FULLERENE AND *QUASI*-FULLERENE CAGES

Recently, Zettl and co-workers [23,190] isolated a new solid fullerene, C_{36}, and from its solid-state ^{13}C NMR concluded a D_{6h} cage structure, known from previous topological studies [26]. The discovery is in agreement with expec-

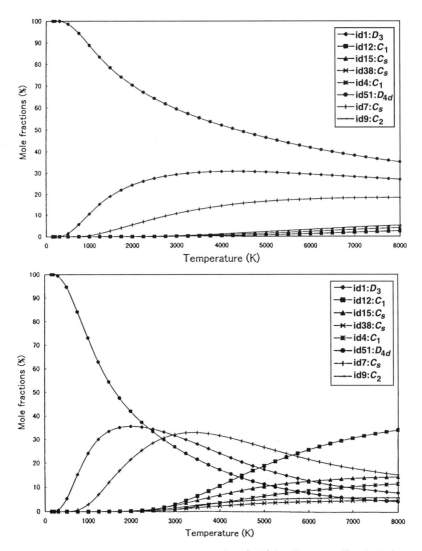

Figure 6.5 The computed relative concentrations [132] for C_{32}. *Top*: Simple Boltzmann factors, Eq. (6.7). *Bottom*: The rigorous w_i terms from Eq. (6.6).

tations of an experimental threshold for observed fullerenes, established from a gas-phase ion chromatography technique by von Helden et al. [191,192]. The fraction of fullerene isomers was observed close to 5% for C_{32}, about 1% for C_{30}, and virtually no cages below $n = 30$. In fact, even synthetic routes toward smaller fullerenes have been studied [193].

C_{36} represents an interesting challenge to computations [149,194–197]. There are just 15 conventional fullerene cages for C_{36} [26]—all being non-IPR structures, of course. In order to reduce the number of pentagon–pentagon junctions, one can consider four-membered and/or seven-membered rings too. Then, Euler's network closure requirement takes the form

$$2n_4 + n_5 - n_7 = 12 \qquad (6.8)$$

where n_i denotes the number of rings with i vertices. This interest in C$_{36}$ *quasi*-fullerenes has been supported by our recent finding [132] that the C$_{32}$ system has a *quasi*-fullerenic ground state, containing two squares. The cages were generated by the topological procedure [63,131,132] based on the generalized Stone–Wales rearrangements. If the topological search is limited to conventional fullerenes and to *quasi*-fullerenes with one or two squares (no heptagon), one heptagon (no square), or one square and one heptagon, the total number of generated cages is 598. Among the twelve C$_{36}$ cages lowest in energy, five are *quasi*-fullerenes.

The molecular geometries are optimized with the SAM1 semiempirical method and the energetics was further checked by density-functional computations at the B3LYP/6-31G* level, that is, the B3LYP/6-31G*//SAM1 approach. Although the topological heuristic program supplied 598 cages, we can ignore the structures especially high in potential energy. Experience says that for higher fullerenes energy handicap as high as 200 kJ/mol can in some cases be overcompensated by a convenient entropy term. We recently reported [69] such a case with one C$_{90}$ cage. However, it can be expected that those entropy effects are somehow proportional to the system dimension. Hence, only the structures with separation energy less than 150 kJ/mol should have a chance to really count in the C$_{36}$ equilibrium isomeric mixture. This bracketing leads to about twelve low-energy structures. Among the twelve selected low-energy cages, seven are the conventional fullerenes. It is convenient to classify the cages by a ring index, a vector containing the counts n_i of the rings considered: n_4, n_5, n_6, n_7. Out of the remaining five *quasi*-fullerenes, three isomers are of the 1,10,9,0 type (one square), one is the 1,11,7,1 type (one square, one heptagon), and one is the 2,11,7,0 type (two squares, no heptagon).

It turns out that in the C$_{36}$ system there are a few cases of considerable differences between the SAM1 and B3LYP/6-31G* approaches. This methodological effect, however, does not influence the selection of the system ground state, which is always represented by the conventional fullerene with the D_{2d} symmetry. There is one high-symmetry species, a conventional D_{6h} fullerene, in the selected 12-membered isomeric set. In fact, the D_{6h} point group is a topological symmetry; however, our more flexible symmetry diagnostic tool [49] gives only a C_{6v} or even C_{2h} symmetry. A closer inspection suggests that we are actually dealing with a Jahn-Teller distortion owing to the degenerate frontier orbitals, although the symmetry distortion is rather small, as can also be seen in Figure 6.6. While in the SAM1 and HF/4-31G approaches the structure is relatively high in energy, it drops in the B3LYP/6-31G*//SAM1 treatment— namely, to only 23 kJ/mol above the ground state. Hence, the structure becomes the second lowest in the B3LYP/6-31G* energy. The energy reduction, however,

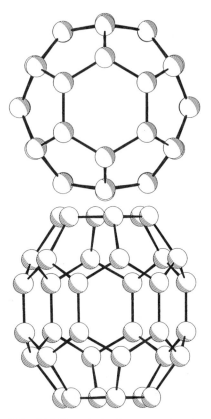

Figure 6.6   Two views of the SAM1 optimized structure [197] of the C_{36} high-temperature prevailing cage.

does not stop here: If we now perform the complete B3LYP/6-31G* geometry optimization, the D_{6h}–D_{2d} separation energy decreases to a mere 15 kJ/mol.

There are other interesting low-energy structures with C_{2v} and C_2 symmetries. On the other hand, the highest member of the twelve-membered set is a C_1 conventional fullerene, being in any considered method more than 165 kJ/mol above the ground state. The lowest *quasi*-fullerene (a C_s symmetry) comes rather high, at least 56 kJ/mol above the ground state. Hence, energetics does not favor *quasi*-fullerenes for $n = 36$, though it does [132] for $n = 32$.

In any case, fullerenes are formed at high temperatures and, thus, entropy effects are possibly important and should be considered within available computational resources. It turns out that in the most sophisticated computational approximation used here, B3LYP/6-31G*, which covers only a part of the correlation energy, just two structures dominate in the region of higher temperatures—the conventional fullerenes D_{6h} and D_{2d} (see Fig. 6.7). Although the D_{2d} ground state has to prevail at low temperatures, the stability order is reversed at moderate temperatures. Although the B3LYP/6-31G* treatment is the

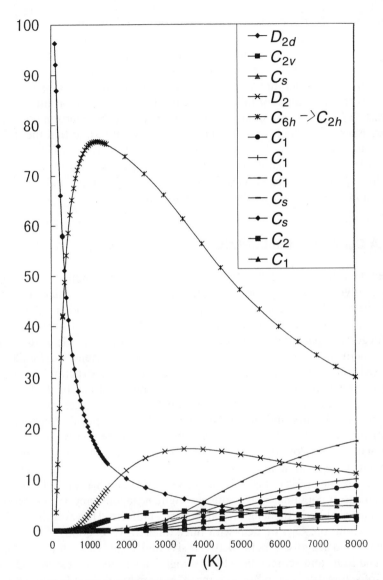

Figure 6.7 The B3LYP/6-31G*//SAM1 computed relative concentrations [197] of the 12 lowest isomers of C$_{36}$.

most sophisticated among the methods applied here, it is still not really clear [198] how good this treatment is for description of the thermochemistry of fullerenes. Hence, without a comparison with experimental data, it would be difficult to decide which method gives the best description of C$_{36}$. However, the solid-state NMR by Piskoti et al. [190] proves the D_{6h} structure. Hence, our B3LYP/6-31G* treatment with two B3LYP/6-31G* fully optimized structures gives the best agreement with experiment, as it predicts the highest relative population of the D_{6h} cage, more than two or even three times higher than the population of the second most abundant isomer. The theoretical prediction, however, is based on the presumption of interisomeric thermodynamic equilibrium. Moreover, an effective synthetic temperature is unclear.

Overall, the computational treatment supplies several interesting conclusions on C_{36}. It reveals that *quasi*-fullerenes are not particularly important for C_{36} (in contrast to C_{32}). The computations also indicate that the interisomeric separation energies can in some cases be quite method sensitive and that the B3LYP/6-31G* treatment is the best affordable method. Finally, C_{36} belongs to the family of systems with an important role of entropy. The entropy contribution actually explains why the species, which is not the system ground state, is still the most abundant isomer isolated experimentally.

6.6   A C_{48} *QUASI*-FULLERENE

C_{48} can be represented [199–201] by one of the thirteen Archimedean or semiregular polyhedra, the so-called truncated cuboctahedron or greater rhombicuboctahedron [201]. Its structure is composed of four-, six-, and eight-membered rings and exhibits a high topological symmetry, O_h (see Fig. 6.8). It belongs to the category of *quasi*-fullerenes, and actually it is a nice example of the species. Its systematic name [6,202] reads: (4,6,8)[48-O_h]*quasi*-fullerene. Taylor [203] has proposed four-membered rings as a possible significant structural pattern for some fullerene structures, and C_{48} serves as a model structure. In fact, four-membered rings represent [76] an important alternative structural pattern for B–N heterofullerenes and nanotubes, and such boron–nitride fullerenes with four-membered cycles have indeed been observed quite recently [204].

Dunlap and Taylor [199] published a density functional computation of C_{48} and pointed out some similarities with C_{60}. The system was also computed [200] by several semiempirical methods (MNDO, AM1, and SAM1). The truncated cuboctahedron contains twelve squares, eight hexagons, and six octagons. In the geometrical body the length of all edges is uniform. However, in its chemical representation, C_{48}, we deal with three different C–C bonds with generally different bond lengths: a bond between six- and eight- (6,8), four- and eight- (4,8), and four- and six-membered (4,6) rings. This frame still exhibits O_h point group of symmetry (Fig. 6.8—note that the diversity of the 4,6- and 6,8-bonds is clearly visible when looking along a threefold axis of symmetry). Hence, we deal with a similar type of relationship as between the (unique) geometrical body of the truncated icosahedron and C_{60} with a ratio of its 5,6- and 6,6-bonds generally different from one (and actually varying [2] from one quantum-chemical geometry optimization to another).

The full geometry optimization and harmonic vibrational analysis show [200] that the C_{48} *quasi*-fullerene is indeed a local energy minimum with O_h or a near-O_h symmetry. The O_h symmetry has, of course, profound consequences for the C_{48} IR vibrational spectrum. Of the 138 vibrational frequencies only

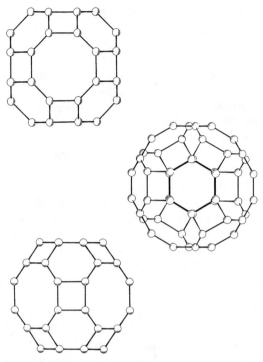

Figure 6.8   Three views [200] of the AM1 optimized $(4,6,8)[48\text{-}O_h]quasi$-fullerene C_{48}.

those belonging to just one class, T_{1u}, can be IR active. This drastic spectral reduction resembles the conditions in C_{60} with the four computationally predicted [205–211] IR lines used [212] in the final, culmination stage of the fullerene synthesis. The computed bond lengths [200] allow us to speak of double (6,8) and single (4,8; 4,6) bonds, this again resembling the C_{60} picture.

However, in order to place the new structural pattern into the fullerene context, we should compare the C_{48} $quasi$-fullerene energetics with a C_{48} pentagon–hexagon structure. There are 199 C_{48} fullerenes [26,213]. A structure of C_2 symmetry was selected [200] as a reference. The $O_h\text{-}C_2$ separation energy somewhat depends on the method used; however, it is consistently large. For example, the SAM1 method gives the $O_h\text{-}C_2$ separation as 1500 kJ/mol. Even though the C_{48} $quasi$-fullerene does not emerge from the computations as particularly stable, it itself does not contradict the possible significance of four-membered cycles for some fullerene-related species, for example, C_{32} [132].

6.7 C₇₆ IPR ISOMERS

The system was isolated by Ettl et al. [33] and since then it has been receiving constant attention [34–42]. In fact, C_{76} is the smallest fullerene that allows for cage isomerism of the IPR structures, though modest. There are just two different IPR structures [26], namely, with the topological symmetries D_2 and T_d. However, the T_d topological symmetry undergoes a reduction as the structure exhibits degenerate, partially filled frontier orbitals. Thus, it has to be distorted according to the Jahn-Teller theorem, lowering both symmetry and energy. In the AM1 calculations [214,215] this relaxation process ends in a D_{2d} symmetry (though other quantum-chemical methods could predict other symmetry subgroups). Moreover, the D_{2d} isomer is still located about 108 kJ/mol above the D_2 ground state on the AM1 potential hypersurface. The separation energy derived directly from the heat of formation at room temperature $\Delta H^\circ_{f,298}$ is 103 kJ/mol. Interestingly, the geometrical distortion is quite small. This distortion can be measured by the rotational constants (identical for a spherical top like a tetrahedral T_d species). Even after the symmetry relaxation the rotational constants remain almost identical for the D_{2d} structure: 0.00173, 0.00173, and 0.00174 cm^{-1}. However, for the other species, D_2, the rotational constants are distinctly different: 0.00194, 0.00171, and 0.00156 cm^{-1}. All computed vibrational frequencies of both structures are real. Hence, we really deal with local energy minima after the symmetry relaxation (the feature is not trivial especially if the Jahn-Teller effect is involved).

From the relative-stability point of view, the D_2 structure is actually in a triply convenient situation. It is lower in energy, it is enhanced by its chirality factor of 2, and its low vibrational frequencies are lower than those of the D_{2d} isomer. Owing to this triple coincidence, it must be prevailing even at very high temperatures. For example, even at a temperature of 4000 K the ground-state structure represents more than 95% of the two-isomer equilibrium mixture. Thus, the computations clearly (and correctly) point out that the D_{2d} C$_{76}$ species can hardly be observed.

6.8 C$_{78}$ IPR ISOMERS

Conditions are very much different if we move from C$_{76}$ to C$_{78}$. The IPR cage isomerism of C$_{78}$ is well known [43–46,87,216] not only from computations but also from experiments. In experiments, at first only two isomers were observed by Diederich et al. [43]: D_3 and C_{2v} symmetry in a ratio 1:5, the latter species being also labeled as $C_{2v}(I)$. Later on, Kikuchi et al. [44] reported three isomers of C$_{78}$ with the symmetries D_3, $C_{2v}(I)$, and $C_{2v}(II)$ and in a ratio of 2:2:5. The challenging observations were soon completed with computations [46] of the relative stabilities. In fact, it was only the third case of such computations in the fullerene field, following two model computations without experimental verification [217,218].

The computations on the C$_{78}$ IPR system were performed [45,46,216] with the semiempirical MNDO, AM1, and PM3 methods. There are [26] five IPR isomers of C$_{78}$ and each of the topologically possible structures indeed produced a local energy minimum in the geometry optimizations. Their symmetries derived from the MNDO optimizations are $D_{3h}(I)$, $D_{3h}(II)$, D_3, $C_{2v}(I)$, and $C_{2v}(II)$. In all three semiempirical methods employed, the $C_{2v}(II)$ structure represents the lowest (global) energy minimum (in fact, in agreement with the MM3 force-field results).

According to the computations, the global minimum represents more than 50% of the equilibrium mixture even at a high temperature of 4000 K. On the other hand, the relative population of the $D_{3h}(I)$ structure is always practically negligible. The remaining three structures exhibit comparable relative stabilities, and in the high-temperature limit this three-membered group represents nearly 50% of the equilibrium isomeric mixture. The picture is basically the same for the three semiempirical energetics considered. Fowler et al. [87] established that the structures $D_{3h}(I)$, $D_{3h}(II)$, D_3, $C_{2v}(I)$, and $C_{2v}(II)$ should exhibit 8, 8, 13, 21, and 22 lines, respectively, in their ^{13}C NMR spectra. Diederich et al. [43] isolated two C_{78} isomers that exhibited 13 and 21 NMR lines, respectively, and were therefore identified as the D_3 and $C_{2v}(I)$ structures. The presence of only two isomers (rather than of five) was rationalized by kinetics. Two product channels were considered, the minor one leading to the D_3 species and the major one to the $C_{2v}(I)$ species. A kinetic control, of course, reduces applicability of the equilibrium thermodynamical treatment.

This, however, was not the end of the story, as later on Kikuchi et al. [44] reported their observation of the D_3, $C_{2v}(I)$, and $C_{2v}(II)$ isomers of C_{78} in a 2:2:5 ratio. This second experiment is actually qualitatively more consistent with the computations than the 1:5:0 ratio [43]. However, in order to get even a quantitative agreement, we have to go to relatively very high temperatures. Still, it should be realized that a relatively small change in the observed ratio can scale down the anticipated temperatures considerably. Moreover, there is certainly room for further computational activities, for example, on the system energetics. In addition to that computational need, there have also been additional experimental results reported recently [219–221]. In particular, the independent ^{3}He NMR spectroscopy data supplied by Saunders and co-workers [5,56,222] could support a more ample isomerism.

6.9 C$_{80}$ IPR ISOMERS

This system of seven topologically possible IPR structures [26] serves a very instructive example of the thermodynamic isomeric interplay. The isomers have in the literature been conventionally coded [47–50,67] by letters rather than by a numerical code, **A–G** (see Fig. 6.9). Of the seven IPR C_{80} structures, the species **B** exhibits a high topological symmetry, I_h, the same as seen in the icosahedral C_{60}. However, according to the concept of the Goldberg polyhedra [80,135,136], based on topological duals, it has to be an electronic open shell. Consequently, it has to undergo a Jahn-Teller distortion toward lower energy and lower symmetry. Similar to the distortions of the topologically tetrahedral T_d C_{76} case, we can also check the Jahn-Teller deformation on the rotational

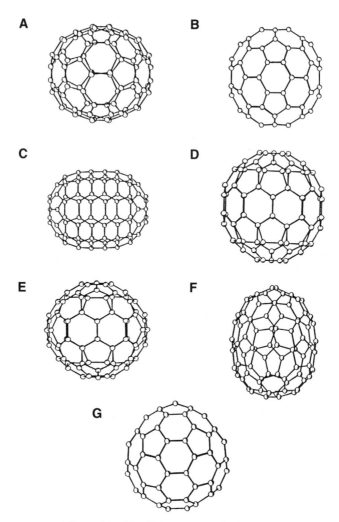

Figure 6.9   The IPR structures [49] of C_{80}.

constants. The distortions are again quite small as the rotational constants of the AM1 fully optimized species **B** read [49]: 0.00155, 0.00156, and 0.00157 cm^{-1}, that is a near-spherical top. The symmetry of the **B** structure is lowered to D_2 in the relaxation process. Nevertheless, the particular species still exhibits the highest energy in the computed IPR set. The computations (at both room and absolute zero temperature) point out the **C** isomer (D_{5d} symmetry) as the system ground state, being followed by the **A** species of D_2 symmetry.

Figure 6.10 presents [67] the temperature dependencies of the relative concentrations, w_i, for the seven-membered mixture under the thermodynamic-equilibrium conditions. The dependencies are reported for the HF/4-31G

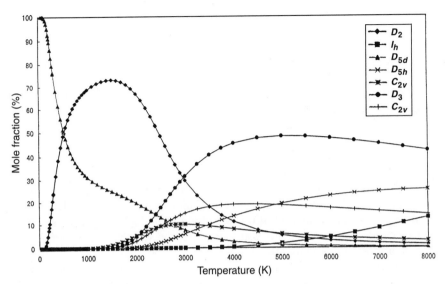

Figure 6.10 The HF/4-31G computed [67] relative concentrations of the C_{80} isomers from Figure 6.9 (topological symmetries indicated).

energetics (Fig. 6.10 refers to the topological symmetries of the isomers). As always at very low temperatures, the ground-state structure, $C-D_{5d}$ symmetry, has to be prevailing, and the relative concentrations of other structures simply have to obey the $\Delta H^\circ_{f,0,\,rel}$ order. This low-temperature region, however, has no relevancy to fullerene synthesis. At higher temperatures the pre-exponential factors in the master equation (6.6) become increasingly important (while the importance of the exponential part is gradually decreasing at the same time). In fact, at a relatively low temperature the **A** (D_2 symmetry) species reaches equimolarity with the **C** species. Beyond this temperature point of the two-component equimolarity, the **A** species of D_2 symmetry is always more populated (although not being the system ground state). Interestingly, at still considerably higher temperatures yet another species becomes the most populated.

The separation energies were computed [49,67] in the SAM1, AM1, HF/STO-3G, HF/3-21G, and HF/4-31G approaches. The results agree reasonably well so that the agreement between various types of computations supports reliability of the predicted relative concentrations. In fact, the computational prediction [49] could soon be critically tested against observations. Hennrich et al. [50] presented NMR spectra of C_{80} consistent with the **A** (D_2 symmetry) species. According to the HF/4-31G computations [67,223], the two-component equimolarity point between the important **A** and **C** structures is reached already at 517 K. At a temperature of some 1500 K (which probably represents a reasonable lower bound of temperatures for the fullerene synthesis) the **A** structure of D_2 symmetry forms about 73% of the equilibrium isomeric mixture. In fact, the **A** isomer reaches its relative concentration maximum at a temperature of 1497 K in the HF/4-31G energetics. Thus, the computations and observations single out as the most populated species a structure that is not the ground state of the system but is still preferred by the Gibbs function. This interesting thermodynamical event has been known with other chemical systems [30,31] too, but it is relatively frequent for isomeric fullerenes owing to the extremely high temperatures needed for their preparation.

6.10 C_{82} IPR ISOMERS

^{13}C NMR spectra for C_{82} were reported by Kikuchi et al. [44], and that time three symmetry species (C_2, C_{2v}, C_{3v}) were found with a ratio of 8:1:1. Later on, however, the original interpretation was modified [4,24,25]; in the newer interpretation only C_2 symmetry is considered for the C_{82} species (while the two former minor components are supposed to be other fullerenes). In fact, it is even suggested [24] that C_{82} has at least two isomers and both of them have C_2 symmetry. There has been constant research interest in the C_{82} system [51–56].

There are already nine C_{82} cage topologies [26] that satisfy the IPR condition. Their relative stabilities at high temperatures were evaluated [55] with the AM1 quantum-chemical semiempirical method. The cage separation energies are rather similar, giving the following energy order at the absolute zero temperature: C_2, C_2, C_s, C_s, C_2, C_s, C_1, C_s, and C_s. In particular, the two lowest structures are of C_2 symmetry and are separated by about 17 kJ/mol only.

At very low temperatures the ground-state structure of a C_2 symmetry is, of course, dominant. At moderate temperatures the other low-energy C_2 isomer represents the second most populated structure. However, its molar fraction exhibits a temperature maximum of about 20% and then decreases. At very high temperatures some other structures show significant populations, including the third C_2 isomer. For a more quantitative correlation of computations and observations, some information on the temperature history of the particular sample would be needed. Any information on a representative reaction temperature during the fullerene synthesis is still virtually missing. However, the computations reveal that the group of C_2 species always represents more than 50% of the equilibrium isomeric mixture, for example, 84% at 1500 K. This computational finding agrees well with the updated interpretation [4,24,25] of the available NMR spectra as it allows for both one or two C_2 structures. Incidentally, with some difficulty, the computations could to some extent also support the original NMR interpretation (C_2, C_{2v}, C_{3v} in a ratio of 8:1:1).

6.11 C_{84} IPR ISOMERS

C_{84} fullerenes are special because they are relatively abundant and, thus, they ere among the very first fullerene species observed [57] immediately after the synthesis of C_{60} and C_{70}. Their observations have constantly been complemented with theoretical studies [60,61,75,86,90,95,225–230] and, in turn, there is now a sophisticated and very productive isolation project going on—already close to some 10 identified C_{84} isomers [224]. Hence, a still more sophisticated wave of the C_{84} computations is to be expected also. However, for C_{84} altogether 24 IPR structures are to be considered and this isomeric count can be quite demanding on computational resources. Kikuchi et al. [44] concluded from ^{13}C NMR spectra the presence of two major isomers of D_2 and D_{2d} symmetry. Saunders and co-workers [5,56,222] with their [3]He NMR technique could visualize up to nine C_{84} isomers. The two major isomers were separated recently by Dennis et al. [62], and their separation work has been continuing [224].

Bakowies et al. [228] computed all the C_{84} IPR isomers at the MNDO semiempirical level. Their study was soon completed [61] with the computational evaluation of the temperature dependencies of the relative concentrations in the isomeric set. In fact, the C_{84} isomeric system is an interesting case where the lowest-energy cage is not the most populated species at the conditions of the fullerene synthesis. The two species reported by Kikuchi et al. [44] are conventionally labeled [228] **22** (D_2 symmetry) and **23** (D_{2d} symmetry). In the MNDO computations [228] the lowest isomer **23** is located only about 2 kJ/mol below structure **22**. On the other hand, the structure highest in energy (also with a D_2 symmetry) is located about 240 kJ/mol above the lowest minimum.

The lowest-energy minimum, **23**, is actually the most populated species only at very low temperatures, below room temperature. The **22/23** two-component equimolarity point is found at a temperature of 276 K only. After this relative stability interchange, structure **22** (slightly higher in potential energy) becomes the most populated isomer for a wide temperature interval. Moreover, the **22/23** ratio is almost constant, being roughly 2:1 anywhere above 500 K; for example, the ratio of the two isomers is 58.2:41.7, 41.4:20.6, and 19.0:8.84 at temperatures of 500, 2000, and 4000 K, respectively. One could interpret [92] this 2:1 ratio in terms of the chirality partion function from Eq. (6.6). There are several other species exhibiting a significant or nonnegligible mole fraction at very high temperatures. This computed feature [61] has recently been confirmed by the observations of Saunders and co-workers [5,56,222]. For example, if we select a threshold w_i value of 10%, the computations can yield three additional isomers of this relatively high stability. The agreement of the computational predictions with newer experiments is comprehensively analyzed in Ōsawa et al. [63]. Competition between some of the species creates an interesting temperature dependency with a local maximum (also shown for C_{80} in Fig. 6.10). In fact, even in very complex fullerene mixtures we should be able, at least for some specific cases, to suggest an optimal region of temperatures where the production yield should be maximized. Although any temperature history of the sample used for the NMR spectroscopy by Kikuchi et al. [44] is virtually unknown, the computed data can match their observed ratio of 2:1 in a very wide temperature interval. Hence, the C_{84} system represents another instructive example of an essential theory–experiment agreement, although further developments are to be expected [62,224].

6.12 C_{86} IPR ISOMERS

As soon as the first isomeric fullerene systems like C_{76}, C_{78}, C_{80}, C_{82}, or C_{84} were successfully computed, it had become rather clear that the interisomeric thermodynamic equilibrium presumption was applicable. This conceptual finding has been somewhat surprising and could be interpreted through a cooperative cancellation of several effects. Anyhow, still higher fullerenic systems should be submitted for computations as long as some observations are available. At present, the observations extend [4,24,25,70] to at least C_{96} (and also include C_{110}). C_{86} is the next system in the row to be computed [64–67]. In contrast to the 24 IPR isomers for C_{84}, there are only 19 topologically different C_{86} IPR structures (Fig. 6.11). They were computed primarily with the SAM1 method [176]. However, the AM1, HF/STO-3G, and STO/3-21G treatments also consistently produce a C_2 species (numbered **17** in Fig. 6.11) as the ground state of the C_{86} IPR set.

Figure 6.12 shows the temperature development of the computed relative stabilities in the C_{86} IPR set. At very low temperatures the ground-state structure is, of course, prevailing; however, its mole fraction decreases relatively fast. At first, a C_s species reaches close to 20% (it has its population maximum at a temperature of 1503 K). There is a third significant species (C_3 symmetry). Although it is relatively high in energy, its entropy contribution still allows for an equimolarity (about 29% each) with the ground state at a temperature of 1739 K. Let us mention, however, that at this temperature two additional species already exhibit relative concentrations of about 10%. Achiba et al. [24] found from ^{13}C NMR spectra the presence of two C_2 isomers in their C_{86} sample; the relative abundance of the isomers was about 4:1. The computations [64–67] suggest C_2 and C_s structures around 1500 K. At higher temperatures, the computations predict a coexistence of more than two isomers. It is, however, true that one computed symmetry is C_s, not C_2, and this minor difference between theory and experiment has not been clarified yet.

6.13 C_{88} IPR ISOMERS

After the small drop in the IPR structures when going from C_{84} to C_{86}, the C_{88} IPR system is already quite extended. There are 35 topologically different C_{88} IPR structures (Fig. 6.13). Nevertheless, this system has been computed and reasonably understood [65–67,231], again primarily using the SAM1 semiempirical method.

As already mentioned (and demonstrated with, e.g., the topologically icosahedral structure for C_{80}), the original topological symmetries represent a kind

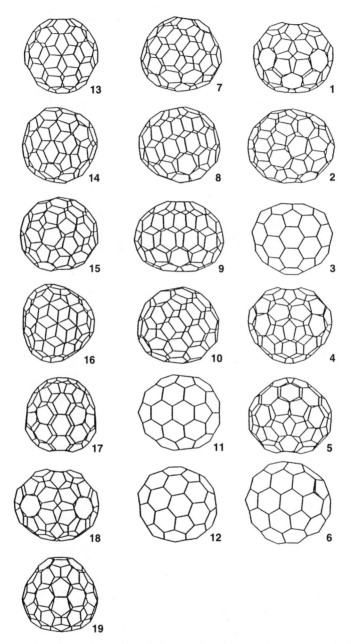

Figure 6.11   The IPR structures [64] of C$_{86}$; numbering according to system in Ref. 130.

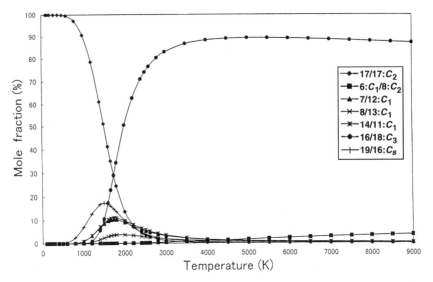

Figure 6.12 The SAM1 computed [64] relative concentrations of the C_{86} isomers from Figure 6.11; numbering according to systems in Refs. 26 and 130.

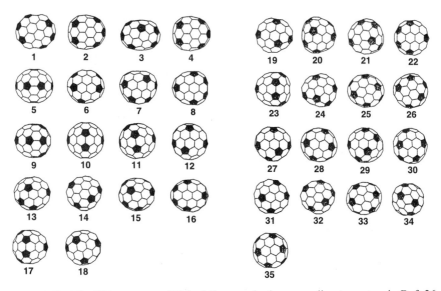

Figure 6.13 The IPR structures [264] of C_{88}; numbering according to system in Ref. 26.

of symmetry upper bound. The true symmetries after quantum-chemical geometry optimizations can be the same or lower (not higher). In particular, during quantum-chemical calculations the topological symmetry can be lowered owing to Jahn-Teller and pseudo-Jahn-Teller effects. In addition to those specific physical reasons, it can also be reduced simply owing to general energy reasons. Hence, only after quantum-chemical geometry optimizations can the symmetry properties be clarified. Moreover, such symmetry reductions are not rare for fullerene and other cages and can frequently happen even without a Jahn-Teller effect. For example, the icosahedral B$_{32}$ cage (a topological boron analogy of C$_{60}$) actually exhibits [232] imaginary frequencies when computed *ab initio*. Similarly, hexa-anion of C$_{60}$ may seem to be icosahedral (as the degenerate LUMO is just filled), but actually its symmetry relaxes with a considerable energy gain [233] of some 60–150 kJ/mol. With these examples in mind, it is not surprising that the C$_{88}$ system at the SAM1 computational level in five cases falls below the topological upper symmetry limit. In four cases the starting topological symmetry itself is so low that it does not allow for degenerate representations. Hence, we deal with only one pure Jahn-Teller distortion (namely, from T to D_2).

The system was computed [231] at four different levels of theory—SAM1, HF/STO-3G, HF/3-21G, and HF/4-31G. The computations consistently predict a species of C_s symmetry as the ground state of the C$_{88}$ IPR set. The IPR cage highest in energy is located about 300 kJ/mol above this system ground state. Owing to the temperature interplay described by Eq. (6.6), the species, which is the second lowest in energy and has a C_2 symmetry, exhibits a fast relative stability increase with a temperature maximum (Fig. 6.14). The maximum molar fraction is about 23% and happens at a temperature of about 1470 K in the SAM1 computations (or already at a temperature of 1270 K, and amounts to about 30%, in the HF/4-31G approach). The third lowest species, again having a C_2 symmetry, has a temperature profile with a maximum also, although rather modest. There is, however, a structure with a steady increase which eventually becomes dominant—a C_1 symmetry, in fact reduced from a C_2 topological expectation. The isomer is relatively rich in potential energy— over 100 kJ/mol above the ground state. It has an equimolarity point with the ground-state structure that comes in the HF/4-31G computations at a temperature of about 1990 K with 29% for each species; see Fig. 6.14. In the SAM1 energetics the equimolarity is reached at somewhat higher temperature, at about 2240 K with about 26% for each of the two isomers. For example, at a temperature of 3000 K the C_1 relative concentration amounts already to 59% and 47% in the HF/4-31G and SAM1 energetics, respectively.

The ^{13}C NMR spectrum for C$_{88}$ in solution is available from Achiba et al. [24] and is consistent with an isomer of C_2 symmetry as a major structure. Although our computations point out even two C_2 structures, their temperature maxima appear at rather low temperatures when the ground-state structure still has a strong population and should be seen in the experiment also. However, if we consider temperatures above 2240 and 1990 K according to the SAM1 and

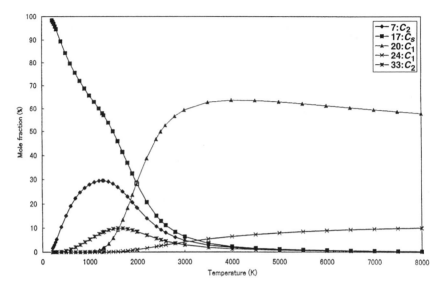

Figure 6.14 The computed [67,231] (HF/4-31G energetics) relative concentrations of the C_{88} isomers from Fig. 6.13.

HF/4-31G predictions, respectively, we shall deal with the fast increasing C_1 structure as the most populated and eventually the dominant one.

As already mentioned, the structure exhibits a C_2 topological symmetry, which is reduced to C_1 only in the quantum-chemical optimization. Such symmetry reductions can bring rather small coordinate distortions. For example, the topologically icosahedral C_{80} cage exhibits [49] Jahn-Teller geometry distortions on the order of 10^{-3} or 10^{-2} Å. Hence, we can speak of a near-icosahedral structure in this case. The C_1 C_{88} structure also exhibits in our search a C_2 symmetry for $\varepsilon > 4 \times 10^{-2}$ Å. However, the vibrational amplitudes of C_{60} from electron diffraction, for example, have still larger values. The same is true, for example, for the vibrational amplitudes of kekulene at room temperature [231]. The vibrational amplitude data suggest that we can actually consider the C_1 structure as having a near C_2 symmetry. This argument would produce an agreement between computation and experiment. Moreover, the theory–experiment comparison also suggests that the measured sample originated at temperatures somewhat over 2000 K.

6.14 C_{90} IPR ISOMERS

C_{90} is another system where the treatment based on the interisomeric thermodynamic equilibrium presumption reveals rich and unexpected internal relationships. The computational treatment is given in reports [68,69]. The C_{90}

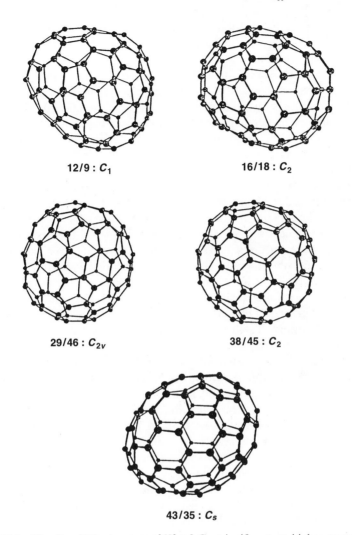

12/9 : C_1 16/18 : C_2

29/46 : C_{2v} 38/45 : C_2

43/35 : C_s

Figure 6.15 The five IPR structures [69] of C$_{90}$ significant at higher temperatures; numbering according to systems Ref. 130/Ref. 26.

system is not small—there are already 46 topologically different C$_{90}$ IPR structures [26]. The computations select one structure with a C_2 symmetry (labeled **38** and **45** according to the numbering systems of Refs. 130 and 26, respectively; see Fig. 6.15) as the system ground state. Interestingly, there is an exception at the HF/3-21G level. The D_{5h} structure (only one such high-symmetry species in the whole set) comes as practically isoenergetic with the **38/45** structure (see Fig. 6.4), but this event is likely to be just a computational artifact.

According to the experimental results reported by Achiba et al. [24], altogether five C$_{90}$ species were identified from ^{13}C NMR spectra. The observed

C_{90} species were in fact distributed in three HPLC fractions: one C_{2v}, three C_2, and one C_1. The interpretation is actually still open [234] to a further adjustment, and more recent surveys [4,25] do not mention the C_1 structure and give two C_2 structures instead. Unless the final interpretation is published, its best available approximation [234] is the original report [24]. The symmetries of the five SAM1 lowest-energy structures (see Fig. 6.15, though only the structures really relevant at higher temperatures are shown there) are C_2 (**38/45**), C_{2v} (**29/46**), C_s (**43/35**), D_{5h} (**19/1**), and C_1 (**22/30**). In other words, at least two of the structures suggested by the potential energy screening do not appear in the experiment [24]. The computation–observation agreement, however, becomes still even worse if we look into the NMR pattern of the computed C_{2v} structure. The last mentioned structure gives 24 lines, 3 of them weaker. The C_{2v} species considered in the experiment is different and exhibits 25 lines, 5 of them weaker.

At best, we could speak of a partial agreement with the experiment at this intermediate stage. In order to get the final picture, however, we really have to investigate possible thermal effects on the relative stabilities in this large isomeric set. Actually, it has been the largest isomeric set ever treated by the combined computational treatment [30] in the literature, though still bigger ones are being processed [235] in our laboratory. After applying the combined treatment, it turns out that there are just five structures that exhibit a significant population in the high-temperature region, relevant to the fullerene synthesis (Fig. 6.16). In addition to the three structures lowest in potential energy (C_2 **38/45**, C_{2v} **29/46**, C_s **43/35**), two high-energy species also exhibit significant populations: C_2 **16/18** ($\Delta E_r = 130$ kJ/mol) and C_1 **12/9** ($\Delta E_r = 202$ kJ/mol). As always, at very low temperatures the ground-state structure has to be the dominant species; however, its decrease of relative stability with increasing temperature found for this system is considerable. According to the SAM1 computations [68,69], the ground-state isomer drops to two-component equimolarity with the **16/18** structure of C_2 symmetry at a temperature of 2012 K. Two other structures, C_s **43/35** and C_{2v} **29/46**, show moderate maxima close to 1500 K. The last structure of Fig. 6.15, C_1 **12/9**, becomes quite important at very high temperatures, though its population is still found too low around 1000 K. Figure 6.16 also screens several less-populated structures; it is of interest that the species with the highest temperature maximum in this second group, **22/30**, also has a C_1 symmetry, and the same is even true for the second-most populated species in the second group, **32/32**. Let us also mention that the results from the SAM1 and HF/4-31G energetics are not particularly different.

We do not know what could be a temperature interval in which the C_{90} sample, examined in the observations [24], was actually synthesized. We may expect the temperatures somewhere beyond 800–1000 K. Our computations predict that at elevated temperatures we primarily deal with the five structures of the following symmetries: C_2, C_2, C_s, C_{2v}, and C_1. This SAM1 high-temperature set compares better with the conclusion [24] from the ^{13}C NMR spectra: C_{2v}, three times C_2, C_1. However, we have to remember that in the computations and experiment we deal with two different C_{2v} isomers with different NMR

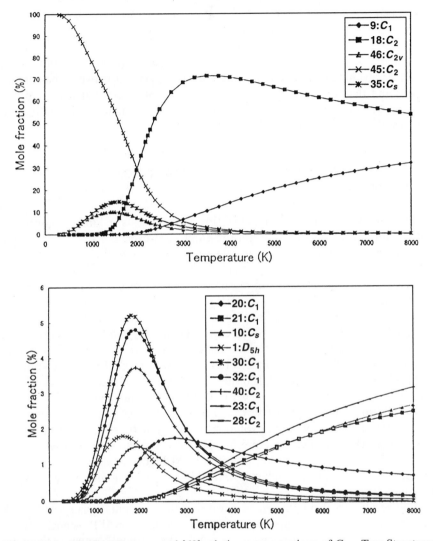

Figure 6.16   The SAM1 computed [69] relative concentrations of C_{90}. *Top*: Structures from Fig. 6.15. *Bottom*: Some less-populated species. Numbering according to system in Ref. 26.

patterns. The experimental conclusion is based on an NMR spectrum consisting [24] of exactly 70 lines, five of them weaker: Those are the primarily observed facts. Any IPR C_{90} isomer of C_2 symmetry always exhibits 45 lines (none weaker). There are C_{2v} C_{90} structures with exactly 25 lines (5 of them weaker). This is the essence of the interpretation [24] of the observed facts. However, in the computations, we deal with a different C_{2v} structure—only 24 NMR lines (3 of them weaker). Moreover, we have to consider still another species, not mentioned in the experimental interpretation, a C_s structure with 46 lines (2 of them weaker). If we now combine together the NMR patterns of the two structures predicted by the computations, C_{2v} and C_s, we also get exactly 70 lines (and exactly 5 of them weaker). With this alternative interpretation of the NMR spectrum [24] we finally reach complete agreement between the computations and the observations. Let us note for completeness that the C_s species is somewhat more populated than the C_{2v} one, which actually should help to equalize the intensities of their NMR lines.

6.15   THE FIRST EXAMPLES OF IMPORTANT NON-IPR HIGHER FULLERENES

In this chapter we have treated the IPR structures as a kind of governing pattern for higher fullerenes. Future research, however, is likely to pay some attention to non-IPR structures in selected systems as well as to some *quasi-fullerenes* (with heptagons in particular). There is a relatively old report of observations of possible but still mysterious isomers of C_{60} and C_{70} presented by Anacleto et al. [236]. It there are really some isomeric cages for the two most common fullerenes, they could be some non-IPR structures (as also indicated in computations [237] of related silicon cages). Hence, we can hardly consider the empirical IPR rule as a law without exceptions. In fact, it was already reported [28] for Ca@C_{72} that the rule indeed did not always work. Even if we consider [235] pristine C_{72}, with no metals inside (Fig. 6.17), we can get some significant non-IPR structures. For C_{72}, not yet isolated, there is just one IPR cage, actually exhibiting a high symmetry, D_{6d}. Figure 6.18 gives the SAM1 computations [235] of the temperature evolution of the isomeric relative stabilities for the system. The IPR D_{6d} structure, in spite of being the system ground state, decreases its relative population pretty fast. At higher temperatures the equilibrium isomeric mixture is in fact governed by a C_2 isomer with one pentagon–pentagon junction.

This result can also be taken as a warning that computations should check the thermal effects on relative stabilities of selected non-IPR isomers in the whole series C_{76}–C_{90}, just done on the purely IPR ground. Actually, there are already some indications that the C_{72} non-IPR stabilomer is not an isolated event. Additional interesting results in this respect are in progress [235], starting with C_{74}. Incidentally, the computational findings also indicate a need for an independent check of the NMR assignments through X-ray structural analysis. Clearly, higher fullerenes are still on the drawing board.

6.16   CONCLUDING REMARKS

This chapter shows the ongoing productive interaction between theory and experiment in fullerene research. For higher fullerenes the mutual interaction is

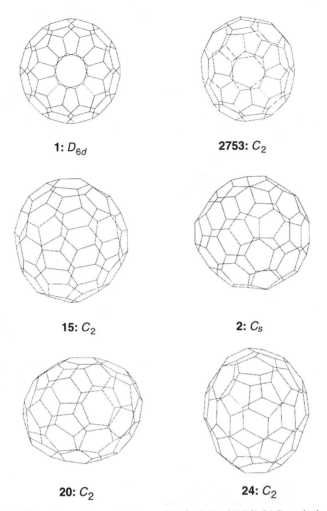

1: D_{6d} **2753:** C_2

15: C_2 **2:** C_s

20: C_2 **24:** C_2

Figure 6.17 Selected low-energy structures [235] of C_{72} (HF/3-21G optimized); there is only one IPR structure, D_{6d}.

indeed essential, given the complexity of the problem. The considerable thermal effects on the relative isomeric populations revealed by the quantum-chemical computations result from a complex interplay between rotational, vibrational, and potential energy terms, chirality factors, and so on. Such effects would never be seen if only energetics is considered and entropy neglected (i.e., the simple Boltzmann factors from Eq. (6.7) instead of the proper terms of Eq. (6.6)). The treatment, however, is built on the presumption of the interisomeric thermodynamic equilibrium. We do not know yet to what degree this presumption is satisfied in reality. We can only acknowledge that the thermodynamic-equilibrium

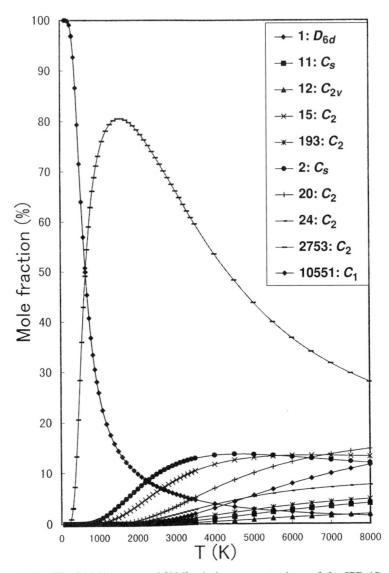

Figure 6.18 The SAM1 computed [235] relative concentrations of the IPR (D_{6d}) and selected non-IPR isomers of C_{72}.

treatment has already produced reasonable computation–observation agreement for the isomeric systems from C_{76} to C_{90}, with no serious failure. This relatively large tested set supports the belief in a still wider applicability of the equilibrium treatment, even up to C_{110}. There are also other aspects of the combined treatment that should be studied further: among them anharmonicity of vibrations, contributions of electronically excited states, and reliability of interisomeric energetics. Future research will organically and simultaneously consider the IPR, non-IPR, and *quasi*-fullerene structures. Once the thermodynamic treatment is completely exposed, interest will gradually shift toward complex kinetic schemes and to fullerene sets with variable stoichiometry.

ACKNOWLEDGMENTS

The reported research has in part been supported by the Japan Science and Technology Corporation (JST) and the Ministry of Education, Science and Culture in Japan. P. D. thanks the Science and Technology Agency of Japan for granting him a visiting-researcher fellowship, and X. Z. thanks the Computer and Communication Foundation (NEC) for granting him a Ph.D. fellowship. The authors also wish to thank the following organizations for kindly permitting the reprinting of copyrighted material: Elsevier Scientific Publishing Company; Marcel Dekker, Inc.; and The Electrochemical Society, Inc.

REFERENCES

1. M. S. Dresselhaus, G. Dresselhaus, and P. C. Eklund, *Science of Fullerenes and Carbon Nanotubes*. Academic Press, San Diego, 1996.

2. J. Cioslowski, *Electronic Structure Calculations on Fullerenes and Their Derivatives*. Oxford University Press, Oxford, 1995.

3. K. Jinno and Y. Saito, *Adv. Chromatogr.* **1996**, *36*, 65.

4. Y. Achiba, K. Kikuchi, Y. Aihara, T. Wakabayashi, Y. Miyake, and M. Kainosho, Trends in Large Fullerenes: Are They Balls or Tubes, in *The Chemical Physics of Fullerenes 10 (and 5) Years Later*, W. Andreoni (ed.). Kluwer Academic Publishers, Dordrecht, 1996, p. 139.

5. M. Saunders, R. J. Cross, H. A. Jiménez-Vázquez, R. Shimshi, and A. Khong, *Science* **1996**, *271*, 1693.

6. *The Chemistry of Fullerenes*, R. Taylor (ed.). World Scientific, Singapore, 1995, p. 1.

7. K. Heinzinger and Z. Slanina, *MPG-Spiegel* No. 2 (1993), 11.

8. J. Baggott, *Perfect Symmetry. The Accidental Discovery of Buckminsterfullerene*. Oxford University Press, Oxford, 1994.

9. H. Aldersey-Williams, *The Most Beautiful Molecule. An Adventure in Chemistry*. Aurum Press, London, 1995.

10. Z. Slanina, *Chem. Intell.* **1998**, *4*, No. 2, 52.

11. J. Mattauch, H. Ewald, O. Hahn, and F. Strassmann, *Z. Phys.* **1943**, *20*, 598.

12. R. E. Honig, *J. Chem. Phys.* **1954**, *22*, 126.

13. J. Drowart, R. P. Burns, G. DeMaria, and M. G. Inghram, *J. Chem. Phys.* **1959**, *31*, 1131.

14. H. Hintenberger, J. Franzen, and K. D. Schuy, *Z. Naturforsch.* **1963**, *18a*, 1236.

15. K. S. Pitzer and E. Clementi, *J. Am. Chem. Soc.* **1959**, *81*, 4477.

16. S. J. Strickler and K. S. Pitzer, Energy Calculations for Polyatomic Carbon Molecules, in *Molecular Orbitals in Chemistry, Physics, and Biology*, B. Pullman and P.-O. Löwdin (eds.). Academic Press, New York, 1964.

17. R. Hoffmann, *Tetrahedron* **1966**, *22*, 521.

18. Z. Slanina, *Radiochem. Radioanal. Lett.* **1975**, *22*, 291.

19. D. E. H. Jones, *New Sci.* **1966**, *32*, 245.

20. E. Ōsawa, *Kagaku* **1970**, *25*, 854; *Chem. Abstr.* **1971**, *74*, 75698v.

21. D. A. Bochvar and E. G. Gal'pern, *Dokl. Akad. Nauk SSSR* **1973**, *209*, 610.

22. R. A. Davidson, *Theor. Chim. Acta* **1981**, *58*, 193.

23. C. Piskoti and A. Zettl, *A New Carbon-Cage Solid: C_{36}*. IWEPNM, Kirchberg/Tyrol, Austria, 1998, p. 33.

24. Y. Achiba, K. Kikuchi, Y. Aihara, T. Wakabayashi, Y. Miyake, and M. Kainosho, Higher Fullerenes: Structure and Properties, in *Science and Technology of Fullerene Materials*, P. Bernier, D. S. Bethune, L. Y. Chiang, T. W. Ebbesen, R. M. Metzger, and J. W. Mintmire (eds.). Materials Research Society, Pittsburgh, 1995.

25. Y. Achiba, *Kagaku* **1997**, *52*, No. 5, 15.

26. P. W. Fowler and D. E. Manolopoulos, *An Atlas of Fullerenes*. Clarendon Press, Oxford, 1995.

27. H. W. Kroto, *Nature* **1987**, *329*, 529.

28. K. Kobayashi, S. Nagase, M. Yoshida, and E. Ōsawa, *J. Am. Chem. Soc.* **1997**, *119*, 12693.

29. Z. Slanina, X. Zhao, and E. Ōsawa, Computing Temperature-Dependent Relative Stabilities of Isomeric Fullerenes and Quasi-Fullerenes, in *Recent Advances in the Chemistry and Physics of Fullerenes and Related Materials*, Vol. 6, K. M. Kadish and R. S. Ruoff (eds.) The Electrochemical Society, Pennington, NJ, 1998, p. 19.

30. Z. Slanina, *Contemporary Theory of Chemical Isomerism*. Kluwer, Dordrecht, 1986.

31. Z. Slanina, *J. Mol. Struct. (Theochem)* **1990**, *65*, 143.

32. W. J. Hehre, L. Radom, P. R. von Schleyer, and J. A. Pople, *Ab Initio Molecular Orbital Theory*. Wiley, New York, 1986.

33. R. Ettl, I. Chao, F. N. Diederich, and R. L. Whetten, *Nature* **1991**, *353*, 149.

34. D. E. Manolopoulos, *J. Chem. Soc. Faraday Trans.* **1991**, *87*, 2861.

35. H. P. Cheng and R. L. Whetten, *Chem. Phys. Lett.* **1992**, *197*, 44.

36. Q. Li, F. Wudl, C. Thilgen, R. L. Whetten, and F. Diederich, *J. Am. Chem. Soc.* **1992**, *114*, 3994.

37. S. Hino, K. Matsumoto, S. Hasegawa, H. Inokuchi, T. Morikawa, T. Takahashi, K. Seki, K. Kikuchi, S. Suzuki, I. Ikemoto, and Y. Achiba, *Chem. Phys. Lett.* **1992**, *197*, 38.

38. G. Orlandi, F. Zerbetto, P. W. Fowler, and D. E. Manolopoulos, *Chem. Phys. Lett.* **1993**, *208*, 441.

39. J. M. Hawkins and A. Meyer, *Science* **1993**, *260*, 1918.

40. S. J. Austin, P. W. Fowler, G. Orlandi, D. E. Manolopoulos, and F. Zerbetto, *Chem. Phys. Lett.* **1994**, *226*, 219.

41. R. H. Michel, H. Schreiber, R. Gierden, F. Hennrich, J. Rockenberger, R. D. Beck, M. M. Kappes, C. Lehner, P. Adelmann, and J. F. Armbruster, *Ber. Bunsenges. Phys. Chem.* **1994**, *98*, 975.

42. R. H. Michel, M. M. Kappes, P. Adelmann, and G. Roth, *Angew. Chem. Int. Ed. Engl.* **1994**, *33*, 1651.

43. F. Diederich, R. L. Whetten, C. Thilgen, R. Ettl, I. Chao, and M. M. Alvarez, *Science* **1991**, *254*, 1768.

44. K. Kikuchi, N. Nakahara, T. Wakabayashi, S. Suzuki, H. Shiromaru, Y. Miyake, K. Saito, I. Ikemoto, M. Kainosho, and Y. Achiba, *Nature* **1992**, *357*, 142.

45. D. Bakowies, A. Geleßus, and W. Thiel, *Chem. Phys. Lett.* **1992**, *197*, 324.

46. Z. Slanina, J.-P. François, D. Bakowies, and W. Thiel, *J. Mol. Struct. (Theochem)* **1993**, *279*, 213.

47. S. J. Woo, E. Kim, and Y. H. Lee, *Phys. Rev. B* **1993**, *47*, 6721.

48. K. Nakao, N. Kurita, and M. Fujita, *Phys. Rev. B* **1994**, *49*, 11415.

49. M.-L. Sun, Z. Slanina, S.-L. Lee, F. Uhlík, and L. Adamowicz, *Chem. Phys. Lett.* **1995**, *246*, 66.

50. F. H. Hennrich, R. H. Michel, A. Fischer, S. Richard-Schneider, S. Gilb, M. M. Kappes, D. Fuchs, M. Bürk, K. Kobayashi, and S. Nagase, *Angew. Chem. Int. Ed. Engl.* **1996**, *35*, 1732.

51. K. Kikuchi, N. Nakahara, T. Wakabayashi, M. Honda, H. Matsumiya, T. Moriwaki, S. Suzuki, H. Shiromaru, K. Saito, K. Yamauchi, I. Ikemoto, and Y. Achiba, *Chem. Phys. Lett.* **1992**, *188*, 177.

52. S. Nagase, K. Kobayashi, T. Kato, and Y. Achiba, *Chem. Phys. Lett.* **1993**, *201*, 475.

53. G. Orlandi, F. Zerbetto, and P. W. Fowler, *J. Phys. Chem.* **1993**, *97*, 13575.

54. S. Nagase and K. Kobayashi, *Chem. Phys. Lett.* **1993**, *214*, 57.

55. Z. Slanina, S.-L. Lee, K. Kobayashi, and S. Nagase, *J. Mol. Struct. (Theochem)* **1995**, *339*, 89.

56. M. Saunders, H. A. Jiménez-Vázquez, R. J. Cross, W. E. Billups, C. Gesenberg, A. Gonzalez, W. Luo, R. C. Haddon, F. Diederich, and A. Herrmann, *J. Am. Chem. Soc.* **1995**, *117*, 9305.

57. F. Diederich, R. Ettl, Y. Rubin, R. L. Whetten, R. Beck, M. Alvarez, S. Anz, D. Sensharma, F. Wudl, K. C. Khemani, and A. Koch, *Science* **1991**, *252*, 548.

58. R. D. Beck, P. S. John, M. M. Alvarez, F. Diederich, and R. L. Whetten, *J. Phys. Chem.* **1991**, *95*, 8402.

59. K. Kikuchi, N. Nakahara, M. Honda, S. Suzuki, K. Saito, H. Shiromaru, K. Yamauchi, I. Ikemoto, T. Kuramochi, S. Hino, and Y. Achiba, *Chem. Lett.* **1991**, 1607.

60. K. Raghavachari and C. M. Rohlfing, *J. Phys. Chem.* **1991**, *95*, 5768.

61. Z. Slanina, J.-P. François, M. Kolb, D. Bakowies, and W. Thiel, *Fullerene Sci. Technol.* **1993**, *1*, 221.

62. T. J. S. Dennis, T. Kai, T. Tomiyama, and H. Shinohara, *Chem. Commun.* **1998**, 619.

63. E. Ōsawa, H. Ueno, M. Yoshida, Z. Slanina, X. Zhao, M. Nishiyama, and H. Saito, *J. Chem. Soc. Perkin Trans. 2* **1998**, 943.

64. Z. Slanina, S.-L. Lee, M. Yoshida, and E. Ōsawa, *Chem. Phys.* **1996**, *209*, 13.

65. Z. Slanina, S.-L. Lee, M. Yoshida, and E. Ōsawa, Computations of 19 IPR Isomers of C_{86} and 35 IPR Isomers of C_{88}, in *Physics and Chemistry of Fullerenes and Their Derivatives*, H. Kuzmany, J. Fink, M. Mehring, and S. Roth (ed.). World Scientific, Singapore, 1996, p. 389.

66. Z. Slanina, S.-L. Lee, M. Yoshida, and E. Ōsawa, SAM1 Computations on 19 IPR Isomers of C_{86} and 35 IPR Isomers of C_{88}, in *Recent Advances in the Chemistry and Physics of Fullerenes and Related Materials*, Vol. 3, K. M. Kadish and R. S. Ruoff (eds.). The Electrochemical Society, Pennington, NJ, 1996, p. 967.

67. Z. Slanina, S.-L. Lee, and L. Adamowicz, *Int. J. Quantum Chem.* **1997**, *63*, 529.

68. Z. Slanina, X. Zhao, S.-L. Lee, and E. Ōsawa, in *Recent Advances in the Chemistry and Physics of Fullerenes and Related Materials*, Vol. 4, K. M. Kadish and R. S. Ruoff (eds.). The Electrochemical Society, Pennington, NJ, 1997.

69. Z. Slanina, X. Zhao, S.-L. Lee, and E. Ōsawa, *Chem. Phys.* **1997**, *219*, 193.

70. R. Mitsumoto, H. Oji, Y. Yamamoto, K. Asato, Y. Ouchi, H. Shinohara, K. Seki, K. Umishita, S. Hino, S. Nagase, K. Kikuchi, and Y. Achiba, *J. Phys. IV* **1997**, *7*, C2–525.

71. Z. Slanina, Ph.D. Thesis, Czechoslovak Academy of Science, Prague, 1974.

72. R. F. Curl, *Philos. Trans. R. Soc. London A* **1993**, *343*, 19.

73. G. E. Scuseria, personal communication (1993).

74. C. Z. Wang, B. L. Zhang, K. M. Ho, and X. Q. Wang, *Int. J. Mod. Phys. B* **1993**, *7*, 4305.

75. D. Bakowies and W. Thiel, *J. Am. Chem. Soc.* **1991**, *113*, 3704.

76. M.-L. Sun, Z. Slanina, and S.-L. Lee, *Chem. Phys. Lett.* **1995**, *233*, 279.

77. Z. Slanina and E. Ōsawa, *Fullerene Sci. Technol.* **1997**, *5*, 167.

78. P. W. Fowler and J. Woolrich, *Chem. Phys. Lett.* **1986**, *127*, 78.

79. P. W. Fowler, *Chem. Phys. Lett.* **1986**, *131*, 444.

80. P. W. Fowler and J. I. Steer, *J. Chem. Soc. Chem. Commun.* **1987**, 1403.

81. P. W. Fowler, J. E. Cremona, and J. I. Steer, *Theor. Chim. Acta* **1988**, *73*, 1.

82. A. Ceulemans and P. W. Fowler, *Phys. Rev. A* **1989**, *39*, 481.

83. A. Ceulemans and P. W. Fowler, *J. Chem. Phys.* **1990**, *93*, 1221.

84. P. W. Fowler, *J. Chem. Soc. Faraday Trans.* **1990**, *86*, 2073.

85. P. Fowler, *Nature* **1991**, *350*, 20.

86. P. W. Fowler, *J. Chem. Soc. Faraday Trans.* **1991**, *87*, 1945.

87. P. W. Fowler, R. C. Batten, and D. E. Manolopoulos, *J. Chem. Soc. Faraday Trans.* **1991**, *87*, 3103.

88. A. Ceulemans and P. W. Fowler, *Nature* **1991**, *353*, 52.

89. P. W. Fowler and D. E. Manolopoulos, *Nature* **1992**, *355*, 428.

90. P. W. Fowler, D. E. Manolopoulos, and R. P. Ryan, *J. Chem. Soc. Chem. Commun.* **1992**, 408.

91. P. W. Fowler, *J. Chem. Soc. Perkin Trans. 2* **1992**, 145.

92. P. W. Fowler, D. E. Manolopoulos, and R. P. Ryan, *Carbon* **1992**, *30*, 1235.

93. D. E. Manolopoulos, D. R. Woodall, and P. W. Fowler, *J. Chem. Soc. Faraday Trans.* **1992**, *88*, 2427.

94. P. W. Fowler and V. Morvan, *J. Chem. Soc. Faraday Trans.* **1992**, *88*, 2631.

95. D. E. Manolopoulos and P. W. Fowler, *J. Chem. Phys.* **1992**, *96*, 7603.

96. P. W. Fowler, D. E. Manolopoulos, D. B. Redmond, and R. P. Ryan, *Chem. Phys. Lett.* **1993**, *202*, 371.

97. D. E. Manolopoulos and P. W. Fowler, *Chem. Phys. Lett.* **1993**, *204*, 1.

98. S.-L. Lee, *Theor. Chim. Acta* **1992**, *81*, 185.

99. A. J. Stone and D. J. Wales, *Chem. Phys. Lett.* **1986**, *128*, 501.

100. C. Coulombeau and A. Rassat, *J. Chim. Phys.* **1991**, *88*, 173.

101. W. N. Lipscomb and L. Massa, *Inorg. Chem.* **1992**, *31*, 2297.

102. A. C. Tang, Q. S. Li, C. W. Liu, and J. Li, *Chem. Phys. Lett.* **1993**, *201*, 465.

103. D. E. Manolopoulos, J. C. May, and S. E. Down, *Chem. Phys. Lett.* **1991**, *181*, 105.

104. X. Liu, D. J. Klein, T. G. Schmalz, and W. A. Seitz, *J. Comput. Chem.* **1991**, *12*, 1252.

105. X. Liu, D. J. Klein, W. A. Seitz, and T. G. Schmalz, *J. Comput. Chem.* **1991**, *12*, 1265.

106. X. Liu, T. G. Schmalz, and D. J. Klein, *Chem. Phys. Lett.* **1992**, *188*, 550.

107. D. E. Manolopoulos, *Chem. Phys. Lett.* **1992**, *192*, 330.

108. X. Liu, T. G. Schmalz, and D. J. Klein, *Chem. Phys. Lett.* **1992**, *192*, 331.

109. M.-L. Sun, M.A. Thesis, National Chung-Cheng University, Chia-Yi, 1995.

110. P. W. Fowler, C. M. Quinn, and D. B. Redmond, *J. Chem. Phys.* **1991**, *95*, 7678.

111. S. Wei, Z. Shi, and A. W. Castleman Jr., *J. Chem. Phys.* **1991**, *94*, 3268.

112. D. J. Klein, W. A. Seitz, and T. G. Schmalz, *J. Phys. Chem.* **1993**, *97*, 1231.

113. D. J. Klein, *Chem. Phys. Lett.* **1994**, *217*, 261.

114. A. T. Balaban, D. J. Klein, and C. A. Folden, *Chem. Phys. Lett.* **1994**, *217*, 266.

115. J. R. Dias, *Chem. Phys. Lett.* **1993**, *209*, 439.

116. A. T. Balaban, X. Liu, D. J. Klein, D. Babic, T. G. Schmalz, W. A. Seitz, and M. Randić, *J. Chem. Inf. Comput. Sci.* **1995**, *35*, 396.

117. K. Balasubramanian, *Chem. Phys. Lett.* **1991**, *182*, 257.

118. K. Balasubramanian, *Chem. Phys. Lett.* **1991**, *183*, 292.

119. K. Balasubramanian, *J. Chem. Inf. Comput. Sci.* **1992**, *32*, 47.

120. K. Balasubramanian, *Chem. Phys. Lett.* **1992**, *198*, 577.

121. K. Balasubramanian, *Chem. Phys. Lett.* **1992**, *197*, 55.

122. K. Balasubramanian, *J. Mol. Spectrosc.* **1992**, *157*, 254.

123. K. Balasubramanian, *Chem. Phys. Lett.* **1993**, *201*, 306.

124. K. Balasubramanian, *Chem. Phys. Lett.* **1993**, *206*, 210.

125. K. Balasubramanian, *Chem. Phys. Lett.* **1993**, *202*, 399.

126. K. Balasubramanian, *J. Phys. Chem.* **1993**, *97*, 4647.

127. K. Balasubramanian, *J. Phys. Chem.* **1993**, *97*, 6990.

128. M. Yoshida and E. Ōsawa, *Bull. Chem. Soc. Jpn.* **1995**, *68*, 2073.

129. M. Yoshida and E. Ōsawa, *Bull. Chem. Soc. Jpn.* **1995**, *68*, 2083.

130. M. Yoshida, Ph.D. Thesis, Toyohashi University of Technology, Toyohashi, 1996.

131. H. Ueno, M.A. Thesis, Toyohashi University of Technology, Toyohashi, 1997.

132. X. Zhao, H. Ueno, Z. Slanina, and E. Ōsawa, C_{32}: An Example of Low Energy Cages with Four-Membered Rings, in *Recent Advances in the Chemistry and Physics of Fullerenes and Related Materials*, Vol. 5, K. M. Kadish and R. S. Ruoff (eds.). The Electrochemical Society, Pennington, NJ, 1997, p. 155.

133. Z. Slanina, F. Uhlík, and L. Adamowicz, *J. Radioanal. Nucl. Chem.* **1997**, *219*, 69.

134. S. Hobday and R. Smith, *J. Chem. Soc. Faraday Trans.* **1997**, *93*, 3919.

135. M. Goldberg, *Tohoku Math. J.* **1934**, *40*, 226.

136. M. Goldberg, *Tohoku Math. J.* **1937**, *43*, 104.

137. D. Bakowies, M. Bühl, and W. Thiel, *J. Am. Chem. Soc.* **1995**, *117*, 10113.

138. D. York, J. P. Lu, and W. Yang, *Phys. Rev. B* **1994**, *49*, 8526.

139. J. P. Lu and W. Yang, *Phys. Rev. B* **1994**, *49*, 11421.

140. E. Ōsawa, Z. Slanina, K. Honda, and X. Zhao, Catalytic Mechanism in the Stone-Wales Rearrangement, in *Recent Advances in the Chemistry and Physics of Fullerenes and Related Materials*, Vol. 5, K. M. Kadish and R. S. Ruoff (eds.). The Electrochemical Society, Pennington, NJ, 1997, p. 138.

141. E. Ōsawa, Z. Slanina, and K. Honda, *Mol. Mat.* **1998**, *10*, 1.

142. T. Braun and H. Rausch, *Anal. Chem.* **1995**, *67*, 1512.

143. B. R. Eggen, M. I. Heggie, G. Jungnickel, C. D. Latham, R. Jones, and P. R. Briddon, *Science* **1996**, *272*, 87.

144. B. R. Eggen, M. I. Heggie, G. Jungnickel, C. D. Latham, R. Jones, and P. R. Briddon, *Fullerene Sci. Technol.* **1997**, *5*, 727.

145. T. P. Martin, N. Malinowski, U. Zimmermann, U. Näher, and H. Schaber, *J. Chem. Phys.* **1993**, *99*, 4210.

146. U. Zimmermann, N. Malinowski, U. Näher, S. Frank, and T. P. Martin, *Phys. Rev. Lett.* **1994**, *72*, 3542.

147. U. Zimmermann, A. Burkhardt, T. P. Martin, and N. Malinowski, in *The Chemical Physics of Fullerenes 10 (and 5) Years Later*, W. Andreoni (ed.). Kluwer Academic Publishers, Dordrecht, 1996.

148. F. Tast, N. Malinowski, S. Frank, M. Heinebrodt, I. M. L. Billas, and T. P. Martin, *Z. Phys. D* **1997**, *40*, 351.

149. Z. Slanina, X. Zhao, and E. Ōsawa, *Chem. Phys. Lett.* **1998**, *290*, 311.

150. Z. Slanina, F. Uhlík, H. Ueno, X. Zhao, and E. Ōsawa, *Fullerene Sci. Technol.* **2000**, *8*.

151. S. W. McElvany, J. H. Callahan, M. M. Ross, L. D. Lamb, and D. R. Huffman, *Science* **1993**, *260*, 1632.

152. J.-P. Deng, D.-D. Ju, G.-R. Her, C.-Y. Mou, C.-J. Chen, Y.-Y. Lin, and C.-C. Han, *J. Phys. Chem.* **1993**, *97*, 11575.

153. Z. Slanina and S.-L. Lee, *J. Mol. Struct. (Theochem)* **1994**, *304*, 173.

154. M.-L. Sun, Z. Slanina, and S.-L. Lee, *Fullerene Sci. Technol.* **1995**, *3*, 627.

155. T. Martin, U. Obst, and J. Rebek, Jr., *Science* **1998**, *281*, 1842.

156. Z. Slanina, S. J. Kim, K. Fox, F. Uhlík, and A. Hinchliffe, Dimers in Earth's and Planetary Atmospheres: The $(H_2O)_2$, $(N_2)_2$, $N_2–O_2$, $(O_2)_2$, $(O_3)_2$, $(CO_2)_2$, $(H_2)_2$, and $Ar–N_2$ Cases, in *Molecular Complexes in Earth's, Planetary, Cometary, and Interstellar Atmospheres*, A. A. Vigasin and Z. Slanina (eds.). World Scientific, Singapore, 1998, p. 100.

157. G. Pólya, *C. R. Acad. Sci.* **1935**, *201*, 1167.

158. G. Pólya, *Acta Math.* **1937**, *68*, 145.

159. S. Fujita, *Symmetry and Combinatorial Enumeration in Chemistry*. Springer-Verlag, Berlin, 1991.

160. L. A. Curtiss and K. Raghavachari, Calculation of Accurate Bond, Energies, Electron Affinities, and Ionization Energies, in *Quantum Mechanical Electronic Structure Calculations with Chemical Accuracy*, S. R. Langhoff (ed.). Kluwer, Dordrecht, 1995, p. 139.

161. K. Raghavachari and L. A. Curtiss, Accurate Theoretical Studies of Small Elemental Clusters, in *Quantum Mechanical Electronic Structure Calculations with Chemical Accuracy*, S. R. Langhoff (ed.). Kluwer, Dordrecht, 1995, p. 173.

162. J. M. Schulman and R. L. Disch, *J. Chem. Soc. Chem. Commun.* **1991**, 411.

163. H. D. Beckhaus, C. Ruchardt, M. Kao, F. Diederich, and C. S. Foote, *Angew. Chem. Int. Ed. Engl.* **1992**, *31*, 63.

164. W. V. Steele, R. D. Chirico, N. K. Smith, W. E. Billups, P. R. Elmore, and A. E. Wheeler, *J. Phys. Chem.* **1992**, *96*, 4731.

165. T. Kiyobayashi and M. Sakiyama, *Fullerene Sci. Technol.* **1993**, *1*, 269.

166. H. P. Diogo, M. E. Minas da Pielade, T. J. S. Dennis, J. P. Hare, H. W. Kroto, R. Taylor, and D. R. M. Walton, *J. Chem. Soc. Faraday Trans.* **1993**, *89*, 3541.

167. H.-D. Beckhaus, S. Verevkin, C. Rüchardt, F. Diederich, C. Thilgen, H.-U. ter Meer, H. Mohn, and W. Müller, *Angew. Chem. Int. Ed. Engl.* **1994**, *33*, 996.

168. T. Clark, *A Handbook of Computational Chemistry. A Practical Guide to Chemical Structure and Energy Calculations*. Wiley, New York, 1985.

169. D. B. Boyd, *Rev. Comput. Chem.* **1990**, *1*, 321.

170. J. J. P. Stewart, *Rev. Comput. Chem.* **1990**, *1*, 45.

171. M. C. Zerner, *Rev. Comput. Chem.* **1991**, *2*, 313.

172. Z. Slanina, S.-L. Lee, and C.-H. Yu, *Rev. Comput. Chem.* **1996**, *8*, 1.

173. M. J. S. Dewar and W. Thiel, *J. Am. Chem. Soc.* **1977**, *99*, 4899.

174. M. J. S. Dewar, E. G. Zoebisch, E. F. Healy, and J. J. P. Stewart, *J. Am. Chem. Soc.* **1985**, *107*, 3902.

175. J. J. P. Stewart, *J. Comput. Chem.* **1989**, *10*, 209.

176. M. J. S. Dewar, C. Jie, and J. Yu, *Tetrahedron* **1993**, *49*, 5003.

177. *AMPAC 6.0*. Semichem, Shavnee, KS, 1997.

178. J. J. P. Stewart, *MOPAC 5.0*. QCPE 455, Indiana University, 1990.

179. M. J. Frisch, G. W. Trucks, H. B. Schlegel, P. M. W. Gill, B. G. Johnson, M. A. Robb, J. R. Cheeseman, T. Keith, G. A. Petersson, J. A. Montgomery, K. Raghavachari, M. A. Al-Laham, V. G. Zakrzewski, J. V. Ortiz, J. B. Foresman, J. Cioslowski, B. B. Stefanov, A. Nanayakkara, M. Challacombe, C. Y. Peng, P. Y. Ayala, W. Chen, M. W. Wong, J. L. Andres, E. S. Replogle, R. Gomperts, R. L. Martin, D. J. Fox, J. S. Binkley, D. J. Defrees, J. Baker, J. P. Stewart, M. Head-

Gordon, C. Gonzalez, and J. A. Pople, *Gaussian 94, Revision E.2*. Gaussian, Inc., Pittsburgh, 1995.

180. W. J. Hehre, L. D. Burke, and A. J. Schusterman, *Spartan*. Wavefunction, Inc., Irvine, CA, 1993.

181. M. Häser, J. Almlöf, and G. E. Scuseria, *Chem. Phys. Lett.* **1991**, *181*, 497.

182. J. A. Pople, R. Krishnan, H. B. Schlegel, D. DeFrees, J. S. Binkley, M. J. Frisch, R. F. Whiteside, R. F. Hout, and W. J. Hehre, *Int. J. Quantum Chem. Quantum Chem. Symp.* **1981**, *15*, 269.

183. E. F. Healy and A Holder, *J. Mol. Struct.* **1993**, *281*, 141.

184. J. A. Pople, A. P. Scott, M. W. Wong, and L. Radom, *Isr. J. Chem.* **1993**, *33*, 345.

185. C. W. Bauschlicher, Jr. and H. Partridge, *J. Chem. Phys.* **1995**, *103*, 1788.

186. Z. Slanina, F. Uhlík, and M. C. Zerner, *Rev. Roum. Chim.* **1991**, *36*, 965.

187. Z. Slanina, *Int. Rev. Phys. Chem.* **1987**, *6*, 251.

188. Z. Slanina, *Theor. Chim. Acta* **1992**, *83*, 257.

189. Z. Slanina and L. Adamowicz, *Thermochim. Acta* **1992**, *205*, 299.

190. C. Piskoti, J. Yarger, and A. Zettl, *Nature* **1998**, *393*, 771.

191. G. von Helden, M. T. Hsu, P. R. Kemper, and M. T. Bowers, *J. Chem. Phys.* **1991**, *95*, 3835.

192. G. von Helden, N. G. Gotts, and M. T. Bowers, *Nature* **1993**, *363*, 60.

193. F. Wahl, J. Wörth, and H. Prinzbach, *Angew. Chem. Int. Ed. Engl.* **1993**, *32*, 1722.

194. L. D. Book, C. Xu, and G. E. Scuseria, *Chem. Phys. Lett.* **1994**, *222*, 281.

195. S. G. Louie, *Nature* **1996**, *384*, 612.

196. J. C. Grossman, M. Côté, S. G. Louie, and M. L. Cohen, *Chem. Phys. Lett.* **1998**, *284*, 344.

197. Z. Slanina, F. Uhlík, H. Ueno, X. Zhao, and E. Ōsawa, *Fullerene Sci. Technol.* **2000**, *8*.

198. K. Raghavachari, Stabilities of Medium-Sized Carbon Clusters: Relative Energies of the Isomers of C_{20}, in *Recent Advances in the Chemistry and Physics of Fullerenes and Related Materials*, Vol. 3, K. M. Kadish and R. S. Ruoff (eds.). The Electrochemical Society, Pennington, NJ, 1996, p. 894.

199. B. I. Dunlap and R. Taylor, *J. Phys. Chem.* **1994**, *98*, 11018.

200. Z. Slanina and S.-L. Lee, *Fullerene Sci. Technol.* **1995**, *3*, 151.

201. I. Hargittai and M. Hargittai, *Symmetry: A Unifying Concept*. Shelter Publications, Bolinas, CA, 1994, p. 99.

202. R. Taylor, *Chem. Intell.* **1995**, *1*, No. 3, 48.

203. R. Taylor, *J. Chem. Soc. Chem. Commun.* **1994**, 1629.

204. D. Goldberg, Y. Bando, O. Stéphan, and K. Kurashimna, *Appl. Phys. Lett.* **1998**, *73*, 2441.

205. Z. C. Wu, D. A. Jelski, and T. F. George, *Chem. Phys. Lett.* **1987**, *137*, 291.

206. V. Elser and R. C. Haddon, *Nature* **1987**, *325*, 792.

207. R. E. Stanton and M. D. Newton, *J. Phys. Chem.* **1988**, *92*, 2141.

208. D. E. Weeks and W. G. Harter, *Chem. Phys. Lett.* **1988**, *144*, 366.

209. D. E. Weeks and W. G. Harter, *J. Chem. Phys.* **1989**, *90*, 4744.

210. J. M. Rudziński, Z. Slanina, E. Ōsawa, and T. Iizuka, Carbon Atom Clusters in Gas Phase: A Theoretical Study, in *8th IUPAC Conference on Physical Organic Chemistry*, Tokyo, 1986, p. 262.

211. Z. Slanina, J. M. Rudziński, M. Togasi, and E. Ōsawa, *J. Mol. Struct.* (*Theochem*) **1989**, *61*, 169.

212. D. R. Huffman and W. Krätschmer, *Mat. Res. Soc. Proc.* **1991**, *206*, 601.

213. D. Babić, D. J. Klein, and C. H. Sah, *Chem. Phys. Lett.* **1993**, *211*, 235.

214. Z. Slanina, M.-L. Sun, S.-L. Lee, and L. Adamowicz, Fullerenes as Multi-Isomeric Mixtures: Computational Studies of the C_{76} and C_{80} Cases, in *Recent Advances in the Chemistry and Physics of Fullerenes and Related Materials*, Vol. 2, K. M. Kadish and R. S. Ruoff (eds.) The Electrochemical Society, Pennington, NJ, 1995, p. 1138.

215. Z. Slanina and S.-L. Lee, C_{76}, C_{80}, C_{82}, C_{86} and C_{88}, Computations of Multi-Isomeric Fullerenes, in *HPC-ASIA '95*, Electronic Proceedings. National Center for High-Performance Computing, Hsinchu, Taiwan, 1995, ch 090.

216. D. Bakowies and W. Thiel, *Chem. Phys.* **1991**, *151*, 309.

217. Z. Slanina, L. Adamowicz, D. Bakowies, and W. Thiel, *Thermochim. Acta* **1992**, *202*, 249.

218. Z. Slanina and L. Adamowicz, *Fullerene Sci. Technol.* **1993**, *1*, 1.

219. R. Taylor, G. J. Langley, A. G. Avent, T. J. S. Dennis, H. W. Kroto, and D. R. M. Walton, *J. Chem. Soc. Perkin Trans. 2* **1993**, 1029.

220. T. Wakabayashi, K. Kikuchi, S. Suzuki, H. Shiromaru, and Y. Achiba, *J. Phys. Chem.* **1994**, *98*, 3090.

221. M. Benz, M. Fanti, P. W. Fowler, D. Fuchs, M. M. Kappes, C. Lehner, R. H. Michel, G. Orlandi, and F. Zerbetto, *J. Phys. Chem.* **1996**, *100*, 13399.

222. R. J. Cross, M. Saunders, A. Khong, R. Shimshi, and S. Uljon, Putting Atoms into Fullerenes, in *Recent Advances in the Chemistry and Physics of Fullerenes and Related Materials*, Vol. 4, K. M. Kadish and R. S. Ruoff (eds.). The Electrochemical Society, Pennington, NJ, 1997, p. 429.

223. S.-L. Lee, M.-L. Sun, and Z. Slanina, *Int. J. Quantum Chem. Quantum Chem. Symp.* **1996**, *30*, 355.

224. T. J. S. Dennis, personal communication (1998).

225. F. Negri, G. Orlandi, and F. Zerbetto, *Chem. Phys. Lett.* **1992**, *189*, 495.

226. K. Raghavachari, *Chem. Phys. Lett.* **1992**, *190*, 397.

227. X.-Q. Wang, C. Z. Wang, B. L. Zhang, and K. M. Ho, *Phys. Rev. Lett.* **1992**, *69*, 69.

228. D. Bakowies, M. Kolb, W. Thiel, S. Richard, R. Ahlrichs, and M. M. Kappes, *Chem. Phys. Lett.* **1992**, *200*, 411.

229. Z. Slanina and S.-L. Lee, *J. Mol. Struct.* (*Theochem*) **1995**, *333*, 153.

230. Z. Slanina and S.-L. Lee, *NanoStruct. Mat.* **1994**, *4*, 39.

231. Z. Slanina, F. Uhlík, M. Yoshida, and E. Ōsawa, *Fullerene Sci. Technol.* **2000**, *8*.

232. Z. Slanina, M.-L. Sun, and S.-L. Lee, *NanoStruct. Mat.* **1997**, *8*, 623.

233. Z. Slanina, S.-L. Lee, F. Uhlík, and L. Adamowicz, B_{32} and C_{60}^{6-}: Two Cases of Lost Icosahedral Symmetry, in *Physics and Chemistry of Fullerenes and Their*

Derivatives, H. Kuzmany, J. Fink, M. Mehring, and S. Roth (eds.). World Scientific, Singapore, 1996, p. 385.

234. Y. Achiba, personal communication (1998).

235. Z. Slanina, X. Zhao, and E. Ōsawa, (*to be published*).

236. J. F. Anacleto, H. Perreault, R. K. Boyd, S. Pleasance, M. A. Quilliam, P. G. Sim, J. B. Howard, Y. Makarovsky, and A. L. Lafleur, *Rapid Commun. Mass. Spectrom.* **1992**, *6*, 214.

237. Z. Slanina, S.-L. Lee, K. Kobayashi, and S. Nagase, *J. Mol. Struct.* (*Theochem*) **1994**, *312*, 175.

CHAPTER 7

POLYMER DERIVATIVES OF FULLERENES

LONG Y. CHIANG and LEE Y. WANG

7.1 INTRODUCTION

Fullerene molecules were found to be highly reactive toward various organic reagents. A great number of chemical reactions involving fullerene molecules were developed in the past six years leading to discovery of various synthetic methods for the mono or multiple additions of alkyl, aryl, organoamino, hydroxy, halide, organoether, and other functional groups on fullerenes [1; see Chapter 10 in this volume]. Extension of these basic organic reactions into polymerization chemistry results in synthetic strategies for the design and preparation of fullerene-derived macromolecules. The purpose of incorporating fullerenes, the related truncated-cage molecules, and the derived fullerides in the chemical structure of polymer materials is often associated with their unique physical characteristics, including energy excitation, optical absorption, fluorescence, intrinsic electron or charge transportation, and photoinduced electron-transfer properties. Most of these important chemical and physical characteristics of C_{60} arise from the high reactivity of pyracylene olefins and its high electronegativity, respectively. It is interesting to note that a moderate derivatization of C_{60} did not lead to a great suppression of the electron- or radical-absorbing capability of the fullerene cage itself. The remaining unfunctionalized olefins on the cage surface are able to retain an appreciable degree of π-conjugation.

Fullerene molecules may be incorporated in the polymer as a molecular disperse entity, aggregated particles, chemically attached segments in the main chains of a polymer (1), grafting components on the side chains of a polymer (2), a molecular core of a starburst polymer (3) and a dendritic polymer (4), or crosslinkers in the polymer network (5), as depicted in Figure 7.1. Chemical functionalization of the fullerenic double bonds via polymer attachments con-

Fullerenes: Chemistry, Physics, and Technology, Edited by Karl M. Kadish and Rodney S. Ruoff.

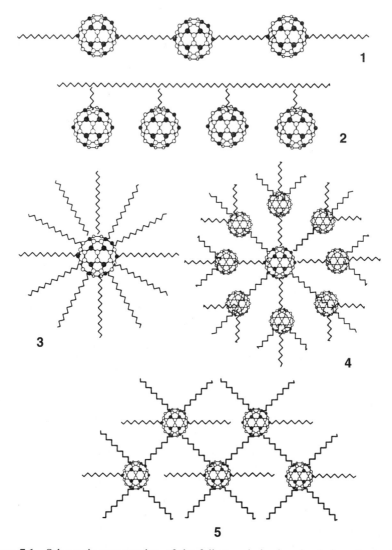

Figure 7.1 Schematic presentation of the fullerene-derived main-chain polymer **1**, the fullerene-grafted side-chain polymer **2**, the fullerene-derived starburst polymer **3** and dendritic polymer **4**, and the fullerene-crosslinked polymer network **5**.

verts the olefinic carbons into the tertiary sp^3 carbons. That effects a local disruption of the olefin conjugation on the cage surface. Thus, shift of the optical absorption bands and the redox potentials of the resulting polymer products were expected. Owing to the nature of the polypyracylene-containing structure of fullerenes, it is difficult to synthesize monodisperse polyfunctional fullerenes; a mixture of regioisomer products with a distribution of the number of functional addends is normally obtained. Chemically, polyfunctional fullerenes may serve as precursors for the production of technologically important dendritic polymers and networks. Therefore, successful synthesis of polyfunctionalized fullerenes with a well-defined, narrow distribution of the number of addends per C_{60} is desirable.

Several synthetic methods reported recently for the attachment of C_{60} onto the functional side groups or end groups of a polymer chain can be categorized, according to their reaction nature, into anionic, cationic, free radical, condensation, hydroamination, cycloazide, and olefin cyclolyzation approaches. The reaction mechanism involved in these synthetic approaches using polyfunctional polymers may bear a close resemblance to that of C_{60} reacting with the corresponding molecular reagents.

7.2 ANIONIC REACTION ON FULLERENE

High reactivity of C_{60} molecules toward certain nucleophiles, such as organolithium or Grignard reagents, has been demonstrated to be advantageous in developing an effective method for multiple nucleophilic additions of these reagents onto fullerenes [2,3]. However, the reaction of C_{60} with methyllithium as a nucleophilic reagent was not selective and controllable. Multiple additions of reagents per C_{60} cage were commonly observed in reactions at ambient temperatures. The low controllability of reaction progress may be due to the fact that early reacted fullerene intermediates exhibited a higher reactivity than that of C_{60} itself. That allowed continuing reactions of the reacted fullerene molecules with more equivalents of the reagent, leaving the unreacted ones untouched unless a substantial excess of reagent was applied. Thus, a broad distribution of the number of addends per C_{60} in the chemical composition of products was obtained normally. An example was given by the treatment of C_{60} with living polystyrene carbanion (PS^-). Products of $C_{60}(PS)_x$ (**6**, Scheme 1) where x ranges from 1 to 10 depending on the initial ratio of PS carbanion used, containing a variable number of polystyrene addends on each C_{60} were reported [4].

Distribution of the number of organo-addends per C_{60} can be narrowed to only 1–2 corresponding to a mixture of monopolystyrene- and dipolystyrene-substituted fullerenes if the reaction conditions were modified to allow a homogeneous contact of reagents with the C_{60} molecules. Condition modifications included the use of an appropriate quantity of relatively polar tetrahydrofuran solvent, during the reaction of polystyryl carbanions with C_{60}, enough to break up the chain-end associative aggregations of polystyryl carbanions, which were formed during the prior living polymerization of styrene in nonpolar hydrocarbon solvents. The homogeneous dispersion of polystyryl carbanions in tetrahydrofuran (THF) prevented increase of the local polystyryl carbanion concentration in an associative form at the C_{60} sites and thus minimized the possible multiple addition on C_{60}. In addition, a low reaction temperature at $-78\,^{\circ}C$ was applied in order to prevent the possible chain transfer to THF and

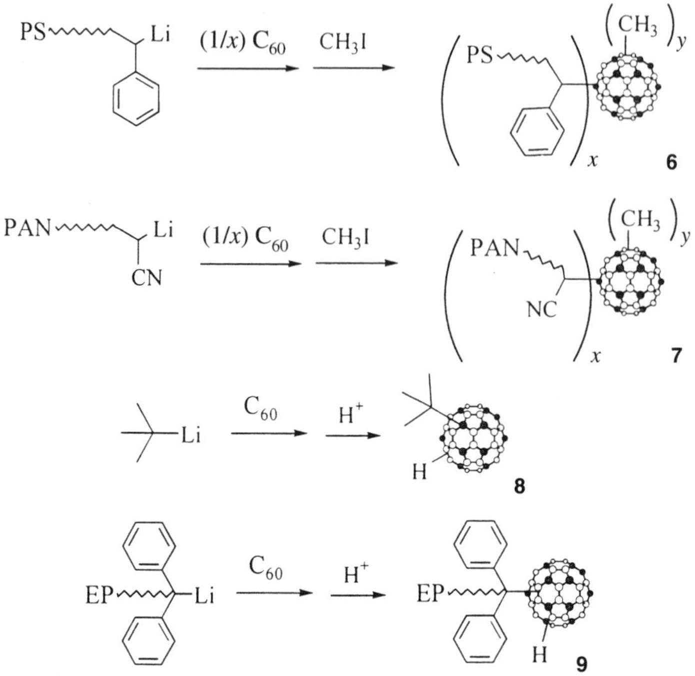

Scheme 1 Synthesis of the polymer-bound fullerene derivatives using either molecular or polymeric lithiated carbanions as nucleophilic reagents in the addition reaction of C_{60}.

to decrease the nucleophilicity of polystyryl carbanions for maximizing the yield of monopolystyrene-substituted fullerenes [5]. A similar synthetic approach was also reported using the living butyl-terminated polyacrylonitrile (PAN) carbanions in the reaction with C_{60} under heterogeneous conditions, leading to soluble polyacrylonitrilated C_{60} products (**7**, Scheme 1) containing multiple PAN arms on each fullerene cage [6].

One interesting anionic approach developed for allowing perfect control of the number of polymer chains being added onto each fullerene molecule was demonstrated by Ederle and Mathis [7] using anionic styrene polymerization as a model. The polymerization was carried out in both polar and nonpolar solvents. The approach resulted in products with a well-defined composition of grafted polystyryl arms per C_{60}. Apparently, variation of the solvent polarity significantly shifted the chemical characteristics of polystyryl carbanions toward their nucleophilic attack on the fullerene molecules. When a nonpolar solvent of toluene was used, a maximum number of grafted chains per C_{60} was found to be 3 and 6 for the stable diphenylalkyl carbanion and the reactive living polystyryl and polyisoprene carbanions, respectively. Conversely, with the application of a polar solvent, such as THF, an electron-transfer process

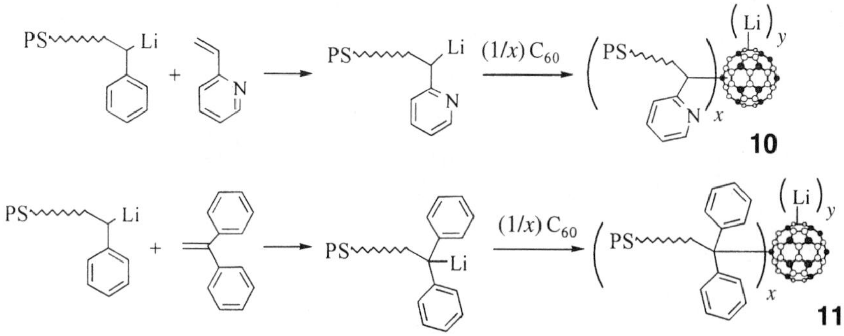

Scheme 2 Synthesis of the polymer-bound fullerene derivatives using polystyryl-2-vinylpyridyl and polystyryl-diphenylethyl carbanions as nucleophilic reagents for the addition reaction of C_{60}.

occurring from the carbanion to the fullerene molecule forming the dianionic C_{60} was found to dominate in the initial reaction mechanism prior to the beginning nucleophilic grafting of polystyrene onto C_{60}. Formation of the monoanionic C_{60}^- and the dianionic C_{60}^{-2} species was readily detectable in their near-infrared spectra, showing an absorption peak at 1075 and 950 nm, respectively [8]. In this case, the number of polymer chain grafts was determined to be 4, 3, and none using styryl, 2-vinylpyridyl (forming **10**, Scheme 2), and diphenylethyl carbanion (forming **11**, Scheme 2), respectively. Capping of a living polystyryl anion with 2-vinylpyridine and diphenylethylene decreased the nucleophilic reactivity of the resulting anionic intermediates toward C_{60}, owing to extended delocalization of negative charges, as shown in Scheme 2. Finally, the maximum number of polymer chains grafted onto each C_{60} using polystyryl-2-vinylpyridyl and polystyryl-diphenylethyl carbanion as a nucleophile in toluene, leading to products of **10** and **11**, respectively, was resolved to be 3 [9]. Similar electron-transfer processes from the terminal anion of a growing polymer chain to the fullerene molecule forming a dianionic C_{60} were also observed if monofunctional living poly(ethylene oxide) anions ($PEO^- - K^+$) or poly(styryl-ethylene oxide) anions were used as the nucleophilic grafting reagents [10].

It is possible to synthesize star-shaped macromolecules derived from C_{60} containing more than six polymer arms or heteroarms. Evidently, hexanionic hexa(polystyryl)fullerenes were found to be reactive enough for initiating polymerization of styrene and methylmethacrylate monomers, as shown in Scheme 3 [11]. However, only one additional polystyrene arm can be attached on hexa(polystyryl)fullerenes, forming protonated hepta(polystyryl)fullerenes **12** after acidification. In the case of methylmethacrylate reactions, a maximum of two additional PMMA arms were added on hexa(polystyryl)fullerenes, giving the corresponding protonated hexa(polystyryl)-di(polymethylmethacrylate)-fullerenes **13**.

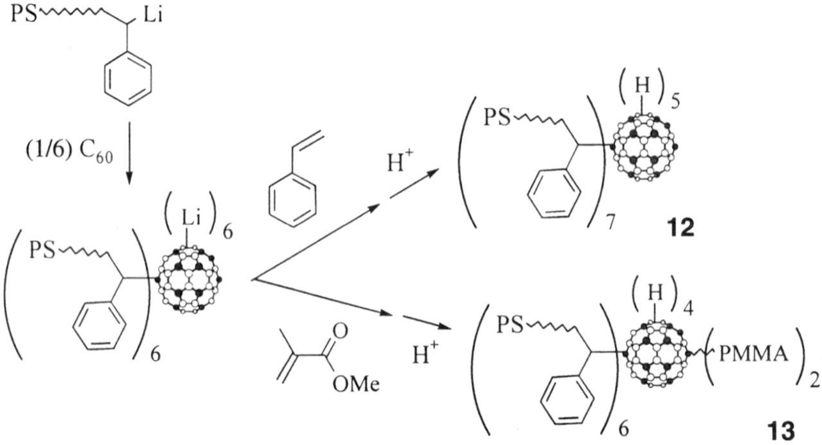

Scheme 3 Polymerization of styrene and methylmethacrylate monomers initiated by hexanionic hexa(polystyryl)fullerenes, producing hepta(polystyryl)fullerenes **12** and hexa(polystyryl)-di(polymethylmethacrylate)fullerenes **13** after protonation.

Monoaddition of an organolithium reagent onto C_{60} was possible by the use of an equivalent amount of *tert*-butyllithium, forming a product of *t*-butylhydro[60]fullerene **8** [12,13]. However, the reaction is expected to be accompanied by a number of minor products of di(*t*-butyl)fullerene and tri(*t*-butyl)fullerene regioisomers. In the presence of highly reactive organolithium reagents, stoichiometrically controlled addition on each C_{60} cage is rather difficult to achieve unless the reactive anionic center is located on a polymer chain. The hosting polymer chain may provide enough space separation or hindrance between two carbanion centers that minimizes the chance of more than one carbanion reacting with a fullerene molecule.

In fact, most synthetic procedures reported for C_{60} grafting on polymers were carried out by this approach. An early example was given by the reaction of C_{60} with the surface lithiated polyethylene (PE) film, producing the surface of polyethylene (**9**, Scheme 1) chemically bound with fullerene molecules. The polyethylene film was composed of high-density polyethylene and a fraction of ethylene oligomers (3% by weight) containing a terminal diphenylethyl group, which was prepared by anionic living oligomerization of ethylene using *n*-BuLi–*N,N,N',N'*-tetramethylethylene-diamine (TMEDA) as an initiator under a monomer pressure of 207 kPa [14]. The dibenzyl proton of the terminal diphenylethyl functional group is weakly acidic. It was deprotonated by the treatment of a diphenylethyl group-containing polyethylene film with *n*-BuLi–TMEDA to yield an anionic surface.

A closely related reaction was applied using polylithiated polyisoprene and polybutadiene as the active polymer matrix. Lithiation of the allylic carbons along the main chain of polyisoprene and polybutadiene was performed with a stoichiometric amount of *sec*-BuLi–TMEDA reagent for the deprotonation of allylic protons [15]. Reaction of C_{60} with polylithiated polyisoprene and polybutadiene yielded the corresponding fullerene-grafted polyolefins. Lithiation of the α-carbons along the main chain of poly(*N*-vinylcarbazole) [16] and the benzyl carbons of poly(*p*-bromostyrene) [17,18] was also reported prior to the reaction with C_{60} molecules.

7.3 FREE RADICAL REACTION ON FULLERENE

The first radical copolymerization of C_{60} and *p*-xylene was demonstrated by treating C_{60} in toluene at $-78\,°C$ with *p*-xylene diradicals, which were generated upon pyrolysis of sublimed paracyclophane at $650\,°C$ [19]. It was known that fullerene molecules exhibit a high intrinsic electronegativity. The electron affinity of C_{60} was found to be 2.65 eV [20]. Electron reduction potentials of the C_{60} molecule have been investigated extensively by studies of cyclic voltammetry [21–27]. Six reversible consecutive electron reductions were achieved in solution, generating a stable hexa-anionic fullerene intermediate, C_{60}^{-6} [27]. These electrochemical data indicated the capability of fullerene molecules in absorbing multiple electrons sequentially.

High electrophilicity of C_{60} even allowed multiple additions of the electron-deficient organoradicals, such as benzyl and *tert*-butyl radicals, to the fullerenic double bounds [28–30]. In principle, absorption of an odd number of electrons or radicals may result in the existence of a residual spin on the cage surface that is readily detectable by the electron spin resonance (ESR) spectroscopic measurements. Rate constants for the addition reaction of the alkyl radicals onto C_{60} were reported to be 10^7–10^9 $m^{-1}s^{-1}$ [31–34], much higher (10^5 times) than those for the propagation of polymer or oligomer chain-end radicals with monomers during the free radical polymerization of styrene and methylmethacrylates. Therefore, it was proposed that the radical initiators were consumed in the early stage of polymerization by the multiple radical initiator additions onto C_{60}, leading to formation of the highly substituted C_{60} derivatives. Polymerization of styrene and methylmethacrylates began to proceed when the rate of addition of radical initiators to the highly substituted C_{60} derivatives became relatively slow as compared with that of the polymer radical propagation with monomers [35].

Free radical copolymerization of styrene or methylmethacrylates and C_{60} was normally carried out using azobisisobutyronitrile (AIBN) as an initiator at $65\,°C$ in either benzene or 1,2-dichlorobenzene [36–39]. One set of the reported results was summarized, as shown in Table 7.1 [38]. The C_{60} content of the polymer in a weight percentage as a function of molecular weight was estimated using the gel permeation chromatography (GPC) measurements with a dual detection system consisting of a differential refractive index (RI) detector and an ultraviolet (UV) detector fixed at 350 nm. The ratio of the integrated UV signal versus the ΔRI signal was assumed to be proportional to the weight

TABLE 7.1 **Copolymerization Data**[a] **of Styrene and C$_{60}$ at 65 °C Using AIBN as an Initiator**

(a) Polymerization of Styrene Monomer Alone

AIBN (mmol)	Yield (%)	M_n (kg/mol)	M_w/M_n
0.347	39.1	10.8	1.7
0.270	35.5	12.9	1.7
0.210	34.0	13.8	1.7
0.140	28.8	17.3	1.8
0.079	24.3	25.1	17

(b) Polymerization of Styrene Monomer in the Presence of C$_{60}$ in a Constant Concentration

AIBN (mmol)	Yield (%)	M_n (kg/mol)	M_w/M_n	wt% of C$_{60}$
0.347	25.7	12.9	1.8	2.2
0.270	14.4	13.8	1.8	3.7
0.210	3.8	8.5	1.5	11.4
0.140	0.58	2.4	1.6	41.1
0.079	0.83	1.2	2.5	56.9

(c) Polymerization of Styrene Monomer and C$_{60}$ with a Constant Concentration of AIBN

C$_{60}$ (mmol)	Yield (%)	M_n (kg/mol)	M_w/M_n	wt% of C$_{60}$
0	40.5	8.3	1.8	0
0.014	36.9	6.0	1.8	0.5
0.028	32.6	12.5	1.7	1.2
0.040	25.7	12.9	1.8	2.2
0.055	11.8	10.8	1.7	6.1

[a] The data were summarized from those in Ref. 38.

percentage of C$_{60}$. Thus, a formula of wt% C$_{60}$ = κ (UV area)/(ΔRI area), where κ is the proportionality constant, was applied [38]. However, the formula can only be used as an approximate routine since the addition of multiple alkyl groups and free radicals to C$_{60}$ may alter its photophysical properties, such as the ultraviolet/visible (UV/VIS) absorption, in a certain degree. Both the polymer yield and the molecular weight (M_n) of polymer increased with the molarity increase of AIBN initiator applied.

The weight percent of C$_{60}$ in the polymer was increased at the expense of the polymer yield and molecular weight. The polydispersity (M_w/M_n) of the C$_{60}$-containing polystyrenes was found to be rather narrow as 1.5–1.8, in a nearly identical range as that of the sytrene homopolymer prepared under similar reaction conditions. Other AIBN-initiated copolymerizations of C$_{60}$ with 4-vinylbenzoic acid, 2-vinylpyridine, and 4-vinylpyridine forming gel-like polymers were also reported [40].

A complicate radical polymerization pathway was proposed based on the elucidation of the polymerization data in different studies. A short induction period was observed before polymerization began. The reaction was accompanied by sizable amounts of a low molecular weight linear polymer [39]. The fraction of C_{60}-containing polystyrenes was detected at the higher end molecular weight distribution of the polymers as shown in the molar chromatograms of the polymerization products. Multiple C_{60} units were counted in a high molecular weight polystyrene. In the case of MMA polymerization, only one C_{60} unit, on average, was estimated in each high molecular weight PMMA polymer molecule, forming a star-like structure. An average branch number of five per C_{60} over the entire molecular weight distribution of PMMA polymers was determined based on the calculation from a random branching model of Zimm and Stockmayer [39].

As far as the free radical polymerization mechanism was concerned, a series of nonselective radical addition and radical coupling reactions involving the C_{60} molecule or its intermediates were proposed. The plausible reaction pathway and most of the possible free radical intermediates produced during copolymerization of C_{60} with styrene and methylmethacrylate monomers are summarized in Scheme 4. A competitive reaction of the initiating radicals (R.) toward either monomers or the fullerene molecules determined the outcome of reactions at the early stage of polymerization. Since the C_{60} molecule exhibits a much faster reaction rate toward the initiating radicals than the propagation rate of monomers, accessibility of monomers to the site of initiating and propagating radicals becomes critical in monitoring the progress and results of the copolymerization. Therefore, variation of either the C_{60} concentration or the amount of AIBN used in polymerization of styrene led to subtle change of the yield and molecular weight of the C_{60}-containing polystyrenes obtained, as shown in Table 7.1, consistent with the proposed radical reaction pathways.

Facile radical couplings between the polymer radicals and fullerene radical intermediates provided a rationalization for formation of the star-shaped high molecular weight polystyrenes. In this case, an even number of polymer arms per C_{60} becomes a preferential chemical composition of the C_{60}-containing products. It was concluded by in situ polymerization studies of methylmethacrylate in an *ESR* tube that no signal contributed from the fullerenic radical intermediates was detected [41]. The dimerization of fullerene derivatives was suggested to be unlikely owing to the fact that the fullerenic radical is not delocalized along the cage surface, but rather is localized around the vicinity of a bulky substituent, as shown in **14** [42,43]. The bulky substituent may hinder the process of dimerization of fullerene derivatives.

Phenomenon of the even-numbered addition of polymeric radical species onto C_{60} was further investigated by allowing exposure of the polystyryl radical intermediates to an excess of C_{60}. The approach minimized the possibility of multiple polymer-chain additions onto one fullerene molecule. Instead of producing the *ESR*-active, monosubstituted fullerene products, an appreciable yield (60–80%) of well-defined di-substituted polystyryl derivatives of C_{60},

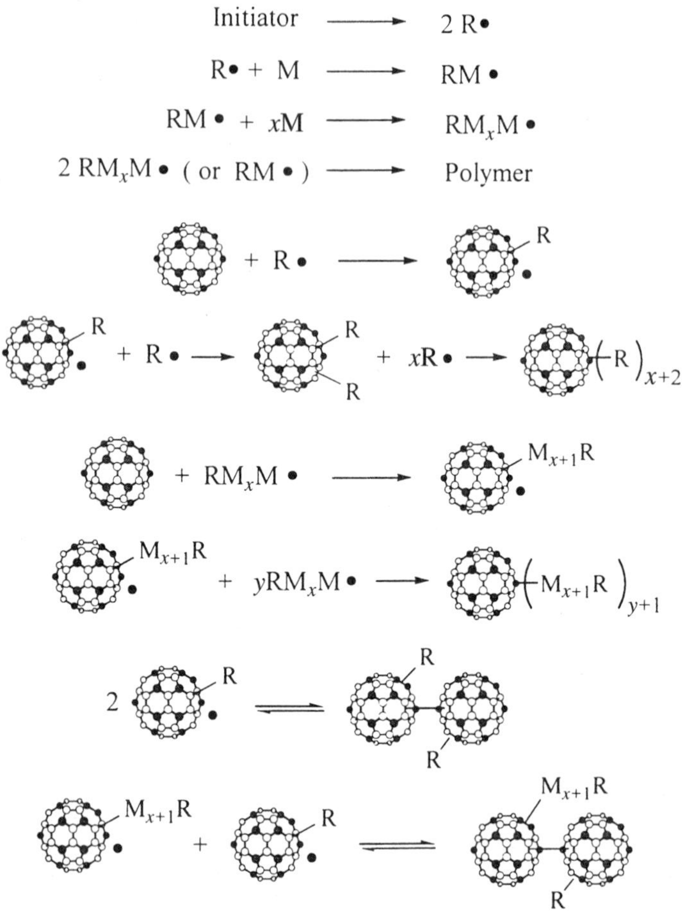

Scheme 4 The plausible reaction mechanism and several possible free radical intermediates produced during the copolymerization of C_{60} with either styrene or methylmethacrylate monomers.

$C_{60}(PS)_2$, was obtained by assumption of a radical mechanism [44]. That provided clear evidence of a preferential facile radical coupling between the monopolystyryl C_{60} radical, $(PS)C_{60}^{\bullet}$, and the polystyryl radical, $PS^{\bullet}$. Polystyryl radicals were thermally regenerated at 125 °C via a deprotection reaction of the 2,2,6,6-tetramethyl-1-piperidinyloxy (TEMPO) polystyrene precursor, prepared under a living free radical polymerization process.

Finally, direct incorporation of C_{60} molecules into a polymer matrix via the free radical method was demonstrated using polycarbonate radicals, generated by either the UV- or AIBN-initiated chain-incision of the parent polycarbonate [45]. However, the approach produced the C_{60}-containing polymer materials with a poorly defined structure.

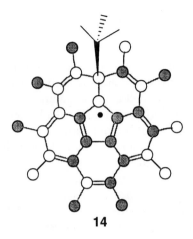

14

7.4 CATIONIC REACTION ON FULLERENE

[60] Fullerene molecules exhibit low susceptibility toward electron oxidation. The oxidation wave of C_{60} at $E_{ox} = +1.30$ V versus Fc/Fc^+ obtained in the cyclic voltammetry study was found to be rather unstable in a neutral organic solution at temperatures from ambient down to $-15\,°C$ [26], indicating its inherent inactivity to multielectron oxidation. The second ionization energy of C_{60} was found to be high at 8.8 eV [46]. Only a relatively short-lived cationic radical of C_{60} was detectable in $(TBA)PF_6$–tetrachloroethane solution (0.1 M) at ambient temperatures, showing a chemically reversible, one-electron oxidation wave at $+1.26$ V versus Fc/Fc^+ [47]. The lifetime of this radical was estimated to be >0.5 min. Apparently, the high electron-ionization threshold (>7.61 V) [48] of the neutral C_{60} molecules greatly hindered the attempt in generating a higher oxidation state of C_{60}. The observed high electron affinity of fullerenes over its electron-donating ability leads to the assumption that this class of materials is more susceptible to the nucleophilic additions than to the electrophilic additions.

Difficulty in detecting the oxidized C_{60} intermediates was presumably owing to the low chemical stability of the resulting cationic radical. Stability of the cationic C_{60} radical can be significantly enhanced in a strong acid medium. Evidently, a respectably intensity of spins, attributed to the cationic C_{60} radical, was detected in the *ESR* spectroscopic studies of C_{60} molecules dissolving in fuming sulfuric acid (H_2SO_4–SO_3) [49–51]. Despite the resistance of C_{60} to the electron oxidation in organic solvents, the conjugated fullerene structure was found to be susceptible to attack by certain electrophilic reagents. Early examples using the electrophilic approach for the synthesis of fullerene derivatives were given by the nitronium (NO_2^+) reaction of C_{60} molecules in the presence of organocarboxylic acids [52–54], yielding several water-soluble polyhydroxylated C_{60} derivatives after hydrolysis of the fullerenic ester intermediates, and the arene addition of benzene and toluene in the presence of aluminum chloride [55].

One example of using the cationic method for grafting C_{60} molecules onto a polymer chain was reported [56]. The C_{60} grafting reaction involved the aluminum chloride ($AlCl_3$)-activated cationic addition reaction of the carbazole moiety of poly(9-vinylcarbazole), denoted as PVK, onto C_{60}. The experiment was carried out in CS_2 at ambient temperatures for 2 h using the quantity of $AlCl_3$ in 1–5 times the fullerene mass [56]. Structural elucidation of the resulting polymer products was made by utilizing a model reaction of C_{60} with 9-vinylcarbazole (EK) performed under similar conditions as reported [56]. Spectroscopic measurements of the C_{60}–EK products revealed the addition reaction of C_{60} occurred at the C_2–C_7, C_3–C_6, and C_4–C_5 positions of the carbazole moiety in a relative substitution ratio of 25%, 47%, and 84%, respectively, instead of exclusively at the most reactive C_3–C_6 carbons. The rational explanation was postulated that three possible double substitutions across the C_4 and C_5, C_3 and C_4, or C_2 and C_3 occurred. That led to a stable 2-carbon bridge across the substituted six-membered ring of the fullerene, creating additional six-membered rings such as the one shown in **17** (Scheme 5). The cationic reaction mechanism using aluminum chloride as a catalyst was then suggested to involve the initial cationic activation of a fullerenic double bond by Lewis

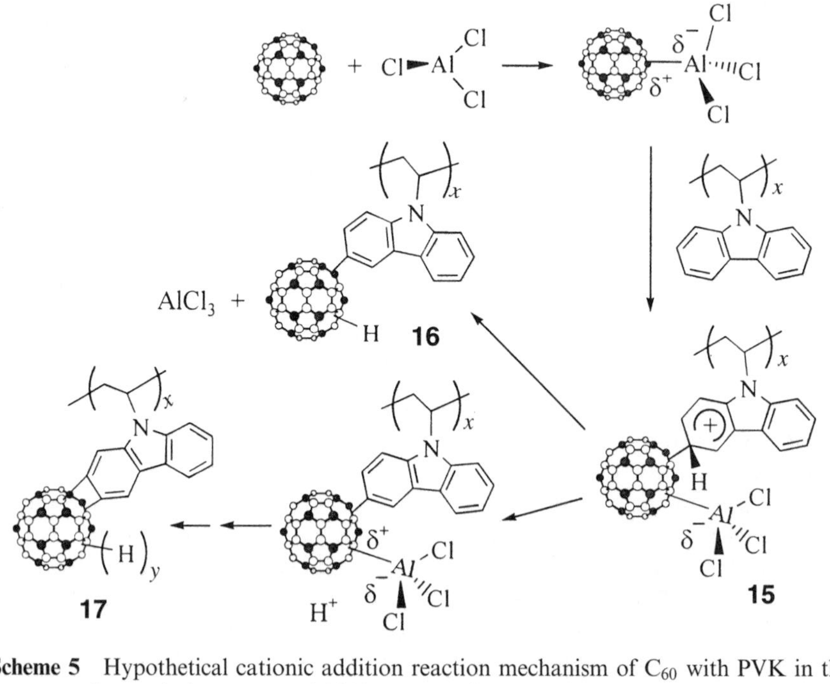

Scheme 5 Hypothetical cationic addition reaction mechanism of C_{60} with PVK in the presence of a catalytic amount of $AlCl_3$, as described in Ref. 56.

acid prior to the nucleophilic attack of the arene moiety of PVK, forming the C_{60}-grafted PVK cation intermediate **15** as shown in Scheme 5 [56]. Release of aluminum chloride from **15** produced the C_{60}-grafted PVK **16**. However, the subsequent second intramolecular cationic arene addition onto the fullerene cage in **15** resulted in formation of the C_{60}-grafted PVK products **17**, containing doubly substituted fused rings.

7.5 CONDENSATION REACTION ON FULLERENE

Fullerene derivatives containing versatile polar functions, such as hydroxy and amino groups, are favorable intermediates for the preparation of ester, amide, urethane, and urea-based polymer materials. Attachment of multifunctional segments covalently bound on the fullerene shell often increases solubility of the products, making processing friendly, and extends the three-dimensional intermolecular interaction among functional groups suitable for the design of interactive polymers. Incorporation of the C_{60} cages hanging as pendent groups along the main chain of organic polymers has been demonstrated by the polycondensation reaction of phenoxy-substituted methanofullerenes, $(HOC_6H_4)_2C_{61}$, with sebacoyl chloride (1.0 equiv.) in dry nitrobenzene at 143 °C for 22 h to give the corresponding polyester **18** in 61% yield, as shown in Scheme 6 [57]. A similar reaction was carried out by reacting bis(phenoxy)-methanofullerenes with hexamethylene diisocyanate o-dichlorobenzene in the presence of 1,4-diazabicyclo[2.2.2]octane (DABCO) to afford the corresponding polyurethane **19** as an insoluble, brown powder in 60% yield. This type of polymer can be classified as a fullerene-containing necklace polymer.

Other fullerene-derived polycondensation polymers **20**, poly(4,4′-didiphenyl-methanofullerene amines), were synthesized by reacting 4,4′-difluorodiphenyl fulleroid, $(FC_6H_4)_2C_{61}$, with aromatic diamine sulfone in dimethylsulfoxide (DMSO) at 100 °C. Using similar reagents, the reaction of $(FC_6H_4)_2C_{61}$ with aromatic dialcohol sulfone in a solvent mixture of toluene and 1-methyl-2-pyrrolidinone (NMP) in the presence of K_2CO_3 at 140–170 °C afforded poly(4,4′-didiphenyl-methanofullerene ethers) in 75–90% yield [58]. One potentially interesting polycondensation was demonstrated using structurally well-defined 1,2:18,36-[60]fullerenobisacetic acid **21** as a linker of amide copolymers. Compound **21** was prepared by hydrogenolysis, with NaH–MeOH as a reagent, of its ester precursor, which was synthesized by the treatment of C_{60} with two equivalents of sulfonium ylide, $Me_2S^+-CH_2^--CO_2Et$, in 17% yield after chromatographic isolation. Reaction of **21** with a mixture of isophthalic acid and diaminodiphenyl ether in the presence of triphenyl phosphite $(PhO)_3P$, LiCl, NMP, and pyridine resulted in the amide copolymer products of poly(1,4-oxybisphenylene fullerenobisacetamide-co-1,4-oxybisphenylene isophthalamide), consisting of the difunctionalized C_{60} cages interlinked in the main chain of copolymers [59]. These copolymers are true examples of the fullerene-containing main-chain polymer **1**.

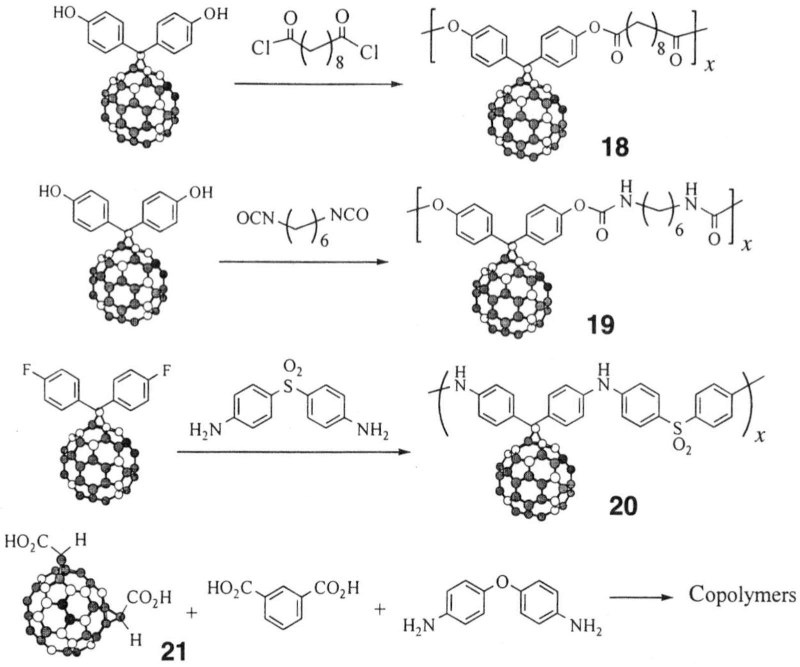

Scheme 6 Polycondensation reactions of the difunctional C_{61} and C_{62} derivatives with conventional diamino and diisocyanide reagents, leading to "necklace" polymers with the C_{60} cages hanging as pendent groups.

Direct grafting of methanofullerene cages onto a polymer chain was made by a condensation reaction of one acid moiety of methano[60]fullerene dicarboxylate, using 1-ethyl-3-(dimethylaminopropyl)carbodiimide (EDAC) as a coupling agent, with an ester functional group of poly(propionylethylenimine-co-ethylenimine) copolymers, which were the products of partially hydrolyzed poly(propionylethylenimine) [60]. The condensation reaction resulted in a highly water-soluble pendant [60]fullerene–poly(propionylethylenimine-co-ethylenimine) polymer.

It has been illustrated that incorporating polyol functional groups onto a ball-shaped C_{60} cage forming polyhydroxylated C_{60} molecules ([60]fullerenols) enriches their utilization as a spherical molecular core in the design of star-shaped polymers [61,62] and as a free radical scavenger in a physiological medium [63–65]. In the synthesis of fullerenol-derived star-shaped polymers, the presynthesized functional polymer or oligomer arms were attached onto the fullerenol molecules via the condensation reaction. The reaction was accomplished on the basis of the nucleophilic attack of fullerenolic hydroxy groups toward isocyanate groups. The chemical reactivity of tertiary fullerenolic hydroxy functional groups is rather limited toward the nucleophilic substitution of a number of electron-withdrawing leaving groups under basic conditions. They

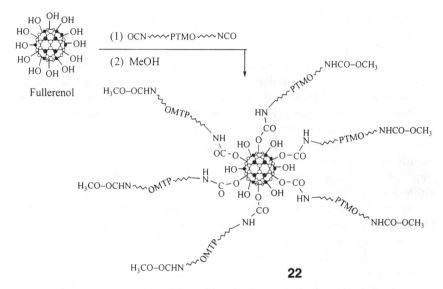

Scheme 7 Polycondensation reactions of dodecahydroxylated C_{60} (fullerenol) with the diisocyanated urethane poly(tetramethylene oxide) (PTMO) prepolymers, producing the star-shaped polymers containing a C_{60} cage as a molecular core.

are, however, moderately reactive with isocyanate groups forming the corresponding urethane moieties. Thus, the condensation reaction of fullerenols was carried out in a dilute solution, to prevent the possible crosslinking side reactions, using diisocyanated poly(tetramethylene oxide) (PTMO) prepolymers as star-arms, as shown in Scheme 7. Fullerenol was synthesized by hydrolysis of its hexacyclosulfated fullerene precursors and characterized to possess an average of 12 hydroxy groups per C_{60} [66,67]. An improved method for the preparation of hexacyclosulfated fullerene precursors and fullerenol was reported using P_2O_5 as an oxidation promoter in the electron-oxidation processes of C_{60} [68]. In the case of C_{70} molecules, the polyhydroxylation reaction of C_{70} under similar conditions and reagents gave [70]fullerenol products consisting of a chemical composition of $C_{70}(OH)_{18}$ [69]. Synthesis of diisocyanated urethane polyether prepolymers was performed by treating poly(tetramethylene oxide) glycol (PTMO) with 4,4′-methane diphenyldiisocyanate (MDI, 2.0 equiv.) in CHCl$_3$ at 60 °C under N$_2$. The resulting fullerenol-derived star-shaped polymers **22** (Scheme 7) were found to be highly viscous and soluble in common organic solvents. It is interesting to note that the macromolecule **22** consists of a hydrophilic core region and multiple hydrophobic polymer arms resembling an amphiphilic core–shell molecule. The molecular mass of **22** was determined by GPC measurements, which were calibrated by linear polystyrene standards and two star-shaped model compounds. Consequently, the average molecular mass of **22** was found to agree well with that of a fullerenol-derived polymer having six linear urethane-connected poly(tetramethylene oxide) arms [61,70]. Its poly-dispersity of 1.45 was notably smaller than that (polydispersity 2.11) of the parent urethane polyether arm, consistent with the theoretical prediction [71].

Similar polycondensation reactions were applied in the preparation of fullerenol-hypercrosslinked poly(urethane-ether) elastomers using $C_{60}(OH)_{12}$ as a crosslinker [72,73]. Treatment of $C_{60}(OH)_{12}$ with the diisocyanated urethane poly(tetramethylene oxide) (PTMO) prepolymers in a concentrated DMF so-lution, using the same molarity of reactive isocyanate groups in prepolymers as that of hydroxy groups in fullerenol, gave the fullerenol-derived elastomer **23** resembling the schematic polymer network **5** shown in Figure 7.1. During the preparation of C_{60}-derived elastomers, rapid reaction of the water molecules with isocyanate groups produced a reactive aniline moiety, which was capable of attacking further on a second isocyanate group with the release of CO_2. Therefore, the presence of a limited quantity of water molecules in the reaction medium brought about a slight variation in the final structure of the crosslinked network, in that it changed the thermal mechanical properties of products [74]. Thus, it is important to dry fullerenol rigorously prior to the isocyanate–hydroxy polycondensation reaction. Gradual increases of the solution viscosity and the degree of gelling were detected as crosslinking proceded. By assuming that the chemical reactivity of isocyanate functions toward fullerenol is similar to that in the synthesis of its star polymer analog **22**, the number of polymer arms chemically attached on each fullerenol molecule in the polymer network **23** could be correlated and estimated to be roughly 5–6. In principle, a high concentration of crosslinking sites on each fullerene cage should allow multiple chemical bondings and interconnections of tightly packed polymer chains on fullerene cages within the network matrix, which should produce thermoset-ting polymer materials exhibiting greatly enhanced physical and mechanical properties.

7.6 HYDROAMINATION REACTION ON FULLERENE

Several fullerene-grafted side-chain polymers were reported. One of the most commonly used synthetic methods for the fullerene grafting or end-capping was based on the nucleophilic hydroamination addition of the primary and sec-ondary amines onto fullerenic olefins [75,76]. In this instance, functionalized polymers containing primary amino groups in the attached side chains, such as poly(ethylenimine), reacted readily with C_{60} in toluene, forming the corre-sponding insoluble polymers [77]. The grafting reaction of C_{60} onto poly[4-[[2-aminoethylimino]methyl]styrene] gave a soluble fullerene-containing polymer **24**, as shown in Scheme 8. In both cases, the fullerene content in the polymer products was determined by the relative intensity of the characteristic UV/VIS absorption peaks of monofunctionalized organofullerenes at 308(sh), 326, and 435 nm [78].

Reaction of amino-terminated polystyrenes with C_{60} was recognized as a

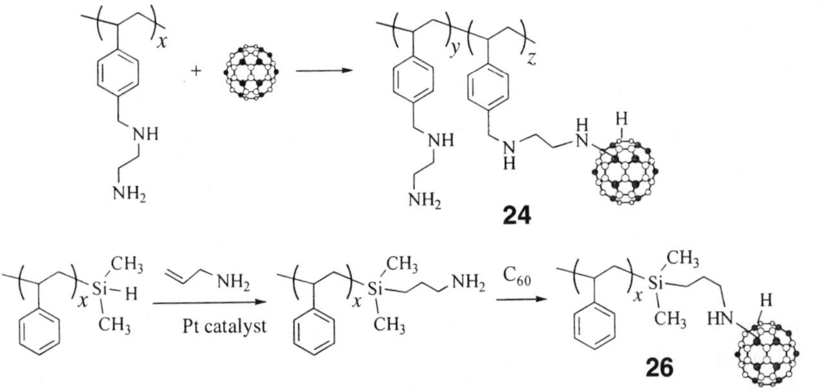

Scheme 8 Grafting the C_{60} molecules onto functionalized polymers containing primary amino groups in the attached side chains via the hydroamination reaction on fullerenes.

well-controlled method for the synthesis of C_{60} end-capped polystyrenes **25** [79]. Retention of the most fullerenic structural integrity in monopolystyryl adduct of fullerene allows preservation of the fundamental electro-optical properties of fullerene, which may partially be lost upon the multiple-addition reaction with addends, such as amines. Amino-terminated polystyrenes were prepared by the living anionic polymerization of styrene followed by the subsequent end-capping addition of dimethylchlorosilane and hydrosilation with allylamine in the presence of Pt catalyst, as shown in Scheme 8. Most of the hydroamination method could lead to a partial crosslinking at the fullerene center since multiple organoamine additions on C_{60} are possible. Application of excess C_{60} in the hydroamination was necessary to minimize formation of products containing a higher degree of substitution than monoaddition on fullerene. Direct mixing of the fullerene molecules with the amine-containing polymers often led to polymer products with a poorly defined structure [80]. In the case of the fullerenic hydroamination with O,O'-bis(2-aminopropyl)polypropylene glycol (Jeffamine) using a Jeffamine/C_{60} ratio of 1:1, the reaction produced complex products exhibiting a chromatographic polymodal distribution, corresponding to a mixture of polymers each with possibly a linear, branched, star-shaped, or cyclic structure [81]. At a higher Jeffamine/C_{60} ratio of 50:1 or 100:1, soluble polymer products exhibiting a narrow molecular mass distribution were obtained within a reaction period of 30 h. A prolonged reaction for a period of 3 weeks resulted in a phase separation of sol–gel fractions from the solution mixture owing to the occurrence of crosslinking at the fullerene site.

Polyhydroamination of C_{60} with either 1,4-butanediamine or 1,6-hexanediamine afforded the corresponding poly(diaminoalkyl) C_{60} derivatives, in 85–94% yield, which were utilized as reagents in the preparation of polyamides with dicarboxylic acid derivatives [82]. Other examples involving hydroamination of C_{60} were given by the fabrication of a self-assembled monolayer

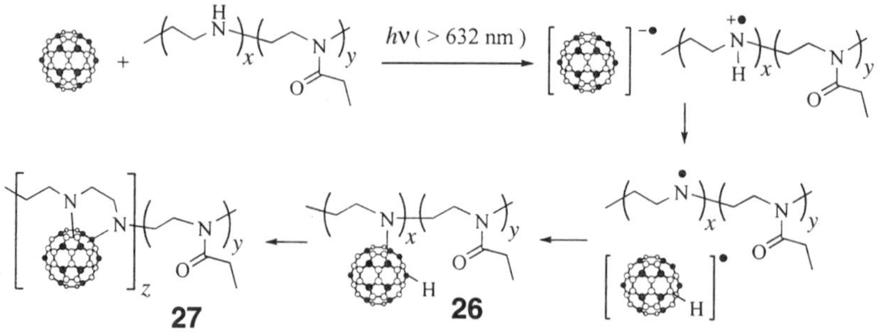

Scheme 9 Photoinduced grafting of the C_{60} molecules onto the functionalized polymers containing secondary amino groups in the main chain.

(SAM) of covalently bonded fullerenes on the $(MeO)_3Si(CH_2)_3NH_2$-modified oxide surfaces [83] or on the $HS(CH_2)_2NH_2$-modified gold surfaces [84].

Reaction of C_{60} with secondary organoamines was a rather slow process compared with that using primary organoamines. Interestingly, efficiency of this reaction was improved by a photoinduced process that facilitates the addition reaction of secondary organoamines onto C_{60}, leading to products of pendant [60]fullerene–poly(propionylethylenimine-*co*-ethylenimine) [85]. The photochemical reaction between C_{60} and the secondary amine moieties of the aminopolymers was found to follow a sequential photoinduced electron-transfer–proton-transfer mechanism, as shown in Scheme 9. The starting material of poly(propionylethyleneimine-*co*-ethyleneimine) was prepared by partial hydrolysis of poly(propionylethyleneimine), which was synthesized by cationic polymerization of 2-ethyl-2-oxazoline in the presence of BF_3–Et_2O in CH_3CN at 80 °C for 24 h. As the photoinduced reaction progresses, the intrapolymeric electron-transfer–proton-transfer processes benefit from the availability of amine units in the immediate vicinity of the excited monofunctionalized C_{60} chromophores **26**. The nucleophilic addition of secondary amines onto C_{60} thus yields exclusively the diamino adducts of C_{60} grafted on the polymer chain. The lack of fullerenic proton detection in the ^1H NMR spectrum of the isolated fractions revealed the occurrence of effective dehydrogenation of fullerene moieties, leading to products of fullerenylated poly(propionylethylenimine-*co*-ethylenimine) **27**.

7.7 CYCLOAZIDE REACTION ON FULLERENE

Reaction of C_{60} with organic azides under elevated temperatures was found to be an efficient process in giving products of fullerene derivatives containing an expanded cage of C_{60} as $C_{60}N$ [86]. The possible reaction mechanism involves either the [2+3] cycloaddition of an azide functional group ($-N_3$) with a fullerenic olefin followed by thermal cleavage of the cycloazide moiety of resulting

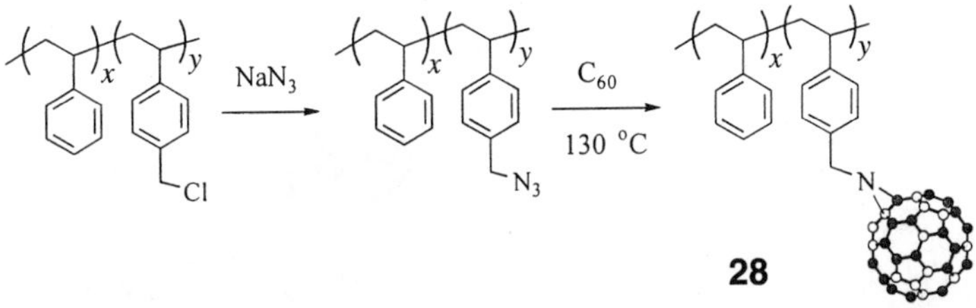

Scheme 10 Grafting the C_{60} molecules as pendent groups onto a polystyrene chain using the corresponding polystyryl azide intermediate as the nitrene-like precursor.

cycloazidofullerene intermediates with the release of N_2 or the direct thermal cleavage of an azide functional group into a reactive nitrene intermediate prior to its attack on a fullerenic olefin. The cycloazide chemistry provides better control of the number of functional addends on each fullerene molecule with the monoadduct of C_{60} as the primary product [87]. Monoaddition of a cyclo-azide moiety onto C_{60} was used for synthesis of the $C_{60}N$ derivatives containing a dendritic substituent with a well-defined structure [88].

Thermal decomposition of the cycloazide functional group chemically attached on a C_{60} molecule forming the corresponding $C_{60}N$ cage was also an efficient, clean process for grafting the fullerene onto a polymer chain. An example was demonstrated by the reaction of C_{60} with azido-substituted poly(styrene-co-methylstyrene) at reflux temperatures of chlorobenzene to give the grafted styrene copolymer **28**, as shown in Scheme 10 [89]. The intermediate polymer of azido-substituted poly(styrene-co-methylstyrene) was synthesized via the replacement of a benzyl halide precursor by sodium azide. This azide reaction with C_{60} was found to proceed primarily through monoaddition, with little or no triaddition and higher addition products. Therefore, the possibility of crosslinking at the fullerene center can be ruled out. Similar azide chemistry was also applied in the preparation of the C_{60} end-capped polystyrene star polymers [90]. The synthetic sequence involved a substitution reaction of benzyl chlorides of a hexa-arm hexachloropolystyrene star polymer with trimethylsilyl azide to give the corresponding hexa-azide star polymer prior to its thermal reaction with C_{60}.

7.8 OLEFIN CYCLIZATION REACTION ON FULLERENE

The Diels–Alder cycloaddition of fullerenes with bis-dienes was investigated extensively and found to be one of the most versatile synthetic methods for the functionalization of fullerene molecules [75,91–93]. The high electronegativity of the fullerene molecules makes them act as electron-deficient dienophiles in reacting with electron-rich organodienes. The Diels–Alder reaction of C_{60} with disubstituted cyclohexadiene and anthracene produced the corresponding disubstituted bicyclohexene adduct and bicycloanthracene adduct of C_{60} as

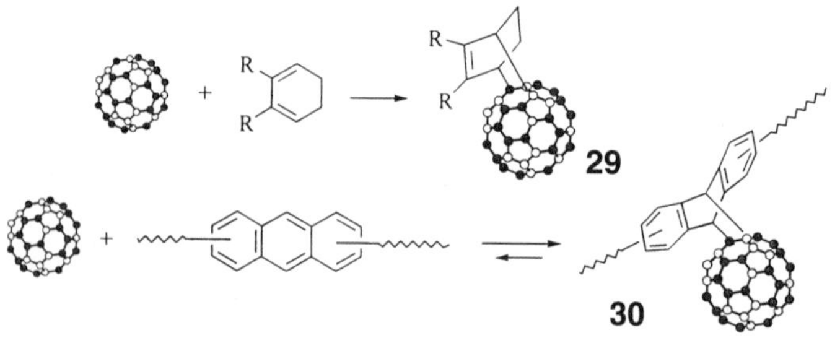

Scheme 11 The Diels–Alder cycloaddition of electron-deficient fullerene dienophiles with electron-rich organic bis-dienes or aromatics, which were chemically attached on a polymer chain.

compounds **29** and **30**, respectively, as shown in Scheme 11. However, uses of the Diels–Alder methodology in the polymerization reaction for grafting the C_{60} molecules onto the polymer chain may be limited by low availability of the diene-containing polymer precursors. Many of the Diels–Alder products were, in fact, thermally degradable with a structural reversal back to their starting fullerene and diene-containing organics. Utilization of this thermal reversibility characteristic in the preparation of thermoreversible polymers was reported recently [94]. In that case, a compound of bisanthracene-decane ether was used as the source of diene, which polymerized readily with an equal molarity of C_{60} in toluene at room temperature. The resulting bicycloanthracene-bound fullerene-containing polymers resembling structure **30** were found to begin disconnecting at temperatures as low as 50 °C, with complete depolymerization occurring at temperatures above 70 °C. Typically, heating of the Diels–Alder polymer products at 75 °C for a period of 20 min resulted in complete reversion of the polymer structure back to its monomer and the C_{60} molecules without the occurrence of obvious side reactions.

7.9 POTENTIAL APPLICATIONS

The high chemical reactivity of the C_{60} molecules allows the synthesis of an enormous number of fullerene-derived polymers suitable for materials and biological applications. Potential uses of molecular fullerene derivatives in a biomedium have been reviewed [95]. Certain biological effects originating from the application of molecular fullerene adducts may be simulated by using its polymer analog. Synthesis of fullerene-containing water-soluble polymers and hydrophilic polymers will be beneficial to the study of their biological significance in physiological medium. Water-soluble derivatives or conjugates of fullerenes were noted to be effective in the inhibition of HIV protease [96], in the generation of singlet oxygen for the DNA cleavage [97,98], and in the photodynamic effect on induced tumor necrosis [99]. Recently, hydrophilic fullerenols [63] and bis(dicarboxylated) dimethanofullerenes [100] were reported to exhibit potent free radical scavenging activities against the reactive oxygen radical species (ROS) under physiologic conditions. Related therapeutic investigations of fullerenol utilizing its free radical scavenging activities included inhibition on the H_2O_2-elicited oxidative damage of brain neurons [64], removal of superoxide radicals in the whole blood associated with the gastric cancer [101], attenuation of the exsanguination-induced bronchoconstriction [65], and antiproliferation of the vascular smooth muscle cells [102]. Combination of discovered biological functions and the variation of polymer structures may allow one to design effective biopolymers for medical uses in the future.

Fullerene molecules are rich in electronic properties. Potential applications of fullerene-containing polymers include conductive materials, optical devices, chemical sensors, electroluminescent cells [103], polymer grid triodes [104], and photolithography [105]. Enhanced optical limiting properties were detected on polystyrene-bound C_{60} solids [106] and poly(3-octylthiophene) solid films sensitized with methanofullerene [107]. One of the most studied physical properties of the C_{60}-containing or C_{60}-doped polymer matrix may be the photoinduced charge transfer from the polymer to the C_{60} molecule. Electron transfer from potent molecular donors to C_{60} resulted in the anionic C_{60} species exhibiting superconductivity, semiconductivity, or optical properties. The electron transfer from weak to moderate organic or polymeric electron donors, such as extended and conjugated polymers, to the fullerene molecule and its derivatives can be made under photoactivation. The photoinduced formation of exciplexes of the singlet excited state of C_{70} molecules with N,N'-diethylaniline in methylcyclohexane at room temperature was reported [108]. The photoinduced electron transfer from highly conjugated polymers having a poly(phenylene vinylene) [109–111], poly(3-alkylthiophene) [112,113], and polyacetylene [114] backbone to the C_{60} molecule was shown to be extremely efficient. These charge-transfer phenomena have been applied in studies on the quantum efficiency of C_{60}-containing polyalkylthiophene as a dry-cell medium in the fabrication of Schottky diode solar cell devices [115,116]. However, the main problem associated with these materials was the readily occurring phase separation or aggregation of the C_{60} molecules within the devices. Therefore, the use of functionalized fullerenes may improve their compatibility with the hosting polymer matrix and circumvent the phase-separation problem. For example, the donor–acceptor heterojunctions of phenyl(methylcaprolate)methanofullerenes and poly(2-methoxy-5-(2'-ethylhexyloxy)–(4-phenylene vinylene) (MEH–PPV) displayed an enhanced efficiency of the polymer photovoltaic cells [117,118]. The direct grafting of C_{60} molecules onto the polymer chain of conjugated polymers, including polythiophene, poly(phenylene vinylene), polyaniline, polypyrrole, and the analogous conjugated polymer derivatives becomes a feasible approach to polymer modification for further improving the electron-transfer efficiency of solar cell devices. This structural modification makes the electron-transfer process from the polymer donor chain to the fullerene moieties of the side chains tunable in an intrapolymer fashion. The design and synthesis of

C_{60}-grafted conjugate polymers with well-defined structure is conceivable by utilization of the anionic, cationic, free radical, condensation, hydroamination, cycloazide, or olefin cyclolyzation reaction on fullerenes as a method of fullerene grafting. At least two C_{60}-grafted polymers of aminopolymer [119] and poly(N-vinylcarbazole) [120] were shown to exhibit photoinduced electron-transfer properties. These preliminary accounts of research and much of the C_{60} chemistry described in this chapter illustrate the early approaches to engineering C_{60}-containing polymer materials by a combination of the electronic properties of fullerides with a variety of fullerene functionalization methods available for the preparation of innovative materials.

REFERENCES

1. R. Taylor and D. R. M. Walton, *Nature* **1993**, *363*, 685.

2. F. Wudl, A. Hirsch, K. Khemani, T. Suzuki, P. M. Allemand, A. Koch, H. Eckert, G. Srdanov, and H. Webb, *ACS Symp. Ser.* **1992**, *481*, 161.

3. A. Hirsch, A. Soi, and H. R. Karfunkel, *Angew. Chem. Int. Ed. Engl.* **1992**, *31*, 766.

4. E. T. Samulski, J. M. DeSimmone, M. O. Hunt, Jr., Y. Z. Menceloglu, R. C. Jarnagin, G. A. York, K. B. Labat, and H. Wang, *Chem. Mater.* **1992**, *4*, 1153.

5. G. D. Wignall, K. A. Affholter, G. J. Bunick, M. O. Hunt, Jr., Y. Z. Menceloglu, J. M. DeSimone, and E. T. Samulski, *Macromolecules* **1995**, *28*, 6000.

6. Y. Chen, W. S. Huang, Z. E. Huang, R. F. Cai, H. K. Yu, S. M. Chen, and X. M. Yan, *Eur. Polym. J.* **1997**, *33*, 823.

7. Y. Ederle and C. Mathis, *Macromolecules* **1997**, *30*, 2546.

8. M. M. Khaled, R. T. Carlin, P. C. Trulove, G. R. Eaton, and S. S. Eaton, *J. Am. Chem. Soc.* **1994**, *116*, 3465.

9. Y. Ederle and C. Mathis, *Fullerene Sci. Technol.* **1996**, *4*, 1177.

10. Y. Ederle, C. Mathis, and R. Nuffer, *Synth. Met.* **1997**, *86*, 2287.

11. Y. Ederle and C. Mathis, *Macromolecules* **1997**, *30*, 4262.

12. A. Hirsch, A. Soi, and H. R. Karfunkel, *Angew. Chem. Int. Engl. Ed.* **1992**, *31*, 766.

13. P. J. Fagan, P. J. Krusic, D. H. Evans, S. A. Lerke, and E. Johnston, *J. Am. Chem. Soc.* **1992**, *114*, 9697.

14. D. E. Bergbreiter and H. N. Gray, *J. Chem. Soc. Chem. Commun.* **1993**, 645.

15. L. Dai, A. W. H. Mau, H. J. Griesser, T. H. Spurling, and J. W. White, *J. Phys. Chem.* **1995**, *99*, 17302.

16. Y. Chen, Z. E. Huang, and R. F. Cai, *J. Polym. Sci. B, Polym. Phys.* **1996**, *34*, 631.

17. Y. Chen, Z. E. Huang, R. F. Cai, S. Q. Kong, S. Chen, Q. Shao, X. Yan, F. Zhao, and D. Fu, *J. Polym. Sci. A Polym. Chem.* **1996**, *34*, 3297.

18. Y. Chen, Z. E. Huang, R. F. Cai, B. C. Yu, W. Ma, S. Chen, Q. Shao, X. Yan, and Y. Huang, *Eur. Polym. J.* **1997**, *33*, 291.

19. D. A. Loy and R. A. Assink, *J. Am. Chem. Soc.* **1992**, *114*, 3977.

20. L. S. Wang, J. Conceicao, C. Jin, and R. E. Smalley, *Chem. Phys. Lett.* **1991**, *182*, 5.

21. R. E. Haufler, J. Conceicao, L. P. F. Chibante, Y. Chai, N. E. Byrne, S. Flanagan, M. M. Haley, S. C. O'Brien, C. Pan, Z. Xiao, W. E. Billups, M. A. Ciufolini, R. H. Hauge, J. L. Margrave, L. J. Wilson, R. F. Curl, and R. E. Smalley, *J. Phys. Chem.* **1990**, *94*, 8634.

22. P. M. Allemand, A. Koch, F. Wudl, Y. Rubin, F. Diederich, M. M. Alvarez, S. J. Anz, and R. L. Whetten, *J. Am. Chem. Soc.* **1991**, *113*, 1050.

23. D. M. Cox, S. Bahal, M. Disko, S. M. Gorun, M. Greaney, C. S. Hsu, E. B. Kollin, J. Millar, J. Robbins, R. D. Sherwood, and P. Tindall, *J. Am. Chem. Soc.* **1991**, *113*, 2940.

24. C. Jehoulet, A. J. Bard, and F. Wudl, *J. Am. Chem. Soc.* **1991**, *113*, 5456.

25. D. Dubois, K. M. Kadish, S. Flanagan, R. E. Haufler, L. P. Chibante, and L. J. Wilson, *J. Am. Chem. Soc.* **1991**, *113*, 4364.

26. D. Dubois, K. M. Kadish, S. Flanagan, and L. J. Wilson, *J. Am. Chem. Soc.* **1991**, *113*, 7773.

27. Q. Xie, E. Perez-Cordero, and L. Echegoyen, *J. Am. Chem. Soc.* **1992**, *114*, 3978.

28. P. J. Krusic, E. Wasserman, P. N. Keizer, J. R. Morton, and K. F. Preston, *Science* **1991**, *254*, 1183.

29. P. J. Krusic, E. Wasserman, B. A. Parkinson, B. Malone, E. R. Holler, Jr., P. N. Keizer, J. R. Morton, and K. F. Preston, *J. Am. Chem. Soc.* **1991**, *113*, 6274.

30. C. N. McEwen, R. G. McKay, and B. S. Larson, *J. Am. Chem. Soc.* **1992**, *114*, 4412.

31. N. D. Dimitrijevic, P. V. Kamat, and R. W. Fessenden, *J. Phys. Chem.* **1993**, *97*, 615.

32. D. M. Guldi, H. Hungerbuhler, E. Janata, and K. D. Asmus, *J. Chem. Soc. Chem. Commun.* **1993**, 84.

33. D. M. Guldi, H. Hungerbuhler, E. Janata, and K. D. Asmus, *J. Phys. Chem.* **1993**, *97*, 11258.

34. M. Walbiner and H. Fischer, *J. Phys. Chem.* **1993**, *97*, 4880.

35. A. G. Camp, A. Lary, and W. T. Ford, *Macromolecules* **1995**, *28*, 7959.

36. C. E. Bunker, G. E. Lawson, and Y. P. Sun, *Macromolecules* **1995**, *28*, 3744.

37. N. Arsalani and K. E. Geckeler, *Fullerene Sci. Technol.* **1996**, *4*, 897.

38. T. Cao and S. E. Webber, *Macromolecules* **1996**, *29*, 3826.

39. W. T. Ford, T. D. Graham, and T. H. Mourey, *Macromolecules* **1997**, *30*, 6422.

40. P. L. Nayak, S. Alva, K. Yang, P. K. Dhal, J. Kumar, and S. K. Tripathy, *Macromolecules* **1997**, *30*, 7351.

41. K. Kirkwood, D. Stewart, and C. T. Imrie, *J. Polym. Sci., A. Polym. Chem.* **1997**, *35*, 3323.

42. J. R. Morton, K. F. Preston, P. J. Krusic, and E. Wasserman, *J. Chem. Soc. Perkin Trans.* **1992**, *2*, 1426.

43. J. R. Morton, K. F. Preston, P. J. Krusic, S. A. Hill, and E. Wasserman, *J. Am. Chem. Soc.* **1992**, *114*, 5454.

44. H. Okamura, T. Terauchi, M. Minoda, T. Fukuda, and K. Komatsu, *Macromolecules* **1997**, *30*, 5279.

45. B. Z. Tang, S. M. Leung, H. Peng, N. T. Yu, and K. C. Su, *Macromolecules* **1997**, *30*, 2848.

46. Y. Achiba, T. Nakagawa, Y. Matsui, S. Suzuki, H. Shiromaru, K. Yamauchi, K. Nishiyama, M. Kainosho, H. Hoshi, Y. Maruyama, and T. Mitani, *Chem. Lett.* **1991**, 1233.

47. Q. Xie, F. Arias, and L. Echegoyen, *J. Am. Chem. Soc.* **1993**, *115*, 9818.

48. D. L. Lichtenberger, K. W. Nebesny, C. D. Ray, D. R. Huffman, and L. D. Lamb, *Chem. Phys. Lett.* **1991**, *176*, 203.

49. S. G. Kukolich and D. R. Huffman, *Chem. Phys. Lett.* **1991**, *182*, 263.

50. H. Thomann, M. Bernardo, and G. P. Miller, *J. Am. Chem. Soc.* **1992**, *114*, 6593.

51. G. P. Miller, C. S. Hsu, H. Thomann, L. Y. Chiang, and M. Bernardo, *Mat, Res. Symp. Proc.* **1992**, *247*, 293.

52. L. Y. Chiang, J. W. Swirczewski, S. Cameron, C. Hsu, S. K. Chowdhury, and K. Creegan, *J. Chem. Soc. Chem. Commun.* **1992**, 1791.

53. L. Y. Chiang, R. B. Upasani, and J. W. Swirczewski, *J. Am. Chem. Soc.* **1992**, *114*, 10154.

54. C. Yu, J. B. Bhonsle, L. Y. Wang, J. G. Lin, B. J. Chen, and L. Y. Chiang, *Fullerene Sci. Technol.* **1997**, *5*, 1407.

55. G. A. Olah, I. Bucsi, C. Lambert, R. Aniszfeld, N. J. Trivedi, D. K. Sensharma, and G. K. S. Prakash, *J. Am. Chem. Soc.* **1991**, *113*, 9387.

56. R. N. Thomas, *J. Polym. Sci. A Polym. Chem.* **1994**, *32*, 2727.

57. S. Shi, K. C. Khemani, C. Li, and F. Wudl, *J. Am. Chem. Soc.* **1992**, *114*, 10656.

58. M. Berrada, Y. Hashimoto, and S. Miyata, *Chem. Mater.* **1994**, *6*, 2023.

59. J. Li, T. Yoshizawa, M. Ikuta, M. Ozawa, K. Nakahara, T. Hasegawa, K. Kitazawa, M. Hayashi, K. Kinbara, M. Nohara, and K. Saigo, *Chem. Lett.* **1997**, 1037.

60. Y. P. Sun, B. Liu, and D. K. Moton, *Chem. Commun.* **1996**, 2699.

61. L. Y. Chiang, L. Y. Wang, S. M. Tseng, J. S. Wu, and K. H. Hsieh, *J. Chem. Soc. Chem. Commun.* **1994**, 2675.

62. L. Y. Chiang, L. Y. Wang, S. M. Tseng, J. S. Wu, and K. H. Hsieh, *Synth. Met.* **1995**, *70*, 1477.

63. L. Y. Chiang, F. J. Lu, and J. T. Lin, *J. Chem. Soc. Chem. Commun.* **1995**, 1283.

64. M. C. Tsai, Y. H. Chen, and L. Y. Chiang, *J. Pharmacy Pharmacol.* **1997**, *49*, 438.

65. Y. L. Lai and L. Y. Chiang, *J. Autonomic Pharmacol.* **1997**, *17*, 229.

66. L. Y. Chiang, L. Y. Wang, J. W. Swirczewski, S. Soled, and S. Cameron, *J. Org. Chem.* **1994**, *59*, 3960.

67. L. Y. Chiang, J. B. Bhonsle, L. Y. Wang, S. F. Shu, T. M. Chang, and J. R. Hwu, *Tetrahedron* **1996**, *52*, 4963.

68. B. H. Chen, J. P. Huang, L. Y. Wang, J. T. Shiea, T. L. Chen, and L. Y. Chiang, *J. Chem. Soc. Perkin Trans. 1* **1998**, 1171.

69. B. H. Chen, J. P. Huang, L. Y. Wang, J. Shiea, T. L. Chen, and L. Y. Chiang, *Synth. Commun.* **1998**, *28*, 3515.

70. S. M. Tseng, L. Y. Wang, K. H. Hsieh, W. B. Liau, and L. Y. Chiang, *Fullerene Sci. Technol.* **1997**, *5*, 1313.

71. T. A. Orofino, *Polymers* **1961**, *2*, 295.

72. L. Y. Chiang, L. Y. Wang, and C. S. Kuo, *Macromolecules* **1995**, *28*, 7574.

73. L. Y. Chiang and L. Y. Wang, *Trends Polym. Sci.* **1996**, *4*, 298.

74. L. Y. Chiang, L. Y. Wang, J. S. Wu, C. S. Kuo, S. M. Tseng, K. H. Hsieh, and W. B. Liau, *J. Polym. Res.* **1996**, *3*, 1.

75. F. Wudl, A. Hirsch, K. C. Khemani, T. Suzuki, P. M. Allemand, A. Koch, H. Eckert, G. Srdanov, and H. M. Webb, *ACS Symp. Ser.* **1992**, *481*, 161.

76. A. Hirsch, Q. Li, and F. Wudl, *Angew. Chem. Int. Ed. Engl.* **1991**, *30*, 1309.

77. K. E. Geckeler and A. Hirsch, *J. Am. Chem. Soc.* **1993**, *115*, 3850.

78. T. Suzuki, Q. Li, K. C. Khemani, F. Wudl, and O. Almarson, *Science* **1991**, *254*, 1186.

79. C. Weis, C. Friedrich, R. Mulhaupt, and H. Frey, *Macromolecules* **1995**, *28*, 403.

80. A. O. Patil, G. W. Schriver, B. Carstensen, and R. D. Lundberg, *Polym. Bull.* **1993**, *30*, 187.

81. N. Manolova, I. Rashkov, F. Beguin, and H. V. Damme, *J. Chem. Soc. Chem. Commun.* **1993**, 1725.

82. J. R. Hwu, T. Y. Kuo, T. M. Chang, H. V. Patel, and K. T. Yong, *Fullerene Sci. Technol.* **1996**, *4*, 407.

83. S. Shi, K. C. Khemani, C. Li, and F. Wudl, *J. Am. Chem. Soc.* **1992**, *114*, 10656.

84. W. B. Calwell, K. Chen, C. A. Mirkin, and S. J. Babinec, *Langmuir* **1993**, *9*, 1945.

85. Y. P. Sun, B. Liu, and G. E. Lawson, *Photochem. Photobiol.* **1997**, *66*, 301.

86. M. Prato, Q. C. Li, F. Wudl, and V. Lucchini, *J. Am. Chem. Soc.* **1993**, *115*, 1148.

87. M. Takeshita, T. Suzuki, and S. Shinkai, *J. Chem. Soc. Chem. Commun.* **1994**, 2587.

88. C. J. Hawker, K. L. Wooley, and M. J. Frechet, *J. Chem. Soc. Chem. Commun.* **1994**, 925.

89. C. J. Hawker, *Macromolecules* **1994**, *27*, 4836.

90. E. Cloutet, Y. Gnanou, J. L. Fillaut, and D. Astruc, *Chem. Commun.* **1996**, 1565.

91. M. F. Meidine, R. Roers, G. J. Langley, A. G. Avent, A. D. Darwish, S. Firth, H. W. Kroto, R. Taylor, and R. M. Walton, *J. Chem. Soc. Chem. Commun.* **1993**, 1342.

92. Y. Rubin, S. Khan, D. Freedberg, and C. Yeretzian, *J. Am. Chem. Soc.* **1993**, *118*, 344.

93. A. Herrmann, F. Diederich, C. Thilgen, H. ter Meer, and G. Muller, *Helv. Chim. Acta* **1994**, *77*, 1689.

94. B. Nie and V. M. Rotello, *Macromolecules* **1997**, *30*, 3949.

95. A. W. Jensen, S. R. Wilson, and D. I. Schuster, *Bioorg. Med. Chem.* **1996**, *4*, 767.

96. S. H. Friedman, D. L. DeCamp, R. P. Sijbesma, G. Srdanov, F. Wudl, and G. L. Kenyon, *J. Am. Chem. Soc.* **1993**, *115*, 6506.

97. H. Tokuyama, S. Yamago, and E. Nakamura, *J. Am. Chem. Soc.* **1993**, *115*, 7918.

98. Y. N. Yamakoshi, T. Yagami, K. Fukuhara, S. Sueyoshi, and N. Miyata, *J. Chem. Soc. Chem. Commun.* **1994**, 517.

99. Y. Tabata, Y. Murakami, and Y. Ikada, *Jpn. J. Cancer Res.* **1997**, *88*, 1108.

100. K. Okuda, T. Mashino, and M. Hirobe, *Bioorg. Med. Chem. Let.* **1996**, *6*, 539.

101. L. Y. Chiang, F. J. Lu, and J. T. Lin, *Proc. Electrochem. Soc.* **1995**, *95-10*, 699.

102. H. C. Huang, L. H. Lu, and L. Y. Chiang, *Proc. Electrochem. Soc.* **1996**, *96-10*, 403.

103. Q. Zheng, R. Sun, X. Zhang, T. Masuda, and T. Kobayashi, *Jpn. J. Appl. Phys.* **1997**, *36*, 1675.

104. J. McElvain, M. Keshavarz, H. Wang, F. Wudl, and A. J. Heeger, *J. Appl. Phys.* **1997**, *81*, 6468.

105. Y. Tajima, Y. Tezuka, T. Ishii, and K. Takeuchi, *Polym. J.* **1997**, *29*, 1016.

106. Y. Kojima, T. Matsuoka, H. Takahashi, and T. Kurauchi, *Macromolecules* **1995**, *28*, 8868.

107. M. Cha, N. S. Sariciftci, A. J. Heeger, J. C. Hummelen, and F. Wudl, *Appl. Phys. Lett.* **1995**, *67*, 3850.

108. Y. Wang, *J. Chem. Phys.* **1992**, *96*, 764.

109. N. S. Saricifti, L. Smilowitz, A. J. Heeger, and F. Wudl, *Science* **1992**, *258*, 1474.

110. S. Morita, S. Kiyomatsu, X. H. Yin, A. A. Zakhidov, T. Noguchi, T. Ohnishi, K. Yoshino, *J. Appl. Phys.* **1993**, *74*, 2860.

111. S. V. Frolov, P. A. Lane, M. Ozaki, K. Yoshino, and Z. V. Vardeny, *Chem. Phys. Lett.* **1998**, *286*, 21.

112. S. Morita, A. A. Zakhidov, and K. Yoshino, *Solid State Commun.* **1992**, *82*, 249.

113. K. Yoshino, X. H. Yin, S. Morita, T. Kawai, and A. A. Zakhidov, *Solid State Commun.* **1993**, *85*, 85.

114. K. Yoshino, T. Akashi, K. Yoshimoto, S. Morita, R. Sugimoto, and A. A. Zakhidov, *Solid State Commun.* **1994**, *90*, 41.

115. C. S. Kuo, F. G. Wakim, S. K. Sengupta, and S. K. Tripathy, *Solid State Commun.* **1993**, *87*, 115.

116. S. Morita, A. A. Zakhidov, and K. Yoshino, *Jpn. J. Appl. Phys.* **1993**, *32*, 873.

117. G. Yu, J. Gao, J. C. Hummelen, F. Wudl, and A. J. Heeger, *Science* **1995**, *270*, 1789.

118. B. Kraabel, J. C. Hummelen, D. Vacar, D. Moses, N. S. Sariciftci, A. J. Heeger, and F, Wudl, *J. Chem. Phys.* **1996**, *104*, 4267.

119. Y. P. Sun, C. E. Bunker, and B. Liu, *Chem. Phys. Lett.* **1997**, *272*, 25.

120. Y. Chen, Z. E. Huang, R. Cai, D. Fan, X. Hou, X. Yan, S. Chen. W. Jin, D. Pan, and S. Wang, *J. Appl. Polym. Sci.* **1996**, *61*, 2185.

CHAPTER 8

ENDOHEDRAL METALLOFULLERENES: PRODUCTION, SEPARATION, AND STRUCTURAL PROPERTIES

HISANORI SHINOHARA

8.1 HISTORICAL PERPECTIVE

Just after the discovery of the "magic number" C_{60} in a laser-vaporized cluster beam mass spectrum by Smalley, Kroto, and co-workers [1], the same research group also found a magic number feature due to LaC_{60} in a mass spectrum prepared by laser vaporization of a $LaCl_2$-impregnated graphite rod [2]. They observed a series of C_n^+ and LaC_n^+ ion species with LaC_{60}^+ as a magic number ion in the mass spectrum (Fig. 8.1) and concluded that a La atom was encaged within the (then hypothetical) soccer-ball-shaped C_{60}.

Shortly after this proposal, however, the Exxon research group [3] claimed that the magic number feature of LaC_{60} in mass spectra was strongly affected by laser ionization conditions and that the La atom was not encaged within C_{60}. After this controversy, further evidence that metal atoms are encaged in C_{60} was reported by the Rice group, showing that LaC_{60}^+ ions did not react with H_2, O_2, NO, and NH_3 [4]. This suggests that reactive metal atoms are protected from the surrounding gases and are indeed trapped inside the C_{60} cage.

The presence of the soccer ball C_{60} was amply demonstrated in 1990 by a simple experiment done by Krätschmer, Huffman, and co-workers [5,6]. They succeeded in producing macroscopic quantities of soccer-ball-shaped C_{60} by using resistive heating of graphite rods under a He atmosphere [5,6]. The resistive heating method was then taken over by the so-called contact arc method [7].

The macroscopic quantities of endohedral metallofullerenes were first ob-

---

Fullerenes: Chemistry, Physics, and Technology, Edited by Karl M. Kadish and Rodney S. Ruoff.
0-471-29089-0 Copyright © 2000 John Wiley & Sons, Inc.

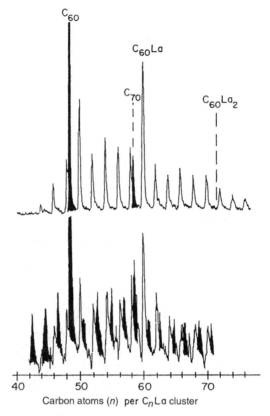

Figure 8.1 Laser-vaporized cluster beam mass spectrum of La–carbon clusters. LaC$_{60}$ is seen as a magic number ion.

tained by the Rice group [8]. They used a high-temperature laser vaporization of La$_2$O$_3$/graphite composite rods and the corresponding contact arc technique as well to produce various sizes of lanthanofullerenes. Contrary to expectation, only the La@C$_{82}$ fullerene survived in solvents and was extracted by toluene, even though La@C$_{60}$ and La@C$_{70}$ were also seen in mass spectra of the sublimed film from soot. Thus, the major lanthanofullerene that is air stable is La@C$_{82}$, and La@C$_{60}$ and La@C$_{70}$ are somehow unstable in air and in solvents.

Thy symbol @ is used to indicate that atoms listed to the left of the @ symbol are encaged in the fullerenes. For example, a C$_{60}$-encaged metal species (M) is then written as M@C$_{60}$ [8]. The corresponding IUPAC nomenclature is different from the conventional M@C$_{60}$ representation. It is recommended by IUPAC that La@C$_{82}$ should be called [82]fullerene-*incar*-lanthanum and should be written iLaC$_{82}$ [9]. However, throughout this chapter the conventional @ description is used for endohedral nature unless otherwise noted.

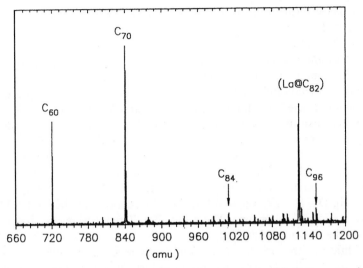

Figure 8.2 FT-ICR mass spectrum of hot toluene extract of fullerene materials produced by high-temperature laser vaporization of a 10% La_2O_3/graphite composite rod.

8.2 ENDOHEDRAL METALLOFULLERENES: A NEW CLASS OF FULLERENE-BASED MATERIALS

Figure 8.2 shows a Fourier transform ion cyclotron resonance (FT-ICR) mass spectrum of hot toluene extract of fullerene materials produced by laser vaporization of a 10% La_2O_3/graphite composite rod [8]. The La@C_{82} metallofullerene is seen and La@C_{60} and La@C_{70} are completely absent in the mass spectrum. The speciality of the La@C_{82} fullerene was confirmed by Whetten and co-workers [10]. They also observed, at relatively high loading ratios of La_2O_3 in composite rods, the La_2@C_{80} produced by the Krätschmer–Huffman method and found it to be another solvent-extractable lanthanofullerene [10,11].

Important information on the electronic structure of La@C_{82} was provided by the IBM Almaden research group. The charge state of the encaged La atom was studied by Johnson et al. [12] using electron spin resonance (ESR). The ESR hyperfine splitting (hfs) analysis of La@C_{82} revealed that the La atom is in the 3+ charge state and that the formal charge state of La@C_{82} is written as La^{3+}@C_{82}^{3-}: Three outer electrons of La are transferring to the C_{82} cage [13].

Several other research groups have extended their works to endohedral yttrium compounds. The Rice–Minnesota University [14] and Nagoya–Mie University [15] research groups also reported solvent-extractable Y@C_{82} and Y_2@C_{82} fullerenes and observed the ESR hfs of Y@C_{82}. From the hfs analyses, both groups concluded that the charge state of the Y atom is 3+ and that a similar intrafullerene electron transfer was taking place in Y@C_{82} as in

La@C_{82}. These results were also confirmed by the NRL group [16]. In addition, they reported the production of mixed dimetallofullerenes like (LaY)@C_{80}. McElvany [17] also reported the production of a series of yttrium fullerenes, Y_m@C_n including Y@C_{82} in the gas phase, by direct laser vaporization of samples containing graphite, yttrium oxide, and fullerenes.

Scandium metallofullerenes were also produced in macroscopic quantity and solvent-extracted by Shinohara et al. [18] and Yannoni et al. [19]. The Sc fullerenes exist in extracts as a variety of species (mono-, di-, tri-, and tetra-scandium fullerenes), typically Sc@C_{82}, Sc_2@C_{74}, Sc_2@C_{82}, Sc_2@C_{84}, and Sc_3@C_{82}. It was found that Sc_3@C_{82} was also an ESR-active species, whereas di- and tetrascandium fullerenes like Sc_2@C_{84} and Sc_4@C_{82} were ESR-silent.

The formation of lanthanide metallofullerenes R@C_{82} (R = Ce, Pr, Nd, Sm, Eu, Gd, Tb, Dy, Ho, Er, Yb, and Lu) was also reported by the UCLA [20] and SRI International [21] groups. These metallofullerenes were also based domi-nantly on the C_{82} fullerene.

In addition to group 3 (Sc, Y, La) and lanthanide metallofullerenes, group 2 metal atoms (Ca, Sr, Ba) were also found to form endohedral metallofullerenes, which have been produced and isolated in milligram quantity [22–28]. These metal atoms have been encaged not only by C_{82} and C_{84} but also such higher fullerenes as C_{72}, C_{74}, and C_{80}. In a similar context, La_2@C_{72} has also been produced and isolated [29,30].

Other important endohedral fullerenes that have been produced are Ca@C_{60} [31–35] and U@C_{60} [31]—C_{60}-based metallofullerenes. The Ca@C_{60} and U@C_{60} fullerenes are unique metallofullerenes, in which Ca and U atoms are encaged by C_{60}, that are quite different from group 3 and lanthanide R@C_{82} type metallofullerenes. An *ab initio* self-consistent field (SCF) Hartree–Fock calculation indicates that the Ca ion in Ca@C_{60} is at the off-center position displaced by 0.7 Å from the center and that the electronic charge of Ca is 2+ [32,36]. A similar theoretical prediction has been made on Sc@C_{60} by Scuseria and co-workers [37]. Metallofullerenes based on C_{60} have been known to be unstable in air and in normal fullerene solvents such as toluene and carbon disulfide. Because none of the C_{60}-based metallofullerenes have been isolated so far, unlike the R@C_{82} type fullerenes, many of the properties including their structures still remain unknown (cf. note added in proof). The metal atoms, which have been reported to form endohedral metallofullerenes, are shown in Table 8.1.

8.3 PRODUCTION AND EXTRACTION FROM SOOT

8.3.1 Production of Metallofullerenes

Endohedral metallofullerenes have been synthesized in several ways, similar to the production of empty fullerenes involving the generation of a carbon-rich vapor or plasma in He or Ar gas atmosphere. Two methods have been routinely

TABLE 8.1   A Part of the Periodic Table Showing the Elements Reported to Form Endohedral Metallofullerenes and Isolated as Purified Materials (as of August 1999)

	1	2	3	4	5	6	7	8
1	H							
2	Li	Be						
3	Na	Mg						
4	K	Ca	Sc	Ti	V	Cr	Mn	Fe
5	Rb	Sr	Y	Zr	Nb	Mo	Tc	Ru
6	Cs	Ba	La	Hf	Ta	W	Re	Os
7	Fr	Ra	Ac					

La	Ce	Pr	Nd	Pm	Sm	Eu	Gd	Tb	Dy	Ho	Er	Tm	Yb	Lu

used to date for preparing macroscopic amounts of metallofullerenes: the high-temperature laser vaporization or the laser-furnace method [7,38] and the standard direct-current arc discharge method [5,6]. Both methods simultaneously generate a mixture of hollow fullerenes (C_{60}, C_{70}, C_{76}, C_{78}, C_{84}, etc.) as well as metallofullerenes.

There is a less common but important method called the metal ion implantation technique for endofullerene production. This technique has been used to produce alkaline metallofullerenes [39–41] and nitrogen atom endofullerenes [42–44], such as $Li@C_{60}$ and $N@C_{60}$, respectively. However, at present the isolation and structural characterization of the metallofullerenes prepared by the above methods have not yet been performed, due to insufficiency of the materials produced by the implantation and the plasma techniques, although partial high-performance liquid chromatography (HPLC) separation on $Li@C_{60}$ has been reported [45].

In the laser furnace method (Fig. 8.3), a target composite rod or disk for laser vaporization, which is composed of metal oxides/graphite with a high-strength pitch binder, is placed in a furnace at 1200 °C [7]. A frequencey-doubled Nd:YAG laser is focused onto the target rod, which is normally rotating/translating to ensure a fresh surface, in an Ar gas flow (100–200 torr) condition. Metallofullerenes and empty fullerenes are produced by the laser vaporization and then flow down the tube with the Ar gas carrier, and finally are trapped on the quartz tube wall near the end of the furnace. To produce fullerenes and metallofullerenes a temperature above 800 °C was found to be necessary; below this critical temperature no fullerenes are produced [7,46,47]. Relatively slow thermal annealing processes are required to form fullerenes and metallofullerenes. The laser furnace method is suited for study of the growth mechanism of fullerenes and metallofullerenes [7,38,48–51]. The laser furnace method also

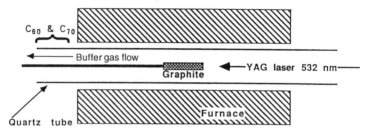

Figure 8.3 Schematic diagram of laser furnace apparatus to produce fullerenes and metallofullerenes by laser vaporization of a rotating graphite target in a tube furnace with a flowing argon carrier gas.

becomes an efficient production method for single-wall carbon nanotubes when Ni/Co or Ni/Fe binary metal is stuffed with graphite powder for target composite rods [52].

Figure 8.4 represents a large-scale direct current (dc) arc discharge apparatus for the production of metallofullerenes developed and installed at Nagoya [25,53–55]. The arc generator consists of a production and a collection chamber, which is equipped with an anaerobic sampling and collection mechanism of raw soot containing metallofullerenes [56,57]. The anaerobic sampling of the soot is preferred to conventional collection under ambient conditions because some of the metallofullerenes in primary soot are air (moisture) sensitive and may be subject to degradation during the soot handling.

Metal oxide/graphite composite rods, for example, La_2O_3 to prepare $La@C_{82}$, are normally used as positive electrodes (anodes) after a high-temperature (above ~1600 °C) heat treatment, where the composite rods are cured and carbonized [8]. At such high temperatures, various metal carbides in the phase of MC_2 are formed in the composite rods [58], which is crucial to an efficient production of endohedral metallofullerenes. For example, the yield of $La@C_{82}$ is increased by a factor of 10 or more when LaC_2-enriched composite rods are used for the arc generation of soot, instead of using La_2O_3 as a starting material for composite rods [56]. The rods (20 mm diameter × 500 mm long) are arced in the direct current (300–500 A) spark mode under 50–100 torr He flow conditions (Fig. 8.1). The soot so produced is collected under totally anaerobic conditions to avoid unnecessary degradation of the metallofullerenes produced during the soot collection and handling.

The production of scandium fullerenes, $Sc_n@C_{82}$ ($n = 1$–4), is particularly interesting, because scandium fullerenes exist in solvent extract as mono-, di- and triscandium fullerenes [18,19], which is quite unique when compared to other rare earth metallofullerenes. Recently, a tetrascandium fullerene, $Sc_4@C_{82}$, has also been produced and isolated [59]. Production of the mono-, di-, tri-, and tetrascandium fullerenes was found to be sensitive to the mixing ratio of scandium and carbon atoms in the composite rods; the relative abundance of di-, tri-, and tetrascandium fullerenes increases as the carbon/scandium ratio decreases. It was observed that the major scandium fullerene

Fullerene Production Chamber

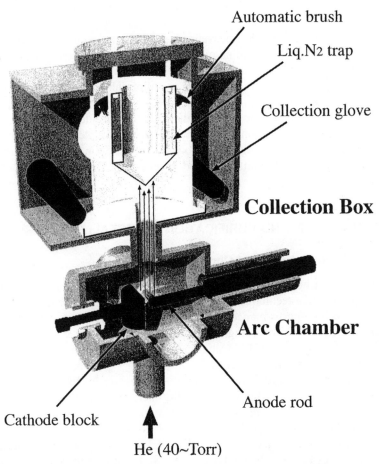

Automatic brush

Liq.N2 trap

Collection glove

Collection Box

Arc Chamber

Anode rod

Cathode block

He (40~Torr)

Figure 8.4 A dc arc discharge apparatus with an anaerobic collection and sampling mechanism. The produced metallofullerene-containing soot is effectively trapped by the liquid N_2 trap in the collection chamber. Typical arc discharge conditions: 40–100 torr He flow, 200–500 A, and 25–30 V.

produced was $Sc_2@C_{84}$ over a wide range of carbon/scandium mixing ratios (10–100) [18,60].

8.3.2 Extraction from Primary Soot

The so-called solvent extraction method by toluene or carbon disulfide is the most common and frequently used method, where metallofullerenes and hollow fullerenes are dissolved in solvents. Soxhlet extraction in a hot solvent or ultrasonic extraction is normally employed to increase the solvent extraction efficiency. However, in many cases, the toluene or CS_2 extraction is not sufficient, since in many cases nearly half of the metallofullerene still remains in the residual soot even after the CS_2 extraction. It has been found that metallofullerenes are further extracted from the residual soot by such solvents as pyridine [61] and 1,2,4-trichlorobenzene [62]. The metallofullerenes were found to be concentrated in this pyridine or trichlorobenzene extracted fraction.

In the vacuum sublimation method [8,11,63], as in the case of empty fullerenes [6,64–67], the raw soot containing metallofullerenes is heated in He gas or in vacuum up to 400 °C, where metallofullerenes such as La@C_{82} and Y@C_{82} start to sublime. The metallofullerenes then condense in a cold trap, leaving the soot and other nonvolatiles behind in the sample holder. Extraction by sublimation has an advantage over solvent extraction in obtaining "sovlent-free" extracts, whereas the latter method is suited for large-scale extraction of metallofullerenes.

8.4 SEPARATION AND PURIFICATION OF METALLOFULLERENES

8.4.1 Purification by Liquid Chromatography

As in the case of hollow fullerenes [64,68,69], liquid chromatography (LC) is the main purification technique for metallofullerenes. One of the most powerful LC methods is high-performance liquid chromatography (HPLC), which allows separation of fullerenes according to their molecular weight, size, shape, or other parameters [70–74]. The HPLC technique even allows separation of structural isomers of various matallofullerenes.

The purification of endohedral metallofullerenes via HPLC had been difficult, mainly because the content of metallofullerenes in raw soot is normally very limited and, furthermore, the solubility in normal HPLC solvents was even lower than that of various higher fullerenes. It took almost two years for metallofullerenes to be completely isolated by the HPLC method [60,75] since the first extraction of La@C_{82} by the Rice group [8]. Following these first isolations of metallofullerenes, isolations with different HPLC columns were also reported [62,76]. The success of purification/isolation was a big breakthrough for the further characterization of metallofullerenes.

8.4.2 HPLC Purification of Metallofullerenes

Scandium metallofullerenes are interesting in terms of separation and purification, because as described in Section 8.3.1 scandium fullerenes appear as mono-, di-, tri-, and tetrascandium fullerenes with several structual isomers that can be separated completely by HPLC.

For example, the scandium fullerenes, such as Sc@C_{82}, Sc_2@C_{84}, and Sc_3@C_{82}, were separated and isolated from various hollow (C_{60}–C_{110}) full-

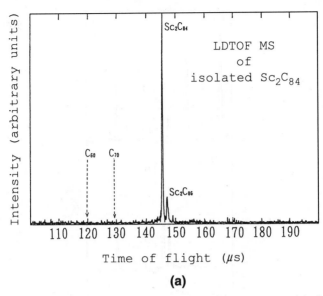

Figure 8.5 The first mass spectra of isolated metallofullerenes: (a) $Sc_2@C_{84}$ and (b) $La@C_{82}$.

erenes by the so-called two-stage HPLC method [57,60,75]. The two-stage HPLC method uses two complementary HPLC columns that have different types of fullerene adsorption mechanisms and are suited for complete separation of the metallofullerenes. The two-stage HPLC method was first successfully applied to the isolation of several discandium fullerenes including $Sc_2@C_{74}$, $Sc_2@C_{82}$, and $Sc_2@C_{84}$ [60] as shown in Figure 8.5a. Kikuchi et al. [75] employed a similar method for the first isolation of $La@C_{82}$ (Fig. 8.5b). To simplify separation, an automated HPLC separation of some endohedral metallofullerenes was employed by Stevenson et al. [77,78].

In the first HPLC stage, the toluene solution of the extracts was separated by a preparative recycling HPLC system (Japan Analytical Industry LC-908-C60) with a Trident-Tri-DNP column (Buckyclutcher I, 21 × 500 mm: Regis Chemical) or a PBB (pentabromobenzyl) column (20 × 250 mm, Nacalai Tesque) with CS_2 eluent. In this HPLC process, the scandium fullerene-containing fractions were separated from other fractions including C_{60}, C_{70}, and higher fullerenes (C_{76}–C_{110}). Complete purification and isolation of various scandium fullerenes was performed in the second HPLC stage by using a Cosmosil Buckyprep column (20 × 250 mm, Nacalai Tesque) with a 100% toluene eluent. Figure 8.6 shows the first and the second HPLC stages of the purification of $Sc@C_{82}$ as an example. The overall extraction/separation scheme is shown in Figure 8.7.

It has been found [61,62] that most of the monometallofullerenes, $R@C_{82}$, have at least two types of structural isomers (I and II), which can be separated

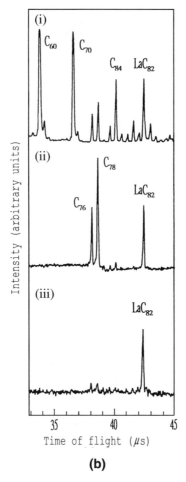

Figure 8.5 (continued)

by the two-stage HPLC technique. The retention times of isomers I are normally shorter than those of isomers II on Buckyprep columns and, in general, isomers I are much more stable than isomers II in air or in various solvents. In some cases, several isomers have been found for a metallofullerene. For example, a monocalcium fullerene, $Ca@C_{82}$, has four isomers (I–IV) that have been produced, isolated, and characterized [22]. Unless otherwise noted, we will discuss the structural and electronic properties of major isomers. The numbering of isomers of metallofullerenes has been done according to the elution order on Buckyprep columns [22].

Normally, the isolation of various metallofullerenes is confirmed by laser-desorption time-of-flight (LDTOF) mass spectrometry. For ESR-active metallofullerenes, the observation of the corresponding hyperfine structures can further confirm the identification and isolation.

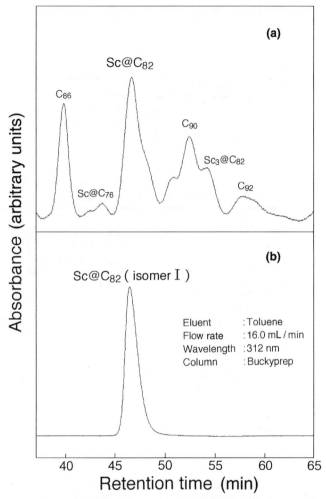

Figure 8.6 (a) An overall high-performance liquid chromatography (HPLC) spectrum for the fraction that contains various scandium metallofullerenes in the first HPLC stage. (b) An isolated HPLC spectrum for $Sc@C_{82}$ (I) after the second HPLC stage. The experimental conditions are presented in the lower right of the figure.

8.5 STRUCTURAL PROPERTIES

8.5.1 Endohedral or Exohedral?

Since the earlier studies on production and solvent extraction of metallofullerenes such as $La/Y/Sc@C_{82}$, there has been controversy over whether or not the metal atom is really trapped inside the fullerene cage [8,13]. In the gas phase, the stability of endohedral metallofullerenes had been studied by laser

Separation Scheme of Sc@C$_{82}$ (isomer I)

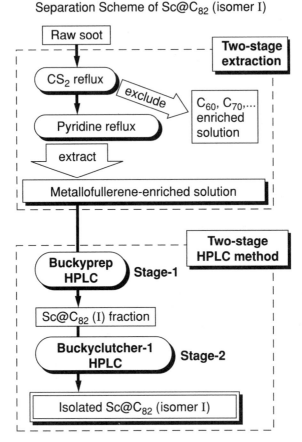

Figure 8.7 Typical extraction and separation/isolation scheme of endohedral metallo-fullerene. Two-stage extraction includes CS$_2$ and pyridine refluxes. The pyridine extract of the residual soot after CS$_2$ extraction is a metallofullerene-enriched extract. The separation and isolation are accomplished by the two-stage HPLC method.

photofragmentation for La@C$_{82}$ and Sc$_2$@C$_{84}$ [79,80], collisional fragmentation with atomic and molecular targets for La@C$_{82}$ and Gd@C$_{82}$ [81], and fragmentaion induced by surface impact for La$_2$@C$_{80}$ [11], La@C$_{82}$, La$_2$@C$_{100}$ [82], Ce@C$_{82}$, and Ce$_2$@C$_{100}$ [83].

Although the most extensive fragmentation was observed in the laser photo-fragmentation, the general tendency of the fragmentations induced by the three excitations was similar to each other: The main fragments from La@C$_{82}$ were C$_2$-loss species such as La@C$_{80}$, La@C$_{78}$, and La@C$_{76}$, and the empty C$_{82}$ fragment was not observed. The results suggest the endohedral nature of La@C$_{82}$, since exohedral La(C$_{60}$) [84] and Fe(C$_{60}$) [85] prepared by gas phase reactions gave C$_{60}$ as the main product upon collisional fragmentation against rare-gas targets.

However, in the solid state, the evidence for the endohedral nature of the

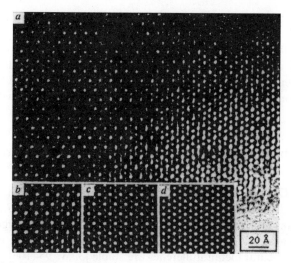

Figure 8.8 High-resolution TEM image of a $Sc_2@C_{84}$ crystal, taken along the [0001] direction. Inset (b) is a Fourier-filtered image that brings out the periodicities present in the original image. Insets (c) and (d) are simulated images of 67 Å thick $Sc_2@C_{84}$ and C_{84} crystals, respectively.

metallofullerenes had not been convincing. For example, two contradictory results were reported for extended X-ray absorption fine structure (EXAFS) experiments on the unpurified extracts of $Y@C_{82}$ (i.e., a mixture of empty fullerenes and metallofullerenes). Soderholm et al. [86] reported that the yttrium atom is exohedrally attached from the outside to the C_{82} cage, whereas Park et al. [87] reported an endohedral nature for $Y@C_{82}$. The first nearest neighbor C–Y distances were 0.253 ± 0.002 nm and 0.24 nm in the former and latter experiments, respectively. Kikuchi et al. [88] performed an EXAFS experiment on a purified $La@C_{82}$ powder material and reported that the first and second nearest neighbor C–La distances are 0.247 ± 0.002 and 0.294 ± 0.007 nm, respectively.

Major experimental evidence had indicated, however, the endohedral nature of the metallofullerenes: The IBM Almaden group reported a high-resolution transmission electron microscopy (HRTEM) experiment on a purified $Sc_2@C_{84}$ material, which suggests that the two scandium atoms are encapsulated in the C_{84} cage [89]. Figure 8.8 shows the HRTEM image of $Sc_2@C_{84}$, suggesting the endohedral nature of the metallofullerene. The UCLA group reported a high-energy collision experiment on $La_2@C_{80}$ against silicon surfaces and found that no collision fragments such as La atoms and C_{80} were observed, also suggesting an endodedral structure of $La_2@C_{80}$ [10]. The Tohoku–Nagoya group reported a series of ultrahigh vacuum scanning tunneling microscopy (UHV-STM) studies on $Sc_2@C_{84}$ and $Y@C_{82}$ adsorbed on clean silicon and copper surfaces, respectively [90–92]. All of the obtained STM images showed spherical shape, which indicates that the metal atoms are encapsulated in the fullerene cages. Gimzewski [93] also studied $Sc_2@C_{84}$ molecules deposited from a CS_2 solution onto Au(110) by STM and obtained some internal structure on the top part of the images.

Although the above experimental results strongly suggest an endohedral nature for the metallofullerenes, the final confirmation of the endohedral nature and detailed endohedral structures of the metallofullerenes have been obtained by synchrotron X-ray diffraction measurements on purified powdered samples [94].

8.5.2 Endohedral Structures as Determined by Synchrotron X-Ray Diffraction

8.5.2.1 $Y@C_{82}$ *Monometallofullerene* Previous experimental evidence including extended X-ray absorption fine structure (EXAFS) [87,88] and high-resolution transmission electron microscopy (HRTEM) [89] suggested that the metal atoms are inside the fullerenes. Theoretical calculations also indicated that such endohedral metallofullerenes are stable [32,37,95–105]. However, the first conclusive experimental evidence of the endohedral nature of a metallofullerene, $Y@C_{82}$, was obtained by a synchrotron X-ray diffraction study. The result indicated that the yttrium atom is encapsulated within the C_{82} fullerene and is strongly bound to the carbon cage [94].

The synchrotron radiation X-ray powder experiment with an imaging plate detector was employed to collect an X-ray powder pattern with good counting statistics [94]. The space group was assigned to $P2_1$, which is monoclinic for $Y@C_{82}$. The experimental data were analyzed in an iterative way using a combination of Rietveld analysis [106] and the maximum entropy method (MEM) [107,108]. The MEM can produce an election density distribution map from a set of X-ray structure factors without using any structural model. Using MEM analysis [109,110], the R_I become as low as 1.5% for $Y@C_{82}$.

A series of iterative steps involving the Rietveld analysis using the revised structural model based on the previous MEM map and the MEM analysis were carried out until no significant improvement was obtained. Eventually, the R_I factor improved from 14.4% to 5.9% ($R_{WP} = 3.0\%$). In Figure 8.9, the best fit of the Rietveld analysis of $Y@C_{82}$ is shown. To visualize the endohedral nature of $Y@C_{82}$, the MEM electron density distribution of $Y@C_{82}$ is shown in Figure 8.10. There exist remarkably high densities just inside the C_{82} cage (Fig. 8.10). The density maximum at the interior of the C_{82} cage corresponds to the yttrium atom, indicating the endohedral structure of the metallofullerene.

There are many local maxima along the cage in $Y@C_{82}$, whereas the electron density of the C_{82} cage is relatively uniform. This suggests that in $Y@C_{82}$ the rotation of the C_{82} cage is very limited around a certain axis even at room temperature, while that in C_{82} is almost free. The MEM electron density map further reveals that the yttrium atom does not reside in the center of the C_{82} cage but is very close to the carbon cage, as suggested theoretically [37,101,103–105,111]. ESR [14,15] and theoretical [105,112] studies suggest the presence of a strong charge-transfer interaction between the Y^{3+} ion and the

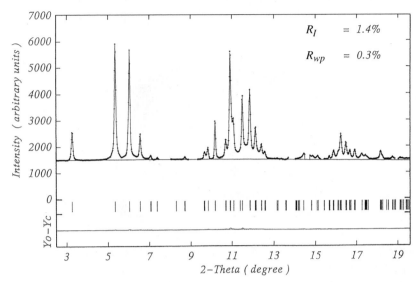

$R_I = 1.4\%$

$R_{wp} = 0.3\%$

Figure 8.9 Synchrotron X-ray diffraction patterns and the corresponding fitting results of Y@C_{82} based on the calculated intensities from the MEM electron density.

C_{82}^{3-} cage, which may cause the aspherical electron density distribution of atoms. The Y–C distance calculated from the MEM map is 2.9(3) Å, which is slightly longer than a theoretical prediction of 2.55–2.65 Å [105]. The X-ray study also reveals that the Y@C_{82} molecules are aligned along the [001] direction in a head-to-tail ($\cdots$ Y@C_{82} $\cdots$ Y@C_{82} $\cdots$ Y@C_{82} $\cdots$) order in the crystal, indicating the presence of strong dipole–dipole and charge-transfer interactions among the Y@C_{82} fullerenes. The ability to achieve molecular alignment of the

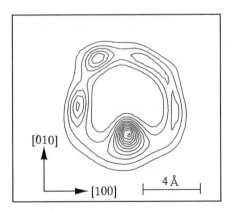

Figure 8.10 The MEM electron density distributions of Y@C_{82} for the (001) section. The counter lines are drawn from $0.0e$ [Å^{-3}] with a $0.5e$ [Å^{-3}] step width. The high density position corresponds to Y atom. The endohedral nature of Y@C_{82} is evident.

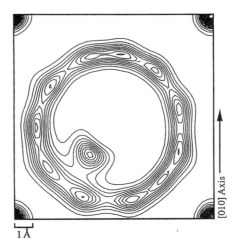

Figure 8.11 The MEM electron density distribution of Sc@C$_{82}$ for the (100) section. The contour lines are drawn with 0.3e [Å^{-3}] intervals.

endohedral metallofullerene in a certain direction in the crystal or clean surfaces may, in future, lead to materials with novel solid-state properties.

8.5.2.2 Sc@C$_{82}$ The endohedral structure of Sc@C$_{82}$ was also studied by synchrotron X-ray diffraction (XRD) with the MEM analysis [113]. The Sc@C$_{82}$ crystal includes solvent toluene molecules and has a $P2_1$ space group, as in the Y@C$_{82}$ case. The MEM electron charge density distribution of Sc@C$_{82}$ is shown in Figure 8.11. The Sc atom is not the center of the fullerene but is close to one of the six-membered rings of the cage. The nearest neighbor Sc–C distance estimated from the MEM map is 2.53(8) Å, which is close to a theoretical value, 2.52–2.61 Å [105]. There are nine IPR-satisfying structural isomers for C$_{82}$. These are $C_2(a)$, $C_2(b)$, $C_2(c)$, C_{2v}, $C_s(a)$, $C_s(b)$, $C_s(c)$, $C_{3v}(a)$, and $C_{3v}(b)$. The present result indicates that the carbon cage of Sc@C$_{82}$ has C_{2v} symmetry (Fig. 8.11).

There has been controversy on whether the encaged Sc atom has a divalent state or a trivalent state [15,18,105,114–116]. The synchrotron X-ray result shows that the number of electrons around the Sc atom is 18.8e, which indicates that the Sc atom in the cage is in a divalent state Sc^{2+}@C$_{82}$$^{2-}$.

8.5.2.3 La@C$_{82}$ The La@C$_{82}$ metallofullerene was the first endohedral metallofullerene to be macroscopically produced and solvent extracted in 1991 [8]. Suematsu et al. [117] first reported the crystal structure of La@C$_{82}$ precipitated from CS$_2$ solution via synchrotron X-ray powder diffraction. The composition is expressed by La@C$_{82}$(CS$_2$)$_{1.5}$. The crystal has cubic structure. The results suggest a molecular alignment in the unit cell, in which the molecules align in the [111] direction with the molecular axis orienting in the same [111] direction. Watanuki et al. [118,119] performed synchrotron XRD measurements on solvent-free powder samples of La@C$_{82}$ and concluded that the

(100) PLANE, SECTION = 22, SINGLE UNIT CELL

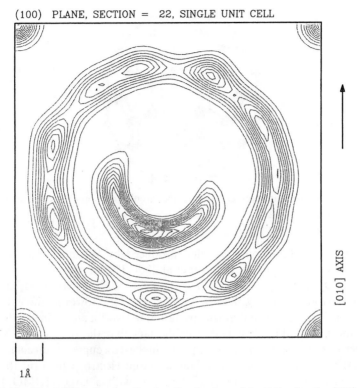

[010] AXIS

1Å

Figure 8.12 The MEM electron density distribution of La@C$_{82}$ for the (100) section. The contour lines are drawn with $0.3e$ [Å^{-3}] intervals.

major part of the crystal has a face-centered cubic (fcc) lattice. Their results strongly suggest the endohedral nature of La@C$_{82}$, where the La atom is displaced from the center of the C$_{82}$ cage by 1.9 Å.

It it was not until quite recently, however, that the detailed endohedral structure of La@C$_{82}$ was revealed experimentally [120]. The electron density distribution of La@C$_{82}$ based on the MEM analysis of the powder XRD data is presented in Figure 8.12. The result shows that the La atom is encapsulated by the C_{2v} isomer of C$_{82}$ as in Sc@C$_{82}$ described earlier. As is seen from the figure, the La atom is not at rest in the cage but rather is in a floating motion along the nearest six-membered ring at room temperature. The result is much different from the Sc@C$_{82}$ and Y@C$_{82}$ cases in which Sc and Y atoms are at standstill in the cage even at room temperature. The light metal atom Sc seems to be more strongly bound to the fullerene cage than the heavy La atom.

8.5.2.4 Dynamic Motion of Metal Atoms Within the Cage   Intrafullerene metal motions have been theoretically predicted by Andreoni and Curioni [103,104,121,122] on La@C$_{60}$ and La@C$_{82}$ on the bases of molecular dynamic simulations. Experimentally, intrafullerene metal motion has been reported on Sc$_2$@C$_{84}$ [123] and on La$_2$@C$_{80}$ [124,125].

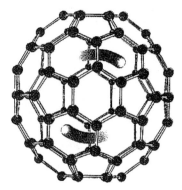

Figure 8.13 A schematic representation of the circular motion of two La atoms inside the I_h-C_{80} cage.

$La_2@C_{80}$ metallofullerene was first produced by Whetten and co-workers [10] and was first isolated by Kikuchi et al. [88]. The empty C_{80} has seven IPR structures (D_2, D_{5d}, C_{2v}, $C_{2v'}$, D_3, D_{5h}, and I_h). A ^{13}C NMR study indicated that the most abundant C_{80} has D_2 symmetry [126]. However, theoretical calculations [127] have shown that encapsulation of two La atoms inside the I_h-C_{80} cage is most favorable. This is due to the fact that the I_h-C_{80} cage has only two electrons in the fourfold degenerate highest occupied molecular orbital (HOMO) level and can accommodate six more electrons to form the stable closed-shell electronic state of $(La^{3+})@C_{80}{}^{6-}$ with a large HOMO–LUMO (lowest unoccupied molecular orbital) gap.

On the bases of ^{13}C NMR and ^{139}La NMR results, Akasaka et al. [124,125] suggested the presence of a circular motion for the encaged La atoms in the C_{80} cage. The metal motion is schematically depicted in Figure 8.13, in which two La atoms can circuit the inside of the spherical I_h-C_{80} cage. The energy barrier for the circuit of the metal cations is very small (ca. 5 kcal/mol). The dynamic behavior of metal atoms should also be reflected in the ^{139}La NMR linewidth, since circulation of two La^{3+} cations produces a new magnetic field inside the cage. Such a linewidth broadening was actually observed with increasing temperature from 305 to 363 K [125].

Similar but much restricted intrafullerene dynamics of encaged metal ions has been reported by Miyake et al. [123] on $Sc_2@C_{84}$. They observed a single ^{45}Sc NMR line, indicating that two Sc atoms in the cage are equivalent. However, in contrast to the $La_2@C_{80}$ case, the internal rotation is hindered by a large barrier of approximately 50 kcal/mol [128].

The intrafullerene dynamics of the Ce atom in the C_{82} cage was also studied by time-differential perturbed angular correlation measurements [129]. The observed angular correlations show the presence of two different chemical species of $Ce@C_{82}$. The data at low temperatures reveal that Ce stays at a certain site for one of the species, whereas for the other the atom has an intramolecular dynamic motion.

8.5.2.5 A Dimetallofullerene: $Sc_2@C_{84}$ Various metallofullerenes supposed to encapsulate two or three metal atoms within fullerene cages, such as
$La_2@C_{80}$ [10,88,124,125,130–132], $La_2@C_{72}$ [29,30,133,134], $Y_2@C_{82}$ [14,15],
$Sc_2@C_{74}$ [60,135], $Sc_2@C_{82}$ [15,19], $Sc_2@C_{84}$ [15,18,19,62,89,136,137],
$Sc_3@C_{82}$ [18,19,57,77,78,133,138–140], and $Er_2@C_{82}$ [141–143] have been successfully synthesized and purified. Among them the scandium dimetallofullerenes, $Sc_2@C_{84}$, are especially interesting, because three structural isomers have
been found and isolated so far [144].

Ab initio theoretical studies [145,146] and the experimental results on
$Sc_2@C_{84}$ including scanning tunneling microscopy [91,135], transmission electron microscopy [89], and ^{13}C-NMR [144] have suggested an endohedral nature. Similar to the monometallofullerenes, a synchrotron powder X-ray study
is reported on $Sc_2@C_{84}$ (isomer III) with the Rietveld/MEM analysis [137].
Figure 8.14 shows a three-dimensional MEM electron density distribution of
$Sc_2@C_{84}(III)$. Six- and five-membered rings are seen on the surface of the
electron density distribution. There are 24 IPR-satisfying structural isomers for
C_{84} [147]. The result reveals that two Sc atoms are encapsulated within the
carbon cage and the $Sc_2@C_{84}(III)$ molecule has D_{2d} symmetry out of the 24
IPR isomers. The Sc atoms are aligned along the C_2 axis of a D_{2d}-$Sc_2@C_{84}$
molecule. The observed Sc–Sc and the nearest Sc–C distances are 3.9(1) and

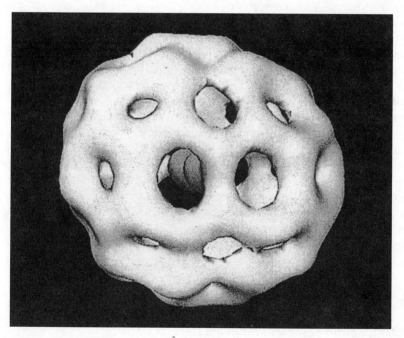

Figure 8.14 The Equicontour 1.3e [Å^{-3}] density map of the MEM charge density for
the $Sc_2@C_{84}$ molecule. The C_2 axis connecting two Sc atoms is just off the axis perpendicular to the plane.

2.4(2) Å, respectively. The D_{2d}-C_{84} was also found to be a major isomer for the empty C_{84} [148].

The number of electrons around each maximum inside the cage is 18.8, which is very close to that of a divalent scandium ion Sc^{2+} (19.0). A theoretical study predicted that the formal electronic structure of $Sc_2@C_{84}$ is well represented by $(Sc^{2+})_2@C_{84}{}^{4-}$, where two 4s electrons of each Sc atom are transferring to the C_{84} cage [145]. The positive charge of the Sc atom from the MEM charge density is +2.2, which is in good accordance with the theoretical value.

The charge density of the present $Sc_2@C_{84}$ molecule shows a distinct feature that is in very sharp contrast to that of $Y@C_{82}$ [94]. The 5,6-ring structure of C_{84} can clearly be seen although the present experiment was performed at room temperature, whereas only averaged charge density of C_{82} was obtained for $Y@C_{82}$ (cf. Fig. 8.10). This suggests that rotations of the C_{82} cage are virtually quenched even at room temperature in a $Sc_2@C_{84}$ solid, while the $Y@C_{82}$ molecule is still in a hindered rotation.

8.5.3 Crystal Structures

Microcrystals of a metallofullerene grown from solution and those that are solvent-free generally have different crystal structures. For example, solvent-free $La@C_{82}$ crystals prepared by sublimation have fcc structure [118,119], whereas toluene- and carbon disulfide-containing $La@C_{82}$ crystals exhibit monoclinic [120] and cubic [117] structures, respectively. Crystals of $Y@C_{82}$ [94], $Sc@C_{82}$ [137], and $Sc_2@C_{84}$ [113] grown from toluene solutions also have monoclinic structures. Furthermore, crystals of empty higher fullerenes such as C_{76} and C_{82} grown from toluene solutions show monoclinic structures [149]. The observed crystal data and structural parameters of $La@C_{82}$, $Y@C_{82}$, $Sc@C_{82}$, and $Sc_2@C_{84}$ are listed in Table 8.2.

One of the important findings in the crystal structure of the mono-metallofullerenes is the alignment of molecules in a certain direction in the crystal. For example, the $Y@C_{82}$ molecules are aligned along the [001] direction in a head-to-tail ($\cdots Y@C_{82} \cdots Y@C_{82} \cdots Y@C_{82} \cdots$) order in the crystal, indicating the presence of strong dipole–dipole and charge-transfer interactions among the $Y@C_{82}$ fullerenes [94]. The crystal data in Table 8.2 also support this idea since the structure parameter c of $Y@C_{82}$ (11.265 Å) is shorter than that of C_{82} (11.383 Å), whereas the parameter a of $Y@C_{82}$ is much longer than that of C_{82}, indicating a shrinkage in crystal packing in the [001] direction.

8.6 ELECTRONIC AND VIBRATIONAL PROPERTIES

8.6.1 Electron Transfer Within the Carbon Cage

Group 3 metallofullerenes exhibit ESR hyperfine structures (hfs), which give important information on the electronic structures of the metallofullerenes.

TABLE 8.2 Structural Data for La@C_{82}, Y@C_{82}, Sc@C_{82}, and Sc$_2$@C_{84} Determined by Synchrotron X-Ray Diffraction

Parameter	Sc@C_{82}	Y@C_{82}	La@C_{82}[a]	Sc$_2$@C_{84}
Lattice type[b]	Monoclinic	Monoclinic	Monoclinic Cubic fcc	Monoclinic
Space group	$P2_1$	$P2_1$	$P2_1$ (monoclinic) $I\bar{4}3d$ (cubic) $Fm\bar{3}m^c$ (fcc)	$P2_1$
Lattice parameters	$a = 18.362(1)$ Å $b = 11.2490(6)$ Å $c = 11.2441(7)$ Å $\beta = 107.996(9)°$	$a = 18.401(2)$ Å $b = 11.281(1)$ Å $c = 11.265(1)$ Å $\beta = 108.07(1)°$	$a_0 = 15.78$ Å (fcc) $a_0 = 25.72 \pm 0.007$ Å (cubic) $a = 18.3345(8)$ Å $b = 11.2446(3)$ Å $c = 11.2320(3)$ Å $\beta = 107.970(6)°$ (monoclinic)	$a = 18.312(1)$ Å $b = 11.2343(6)$ Å $c = 11.2455(5)$ Å $\beta = 107.88(1)°$
Nearest metal–carbon distance	2.53(8) Å	2.9(3) Å	2.55(7) Å	2.4(2) Å

[a] Crystals grown from CS_2 solution have a cubic lattice, whereas solvent-free crystals have a fcc lattice.
[b] All crystals grown from toluene solution have monoclinic lattice.
[c] Patterson symmetry.

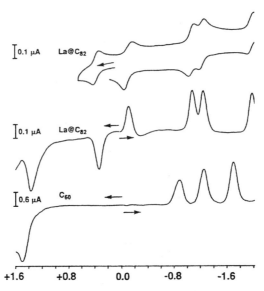

Figure 8.15 Cyclic voltammogram of La@C$_{82}$ and differential pulse voltammograms of La@C$_{82}$ and C$_{60}$ in o-dichlorobenzene.

Typical ESR-active endohedrals are La@C$_{82}$, Y@C$_{82}$, and Sc@C$_{82}$. The ESR hfs of a metallofullerene was first observed in La@C$_{82}$ by the IBM Almaden group [12] and was discussed within the framework of an intrafullerene electron transfer. The observation of eight equally spaced lines provides evidence of isotropic electron–nuclear hyperfine coupling (hfc) to ^{139}La with nuclear spin quantum number $\frac{7}{2}$. The observed electron g-value of 2.0010, close to that measured for the C$_{60}$ radical anion [150,151], indicates that a single unpaired electron resides in the lowest unoccupied molecular orbital (LUMO) of the carbon cage. They also observed ^{13}C hyperfine satellites, providing that the unpaired electron couples to both the ^{139}La and the carbon atoms. The observed hfc (1.2 G) is very small compared to that (50 G) measured for La^{2+} substituted in CaF$_2$ [152]. Therefore, it was concluded that the La atom in the C$_{82}$ cage must be in the 3+ state, which gives a formal charge state of La^{3+}@C$_{82}$$^{3-}$. Ultraviolet photoelectron spectroscopy (UPS) [153,154] and very recent X-ray diffraction [120] results strongly support this conclusion.

ESR hfs obtained for Y@C$_{82}$ indicated that the yttrium in C$_{82}$ is also in the 3+ oxidation state [14,15]. However, there has been controversy as to whether Sc@C$_{82}$ has a 3+ or 2+ charge state as described in Section 8.5.2 [18,105,114]. Recent X-ray diffraction [120] and UPS [155] results indicate a 2+ state leading to Sc^{2+}@C$_{82}$$^{2-}$. Theoretical calculations suggest that the electronic structure of Sc@C$_{82}$ is well represented by Sc^{2+}@C$_{82}$$^{2-}$ [105,111,156].

The linewidths of the ESR hfs of La@C$_{82}$, Sc@C$_{82}$, and Gd@C$_{82}$ have been discussed thoroughly by Kato et al. [138,157–159] in terms of the spin–rotation coupling interaction. Dinse and co-workers [114,115,160] investigated the temperature dependence of the ESR linewidths of La@C_{82}, La@C_{90}, and Sc@C_{82} in different solvents and obtained information on the nuclear quadrupole interactions in these metallofullerenes. Dunsch and co-workers [161–163] studied ^{13}C satellite structures of M@C_{82} (M = Sc, Y, La) in detail and reported that the manifold of ^{13}C hfc constants could be interpreted by the calculated spin density distributions.

ESR spectra of La@C_{82}, Y@C_{82}, Ho@C_{82}, and Tm@C_{82} taken from the solid soot extract were reported by Bartl et al. [161,164–166] and reported low resolved but split hyperfine structure, indicating that the metal atoms exist in ionic form in the fullerene cage and also in the solid state. The research group also reported [167] the principal values of the hyperfine tensor **A** and the relative orientation of **g** and **A** tensors of M@C_{82} (M = Sc, Y, La) by applying three- and four-pulse electron spin echo envelope modulation (ESEEM) techniques.

8.6.2 Electrochemistry of Metallofullerenes

Further electronic properties of endohedral metallofullerenes, such as reduction/oxidation (redox) properties, have also been investigated electrochemically by using cyclic voltammetry (CV). Suzuki et al. [168] measured cyclic voltammograms on La@C_{82} and found unusual redox properties of the metallofullerene, which differ significantly from those of empty fullerenes. Figure 8.15 shows the CV results for La@C_{82}. The first reversible oxidation potential is approximately equal to that of ferrocene, indicating that La@C_{82} is a moderate electron donor. The first reduction and oxidation potentials indicate that it should form both cationic and anionic charge-transfer complexes. In addition, La@C_{82} is a stronger electron acceptor than empty fullerenes such as C_{60}, C_{70}, and C_{84}. A schematic energy level diagram of La@C_{82} is presented in Figure 8.16. The CV results on La@C_{82} revealed that La@C_{82} is a good electron donor as well as a good electron acceptor and that at least five electrons can be transferred to the C_{82} cage while maintaining the 3+ charge state of the encaged La atom, that is, $(La^{3+}@C_{82}{}^{3-})^{5-}$. The CV measurements indicate that the oxidation state of the yttrium atom in Y@C_{82} [169] is close to that of La@C_{82}, likely 3+. The electrochemistry of Y@C_{82} is almost identical to that of La@C_{82} [140,169]. Other monometallofullerenes such as Y@C_{82}, Ce@C_{82}, and Gd@C_{82} have a similar tendency in their redox properties [169]. Anderson et al. [139] reported CV data on Sc_3@C_{82}. The CV measurements were also done on other lanthanide fullerenes: Pr@C_{82}, Nd@C_{82}, Tb@C_{82}, Dy@C_{82}, Ho@C_{82}, Er@C_{82}, and Lu@C_{82} by Wang et al. [170].

Dunsch and co-workers [166,171–173] studied electron transfers in metallofullerenes by CV coupled with in situ ESR experiments. The electron transfer to the endohedral La@C_{82} molecule studied by this method gives evidence of a charge in the electronic state of the fullerene; the electrochemical reaction in the anodic scan causes the formation of $La^{3+}@C_{82}{}^{3-}$ and during the cathodic scan

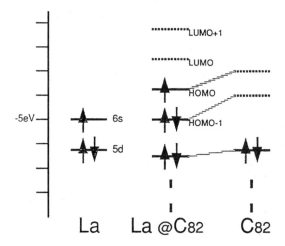

Figure 8.16   Schematic energy level diagram of La@C_{82}.

the spin concentration decreases because the La^{3+}@C_{82}^{2-} structure formed by reduction is not paramagnetic.

8.6.3   Bulk Electronic Properties

The electronic properties of several metallofullerenes in the solid state have been studied by ultraviolet photoemission spectroscopy (UPS) [153,154,174,175]. Figure 8.17 is a UPS spectrum of La@C_{82} [154]. The first (small) peak is 0.64 eV below the Fermi level corresponding to the singly occupied molecular orbital (SOMO) level of La@C_{82}, where the spectral onset is 0.35 eV from the Fermi level. Hino et al. [153] also observed peaks at 0.9 and 1.6 eV below the Fermi level in their UPS measurement on La@C_{82}, which are absent in the corresponding empty C_{82} spectrum. The peaks at 1.6 and 0.9 eV correspond to electronic transitions from the La atom to the HOMO (highest occupied molecular orbital) and HOMO-1 levels of La@C_{82}, respectively (Fig. 8.17). Since the observed intensity ratio of the two peaks is about 2:1, they concluded that three electrons of the La atom are transferred to the fullerene cage, that is, La^{3+}@C_{82}^{3-}, which is consistent with the results obtained by ESR hyperfine splittings.

Pichler et al. [174,175] studied the valency of the Tm ion in the endohedral Tm@C_{82} fullerene by UPS and X-ray photoelectron spectroscopy (XPS). The similarity of the Tm 4d core level photoemission spectrum to that calculated for Yb^{3+} suggests a 4f^{13} ground-state configuration for the Tm ion. The real part of the optical conductivity of the C_{3v} and C_s isomers is shown in Figure 8.18 [175]. The energy gap transitions of the two isomers are different, where the onsets of the isomers are 0.6 eV (C_s) and 0.8 eV (C_{3v}). The UPS measurements

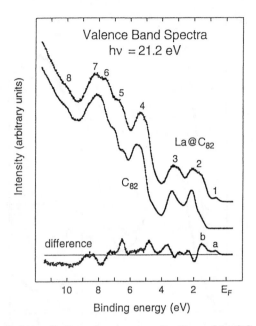

Figure 8.17 Comparison of valence band spectra for C_{82} and La@C_{82} measured at 21.2 eV. Feature 1 is attributed to SOMO-derived levels centered 0.64 eV below the Fermi level with an onset at 0.35 eV.

on Sc_2@C_{84}(III) [136], Gd@C_{82} [176], and Sc@C_{82}(I) [155] were also reported in which the band gaps of these metallofullerenes were obtained.

8.6.4 Vibrational Structures of Metallofullerenes

Vibrational structures of some metallofullerenes have been studied by IR and Raman spectroscopy [75,177–181]. Some of the vibrational absorption lines of Sc_2@C_{84}(III) are strongly enhanced if compared to the spectrum of the empty cage [177,178]. With decreasing temperature, a dramatic narrowing of the lines was observed. The linewidth shows an Arrhenius-like behavior between 200 and 300 K, provided that the main contribution to it comes from rotational diffusion.

Lebedkin et al. [179,180] reported vibrations due to the encapsulated metal ions in the cage for M@C_{82} (M = La, Y, Ce, Gd) based on IR and Raman measurements. Figure 8.19 shows the Raman spectra of M@C_{82} (M = La, Y, Ce, Gd) [180]. The peaks around 150 cm^{-1} can be attributed to internal vibrational modes, most probably metal-to-cage vibrations. Almost all peaks are observed at similar positions. These peaks were strongly broadened when the samples were exposed to air. This result is in agreement with a near-edge X-ray absorption fine-structure (NEXAFS) study [181], where the pronounced effect of air on the spectra of La@C_{82} films could be reduced only by heating them

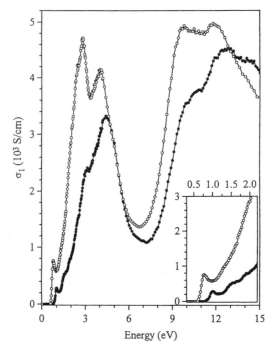

Figure 8.18 Real part of the optical conductivity (σ_1) of the C_{3v} isomer (●) and C_s isomer (○) of Tm@C_{82}. The inset shows the region of the energy gap on an extended scale.

to 600 °C. The far-infrared (FIR) spectra of M@C_{82} (M = La, Y, Ce) [180] support the picture derived from the Raman measurements. The metal-dependent FIR peaks between 150 and 200 cm^{-1} correspond well to their Raman counterparts.

More detailed IR and Raman measurements were reported by Dunsch and co-workers [182] on the three isomers of Tm@C_{82} [172,183]. The frequency of the metal-to-cage vibration is little affected by the fullerene cage isomerization but is strongly dependent on the kind of metal ion inside the fullerene. The metal-to-cage peaks (116 cm^{-1}) are almost invariant among the isomers but are much smaller than those of M@C_{82} (M = La, Y, Ce, Gd) described earlier. It was proposed that the metal-to-cage vibration is sensitive to the charge state of the encaged metal atom; the frequency of this vibration for the trivalent met-allofullerenes, M@C_{82} (M = La, Y, Ce, Gd), is much higher than those of the divalent metallofullernes such as Tm@C_{82} and Eu@C_{74} [181]. This is expected if the metal–cage bonding is basically electrostatic [94,184] and the metals bear the same charge.

Inelastic neutron scattering (INS) results for La@C_{82} and Y@C_{82} [179,180] also presented evidence for the metal-to-cage vibration at 180, 150, and 85 cm^{-1}. An interval from 100 to 200 cm^{-1} in the Raman and FIR spectra of

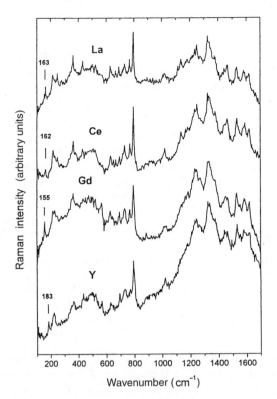

Figure 8.19 Raman spectra of M@C_{82} (M = La, Ce, Gd, Y) metallofullerenes. The spectra are shifted for clarity.

M@C_{82} (M = divalent or trivalent metal atoms) can be regarded as a "metal-fingerprint" range.

8.7 PROSPECTS

Endohedral metallofullerenes are novel forms of fullerene-related materials, which were obtained in their purified form in January 1993. Since then, as seen in this chapter, numerous investigations in various aspects of the metallofullerenes have been intensively done by many research groups. These studies have revealed unique structural and novel electronic properties of the metallofullerenes. Even so, there still remain many potentially important and intriguing topics. For instance, the solid-state properties of most of the metallofullerenes, such as electric conductivity and magnetic behavior, are not well known. Only limited studies have been reported. Organic functionalizations of the metallofullerenes will be an important direction to synthesize further new materials based on endohedral metallofullerenes, because metallofullerenes are generally more reactive, either thermally or photochemically, than the corresponding empty fullerenes [124,132,185].

Metallofullerenes will also become an important nanostructured material for future nanoscaled-electronic devices, because the band gaps of endohedral metallofullerenes, for example, can be varied between 1.0 and 0.2 eV depending on the fullerene size, the kind of metal atom(s), and the number of metal atoms encapsulated.

Physiological and medical applications of the endohedral metallofullerenes will be important in relation to tracer chemistry and contrast agents in biological systems, and definitely await future studies. In any case, the endohedral metallofullerenes will continue to tantalize physicists, chemists, and materials scientists for years to come.

Note Added in Proof

Very recently, the first $M@C_{60}$ type metallofullerenes were purified and isolated for $Er@C_{60}$ [186] and $Eu@C_{60}$ [187].

ACKNOWLEDGMENTS

The author is indebted for the synchrotron X-ray collaboration with Professor M. Takata, Dr. Eiji Nishibori, and Professor M. Sakata (Nagoya University). The author thanks Drs. T. J. S. Dennis, C. R. Wang, T. Tomiyama, and T. Sugai (Nagoya University) for their help. The present study has been supported by a Future Programme "Advanced Processes for New Carbon Nano-Materials" sponsored by the Japan Society for Promotion of Science.

REFERENCES

1. H. Kroto, J. R. Heath, S. C. O'Brien, R. F. Curl, and R. F. Smalley, *Nature* **1985**, *318*, 162.

2. J. Heath, S. C. O'Brien, Q. Zhang, Y. Liu, R. F. Curl, H. W. Kroto, F. K. Tittel, and R. E. Smalley, *J. Am. Chem. Soc.* **1985**, *107*, 7779.

3. D. M. Cox, D. J. Trevor, K. C. Reichmann, and A. Kaldor, *J. Am. Chem. Soc.* **1986**, *108*, 2457.

4. F. D. Weiss, J. L. Elkind, S. C. O'Brien, R. F. Curl, and R. E. Smalley, *J. Am. Chem. Soc.* **1988**, *110*, 4464.

5. W. Kraetschmer, K. Fostiropoulos, and D. R. Huffman, *Chem. Phys. Lett.* **1990**, *170*, 167.

6. W. Kraetschmer, K. Fostiropoulos, L. D. Lamb, and D. R. Huffman, *Nature* **1990**, *347*, 354.

7. R. E. Haufler, Y. Chai, L. P. F. Chibante, J. Conceicao, C. Jin, L. S. Wang, S. Maruyama, and R. E. Smalley, "Carbon Arc Generation of C_{60}" in *Cluster-*

Assembled Materials, Vol. 206, R. S. Averback, J. Bernhoc, and D. L. Nelson (eds.). Materials Research Society, Pittsburgh, 1991, pp. 627–637.

8. Y. Chai, T. Guo, C. Jin, R. E. Haufler, L. P. F. Chibante, J. Fure, L. Wang, J. M. Alford, and R. E. Smalley, *J. Phys. Chem.* **1991**, *95*, 7564.

9. E. W. Godly and R. Taylor, *Pure Appl. Chem.* **1997**, *69*, 1411.

10. M. M. Alvarez, E. G. Gillan, K. Holczer, R. B. Kaner, K. S. Min, and R. L. Whetten, *J. Phys. Chem.* **1991**, *95*, 10561.

11. C. Yeretzian, K. Hansen, M. M. Alvarez, K. S. Min, E. G. Gillan, K. Holczer, R. B. Kaner, and R. L. Whetten, *Chem. Phys. Lett.* **1992**, *196*, 337.

12. R. D. Johnson, M. S. de Vries, J. Salem, D. S. Bethune, and C. S. Yannoni, *Nature* **1992**, *355*, 239.

13. D. S. Bethune, R. D. Johnson, J. R. Salem, M. S. de Vries, and C. S. Yannoni, *Nature* **1993**, *336*, 123.

14. J. H. Weaver, Y. Chai, G. H. Kroll, C. Jin, T. R. Ohno, R. E. Haufler, T. Guo, J. M. Alford, J. Conceicao, L. P. F. Chibante, A. Jain, G. Palmer, and R. E. Smalley, *Chem. Phys. Lett.* **1992**, *190*, 460.

15. H. Shinohara, H. Sato, Y. Saito, M. Ohkohchi, and Y. Ando, *J. Phys. Chem.* **1992**, *96*, 3571.

16. M. M. Ross, H. H. Nelson, J. H. Callahan, and S. W. McElvany, *J. Phys. Chem.* **1992**, *96*, 5231.

17. S. W. McElvany, *J. Phys. Chem.* **1992**, *96*, 4935.

18. H. Shinohara, H. Sato, M. Ohchochi, Y. Ando, T. Kodama, T. Shida, T. Kato, and Y. Saito, *Nature* **1992**, *357*, 52.

19. C. S. Yannoni, M. Hoinkis, M. S. de Vries, D. S. Bethune, J. R. Salem, M. S. Crowder, and R. D. Johnson, *Science* **1992**, *256*, 1191.

20. E. Gillan, C. Yeretzian, K. S. Min, M. M. Alvarez, R. L. Whetten, and R. B. Kaner, *J. Phys. Chem.* **1992**, *96*, 6869.

21. L. Moro, R. S. Ruoff, C. H. Becker, D. C. Lorents, and R. Malhotra, *J. Phys. Chem.* **1993**, *97*, 6801.

22. Z. Xu, T. Nakane, and H. Shinohara, *J. Am. Chem. Soc.* **1996**, *118*, 11309.

23. T. J. S. Dennis and H. Shinohara, *Chem. Phys. Lett.* **1997**, *278*, 107.

24. T. J. S. Dennis and H. Shinohara, "Production and Isolation of Barium-Based Metallofullerenes" in *Fullerenes: Recent Advances in the Chemistry and Physics of Fullerenes and Related Materials*, Vol. 4, K. Kadish and R. Ruoff (eds.). The Electrochemical Society, Pennington, NJ, 1997, pp. 507–515.

25. T. J. S. Dennis and H. Shinohara, *Appl. Phys. A* **1998**, *66*, 243.

26. T. J. S. Dennis and H. Shinohara, *J. Chem. Soc. Chem. Commun.* **1998**, 883.

27. S. M. Wan, H.-W. Zhang, T. S. C. Tso, K. P. Kwong, T. Wong, R. S. Ruoff, M. Inakuma, and H. Shinohara, "The Isolation of Endohedral Metallofullerenes with Octadecylsilica Columns" in *Fullerenes: Recent Advances in the Chemistry and Physics of Fullerenes and Related Materials*, Vol. 4, K. Kadish and R. Ruoff (eds.). The Electrochemical Society, Pennington, NJ, 1997, pp. 490–506.

28. S. M. Wan, H.-W. Zhang, T. Nakane, Z. Xu, M. Inakuma, H. Shinohara, K. Kobayashi, and S. Nagase, *J. Am. Chem. Soc.* **1998**, *120*, 6806.

29. D. S. Bethune, P. H. M. van Loosdrecht, C.-H. Kiang, R. Beyers, J. R. Salem,

M. S. de Vries, R. D. Johnson, P. Burbank, T. Glass, S. Stevenson, Z. Sun, and H. D. Dorn, *Materials Research Society Fall Meeting Abstract* **1994**, 255.

30. S. Stevenson, P. Burbank, K. Harich, Z. Sun, H. C. Dorn, P. H. M. van Loosdrecht, M. S. de Vries, J. R. Salem, C.-H. Kiang, R. D. Johnson, and D. S. Bethune, *J. Phys. Chem.* **1998**, *102*, 2833.

31. T. Guo, M. D. Diener, Y. Chai, M. J. Alford, R. E. Haufler, S. M. McClure, T. Ohno, J. H. Weaver, G. E. Scuseria, and R. E. Smalley, *Science* **1992**, *257*, 1661.

32. L. S. Wang, J. M. Alford, Y. Chai, M. Diener, J. Zang, S. M. McClure, T. Guo, G. E. Scuseria, and R. E. Smalley, *Chem. Phys. Lett.* **1993**, *207*, 354.

33. L. S. Wang, J. M. Alford, Y. Chai, M. Diener, and R. E. Smalley, *Z. Phys. D.* **1993**, *26*, S297.

34. Y. Wang, D. Tomanek, and R. Ruoff, *Chem. Phys. Lett.* **1993**, *208*, 79.

35. Y. Kubozono, T. Ohta, T. Hayashibara, H. Maeda, H. Ishida, S. Kashino, K. Oshima, H. Yamazaki, S. Ukita, and T. Sogabe, *Chem. Lett.* **1995**, 457.

36. G. E. Scuseria, *J. Chem. Phys.* **1992**, *97*, 7528.

37. T. Guo, G. K. Odom, and G. E. Scuseria, *J. Phys. Chem.* **1994**, *98*, 7745.

38. Z. C. Ying, C. Jin, R. L. Hettich, A. A. Puretzky, R. E. Haufler, and R. N. Compton, "Production and Characterization of Metallofullerene 'Superatoms'" in *Fullerenes: Recent Advances in the Chemistry and Physics of Fullerenes and Related Materials*, Vol. 1, K. Kadish and R. Ruoff (eds.). The Electrochemical Society, Pennington, NJ, 1994, pp. 1402–1412.

39. R. Tellgmann, N. Krawez, S.-H. Lin, I. V. Hertel, and E. E. B. Campbell, *Nature* **1996**, *382*, 407.

40. R. Tellgmann, N. Krawez, I. V. Hertel, and E. E. B. Campbell, *Fullerenes and Fullerene Nanostructures*, H. Kuzmany, J. Fink, M. Mehring, and S. Roth (eds.). World Scientific, London, 1996, pp. 168–172.

41. E. E. B. Campbell, R. Tellgmann, N. Krawez, and I. V. Hertel, *J. Phys. Chem. Solids* **1997**, *58*, 1763.

42. T. A. Murphy, T. Pawlik, A. Weidinger, M. Hoehne, R. Alcala, and J. M. Spaeth, *Phys. Rev. Lett.* **1996**, *77*, 1075.

43. C. Knapp, K.-P. Dinse, B. Pietzak, M. Waiblinger, and A. Weidinger, *Chem. Phys. Lett.* **1997**, *272*, 433.

44. H. Mauser, N. van E. Hommes, T. Clark, A. Hirsch, B. Pietzak, A. Weidinger, and L. Dunsch, *Angew. Chem. Int. Ed.* **1997**, *36*, 2835.

45. N. Krawez, R. Tellgmann, A. Gromov, W. Kraetschmer, and E. E. B. Campbell, in *Molecular Nanostructures*, H. Kuzmany, J. Fink, M. Mehring, and S. Roth (eds.). World Scientific, London, 1998, pp. 184–188.

46. S. Suzuki, D. Kasuya, T. Suganuma, H. Shiromaru, Y. Achiba, and H. Kataura, in *Fullerenes: Recent Advances in the Chemistry and Physics of Fullerenes and Related Materials*, Vol. 4, K. Kadish and R. Ruoff (eds.). The Electrochemical Society, Pennington, NJ, 1997, pp. 485–489.

47. T. Wakabayashi, D. Kasuya, H. Shiromaru, S. Suzuki, K. Kikuchi, and Y. Achiba, *Z. Phys. D* **1997**, *40*, 414.

48. R. F. Curl and R. E. Smalley, *Sci. Am.* **1991**, *265*, 54.

49. T. Wakabayashi and Y. Achiba, *Chem. Phys. Lett.* **1992**, *190*, 465.

50. R. E. Smally, *Acc. Chem. Res.* **1992**, *25*, 98.

51. T. Wakabayashi, K. Kikuchi, H. Shiromaru, S. Suzuki, and Y. Achiba, *Z. Phys. D* **1993**, *26*, S258.

52. A. Thess, R. Lee, P. Nikolaev, H. Dai, P. Petit, J. Robert, C. Xu, Y. H. Lee, S. G. Kim, A. G. Rinzler, D. T. Cobert, G. E. Scuseria, D. Tomanek, J. E. Fisher, and R. E. Smalley, *Science* **1996**, *273*, 483.

53. H. Shinohara, M. Takata, M. Sakata, T. Hashizume, and T. Sakurai, *Mater. Sci. Forum* **1996**, *232*, 207.

54. T. Nakane, Z. Xu, E. Yamamoto, T. Sugai, T. Tomiyama, and H. Shinohara, *Fullerene Sci. Technol.* **1997**, *5*, 829.

55. H. Shinohara, in *Advances in Metal and Semiconductor Clusters*, Vol. 4, M. Duncan (ed.). JAI Press, New York, 1998, pp. 205–226.

56. S. Bandow, H. Shinohara, Y. Saito, M. Ohkohchi, and Y. Ando, *J. Phys. Chem.* **1993**, *97*, 6101.

57. H. Shinohara, M. Inakuma, N. Hayashi, H. Sato, Y. Saito, T. Kato, and S. Bandow, *J. Phys. Chem.* **1994**, *98*, 8597.

58. G. Adachi, N. Imanaka, and Z. Fuzhong, "Chap. 99, Rare Earth Carbides" in *Handbook on the Physics & Chemistry of Rare Earths*, Vol. 15, K. A. Gschneider, Jr. and L. Eyring (eds.). Elsevier Amsterdam, 1991, p. 61.

59. K. Kuroki, T. Kuriyama, M. Inakuma, and H. Shinohara, unpublished results.

60. H. Shinohara, H. Yamaguchi, N. Hayashi, H. Sato, M. Ohkohchi, Y. Ando, and Y. Saito, *J. Phys. Chem.* **1993**, *97*, 4259.

61. M. Inakuma, M. Ohno, and H. Shinohara, "ESR Studies of Endohedral Mono-Metallofullerenes" in *Fullerenes: Recent Advances in the Chemistry and Physics of Fullerenes and Related Materials*, Vol. 2, K. Kadish and R. Ruoff (eds.). The Electrochemical Society, Pennington, NJ, 1995, pp. 330–342.

62. K. Yamamoto, H. Funasaka, T. Takahashi, and T. Akasaka, *J. Phys. Chem.* **1994**, *98*, 2008.

63. M. D. Diener, C. A. Smith, and D. K. Veirs, *Chem. Mater.* **1997**, *9*, 1773.

64. R. Taylor, J. P. Hare, A. K. Abdul-Sada, and H. W. Kroto, *J. Chem. Soc. Chem. Commun.* **1990**, 1423.

65. J. Abrefah, D. R. Olander, M. Balooch, and W. J. Siekhaus, *Appl. Phys. Lett.* **1992**, *60*, 1313.

66. C. Pan, M. P. Sampson, Y. Chai, R. H. Hauge, and J. L. Margrave, *J. Phys. Chem.* **1991**, *95*, 2944.

67. R. D. Averitt, J. M. Alford, and N. J. Halas, *Appl. Phys. Lett.* **1994**, *65*, 374.

68. H. Ajie, M. M. Alvarez, S. J. Anz, R. D. Beck, F. Diederich, K. Fostiropoulos, D. R. Hufman, W. Kraetschmer, Y. Rubin, K. E. Schriver, D. Sensharma, and R. L. Whetten, *J. Phys. Chem.* **1990**, *94*, 8630.

69. W. A. Scrivens, P. V. Bedworth, and M. J. Tour, *J. Am. Chem. Soc.* **1992**, *114*, 7917.

70. K. Kikuchi, N. Nakahara, M. Honda, S. Suzuki, K. Saito, H. Shiromaru, K. Yamauchi, I. Ikemoto, T. Kuromachi, S. Hino, and Y. Achiba, *Chem. Lett.* **1991**, 1607.

71. K. Kikuchi, N. Nakahara, T. Wakabayashi, M. Honda, H. Matsumiya, T. Moriwaki, S. Suzuki, H. Shiromaru, K. Saito, K. Yamauchi, I. Ikemoto, and Y. Achiba, *Chem. Phys. Lett.* **1992**, *188*, 177.

72. R. C. Klute, H. C. Dorn, and H. M. McNair, *J. Chromatogr. Sci.* **1992**, *30*, 438.

73. M. S. Meier and J. P. Selegue, *J. Org. Chem.* **1992**, *57*, 1924.

74. K. Jinno and Y. Saito, *Adv. Chromatogr.* **1996**, *36*, 65.

75. K. Kikuchi, S. Suzuki, Y. Nakao, N. Nakahara, T. Wakabayashi, H. Shiromaru, I. Saito, I. Ikemoto, and Y. Achiba, *Chem. Phys. Lett.* **1993**, *216*, 67.

76. M. Savina, G. Martin, J. Xiao, N. Milanovich, M. E. Meyerhoff, and A. H. Francis, "Chromatographic Isolation and EPR Characterization of La@C_{82}" in *Fullerenes: Recent Advances in the Chemistry and Physics of Fullerenes and Related Materials*, Vol. 1, K. Kadish and R. Ruoff (eds.). The Electrochemical Society, Pennington, NJ, 1994, pp. 1309–1319.

77. S. Stevenson, H. C. Dorn, P. Burbank, K. Harich, J. Haynes, C. H. Klang, J. R. Salem, M. S. de Vries, P. H. M. van Loosdrecht, R. D. Johnson, C. S. Yannoni, and D. S. Bethune, *Anal. Chem.* **1994**, *66*, 2675.

78. S. Stevenson, H. C. Dorn, P. Burbank, K. Harich, Z. Sun, C. H. Klang, J. R. Salem, M. S. de Vries, P. H. M. van Loosdrecht, R. D. Johnson, C. S. Yannoni, and D. S. Bethune, *Anal. Chem.* **1994**, *66*, 2680.

79. T. Wakabayashi, H. Shiromaru, S. Suzuki, K. Kikuchi, and Y. Achiba, *Surf. Rev. Lett.* **1996**, *3*, 793.

80. S. Suzuki, Y. Kojima, H. Shiromaru, Y. Achiba, T. Walabayashi, R. Tellgmann, E. E. B. Campbell, and I. V. Hertel, *Z. Phys. D* **1997**, *40*, 410.

81. D. C. Lorents, D. H. Yu, C. Brink, N. Jensen, and P. Hvelplund, *Chem. Phys. Lett.* **1995**, *236*, 141.

82. R. D. Beck, P. Weis, J. Rockenberger, R. Michel, D. Fuchs, M. Benz, and M. M. Kappes, *Surf. Sci. Lett.* **1996**, *3*, 881.

83. R. D. Beck, P. Weis, J. Rockenberger, and M. M. Kappes, *Surf. Sci. Lett.* **1996**, *3*, 771.

84. Y. Huang and B. S. Freiser, *J. Am. Chem. Soc.* **1991**, *113*, 9418.

85. L. M. Roth, Y. Huang, J. T. Schwedler, C. J. Cassady, D. Ben-Amotz, B. Kahr, and B. S. Freiser, *J. Am. Chem. Soc.* **1991**, *113*, 6298.

86. L. Soderholm, P. Wurz, K. R. Lykke, D. H. Parker, and F. W. Lytle, *J. Phys. Chem.* **1992**, *96*, 7153.

87. C. H. Park, B. O. Wells, J. Dicarlo, Z. X. Shen, J. R. Salem, D. S. Bethune, C. S. Yannoni, R. D. Johnson, M. S. de Vries, C. Booth, F. Bridges, and P. Pianetta, *Chem. Phys. Lett.* **1993**, *213*, 196.

88. K. Kikuchi, Y. Nakao, Y. Achiba, and M. Nomura, "Endohedral Metallofullerenes, LaC_{82} and La$_2$$C_{80}$" in *Fullerenes: Recent Advances in the Chemistry and Physics of Fullerenes and Related Materials*, Vol. 1, K. Kadish and R. Ruoff (eds.). The Electrochemical Society, Pennington, NJ, 1994, pp. 1300–1308.

89. R. Beyers, C.-H. Kiang, R. D. Johnson, J. R. Salem, M. S. de Vries, C. S. Yannoni, D. S. Bethune, H. C. Dorn, P. Burbank, K. Harich, and S. Stevenson, *Nature* **1994**, *370*, 196.

91. H. Shinohara, N. Hayashi, H. Sato, Y. Saito, X. D. Wang, T. Hashizume, and T. Sakurai, *J. Phys. Chem.* **1993**, *97*, 13438.

92. T. Sakurai, X. D. Wang, Q. K. Xue, Y. Hasegawa, T. Hashizume, and H. Shinohara, *Prog. Surf. Sci.* **1996**, *51*, 263.

93. J. K. Gimzewski, "Scanning Tunneling and Local Probe Studies of Fullerenes" in *The Chemical Physics of Fullerenes 10 (and 5) Years Later*, Vol. 316, W. Andreoni (ed.). Kluwer, Dordrecht, 1996, pp. 117–136.

94. M. Takada, B. Umeda, E. Nishibori, M. Sakata, Y. Saito, M. Ohno, and H. Shinohara, *Nature* **1995**, *377*, 46.

95. A. Rosen and B. Waestberg, *J. Am. Chem. Soc.* **1988**, *110*, 8701.

96. A. Rosen and B. Waestberg, *Z. Phys. D* **1989**, *12*, 387.

97. J. Cioslowski and E. D. Fleishcmann, *J. Chem. Phys.* **1991**, *94*, 3730.

98. A. H. H. Chang, W. C. Ermler, and R. M. Pitzer, *J. Chem. Phys.* **1991**, *94*, 5004.

99. D. E. Manolopoulos and P. W. Fowler, *Chem. Phys. Lett.* **1991**, *187*, 1.

100. D. E. Manolopoulos and P. W. Fowler, *J. Chem. Soc. Faraday Trans.* **1992**, *88*, 1225.

101. K. Laasonen, W. Andreoni, and M. Parrinello, *Science* **1992**, *258*, 1916.

102. S. Saito and S. Sawada, *Chem. Phys. Lett.* **1992**, *198*, 466.

103. W. Andreoni and A. Curioni, *Phys. Rev. Lett.* **1996**, *77*, 834.

104. W. Andreoni and A. Curioni, "Theory of (La, Y) Endohedrally Doped Fullerenes: The Past and the Present" in *The Chemical Physics of Fullerenes 10 (and 5) Years Later*, W. Andreoni (ed.). Kluwer, Dordrecht, 1996, pp. 183–196.

105. S. Nagase and K. Kobayashi, *Chem. Phys. Lett.* **1993**, *214*, 57.

106. H. M. Rietveld, *J. Appl. Crystallogr.* **1968**, *2*, 65.

107. D. M. Collins, *Nature* **1982**, *298*, 49.

108. G. Bricogne, *Acta Crystallogr. Sec. A* **1988**, *44*, 517.

109. M. Sakata and M. Sato, *Acta Crystallogr. Sec. A* **1990**, *46*, 263.

110. S. Kumazawa, Y. Kubota, M. Tanaka, M. Sakata, and Y. Ishibashi, *J. Appl. Crystallogr.* **1993**, *26*, 453.

111. S. Nagase and K. Kobayashi, *Chem. Phys. Lett.* **1994**, *228*, 106.

112. J. Schulte, M. C. Boehm, and K. P. Dinse, *Chem. Phys. Lett.* **1996**, *259*, 48.

113. M. Takada, E. Nishibori, B. Umeda, M. Sakata, M. Inakuma, and H. Shinohara, *Chem. Phys. Lett.* **1998**, *298*, 79.

114. M. Ruebsam, P. Schweitzer, and K. P. Dinse, "EPR-Investigation of Metallo-Endofullerenes" in *Fullerenes and Fullerene Nanostructures*, H. Kuzmany, J. Fink, M. Mehring, and S. Roth (eds.). World Scientific, London, 1996, pp. 173–177.

115. M. Ruebsam, C. P. Knapp, P. Schweitzer, and K. P. Dinse, "EPR-Investigation of Metallo-Endofullerenes in Solution" in *Fullerenes: Recent Advances in the Chemistry and Physics of Fullerenes and Related Materials*, Vol. 3, K. Kadish and R. Ruoff (eds.). The Electrochemical Society, Pennington, NJ, 1996, pp. 602–608.

116. J. Schulte, M. C. Boehm, and K. P. Dinse, "Electronic Structure of Sc@C_{82}" in *Molecular Nanostructures*, H. Kuzmany J. Fink, M. Mehring, and S. Roth (eds.). World Scientific, London, 1998, pp. 189–192.

117. H. Suematsu, Y. Murakami, H. Kawata, Y. Fujii, N. Hamaya, K. Kikuchi, Y. Achiba, and I. Ikemoto, *Mat. Res. Soc. Symp. Proc.* **1994**, *349*, 213.

118. T. Watanuki, A. Fujiwara, K. Ishii, H. Suematsu, H. Nakao, Y. Fujii, H. Kawada, Y. Murakami, Y. Nakao, K. Kikuchi, Y. Achiba, and Y. Maniwa, *Photon Factory Report* **1995**, *13*, 333.

119. T. Watanuki, A. Fujiwara, K. Ishii, T. Shibata, H. Suematsu, H. Nakao, Y. Fujii, H. Kawada, Y. Murakami, K. Kikuchi, Y. Achiba, and Y. Maniwa, *Photon Factory Report* **1996**, *14*, 403.

120. E. Nishibori, M. Sakata, M. Takata, M. Hasegawa, and H. Shinohara, *Chem. Phys. Lett.* submitted, 2000.

121. W. Andreoni and A. Curioni, "Structure and Dynamics of Metallofullerenes: New Insights" in *Fullerenes: Recent Advances in the Chemistry and Physics of Fullerenes and Related Materials*, Vol. 4, K. Kadish and R. Ruoff (eds.). The Electrochemical Society, Pennington, NJ, 1997, pp. 516–522.

122. W. Andreoni and A. Curioni, *Appl. Phys. A* **1998**, *66*, 299.

123. Y. Miyake, S. Suzuki, Y. Kojima, K. Kikuchi, K. Kobayashi, S. Nagase, M. Kainosho, Y. Achiba, Y. Maniwa, and K. Fisher, *J. Phys. Chem.* **1996**, *100*, 9579.

124. T. Akasaka, S. Nagase, K. Kobayashi, T. Suzuki, T. Kato, K. Kikuchi, Y. Achiba, K. Yamamoto, H. Funasaka, and T. Takahashi, *Angew. Chem. Int. Ed. Engl.* **1995**, *34*, 2139.

125. T. Akasaka, S. Nagase, K. Kobayashi, M. Waelchli, K. Yamamoto, H. Funasaka, M. Kako, T. Hoshino, and T. Erata, *Angew. Chem. Int. Ed. Engl.* **1997**, *36*, 1643.

126. F. H. Hennrich, R. H. Michel, A. Fisher, S. Richard-Schneider, S. Gilb, M. M. Kappes, D. Fuchs, M. Buerk, K. Kobayashi, and S. Nagase, *Angew. Chem. Int. Ed. Engl.* **1996**, *35*, 1732.

127. K. Kobayashi, S. Nagase, and T. Akasaka, *Chem. Phys. Lett.* **1995**, *245*, 230.

128. S. Nagase and K. Kobayashi, *Chem. Phys. Lett.* **1994**, *231*, 319.

129. W. Sato, K. Sueki, K. Kikuchi, K. Kobayashi, S. Suzuki, Y. Achiba, H. Nakahara, Y. Ohkubo, F. Ambe, and K. Asai, *Phys. Rev. Lett.* **1998**, *80*, 133.

130. T. Suzuki, Y. Maruyama, T. Kato, K. Kikuchi, Y. Nakao, Y. Achiba, K. Kobayashi, and S. Nagase, *Angew. Chem.* **1995**, *107*, 1228.

131. T. Suzuki, Y. Maruyama, T. Kato, K. Kikuchi, Y. Nakao, Y. Achiba, K. Kobayashi, and S. Nagase, *Angew. Chem. Int. Ed. Engl.* **1995**, *34*, 1094.

132. T. Akasaka, T. Kato, K. Kobayashi, S. Nagase, K. Yamamoto, H. Funasaka, and T. Takahashi, *Nature* **1995**, *374*, 600.

133. P. H. M. van Loosdrecht, R. D. Johnson, D. S. Bethune, H. C. Dorn, P. Burbank, and S. Stevenson, *Phys. Rev. Lett.* **1994**, *73*, 3415.

134. D. S. Bethune, "Adding Metal to Carbon: Production and Properties of Metallofullerenes and Single-Layer Nanotubes" in *The Chemical Physics of Fullerenes 10 (and 5) Years Later*, W. Andreoni (ed.). Kluwer, Dordrecht, 1996, pp. 165–181.

135. X. D. Wang, T. Hashizume, Q. Xue, H. Shinohara, Y. Saito, Y. Nishina, and T. Sakurai, *Chem. Phys. Lett.* **1993**, *216*, 409.

136. T. Takahashi, A. Ito, M. Inakuma and H. Shinohara, *Phys. Rev. B* **1995**, *52*, 13812.

137. M. Takada, E. Nishibori, B. Umeda, M. Sakata, E. Yamamoto, and H. Shinohara, *Phys. Rev. Lett.* **1997**, *78*, 3330.

138. T. Kato, S. Bandow, M. Inakuma, and H. Shinohara, *J. Phys. Chem.* **1995**, *99*, 856.

139. M. R. Anderson, H. C. Dorn, S. Stevenson, P. M. Burbank, and J. R. Gibson, *J. Am. Chem. Soc.* **1997**, *119*, 437.

140. M. R. Anderson, H. C. Dorn, P. M. Burbank, and J. R. Gibson, "Voltammetric Studies of Mn@C_{82}" in *Fullerenes: Recent Advances in the Chemistry and Physics of Fullerenes and Related Materials*, Vol. 4, K. Kadish and R. Ruoff (eds.). The Electrochemical Society, Pennington, NJ, 1997, pp. 448–456.

141. H. C. Dorn, S. Stevenson, P. Burbank, Z. Sun, T. Glass, K. Harich, P. H. M. van Loosdrecht, R. D. Johnson, R. Beyers, J. R. Salem, M. S. de Vries, C. S. Yannoni, D. S. Bethune, and C. H. Kiang, "Endohedral Metallofullerenes: Isolation and Characterization" in *Science and Technology of Fullerenes*, P. Bernier, D. S. Bethune, L. Y. Chiang, T. W. Ebbesen, R. M. Metzger, and J. W. Mintmire (eds.). Materials Research Society, Pittsburgh, 1995, pp. 123–135.

142. R. M. Macfarlane, G. Wittmann, P. H. M. van Loosdrecht, M. S. de Vries, D. S. Bethune, S. Stevenson, and H. C. Dorn, *Phys. Rev. Lett.* **1997**, *79*, 1397.

143. X. Ding, J. M. Alford, and J. C. Wright, *Chem. Phys. Lett.* **1997**, *269*, 72.

144. E. Yamamoto, M. Tansho, T. Tomiyama, H. Shinohara, H. Kawahara and Y. Kobayashi, *J. Am. Chem. Soc.* **1996**, *118*, 2293.

145. S. Nagase and K. Kobayashi, *Chem. Phys. Lett.* **1997**, *276*, 55.

146. S. Nagase, K. Kobayashi, and T. Akasaka, *Bull. Chem. Soc. Jpn.* **1996**, *69*, 2131.

147. D. E. Manolopoulos and P. W. Fowler, *J. Chem. Phys.* **1992**, *96*, 7603.

148. K. Kikuchi, N. Nakahara, T. Wakabayashi, S. Suzuki, H. Shiromaru, Y. Miyake, K. Saito, I. Ikemoto, M. Kainosho, and Y. Achiba, *Nature* **1992**, *357*, 142.

149. H. Kawata, Y. Fujii, H. Nakao, Y. Murakami, T. Watanuki, H. Suematsu, K. Kikuchi, Y. Achiba, and I. Ikemoto, *Phys. Rev. B* **1995**, *51*, 8723.

150. P. M. Allemand, G. Srdanov, A. Koch, K. Khemani, F. Wudl, Y. Rubin, F. Diederich, M. M. Alvarez, S. J. Anz, and R. L. Whetten, *J. Am. Chem. Soc.* **1991**, *113*, 2780.

151. P. J. Krusic, E. Wasserman, B. A. Parkinson, B. Malone, and E. R. Holler, *J. Am. Chem. Soc.* **1991**, *113*, 6274.

152. O. Pilla and H. Bill, *J. Phys. C Solid State Phys.* **1984**, *17*, 3263.

153. S. Hino, H. Takahashi, K. Iwasaki, K. Matsumoto, T. Miyazaki, S. Hasegawa, K. Kikuchi, and Y. Achiba, *Phys. Rev. Lett.* **1993**, *71*, 4261.

154. M. D. Poirier, M. Knupfer, J. H. Weaver, W. Andreoni, K. Laasonen, M. Parrinello, D. S. Bethune, K. Kikuchi, and Y. Achiba, *Phys. Rev. B* **1994**, *49*, 17403.

155. S. Hino, K. Umishita, K. Iwasaki, T. Miyamae, M. Inakuma, and H. Shinohara, *Chem. Phys. Lett.* **1999**, *300*, 145.

156. S. Nagase and K. Kobayashi, *J. Chem. Soc. Chem. Commun.* **1994**, 1837.

157. T. Kato, S. Suzuki, K. Kichuchi, and Y. Achiba, *J. Phys. Chem.* **1993**, *97*, 13425.

158. T. Kato, T. Suzuki, K. Yamamoto, H. Funasaka, and T. Takahashi, "An ESR Detection of the Spin Multiplet of Gd@C_{82}" in *Fullerenes: Recent Advances in the Chemistry and Physics of Fullerenes and Related Materials*, Vol. 2, K. Kadish and R. Ruoff (eds.). The Electrochemical Society, Pennington, NJ, 1995, pp. 733–739.

159. T. Kato, T. Akasaka, K. Kobayashi, S. Nagase, K. Yamamoto, H. Funasaka, and T. Takahashi, *Appl. Magn. Reson.* **1996**, *11*, 293.

160. M. Ruebsam, M. Plueschau, P. Schweitzer, K. P. Dinse, D. Fuchs, H. Rietschel, R. H. Michel, M. Benz, and M. M. Kappes, "2D EPR Investigation of Endohedral La@C_{82} in Solution" in *Physics and Chemistry of Fullerenes and Derivatives*, H. Kuzmany, J. Fink, M. Mehring, and S. Roth (eds.). World Scientific, London, 1995, pp. 117–121.

161. A. Bartl, L. Dunsch, J. Froehner, and U. Kirbach, *Chem. Phys. Lett.* **1994**, *229*, 115.

162. A. Bartl, L. Dunsch, U. Kirbach, and G. Seifert, *Synth. Met.* **1997**, *86*, 2395.

163. G. Seifert, A. Bartl, L. Dunsch, A. Ayuela, and A. Rockenbauer, *Appl. Phys. A* **1998**, *66*, 265.

164. A. Bartl, L. Dunsch, U. Kirbach, and B. Schandert, *Synth. Met.* **1995**, *70*, 1365.

165. A. Bartl, L. Dunsch, and U. Kirbach, *Solid State Commun.* **1995**, *94*, 827.

166. A. Bartl, L. Dunsch, and U. Kirbach, *Appl. Magn. Reson.* **1996**, *11*, 301.

167. S. Knorr, A. Grupp, M. Mehring, U. Kirbach, A. Bartl, and L. Dunsch, *Appl. Phys. A* **1998**, *66*, 257.

168. T. Suzuki, Y. Maruyama, T. Kato, K. Kikuchi, and K. Achiba, *J. Am. Chem. Soc.* **1993**, *115*, 11006.

169. T. Suzuki, K. Kikuchi, F. Oguri, Y. Nakao, S. Suzuki, Y. Achiba, K. Yamamoto, H. Funasaka, and T. Takahashi, *Tetrahedron* **1996**, *52*, 4973.

170. W. Wang, J. Ding, S. Yang, and X. Y. Li, "Electrochemical Properties of 4F-Block Metallofullerenes" in *Fullerenes: Recent Advances in the Chemistry and Physics of Fullerenes and Related Materials*, Vol. 4, K. Kadish and R. Ruoff (eds.). The Electrochemical Society, Pennington, NJ, 1997, pp. 417–428.

171. L. Dunsch, A. Bartl, U. Kirbach, and J. Froehner, "ESR-Spectroscopic Studies of Endohedral Fullerenes Interaction in Solution" in *Fullerenes: Recent Advances in the Chemistry and Physics of Fullerenes and Related Materials*, Vol. 2, K. Kadish and R. Ruoff (eds.). The Electrochemical Society, Pennington, NJ, 1995, pp. 182–190.

172. L. Dunsch, P. Kuran, U. Kirbach, and D. Scheller, "The Isomerization of Endohedral Fullerenes and Its Influence on the Electronic Properties of These Structures" in *Fullerenes: Recent Advances in the Chemistry and Physics of Fullerenes and Related Materials*, Vol. 4, K. Kadish and R. Ruoff (eds.). The Electrochemical Society, Pennington, NJ, 1997, pp. 523–536.

173. A. Petr, L. Dunsch, and A. Neudeck, *J. Electroanal. Chem.* **1996**, *412*, 153.

174. T. Pichler, M. S. Golden, M. Knupfer, J. Fink, U. Kirbach, P. Kuran, and L. Dunsch, *Phys. Rev. Lett.* **1997**, *79,*, 3026.

175. T. Pichler, M. Knupfer, M. S. Golden, T. Boeske, J. Fink, U. Kirbach, P. Kuran, L. Dunsch, and Ch. Jung, *Appl. Phys. A* **1998**, *66*, 281.

176. S. Hino, K. Umishita, K. Iwasaki, T. Miyazaki, T. Miyamae, K. Kikuchi, and Y. Achiba, *Chem. Phys. Lett.* **1997**, *281*, 115.

177. T. Pichler, H. Kuzmany, E. Yamamoto, and H. Shinohara, "Vibrational Structure of Sc_2@C_{84} Analysed by IR Spectroscopy" in *Fullerenes and Fullerene Nanostructures*, H. Kuzmany, J. Fink, M. Mehring, and S. Roth (eds.). World Scientific, London, 1996, pp. 178–181.

178. M. Hulman, T. Pichler, H. Kuzmany, F. Zerbetto, E. Yamamoto, and H. Shinohara, *J. Mol. Struct.* **1997**, *408/409*, 359.

179. S. Lebedkin, B. Renker, R. Heid, H. Rietschel, and H. Schober, "The Vibrational Spectroscopy of M@C_{82} (M = La, Y): Raman, Far-Infrared, and Neutron Inelastic Scattering Results" in *Molecular Nanostructures*, H. Kuzmany, J. Fink, M. Mehring, and S. Roth (eds.). World Scientific, London, 1998, pp. 203–206.

180. S. Lebedkin, B. Renker, R. Heid, H. Schober, and H. Rietschel, *Appl. Phys. A* **1998**, *66*, 273.

181. M. Buerk, M. Schmidt, T. R. Cummins, N. Nuecker, J. F. Ambruster, S. Schuppler, D. Fuchs, P. Adelmann, R. H. Michel, and M. M. Kappes, "Doping of Higher Fullerenes: New Aspects" in *Fullerenes and Fullerene Nanostructures*, H. Kuzmany, J. Fink, M. Mehring, and S. Roth (eds.). World Scientific, London, 1996, pp. 196–199.

181. L. Dunsch, P. Kuran, and M. Krause, "Vibrational Characterization of Three Tm@C_{82} Isomers" in *Fullerenes: Recent Advances in the Chemistry and Physics of Fullerenes and Related Materials*, Vol. 6, K. Kadish and R. Ruoff (eds.). The Electrochemical Society, Pennington, NJ, 1998, pp. 1031–1038.

182. L. Dunsch, D. Eckert, J. Froehner, A. Bartl, P. Kuran, M. Wolf, and K. H. Mueller, "Magnetic Properties of Endo- and Exohedral Rare—Earth Fullerenes" in *Fullerenes: Recent Advances in the Chemistry and Physics of Fullerenes and Related Materials*, Vol. 6, K. Kadish and R. Ruoff (eds.). The Electrochemical Society, Pennington, NJ, 1998, pp. 955–966.

183. U. Kirbach and L. Dunsch, *Angew. Chem. Int. Ed. Engl.* **1996**, *35*, 2380.

184. K. Kobayashi and S. Nagase, *Chem. Phys. Lett.* **1998**, *282*, 325.

185. T. Akasaka, S. Nagase, K. Kobayashi, T. Suzuki, T. Kato, K. Yamamoto, H. Funasaka, and T. Takahashi, *J. Chem. Soc. Chem. Commun.* **1995**, 1343.

186. T. Ogawa, T. Sugai, and H. Shinohara, *J. Am. Chem. Soc.* **2000**, *122*, 3538.

187. T. Inoue et al. *Chem. Phys. Lett.* **2000**, *316*, 381.

CHAPTER 9

ENDOHEDRAL METALLOFULLERENES: THEORY, ELECTROCHEMISTRY, AND CHEMICAL REACTIONS

SHIGERU NAGASE, KAORU KOBAYASHI, TAKESHI AKASAKA, and TAKATSUGU WAKAHARA

9.1 INTRODUCTION

Since the first proposal in 1985 [1] immediately after the discovery of the special stability of C_{60} [2], encapsulation of one or more metal atoms inside hollow fullerene cages (endohedral metallofullerenes) has long attracted special attention because it could lead to new spherical molecules with novel properties unexpected for empty fullerenes. In 1991 Smalley and co-workers showed for the first time that several lanthanum-containing fullerenes can be produced by laser vaporization of graphite rods impregnated with the appropriate metal oxides, and extraction with toluene yields mostly $La@C_{82}$ as a stable compound (the symbol @ denotes that La is located inside C_{82}) [3]. Triggered by this success, great efforts have been made for the production and characterization of endohedral metallofullerenes. Up to now it has been demonstrated that group 2 (M = Ca, Sr, Ba) and group 3 metals (M = Sc, Y, La), and most lanthanide metals can be trapped inside the higher fullerenes such as C_{80}, C_{82}, and C_{84} (much less abundant than C_{60} and C_{70}) to form soluble and relatively stable endohedral metallofullerenes ($M_m@C_n$) [4].

Because of the difficulty in producing pure samples in large quantities, the experimental characterization of endohedral metallofullerenes has been hindered. Recent important progress is marked by the successful isolation and purification of metallofullerenes such as $Sc@C_{82}$ [5], $Y@C_{82}$ [6], $La@C_{82}$ [7], $Gd@C_{82}$ [8], $La_2@C_{80}$ [9], $Sc_2@C_{84}$ [10], and $Sc_3@C_{82}$ [11] in macroscopic quantities. This has made it possible to investigate the interesting electronic properties and chemical reactivities [4]. As shown in recent reviews [12], the

Fullerenes: Chemistry, Physics, and Technology, Edited by Karl M. Kadish and Rodney S. Ruoff.
0-471-29089-0 Copyright © 2000 John Wiley & Sons, Inc.

determination of cage structures and metal positions is currently of central interest since it is an important clue in disclosing the formation mechanism and developing new routes to bulk production of endohedral metallofullerenes. It has generally been accepted that extractable metallofullerenes take on endohedral structures. However, a definitive structure proof has to be performed for each of the metallofullerenes. It is also of current interest whether metal atoms are rigidly attached to fullerene cages or move about freely.

There are a huge number of isomers for fullerenes consisting of pentagonal and hexagonal carbon rings. Only the isomers that satisfy the so-called isolated pentagon rule (IPR) [13] have been isolated and experimentally characterized [14]. Therefore, it has often been assumed that metals are encapsulated inside the most stable or abundant fullerene isomers. However, it is found that the cage structures of metallofullerenes do not necessarily coincide with those of the most stable fullerene isomers. In addition, unconventional cage structures are significantly stabilized by endohedral metal-doping, which violate the isolated pentagon rule or contain heptagonal rings. These enrich the chemistry of endohedral metallofullerenes.

In this chapter, recent progress is summarized in the theoretical and experimental studies of endohedral monometallofullerenes and dimetallofullerenes. Although emphasis may mainly be put on our recent studies, the present purposes are to disclose (1) how metals are trapped inside fullerene cages, (2) what the cage structures are, (3) what the electronic states of metals and cages are, (4) how the electronic properties and chemical reactivities of empty fullerenes change upon endohedral metal-doping, and (5) how the electronic properties of endohedral metallofullerenes change upon substitutional doping. It is hoped that these will be of great help in understanding the growth mechanism and developing new routes to bulk production as well as in the applications of endohedral metallofullerenes in several fields as new molecules with novel properties.

9.2   MONOMETALLOFULLERENES (M@C_{82}, M = Ca, Sc, Y, La, AND LANTHANIDES)

Almost without exception, it has been observed that M@C_n is most abundantly extracted when $n = 82$ (a special magic number) as a representative and stable monometallofullerene.

9.2.1   Cage Structures and Metal Positions

For the C_{82} fullerene, there are nine distinct isomers that satisfy the isolated pentagon rule [15], as shown in Figure 9.1. The measurement of the ^{13}C NMR spectrum showed that only three isomers with C_2, C_{2v}, and C_{3v} symmetries were produced in a ratio of 8:1:1 [16]. According to theoretical calculations, the $C_2(a)$ isomer is most stable [17]. However, it was found that the isomers with C_{2v} and C_{3v} symmetries do not correspond to energy minima, but distort to

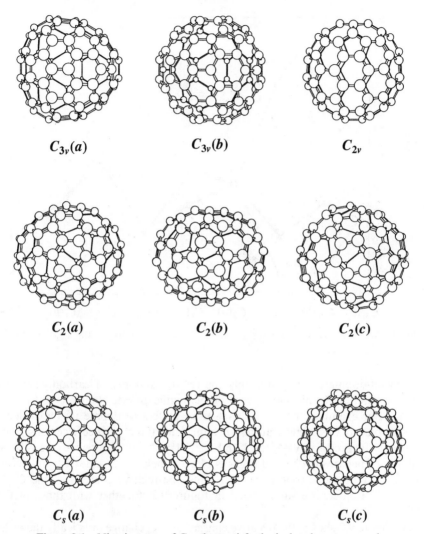

Figure 9.1 Nine isomers of C_{82} that satisfy the isolated pentagon rule.

have lower C_2 and C_s symmetries, respectively, because of energetic reasons and Jahn-Teller distortion. This contradicted the assignment of the observed NMR lines. However, a recent reinvestigation of the NMR lines with a newly purified C_{82} sample shows that one isomer with C_2 symmetry is produced while the two minor species are supposed to be other fullerenes [18]. The most abundant production of the $C_2(a)$ isomer is also confirmed by calculations with temperature effects [19].

It has long been an open question in which isomers of C_{82} a metal atom (M) is encapsulated. It was often assumed that M was encapsulated inside the most stable $C_2(a)$ isomer [17,20–23]. However, it is important to note that endohe-

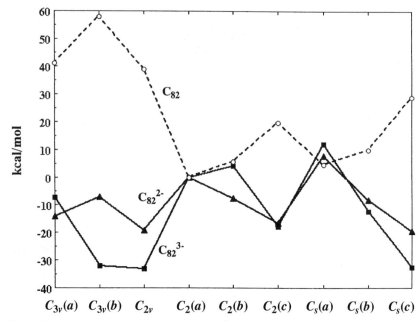

Figure 9.2 Energies of C_{82}, $C_{82}{}^{2-}$, and $C_{82}{}^{3-}$ isomers relative to the $C_2(a)$ isomer.

dral metal-doping changes not only the relative stability of carbon cages but also the inner electronic states, because of significant electron transfer from M to C_{82}. For example, the Sc $(3d^1 4s^2)$ atom prefers to donate its two valence 4s electrons to C_{82} while transfer of almost all three valence electrons is increasingly favored for Y $(4d^1 5s^2)$ and La $(5d^1 6s^2)$ [20]. Therefore, the relative stability of negatively charged cages is of great help as an index for predicting the most favorable carbon cage [12]. Thus, the relative energies of the $C_{82}{}^{2-}$ and $C_{82}{}^{3-}$ isomers are summarized in Figure 9.2, together with those of C_{82} [24].

As Figure 9.2 shows, the negative charges on C_{82} change drastically the order of the relative stability of the cage isomers. The $C_2(a)$ cage is most stable for C_{82}. However, it is highly destabilized upon accepting electrons and becomes the second and third most unstable for $C_{82}{}^{2-}$ and $C_{82}{}^{3-}$, respectively. Instead, the C_{2v}, $C_s(c)$, $C_2(c)$, and $C_{3v}(a)$ cages (which are highly unstable for C_{82}) become stable for $C_{82}{}^{2-}$, while the C_{2v}, $C_{3v}(b)$, and $C_s(c)$ cages are highly stabilized for $C_{82}{}^{3-}$. In this context, it is interesting that two or three isomers are extractable for M@C_{82} [25–27]. The C_{2v} cage is most stable for both $C_{82}{}^{2-}$ and $C_{82}{}^{3-}$.

The structures of La@C_{82} optimized by placing La inside the C_{2v}, $C_{3v}(b)$, and $C_s(c)$ cages (most stable for $C_{82}{}^{3-}$) are shown as **1–3** in Figure 9.3, respectively [24]. For comparison, the structure optimized with the $C_2(a)$ cage (most stable for C_{82}) is also shown as **4** in Figure 9.3. Endohedral structures **1–**

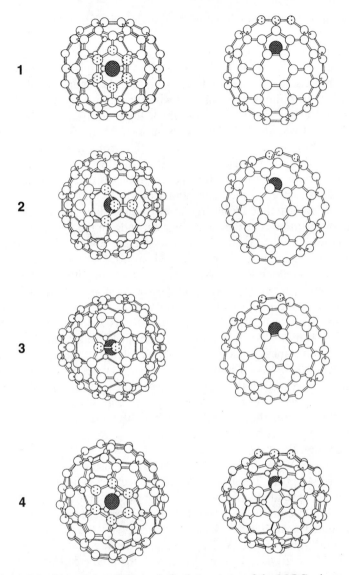

Figure 9.3   Two views of the optimized structures of the M@C$_{82}$ isomers.

4 have C_{2v}, C_{3v}, C_s, and C_2 symmetry, respectively. In all these cases, almost three valence electrons on La are transferred onto C$_{82}$; the electronic structures are formally described as La^{3+}C$_{82}{}^{3-}$, despite significant back electron transfer from C$_{82}$ to the d orbitals on La. As Figure 9.3 shows, the La atom is located at the off-center position of C$_{82}$, and the favorable positions differ significantly for **1–4**. As will be discussed in Section 9.4, the La atom is located near the minimum of the electrostatic potential of the C$_{82}{}^{3-}$ cages; electrostatic interactions play a dominant role both in stabilizing endohedral structures and in determining metal positions.

As expected from the relative energies of the $C_{82}{}^{3-}$ cages in Figure 9.2, **1–3** were calculated to be 29–34 kcal/mol more stable than **4** (encapsulation of La inside the most stable $C_2(a)$ isomer of C_{82}) [24]. Structure **1** is 4–5 kcal/mol more stable than **2** and **3**, agreeing well with the relative energies of $C_{82}{}^{3-}$. The same was calculated for $Y@C_{82}$ (in which three electrons are transferred to C_{82}). Structure **1** was also calculated to be most stable for $Sc@C_{82}$, although the second and third most stable structures differ from those of $La@C_{82}$ and $Y@C_{82}$, reflecting the two-electron transfer from Sc to C_{82} (see the relative energies of $C_{82}{}^{2-}$ in Figure 9.2). It has been predicted that **1** with C_{2v} symmetry is the likeliest structure of the main isomer of $M@C_{82}$ [24].

It can be argued that the energetic advantage does not necessarily mean abundant production since the abundance depends on the growth processes and kinetics. However, the structural features of **1** are noteworthy. As Figure 9.3 shows, **1** is most highly stabilized when M approaches the center of one hexagonal ring in C_{82} along the C_2 axis. This is consistent with the extended X-ray absorption fine structure (EXAFS) studies of $Y@C_{82}$ [28] and $La@C_{82}$ [29], which show that the number of the nearest neighbor carbons is 6. The calculated distances between M and hexagonal carbons are 2.497 and 2.498 Å for $Sc@C_{82}$, 2.464 and 2.477 Å for $Y@C_{82}$, and 2.648 and 2.661 Å for $La@C_{82}$ [24]. The Y–C distances are in good agreement with the average value of 2.4 Å evaluated from the EXAFS study [28]. In a synchrotron X-ray powder diffraction study of $Y@C_{82}$ [30], the Y–C distances were once estimated to be 2.9 Å. However, a recent value of 2.47 Å [31], refined with a high resolution, agrees almost perfectly with the calculated values, although the cage structure and symmetry have not yet been resolved.

Very recently, the cage structure and symmetry of $Sc@C_{82}$ have been determined from the X-ray powder diffraction study using a maximum entropy method (MEM) [32a]. The obtained MEM electron densities reveal that the Sc atom is encapsulated inside the C_{2v} cage of C_{82} and is located at the off-centered position adjacent to a hexagonal ring, as predicted theoretically [24]. The determined nearest Sc–C distance of 2.53 ± 0.08 Å agrees with the predicted average value of 2.50 Å. The net charge density of +2.2 on Sc estimated from the MEM analysis is also consistent with the predicted electronic structure of $Sc^{2+}C_{82}{}^{2-}$. In addition, the C_{2v} structure of $La@C_{82}$ is also verified from the recent ^{13}C NMR study of $La@C_{82}$ in its anionic form [32b].

Because of the difficulty in preparing single crystals even for the most abundant $M@C_{82}$, ^{13}C NMR probes are at present most helpful for the determination of cage structures and symmetry. However, the paramagnetic nature prevents NMR measurements. For this reason, the structural determination of monometallofullerenes has been delayed. In this context, the recent successful isolation and purification of the $Ca@C_n$ isomers [33] are noteworthy as the first example of group 2 metal-containing fullerenes for which diamagnetic properties are expected. In $Ca@C_{82}$, two electrons are transferred from Ca to C_{82} [34].

As is expected from the relative stability of the C$_{82}{}^{2-}$ isomers in Figure 9.2, encapsulations of a Ca atom inside the C_{2v}, $C_s(c)$, $C_2(c)$, and $C_{3v}(a)$ cages are calculated to be most stable and close in energy [34]. This is consistent with the fact that four isomers are extractable and isolated for Ca@C$_{82}$ [33]. In this context, it is interesting that three stable isomers are isolated for Tm@C$_{82}$, which are expected to have an electronic structure of Tm^{2+}C$_{82}{}^{2-}$ [27]. For one of the four isomers isolated for Ca@C$_{82}$, the ^{13}C NMR spectrum is successfully measured and shows a total of 41 distinct ^{13}C NMR lines with equal intensity [33b]. This agrees with the endohedral structure optimized with the $C_2(c)$ cage of C$_{82}$, which has 41 nonequivalent carbons [34]. It is expected that NMR probes are also successful for the remaining isomers.

It appears that cage-like structures with M inside (or outside) are first formed through several steps. In many cases, the initial cages do not correspond to stable structures. However, they can isomerize to find stable fullerene cages during annealing processes, because the overall process is highly exothermic. The M atom wrapped in the C_{2v} cage is stabilized with the binding energies (relative to M + C$_{82}$) of 77, 106, and 115 kcal/mol for M = Sc, Y, and La, respectively [24]. These values are much larger than that of 39 kcal/mol for Ca@C$_{82}$ [34], agreeing with the fact that group 3 metals are more abundantly encapsulated than group 2 metals.

9.2.2 Electronic Structures

Electron paramagnetic resonance (EPR) spectroscopy provides useful information on electronic structures even with extracts containing a small amount of endohedral metallofullerenes. The first successful measurement of the EPR spectrum was achieved for La@C$_{82}$ [35]. An important observation is that the ^{139}La hyperfine coupling constant is very small and the g value is close to the free-spin value of 2.0023 (similar to those found for fullerene anion radicals). It was suggested that these EPR features result from the electronic structure described as La^{3+}C$_{82}{}^{3-}$. Very small hyperfine coupling constants and near free-spin g values were also observed for Sc@C$_{82}$ [26,36] and Y@C$_{82}$ [26,37]. Thus, it was believed that the Sc atom as well as the Y atom has a 3+ charge inside C$_{82}$.

It has been pointed out that the EPR data cannot always be direct evidence for M^{3+}C$_{82}{}^{3-}$ [20]. The electronic structures depend on the types of entrapped metals. As is apparent from the compact and low-lying d orbital of Sc shown in Figure 9.4 [38], the Sc (3d^{1}4s^2) atom donates only its two 4s valence electrons to C$_{82}$, leaving behind its 3d electron on Sc. However, transfer of all three valence electrons is increasingly favored upon going to Y@C$_{82}$ and La@C$_{82}$ because of the diffuse and higher-lying d orbitals, as originally interpreted from the EPR data. The electronic structure of La^{3+}C$_{82}{}^{3-}$ is also calculated with the La atom placed even inside the much more unstable C_{3v} cage of C$_{82}$ [21,39], while it is supported from the observations of the ultraviolet photoelectron spectrum [40], quadruple coupling constant [41,42], and effective magnetic

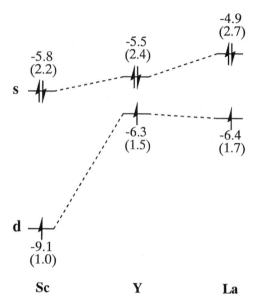

Figure 9.4 The energy levels (eV) and sizes (Å) of atomic orbitals of Sc ($3d^1 4s^2$), Y ($4d^1 5s^2$), and La ($5d^1 6s^2$). The orbital sizes measured by the $\langle r \rangle$ values are in parentheses.

moment [8b]. On the other hand, the difference in the electronic structures of $Sc^{2+}C_{82}{}^{2-}$ and $La^{3+}C_{82}{}^{3-}$ is well reflected in the differences in the anisotropic components of the hyperfine and g tensors [43] and in the ultraviolet–visible (UV/VIS) absorption spectra [5].

Encapsulation of the still heavier 4f block lanthanides (14 atoms following La in the periodic table) inside the C_{2v} cage of C_{82} has been theoretically investigated [24], since their electronic states have been controversial [44–46]. As representative example, the first (Ce, Pr, and Nd), middle (Eu and Gd), and last (Yb and Lu) members in the series were considered for $M@C_{82}$. When M = Ce ($4f^1 5d^1 6s^2$), Pr ($4f^3 6s^2$), Nd ($4f^4 6s^2$), and Gd ($4f^7 5d^1 6s^2$), a total of three electrons (6s and 5d electrons on Ce and Gd while 6s and 4f on Pr and Nd) is transferred to C_{82}, as shown in Figure 9.5a. Consequently, the lowest unoccupied molecular orbital (LUMO) and LUMO+1 of C_{82} are doubly and singly occupied (see Fig. 9.5b), respectively, as in the cases of M = Y and La. Therefore, the UV/VIS spectra observed for M = Ce, Pr, Nd, and Gd have a strong resemblance to those for M = Y and La. From this resemblance, it has been concluded that the lanthanide atoms have a formal charge of 3+ in $M@C_{82}$ [44–46]. As Figure 9.5c shows, however, each of the occupied orbitals of C_{82} has a very small tail on the 5d orbitals as a result of slight back transfer. The tail density is at most only 0.02, but the total sum of the mist densities from 200 occupied orbitals of C_{82} leads to a large 5d population of 0.6–0.7 on M. Therefore, the formal charge on Ce, Pr, Nd, and Gd should be recognized as 2+, as also emphasized for dimetallofullerenes ($M_2@C_{80}$) [47]. It is important

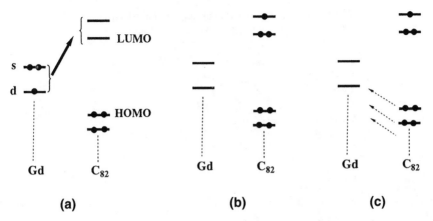

Figure 9.5 Electron transfer between Gd and C$_{82}$.

to note that the spectroscopic data reflect only the orbital picture in Figure 9.5b.

On the other hand, Eu ($4f^7 6s^2$) and Yb ($4f^{14} 6s^2$) donate only two 6s valence electrons to C$_{82}$, reflecting the stable f^7 and f^{14} electron configurations [24]. This allows the formal charge to be described as 2+ in M@C$_{82}$, though approximately 0.2 electrons are back-transferred from C$_{82}$. The last lanthanide member, Lu ($4f^{14} 5d^1 6s^2$), has the same valence electron configuration as Sc, Y, and La. In Lu@C$_{82}$, however, one 5d electron and one 6s electron are transferred, a large amount of 6s electrons remaining on Lu because of the large relativistic contraction and stabilization of the 6s orbital.

9.2.3 Electronic Properties

Ionization potentials (I_p) and electron affinities (E_a) play an important role in characterizing the electronic properties and reactivities of endohedral metallofullerenes. It is interesting to compare these values with those of typical fullerenes such as C$_{60}$ and C$_{70}$ (whose properties have widely been investigated). However, no direct experimental data are available for endohedral metallofullerenes [48]. Therefore, the I_p and E_a values calculated are summarized in Table 9.1 together with those of C$_{60}$ and C$_{70}$ [49,50]. The calculated I_p and E_a values of 7.78 and 2.57 eV for C$_{60}$ agree well with the available experimental values of 7.64 eV [51] and 2.65 eV [52], respectively.

It was suggested early that metallofullerenes have lower ionization potentials than empty fullerenes [53]. This can be quantitatively confirmed in Table 9.1. The I_p values for M@C$_{82}$ are 1.2–1.6 eV smaller than those for C$_{60}$ and C$_{70}$. In addition, the E_a values for M@C$_{82}$ are 0.4–0.7 eV larger than those for C$_{60}$ and C$_{70}$. These indicate that endohedral metallofullerenes can act as stronger electron donors as well as stronger electron acceptors. As discussed later, this is consistent with the shifts of the first oxidation and reduction potential peaks measured in solution for M@C$_{82}$, relative to those of C$_{60}$ and C$_{70}$.

TABLE 9.1 Ionization Potentials (I_p) and Electron Affinities (E_a) Calculated for M@C$_{82}$ and Fullerenes

Compound	I_p (eV)	E_a (eV)
Sc@C$_{82}$	6.45	3.08
Y@C$_{82}$	6.22	3.20
La@C$_{82}$	6.19	3.22
Ce@C$_{82}$	6.46	3.19
Eu@C$_{82}$	6.49	3.22
Gd@C$_{82}$	6.25	3.20
C$_{60}$	7.78	2.57
C$_{70}$	7.64	2.69

As Table 9.1 shows, endohedral metallofullerenes have rather similar ionization potentials. For example, the difference in I_p between Sc@C$_{82}$ and La@C$_{82}$ is only 0.3 eV, although the original LUMO level (occupied in Sc@C$_{82}$) of C$_{82}$ is responsible for the ionization of Sc@C$_{82}$ while the original LUMO+1 level (singly occupied in La@C$_{82}$) of C$_{82}$ is responsible for that of La@C$_{82}$. This is because the energy gap between the original LUMO and LUMO+1 of C$_{82}$ is only 0.6 eV. Similar trends are also seen for E_a.

A remarkable finding is that the net charge on the M atom in M@C$_{82}$ is little changed even when M@C$_{82}$ loses or accepts an electron [49,50]. This allows the formal view that electron removal and gain take place on the C$_{82}$ cage, and M acts as a positive core as in a sort of "superatom." In an attempt to confirm this view, the I_p and E_a values of C$_{82}^{2-}$ and C$_{82}^{3-}$ were calculated by placing a point charge of 2+ or 3+ instead of the M atom [49,50]. These model calculations give the I_p and E_a values that agree well (in both magnitude and order) with those calculated for M@C$_{82}$; note that electron attachment to C$_{82}^{2-}$ and C$_{82}^{3-}$ is not significantly bound or unbound without the point charge. This supports again the view that endohedral metallofullerenes are composed of a positively charged core-metal and a negatively charged cage, which are mostly stabilized by electrostatic interactions. Thus, it may be expected that endohedral metallofullerenes such as Y@C$_{82}$ and La@C$_{82}$ behave like a Li atom, while Sc@C$_{82}$ (in which the C$_{82}$ cage has formally closed-shell character) behaves like a He atom. In this context, a recent STM study is interesting because it shows that Y@C$_{82}$ forms dimers and trimers by a covalent interaction along the step edge of the Cu(111) clean surface in a way that Li atoms form the Li$_2$ and Li$_3$ molecules [54].

9.2.4 Electrochemical Properties

To provide insight into the electronic structures of endohedral monometallofullerenes, the reduction and oxidation potentials have been measured. As summarized in Table 9.2, the redox properties of the major isomers of Y@C$_{82}$ [6],

TABLE 9.2 Redox Potentials (in volts) of Metallofullerenes and Empty Fullerenes

Compound	$^{ox}E_2$	$^{ox}E_1$	$^{red}E_1$	$^{red}E_2$	$^{red}E_3$	$^{red}E_4$	$^{red}E_5$	$^{red}E_6$
Y@C$_{82}$	+1.07	+0.10	−0.37	−1.34		−2.22	−2.47	
La@C$_{82}$	+1.07	+0.07	−0.42	−1.37	−1.53	−2.26	−2.46	
La@C$_{82}$ (minor)	+1.08	−0.07	−0.47	−1.40		−2.01	−2.41	
Ce@C$_{82}$	+1.08	+0.08	−0.41	−1.41	−1.53	−1.79	−2.25	−2.50
Pr@C$_{82}$	+1.08	+0.07	−0.39	−1.35	−1.46	−2.21	−2.48	
Gd@C$_{82}$	+1.08	+0.09	−0.39	−1.38		−2.22		
C$_{60}$		+1.21	−1.12	−1.50	−1.95	−2.41		
C$_{70}$		+1.19	−1.09	−1.48	−1.87	−2.30		
C$_{76}$		+0.81	−0.94	−1.26	−1.72	−2.13		
C$_{82}$		+0.72	−0.69	−1.04	−1.58	−1.94		
C$_{86}$		+0.73	−0.58	−0.85	−1.60	−1.96		

La@C$_{82}$ [55], Ce@C$_{82}$ [56], Pr@C$_{82}$ [57], and Gd@C$_{82}$ [58] differ significantly from those of empty fullerenes. The reduction of the monometallofullerenes showed five reversible waves by cyclic voltammetry (CV), even though three electrons are already transferred on the C$_{82}$ cage. Monometallofullerenes are stronger electron acceptors than empty fullerenes.

By assuming that all reversible redox processes observed involve single electron transfer, the electrochemical behaviors of La@C$_{82}$ have been explained in the following way [56]: (1) The removal of the unpaired electron corresponds to the first oxidation process, the resultant La@C$_{82}^+$ having no unpaired electron. (2) The first reduction is relatively facile since the singly occupied molecular orbital (SOMO) is filled up to afford the closed-shell species La@C$_{82}^-$. (3) The low-lying HOMO−1 could be responsible for the irreversible formation of La@C$_{82}^{2+}$. (4) Since calculations show that the LUMO and LUMO+1 of La@C$_{82}$ derive from C$_{82}$ (not from 5d and 6s atomic orbitals of the La atom) [20], the C$_{82}$ cage can carry five extra electrons in solutions.

One reversible oxidation and four reversible reductions were observed for Y@C$_{82}$. The peak current intensity of the second reduction is twice that of each of the other redox peaks; the shape of the current voltage curve suggests a simultaneous two-electron transfer, not an overlap of two one-electron transfers. The remarkably small potential difference between the first oxidation and reduction suggests that the original LUMO+1 of C$_{82}$ is singly occupied, as proposed for La@C$_{82}$. The oxidation state of the Y atom is, therefore, close to that of the La atom. The first oxidation and reduction potentials of Y@C$_{82}$ are anodically shifted, relative to those of La@C$_{82}$. These relatively small shifts are in agreement with the fact that the I_p and E_a values of Y@C$_{82}$ are close to those of La@C$_{82}$, respectively, as shown in Table 9.1.

As observed for La@C$_{82}$ and Y@C$_{82}$, the first redox processes of Ce@C$_{82}$, Pr@C$_{82}$, and Gd@C$_{82}$ are reversible, and their potentials are very close to those of La@C_{82} and Y@C_{82}. The second and third reductions of Gd@C_{82} are two-electron processes, but the latter reduction is irreversible. The second and third reductions of Ce@C_{82} are irreversible. Interestingly, after one reversible and two irreversible reductions, three other reversible reductions were observed for Ce@C_{82}.

The difference between the first oxidation and reduction potentials is very small, probably because of the open-shell electronic structure. Monometallofullerenes act as strong electron donors as well as powerful electron acceptors, compared with empty fullerenes such as C_{60}, C_{70}, C_{76}, C_{78}, and C_{86}. The electronic properties of Ce@C_{82}, Pr@C_{82}, and Gd@C_{82} are very similar to those of La@C_{82} and Y@C_{82}, although the Ce, Pr, and Gd atoms have 4f electrons. This suggests that these 4f electrons do not play an important role in monometallofullerene chemistry as seen in organic and inorganic lanthanide chemistry.

The redox property has also been measured for the minor isomer of La@C_{82}, as shown in Table 9.2 [59]. Its CV data show one reversible oxidation and four reversible reductions as well as an irreversible oxidation. The first oxidation potential shifts negatively relative to that of the major isomer. This is in agreement with the observation that the minor isomer is more sensitive to oxygen. The minor isomer has a small potential difference between the first oxidation and reduction. The observed differences between major and minor isomers could be attributed to the difference in the C_{82} cages, but the actual cage structure is expected to be verified in a future experimental study.

9.2.5 Reactivities

A huge number of experimental studies have been performed to functionalize empty fullerenes such as C_{60} and C_{70} to understand the basic chemical properties and obtain new derivatives with interesting material, catalytic, or biological properties [60]. A new procedure to functionalize C_{60} and C_{70} by the addition of silicon and germanium compounds has been developed [61–68]. It is an interesting challenge to disclose how the reactivities of empty fullerenes are modified upon endohedral metal-doping [69]. Thus, the first exohedral functionalization was carried out for La@C_{82} with 1,1,2,2-tetramesityl-1,2-disilirane [70], as shown in Table 9.3.

La@C_{82}   **Disilirane**

TABLE 9.3   Reactivities, Oxidation ($^{ox}E_1$) and Reduction ($^{red}E_1$) Potentials, and HOMO–LUMO Levels of Metallofullerenes and Empty Fullerenes

| Compound | Reactivity | | $^{ox}E_1$ (V) | $^{red}E_1$ (V) | HOMO (eV) | LUMO (eV) |
	$h\nu$	Heat				
La@C$_{82}$	Yes	Yes	+0.07	−0.42		
Pr@C$_{82}$	Yes	Yes	+0.07	−0.39		
Gd@C$_{82}$	Yes	Yes	+0.09	−0.39		
La$_2$@C$_{80}$	Yes	Yes	+0.56	−0.31	−7.18	−2.55
Sc$_2$@C$_{84}$	Yes	No	+0.53	−0.97	−6.77	−1.38
C$_{60}$	Yes	No	+1.21	−1.12	−8.33	−0.63
C$_{70}$	Yes	No	+1.19	−1.09	−7.97	−0.84
C$_{76}$	Yes	No	+0.81	−0.94	−7.42	−1.13
C$_{78}$	Yes	No	+0.85	−0.73	−7.44	−1.49
C$_{82}$	Yes	No	+0.72	−0.69	−7.36	−1.52
C$_{84}$	Yes	No	+1.04	−0.75	−7.79	−1.34

The photochemical reaction was first tested. It was verified from the laser-desorption time of flight (LDTOF) and fast atom bombardment (FAB) mass spectra that the reaction leads to its 1:1 adduct; no molecular ion peaks ascribable to multiple-addition products were observed. It is noteworthy that La@C$_{82}$ can also be functionalized, as in the case of empty fullerenes. In addition, an interesting finding is that La@C$_{82}$ reacts thermally with disilirane, affording the 1:1 adduct. This is in sharp contrast to the fact that empty fullerenes react with disilirane only in a photochemical way [63,64]. Apparently, the facile thermal addition of disilirane to La@C$_{82}$ is due to the stronger electron acceptor and donor properties (see Tables 9.1 and 9.3). The EPR spectra measured during the reaction reveal the formation of at most two regioisomers with a different La isotropic splitting. This suggests that the regioselectivity as well as the reactivity of empty fullerenes could be controlled by endohedral metal-doping.

To see how the reactivity changes when a different metal is inside the cage, the reactions of Pr@C$_{82}$ [57] and Gd@C$_{82}$ [58] with disilirane were also investigated. It is found that Pr@C$_{82}$ and Gd@C$_{82}$ are functionalized both photochemically and thermally, as in the case of La@C$_{82}$. This is not surprising since the I_p and E_a values of Gd@C$_{82}$ are comparable to those of La@C$_{82}$, as shown in Tables 9.1 and 9.3. Also successful have been the functionalizations of La@C$_{82}$ by digermirane [71], diphenyldiazomethane [72], and diazirine [73].

It is expected that these successful exohedral derivatizations of endohedral metallofullerenes trigger much more work and constitute an important stepping stone on the way to new practical applications since they can also serve as precursors of new types of molecules. In this context, it should be emphasized that the present derivatizations differ significantly from those of endohedral noble gas fullerenes such as He@C$_{60}$ and He@C$_{70}$ [74]. In the chemical derivatizations of La@C_{82}, Pr@C_{82}, and Gd@C_{82}, the degree of the electron transfer from the electropositive metal to the fullerene cage plays a crucial role and thereby tunes the electronic properties of the fullerene cage, unlike the noble gas case.

9.3   DIMETALLOFULLERENES (La$_2$@C$_{80}$ AND Sc$_2$@C$_{84}$)

Two metal atoms can even be encapsulated inside the higher fullerene cages to form extractable and air-stable endohedral compounds. Among these, La$_2$@C_{80} and Sc$_2$@C_{84} have attracted special attention as representative and abundant dimetallofullerenes.

9.3.1   Structures and Electronic States

We first discuss La$_2$@C_{80}. As Figure 9.6 shows, there are seven isomers for the C_{80} fullerene, which satisfy the isolated pentagon rule [15]. It has been calculated that the D_2 and D_{5d} isomers are most stable while the I_h isomer is most unstable; the former two isomers close in energy are approximately 52 kcal/mol more stable than the latter [75]. The HOMO–LUMO gaps suggest that the D_2 isomer is less reactive and perhaps more air stable than D_{5d}. The isomer distribution of C_{80} was an open question since C_{80} was a "missing" fullerene between C_{60} and C_{96} because of the difficulty in separation [18]. However, the high abundance ($>90\%$) of the D_2 isomer has recently been verified from the successful isolation and analysis of the observed ^{13}C NMR lines [76].

It seems plausible to assume that two La atoms are encapsulated inside the most abundant D_2 or the isoenergetic D_{5d} isomer of C_{80}. However, it has been found that encapsulation of two La atoms inside the most unstable I_h cage is most favorable; it is 63 and 79 kcal/mol more stable than encapsulation inside the D_2 and D_{5d} cages, respectively [75]. This is because the I_h cage of C_{80} has only two electrons in the fourfold degenerate HOMOs and can accommodate six more electrons to form a stable closed-shell electronic state of $(La^{3+})_2C_{80}{}^{6-}$ with a large HOMO–LUMO gap. The most stable endohedral structure (**5**) optimized with the I_h cage is shown in Figure 9.7 [75]. It has D_{2h} symmetry in which two La atoms are located equivalently along a C_2 axis with a long La–La distance of 3.655 Å, facing the hexagonal ring of the C_{80} cage with the La–C distances of 2.567–2.598 Å.

It may be argued that the energetic advantage does not necessarily mean abundant production. However, it has finally been verified form the ^{13}C and ^{139}La NMR studies that two La atoms are indeed encapsulated equivalently inside the I_h cage [77]. This has also been confirmed by the skeletal vibration mode observed in the Fourier transform infrared (FTIR) study [78]. These again give support to the view that an energetically stable structure with a large HOMO–LUMO gap is most abundantly produced and isolated, as is so for empty fullerenes.

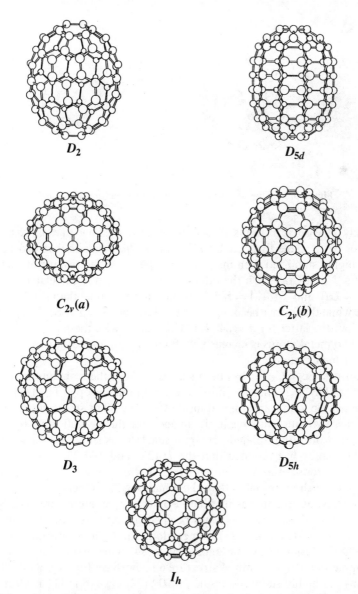

Figure 9.6 Seven isomers of C$_{80}$ that satisfy the isolated pentagon rule.

A minor isomer of La$_2$@C$_{80}$ has recently been extracted and isolated [79]. From the ^{13}C NMR measurement and theoretical calculations, it is suggested that two La atoms are encapsulated inside the D_{5h} cage of C$_{80}$ (the second most stable for C$_{80}{}^{6-}$).

Ion mobility measurements suggest that La(La@C$_{80}$) with one La atom outside the cage is formed because of the electrostatic repulsion between the

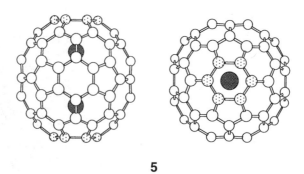

5

Figure 9.7   Two views of the optimized structure of $La_2@C_{80}$.

positively charged La atoms [80,81]. Calculations show that $La_2@C_{80}$ is 68 kcal/mol more stable than $La(La@C_{80})$ [47]. This suggests that two La atoms can be highly stabilized even inside C_{80} by the large electrostatic attraction between La^{3+} and $C_{80}{}^{6-}$. On the other hand, $M_2@C_n$ was calculated to be only 29 (M = La) and 51 (M = Sc) kcal/mol more stable than $M(M@C_n)$ for $n = 74$, while the former becomes 75 (M = La) and 70 (M = Sc) kcal/mol more stable than the latter for $n = 84$ [4b]. This agrees with the fact that encapsulation of two metal atoms is increasingly favored with an increase in n (especially for larger metal atoms).

We turn to $Sc_2@C_{84}$. It has been suggested from the scanning tunneling microscopy (STM) [82] and transmission electron microscopy (TEM) [83] studies that the two Sc atoms are trapped inside C_{84}. The STM study shows that there are at least two different sizes in the image. For the C_{84} fullerene, there are 24 isomers that obey the isolated pentagon rule [15], as shown in Figure 9.8. The ^{13}C NMR study has revealed that the $D_2(22)$ and $D_{2d}(23)$ isomers are most abundantly produced as a 2:1 mixture [16]. These two isomers are very close in energy and much more stable than the remaining 22 isomers.

An interesting question is in which isomer two Sc atoms are encapsulated. From theoretical calculations, it has been suggested that they are equivalently encapsulated inside the $D_{2d}(23)$ isomer along the C_2 axis bisecting the double bounds at the fusion of two hexagonal rings [84]. This endohedral structure has D_{2d} symmetry and its optimized structure **6** is shown in Figure 9.9 [85]. The two Sc atoms are highly stabilized inside the $D_{2d}(23)$ cage; **6** is 111 kcal/mol more stable than $2Sc + C_{84}$ [85]. This stabilization is mostly electrostatic and due to polarization. The ^{13}C NMR spectrum observed recently for the most abundant isomer of $Sc_2@C_{84}$ shows a total of 11 distinct ^{13}C NMR lines (10 lines of nearly equal intensity and 1 of half intensity) [86]. This agrees perfectly with structure **6** with 11 nonequivalent carbons $(10(\times8) + 1(\times4)$ carbons). In addition, only one single ^{45}Sc NMR signal observed confirms that two Sc atoms are equivalent [87].

The distance between the Sc atoms is greatly increased to 4.029 Å by approximately 1.4 Å from the distance in the ground state of Sc_2 [88], and each Sc

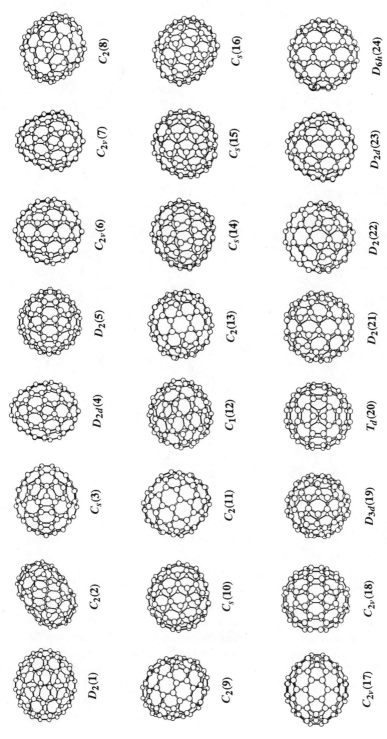

$C_2(8)$ $C_{2v}(7)$ $C_{2v}(6)$ $D_2(5)$ $D_{2d}(4)$ $C_s(3)$ $C_2(2)$ $D_2(1)$

$C_s(16)$ $C_s(15)$ $C_s(14)$ $C_2(13)$ $C_1(12)$ $C_2(11)$ $C_s(10)$ $C_2(9)$

$D_{6h}(24)$ $D_{2d}(23)$ $D_2(22)$ $D_2(21)$ $T_d(20)$ $D_{3d}(19)$ $C_{2v}(18)$ $C_{2v}(17)$

Figure 9.8 Twenty-four isomers of C_{84} that satisfy the isolated pentagon rule.

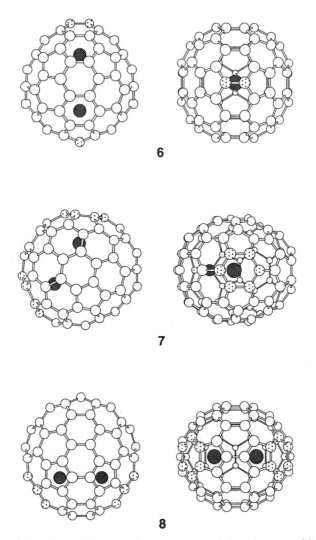

Figure 9.9   Two views of the optimized structures of three isomers of $Sc_2@C_{84}$.

is only 2.358 Å apart from the nearest neighbor carbon. This indicates that two
Sc atoms are not encapsulated as a dimer. The Sc–Sc and Sc–C distances agree
well with the values of Sc–Sc = 3.9 ± 0.1 Å and Sc–C = 2.4 ± 0.2 Å evaluated
from the synchrotron X-ray powder diffraction study [89]. The long Sc–Sc distance is not surprising since almost four electrons are transferred from the
bonding orbitals of Sc_2 to C_{84}. The large red shifts of UV/VIS absorption peaks
observed for $Sc_2@C_{84}$ relative to C_{84} [90] are well explained as a result of the
electron transfer. The formal electronic structure of $(Sc^{2+})_2C_{84}{}^{4-}$ is also consistent with the X-ray photoemission spectrum [91] as well as the net charge

density of +2.2 on Sc obtained from the X-ray powder diffraction study [89]. The calculated singlet state agrees with the EPR study, which shows no para-magnetic signal [90].

From the MEM analysis of the powder diffraction data of Sc$_2$@C$_{84}$ [89], it has been claimed that the C–C double bonds that two Sc atoms face are unusually stretched to 1.9 ± 0.1 Å, unlike the predicted distance of 1.434 Å [85] (only ca. 0.1 Å longer than that of the D_{2d}(23) cage of C$_{84}$). This discrepancy should be resolved in a future study. However, it has been pointed out that such a D_{2d} structure with unusually long C–C bonds is 93 kcal/mol less stable and does not correspond to an energy minimum [12a]. In addition, the structure is not consistent with the ^{13}C NMR data [86]: the chemical shifts observed for Sc$_2$@C$_{84}$ are all in the range of 134–148 ppm and differ only by 4.9 ppm from those of C$_{84}$. In this context, it is interesting that similar long C–C distances of 1.9 ± 0.15 Å are also observed for the polymeric RbC$_{60}$ in the powder X-ray diffraction study [92]. However, these long distances conflict with the shorter distances obtained recently in the X-ray crystallographic study of the C$_{60}$ dimer [93] as well as in the extensive calculations of related compounds [94]. This may suggest that the distances of interacting C–C bonds tend to be significantly overestimated in powder X-ray diffraction studies.

In contrast to the empty case, it was calculated that encapsulation of two Sc atoms inside the D_2(22) cage of C$_{84}$ (close in energy to D_{2d}(23)) is 11.6 kcal/mol less stable than structure **6**. More stable endohedral structures were found by placing two Sc atoms inside the considerably unstable C_s(10) and C_{2v}(7) cages of C$_{84}$. This is because these two cages are highly stabilized upon accepting four electrons from two Sc atoms [85]. As Figure 9.9 shows, the optimized structures **7** and **8** have C_s and C_{2v} symmetry, respectively; these are only 5.2 kcal/mol less stable and 2.6 kcal/mol more stable than structure **6**, respectively [85].

Recently, two additional isomers have successfully been isolated from a purified sample of Sc$_2$@C$_{84}$ using a two-stage high-performance liquid chromatography (HPLC) method [86]. The ^{13}C NMR spectrum observed for one of these two isomers shows a total of 46 NMR lines (38 + 8(half intensity)) [86b]. This agrees completely with structure **7** with 46 nonequivalent carbons (38($\times$2) + 8($\times$1) carbons). The observation of two ^{45}Sc NMR peaks [87] confirms that two Sc atoms are nonequivalent, as shown in Figure 9.9. Unfortunately, these ^{13}C and ^{45}Sc NMR studies suggest nothing about the metal positions, because all positions on the C_s plane in **7** satisfy the observed NMR spectra. The metal positions predicted for **7** in Figure 9.9 are expected to be determined in a future X-ray study. It is also expected to be verified that the remaining isomer has structure **8** with C_{2v} symmetry. These will support that energetically stable structures are most abundantly produced. The HOMO and LUMO levels are −6.8 and −1.4 eV for **6**, −6.4 and −1.9 eV for **7**, and −6.1 and −1.4 eV for **8**, respectively [85]. These suggest that isomers **7** and **8** of Sc$_2$@C$_{84}$ are more reactive than **6**.

It was recently suggested that the remaining isomer has C_{2v} symmetry but its endohedral structure originates from the $C_{2v}(17)$ cage (not $C_{2v}(7)$), since a total of 24 ^{13}C NMR lines could be observed [86b], while structure **8** (C_{2v}) in Figure 9.9 has 23 nonequivalent carbons. However, the endohedral structure optimized with the $C_{2v}(17)$ cage is 27–29 kcal/mol less stable than **8** [85]. In addition, it has an unusually small HOMO–LUMO gap due to the remarkably high-lying HOMO level and is thereby highly unstable toward air and moisture. This may conflict with the fact that the remaining isomer differs a little in reactivity from **6**. One of the observed 24 lines might be ascribed to impurities, although the discrepancy should be resolved in a future study.

Unless M = La, extracting and isolating $M_2@C_{80}$ in macroscopic quantities has been unsuccessful [4]. To check whether the endohedral structure with the I_h cage is always the most stable, encapsulation of Sc and Y atoms (the lighter members in group 3) was tested [47]. For $Y_2@C_{80}$, it was calculated that almost three valence electrons are transferred from Y to form a closed shell on C_{80}, as in $La_2@C_{80}$. In $Sc_2@C_{80}$, however, only two valence electrons were transferred from Sc. Consequently, one of the fourfold degenerate HOMOs of the I_h cage remained to be unfilled and lie as low as −3.7 eV. Such two-electron transfer from Sc has been established for $Sc@C_{82}$ [20,24,32], $Sc@C_{60}$ [95], and $Sc_2@C_{84}$ [84,85,89]. However, it is quite unfavorable for $Sc_2@C_{80}$, because a low-lying unfilled orbital remains on C_{80}, which leads to a high reactivity. Rather, two Sc atoms were stabilized inside the D_{5h} cage of C_{80} because of the two-electron transfer from Sc. However, the C_{80}-derived HOMO and LUMO levels were 0.8 eV higher and 1.7 eV lower than those of $La_2@C_{80}$, respectively, suggesting again a higher reactivity of $Sc_2@C_{80}$. On the other hand, encapsulation of two La atoms inside the D_{5h} cage of C_{80} was 36 kcal/mol less stable than that inside the I_h cage. Thus, metals that can fill up the fourfold degenerate HOMOs of the I_h cage of C_{80} are most suitable for $M_2@C_{80}$. Encapsulation of such metals is expected to be most abundantly produced and extracted [47].

The still heavier metals, the first (Ce and Pr), middle (Eu and Gd), and last (Yb and Lu) members in the lanthanide series, were considered for $M_2@C_{80}$ [47]. When M = Ce ($4f^1 5d^1 6s^2$) and Pr ($4f^3 6s^2$), the fourfold degenerate HOMOs of the I_h cage of C_{80} were filled up as a result of three-electron transfer from each metal. When M = Gd ($4f^7 5d^1 6s^2$), they were not completely filled up. However, the ferromagnetic coupling between two unpaired $4f^7$ sets is interesting since a total spin of $S = 7$ (14 parallel spins on Gd_2) is the highest known for endohedral metallofullerenes. On the other hand, Eu ($4f^7 6s^2$), Yb ($4f^{14} 6s^2$), and Lu ($4f^{14} 5d^1 6s^2$) prefer to donate only two valence electrons to C_{80}, as does Sc, because of the stable f^7 and f^{14} electron configurations and the large relativistic contraction and stabilization of the 6s orbital of Lu. These again suggest that the stability and electronic state of $M_2@C_{80}$ depend strongly on metal atoms.

Unlike the case of $Sc_2@C_{84}$, $La_2@C_{84}$ has not been isolated. It is calculated with the $D_{2d}(23)$ cage that $La_2@C_{84}$ has higher HOMO and lower LUMO levels (thereby more reactive) than $Sc_2@C_{84}$ [4a].

Up to now, little has been known about the features of chemical bonding in endohedral metallofullerenes. In an attempt to disclose the general features, a topological analysis of the electron density distribution $\rho(r)$ and its Laplacian $\nabla^2\rho(r)$ has been carried out for the main isomer (**6**) of Sc$_2$@C$_{84}$ [96], according to the "atoms in molecules (AIM)" theory [97]. As widely demonstrated, the analysis provides important information on the distribution and nature of chemical bonds between the atoms in molecules [97]. Thus, the relevant bond critical points (r_b) between two atoms were located, at which the gradient of $\rho(r)$ vanishes ($\nabla\rho(r_b) = 0$) and the electron density is minimum along the bond and maximum in the other two directions. The Laplacian $\nabla^2\rho(r)$ identifies whether the electron density is locally concentrated (a negative value) or depleted (a positive value). For covalent (or polar) bonds arising from electron sharing, $\nabla^2\rho(r_b)$ becomes negative and large in magnitude as a result of electron concentration, and then $\rho(r_b)$ provides a measure of the strength of covalent bonding. For ionic bonds, $\nabla^2\rho(r_b)$ is positive and $\rho(r_b)$ is relatively low in value because of the contraction of electrons toward the interacting atoms, reflecting a closed-shell interaction. Bond paths of maximum electron density through r_b between two atoms are images of chemical bonds, the existence being necessary and sufficient for the two atoms to be bonded to (or interact with) one another [98]. Information on the shape of the electron density distribution along the bond is given by the principal curvatures ($\lambda_1 < \lambda_2 < \lambda_3$) of $\rho(r_b)$. The ratio of the two negative curvatures ($\lambda_1/\lambda_2 - 1$), called the bond ellipticity, provides a measure of the deviation of the electron density distribution from cylindrical symmetry and thus monitors the amount of π character in the case of double bonds [97]. In the following discussion, the $\rho(r)$ and $\nabla^2\rho(r)$ values are given in atomic units (au).

For structure **6** of Sc$_2$@C$_{84}$ in Figure 9.9, a total of 126 linkages are drawn between pairs of carbon atoms. Accordingly, 126 bond paths passing through a bond critical point (r_b) were fond between the carbon atoms, as shown in Figure 9.10a. Most of the bond paths were calculated to be slightly longer than the corresponding interatomic distances, reflecting that the C–C bonds are curved toward the outside of the C$_{84}$ cage. The electron densities $\rho(r_b)$ at each

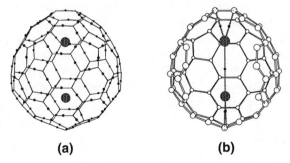

(a)　　　　　　**(b)**

Figure 9.10　The molecular graphs of structure **6** of Sc$_2$@C$_{84}$. Bond paths (lines) and bond critical points (dots) for C–C bonds are shown in (a) while those for Sc–C and Sc–Sc bonds are in (b).

critical point are 0.27–0.34; the Laplacian $\nabla^2\rho(r_b)$ values are highly negative in the range from -0.64 to -1.00. These large $\rho(r_b)$ and negative $\nabla^2\rho(r_b)$ values show that the C–C bonds are strongly covalent and compare well with those of $\rho(r_b) = 0.32$ and $\nabla^2\rho(r_b) = -0.89$ for the C–C bonds in benzene. The bond ellipticities of 0.14–0.22 are also comparable to that of 0.21 for benzene, suggesting that the C–C bonds preserve π-bond character even after receiving a total of four electrons onto the C_{84} cage.

Upon going from C_{84} to $Sc_2@C_{84}$, the $\rho(r_b)$ and $\nabla^2\rho(r_b)$ values for the double bonds (at the top and bottom of the molecules) interacting with Sc atoms are most changed; for other bonds, the changes are at most 0.03 for $\rho(r_b)$ and 0.15 for $\nabla^2\rho(r_b)$. The largest changes for the double bonds are due to the fact that the distances are increased by 0.095 Å from 1.339 (C_{84}) to 1.434 ($Sc_2@C_{84}$) Å, as a result of four-electron transfer from two Sc atoms to the degenerated LUMO of C_{84} with antibonding character over the double bonds. As measured by the bond ellipticities (0.24 for C_{84} versus 0.19 for $Sc_2@C_{84}$), however, the bonds have considerable π-bond character even in $Sc_2@C_{84}$.

As Figure 9.10b shows, a total of four bond paths are found between Sc and the double bond carbons discussed above. In other words, the Sc atoms interact only with the double bonds. The $\rho(r_b)$ and $\nabla^2\rho(r_b)$ values are 0.05 and 0.17, respectively. These very low $\rho(r_b)$ and positive $\nabla^2\rho(r_b)$ values verify that the Sc–C bonds are highly ionic, as anticipated from the electronic structure of $(Sc^{2+})_2C_{84}^{4-}$. This is consistent with the fact that the binding energies of endohedral metallofullerenes are dominated by electrostatic interactions [4b]. The delocalized electron interaction between Sc and C is reflected in the high bond ellipticity of 0.81.

As Figure 9.10b shows, a bond path is also located between two Sc atoms; the $\rho(r_b)$ and $\nabla^2\rho(r_b)$ values are low (0.02) and slightly negative (-0.01), respectively. This suggests that there is a small covalent interaction between two Sc atoms. Because of this small interaction, a teardrop-like electron density distribution is produced between two Sc atoms (see Figure 9.11), as found in the X-ray powder study [89]. Thus, no full support can be given to the view that the teardrop feature may be a result of the Sc–Sc stretching vibrations [89]. It is rather expected that the thermal motion of the Sc atoms along the double bonds will be verified in a future experimental study [99].

As found for $Sc_2@C_{84}$, metal atoms interact with specific cage carbons. Metal–carbon bonds are highly ionic with large bond ellipticities. Despite significant electron transfer from metals to carbon cages, carbon–carbon bonds are strongly covalent with π-bond character. The interaction between two Sc atoms is very weak with covalent character.

9.3.2 Reactivities

It is of great interest how the reactivities change when two metals are encapsulated inside fullerene cages. Thus, the reactions with disilirane have been investigated [100]. As shown in Table 9.3, both $La_2@C_{80}$ and $Sc_2@C_{84}$ (D_{2d}) can

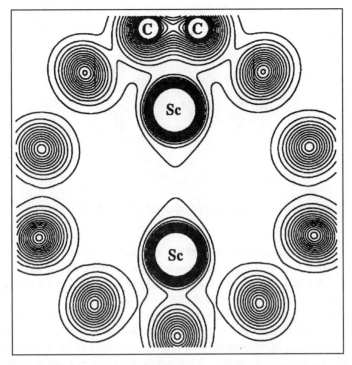

Figure 9.11 The electron density $\rho(r)$ map of structure **6** of $Sc_2@C_{84}$ plotted up to 0.46 au with 20 contour lines in the plane containing two double-bond carbons and Sc atoms.

be functionalized by the photochemical addition of disilirane. These are the first examples of the derivatization of endohedral dimetallofullerenes. An interesting finding is that $La_2@C_{80}$ reacts also thermally with disilirane to provide a 1:1 adduct(s). This is because the low-lying La-derived LUMO level of $La_2@C_{80}$ is 1–2 eV lower than those of the empty fullerenes, being consistent with the fact that $La_2@C_{80}$ has a lower reduction potential [9]. On the other hand, the thermal addition of disilirane to $Sc_2@C_{84}$ (D_{2d}) is suppressed, as in empty fullerenes, because of the higher-lying LUMO level. This is consistent with the higher reduction potential of $Sc_2@C_{84}$. These results strongly suggest that the thermal additions are initiated by electron transfer from disilirane to endohedral metallofullerenes with strong electron-accepting power, regardless of the number of metals.

9.4 THE MOTION OF ENCAPSULATED METAL ATOMS

Molecular rotation within a cage is expected to be of great help in designing functional molecular devices [101]. Thus, it is currently of growing interest

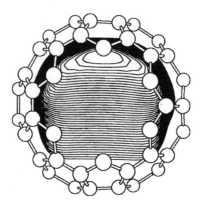

Figure 9.12   The electrostatic potential map inside $C_{82}{}^{3-}$ (C_{2v}).

whether encapsulated metals are rigidly attached to cage carbons or move about. To provide insight into the nature of the inside of fullerene cages, the electrostatic potentials inside the C_{80} (I_h), C_{82} (C_{2v}), and C_{84} ($D_{2d}(23)$) cages have been investigated [4b,24,34,99].

An interesting finding is that the values of the electrostatic potentials are all positive (i.e., destabilization for cationic species) at any positions inside the cages, reflecting that the probability of finding electrons inside is much smaller than that of finding them outside. Such positive values are not specific to the higher fullerenes but rather common to neutral carbon cages, as also found for C_{60} and C_{70} [102]. In other words, the inside of fullerene carbon cages is originally suitable for the accommodation of anionic or neutral species. Accordingly, it is calculated that anionic species are highly stabilized in carbon fullerene cages [103], unlike silicon fullerene cages [104].

However, the situation is drastically changed as electrons are transferred onto carbon cages [4b,24,34,99]. To make this clearer, the electrostatic potentials calculated for $C_{82}{}^{3-}$ (C_{2v} in Fig. 9.1) and $C_{84}{}^{4-}$ ($D_{2d}(23)$ in Fig. 9.8) are shown in Figures 9.12 and 9.13, respectively. These electrostatic potentials provide highly negative values (-221 kcal/mol for $C_{82}{}^{3-}$ and -275 kcal/mol for $C_{84}{}^{4-}$ even at the center of the cages) and are very suitable for the accommodation of cationic species. This is why encapsulated metals do not preserve a neutral state but prefer highly cationic states. Obviously, energy cost in electron transfer from metals to cages is smaller for metals with lower I_p and cages with higher E_a; it is overcome enough by resultant electrostatic attractions. These attractions are highly enhanced with an increase in the electron transfer. Thus, it is not surprising that the atoms encapsulated up to now are mostly group 3 and lanthanide metals for which multiple ionizations are facile.

The positively charged metals should be most highly stabilized at the minima of the electrostatic potentials. Thus, the metal positions calculated for La@C_{82} and Sc$_2$@C_{84} in Figures 9.3 and 9.9 correspond to the minima of the electrostatic potentials in Figures 9.12 and 9.13, respectively. This again confirms that

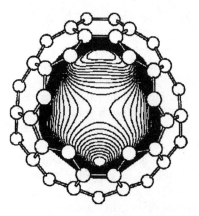

Figure 9.13   The electrostatic potential map inside C_{84}^{4-} $(D_{2d}(23))$.

electrostatic interactions play a dominant role in both stabilizing endohedral metallofullerenes and determining metal positions.

As Figures 9.12 and 9.13 show, the electrostatic potentials are very flat around the minima. Accordingly, for example, the actual calculations of $Sc_2@C_{84}$ show that the energy changes are only approximately 1–2 kcal/mol even when each Sc atom moves off the minima by approximately 1 Å along the double bonds in C_{84}, although further large displacement leads to great destabilization [99]. These small energy changes suggest that metals are not still but oscillate with considerable amplitude at room temperature [105].

Much more interesting is the case of $C_{80}^{6-}(I_h)$. As Figure 9.14 shows, the electrostatic potential map shows almost concentric circles with no clear minima [99], reflecting the round cage structure. The actual calculations in $La_2@C_{80}$ show that the overall barrier is approximately 5 kcal/mol even if two La atoms make the circuit of the inside of the cage [99,106]. Accordingly, it is

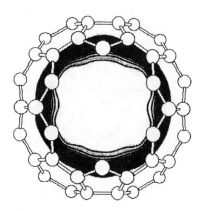

Figure 9.14   The electrostatic potential map inside C_{80}^{6-} (I_h).

measured from the temperature-dependent ^{139}La NMR study that the $\Delta H^{\ddagger}$ value for the rotation is 5.6 kcal/mol [107]. This small energy barrier is interesting since it can make the metals stop with a decrease in temperature and circulate at high speed with an increase in temperature. Such temperature control of metal motion inside fullerene cages may lead to an interesting "molecular device" with new electronic or magnetic properties.

The circuit of two La^{3+} atoms (with a large moment of inertia) inside C_{80} produces a magnetic field at the positions of La. In the ^{139}La NMR study, this should be reflected in the nuclear magnetic relaxation rate (hence, the linewidth). Ordinarily, interaction between La nuclear spins and the induced magnetic field (the so-called spin–rotation interaction) does not contribute significantly to the relaxation time in solutions, since molecular rotation is dominantly quenched by neighbors. In $La_2@C_{80}$, however, the rotation of two La atoms can be preserved due to the unique cage protection so that the spin–rotation interaction has a drastic effect on the relaxation process. Relaxation by the spin–rotation interaction leads to an increase in linewidth with increasing temperature, unlike other types of relaxation [108]. By observing a large broadening of the ^{139}La NMR linewidth with increasing temperature, it has been found that the La atoms are circulating inside the round cage even at room temperature, as shown in Figure 9.15 [77]. This is the first experimental evidence for the circular motion of metal atoms in endohedral metallofullerenes.

The small rotational barriers suggest that two La atoms can stop at the most stable positions with a decrease in temperature. Therefore, the ^{13}C NMR measurements were carried out at lower temperatures. By decreasing temperature to 258 K, broad ^{13}C NMR signals were observed, which may be ascribed to the overlap of 13 NMR lines [77], as expected from the static D_{2h} structure (Fig. 9.7) with 13 nonequivalent carbons.

It is expected that the dynamic behaviors of encapsulated metals will extend the research area of endohedral metallofullerenes [99]. It is interesting to control the rotating charges using $Ce_2@C_{80}$ and $Pr_2@C_{80}$: these metal atoms are

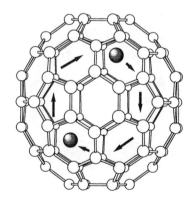

Figure 9.15 The rapid circular motion of two La atoms in $La_2@C_{80}$.

less positively charged (formally, 2+). Also interesting is the circular rotation of high spins in $Gd_2@C_{80}$.

9.5 UNCONVENTIONAL CAGE STRUCTURES OF ENDOHEDRAL METALLOFULLERENES

For the endohedral metallofullerenes extracted and isolated up to now, it has been found that the carbon cages consist of only pentagons and hexagons and satisfy the isolated pentagon rule (IPR). However, it is an important question whether metals can stabilize unconventional cage structures during growth and annealing processes, as summarized in a recent review [12c].

9.5.1 Ca@C$_{72}$

For the C_{72} fullerene, there is only one IPR-satisfying isomer of D_{6d} symmetry [15]. The endohedral structure (**9**) of Ca@C$_{72}$ optimized with this D_{6d} cage is shown in Figure 9.16, which has C_s symmetry. To investigate whether the endohedral structure is most stable, a total of 431,240 closed cage structures composed of pentagons, hexagons, and one heptagon were generated for C_{72}; among these, 11,190 structures correspond to conventional fullerenes consisting of only pentagons and hexagons. By searching a huge number of generated structures, it has been found that there are much more stable endohedral structures [109].

As Figure 9.16 shows, structures **10** (C_2) and **11** (C_{2v}) contain a pair of adjacent pentagons. Nevertheless, **10** is 37 kcal/mol more stable than **9**, while **11** is 36 kcal/mol more stable than **9**. These are remarkable as the first examples invalidating IPR established in fullerene chemistry. Upon optimization without Ca, the carbon cage of **10** becomes 10 kcal/mol less stable than the IPR-satisfying D_{6d} cage, because it has a pair of adjacent pentagons. Interestingly, the non-IPR carbon cage of **11** is 8 kcal/mol more stable than the D_{6d} cage even after optimization without Ca. This IPR violation for the empty case is also noteworthy since the structure and stability of C_{72} have long been controversial [15,110]. As Figure 9.17 shows, the carbon cage of **11** is formally obtainable by adding a C_2 fragment onto a hexagonal ring in the belt of the IPR-satisfying C_{70} (D_{5h}).

Structure **12** (C_s) in Figure 9.16 contains one heptagonal ring (surrounded by two groups consisting of two and three adjacent pentagons), thereby not obeying the traditional definition of fullerenes. However, **12** was calculated to be 18 kcal/mol more stable than **9**. This is the first example of a heptagon-containing structure that is more stable than the conventional fullerene structure. A stable structure with one heptagonal ring has been calculated for C_{62} [111]; however, it should be noted that C_{62} has no IPR-satisfying structure. Upon optimization without Ca, the heptagon-containing cage of **12** becomes 26 kcal/mol less stable than the IPR-satisfying D_{6d} cage. On the other hand,

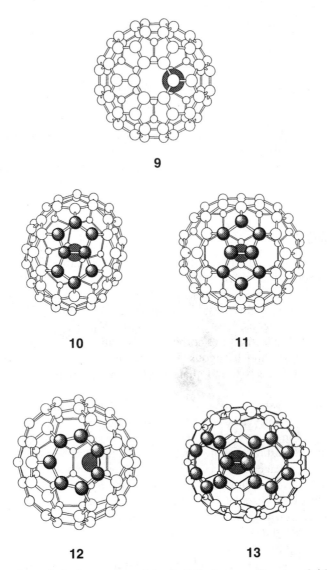

Figure 9.16   The optimized structures of the Ca@C_{72} isomers. Structure **9** (C_s) satisfies the IPR while **10** (C_2) and **11** (C_{2v}) contain a pair of adjacent pentagons. Structures **12** (C_s) and **13** (C_s) have one and two heptagonal rings, respectively.

structure **13** in Figure 9.16 possesses two heptagonal rings. Nevertheless, **13** is 2 kcal/mol more stable than **9**. Without Ca, the two-heptagon-containing cage is 24 kcal/mol less stable than the D_{6d} cage; it is formally obtainable by adding a C_2 fragment on C_{70} (D_{5h}), as shown in Figure 9.17, and is 2 kcal/mol more stable than the one-heptagon-containing cage of **12**. As is obvious from these

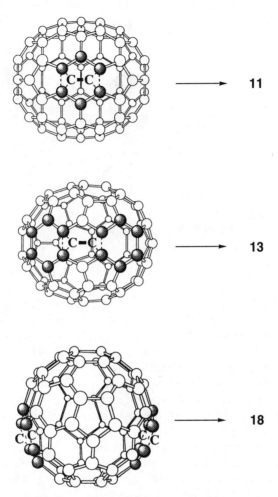

Figure 9.17 Formal addition of one or two C_2 fragments on C_{70} (D_{5h}) leads to the cage of **11**, **13**, and **18**.

results, heptagon-containing structures as well as non-IPR structures can be highly stabilized by metal-doping.

The binding energies (relative to Ca + C_{72}) are 36 (**10**), 16 (**11**), 33 (**12**), and 15 (**13**) kcal/mol. These confirm that a Ca atom can be endohedrally strongly bound to the non-IPR and heptagon-containing C_{72} cages. In contrast, encapsulation of Ca inside the IPR-satisfying D_{6d} cage leading to structure **9** was calculated to be 12 kcal/mol less stable than Ca + C_{72}. This reflects the leapfrog structure of the D_{6d} cage with a large HOMO–LUMO gap [15]. Two electrons are transferred in Ca@C_{72} from Ca to C_{72}. However, a high energy cost is required for the D_{6d} cage because of the high-lying LUMO level. It seems un-

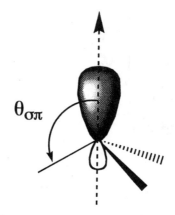

Figure 9.18 Pyramidalization angle.

likely that a Ca atom is wrapped in the IPR-satisfying C_{72} cage during annealing rearrangements.

Recently, $Ca@C_{72}$ has been isolated and purified [112a], for which a ^{13}C NMR study is in progress [113]. It is expected to be verified that $Ca@C_{72}$ takes unconventional structures such as **10** and **11**. Since only trace mass spectral evidence has been available for C_{72}, the extraction and isolation of $Ca@C_{72}$ are apparently owing to Ca-mediated stabilization.

The HOMO levels of **10**, **11**, **12**, and **13** are -7.0, -6.4, -7.4, and -6.6 eV, respectively. These are 0.8–1.8 eV lower than that of -5.6 eV for **9**, suggesting that the unconventional structures are less reactive especially toward oxygen than the IPR-satisfying **9**.

The local strain inherent in closed-cage structures also plays an important role in determining the reactivities. In this context, the pyramidalization angles from the π-orbital axis vector (POAV) analysis, $\theta_p = (\theta_{\sigma\pi} - 90)°$ (for $\theta_{\sigma\pi}$, see Fig. 9.18), provide a useful index of the local strain at each carbon atom: an increase in θ_p leads to a high local strain (a high curvature on the cage) [114]. The calculated pyramidalization angles are given in Table 9.4 [12c]. The maximum value (θ_p^{max}) for the IPR-satisfying C_{72} cage is 12.3° and highly increased to 14.2° upon endohedral Ca-doping. As Table 9.4 shows, the θ_p^{max} values for

TABLE 9.4 θ_p^{max} in C_{72} and $Ca@C_{72}$

Structure	C_{72}	$Ca@C_{72}$
9	12.3°	14.2°
10	16.0°	16.6°
11	18.3°	15.1°
12	16.1°	16.4°
13	16.2°	15.1°

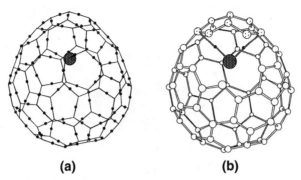

(a)   **(b)**

Figure 9.19 The molecular graphs of structure **10** of Ca@C$_{72}$. Bond paths (lines) and bond critical points (dots) for C–C bonds are shown in (a) while those for Ca–C bonds are in (b).

the non-IPR and heptagon-containing C$_{72}$ cages are larger because of the defects and become rather smaller (or change little) upon endohedral Ca-doping. However, the resultant values of 15.1°–16.6° are larger than those of 11.0°–12.1° for four stable isomers of Ca@C$_{82}$. This is consistent with the fact that Ca@C$_{72}$ is more reactive and less abundantly produced than Ca@C$_{82}$ [33].

To provide insight into the nature of the bonding in the non-IPR isomers **10** and **11** of Ca@C$_{72}$, the topological analysis of electron density $\rho(r)$ and its Laplacian $\nabla^2\rho(r)$ has been carried out [96]. As Figures 9.19a and 9.20a show, 108 bond paths and bond critical points (r_b) were found between the carbon atoms, as expected from the carbon networks. The $\rho(r_b)$ and $\nabla^2\rho(r_b)$ values, which are comparable to those of Sc$_2$@C$_{84}$, suggest that there are strong covalent interactions between the carbon atoms; even for the C–C bonds in adjacent pentagons, the $\rho(r_b)$ values are large (0.28–0.30 for **10** and 0.28–0.31 for **11**) while the $\nabla^2\rho(r_b)$ values are highly negative (−0.66 to −0.79 for **10** and

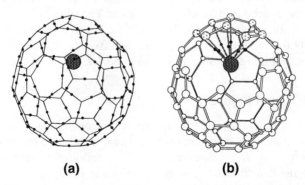

(a)   **(b)**

Figure 9.20 The molecular graphs of structure **11** of Ca@C$_{72}$. Bond paths (lines) and bond critical points (dots) for C–C bonds are shown in (a) while those for Ca–C bonds are in (b).

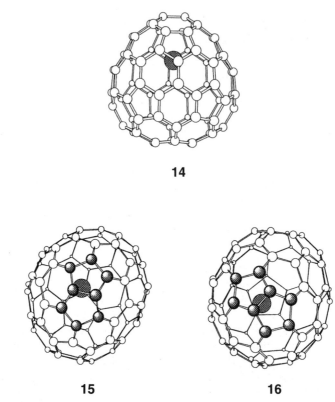

Figure 9.21 The optimized structures of the Ca@C$_{74}$ isomers. Structure **14** (C_{2v}) satisfies the IPR while **15** (C_1) and **16** (C_1) contain a pair of adjacent pentagons.

-0.69 to -0.82 for **11**). As Figures 9.19b and 9.20b show, two and eight bond paths were found between Ca and C for **10** and **11**, respectively; Ca interacts with specific cage carbons in adjacent pentagons. The $\rho(r_b)$ values for the Ca–C bonds are 0.01 for **10** and 0.02 for **11**, while the $\nabla^2\rho(r_b)$ values are 0.07 for **10** and 0.08–0.09 for **11**. These low $\rho(r_b)$ and positive $\nabla^2\rho(r_b)$ values suggest that the Ca–C bonds are highly ionic, as found in Sc$_2$@C$_{84}$.

9.5.2   Ca@C$_{74}$ and Sc$_2$@C$_{74}$

In an attempt to generalize the appearance of unconventional structures, endohedral Ca-doping was investigated for C$_{74}$. The C$_{74}$ fullerene has only one IPR-satisfying isomer of D_{3h} symmetry [15]. The endohedral structure (**14**) of Ca@C$_{74}$ optimized with this D_{3h} isomer is shown in Figure 9.21, which has C_{2v} symmetry. Although C$_{74}$ itself has not yet been successfully isolated and remains as a "missing" fullerene between C$_{60}$ and C$_{94}$ [112b], Ca@C$_{74}$ has recently been isolated and purified [112a].

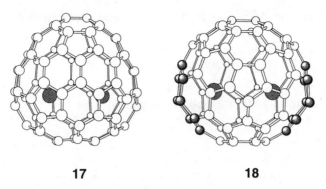

17   **18**

Figure 9.22   The optimized structures of the $Sc_2@C_{74}$ isomers. Structure **17** (C_{2v}) satisfies the IPR while **18** (C_{2v}) contains two pairs of adjacent pentagons.

A total of 615,576 closed-cage structures were generated for C_{74}, which are composed of pentagons, hexagons, and one heptagon [12c]. Two electrons are transferred in $Ca@C_{74}$ from Ca to C_{74}. Therefore, the relative energies of the C_{74}^{2-} isomers are of great help in predicting correctly the most favorable carbon cages, as already demonstrated for several examples. From the 615,576 structures, a search was made for two cages stable for C_{74}^{2-}, which contain a pair of adjacent pentagons. The endohedral structures (**15** and **16**) of $Ca@C_{74}$ optimized with these cages are shown in Figure 9.21 and have C_1 symmetry. However, **15** and **16** were calculated to be 16 and 18 kcal/mol less stable than **14**, respectively. This suggests that the IPR-satisfying **14** is the most probable structure for $Ca@C_{74}$.

Therefore, encapsulation of two Sc atoms was next tested. The endohedral structure (**17**) of $Sc_2@C_{74}$ optimized with the IPR-satisfying C_{74} cage is shown in Figure 9.22 and has C_{2v} symmetry [85]. Since a total of four electrons are transferred to C_{74} in $Sc_2@C_{74}$, a stable cage for C_{74}^{4-} was selected from the 615,576 structures of C_{74}, which contains two pairs of adjacent pentagons. The endohedral structure (**18**) of $Sc_2@C_{74}$ optimized with this defect cage is also shown in Figure 9.22. It is remarkable that **18** is only 1.5 kcal/mol less stable than the IPR-satisfying **17**. It is interesting that the carbon cage of **18** is also formally obtainable by adding two C_2 fragments onto two different hexagonal rings in C_{70} (D_{5h}), as shown in Figure 9.17. The HOMO and LUMO levels of -6.2 and -1.5 eV for **18** are 0.4 eV lower and 0.6 eV higher than those of **17**, respectively. This clearly indicates that **18** is much less reactive than **17**. It has been suggested that $Sc_2@C_{74}$ is an interesting target for the experimental study of unconventional structures [12c].

With the availability of a purified sample of $Sc_2@C_{74}$, the ^{13}C NMR spectrum has very recently been measured and consists of a total of 39 lines from 130 to 160 ppm [115]. Generally, the carbon atoms in IPR-satisfying fullerene cages exhibit the ^{13}C chemical shifts in the range of 130–151 ppm. Therefore, the two highly deshielded lines at 161.11 and 160.66 ppm imply the presence of adjacent pentagons. According to theoretical calculations [16c], however, the non-IPR C_s structure proposed in the NMR study does not correspond to an energy minimum. It is an interesting subject to find an unconventional structure that is consistent with the observed ^{13}C NMR spectrum.

9.5.3   Endohedrally Metal-Doped Heterofullerenes

Considerable interest is currently directed toward the variation of the chemical and physical properties of endohedral metallofullerenes to extend their applications. For this purpose, an attempt is made to substitute one or more cage carbons by heteroatoms. Such substitutional doping has progressed considerably for empty fullerenes. Most interesting is the recent preparation of azafullerene ions such as $C_{59}N^+$ and $C_{69}N^+$ [116–118]. However, the resultant $C_{59}N$ and $C_{69}N$ are highly reactive because of the strong radical character of the cages. For this reason, only the dimers or derivatives have been isolated and their properties have been characterized [119]. It is an interesting challenge to prepare a new class of heterofullerenes whose properties are significantly enhanced by a combination of endohedral and substitutional doping. Very recently, it has been found that La-doped azafullerene ions such as $La_2@C_{79}N^+$ and $La@C_{81}N^+$ are efficiently formed by fast atom bombardment mass fragmentation of the reaction product of benzyl azide with $La_2@C_{80}$ and LaC_{82} [120].

$La_2@C_{79}N^+$ and $La@C_{81}N^+$ are isoelectronic with $La_2@C_{80}$ and $La@C_{82}$, respectively. Calculations of $La_2@C_{79}N^+$ show that almost all three valence electrons are transferred from each La to the LUMO, LUMO+1, and LUMO+2 of the azafullerene cage, as in $La_2@C_{80}$. Consequently, $La_2@C_{79}N^+$ has a closed-shell electronic structure. Acceptance of an electron leads to the $La_2@C_{79}N$ radical. An important finding is that the electron is not distributed on the cage but is localized on each La with a spin density of 0.5; two La atoms can serve as a spin-absorbent. The most stable optimized structure of $La_2@C_{79}N$ is shown in Figure 9.23, which has C_s symmetry. The La–La distance of 3.585 Å is shorter than those of 3.655 and 3.622 Å for $La_2@C_{80}$ and $La_2@C_{79}N^+$, respectively, as a result of the acceptance of an electron into the La–La bonding orbital and the decreased positive charge on La. In $La_2@C_{79}N$, the nitrogen atom occupies the position that is 4.518 Å away from each La. On the other hand, three electrons are transferred in $La@C_{81}N^+$ from La to the LUMO and LUMO+1 of the azafullerene cage, as in $La@C_{82}$. However, the singly occupied LUMO+1 level is filled up in $La@C_{81}N$ by accepting an electron. Thus, the cage atoms of both $La_2@C_{79}N$ and $La@C_{81}N$ have no significant radical character. This is in sharp contrast with the high spin localization onto the carbons near nitrogen in $C_{59}N$ and $C_{69}N$, which leads to facile dimerization and abstraction [119].

Because of the unique stability and properties, it is interesting to prepare and isolate $La_2@C_{79}N$ and $La@C_{81}N$ in macroscopic amounts in order to develop a new class of cage molecules promising for material applications.

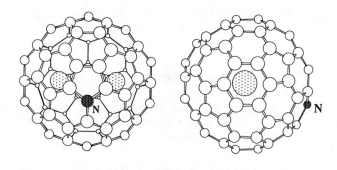

19

Figure 9.23 Two views of the optimized structure (C_s) of $La_2@C_{79}N$.

9.6 CONCLUSION

The fundamental studies of endohedral metallofullerenes have greatly progressed in the last few years through a close interplay of theoretical calculations
and experiments. Almost without exception, it has been found up to now that
energetically stable structures with a large HOMO–LUMO gap are abundantly
produced. The cage structures do not necessarily coincide with those of abundant fullerenes because of important electron transfer from metals to fullerenes.
In addition, new cage structures that violate the isolated pentagon rule and
contain heptagonal rings or heteroatoms can be stabilized by metal-doping,
enriching the research area of endohedral metallofullerenes. Electrostatic interactions play an important role in determining the cage structures and metal
positions. It is very likely that the endohedral structures are formed during the
annealing process. In conclusion, endohedral metallofullerenes differ significantly in electronic properties and reactivities from empty fullerenes. It is hoped
that the intersting differences constitute an important stepping stone on the way
to material, catalytic, and biological applications.

REFERENCES

1. J. R. Heath, S. C. O'Brein, Q. Zhang, Y. Liu, R. F. Curl, H. W. Kroto, F. K.
   Tittle, and R. E. Smalley, *J. Am. Chem. Soc.* **1985**, *107*, 7779.

2. H. Kroto, J. R. heath, S. C. O'Brien, R. E. Curl, and R. E. Smally, *Nature* **1985**,
   318, 162.

3. Y. Chai, T. Guo, C. Jin, R. E. Haufler, L. P. F. Chibante, J. Fure, L. Wang, J. M.
   Alford, and R. E. Smalley, *J. Phys. Chem.* **1991**, *95*, 7564.

4. For recent reviews, see: (a) D. S. Bethune, R. D. Johnson, J. R. Salem, M. S. de
   Vries, and C. S. Yannoni, *Nature* **1993**, *366*, 123; (b) S. Nagase, K. Kobayashi, and
   T. Akasaka, *Bull. Chem. Soc. Jpn.* **1996**, *69*, 2131.

5. M. Inakuma, M. Ohno, and H. Shinohara, "ESR Studies of Endohedral Mono-metallofullerenes" in *Recent Advances in the Chemistry and Physics of Fullerenes and Related Materials*, K. M. Kadish and R. S. Ruoff (eds.). The Electrochemical Society, Pennington, NJ, 1995, p. 330.

6. K. Kikuchi, Y. Nakao, S. Suzuki, Y. Achiba, T. Suzuki, and Y. Maruyama, *J. Am. Chem. Soc.* **1994**, *116*, 9367.

7. (a) K. Kikuchi, S. Suzuki, Y. Nakao, H. Nakahara, T. Wakabayashi, H. Shiromaru, K. Saito, I. Ikemoto, and Y. Achiba, *Chem. Phys. Lett.* **1993**, *216*, 67; (b) K. Yamamoto, H. Funasaka, T. Takahashi, and T. Akasaka, *J. Phys. Chem.* **1994**, *98*, 208.

8. (a) K. Kikuchi, K. Kobayashi, S. Sueki, S. Suzuki, H. Nakahara, Y. Achiba, K. Tomura, and M. Katada, *J. Am. Chem. Soc.* **1994**, *116*, 9775; (b) H. Funasaka, K. Sugiyama, K. Yamamoto, and T. Takahashi, *J. Phys. Chem.* **1995**, *99*, 1826; (c) H. Funasaka, K. Sakurai, Y. Oda, K. Yamamoto, and T. Takahashi, *Chem. Phys. Lett.* **1995**, *232*, 273.

9. T. Suzuki, Y. Maruyama, T. Kato, K. Kikuchi, Y. Nakao, Y. Achiba, K. Kobayashi, and S. Nagase, *Angew. Chem. Int. Ed. Engl.* **1995**, *34*, 1094.

10. H. Shinohara, H. Yamaguchi, N. Hayashi, H. Sato, M. Ohkohchi, Y. Ando, and Y. Saito, *J. Phys. Chem.* **1993**, *97*, 4259.

11. H. Shinohara, M. Inakuma, N. Hayashi, H. Sato, Y. Saito, T. Kato, and S. Bandow, *J. Phys. Chem.* **1994**, *98*, 8597.

12. (a) S. Nagase, K. Kobayashi, and T. Akasaka, *J. Mol. Struct.* (*Theochem*) **1997**, *398*, 221; (b) S. Nagase, K. Kobayashi, and T. Akasaka, *J. Comput. Chem.* **1998**, *19*, 232; (c) S. Nagase, K. Kobayashi, and T. Akasaka, *J. Mol. Struct.* (*Theochem*), **1999**, *461&462*, 97.

13. (a) H. W. Kroto, *Nature* **1987**, *329*, 529; (b) T. Schmaltz, W. A. Seitz, D. J. Klein, and G. E. Hite, *J. Am. Chem. Soc.* **1988**; *110*, 1113.

14. Y. Achiba, K. Kikuchi, Y. Aihara, T. Wakabayashi, Y. Miyake, and M. Kainosho, "Trends in Large Fullerenes: Are They Balls or Tubes" in *The Chemical Physics of Fullerenes 10 (and 5) Years Later*, W. Andreoni (ed.). Kluwer Academic Publishers, Dordrecht, 1996, p. 139.

15. P. W. Fowler and D. E. Manolopoulos, *An Atlas of Fullerenes*. Clarendon, Oxford, 1995.

16. K. Kikuchi, H. Nakahara, T. Wakabayashi, S. Suzuki, H. Shiromaru, Y. Miyake, K. Saito, I. Ikemoto, M. Kainosho, and Y. Achiba, *Nature* **1992**, *357*, 142.

17. S. Nagase, K. Kobayashi, T. Kato, and Y. Achiba, *Chem. Phys. Lett.* **1993**, *201*, 475.

18. Y. Achiba, K. Kikuchi, Y. Aihara, T. Wakabayashi, Y. Miyake, and M. Kainosho, *Mat. Res. Soc. Symp. Proc.* **1995**, *359*, 3.

19. Z. Slanina, S.-L. Lee, K. Kobayashi, and S. Nagase, *J. Mol. Struct.* (*Theochem*) **1995**, *339*, 89.

20. S. Nagase and K. Kobayashi, *Chem. Phys. Lett.* **1993**, *214*, 57.

21. D. M. Poirier, M. Knupfer, J. H. Weaver, W. Andreoni, K. Laasonen, M. Parrinello, D. S. Bethune, K. Kikuchi, and Y. Achiba, *Phys. Rev. B* **1994**, *49*, 17403.

22. W. Andreoni and A. Curioni, *Phys. Rev. Lett.* **1996**, *77*, 834.

23. (a) J. Schulte, M. C. Böhm, and K.-P. Dinse, *Chem. Phys. Lett.* **1996**, *259*, 48; (b) J. Schulte, M. C. Böhm, and K.-P. Dinse, *J. Mol. Struct. (Theochem)* **1998**, *427*, 279.

24. K. Kobayashi and S. Nagase, *Chem. Phys. Lett.* **1988**, *282*, 325.

25. M. Hoinkis, C. S. Yannoni, D. S. Bethune, J. R. Salem, R. D. Johnson, M. S. Crowder, and M. S. de Vries, *Chem. Phys. Lett.* **1992**, *198*, 461.

26. S. Suzuki, S. Kawata, H. Shiromaru, K. Yamauchi, K. Kikuchi, T. Kato, and Y. Achiba, *J. Phys. Chem.* **1992**, *96*, 7159.

27. U. Kirbach and L. Dunsch, *Angew. Chem. Int. Ed. Engl.* **1996**, *35*, 2380.

28. C.-H. Park, B. O. Wells, J. DiCarlo, Z.-X. Shen. J. R. Salem, D. S. Bethune, C. S. Yannoni, R. D. Johnson, M. S. de Vries, C. Booth, F. Bridges, and P. Pienetta, *Chem. Phys. Lett.* **1993**, *213*, 196.

29. M. Nomura, Y. Nakao, K. Kikuchi, and Y. Achiba, *Physica B* **1995**, *208&209*, 539.

30. M. Takata, B. Umeda, E. Nishibori, M. Sakata, Y. Saito, M. Ohno, and H. Shinohara, *Nature* **1995**, *377*, 46.

31. M. Takata, E. Nishibori, M. Sakata, Y. Saito, and H. Shinohara, "The Endohedral Nature of The Metallofullerene Y@C$_{82}$" in *Fullerenes and Fullerene Nanostructures*, H. Kuzmany, J. Fink, M. Mehring, and S. Roth (eds.). World Scientific, Singapore, 1996, p. 155.

32. (a) E. Nishibori, M. Takata, M. Sakata, M. Inakuma, and H. Shinohara, *Chem. Phys. Lett.* **1998**, *298*, 79; (b) T. Akasaka, T. Wakahara, S. Nagase, K. Kobayashi, M. Wälchli, K. Yamamoto, M. Kondo, S. Shirakura, S. Shingo, Y. Maeda, T. Kato, M. Kako, Y. Nakadaira, X. Gao, E. V. Caemelbecke, and K. M. Kadish, submitted.

33. (a) Z. Xu, T. Nakane, and H. Shinohara, *J. Am. Chem. Soc.* **1996**, *118*, 11310; (b) T. J. S. Dennis and H. Shinohara, *Appl. Phys. A*, **1998**, *66*, 243.

34. K. Kobayashi and S. Nagase, *Chem. Phys. Lett.* **1997**, *274*, 226.

35. R. D. Johnson, M. S. de Vries, J. Salem, D. S. Bethune, and C. S. Yannoni, *Nature* **1992**, *355*, 239.

36. (a) H. Shinohara, H. Sato, M. Ohkohcni, Y. Ando, T. Kodama, T. Shida, T. Kato, and Y. Saito, *Nature* **1992**, *357*, 52; (b) C. S. Yannoni, M. Hoinkis, M. S. de Vries, D. S. Bethune, J. R. Salem, M. S. Crowder, and R. D. Johnson, *Science* **1992**, *256*, 1191.

37. (a) J. H. Weaver, Y. Chai, G. H. Kroll, C. Jin, T. R. Ohno, R. E. Haufler, T. Guo, J. M. Alford, J. Conceicao, L. P. F. Chibante, A. Jain, G. Palmer, and R. E. Smalley, *Chem. Phys. Lett.* **1992**, *190*, 460; (b) H. Shinohara, H. Sato, Y. Saito, M. Ohkohchi, and Y. Ando, *J. Phys. Chem.* **1992**, *96*, 3571.

38. J. P. Desclaux, *At. Data Nucl. Data Tables* **1973**, *12*, 311.

39. K. Laasonen, W. Andreoni, and M. Parrinello, *Science* **1992**, *258*, 1916.

40. S. Hino, H. Takahashi, K. Iwasaki, K. Matumoto, T. Miyazaki, H. Hasegawa, K. Kikuchi, and Y. Achiba, *Phys. Rev. Lett.* **1993**, *71*, 4261.

41. M. Rübsam, M. Plüschau, P. Schweitzer, K.-P. Dinse, D. Fuchs, H. Rietschel, R. H. Michel, M. Benz, and M. M. Kappes, *Chem. Phys. Lett.* **1995**, *240*, 615.

42. M. Rübsam, M. Plüschau, J. Schutle, K.-P. Dinse, D. Fuchs, H. Rietschel, R. H. Michel, M. Benz, and M. M. Kappes, "One- and Two-dimensional EPR Investi-

gation of Metallo-endofullerenes in solution" in *Recent Advances in the Chemistry and Physics of Fullerenes and Related Materials*, K. M. Kadish and R. S. Ruoff (eds.) The Electrochemical Society, Pennington, NJ, 1995, p. 724.

43. T. Kato, S. Suzuki, K. Kikuchi, and Y. Achiba, *J. Phys. Chem.* **1993**, *51*, 13425.

44. J. Ding, L.-T. Weng, and S. Yang, *J. Phys. Chem.* **1996**, *100*, 11120.

45. J. Ding and S. Yang, *J. Am. Chem. Soc.* **1996**, *118*, 11254.

46. J. Ding, N. Lin, L.-Y. Weng, N. Cue, and S. Yang, *Chem. Phys. Lett.* **1996**, *261*, 92.

47. K. Kobayashi and S. Nagase, *Chem. Phys. Lett.* **1996**, *262*, 227.

48. For the values ($I_p = 6.2$–6.4 eV and $E_a = 2.7$–3.3 eV) for La@C_n ($n = 60, 74, 82$) bracketed from charge exchange reactions, see: R. L. Hettich, Z. C. Ying, and R. N. Compton, "Structural Determination and Ionic Properties of Endoehedral Lanthanum Fullerenes" in *Recent Advances in the Chemistry and Physics of Fullerenes and Related Materials*, K. M. Kadish and R. S. Ruoff (eds.). The Electrochemical Society, Pennington, NJ, 1995, p. 1457. For the values ($E_a = 3.2$–3.3 eV) for Gd@C_{82} ($n = 74, 76, 78, 80$, and 82) from Knudsen cell mass spectrometry, see: O. V. Boltalina, I. N. Ioffe, I. D. Sorokin, and L. N. Sidorov, *J. Phys. Chem. A* **1997**, *101*, 9561.

49. S. Nagase and K. Kobayashi, *Chem. Phys. Lett.* **1994**, *228*, 106.

50. S. Nagase and K. Kobayashi, *J. Chem. Soc. Chem. Commun.* **1994**, 1837.

51. D. L. Lichtenberger, M. E. Jatcko, K. W. Nebesny, C. D. Ray, D. R. Huffman, and L. D. Lamp, *Chem. Phys. Lett.* **1992**, *198*, 459.

52. L.-S. Wang, J. Conceicao, C. Jin, and R. E. Smalley, *Chem. Phys. Lett.* **1991**, *182*, 5.

53. (a) D. M. Cox, D. J. Trevor, K. C. Reichmann, and A. Kaldor, *J. Am. Chem. Soc.* **1986**, *108*, 2457; (b) D. M. Cox, K. C. Reichmann, and A. Kaldor, *J. Chem. Phys.* **1988**, *88*, 1588.

54. H. Shinohara, M. Inakuma, M. Kishida, S. Yamazaki, T. Hashizume, and T. Sakurai, *J. Phys. Chem.* **1995**, *99*, 13769.

55. T. Suzuki, Y. Maruyama, T. Kato, K. Kikuchi, and Y. Achiba, *J. Am. Chem. Soc.* **1993**, *115*, 11006.

56. T. Suzuki, K. Kikuchi, F. Oguri, Y. Nakao, S. Suzuki, Y. Achiba, K. Yamamoto, H. Funasaka, and T. Takahashi, *Tetrahedron* **1996**, *52*, 4973.

57. (a) T. Akasaka, S. Okubo, T. Wakahara, K. Yamamoto, T. Kato, T. Suzuki, S. Nagase, and K. Kobayashi, "Isolation and Characterization of A Pr@C_{82} Isomer" in *Recent Advances in the Chemistry and Physics of Fullerenes and Related Materials*, K. M. Kadish and R. S. Ruoff (eds.). The Electrochemical Society, Pennington, NJ, 1999, p. 771; (b) T. Akasaka, S. Okubo, M. Kondo, Y. Maeda, T. Wakahara, T. Kato, T. Suzuki, K. Yamamoto, K. Kobayashi, and S. Nagase, *Chem. Phys. Lett.*, **2000**, *319*, 153.

58. T. Akasaka, S. Nagase, K. Kobayashi, T. Suzuki, T. Kato, K. Yamamoto, H. Funasaka, and T. Takahashi, *J. Chem. Soc. Chem. Commun.* **1995**, 1343.

59. K. Yamamoto, H. Funasaka, T. Takahashi, T. Akasaka, T. Suzuki, and Y. Maruyama, *J. Phys. Chem.* **1994**, *98*, 12831.

60. For recent reviews, see: (a) A. Hirsch, *The Chemistry of the Fullerenes*. Thieme, New York, 1994; (b) *The Chemistry of Fullerenes*, R Taylor (ed.). World Scientific, London, 1995.

61. T. Akasaka, W. Ando, K. Kobayashi, and S. Nagase, *J. Am. Chem. Soc.* **1993**, *115*, 1605.

62. T. Akasaka, W. Ando, K. Kobayashi, and S. Nagase, *Fullerene Sci. Technol.* **1993**, *1*, 339.

63. T. Akasaka, W. Ando, K. Kobayashi, and S. Nagase, *J. Am. Chem. Soc.* **1993**, *115*, 10366.

64. T. Akasaka, W. Ando, K. Kobayashi, and S. Nagase, *J. Am. Chem. Soc.* **1994**, *116*, 2627.

65. T. Suzuki, Y. Maruyama, T. Akasaka, W. Ando, K. Kobayashi, and S. Nagase, *J. Am. Chem. Soc.* **1994**, *116*, 1359.

66. T. Akasaka, W. Ando, K. Kobayashi, and S. Nagase, *Trans, Mat. Res. Soc. Jpn.* **1994**, *14B*, 1091.

67. T. Akasaka, W. Ando, K. Kobayashi, and S. Nagase, (1994) p. 723, "Chemical Derivatization of Fullerenes with Organosilicon Compounds"; (1995) p. 1125, "Chemical Derivatization of Fullerenes: [4 + 3] Cycloaddition of C_{60}" in *Recent Advances in the Chemistry and Physics of Fullerenes and Related Materials*, K. M. Kadish and R. S. Ruoff (eds.). The Electrochemical Society, Pennington, NJ.

68. T. Akasaka, E. Mitsuhida, W. Ando, K. Kobayashi, and S. Nagase, *J. Chem. Soc. Chem. Commun.* **1995**, 1529.

69. For a recent review, see: T. Akasaka, S. Nagase, and K. Kobayashi, *J. Synth. Org. Chem. Jpn.* **1996**, *54*, 580.

70. T. Akasaka, T. Kato, K. Kobayashi, S. Nagase, K. Yamamoto, H. Funasaka, and T. Takahashi, *Nature* **1995**, *374*, 600.

71. T. Akasaka, T. Kato, K. Kobayashi, S. Nagase, K. Yamamoto, H. Funasaka, and T. Takahashi, *Tetrahedron* **1996**, *52*, 5015.

72. T. Suzuki, Y. Maruyama, T. Kato, T. Akasaka, K. Kobayashi, S. Nagase, K. Yamamoto, H. Funasaka, and T. Takahashi, *J. Am. Chem. Soc.* **1995**, *117*, 9606.

73. T. Akasaka, Y. Niino, S. Okubo, T. Wakahara, M. T. Liu, T. Kato, S. Nagase, K. Kobayashi, and K. Yamamoto, to be published.

74. (a) A. B. Smith, R. M. Strongin, L. Brard, W. J. Romanow, M. Saunders, H. A. Jiménez-Vázquez, and R. J. Cross, *J. Am. Chem. Soc.* **1994**, *116*, 10831; (b) M. Saunders, H. A. Jiménez-Vázquez, R. J. Cross, and R. J. Poreda, *Science* **1993**, *259*, 1428; (c) M. Saunders, H. A. Jiménez-Vázquez, R. J. Cross, S. Mroczkowski, D. I. Freedburg, and F. A. L. Anet, *Nature* **1994**, *367*, 256; (d) M. Saunders, H. A. Jiménez-Vázquez, R. J. Cross, S. Mroczkowski, M. L. Cross, D. E. Giblin, and R. J. Poreda, *J. Am. Chem. Soc.* **1994**, *116*, 2193; (e) M. Saunders, H. A. Jiménez-Vázquez, B. W. Bangerter, R. J. Cross, S. Mroczkowski, D. I. Freedburg, and F. A. Anet, *J. Am. Chem. Soc.* **1994**, *116*, 3621; (f) M. Saunders, H. A. Jiménez-Vázquez, R. J. Cross, W. E. Billups, C. Gesenberg, A. Gonzalez, W. Luo, R. C. Haddon, F. Diederich, and A. Hermann, *J. Am. Chem. Soc.* **1995**, *117*, 9305.

75. K. Kobayashi, S. Nagase, and T. Akasaka, *Chem. Phys. Lett.* **1995**, *245*, 230.

76. F. H. Hennrich, R. H. Michel, A. Fischer, S. Richard-Schneider, S. Gilb, M. M. Kappes, D. Fuchs, M. Bürk, K. Kobayashi, and S. Nagase, *Angew. Chem. Int. Ed. Engl.* **1996**, *35*, 1732.

77. T. Akasaka, S. Nagase, K. Kobayashi, M. Wälchli, K. Yamamoto, H. Funasaka, M. Kato, T. Hoshino, and T. Erata, *Angew. Chem. Int. Ed. Engl.* **1997**, *36*, 1643.

78. M. Moriyama, T. Sato, A. Yabe, T. Akasaka, T. Wakahara, S. Nagase, K. Kobayashi, and K. Yamamoto, *Chem. Lett.*, in press.

79. (a) K. Yamamoto, T. Ishiguro, K. Sakurai, H. Funasaka, and T. Akasaka, "Isolation and Characterization of $La_2@C_{80}$ Isomers" in *Recent Advances in the Chemistry and Physics of Fullerenes and Related Materials*, K. M. Kadish and R. S. Ruoff (eds.). The Electrochemical Society, Pennington, NJ, 1997, p. 743; (b) K. Yamamoto, T. Akasaka, M. Okamura, S. Nagase, and K. Kobayashi, in preparation.

80. (a) K. B. Shekimov and M. F. Jarrold, *J. Am. Chem. Soc.* **1995**, *117*, 6404; (b) K. B. Shekimov and M. F. Jarrold, *J. Am. Chem. Soc.* **1996**, *118*, 1139.

81. K. B. Shelkimov, D. E. Clemmer, and M. F. Jarrold, *J. Chem. Soc. Dalton Trans.* **1996**, 567.

82. H. Shinohara, N. Hayashi, H. Sato, Y. Saito, X.-D. Wang, T. Hashizume, and T. Sakurai, *J. Phys. Chem.* **1993**, *97*, 13438.

83. R. Beyers, C.-H. Kiang, R. D. Johnson, J. R. Salem, M. S. de Vries, C. S. Yannoni, D. S. Bethune, H. C. Dorn, P. Burbank, K. Harich, and S. Stevenson, *Nature*, **1994**, *370*, 196.

84. S. Nagase and K. Kobayashi, *Chem. Phys. Lett.* **1994**, *231*, 319.

85. S. Nagase and K. Kobayashi, *Chem. Phys. Lett.* **1977**, *276*, 55.

86. (a) E. Yamamoto, M. Tansho, T. Tomiyama, H. Shinohara, H. Kawahara, and Y. Kobayashi, *J. Am. Chem. Soc.* **1996**, *118*, 2293; (b) H. Shinohara, private communication.

87. Y. Miyake, S. Suzuki, Y. Kojima, K. Kikuchi, K. Kobayashi, S. Nagase, M. Kainosho, Y. Achiba, Y. Maniwa, and K. Fisher, *J. Phys. Chem.* **1996**, *100*, 9579.

88. (a) L. B. Knight, A. J. Mckinley, R. M. Babb D. W. Hill, and M. D. Morse, *J. Chem. Phys.* **1993**, *99*, 7376; (b) H. Åkeby, L. G. M. Pettersson, and P. E. M. Siegbahn, *J. Chem. Phys.* **1992**, *97*, 1850; (c) I. Pápai and M. Castro, *Chem. Phys. Lett.* **1997**, *267*, 551.

89. M. Takata, E. Nishibori, B. Umeda, M. Sakata, E. Yamamoto, and H. Shinohara, *Phys. Rev. Lett.* **1997**, *78*, 3330.

90. H. Shinohara, H. Yamaguchi, N. Hayashi, H. Sato, M. Ohkohchi, Y. Ando, and Y. Saito, *J. Phys. Chem.* **1993**, *97*, 4259.

91. T. Takahashi, A. Ito, M. Inakuma, and H. Shinohara, *Phys. Rev. B* **1995**, *52*, 13812.

92. P. W. Stephens, G. Bortel, G. Faigel, M. Tegze, A. Janossy, S. Pekker, G. Oszlanyi, and L. Forro, *Nature* **1994**, *370*, 636.

93. G. Wang, K. Komatsu, Y. Murata, and M. Shiro, *Nature* **1997**, *387*, 583.

94. C. H. Choi and M. Kertesz, *Chem. Phys. Lett.* **1998**, *282*, 318.

95. T. Guo, G. K. Odom, and G. E. Scuseria, *J. Phys. Chem.* **1994**, *98*, 7745.

96. K. Kobayashi and S. Nagase, *Chem. Phys. Lett.*, **1999**, *302*, 312.

97. (a) R. F. W. Bader, *Atoms in Molecules—A Qunatum Theory*, Oxford University Press, Oxford, 1990; (b) R. F. W. Bader, *Chem. Rev.* **1991**, *91*, 893.

98. R. F. W. Bader, *J. Phys. Chem. A* **1998**, *102*, 7314.

99. K. Kobayashi, S. Nagase, and T. Akasaka, *Chem. Phys. Lett.* **1996**, *261*, 502.

100. T. Akasaka, S. Nagase, K. Kobayashi, T. Suzuki, T. Kato, K. Kikuchi, Y. Achiba, K. Yamamoto, H. Funasaka, and T. Takahashi, *Angew. Chem. Int. Ed. Engl.* **1995**, *34*, 2139.

101. T. C. Bedard and J. S. Moore, *J. Am. Chem. Soc.* **1995**, *117*, 10662.

102. (a) J. Cioslowski and A. Nanayakkara, *J. Chem. Phys.* **1992**, *96*, 8354; (b) J. Cioslowski, *Electronic Structure Calculations on Fullerenes and Their Derivatives*, Oxford University Press, New York, 1995, Chap. 8.

103. J. Cioslowski and E. D. Fleischmann, *J. Chem. Phys.* **1991**, *94*, 3730.

104. (a) S. Nagase and K. Kobayashi, *Chem. Phys. Lett.* **1991**, *187*, 291; (b) S. Nagase, *Pure Appl. Chem.* **1993**, *65*, 675.

105. However, the internal rotation is hindered by a large barrier of ca. 50 kcal/mol [84]. On the other hand, two Sc atoms in $Sc_2@C_{84}$ (C_s) can rotate inside the cage at high temperatures because of a small barrier of ca. 25 kcal/mol. This is confirmed by observing that two ^{45}Sc NMR signals coalesce at 383 K [87].

106. As Figure 9.7 shows, $La_2@C_{80}$ is most stable when each La is directed to the center of the hexagonal ring of C_{80}. It is destabilized by 14.3 kcal/mol when each La is directed to the center of the pentagonal ring. Except for this, La can freely move about with barriers of 4.9–5.5 kcal/mol. Thus, there are many paths for the circuit of La. When each La circles about along the hexagonal rings, the barrier is the smallest (4.9 kcal/mol).

107. T. Akasaka, M. Okamura, S. Nagase, K. Kobayashi, M. Wälchli, K. Yamamoto and T. Kato, to be published.

108. For example, see: R. G. Kidd, Chapter 5, "Quadrupolar and Other Types of Relaxation" in *NMR of Newly Accessible Nuclei*, Vol. 1, P. Laszlo (ed.). Academic Press, New York, 1983, p. 103.

109. K. Kobayashi, S. Nagase, M. Yoshida, and E. Osawa, *J. Am. Chem. Soc.* **1997**, *119*, 12693.

110. K. Raghavachari, *Z. Phys. D* **1993**, *26*, S261.

111. A. Ayuela, P. W. Fowler, D. Mitchell, R. Schmidt, G. Seifert, and F. Zerbetto, *J. Phys. Chem.* **1996**, *100*, 15634.

112. (a) T. S. M. Wan, H.-W. Zhang, T. Nakane, Z. Xu, M. Inakuma, H. Shinohara, K. Kobayashi, and S. Nagase, *J. Am. Chem Soc.* **1998**, *120*, 6806. (b) For the recent successful preparation of the C_{74} anion, see: M. D. Diener and J. M. Alford, *Nature*, **1998**, *393*, 668.

113. T. J. S. Dennis and H. Shinohara, private communication.

114. R. C. Haddon, *Science* **1993**, *261*, 1545.

115. H. C. Dorn, S. Stevenson, P. Burbank, K. Harich, Z. Sun, T. Glass, M. Anderson, D. S. Bethune, and M. Sherwood "Isolation and Structure of $Sc_2@C_{74}$" in *Recent Advances in the Chemistry and Physics of Fullerenes and Related Materials*, K. M. Kadish and R. S. Ruoff (eds.). The Electrochemical Society, Pennington, NJ, 1998, p. 990.

116. J. C. Hummelen, B. Knight, J. Pavlovich, R. Gonzalez, and F. Wudl, *Science* **1995**, *269*, 1554.

117. I. Lamparth, B. Nuber, G. Schick, A. Skiebe, T. Grösser, and A. Hirsch, *Angew. Chem. Int. Ed. Engl.* **1995**, *34*, 2257.

118. J. Averdung, H. Luftmann, I. Schlachter, and J. Mattey, *Tetrahedron* **1995**, *51*, 6977.

119. (a) B. Nuber and A. Hirsch, *J. Chem. Soc. Chem. Commun.* **1996**, 1421. (b) K. Prassides, M. Keshavarz-K, J. C. Hummelen, W. Andreoni, P. Giannozzi, E. Beer,

C. Bellavia, L. Cristofolini, R. Gonzalez, A. Lappas, Y. Murata, M. Malecki, V. Srdanov, and F. Wudl, *Science* **1996**, *271*, 1883. (c) C. Bellavia-Lund, M. Keshavarz-K, T. Collins, and F. Wudl, *J. Am. Chem. Soc.* **1997**, *119*, 8101. (d) A. Gruss, K.-P. Dinse, A. Hirsch, B. Nuber, and U. Reuther, *J. Am. Chem. Soc.* **1997**, *119*, 8728. (e) K. Hasharoni, C. Bellavia-Lund, M. Keshavarz-K., G. Srdanov, and F. Wudl, *J. Am. Chem. Soc.* **1997**, *119*, 11128. (f) M. Bühl, A. Curioni, and W. Andreoni, *Chem. Phys. Lett.* **1997**, *274*, 231. (g) T. Pichler, M. Knupfer, M. S. Golden, S. Haffner, R. Friedlein, J. Fink, W. Andreoni, A. Curioni, M. Keshavarz-K, C. Bellavia-Lund, A. Sastre, J.-C. Hummelen, and F. Wudl, *Phys. Rev. Lett.* **1997**, *78*, 4249.

120. T. Akasaka, S. Okubo, T. Wakahara, K. Kobayashi, S. Nagase, M. Kako, Y. Nakadaira, T. Kato, K. Yamamoto, H. Funasaka, Y. Kitayama, and K. Matsuura, *Chem. Lett.* **1999**, 945.

CHAPTER 10

BIOLOGICAL ASPECTS OF FULLERENES

STEPHEN R. WILSON

10.1 INTRODUCTION

Over the past few years, investigations of fullerene biological applications have greatly increased. Much of the early work has been extensively reviewed [1,2]. Several fundamental physical and chemical properties of fullerenes govern how they may be adapted for biological use and are briefly reviewed here [3].

10.1.1 Solubility

First, fullerene C_{60} itself is not water soluble, although many derivatives certainly can be very water soluble [4]. Studies concerning fullerene C_{60} solubilities reported no solubility in polar solvents such as methanol or water. On the other hand, a water suspension of ^{14}C-labeled C_{60} was stable for long periods and could be delivered to cells [5]. Solubilization of C_{60} using cyclodextrins or calixerenes has also been reported [6], as well as the use of other solubilization methods—detergents, phospholipids, micelles, liposomes, and so on [7]. Fullerene–polyethylene glycol (PEG) derivatization has been especially successful [8]. Monofunctional derivatives of C_{60} containing polar side chains are sometimes quite water soluble. For example, Brettreich and Hirsch [9] have recently reported preparation of the extremely water-soluble fullerene C_{60} dendrimer **1** (solubility ~8.7 mg/mL water at pH = 7.4).

Polyadducts such as the polyhydroxyfullerols $C_{60}(OH)_n$ are often very water soluble and have often been used for biological experiments (see compound **6**, Sections 10.2.2 and 10.8). As a general rule, the greater the number of polar groups added to the fullerene core, the greater the water solubility. On the other hand, many important biological applications also employ lipophilic fullerenes.

A recent report describes the octanol/water partition coefficient (log P) for

Fullerenes: Chemistry, Physics, and Technology, Edited by Karl M. Kadish and Rodney S. Ruoff.
0-471-29089-0 Copyright © 2000 John Wiley & Sons, Inc.

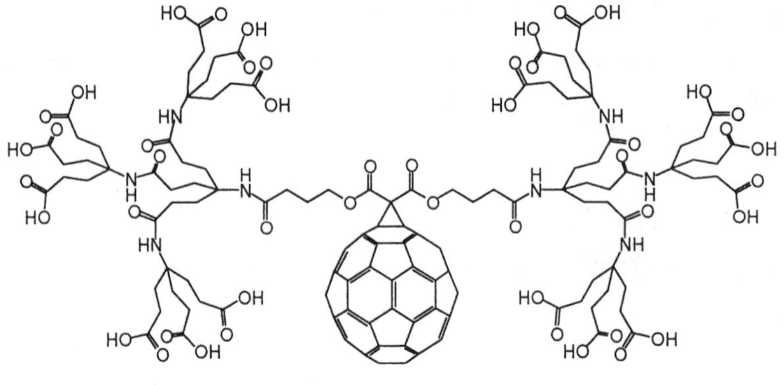

1

several fullerene compounds [2c]. In order to maximize the chances of advantageous adsorption and distribution properties of bioactive compounds, it is generally accepted that there are certain desirable ranges of lipophilicity (log P values of -2 to $+2$). The partition coefficient for C_{60} itself (calculated as log $P = 4.5$) contrasts with a C_3-tris(malonate) (see compound **15**, Section 10.4) with log $P = 0.54$ at pH 7.2.

10.1.2 Redox Behavior

C_{60} may be reversibly reduced by up to six electrons [10]. The electron affinity of C_{60} can be explained qualitatively by considering its numerous pyracylene units, which upon receiving two electrons go from an unstable $4n$ pi system to a stable aromatic $4n + 2$ pi system. Thus, the reduction potential of C_{60} is similar to a quinone and leads to applications in quenching of oxygen radicals [11].

10.1.3 Photophysical Properties

Fullerenes absorb strongly in the ultraviolet (UV) and moderately in the visible regions of the spectrum, form a long-lived triplet state, and sensitize the formation of singlet molecular oxygen with almost 100% efficiency [12a]. Applications that require photosensitizers, such as photodynamic therapy (PDT), are thus a fruitful area of fullerene research. Nonlinear optical properties [12b] observed for fullerenes also point to applications for protection against strong UV irradiation.

10.1.4 Free Radical Trapping

C_{60} has been called a "free radical sponge" [13]. It rapidly reacts with all sorts of reactive free radicals and a number of biomedical applications take advantage of this property.

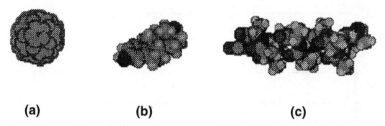

(a)   **(b)**   **(c)**

Figure 10.1   (a) C_{60}. (b) Steroid (premarin, MW $\sim$ 250). (c) Peptide alpha-helix.

10.1.5 Chemical Reactivity

The chemical reactivity of fullerene C_{60} is now fairly well established [3a]. Many derivatives of C_{60} and C_{70} are known, and numerous fullerene compounds are now available for study. This topic is reviewed in detail in Chapter 3.

10.1.6 Shape and Size

C_{60} is an extraordinarily symmetrical molecule (symmetry group I_h) [14]. It is 7.2 Å in diameter and about the size of a steroid or peptide alpha-helix (Fig. 10.1).

If one considers only the molecular weight (MW) as a measure of size, C_{60} seems large (MW = 720), but based on true surface area, the size of fullerene C_{60} is less than one-third that expected for molecules of the same molecular weight. This is undoubtedly due to the unusual close packing of its 60 carbon atoms. Based on calculations of the molecular surface area (Fig. 10.2), a fullerene C_{60} molecule has the effective size of a $C_{16}H_{34}$ hydrocarbon, that is, a 16-carbon compound with MW = 226. Figure 10.2 shows a plot of surface areas and molecular weights for a series of straight-chain hydrocarbons from C_4 to C_{20} [15] compared to fullerene C_{60}. This dramatically demonstrates the poor correlation of the MW of C_{60} in comparison to the surface areas of the other compounds.

The molecular size of a potential drug is an important factor in assessing oral bioavailability. Compounds of lower molecular weight (ideally less than MW = 700) appear to have higher prospects for oral bioavailability. For this type of analysis, the "effective molecular weight" of the fullerene C_{60} core is 226 amu based on data from Figure 10.2.

10.2 ENZYME INHIBITION/RECEPTOR BINDING

10.2.1 An Early Example

In 1996 Tokuyama and co-workers reported one of the first examples of a biological study of a fullerene compound [16]. Fullerene compounds **2a–c** were

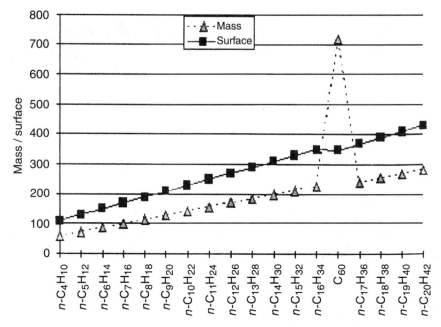

Figure 10.2   Surface areas and molecular weights of straight-chain hydrocarbons versus fullerene-C_{60}.

tested in a series of enzyme assays and found to inhibit cysteine proteases (*m*-calpain and papain) and serine proteases (trypsin, plasmin, and thrombin), especially when exposed to low-energy light.

2a R = COCH$_2$CH$_2$COOH
2b R = H
2c R = COCH$_2$CH$_2$COO–

The same compounds were inactive against cathepsin D, acyl-COA cholesterol acyltransferase, diacylglycerol acyltransferase, HIV-reverse transcriptase (RT), and several steroid biosynthesis enzymes.

10.2.2   HIV Protease

Also in 1993, Friedman and co-workers proposed that a C_{60} molecule could fit snugly into the hydrophobic cavity of HIV protease and inhibit the enzyme

[17]. Using molecular modeling, Friedman fit an energy-minimized C_{60} derivative **3** into the enzyme active site of HIV protease, showing good van der Waals interactions with the hydrophobic surface of the cavity.

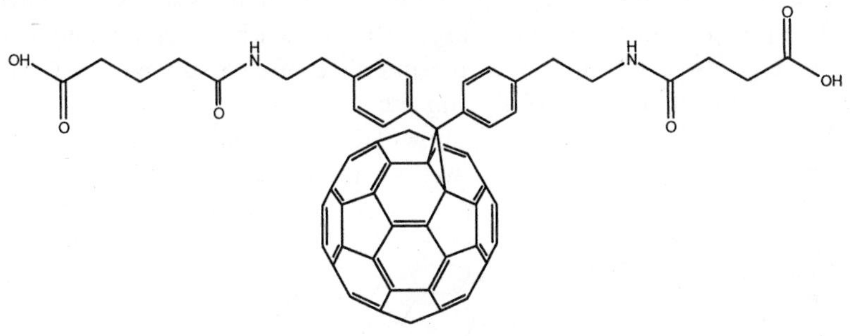

3

Experimentally, compound **3** was shown to be a good competitive inhibitor of HIV-1 protease with a K_i value of 5.3 µM [17b]. For comparison, the best peptide-based protease inhibitors are effective in the low nanomolar range. Binding of fullerene HIV-1 protease inhibitors was optimized by computer by manipulating the structure while computationally predicting increases in hydrophobic desolvation. The affinity and activity of fullerene-based inhibitors were increased by evaluation of hydrophobic contacts of hypothetical derivatives docked into the active site [17c].

Two target molecules were identified by this method and subsequently synthesized. A diphenyl C_{60} alcohol **4** and a diisopropyl cyclohexyl C_{60} alcohol **5** showed improved anti-HIV-1 activities with K_i values of 103 and 150 nM, respectively.

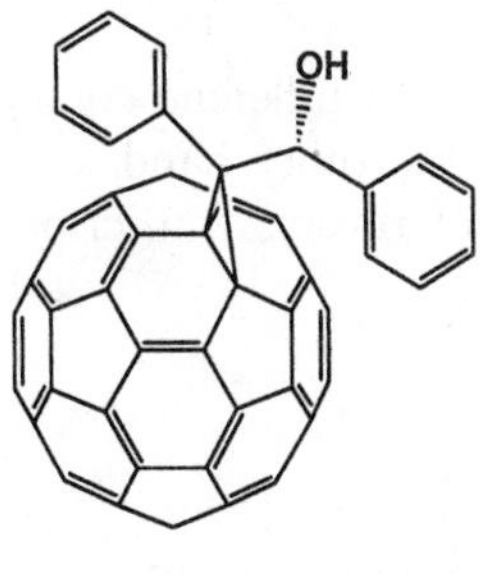

4 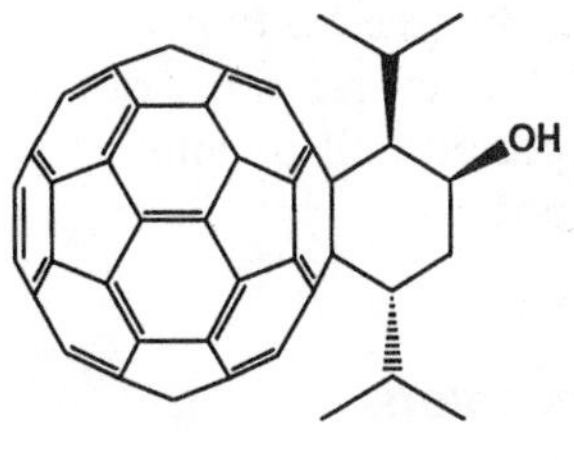   **5**

Because of the success of this approach and consistency of the model with experimental data, other groups have sought to improve on the binding affinity of these compounds in a rational manner. More examples of promising fullerene-based antiviral drugs are discussed in Section 10.3.

10.2.3 Redox Enzymes

The inhibition of glutathione-S-transferase (GST) by C_{60} [18] was recently reported. C_{60} was solubilized with polyvinylpyrrolidone and showed non-competitive inhibition GST activity in mouse liver in an assay using ethacrynic acid. The possible inhibition mechanism was shown by docking studies using SYBYL. C_{60} fits in the hollow at the interface of two subunits of GST and causes allosteric changes in the binding site.

Polyhydroxyfullerol **6** has been shown to inhibit microsomal cytochrome P_{450}-dependent monooxygenases as well as mitochondrial oxidative phosphorylation [19].

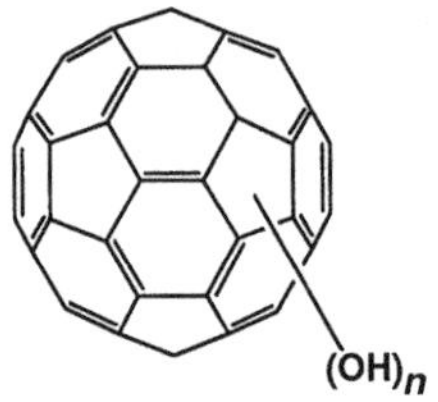

6 (n= **6–15**)

Monooxygenase activities toward benzo[a]pyrene, 7-ethoxycoumarin, aniline, and erythromycin have IC_{50} values of 42, 94, 102, and 349 μM, respectively. Fullerenol **6** exhibited noncompetitive and mixed-type inhibition in benzo[a]-pyrene hydroxylation. Additions of fullerol **6** to rat liver mitochondria resulted in a dose-dependent inhibition of adenosine 5′-diphosphate (ADP)-induced uncoupling and mitochondrial adenosine triphosphatase (ATPase) activity with an IC_{50} value of 7.1 μM.

10.2.4 Reverse Transcriptase

An early report (see Section 10.2.1) suggested that fullerene compounds **2a–c** did not inhibit reverse transcriptase [15]. On the other hand, nonderivatized C_{60} was reported [20] to inhibit the activity of reverse transcriptase with an $IC_{50} \sim 0.3$ μM.

10.2.5 Receptors

Several reports of fullerene compound interaction directly with biological receptors have appeared. In one approach, the Wilson group [21] prepared a fullerene–estrone hybrid compound **7**.

Compound **7** had estrogenic activity, binding to cytosolic estradiol receptor with $K_d \sim 40$ μM. In another report, Toniolo et al. [22] prepared a fullerene peptide compound **8** that is an analog of peptide T. This compound exhibits potent activity in a CD4 receptor-mediated human monocyte chemotaxis assay.

7

8

The effects of C_{60} solubilized with polyvinylpyrrolidine on central nervous system (CNS) receptor response in various tissues has been examined. No effect on receptor responses of several neurotransmitters could be observed in a guinea pig ileum. For example, acetylcholine-, L-isoprenoaline-, and serotonin-mediated responses were unaffected by the fullerene at 4 μM concentration [23].

10.2.6 Direct Binding Information

Computer models for the interaction of fullerenes with peptides, proteins, and DNA have been reported. Calculated binding studies of fullerenes to HIV protease [17a,c] and glutathione-*S*-transferase [42] have appeared. In addition, computed interaction of a fullerene compound with DNA [24] and binding of C_{60} to a specific peptide helix [25a] (Fig. 10.3) show evidence for molecular recognition. A recent report by Mi et al. [25b] describes extensive molecular dynamics simulations for the binding of several anti-HIV fullerene compounds with HIV protease.

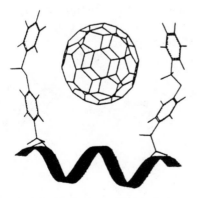

Figure 10.3   Model of helical peptide binding to fullerene-C_{60}.

10.3 ANTICANCER/ANTIVIRAL

10.3.1 Anti-HIV

A water-soluble fullerene derivative was shown to be very active against HIV protease in cell culture [26]. N-tris(hydroxymethyl)propylamido methanofullerene C_{60} derivative **9** was quite active against HIV-l in acutely infected human peripheral blood mononuclear (PBM) cells ($EC_{50} = 2.5$ µM and exhibited no cytotoxicity in two human and one monkey cell system ($LD_{50} > 100$ µM in PBM, CEM (a human leukemia cell line), and Vero cells).

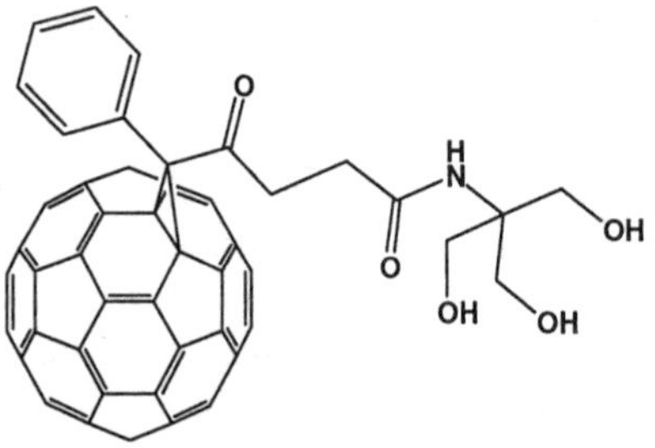

9

Fullerene derivative **3** (see Section 10.2.2) was also tested against HIV-1 chronically infected cells. In PBM cells infected with HIV type ILA-l, the antiviral activity (EC_{50}) was found to be 7.3 µM. Since many anti-HIV agents cause the development of drug-resistant viral strains, it was interesting to note that fullerene compound **3** was shown to be equipotent against both azidothymidine (AZT)-susceptible, and AZT-resistant HIV-1, indicating no cross-resistance between **3** and AZT.

Eleven new C_{60} and C_{70} derivatives were synthesized at New York University by Wilson and Schuster [27] and were tested as dimethylsulfoxide (DMSO)/water emulsions against human PBM cells infected with HIV type ILA-l. All but one of these compounds showed antiviral activity (EC_{50}) in the low micromolar range. The three most active were compounds **10**, **11**, and **12** (EC_{50} = 0.88. 2.9, and 2.2 µM, respectively).

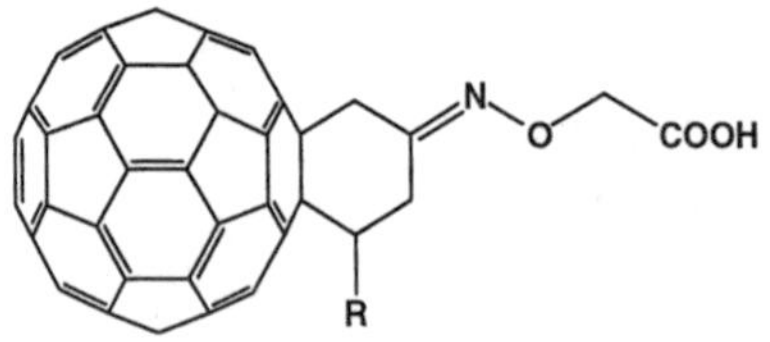

10 R = H
11 R = OCH$_3$

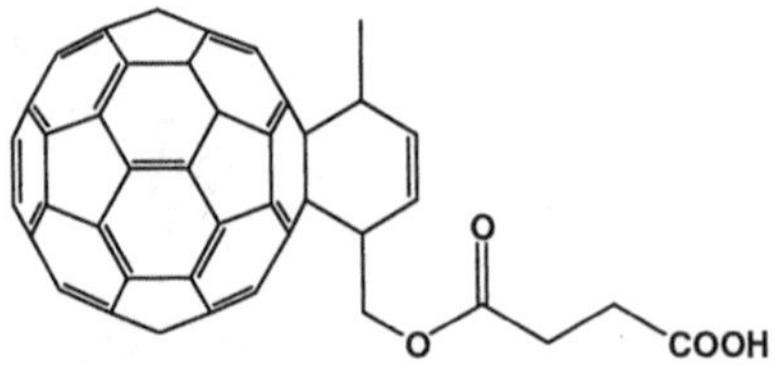

12

All but one test compound demonstrated no toxicity in PBM and Vero cells (IC_{50}, > 100 μM).

10.3.2 Other Antiviral Action

Direct inactivation of viruses with C_{60} itself has also been reported [28]. Photodynamic reactions induced by singlet oxygen-generating agents are known to inactivate enveloped viruses. Such viruses in blood are inactivated by C_{60} in the presence of O_2 and light. C_{60} was shown to mediate the inactivation of viruses from two different families—Semliki Forest virus (SFV, Togaviridae) and vesicular stomatitis virus (VSV, Rhabdoviridae).

Nonderivatized C_{60} has also been reported [29] to inhibit the replication of simian immunodeficiency virus (SIV) and Moloney murine leukemia virus (M-MuLV) in vitro.

Direct virucidal action of several fullerene compounds was also reported by Schuster et al. [27].

10.3.3 Anticancer Fullerenes

Cell division (mitosis) is critically dependent on chromosome rearrangements that involve structures called microtubules. These rigid tubulin polymers are ideal targets for anticancer drugs, since cancer cells are dividing rapidly while normal cells are not. Effects of the water-soluble fullerene compound $C_{60}(OH)_{24}$ **13** on microtubule assembly have been observed [30].

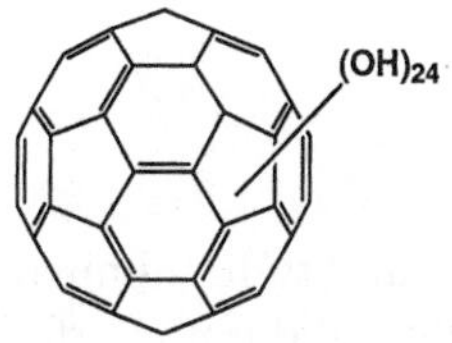

13

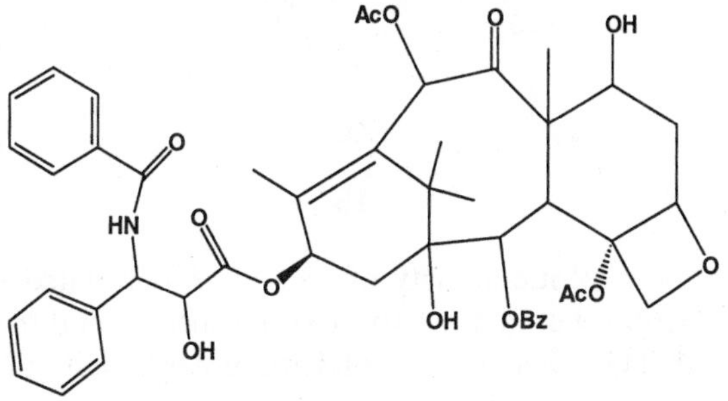

14

Microtubules, isolated by three cycles of temperature-dependent assembly and disassembly, were incubated in reassembly buffer with or without $C_{60}(OH)_{24}$ **13**. The rate and extent of microtubule formation could be determined by transmission electron microscopy and turbometric measurements—compound **13** strongly blocked assembly. In addition, compound **13** affected the growth kinetics of human lymphocyte cultures and HEP-2 epidermal carcinoma cell cultures. These results showed that compound **13** inhibits cell growth via inhibition of mitotic spindle formation in a manner similar to the important cancer drug taxol **14**.

Liposomes containing 1.09×10^{-4} M C_{60} were reported to have anticancer effects on human cervical cancer cells [31]. Discussion of the use of fullerene compounds as anticancer agents using photodynamic therapy (PDT) is deferred to Section 10.6.

10.4 CELL-SIGNALING—APOPTOSIS

Transforming growth factor β (TGF-β) is a representative member of a superfamily of regulatory proteins involved in signal transduction [32]. It has been shown to induce apoptosis in normal hepatocytes and hepatoma cells both in vivo and in vitro. While the mechanism by which TGF-β induces apoptosis is not clear, its effects are mediated by antioxidants such as cysteine or ascorbic acid. Modification of apoptotic signaling by TGF-β in human hepatoma cells by a fullerene compound has been observed [33]. The free radical scavenger compounds—carboxyfullerene–C_{60} derivatives **15** and **16**—were found to prevent TGF-β-induced apoptosis in TGF-β treated human hepatoma HEP-3B cells.

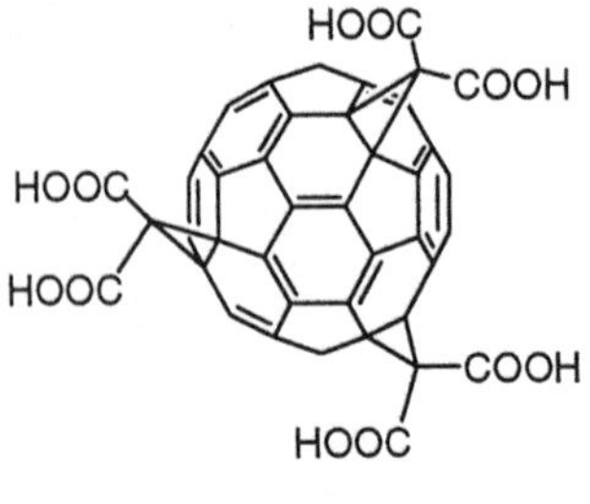

15 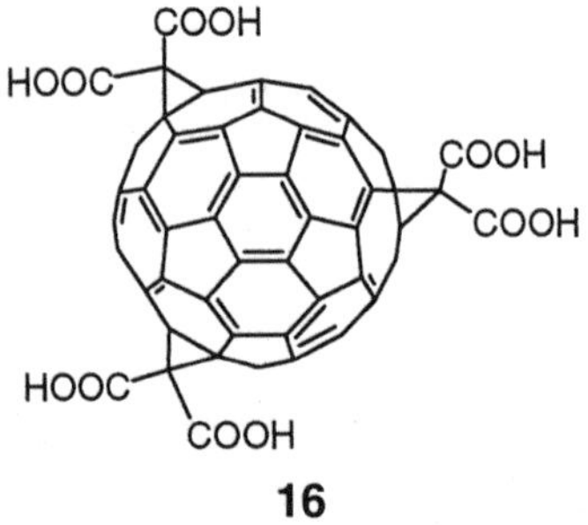   **16**

Anti-apoptotic activity of **15** and **16** correlated with their ability to eliminate TGF-β-generated reactive oxygen species, and the compounds did not interfere with TGF-β-activated PM-1 promoter activity in the HEP-3B cells. The C_3 regioisomer **15** was more potent in protecting cells from apoptosis than the D_3 isomer **16**.

In a recent report [34] C_{60} was used as a tool to establish some facts about Fas (a proto-oncogene) apoptotic signaling. The lipophilic second messenger ceramide has been suggested to have a role in mediating apoptotic signal for Fas. Several experiments showed this was not the case, including the use of C_{60} to antagonize ceramide-triggered but not Fas-triggered apoptosis.

Dugan et al. [35] studied the effects of fullerene compounds on apoptotic death of culture neurons. Polyhydroxy fullerols **6** and **13** decreased excitotoxic neuronal death by 80% following N-methyl-D-aspartate (NMDA) treatment.

Other examples of the effects of fullerene compounds **15** and **16** on apoptosis are discussed in Section 10.8, which covers antioxidants.

C_{60}, solubilized with polyvinylpyrrolidine, promoted cell differentiation (3.2 times control) and chondriogenesis of LB (a B lymphoblast cell line) cells at 40 $\mu g/mL$ in a concentration-dependent process [7i].

10.5 DNA AND GENOMIC APPLICATIONS

Water-soluble compounds **2a** and **2c** were found to cleave deoxyribonucleic acid (DNA) when incubated with supercoiled pBR322 DNA and irradiated with light [36]. More importantly, a hybrid DNA–fullerene compound **17** was shown to specifically cleave a 182 base-pair fragment at guanine residues upon exposure to light.

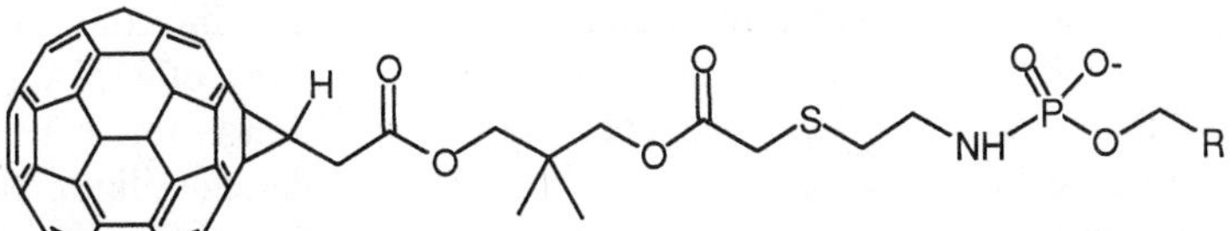

17 R = T(3')CTTTCCTCTTCTT(5')

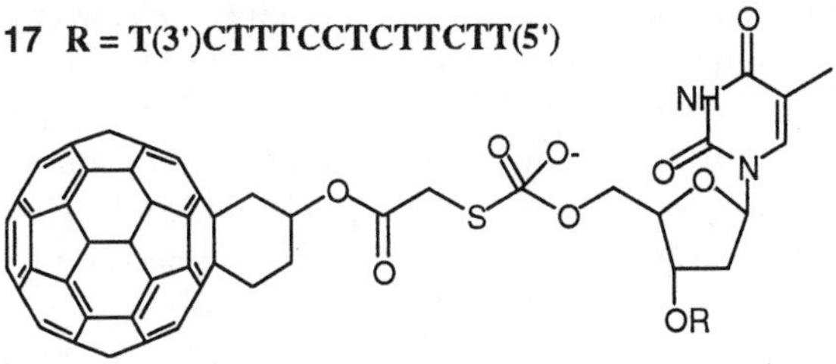

18 (R = C(5')TAACGACAATATGTACAAGCCTAATTGTGTAGCATCT

It was initially presumed that the mechanism for scission of oligonucleotides involves photoexcitation of the fullerene group followed by sensitized formation of $^{1}O_{2}$, which would then cleave the oligonucleotide.

Boutorine et al. [37] described the synthesis of a fullerene–oligonucleotide **18** that could bind single- or double-stranded DNA as well as other forms. Again, the conjugate cleaved specifically at guanine residues proximal to the fullerene moiety upon exposure to light.

In a similar study, another fullerene–oligonucleotide **19** also was shown to cut DNA at guanine residues near the fullerene terminus of the hybrid [24]. The possible intermediacy of singlet oxygen was examined in this case by comparing the reactivity of the fullerene–oligonucleotide with a similarly linked eosin-oligonucleotide (eosin is a compound known to sensitize singlet oxygen formation). Since singlet molecular oxygen has a significantly longer lifetime in

19 R = oligo-1
20 R = oligo-2
21 R = anti-luciferase

D_2O instead of H_2O ($\tau_{1/2}$ in $H_2O = 2$ μs; $\tau_{1/2}$ in $D_2O = 20$ μs [38]) when the PDT experiment was carried out in D_2O [39], eosin–oligonucleotide cleavage was more efficient, presumably due to the longer singlet oxygen half-life and consistent with a singlet oxygen mechanism. On the other hand, the fullerene–oligonucleotide hybrid exhibited the *same* rate of cleavage of DNA in both solvents! Also, a singlet oxygen quencher, sodium azide, was found to inhibit the eosin–oligonucleotide DNA cleavage, but not the fullerene–oligonucleotide cleavage. Taken together, these results suggest that the fullerene–oligonucleotide cleavage does not involve the singlet oxygen mechanism, but rather some other mechanism must be involved, such as a direct photoinduced electron transfer from guanosine residues to the fullerene.

Antisense oligonucleotides [40] tethered to C_{60} were investigated as a means to enhance oligo transport into cells. A nucleophilic sulfhydryl or amino group was introduced at the 3′- or 5′-terminus of the oligonucleotide and a C_{60} moiety with an alkylating group or carboxyl group was attached via alkylation or acylation of the nucleophilic residue to give **20** and **21**. The hybrids were purified by electrophoresis, and the electrophoretic mobility, UV spectra, and other properties indicated that **17** and **18** formed aggregates from tens to several hundreds of nanometers in diameter. The C_{60}–oligonucleotide 14-*mer* **20** formed a duplex with the complementary DNA sequences and triplexes with corresponding double-stranded DNAs and showed light-induced guanine-specific cleavage of targets.

Direct interaction of a fullerene–anti-luciferase oligonucleotide conjugate **21** with *J. Jian Jurkat* cells expressing luciferase was also studied. The antisense biological activity of the fullerene conjugate **21** did not exceed that of the non-conjugated oligonucleotide, even when the conjugate was adsorbed on the hydrophobic nanoparticles [41].

Another type of experiment was the use of C_{60} solubilized with polyvinyl-pyrrolidone, which, on irradiation, showed mutagenicity and DNA damage consistent with 8-OH-dG formation and DNA-strand scission [42].

Studies of mutagenicity and genotoxicity of C_{60} examined mutagenicity and lipid peroxidation with and without irradiation [43]. Zakharenko et al. [44] estimated the genotoxicity of fullerene C_{60} and fullerol compounds by assaying for mutations in *Escherichia coli* and larvae and found that C_{60} and a poly-hydroxylated derivative were essentially nongenotoxic.

The preparation of DNA–fullerene hybrid materials using a cationic C_{60} compound allows transmission electron microscopy (TEM) imaging of DNA without the use of heavy metals. The cationic compound $C_{60}–N, N$-dimethyl pyrrolidinium iodide salt binds to DNA and provides excellent contrast for imaging individual DNA molecules [45].

A C_{60} derivative was linked to a gold surface and DNA was shown to bind and be cleaved with the gold-linked fullerene [46]. Site-specific DNA cleavage by incorporation of C_{60} in certain cationic monolayers was observed. Other examples of bioreactive fullerene surfaces are discussed in Section 10.13.

10.6 PHOTODYNAMIC THERAPY (PDT)

As mentioned in Section 10.1.4, C_{60} and C_{70} are efficiently converted to their triplet excited states on UV and visible irradiation, and the triplet states efficiently sensitize the formation of $^1O_2$. For this reason, one of the earliest suggestions for possible biological activity of fullerenes was the potential for photodynamic damage to biological systems [47]. The therapeutic application of PDT has been the topic of considerable excitement recently, since the 1995 approval of the first PDT drug Photofrin **22** [48].

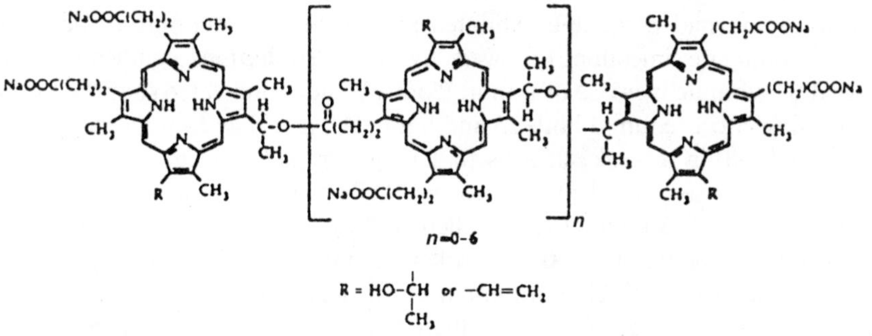

$$n = 0-6$$

$$R = HO-\overset{|}{\underset{CH_3}{CH}} \text{ or } -CH=CH_2$$

22

Photofrin is a porphyrin photosensitizer and its application in PDT is based on two properties: (1) sensitization of $^1O_2$ at wavelengths of light that penetrate tissues and (2) localization of the very hydrophobic drug in tumor cells, which are generally hydrophobic tissues. From a practical standpoint, since C_{60} and C_{70} do not absorb effectively at long wavelengths, their use as in vivo sensitizers may be limited. A potential solution to this problem has been reported by Cheng et al. [49], which involves linking a C_{60} moiety to a light-harvesting "antenna" such as a porphyrin.

Nonetheless, fullerene compound **2a** was shown to be cytotoxic to cells upon incubation with HeLa S3 cells and irradiation with light [16]. The same mixture of fullerenes and cells was unaffected when not exposed to light. A reference compound lacking the fullerene moiety but possessing a side chain similar to that of **2a** was inactive when incubated with the cells and exposed to the same light source. These results were explained by the formation of $^1O_2$ within the cells.

An interesting study involved reaction of C_{60} with polyethylene glycols (PEGs) having terminal primary amino groups to form water-soluble PEG–C_{60} conjugates **23** and **24**. These conjugates showed strong cytotoxicity to L929 cells upon visible light irradiation as a result of superoxide formation [50]. Cytochrome-c was reduced when irradiated in the presence of the PEG–fullerene derivative, implying generation of superoxide. When superoxide dismutase (SOD) was added, reduction of cytochrome-c was suppressed.

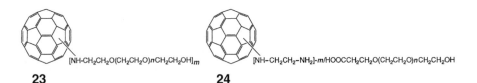

23   24

A study by the same group, using PEG–C_{60} compounds with animals gave encouraging results [51,52]. The C_{60}–PEG conjugate **23**, upon intravenous injection into tumor-bearing mice, accumulated in the tumor to a greater extent than in the normal tissue and exhibited prolonged C_{60} retention at the tumor tissue. Conjugate injection followed by light irradiation significantly suppressed the volume increase of tumor. Control studies indicated that the tumor-suppressing effect required both **23** and light. The fact that **23** was cleared from the animal within a week suggests that long-term side effects associated with photodynamic use of the FDA-approved PDT agent, photophrin, may be substantially reduced using analogous fullerene drugs.

In another report, Li and co-workers reported that when a mixture of liposome-encapsulated C_{60} and human cervical cancer cells was irradiated for a short period of time [29], the irradiated cells showed a twofold *increase* in proliferation relative to the unirradiated cells with longer irradiation times.

Another study of the PDT action of C_{60}-containing liposomes on cancer cells was reported [53]. After the fusion between the C_{60}–phosphatidylcholine liposome (C_{60} concentration ca. 20 mg/L) and HeLa cells, the preparation was irradiated with a tungsten-halogen lamp (4000 nm, 30 min). Most of the cells were killed as shown by a MTT (3-(4,5-dimethylthiazol-2-yl)-2,5-diphenyl-tetrazolium bromide) assay. Biochemical tests showed that photoexcited C_{60} decreased the sulfhydryl content of the membrane protein and increased the malondialdehyde (MDA) level by peroxidation of the membrane lipid.

A further report of tumor targeting of the PEG-modified fullerene compound **23** and its photodynamic effects was reported. Conjugate **23** showed photodynamic activity against fibrosarcoma tumors in mice and effects on the erythrocyte membrane [54,55]. Distinct damage to important amino acids of membrane proteins, such as valine and tryptophan, and oxidation of membrane protein sulfhydryl groups and peroxidation of unsaturated fatty acids were observed. NaN_3 and SOD were used as inhibitors to demonstrate an effect of "active oxygen species" such as oxygen free radicals $O^{\cdot-}$ and $OH\cdot$.

A study of the effect of substitution on the fullerene core on singlet oxygen production efficiency was reported recently. Singlet molecular oxygen from a series of fullerene derivatives was measured as a function of sequential functionalization of the core [56], by its characteristic 1268 nm emission. The efficiency was independent of the kind of addends but decreased with an increase in the number of addends, as might be expected as the extent of fullerene-type conjugation was decreased and the benzenoid character increased.

10.7 ELECTRON-TRANSFER AND MEMBRANE EFFECTS

C_{60} readily accepts electrons as mentioned earlier [57]. This has led to studies on possible electron-transfer abilities of fullerenes in biological systems. In a report by Hwang and Mauzerall in 1993 [58], C_{70} was shown to mediate electron transport through a lipid bilayer system. A C_{70}-containing lipid bilayer was prepared covering a 1.5 mm hole in a thin Teflon sheet. On one side of the lipid bilayer was an aqueous compartment containing ascorbate as the electron donor, and on the other side, an aqueous compartment containing anthraquinone 2-sulfonate as the electron acceptor. On irradiation, a current was observed between the aqueous compartments. In the absence of C_{70} or in the presence of oxygen no current was observed. Two possible mechanisms were suggested. One mechanism is electron transport via electron hopping between fullerene molecules in the membrane. Alternatively, current could result by reduction of C_{70} near the boundary, diffusion of the C_{70}-radical anion across the bilayer, and release of the electron on the other side of the membrane.

An application of fullerene redox chemistry to biosensor technology has been reported [59]. A supported bilayer lipid membrane containing C_{60} was prepared and light-induced voltage and current were generated. Using such membranes it was possible to construct sensor probes and investigate redox reactions and light-induced electron transfer across the lipid bilayer. The C_{60}-containing bilayer lipid membrane was shown to be a light-sensitive diode capable of photoinduced charge separation and was shown to be useful for electrochemical biosensor electronics device development.

The Diederich group has extended this concept to the formation of glucosylated fullerene layers that have prospects for applications as bisensors [60]. Stable Langmuir–Blodgett films of fullerene–glycodendron conjugates were prepared. The amphiphilic C_{60}–dendrimer conjugates form stable, ordered monomolecular Langmuir layers at the air–water interface and could have potential applications as glycoprotein biosensors.

Self-organized fullerene–lipid films containing a C_{60} derivative with a lipid chain addend were studied by differential scanning calorimetry and other methods [61].

Finally, a study of the structural and dynamic effects of lipophilic fullerene derivatives such as **25** in phospholipid bilayers reveals novel long-range ordering and formation of rod-like structures in the membrane [62].

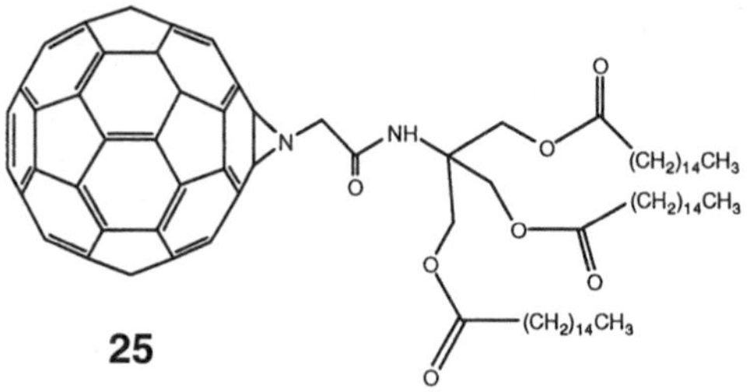

25

10.8 ANTIOXIDANT PROPERTIES

Since 1994, Chiang and co-workers have reported many interesting biological effects of water-soluble polyhydroxylated fullerols **6** [11]. Examples of **6** as a potent antioxidant in vitro and in vivo are of its ability to scavenge intermediate free radicals [63]. Polyhydroxylated fullerols also have excellent efficiency in eliminating superoxide radicals ($O_2^{\cdot-}$), generated by xanthine/xanthine oxidase.

In another report, a water-soluble fullerene, bis(malonate) derivative $C_{62}(COOH)_4$ **26**, decreased the toxicity of active oxygen generators, paraquat and heteropentalene, in *E. coli* [64]. The possible mechanism is one in which $C_{62}(COOH)_4$ quenches radicals or superoxide. $C_{62}(COOH)_4$ thus can affect active oxygen toxicity in vivo.

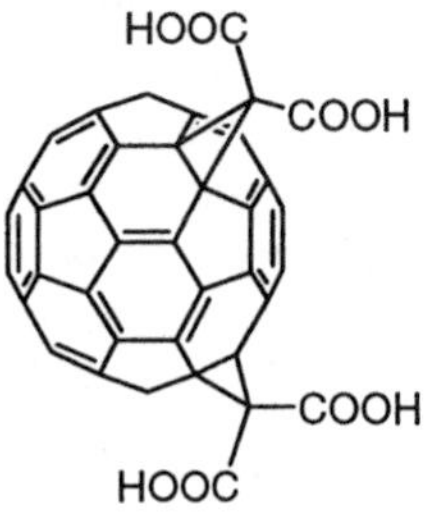

26

The biological activities of water-soluble iron complexes of C_{60} derivatives have been evaluated. These compounds—basically C_{60}–EDTA hybrids—affect reactive oxygen species generated by granulocytes and mononuclear phagocytes in whole blood [65].

Water-soluble fullerene compounds attenuate bronchoconstriction in guinea pigs [66]. Fullerol **6** normally does not cause significant alteration in respiratory function except at high doses (>2 mg/kg). It was suspected that oxygen radicals play an important role in certain types of airway constriction. This study investigated the effects of antioxidants in guinea pigs. Beginning 15 min after induction of airway constriction (exsanguination), there was gradual increase in airway constriction with time. Administration of either fullerol **6** or the drug deferoxamine significantly ameliorated bronchoconstriction effects.

In a related study, fullerol **6** was also shown to attenuate bronchoconstriction induced by a xanthine/xanthine oxidase system [67]. The intratracheal instillation of xanthine/xanthine oxidase caused a marked decrease in dynamic respiratory compliance—airway constriction—which was significantly reduced by fullerol **6**, but not by other agents or receptor antagonists.

In a similar way, the C_{60}–dimalonic acid **26** was shown to affect active oxygen toxicity [68]. The water-soluble compound **26** quenches superoxide radical ($O_2^{-\cdot}$), generated from the xanthine/xanthine oxidase system, oxidizes reduced cytochrome-c in the presence of hydrogen peroxide, and inhibits glutathione reductase activity.

A recent series of papers reveals what may be an important effect of fullerene compounds on nitric oxide signaling pathways [69]. A fullerene derivative, C_{60}–monomalonic acid **27**, caused an inhibitory effect on nitric oxide-dependent relaxation of aortic smooth muscle [70].

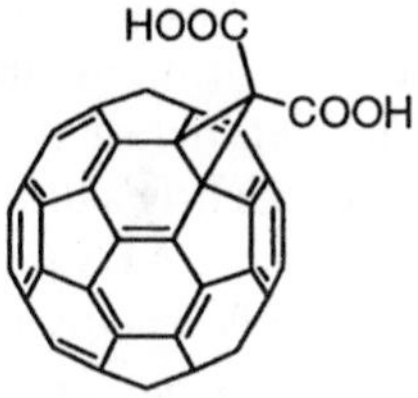

27

The effects of C_{60}–monomalonic acid derivative **27** on several tissues were studied: endothelium-containing or denuded rabbit aorta, guinea pig trachea and ileum, and rat stomach, vas deferens, and uterus. For example, a nitric oxide-generating agent, S-nitroso-N-acetylpenicillamine, caused the relaxation of aorta without endothelium in a concentration-dependent manner. Compound **27** at 10^{-5} M significantly reduced the maximal response of the relaxation induced by acetylcholine in endothelium-containing rabbit thoracic aorta. The acetylcholine-induced relaxation was recovered in the presence of superoxide dismutase (SOD, 250 units/mL). This inhibitory effect of **27** was masked in the presence of SOD. Sodium nitroprusside-induced relaxation was not affected by either **31** or SOD. In the other tissues, neither **27** nor C_{60} had an effect on the responses induced by any agonist. These observations indicate that

27 inhibits the endothelium-dependent relaxation induced by acetylcholine but does not affect the agonist-induced contractile response of smooth muscle.

Continued studies [71] of this compound showed that **27** produced an augmentation of phenylephrine-induced muscle tone and reduced both the acetylcholine-induced maximum relaxation and the amplitude of substance P $(10^{-8}$ M)-induced relaxation in endothelium-containing thoracic rabbit aorta. The acetylcholine- and substance P-induced relaxation was restored in the presence of SOD (250 units/mL).

Another report studied the active oxygen radical scavenging ability of water-soluble fullerols [72] using electron spin resonance (ESR). Fullerol **6** had excellent scavenging ability for both hydroxyl radical and superoxide anion as shown by ESR. A similar study employed ESR to examine fullerene reaction with photogenerated radicals [73].

Fullerene antioxidants also show promise as neuroprotectants in cells and whole animals. Dugan et al. [35] reported that the free radical scavenger fullerol **6** reduced excitotoxic and apoptotic death of cultured cortical neurons. In cortical cell cultures exposed to excitotoxic and apoptotic injuries, two other polyhydroxylated C_{60} fullerol derivatives, $(C_{60}(OH)_n, n = 12)$ and $(C_{60}(OH)_nO_m, n = 18–20, m = 3–7)$ (cf. fullerol **6**) demonstrated excellent antioxidant capabilities when tested by ESR with a spin-trapping agent and a hydroxyl radical-generating system. These water-soluble agents decreased (by 80%) excitotoxic neuronal cell death following brief exposure to cell-death promoters, such as NMDA (N-methyl-D-asportate).

The same group reported similar studies [74] using carboxyfullerenes **15** and **16**, which reduced apoptotic neuronal death in cultured neurons and also demonstrated protection against neurodegeneration in a mouse model of amyotrophic lateral sclerosis (ALS or Lou Gehrig's disease). This experiment showed delayed onset of ALS symptoms in the FALS mouse (a mutant that lacks the SOD gene) and allowed the carboxyfullerene-treated mice to live 8 days longer (~138 days) compared to an untreated group (130 days).

The hippocampus has been used as a model to study the role of the free radical trapper fullerol **6** in prevention of hydrogen peroxide- and cumene hydroperoxide-elicited damage in hippocampus tissue slices from rat. Fullerol **6** prevented hydrogen peroxide- and cumene hydroperoxide-elicited changes [75].

10.9   RADIOTHERAPY, TRACER STUDIES, AND MEDICAL IMAGING

10.9.1   Radiocarbons

Samples of C_{60} have been enriched in both radioisotopic carbon variants (^{11}C and ^{14}C). The production of ^{11}C-fullerenes for use as possible radiotracers was accomplished recently [76]. A complete complement of ^{11}C-radiolabeled fullerenes C_n ($n = 60, 70, 76, 78,$ and 84) was produced, purified by high-performance liquid chromatography (HPLC), and characterized using optical absorption spectroscopy and an ultrasensitive counting technique for radioactivity measurement. This successful labeling of fullerenes with ^{11}C provides possible applications as short half-life tracers.

Several years ago, Scrivens et al. [77] reported the preparation of ^{14}C-enriched C_{60} by arc production of fullerenes from ^{14}C-enriched graphite. They also studied cellular uptake of underivatized ^{14}C-C_{60} using a fine aqueous suspension and monitored its uptake by cells. This suspension was taken up by human keratinocytes but did not appear to affect the proliferation rate of cultured human keratinocytes and human fibroblasts. An alternative approach to ^{14}C-C_{60} was reported by the Wilson group, using a ^{14}C-enriched steroid as the carbon source [78].

10.9.2 Metallofullerene Tracers

Since the first discovery of fullerenes, the hole has been viewed as a place to encapsulate atoms. One good prospect for fullerene bioapplication involves placing radioactive metal atoms safely inside the fullerene cage, leading to products with the potential for radiotherapy or medical imaging. Once inside, the toxic heavy metal would have no chance for escape.

Several applications of endohedral metallofullerenes in medicine have already been discovered. The most comprehensive is a report on the in vivo biological behavior of a water-soluble ^{166}Ho metallofullerene [79]. Neutron irradiation of purified holmium metallofullerenes, Ho@C_{82} and Ho$_2$@C_{24}, and a water-soluble metallofullerene derivative Ho$_2$@C_{84}(OH)$_n$ has been performed using a thermal neutron source. Properties of several endohedral radioactive elements have been studied in anticipation of comprehensive applications in nuclear medicine.

The encapsulation of a metal important for magnetic resonance imaging (MRI)—the relaxation agent Gd—was also reported by the same group [80]. A Japanese patent application reports a similar synthesis and proposed use of Gd@C_n compounds [81].

10.10 IMMUNOLOGY OF FULLERENES

Immunological studies of C_{60} itself are limited. Several reports on industrial toxicology show fullerenes have virtually no inflammatory effect in mice and rats and do not elicit an immune response [82].

A more recent paper compares the immunological effects of C_{60} with quartz particles [83]. Either fullerene soot or C_{60} was incubated with bovine alveolar and HL-60 macrophages to test cytotoxicity, lysosomal damage, cytokine production, and the formation of reactive oxygen species. The results were compared to quartz. C_{60} and soot were not cytotoxic to the cells and C_{60} induced a lower chemotaxis than quartz. Both induced the production of tumor necrosis factor and interleukin-8. Stimulation of the release of superoxide and other reactive oxygen species was not observed by either C_{60} or soot [84].

The true antigenicity of the C_{60} molecule and fullerene compounds was recently determined by the preparation of fullerene-specific antibodies using similar techniques to the preparation of antibodies to steroids by Erlanger, Wilson, and co-workers [85]. An important principle in immunology—clonal selection theory—states that antigens elicit the production of antibodies by selecting for specific antibody-producing cells already present in the "immune repertoire" of immunized animals [86]. Although there is debate about the size of the available repertoire [87,88], immunologists usually assume that it is diverse enough to produce antibodies to any molecule. The question arises as to whether the immune repertoire is "complete" enough to recognize and respond to the unprecedented buckminsterfullerene structure and, particularly, if the usual immune response mechanism can function. This requires that a fullerene–protein conjugate must be processed by B cells, and then the processed peptides and fullerene–peptide hybrids displayed on the B cell surface for recognition by T cells and production of high-affinity IgG-type antibodies to the fullerene sphere.

Fullerene derivatives **12** and **28** were conjugated with bovine and rabbit serum albumin (BSA and RSA) and bovine thyroglobulin (TG) [89].

28

Immunization of mice with a C_{60} fullerene–thyroglobulin conjugate (compound **12**–TG) produced a polyclonal immune response comprised of IgG antibodies specific for C_{60} fullerenes and a subpopulation that cross-reacted with a C_{70} fullerene derivative. Detection was possible by enzyme-linked immunoadsorbant assay (ELISA) using an IgG-specific second antibody showing that the antibodies raised were of the IgG isotype.

These results establish that derivatization of TG by a fullerene molecule did not prevent internal processing by B cells, and subsequent peptide display to T cells occurs, presumably by the process of linked recognition [90,91].

In contrast to the design of fullerene compounds that could bind to known enzyme or receptor sites (cf. Section 10.2.6), the direct formation of fullerene antibodies uses the immune system to create recognition sites for fullerenes. What types of possible interactions with the combining sites of IgG antibodies are possible? We might expect the interactions of a C_{60} with the peptide chains at the antibody combining site to involve hydrophobic groups. Other possibilities include interaction with $-NH_2$ or $-SH$ side chains or possibly charge-transfer interactions. A picture of binding will emerge from the gene sequence of fullerene monoclonal antibodies currently being produced, as well as future protein X-ray crystallography. In a subsequent *Proceedings of the National*

Academy of Science commentary on anti-fullerene antibodies, a well-known immunologist discussed insights into the origins of immune recognition [92] revealed from C_{60} antibodies [85].

10.11 CARBON NANOTUBES

Recent interest in the tubular forms of fullerene called nanotubes has resulted in many proposed applications for these remarkable materials [93]. A few applications of carbon nanotubes to biological systems have already appeared. First, a patent application proposes the use of functionalized nanotubes for biocatalysis and for enzyme immobilization [94]. Another describes the use of luminescence nanotube arrays for detection of biomolecules [95].

Two interesting papers report the encapsulation of proteins inside carbon nanotubes [96,97]. Ends were cut from 2–8 nM nanotubes and the resulting "pipes" were filled with the enzymes Zn_2Cd_2–metallothionein, cytochromes-*c* and -*c*3, and β-lactamase, as determined by TEM and enzyme assay techniques. Enzyme activity appeared largely unaffected, implying major enzyme conformational changes did not occur—in other words, the nanotube acts as a "host" for the biomolecule.

While much of the chemistry and biology of nanotubes is still virgin territory, there have been some studies of large-diameter carbon fibers. Carbon fibers are used commercially for their thermal resistance, excellent electrical conductivity, flexibility, and high tensile strength [98a]. Inhalation toxicity studies with respirable-sized carbon fibers show some pulmonary biochemical and cellular effects [98b]. Following a 1–5 day exposure period, cells and fluids from carbon fiber-exposed animals were recovered and various biochemical values were measured. A 5 day exposure to respirable carbon fibers at concentrations of 50 and 120 mg/L produced dose-dependent, transient inflammatory responses in the lungs of exposed rats. No significant differences in the morphology or in vitro phagocytic capacities of macrophages were observed.

10.12 METABOLISM, EXCRETION, AND TOXICITY

One of the first studies of the biodistribution of fullerene compounds was the in vivo behavior of the ^{13}C-radiolabeled monosubstituted fullerene compound **29** [99].

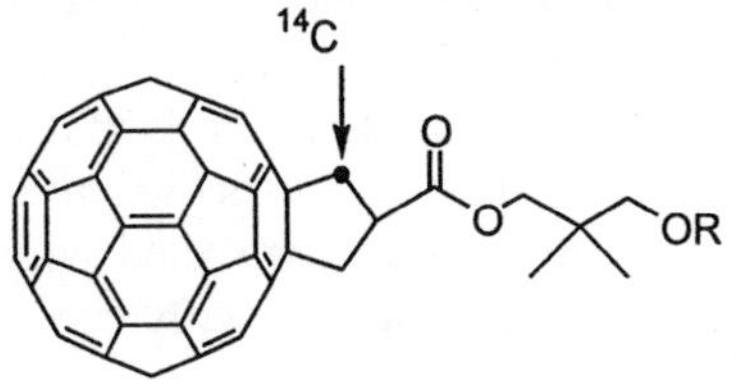

29   R = COCH₂CH₂COOH

When administered orally to rats, this compound was not efficiently absorbed (only 4%) and was excreted in feces. When injected intravenously, the compound rapidly distributed in tissues, and most of the compound was retained in the animal for at least a week. Uptake was primarily in the liver and kidneys after 30 h, and the amount of compound remaining decreased to less than 2% after 7 days. No acute toxicity was observed after intravenous (IV) injection; this compound also appeared to cross the blood–brain barrier to a small extent (~1%). The in vivo behavior of this compound is consistent with its observed lipophilicity.

Using ^{14}C-labeled C_{60} [100], the Tour group studied tissue uptake for C_{60} itself as well as a water-soluble compound, the quaternary ammonium fullerene **30**.

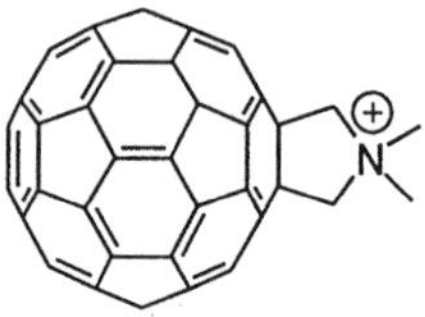

30

Both compounds cleared rapidly from circulation and distributed to liver, spleen, muscle, skin, and lung. The more lipophilic C_{60} distributed mostly to the liver and remained there for 120 h, whereas more polar compound **30** distributed to liver only half the extent of the native fullerene.

In vivo studies of polyhydroxyfullerene compounds employing endohedral metals as radiotracers was reported by Wilson and co-workers [101]. A water-soluble higher fullerene C_{84} derivative $M@C_{84}(OH)_n$ **31**, encapsulating a therapeutic radiometal, was administered intraperitoneally to rats.

$$M@C_{84}(OH)_n$$

31

Gamma-camera imaging of the rats showed an even distribution of the fullerene compound throughout the blood pool with no selective tissue localization for up to 48 h. The compound was excreted in the urine and distributed in organs [80].

The acute and subacute toxicities of water-soluble polyalkylsulfonated fullerols were studied [102]. Polyalkylsulfonated C_{60} **32** is a highly water-soluble antioxidant.

A 50 mg/mL aqueous solution was used in rats in a single-dose acute toxicity study or a 12 day subcutaneous study. No rats died after oral administration. In an LD_{50} intraperitoneal injection study, rats died within 30 h with a dose of approximately 600 mg/kg. Rats injected with the compound immediately eliminated it through the kidney.

32

Compound **2** has been shown to have antiviral activity as discussed in Section 10.3. This compound was shown to be well tolerated in mice after intraperitoneal administration. A study was undertaken to develop HPLC methodology to measure compound **2** in blood and to characterize the preclinical pharmacokinetics in rats [103]. Following IV administration of the fullerene derivative at a dose of 15 mg/kg of body weight, the concentration of **2** in plasma declined with a half-life of 6.8 h. Urine samples obtained 24 h after intravenous administration did not contain detectable levels of the compound, indicating the absence of a significant renal clearance mechanism.

A recent study of fullerol **6** showed the compound suppressed microsomal cytochrome P450-dependent monooxygenases and inhibited mitochondrial oxidative phosphorylation [104]. The acute toxicity of fullerol **6** was estimated to be $LD_{50} = 1.2$ g/kg.

Finally, an interesting report has just appeared showing that several species of fungi can grow in fullerenes as the sole source of carbon [105].

10.13 COMMERCIAL PROSPECTS

Only a few thousand fullerene compounds have been reported. For comparison, the total number of known organic compounds in the *Chemical Abstracts Registry File* number more than 12 million. In fact, since fullerene C_{60} only became commercially available about seven years ago, it should be considered impressive that many studies have already been published on the effects of C_{60} compounds as drugs.

At the time of this writing there are no fullerene drugs on the market. Of the several interesting lead compounds discussed earlier, neuroprotectant compound **15** may be the furthest toward commercialization [74]. Compound **15** (discussed in Section 10.8 and chap 11) is a drug that could be useful against ALS, also known as Lou Gehrig's disease, for which there is no cure. A patent was filed for the antioxidant neuroprotectant ALS action mentioned above [106].

There are several other possible fullerene-based biomedical products. A patent application has been issued for fullerene-coated cell-culture glassware

[107]. It was reported that DNA and proteins bound to surfaces in the following order: nitrocellulose > fullerene-coated glass > glass > polystyrene [108].

Another potential application takes advantage of the general antiviral activity of fullerenes. Fullerene-impregnated, medical membranes such as condoms and surgical gloves were reported. Langmuir–Blodgett coating of C_{60} onto latex rubber affords rubber with therapeutic antiviral properties [109]. Another report described favorable blood contact properties of surface-immobilized C_{60} [110].

Taking advantage of the UV-absorbing ability of C_{60}, sunscreen cosmetics containing fullerenes with UV-protection effects using oils as dissolving agents have been prepared. For example, cosmetics have been prepared as oil–water emulsions containing C_{70}, cetyl 2-ethylhexanoate, glyceryl tri(2-ethylhexanoate), pentaerythritol tetra(2-ethylhexanoate), and other ingredients [111].

A patent application reported the use of water-soluble fullerenes as photosensitizers for photodynamic therapy [112]. Effective photosensitizers for photodynamic therapy were prepared by chemically modifying C_{60} with water-soluble polymers.

The two major research groups involved in antioxidant and free radical applications of fullerenes have filed patents relating to fullerene derivatives, such as fullerols, which are capable of scavenging free radicals [113]. The fullerol was shown to decrease the free radical content by up to 86% in human blood samples from patients with acute pancreatitis or gastric cancer.

ACKNOWLEDGMENTS

I would like to thank the NSF and PRF for support of the New York University fullerene group. Thanks also to my colleague, David Schuster, for his comments on the manuscript.

REFERENCES

1. (a) A. W. Jensen, S. R. Wilson, and D. I. Schuster, *Biorg. Med. Chem.* **1996**, *4*(6), 767. (b) T. Da Ros and M. Prato, *Chem Commun.* **1999**, 663.

2. (a) C. L. Hill, *Proc. Electrochem. Soc.* **1994**, *94*(24), 659. (b) C. L. Hill, *Proc. Electrochem. Soc.* **1994**, *95*, 683. (c) S. R. Wilson, *Proc. Electrochem. Soc.* **1997**, *97*(42), *322*. (d) [CAS 126:101484] X. Wan and J. Li, *Daxue Huaxue* **1996**, *11*(3), 25. (e) S. Nagase, K. Kobayashi, and T. Akasaka, *Bull. Chem. Soc. Jpn.* **1996**, *69*(8), 2131. (f) [CAS 125:57997], Y. Rubin, *NATO ASI Ser. Ser. E* **1996**, *316*, 295. (g) Z. Xu, Z. Y. Suo, X. W. Wei, and D. X. Zhu, *Prog. Biochem. Biophys.* **1998**, *25*(2), 130.

3. (a) A. Hirsch, *The Chemistry of the Fullerenes.* George Thieme Verlag, New York, 1994. (b) R. Taylor, *The Chemistry of the Fullerenes.* World Scientific, Singapore, 1995.

4. (a) R. S. Ruoff, D. S. Tse, R. Maihotra, and D. C. Lorents, *J Phys. Chem.* **1993**, *97*, 3379. (b) G. V. Andrievsky, M. V. Kosevich, O. M. Vovk, V. S. Shelkovsky, and L. A. Vashchenko, *Proc. Electrochem. Soc.* **1995**, *95*(10), 1591.

5. R. Bullard-Dillard, K. Creek, W. Scrivens, and J. Tour, *Bioorg. Chem.* **1996**, *24*, 376.

6. (a) T. Andersson, K. Nusson, M. Sundahi, O. Wesunan, and O. Wennersirom, *J Chem. Soc. Chem. Commun.* **1992**, 604. (b) M. Sundahl, T. Andersson, K. Nusson, O. Wennentrom, and O. Wesunan, *Synth. Meth.* **1993**, *55*, 3252. (c) J. L. Atwood, O. A. Koulsantonis, and C. L. Raston, *Nature* **1994**, *368*, 229.

7. (a) F. Moussa, F. Trivin, R. Ceolin, M. Hadchouel, P.-Y. Sizaret, V. Greugny, C. Fabre, A. Rassat, and H. Szwarc, *Fullerene Sci. Technol.* **1996**, *4*, 21. (b) J. L. Garaud, J. M. Janot, G. Miquel, and P. Seta, *J. Membr. Sci.* **1994**, *91*, 259. (c) K. C. Hwang and D. Mauzerall, *J Am. Chem. Soc.* **1992**, *114*, 9705. (d) R. V. Bensasson, B. Bienvenue, M. Dellinger, S. Leach, and P. Seta, *J. Phys. Chem.* **1994**, *98*, 3492. (e) H. Hungerbuhler, D. M. Guldi, and K.-D. Asmus, *J Am. Chem. Soc.* **1993**, *115*, 3386. (f) Y. N. Yamakoshi, T. Yagami, K. Fukuhara, S. Sueyoshi, and N. Miyata, *J. Chem. Soc. Chem. Commun.* **1994**, 517. (i) T. Tsuchiya, Y. N. Yamakoshi, and N. Miyata, *Biochem. Biophys. Res. Comnum.* **1995**, *206*, 885.

8. N. V. Katre, *Adv. Drug Delivery Res.* **1993**, *10*, 91.

9. M. Brettreich and A. Hirsch, *Tetrahedron Lett.* **1998**, *27*, 2731.

10. Q. Xie, B. Perez-Cordero, and L. Echegoyen, *J Am. Chem. Soc.* **1992**, *114*, 3978.

11. L. Y. Chiang, L.-Y. Wang, J. W. Swirczewski, S. Soled, and S. Cameron, *J. Org. Chem.* **1994**, *59*, 3960.

12. (a) J. W. Aborgast, A. P. Darmanyan, C. S. Foote, F. N. Diederich, R. L. Whetten, and Y. Rubin *J. Phys. Chem.* **1991**, *95*, 11. (b) B. Ma, J. E. Riggs, and Y.-P. Sun, *J. Phys. Chem.* **1998**, *102*, 5999.

13. F. N. Tebbe, R. L. Harlow, D. B. Chase, L. E. Firment, E. R. Holler, B. S. Malone, P. J. Krusic, and E. Wasserman, *J. Am. Chem. Soc.* **1991**, *113*, 9900.

14. J. Baggott, *Perfect Symmetry.* Oxford University Press, Oxford, 1994.

15. I. Tunion, E. Silla, and J. L. Pascual-Ahuir, *Prot. Eng.* **1992**, *5*, 715.

16. E. Nakamura, H. Tokuyama, S. Yamago, T. Shiraki, and Y. Sugiura, *Bull. Chem. Soc. Jpn.* **1996**, *69*, 2143.

17. (a) S. H. Friedman, D. L. DeCamp, R. P. Sijbesma, G. Srdanov, F. Wudl, and G. L. Kenyon, *J. Am. Chem. Soc.* **1993**, *115*, 6506. (b) R. P. Sijbesma, G. Srdanov, F. Wudl, J. A. Castoro, C. Wilkins, S. H. Friedman, D. L. DeCamp, and G. L. Kenyon, *J. Am. Chem. Soc.* **1993**, *115*, 6510. (c) S. H. Friedman, P. S. Ganpathi, Y. Rubin, and G. L. Kenyon, *J. Med. Chem.* **1998**, *41*, 2424.

18. N. Miyata, Y. Yamakoshi, H. Inoue, M. Kojima, K. Takahashi, and N. Iwata, *Proc. Electrochem. Soc.* **1998**, *98–8*, 1227.

19. T.-H. Ueng, J.-J. Kang, H.-W. Wang, Y.-W. Cheng, and L. Y. Chiang, *Toxicology Lett.* **1997**, *93*, 29.

20. J. Nacsa, J. Segesdi, A. Gyuris, T. Braun, H. Rausch, A. Buvari-Barcza, L. Barcza, J. Minarovits, and J. Molnar, *Fullerene Sci. Technol.* **1997**, *5*(5), 969.

21. S. R. Wilson, Q. Lu, J. Cao, H. Zhao, Y. Wu, and D. I. Schuster, *Proc. Electrochem. Soc.* **1995**, *95*(10), 1179.

22. C. Toniolo, A. Bianco, M. Maggini, G. Scorrano, M. Prato, M. Marastoni, R. Tomatis, S. Spisani, G. Palo, and E. Blair, *J. Med. Chem.* **1994**, *37*, 4558.

23. M. Satoh, K. Matsuo, Y. Takanashi, and I. Takayagi, *Gen. Pharmacol.* **1995**, *26*, 1533.

24. Y.-Z. An, C.-H. B. Chen, J. L. Anderson, D. S. Sigman, C. Foote, and Y. Rubin, *Tetrahedron* **1996**, *52*, 5179.

25. (a) A. Bianco, F. Gasparrini, M. Maggini, D. Misiti, A. Polese, M. Prato, G. Scorrano, C. Toniolo, and C. Villani, *J. Am. Chem. Soc.* **1997**, *119*, 7550. (b) H. Mi, M. E. Tuckerman, D. I. Schuster, and S. R. Wilson, *Proc. Electrochem. Soc.* **1999**, *99–12*, 256.

26. R. F. Schinazi, C. Bellavia, R. Gonzalez, C. L. Hill, and F. Wudl, *Proc. Electrochem. Soc.* **1995**, *95*(10), 696.

27. D. I. Schuster, S. R. Wilson, and R. F. Schinazi, *Bioorg. Med. Chem. Lett.* **1996**, *6*, 1253.

28. F. Kaesermann and C. Kempf, *Antiviral Res.* **1997**, *34*, 65.

29. J. Nacsa, J. Segesdi, A. Gyuris, T. Braun, H. Rausch, A. Buvari-Barcza, L. Barcza, J. Minarovits, and J. Molnar, *Fullerene Sci. Technol.* **1997**, *5*(5), 969.

30. J. Simic-Krstic, *Arch. Oncol.* **1997**, *5*(1), 143.

31. L. Wenzhu, K. Qian, W. Huang, Z. Zhang, and W. Chen, *Chin. Phys. Lett.* **1994**, *11*, 207.

32. (a) M. P. Mattson, *Neuroprotective Signal Transduction* (*Contemporary Neuroscience*). Humana Press, Totowa, NJ, 1998. (b) H. J. Forman, and E. Cadenas, *Oxidative Stress and Signal Transduction*. Chapman and Hall, New York, 1997.

33. Y. L. Huang, C. K. F. Shen, T. Y. Luli, H. C. Yang, K. C. Hwang, and C. K. Chou, *Eur. J. Biochem.* **1998**, *254*(1), 3843.

34. S.-C. Hsu, C.-C. Wu, T.-Y. Luh, C.-K. Chu, S.-H. Han, and M.-Z. Lai, *Blood* **1998**, *91*, 2658.

35. L. L. Dugan, J. K. Gabrielson, S. P. Yu, T.-S. Lin, and D. W. Choi, *Neurobiol. Dis.* **1996**, *3*, 129.

36. H. Tokuyama, S. Yamada, E. Nakamura, T. Shiraki, and Y. Sugiura, *J. Am. Chem. Soc.* **1993**, *115*, 7918.

37. A. S. Boutorine, H. Tokuyama, M. Takasugi, H. Isobe, E. Nakamura, and C. Helene, *Angew. Chem. Int. Ed. Engl.* **1994**, *33*, 2462.

38. N. J. Turro, *Modern Molecular Photochemistry*. Benjamin/Cummings, Menlo Park, CA, 1978.

39. P. B. Merkel and D. R. Keanis, *J. Am. Chem. Soc.* **1972**, *94*, 7244.

40. (a) C. A. Stein and A. M. Krieg, *Applied Antisense Oligonucleotide Technology*. Wiley-Liss, New York, 1998. (b) C. Lichenstein and W. Nellen, *Antisense Technology: A Practical Approach*. Oxford University Press, New York, 1997.

41. A. S. Boutorine, O. Balland, H. Tokuyama, M. Takasugi, H. Isobe, E. Nakamura, and C. Helene, *Proc. Electrochem. Soc.* **1997**, *97*(42), 186.

42. N. Miyata and Y. Yamakoshi, *Proc. Electrochem. Soc.* **1997**, *97*(42), 345.

43. N. Sera, H. Tokiwa, and N. Miyate, *Carcinogenesis* **1996**, *17*, 2163.

44. L. P. Zakharenko, I. K. Zakharov, S. N. Lunegov, and A. A. Nikifornov, *Dokl. Biol. Sci.* **1994**, *335*, 261.

45. A. M. Cassell, W. A. Scrivens, and J. M. Tour, *Angew. Chem. Int. Ed.* **1998**, *37* 1528.

46. N. Higashi, T. Inoue, and M. Niwa, *Chem Commun* **1997**, 1507.

47. (a) J. W. Aborgast, A. P. Darmanyan, C. S. Foote, F. N. Diederich, R. L. Whetten, and Y. Rubin, *J. Phys. Chem.* **1991**, *95*, 11. (b) C. S. Foote, *ACS Symp. Ser.* **1995**, *616*, 17.

48. (a) J. G. Levy, *Tibtech* **1995**, 1314. (b) "A Ray of Hope for Cancer Patients," *Business Week*, June 10, 1996.

49. P. Cheng, S. R. Wilson, and D. I. Schuster, *Chem. Commun.* **1999**, 89 and references cited.

50. N. Nakajima, C. Nishi, F.-M. Li, and Y. Ikada, *Fullerene Sci. Technol.* **1996**, *4*(1), 1.

51. Y. Tabata, Y. Murakarni, and Y. Ikada, *Fullerene Sci. Technol.* **1997**, *5*(5), 989.

52. Y. Tabata, Y. Muramani, and Y. Ikada, *Jpn. Cancer Res.* **1997**, *88*, 1108.

53. K. Qian, Q. Yan, W. Huang, and W. Li, *Shengwu Hwaxue Yu Shengwu Wuli Jinzhan* **1997**, *24*(3), 237.

54. Y. Tabata, Y. Murakami, and Y. Ikada, *Proc. Int. Symp. Controlled Release Bioact. Mater.* **1997**, *24*, 122. [CAS 127:119061]

55. W.-D. Huang, K.-X. Qian, and W.-Z. Li, *Shengwu Huaxue Zazhi* **1996**, *12*(4), 483. [CAS 125:189625]

56. T. Hamano, K. Okuda, T. Mashino, M. Hirobe, K. Arakane, A. Ryu, S. Mashiko, and T. Nagano, *Chem. Commun.* **1997**, 21.

57. N. Martin, L. Sanchez, B. Illescas, and I. Perez, *Chem Rev.* **1998**, *98*, 257.

58. K. C. Hwang and D. Mauzerall, *Nature* **1993**, *361*, 138.

59. H. T. Tien, L.-G. Wang, X. Wang, and A. L. Ottova, *Bioelectrochem. Bioenerg.* **1997**, *42*(2), 161.

60. F. Cardullo, F. Diederich, L. Echegoyen, T. Habicher, N. Jayaraman, R. M. Leblanc, J. F. Stoddart, and S. Wang, *Langmuir* **1998**, *14*(8), 1955.

61. H. Murakami, Y. Watanabe, and N. Nakashima, *J. Am. Chem. Soc.* **1996**, *118*, 4484.

62. M. Hetzer, S. Bayerl, X. Camps, O. Vostrowsky, A. Hirsch, and T. M. Bayerl, *Adv. Mater.* **1997**, *9*(11), 913.

63. L. Y. Chiang, F.-J. Lu, and J.-T. Lin, *J. Chem. Soc. Chem. Commun.* **1995**, 1283.

64. K. Okuda, M. Hirobe, M. Mochizuki, and T. Mashino, *Proc. Electrochem. Soc.* **1997**, *97*(42), 337.

65. Y. I. Pukhova, G. N. Churilov, V. G. Isakova, A. Ya. Korets, and Y. N. Titarenko, *Dokl. Akad. Nauk* **1997**, *355*(2), 269. [CAS 128:123774]

66. Y.-L. Lai and L. Y. Chiang, *J. Auton. Pharmacol.* **1997**, *17*(4), 229.

67. Y.-L. Lai, W.-Y. Chiou, and L. Y. Chiang, *Fullerene Sci. Technol.* **1997**, *5*(5), 1057.

68. K. Okuda, T. Mashino, and M. Hirobe, *Proc. Electrochem. Soc.* **1997**, *96*(10), 411.

69. J. Lincoln, H. V. Hoyle, and G. Burnstock, *Nitric Oxide in Health and Disease.* Cambridge University Press, Cambridge, 1997.

70. M. Satoh, K. Matsuo, H. Kiriya, T. Mashino, M. Hirobe, and I. Takayanagi, *Gen. Pharmacol.* **1997**, *29*(3), 345.

71. M. Satoh, K. Matsuo, K. Hitoshi, T. Mashino, T. Nagano, M. Hirobe, and I. Takayanagi, *Eur. J. Pharmacol.* **1997**, *327*(2/3), 175.

72. (a) Y. Zhu, D. Sun, Guizhen, Liu, Z. Liu, R. Zhan, S. Liu, and X. Gaodeng, *Huaxue Xuebao* **1996**, *17*(7), 1127. (b) D. Sun, Y. Zhu, Z. Liu, G. Liu, X. Guo, R. Zhan, and S. Liu, *Chin. Sci. Bull.* **1997**, *42*(9), 748.

73. Y. Yamakoshi, S. Sueyoshi, K. Fukuhara, and N. Miyata, *J. Am. Chem. Soc.* **1998**, *120*, 12363.

74. L. L. Dugan, D. M. Turetsky, C. Du, D. Lobner, M. Wheeler, C. R. Almli, C. K.-F. Shen, T.-Y. Luh, D. W. Choi, and T.-S. Lin, *Proc. Natl. Acad. Sci. USA* **1997**, *94*, 9434.

75. M.-C. Tsai, Y. H. Chen, and L. Y. Chiang, *J. Pharm. Pharmacol.* **1997**, *49*, 438.

76. M. G. Mitch, L. R. Karam, B. M. Coursey, and I. Sagdeev, *Fullerene Sci. Technol.* **1997**, *5*(5), 855.

77. W. A. Scrivens, J. M. Tour, K. E. Creek, and L. Pirisi, *J. Am. Chem Soc.* **1994**, *116*, 4517.

78. S. R. Wilson and E. Chin, *Fullerene Sci. Technol.* **1996**, *4*(1), 43.

79. T. P. Thrash, D. W. Cagle, M. Alford, G. J. Ehrhardt, J. C. Lattimer, and L. J. Wilson, *Proc. Electrochem. Soc.* **1997**, *97*, 14349.

80. D. W. Cagle, J. M. Alford, J. Tien, and L. J. Wilson, *Proc. Electrochem. Soc.* **1997**, *97*(14), 361.

81. H. Shinohara, K. Yogi, and J. Nakamura, *Jpn. Pat. Appl.* 94–285395. [CAS 125:136631]

82. M. A. Nelson, F. E. Domann, G. T. Bowdon, S. B. Hooser, Q. Fernando, and D. E. Carter, *Toxcol. Ind. Health* **1993**, *9*(4), 623.

83. T. Baierl, E. Drosselmeyer, A. Seidel, and S. Hippeli, *Exp. Toxicol. Pathol.* **1996**, *48*(6), 508.

84. T. Baierl and A. Seidel, *Fullerene Sci. Technol.* **1996**, *4*(5), 1073.

85. B.-X. Chen, S. R. Wilson, M. Das, D. J. Coughlin, and B. F. Erlanger, *Proc. Natl. Acad. Sci. USA* **1998**, *95*, 10809.

86. N. K. Jerne, *Ann. Immunol. (Inst. Pasteur)* **1974**, *125C*, 373.

87. K. Rajewsky, I. Forster, and A. Kumano, *Science* **1987**, *238*, 1088.

88. R. E. Langman, *Int. J. Clin. Lab. Res.* **1992**, *22*, 63.

89. B. F. Erlanger and E. Brand, *J. Am. Chem. Soc.* **1951**, *73*, 4026.

90. C. A. Janeway, Jr., and P. Travers, *Immunobiology.* Garland Publishing, New York, 1996.

91. (a) A. Johnstone and R. Thorpe, *Immunochemistry in Practice.* Blackwell Scientific Publications, Oxford, 1982. (b) J. H. Arevolo, C. A. Hassig, M. J. Stura, M. J. Taussig, and S. R. Wilson, *J. Mol. Biol.* **1994**, *241*, 663.

92. D. Izhaky and I. Pecht, *Proc. Natl. Acad. Sci. USA* **1998**, *95*, 11509.

93. (a) B. I. Yacobson and R. E. Smalley, Am. Sci. **1997**, *85*, 324. (b) M. S. Dresselhaus, G. Dresselhaus, and P. Ekland, *Science of Fullerenes and Carbon Nanotubes.* Academic Press, New York, 1996.

94. A. Fischer, R. Hoch, D. Moy, M. Lu, M. Martin, C. M. Niu, N. Ogata, H. Tennent, L. Dong, J. Sun, L. Helms, F. Jameison, P. Liang, and D. Simpson, *PCT Int. Appl.* **1997**, WO 9732571 A1. [CAS 127:283826]

95. R. J. Massey, M. T. Martin, L. Dong, M. Lu, A. Fischer, F. Jameison, P. Liang, R. Hoch, and J. Leland, *PCT Int. Appl.* **1997**, WO 9733176 A1.

96. J. J. Davis, M. L. H. Green, A. O. Hill, Y. C. Leung, P. J. Sadler, J. Sloan, A. V. Xavier, and S. C. Tsang, *Inorg. Chim. Acta* **1998**, *272*, 261.

97. J. Cook, J. Sloan, and M. L. H. Green, *Fullerene Sci. Technol.* **1997**, *5*(4), 695.

98. (a) *Carbon Fibers*, J.-B. Donnet, T. K. Want, J. C. Peng, and M. Peng (eds.). Marcel Dekker, New York, 1998. (b) D. B. Warheit, J. F. Hansen, M. C. Carakostas, and M. A. Hartsky, *Ann. Occup. Hyg.* **1994**, *38* (Suppl. 1), 769.

99. S. Yamago, H. Tokuyama, E. Nakamura, K. Kikuchi, S. Kananishi, K. Seuki, H. Nakahara, S. Enomoto, and F. Ambe, *Chem. Biol.* **1995**, *2*, 385.

100. R. Bullard-Dillard, K. Creek, W. Scrivens, and J. Tour, *Bioorg. Chem.* **1996**, *24* 376.

101. D. W. Cagle, S. J. Kennel, J. M. Alford, and L. J. Wilson, *Proc. Natl. Acad. Sci. USA* **1999**, *96*, 5182.

102. H. H. C. Chen, C. Yu, T. H. Ui, S. Chen, B. J. Cren, and K. J. Huang, *Ann. Toxicol. Pathol.* **1998**, *26*, 43.

103. P. Rajagopalan, F. Wudl, R. F. Schinazi, and F. D. Boudinoti, *Antimicrobial Agents Antiviral Therapy* **1996**, *40*, 2262.

104. T.-H. Ueng, J.-J. Kang, H.-W. Wang, Y.-W. Cheng, and L. Y. Chiang, *Toxicol. Lett.* **1997**, *93*, 29.

105. L. K. Panina, V. E. Kurochkin, E. V. Bogomolvova, A. A. Evstrapov, and N. G. Spitsing, *Dokl. RAS* **1997**, *351*, 275.

106. D. W. Choi, L. Dugan, T.-S. T. Lin, and T.-Y. Luh, *PCT Int. Appl.* **1997**, WO 9746227. [CAS 128:70789]

107. R. C. Richmond and U. J. Gibson, *PCT Appl.* **1992**, WO 9400552. [CAS 120:158195y]

108. R. C. Richmond and U. J. Gibson, *Proc. Electrochem. Soc.* **1995**, *95*(10), 684.

109. M. C. Petty, M. R. Bryce, and G. Williams, *PCT Appl* [CAS 123:266208]

110. L. M. Lander, W. J. Brittain, and E. A. Vogler, *Langmuir* **1995**, *11*(1), 375.

111. K. Miyazawa, F. Matsuzaki, T. Hariki, and M. Yamaguchi, *Jpn. Kokai Tokkyo Koho* **1997**, JP 09278625. [CAS 127:362482]

112. Y. Ikada, Y. Tabata, and N. Nakajima, *Jpn. Kokai Tokkyo Koho* **1997**, JP 09235235. [CAS 127:268068]

113. L. Y. Chiang, *Eur. Pat. Appl.* **1997**, EP 770577. [CAS 127:34015]

CHAPTER 11

CARBOXYFULLERENES AS NEUROPROTECTIVE ANTIOXIDANTS

LAURA L. DUGAN, EVA LOVETT, SARAH CUDDIHY, BEI-WEN MA, TIEN-SUNG LIN, and DENNIS W. CHOI

11.1 INTRODUCTION

Free-radical-mediated oxidative injury has been implicated in the pathogenesis of an extensive list of human diseases ranging from cancer to tissue ischemia to complications of infectious diseases. The contribution of oxidative damage to neurological conditions may be especially prominent for a number of reasons, including the reliance of the brain on aerobic metabolism, its rich content of unsaturated fatty acids—targets of lipid peroxidation—and because the nervous system, unlike many other tissues, has limited ability to regenerate damaged tissue. Free-radical-mediated injury may occur during acute insults such as stroke, head trauma, and spinal cord injury, and in chronic neurodegenerative conditions, such as Parkinson's disease and Alzheimer's dementia. Supporting a significant role for oxidative injury in central nervous system (CNS) disease, recent human trials of vitamin E in Alzheimer's patients found that treatment delayed the onset and slowed the progression of the disease.

Evidence from studies in cell culture and animal models of disease suggest that reactive oxygen species—superoxide ($O_2^{\cdot-}$) and hydroxyl ($\cdot OH$) radicals, and the nonradical molecules H_2O_2 and hypochlorous acid—are major contributors to oxidative injury in mammals. In addition, nitric oxide ($NO\cdot$), a biologically generated free radical gas with limited intrinsic reactivity, can combine with $O_2^{\cdot-}$ to generate the reactive oxidant, peroxynitrite ($ONOO^-$). These reactive oxygen species may then cause oxidative damage to cellular components, such as peroxidation of cell membrane lipids, oxidation and fragmentation of DNA, inactivation of transport proteins, and inhibition of energy production by mitochondria [1]. Furthermore, while nitric oxide and hypochlorous acid appear to be generated by a restricted number of cell types, $O_2^{\cdot-}$

Fullerenes: Chemistry, Physics, and Technology, Edited by Karl M. Kadish and Rodney S. Ruoff.
0-471-29089-0 Copyright © 2000 John Wiley & Sons, Inc.

and H_2O_2 are ubiquitous "by-products" of many biological processes, including energy metabolism and macromolecule synthesis.

Two processes have emerged as important contributors to tissue injury during insults to the CNS: glutamate receptor-mediated excitotoxicity and apoptotic cell death. Reactive oxygen species have been implicated in both cell death pathways. The concept that oxidative stress may mediate a component of excitotoxic neuronal damage is supported by the ability of free radical scavengers such as the 21-aminosteroid "lazaroids" [2], the vitamin E analog, trolox [3], and spin-trapping agents such as PBN [4] to reduce excitotoxic neuronal death. Furthermore, several laboratories have directly demonstrated free radical formation as a consequence of glutamate receptor overstimulation, using electron paramagnetic resonance (EPR) [5,6] spectroscopy or fluorescence microscopy with oxidation-sensitive dyes [6,7].

Apoptosis may mediate delayed neuronal degeneration after ischemia-reperfusion [8] and has been proposed to occur in certain neurodegenerative diseases, such as Huntington's disease, Alzheimer's dementia, and amyotrophic lateral sclerosis (ALS) [9]. Neuronal apoptosis initiated by removal of growth factors induces a burst of reactive oxygen species [10], and antioxidants block apoptosis in several model systems [10,11]. In addition, the anti-apoptosis protein Bcl-2 may mediate its cytoprotective effects by altering cellular free radical homeostasis [12,13].

In light of the multiple injury processes to which free radicals contribute, we have continued to explore antioxidant agents as broadly effective therapeutic strategies for CNS injury and disease. In the past several years, we have focused on developing water-soluble derivatives of C_{60} as biologically effective antioxidants based on the premise that the potent innate antioxidant properties of C_{60} could be harnessed for use in biological systems by the addition of functional groups that enhance its water solubility.

11.2 STRUCTURES OF MALONIC ACID C_{60} DERIVATIVES

We generated 2,2-fulleromalonic acid derivatives of C_{60} (carboxyfullerenes) by the method of Hirsch, 1994 [14], as modified [15]. C_{60} (>99.95%) was obtained from MER Corp. (Tucson, AZ, USA). Other reagents were purchased from Sigma, Aldrich, or other standard suppliers. Malonic C_{60} esters were synthesized and then chromatographed on silica gel. Fractions were evaluated for structure and purity by ^{13}C NMR, and confirmed by reverse-phase high-performance liquid chromatography (HPLC) analysis. Pure esters were hydrolyzed to acids, and purity and identity of the resulting malonic acids were confirmed by ^{13}C NMR, HPLC, and mass spectrometry/mass spectrometry (MS/MS) [16]. This strategy provided four main tris(malonic) acid C_{60} isomers, C_3, D_3 (Fig. 11.1), referred to by their molecular symmetry, and compounds E and G. The latter two isomers appear to have asymmetric additions of malonic acid to the C_{60} sphere. All four tris adducts are highly soluble in water (up to ~100 mM) but

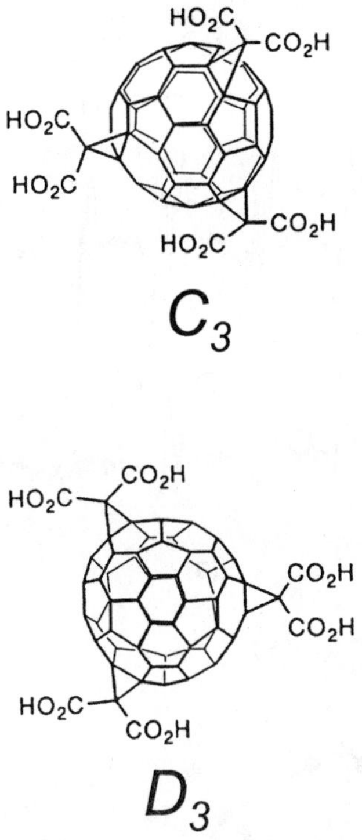

Figure 11.1 Structures of the C_3 and D_3 regioisomers of tris(malonic) acid C_{60}. The malonic acid groups are positioned equatorially in the D_3 compound and are clustered in one hemisphere in the C_3 compound [14].

also retain some of lipophilic character of native C_{60}. The C_3 isomer enters lipid bilayers to a greater extent than D_3 [15].

11.3 ANTIOXIDANT PROPERTIES OF CARBOXYFULLERENES

After generating pure preparations of tris(malonic) acid C_{60} isomers, we next wished to determine whether they retained the potent free radical scavenging capability of native C_{60}. We analyzed superoxide ($O_2^{\cdot-}$) scavenging by C_3 by two complementary methods. Electron paramagnetic resonance (EPR) spectrometry was used to assess $O_2^{\cdot-}$ scavenging by malonic acid C_{60} derivatives as described [15]. Samples in Tris buffer were loaded into a quartz flat cell ($60 \times 10 \times 0.25$ mm) and analyzed on a Bruker 200 X-band spectrometer, with settings: power $= 1.6$ mW, modulation $= 1$ G, field modulation $= 100$ Hz, R.G. $= 3.2 \times 10^5$. Superoxide was generated by the oxidation of 25 μM xanthine $+ 10$ mU xanthine oxidase. C_3 was an effective scavenger of $O_2^{\cdot-}$ as demonstrated by two different assay techniques. Using EPR spectroscopy per-

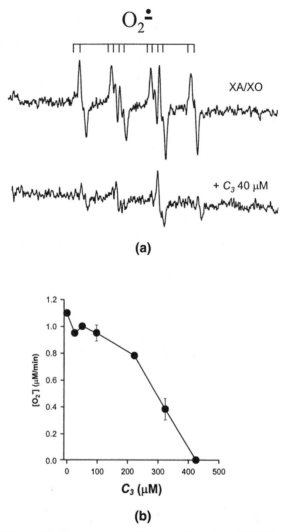

Figure 11.2 C_3 is an effective scavenger of superoxide anion: (a) EPR spectrum of $O_2^{\cdot-}$ (2-4-4-2 hyperfine splitting pattern for $O_2^{\cdot-}$ is identified) generated through oxidation of xanthine by xanthine oxidase, detected by the spin-trapping agent DMPO, with vehicle (water; *top trace*) or C_3 (40 µM; *bottom trace*) added 40 seconds after the reaction was started. Settings for the EPR were as described in the text. C_3 also effectively eliminated $O_2^{\cdot-}$ in a second assay system (b), in which reduction of cytochrome-c by $O_2^{\cdot-}$ (from xanthine/xanthine oxidase) was followed as absorbance at 550 nm.

formed at room temperature with dimethyl-1-pyrroline *N*-oxide (DMPO) as the spin-trapping agent (100 mM), C_3 eliminated the superoxide signal generated by xanthine oxidase/xanthine (Fig. 11.2a). While spin-trapping nitrones or nitroxides are among the few classes of antioxidants capable of reacting with

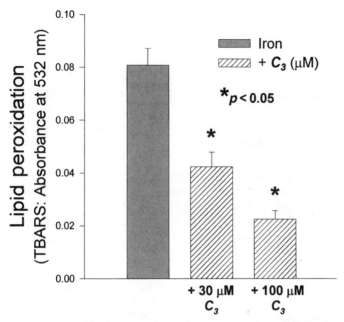

Figure 11.3 Iron-induced lipid peroxidation is blocked by C_3. Production of aldehyde lipid peroxidation products, assayed by the standard TBARS method and detected by their absorbance at 532 nm, is inhibited by C_3. Values are mean μ SEM, $n = 4$, $p < 0.05$ by ANOVA, followed by Student–Neuman–Keuls, representative of two additional experiments.

$O_2^{\cdot-}$, C_3 appeared to be a more avid scavenger of $O_2^{\cdot-}$ than DMPO, as C_3 (40 μM) was able to effectively out-compete DMPO, present at 100 mM, for $O_2^{\cdot-}$.

Superoxide elimination by C_3 was also assessed by a second method, based on the reduction of cytochrome-c (absorbance 550 nm) by $O_2^{\cdot-}$. Production of $O_2^{\cdot-}$ (from xanthine oxidase/hypoxanthine) was determined spectrophotometrically by following absorbance at 550 nm. The reaction gave a linear increase in absorbance for more than 30 minutes. Concentrations of C_3 from 25 to 425 μM were included to estimate a half-inhibitory concentration (Fig. 11.2b).

C_3 also acts as an effective radical scavenger in one-electron Fenton-type reactions, as demonstrated by EPR spectrometry of the reaction of H_2O_2 with iron [15], and in a thiobarbituric acid reactive substances (TBARS) assay for iron-catalyzed lipid peroxidation products (Fig. 11.3). Measurement of TBARS was performed by a modification of the method described [17]. Phosphatidylcholine (18:1, 20:4, Sigma) was added to a test tube, the solvent evaporated, and Tris buffer with 100 μM ferrous and ferric iron (50:50), with or without C_3, was added. Samples were sonicated to form a fine suspension of the lipid in the buffer. The reaction was incubated for 30 min at 37 °C and stopped by addition of cold trichloroacetic acid–thiobarbituric acid solution containing BHT (0.2%)

to halt further peroxidation. Samples were boiled for 10 min, cooled, and the chromagen was analyzed at 532 nm. Iron increased TBARS formation, while C_3 decreased lipid peroxidation.

11.4   NEUROPROTECTIVE PROPERTIES: EXCITOTOXICITY

We then explored the neuroprotective and cytoprotective properties of malonic acid C_{60} derivatives. Neocortical cell cultures were prepared from fetal (E15) Swiss–Webster mice (Simonson) as described previously [18]. Cortical hemispheres were dissected away from the rest of the brain and placed in media stock (MS: Eagle's Minimal Essential Media minus L-glutamine, Gibco 11430-022, with 20 mM glucose, 26.2 mM $NaHCO_3$) for 10 min. The cortices were briefly centrifuged, the trypsin was removed, and the hemispheres were resuspended in plating medium, which consists of MS plus 5% fetal calf serum and 5% horse serum (Hyclone) for pure neuronal cultures (<0.5% astrocytes). For neuron-astrocyte cocultures, the same media, supplemented with L-glutamine (2 mM) was used. After trituration, cell suspensions were diluted and plated onto a preexisting bed of mouse cortical astrocytes in 24-well Primaria culture plates (for mixed cultures), or onto poly-D-lysine (PDL): laminin-coated 24-well plates (for pure neuronal cultures). Cultures were fed biweekly with growth medium (MS with 10% horse serum, 2 mM L-glutamine), until the final feeding at day 11 or 12 in vitro, when cultures were fed with MS supplemented with 2 mM L-glutamine. Cells were used for experiments 14–16 days after plating.

Brief exposure to N-methyl-D-aspartate (NMDA) was carried out as described [18]. The cell culture medium was exchanged twice with N-[2-hydroxyethyl] piperazine-N'-[2-ethane sulfonic acid] (HEPES), bicarbonate-buffered balanced salt solution (HBBSS), and NMDA (200 µM) was added alone or with an antioxidant compound for 10 min. Antioxidants tested included tocopherol, phenyl-t-butyl-nitrone (PBN), trolox, ubiquinone, and the 21-amino steroids (U74500A and U74006F). The stock solution of C_3 was made in dilute NaOH to give a neutral salt solution, stock solutions of ubiquinone and the lazaroids were made in DMSO, and stocks of other drugs were prepared in HBBSS. Exposure to NMDA was terminated by exchanging the medium twice with HBBSS, then twice with MS. Washed controls received HBBSS without drugs. Slowly triggered NMDA or α-amino-3-hydroxy-5-methyl-4-isoxazolepropionic acid (AMPA) receptor-mediated excitotoxicity was induced by the addition of NMDA (15 µM) or AMPA (6 µM plus 10 µM MK-801 to eliminate a secondary contribution of NMDA receptors) with or without C_3 for 24 h. The cells were returned to the 37 °C (5% CO_2) incubator for 24 h, when injury was assessed by determination of lactate dehydrogenate (LDH) release [19].

Coapplication of C_3 with NMDA for 24 h blocked most of the neuronal cell death induced by this low, chronic NMDA exposure (Fig. 11.4a). In addition, in a head-to-head comparison of C_3 with several other benchmark antioxidants,

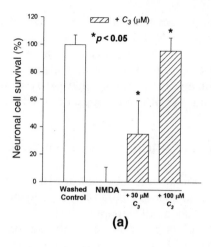

(a)

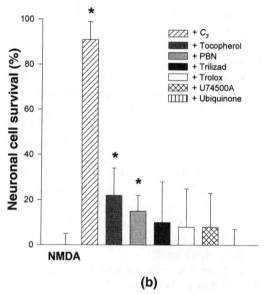

(b)

Figure 11.4 Neuroprotection by tris(malonic) acid C_{60} derivatives. C_3 provides near-complete protection of cortical neurons exposed to NMDA (12.5 μM) for 24 h (a). Values are shown as % cell death, mean ± SEM, $*p < 0.05$ by ANOVA, Student–Neuman–Keuls, $n = 4$. The neuroprotective efficacy of C_3 was compared head-to-head with that of other antioxidant compounds (b) against the fulminant injury produced by treatment of cortical cultures with NMDA, 200 μM for 10 min. The maximal protection achieved for each compound, determined from a full dose–response curve, is shown. Other than C_3, only α-tocopherol and the spin-trap, PBN, were significantly protective. Values are % cell survival, mean ± SEM, $*p < 0.05$ by ANOVA, Student–Neuman–Keuls, $n = 4$–8.

C_3 demonstrated superior potency and efficacy against toxicity initiated by brief (10 min) exposure to high-dose NMDA (Fig. 11.4b). C_3 also provided protection against AMPA-receptor-mediated injury in both neurons [15] and oligodendroglia [20], the cell type responsible for central myelin formation. Other bis-, tris-, and tetramalonic acid C_{60} isomers have shown excellent protective efficacy against NMDA and AMPA receptor excitotoxicity (data not shown).

11.5 NEUROPROTECTIVE PROPERTIES: APOPTOSIS

Several lines of evidence support the idea that apoptosis contributes to brain damage in CNS injury. A number of laboratories have found morphological or biochemical changes associated with apoptosis [21,22], and pharmacological and molecular strategies that target apoptosis limit tissue damage in a number of CNS injury models [8,23,24]. Concensus is emerging that reactive oxygen species contribute importantly to apoptotic cell death. Oxidative stress can trigger apoptosis, and free radical production is an early event in programmed cell death [10]. Antioxidants are neuroprotective against programmed cell death/apoptosis triggered by a wide variety of insults.

To determine whether C_{60} derivatives were effective agents against apoptotic forms of cell death, we tested these compounds in cortical neuronal cultures grown in the absence of glial cells. When these pure neuronal cultures are deprived of serum, they undergo apoptotic death. As previously described, within 3–4 h after removal of serum, neurons generate a substantial burst of reactive oxygen species, and early treatment with antioxidants protected cells from subsequent apoptotic death [10,15]. Using confocal microscopy to evaluate pure neuronal cultures loaded with the oxidation-sensitive fluorescent compound dihydrorhodamine 123, we followed generation of reactive oxygen species over time [6]. Neurons deprived of serum for 4 h (Fig. 11.5) show increased fluorescence, due to dye oxidation, compared to control cultures. If C_3 (10 μM) is included at the time of serum removal, cells fail to exhibit enhanced free radical production and are rescued from apoptotic death (Fig. 11.5) [15]. C_{60} derivatives also blocked apoptosis triggered by other insults, including exposure to the Alzheimer's peptide $A\beta_{1-42}$ [14]. More recently, we have demonstrated neuroprotection by these compounds against toxin-induced death of dopaminergic neurons [25] and growth factor-deprivation of superior cervical ganglion neurons (G. Putcha, E. Johnson, and L. L. Dugan, *unpublished observations*).

11.6 NEUROPROTECTION IN VIVO

We are continuing to explore the ability of the malonic acid C_{60} derivatives to provide neuroprotection in animal models of nervous system disease. FALS mice (SOD1 G93A, Gurney G1 strain) were used for therapeutic trials of C_3. As reported by Gurney and colleagues [26], these animals develop motor

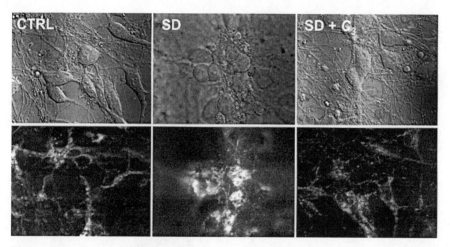

Figure 11.5 Free radical production by cortical neurons in response to removal of growth factor support. Images are differential interference contrast (*top*) or fluorescence (*bottom*) photomicrographs of cultured cortical neurons, using the oxidation-sensitive dye dihydrorhodamine. Cell cultures were loaded with the nonfluorescent compound dihydrorhodamine and washed into medium with or without serum, which provides growth factor support for these cultures. Production of reactive oxygen species was observed as increasing fluorescence intensity. The enhanced fluorescence seen in the center image indicates increased free radical generation by neurons undergoing apoptotic cell death due to growth factor deprivation. Treatment with C_3 decreases free radical formation, blocks the degenerative cell changes associated with apoptosis, and blocks apoptotic neuronal death (images at far right) [15].

symptoms by approximately 90 days of age and are moribund between 125 and 145 days of age. We originally reported that chronic administration of C_3, 15 mg/kg/day, beginning at 9 weeks of age [15], provided a 9 day increase in survival. A subsequent therapeutic trial initiated treatment with 1 mg/kg/day C_3 at 5 weeks of age, using Alzet mini-osmotic pumps (28 day) containing C_3 or saline, which were implanted intraperitoneally. Reimplantation with a fresh pump was performed monthly for the next 3–4 months, with removal of the depleted pump at the time of surgery. This second treatment study in FALS mice produced slightly better protection (Fig. 11.6); both the onset of motor deterioration and death were delayed by ~12 days by this treatment.

In additional ongoing studies in vivo, C_3 infusion also enhanced motor function recovery in rats after cord compression trauma. Rats that received C_3 (3 mg/kg/day) by i.p. Alzet pump for only one week after undergoing impact trauma to their lumbar cord scored approximately 6 points (out of 21) on the BBB locomotor scale (X. Z. Liu, D. W. Choi, and L. L. Dugan, *unpublished observations*). We also have observed decreased brain infarct volume in neonatal rats after hypoxic-ischemic injury (A. D'Costa, D. Holtzman, and L. L. Dugan *unpublished observations*) and in rats in a focal ischemia model of stroke

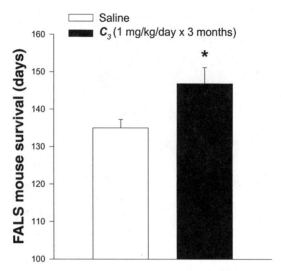

Figure 11.6   G93A FALS mice receiving C_3 1 mg/kg/day by intraperitoneal (i.p.) infusion demonstrated increased survival. Survival of FALS mice treated with C_3 or vehicle (normal saline) is shown. Values are mean (days) $\pm$ SEM, $*p < 0.05$, by t-test.

(Y. Y. He, C. Y. Hsu, L. L. Dugan, *unpublished observations*). Dose-finding studies and exploration of oral administration are ongoing for each of these models.

11.7   DISCUSSION

In 1991, Krusic reported that the native C_{60} fullerene could scavenge more than 30 methyl radicals per C_{60} molecule and termed C_{60} a "radical sponge." Such a multihit small molecule free radical scavenger would be highly desirable when restricted drug delivery, for example, across the blood–brain barrier, would limit drug efficacy. However, C_{60} itself is insoluble in all but a few solvents, primarily toluene. Although many types of water-soluble C_{60} derivative have been synthesized over the past few years, we chose to focus on malonic acid (carboxyl) derivatives of C_{60} for a number of reasons. Unlike most other procedures to derivatize C_{60}, the Hirsch synthesis of malonic acid adducts allows a limited number of isomers with identified structures to be generated and purified. In addition, we predicted that carboxylic acid groups would be well tolerated biologically and would provide good aqueous solubility for the derivatives. Although we have studied all of the tris isomers, we have worked most extensively with the isomer with "C_3" structural symmetry, which also has lipophilic properties that we believe contribute to its neuroprotective efficacy [15].

We generated specific regioisomers of tris(malonic) acid C_{60} and evaluated their free radical scavenging and neuroprotective properties. Using EPR spectroscopy and biochemical assays of free radical scavenging, we found that these derivatives retained the potent free radical scavenging properties reported for the native C_{60} molecule. In cell culture experiments, these agents rescued cortical neurons from a broad range of insults. We studied the C_3 isomer most extensively and found that it inhibited excitotoxic death of cultured cortical neurons induced by 24 h exposure to N-methyl-D-aspartate (NMDA), α-amino-3-hydroxy-5-methyl-4-isoxazolepropionic acid (AMPA). Surprisingly, this derivative blocked even the rapidly triggered neurotoxicity produced by brief exposure to high concentrations of NMDA, a fulminant injury with limited sensitivity to all other types of intervention short of direct blockade of the NMDA receptor. The C_3 isomer also reduced apoptotic neuronal death produced by several insults, including serum deprivation and exposure to the $A\beta_{1-42}$ peptide and has shown robust neuroprotection in a number of other cell culture models of neurological disease including Parkinson's disease. Continuous intraperitoneal infusion of the C_3 derivative in a transgenic mouse model of Lou Gehrig's disease (amyotrophic lateral sclerosis, ALS) carrying the human mutant (G93A) superoxide dismutase gene responsible for a form of familial amyotrophic lateral sclerosis delayed both death and functional deterioration. Preliminary studies in in vivo models of several other neurological diseases have been promising, indicating that polar carboxylic acid C_{60} derivatives have attractive therapeutic properties and may have potential as therapeutic agents in several acute and chronic neurological disease states. Several bis and tris isomers of malonic acid C_{60} were synthesized and found to be extremely potent free radical scavengers by EPR, and by biochemical assays of cytochrome-c reduction and TBARS production. The C_3 regioisomer was studied most extensively and provided robust neuroprotection against excitotoxic injury in cortical neurons and oligodendroglia, and apoptotic neurodegeneration of cortical neurons and mesencephalic neurons. Preliminary experiments also showed the C_3 isomer to be neuroprotective in animal models of several acute and chronic neurological diseases.

Although many types of reactive oxygen species (ROS) can be generated by cells under pathophysiological conditions, $O_2^{\cdot-}$ may be an especially attractive target for antioxidant compounds. Several sources of "pathological" ROS, such as the mitochondrial electron transport chain, and arachidonic acid-metabolizing enzymes may generate primarily $O_2^{\cdot-}$. Superoxide has a relatively long half-life and so might persist long enough to be eliminated by an effective antioxidant. Superoxide may interact with nitric oxide to produce the toxic oxidant, peroxynitrite, or may undergo metal-catalyzed Haber–Weiss reactions to generate hydroxyl radical. A unique feature of the C_{60} derivatives is their ability to eliminate both superoxide anion ($O_2^{\cdot-}$), as well as hydroxyl radical ($\cdot$OH), at low (micromolar) concentrations [15]. We believe that the ability of these molecules to detoxify $O_2^{\cdot-}$ is responsible for the greater degree of neuroprotection they afford compared to other antioxidants we have studied.

We have also expanded the populations of neurons in which neuroprotection by C_3 can be observed. Application of 6-hydroxydopamine to substantia nigra neurons in culture can produce loss of tyrosine hydroxylase positive neurons, and C_3 provides substantial protection against this injury [25]. In addition, nonneuronal CNS cell types can be rescued by C_3. Mature oligodendroglia can also be rapidly killed by excitotoxicity, in a manner very similar to neurons, and can be protected against this injury by C_3 [20]. The concentration at which C_3 was effective in oligodendroglia was actually lower than that required to protect neurons from this same type of injury.

In summary, we have found water-soluble C_{60} derivatives to be broadly effective neuroprotective agents in vitro, blocking excitotoxic, apoptotic, and toxin-induced cell death of neurons and glial cells [15,27]. Furthermore, in ongoing studies, these molecules appear to provide effective neuroprotection in vivo, as well. Our continuing broader goals are to develop these molecules as tools to study the role of specific reactive oxygen species, such as $O_2^{\cdot-}$ and H_2O_2, as both signaling molecules and neurotoxins, to begin to define the structure–function relationships for these promising compounds, and to define their pharmacokinetic and toxicity profiles as necessary steps toward human clinical trials.

REFERENCES

1. B. Halliwell, *J. Neurochem.* **1992**, *59*, 1609.

2. H. Monyer, D. M. Hartley, and D. W. Choi, *Neuron* **1990**, *5*, 121.

3. H. S. Chow, J. J. Lynch, 3rd, K. Rose, and D. W. Choi, *Brain Res.* **1994**, *639*, 102.

4. T. L. Yue, J. L. Gu, P. G. Lysko, H. Y. Cheng, F. C. Barone, and G. Feuerstein, *Brain Res.* **1992**, *574*, 193.

5. M. Lafon-Cazal, S. Pietri, M. Culcasi, and J. Bockaert, *Nature* **1993**, *364*, 535.

6. L. L. Dugan, S. L. Sensi, L. M. T. Canzoniero, S. D. Handran, S. M. Rothman, T.-S. Lin, M. P. Goldberg, and D. W. Choi, *J. Neurosci.* **1995**, *15*, 6377.

7. I. J. Reynolds and T. G. Hastings, *J. Neurosci.* **1995**, *15*, 3318.

8. D. W. Choi, *Curr. Opin. Neurobiol.* **1996**, *6*, 667.

9. P. Desjardins and S. Ledoux, *Metab. Brain Dis.* **1998**, *13*(2), 79.

10. L. J. Greenlund, T. L. Deckwerth, and E. M. Johnson, Jr., *Neuron* **1995**, *14*, 303.

11. R. R. Ratan, T. H. Murphy, and J. M. Baraban, *J. Neurochem.* **1994**, *62*, 376.

12. D. M. Hockenbery, Z. N. Oltvai, X. M. Yin, C. L. Milliman, and S. J. Korsmeyer, *Cell* **1993**, *75*, 241.

13. D. J. Kane, T. A. Sarafian, R. Anton, H. Hahn, E. B. Gralla, J. S. Valentine, T. Ord, and D. E. Bredesen, *Science* **1993**, *262*, 1274.

14. I. Lamparth and A. Hirsch, *J. Chem. Soc. Chem. Commun.* **1994**, 1727.

15. L. L. Dugan, D. M. Turetsky, C. Du, D. Lobner, M. Wheeler, R. Almli, C. K. F. Shen, T. Y. Luh, D. W. Choi, and T. S. Lin, *Proc. Natl. Acad. Sci. USA* **1997**, *94*, 9434.

16. M. Grayson, E. Lovett, L. Dugan, Y. Wang, and M. Gross, *Abs. Am. Soc. Mass Spectrosc.* **1999**, 929.

17. D. K. Anderson, L. L. Dugan, E. D. Means, and L. A. Horrocks, *Brain Res.* **1994**, *637*, 119.

18. L. L. Dugan, V. M. G. Bruno, S. M. Amagasu, and R. G. Giffard, *J. Neurosci.* **1995**, *15*, 4545.

19. J. Koh and D. W. Choi, *J. Neurosci. Meth.* **1987**, *20*, 83.

20. M. P. Goldberg, S. P. Althomsons, T. Chapman, D. W. Choi, and L. L. Dugan, *Neurology* **1998**, *A370* (Suppl. 59), 2.

21. J. P. MacManus, A. M. Buchan, I. E. Hill, I. Rasquinha, and E. Preston, *Neurosci. Lett.* **1993**, *164*, 89.

22. T. Nitatori, N. Sato, S. Waguri, Y. Karasawa, H. Araki, K. Shibanai, E. Kominami, Uchiyama. *J. Neurosci.* **1995**, *15*, 1001.

23. E. M. Johnson, Jr., L. J. S. Greenlund, P. T. Akins, and C. Y. Hsu, *J. Neurotrauma* **1995**, *12*, 843.

24. S. A. Loddick, A. MacKenzie, and N. J. Rothwell, *NeuroReport* **1996**, *7*, 1465.

25. J. Lotharius, L. L. Dugan, and K. L. O'Malley, *J. Neurosci.* **1999**.

26. M. E. Gurney, P. Haifeng, A. Y. Chiu, M. C. Dal Canto, C. Y. Polchow, et al. *Science* **1994**, *264*, 1772.

27. L. L. Dugan, J. K. Gabrielsen, S. P. Yu, T. S. Lin, and D. W. Choi, *Neurobiol. Dis.* **1996**, *3*, 129.

CHAPTER 12

FULLERENES AND FULLERENE IONS IN THE GAS PHASE

DIETHARD K. BOHME, OLGA V. BOLTALINA, and PREBEN HVELPLUND

12.1 INTRODUCTION

This chapter focuses on the gas phase physics and chemistry of fullerenes, fullerene ions, and their derivatives. The production of fullerene ions in the 1980s by laser desorption of graphite and their mass spectrometric analysis provided the first information about the structure and reactivity of fullerenes. But the discovery of a method for preparation and isolation of macroscopic amounts of some fullerenes (C_{60} and C_{70}) in the early 1990s was decisive in revolutionizing the physics and chemistry of fullerenes. Solid fullerene material could now be brought into the gas phase by heating to moderate temperatures ($\sim 500\,°C$) and both positively and negatively charged fullerenes could be produced routinely in various types of ion sources. Singly and multiply charged fullerene ions, both positive and negative, were shown to exhibit a wealth of chemistry in the gas phase at room temperature, ranging from electron transfer to multiple derivatization and ion-induced polymerization. Electron capture and loss processes along with fullerene fragmentation were explored in high-energy collision studies. High-energy collisions with atoms were shown to lead to the remarkable insertion of the atoms into the hollow fullerene cage to form *endohedral* fullerenes. Furthermore, fusion and multifragmentation processes were demonstrated for collisions between two fullerenes or between other clusters and fullerenes. The field of gas phase physics and chemistry involving fullerenes and their ions has indeed experienced spectacular growth in the last decade and has provided exciting opportunities for exploring science, both the known and the completely new.

Fullerenes: Chemistry, Physics, and Technology, Edited by Karl M. Kadish and Rodney S. Ruoff.
0-471-29089-0   Copyright © 2000 John Wiley & Sons, Inc.

12.2 VAPORIZATION PROPERTIES

One of the most striking differences between the two forms of carbon—graphite and fullerene—is the difference in their volatilities. In order to transfer graphite into the gas phase heating up to 2000 °C is needed. By contrast, fullerene can be sublimed at 500 °C. This has prompted studies of the vaporization behavior of the fullerene mixtures and fullerenes that were available in pure forms. Sublimation enthalpy, $\Delta_{sub}H°$ is defined as the enthalpy of reaction (12.1) involving transfer of C_{60} from the crystalline state into the gas phase:

$$C_{60}(cr) = C_{60}(g) \tag{12.1}$$

$$\Delta_{sub}H° = \Delta_r H°(1.1) = \Delta_f H°(C_{60}(g)) - \Delta_f H°(C_{60}(cr)) \tag{12.2}$$

The common way of determining sublimation enthalpy involves measurement of the vapor pressure as a function of temperature, which is followed by the van't Hoff analysis. Most sublimation studies were performed by Knudsen cell mass spectrometry (KCMS) on the pure C_{60} and C_{60}/C_{70} mixtures [1–4]. Several other experimental methods were also applied: effusion torsion (ETM), the Knudsen effusion method (KEM) [5,6], desorption rate measurements [7], and thermogravimetry [8].

12.2.1 [60]Fullerene

Characterization of the purity of fullerene samples is crucial in thermochemical determinations. Besides impurities of the higher fullerenes $(C_{n>70})$ that can eventually be removed during the high-performance liquid chromatography (HPLC) recycling procedure, the presence of solvent traces was found to cause one of the main uncertainties in vapor pressure results. From comparison of the data on evaporation of the sublimed C_{60} sample with that from a sample containing traces of solvent, it was found that at the final stage of evaporation vapor pressures may differ by 30–40%: No increase in the vapor pressure of the unsublimed sample with temperature was observed. Nonvolatile residue (up to 20% of the initial amount) was formed in the cell after evaporation. It was not dissolved in organic solvents and was referred to as amorphous carbon [4] or polymerized carbonaceous material [6]. No serious attempts to characterize this residue were undertaken. In some cases thermal degradation of fullerene samples during evaporation could be avoided, but it was found to depend greatly on the rate of heating: The faster the temperature was raised the smaller the extent of degradation observed (if at all) [9].

In Figure 12.1 data on the vapor pressure of [60]fullerene are presented as a function of temperature. $\Delta_{sub}H°$ values obtained in the KCMS studies are about 20 kJ/mol lower (Table 12.1) than the KEM determination, and the vapor pressure results are higher than other published data (Fig. 12.1). It is probable that the discrepancy can be accounted for by the differing quality of

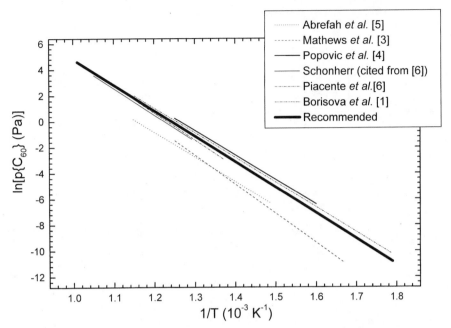

Figure 12.1 Temperature dependence of vapor pressure of [60]fullerene.

the samples that were used by the researchers. Application of the pure sublimed sample of [60]fullerene in the KCMS study [4] helped to avoid complications due to thermal degradation. On the other hand, in KCMS the uncertainty in ionization cross-section values can be a source of significant errors in the absolute vapor pressure data.

Combining the value $\Delta_{sub}H°(C_{60}, 298.15\ \mathrm{K}) = 171 \pm 23$ kJ/mol from Table 12.1 with the solid-state formation enthalpy (2355 ± 15 kJ/mol) [10], one gets the gas-phase formation enthalpy of [60]fullerene as 2526 ± 17 kJ/mol.

12.2.2 [70]Fullerene

Though much fewer sublimation studies were performed on [70]fullerene, similar problems have been encountered by the researchers: thermal degradation of the unsublimed samples and calibration errors [4,5,9,11]. Furthermore, it was found that even sublimed [70]fullerene sample degraded to some extent at elevated temperatures. The temperature equation for vapor pressure of [70]fullerene at 783–904 K is given as

$$\log[p(\mathrm{Pa})] = -9917 \pm 160/T + 11.38 \pm 0.15 \qquad (12.3)$$

Related sublimation enthalpy (190 ± 3 kJ/mol) was recalculated to 298.15 K

TABLE 12.1 Saturated Vapor Pressure and Sublimation Enthalpy of C_{60}

Reference	Year	Temperature interval, K	$\ln[p(Pa)] = -B/T + A$ B, K	$\ln[p(Pa)] = -B/T + A$ A	$\Delta_{sub}H^{\circ}_T$ (kJ/mol)	T (K)	P_{780} (Pa)	$\Delta_{sub}H^{\circ}_{298}$ (kJ/mol) II law[a]	$\Delta_{sub}H^{\circ}_{298}$ (kJ/mol) III law[b]
J. Abrefah et al. [5]	1992	673–873	19095 ± 502	22.08 ± 0.51	159 ± 4	775[c]	0.0908	165 ± 13	180 ± 6
C. K. Mathews et al. [3]	1992	600–800	22512 ± 318	26.66 ± 0.30	181 ± 2	700	0.111	185 ± 11	179 ± 6
A. Popović et al. [4]	1994	625–800	19056 ± 276	24.09 ± 0.32	158 ± 3	702[c]	0.708	162 ± 12	167 ± 6
E. Schonherr[d]	1995	774–953	20412 ± 755	25.01 ± 0.90	170 ± 1	854[c]	0.312	178 ± 10	172 ± 6
V. Piacente et al. [6]	1995	730–990	21078 ± 345	25.97 ± 0.46	175 ± 3	860	0.350	181 ± 2	172 ± 6
D. Yu. Borisova et al. [1]	1998	560–870	19200 ± 300	24.10 ± 0.40	159 ± 3	681[c]	0.597	162 ± 12	168 ± 6
Mean[e]	1999	560–990	19905 ± 1597[e]	24.71 ± 2.40[g]	166 ± 13[e]	765	0.447[f]	171 ± 23	170 ± 9

[a] Recalculated to 298 K in this work except [6] using thermodynamic functions from ref. [6].

[b] Calculated from P_{780} and thermodynamic functions from ref. [6].

[c] Mean temperatures were calculated by averaging 1/T in the temperature interval.

[d] The temperature equation was evaluated in ref. [6] using the vapor pressure points from the unpublished work of Schonherr.

[e] Obtained by averaging data from refs. [1,4,6] and data of Schonherr at mean temperatures.

[f] Evaluated by averaging $\ln[(p(Pa)]_{780}$ from refs. [1,4,6] and data of Schonherr.

[g] Obtained by averaging $\ln[(p(Pa)]_{780}$ from refs. [1,4,6] obtained from $\Delta_{sub}H^{\circ}_T$ and P_{780}.

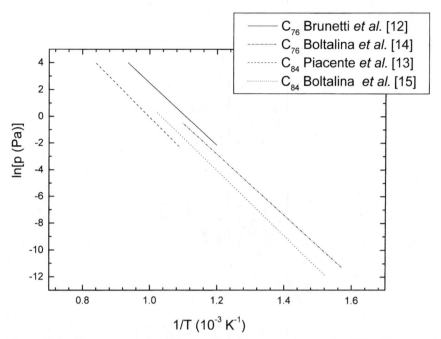

Figure 12.2 Temperature dependence of the vapor pressure for [76]fullerene and [84]fullerene.

using the estimated thermodynamic functions for the solid and gas phase [70]fullerenes yielding the value of 200 ± 6 kJ/mol [9].

12.2.3 Higher Fullerenes

Sublimation studies of the higher fullerenes (C_{76} and C_{84}) were reported by two groups; the effusion torsion method [12,13] and KCMS [14,15] were applied. The results on vapor pressure (shown in Fig. 12.2) cover different temperature intervals, but the sublimation enthalpies are in good agreement (Table 12.2). The temperature equations for [76]fullerene are as follows:

$$\log[p(\text{Pa})] = -9904 \pm 356/T + 10.65 \pm 0.47 \quad (637\text{--}911 \text{ K}) \text{ (Ref. 14)} \quad (12.4)$$

$$\log[p(\text{Pa})] = -10150 \pm 150/T + 11.23 \pm 0.20 \quad (834\text{--}1069 \text{ K}) \text{ (Ref. 12)} \quad (12.5)$$

For [84]fullerene,

$$\log[p(\text{Pa})] = -10570 \pm 234/T + 10.92 \pm 0.28 \quad (658\text{--}980 \text{ K}) \text{ (Ref. 15)} \quad (12.6)$$

$$\log[p(\text{Pa})] = -10950 \pm 300/T + 10.92 \pm 0.30 \quad (920\text{--}1190 \text{ K}) \text{ (Ref. 13)} \quad (12.7)$$

In Figure 12.3a, the sublimation enthalpy of fullerenes is presented as a func-

TABLE 12.2 Sublimation Enthalpy of the Higher Fullerenes

Fullerene	Pure Compound				Fullerene Mixture		
	Temperature Interval (K)	$\Delta_{sub}H^{\circ}{}_{T}$ (kJ/mol)	$\Delta_{sub}H^{\circ}{}_{298}$ (kJ/mol)	Reference	Temperature Interval (K)	$\Delta_{sub}H^{\circ}{}_{T}$ (kJ/mol)	Reference
C_{76}	834–1069	194 ± 4	206 ± 4	12	681–901	218 ± 14	19
	637–911	190 ± 7		14	800–950	196 ± 75	14
C_{78}	—	—	—	—	681–901	220 ± 15	14
	—	—		—	—	—	—
C_{84}	920–1190	210 ± 6	225 ± 6	13	800–950	247 ± 25	19
	658–980	202 ± 4		15	681–901	225 ± 14	15

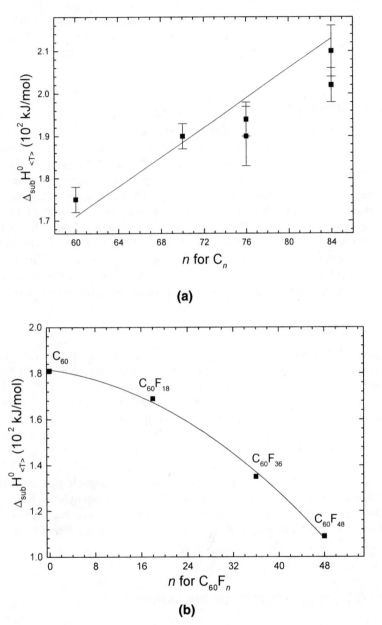

Figure 12.3 Sublimation enthalpy of (a) fullerenes and (b) fluorinated [60]fullerenes as a function of molecular composition.

tion of the number of carbon atoms, showing a gradual increase in $\Delta_{sub}H(C_n)$ with n. It agrees with the earlier qualitative observation that the heavier the fullerene molecule the lower its volatility and the higher the temperature needed to transfer it into the gas phase.

12.2.4   Fullerene Mixtures

The thermodynamic behavior of the binary system C_{60}–C_{70} is still not well established. The total vapor pressure and the partial pressures of [60]fullerene and [70]fullerene were determined over fullerite (fullerene mixture obtained after the Soxhlet procedure) and specially prepared mixtures of pure [60]fullerene and [70]fullerene in a wide range of compositions [16–18]. These studies resulted in the eutectic-type phase diagram with very limited mutual solubility of fullerenes; in close proximity to pure [70]fullerene and [60]fullerene (from 0 to 25 mol%) the components do not form a solid solution [17]. Samples of the fullerene mixtures enriched with the heavier fullerenes ((C_{60} + C_{70}) < 10 mol%) were also shown to be nonideal solid solutions, exhibiting significantly higher partial sublimation enthalpies than those obtained for the pure fullerenes (Table 12.2).

12.2.5   Fullerene Derivatives

Sublimation enthalpies were determined for the series of fluorinated [60]fullerenes with differing fluorine content ($n(F) = 18, 36, 48$) by using KCMS, the effusion torsion method, and the Knudsen effusion method [20]. In Figure 12.3b, the $\Delta_{sub}H^{\circ}$ values are plotted together with data for [60]fullerene, showing the decrease in sublimation enthalpy with the number of added fluorine atoms. This result is consistent with the general trend in volatility of halogenated hydrocarbons: Derivatization with halogen atoms results in much weaker intermolecular interactions in the solid phase and leads to a higher volatility. It is also true for the various products of organic functionalization of fullerenes—methoxylated, perfluoroalkylated, and phenylated fullerenes—although no quantitative data on the vapor pressure have been reported. By contrast, endohedral derivatives with encaged metals were found to be generally less volatile than the parent fullerenes.

12.3   COLLISION STUDIES OF FULLERENES

12.3.1   Fragmentation of C_{60} by Energetic Ions

Since purified C_{60} is now readily available, and since the powder is easy to vaporize [5], C_{60} has been widely used as a target gas in collision studies. The vapor pressure of C_{60} is ~1 mtorr, a typical target pressure at 525 °C. Collision studies have been performed in differentially pumped heated target cells, and in a configuration where an effusive molecular beam has been used as a target. Interaction between electrons or photons and fullerenes will not be discussed in this chapter. The reader is referred to a review article by Scheier et al. [21], concerning electron impact on fullerenes, and to an article by Deng et al. [22], about photoexcitation of C_{60}.

Fragmentation of C_{60} has been studied for a variety of atomic projectile ions with energies ranging from kiloelectron volt (keV) to gigaelectron volt (GeV) energies and with a broad selection of charge states [23–31]. The mechanism of energy transfer to the C_{60} molecule depends primarily on the projectile atomic number charge state and kinetic energy. This dependence is most easily visualized by a comparison to stopping power of the same projectile in a carbon foil. As discussed by Lindhard et al. [32], elastic energy transfer to individual atoms plays a dominant role at low energies, whereas inelastic energy transfer dominates at high energies.

For multiply charged high-energy projectiles such as 625 MeV Xe^{35+} [23], excitation of the giant dipole plasmon resonance is found to be the dominant mechanism of energy transfer to the C_{60} molecule. The measured positive fragment distribution is shown in Figure 12.4 as solid points together with a histogram calculated on the basis of multifragmentation and giant plasmon excitation, which also include bond percolation with the bond-breaking probability as an adjustable parameter. The data show features similar to those obtained for multifragmentation of nuclei bombarded with high-energy protons (e.g., the U-shaped fragment distribution.)

At low collision energies the situation is quite different, as shown in Figure 12.5. When C_{60} is hit by 74 eV Li^+ ions, the fragment spectrum is dominated by C_{60}^+ and its C_2 loss fragments [27]. In this situation the energy transfer is mainly taking place via semielastic collisions between the Li atoms and individual atoms in the cage. As a result of this interaction the C_{60} molecule is heated and evaporative cooling via successive emission of C_2 dimers accounts for the fragmentation pattern.

At medium collision energies (keV and MeV) the interaction results predominantly in electronic excitation of the collision partners. In this case the fragmentation process depends strongly on how electronic excitation couples to vibrational degrees of freedom, resulting in emission of smaller or larger fragment ions with a distribution of charge states. The fragment spectrum shown in Figure 12.6b [24] is typical for medium-energy collisions. C_{60}^+ is formed in distant collisions whereas multiply charged C_{60}^{q+} ions, followed by their C_2 loss peaks, and small fragments C_n^+ ($n < 15$) result from closer collisions. These conclusions are supported by the spectrum shown in Figure 12.6a, where closer collisions have been ensured by requiring double electron loss from the projectile ion as compared to single electron loss, which was required when collecting the spectrum shown in Figure 12.6b.

An extreme situation in atomic collisions occurs when highly charged low-energy ions are used as projectiles. In this case electrons are captured from the

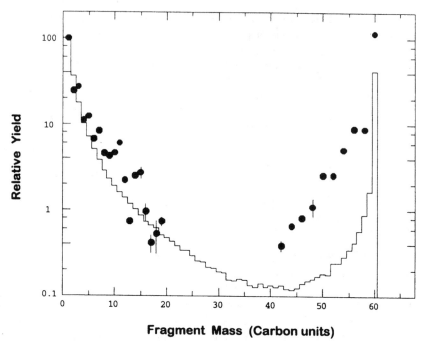

Figure 12.4 Measured mass distribution (solid points) for positive fragments arising from C_{60} bombarded by 625 MeV $^{136}Xe^{35+}$ ions. The histogram is the distribution calculated on the basis of a multifragmentation model (see text). The fragment detection efficiency was taken to be 20%. The error bars are smaller than the points, except when shown. The errors reflect statistical fitting errors and also any ambiguities in m/q values (see Fig. 12.6). (The absence of experimental points for fragment masses between about 20 and about 40 is due to ambiguities of this sort.)

C_{60} target at large impact parameters and the resulting "fragment" spectrum is dominated by multiply charged C_{60}^{q+} ions. Fragmentation of C_{60} has been observed only when coincidence with projectiles, which have captured several electrons, has been required [26]. This type of collision will be discussed further in the Section 12.3.3.

12.3.2 Fullerene Ions as Projectiles in Collisions

The strong C–C bonds give the fullerenes a fantastic resilience that permits them to be vaporized and ionized without being destroyed. C_{60} ions have been prepared in laser desorption [33], electron impact [34], plasma [35], electrospray [36], and sputter ion sources [37] with charge states ranging from +4 to −2. Other fullerenes, endohedrals, and fullerene derivatives have also been ionized and used as projectiles in collision experiments.

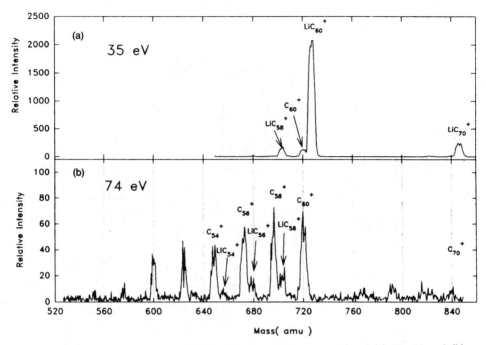

Figure 12.5 Mass spectrum of $Li^+ + C_{60}$ at collision energies of (a) 35 eV and (b) 74 eV.

Fragment size and charge state information concerning products resulting from collisions involving fullerenes have been obtained in various ways. The techniques involved include time-of-flight (TOF) spectrometry [33] and quadrupole mass spectrometry (QMS) [38] including Fourrier transform (FT) analysis. Sector instruments that involve magnetic and electrostatic analyzers have also been used [39]. Ion traps [40] and storage rings [41] are also included in the collection of instruments that have served in fullerene collision investigations. Kinetic energies range from electron volts (eV) in ion traps to kiloelectron volts (keV) in mass spectrometers and accelerators. Since fullerenes are heavy particles compared to most commonly used target gases, the center-of-mass energy E_{cm} is a convenient measure of the collision energy. If the fullerene projectile with mass M_1 and kinetic energy E_k collides with a particle with mass M_2, then

$$E_{cm} = \frac{M_1}{M_1 + M_2} E_k$$

The center-of-mass energies range from a few electron volts for light target gases to kiloelectron volts for heavier target gases. In the special case where C_{60}

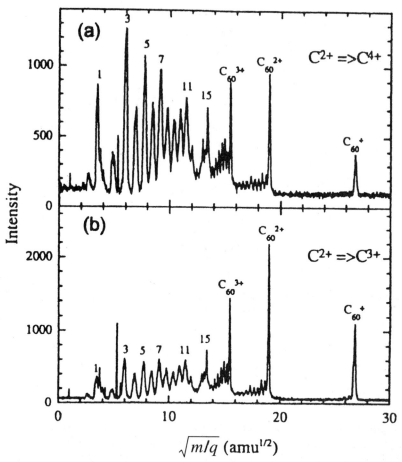

Figure 12.6   TOF spectra for the electron loss cases of a 15.6 MeV C^{2+} projectile.

is used as both projectile and target, $E_{cm} = E_k/2$. For most collision studies involving fullerenes as projectiles, the velocity of the fullerene is about 10^5 m/s.

A typical positive ion fragmentation spectrum is shown in Figure 12.7; the projectile is 8 keV C_{60}^+ and the target gas is O_2 [34]. C_{60}^+ fragments into C_n^+ fullerenes for $n \geq 32$ and even. Below $n = 30$ both even- and odd-numbered fragments appear. These fragments are assumed to take the form of linear chains for $n < 10$ and mono- or polycyclic rings for $n > 10$ [42]. The C_{4n-1} fragment ions form a subgroup of particular intense peaks with C_{15}^+, C_{19}^+, and C_{23}^+ as the most prominent group members. The fragmentation pattern depends on several collision parameters, the most important being the mass of the target atom, the center-of-mass energy, and the internal energy of the fullerene projectile. At center-of-mass energies ~1 keV, the energy transfer is taking

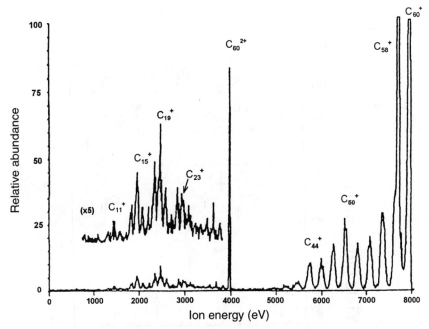

Figure 12.7 The 8 keV collision-induced dissociation/mass analyzed ion kinetic energy spectrum of C_{60} with O_2 target gas. The parent ion (C_{60}^+) appears at approximately 7900 eV and the charge stripping (C_{60}^{2+}) product ion appears at approximately 3950 eV.

place primarily via semielastic collisions between the fullerene carbon atoms and the target atom. The result of such collisions is prompt ejection of smaller or larger clusters of carbon atoms followed by evaporative cooling via C_2 emission of the remaining hot fullerene part and C_3 emission from hot chains or ring fragments. The relative population of the two fragment groups, $n > 30$ and $n < 30$ was found [43] to depend on the target atom mass in a way where larger target atom mass results in a more dominant population of fragment peaks with n values smaller than 30. This finding is in accordance with molecular dynamics simulations [44], which show that, for heavy target atoms, prompt energy transfer to a large group of C atoms may lead to a collective release of this group of atoms. For light target atoms energy is transferred predominantly into heat (vibrational excitation) of the fullerene, and C_2 emission is the dominating decay process. In this case positive fullerene $(n > 30)$ fragment ions dominate the fragment spectrum.

The information obtained from inclusive fragment distributions suffers from the lack of knowledge about fragment partners. Such information has been obtained in a pioneering study by Brink et al. [45], and more recently by Vandenbosch et al. [46], by performing coincidence experiments, where the size distribution of the partners to a particular fragment product ion was registered.

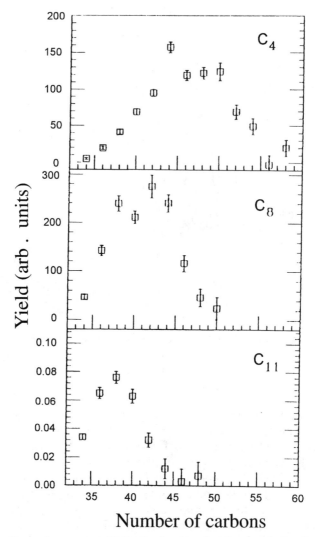

Figure 12.8   Heavy fragment yield distribution in coincidence with C_4, C_8, and C_{11} for 75 keV C_{60} on H_2.

Figure 12.8 shows an exclusive fragment distribution [46], where H_2 target atoms are bombarded by 75 keV C_{60}^- projectiles. C_8^+ ions were detected in coincidence with positive fullerene ions, and it should be noted that the most intense partner peak relates to C_{42}^+, and that a C_{52}^+ peak is absent in the spectrum. This finding tells us that pure binary fragmentation is unlikely to occur. Instead, either multiple fragmentation or, more likely, fragmentation followed by evaporative cooling is the dominant reaction pathway.

The internal temperature of the fullerene projectile is an important collision parameter, which, unfortunately, is very difficult to control or to determine. By the use of electrospray ion sources [47], cold, room-temperature ions have successfully been created for collision experiments. Smalley and co-workers [48] used buffer gas for cooling hot fullerenes and demonstrated how a quantity such as the electron affinity depends on the fullerene temperature. In most experiments, however, a broad projectile temperature distribution is accepted. The maximum fullerene temperatures within this distribution are determined by the activation energy E_a for the decay of the ion in question through the Boltzmann factor $e^{-E_a/kT}$, where T is the fullerene temperature and k is the Boltzmann constant. As an example, C_{60}^- and C_{60}^+ ions, produced with a broad temperature distribution in the ion source, will have temperatures less than 1500 K and 4000 K, respectively, after approximately 100 μs, a typical flight time in most collision experiments. As shown in Figure 12.9, it has been demonstrated by Larsen et al. [43] that the collision-induced positive ion fragment distribution depends strongly on whether C_{60}^+ or C_{60}^- ions are used as projectiles. This difference was related to differences in maximum temperature possessed by cations and anions.

In most experiments, information about neutral fragments is missing entirely. In order to overcome this problem, collisional reionization followed by electrostatic mass dispersion has been attempted by McHale et al. [49]. One of the major problems with this technique is the collisional reionization part of the experiment, where neutral fragmentations pass a second target cell. It is difficult to account for the size-dependent ionization cross section of neutral fragments, and, furthermore, ionization is accompanied by further fragmentation of the fragments. These problems become less severe at higher energies. In Figure 12.10 are shown neutral fragment distributions for 50 keV C_{60}^- and C_{60}^+ in collisions with He [50]. H_2 has been used as a reionization gas in the second gas cell. It should be noted that to a first approximation neutral fullerenes result from collisions involving negative fullerenes while small neutral C_n clusters with $n < 20$ result from collisions where positive fullerenes fragment. This observation is in accordance with the concept that the charge stays with the large collision partner when positive fullerenes fragment, while electron loss is the dominating "destruction" process for negative fullerenes.

Fullerenes can act as atom containers since the cavity is large enough to house one or two atoms [51]. Such complexes have been dubbed endohedrals, for which the symbol $A@C_n$ is normally used. Fullerene derivatives are another group of fullerene-based molecules where the extra atoms or molecules are placed on the other surface of the fullerene. Such derivatives are also called exohedral complexes, for which a symbol like $C_{60}H_{36}$ can be used. Many different endo- and exohedral complexes have been produced [52] and characterized, and quite a few have been used as projectile ions in collision studies. An example of the outcome of collisions involving endo- and exohedrals is shown in Figure 12.11 [53]. The fragment distribution for collisions between argon endohedrals ($Ar@C_{60}$) and He atoms demonstrate that substantial fragmentation of the carbon cage can occur without loss of the trapped Ar atom. The

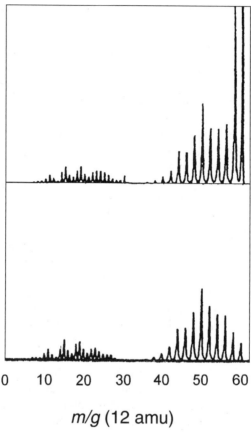

Figure 12.9 (*Top*) Positive ion fragmentation spectrum obtained by collision of C_{60}^+ with Ne at an energy of 50 eV. (*Bottom*) Positive ion fragmentation spectrum obtained by collision of C_{60}^- with Ne at an energy of 50 keV.

smallest endohedral fullerene in this example is Ar@C_{48}, indicating that even smaller fullerenes cannot encage the Ar atom and hence burst apart. This collision process is an analog to the "shrink wrap" model proposed by Smalley and co-workers [54] as an explanation of fragment distributions resulting from laser vaporization of endohedral complexes. The fragment distribution for the exohedral complex $C_{60}F$ shown in Figure 12.11b [55] is different from the distribution shown in Figure 12.11a in the sense that, with the exception of a $C_{58}F$ "peak," only fullerene fragments are observed. This qualitative difference between spectra, originating from fragmentation of endohedrals and exohedrals, is believed to be of general validity.

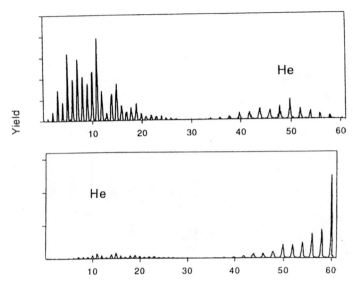

Figure 12.10 (*Top*) The 50 keV C_{60}^{+} on He; neutral fragment distribution. (*Bottom*) The 50 keV C_{60}^{-} on He; neutral fragment distribution.

Formation of fullerene endohedrals, that is, fullerenes encaging atoms in the central cavity, has been observed as a result of C_{60} bombardment by low-energy ions, as shown in Figure 12.5 [27]. The energy variation of the probability for formation of $Ne@C_{60}$ (endohedral Ne) in collisions between Ne and C_{60}^{+} is shown in Figure 12.12 [56]. The observed threshold value at around 10 eV can be accounted for by simple bond-breaking models, as can the decrease in probability at higher energies. In simple terms, one can say that the Ne atom cannot penetrate the fullerene surface if its energy is too low and it passes right through the fullerene if its energy is too high.

As has been demonstrated earlier, fullerenes can be used both as target particles and as projectiles in collision experiments. It is now possible to perform cluster–cluster collisions with a reasonable signal intensity. For a general discussion of cluster–cluster collisions the reader is referred to articles by Schmidt and co-workers. (e.g., see Knospe et al. [57]).

Many reactions known from collisions between nuclei, such as fusion, deep inelastic collisions, and multifragmentation, have been studied [58]. As an example of cluster–cluster collision, fusion, resulting from collisions between two C molecules, attracts special attention [59]. The cross section for this process is shown in Figure 12.13 as a function of E^{-1}, where E is the center-of-mass energy in the collision. The cross section attains its maximum value of around 2×10^{-16} cm^2 at an energy of about 150 eV. The fusion process has been modeled using molecular dynamics simulations, and reasonable agreement with experimental results [59] has been obtained.

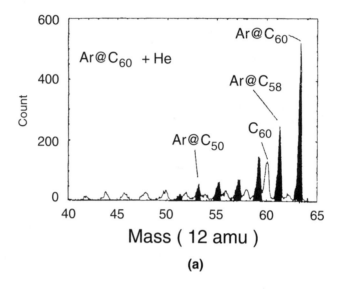

(a)

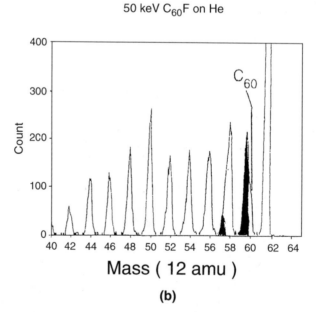

(b)

Figure 12.11   Distribution of positively charged fragments for a) 50 keV Ar@C_{60}^- on He, and b) 50 keV $C_{60}F^-$ on He.

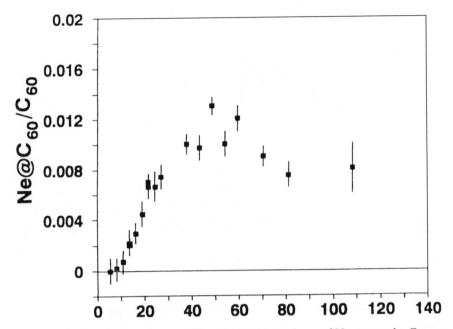

Figure 12.12 Center-of-mass collision energy dependence of Ne capture by C_{60}.

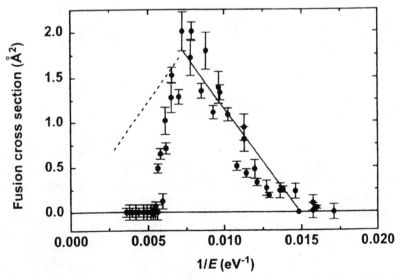

Figure 12.13 Experimental absolute fusion cross sections as a function of $1/E$. The solid line is a least squares fit to the experimental data points in the low-energy range (≤ -130 eV); the dashed line is the expected energy dependence if the instability of the fused compound is due to centrifugal fragmentation.

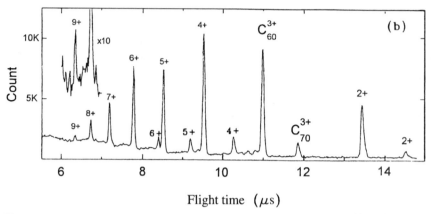

Figure 12.14   Time-of-flight spectrum from impact of v = 0.30 a.u. Bi^{44+} on fullerenes (see text).

12.3.3   Capture and Loss of Electrons in Fullerene Collisions

Multiple electron transfer plays an important role in collisions between multiply charged ions and C_{60} molecules. Selberg et al. [60] have demonstrated that ~18 electrons can be captured by multiply charged projectiles in collisions with C_{60}. The C_{60} molecule acts as an efficient electron donor compared to a heavy atom since electron binding energies are more moderate in C_{60} than in atoms. Jin et al. [61] found that C_{60}^{+q} ions with q up to 9 can be created in collisions between Bi^{44+} and C_{60} at collision energies about 500 keV. The lifetime of C_{60}^{8+} has been estimated to be at least 5 µs, and for ions with lower charge states the lifetime is even longer. Figure 12.14 shows the C_{60} charge state spectrum obtained in these investigations and it is argued that the removal of more than 11 electrons will most probably result in an immediate disruption of the fullerene.

In collisions between fast moderately charged particles and C_{60}, the fullerene might be considered as a "small piece of carbon foil." The thickness of this foil, $(1-2) \times 10^{16}$ atoms/cm², is sufficient to ensure that equilibrium charge state distributions have been attained by the emerging ions. This reasoning has been used by Larsson et al. [62] to explain a connection between energy loss and charge state for 100 keV Ar^{3+} ions, which emerge from the C_{60} cage as Ar^{+}.

Fullerene projectiles, due to their high stability, can capture and lose electrons in collisions with atoms or molecules [63–67]. Cross sections as large as 10^{-14} cm² have been measured for electron capture in $C_{60}^{+2} + C_{60}$ collisions, while typical electron loss cross sections range from 1×10^{-16} to 10×10^{-16} cm² [67]. Electron capture by fluorinated C_{60} anions in collisions with atomic targets have recently been reported [68]. The cross section for this endoergic process was found to be as large as 10^{-16} cm², and it was argued that the capture reaction proceeds via a curve crossing between potential energy curves related to the incoming and outgoing reaction channels.

12.4 POSITIVE-ION CHEMISTRY OF FULLERENES

The positive-ion chemistry of fullerenes received immediate attention upon the first reported isolation of C_{60} [69]. Early studies of this chemistry [70] focused on the gas phase chemical ionization of C_{60} (e.g., in methane or isobutane) [71,72], in part to provide more information on the purity of fullerene fractions. It was found that fullerenes are readily protonated in reactions of type (12.8) with $RH^+ = CH_5^+$, for example, or otherwise derivatized in reactions of type (12.9) with $RH^+ = C_2H_5^+$ or $C_4H_9^+$, for example.

$$RH^+ + C_{60} \rightarrow C_{60}H^+ + R \qquad (12.8)$$

$$RH^+ + C_{60} \rightarrow C_{60}RH^+ \qquad (12.9)$$

The proton affinities of C_{60} and C_{70} have been bracketed with ion/molecule reactions of type (12.8) to be within 205.5 ± 1.5 kcal mol^{-1} and $C_{60}H^+$ has been observed to transfer a proton to C_{70} indicating that $PA(C_{70}) > PA(C_{60})$ [71].

Fascination with the hollow nature of C_{60} led to early experiments that demonstrated the remarkable formation of *endohedral* adducts in high-energy collisions of C_{60}^+ and C_{70}^+ with rare gas atoms as in reactions (12.10a) and (12.10b) with $m = 2, 4, 6, \ldots$ [73–75].

$$C_{60}^+ + M \rightarrow M@C_{60}^+ \qquad (12.10a)$$

$$C_{60}^+ + M \rightarrow M@C_{60-m}^+ + C_m \qquad (12.10b)$$

Similar reactions were demonstrated with C_{60}^{2+} and C_{60}^{3+} and multiple insertion also has been reported [74–77].

Reactions between C_{60} and metal ions $M^+ = Fe^+$, Co^+, Ni^+, Cu^+, Rh^+, La^+, V^+, and VO^+ have been shown to lead to the formation of *exohedral* adducts MC_{60}^+ in addition reactions of type (12.11) occuring at low pressures ($\approx 1 \times 10^{-7}$ torr) and at near-thermal energies [78,79].

$$M^+ + C_{60} \rightarrow C_{60}M^+ \qquad (12.11)$$

Varying amounts of electron transfer were observed to compete with addition. $C_{60}Ni^+$ reacts further to produce the "dumbbell" ion $(C_{60})_2Ni^+$ [79b]. Also, systematic measurements at higher energies have been reported for collisions of C_{60} with the atomic ions $X^+ = Ne^+$, O^+, C^+, N^+, Li^+, Na^+, K^+, B^+, F^+, Fe^+, and W^+ that can give rise to the onset of (endothermic) electron transfer, dissociative electron transfer, and insertion as indicated by reactions (12.12a),

(12.12b), and (12.12c), respectively [27,80]:

$$X^+ + C_{60} \rightarrow C_{60}^+ + X \tag{12.12a}$$

$$X^+ + C_{60} \rightarrow C_{60-2n}^+ + C_{2n} + X \tag{12.12b}$$

$$X^+ + C_{60} \rightarrow X@C_{60}^+ \tag{12.12c}$$

The collision-energy dependence measured for the reaction $Ne^+ + C_{60}$ is shown in Figure 12.15 and indicates an onset for dissociative electron transfer and insertion at collision energies of about 20 and 30 eV, respectively [80f].

12.4.1 Electron-Transfer Reactions

Electron-transfer reactions with buckminsterfullerene have previously been surveyed in considerable detail [81]. Experiments have shown that an electron can be removed from C_{60} in the gas phase at room temperature by cations having a recombination energy greater than the first ionization energy of C_{60}, $IE(C_{60}) = 7.64 \pm 0.02$ eV [82], and by metastable rare-gas atoms with an excitation energy greater than $IE(C_{60})$ in Penning or chemi-ionization reactions [83]. Remarkably, He^+ and Ne^+ can remove *two* electrons from C_{60} at thermal energies in a process apparently analogous to the emission of electrons from metal surfaces by noble-gas ion impact [84].

C_{60}^{n+} cations will remove electrons from neutral molecules with sufficiently low ionization energies. With $C_{60}^+(n = 1)$ the threshold for electron transfer is determined simply by its recombination energy (7.64 ± 0.02 eV) [85]. When $n > 1$, a barrier arises from Coulombic repulsion between the two charged product ions [86]. Thus, the threshold for the occurrence of electron transfer to C_{60}^{2+} (RE = 11.36 ± 0.05 eV) [81] has been determined to be at IE ≤ 9.58 eV [86], while that for electron transfer to C_{60}^{3+} (RE = 15.6 ± 0.5 or 16.6 ± 1 eV) [81,87] has been shown to be at IE ≤ 11.2 eV [81,88]. Triply charged C_{60}^{3+} cations also have been observed to abstract *two* electrons from corannulene and the polycyclic aromatic hydrocarbons (PAHs) anthracene, pyrene, and benzo[rst]pentaphene [89]. These molecules have sufficiently low first and second ionization energies to make double-electron transfer thermodynamically and kinetically favorable.

12.4.2 Addition Reactions

Experimental measurements of gas phase reactions of fullerene cations leading to chemical bond formation or derivatization of the fullerene surface have been surveyed for the first three charge states of C_{60} [90].

C_{60}^+ has been observed to be quite unreactive toward many molecules in helium buffer gas at 0.35 torr [90]. The important exceptions are shown in Figure 12.16, which indicates that C_{60}^+ has been seen to bond to H atoms [86b,91], to strong nucleophiles such as ammonia and saturated amines [92],

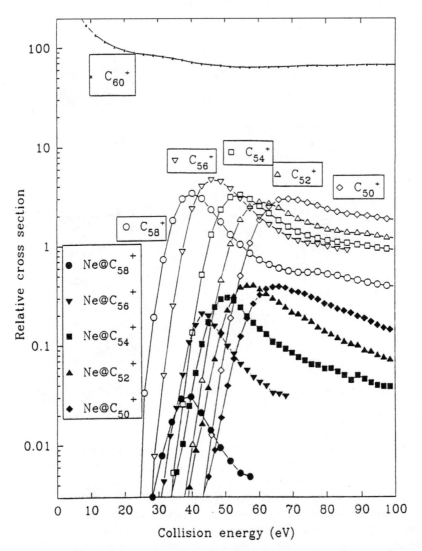

Figure 12.15 Collision energy dependence for all significant product channels observed for the reaction of Ne^+ with C_{60} with an octapole guided-ion beam mass spectrometer.

and to molecules capable of Diels–Alder addition [93]. The reaction of C_{60}^+ with iron pentacarbonyl is the only example of a *bimolecular* derivatization reaction reported to date [94,95]. The difficulty in covalent bonding to C_{60}^+ has been attributed to the distortion of the C_{60} carbon cage required at the C-site of bond formation with the substituent so as to achieve sp^3 hybridization [96]. A remarkable trend is observed for the reactions of C_{60}^+ with amines: Ammonia adds only slowly, but the rate of addition approaches the collision rate with

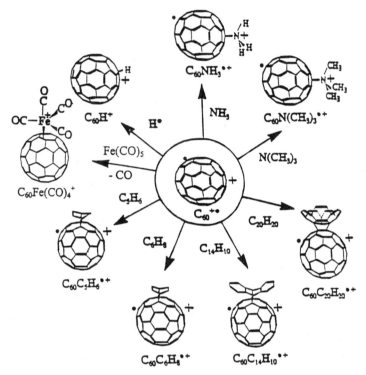

Figure 12.16   An overview of derivatization reactions of C_{60}^{+} observed at a temperature of 294 ± 2 K and a helium pressure of 0.35 ± 0.01 torr.

increasing alkyl substitution. This behavior is consistent with a trend in inductive electron donation (the Lewis basicity of the amine) and a trend in the number of degrees of freedom effective in energy dispersal [92].

C_{60}^{2+} is generally easier to derivatize than C_{60}^{+}. This enhanced reactivity has been attributed to the stronger electrostatic interaction between molecules and C_{60}^{2+} that may serve to overcome the activation barrier associated with the change in hybridization required at the site of bonding [96]. Figure 12.17 indicates derivatization reactions observed with C_{60}^{2+}. The chemistry of C_{60}^{2+} has been explored with H atoms [86b,91], ammonia and aliphatic amines [92], nitriles [97], water, alcohols and ethers [98], aldehydes, ketones, carboxylic acids and esters [99], cyclic aliphatic oxides [100], and unsaturated hydrocarbons [101–103]. Electron transfer is an important competitive channel for some of the observed derivatization reactions and, for molecules with sufficiently low ionization energies, can become the only reaction channel. However, electron transfer with C_{60}^{2+} occurs in the presence of an activation barrier that arises from Coulomb repulsion between the charged product ions [86]. This is evident in Figure 12.18, which shows a delayed onset for competing electron

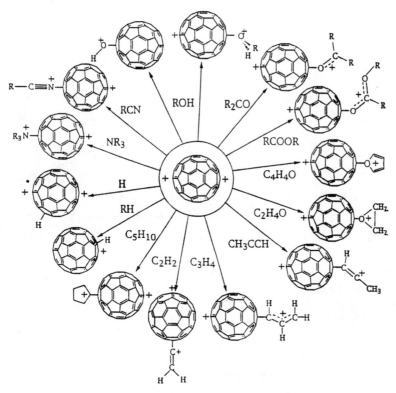

Figure 12.17 An overview of derivatization reactions of C_{60}^{2+} observed at a temperature of 294 ± 2 K and a helium pressure of 0.35 ± 0.01 torr.

transfer at about 9.6 eV that is well below the recombination energy of C_{60}^{2+} (11.36 ± 0.05 eV).

Derivatization reactions also have been observed with C_{60}^{3+}, but electron transfer is even more competitive than with C_{60}^{2+}. For example, although ammonia has been reported to add to C_{60}^{3+}, only electron transfer has been seen with amines [92]. Reactions have been investigated with H atoms [86,91], nitriles [97], water, alcohols and ethers [98], aldehydes, ketones, carboxylic acids and esters [99], cyclic aliphatic oxides [100], and unsaturated hydrocarbons [101–103]. Figure 12.19 provides a summary of addition reactions observed with C_{60}^{3+}. New charge separation channels appear for some of the reactions with C_{60}^{3+}. For example, dissociative electron transfer to produce $C_{60}^{2+}/$ $CH_2NH_2^+$ and C_{60}^{2+}/CH_3^+ is the dominant reaction channel with ethylamine [92]. Also, dissociative addition reactions, such as hydride and hydroxide transfer, become more effective. This is the case with water, some alcohols, and some alkanes. Halide transfer from HCl and HBr also has been observed [104].

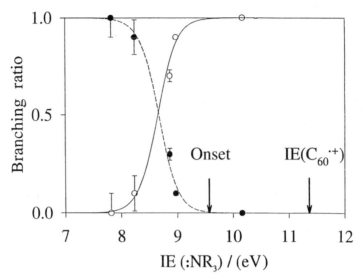

Figure 12.18   Competition between electron transfer (dotted line) and adduct formation (solid line) observed for reactions of C_{60}^{2+} with ammonia and amines at a temperature of 294 ± 2 K and a helium pressure of 0.35 ± 0.01 torr.

12.4.3   Higher-Order Reactions

12.4.3.1   *Cation-Transfer Reactions*   A number of secondary reactions have been observed with derivatized fullerene cations that involve the transfer of a proton, a methyl cation, or the entire ionized substituent. When the substituent contains a hydrogen atom, $C_{60}XH^{n+}$ may undergo partial or complete proton transfer with a suitable proton acceptor. Thus, secondary reactions of singly charged $C_{60}XH^{+}$ ions produce proton-bound adducts of the type $C_{60}X \cdots H^{+} \cdots XH$ with $XH = NH_3$, CH_3NH_2, $C_2H_5NH_2$, and $(CH_3)_2NH$ [92]. Doubly charged derivatized $C_{60}(XH)_m^{2+}$ $(m = 1, 2)$ cations have been observed to undergo proton-transfer reactions of type (12.13) with $XH =$ ammonia, amines, alcohols, carboxylic acids, ketones, and cyanides [105a]. A rationale has been presented for assessing this proton-transfer reactivity in terms of an *apparent* gas phase acidity of the dicationic species $C_{60}(XH)_m^{2+}$ and the first steps have been taken toward establishing a gas phase acidity ladder for derivatized fullerene cations [105a]. Also, the gas phase acidity of $C_{60}H^{2+}$ has been established from the occurrence and nonoccurrence of proton-transfer reactions with this cation [105b].

$$C_{60}(XH)_m^{2+} + XH \rightarrow C_{60}(XH)_{m-1}X^{+} + XH_2^{+} \qquad (12.13)$$

Reactions with methyl formate and methyl acetate have been reported that involve the transfer of a methyl cation according to Eq. (12.14), proceeding in

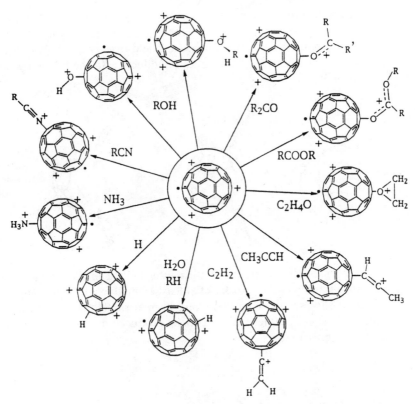

Figure 12.19 An overview of derivatization reactions of C_{60}^{3+} observed at a temperature of 294 ± 2 K and a helium pressure of 0.35 ± 0.01 torr.

competition with adduct formation and proton transfer [99]:

$$C_{60}(XCH_3)^{3+} + XCH_3 \rightarrow C_{60}(X)^{2+} + X(CH_3)_2^{+} \qquad (12.14)$$

Also, several reactions of derivatized fullerene cations have been reported that involve the complete transfer of the ionized substituent according to Eq. (12.15) with R = CN, C_2H_3, C_3H_3, and CH_2CN [97]:

$$C_{60}(NCR)^{3+} + RCN \rightarrow C_{60}^{2+} + (RCN)_2^{+} \qquad (12.15)$$

It has been suggested that these reactions are driven by the stability of the product dimer cation.

12.4.3.2 Surface Derivatization
A number of higher-order reactions have been observed that result in the sequential derivatization of the surface of the fullerene cation to produce "fuzzy balls." H atoms have been observed to add

sequentially to the first three charge states of C_{60}^{n+} to produce $C_{60}H_m^{n+}$ with m up to 10 and constrained only by the hydrogen atom density experimentally available [86b,91]. Nitriles have been observed to add rapidly in numbers equal to the number of charges on C_{60}^{n+} to produce adducts with the following suggested structures [97]:

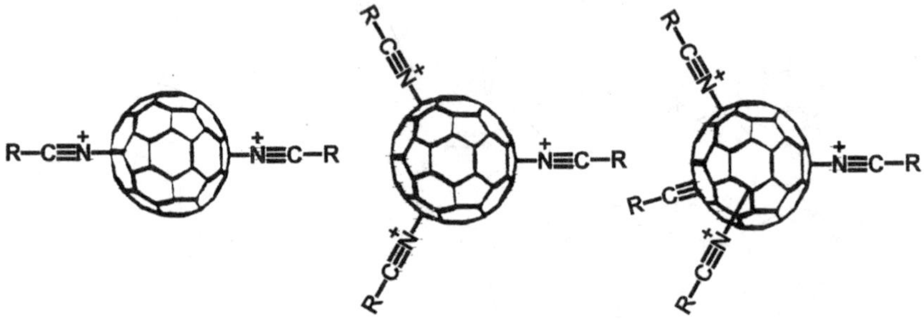

C_{60}^{2+} has been observed to react with methyl nitrite in sequential bimolecular reactions leading to the transfer of at least 15 methoxy groups [106]. The addition kinetics shows periodicities that have been attributed to periodicities in the stabilities of the surface-derivatized cations produced, of which the following structure is representative:

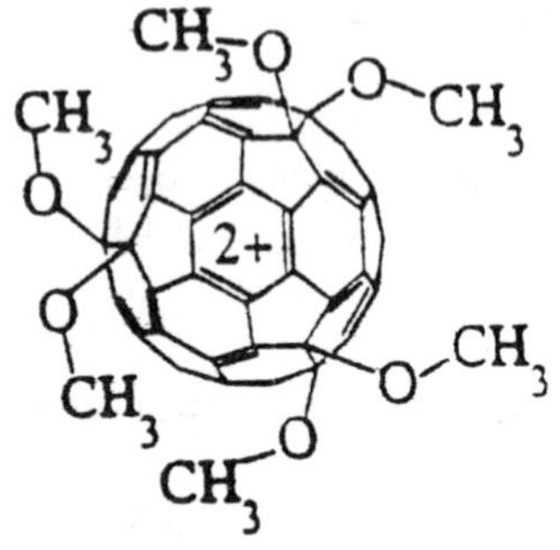

12.4.3.3 Ball-and-Chain Derivatization

Multiple molecular derivatization leading to ball-and-chain formation, as illustrated in reaction (12.16), has been reported for reactions of multiply charged fullerenes, particularly with monomers well known from traditional polymer chemistry. The monomers 1,3-butadiene [101], ethylene oxide [100], 1-butene [102a], allene and propyne [103], and ethylene and acetylene [102b] have all been found to polymerize with C_{60}^{2+}, C_{70}^{2+}, or C_{60}^{3+}. Experimental results obtained for the multicollision-induced dissociation of the product cations $C_{60}(\text{1-butene})_n^{2+}$, $C_{60}(\text{allene})_n^{2+}$, $C_{60}(\text{propyne})_n^{2+}$, and $C_{60}(\text{propyne})_n^{3+}$ are consistent with growth by ball-and-chain propagation in these cases.

$$C_{60}^{2+} \xrightarrow{M} + C_{60}-M^{+} \xrightarrow{nM} + C_{60}-(M)_n-M^{+}$$

$$(12.16)$$

The chemistry of C_{60}^{2+} with allene is particularly interesting because the sequential addition leads to a striking periodicity in reactivity: The even-numbered adducts react about 10 times faster than the odd-numbered adducts (see Fig. 12.20). Also, a striking periodicity in stability is revealed by multi-collision-induced dissociation experiments (see Fig. 12.20). A mechanism has been proposed that involves the alternating formation of *charge-localized even-numbered acyclic allyl cations* that are more reactive and the formation of *charge-delocalized cyclic odd-numbered allyl cations* that are less reactive [103].

12.4.3.4 Influence of Surface Features on the Reactivity of Fullerene Ions

The influence of surface strain on the chemical reactivity of fullerene cations has been explored experimentally with a number of neutral reagents. Early investigations of addition reactions of C_{56}^{+}, C_{58}^{+}, and C_{60}^{+} with NH_3, of addition reactions of C_{56}^{2+}, C_{58}^{2+}, and C_{60}^{2+} with C_2H_4 and CH_3CN, and of hydride transfer reactions of C_{56}^{2+}, C_{58}^{2+}, and C_{60}^{2+} with n-C_4H_{10} revealed a strong dependence of reactivity on the size of the fullerene cation (see Fig. 12.21) [107]. The least-strained C_{60}^{n+} cations were found to be the least reactive as predicted [108]. Systematic studies of addition reactions of C_{56}^{+}, C_{58}^{+}, C_{60}^{+}, C_{70}^{+}, corannulene$^+$, and coronene$^+$ with cyclopentadiene and 1,3-cyclohexadiene indicated that the efficiency of bond formation depends strongly on the curvature of the carbonaceous surface, as illustrated in Figure 12.21 [109].

12.4.3.5 Reactions of Atomic Exohedral Adduct Ions

Some progress has been made in elucidating the influence of C_{60} as a ligand on the reactivity of metal ions and whether the derivatization of C_{60} can be catalyzed by the metal center. Thus, the rate of the bimolecular O-atom transfer reaction of nitrous oxide with Fe^+ is enhanced by a factor of more than 3 in the presence of C_{60} [110]. The addition reaction of CO with Fe^+, which does not occur measurably for the isolated cation, proceeds rapidly in the presence of C_{60} and sequentially leads to the addition of four CO molecules to the Fe center [110]. The exohedral adduct ion $C_{60}Co^+$ has been shown to add ethylene, propene, 1-butene, and isobutene and collision induced dissociation (CID) studies have demonstrated that the alkene groups in the resulting $C_{60}CoC_nH_{2n}^+$ are bound to the metal center [111]. An interesting derivatization sequence occurs in the reaction of $C_{60}Co^+$ with cyclopropane, reaction (12.17a) ($m = 0$–2) and reaction (12.17b) ($m = 0$–3):

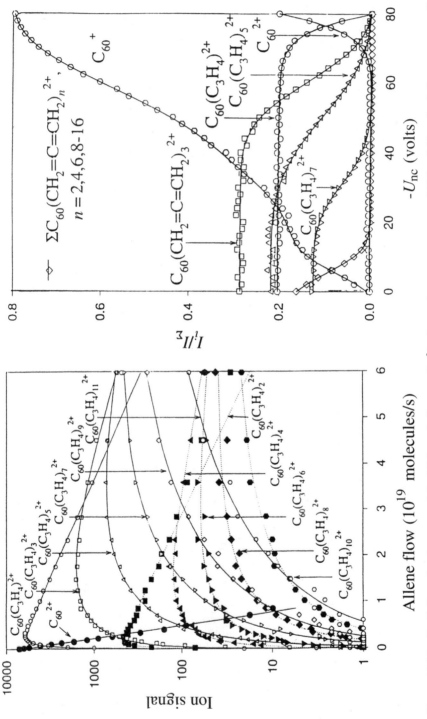

Figure 12.20 (*Left*) Data for the sequential addition of allene to C_{60}^{2+} observed at a temperature of 294 ± 2 K and a helium pressure of 0.35 ± 0.01 torr. (*Right*) measured profiles for the multicollision-induced dissociation of $C_{60}(\text{allene})^{2+}$ at an allene flow of 4.5×10^{19} molecules/s in 10% argon/helium at a total pressure of 0.30 ± 0.01 torr.

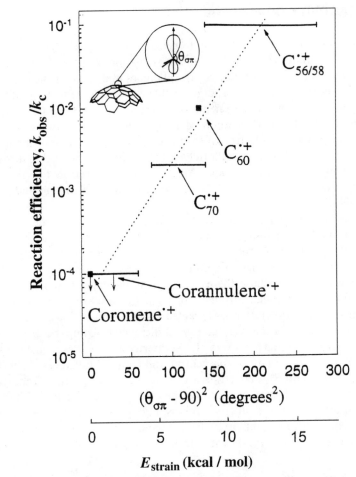

Figure 12.21 A correlation between reaction efficiency, k_{obs}/k_c, with the square of the POAV angle and the strain energy for addition reactions with cyclopentadiene at room temperature and a helium pressure of 0.35 ± 0.01 torr. k_{obs} is the measured rate coefficient and k_c is the collision rate coefficient, which is estimated to be 10^{-9} cm^3 (molecule · s).

$$C_{60}Co(CH_2)_m{}^+ + c\text{-}C_3H_6 \rightarrow C_{60}Co(CH_2)_{m+1}{}^+ + C_2H_4 \qquad (12.17a)$$

$$\rightarrow Co_{60}(CH_2)_m{}^+ + CoC_3H_6 \qquad (12.17b)$$

CID studies suggest that the $C_{60}Co(CH_2)_{m+1}{}^+$ ions involve coordination of Co to a derivatized C_{60} [111]. In related studies, reactions of C_{60} with Fe(benzyne)$^+$ and Fe(biphenylene)$^+$ have been shown to form metallacyclic C_{60} derivatives [112].

12.5   REACTIVITY OF FULLERENE ANIONS

Fullerene anions are remarkably easy to generate in the gas phase by a variety of ionization techniques [113–116]. A rare observation of stable doubly charged anions in the gas phase was reported for fullerenes and some of the derivatives with electronegative addends. The reversible successive reduction of C_{60}^{n-} ($n = 1$–6) in solution was experimentally observed (a reader can be referred to the recent review [117]). Formation of many fullerene derivatives is accompanied by the transfer of electron(s) to the fullerene moiety, one of the famous examples being endohedral metallofullerenes in which a metal donates up to three electrons to the fullerene cage. All these facts allowed P. Fowler to call electron deficiency of fullerenes a "recurring theme in the chemistry and physics of fullerenes" [118].

Studies of the gas phase reactions of fullerene anions can provide invaluable information on the "intrinsic" chemical reactivity of fullerene anions, which is not affected by solvation processes. However, negative ion chemistry of fullerenes has attracted much less attention than studies of reactions of fullerene cations. It was partly because the earlier reports indicated that the gas phase fullerene anions were chemically inert (e.g., in the reaction with D_2O) [119], whereas positive carbon cluster ions reacted rapidly with various gases.

Lower reactivity of negatively charged fullerenes in comparison with their positively charged counterparts was also supported by the theoretical consideration of relative aromaticity of the neutral and charged fullerenes [120]. Bond resonance energy (BRE), defined as a contribution of a given π bond to the aromaticity, was used as a quantitative measure of the kinetic stability of a molecule or an ion. For the fullerene anions, high positive BREs were obtained that indicated their low reactivity, whereas the cations that had large negative BREs, were, on the contrary, shown to be kinetically unstable and very reactive.

12.5.1   Electron Affinity of [60]Fullerene and [70]Fullerene

Electron affinity (EA) represents a measure of ability of a molecule to attach an electron. If the reaction

$$A + e = A^{-}, \qquad \Delta_r H^{\circ}(1) = -EA(A) \qquad (12.18)$$

is exothermic, then the stable anion can be formed and experimentally observed. In the gas phase, charge-transfer reactions are governed by the EAs of reacting species.

The Rice University group first observed an intense C_{60}^{-} peak in the laser desorption (LD) mass spectrum (Fig. 12.22) that was ascribed to the enhanced stability of the proposed cage structure for the 60-carbon atom cluster [121]. The ultraviolet photoelectron spectroscopy (UPS) study of some large carbon clusters showed that the EA values for the clusters C_n ($n = 50$–84) fall within the range of 2.5–3.0 eV, with C_{60} having the lowest EA of 2.6–2.8 eV [122]. For

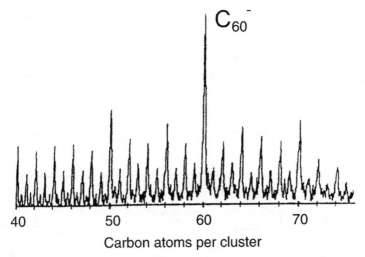

Figure 12.22 Negative-ion laser ionization mass spectrum of carbon clusters, with the prominent C_{60}^- peak at 720 amu published by the Smalley group in 1986 [121].

more refined EA determination the beam of C_{60}^- was cooled by introducing it into the "waiting room"—a chamber filled with the buffer gas. Fullerene anions underwent multiple collisions, thereby losing excess internal energy gained during laser vaporization/ionization. Then the photodetachment threshold of C_{60}^- was measured and this yielded $EA(C_{60}) = 2.65 \pm 0.05$ eV [123]. This value was later corrected to 2.666 ± 0.001 eV in the experiment performed in the ASTRID ion storage ring [124]. This technique retained the C_{60}^- beam in the ion storage ring for times up to a few seconds. Figure 12.23 illustrates the decrease in the excess internal energy with the time after injection of [60]fullerene anions in the storage ring. A similar experiment was performed on the [70]fullerene anion, yielding $EA = 2.676$ eV.

Thermodynamic EA determination of [70]fullerene was performed in the study of electron exchange reaction between C_{60}^- and C_{70},

$$C_{60}^- + C_{70} = C_{70}^- + C_{60} \tag{12.19}$$

using the method of ion molecular equilibrium (IME)—a modification of Knudsen cell mass spectrometry (KCMS) [125]. In addition to the electron impact ionization (EI) mode, which is used in KCMS for mass analysis of neutral species evaporating from the effusion cell, there is also a thermal ionization mode for analysis of thermal ions that are generated inside the cell. Consecutive measurements of the negative and positive ion ratios yield the value of the equilibrium constant, k_p. Gibbs energy, $\Delta_r G^\circ_T$ (Eq. (12.19)), was determined as 7.7 ± 1.4 kJ/mol, and, under the assumption of zero entropy change of the isomolecular reaction (12.19), it leads to $\Delta EA(C_{70}, C_{60}) =$

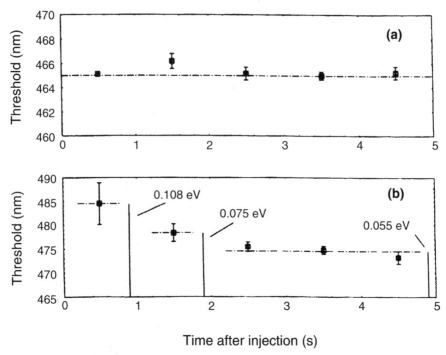

Figure 12.23 Variation in photodetachment thresholds in (a) lower and (b) higher energy thresholds for C_{60}^- with time after injection into the storage ring [124]. The vertical lines in (b) indicate the amount of internal energy possessed by C_{60}^-.

0.08 ± 0.01 eV, the difference being somewhat larger than that found from the direct electron photodetachment measurements for C_{60} and C_{70} ($\Delta EA = 0.01$ eV) [124]. It is likely that this disagreement stems from the difference between adiabatic and vertical ΔEA values (the former is usually derived in equilibrium measurements). The IME result is supported by the relative [60]fullerene and [70]fullerene ion mobility studies [126] that resulted in the same ΔEA value of 0.073 eV. Also very close to the IME result is the ΔEA estimation from the reduction potentials of C_{70} and C_{60} in various solvents: ΔEA falls within the 0.05–0.07 eV range [117].

12.5.2   Two-Electron Addition to Fullerene

A remarkable observation of doubly charged anions was reported for fullerenes. Attachment of two electrons to fullerenes could be initiated under various ionization conditions: laser desorption/ionization (LD) [127], low-energy electron capture [128], collisionally induced dissociative electron capture [129], and electrospray ionization (ES) [130]. Derivatization of fullerenes with electronegative addends facilitates generation of multiply charged anions in the gas phase due to the high electron affinities and the larger size, which allow a better delocalization of the excessive charge on the cage. The latter statement has been illustrated in the calculations on the most stable structures of $C_{60}(CN)_n^{2-}$. When the $C_{60}(CN)_n$ molecule acquires two charges, the structure with the larger separation of the addends is favored. Doubly charged cyanofullerene anions $C_{60}(CN)_n^{2-}$ ($n = 2, 4, 6$) were observed in the negative ion ES mass spectra of freshly prepared solutions of C_{60} and NaCN [131]. Fluorinated fullerene dianions $C_{60}F_n^{2-}$ ($n = 44$–47) were first found in the ES mass spectrum, when ferrocene was used as an electron donor in a solution of $C_{60}F_{48}$ in dichloromethane [130]. Attachment of the low-energy electrons [132] or the dissociative/nondissociative electron attachment to $C_{60}F_{2n-1}^-$ in collisions with gases (Ar, Xe) has also yielded formation of the doubly charged anions [129]. Perfluoroalkylated fullerenes were reported to form doubly charged anions under the negative ion CI mass spectrometry conditions [133].

12.5.3 Electron Exchange Reactions

12.5.3.1 Higher Fullerenes

As the first approximation, fullerenes can be considered as hollow conducting spheres of varying diameters, and simple models may be valid for rough estimation of some properties of fullerenes as a function of their sizes. Using the approach first suggested by F. Smith in 1960 for some conjugated polyaromatics [134] to the fullerenes, EAs were calculated from an equation relating EA and a geometrical size of the sphere (R_N) [135]:

$$EA(C_n) = V_f - \frac{a_0}{2R_N} \times 27.2 \text{ eV} \tag{12.20}$$

where V_f is the work function of the graphite surface and R_N is determined as $R_N = 3.55(N/60)^{1/2}$.

The resulting EA estimates for the fullerene series up to $n(C) = 120$ increase from 2.66 eV for C_{60} to 3.3 eV for C_{120}, asymptotically approaching the work function value of graphite (4.3 eV). A similar approach can be used for estimating ionization energies [136]. In general, this simple model agrees with the experiment that also shows a gradual increase in the EAs with the carbon atom number (Fig. 12.24). However, some experimental findings (e.g., enhanced EA value for [74]fullerene) could not be explained within this model.

EA determination for the higher fullerenes was performed by the IME method [137]. The electron exchange reactions between C_{70}^- and C_{2n} ($2n = 72$–84),

$$C_{70}^- + C_{2n} = C_{2n}^- + C_{70} \tag{12.21}$$

and between C_{84}^- and the heavier fullerenes C_{2n} ($2n = 84$–106),

$$C_{84}^- + C_{2n} = C_{2n}^- + C_{84} \tag{12.22}$$

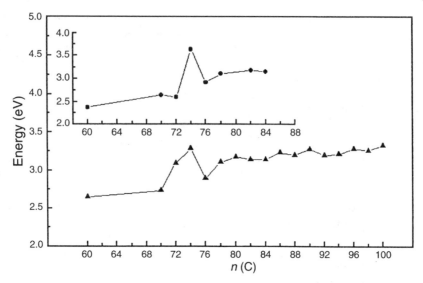

Figure 12.24   Electron affinity of fullerenes. Inset shows theoretical EAs (G. Seifert, *unpublished calculations*).

were studied under equilibrium conditions and then relative values of EA have been anchored to EA(C$_{60}$). The EA data for the higher fullerenes in Table 12.3 do not relate to the single isomeric species—the fullerenes present in the un-purified samples as mixtures of two or more stable isomers (except those for [70]-, [72]-, [74]-, and [76]fullerenes). Interestingly, density-functional theory (DFT) calculations for the two major isomers of [84]fullerene—D_{2d} and D_2—resulted in very close EA values of 3.13 and 3.14 eV, respectively [128], and in excellent agreement with the experimental value for the unpurified [84]fullerene sample (3.15 ± 0.04 eV) [137].

The result on EA(C$_{74}$) is particularly interesting, because up to the present it has not been possible to isolate [74]fullerene from the primary fullerene soot and characterize its structure and properties. Apparently, extraction of [74]fullerene together with the other soluble fullerenes is hindered because of the prominent electron-withdrawing character of [74]fullerene. Irregularity in the EA trend in the fullerene series has also been demonstrated theoretically. (Calculated EA data are shown as an inset in Fig. 12.24).

12.5.3.2 Fullerene Derivatives   Theoretical calculations of electron affinity of endohedral metallofullerenes indicated that electron deficiency of fullerenes is enhanced when a metal is entrapped inside the cage [138]. A metal atom is believed to donate its three or two electrons to the fullerene's lowest unoccupied molecular orbital (LUMO). The EA value estimated from the photoelectron spectrum of Ca@C$_{60}$ was shown to be higher than for [60]fullerene by about 0.35 eV [139]. Fourier transform mass spectrometry (FTMS) has been used for

TABLE 12.3 Electron Affinity of Fullerenes

Molecule	EA (eV)	Method	Reference	Molecule	EA (eV)[a]
C_{60}	2.65 ± 0.05	PES[b]	123	C_{84}	3.15 ± 0.04
	2.666 ± 0.001	PD[c]	124	C_{86}	3.21 ± 0.04
C_{70}	2.676 ± 0.001	PD	124	C_{88}	3.29 ± 0.05
	2.75 ± 0.01	IME	137	C_{90}	3.29 ± 0.05
C_{72}	3.11 ± 0.05	IME	137	C_{92}	3.22 ± 0.05
C_{74}	3.30 ± 0.05	IME	137	C_{94}	3.22 ± 0.05
C_{76}	2.90 ± 0.02	IME	137	C_{96}	3.19 ± 0.05
	2.9	DFT	d	C_{98}	3.28 ± 0.06
C_{78}	3.12 ± 0.03	IME	137	C_{100}	3.33 ± 0.05
C_{80}	3.19 ± 0.05	IME	137	C_{102}	3.39
C_{82}	3.16 ± 0.05	IME	137	C_{104}	3.43
C_{84}	3.14	DFT	128	C_{106}	3.41

[a] Determined by IME [137].

[b] Photoelectron spectroscopy.

[c] Photodetachment threshold of the ions from the storage ring (see text).

[d] Density-functional theory calculation (H.-P. Cheng and R. L. Whetten, *Chem. Phys. Lett.* **1992**, *197*, 44.)

the observation of electron exchange and "bracketing" EAs of some endohedral metallofullerenes [140] and fluorofullerenes [141]. Anions were produced by the slow electron attachment, then were trapped in the Ion Cyclotron Resonance (ICR) cell for times up to a few seconds and were allowed to react with various molecules whose EAs are well established. In the study of $La@C_n$, electron exchange occurred with the molecules whose EAs did not exceed 2.8 eV (EA of fluoranil), and $La@C_n$ did not react with tetracyano-quino-dimethane (TCNQ) (EA = 3.3 eV).

$$La@C_n + A^- = A + La@C_n^- \qquad (12.23)$$

In the IME study [143], equilibrium constants were directly measured for the electron exchange between the anions of hollow fullerenes and endohedral Gd metallofullerenes:

$$Gd@C_n + C_n^- = Gd@C_n^- + C_n \qquad (12.24)$$

yielding the EAs for $Gd@C_n$ (Table 12.4) that were about 0.2–0.3 eV higher than those of the corresponding empty fullerenes. It was shown, however, that the electron-accepting ability of metallofullerenes does not significantly depend on the type of fullerene cage in which the metal atom is entrapped. In case of $Gd@C_n$, EA values fall within the narrow range of 3.25 ± 0.05 eV.

TABLE 12.4   Electron Affinity of Endohedral Metallofullerenes

Molecule	EA (eV)	Method[a]	Reference
$Ca@C_{60}$	3.0	PES	139
$Gd@C_{60}$	2.9 ± 0.2	IME	142
$Gd@C_{74}$	3.24	IME	144
$Gd@C_{76}$	3.2 ± 0.1	IME	144
$Gd@C_{78}$	3.26	IME	144
$Gd@C_{80}$	3.3 ± 0.1	IME	144
$Gd@C_{82}$	3.3 ± 0.1	IME	144
$La@C_{74}$	>2.9	IME	144
$La@C_n$, $n = 60, 70\text{--}84$	$2.7 < EA < 3.3$	IMB[b]	145
$La@C_{60}$	3.8	DFT	146
$La@C_{82}$	3.22	DFT	138
$Y@C_{82}$	3.12	DFT	138
$Sc@C_{82}$	3.08	DFT	138
$Ce@C_{82}$	3.19	DFT	138
$La@C_{82}$	3.35 ± 0.06	Estimate from E_{red}	147
$Y@C_{82}$	3.43 ± 0.06	Estimate from E_{red}	147

[a] Acronyms defined in footnotes to Table 12.3.

[b] Bracketing method.

Table 12.5 summarizes experimental EAs for fluorofullerenes obtained by the "bracketing" technique in the FTMS and IME studies. In FTMS, experiments were performed with the highly fluorinated [60]fullerene and [70]fullerene, and the EA value was "bracketed" between 2.8 and 4.3 eV. Experimentally, the electron exchange reaction was observed as the relative change in the intensity of the reacting species injected in the ion trap with time. Figure 12.25 shows that the relative intensity of the ion with higher EA $(C_{60}F_{44,46,48}{}^-)$ increases with the time of reaction, whereas the intensity of the $NO_3{}^-$ ion with the lower EA decreases.

In the IME experiment, mono- and difluorofullerenes were generated in the solid-state reactions between fullerene and mild fluorinating reagents and gas phase species, both neutral and charged, were equilibrated at the elevated tem-

TABLE 12.5   Electron Affinity of Fluorofullerenes

Molecule	EA (eV)	Method	Reference
$C_{60}F$	2.78	IME	148
$C_{60}F_2$	2.74 ± 0.07	IME	148
$C_{60}F_{36}$	(3.48)	Estimate from E_{red}	149
$C_{60}F_{44,46,48}$	4.06 ± 0.30	IMB[a]	141, 150
$C_{70}F_2$	2.80 ± 0.07	IME	148
$C_{70}F_{52,54}$	4.06 ± 0.30	IMB	141

[a] Bracketing method.

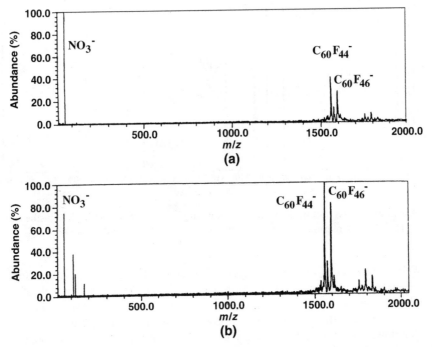

Figure 12.25 Mass spectra showing electron exchange reaction between $C_{60}F_x$ and NO_3^-: reagents in the gas phase (a) before the reaction and (b) after 4 s. Adapted from Ref 141.

peratures. The absolute EAs for fluorofullerenes were evaluated relative to $EA(C_{60})$. A substantial increase in EA is seen with the increase in fluorination degree: from 2.74 eV ($C_{60}F_2$) to 4.05 eV ($C_{60}F_{48}$).

12.5.4 Addition Reactions

The reactivity of anions of [60]fullerene and [70]fullerene under thermal conditions were studied by the flowing afterglow method [151]. Fullerene anions reacted with various gases in the flow tube and the thermalized products were recorded mass spectrometrically. As neither C_{60}^- nor C_{70}^- reacted with water and various strong acids, including fluoroacetic acid ($G_{acid} = 317$ kcal/mol), it was concluded that either fullerene anions have higher acidity or the proton-transfer reactions were not observed due to the kinetic barrier to proton abstraction. Reactions of C_{60}^- and C_{70}^- with NO_2 resulted in the adduct formation, $C_{60}NO_2^-$ and $C_{70}NO_2^-$ (Fig. 12.26). The rate constants were found to be very close and not dependent on the flow tube pressure. No charge transfer was observed, which was a reflection of the fact that the EA value of either fullerene was higher than that of NO_2 (EA = 2.28 eV). Reactions with O_2, NO, and BF_3 were very slow; no adduct formation was mentioned.

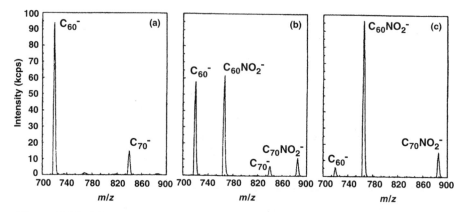

Figure 12.26 Mass spectra showing reactions of C_{60}^- and C_{70}^- with NO_2 in the flow tube [151] with increasing flow rates (STP cm^3/min): (a) 0.04, (b) 0.56, and (c) 2.6.

While addition of molecular oxygen to fullerene anion did not occur under the flow tube conditions, this reaction was observed in the collision cell of the tandem mass spectrometer [152]. O_2 was pressurized in the collision cell up to 1 mtorr, whereas the energy of C_{60}^- produced by chemical ionization was relatively low. The observed mass spectrum is shown in Figure 12.27, the peak due to addition of a single O_2 molecule to fullerene anion is clearly seen at m/z = 752. Similar experiments with NO as a reacting gas led to a different reaction product, that is, formation of fullerene monooxide. CID spectra indicated that negative ions $C_{60}O^-$ and $C_{60}O_2^-$ have remarkably different structures, the latter being supposedly the precursor for the open-cage fullerene structures. Addition of molecular oxygen activates the cage fragmentation to a much more significant extent than addition of a single oxygen atom.

In the negative ion chemical ionization (NICI) mass spectrometry, fullerene anions are efficiently produced. The cross section for C_{70}^- formation was found to be higher than for C_{60}^-. When O_2 was introduced into the CI cell, a series of oxides $C_{60}O_n^-$ ($n = 1$–6) were recorded in the NICI mass spectrum. Mono- and difluorinated [60]fullerene anions were observed when SF_6 was let into the CI cell as a source of fluorine [153].

12.5.5 Fluorine Exchange Reactions

The IME technique was applied for generating $C_{60}F_n^-$ ($n = 1, 2, 4$) and $C_{70+2n}F_2^-$ via the reactions of fullerenes with transition metal fluorides or highly fluorinated fullerenes as fluorinating reagents [154]. This resulted in the determination of the Gibbs energies of the fluorine exchange reactions:

$$2C_{60}F^- = C_{60}F_2^- + C_{60}^- \tag{12.25}$$

$$C_{60}F_4^- + C_{60}^- = 2C_{60}F_2^- \tag{12.26}$$

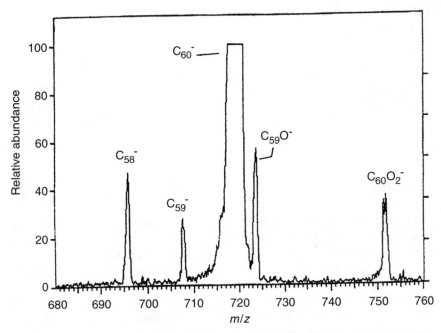

Figure 12.27 Mass spectrum obtained when mass selected C_{60}^- undergoes 110 eV reactive collisions with O_2 (1 mtorr) in the collision cell of the triple quadrupole instrument [152].

$$C_{60}F_2^- + C_{70+2n}^- = C_{60}^- + C_{70+2n}F_2^- \qquad (12.27)$$

$$C_{60}^- + 2MF_{x+1}^- = C_{60}F_2^- + 2MF_x^- \qquad (12.28)$$

where MF_x^- is either FeF_3^- or UF_5^-. These ions were used as anchors for the absolute C–F bond dissociation energy (BDE) determination for $C_{2n}F_{2x}$. BDE is defined as the enthalpy of reaction (12.29) divided by the number of fluorine atoms:

$$C_{2n}F_{2x} = C_{2n} + 2xF \qquad (12.29)$$

$$(C\text{–}F)BDE(C_{2n}F_{2x}) = \Delta_r H(\text{Eq. (12.29)})/2x$$

Table 12.6 presents the derived thermochemical parameters for mono- and difluoro [60]fullerene molecules and anions. The (C–F) BDE estimates for difluorinated higher fullerenes fall within a narrow range, 386 ± 14 kJ/mol. A slightly higher value is obtained for $C_{74}F_2^-$ in accord with the EA trend in the higher fullerene series (Table 12.3). Comparison with the simple fluoroorganic molecules shows that (C–F) BDE of mono- and difluorofullerenes is higher than that of fluorobenzene (290 kJ/mol) and very close to difluoroethane (353 kJ/mol) but much lower than (C–F) BDE in CF_4 (545 kJ/mol).

TABLE 12.6 C–F Bond Dissociation Energies (kJ/mol) for $C_{60}F_n$ ($n = 1, 2$), $C_{70}F_2$, and Their Anions

$C_{60}F$	$C_{60}F_2$	$C_{70}F_2$	$C_{60}F^-$	$C_{60}F_2^-$	$C_{70}F_2^-$	Reference
≥ 362	369	366	388	372	370	154
403	365					155
	306					156
289						157

CONCLUSION

Clearly, there has been significant progress in the elucidation of fundamental physicochemical properties of fullerenes and their positive and negative ions in the gas phase. Such information is useful as a benchmark for the behaviour of these species in the condensed phase. Further advances are expected, particularly as larger fullerenes and endohedral fullerenes or fullerenes derivatized exohedrally become increasingly available for study. For example, it will become increasingly possible to explore the physics and chemistry of metal atoms and organic or organometallic molecules attached to curved carbonaceous surfaces in "gas-phase surface" experiments. Also, research could then be directed to a study of the influence of atoms or molecules trapped inside a fullerene molecule on each other and on the physicochemistry of the immediate "world" outside the carbonaceous surface. "The sky then becomes the limit"!

REFERENCES

1. D. Yu. Borisova, A. A. Mavrin, L. N. Sidorov, E. V. Skokan, J. G. Edwards, F. M. Spiridonov, A. Ya. Borshchevsky, and I. N. Ioffe, *Fullerene Sci. Technol.* **1998**, *6*, 519.

2. M. V. Korobov, E. V. Skokan, and D. Yu. Borisova, *J. Phys. Chem. (Russian)* **1994**, *70*, 999.

3. C. K. Mathews, M. SaiBaba, T. S. Lakshmi Narasimhan, R. Balasubramanian, N. Sivaraman, and P. R. Vasudeva Rao, *J. Phys. Chem.* **1992**, *96*, 3566.

4. A. Popovic, G. Drasic, and J. Marsel, *Rapid Commun. Mass Spectrom.* **1994**, *8*, 985.

5. J. Abrefah, R. D. Olander, M. Balooch, and J. W. Siekhaus, *Appl. Phys. Lett.* **1992**, *60*, 1313.

6. V. Piacente, G. Gigli, P. Scardala, A. Guistini, and D. Ferro, *J. Phys. Chem.* **1995**, *99*, 14052.

7. A. Tokmakoff, A. Haynes, and S. M. George, *Chem. Phys. Lett.* **1991**, *186*, 450.

8. S. Chen, R. A. Kortan, C. R. Haddon, and A. D. Fleming, *J. Phys. Chem.* **1992**, *96*, 1016.

9. V. Piacente, G. Gigli, P. Scardala, A. Guistini, and G. Bardi, *J. Phys. Chem.* **1996**, *100*, 15.

10. V. P. Kolesov, S. M. Pimenova, A. A. Kurskaya, and T. S. Papina, *J. Chem. Thermodynamics* **1996**, *28*, 1121.

11. C. K. Mathews, M. SaiBaba, S. T. Lakshmi Narasimhan, R. Balasubramanian, N. Sivaraman, G. T. Srinivasan, and R. P. Vasudeva Rao, *Fullerene Sci. Technol.* **1993**, *1*, 101.

12. B. Brunetti, G. Gigli, E. Giglio, V. Piacente, and P. Scardala, *J. Phys. Chem. A* **1997**, *101*, 10715.

13. V. Piacente, C. Palchetti, G. Gigli, and P. Scardala, *J. Phys. Chem. A* **1997**, *101*, 4303.

14. O. V. Boltalina, V. Y. Markov, A. Ya. Borshchevsky, L. N. Sidorov, V. N. Bezmelnitsin, A. V. Eletskii, and R. Taylor, *Rapid Commun. Mass Spectrom.* **1998**, *12*, 1028.

15. O. V. Boltalina, V. Y. Markov, A. Ya. Borshchevsky, V. Y. Davydov, L. N. Sidorov, V. N. Bezmelnitsin, A. V. Eletskii, and R. Taylor, *Mendeleev Commun.* **1998**, 141.

16. M. V. Korobov, E. V. Skokan, P. A. Dorozko, L. Homich, and A. Kurskaya, *Recent Advances in the Chemistry and Physics of Fullerenes and Related Materials* **1994**, *1*, 1595.

17. A. Popovic, *Rapid Commun. Mass Spectrom.* **1996**, *10*, 1433.

18. M. SaiBaba, T. S. Lakshmi Narasimhan, R. Balasubramanian, N. Sivaraman, and C. K. Mathews, *J. Phys. Chem.* **1994**, *98*, 1333.

19. M. Moalem, M. Balooch, R. S. Ruoff, and A. V. Hamza, *J. Phys. Chem.* **1995**, *99*, 16736.

20. O. V. Boltalina, V. Y. Markov, A. Ya. Borshchevsky, G. Gigli, and J. Balducci, *J. Phys. Chem.* **1999**, *103*, 3828.

21. P. Scheier, B. Dünser, R. Worgotter, S. Matt, D. Muigg, G. Senn, and T. D. Märk, *Int. Rev. Phys. Chem.* **1996**, *15*, 93.

22. R. Deng, G. Littlefield, and O. Echt, *Z. Phys. D* **1997**, *40*, 355.

23. S. Cheng, H. G. Berry, R. W. Dunford, H. Esbensen, D. S. Gemmel, E. P. Kanter, and T. LeBrun, *Phys. Rev. A* **1996**, *54*, 3182.

24. Y. Nakai, A. Itoh, T. Kambara, Y. Bitoh, and Y. Awaya, *J. Phys. B* **1997**, *30*, 3049.

25. D. C. Lorents, *Comments At. Mol. Phys.* **1997**, *33*, 125.

26. B. Walch, C. L. Cocke, R. Voelpel, and E. Salzborn, *Phys. Rev. Lett.* **1994**, *72*, 1439.

27. Z. Wan, J. F. Christian, Y. Basir, and S. L. Anderson, *J. Chem. Phys.* **1993**, *99*, 5858.

28. S. Martin, L. Chen, A. Denis, and J. Désesquelles, *Phys. Rev. A* **1998**, *57*, 4518.

29. A. Itoh, H. Tsuchida, K. Miyaba, M. Imai, and N. Imanishi, *Nucl. Instrum. Methods B* **1997**, *129*, 363.

30. T. Schlathölter, R. Hoekstra, and R. Morgenstern, *J. Phys. B* **1998**, *31*, 1321.

31. A. Reinköster, U. Werner, and H. O. Lutz, *Europhys. Lett.* **1998**, *43*, 653.

32. J. Lindhard, V. Nielsen, and M. Scharff, *Mat. Fys. Medd. Dan. Vid. Selsk.* **1968**, *36*, 10.

33. R. Ehlich, A. V. Glotov, K. Hansen, and E. E. B. Campbell, *J. Phys. B* **1996**, *29*, 5143.

34. R. J. Doyl and M. M. Ross, *J. Phys. Chem.* **1991**, *95*, 4954.

35. P. Hvelplund, L. H. Andersen, H. K. Haugen, J. Lindhard, D. C. Lorents, R. Malhorat, R. Ruoff, *Phys. Rev. Lett.* **1992**, *69*, 1915.

36. T. Y. Liu, L. L. Shiu, T. Y. Luh, and G. R. Her, *Rapid Commun. Mass Spectrom.* **1995**, *9*, 93.

37. P. Hekånsson, S. Della-Negra, J. P. Mouffron, B. Waast, and P. A. Sullivan, *Nucl. Instrum. Methods B* **1996**, *112*, 39.

38. A. A. Tuinmann, P. Mukherjee, and J. L. Adcock, *J. Phys. Chem.* **1992**, *96*, 7584.

39. D. C. Lorents, D. H. Yu, C. Brink, N. Jensen, and P. Hvelplund, *Chem. Phys. Lett.* **1995**, *236*, 141.

40. D. B. Cameron and J. H. Parks, *Chem. Phys. Lett.* **1997**, *272*, 18.

41. J. U. Andersen, C. Brink, P. Hvelplund, M. O. Larsson, B. B. Nielsen, and H. Shen, *Phys. Rev. Lett.* **1996**, *77*, 399.

42. R. L. Whetten and C. Yeretzian, *Int. J. Mass Spectrom. Ion Processes* **1994**, *138*, 63.

43. M. C. Larsen, P. Hvelplund, M. O. Larsson, and H. Shen, *Eur. Phys. J. D.* **1999**, *5*, 283.

44. R. Ehlich, O. Knospe, and R. Schmidt, *J. Phys. B* **1997**, *30*, 5429.

45. C. Brink, L. H. Andersen, P. Hvelplund, and D. H. Yu, *Z. Phys. D* **1994**, *29*, 45.

46. R. Vandenbosch, B. P. Henry, C. Cooper, M. L. Gardel, J. F. Liang, and D. I. Will, *Phys. Rev. Lett.* **1998**, *81*, 1821.

47. A. Dupont, J. P. Gisselbrecht, E. Leize, L. Wagner, and A. V. Dorsselaer, *Tetrahedron Lett.* **1994**, *35*, 6083.

48. L. S. Wang, J. Conceicao, C. Jin, and R. E. Smalley, *Chem. Phys. Lett.* **1991**, *182*, 5.

49. K. J. McHale, M. J. Polce, and C. Wesdemiotis, *J. Mass Sprectrom.* **1995**, *30*, 33.

50. M. C. Larsen, unpublished data.

51. A. Lahamer, Z. C. Ying, R. E. Haufler, R. L. Hettich, and R. N. Compton, *Adv. Metal Semicond. Clusters* **1998**, *4*, 179.

52. H. Schwarz, T. Weiske, D. K. Bohme, and J. Hrusåk, Exo-and Endohedral Fullerene Complexes in the Gas Phase, in *Buckminsterfullerenes*, W. E. Billups and M. A. Ciufolini (eds.). VCH Publishers, New York **1993**, Chap. 10.

53. C. Brink, P. Hvelplund, H. Shen, H. A. Jiménez-Vázquez, R. J. Cross, and M. Saunders, *Chem. Phys. Lett.* **1998**, *290*, 551.

54. J. L. Elkind, F. D. Weiss, J. M. Alford, R. T. Laaksonen, and R. E. Smalley, *J. Am. Chem. Soc.* **1981**, *74*, 6511.

55. O. V. Boltalina, 1998, unpublished.

56. E. E. Campbell, R. Ehlich, A. Hielscher, J. M. Frazao, and I. V. Hertel, *Z. Phys. D* **1992**, *23*, 1.

57. O. Knospe, A. V. Glotov, G. Seifert, and R. Schmidt, *J. Phys. B* **1996**, *29*, 5163.

58. B. Farizon, M. Farizon, M. J. Gaillard, R. Genre, S. Louc, J. Martin, J. P. Buchet, M. Carré, G. Senn, P. Scheier, and T. D. Märk. *Int. J. Mass Spectrom. Ion Processes* **1997**, *164*, 225.

59. F. Rohmund, E. E. Campbell, O. Knospe, G. Seifert, and R. Schmidt, *Phys. Rev. Lett.* **1996**, *76*, 3289.

60. N. Selberg, A. Bárány, C. Biedermann, C. J. Setterlind, H. Cederquist, A. Langereis, M. O. Larsson, A. Wännström, and P. Hvelplund, *Phys. Rev. A* **1996**, *53*, 874.

61. J. Jin, H. Khemliche, M. H. Prior, and Z. Xie, *Phys. Rev. A* **1996**, *53*, 615.

62. M. O. Larsson, P. Hvelplund, M. C. Larsen, H. Shen, H. Cederquist, and H. T. Schmidt, *Int. J. Mass Sprectrom.* **1998**, *177*, 51.

63. H. Shen, P. Hvelplund, D. Mathur, A. Bàràny, H. Cederquist, N. Selberg, and D. C. Lorents, *Phys. Rev. A* **1995**, *52*, 3847.

64. F. Rohmund and E. E. B. Campbell, *Chem. Phys. Lett.* **1995**, *245*, 237.

65. F. Rohmund and E. E. B. Campbell, *J. Phys. B* **1997**, *30*, 5293.

66. F. Rohmund and E. E. B. Campbell, *Z. Phys. D.* **1997**, *40*, 399.

67. H. Shen, C. Brink, P. Hvelplund, and M. O. Larsson, *Z. Phys. D* **1997**, *40*, 371.

68. O. V. Boltalina, P. Hvelplund, M. C. Larsen, and M. O. Larsson, *Phys. Rev. Lett.* **1998**, *80*, 5101.

69. W. Krätschmer, L. D. Lamb, K. Fostiropoulos and D. R. Huffman, *Nature,* **1990**, *347*, 354.

70. (a) S. W. McElvany and M. M. Ross, *J. Am. Soc. Mass Spectrom.* **1992**, *3*, 268. (b) S. W. McElvany, M. M. Ross and J. H. Callahan, *Acc. Chem. Res.* **1992**, *25*, 162.

71. S. W. McElvany and J. H. Callahan, *J. Phys. Chem.* **1991**, *95*, 6187.

72. D. Schröder, D. K. Bohme, T. Weiske, and H. Schwarz, *Int. J. Mass Spec. Ion Processes* **1992**, *116*, R13.

73. (a) T. Weiske, D. K. Bohme, J. Hrusak, W. Krätschmer and H. Schwarz, *Angew. Chem. Int. Ed. Eng.* **1991**, *30*, 884. (b) T. Weiske, D. K. Bohme and H. Schwarz, *J. Phys. Chem.* **1991**, *95*, 8451.

74. (a) K. A. Caldwell, D. E. Giblin, C. S. Hsu, D. Cox and M. I. Gross, *J. Am. Chem. Soc.* **1991**, *113*, 8519. (b) K. A. Caldwell, D. E. Giblin and M. I. Gross, *J. Am. Chem. Soc.* **1992**, *114*, 3743.

75. M. M. Ross and J. H. Callahan, *J. Phys. Chem.* **1991**, *95*, 5720.

76. (a) T. Weiske, J. Hrusak, D. K. Bohme and H. Schwarz, *Chem. Phys. Letters* **1991**, *186*, 459. (b) T. Weiske, D. K. Bohme and H. Schwarz, *J. Phys. Chem.* **1991**, *95*, 8451.

77. T. Weiske and H. Schwarz, *Angew. Chem., Int. Ed. Eng.* **1992**, *31*, 605.

78. L. S. Roth, Y. Huang, J. T. Schwedler, C. J. Cassidy and B. S. Freiser, *J. Am. Chem. Soc.* **1991**, *113*, 6298.

79. (a) Y. Huang and B. S. Freiser, J. Am. Chem. Soc. **1991**, *113*, 8186. (b) Y. Huang and B. S. Freiser, *J. Am. Chem. Soc.* **1991**, *113*, 9418.

80. (a) Z. Wan, J. F. Christian and S. L. Anderson, J. Chem. Phys. **1992**, *96*, 3344. (b) J. F. Christian, Z. Wan and S. L. Anderson, *Chem. Phys. Lett.* **1992**, *199*, 373. (c) J. F. Christian, Z. Wan and S. L. Anderson, *J. Phys. Chem.* **1992**, *96*, 3574. (d) J. F. Christian, Z. Wan and S. L. Anderson, *J. Phys. Chem.* **1992**, *96*, 10597. (e) J. F. Christian, Z. Wan and S. L. Anderson, *J. Chem. Phys.* **1993**, *99*, 3468. (f) Y. Basir, Z. Wan, J. F. Christian and S. L. Anderson, S. L., *Int. J. Mass Spectrom. Ion Processes*, **1994**, *138*, 173. (g) Y. Basir and S. L. Anderson, S. L., *Chem. Phys. Lett.*, **1995**, *243*, 45.

81. D. K. Bohme, *Int. Rev. Phys. Chem.* **1994**, *13*, 163.

82. D. L. Lichtenberger, M. E. Rempe, and S. B. Gogosha, *Chem. Phys. Lett.*, **1992**, *198*, 454.

83. G. Javahery, S. Petrie, J. Wang, X. Wang and D. K. Bohme, *Int. J. Mass Spectrom. Ion Processes*, **1993**, *125*, R13.

84. G. Javahery, S. Petrie, J. Wang and D. K. Bohme, *Chem. Phys. Lett.* **1992**, *195*, 7.

85. J. A. Zimmerman, J. R. Eyler, S. B. H. Bach and S. W. McElvany, *J. Chem. Phys.* **1991**, 94.

86. (a) S. Petrie, J. Javahery, J. Wang and D. K. Bohme, *J. Phys. Chem.* **1992**, *96*, 6121. (b) S. Petrie, G. Javahery, J. Wang and D. K. Bohme, *J. Am. Chem. Soc.* **1992**, *114*, 6268.

87. P. Scheier, B. Dünser, R. Wörgötter, M. Lezius, R. Robl and T. D. Märk, *Int. J. Mass Spec. Ion Processes* **1994**, *138*, 77.

88. G. Javahery, H. Wincel, S. Petrie and D. K. Bohme, *Chem. Phys. Lett.* **1993**, *204*, 467.

89. G. Javahery, H. Becker, S. Petrie, P. C. Cheng, H. Schwarz, L. T. Scott and D. K. Bohme, *Org. Mass Spectrom.* **1993**, *20*, 1005.

90. D. K. Bohme in *Recent Advances in the Chemistry and Physics of Fullerenes and Related Materials*. R. S. Ruoff and K. M. Kadish (Eds.), Electrochemical Society Proceedings 95-10, 1465; Electrochemical Society, Inc., Pennington, N.J., 1995.

91. S. Petrie, H. Becker, V. I. Baranov and D. K. Bohme, *Int. J. Mass Spectrom. Ion Processes* **1995**, *145*, 79.

92. G. Javahery, S. Petrie, H. Wincel, J. Wang and D. K. Bohme, *J. Am. Chem. Soc.* **1993**, *115*, 5716.

93. H. Becker, G. Javahery, S. Petrie and D. K. Bohme, *J. Phys. Chem.* **1994**, *98*, 5591.

94. Q. Jiao, Y. Huang, S. A. Lee, J. R. Gord and B. S. Freiser, *J. Am. Chem. Soc.* **1992**, *114*, 2726.

95. V. Baranov and D. K. Bohme, *Int. J. Mass Spectrom. Ion Processes* **1997**, *165/166*, 249.

96. S. Petrie and D. K. Bohme, *Can. J. Chem.* **1994**, *72*, 577.

97. G. Javahery, S. Petrie, J. Wang, H. Wincel and D. K. Bohme, *J. Am. Chem. Soc.* **1993**, *115*, 9701.

98. G. Javahery, S. Petrie, H. Wincel, J. Wang and D. K. Bohme, *J. Am. Chem. Soc.* **1993**, *115*, 6295.

99. S. Petrie, G. Javahery, H. Wincel, J. Wang and D. K. Bohme, Int. J. Mass Spectrom. Ion Processes **1994**, *138*, 187.

100. J. Wang, G. Javahery, S. Petrie, A. C. Hopkinson and D. K. Bohme, *Angew. Chem. Int. Ed. Engl.* **1994**, *33*, 206.

101. J. Wang, G. Javahery, S. Petrie and D. K. Bohme, *J. Am. Chem. Soc.* **1992**, *114*, 9665.

102. (a) J. Wang, V. Baranov and D. K. Bohme, *J. Am. Soc. Mass Spectrom.* **1996**, *7*, 261. (b) J. Wang, G. Javahery, V. Baranov and D. K. Bohme, *Tetrahedron* **1996**, *52*, 5191.

103. V. Baranov, J. Wang, A. C. Hopkinson and D. K. Bohme, *J. Am. Chem. Soc.* **1997**, *119*, 2040.

104. V. Baranov and D. K. Bohme, *Chem. Phys. Lett.* **1996**, *258*, 203.

105. (a) S. Petrie, G. Javahery and D. K. Bohme, *Int. J. Mass Spectrom. Ion Processes* **1993**, *124*, 145. (b) S. Petrie, G. Javahery, H. Wincel and D. K. Bohme, *J. Am. Chem. Soc.* **1993**, *113*, 6290.

106. V. Baranov, A. C. Hopkinson and D. K. Bohme, *J. Am. Chem. Soc.* **1997**, *119*, 7055.

107. S. Petrie and D. K. Bohme, *Nature* **1993**, *165*, 426.

108. R. C. Haddon, *Science* **1993**, *261*, 1545.

109. H. Becker, L. T. Scott and D. K. Bohme, *Int. J. Mass Spectrom. Ion Processes* **1997**, *167/168*, 519.

110. V. Baranov and D. K. Bohme, *Int. J. Mass Spectrom. Ion Processes* **1995**, *149/150*, 543.

111. S. Z. Kan, Y. G. Byun and B. S. Freiser, *J. Am. Chem. Soc.* **1994**, *116*, 8815.

112. S. Z. Kan, Y. G. Byun and B. S. Freiser, *J. Am. Chem. Soc.* **1995**, *117*, 1177.

113. H. Ajie, M. M. Alvarez, S. J. Anz, R. D. Beck, F. Diederich, K. Fostiropoulos, D. R. Huffman, W. Kratschmer, Y. Rubin, K. E. Schriver, D. Sensharma, R. L. Whetten, *J. Phys. Chem.* **1990**, *94*, 8630.

114. S. Fujimaki, I. Kudaka, T. Sato, K. Hiraoka, H. Shinohara, Y. Saito, and K. Nojima, *Rapid Commun. Mass Spectrom.* **1993**, *7*, 1077.

115. H. Shinohara, H. Saito, M. Takayama, A. Izuoka, and T. Sugawara, *J. Phys. Chem.* **1991**, *95*, 8449.

116. J. M. Wood, B. Kahr, I. I. Hoke, L. Dejarme, R. G. Cooks, and D. Ben-Amotz, *J. Am. Chem. Soc.* **1991**, *113*, 5907.

117. L. Echegoyen and L. Echegoyen, *Acc. Chem Res.* **1998**, *31*, 593.

118. P. W. Fowler and A. Ceulemans, *J. Phys. Chem.* **1995**, *99*, 508.

119. R. F. Curl and R. E. Smalley, *Science* **1988**, *242*, 1017.

120. J. Aihara, *J. Phys. Chem.* **1995**, *99*, 12739.

121. S. C. O'Brien, J. R. Heath, H. W. Kroto, R. F. Curl, and R. E. Smalley, *Chem. Phys. Lett.* **1986**, *132*, 99.

122. S. H. Yang, C. L. Pettiette, J. Conceicao, O. Cheshnovsky, and R. E. Smalley, *Chem. Phys. Lett.* **1987**, *139*, 233.

123. L. Wang, J. Conceicao, C. Jin, and R. E. Smalley, *Chem. Phys. Lett.* **1991**, *182*, 5.

124. C. Brink, L. H. Andersen, P. Hvelplund, D. Mathur, and J. D. Volstad, *Chem. Phys. Lett.* **1995**, *233*, 52.

125. O. V. Boltalina, A. Y. Borshchevskii, L. N. Sidorov, E. V. Sukhanova, and E. V. Skokan, *Rapid Commun. Mass Spectrom.* **1993**, *7*, 1009.

126. M. E. Burba, S. K. Lim, and A. C. Aldrecht, *J. Phys. Chem.* **1995**, *99*, 11839.

127. P. A. Limbach, L. Schweikhard, K. A. Cowen, M. T. McDermott, A. G. Marshall, and J. V. Coe, *J. Am. Chem. Soc.* **1991**, *113*, 6795.

128. R. N. Compton, A. A. Tuinman, C. E. Klots, M. R. Pederson, and D. C. Patton, *Phys. Rev. Lett.* **1997**, *78*, 4367.

129. O. V. Boltalina, P. Hvelplund, M. C. Larsen, and M. O. Larsson, *Phys. Rev. Lett.* **1998**, *80*, 5101.

130. F. Zhou, G. J. Van Berkel, and B. T. Donovan, *J. Am. Chem. Soc.* **1994**, *116*, 5485.

131. G. Khairallah and P. Barrie, *Chem. Phys. Lett.* **1997**, *268*, 218.

132. C. Jin and R. N. Compton, *Phys. Rev. Lett.* **1994**, *73*, 2821.

133. C. N. McEwen, P. J. Fagan, and P. J. Krusic, *Int. J. Mass Spectrom. Ion Processes* **1995**, *146/147*, 297.

134. F. T. Smith, *J. Chem. Phys.* **1961**, *34*, 793.

135. O. V. Boltalina, L. N. Sidorov, E. V. Sukhanova, and P. Hvelplund, in *Recent Advances in the Chemistry and Physics of Fullerenes and Related Materials*, R. S. Ruoff and K. M. Kadish (eds.), The Electrochemical Society Proceedings, Vol. 94-10, pp. 1607–1618. The Electrochemical Society, Pennington, NJ, 1994.

136. G. Seifert, K. Veitze, and R. Schmidt, *J. Phys. B At. Mol. Opt. Phys.* **1996**, *29*, 5183.

137. O. V. Boltalina, E. V. Dashkova, and L. N. Sidorov, *Chem. Phys. Lett.* **1996**, *256*, 253.

138. S. Nagase and K. Kobayashi, *Chem. Phys. Lett.* **1994**, *228*, 106.

139. L. S. Wang, J. M. Alford, Y. Chai, M. Diener, J. Zhang, S. M. McClure, T. Guo, G. E. Scuseria, and R. E. Smalley, *Chem. Phys. Lett* **1993**, *207*, 354.

140. R. L. Hettich, C. Ying, and R. N. Compton, in *Recent Advances in the Chemistry and Physics of Fullerenes and Related Materials*, R. S. Ruoff and K. M. Kadish (eds.), The Electrochemical Society Proceedings, Vol. 95-10, p. 1457. The Electrochemical Society, Pennington, NJ, 1995.

141. R. Hettich, C. Jin, and R. Compton, *Int. J. Mass. Spectrom. Ion Processes* **1994**, *138*, 263.

142. O. V. Boltalina, in *Recent Advances in the Chemistry and Physics of Fullerenes and Related Materials*, R. S. Ruoff and K. M. Kadish (eds.), The Electrochemical Society Proceedings, Vol. 95-10, p. 1395. The Electrochemical Society, Pennington, NJ, 1995.

143. O. V. Boltalina and L. N. Sidorov, *Mol. Mat.* **1998**, *10*, 71.

144. O. V. Boltalina, I. N. Ioffe, I. D. Sorokin, and L. N. Sidorov, *J. Phys. Chem. A.* **1997**, *101*, 9651.

145. Z. C. Ying, *Recent Advances in the Chemistry and Physics of Fullerenes and Related Materials* **1994**, *1*, 1402.

146. A. Rosen and B. J. Wastberg, *Chem. Phys.* **1989**, *90*, 2525.

147. R. Ruoff, K. Kadish, P. Boulas, and E. C. M. Chen, *J. Phys. Chem.* **1995**, *99*, 8843.

148. O. V. Boltalina, L. N. Sidorov, E. V. Sukhanova, and I. D. Sorokin, *Chem. Phys. Lett.* **1994**, *230*, 567.

149. N. Liu, Y. Morio, F. Okino, H. Tuhara, O. V. Boltalina, and V. K. Pavlovich, *Synth. Metals* **1997**, *86*, 2289.

150. R. L. Hettich, C. Jin, P. F. Britt, A. A. Tuinman, and R. N. Compton, *Mat. Res. Soc. Symp. Proc.* **1994**, *349*, 133.

151. L. S. Sunderlin, J. A. Paulino, J. Chow, B. Kahr, D. Ben-Amotz, and R. R. Squires, *J. Am. Chem. Soc.* **1991**, *113*, 5489.

152. J. H. Callahan, S. W. McElvany, and M. M. Ross, *Int. J. Mass Spectrom. Ion Processes* **1994**, *138*, 221.

153. S. W. McElvany and J. H. Callahan, *J. Phys. Chem.* **1991**, *95*, 6186.

154. O. V. Boltalina, D. B. Ponomarev, A. Y. Borshchevskii, and L. N. Sidorov, *J. Phys. Chem.* **1997**, *101*, 2574.

155. L. P. Malkerova, D. V. Sevast'yanov, A. S. Alikhanyan, S. P. Ionov, and N. G. Spitsyna, *Dokl. Chem.* **1995**, *342*, 142.

156. D. Bakowies and W. Thiel, *Chem. Phys. Lett.* **1992**, *193*, 236.

157. A. Hirsch, H. Mauser, and T. Clark, in *Molecular Nanostructures*, H. Kuzmany, J. Fink, M. Mehring, and S. Roth (eds.). World Scientific Publishing, Singapore, 1997, pp. 55–60.

CHAPTER 13

FULLERENE–SURFACE INTERACTIONS

ALEX V. HAMZA

13.1 INTRODUCTION

This chapter reviews recent work on fullerene–surface interactions. The review excludes theoretical developments, endohedral fullerenes and nanotubes, Langmuir–Blodgett fullerene films, and doped fullerenes (i.e., fullerene superconductors), as these subjects are covered in other chapters. Also excluded are devices fabricated with fullerenes, as these are also topics of other chapters.

There are three recent, excellent reviews on fullerene–surface interactions [1–3]. Thus, this chapter will focus on developments since those reviews were written. This chapter is not intended to be an exhaustive review; therefore, the examples discussed are intended to be illustrative. The author apologizes for the inevitable omissions of important work in this area.

Until recently, the primary interaction between surfaces and fullerenes was thought to be the transfer of charge from substrate to fullerene. To this author's knowledge, a surface interaction involving charge transfer from the fullerene to the surface has not been observed. C_{60} has a large ionization potential (7.6 eV [3]) compared to most surface work functions and a reasonably large electron affinity (2.7 eV [3]). Therefore, C_{60} would not be expected to donate electrons to surfaces. The difficulty of promoting an electron from the valence band to the conduction band defines the binding interaction between fullerene and substrate. For metals, where promotion of an electron to the conduction band is facile, charge transfer is observed. For semiconductors and insulators, where promotion requires surmounting a band gap, charge transfer would be more difficult. At first sight, this picture is not unreasonable. The binding of fullerenes to metals would be stronger than their binding to insulators and semiconductors. In general, fullerenes have very weak binding to insulators. An exception to this

---

Fullerenes: Chemistry, Physics, and Technology, Edited by Karl M. Kadish and Rodney S. Ruoff.
0-471-29089-0   Copyright © 2000 John Wiley & Sons, Inc.

rule occurs with semiconductors. Many semiconductor surfaces have surface dangling bonds. Very strong covalent bonding between substrate and fullerene can occur in this situation. However, if the surface dangling bonds are "healed" by the restructuring of the surface, the binding can be described by (the lack of) charge transfer. Late transition metals and aluminum also interact with fullerenes via (minimal) charge transfer and (more importantly) via covalent interactions. In general, simple metals and noble metals interact with fullerenes via charge transfer (ionic) bonding; insulators and healed semiconductors interact via van der Waals attractions; and unhealed semiconductors, late transition metals, and aluminum interact with fullerenes via covalent interactions.

The search for application of the interesting and unique properties of fullerenes has led to many promising experiments in the surface science of fullerenes. Perhaps, the simplest is the use of fullerenes as novel sources of pure carbon for the growth of existing materials (i.e., SiC [4], TiC [5], and diamond [6,7]). More exciting is the use of the molecular properties of fullerenes to produce novel devices (i.e., the molecular abacus [8] and optical switching [9]). Because of the range of binding strengths of fullerenes to surfaces, fullerenes have already found uses in scientific applications to preserve sensitive surface structures in harsh (i.e., ambient) environments [10–13].

The outline of this chapter is as follows. First, the interaction of metal surfaces and fullerenes are discussed (Section 13.2). The predominantly ionic bonding of fullerenes on noble metals begins the section and the covalent nature of the aluminum–surface and late transition metal–surface interactions with fullerenes ends the section. Second, the interactions of semiconductor surfaces with fullerenes are addressed (Section 13.3). Silicon surfaces and the growth of silicon carbide have received the most attention. In Section 13.4, the interaction of fullerenes with insulators is described. Section 13.5 describes the use of fullerene films to probe surface processes, such as the growth of new surface structures and the interaction of ionic particles with surfaces. The summary section includes a table of all systems discussed in this chapter and their fullerene binding interactions.

13.2 FULLERENE–METAL INTERACTIONS

13.2.1 Silver

The interaction of C_{60} with the Ag(001) surface has received a good deal of attention quite recently. Goldani et al. [14] have used photoelectron spectroscopy and low-energy electron diffraction and Giudice et al. [15] have used scanning tunneling microscopy (STM) to study the morphology and electronic structure of C_{60} on Ag(001). The photoelectron spectroscopy work has shown that at about 100 K a pseudogap opens at the Fermi level due to charge transfer of about three electrons per molecule (see Fig. 13.1). This is a very exciting result since the transfer of three electrons per molecule is the same transfer that

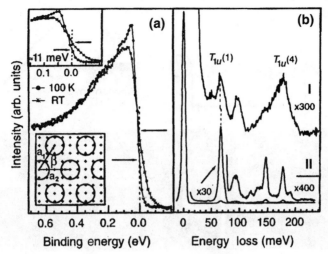

Figure 13.1 (a) Photoemission spectra of one-monolayer (1ML) C_{60}/Ag(100) at 300 K (—*—) and 100 K (—•—). The region near the Fermi level is shown in detail in the top inset, while the inset at the bottom shows the real space ordering of the C_{60} molecules (big circles) on the Ag(100) surface (crosses). $|\mathbf{a}_1| = |\mathbf{a}_2| = 10.42$ Å and $\beta = 56.3°$. (b) High-resolution electron energy loss spectra of 1ML-C_{60}/Ag(100) (I, top) and of a multilayer (II, bottom). Reprinted with permission from Ref. 14.

occurs in superconducting fullerenes. The reduction of the density of states at the Fermi level at 100 K is also consistent with the electronic structure that may be present in superconducting material. This is a tantalizing behavior, though it is not believed that the C_{60}/Ag(001) phase is superconducting. The STM measurement of the C_{60}/Ag(001) phase reveals a $c(6 \times 4)$ structure with a 10.42 Å spacing, much larger than the 10.02 Å spacing of the face-centered cubic (fcc) (111) C_{60} and nearly the spacing of the superconducting Cs_3C_{60}. After annealing to 680 K, the bright and dim C_{60} adsorbates seen in STM were ascribed to different bonding interactions. Also after annealing, intramolecular structure was observed, revealing a nonrotating fullerene and indicating that the charge transfer was thermally activated.

Magnano et al. [16], Purdie et al. [17], and Kovac et al. [18] have extended study on the C_{60}/Ag(110) interaction. The key observation of this work is the population of the π^*-derived lowest unoccupied molecular orbital (LUMO) of C_{60}. The increased density of states observed at the Fermi level is attributed to the LUMO crossing the Fermi level. Previous scanning tunneling spectroscopy (STS) [19] had observed no appreciable Fermi level crossing of the LUMO. The discrepancy is due to the different surface sensitivity of the photoemission and STS measurements. STS probes the top carbon atoms including only carbon nearest-neighbors while photoemission probes all the carbon atoms in the epitaxial layer ($c(4 \times 4)$). The filling of the LUMO points to significant charge transfer in this system.

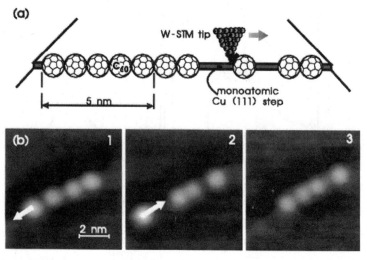

Figure 13.2 (a) Schematic of W-STM tip "pushing" a C_{60} molecule along Cu(111) monoatomic step and (b) sequence of STM images of 4 C60 molecules adsorbed on the lower step on Cu(111). Reprinted with permission from Ref. 8.

In keeping with the charge transfer observed for C_{60}/Ag systems, Caudano et al. [20] have observed a softening (shift to lower frequency) of the "pentagonal pinch" vibrational mode upon C_{60} adsorption on Ag(111) via sum-frequency generation (a nonlinear optical technique). The activity of this mode in sum-frequency generation requires some infrared activity, which is ascribed to charge transfer, as is the softening of the mode.

13.2.2 Copper

One of the most exciting recent developments in fullerene–surface interactions has been the manipulation of individual fullerenes with scanning tunneling microscope tips into a recognizable structure. Gimzewski and co-workers [8] have demonstrated a molecular abacus based on the positioning of C_{60} along monoatomic steps on Cu(111) (see Fig. 13.2). The key interactions responsible for the ability to position C_{60} on Cu(111) are the increased affinity of the C_{60} for step sites, the STM induced formation of kink sites, and the pushing or pulling of the C_{60} by the STM tip. Positioning C_{60} on Cu(111) terrace sites is very tedious since the C_{60} tries to avoid the tip, and motion in many directions is equally likely (i.e., sideways). The step edge keeps the C_{60} moving in a straight line and the STM-induced kink sites hold the C_{60} in place on the step edge until the next tip manipulation procedure, which slides, pulls, or pushes the molecule [21]. Sliding and pulling of C_{60} on Si surfaces was also demonstrated in earlier work [1].

Wang et al. [22] have observed the segregation of C_{60} to domain boundaries in the Cu(111)–(4 × 4) C_{60}/C_{70} phase with STM. Close-packed face-centered cubic (111) C_{60} has a lattice spacing of 10.0 Å, only 2% smaller than 4 times the Cu lattice spacing (10.2 Å); and close-packed face-centered cubic (111) C_{70} has a lattice spacing of 10.63 Å, only 4% larger than 4 times the Cu lattice spacing, if the long axis of the C_{70} molecule is perpendicular to the surface. Mixing of C_{60} and C_{70} would produce an overlayer with very little lattice mismatch; however, the stronger C_{70}–C_{70} interaction leads to segregation of the weaker bound C_{60} to the domain boundaries. An interesting interplay between the adsorbate–adsorbate interaction and the adsorbate–surface interaction produces this structure.

The Cu(110) surface and the Ni(110) surface have very similar atomic structures, which in their clean states do not reconstruct. However, their interactions with C_{60} adsorbates are surprisingly different. Murray et al. [23] have investigated the interaction of C_{60} with these two surfaces using STM and low-energy electron diffraction. On the Cu(110) surface, a distorted hexagonal overlayer, in which every third row of C_{60} molecules is relaxed, is observed. The nearest-neighbor spacing along the rows is 11.2 Å and the nearest-neighbor spacing between rows varies from 9.7 to 11.1 Å depending on which rows are considered. This spacing differs from the face-centered cubic C_{60} (111) spacing of 10.02 Å, since the Cu(110) lattice does not allow for the formation of a simple, undistorted, comensurate hexagonal C_{60} overlayer. The $C_{60}/Ni(110)$ overlayer upon annealing to 575 K forms an overall morphology that is rough compared to Cu(110). The C_{60} molecules are aligned in one-dimensional rows along the $\langle 001 \rangle$ direction. The vertical displacements are 1.3 Å for adjacent rows, which is the monoatomic Ni(110) step height. The spacing along the rows is 10.5 Å, 5% larger than the face-centered cubic $C_{60}(111)$ spacing, but commensurate with three Ni lattice spacings along $\langle 100 \rangle$. The one-dimensional rows are 9.5 Å apart in the plane; but with the vertical displacement, the actual C_{60}–C_{60} spacing is 10.0 Å (the bulk C_{60} distance). This structure can be explained by the surface restructuring via the addition/removal of $\langle 001 \rangle$ Ni rows, which produce the unusual (100) microfacets. These microfacets allow a higher coordination for the C_{60} binding. Why does the Ni(110) surface reconstruct to support the closer packed C_{60} structure and not the Cu(110) surface? The answer lies in the energetics. The bonding between C_{60} and Ni is stronger than the interaction between C_{60} and Cu. Thus, the energy cost to rearrange the Ni surface atoms is more than compensated for by the increase in the Ni–C_{60} binding. Because of the charge transfer, the LUMO of the C_{60} overlaps the Ni d band. The surface rearrangement (more open surface) may also raise the energy of the d band, leading to an even stronger C_{60}–Ni interaction on the reconstructed surface.

Inverse photoemission measurements of the $C_{60}/Cu(111)$ system have been performed by Tsuei and Johnson [24]. They have observed a new state with parabolic dispersion below the vacuum level in the (4×4) C_{60} overlayer. Since the effective mass of this state is close to unity (free electron-like) and the overlayer is metallic, this result suggests that this new state is an image potential state (e.g., see Ref. 25).

13.2.3 Gold

In another exciting development, the electrical properties of a single C_{60} molecule on a Au(110)–(1×2) surface have been measured with a scanning tunneling tip [26,27]. Probing an isolated C_{60} molecule on a Au(110)–(1×2) surface with a STM tip affords the determination of nanoscopic electrical properties. Two important conclusions were revealed. First, the compression of the C_{60} molecule by the tip results in a shifting of the energy levels such that the band gap is reduced. The effect of this induced gap shrinking explains why it is sometimes possible to image bulk insulators with a STM. Second, the measured 54 MΩ resistance of the junction corresponds to the apparent resistance of a single C_{60} molecule. In addition, inelastic scattering of the electrons in the fullerene molecule appears to be very rare at low current intensity in the compressed configuration.

Hunt et al. [28] have measured the coupling of the interaction of fullerene and a metal substrate and subsequently adsorbed fullerene layers. They have concluded that on the Au(110) substrate only the fullerenes directly in contact with the surface display any significant change in their electronic structure. The first monolayer is metallic in electron energy loss measurements, whereas the second monolayer exhibits excitonic features characteristic of a semiconductor. In photoemission measurement, the highest occupied molecular orbitals (HOMOs) of the first monolayer shift 0.5 eV in binding energy and broaden; whereas the second monolayer exhibits the photoemission spectrum of the multilayer. This is also true for a Cs-precovered (4MLs) Au(110) surface at 98 K. At higher temperatures, Cs diffuses into the overlayer.

Inverse photoemission spectroscopy is a powerful tool to probe the unoccupied electron structure of the fullerene–substrate system [2]. Pedio et al. [29] have extended these studies, applying this technique to the C_{60}/Au(110) (2×1) system. There is evidence of shifts in the π-derived LUMOs toward the Fermi level, which is expected due to charge transfer. In the single monolayer, the molecular orbitals broaden. The broadening could be due to a rehybridization of the orbitals with metal d orbitals (i.e., some covalent bonding), or splitting of the degeneracy of the orbitals due to the charge transfer. Upon heating, the surface reconstructs [30] to form a (6×5) structure. The broadening, due to rehybridization or splitting of the degeneracy, is still detectable but reduced on the (6×5) surface, suggesting the dominance of charge transfer over covalent bonding in this case. This was also observed in photoelectron spectroscopy and inverse photoemission by Purdie et al. [17].

In a very interesting spin-off experiment, Hevesi et al. [31] have measured the cross section for the impact-excited electron energy loss channels for a thick

$C_{60}(111)$ overlayer on Au(110). A negative ion resonance evidenced by a dramatic cross-section enhancement at a primary electron energy of $\sim$2.9 eV is observed. The formation of a negative ion state can prolong the effective interaction time of the incident electron with the intramolecular vibration, which is probed by the high-resolution electron energy loss spectroscopy. These results are consistent with a resonant process when the incoming electron is trapped in the third lowest unoccupied molecular orbital band of the C_{60} solid. The third LUMO band is just above the vacuum level. This resonance may also account for the very limited inelastic mean free path (10–20 Å) of low-energy electrons in fullerene-based materials [32].

13.2.4 Aluminum

Binding of fullerenes to the Al(111) and Al(110) surfaces is a very interesting case because of the enhanced strength of the substrate–C_{60} bond relative to the substrate–substrate bond. It also seems clear that more than a charge-transfer model is required to describe the binding of fullerenes to aluminum metal surfaces. Based on photoelectron spectroscopy and X-ray absorption spectroscopy, Maxwell et al. [33–35] have determined that the bonding has covalent character. The HOMO and the LUMO both shift to increase the C_{60} gap, suggesting a rehybridization of the orbitals with the aluminum orbitals. This observation is consistent with early work on aluminum foils [12], where a 0.3 eV shift in the π-derived HOMOs was observed. Also very provocative is the shift of the Al 2p orbital to lower binding energy. Maxwell et al. [33,34] have ascribed this shift to a reconstruction of the aluminum substrate, since the Al 2p shift is only observed on the C_{60}/Al(111)–(6 × 6) surface; whereas Hamza et al. [12] have ascribed the shift to screening of the Al 2p hole by the highly polarizable fullerene. In either case, a change in the substrate binding occurs because of interaction with the fullerene. In an extension of the work of Maxwell and colleagues, Johansson et al. [36] have observed formation of the (6 × 6) overlayer after annealing the C_{60}/Al(111) surface at 490 K and not at room temperature. This observation is consistent with the motion of aluminum substrate atoms, as suggested earlier [12,33]. In addition, the STM measurements showing intramolecular structure indicate that C_{60} is not freely rotating on the surface even at room temperature and showed preferential binding to the lower terrace of step edges. Unlike silicon, where there is also strong binding, the fullerene is mobile on the Al(111) surface, at least until the surface is annealed to 490 K.

Maxwell et al. [34] have measured the work function change for Al(111) and Al(110) upon adsorption of C_{60}. By comparing their measurements with those from other substrates [17,34], they conclude that the work function for a C_{60} layer on a metal surface is to first order due to the fundamental dielectric response of the fullerene overlayer. The chemical binding at the interface contributes only slightly to the observed $\sim$5 eV work function. In the past, change in the work function has often been associated with a change in the surface dipoles. Thus, for surfaces with work functions less than 5 eV, a change on adsorption to 5 eV would be interpreted as a charge transfer from the metal to the fullerene layer. On the other hand, for surfaces with work function greater than 5 eV, a change on adsorption to a work function of 5 eV would be interpreted as a charge transfer from the fullerene layer to the metal, as was proposed for Rh(111) [37]. The latter is unlikely the case, since the observed work function is due to the fundamental dielectric properties of the fullerene and not the interaction of the fullerene with the surface.

In an exciting development, Fasel et al. [38] have demonstrated the power of X-ray photoelectron diffraction to determine the molecular orientation of adsorbed C_{60} on Al(111), Al(001), Cu(111), and Cu(110) surfaces. The strong forward-focusing of the emitted electrons allows a straightforward structural interpretation of the X-ray photoelectron diffraction data. C_{60} adsorbates on Al(111) have a sixfold symmetry with a six-membered ring facing the surface with C–C bonds lying along the $\langle 10\bar{1} \rangle$ direction, which was also observed by STM [36]. In agreement with STM measurements [1], the orientation of C_{60} on Cu(111) is with a six-membered ring nearest the substrate, but with the C–C bonds perpendicular to the $\langle 10\bar{1} \rangle$ direction. On Cu(110) and Al(001), the adsorption is on 5-6 bonds and on edge atoms, respectively (see Fig. 13.3 for

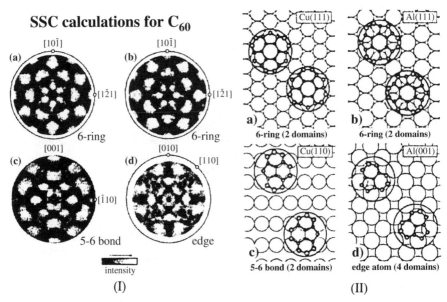

Figure 13.3 (I) Single scattering cluster calculation that reproduces the experimental X-ray photoelectron diffraction patterns for C_{60} on Cu(111), Al(111), Cu(110), and Al(001). (II) Molecular orientations corresponding to these calculations for C_{60} on Cu(111), Al(111), Cu(110), and Al(001). Substrate lattice spacing and C–C distances are properly scaled. For clarity, only the atoms closest to the substrate are shown as black dots. The molecule substrate registry and intermolecular distances are arbitary. Reproduced with permission from Ref. 38.

the molecular orientations). The forward scattering distribution observed in photoelectron diffraction is simpler to relate to atomic geometry than the local density of states, which is determined by scanning tunneling microscopy (see Section 13.3.1). X-ray photoelectron diffraction can give an unambiguous determination of the structure of fullerene absorbates on surfaces.

13.2.5 Transition Metals

Cepek et al. [39] have investigated the binding of C_{60} to Pt(111) and Ni(110) surfaces. These experiments reveal that fullerenes are bound rather strongly to both surfaces. The vibrational modes of one-monolayer (1ML) C_{60} on Pt(111) are, within experimental error, the same as that for the C_{60} multilayer. This result suggests that there is very little charge transfer between C_{60} and Pt. Perhaps the large work function of Pt(111) plays a role in the lack of charge transfer observed, although earlier experiments showed little effect of work function on electron transfer [40] (also see Section 13.2.3). The absence of electronic excitation features in the electron energy loss spectra in the 1–7.5 eV energy range for the 1ML-C_{60}/Pt(111) suggests a rehybridization of the molecular orbitals of C_{60}. The valence band photoemission measurements with broadening of the HOMOs at 2 and 3 eV binding energy support this picture of covalent bonding. In addition, C_{60} does not desorb from this surface even after heating to 1050 K. C_{60} decomposes to a graphitic overlayer at 1050 K, a temperature consistent with previous measurements of the stability of the C_{60} solid [41] and decomposition of C_{60} on Si [42,43].

Previous studies of the C_{60}/Ni(110) surface [40] have shown a strong binding interaction with high-resolution electron energy loss spectroscopy. The strong bonding was attributed to charge transfer because of the softening of the vibrational modes. Perhaps covalent bonding might also lead to a softening of the modes. Cepek et al. [39] have observed fragmentation of C_{60} at 760 K on Ni(110) and formation of a carbidic overlayer with diffusion of carbon into the bulk at >800 K. Surface-catalyzed fullerene fragmentation as seen on Ni(110) and Pt(111) has been observed previously by Hamza et al. [44] on silica and highly oriented pyrolytic graphite with activation energy of ~2.5 eV, which is similar to solid C_{60} decomposition [41]. The thermodynamics of carbide formation can significantly lower the activation barrier for C_{60} decomposition as observed on stainless steel (C_{60} decomposes at 650 K on Ar^+ sputter-cleaned stainless steel [45]).

Norin et al. [46] have speculated on the possibility of producing transition metal fullerides. The high cohesive energy of transition metals makes transition metal fullerides unstable to formation of a two-phase system of transition metal and solid C_{60}. Norin et al. [46] have produced metastable Ti_xC_{60} films by co-sublimation of Ti and C_{60} onto substrates at ~400 K in ultrahigh vacuum. Raman spectroscopy showed a softening of the "pentagonal pinch" mode ($A_g(2)$) upon incorporation of Ti, consistent with a partial charge transfer from Ti to C_{60}. The softening was less than for alkali metal fullerides, consistent with some covalent contribution to the binding. Maximum Ti content was about 5 at% ($Ti_{\sim 3.5}C_{60}$); higher Ti content led to the formation of TiC. The metastable transition metal fulleride was formed in this case because the effects of the high cohesive energy of titanium were minimized by cosublimation of Ti and C_{60}.

13.3   FULLERENE–SEMICONDUCTOR INTERACTIONS

13.3.1   Silicon

In early work on fullerene adsorption on Si substrates, ordered overlayers were not observed. By careful preparation of the overlayer, Klyachko and Chen [47] have been able to study the ordering of C_{60} on anisotropic surfaces, Si(100)–(2 × 1) and Ge(100)–(2 × 1). While the two surfaces and overlayer structures are very similar (the Ge(100)–(2 × 1) lattice is 4% larger than Si(100)–(2 × 1)), the delicate balance between short-range repulsive forces and long-rang attractive forces between adsorbates determines that long-range order is easier to achieve on the Ge(100)–(2 × 1) (i.e., stable) than on the Si(100)–(2 × 1) surface (i.e., unstable). Both Ge(100) and Si(100) heal one of the dangling bonds per surface atom by reconstructing the surface with surface dimers. On both surfaces, C_{60} adsorbs between dimer rows and at the center of four adjacent dimers. Indeed, local ordering observed previously in C_{60}/Si(100)–(2 × 1) was assigned to a (3 × 4) superstructure [1]. The potential barrier on the dimer rows is stronger than the intermolecular forces and confines the molecule to the trough. However, molecular ordering along the trough is dependent on the intermolecular forces. The ideal (3 × 4) structure is shown in Figure 13.4. The nearest-neighbor distance is r_{12} and the next-nearest-neighbor distance is r_{13}. For silicon the ideal (3 × 4) would have r_{12} equal to 9.60 Å and r_{13} equal to 11.52 Å. Since r_{12} is much less than the equilibrium distance, there is a net repulsive force in this configuration, causing the C_{60} molecules to move away from each other along the troughs until the weakened repulsive force between 1–2 pairs is balanced by the attractive force exerted on the 1–3 pair. On the other hand, the ideal (3 × 4) structure on Ge would have r_{12} equal to 10.0 Å and r_{13} equal to 12.0 Å. In this case r_{12} is close to the equilibrium position. Thus, the attraction between 1–3 pairs would cause the structure to compress until the attractive and repulsive forces balance. Therefore, it is difficult to produce a long-range, well-ordered C_{60} overlayer on the Si(100)–(2 × 1) surface.

Yao et al. [48,49] have observed different bonding configurations for C_{60} on Si(100) by observation of different intramolecular structures via scanning tunneling microscopy. They assign the C_{60} adsorbed at room temperature to a "physisorbed" molecule. The C_{60} adsorbed between dimer rows has been observed earlier by many workers [1,47] and is not particularly mobile on the surface. That this molecule should be assigned as a physisorbed state is questionable. The fact that intramolecular structure can be observed suggests that the molecule is not rotating on the surface, as would be expected for a weakly

(A)

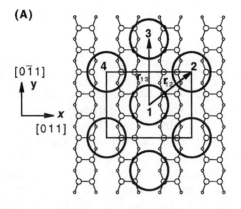

(B)

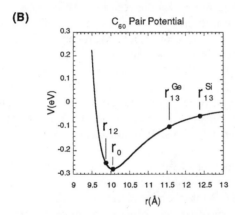

Figure 13.4 (a) The ideal (3×4) C_{60} superlattice, marked by the rectangle, on the (2×1) dimer row surface of Si(100) or Ge(100). The C_{60} are represented by 7 Å large circles scaled in proportion to the substrate lattice. (b) The Girifalco potential for an isolated pair of C_{60} molecules separated by distance r. The positions r_{12} and r_{13} are indicated. Reprinted with permission from Ref. 47.

bound species. This physisorbed molecule does not desorb from the surface; it can only be converted to the other "chemisorbed" states by annealing. The transformation to the chemisorbed state occurs suspiciously close to the temperature at which reaction of C_{60} to form SiC begins [43]. The chemisorbed C_{60} molecules are bound on top of dimer rows and exhibit different intramolecular structure from the C_{60} molecules bound in the trough. The change in intramolecular structure is a change in the local density of states indicative of a change in the bonding of the molecule, perhaps similar to the observations of cage opening on Si(111)–(7×7) [42].

In an extension of the work by Hong et al. [10,11], Ma et al. [50] have shown that the protective properties of a fullerene overlayer are achieved on the Si(110)–(16×2) surface. In addition, the adsorption of C_{60} on Si(110)–(16×2)

is essentially identical to the (111) and (100) surfaces. LeLay et al. [13] have also demonstrated the encapsulation of C_{60} on Si(111)–($\sqrt{3} \times \sqrt{3}$) Rotated 30°-Ag, which nicely preserves the surface reconstruction under ambient conditions. The thermal stability of the encapsulating layer was also studied; upon heating, a replacement reaction takes place between Ag and C_{60}. Ag desorbs at about 850 K and a strongly bound C_{60} monolayer remains on top of the silicon surface.

Dumas et al. [51] have studied the interaction of C_{60} with hydrogen-terminated Si(111)–(1 × 1) substrate. Upon adsorption of C_{60}, the SiH stretching mode keeps its intensity and its orientation perpendicular to the plane of the surface. The binding of C_{60} to this surface is weak (van der Waals) since the dangling bonds are passivated by hydrogen. The width of the elastic scattering peak in the high-resolution electron energy loss spectrum indicates an ordered overlayer. However, the lack of a low-energy electron diffraction pattern suggests that the adsorbed molecules are rotating. An extension of this study was carried out by Sakamoto and co-workers [52,53]. By passivating dangling bonds of the Si(111)–(7 × 7) with atomic hydrogen, a Si(111)–δ(7 × 7) surface can be produced. There is also *no* softening of the vibrational modes of C_{60} on the H-terminated Si(111)–δ(7 × 7) surface, whereas the softening of the vibrational modes on Si(111)–(7 × 7) is observed. The softening is ascribed to a three-electron charge transfer from the silicon dangling bonds to C_{60}. The number of electrons transferred is determined by Raman frequency shifts observed in alkali-doped C_{60} films. It is not clear if a covalent bond will show similar mode softening.

Ever since the structure of fullerenes was proposed and gram quantities of material have become available, researchers have tried to image the cage structure of fullerenes. While the photoelectron diffraction described earlier (see Section 13.2.4) is the most straightforward technique for imaging the fullerene structure, it is a reciprocal-space technique. Scanning tunneling microscopy is a real-space technique with the capability to determine structure on the atomic scale. However, STM images the local density of states, which can be related to atomic structure via first principles calculations. The first such calculation was performed by Yajima and Tsukada [54] for C_{60} on Si(100)–(2 × 1) using the discrete-variational Xα method with a linear combination of atomic orbitals basis. Experimental STM images were reproduced with a strong C_{60}–Si interaction responsible for the observed features (four stripes [55]). For those systems in which an ordered overlayer is produced, photoelectron diffraction will be the method to determine intramolecular structure and orientation. For systems with only local orders, a combination of first principle calculations and STM will be the method to determine intramolecular structure and orientation.

Attempts are now being made to model the behavior of C_{60} on surfaces with molecular dynamics simulations. One such attempt by Katircioglu and Erkoc [56] did predict the decomposition of C_{60} on Si(111) at around 900 K. Unfortunately, the simulations also show the incorporation of C_{60} into the silicon surface at 300, 500, and 700 K, in clear contradiction to experimental observations [1,48]. Clearly more work is needed.

13.3.2 Silicon Carbide

Li et al. [57] have studied the adsorption of C_{60} on 6H-SiC(0001) with STM. The 6H-SiC(0001) surface exhibits a variety of surfaces phases. The $(\sqrt{3} \times \sqrt{3})$ is believed to be C-rich with little corrugation (0.2 Å) and the (3×3) surface is believed to be Si-rich with large corrugation (2.5 Å). The binding to the two investigated 6H-SiC(0001) surfaces is very similar to the binding to Si, in that the intramolecular structure may be observed, suggesting the fullerenes do not rotate on the surface. In addition, the observed intramolecular structure (three stripes in the local density of states) is similar on silicon surfaces and SiC surfaces. Also, the fullerene molecules showed no evidence of surface diffusion even upon heating to 600–700 K as is observed on silicon surfaces [1]. Although the atomic structure of the surface phases of 6H-SiC(0001) is not well known, the STM results suggest that C_{60} binds to silicon sites on both SiC surfaces.

13.3.2.1 Silicon Carbide Growth In earlier work, many researchers [1,43,58–63] have demonstrated that it is possible to grow epitaxial, cubic silicon carbide on silicon substrates using fullerenes as the carbon source. The growth is characterized by diffusion of Si atoms from the substrate to the growing surface. The growth is also shown to be selective; growth is observed on silicon, but not on silicon dioxide between 900 and 1200 K [43,60]. The low temperature at which this growth process occurs makes this an attractive process for the formation of epilayers of this high-temperature, high-power material for device fabrication.

Henke et al. [64] and Geier et al. [65] have extended their work on the growth of epitaxial 3c-SiC with C_{60} on Si(111) to include C_{70} and Si(100). The structural quality of the films was shown to be sensitive to the growth temperature between 1000 and 1250 K, with the best films grown at 1150 K. Interestingly, they show that good epitaxial 3c-SiC films can be grown at lower temperature (1000 K) with C_{70} as the carbon source. This is surprising since the vast majority of the impinging C_{70} are unreactive until a surface temperature of 1150 K [66,67] (see below). The lower reactivity may control the growth rate in such a way as to promote the epitaxial growth.

In recent work, Morse et al. [68] have shown that the mechanical properties of SiC films grown from fullerene precursors are very close to bulk SiC values, suggesting the use of these films in microelectromechanical systems. The mechanical properties (hardness, elastic modulus, and coefficient of friction) were determined via atomic force microscopy based nanoidentation. The measured value of the elastic modulus (310 GPa) suggested that the porosity of the films was 18%. Levinson et al. [4] have shown that the growth of SiC via cosublimation of silicon and C_{60} can produce fully dense films with an elastic modulus of 450 GPa, the value of bulk SiC. In addition, cosublimation of silicon can be used to eliminate interface voids during the growth process of SiC via fullerenes in the absence of cosublimation. If silicon is supplied to the growing film from the substrate via adatom diffusion, voids will form at the interface between silicon and SiC. The adatoms are usually supplied from defects or the free surface or other interfaces. Unfortunately, the cosublimation of silicon during growth removes the selectivity of the growth process. In the absence of cosublimation, C_{60} will react on silicon at low substrate temperature (>900 K), but C_{60} will not react with silicon dioxide until a high temperature is reached (>1200 K). Cosublimation of silicon leads to blanket deposition. Levinson et al. [4] have shown that patterned SiC can still be grown with fullerene precursors and cosublimation of silicon with a lift-off technique.

Balooch and co-workers [66,67] have extended their work on the growth of silicon carbide on silicon with C_{60} to include higher fullerenes. Interestingly, the growth of SiC via C_{84} precursors is very similar to that of C_{60}; however, the reactivity of C_{70} to form SiC is strikingly different. In agreement with earlier STM work [69], C_{70} did not completely react with the silicon substrate until a surface temperature of 1200 K was reached. This was attributed to the stability of the C_{70} molecule oriented "side on" on the surface. Measurements of the angular distribution of scattered fullerenes [67] show that only a trace amount of C_{70} reacts at 950 K on silicon. The trace reaction roughens the surface (by forming SiC) so that the angular distribution of the scattered molecules becomes cosine. The change in the angular distribution accounts for the 20% drop in the scattered C_{70} signal at 950 K and not the reactivity of the molecule [67]. It seems that the "end on" configuration is only sampled rarely during the surface interaction.

Li et al. [57] have shown that SiC does not grow on the 6H-SiC(0001)–($\sqrt{3} \times \sqrt{3}$) (carbon-rich) surface even at surface temperatures of 1300 K. However, on the Si-rich (3 × 3) surface, epixtaxial growth of SiC is observed, in accordance with the observations of Levinson et al. [4] of silicon diffusion in the growth of SiC on silicon (i.e., silicon must be present at the growth surface for growth to continue). Recently, Pascual et al. [70] have observed the (3 × 3) phase after annealing to high temperature (>1150 K) on the SiC(111) surface formed by growth on Si(111) with C_{60}. A second ($2\sqrt{3} \times 2\sqrt{3}$) phase, the origin of which is not yet known, is also observed. Sakamoto et al. [71] have also measured the vibrational modes of the $C_{60}/3c$-SiC(111) system as a function of annealing temperature.

McGinnis et al. [72] have used thin films of C_{60} to serve as a mask for tungsten chemical vapor deposition on silicon substrates at 650 K. Since an unstabilized fullerene film would rapidly sublime at 650 K, the fullerene mask was stabilized by exposure to oxygen and ultraviolet light. The oxygen/ultraviolet exposure very likely polymerized the film [3]. The stabilized mask was completely selective to tungsten deposition. These masks may find application as effective masks at high temperature, because of the ease with which they can be produced.

While not specifically based on a fullerene–surface interaction, Gruen et al. [6] have grown nanocrystalline diamond films on silicon substrates by plasma-enhanced chemical vapor deposition with fullerene precursors. An argon–C_{60} (0.14% atomic carbon content) microwave plasma fired at 1120 K with 1500 W for 1–3 hours leads to a diamond growth rate of 1.2 μm/h. Key here is the growth of a diamond film in the absence of hydrogen and oxygen. It is speculated that the $Ar-C_{60}$ plasma produces C_2, which is incorporated in the growing diamond. Very recently, Erdemir et al. [7] have demonstrated the excellent mechanical properties of these nanocrystalline diamond films especially after laser polishing.

13.3.3 GaAs

Normally, the interaction of fullerenes with GaAs surfaces is so weak that the fullerene–fullerene interaction dominates the observed structure, which is often a fcc (111) surface structure. In the case of the GaAs(001)–(4 × 2) As-rich surface, a second "strained" fullerene structure is observed [73]. The (4 × 2) reconstruction results in either As dimers (As-rich surface) or Ga dimers (Ga-rich surface). C_{60} adsorbs in the troughs between dimer rows as on Si(100)–(2 × 1) and Ge(100)–(2 × 1). On the Ga-rich surface, which is electron deficient and hence has virtually no charge-transfer bonding, C_{60} forms a quasi-close-packed structure. However, on the As-rich (4 × 2) surface, which is an electron-rich surface, a fcc (110) C_{60} phase is observed as with $TiO_2(100)$–(1 × 3) (see below). Interestingly, in this case it is not only the templating of the geometric substrate but also the increased substrate–C_{60} binding (due to charge transfer) and the interaction between surface dipoles that leads to the rarer fcc (110) fullerene surface phase.

13.3.4 Germanium

Dunphy et al. [74] have demonstrated the bidirectional step-flow growth on Ge(100)–(2 × 1) and GaAs(110) of heteroepitaxial C_{60} layers. This bidirectional step-flow growth is a direct consequence of a large out-of-plane lattice mismatch and thus should be common in the growth of large molecules and clusters on monoatomically stepped substrates. C_{60} molecules adsorbed at a step edge protrude well above the step edge's top, creating an inverted step edge. The inverted step edge acts as a nucleation site for subsequent growth on the upper terrace. Thus, there is step flow in the direction of decending steps (normal step-flow growth) and step flow in the direction of ascending steps. These observations may resolve some of the confusion as to whether C_{60} adsorbates bind to upper or lower step edges (i.e., both upper and lower).

13.4 FULLERENE–INSULATOR INTERACTIONS

Murray et al. [75] have investigated the possibility of producing various C_{60} structures by judicious choice of the growth substrate or template. They have chosen a $TiO_2(100)$–(1 × 3) surface to act as a template. The interplay of the intramolecular forces (van der Waals attractions), the fullerene–substrate bonding, and the bonding within the substrate (i.e., the sensitivity to surface reconstruction) will determine the C_{60} overlayer structure. In the case of the $TiO_2(100)$–(1×3) surface, the fcc (110) overlayer of C_{60} is formed. Many previous observations of closed-packed C_{60} overlayers have been the fcc (111) plane (e.g., GaAs [76], mica [77]).

Kuhnke et al. [9,78,79] have investigated the dynamics of the optical properties of C_{60} at a quartz (SiO_2) interface. In general, the optical properties of the isolated fullerene and the fullerene solid are quite similar owing to the relatively weak van der Waals interactions between molecules. On metal surfaces (where charge transfer is observed) and on semiconductor surfaces (where covalent bonding is observed), the optical properties may be significantly altered. On SiO_2 surfaces, the interaction between fullerenes and the substrate is still weaker than the C_{60} intramolecular attraction [80]. Even with this weak interaction, changes in the optical properties are observed due to the change in the site symmetry and the presence of a different dielectric medium. The decay time of the quenching of the second harmonic generation (due to population of the excited state) in C_{60} is nearly three orders of magnitude faster at the SiO_2 interface. This suggests that the excitonic lifetime at the interface is greatly reduced due to the proximity to the dielectric material. Similar behavior has been shown for C_{60}/alkanethiol/Au "sandwiches" [79]. Here, the decay time of the second harmonic quenching increases from 44 ps at 1 nm alkanethiol thickness to 520 ps at 2.4 nm alkanethiol thickness. This also has practical consequences. An optical switch with 40–70 ps switching time could be built based on this phenomenon.

Since the early work of Wang [81], there have been some efforts to characterize the C_{60} enhancement of the photoconductivity of organic polymers and molecules. Recent experiments by Schlebusch et al. [82] using ultraviolet photoelectron spectroscopy and X-ray absorption near-edge spectroscopy have shown that photoconductivity enhancement is possible through an electron-transfer mechanism from excitonic states of the organic photoconductor and the lowest lying unoccupied molecular orbital of C_{60}. While some organic photoconductors and fullerenes readily mix, Cu-phthalocyanine and C_{60} form a stable interface, making this a possible heterojunction for photovoltaic applications.

In a preliminary study, based on the work of Ohno et al. [83], Wang and Wang [84] have observed the growth of amorphous TiC by annealing to 500 K a Ti overlayer on a C_{60} overlayer on Ti metal. Norin et al. [5] have demonstrated a very exciting application of these results by using C_{60} as a source of carbon for growth of epitaxial titanium carbide on MgO(001). Coevaporation of C_{60} and titanium onto a MgO(001) crystal resulted in the epitaxial growth of TiC at a temperature as low as 700 K, hundreds of degrees lower than that in previous reports of epitaxial TiC growth. The TiC films were smooth, adhesive, and electrically conducting. Interestingly, the authors propose that the binding of Ti to C_{60} may facilitate the epitaxy.

13.5 FULLERENES AS PROBES OF SURFACE PROCESSES

The fullerenes are proving very useful in the scientific community as a source of well-controlled thin films. As discussed earlier, fullerene thin films proved useful in encapsulating fragile surface reconstructions. In a recent development, fullerene layers on substrates are functioning as templates for the formation of novel structures.

The highly corrugated surfaces of crystalline overlayers of large molecules could be the template for quantum dots. Klyachko and Chen [85] have explored this possibility with C_{60} overlayers acting as a template for Si nanoclusters. The first striking result here is that the C_{60} overlayer is sufficiently stable to support the growth of clusters. The second is that the Si clusters have enhanced contributions from s states in the Auger electron spectrum, suggesting less sp hybridization in the Si clusters than in bulk silicon. Third, scanning tunneling spectroscopy of the clusters indicate that they have large band gaps, suggestive of quantum confinement for small particles [86]. And the fourth is the tantalizing result of stepped conductance at higher sample bias, suggestive of the formation of a single electron tunneling junction.

Ruan and Chen [87] have discovered a self-limiting growth process of Au clusters on a fullerene layer. This process could potentially be used to fabricate C_{60}-passivated monodisperse metal clusters. Room temperature deposition of Au on a thin C_{60} film grown on a Si(111) substrate leads to the formation of three-dimensional Au clusters on the C_{60} film. As the cluster grows, its facets become binding sites that attract C_{60} molecules, detaching themselves from step edges and diffusing on the surface. The nested clusters have lateral dimensions of 30–70 Å. The binding of C_{60} to Au has been demonstrated to be via charge transfer (i.e., ionic). A small cluster must be large enough that the transfer of electrons to the C_{60} molecules forming the nest does not destabilize the Au cluster. Thus, binding of C_{60} to the Au cluster cannot occur until a certain size is reached. Once that size is reached cluster growth ends because the facets become terminated with C_{60}.

Using GaAs(110) as a template, Biermann et al. [88] have investigated the interaction of Fe with close-packed fcc (111) C_{60} films. The Fe atoms clustered on the surface and exhibited a rehybridization with the π orbitals of the C_{60} substrate in ultraviolet photoelectron spectroscopy. Interesting here is the relatively strong coupling between the fullerene and the iron, but the Fe atoms still minimize the surface energy by forming clusters. The formation of clusters on the fullerite film seems to be a rather general phenomenon.

Another area where fullerenes are finding wide use is in understanding the interaction of ions with surfaces. Work on the radiation damage of fullerenes by energetic ions does not yet reveal whether the damage process is dominated by electronic or nuclear energy transfer. Naturally, the stability of the fullerene to radiation damage is also of fundamental importance in understanding the properties of this carbon allotrope.

TABLE 13.1 Experimental Damage Cross Sections for Various Incident Ions and Energies and Charge States

Ion	Energy (keV)	Experimental Cross Section (cm^2)	S_{elect}[a] $(eV/\mathring{A})$	S_{nucl}[a] $(eV/\mathring{A})$	Reference
H^+	100	6.1×10^{-17}	11.7	0.0104	92
H^+	160	5.0×10^{-17}	11.4	0.0122	93
He^+	300	2.2×10^{-16}	31.4	0.0974	93
Li^+	30	6.9×10^{-16}	15.1	1.8	94
C^+	300	2.05×10^{-15}	54.8	1.84	93
Ar^+	220	3.85×10^{-14}	59.2	32.4	93
Ar^+	2	0.85×10^{-14}	10.1	55.6	95
Xe^+	620	6.0×10^{-13}	223	156	96
Xe^{44+}	44	4.89×10^{-13}	242	168	91
Xe^{44+}	308	1.28×10^{-12}	483	185	91
I^{10+}	55000	3.2×10^{-12}	875	10.4	97

[a]For singly charged ions the nuclear and electronic stopping (S_{elect}, S_{nucl}) is calculated by TRIM [89]. For multiply charged ions the electronic stopping is from Ref. 90. and the nuclear stopping from TRIM.

The response of fullerene films under bombardment of singly and multiply charged ions has been investigated through a broad spectrum of energies (2 keV to 55 MeV) by various groups summarized in Table 13.1. The underlying question is to identify the energy dissipation channel—nuclear or electronic stopping—responsible for the destruction of fullerenes. For the case of energetic protons [92] and swift heavy ions [94], it has been proposed that electronic energy transfer tends to dominate the damage process. Conversely, for the case of light and intermediate ions, nuclear energy-transfer processes dominate fullerene destruction. Interestingly, there is sufficient energy transfer in all cases (both electronic and nuclear) to destroy the fullerene cage. While the fullerenes are damaged with relative ease (compared to graphite and diamond), they can dissipate a substantial amount of nuclear and electronic energy before being destroyed.

Newman et al. [91] have investigated the effects of electronic stopping on the destruction of fullerenes by using slow, highly charged ions. By varying the charge of the incident ion, the electronic stopping can be varied independent of the nuclear stopping to first order. They have found, in agreement with Fink et al. [94], that nuclear stopping is approximately 95% efficient in damaging fullerenes with a near linear dependence on nuclear stopping power; whereas electronic stopping is about 0.1% efficient in damaging fullerenes with a greater than quadratic dependence on electronic stopping power.

Impact of low-energy ions with free fullerene molecules results in the "evaporation" of two carbon atoms or C_2 [98]. The fragment remains otherwise intact.

$$C_{2n} \rightarrow C_{2(n-1)} + 2C \text{ (or } C_2) \tag{13.1}$$

Similar behavior has been observed in surface scattering experiments [44], where $C_{2(n-1)}$ fullerenes (C_{58}) are observed. The surface showed a buildup of carbon. Whether C_1 or C_2 fragments are removed is unknown. In the sputtering of positively charged fullerene and fullerene fragments from fullerene films, again the $C_{2(n-x)}^+$ behavior is observed. See Schenkel et al. [99], for example. In this case, it is also not known if the fullerene becomes charged before, during, or after evaporation of carbon atoms. Newman et al. [91] have shown that for

TABLE 13.2   Summary of results for C_{60}/Surface Systems after Refs. 3 and 34

Substrate	Desorption Temperature T_D (K)	Type of Binding	Mobility at Room Temperature	Reference
C_{60}	500–600	van der Waals		103
Graphite	550–620	van der Waals	Mobile	104, 105
GeS(001)	490	van der Waals	Mobile	2
GaAs(100)–As		Ionic	Mobile	73
GaA(100)–Ga		van der Waals	Mobile	73
GaAs(110)		Weak	Mobile	74
SiO₂	470	Weak	Mobile	106
Aluminum	750[a]	Strong		12
Al(110)	730	Covalent	Mobile	34
Al(111)	730	Covalent	Mobile	34
Ag(110)		Ionic	Mobile	16–18
Ag(111)	770	Ionic	Mobile	107
Au(110)	800	Ionic	Mobile	30, 108
Au(111)	770	Ionic	Mobile	107
Cu(110)	730	Ionic	Mobile	23
Cu(111)		Ionic	Mobile	109
Si(100)	Dissociation at 1070	Covalent	Immobile	106, 109
Si(111)	Dissociation at 1100	Covalent	Immobile	42, 61, 110, 111
Ge(111)	970	Covalent	Immobile	110, 111
Ge(100)			Low mobility	74
Stainless steel	Dissociation at 650			45
Ni(110)	Dissociation at 760	Covalent	Low mobility	39, 40
Pt(111)	Dissociation at 1050	Covalent	Low mobility	39

[a] No desorption was observed; carbon was removed from the surface as measured by Auger electron spectroscopy.

doubly charged parent ions the evaporation process involves C_1, C_2, and even higher C_x $(x > 2)$ carbon fragments, as opposed to sequential C_2 evaporation observed for charged fullerenes in the gas phase [100–102]. There is still work to be done to fully understand the fullerene fragmentation process.

13.6 SUMMARY

In their recent paper, Maxwell et al. [34] have assembled a summary of many C_{60}–surface interactions. Their summary is supplemented with additional information and presented in Table 13.2. The fullerene/surface systems are grouped roughly in order of the strength of the fullerene–substrate interaction. The bonding is characterized as weak (predominantly van der Waals interactions), ionic (intermediate strength, predominantly charge-transfer complexes), or covalent (strong interaction with rehybridization of the molecular orbitals). The desorption or dissociation temperature is also listed. The mobility of the fullerene at room temperature is a measure of the corrugation of the binding potential. Generally, as the binding becomes stronger, the mobility decreases.

As more of the surface properties of fullerenes become known, application of this material to devices becomes more promising.

ACKNOWLEDGMENTS

The author would like to thank Dr. Mehdi Balooch and Dr. Thomas Schenkel for helpful comments during the preparation of the manuscript for this chapter. This work was performed under the auspices of the U.S. Department of Energy at Lawrence Livermore National Laboratory under contract number W-7405-ENG-48.

REFERENCES

1. T. Sakurai, X.-D. Wang, Q. K. Xue, Y. Hasegawa, T. Hashizume, and H. Shinohara, *Prog. Surf Sci.* **1996**, *51*, 263.

2. G. Gensterblum, *J. Electron Spectrosc. Relat. Phenom.* **1996**, *81*, 89.

3. M. S. Dresselhaus, G. Dresselhaus, and P. C. Eklund, *Science of Fullerenes and Carbon Nanotubes.* Academic Press San Diego, 1996.

4. J. A. Levinson, A. V. Hamza, E. S. G. Shaqfeh, and M. Balooch, *J. Vacuum Sci. Technol.* **1998**, *A16*, 2385.

5. L. Norin, S. McGinnis, U. Jannsson, and J.-O. Carlsson, *J. Vacuum Sci. Technol.* **1997**, *A15*, 3082.

6. D. M. Gruen, S. Liu, A. Krauss, J. Luo, and X. Pan, *Appl. Phys. Lett.* **1994**, *64*, 1502.

7. A Erdemir, M. Halter, G. R. Fenske, A Krauss, and D. M. Gruen, *Surf. Coating Technol.* **1997**, *94(5)*, 537.

8. M. T. Cuberes, R. R. Schlitter, and J. K. Gimzewski, *Appl. Phys. Lett.* **1996**, *69*, 3016.

9. K. Kuhnke, R. Becker, and K. Kern, *Chem. Phys. Lett.* **1996**, *257*, 569.

10. H. Hong, W. E. McMahon, P. Zschack, D.-S. Lin, R. D. Aburano, H. Chen, and T.-C. Chiang, *Appl. Phys. Lett.* **1992**, *61*, 3127.

11. H. Hong, R. D. Aburano, E. S. Hirshorn, P. Zschack, H. Chen, and T.-C. Chiang, *Phys. Rev. B* **1993**, *47*, 6450.

12. A. V. Hamza, J. Dykes, W. D. Mosley, L. Dinh, and M. Balooch, *Surf. Sci.* **1994**, *318*, 368.

13. G. LeLay, M. Göthelid, V. Yu. Aristov, A. Cricenti, M. C. Håkansson, C. Giammichele, P. Perfetti, J. Avila, and M. C. Asensio, *Surf. Sci.* **1997**, *377–379*, 1061.

14. A. Goldani, C. Cepek, E. Magnano, A. D. Laine, S. Vandre, and M. Sancrotti, *Phys. Rev. B* **1998**, *58*, 2228.

15. E. Giudice, E. Magnano, S. Rusponi, C. Boragno, and U. Valbusa, *Surf. Sci.* **1998**, *405*, L561.

16. E. Magnano, S. Vandre, C. Cepek, A. Goldani, A. D. Laine, G. M. Curro, A. Santaniello, and M. Sancrotti, *Surf. Sci.* **1997**, *377–379*, 1066.

17. D. Purdie, H. Bernhoff, and B. Reihl, *Surf. Sci.* **1996**, *364*, 279.

18. J. Kovac, G. Scarel, O. Sakho, and M. Sancrotti, *J. Electron Spectrosc. Relat. Phenom.* **1995**, *72*, 71.

19. T. David, J. K. Gimzewski, D. Purdie, B. Reihl, and R. R. Schlitter, *Phys. Rev. B* **1994**, *50*, 5810.

20. Y. Caudano, A. Peremans, P. A. Thiry, P. Dumas, and A. Tadjeddine, *Surf. Sci.* **1997**, *377–379*, 1071.

21. H. Tang, M. T. Cuberes, C. Joachim, and J. K. Gimzewski, *Surf. Sci.* **1997**, *386*, 115.

22. X.-D. Wang, T. Hashizume, V. Yu. Yurov, Q. K. Xue, H. Shinohara, Y. Kuk, Y. Nishina, and T. Sakurai, *Z. Phys. Chem.* **1997**, *202*, 117.

23. P. W. Murray, M. O. Pedersen, E. Laegaard, I. Stensgaard, and F. Besenbacher, *Phys. Rev. B* **1997**, *55*, 9360.

24. K. D. Tsuei and P. D. Johnson, *Solid State Commun.* **1997**, *101*, 337.

25. A. V. Hamza and G. D. Kubiak, *J. Vacuum Sci. Technol. A* **1990**, *8*, 2687.

26. C. Joachim, J. K. Gimzewski, R. R. Schlitter, and C. Chavy, *Phys. Rev. Lett.* **1995**, *74*, 2102.

27. C. Joachim and J. K. Gimzewski, *Europhys. Lett.* **1995**, *30*, 409.

28. M. R. C. Hunt, P. Rudolf, and S. Modesti, *Phys. Rev. B* **1997**, *55*, 7882.

29. M. Pedio, M. L. Grilli, C. Ottaviani, M. Capozi, C. Quaresima, P. Perfetti, P. A. Thiry, R. Caudano, and P. Rudolf, *J. Electron Spectrosc. Relat. Phenom.* **1995**, *76*, 405.

30. J. K. Gimzewski, S. Modesti, and R. R. Schlitter, *Phys. Rev. Lett.* **1994**, *72*, 1036.

31. K. Hevesi, C.-A. Fustin, P. Rudolf, L.-M. Yu, B.-Y. Han, G. Gensterblum, P. A. Thiry, J.-J. Pireaux, and R. Caudano, *J. Electron Spectrosc. Relat. Phenom.* **1995**, *76*, 115.

32. G. K. Wertheim, D. N. E. Buchanan, E. E. Chaban, and J. E. Rowe, *Solid State Commun.* **1992**, *83*, 785.

33. A. J. Maxwell, P. A. Brühwiler, S. Andersson, D. Arvanitis, B. Hernnäs, O. Karis, D. C. Mancini, and N. Martensson, S. M. Gray, M. K.-J. Johansson, and L. S. O. Johansson, *Phys. Rev. B* **1995**, *52*, R5546.

34. A. J. Maxwell, P. A. Brühwiler, D. Arvanitis, J. Hasselström, M. K.-J. Johansson, and N. Martensson, *Phys. Rev. B* **1998**, *57*, 7312.

35. A. J. Maxwell, P. A. Brühwiler, D. Arvanitis, J. Hasselström, and N Martensson, *Phys. Rev. Lett.* **1997**, *79*, 1567.

36. M. K.-J. Johansson, A. J. Maxwell, S. M. Gray, P. A. Brühwiler, and L. S. O. Johansson, *Surf. Sci.* **1998**, *397*, 314.

37. L. Q. Jiang and B. E. Koel, *Chem. Phys. Lett.* **1994**, *223*, 69.

38. R. Fasel, P. Aebi, R. G. Agostino, D. Naumovic, J. Osterwalder, A. Santaniello, and L. Schlapbach, *Phys. Rev. Lett.* **1996**, *76*, 4733.

39. C. Cepek, A. Goldani, and S. Modesti, *Phys. Rev. B* **1996**, *53*, 7466.

40. M. R. C. Hunt, S. Modesti, P. Rudolf, and R. E. Palmer, *Phys. Rev. B* **1995**, *51*, 10039.

41. S. D. Liefer, D. G. Goodwin, M. S. Anderson, and J. R. Anderson, *Phys. Rev. B* **1995**, *51*, 9973.

42. M. Balooch and A. V. Hamza, *Appl. Phys. Lett.* **1993**, *63*, 150.

43. A. V. Hamza, M. Balooch, and M. Moalem, *Surf. Sci.* **1994**, *317*, L1129.

44. A. V. Hamza, M. Balooch, M. Moalem, and D. R. Olander, *Chem. Phys. Lett.* **1994**, *228*, 117.

45. A. V. Hamza and M. Balooch, unpublished results.

46. L. Norin, U. Jansson, C. Dyer, P. Jacobsson, and S. McGinnis, *Chem. Mater.* **1998**, *10*, 1184.

47. D. Klyachko and D. M. Chen, *Phys. Rev. Lett.* **1995**, *75*, 3693.

48. X. Yao, T. G. Ruskell, R. K. Workman, D. Sarid, and D. Chen, *Surf. Sci.* **1996**, *367*, L85.

49. X. Yao, T. G. Ruskell, R. K. Workman, D. Sarid, and D. Chen, *Surf. Sci.* **1996**, *366*, L743.

50. Y.-R. Ma, P. Moriarty, M. D. Upward, and P. H. Beton, *Surf. Sci.* **1998**, *397*, 421.

51. P. Dumas, M. Gruyters, P. Rudolf, Y. He, L.-M. Yu, G. Gensterblum, R. Caudano, and Y. J. Chabal, *Surf. Sci.* **1996**, *368*, 330.

52. K. Sakamoto, T. Wakita, M. Harada, and A. Kasuya, *Surf. Sci.* **1998**, *402–404*, 523.

53. S. Suto, K. Sakamoto, T. Wakita, C.-W. Hu, and A. Kasuya, *Phys. Rev. B* **1997**, *56*, 7439.

54. A. Yajima and M. Tsukada, *Surf. Sci.* **1996**, *366*, L715.

55. T. Hashizume, X.-D. Wang, Y. Nishina, H. Shinohara, Y. Saito, Y. Kuk, and T. Sakurai, *Jpn. J. Appl. Phys.* **1993**, *31*, L880.

56. S. Katircioglu and S. Erkoc, *Surf. Sci.* **1997**, *383*, L775.

57. L. Li, Y. Hasagawa, H. Shinohara, and T. Sakurai, *J. Phys. IV* **1996**, *6(5)*, 173.

58. M. Balooch and A. V. Hamza, *J. Vacuum Sci. Technol. B* **1994**, *12*, 3218.

59. D. Chen, R. Workman, and D. Sarid, *Electron. Lett.* **1994**, *30*, 1007.

60. D. Chen, R. Workman, and D. Sarid, *Surf. Sci.* **1995**, *344*, 23.

61. S. Henke, M. Philipp, B. Rauschenbach, and B. Stritzker, in *Physics and Chemistry of Fullerenes and Derivatives*, H. Kuzmany, J. Fink, M. Mehring, and S. Roth (eds.). World Scientific, Singapore, 1995, p. 81.

62. S. Henke, B. Stritzker, and B. Rauschenbach, *J. Appl. Phys.* **1995**, *78*, 2070.

63. L. Moro, A. Paul, D. C. Lorents, R. Malhotra, and R. Ruoff, *Appl. Surf. Sci.* **1997**, *119*, 76.

64. S. Henke, M. Philipp, B. Rauschenbach, and B. Stritzker, *Mater. Sci. Eng. B Solid State Mater. Adv. Technol.* **1994**, *36*, 291.

65. S. Geier, M. Zeitler, K. Helming, M. Philip, S. Henke, B. Rauschenbach, and B. Stritzker, *Appl. Phys. A Mater. Sci. Processing* **1997**, *64*, 139.

66. M. Moalem, M. Balooch, A. V. Hamza, and R. S. Ruoff, *J. Phys. Chem.* **1995**, *99*, 16736.

67. A. V. Hamza and M. Balooch, *J. Phys. Chem. Solids* **1997**, *58*, 1973.

68. K. Morse, T. P. Weihs, A. V. Hamza, M. Balooch, Z. Jiang, and D. B. Bogy, *J. Tribology Trans. ASME* **1997**, *119*, 26.

69. X.-D. Wang, T Hashizume, H. Shinohara, Y. Saito, Y. Nishina, and T. Sakurai, *Phys. Rev. B* **1993**, *47*, 15923.

70. J. I. Pascual, J. Gomez-Herrero, and A. M. Baro, *Surf. Sci.* **1998**, *397*, L267.

71. K. Sakamoto, T. Suzuki, M. Harada, T. Wakita, S. Suto, and A. Kasuya, *Phys. Rev. B* **1998**, *57*, 9003.

72. S. McGinnis, L. Norin, U. Jansson, and J.-O. Carlsson, *Appl. Phys. Lett.* **1997**, *70*, 586.

73. T. Sakurai, Q. Xue, T. Hashizume, and Y. Hasagawa, *J. Vacuum Sci. Technol. B* **1997**, *15*, 1628.

74. J. C. Dunphy, D. Klyachko, H. Xu, and D. M. Chen, *Surf. Sci.* **1997**, *383*, L760.

75. P. W. Murray, J. K. Gimzewski, R. R. Schlitter, and G. Thornton, *Surf. Sci.* **1996**, *367*, L79.

76. Y. Z. Li, J. C. Patrin, M. Chander, J. H. Weaver, L. P. F. Chibante, and R. E. Smalley, *Science* **1991**, *252*, 547.

77. D. Schmicker, S. Schmidt, J. G. Skofronick, J. P. Toennies, and R. Vollmer, *Phys. Rev. B* **1991**, *44*, 10995.

78. K. Kuhnke, R. Becker, H. Berger, and K. Kern, *J. Appl. Phys.* **1996**, *79*, 3781.

79. K. Kuhnke, R. Becker, and K. Kern, *Surf. Sci.* **1996**, *377–379*, 1056.

80. M. Moalem, M. Balooch, A. V. Hamza, W. J. Siekhaus, and D. R. Olander, *J. Chem. Phys.* **1993**, *99*, 4855.

81. Y. Wang, *Nature* **1992**, *365*, 585.

82. C. Schlebusch, B. Kessler, S. Cramm, and W. Eberhardt, *Synth. Metals* **1996**, *77*, 151.

83. T. R. Ohno, Y. Chen, S. E. Harvey, G. H. Kroll, P. J. Benning, J. H. Weaver, L. P. F. Chibante, and R. E. Smalley, *Phys. Rev. B* **1993**, *47*, 2389.

84. W. H. Wang and W. K. Wang, *J. Appl. Phys.* **1996**, *79*, 149.

85. D. Klyachko and D. M. Chen, *J. Vacuum Sci. Technol. B* **1997**, *15*, 1295.

86. T. Takagahara and K. Takeda, *Phys. Rev. B* **1992**, *46*, 15578.

87. L. Ruan and D. M. Chen, *Surf. Sci.* **1997**, *393*, L113.

88. M. Biermann, B. Kessler, S. Krummacher, and W. Eberhardt, *Solid State Commun.* **1995**, *95*, 1.

89. J. F. Zeigler, J. P Biersack, and V. Littmark, *The Stopping and Range of Ions in Solids.* Pergamon, New York, 1985.

90. T. Schenkel, A. V. Hamza, A. V. Barnes, and D. H. Schneider, *Phys. Rev. A* **1997**, *56*, R1701.

91. M. Newman, A. V. Hamza, T. Schenkel, A. V. Barnes, G. Machicoane, M. Hattass, T. Niedermayr, and D. H. Schneider, to be published.

92. R. G. Musket, R. A. Hawley-Fedder, and W. L. Bell, *Radiat. Effects Defects Solids* **1991**, *118*, 225.

93. J. Kastner, H. Kuzmany, and L. Palmetshofer, *Appl. Phys. Lett.* **1994**, *65*, 543.

94. D. Fink, R. Klett, P. Szimkoviak, J. Kastner, L. Palmetshofer, L. T. Chadderton, L. Wang, and H. Kusmany, *Nucl. Instrum. Methods Phys Res. B* **1996**, *108*, 114.

95. A. Hoffman, P. J. K. Paterson, S. T. Johnston, and S. Prawer, *Phys. Rev. B* **1996**, *53*, 1573.

96. S. Prawer, K. W. Nugent, S. Biggs, D. McCulloch, W. H. Leong, A. Hoffman, and R. Kalish, *Phys. Rev. B* **1995**, *52*, 841.

97. R. M. Papaléo, A. Hallén, J. Eriksson, G. Brinkmalm, P. Demirev, P. Håkansson, and B. U. R. Sundqvist, *Nucl. Instrum. Methods B* **1994**, *91*, 124.

98. H. W. Kroto, A. W. Allaf, and S. P. Balm, *Chem. Rev.* **1991**, *91*, 198.

99. T. Schenkel, A. V. Barnes, A. V. Hamza, and D. H. Schneider, *Eur. Phys. D* **1998**, *1*, 297.

100. P. Scheier, B. Dünser, R. Wörgötter, D. Muigg, S. Matt, O. Echt, M. Foltin, and T. D. Märk, *Phys. Rev. Lett.* **1996**, *77*, 2654.

101. P. Scheier, B. Dünser, and T. D. Märk, *Phys. Rev. Lett.* **1995**, *74*, 3368.

102. B. Dünser, O. Echt, P. Scheier, and T. D. Märk, *Phys. Rev. Lett.* **1997**, *79*, 3861.

103. A. V. Hamza and M. Balooch, *Chem. Phys. Lett.* **1992**, *198*, 603.

104. P. A. Brühwiler, A. J. Maxwell, P. Baltzer, S. Andersson, D. Arvanitis, L. Karlsson, and N. Martensson, *Chem. Phys. Lett.* **1997**, *279*, 85.

105. A. V. Hamza and M. Balooch, unpublished results.

106. A. V. Hamza and M. Balooch, *Chem. Phys. Lett.* **1993**, *201*, 404.

107. E. I. Altman and R. J. Colton, *Surf. Sci.* **1993**, *295*, 13.

108. S. Modesti, R. R. Schlittler, and J. K. Gimzewski, *Surf. Sci.* **1995**, *331–333*, 1129.

109. T. Hashizume and T. Sakurai, *J. Vacuum Sci. Technol. B* **1994**, *12*, 1992.

110. D. M. Chen, H. Xu, W. N. Creager, and P. Burnett, *J. Vacuum Sci. Technol. B* **1994**, *12*, 1910.

111. H. Xu, D. M. Chen, and W. N. Creager, *Phys. Rev. B* **1994**, *50*, 8454.

CHAPTER 14

STRUCTURES OF FULLERENE-BASED SOLIDS

KOSMAS PRASSIDES and SERENA MARGADONNA

14.1 INTRODUCTION

The isolation and preparation ten years ago of the fullerenes, a set of hollow, closed-cage molecules consisting purely of carbon, in bulk quantities sparked off a truly remarkable interdisciplinary research activity, encompassing diverse fields of chemistry, physics, and materials science. The structural and dynamical properties of pristine fullerene solids posed many challenging questions to modern day experimental and theoretical science. In addition, many atomic and molecular species were found to readily intercalate in the large interfullerene voids present in the solids, with or without charge transfer. Fullerene solids appear to act as "molecular sponges" with almost all atoms or molecules that come into contact with them, being readily trapped in their lattice and resulting in a series of molecular adducts. Charge transfer to the fullerene units leads to salt or donor–acceptor adduct formation. Finally, notwithstanding their stability, which led to their original detection, the fullerenes readily participate in a plethora of chemical reactions and undergo facile functionalization. Of particular interest have been reports of the synthesis of polymeric forms of C_{60}. The focus of solid state and materials science research has gradually shifted away from pristine samples toward fullerene derivatives, which display magnetic and superconducting properties. In addition, the formation and properties of other curved forms of carbon, like nanotubes and buckyonions is currently attracting considerable interest, culminating with the bulk production of single-wall nanotubes with nearly uniform diameters, that self-organize into "ropes" by condensation of laser-vaporized carbon–nickel–cobalt mixtures at high temperatures.

This chapter focuses principally on the structural properties of pristine fullerenes in the solid state, drawing from the results of a number of experimental

Fullerenes: Chemistry, Physics, and Technology, Edited by Karl M. Kadish and Rodney S. Ruoff.
0-471-29089-0   Copyright © 2000 John Wiley & Sons, Inc.

techniques. The structural phase transitions accompanying orientational ordering of the molecules are sensitively related to deviations from isotropy arising through the presence of more than one type of carbon–carbon bonds, which give rise to an anisotropic π electronic distribution on the fullerene surface. Reduction in molecular symmetry is in general accompanied by a much more complicated behavior in the solid state. In addition, some of the salient features of fullerene intercalation chemistry as well as the properties of the resulting materials for which a vigorous research activity is still maintained will be described, with emphasis on their superconducting and magnetic properties. Bridged fullerene structures were considered for a while to be quite exotic, especially as notions of the special stability associated with the parent fullerene molecule, C_{60}, prevailed. However, the family of bridged dimeric and polymeric fullerene intercalated compounds has now grown considerably and a detailed account of their structural properties is included in this chapter. Finally, as considerable excitement has been generated by the chemical synthesis and purification in bulk quantities of the first nitrogen-substituted fullerenes, which afford excellent opportunities to explore the properties of the condensed phases of heterofullerenes and their chemical derivatives, the field of heterofullerene research is reviewed.

14.2 SOLID C_{60}

Most of the early research activity in the field concentrated on the physical, chemical, and structural properties of the parent C_{60} fullerene together with those of its polymeric forms and many derivatives. As a result, there is a considerable knowledge accumulated and a widespread agreement on the structural and dynamical properties.

Early X-ray diffraction work [1,2] revealed that at room temperature the C_{60} crystalline powder consisted of spheroidal molecules of diameter 7.1 Å, forming a random mixture of hexagonal close-packed (hcp) and face-centered cubic (fcc) arrays. However, careful elimination, by sublimation, of solvent molecules trapped in interstitial cavities leads only to a fcc crystal structure. The possible space groups consistent with the icosahedral molecular symmetry were discussed by Fleming et al. [3]. The C_{60} molecule takes the form of a truncated icosahedron and has molecular point group symmetry $m\bar{3}5(I_h)$ (Fig. 14.1) [4]. All carbon atoms are equivalent and there are two types of C–C bonds: 30 short "double bonds" (6:6) that fuse two hexagons and 60 long "single bonds" that fuse a hexagon to a pentagon (6:5). Pseudohexagonal faces are centers of threefold rotation symmetry, pentagonal faces are centers of fivefold rotational symmetry, edges between two hexagons are centers of twofold rotational symmetry, and the entire molecule is symmetric under inversion. Periodic translational symmetry is incompatible with fivefold rotational symmetry, and the maximal subgroup consistent with crystalline symmetry is $m\bar{3}$. This means that when a static C_{60} molecule is placed in a cubic crystalline lattice the icosahedral

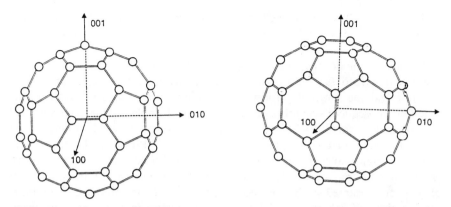

Figure 14.1 The two "standard orientations" of the C$_{60}$ molecule related by a 90° rotation about any Cartesian axis. The Cartesian axes in a cubic crystal pass through three orthogonal twofold molecular axes [4]. Reprinted from *J. Phys. Chem. Solids*, **53**, P. A. Heiney, p. 1334, 1992, with permission from Elsevier Science.

symmetry is broken, increasing the number of inequivalent carbon atoms to three and the number of positional parameters required to describe the molecule to eight. In the simplest possible crystalline structure, all molecules in the lattice have the same orientation. For this case, the relative orientation of a truncated icosahedron placed at the origin of a cubic lattice is such that the [100] axes pass through three twofold molecular axes (hexagon–hexagon edges), and the four [111] axes pass through hexagonal faces; this is referred to as the "standard orientation" (Fig. 14.1). There are actually two such standard orientations of the molecule, related by a $\pi/2$ rotation about any of the [100] axes, or equivalently by a reflection through a (100) plane. In a fcc lattice, four C$_{60}$ molecules are placed at the corners and at the midpoints of the faces of the cubic cell and if they all have the same orientation, then the crystal symmetry is $Fm\bar{3}$; on the other hand, if the molecules are orientationally disordered, then the crystal symmetry is raised to $Fm\bar{3}m$. Powder neutron [5] and X-ray [2] diffraction measurements proved that the space group is $Fm\bar{3}m$, with all four quasispherical C$_{60}$ units being symmetry equivalent (the cubic lattice constant at 290 K is 14.1569(5) Å). Subsequently, a variety of different experimental techniques, including nuclear magnetic resonance (NMR) [6,7], quasielastic neutron scattering [8], and muon spin relaxation (μSR) [9] were employed to study the orientational disorder. These measurements showed that the fcc phase could be considered as a plastic "rotator" phase in which the C$_{60}$ molecules rotate almost freely, effectively randomly with no time-averaged preferred orientation in space. Rotational correlation times τ_{rot} of 9–12 ps at room temperature are only three to four times longer than expected for unhindered gas phase rotation and faster than any other known solid state rotor [8,10]; furthermore, C$_{60}$ reorients even faster in the plastic phase than in tetrachloroethane solution ($\tau \approx 15$ ps). Thus, the high-temperature description of pristine C$_{60}$ that has emerged is of a prototypical plastic crystal (cf. adamantane and norbornane) with a well-defined translational order and a smooth rotational potential with many shallow minima, leading to continuous small-angle molecular motions.

Within this model the electronic density surface of each fullerene molecule could be described as a uniform spherical shell with the diameter of the carbon skeleton and the molecular form factor being simply the zero-order Bessel function. Subsequent synchrotron X-ray studies of single crystals have revealed the existence of small deviations from perfect sphericity [11]. The single crystal diffraction data analysis employed the technique of Press and Hüller [12], in which, for rigid molecules and for atoms confined to a spherical shell, an orientational scattering density function can be defined in terms of an infinite sum of symmetry-adapted surface harmonic (SASH) functions obtained from linear combinations of spherical harmonics. The first term in the expansion corresponds to the ideal spherical shell, while the higher-order SASH terms modulate the deviations from ideality, but with only the coefficients, which transform as the totally symmetric representation of the point group of the site symmetry of the shell of atoms being nonzero. The analysis of Chow et al. [11] revealed for C_{60} at room temperature an electron density deficiency ($\sim16\%$) present along the cube diagonal $\langle111\rangle$ directions and an electron excess ($\sim10\%$) along the $\langle110\rangle$ directions (Fig. 14.2). This lack of uniformity of the electron density on the spherical shell proved that actually the molecules are not "pure" free rotors, but they experience a crystal field of cubic symmetry that generates an orientational probability distribution function. Various theoretical studies have attempted to model the C_{60} orientational potential in the high-temperature fcc

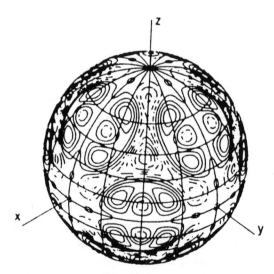

Figure 14.2 The orientation distribution function for C_{60} at room temperature viewed down one of the $\langle111\rangle$ directions. The zero-order term is omitted to emphasize differences from spherical symmetry. Solid (dashed) contours indicate excess (deficit) carbon density with respect to the zero-order term. Contour intervals are ±0.024 [11]. Reprinted with permission from P. C. Chow, 1992, *Phys. Rev. Lett.*, **69**.

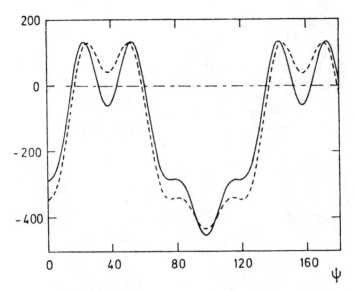

Figure 14.3 Crystal field potential, $V(\psi)$, for the orientationally disordered phase of C$_{60}$. Dotted line, $a = 14.15$ Å; solid line, $a = 14.10$ Å [15]. Reprinted from *J. Chem. Phys.*, **101**, Lamoen and Michel, p. 1440, 1994, with permission from AIP.

phase [13–15]. Lamoen and Michel modeled the orientational potential employing a Born–Mayer interaction for the repulsive part and a van der Waals attraction between carbon atoms; they also included other interaction centers along the (6:6) bonds and at the centers of the (6:5) bonds. The orientational intermolecular potential so obtained was then expanded in terms of symmetry adapted surface functions and the calculated coefficients were in good agreement with those deduced from the synchrotron diffraction experiments. Plotting the calculated crystal field per molecule as a function of the rotation angle ψ (the molecule being rotated counterclockwise around the [111] axis away from the standard orientation), they found the existence of an absolute minimum at $\psi = 98°$ and a secondary minimum at $\psi = 38°$ that becomes deeper and deeper with lattice contraction (Fig. 14.3). More recently, Launois et al. [16] measured the X-ray diffuse scattering of a C$_{60}$ single crystal at room temperature. The diffuse scattering assumes the form of two broad halos at wave vectors of about 3.3 and 5.3 Å^{-1} and a rich azimuthal modulation of the halos revealed a strong intermolecular correlation. The Lamoen–Michel model seemed to fit satisfactorily the relative intensities of the diffuse scattering features. The positions of the diffuse scattering maxima, corresponding to competing instabilities, are found to be independent of the model of intermolecular interaction, but they are related to the shape of the large and highly symmetrical C$_{60}$ molecule.

As the temperature is lowered, the rotational properties of the C$_{60}$ molecules become more complicated. In analogy with adamantane, which shows a phase transition at 208.6 K from a high-temperature fcc $Fm\overline{3}m$ to an orientationally ordered tetragonal $P42_1c$ structure [17], differential scanning calorimetry (DSC)

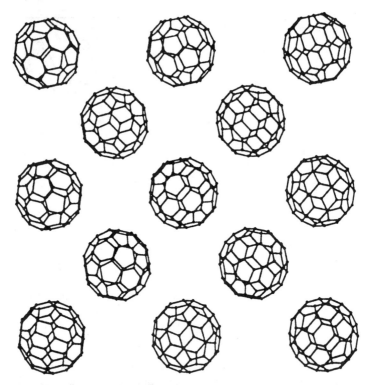

Figure 14.4   Basal plane projection of the low-temperature structure of C_{60} (space group $Pa\bar{3}$).

measurements of C_{60} [18] first established the existence of a phase transition near 250 K that was subsequently confirmed by means of X-ray diffraction [2]. New reflections appear indexing on a primitive cubic unit cell. The results were interpreted as manifesting the orientational ordering of the four C_{60} molecules in the unit cell, which ceased to be symmetry equivalent, as orientational long-range order developed. Sachidanandam and Harris [19] proposed the cubic $Pa\bar{3}$ space group for the low-temperature crystal structure of C_{60}, starting from an ideal $Fm\bar{3}$ space group with the molecules in their standard orientation. The equilibrium configuration was found for at $\psi \approx 94°$. High-resolution neutron diffraction measurements on a highly crystalline, solvent-free C_{60} sample were employed for the structure determination at 5 K [20]. Rietveld profile refinement confirmed the $Pa\bar{3}$ space group ($a = 10.0408(1)\,\text{Å}$) and resulted in an improved value for the rotation angle about [111] of ∼98° (Fig. 14.4), reminiscent of the miminum found by Lamoen and colleagues for the high-temperature fcc phase. Closer inspection of the relative orientations of neighboring C_{60} molecules (of S_6 local symmetry) in the lattice (Fig. 14.5) readily reveals the reason for which early theoretical attempts to model the intermolecular interactions had failed. C_{60} is indeed a quasispherical molecule and, consequently,

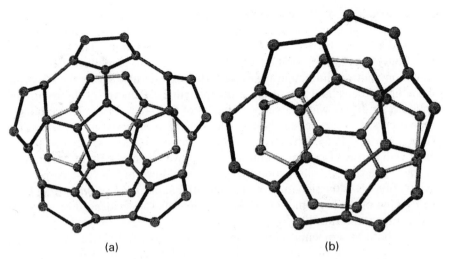

Figure 14.5 Intermolecular C$_{60}$–C$_{60}$ nearest contact in the primitive cubic structure, as viewed along the (110) center-to-center direction for rotation angles about [111]: (a) $\psi = 98°$ and (b) $\psi = 38°$.

the intermolecular energy is dominated by the van der Waals contributions. However, a fundamental difference from graphite is the presence of isolated pentagonal units on the surface of the sphere. The result is that the two types of C–C bonds lead to an anisotropic electronic distribution. Thus, bonding models developed for the graphitic sheets ought to be modified accordingly in order to be applicable to the fullerenes. Figure 14.5a shows the intermolecular C$_{60}$–C$_{60}$ nearest contact, as viewed along the center-to-center direction; a high π-order 6:6 ring fusion is positioned almost parallel ($\sim$179.6°) to the pentagonal face, which is the most "electron-deficient" region of the surface of the fullerene cage, leading to a subtle optimization of the molecular electrostatic potential. The seemingly arbitrary rotation angle of $\psi = 98°$ in $Pa\overline{3}$ thus ensures all 12 nearest-neighbor interactions are optimized with 6 of the 30 short bonds and 6 of the 12 pentagonal faces of each molecular unit taking part in the inter-molecular interactions. The C–C bonds range from 1.379(10) Å to 1.485(10) Å with average values for the 6:6 ring fusions of 1.404(10) Å and for the 6:5 ring fusions of 1.448(10) Å, in excellent agreement with the values derived from NMR data of 1.400(15) Å and 1.450(15) Å, respectively. Subsequent theoretical calculations based on this model for the intermolecular interactions [21,22] took into account the electrostatic interaction by assigning different partial bond charges to the high ($q \approx -0.54e$) and low ($q \approx 0.27e$) π-order C–C bonds. The low-temperature global energy minimum was found to be the experimentally observed $Pa\overline{3}$ crystal structure with a rotation angle of $\sim$98.7° [21]. Alter-natively, Sprik et al. [22] arrive at a similar global minimum by assigning a negative charge of $-0.35e$ to the high π-order bonds and positive charges (0.175e) to the carbon atoms. NMR studies [7] of the rotational dynamics of

C_{60} revealed that at the phase transition there is a dramatic change in the nature of the dynamics: the molecular motion in the low-temperature phase was described as jumps between symmetry-equivalent orientations. The temperature evolution of the lattice constant of the cubic unit cell, as derived from high-resolution neutron diffraction measurements [5,23], clearly shows the signature of the first-order phase transition, accompanying the change in the rotational dynamics at 260 K where a sudden change of 0.344(8)% in the lattice dimensions occurs. Coexistence of the two phases is also observed at T_c, where the fcc/pc ratio is ~45:55. The lattice constant smoothly decreases below 260 K, albeit with a distinct curvature until, at 90 K, a well-defined cusp in its variation appears. The smooth contraction then continues down to 5 K. The cusp in the vicinity of 85 K is the signature of an orientational glass transition, arising from the freezing of the jump motion of the C_{60} molecules, as evidenced by a variety of experimental results [24].

Since both experimental diffraction [20] and theoretical [21,22] work reveal the crucial role that the electronic anisotropy of C_{60} plays in the intermolecular interactions, an attempt was made to refine the dynamic model proposed by the NMR studies by searching for statistically preferred orientations close in energy to the global minimum configuration of Figure 14.5a. For instance, a 60° rigid rotation about the [111] direction, corresponding to $\psi \approx 38°$, should lead to a local minimum in energy [5] since the intermolecular C_{60}–C_{60} nearest contact is now characterized by a high π-order 6:6 ring fusion facing (~159°) a hexagonal face of its neighbor (Fig. 14.5b). It is worthy to remember at this point that both configurations were also minima for the orientational potential of the fcc phase. The theoretical calculations of Lu et al. [21] have then confirmed this model for the $Pa\bar{3}$ phase, estimating that a ~100 meV energy difference exists between the two configurations with the height of the hindrance potential estimated to be ~300 meV. Thus, assuming 60° hops about the [111] direction, the primitive cubic phase was modeled by two energetically favorable orientations coexisting, with their fractions refined upon fitting the diffraction data in the 5–260 K temperature range. The fraction of the 98° phase is found to increase smoothly down to 90 K and then to remain constant at a value of $p = 0.835(4)$ down to 5 K (Fig. 14.6). The low-temperature structure of C_{60} is then best described below 90 K as an orientational glass; a finite amount of orientational disorder is frozen in, as the thermal energies become much smaller than the rotational barriers. In the intermediate temperature (90–260 K) regime, the C_{60} molecules appear to execute jump rotations, not only among symmetry-equivalent positions (e.g., 72° jumps about the [110] direction) as implied by NMR, but also among inequivalent nearly degenerate positions (36° random jumps about the [110] directions that are straddled by 6:6 fusions) [25]. An estimate of the energy difference, between the two nearly degenerate configurations ($\psi = 98°$ and 38°) was obtained by both neutron diffraction [5,23] and thermal conductivity [26] studies agreeing on a value of $\Delta E \approx 11.9$ meV. The coexistence of the two preferred orientations in the primitive cubic structure was also confirmed by a single crystal X-ray diffraction study in which the

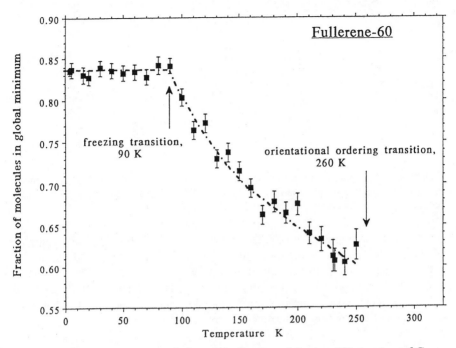

Figure 14.6 Temperature evolution of the fraction of the $\psi = 98°$ structure of C$_{60}$.

intensities of some Bragg reflections were monitored from room temperature down to 23 K [27,28]. This model was revisited by Blaschko et al. [29] in an elastic diffuse neutron scattering investigation in which they confirm the existence of the two orientations in the $Pa\bar{3}$ structure but with slightly different orientational angles $\psi \approx 102°$ and $\psi \approx 42°$. Finally, the performance by the molecules of small angular oscillations [30] about their equilibrium orientations should be taken into account to explain the presence of diffuse scattering (peaking at $Q \approx 3.5$ and $5.8\,Å^{-1}$) even at 5 K. Using a Monte Carlo simulation, Copley et al. [31] find that a root mean square (rms) librational amplitude of $1.9°$ and a fraction $p \approx 0.833$ of the majority phase can account well for both the observed Bragg and diffuse scattering at 5 K.

Of particular interest have been reports of the synthesis of polymeric forms of C$_{60}$. At ambient pressure, polymerization products, insoluble in the common organic solvents in which fullerenes dissolve, have been identified when C$_{60}$ is exposed to a modest flux of visible or ultraviolet light in the temperature range $250 < T < 373$ K [32]. The mechanism proposed for this photopolymerization reaction is a "[2+2] cycloaddition," which results in the formation of four-membered carbon rings, fusing together adjacent molecules. Photopolymerization is inhibited below the orientational ordering transition of C$_{60}$ as the probability of short carbon–carbon bonds coming to close intermolecular contact diminishes. In addition, different types of polymeric structures have been iden-

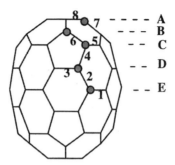

Figure 14.7 Schematic diagram of the C_{70} molecule. The circles label the five symmetry inequivalent types of C atoms. The eight different C–C bonds are consecutively numbered from the equator toward the poles.

tified under the application of high pressure. An orthorhombic body-centered phase comprised of one-dimensional C_{60} chains was found at 300 °C and 8 GPa. At higher temperatures and lower pressures, two other phases made of two-dimensional polymerized layers were identified: a pseudo-body-centered tetragonal phase containing quadratic polymerized layers and a rhombohedral phase with ABC packing of hexagonal polymerized layers [33,34].

14.3 SOLID C_{70} AND HIGHER FULLERENES

The insertion of 10 additional atoms into C_{60} in the form of an equatorial belt reduces the symmetry of the resulting fullerene C_{70} to D_{5h}, increases the number of sets of inequivalent C atoms to five ($n_1:n_2:n_3:n_4:n_5 = 10:10:20:20:10$, as indicated in Fig. 14.7) [35], and requires the values of 12 geometrical parameters to define the structure. High-Q pulsed neutron powder diffraction measurements in the temperature range 20–300 K [36] were used to determine the intra- and intermolecular structures of pristine C_{70}. Employing a D_{5h} molecular model for the intramolecular structure and a free-form Monte Carlo solution for the intermolecular pair-correlation function, values for the eight symmetry-inequivalent C–C bond lengths were extracted. These values are shown in Table 14.1 together with the dimensions of the long d_1 (distance between the planes of the two pentagonal end faces) and the short d_2 (diameter of the circle passing through the 10 equatorial atoms) axes of the molecule (Fig. 14.7). The five hexagonal faces around the equator comprise two types of bond lengths, which are essentially equal ($\sim$1.42 Å) and are joined together by the longest C–C bonds of the molecule ($\sim$1.48 Å). This is consistent with the chemical inertness exhibited by the equatorial region of the molecule. On the other hand, the polar regions exhibit bond length patterns reminiscent of C_{60} and mimic its chemical behavior, essentially acting as electron-deficient olefins. Little change in the *intra*molecular structure is evident between 20 and 300 K with the long and short molecular axes remaining essentially constant at 7.96 and 7.12 Å,

TABLE 14.1   C–C Bond Distances (in Å) in C$_{70}$ at 20 K

Bond Type[a]	20 K
Hexagon–hexagon (**1**)	1.479(7)
Hexagon–pentagon (**2**)	1.417(6)
Within pentagon (**3**)	1.438(4)
Within pentagon (**4**)	1.46(1)
Pentagon–pentagon (**5**)	1.376(7)
Within pentagon (**6**)	1.449(6)
Pentagon–pentagon (**7**)	1.385(7)
Within pentagon (**8**)	1.459(4)
d_1	7.960(6)
d_2	7.123(7)

[a] d_1 is the distance between the planes of the two pentagonal end faces and d_2 is the diameter of the equatorial circle. The bold numbers in parentheses refer to the C–C bond types shown in Figure 14.7 from the equator toward the poles of the C$_{70}$ molecule.

respectively. These results are in excellent agreement with both theoretical calculations [37] and experimental results obtained from gas phase electron diffraction [38] but differ from earlier low-resolution work [39], which conjectured the presence of a pinching deformation in the equatorial region of the molecule.

Early X-ray diffraction and electron microscopy work [40] had revealed the presence in solid C$_{70}$ of more defects than in C$_{60}$ and the coexistence of cubic close-packed (ccp) and hexagonal close-packed (hcp) crystalline phases. Moreover, the minority hcp phase proved harder to eliminate than in the case of C$_{60}$, and most early reported studies were performed on mixed phase materials. In addition, trapped solvent molecules in the interstitial spacing of solid C$_{70}$ were more tenaciously retained, even after repeated sublimations. Structural studies on single crystals of the least stable minority hcp structural modification [41] revealed two phase transitions on cooling: first to a deformed hcp structure, followed by a transition to a monoclinic structure.

At high temperatures, both neutron [42,43] and X-ray [44] diffraction reveal that the most stable structural modification of solid C$_{70}$ is fcc (space group $Fm\bar{3}m$). The cubic lattice constant ($a = 14.976(7)$ Å at 430 K) scales with the long molecular axis, as the molecules tumble fast enough to average out the anisotropy. On cooling, the first evidence of extra reflections, not allowed in the fcc space group, appears at ~300–280 K when a small peak starts growing on the low d-spacing side of the $(220)_c$ reflection. On further cooling, new peaks grow at the expense of the fcc reflections. The diffraction profile develops gradually down to 200 K. Little change occurs below this temperature down to 5 K. The observed diffraction profiles in the 280–200 K temperature range can be rationalized by considering the coexistence of a fcc and a rhombohedral ($a_{rhomb} \approx a_{cub}$, $\alpha \approx 85.6°$, space group $R\bar{3}m$) phase; in the latter, the long mo-

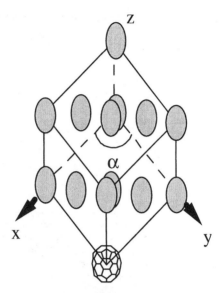

Figure 14.8 Schematic diagram of the rhombohedral unit cell of solid C_{70} ($\alpha \approx 85.6°$).

lecular axis aligns toward the unit cell diagonal (Fig. 14.8). What appears remarkable is that a fraction of the high-symmetry disordered phase is still present even at very low temperatures. Below 180 K, the neutron diffraction profile is consistent with the existence of an orientationally ordered structure with monoclinic symmetry. These results are in agreement with the predictions of computer simulations [45] and the high-resolution X-ray [44] powder diffraction measurements; the latter clearly identified the low-temperature structure as monoclinic (possible space groups $C2$, Cm, $P2_1$, or Pm) (Fig. 14.9). The X-ray study finds that the fcc → rhombohedral phase transition occurs at 345 K and the rhombohedral → monoclinic one at 295 K on cooling. It is noteworthy, however, that on heating, even though the rhombohedral → fcc transition again occurs at 280 K, the low-temperature structure does not disappear until the temperature is raised to 340–350 K, implying a ~70 K hysteresis effect. The structural properties of C_{70} are thus sensitive to the stresses accompanying the transitions and the high concentration of defects, associated with static and dynamic disorder, that lead to extensive "undercooling" for the fcc phase. The hindrance potential barrier associated with the microstrains and defects appears to be of the order of 350 K.

The temperature dependence of the structural results show that at high temperature the structure is fcc in agreement with the quasi-isotropic motion of the ellipsoidal molecules. The appearance of the rhombohedral phase on cooling is the result of a molecular orientational ordering with the molecules preferentially orienting along the unit cell diagonal. This is reflected in both an abrupt expansion along $\langle 111 \rangle$ (~1.9 Å) as the long axis of the molecule is now pointing toward this direction and an abrupt contraction along the close-packing $\langle 110 \rangle$

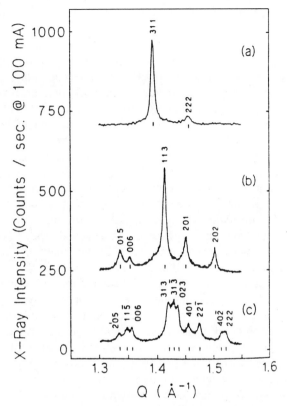

Figure 14.9 X-ray diffraction pattern of solid C$_{70}$ at (a) 440 K (fcc structure), (b) 300 K (rhombohedral structure), and (c) 15 K (monoclinic structure) [44]. Reprinted from *Chem. Phys.*, **101**, G. B. M. Vaughan, p. 602, 1993, with permission from Elsevier Science.

direction ($\sim$0.4 Å), reflecting their smaller size in the equatorial plane. A small change in volume is observed at the phase transition, while the rhombohedral angle α does not change with temperature. On further cooling the complete ordering of the molecules stabilizes a monoclinic strucure.

Modest pressure at ambient temperature is also enough to cause an irreversible ordering transition at $\sim$0.35 GPa [42], the resulting rhombohedral structure ($\alpha \approx 85.4(1)°$) being slightly less compressible ($\kappa = 4.0(1.4) \times 10^{-2}$ GPa^{-1}) than the simple cubic structure of C$_{60}$ ($\kappa = 5.5(0.6) \times 10^{-2}$ GPa^{-1}). This may be related to the anisotropy of the C$_{70}$ molecules compared to the quasispherical C$_{60}$ units, which can pack much more efficiently. No fcc fraction is present above 1 GPa. When pressures higher than 6.5 GPa were applied, a progressive reduction in the intensities of the 311, 31$\bar{1}$, 11$\bar{3}$, and 22$\bar{2}$ reflections was observed and the solid becomes more compressible. This may be associated with a distortion of the rhombohedral unit cell, possibly driven by a reduced shear strength arising from the anisotropic structure of the molecules. Further increase of pressure to 11 GPa results in the appearance of an amorphous carbon phase; complete collapse of the Bragg intensity occurs by 18 GPa with amorphous carbon present to 25 GPa.

In an attempt to investigate the possibility of polymerization in C_{70} suggested by the amorphization above 18 GPa, more recent studies focused on the high-temperature and high-pressure part of the phase diagram. Iwasa et al. [46] submitted both hcp and fcc C_{70} to temperatures up to 800 °C at 5 GPa. They observed clear signs of polymerization after the experiment by both X-ray diffraction and infrared (IR) spectroscopy, while Premila et al. [47] found dimerization of C_{70} after high-pressure, high-temperature (HPHT) treatment ($P < 7.5$ GPa and $T < 750$ °C). In both cases, the higher pressure needed for polymerization than that of C_{60} is due to the decrease in reactivity, as the 6:6 bonds taking part in the [2+2] cycloaddiction reaction are only localized in the polar part of the molecule. At higher pressures, the phase diagram becomes even more complicated with the possibility of a second type of polymerization, like the formation of a three-dimensional (3D) polymerized tetragonal phase that has been found at $P > 9.5$ GPa ($T < 1770$ K) [34,48].

Until recently, it has been difficult to investigate the solid state properties of higher fullerenes C_n ($n > 70$), despite their potentially interesting physical and chemical properties. This was mainly a consequence of the difficulties associated with separation and purification of these molecules. Furthermore, many higher fullerenes exist in several isomeric forms, which satisfy the isolated pentagon rule and need to be separated from one another to allow investigation. Such separation has been achieved only recently.

C_{76} exists in two optical isomeric forms with D_2 symmetry (Fig. 14.10a) and has an approximately ellipsoidal shape with its long and short axes 4.39, 3.79, and 3.32 Å, similar to C_{70}. The enantiomeric mixture when cocrystallized with toluene adopts in the solid state a monoclinic structure with a space group $P2_1$ [49]. When solvated, C_{76} forms unusually systematic twins resulting in a tenfold symmetry in the diffraction pattern. In this as well as in the pentane solvated C_{60} and C_{70} that also show very similar twinning, the internal molecular symmetry does not play any major role in the crystallization. All three fullerenes when solvated adopt a very similar monoclinic structure in which the character of each fullerene practically disappears, implying a dynamically rotating state. When solvent free, the enantiomeric mixture of C_{76} adopts a fcc structure, suggesting orientationally disordered molecules. X-ray diffraction measurements did not detect any phase transition in the temperature range from 300 to 7.6 K [49]. The C_{76} molecules remain disordered at all temperatures with neither an orientational ordering transition (cf. C_{60}) nor a phase transition to a low-symmetry structure (cf. C_{70}), implying a statically disordered molecular arrangement at the fcc sites. Furthermore, a ^{13}C NMR study of the C_{76} crystal shows that the molecule is rotating rapidly at room temperature but gradually slows down as the temperature decreases but without a drastic change that may suggest any phase transition [50]. This is in contrast with theoretical results obtained on a simulated crystal of one-handed chiral C_{76} that showed a phase transition at 200 K [51]. This disagreement may be related to the fact that in all

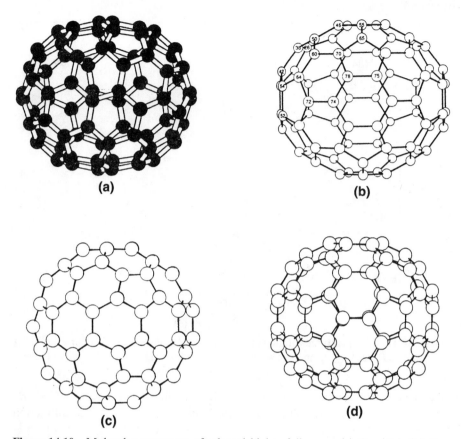

Figure 14.10   Molecular structures of selected higher fullerenes: (a) C_{76} (D_2), (b) C_{78} (C'_{2v}), (c) C_{84} (D_{2d}), and (d) C_{84} (D_2) isomers.

experimental work the sample is a mixture of right-handed and left-handed chiral molecules and to the effect of crystallinity and the presence of impurities. In the case of C_{78}, the isolated pentagon rule excludes all but five isomers. Two of these possess C_{2v} symmetry but differ in the number of inequivalent carbon atoms and theoretical calculations find the one labeled C'_{2v} to be most stable (Fig. 14.10b) [52]. X-ray diffraction measurements on an isomer-pure solid C'_{2v} C_{78} sample showed similar behavior to that of C_{76}, as C_{78} also adopts a fcc structure with rotationally disordered molecules in the temperature range between 300 and 100 K.

C_{84} is the most abundant fullerene after C_{60} and C_{70}. Among the 24 possible structural isomers, it is usually isolated as a mixture of only those with D_2 and D_{2d} symmetry (Fig. 14.10c,d) in a 2:1 abundance ratio, respectively. A remarkable feature of the D_{2d} isomer is its quasispherical shape (aspect ratio = 1.06) [53] and its low-lying doubly degenerate lowest unoccupied molecular orbital (LUMO) [54], which may lead to facile reaction with electron donors. Indeed, it has been found that a potassium intercalation compound of stoichiometry

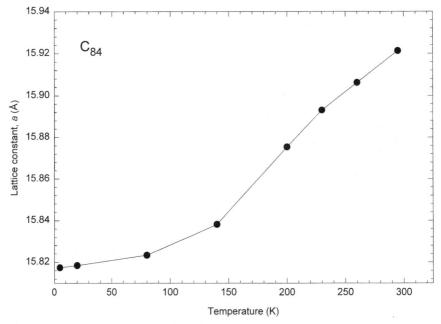

Figure 14.11 Temperature evolution of the cubic lattice constant, a, of sublimed solid C_{84} (mixture of D_2 and D_{2d} isomers in a 2:1 abundance ratio). The line is a guide to the eye.

$K_{8+x}C_{84}$ can be isolated after doping with potassium to saturation [55]. A D_2 and D_{2d} isomer mixture solid sample was studied by X-ray powder diffraction (Fig. 14.11). It adopts a fcc crystal structure at all temperatures down to 5 K ($a = 15.817(4)$ Å at 20 K), in analogy with C_{76} and C_{78} [56]. The absence of orientational ordering effects could be due to the coexistence of the two isomers, as static disorder is dominant at all temperatures. The D_{2d} isomer has been isolated and studied by single crystal X-ray diffraction in the crystalline adduct η^2-$C_{84}Ir(CO)Cl(PPh_3)_2$ [57]. Afterward, chromatographic techniques were employed to successfully separate the two isomers [58] and it became possible to study isolated D_2 and D_{2d} in pure form. In both cases, the diffraction profiles show phase transitions to low-symmetry orientationally ordered phases on cooling, in analogy with C_{70}. In the case of C_{84}, the isomer mixture and the two isolated isomers behave differently with temperature, exemplifying the necessity of working on pure isomeric forms for a proper understanding of the chemical, physical, and structural properties of the higher fullerenes.

14.4 INTERCALATED FULLERIDES

14.4.1 Alkali Fullerides

Intercalation of solid C_{60} with electron donors, like the alkali metals, results in a wealth of intercalated fulleride salts with stoichiometries A_xC_{60}, where x can be as low as 1 (Cs_1C_{60}) or as high as 12 ($Li_{12}C_{60}$). This reflects the ease of reduction of C_{60}, especially to oxidation states ranging from -1 to -6. Most prominent among these salts are those with composition A_3C_{60}, which are metallic and become superconducting at critical temperatures, T_c, as high as 33 K (at ambient pressure) for $RbCs_2C_{60}$ [59], at present surpassed only by the high T_c superconducting cuprates. The appearance of superconductivity in alkali fullerides has led to extensive efforts in attempting to understand their electronic, structural, and dynamic properties and to elucidate the origin of their high T_c and the mechanism responsible for superconducting pair formation. The fullerides are potentially simpler materials than the high-T_c cuprates and thus perhaps more easily amenable to theoretical modeling; still they do pose many fresh challenges, principally arising from the comparable magnitudes of the Fermi and phonon energies and the role of electronic correlations. Theoretical treatments of superconductivity in the fullerides consider purely electronic [60] as well as phonon-induced pairing, involving either low-energy intermolecular [61] or high-energy intramolecular modes [62,63]. There is extensive evidence from a plethora of experimental techniques (neutron and Raman spectroscopy, heat capacity, optical reflectivity, isotope effect, and resistivity measurements) for a traditional weak-coupling phonon-mediated mechanism of superconductivity with an electron–phonon coupling strength distributed over a wide range of energies (33–195 meV) [64].

The structure of pristine fcc C_{60} is characterized by the presence of two types of unoccupied interstitial holes with high symmetry: the smaller tetrahedral, T (two per C_{60} unit with a radius of 1.12 Å), and the larger octahedral, O (one per C_{60} unit with radius of 2.06 Å), sites. If we disregard the orientational order of the C_{60} units, four types of crystalline structures of intercalated fullerides may be realized, depending on the relative occupancy of the T and O sites: (1) the rock-salt type (AC_{60}) in which one alkali metal ion is accommodated in every O site, (2) the antifluorite type (A_2C_{60}) in which both T sites are occupied, (3) the cryolite type (A_3C_{60}) in which all T and O sites are fully occupied, and (4) the zinc blende type (AC_{60}) in which the alkali metal ions occupy half the available T sites.

The structure of the parent superconducting fullerides A_3C_{60} (A = K, Rb) [65] is based on the cryolite structural type (Fig. 14.12). It is face-centered cubic (fcc) with the C_{60}^{3-} ions adopting two different molecular orientations related by a 90° rotation about the $\langle 001 \rangle$ or equivalently by 44° 23′ about the $\langle 111 \rangle$ crystal axis (merohedral molecular disorder) and performing small-amplitude librational motion up to 700 K. The effect of the orientational disorder is to raise the crystal symmetry from $Fm\overline{3}$ (compatible with perfect orientational ordering of the molecules) to $Fm\overline{3}m$. As the ionic size of both K$^+$ and Rb$^+$ is larger than that of the empty T sites, rotational motion of the C_{60}^{3-} ions is restricted and the crystal structure can be understood in terms of the strong repulsive interactions between the alkali ions and the fulleride units. Thus, the latter are forced to expose toward the tetrahedral alkali ions the part of their quasispherical shape with the largest surface area, namely, the hexagonal faces, thus leading to maximization of the A$^+$–C(C_{60}) distances. The resulting alkali

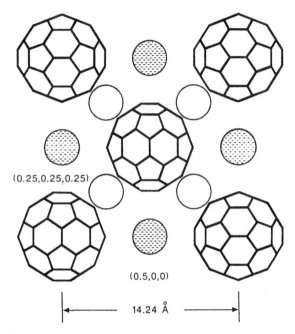

(0.25,0.25,0.25)

(0.5,0,0)

14.24 Å

Figure 14.12 The structure of K_3C_{60}. The open and hatched spheres represent the potassium at the tetrahedral and octahedral sites, respectively [65]. Reprinted from *Nature*, **351**, P. W. Stephens, pp. 632–634, 2000.

metal coordination geometry is to four hexagonal faces of neighboring fulleride ions (T site, coordination number 24) and to six hexagon–hexagon (6:6) fusions (octahedral site, coordination number 12). The control of the structural properties of A_3C_{60} by the tetrahedral ions is manifested in the expanded dimensions of the fcc unit cell ($a = 14.240\,\text{Å}$ for K_3C_{60}, $a = 14.157\,\text{Å}$ for C_{60} at room temperature) and the extreme robustness of the structure, which shows no phase change either to an orientational liquid phase at high temperatures or to ordering of the standard orientations at low temperatures. Experimental evidence derived from the relationship between lattice dimensions (i.e., interfullerene spacing) and superconducting transition temperatures in A_3C_{60} (Fig. 14.13) both at ambient [66] and at high pressures [67] is consistent with T_c being modulated by the density of states at the Fermi level, $N(\varepsilon_F)$. As the interfullerene spacing increases, the overlap between the molecules decreases; this leads to a reduced bandwidth and, for fixed band filling, to an increased density of states. From a Bardeen-Cooper-Schrieffer (BCS) type relationship,

$$T_c \propto (\hbar\omega_{ph}) \exp[-1/N(\varepsilon_F)V] = (\hbar\omega_{ph}) \exp(-1/\lambda) \qquad (14.1)$$

the observed high T_c may be understood in terms of a high average phonon frequency $\hbar\omega_{ph}$, resulting from the light carbon mass and the large force constants associated with the intramolecular modes, and a high density-of-states

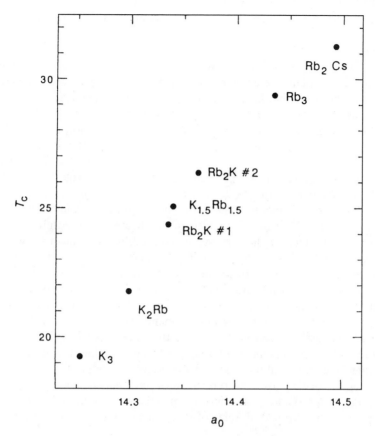

Figure 14.13 Variation of T_c (K) with lattice parameter a_0 (Å) for various compositions of A_3C_{60} [66]. Reprinted from *Nature*, **352**, R. M. Fleming, pp. 140–141, 2000.

(DOS) at ε_F, resulting from the weak intermolecular interactions and strongly scattering intramolecular modes.

Despite the apparent simplicity of the structure of A_3C_{60} fullerides, some important issues are not as yet fully resolved. One feature of the structural refinements of both synchrotron X-ray and neutron powder diffraction data is the large isotropic temperature factor found for the octahedral ion ($\sim$16 Å^2), which implies the existence of substantial static and/or dynamic disorder and may signify strong displacements away from the center of the O site [65,68]. Extended X-ray absorption fine structure (EXAFS) studies [69,70] have indeed repeatedly pointed toward the existence of such displacements at the local level. In addition, the accepted fcc crystal structure is consistent with the presence of only two inequivalent alkali ions. However, ^{39}K and ^{87}Rb NMR measurements found, in both K_3C_{60} and Rb_3C_{60}, that while at high temperatures two resonances (T and O) with an intensity ratio 2:1 are present, on cooling, the peak associated with the tetrahedrally coordinated ions splits into two components (T and T') [71–74]. As the NMR technique sensitively probes the local structure, such observations raise the question of the presence of structural distortions or defects, not detected by bulk diffraction measurements. Various possibilities to explain the appearance of the new T' line at low temperatures have been discussed in the literature. These include the existence of local misorientations of C_{60}^{3-} ions, displacement of the octahedral ions off the center of the octahedral interstices, molecular Jahn-Teller distortions, and structural instabilities like charge density waves. An additional possibility includes the presence of a small concentration x of tetrahedral vacancies. This acquires additional significance for electronic and conducting properties, as it has been argued that stoichiometric A_3C_{60} fullerides should be insulating, and their metallic and superconducting behavior could only arise through the effects of nonstoichiometry [75]. Careful analysis of synchrotron X-ray diffraction data has provided support for the presence of such tetrahedral vacancies with x of the order of 0.07 [76]. Then the alkali T' sites may be interpreted as representing the nearest neighbors of these vacancies without the necessity of off-center displacements of the octahedral ions (Fig. 14.14) [77] and the collapse of the two lines into one at high temperature may be understood in terms of vacancy diffusion [78]. Nonetheless, universal agreement has not been reached and while it is clear that excess electron density exists at the T' sites, spin-echo double resonance NMR measurements find no site-to-site correlation with the T' sites occupying the tetrahedral interstices in a completely random fashion [79].

The control of the structural properties of intercalated fullerides by the ion occupying the tetrahedral interstices is immediately manifested when Na^+ is introduced in the tetrahedral holes of a ccp array of C_{60}^{3-} ions. As has been

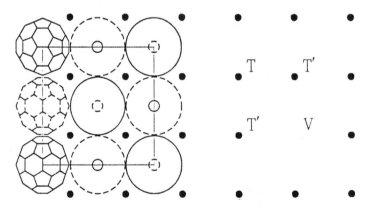

Figure 14.14   Crystal structure of Rb_3C_{60}. Big circles denote C_{60} molecules randomly distributed between the two orientations shown; small circles denote octahedral cations (solid lines: $z = 0$, dashed lines: $z = 1/2$). Filled circles denote tetrahedral cations ($z = 1/4, 3/4$). V is a tetrahedral vacancy with T and T' sites corresponding to the alkali NMR lines [77]. Reprinted from *Europhys. Lett.*, **41**, G. M. Bendele, p. 554, 1998, with permission from Elsevier Science.

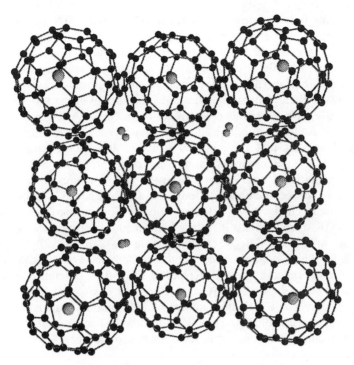

Figure 14.15 Unit-cell basal plane projection of the primitive cubic structure of $Na_2A'C_{60}$ ($A' = Rb, Cs$).

shown by a high-resolution study of lightly doped Na_xC_{60} intercalates [80], the C_{60}^{3-} ions are no longer necessarily confined to their two standard orientations found in K_3C_{60}. As the Na^+ ionic radius (0.95 Å) is smaller than the size of the hole, there is evidently enough space for the C_{60}^{3-} ions to rotate in such a way as to optimize both the attractive $Na^+-C_{60}^{3-}$ interactions and the $C_{60}^{3-}-C_{60}^{3-}$ contacts. The resulting structure is experimentally found to be fcc, but to differ from that of merohedrally disordered fullerides with respect to the orientational state of the C_{60}^{3-} ions, which are now quasispherically disordered. On cooling, a phase transition occurs in the vicinity of room temperature to a primitive cubic structure (space group $Pa\bar{3}$) (Fig. 14.15) for both superconducting Na_2CsC_{60} [81] and Na_2RbC_{60} [82]. Quite remarkably, optimization in the primitive cubic structure occurs for the same orientational state found in solid C_{60}, namely, an anticlockwise rotation about $\langle 111 \rangle$ of $\sim 98°$. For this setting, the $C_{60}^{3-}-C_{60}^{3-}$ contacts are optimized as 6:6 fusions "nest" over pentagonal faces of neighboring ions and at the same time, the $Na^+-C_{60}^{3-}$ coordination is highly optimal, as each ion presents two hexagonal faces and six 6:6 fusions to its eight neighboring Na^+ ions, reducing their coordination number to 12. Optimal coordination is also preserved for the ions (Cs^+ or Rb^+) occupying the octahedral holes, as they now coordinate to six hexagon–pentagon (6:5) fusions of neighboring fulleride ions. The primitive cubic family of $Na_2(A,A')C_{60}$ salts displays a much steeper rate of change of T_c with interfullerene spacing than that exhibited by merohedrally disordered fcc fullerides (Fig. 14.13), while at the same time, the effects of physical [83] and chemical [84] pressure on the superconducting properties are not identical, and chemical pressure suppresses T_c much faster than physical pressure does. The origin of the faster depression of T_c with decreasing interfullerene separation in $Na_2(A,A')C_{60}$ is of particular interest, as it could potentially lead to the attainment of much higher T_c. Empirically, it appears that the modified structure and intermolecular potential sensitively affect the electronic and conducting properties [81,83,84]. The changed orientational state of the ions in the $Pa\bar{3}$ structures could affect electron hopping between neighboring ions and, as a consequence, might lead to a modified rate of change of $N(\varepsilon_F)$, and hence of T_c with interfullerene spacing. Indeed, NMR measurements [85] have confirmed that the values of $N(\varepsilon_F)$ for Na_2RbC_{60} and Na_2KC_{60} (~ 7 states/eV-spin-C_{60}) are strongly suppressed. However, both high-pressure susceptibility measurements on Na_2CsC_{60} and the temperature dependence of its EPR spectra show little difference in the a dependence of $N(\varepsilon_F)$ between Na_2CsC_{60} ($Pa\bar{3}$) and K_3C_{60} and Rb_3C_{60} ($Fm\bar{3}m$) [86–88], in agreement with the findings of tight-binding calculations [89]. In fact, the dependence of $N(\varepsilon_F)$ on the interfullerene separation d can be described by a single functional form for both structural types [88]:

$$N(\varepsilon_F)^{-1} = (10^4 d)\exp(-d/0.25) - 0.03 \, \text{eV-spin-}C_{60} \qquad (14.2)$$

A similar conclusion has also come from ^{13}C NMR Knight shift measurements, which find similar electronic states around room temperature for both fcc and simple cubic superconductors, ruling out different $N(\varepsilon_F)$–a relations [90]. The apparent contradictions observed for the properties of the sodium intercalated phases can now be resolved, as recent X-ray and neutron diffraction, ^{13}C and ^{23}Na NMR, and susceptibility measurements have revealed a more complicated situation than hitherto appreciated (vide infra).

Further deviations in the structural and electronic properties are encountered when the small weakly electropositive Li^+ ions are introduced in the tetrahedral holes. Li_2CsC_{60} and Li_2RbC_{60} are not superconducting, despite the apparent correct band filling. The nonsuperconducting properties observed for Li_2CsC_{60} can be related to the enhanced interactions between Li and C_{60} that modify the electronic structure at the Fermi level through partial hybridization of carbon p_z and metal 2s orbitals [91]. Crystallographic analysis of Li_2CsC_{60} has revealed that a fcc structure ($Fm\bar{3}m$) comprising spherically disordered C_{60}^{3-} ions is adopted even at low temperatures. A symmetry-adapted spherical harmonics (SASH) analysis of the disordered C_{60}^{3-} ions reveals excess carbon density along the $\langle 111 \rangle$ direction (Fig. 14.16), indicative of the Li^+–C interaction, which has not been encountered in any of the superconducting alkali metal fullerides [91]. As a result, the formal charge of the C_{60}^{n-} ions is less than 3, implying a less than half-full t_{1u} band. Raman measurements have established a

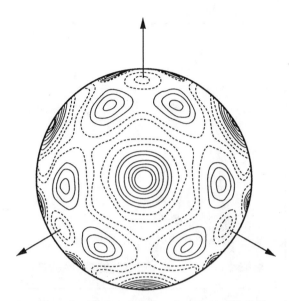

Figure 14.16 Orientation distribution function for C_{60} in Li_2CsC_{60} viewed down the $\langle 111 \rangle$ direction. The zero-order term is omitted to emphasize differences from spherical symmetry. Solid and dashed contours indicate excess and deficit, respectively, carbon density with respect to the zero-order term. Contour intervals are ± 0.128.

value of the effective doping level, $n \approx 2.5$ [92]. The strong Li^+–C interaction controls the structural properties of Li_2CsC_{60}, which remains strictly cubic at all temperatures, under different heating/cooling protocols, and up to a pressure of 6 GPa [93].

In order to overcome the Li^+–C interactions and achieve half-filling ($n = 3$) of the t_{1u} band, the series of fulleride salts, Li_xCsC_{60} ($1.5 \leq x \leq 6$) was successfully synthesized. By adjusting the Li content, x, the electron transfer from the alkali metals to C_{60} and the filling level of the conduction band was controlled. Half-filling was achieved for the composition Li_xCsC_{60}, which is a bulk superconductor with $T_c = 10.5$ K (Fig. 14.17) [92]. The structural properties of Li_xCsC_{60} were studied by high-resolution synchrotron X-ray diffraction [94]. At low temperatures, it is primitive cubic (space group $Pa\bar{3}$), isostructural with the Na_2AC_{60} phases with orientationally ordered C_{60}^{3-} ions in the unit cell. The tetrahedral and octahedral interstices are occupied by Li and Cs, respectively, while the excess Li (one per C_{60} unit) is disordered at the corners of a $(Li_8)_{1/8}$ cube with an edge length of ~3.4 Å, centered at the octahedral sites (Fig. 14.18). No anomalous close contacts between either the alkali metal ions or between Li^+ and the C_{60} units are encountered. The observed geometry of the Li defect is consistent with the existence of additional overdoped Li_xCsC_{60} phases with x between 4 and 6. On heating to room temperature, a phase transition occurs to a face centered cubic structure (space group $Fm\bar{3}m$), which contains orientationally disordered C_{60}^{3-} ions. The geometry of the Li^+ defect is unaffected by

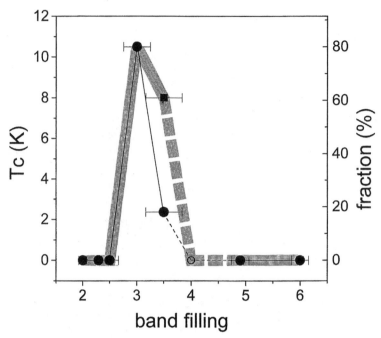

Figure 14.17   Dependence of T_c (squares) and superconducting fraction (solid circles) of the Li_xCsC_{60} samples on the estimated band filling of the C_{60} t_{1u} band.

this phase change. These results provide clear evidence of the importance of the Li^+–C interactions in sensitively controlling the structural, conducting, and electronic properties of the family of lithium intercalated fullerides.

The fcc structure of C_{60} can be modified into additional structural types when more than three ions per C_{60} are incorporated. The adopted structural types in A_xC_{60} are dependent on both the nature of A and the doping level x. For instance, when A = K, Rb, or Cs, body-centered cubic (bcc) phases (space group $Im\bar{3}$) form with stoichiometry A_6C_{60} with the cations now occupying six symmetry-equivalent tetrahedral sites [95]. Complete filling of the t_{1u} band results in insulating behavior. Line phases also exist for $x = 4$; the resulting K_4C_{60} and Rb_4C_{60} phases adopt body-centered tetragonal (bct) structures (space group $I4/mmm$), which derive from the bcc structure by creating two cation vacancies that order along the c direction [96,97]. Cs_4C_{60} is further distorted to give a body-centered orthorhombic (bco) structure (space group $Immm$) [98]. While simple band structure considerations would predict these phases to be metallic, experimentally they are found to be insulating, most likely because of the combined effect of electron correlation and electron–phonon interaction on the narrow t_{1u}-derived conduction band [99–101]. Changing the cation to Na also leads to different structural types. Na intercalation has been shown to be possible up to $x = 11$. In Na_2C_{60}, the Na ions

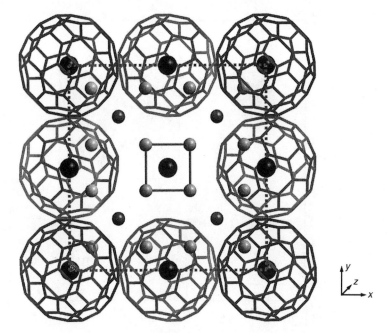

Figure 14.18 Unit-cell basal plane projection of the primitive cubic Li_3CsC_{60} structure at low temperatures. The C_{60} units at the center of the (110) face have been removed for clarity.

reside only in the tetrahedral sites [102]. Na_6C_{60} is a particularly stable composition, which shows the novel feature of tetrahedral Na_4 clusters with Na–Na separations of ∼3.2 Å accommodated in the large octahedral sites [102,103], while the structure remains fcc. $Na_{10}C_{60}$ adopts a similar structure but now both cube clusters and the octahedral sites are occupied (Na–Na separation ≈ 2.8 Å) [104].

14.4.2 Alkaline- and Rare-Earth Fullerides

Alkaline-earth and lanthanide fulleride chemistry is not as widely established as that of the alkali metals. Still a broad range of compositions is gradually becoming available, exhibiting interesting structural and electronic properties. For instance, synthesis of mixed alkali–alkaline-earth fullerides can be employed to increase the effective doping level to higher values than 3 for the family of fcc salts with stoichiometry $M_{3-y}Ba_yC_{60}$ ($0.2 < y < 2$, M = K, Rb, Cs) [105]. In this case, superconductivity disappeared for $n = 5$. An interesting observation that may prove of relevance to the electronic properties of these systems comes from the analysis of high-resolution powder neutron diffraction data [106] on the Cs_2BaC_{60} member of the above series, which is found not to be strictly isostructural with the merohedrally disordered fullerides (space group

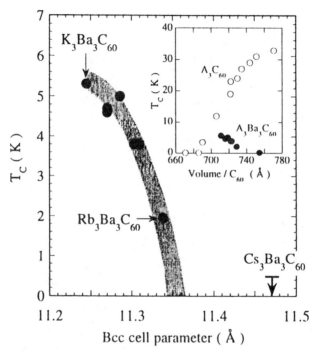

Figure 14.19 Relation between T_c and the cell parameter for $A_3Ba_3C_{60}$ salts. The inset shows a comparison with the behaviour of A_3C_{60} salts [109]. Reprinted with permission from Iwasa et al., 1998, *Phys. Rev. B*, **57**.

$Fm\bar{3}m$) as proposed by synchrotron X-ray diffraction work [107], but rather to possess a different orientational ordering scheme (space group $Fm\bar{3}$) with one of the two "standard orientations" of C_{60} predominating. Mixed compositions have also been synthesized for higher doping concentrations of alkaline earths, resulting in doping levels as high as $n = 9$, which correspond to half-filling of the triply degenerate LUMO+1 t_{1g} state of C_{60}. Prominent among these are the superconducting compositions $A_3Ba_3C_{60}$ (A = K, Rb, Cs) [108,109]. These are isostructural with the A_6C_{60} salts, adopting bcc structures (space group $Im\bar{3}$). However, in sharp contrast with the behavior of the alkali fulleride superconductors, T_c decreases with increasing lattice constant with the maximum value of 5.4 K attained for $K_3Ba_3C_{60}$ (Fig. 14.19).

Binary alkaline-earth fullerides have also been synthesized. Calcium forms a primitive cubic salt with stoichiometry Ca_5C_{60}, which becomes superconducting at 8.4 K [110]. On the other hand, strontium and barium metals form stable compositions AE_xC_{60} for x = 3, 4, and 6. AE_3C_{60} (AE = Ba, Sr) are insulating and adopt the A15 structural type (space group $Pm\bar{3}n$) with lattice constants of 11.34 and 11.14 Å, respectively [111,112]. It is noteworthy that this structural type is also adopted by Cs_3C_{60}, which is superconducting at ~40 K at elevated pressures, the highest reported T_c for a fulleride superconductor [113]. While

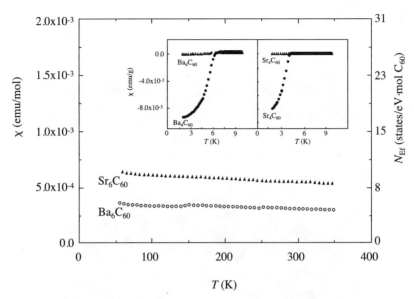

Figure 14.20 Magnetic susceptibility of Ba_6C_{60} and Sr_6C_{60}. The inset shows the temperature dependence of the magnetic susceptibility (5 G) for Ba_6C_{60}, Sr_6C_{60}, Ba_4C_{60}, and Sr_4C_{60}.

the Ba_3C_{60} salt is single phase, in Sr_3C_{60}, there is a coexisting fcc phase with $a = 14.144$ Å. The AE_6C_{60} (AE = Ba, Sr) phases have proved more controversial. Early work established their structures correctly as bcc (space group $Im\bar{3}$) [111,114] but proposed they were superconducting with $T_c = 4$ K for Sr_6C_{60} and 6.5 K for Ba_6C_{60}. This was disputed in later work [115,116]. It has now been unambiguously established that both Sr_6C_{60} and Ba_6C_{60} are metallic, but not superconducting (Fig. 14.20) [117]. Their density of states at the Fermi level is found to increase with decreasing interfullerene separation, in agreement with theoretical results, which find strong hybridization between the alkaline-earth d and s orbitals and C_{60} $p\pi$ orbitals [118,119]. Photoemission studies also reveal a hybrid band with incomplete charge transfer from the alkaline-earth metals to C_{60} [120]. The nature of the bulk superconducting phases in the Ba–C_{60} and Sr–C_{60} systems has now been established unambiguously to be the orthorhombic (space group $Immm$) phases Ba_4C_{60} and Sr_4C_{60} (Fig. 14.21) [121]. High-resolution X-ray diffraction has revealed highly anisotropic phases ($a = 11.601$ Å, $b = 11.184$ Å, $c = 10.884$ Å at 5 K for Ba_4C_{60}) that show no thermal expansion along the short c orthorhombic axis ($\delta c/c = -0.006(7)\%$ between 5 and 295 K), the signature of strong hybridization between metal and carbon orbitals. In sharp contrast to the alkali fulleride A_4C_{60} phases whose metallic state suffers from various instabilities, superconductivity is encountered in AE_4C_{60} away from nominal half-filling of the t_{1g}-derived conduction band and for noncubic crystal structures.

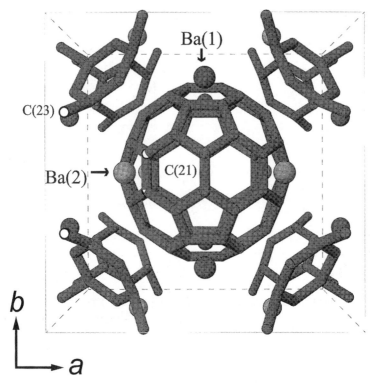

Figure 14.21   Projection of the body-centered orthorhombic structure of Ba_4C_{60} on the [110] basal plane. The two sets of crystallographically distinct barium ions, Ba(1) (*m2m* site) and Ba(2) (*2mm* site), are depicted as dark and light grey spheres, respectively. The hexagon C(21) and pentagon C(23) atoms, which are in close contact with Ba(2), are depicted as white spheres.

Finally, intercalation of C_{60} with the rare earths Yb and Sm has so far led to the isolation of the superconducting phases $Yb_{2.75}C_{60}$ ($T_c = 6$ K) [122,123] and $Sm_{2.75}C_{60}$ ($T_c = 8$ K) [124]. These adopt structures in which the rare-earth ions occupy off-centered interstitial sites in the face-centered cubic C_{60} lattice, leaving eight ordered tetrahedral sites vacant in an orthorhombic unit cell, which is doubled in each direction ($a = 28.17(5)$ Å, $b = 28.07(3)$ Å, $c = 28.27(3)$ Å for $Sm_{2.75}C_{60}$).

14.5   POLYMERIZED FULLERIDES

14.5.1 AC_{60} (A = K, Rb, Cs) Salts

Fullerides with AC_{60} (A = K, Rb, Cs) stoichiometry have attracted particular interest because of their rich phase diagram and complicated structural, electronic, conducting, and magnetic properties. The existence of a crystal phase with stoichiometry AC_{60} was first indicated by Raman [125], photoemission [126], and X-ray diffraction [127] studies. Above 400 K, the AC_{60} salts showed unsurprising behavior, adopting a fcc rock-salt structure with the A^+ cations accommodated in the octahedral interstices [127,128]. Below 400 K, the most stable form was originally shown to adopt orthorhombic crystal symmetry [128–132] with pronounced quasi-one-dimensional character and unusually close contacts between the fulleride ions; partial covalent bonding between fullerenes along the chains and their consequent molecular deformation from icosahedral symmetry have been postulated to account for these observations. Structural characterization was originally achieved by synchrotron X-ray powder diffraction [129] studies, which established a cell of dimensions $a = 9.11$ Å, $b = 10.10$ Å, and $c = 14.21$ Å (space group *Pmnn*) for RbC_{60} at room temperature with fulleride linkages along the polymer chains achieved by a [2+2] cyclo-addition mechanism, in analogy with the photopolymerization reaction of pristine C_{60}. The contraction along the polymerization a axis is ~0.9 Å, compared to the center-to-center interfullerene distances of about 10 Å, encountered in monomeric fullerenes and their derivatives. However, subsequent single crystal X-ray diffraction and diffuse scattering studies on KC_{60} and RbC_{60} [133] revealed that they possess different chain orientations about their axes, which are described by distinct space groups *Pmnn* (orthorhombic) and $I2/m$ (monoclinic), respectively (Fig. 14.22).

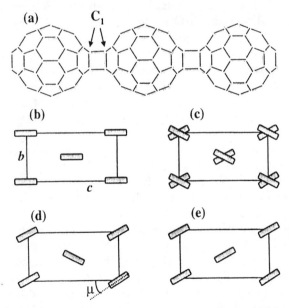

Figure 14.22 (a) Linear polymer chains formed by [2+2] cycloaddition in AC_{60} salts. Schematic drawing of chain orientations for (b) ordered *Immm*, (c) disordered *Immm*, (d) *Pmnn*, and (e) $I2/m$. The shaded bars indicate the orientation of the covalently bonded polymer chains along a by projection on the crystallographic bc plane [133]. Reprinted with permission from P. Launois, 1998, *Phys. Rev. Lett.*, **81**.

Remarkably, these polymer phases were found to be air stable, in contrast to the vast majority of fulleride salts, which display extreme air sensitivity [131]. In addition, they showed metallic behavior, but in common with many other low-dimensional conductors, a metal–insulator transition occurs at low temperature [128]. The transition to the insulating state in the vicinity of 50 K [134] is accompanied by a magnetic transition. Thus, the temperature evolution of the static susceptibility, χ_s, as measured by electron paramagnetic resonance (EPR), revealed a sharp decrease to zero in the vicinity of 50 K, attributed to a spin density wave (SDW) instability of the conducting C_{60}^- linear chains [128]. The origin and true nature of the low-temperature magnetic phase remained the subject of controversy for a while. Zero-field muon spin relaxation (ZF-μSR) studies of RbC_{60} and CsC_{60} samples [135–137] revealed frozen electronic moments in the low-temperature phase, but spontaneous Larmor frequencies that would have unambiguously implied long-range magnetic order (LRO), associated with antiferromagnetism (AF) or spin density wave (SDW) formation, were absent. The muons experienced a local magnetic field that peaked close to zero and with large spatial inhomogeneities, which may be due to a number of physical factors, including a wide distribution of chain $(C_{60}^-)_n$ lengths and orientational disorder effects. Strong evidence for the presence of an antiferromagnetically ordered ground state came from the observation in both RbC_{60} and CsC_{60} of antiferromagnetic resonance at high frequencies (75, 150, and 225 GHz) [138]. The sublattice magnetization was found to be independent of applied magnetic field up to at least 8 T, while magnetic fluctuations were present between 35 and 50 K. Comparison with the SDW system, $(TMTSeF)_2PF_6$ implied that the fulleride polymers are also quasi-1D SDW systems with 3D ordering at low temperatures. However, later work [139] cautioned that unequivocal proof for an antiferromagnetic ground state cannot be established, as the data could also be interpreted in terms of a spin-glass model and final confirmation should await single crystal antiferromagnetic resonance (AFMR) studies.

Direct evidence for the presence of sp^3 hybridized carbon atoms in the AC_{60} polymeric structures has come from magic-angle spinning (MAS) ^{13}C NMR spectroscopy [140,141]. The results at low temperature show the presence of a line at 50 ppm with a long spin lattice relaxation time of about 1.3 s, consistent with polymerization along the a axis of the monoclinic unit cell. Another piece of direct spectroscopic identification of the intermolecular bridging C–C modes in polymeric AC_{60} (A = Rb, Cs) has been achieved by neutron inelastic scattering measurements [142–144]. The generalized phonon density of states, $G(\omega)$, of monoclinic CsC_{60} shows a broad peak between 11 and 23 meV, the signature of interfullerene C–C bond formation, as in the absence of covalent contributions to the interfullerene bonding the intermolecular interactions are dominated only by van der Waals forces, resulting in very soft interball modes that do not extend beyond 8 meV. These modes disappear abruptly as the transition to the high-temperature rock-salt structure occurs. Adams et al. [145] have developed a first-principles quantum molecular dynamics model to de-

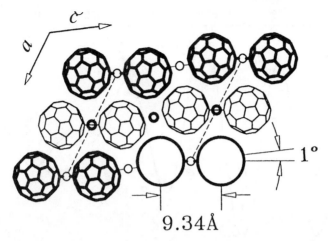

Figure 14.23 Monoclinic structure of the dimer phase of RbC_{60}. Bold and normal fullerenes and alkali ions are at $y = 0$ and $y = 0.5$, respectively [148]. Reprinted with permission from G. Oszlanyi, 1996, *Phys. Rev. B*, **54** and Pekker, 1998, *Chem. Phys. Lett.*, **282**.

scribe the vibrational properties of polymerized fullerenes. Rigid ball–ball chain modes are predicted to occur in the energy range 10–20 meV in good agreement with the measured low-temperature neutron spectra.

The polymer phase of the AC_{60} salts has dominated the physical studies on these systems. However, depending on the cooling rate from the high-temperature rock-salt fcc phase, various other metastable phases are encountered. For instance, a medium quenching rate to just below room temperature leads to the stabilization of a low-symmetry insulating structure that has been characterized by synchrotron X-ray diffraction studies to comprise $(C_{60}^-)_2$ dimers, bridged by single C–C bonds [146–148]. The RbC_{60} dimer phase crystallizes in the monoclinic space group $P2_1/a$ with lattice parameters $a = 17.141(5)\,\text{Å}$, $b = 9.929(5)\,\text{Å}$, $c = 19.277(5)\,\text{Å}$, $\beta = 124.4°$ at 220 K (Fig. 14.23) [148] and transforms back to the polymer phase on heating. The chain axis is now along b and is somewhat longer than that found in the RbC_{60} polymer, implying a smaller contraction of $\sim 0.7\,\text{Å}$ from the interfullerene separations found in monomeric fullerenes. This is consistent with the presence of a single bridging C–C bond, instead of the four-membered rings of the polymers. Unfortunately, powder X-ray data do not allow refinement of the bridging C–C bond, which was constrained in the course of the refinements to $1.54\,\text{Å}$ [148], but neutron inelastic measurements clearly revealed a different form of the intermolecular phonon DOS (PDOS) with sharp pronounced peaks observed at 10.4, 12.2, and 13 meV [142]. It is of interest to note that a microscopic theory of the phase changes in RbC_{60} [149], recently developed by studying the orientation dependence of the crystal and molecular field, identified additional channels of distortion leading to both polymerization and dimerization (Fig. 14.24).

Another metastable conducting phase of RbC_{60} and CsC_{60} was reported

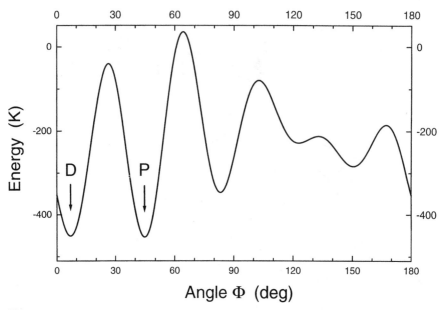

Figure 14.24   The molecular field when the C_{60} molecule is rotated away from the orthorhombic starting orientation about the [110] axis of the cubic system. The developing minima at 45° and 9° correspond to competitive polymer and dimer phases, respectively.

with very rapid quenching by immersing the high-T sample in liquid nitrogen [150,151]. In this case, the 3D isotropic character of the structure is retained while orientational ordering of the C_{60}^{-} ions leads to a simple cubic structure (space group $Pa\bar{3}$) with $a = 13.9671(3)$ Å at 4.5 K. Upon heating, it returns to the most stable polymer phase with the following sequence of transformations: quenched 3D primitive cubic phase → dimer phase → 3D primitive cubic phase (again) → 3D fcc phase (the same orientational ordering transition encountered for C_{60} and Na_2AC_{60}) → 1D polymer phase → 3D fcc phase (again). The observed transformations can be understood as shown in the schematic phase diagram in Figure 14.25, derived from the combined results of powder X-ray and neutron diffraction, EPR, and DTA measurements. The transformations from the low-temperature primitive cubic to the dimer phase and from the midtemperature fcc to the 1D polymer phase are nonequilibrium processes, while the rest are equilibrium processes. The heat absorption at the 3D primitive cubic → 3D fcc phase is approximately 3 J/g both for RbC_{60} and CsC_{60}, the value straddling those of 1–2 J/g for Na_2AC_{60} and 9 J/g for C_{60}.

A final comment concerns the spin susceptibilities measured by EPR for the 1D polymer and the quenched 3D phases. The observed Pauli-like susceptibilities imply that both phases are conducting. However, while the quasi-1D polymer phase shows a metal–insulator instability, the primitive cubic phase reveals contrasting electronic behavior on the local scale, as revealed by ^{13}C

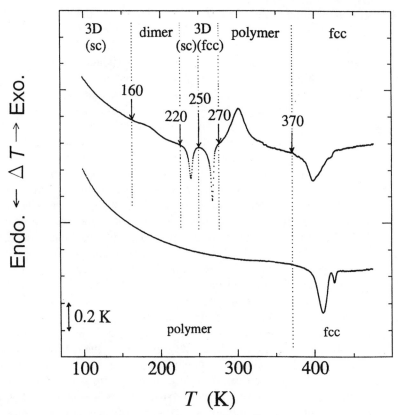

Figure 14.25 Schematic energy diagram of the CsC_{60} phases, depending on the cooling rate from the high-temperature fcc phase. Complementary DTA and EPR data are included to emphasize the phase transformations and accompanying changes in dimensionality.

and ^{133}Cs NMR measurements [152]. A partial spin gap opens below $T < 50$ K, ascribed to the occurrence of localized spin singlets on a small fraction of the C_{60}^- ions.

14.5.2 $Na_2A'C_{60}$ (A' = K, Rb, Cs) Salts

As discussed earlier, $Na_2A'C_{60}$ salts, incorporating the small Na^+ ion in the tetrahedral holes, are fcc at high temperatures, containing orientationally disordered C_{60}^{3-} ions. On cooling, a phase transition to an orientationally ordered primitive cubic phase (Fig. 14.15), isostructural with pristine C_{60}, has been identified [81,82]. In addition, structural studies on Na_2RbC_{60} at elevated pressures [153] have revealed the existence of a symmetry-lowering transition to a phase with unusually short C_{60}^{3-}–C_{60}^{3-} distances ($a = 9.35$ Å), reminiscent of the situation encountered in polymerized $A'C_{60}$ salts ($a = 9.11$ Å). As

Na_2RbC_{60} has an anomalously low T_c (3.5 K) [154] and the results of many physical measurements [85,86] have been incompatible with the existence of a single high-symmetry low-temperature phase, the question of the true ground state structure of Na_2RbC_{60} and its analogs is of particular importance. The major breakthrough in understanding its structural and physical properties was achieved when special attention was paid to the heating and cooling protocols in the course of experimental studies. Thus, it was found that, if special care was taken to cool the sample very slowly down to 180–200 K (with cooling rates as slow as 1 K/h), the ground state of Na_2RbC_{60} at ambient pressure and low temperatures is not the ordered $Pa\bar{3}$ phase believed up to then; instead, slow cooling was found to mimic the effect of applied pressure and to lead to the stabilization of a phase with crystal symmetry lower than cubic and with a short interball center-to-center distance of ~9.38 Å, which is essentially insensitive to cooling [155,156]. The cubic phase is metastable and does survive for relatively rapid cooling rates. These results provided an unexpected dimension to the behavior of fulleride salts, pointing to a richer abundance at ambient pressure of bridged fullerene structures than hitherto appreciated and were of value in rationalizing many of the puzzling features associated with the properties of sodium (and lithium) C_{60} salts.

Comprehensive studies of the structural, electronic, and conducting properties of the low-temperature phase in sodium intercalated salts have now been undertaken. The determination of the crystal structure of the parent Na_2RbC_{60} salt was accomplished by Rietveld refinement of synchrotron X-ray powder diffraction data [157]. The unit cell of the low-temperature structure is derived from the primitive cubic room temperature structure by a ~0.6 Å contraction along one of the cubic face diagonals and a ~0.4° tilt of one of the axes, resulting in monoclinic (space group $P2_1/a$) lattice parameters of $a = 13.711$ Å, $b = 14.554$ Å, $c = 9.373$ Å, $\beta = 133.53°$ at $T = 180$ K. The relative orientation of the fulleride neighboring ions in the unit cell can be described by a rotation angle ψ about the short c axis of ~82° (where $\psi = 90°$ would describe the on-ball double C–C bond associated with the contact C atom pointing along the b axis). At this angle, the relative orientation of the molecules is such that pairs of C atoms located on the c axis from neighboring fullerenes are brought into close proximity. The resulting interfullerene contacts (Fig. 14.26) can be described as single C–C bonds, leading to a quasi-one-dimensional polymeric structure. The bridging geometry of the fulleride ions was characterized more accurately by high-resolution neutron powder diffraction [158]. The results obtained were in excellent agreement with those of the synchrotron X-ray experiments with a deep minimum in the R-factors of the Rietveld refinements again found at an angle $\psi = 82° \pm 1°$. Following this, an exhaustive series of Rietveld refinements in which the molecular orientation and geometry of C_{60} units were systematically varied (Fig. 14.27) was performed. The intermolecular C4–C4 bond converged to a value of 1.70(7) Å. At the same time, the C4–C4 bond was inclined by ~7.7° from the c axis within the bc plane. The on-ball C–C bond distances converged to values of C4–C3 = 1.54(5) Å, C4–C12 = 1.50(5) Å, and

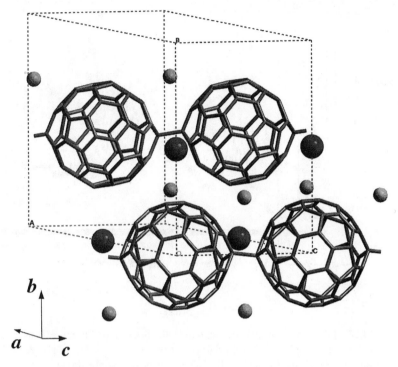

Figure 14.26 Crystal structure of the Na_2RbC_{60} polymer.

C4–C14 = 1.42(8) Å (Fig. 14.28). Direct spectroscopic identification of the bridging C–C modes in polymeric Na_2RbC_{60} has been achieved by neutron inelastic scattering measurements [159]. The generalized phonon density of states, $G(\omega)$, was measured at temperatures of 200 and 320 K. A broad peak between 8 and 25 meV is strongly suppressed in intensity on going through the transition, coinciding with the intermolecular region of vibrational spectra of fullerenes and their derivatives.

The origin of the differing structures adopted by RbC_{60} (chains bridged by two C–C bonds) and Na_2RbC_{60} (chains bridged by one C–C bond) is of considerable interest [157,158]. The microscopic theory of the phase changes in RbC_{60} [149] described earlier has identified two competing orientations (Fig. 14.24) for the low-temperature structure when rotations Φ about the cubic [110] axis are considered, one corresponding to the formation of a doubly bonded polymer at $\Phi = 45°$ and the other to the formation of a singly bonded dimer (or polymer) at $\Phi = 9°$. The observed behavior is strongly associated with the different charged states of the fulleride ions. In C_{60}^- salts, the presence of a single unpaired electron per ball allows only the formation of dimers, with polymer formation occurring via a different mechanism, namely, a [2+2] cycloaddition reaction. However, in the case of C_{60}^{3-} salts, the availability of additional

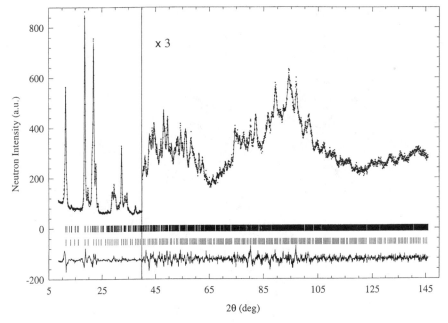

Figure 14.27 Observed (points), calculated (solid line), and difference (lower panel) plots for the final two-phase Rietveld refinement of the high-resolution powder neutron diffraction data of Na_2RbC_{60} at 2.5 K. Reflection positions are shown as tick marks. The tick marks show the reflection positions for the monoclinic (top set) and primitive cubic (bottom set) cells.

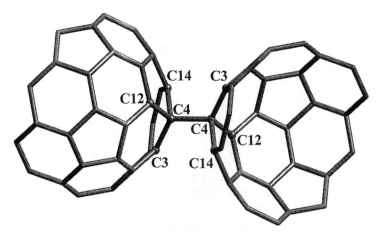

Figure 14.28 Geometry of the frontier (C3, C12, C14) and bridging (C4) carbon atoms on adjacent fullerenes in Na_2RbC_{60}, showing the optimum orientation and distortion of the molecules along the direction of polymerization (c axis).

electrons allows the bridging of each fulleride unit with two nearest neighbors through single C–C bond formation, leading to a polymeric structure. Comparative local density approximation (LDA) and quantum chemical studies of the stabilities of a variety of bridging structures for C_{60}^- and C_{60}^{3-} have confirmed these arguments [160,161]. In addition, steric factors appear to disfavor the [2+2] cycloaddition route for polymer formation in Na_2RbC_{60}. If Na_2RbC_{60} were to adopt a structure similar to that of the RbC_{60} polymer, Na^+ would encounter severe steric crowding, which is relieved in the case of the singly bonded polymer.

An additional difference between the polymer phases of Na_2RbC_{60} and RbC_{60} arises from the different precursor phases. The RbC_{60} polymer is formed at $T < 400$ K from the orientationally disordered fcc rock-salt (space group $Fm\bar{3}m$) structure, while the Na_2RbC_{60} polymer forms from the orientationally ordered primitive cubic $Pa\bar{3}$ phase at much lower temperature, $\sim$250 K. This has been demonstrated unambiguously by variable temperature solid state ^{13}C and ^{23}Na NMR spectroscopic measurements [156]. In brief, the magic-angle spinning ^{13}C NMR spectrum at 333 K displays a single sharp peak at δ 189, consistent with the fast rotational motion in the fcc phase. On cooling, a second peak appears gradually at δ 186, with an intensity growing at the expense of the first resonance, which becomes less intense until it eventually disappears. It can be assigned to the primitive cubic phase and remains sharp at 253 K, consistent with rapid large-amplitude librational motions of the fulleride ions about their equilibrium orientations. Slow cooling to 213 K leads to a reduction in intensity of the ^{13}C resonance (to $\sim$50% of its original value) and the appearance of numerous new features over the chemical shift range δ 0–250, implying low crystalline symmetry and considerable distortion of the C_{60}^{3-} ions. These results are mirrored by those obtained from static ^{23}Na NMR spectroscopy. So quasi-free molecular rotations in the fcc precursor phase of RbC_{60} allow frequent optimal contact between fullerenes for polymer formation through [2+2] cycloaddition. This contrasts with the extremely slow kinetics (vide infra) exhibited by the monomer → polymer transformation in Na_2RbC_{60}. The primitive cubic precursor is characterized by fullerene orientations in which 6:6 carbon bonds nest over pentagonal faces of neighboring molecules. Carbon bonds can form during orientational jumps, as the fullerenes perform small-amplitude librational motion [162], but the resulting kinetics are extremely slow and the transformation does not go to completion with the monomer phase surviving to low temperatures. One can also conclude that formation of additional C–C bridging bonds from the orientationally ordered cubic phase will be even slower.

The monomer → polymer transformation in Na_2RbC_{60} has been studied in detail by temperature-dependent synchrotron X-ray and neutron powder diffraction [159,163,164] during both slow cooling and heating protocols. On slow cooling, the fraction of polymeric phase starts to become significant at a temperature of 250 K ($\phi \approx 10\%$) increasing up to $\phi \approx 53\%$ on cooling to 200 K. After remaining for 7 hours at 200 K, ϕ only increases by $\sim$7% with its time evolution approximately described by a linear law, $\phi(t) = 53.0(2) + 0.018(1) \times t$ (min^{-1}). The extremely slow kinetics of the polymerization reaction are clearly evident with a complete transformation at this rate at 200 K requiring ∼44 hours. On heating, hysteretic behavior is observed, with the monoclinic phase disappearing at ∼270 K.

The formation of the polymer has also been explored in other ternary and quaternary sodium fullerides with stoichiometry $Na_2Rb_{1-x}Cs_xC_{60}$ $(0 \leq x \leq 1)$ and Na_2KC_{60}. Figure 14.29 shows the synchrotron X-ray diffraction patterns of a series of such salts at ambient pressure, obtained after slow cooling at 200 K

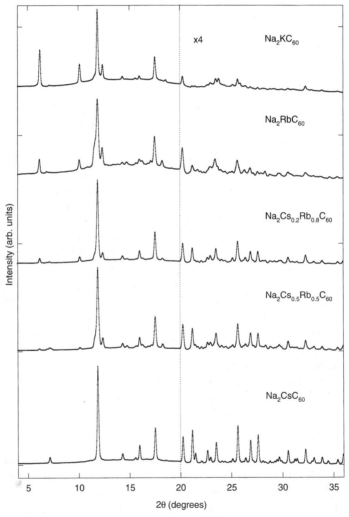

Figure 14.29 Synchrotron X-ray diffraction data of some sodium fullerides after slow cooling ($\lambda = 0.873\,\text{Å}$, $T = 200$ K).

[165]. At present, all the fullerides studied are found to polymerize, except the unique case of Na_2CsC_{60}, which remains strictly cubic under a variety of cooling treatments. It thus appears that polymer formation occurs only below a critical interball separation. This can be confirmed by studying the pressure dependence of the structure of solid Na_2CsC_{60} [163,164,166]. Synchrotron X-ray powder diffraction experiments have been performed on Na_2CsC_{60} at ambient temperature up to 8.25 GPa. When a pressure of 0.74 GPa is reached, the polymeric phase appears and coexists with the cubic one up to 1.05 GPa. The pressure–volume curve for the monoclinic phase up to ~8.25 GPa has been fitted with the Murnaghan equation of state (EOS): $P = (K_o/K_o')[(V_o/V)^{K_o'} - 1]$, where K_o is the average bulk modulus, $K_o'(= dK_o/dp)$ is its pressure derivative, and V_o is the unit cell volume at zero pressure. K_o for the polymer was found to be 28(1) GPa (compressibility, $\kappa = 4.0(1) \times 10^{-2}$ GPa^{-1}) with a pressure derivative $K_o' = 11(1)$. Substantial anisotropy in the compressibility also exists, with the structure being least compressible along c—the polymeric chain—and most compressible along b—the interchain direction. Similar behavior to Na_2CsC_{60} is also exhibited by the superconducting fulleride Li_3CsC_{60}, which polymerizes at high pressure to afford an isostructural polymer structure with the excess Li disordered over the corners of a $(Li_{1/8})_8$ cluster, centered at the Cs^+ site (Fig. 14.30) [167].

Finally, one would like to know whether the polymeric phase of Na_2RbC_{60} is superconducting; Direct current (dc) magnetization measurements for a rapidly cooled sample show bulk superconductivity with $T_c = 3.8$ K. However, if the sample is kept at 180 K for ~10 hours to ensure conversion to the polymer before further cooling to 2 K, there is little change in T_c (~3.6 K) but a drastic decrease in the superconducting fraction (to ~25% of the original value), consistent with two coexisting fractions: one superconducting (cubic) and one non-superconducting (monoclinic). Similar experiments on Na_2CsC_{60} ($T_c = 12$ K) reveal no dependence of the magnetization on thermal history, in agreement with the nonappearance of polymeric Na_2CsC_{60} in the diffraction experiments. Zhu [153] had suggested that pressure-polymerized $Na_2A'C_{60}$ may be superconducting, based on the observation of superconductivity in pressurized Na_2CsC_{60} [83]. The observation that superconductivity in Na_2RbC_{60} and Na_2CsC_{60} is confined only to the isotropic 3D phases does not support that conjecture. However, the connection of the suppressed value of T_c in quenched Na_2RbC_{60} with the presence of the monoclinic phase is intriguing. Figure 14.31 depicts the relationship between T_c and lattice parameter a for a variety of fulleride salts at both ambient and elevated pressures [168]. The high-pressure results define branches in this diagram with slopes about five to six times smaller than that encountered in the quaternary fullerides, $Na_2Rb_{1-x}Cs_xC_{60}$ ($0 \leq x \leq 1$) at ambient pressure [84]. Thus, while in the fcc fullerides the effects of chemical and physical pressure on the superconducting properties differ little, Na-containing primitive cubic fullerides show a clear differentiation, with chemical substitution leading to a much faster suppression of T_c than application of pressure. The pressure dependence of the superconducting properties

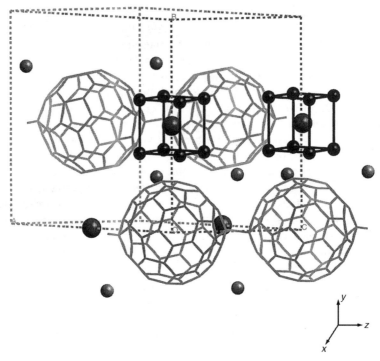

Figure 14.30 The crystal structure of polymerized Li_3CsC_{60}. The fullerenes form polymeric chains connected via single C–C bonds. The Cs^+ and Li^+ ions in the tetrahedral site are depicted as large dark and small light spheres, respectively, while the Li defects are shown as small dark spheres. The defect around the $(1/2, 0, 0)$ position has been removed for clarity.

of individual members of the $Na_2Rb_{1-x}Cs_xC_{60}$ ($0 \leq x \leq 1$) family differs very little from the phenomenology established for the fcc fullerides in which the larger alkali ions K^+, Rb^+, and Cs^+ reside in the tetrahedral interstices. Thus, the faster depression of T_c with interfullerene separation at ambient pressure, as x varies in $Na_2Rb_{1-x}Cs_xC_{60}$, appears to be metal specific. Its origin still poses intriguing questions and has yet to be explained unambiguously. While the conjecture that the dependence of T_c on interfullerene separation is generally much steeper in the primitive cubic structure than in the fcc one can now be discarded, a number of additional possible explanations still remain. These include (1) the existence of an as yet unidentified low-symmetry distortion of the primitive cubic structure; (2) a very sensitive modulation of the degree of electron transfer between Na and C_{60} by the interfullerene separation, which may lead to deviations from half-filling of the conduction band and rapid suppression of superconductivity, in analogy with the situation encountered in nonsuperconducting Li_2CsC_{60} [91]; and (3) the presence of intergrowths, or coexistence at the microscopic level, of superconducting cubic and nonsuper-

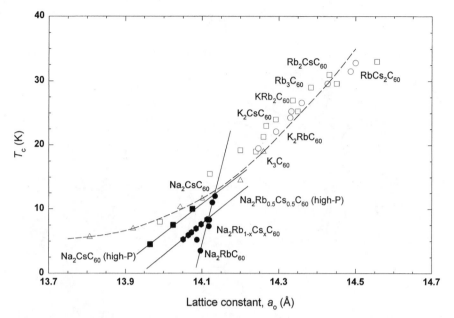

Figure 14.31 Relationship between the superconducting transition temperature, T_c, and the cubic lattice constant, a, of fulleride salts over a wide range of values for a.

conducting polymer domains, with the size of the latter growing as the lattice constant decreases with decreasing x in $Na_2Rb_{1-x}Cs_xC_{60}$.

The temperature dependence of the ESR susceptibility of the Na_2RbC_{60} polymer has revealed that it is conducting down to 4 K [169,170]. The results exclude the possibility of the existence of a ground state instability in this polymer and the formation of either a SDW or a CDW state. However, when the interchain separation is increased by forming the quaternary salt $Na_2Rb_{0.3}Cs_{0.7}C_{60}$, the system becomes more one-dimensional and a pseudogap of magnetic origin opens at the Fermi level below 45 K with three-dimensional magnetic ordering setting in below $T_N \approx 15$ K, as confirmed by the observation of an antiferromagnetic resonance at high field [171].

14.5.3 Na₄C₆₀ Salt

The crystal structures adopted by alkali fulleride salts, A_xC_{60}, have been known to depend sensitively on the value x as well as on the type of A used. For instance, in addition to the well-characterized merohedrally disordered fcc phases, A_3C_{60}, body-centered orthorhombic (bco), Cs_4C_{60} [98], and body-centered tetragonal (bct), A_4C_{60} (A = K, Rb), phases have been isolated [96]. This trend is followed when A = Na. Thus, an isostructural bct phase can also be isolated for Na_4C_{60} at high temperatures (space group $I4/mmm$, $a = 11.731$ Å, $c = 10.438$ Å at 550 K) [172]. The interfullerene distances are

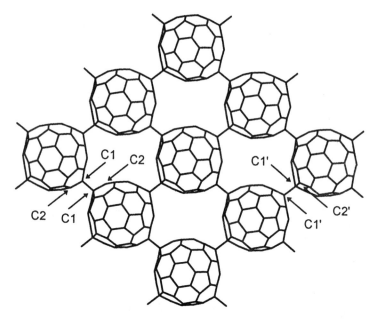

Figure 14.32   Polymer plane of C_{60} showing the near-neighbor molecular orientations and distortions in Na_4C_{60} [161,173]. Reprinted from *Phys. Rev. Lett.*, **81**, Launois, p. 4420, 1998, with permission from AIP.

~9.80 Å, implying the presence of $C_{60}{}^{4-}$ monomers in the structure. However, cooling below ~500 K leads to a reversible structural transition to a low-symmetry phase, which can best be described as comprising a two-dimensional network of $C_{60}{}^{4-}$ units, each bridged by four single C–C bonds to its neighbors within each plane (Fig. 14.32) [173]. Rietveld refinements of synchrotron X-ray powder diffraction data led to the structural characterization of this polymeric phase, which crystallizes in the monoclinic space group $I2/m$ with lattice dimensions $a = 11.235(5)$ Å, $b = 11.719(5)$ Å, $c = 10.276(5)$ Å, $\beta = 96.16(5)°$ at room temperature [173]. The nearest interfullerene distances within the plane are 9.28 Å, while they are 9.93 Å out of the plane. The short 9.28 Å distance is comparable to those found in all other cases in which single C–C bonds bridge neighboring fullerene units (dimerized AC_{60}, $Na_2A'C_{60}$, $(C_{59}N)_2$). The polymerization enthalpy has been reported as 50 J/g, which is almost four times as large as that measured for the singly bonded RbC_{60} dimer, 13 J/g [172]. Finally, of particular importance is the observation that the Na_4C_{60} solid is metallic in both its polymeric and monomeric forms, in contrast to the monomeric bct analogs, A_4C_{60}, formed by the large alkali metals K and Rb, which are insulating. The unique behavior of Na_4C_{60} has been rationalized within a Mott–Hubbard picture; hence, the short interfullerene distances lead to a reduced value of the on-site correlation energy, U, and to an increased value of the conduction bandwidth, W, thus shifting the (U/W) ratio below a critical value for a correlated metal-to-insulator transition [172].

14.6 HETEROFULLERENES AND THEIR DERIVATIVES

The simplest variant of fullerene clusters consists of substituting one of the carbon atoms on their skeleton by an aliovalent atom to afford on-ball doped molecules. Such systems are expected to display significantly perturbed structural and electronic character when compared to the parent fullerene precursors [174]. The first bulk synthesis of a heterofullerene, namely, the azafullerene with stoichiometry $C_{59}N$, was achieved by a three-step reaction sequence, starting from C_{60} [175,176]. As a result of the trivalency of nitrogen, compared to the tetravalency of carbon, nitrogen substitution of a single carbon atom on the C_{60} skeleton leads to the azafullerene radical $C_{59}N^{\cdot}$, which is isoelectronic with the $C_{60}^{\cdot -}$ radical anion—a process somewhat similar to doping silicon with phosphorus. $C_{59}N^{\cdot}$ is found to rapidly dimerize to yield the $(C_{59}N)_2$ dimer. If the synthetic procedure is somewhat modified to involve the presence of a reducing agent stable to strong acid conditions, dimerization is prevented, and the monomeric hydroazafullerene $C_{59}HN$ can be isolated [177].

The introduction of the nitrogen atom in the fullerene cluster strongly perturbs the electronic and geometric character of the parent C_{60} molecule, resulting in a very reactive $C_{59}N$ radical. Theoretical calculations [178] find that the $C_{59}HN$ molecule is particularly stable (binding energy ≈ 72 kcal/mol), while the $(C_{59}N)_2$ dimer adopts a closed structure (binding energy ≈ 18 kcal/mol) in which the nitrogens are *trans* to each other with an intermolecular C–C bond, formed by C atoms neighboring the N atoms on each monomer, of length 1.61 Å. Experimentally, $(C_{59}N)_2$ is found to be a diamagnetic, insulating solid. Direct observation of the $C_{59}N$ radicals has been achieved by light-induced and thermal homolytic ESR studies [179–181]. The ESR spectrum of $(C_{59}N)_2$ shows three sharp (70 mG) lines, typical of a single ^{14}N hyperfine splitting with a hyperfine coupling constant of 3.73 G. The g-value of the $C_{59}N$ radical is 2.0013(2), higher than that reported for the Jahn-Teller distorted C_{60} radical anion, 1.9991. In addition, Raman spectroscopic studies have led to the identification of the interfullerene C–C vibrational modes that occur at 82, 103, and 111 cm^{-1} [182].

Solid $(C_{59}N)_2$ tenaciously retains solvent molecules, which, however, can be removed by sublimation. Synchrotron X-ray diffraction measurements on sublimed $(C_{59}N)_2$ provide unambiguous evidence for molecular dimerization, with the solid adopting a monoclinic cell (space group $P2_1/a$) with lattice dimensions $a = 17.25$ Å, $b = 9.96$ Å, $c = 19.44$ Å, and $\beta = 124.3°$ (Fig. 14.33) [183]. This is reminiscent of the situation encountered for the metastable phase of RbC_{60} in which the C_{60}^- anions form dimeric units. The intradimer center-to-center separations in $(C_{59}N)_2$ are ~ 9.41 Å, comparable to the distances found in other fullerene systems bridged by single C–C bonds like the isoelectronic $(C_{60})_2^{2-}$ dimers in RbC_{60}, the $(C_{60}^{3-})_n$ polymer in Na_2RbC_{60} and the two-dimensional $(C_{60}^{4-})_n$ polymer in Na_4C_{60}. All other distances are considerably larger with the average distance being 9.97 Å, only marginally smaller than that in orientationally disordered C_{60} (10.02 Å).

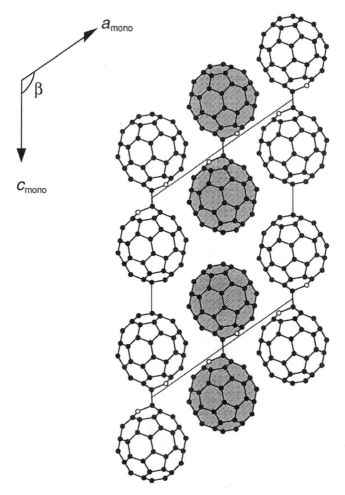

Figure 14.33 Unit-cell basal plane projection of the structure of $(C_{59}N)_2$ down the b axis. Only one of the two merohedral images is shown. The open circles label the N atoms. Shaded molecules lie at $b = 1/2$.

The diversity of bonding interactions present in the $(C_{59}N)_2$ solid has been probed by pressure tuning of the intermolecular distances in its hexagonal form [184]. While no pressure-induced phase transition was evident, the (c/a) ratio increased gradually with increasing P from 1.623 at ambient pressure to the ideal hcp value of 1.633 at ~6.5 GPa; in this pressure range, the interdimer center-to-center distances are compressed to values comparable to the intradimer distances. The effect of high pressure, up to 10 GPa, on the phonon spectra of azafullerene, $(C_{59}N)_2$, at room temperature has also been investigated by Raman spectroscopy. The pressure dependence of the selected phonon frequencies exhibits reversible changes in their pressure coefficient in the same pressure region, namely, 6.0 ± 0.5 GPa [185]. The azafullerene molecular units survive intact to pressures as high as 22 GPa and the solid is somewhat less compressible (atmospheric-pressure isothermal bulk modulus, $K_0 = 21.5(8)$ GPa) than pristine C_{60}, comprising stiffer nearest-neighbor bonds [184]. Finally, a study of the electronic structure by electron energy loss and photoemission spectroscopy [186] revealed little mixing between the N and C electronic states with strong localization of the excess electron at the N atom. Thus, the chemical picture of $(C_{59}N)_2$ that has emerged thus far is consistent with the presence of a triply coordinated N atom with a lone pair in each azafullerene unit.

The measured redox potentials by cyclic voltammetry imply that both $(C_{59}N)_2$ and $C_{59}HN$ should be more easily reduced than C_{60}. Indeed, $(C_{59}N)_2$ solid is easily reduced by alkali metals to afford the monomeric $C_{59}N^{6-}$ azafulleride ion, which has been isolated and structurally characterized as the $A_6C_{59}N$ (A = K, Rb) salts [187,188]. It is instructive to compare the properties of this salt with those of its C_{60} analog, K_6C_{60}. C_{60} has a triply degenerate t_{1u} lowest unoccupied molecular orbital (LUMO), which, when half-filled with electrons stemming from alkali metal atoms, gives rise to the metallic A_3C_{60} salts with half-filled electronic bands. When the t_{1u} level is completely filled, the stoichiometry requires six alkali metal ions, the resistivity reaches a maximum, and the resulting insulating A_6C_{60} salts adopt a body-centered cubic (bcc) structure. In the case of the $C_{59}N$ molecule, as a result of the symmetry lowering, the t_{1u} triplet LUMO is no longer degenerate:

$$\text{---} \quad \text{LUMO} \qquad \text{--}\!\!+\!\!\text{--} \quad \text{LUMO}$$

$$C_{60} \qquad\qquad\qquad C_{59}N$$

Complete filling of the levels should require, in principle, only five alkali metal atoms. Thus, reduction of $C_{59}N$ with six electrons necessitates population of the LUMO+1 state:

$$\text{---} \quad \text{LUMO+1} \qquad + \quad \text{LUMO+1}$$

$$\text{⧺ ⧺ ⧺} \quad \text{LUMO} \qquad \text{⧺⧻⧺} \quad \text{LUMO}$$

$$C_{60}{}^{6-} \qquad\qquad\qquad C_{59}N^{6-}$$

The resulting $A_6C_{59}N$ salts adopt bcc structures (space group $Im\bar{3}$), essentially isostructural with A_6C_{60}, but with somewhat smaller lattice constants (Fig. 14.34). The center-to-center distance between neighboring azafulleride ions at ambient temperature is 9.83 Å for $K_6C_{59}N$ and 10.00 Å for $Rb_6C_{59}N$. Band structure calculations suggest the likelihood of metallic behavior, if the effect of spin polarization on the half-filled band is ignored. The measured four-probe conductivities of pressed pellets of these azafullerides at room temperature range from $\sigma \approx 0.1$ to $1 \ \Omega^{-1} \ cm^{-1}$, implying significantly increased values of

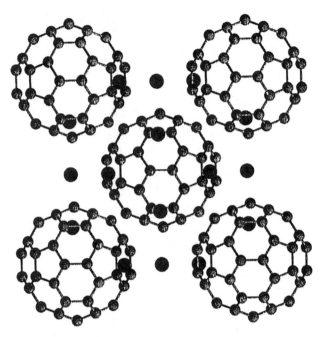

Figure 14.34   Basal plane projection of the body-centered cubic structure of $A_6C_{59}N$ salts.

conductivity than those of the wide gap insulators A_6C_{60} and $(C_{59}N)_2$, despite the expected effects of grain boundaries. The values are also somewhat larger than those in C_{60}^- salts, like $(TDAE)C_{60}$. Superconducting quantum interference device (SQUID) magnetic measurements reveal that the temperature dependence of the susceptibility, χ, can be well described by the expression $\chi = \chi_0 + C/(T - \Theta)$, that is, a temperature-independent diamagnetic contribution and a Curie–Weiss term. No large Pauli paramagnetic term, which will signify metallic behavior, is evident unless the density of states at the Fermi level is much smaller than those encountered in metallic alkali fullerides, and the corresponding temperature-independent Pauli term is dwarfed by the large value of the core diamagnetism. The Curie–Weiss term implies antiferromagnetically coupled localized spins on the $(C_{59}N)^{6-}$ ions with values of $\Theta \approx -(5-7)$ K; the Curie constants are $C \approx 0.03$ emu $\cdot$ K/mol, corresponding to a magnetic moment of $\sim 0.5\ \mu_B$ per $(C_{59}N)^{6-}$ ion, much reduced from the value expected for a single unpaired electron per ball. The origin of this reduction is unclear but it might be related to peculiarities of the electronic structure of the $A_6C_{59}N$ salts, associated with the substantial localized distortion in the vicinity of the N atom. Alternatively, the paramagnetic term may reflect the presence of large concentrations ($\sim 10\%$) of defects or impurities in the solid; this argument, however, is weakened by the observation that we obtain similar values for the magnetic moments for all $A_6C_{59}N$ salts studied. In any event, the

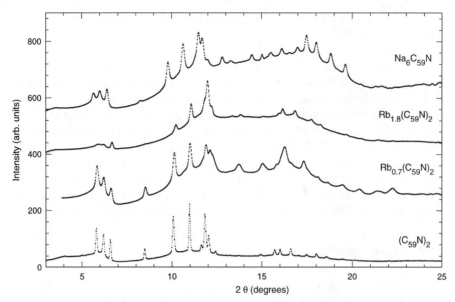

Figure 14.35 Synchrotron X-ray diffraction profiles of $Na_6C_{59}N$, $Rb_{1.8}(C_{59}N)_2$, $Rb_{0.7}(C_{59}N)_2$, and hexagonal $(C_{59}N)_2$ at ambient temperature ($\lambda = 0.873$ Å).

discovery that azafullerene can give rise to a series of alkali metal intercalates opens the way to the synthesis of materials with electronic properties potentially as rich as those of their fullerene antecedents.

The above data on saturation-doped $A_6C_{59}N$ salts clearly establish that the $(C_{59}N)_2$ dimer reverts to a monomer upon reaction with excess alkali metals. However, the electrochemical data [175] on $(C_{59}N)_2$ show that the addition of the first three electrons to the dimer gives rise to double waves in cyclic voltametric experiments, implying the survival of the bridging C–C bond upon reduction. Attempts to explore the solid state intercalation chemistry of $(C_{59}N)_2$ at lower doping levels led to the synthesis and characterization [188] of salts with stoichiometry $Rb_{2x}(C_{59}N)_2$ with $x = 0.35$ and 0.9. The resulting crystalline products were studied by synchrotron X-ray diffraction (Fig. 14.35), revealing hexagonal structures (space group $P6_3/mmc$) with the intercalated Rb essentially residing in the octahedral interstices. The lattice dimensions show a slight contraction upon intercalation ($a = 9.87$ Å, $c = 16.12$ Å for $Rb_{0.7}(C_{59}N)_2$ and $a = 9.81$ Å, $c = 16.06$ Å for $Rb_{1.8}(C_{59}N)_2$) relative to pristine $(C_{59}N)_2$, consistent with charge transfer from Rb to $(C_{59}N)_2$. The dc magnetization measurements reveal the presence of a Curie-like paramagnetic component superimposed on a temperature-independent background.

Finally, examination of the as-prepared azafullerene solid, $(C_{59}N)_2$, under the electron microscope reveals a remarkable self-assembly of the quasispherical nitrogen-substituted fullerene units on the mesoscopic length scale into large (micrometer-size) hollow spherical particles [189]. Unlike the curling of extended graphitic networks, which occurs only under intense electron-beam irradiation, the mesoscopic spherical structures of azafullerene can be synthesized chemically under mild conditions. The origin of the spherical, hollow particle morphology is of interest, as it has no precedent among other all-carbon molecular fullerene solids. It may arise from the self-assembly of smaller spherical particles as a result of rapid solvent evaporation, analogous to spherical particles resulting from spray-drying procedures or from flash-evaporation of foams. It is remarkable that the small azafullerene balls form a hierarchical mesoscopic structure that is itself spherical and tantalizing to contemplate the possibility that the holey nature of the balls is a consequence of the aspherical nature of the polar molecule itself. The ability to synthesize novel macroscopic forms of fullerene-based solids could lead to important materials applications.

14.7 CONCLUSION

Work on solid fullerenes and their numerous derivatives continues unabated to produce novel and unexpected results. We attempted to deal in the present chapter with many of the aspects of fullerene solid state chemistry and physics in some detail. We hope that the breadth and diversity of the examples chosen to illustrate the advances in this field have been able to testify to the vigor and the continued vitality of the subject.

REFERENCES

1. W. Krätschmer, L. D. Lamb, K. Fostiropoulos, and D. R. Huffman, *Nature* **1990**, *347*, 354.

2. P. A. Heiney, J. E. Fischer, A. R. McGhie, W. J. Romanow, A. M. Denenstein, J. P. McCauley, A. B. Smith, and D. E. Cox, *Phys. Rev. Lett.* **1991**, *66*, 2911.

3. R. Fleming, B. Hessen, T. Siegrist, A. R. Cortan, P. Marsh, R. Tycko, G. Dabbagh, and R. C. Haddon, in *Synthesis Properties and Chemistry of Large Carbon Clusters*. American Chemical Society Symposium Series Vol. 481. American Chemical Society, Washington, DC, 1991, p. 25.

4. P. A. Heiney, *J. Phys. Chem. Solids* **1992**, *53*, 1333.

5. W. I. F. David, R. M. Ibberson, T. J. S. Dennis, J. P. Hare, and K. Prassides, *Europhys. Lett.* **1992**, *18*, 225.

6. C. S. Yannoni, R. D. Johnson, G. Meijer, D. S. Bethune, and J. R. Salem, *J. Phys. Chem.* **1991**, *95*, 9.

7. R. D. Johnson, C. S. Yannoni, H. C. Dorn, J. R. Salem, and D. S. Bethune, *Science* **1992**, *255*, 1235.

8. D. A. Neumann, J. R. D. Copley, R. L. Cappelletti, W. A. Kamitakahara, R. M. Lindstrom, K. M. Creegan, D. M. Cox, W. J. Romanow, N. Coustel, J. P. McCauley, N. C. Maliszewskyj, J. E. Fischer, and A. B. Smith, *Phys. Rev. Lett.* **1991**, *67*, 3808.

9. R. F. Kiefl, J. W. Schneider, A. MacFarlane, K. Chow, T. L. Duty, T. L. Estle, B. Hitti, R. L. Lichti, E. J. Ansaldo, C. Schwab, P. W. Percival, G. Wei, S. Wlodek, K. Kojima, W. J. Romanow, J. P. McCauley, N. Coustel, J. E. Fischer, and A. B. Smith, *Phys. Rev. Lett.* **1992**, *68*, 1347.

10. R. Tycko, G. Dabbagh, R. M. Fleming, R. C. Haddon, A. V. Makhija, and S. M. Zahurak, *Phys. Rev. Lett.* **1991**, *67*, 1886.

11. P. C. Chow, X. Jiang, G. Reiter, P. Wochner, S. C. Moss, J. D. Axe, J. C. Hanson, R. K. McMullan, R. L. Meng, and C. W. Chu, *Phys. Rev. Lett.* **1992**, *69*, 2943.

12. W. Press and A. Hüller, *Acta Crystallogr. A* **1971**, *29*, 252.

13. K. H. Michel, J. R. D. Copley, and D. A. Neumann, *Phys. Rev. Lett.* **1992**, *68*, 2929.

14. D. Lamoen and K. H. Michel, *Z. Phys. B* **1993**, *92*, 323.

15. D. Lamoen and K. H. Michel, *J. Chem. Phys.* **1994**, *101*, 1435.

16. P. Launois, S. Ravy, and R. Moret, *Phys. Rev. B* **1997**, *55*, 2651; P. Launois, S. Ravy, and R. Moret, *Int. J. Mod. Phys. B* **1999**, *13*, 253.

17. C. E. Nordmann and D. L. Schmitkons, *Acta Crystallogr.* **1965**, *18*, 764; J. P. Amourex, M. Bée, and J. C. Damien, *Acta Crystallogr. B* **1980**, *36*, 2633.

18. A. Dworkin, H. Szwarc, S. Leach, J. P. Hare, T. J. S. Dennis, H. W. Kroto, R. Taylor, and D. R. M. Walton, *C.R. Acad. Sci. Paris II* **1991**, *312*, 979.

19. R. Sachidanandam and A. B. Harris, *Phys. Rev. Lett.* **1991**, *67*, 1467.

20. W. I. F. David, R. M. Ibberson, J. C. Matthewman, K. Prassides, T. J. S. Dennis, J. P. Hare, H. W. Kroto, R. Taylor, and D. R. M. Walton, *Nature* **1991**, *353*, 147.

21. P. Lu, X. P. Li, and R. M. Martin, *Phys. Rev. Lett.* **1992**, *68*, 1551; X. P. Li, J. P. Lu, and R. M. Martin, *Phys. Rev. B* **1992**, *46*, 4300.

22. M. Sprik, A. Cheng, and M. L. Klein, *J. Phys. Chem.* **1992**, *95*, 6762.

23. W. I. F. David, R. M. Ibberson, and T. Matsuo, *Proc. R. Soc. A* **1993**, *442*, 129.

24. C. Meingast and F. Gugenberger, *Mod. Phys. Lett. B* **1993**, *7*, 1703.

25. W. I. F. David, R. M. Ibberson, T. J. S. Dennis, J. P. Hare, and K. Prassides, *Europhys. Lett.* **1992**, *18*, 735.

26. R. C. Yu, N. Tea, M. B. Salamon, D. Lorents, and R. Malhotra, *Phys. Rev. Lett.* **1992**, *68*, 2050.

27. H. B. Burgi, E. Blanc, D. Schwarzenbach, S. Z. Liu, Y. J. Lu, M. M. Kappes, and J. A. Ibers, *Angew. Chem.* **1992**, *31*, 640.

28. R. Moret, P. Launois, and S. Ravy, *Fullerene Sci. Technol.* **1996**, *4*, 1287.

29. O. Blaschko, G. Krexner, C. Maier, and R. Karawatzki, *Phys. Rev. B* **1997**, *56*, 2288.

30. D. A. Neumann, J. R. D. Copley, D. Reznik, W. A. Kamitakahara, J. J. Rush, R. L. Paul, and R. M. Lindstrom, *J. Phys. Chem. Solids* **1993**, *54*, 1699.

31. J. R. D. Copley, D. A. Neumann, R. L. Cappelletti, and W. A. Kamitakahara, *J. Phys. Chem. Solids* **1992**, *53*, 1353.

32. A. M. Rao, P. Zhou, K. A. Wang, G. T. Hager, J. M. Holden, Y. Wang, W. T. Lee, X. X. Bi, P. C. Eklund, D. S. Cornett, M. A. Duncan, and I. J. Amster, *Science* **1993**, *259*, 955.

33. M. Nunez-Regueiro, L. Marques, J. L. Hodeau, and M. Perroux, *Phys. Rev. Lett.* **1995**, *74*, 278; Y. Iwasa, T. Arima, R. M. Fleming, T. Siegrist, O. Zhou, R. C.

Haddon, L. J. Rothberg, K. B. Lyons, H. L. Carter, A. F. Hebard, R. Tycko, G. Dabbagh, J. J. Krajewski, G. A. Thomas, and T. Yagi, *Science* **1994**, *264*, 1570; C. Xu and G. E. Scuseria, *Phys. Rev. Lett.* **1995**, *74*, 274; L. Marques, J. L. Hodeau, M. Nunez-Regueiro, and M. Perroux, *Phys. Rev. B* **1996**, *12*, 633.

34. B. Sundqvist, *Adv. Phys.* **1999**, *48*, 1.

35. R. Taylor, J. P. Hare, A. K. Abdul-Sada, and H. W. Kroto, *J. Chem. Soc. Chem. Commun.* **1990**, 1423.

36. A. V. Nikolaev, T. J. S. Dennis, K. Prassides, and A. K. Soper, *Chem. Phys. Lett.* **1994**, *223*, 143.

37. J. Baker, P. W. Fowler, P. Lazzaretti, M. Malagoli, and R. Zanasi, *Chem. Phys. Lett.* **1991**, *184*, 182; K. Raghavachari and C. M. Rohlfing, *J. Phys. Chem.* **1991**, *95*, 5768; F. Negri, G. Orlandi, and F. Zerbetto, *J. Am. Chem. Soc.* **1991**, *113*, 6037; G. E. Scuseria, *Chem. Phys. Lett.* **1991**, *180*, 451; C. Z. Wang, C. H. Xu, and K. H. Ho, *Phys. Rev. B* **1992**, *46*, 9761; W. Andreoni, F. Gygi, and M. Parrinello, *Chem. Phys. Lett.* **1992**, *189*, 241.

38. K. Hedberg, L. Hedberg, M. Buhl, D. S. Bethume, C. A. Brown, and R. D. Johnson, *J. Am. Chem. Soc.* **1997**, *119*, 5314.

39. D. R. McKenzie, C. A. Davis, D. J. H. Cockayne, D. A. Muller, and A. M. Vassallo, *Nature* **1992**, *355*, 622.

40. G. B. M. Vaughan, P. A. Heiney, J. E. Fischer, D. E. Luzzi, D. A. Ricketts-Foot, A. R. McGhie, Y.-W. Hui, A. L. Smith, D. E. Cox, W. J. Romanow, B. H. Allen, N. Coustel, J. P. McCauley, and A. B. Smith, *Science* **1991**, *254*, 1350.

41. M. A. Verheijen, H. Meekes, G. Meijer, P. Bennema, J. L. de Boer, S. van Smaalen, G. van Tendeloo, S. Amelinckx, S. Muto, and J. van Landuyt, *Chem. Phys.* **1992**, *166*, 287; G. van Tendeloo, S. Amelinckx, J. L. de Boer, S. van Smaalen, M. A. Verheijen, H. Meekes, and G. Meijer, *Europhys. Lett.* **1993**, *21*, 329.

42. C. Christides, I. M. Thomas, T. J. S. Dennis, and K. Prassides, *Europhys. Lett.* **1993**, *22*, 611.

43. C. Christides, T. J. S. Dennis, K. Prassides, R. L. Cappelletti, D. A. Neumann, and J. R. D. Copley, *Phys. Rev. B* **1994**, *49*, 2897.

44. G. B. M. Vaughan, P. A. Heiney, D. E. Cox, J. E. Fischer, A. R. McGhie, A. L. Smith, R. M. Strongin, M. A. Cichy, and A. B. Smith, *Chem. Phys.* **1993**, *178*, 599.

45. M. Sprik, A. Cheng, and M. L. Klein, *Phys. Rev. Lett.* **1992**, *69*, 1660.

46. Y. Iwasa, T. Furudate, T. Fukawa, T. Ozaki, T. Mitagi, T. Yagi, and T. Arima, *Appl. Phys. A* **1997**, *64*, 251.

47. M. Premila, C. S. Sundar, P. Ch. Sahu, A. Bharathi, Y. Hariharan, D. V. S. Muthu, and A. K. Sood, *Solid State Commun.* **1997**, *104*, 237.

48. V. D. Blank, N. R. Serebryanaya, G. A. Dubitsky, S. G. Buga, V. N. Denisov, B. N. Mavrin, A. N. Ivlev, S. N. Sulyanov, and N. A. Lvova, *Phys. Lett. A* **1998**, *248*, 415.

49. H. Kawada, Y. Fujii, H. Nakao, Y. Murakami, T. Watanuki, H. Suematsu, K. Kikuchi, Y. Achiba, and I. Ikemoto, *Phys. Rev. B* **1995**, *51*, 8723.

50. Y. Maniwa and K. Kume, *Phys. Rev. B* **1996**, *53*, 14196.

51. T. Nishikawa and K. Yokoi, *J. Phys. Soc. Jpn.* **1998**, *67*, 899.

52. J. C. Niles and X. Q. Wang, *J. Chem. Phys.* **1995**, *103*, 7040.

53. X. Q. Wang, C. Z. Wang, B. L. Zhang, and K. M. Ho, *Chem. Phys. Lett.* **1993**, *207*, 349; S. Saito, S. Sawada, N. Hamada, and A. Oshiyama, *Mater. Sci. Eng. B* **1993**, *19*, 105.

54. S. Negase and K. Kobayashi, *Chem. Phys. Lett.* **1994**, *231*, 319.

55. K. M. Allen, T. J. S. Dennis, M. J. Rosseinsky, and H. Shinohara, *J. Am. Chem. Soc.* **1998**, *120*, 6681.

56. S. Margadonna, C. M. Brown, T. J. S. Dennis, A. Lappas, P. Pattison, K. Prassides, and H. Shinohara, *Chem. Mater.* **1998**, *10*, 1742.

57. A. L. Balch, A. S. Ginwalla, J. W. Lee, B. C. Noll, and M. Olmstead, *J. Am. Chem. Soc.* **1994**, *116*, 2227.

58. T. J. S. Dennis, K. Tsutomu, T. Tomiyama, and H. Shinohara, *Chem. Commun.* **1998**, 619.

59. K. Tanigaki, T. W. Ebbesen, S. Saito, J. Mizuki, J. S. Tsai, Y. Shimakawa, Y. Kubo, and S. Kuroshima, *Nature* **1991**, *352*, 222.

60. S. Chakravarty, M. P. Gelfand, and S. Kivelson, *Science* **1991**, *254*, 970.

61. O. V. Dolgov and I. I. Mazin, *Solid State Commun.* **1992**, *81*, 935.

62. C. M. Varma, J. Zaanen, and K. Raghavachari, *Science* **1991**, *254*, 989.

63. M. Schlüter, M. Lannoo, M. Needels, G. A. Baraff, and D. Tománek, *Phys. Rev. Lett.* **1992**, *68*, 526.

64. O. Gunnarsson, *Rev. Mod. Phys.* **1997**, *69*, 575.

65. P. W. Stephens, L. Mihaly, P. L. Lee, R. L. Whetten, S. M. Huang, R. Kaner, F. Deiderich, and K. Holczer, *Nature* **1991**, *351*, 632.

66. R. M. Fleming, A. P. Ramirez, M. J. Rosseinsky, D. W. Murphy, R. C. Haddon, S. M. Zahurak, and A. V. Makhija, *Nature* **1991**, *352*, 787.

67. G. Sparn, J. D. Thompson, R. L. Whetten, S. M. Huang, R. B. Kaner, F. Diederich, G. Grüner, and K. Holczer, *Phys. Rev. Lett.* **1992**, *68*, 1228.

68. K. M. Allen, W. I. F. David, J. M. Fox, R. M. Ibberson, and M. J. Rosseinsky, *Chem. Mater.* **1995**, *7*, 764.

69. G. Nowitzke, G. Wortmann, H. Werner, and R. Schlogl, *Phys. Rev. B* **1996**, *54*, 13230.

70. I. Hirosawa, H. Kimura, J. Mizuki, and K. Tanigaki, *Phys. Rev. B* **1995**, *51*, 3038.

71. R. E. Walstedt, D. W. Murphy, and M. J. Rosseinsky, *Nature* **1993**, *362*, 611.

72. G. Zimmer, M. Helmle, M. Mehring, F. Rachdi, J. Reichenbach, L. Firlej, and P. Bernier, *Europhys. Lett.* **1993**, *24*, 59.

73. H. Alloul, K. Holczer, Y. Yoshinari, and O. Klein, *Physica C* **1994**, *235*, 2509.

74. G. Zimmer, K. F. Their, M. Mehring, F. Rachdi, and J. E. Fischer, *Phys. Rev. B* **1996**, *53*, 5620.

75. R. W. Lof, M. A. VanVeenendaal, B. Koopmans, H. T. Jonkman, and G. A. Sawatzky, *Phys. Rev. Lett.* **1992**, *68*, 3924.

76. J. E. Fischer, G. Bendele, R. Dinnebier, P. W. Stephens, C. L. Lin, N. Bykovetz, and Q. Zhu, *J. Phys. Chem Solids* **1995**, *56*, 1445.

77. G. M. Bendele, P. W. Stephens, and J. E. Fischer, *Europhys. Lett.* **1998**, *41*, 553.

78. M. Apostol, C. Goze, F. Rachdi, M. Mehring, and J. E. Fischer, *Solid State Commun.* **1996**, *98*, 253.

79. C. H. Pennington, C. Hahm, V. A. Stenger, K. Gorny, C. H. Recchia, J. A. Martindale, D. R. Buffinger, and R. P. Ziebarth, *Phys. Rev. B* **1996**, *54*, R6853.

80. T. Yildirim, J. E. Fischer, A. B. Harris, P. W. Stephens, D. Liu, L. Brard, R. M. Strongin, and A. B. Smith, *Phys. Rev. Lett.* **1993**, *71*, 1383.

81. K. Prassides, C. Christides, I. M. Thomas, J. Mizuki, K. Tanigaki, I. Hirosawa, and T. W. Ebbesen, *Science* **1994**, *263*, 950.

82. K. Kniaz, J. E. Fischer, Q. Zhu, M. J. Rosseinsky, O. Zhou, and D. W. Murphy, *Solid State Commun.* **1993**, *88*, 47.

83. J. Mizuki, M. Takai, H. Takahashi, N. Mori, I. Hirosawa, K. Tanigaki, and K. Prassides, *Phys. Rev. B* **1994**, *50*, 3466.

84. T. Yildirim, J. E. Fischer, R. Dinnebier, P. W. Stephens, and C. L. Lin, *Solid State Commun.* **1995**, *93*, 269.

85. Y. Maniwa, T. Saito, K. Kume, K. Kikuchi, I. Ikemoto, S. Suzuki, Y. Achiba, I. Hirosawa, M. Kosaka, and K. Tanigaki, *Phys. Rev. B* **1995**, *52*, R7054.

86. K. Tanigaki and K. Prassides, *J. Mater. Chem.* **1995**, *5*, 1515.

87. P. Petit and J. Robert, *Appl. Magn. Reson.* **1996**, *11*, 183.

88. P. Petit, J. Robert, T. Yildirim, and J. E. Fischer, *Phys. Rev. B* **1998**, *57*, 1226.

89. N. Laouini, O. K. Andersen, and O. Gunnarsson, *Phys. Rev. B* **1995**, *51*, 17446.

90. Y. Maniwa, D. Sugiura, K. Kume, K. Kikuchi, S. Suzuki, Y. Achiba, I. Hirosawa, K. Tanigaki, H. Shimoda, and Y. Iwasa, *Phys. Rev. B* **1996**, *54*, R6861.

91. I. Hirosawa, K. Prassides, J. Mizuki, K. Tanigaki, A. Gevaert, A. Lappas, and J. C. Cockcroft, *Science* **1994**, *264*, 1294.

92. M. Kosaka, K. Tanigaki, K. Prassides, M. Margadonna, A. Lappas, C. M. Brown, and A. N. Fitch, *Phys. Rev. B* **1999**, 59, R6628.

93. S. Margadonna, C. M. Brown, K. Prassides, A. N. Fitch, K. D. Knudsen, T. Le Bihan, M. Mezouar, I. Hirosawa, and K. Tanigaki, *Int. J. Inorg. Mater.* **1999**, *1*, 157.

94. S. Margadonna, K. Prassides, A. N. Fitch, M. Kosaka, and K. Tanigaki, *J. Am. Chem. Soc.* **1999**, *121*, 6318.

95. O. Zhou, J. E. Fischer, N. Coustel, S. Kycia, Q. Zhu, A. R. McGhie, W. J. Romanow, J. P. McCauley, A. B. Smith, and D. E. Cox, *Nature* **1991**, *351*, 462.

96. R. M. Fleming, M. J. Rosseinsky, A. P. Ramirez, D. W. Murphy, J. C. Tully, R. C. Haddon, T. Siegrist, R. Tycko, S. H. Glarum, P. Marsh, G. Dabbagh, S. M. Zahurak, A. V. Makhija, and C. Hampton, *Nature* **1991**, *352*, 701.

97. C. A. Kuntscher, G. M. Bendele, and P. W. Stephens, *Phys. Rev. B* **1997**, *55*, R3366.

98. P. Dahlke, P. F. Henry, and M. J. Rosseinsky, *J. Mater. Chem.* **1998**, *8*, 1571.

99. R. F. Kiefl, T. L. Duty, J. W. Schneider, A. MacFarlane, K. Chow, J. W. Elzey, P. Meudels, G. D. Morris, J. H. Brewer, E. J. Ansaldo, C. Niedermayer, D. R. Noakes, C. E. Stronach, B. Hitti, and J. E. Fischer, *Phys. Rev. Lett.* **1992**, *69*, 2005.

100. P. J. Benning, F. Stepniak, D. M. Poirier, J. L. Martins, J. H. Weaver, L. P. F. Chibante, and R. E. Smalley, *Phys. Rev. B* **1993**, *47*, 13843.

101. M. Knupfer and J. Fink, *Phys. Rev. Lett.* **1997**, *79*, 2714.

102. M. J. Rosseinsky, D. W. Murphy, R. M. Fleming, R. Tycko, A. P. Ramirez, T. Siegrist, G. Dabbagh, and S. E. Barrett, *Nature* **1992**, *356*, 416.

103. W. Andreoni, P. Gianozzi, and M. Parrinello, *Phys. Rev. Lett.* **1994**, *72*, 848.

104. T. Yildirim, O. Zhou, J. E. Fischer, N. Bykovetz, R. A. Strongin, M. A. Cichy, A. B. Smith, C. L. Lin, and R. Jelinek, *Nature* **1992**, *360*, 568.

105. T. Yildirim, L. Barbedette, J. E. Fischer, C. L. Lin, J. Robert, P. Petit, and T. T. M. Palstra, *Phys. Rev. Lett.* **1996**, *77*, 167.

106. A. C. Duggan, J. M. Fox, P. F. Henry, S. J. Heyes, D. E. Laurie, and M. J. Rosseinsky, *Chem. Commun.* **1996**, 1191

107. T. Yildirim, L. Barbedette, J. E. Fischer, G. M. Bendele, P. W. Stephens, C. L. Lin, C. Goze, F. Rachdi, J. Robert, P. Petit, and T. T. M. Palstra, *Phys. Rev. B* **1996**, *54*, 11981.

108. Y. Iwasa, H. Hayashi, T. Furudate, and T. Mitani, *Phys. Rev. B* **1996**, *54*, 14960.

109. Y. Iwasa, M. Kawaguchi, H. Iwasaki, T. Mitani, N. Wada, and T. Hasegawa, *Phys. Rev. B* **1998**, *57*, 13395.

110. A. R. Kortan, N. Kopylov, S. Glarum, E. M. Gyorgy, A. P. Ramirez, R. M. Fleming, F. A. Thiel, and R. C. Haddon, *Nature* **1992**, *355*, 529.

111. A. R. Kortan, N. Kopylov, R. M. Fleming, O. Zhou, F. A. Thiel, R. C. Haddon, and K. M. Rabe, *Phys. Rev. B* **1993**, *47*, 13070.

112. A. R. Kortan, N. Kopylov, E. Ozdas, A. P. Ramirez, R. M. Fleming, and R. C. Haddon, *Chem. Phys. Lett.* **1994**, *223*, 501

113. T. T. M. Palstra, O. Zhou, Y. Iwasa, P. E. Sulewski, R. M. Fleming, and B. R. Zegarski, *Solid State Commun.* **1994**, *93*, 327.

114. A. R. Kortan, N. Kopylov, S. Glarum, E. M. Gyorgy, A. P. Ramirez, R. M. Fleming, O. Zhou, F. A. Thiel, P. L. Trevor, and R. C. Haddon, *Nature* **1992**, *360*, 566.

115. M. Baenitz, L. Heinze, L. Luders, H. Werner, R. Schlogl, M. Weiden, G. Sparn, and F. Steglich, *Solid State Commun.* **1995**, *96*, 539.

116. M. Kraus, M. Kanowski, M. Baenitz, H. Werner, R. Schlogl, E. W. Scheidt, H. M. Vieth, and K. Luders, *Fullerene Sci. Technol.* **1995**, *3*, 115.

117. B. Gogia, K. Kordatos, H. Suematsu, K. Tanigaki, and K. Prassides, *Phys. Rev. B* **1998**, *58*, 1077.

118. S. C. Erwin, and M. R. Pederson, *Phys. Rev. B* **1993**, *47*, 14657.

119. S. Saito and A. Oshiyama, *Phys. Rev. Lett.* **1993**, *71*, 121.

120. T. Schedel-Niedrig, M. C. Bohm, H. Werner, J. Schulte, and R. Schlogh, *Phys. Rev. B* **1997**, *55*, 13542.

121. C. M. Brown, S. Taga, B. Gogia, K. Kordatos, S. Margadonna, K. Prassides, Y. Iwasa, K. Tanigaki, A. N. Fitch, and P. Pattison, *Phys. Rev. Lett.* **1999**, *83*, 2258.

122. E. Ozdas, A. R. Kortan, N. Kopylov, A. P. Ramirez, T. Siegrist, K. M. Rabe, H. E. Bair, S. Schuppler, and P. H. Citrin, *Nature* **1995**, *375*, 126.

123. P. H. Citrin, E. Ozdas, S. Schuppler, A. R. Kortan, and K. B. Lyons, *Phys. Rev. B* **1997**, *56*, 5213.

124. X. H. Chen and G. Roth, *Phys. Rev. B* **1995**, *52*, 15534.

125. J. Winter and H. Kuzmany, *Solid State Commun.* **1992**, *84*, 935.

126. D. M. Poirier, T. R. Ohno, G. H. Kroll, P. J. Benning, F. Stepniak, J. H. Weaver, L. P. F. Chibante, and R. E. Smalley, *Phys Rev. B* **1993**, *47*, 9870.

127. Q. Zhu, O. Zhou, J. E. Fischer, A. R. McGhie, W. J. Romanow, R. M. Strongin, M. A. Cichy, and A. B. Smith, *Phys. Rev. B* **1993**, *47*, 9870.

128. O. Chauvet, G. Oszlanyi, L. Forro, P. W. Stephens, M. Tegze, G. Faigel, and A. Janossy, *Phys. Rev. Lett.* **1994**, *72*, 2721.

129. P. W. Stephens, G. Bortel, G. Faigel, M. Tegze, A. Janossy, S. Pekker, G. Oszlanyi, and L. Forro, *Nature* **1994**, *370*, 636.

130. S. Pekker, L. Forro, L. Mihaly, and A. Janossy, *Solid State Commun.* **1994**, *90*, 349.

131. S. Pekker, A. Janossy, L. Mihaly, O. Chauvet, M. Carrard, and L. Forro, *Science* **1994**, *265*, 1077.

132. G. Faigel, G. Bortel, M. Tegze, L. Granasy, S. Pekker, G. Oszlanyi, O. Chauvet, G. Baumgartner, L. Forro, P. W. Stephens, L. Mihaly, and A. Janossy, *Phys. Rev. B* **1995**, *52*, 3199.

133. P. Launois, R. Moret, J. Hone, and A. Zettl, *Phys. Rev. Lett.* **1998**, *81*, 4420.

134. F. Bommeli, L. Degiorgi, P. Wachter, O. Legeza, A. Janossy, G. Oszlanyi, O. Chauvet, and L. Forro, *Phys. Rev. B* **1995**, *51*, 14794.

135. Y. J. Uemura, K. Kojima, G. M. Luke, W. D. Wu, G. Oszlanyi, O. Chauvet, and L. Forro, *Phys. Rev. B* **1995**, *52*, R6991.

136. W. A. MacFarlane, R. F. Kiefl, S. Dunsiger, J. E. Sonier, and J. E. Fischer, *Phys. Rev. B* **1995**, *52*, R6995.

137. L. Cristofolini, A. Lappas, K. Vavekis, K. Prassides, R. DeRenzi, M. Ricco, A. Schenck, A. Amato, F. N. Gygax, M. Kosaka, and K. Tanigaki, *J. Phys. Condens. Matter* **1995**, *7*, L567.

138. A. Janossy, N. Nemes, T. Feher, G. Oszlanyi, G. Baumgartner, and L. Forro, *Phys, Rev. Lett.* **1997**, *79*, 2718.

139. M. Bennati, R. G. Griffin, S. Knorr, A. Grupp, and M. Mehring, *Phys. Rev. B* **1998**, *58*, 15603.

140. T. Kalber, G. Zimmer, and M. Mehring, *Z. Phys. B* **1995**, *97*, 1.

141. H. Alloul, V. Brouet, E. Lafontaine, L. Malier, and L. Forro, *Phys. Rev. Lett.* **1996**, *76*, 2922.

142. B. Renker, H. Schober, F. Gompf, R. Heid, and E. Ressouche, *Phys. Rev. B* **1996**, *53*, R14701.

143. L. Cristofolini, C. M. Brown, A. J. Dianoux, M. Kosaka, K. Prassides, K. Tanigaki, and K. Vavekis, *Chem. Commun.* **1996**, 2465.

144. H. M. Guerrero, R. L. Cappelletti, D. A. Neumann, and T. Yildirim, *Chem. Phys. Lett.* **1998**, *297*, 265.

145. G. B. Adams, J. B. Page, O. F. Sankey, and M. O'Keeffe, *Phys. Rev. B* **1994**, *50*, 17471.

146. Q. Zhu, D. E. Cox, and J. E. Fischer, *Phys. Rev. B* **1995**, *51*, 3966.

147. G. Oszlanyi, G. Bortel, G. Faigel, M. Tegze, L. Granasy, S. Pekker, P. W. Stephens, G. Bendele, R. Dinnebier, G. Mihaly, A. Janossy, O. Chauvet, and L. Forro, *Phys. Rev. B* **1995**, *51*, 12228.

148. G. Oszlanyi, G. Bortel, G. Faigel, L. Granasy, G. Bendele, P. W. Stephens, and L. Forro, *Phys. Rev. B* **1996**, *54*, 11849.

149. A. V. Nikolaev, K. Prassides, and K. H. Michel, *J. Chem. Phys.* **1998**, *108*, 4912.

150. M. Kosaka, K. Tanigaki, T. Tanaka, T. Take, A. Lappas, and K. Prassides, *Phys. Rev. B* **1995**, *51*, 12018.

151. A. Lappas, M. Kosaka, K. Tanigaki, and K. Prassides, *J. Am. Chem. Soc.* **1995**, *117*, 7560.

152. V. Brouet, H. Alloul, F. Quere, G. Baumgartner, and L. Forro, *Phys. Rev. Lett.* **1999**, *82*, 2131.

153. Q. Zhu, *Phys. Rev. B* **1995**, *52*, R723.

154. K. Tanigaki, I. Hirosawa, T. E. Ebbesen, J. Mizuki, Y. Shimakawa, Y. Kubo, J. S. Tsai, and S. Kuroshima, *Nature* **1992**, *356*, 419.

155. K. Prassides, K. Vavekis, K. Kordatos, K. Tanigaki, G. M. Bendele, and P. W. Stephens, *J. Am. Chem. Soc.* **1997**, *119*, 834.

156. L. Cristofolini, K. Kordatos, G. A. Lawless, K. Prassides, K. Tanigaki, and M. P. Waugh, *Chem. Commun.* **1997**, 375.

157. G. M. Bendele, P. W. Stephens, K. Prassides, K. Vavekis, K. Kordatos, and K. Tanigaki, *Phys. Rev. Lett.* **1998**, *80*, 736.

158. A. Lappas, C. M. Brown, K. Kordatos, E. Suard, K. Tanigaki, and K. Prassides, *J. Phys. Condens. Matter* **1999**, *11*, 371.

159. K. Prassides, C. M. Brown, S. Margadonna, K. Kordatos, K. Tanigaki, E. Suard, A. J. Dianoux, and K. Knudsen, *J. Mater. Chem.* **2000**, in press.

160. T. Ogitsu, S. Margadonna, K. Prassides, K. Tanigaki, K. Kusakabe, and S. Tsuneyuki, in *Recent Advances in the Chemistry and Physics of Fullerenes and Related Materials*, Vol. 6, K. M. Kadish and R. S. Ruoff (eds.). The Electrochemical Society, Pennington, NJ, 1998, p. 666.

161. S. Pekker, G. Oszlanyi, and G. Faigel, *Chem. Phys. Lett.* **1998**, *282*, 435.

162. C. Christides, K. Prassides, D. A. Neumann, J. R. D. Copley, J. Mizuki, K. Tanigaki, I. Hirosawa, and T. W. Ebbesen, *Europhys. Lett.* **1993**, *24*, 755.

163. S. Margadonna, C. M. Brown, A. Lappas, K. Kordatos, K. Tanigaki, and K. Prassides, in *Electronic Properties of Novel Materials*, H. Kuzmany, J. Fink, M. Mehring, and G. Roth (eds.). World Scientific Publishing, Singapore, 1998, p. 327.

164. S. Margadonna, C. M. Brown, A. Lappas, K. Kordatos, K. Tanigaki, and K. Prassides, in *Recent Advances in the Chemistry and Physics of Fullerenes and Related Materials*, Vol. 6, K. M. Kadish and R. S. Ruoff (eds.). The Electrochemical Society, Pennington, NJ, 1998, p. 642.

165. K. Prassides, K. Tanigaki, and Y. Iwasa, *Physica C* **1997**, *282*, 307.

166. S. Margadonna, C. M. Brown, A. Lappas, K. Prassides, K. Tanigaki, K. D. Knudsen, T. Le Bihan, and M. Mézouar, *J. Solid State Chem.* **1999**, *145*, 471.

167. S. Margadonna, K. Prassides, K. D. Knudsen, M. Hanfland, M. Kosaka, and K. Tanigaki, *Chem. Mater.* **1999**, *11*, 2960.

168. C. M. Brown, T. Takenobu, K. Kordatos, K. Prassides, Y. Iwasa, and K. Tanigaki, *Phys. Rev. B* **1999**, *59*, 4439.

169. D. Arcon, K. Prassides, S. Margadonna, A. L. Maniero, L. C. Brunel, and K. Tanigaki, *Phys. Rev. B* **1999**, *60*, 3856.

170. F. Simon and L. Forro, personal communication.

171. D. Arcon, K. Prassides, A. L. Maniero, and L. C. Brunel, *Phys. Rev. Lett.* **2000**, *84*, 562.

172. G. Oszlanyi, G. Baumgartner, G. Faigel, L. Granasy, and L. Forro, *Phys. Rev. B* **1998**, *58*, 5.

173. G. Oszlanyi, G. Baumgartner, G. Faigel, and L. Forro, *Phys. Rev. Lett.* **1997**, *78*, 4438.

174. W. Andreoni, F. Gygi, and M. Parinello, *Chem. Phys. Lett.* **1992**, *190*, 159.

175. J. C. Hummelen, B. Knight, J. Pavlovich, R. Gonzalez, and F. Wudl, *Science* **1995**, *269*, 1554.

176. B. Nuber and A. Hirsch, *Chem. Commun.* **1996** 1421.

177. M. Keshavarz-K., R. Gonzalez, R. G. Hicks, G. Srdanov, V. I. Srdanov, T. G. Collins, J. C. Hummelen, C. Bellavia-Lund, J. Pavlovich, F. Wudl, and K. Holczer, *Nature* **1996**, *383*, 147.

178. W. Andreoni, A. Curioni, K. Holczer, K. Prassides, M. Keshavarz-K., J. C. Hummelen, and F. Wudl, *J. Am. Chem. Soc.* **1996**, *118*, 11335.

179. K. Hasharoni, C. Bellavia-Lund, M. Keshavarz-K., G. Srdanov, and F. Wudl, *J. Am. Chem. Soc.* **1997**, *119*, 11128.

180. A. Gruss, K. P. Dinse, A. Hirsch, B. Nuber, and U. Reuther, *J. Am. Chem. Soc.* **1997**, *119*, 8728.

181. F. Simon, D. Arcon, N. Tagmatarchis, S. Garaj, L. Forro, and K. Prassides, *J. Phys. Chem. A* **1999**, *103*, 6969.

182. H. Kuzmany, W. Plank, J. Winter, O. Bubay, N. Tagmatarchis, and K. Prassides, *Phys. Rev. B* **1999**, *60*, 1005.

183. C. M. Brown, L. Cristofolini, K. Kordatos, K. Prassides, C. Bellavia, R. Gonzalez, M. Keshavarz-K., F. Wudl, A. K. Cheetham, J. P. Zhang, A. N. Fitch, and P. Pattison, *Chem. Mater.* **1996**, *8*, 2548.

184. C. M. Brown, E. Beer, C. Bellavia, L. Cristofolini, R. Gonzalez, M. Hanfland, D. Häusermann, M. Keshavarz-K., K. Kordatos, K. Prassides, and F. Wudl, *J. Am. Chem. Soc.* **1996**, *118*, 8715.

185. J. Arvanitidis, K. Papagelis, K. P. Meletov, G. A. Kourouklis, S. Ves, K. Kordatos, F. Wudl, and K. Prassides, *Phys. Rev. B* **1999**, *59*, 3180.

186. T. Pichler, M. Knupfer, M. S. Golden, J. Fink, J. Winter, M. Haluska, H. Kuzmany, M. Keshavarz-K., C. Bellavia-Lund, A. Sastre, J. C. Hummelen, and F. Wudl, *Appl. Phys. A* **1997**, *64*, 301.

187. K. Prassides, M. Keshavarz-K., J. C. Hummelen, W. Andreoni, P. Giannozzi, E. Beer, C. Bellavia, L. Cristofolini, R. González, A. Lappas, Y. Murata, M. Malecki, V. Srdanov, and F. Wudl, *Science* **1996**, *271*, 1833.

188. K. Kordatos, C. M. Brown, K. Prassides, C. Bellavia-Lund, P. de la Cruz, F. Wudl, J. D. Thompson, and A. N. Fitch, *Molecular Nanostructures*, in H. Kuzmany, J. Fink, M. Mehring, and G. Roth (eds.). World Scientific Publishing, Singapore, 1997, p. 73.

189. K. Prassides, M. Keshavarz-K., E. Beer, C. Bellavia, R. González, Y. Murata, F. Wudl, A. K. Cheetham, and J. P. Zhang, *Chem. Mater.* **1996**, *8*, 2405.

CHAPTER 15

FULLERENES UNDER HIGH PRESSURE

BERTIL SUNDQVIST

15.1 INTRODUCTION

The structures and properties of solid fullerenes at atmospheric pressure have been discussed by many authors, for example, by Prassides and Margadonna in Chapter 14 in this volume and by Dresselhaus et al. [1]. In contrast, information on the properties and phases of fullerenes under high pressure is still available mainly in the form of original research papers, as short sections in general reviews, or as short reviews on specialized topics. Only two more comprehensive reviews have appeared, one focusing on the polymeric high-pressure phases of C_{60} [2] and the other covering mostly high-pressure work carried out on fullerenes and their derivatives and compounds before the middle of 1997 [3]. This small number of reviews is rather surprising, because high pressure has very interesting effects on both the structures and the physical properties of solid fullerenes. To remedy this situation the present chapter will discuss the effects of high pressure on fullerenes in a wide sense, including both pure and doped materials as well as the polymeric phases that may be obtained by various methods, but excluding the effects of pressure on the superconducting properties of doped fullerenes, which are discussed in Chapter 16 of this volume.

The subject to be discussed is a vast one and the material can be organized in many ways. Sections 15.2 and 15.3 deal with materials based on individual molecules (monomeric materials) and based on polymeric structures, respectively. The structural phase diagram of C_{60} is much better defined than those of C_{70} and other higher fullerenes because of the high symmetry of the molecule. Because C_{60} is also easiest to produce in pure form, most high-pressure studies so far have been carried out on this material. The discussion here will therefore

Fullerenes: Chemistry, Physics, and Technology, Edited by Karl M. Kadish and Rodney S. Ruoff.
0471-29089-0   Copyright © 2000 John Wiley & Sons, Inc.

concentrate on C_{60}, with shorter discussions devoted to the properties and phases of C_{70} and of doped, intercalated, or functionalized fullerenes. No introductory information on the structures and phases of fullerenes at atmospheric pressure will be given here, since excellent reviews are already available [1] (see also Chapter 14 in this volume). "High" pressure will usually be defined as pressure above 100 MPa (1 kbar), although in some cases, data obtained at lower pressures will be discussed. High-pressure techniques will not be discussed in detail. When, in some cases, mention will be made of particular devices or techniques, the interested reader is referred to any one of several excellent recent specialized books [4] for more information.

15.2   MOLECULAR FULLERENES UNDER PRESSURE

15.2.1   Interactions Between Fullerenes and Pressure Media

15.2.1.1   Intercalation of Solid Fullerenes Under Pressure   In high-pressure studies the applied pressure, p, must be transmitted to a specimen through a medium, which may be a solid, a liquid, or a gas. In this process reactions may occur between fullerenes and their high-pressure environment. With fluid media, the most common type of reaction is intercalation of fluid into the fullerene lattice. Because of the large size of the fullerene molecules, the interstitial voids in the lattice are large enough to accommodate the atoms or molecules of many commonly used gaseous pressure-transmitting media. If such a gas is used to transmit pressure to a fullerene specimen, care must be taken to ensure that any effects observed in the experiment are indeed due to the applied pressure and not to unintentional intercalation of the gas. Noble gases like He are in many ways "ideal" pressure media. They are inert, insulating, optically transparent, and hydrostatic to very high pressures [4], and many early experiments were therefore carried out under He gas. The results from these, however, must be evaluated with some care, due to the possibility of intercalation.

The intercalation into C_{60} and C_{70} of many fluids, such as the noble gases [5–10], O_2 [11,12], N_2 [5], H_2 [11,13], CH_4 [14], and CO [15,16], as well as H_2O [17], has been well studied, and a strong correlation has been found between the atomic or molecular dimensions of the intercalant and the ease of penetration into the fullerene lattice. This is well illustrated by the intercalation kinetics of noble gases into C_{60}, as reviewed by Kwei et al. [8]. At room temperature (RT), He (atomic radius 0.933 Å) diffuses into the fullerene lattice on a time scale of minutes while Ar (radius 1.54 Å) does not intercalate measurably into C_{60} even over six days at 0.6 GPa. Ne, which has an intermediate atomic radius of 1.12 Å, diffuses into C_{60} with a time constant of about 1 h in face-centered cubic (fcc) C_{60} and 5 h in simple cubic (sc) C_{60}, as shown in Figure 15.1. The observed difference in diffusion rate between the two phases indicates that molecular rotation is very important for the diffusion of intercalants, and it has indeed been

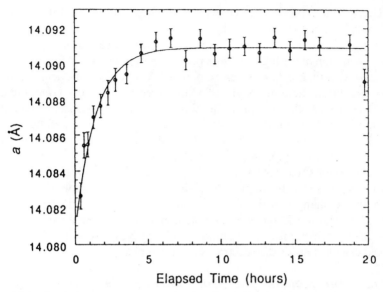

Figure 15.1   Lattice parameter versus time during intercalation of Ne into C_{60} at 207 MPa. Reprinted from Ref. 8, p. 250, by courtesy of Marcel Dekker, Inc.

observed [18] that even He does not diffuse into C_{60} in any significant amount at very low temperatures, $T < 180$ K, where molecular ratcheting between orientational states is very slow. Conversely, heating above RT significantly increases the diffusion rate, and above 200 °C most noble gases diffuse rapidly into both C_{60} [9] and C_{70} [10] at 170 MPa, forming structurally ordered phases of the type ArC_{60}. While the C_{60} compounds are metastable under ambient conditions, Ar diffuses out of the ArC_{70} lattice with a time constant of about 48 h, and XeC_{70} is the only stable C_{70} compound [10]. Under similar p–T conditions, CH_4 [14], CO [15], and H_2O [17] also intercalate into C_{60}. Hydrogen easily intercalates into C_{60} at RT [11,13], as might be expected. In this case, high-pressure treatment at high temperature (620 K) gives a reaction between the C_{60} host and intercalated hydrogen, transforming the material into an intercalation compound of H_2 in $C_{60}H_x$ [13]. In addition to pressure, intercalant size, and rotational motion in the host lattice, other parameters also seem to be important in determining diffusion rates, since N_2 and O_2 have practically identical molecular dimensions but very different diffusion rates. While O_2 easily diffuses into the C_{60} lattice [11,12], N_2 intercalates very slowly and only to a very small extent [5,12]. Again, no significant differences seem to exist between the intercalation behaviors of C_{60} and C_{70} [11].

Other materials can also intercalate into C_{60}, and several experiments have been carried out on the intercalation of alkali metals into both C_{60} [19] and carbon nanotubes [20] under pressure. The resulting compounds are claimed to be very highly doped, with Li concentrations up to near $Li_{30}C_{60}$ and Li_2C, respectively, but little is still known about these materials. Intercalation of alkali metals seems to close the lattice for gas intercalation, since at RT, Ne does not penetrate into Na_2CsC_{60} even after several weeks at 0.5 GPa [21].

The discussion of the high-pressure properties of intercalated fullerenes will be integrated with that of their fullerene parents. Although no particular emphasis will be placed on these materials, they are potentially of very large interest in themselves from both an application point of view (as, e.g., gas storage materials) and from the point of view of basic science. For example, Holleman et al. [16] studied the evolution of the dynamics of CO molecules in the octahedral voids [1] (see also Chapter 14 in this volume) of a C_{60} lattice, both as a function of T and as the lattice was compressed up to 3.2 GPa at RT. The infrared (IR) rotovibrational spectrum of this "stick-in-a-box" system evolved very clearly with both the changes in the size and shape of the "box" under pressure [16] and with the changes in the effective potential at the rotational fcc–sc transition of the host [15,16], and this evolution could be reproduced by quantum mechanical calculations of the spectra using reasonable potentials.

15.2.1.2 *Formation of Endohedral Compounds*

Quasispherical fullerenes like C_{60} and C_{70} consist of a hollow shell, which can be used as a container for other atoms or molecules. Such endohedral compounds should have very interesting physical and/or chemical properties [1], and a number of methods have been devised to produce materials of this type. The standard fullerene production method in which graphite rods are evaporated in a noble gas environment can be used to produce both metal-filled fullerenes, if mixed metal–graphite rods are used, and small amounts of noble-gas-filled fullerenes [1], but endohedral compounds can also be formed by irradiating thin C_{60} layers by high-energy ions [22]. A third method of particular interest here is to produce endohedral compounds by treating fullerenes at high temperatures under high gas pressure. Saunders et al. [23] compressed solid C_{60} in noble gases near 0.3 GPa at temperatures near 925 K and found that about 0.1% of the fullerene molecules captured noble gas guest atoms. He, Ne, Ar, and Kr all formed similar concentrations of endohedral compounds while the larger Xe molecule did not enter the C_{60} cage. At the high temperatures used, C–C bonds are believed to break spontaneously to open transient windows on the C_{60} molecules, allowing noble gas atoms to enter and get trapped inside. After cooling and depressurization the compounds seem to remain stable for long times under ambient conditions. Little work has been carried out in this area, but Saunders et al. [24] have discussed the method used, the properties of endohedral noble gas fullerenes and their chemical derivatives, and their possible applications. A very recent publication even reports the formation of artificial molecules of Ne_2 inside C_{70} molecules [25], and experiments on multiwalled carbon nanotubes [26] showed that at 0.17 GPa and 650 °C, Ar gas can also penetrate into the central cavity of these and remain trapped there for several months under ambient conditions. However, no high-pressure studies have been carried out on endohedral compounds and they will not be discussed further here.

15.2.2   Phase Diagrams of Molecular Fullerenes

15.2.2.1   Phase Diagrams of C_{60} and C_{60}-Based Compounds

Orientationally Ordered sc Phase   At atmospheric pressure molecular C_{60} has three structurally different phases [1] (see also Chapter 14 in this volume): a rotationally disordered face-centered cubic (fcc) phase at high temperatures, a primitive cubic or simple cubic (sc) phase with a temperature-dependent degree of orientational order below $T_o = 260$ K, and a sc phase with frozen-in orientational disorder ("orientational glass") below the glassy crystal transition temperature (or glass transition temperature) at $T_g \approx 90$ K. Both T_o and T_g increase with increasing p. Because an understanding of the orientational structure and its evolution with p and T in the sc phase will be useful in discussions of the pressure dependence of both the phase transitions and of the structures of the high-T and low-T phases, the discussion will start with this intermediate phase.

In solid C_{60} there are two possible orientational states in the sc phase at zero pressure. These will be denoted by the descriptive notations P (pentagon) and H (hexagon) orientations, referring to the orientation of a double bond on one molecule relative to the geometric structure of its neighbor [1] (see also Chapter 14 in this volume). Well below the transition temperature T_o the relative fraction f of P-oriented molecules follows a thermal distribution:

$$f(T) = 1/[1 + \exp(-\Delta/k_B T)] \qquad (15.1)$$

Here Δ is the difference in energy between the two states. At zero pressure, the P-oriented state has a slightly lower energy and is the preferred low-pressure orientation. However, the H state has a slightly smaller molecular volume and a simple thermodynamic argument shows that it should be favored at high p [27]. Δ is very small [1] (see also Chapter 14 in this volume), and f is therefore very sensitive to changes in both p and T. David and Ibberson [27] used neutron diffraction to show that, at 150 K, the relative energies of the P and H states were equal at a pressure of 191 MPa, as shown in Figure 15.2. Assuming Δ to be a linear function of p, Eq. (15.1) can be used to calculate an approximate value for f in the sc phase at any T or p as shown in Figure 15.3, which clearly indicates that above about 200 MPa the H orientation is always preferred. At zero pressure there is a strong orientational disorder close to T_o. Although the orientational order improves on cooling, it is impossible to obtain a fully orientationally ordered state because of the glass transition near 90 K, below which no further molecular reorientation is possible, and below which $f \approx 0.83$. At sufficiently high pressures, on the other hand, an H-oriented state should exist at low temperatures as suggested by Sundqvist et al. [28]. Later studies by Raman spectroscopy [29], neutron diffraction [30,31] and IR spectroscopy on CO molecules [16] trapped in a C_{60} lattice have verified that such an H-oriented ordered phase indeed exists. Conversely, if the C_{60} lattice is expanded by intercalation, the resulting increase in Δ will drive the orientational order further toward a P-oriented state, and available data [32] are compatible

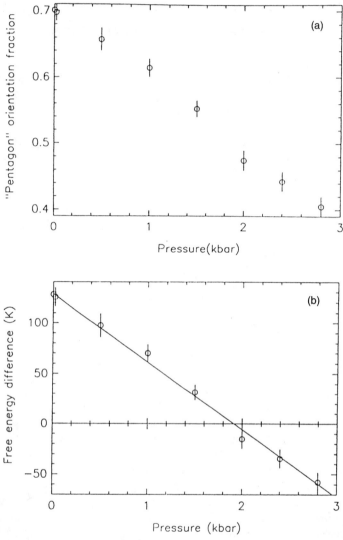

Figure 15.2 (a) Fractional occupation of P-oriented state versus pressure at 150 K. (b) Difference in free energy between the P and H states at 150 K. Reprinted with permission from Ref. 27. Copyright © 1998 IOP Publishing Ltd.

with $f > 0.9$ at low T. The structural properties of C_{60} often seem to be determined primarily by the intermolecular distance. While David and Ibberson [27] found $\Delta = 0$ near 191 GPa at 150 K, studies carried out near 90–100 K [18,29,31,33] consistently indicate lower equilibrium pressures near 0.15 GPa, possibly because orientational equilibrium occurs at the particular lattice *volume* corresponding to 191 MPa and 150 K such that the "equilibrium pressure"

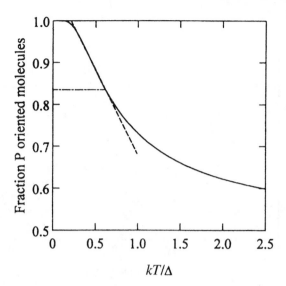

Figure 15.3   Relative fraction f of P-oriented molecules calculated from Eq. (15.1) versus temperature T. Dashed line is a linear fit to the low-temperature behavior; horizontal dot-dashed line shows the actual maximum fraction observed below T_c. Reprinted with permission from Ref. 46. Copyright © 1996 American Physical Society.

depends slightly on temperature. (The results in Fig. 15.3 were calculated taking this effect into account.)

The form of Eq. (15.1) shows f to vary smoothly with T and p. In principle, f can never be zero, and no "transition" to a pure H-oriented state should exist. However, rather sharp changes have been observed in a number of physical properties, such as the bulk modulus and the Raman and IR spectra, and in structural studies, at pressures corresponding to values of f in the range 0.02–0.10 [3]. Although the shape of $f = f(T)$ implies that any property sensitive to the orientational structure should show a fairly sharp change in its T dependence near $kT \approx 0.25\Delta$, this is probably not sufficient to explain the anomalous changes observed. A more probable explanation can be found from the change in intermolecular potential brought about by the molecular reorientation. Many calculations have been carried out to reproduce the phase transition behavior and the basic physical properties of C_{60} [1,3] (see also Chapter 14 in this volume). The interaction potentials used usually consist of two parts, a basic molecular potential similar to a van der Waals interaction plus an electrostatic potential modeling the charge distribution on the molecule, essential to reproduce the orientational ordering in the sc phase. Good overviews of the models used are given in two recent studies [34,35]. In most cases, calculations of the relative energies of different orientational states for one particular molecule assume that all 12 neighboring molecules are P oriented, as ideally expected under ambient pressure at low T. As shown by the dotted curve in Figure 15.4

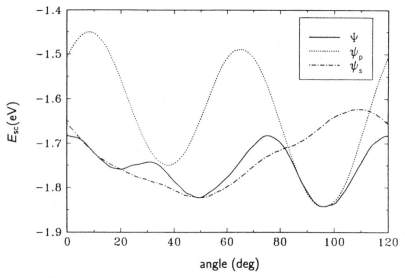

Figure 15.4   Intermolecular energy E_m as a function of rotation around the $\langle 111 \rangle$ axis. Dotted curve: all neighbors P oriented; dot-dash curve: all neighbors H oriented; solid curve: coherent, symmetry-preserving rotation of all molecules. Reprinted with permission from Ref. 36. Copyright © 1994 American Physical Society.

the molecule studied can then be either P oriented, and thus in a global energy minimum, or H oriented, in a local energy minimum [36]. The energy difference between these states is small, and the threshold energy is relatively high [1] (see also Chapter 14 in this volume). However, if all neighbors are assumed to be H oriented (dashed curve in Fig. 15.4) only one single minimum, corresponding to the H orientation, is obtained. Orientational defects are thus allowed in a preferably P-oriented phase but not in an H-oriented one. This indicates that a sharp final transition into a fully H-oriented state might actually occur, since below some critical value of f the intermolecular potential will force all remaining P-oriented molecules into the H orientation. Once formed, such an H-oriented state should be very stable and might even remain stable down to the equilibrium pressure between the P and H orientations, that is, about 0.2 GPa. However, no evidence for such hysteretic behavior has been observed. Another possible mechanism has been suggested by Pintschovius et al. [31]. Assuming Δ to be a linear function of f and taking into account that the P- and H-oriented states have different compressibilities (see below), they used an expression similar to Eq. (15.1) to calculate $f = f(p)$ and found a rather sharp final transition to an H-oriented high-pressure state.

Intercalation of C_{60} by gases [8] or the addition of small side groups to the molecules [37] seem to have relatively small effects on the orientational structure at zero pressure. The expansion of the lattice on intercalation, equivalent to a negative pressure, may improve the low-T orientational order, as mentioned above, and intercalation will probably also change the detailed evolution of the orientational structure under pressure. However, no study of this effect has yet been carried out. On the other hand, the addition of side groups to form $C_{60}O$, $C_{61}H_2$, and so on make molecular rotation more difficult and, in particular, limits the orientation of the added side group by forcing it to point toward any one of the octahedral voids in the lattice, changing the population of allowed orientational states from that of pristine C_{60}. Again, no structural studies under pressure have been reported.

Glass Transition Line and the Glassy Crystal Phase   The pressure dependence of the glass transition temperature T_g has been studied by several methods. Andersson et al. [33,38] give a value $dT_g/dp = +62$ K/GPa for the slope of this line based on thermal conductivity studies, and similar values have been found in other types of studies [18,31]. Care must be taken when studying this line, however. Because the orientational structure varies with T and p in the sc phase at $T > T_g$, different orientational structures are frozen-in on cooling through T_g at different pressures. Once the sample has been cooled through T_g, however, no further structural evolution can take place. If the pressure is changed while in the glassy phase, very interesting relaxation effects can therefore be observed when heating back through the glass transition. When molecular reorientation is enabled by the rise in T, molecules try to reorient from the frozen-in structure to the "equilibrium" orientational structure at T_g at the new pressure. Depending on thermal history, the observed T_g can be shifted significantly from the value observed when heating and cooling under isobaric conditions [33,38], mainly because the reorientation changes the molecular volume. If pressure has been increased at low T, the molecules reorient from the P to the H orientation. Because the latter has a smaller molecular volume, more space becomes available for relaxation, and the glass transition occurs at lower temperatures than expected. This effect explains the small $dT_g/dp = 35$ K/GPa observed by Wolk et al. [29] in an experiment where the sample was always cooled through T_g at zero pressure, then heated through T_g at elevated pressure. Conversely, in an H $\rightarrow$ P reorientation the final state has a larger molecular volume, the relaxation becomes frustrated, and the transition shifts to higher temperatures. The relaxation behavior of C_{60} near T_g has been studied through measurements of the thermal conductivity [33,38], the thermal expansion [18], and the compressibility [31]. The region close to the glass transition line in C_{60} is potentially a very interesting part of the phase diagram for further study of relaxation in orientationally disordered materials.

In principle, the slope dT_g/dp can be estimated from the standard expression [39] $dT_g/dp = T_g V_g \Delta\alpha/\Delta c_p$, where $\Delta\alpha$ and Δc_p are the step changes in the thermal expansivity α and the specific heat capacity c_p at the transition. For C_{60} this formula predicts a negative value for dT_g/dp because the P-H molecular reorientation with T in the sc phase, together with the different molecular volumes of the two states, leads to a negative $\Delta\alpha$ at the transition. A positive slope in better agreement with experiment can be found if $\Delta\alpha$ is corrected for this effect [3].

A linear extrapolation to RT shows that the glass transition should occur at an approximate pressure of 3.3 GPa, well above the range where anomalies have been observed in various physical properties (see above and in Ref. 3). Since glass transition lines usually curve away from the temperature axis, the glass transition at RT probably occurs at even higher pressures. While no experiments have detected any unambiguous anomalies in this range, it should be noted that such a glass transition is probably difficult to observe. At these pressures the C_{60} lattice should be almost perfectly H oriented, and at the glass transition the molecules stop jumping between equivalent H orientational states.

As discussed above, modified fullerenes have phase diagrams similar to that of C_{60}. For $C_{61}H_2$ T_g is rather high, about 140 K [37], because of the higher barrier for rotation. Compression measurements [40] on deuterated $C_{61}H_2$ show an approximate dT_g/dp near 100 K/GPa, indicating much stronger pressure effects on T_g than in pure C_{60}.

Returning to pure C_{60}, the actual orientational structure of the low-T glassy crystal phase can be varied at will by choosing a suitable pressure at which to cool the sample, but once cooled through T_g, the orientational structure should be fixed. Although few studies have been carried out in this phase, little probably happens to the structure as pressure is increased except that the molecules will be pushed closer together. Above some pressure, on the order of 4–6 GPa, the molecules will be close enough together that polymerization may occur at RT (see discussion of polymeric states in Section 15.3). Whether different frozen-in orientational states will give discernible differences in polymer structures or polymerization rates, though, has never been investigated.

fcc–sc Transition and the High-Temperature fcc Phase The pressure dependence of the fcc–sc transition temperature of C_{60} has been well studied by many groups, and since the transition is of first order, many different methods can be used to observe it [3]. The first experiments were carried out soon after the discovery of the sc phase [41,42] using differential thermal analysis in a He atmosphere. Unfortunately, as discussed above, He gas easily intercalates in C_{60}, and the results obtained in these early studies are not representative for the behavior of pure C_{60}. This was first observed by Samara et al. [5], who found large differences in dT_o/dp depending on the pressure-transmitting media used. This is very clearly illustrated in Figure 15.5, showing the very different slopes obtained in He and in N_2, respectively. Table 15.1 summarizes present knowledge by giving selected data for the slope dT_o/dp as obtained by various methods in different pressure media. (Data from a larger number of studies have been collected elsewhere [3].) The values in the top group have been obtained in nonintercalating media and should be intrinsic to C_{60}. The average slope is $dT_o/dp = 160$ K/GPa, significantly higher than for dT_g/dp but in good agreement with values estimated from the Clausius–Clapeyron relation. The slope

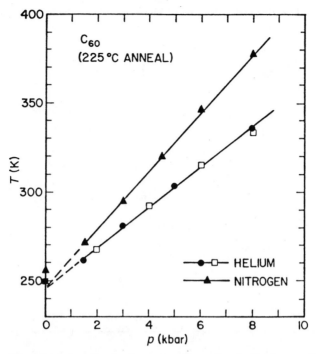

Figure 15.5 Phase boundary for the fcc-sc transformation in C_{60} in two different pressure media, He (intercalating) and N_2 (nonintercalating). Reprinted with permission from Ref. 5. Copyright © 1993 American Physical Society.

has also been calculated by several groups as tests of theoretical models of the types described above for the lattice potential of C_{60}, with variable degrees of success. The most successful calculation is probably that of Lamoen and Michel [47], who found $dT_o/dp = 182$ K/GPa.

When noble gases are intercalated into C_{60} the composite lattice becomes less compressible because of the increasing repulsive interactions between the molecules. However, the orientational interaction between molecules seems to be little influenced by the presence of the noble gas atoms, and at RT the rotational transition still occurs at the same intermolecular distance as for pure C_{60} [6], as shown in Figure 15.6. Because of the different compressibilities, different slopes dT_o/dp will thus be obtained in different gases. In He, which is the most investigated intercalating medium, the average slope is $dT_o/dp = 112$ K/GPa (Table 15.1). Thus, a significantly higher pressure is needed at RT to stop the molecular rotation in He compared to nonintercalating media. The variation in transformation pressure with intercalation was verified by Schirber et al. [6] in a very elegant experiment, based on the fact that Ne intercalates relative slowly. A C_{60} sample was rapidly compressed by Ne gas until it transformed into the sc phase near 0.275 GPa. Over the next hour Ne diffused into the sample, increasing the lattice parameter and causing a transformation back to the fcc

TABLE 15.1   Experimentally Determined Slopes dT_o/dp for the fcc–sc Phase Transition Line in Pristine and Intercalated C_{60}

Pressure Medium	dT_o/dp (K/GPa)	Reference
	Pristine C_{60}	
Ar	141	6
Ar	174.2	18
Ar	170	31
N_2	164	5
Pentane	159	5
Isopentane	160	43
Oil	150	44
C_{60}	163	45,46
	Intercalated C_{60}	
He	117	41
He	104	42
He	109	5
He	111	6
He	119.5	18
Ne	118	6

phase (cf. Fig. 15.6). While the light noble gases diffuse out of the C_{60} lattice at zero pressure, other gases are able to form intercalation compounds that are stable at zero pressure for extended periods. In metastable compounds like ArC_{60} [9], as well as in oxygen intercalated C_{60} [12], T_o has been observed to be depressed by several kelvins, as would be expected for an expanded lattice ("negative pressure" effect).

The change in volume at the transition is very close to 1% at zero pressure [1], and two groups have observed smaller changes in volume under pressure [18,46]. However, recent accurate measurements [31] show no such difference. More complicated anomalies have also been observed at the transition. Samara et al. [42] found that differential thermal analysis (DTA) transition peaks evolved from slightly anisotropic at zero pressure to having definite shoulders at the highest pressures and suggested that two rotational states, hindered and unhindered rotation, might exist near T_o. Double peaks have also been observed at zero pressure by modulated differential scanning calorimetry [48]. An intermediate state might thus exist close to T_o.

Molecular rotation in fcc C_{60} is not completely free, and at low temperatures there is a strong correlation between the rotation of neighboring molecules such that large corotating clusters, with dimensions up to 40 Å, exist [49]. In principle, very little should happen to this structure under pressure except for a slowing down of the rotational motion because of the increasing intermolecular interactions, except that near the fcc–sc boundary there should always be a zone

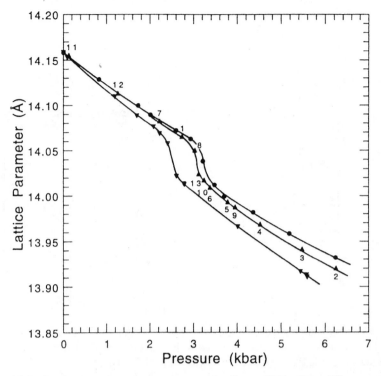

Figure 15.6 Lattice parameter versus pressure for C_{60} at RT in three different pressure media, from top down He, Ne, and Ar. Number indicates order of measurements. Reprinted with permission from Ref. 6. Copyright © 1995 American Physical Society.

with correlated molecular rotation. However, very recent resistivity studies [44] below 350 K show a repeatable cusp structure in the conductivity versus pressure at a weakly temperature-dependent pressure close to 0.2 GPa. This corresponds closely to the pressure at which $\Delta = 0$ in the sc phase, suggesting the possibility that even the correlated molecular rotation in the fcc phase might be modulated by the pressure dependence of the intermolecular orientational potential.

At very high temperatures molecular C_{60} decomposes into amorphous carbon. While isolated C_{60} molecules seem to be marginally stable to about 2000 K [50], solid C_{60} breaks down at temperatures just below 1000 K [51] by a thermally activated process. At very low pressures C_{60} sublimes on heating without forming a liquid phase. The possible existence of a liquid phase under pressure has been the subject of some discussion [3,52–54], but since all theoretical predictions agree that such a phase should only exist above 1800 K, where the molecule is no longer stable, the "normal" liquid phase of atomic carbon [55] probably remains the only liquid phase of carbon.

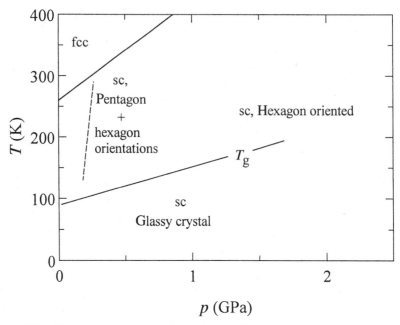

Figure 15.7   Pressure–temperature phase diagram for molecular C_{60}. The slopes of the phase lines have been calculated as discussed in the text. The dashed line indicates the line along which the P and H orientations have equal free energies.

Phase Diagram for C_{60} Figure 15.7 shows the pressure–temperature phase diagram of pure C_{60} as drawn from the information given in this section. Little is known about the high-temperature behavior, particularly about the breakdown of C_{60} into amorphous carbon. The slope of the fcc–sc phase line has been taken as the average value for pure C_{60} from Table 15.1, while the slope of the glass transition line has been taken from Andersson et al. [33,38]. Because no clear way has been found to illustrate the detailed evolution of the orientational structure in the sc phase with pressure and temperature, only the "equilibrium" coordinates, where $\Delta = 0$ and the P and H orientations are equally probable, are shown by the dashed, nearly vertical line near 0.2 GPa. The resistance anomalies observed in the fcc phase [44] fall very close to an extrapolation of this line.

15.2.2.2   Phase Diagram of C_{70} The phase diagram for C_{70} is more complicated than that of C_{60} because of the additional possibilities arising from the molecular anisotropy. It should contain at least three phases: a high-T phase with freely rotating, quasispherical molecules, an intermediate phase with uniaxially rotating molecules, and a low-T phase with nonrotating or ratcheting molecules. An additional difficulty is that C_{70} has two competing structures at high temperatures [1], the hexagonal close-packed (hcp) structure and the fcc one. These differ only in their stacking sequences, and because of the weak intermolecular interactions both types of crystals usually contain a sufficient amount of stacking faults to be considered mixed phases. On cooling, the fcc and hcp structures evolve through different structural sequences, but because most crystals are mixed, several transitions can often be observed. The identification of the various transitions and phases is still somewhat controversial [56,57], especially since impurities such as C_{60} can drastically change the transition behavior [57,58]. Recent work [59] suggests that the fcc phase is the most stable phase below 600 K and the hcp phase above, but this remains to be verified. However, after high-pressure treatment, C_{70} always returns to the fcc structure at zero pressure, irrespective of the initial (pretreatment) structure [60], suggesting that the fcc phase is indeed the most stable. In the fcc-based structural sequence, fcc C_{70} transforms on cooling into a rhombohedral (rhh) phase with uniaxially rotating molecules oriented along the $\langle 111 \rangle$ axis near 355 K, and then to a monoclinic (mcc), orientationally ordered phase below 280 K [57].

The most complete study of the phase diagram of molecular C_{70} has been carried out by Kawamura et al. [61] using X-ray diffraction (XRD) studies as a function of T. Their results are shown in Figure 15.8. The initial structure at RT was rhh, and all samples transformed to fcc on heating under pressure. (Samples intially in the fcc phase at RT transformed into rhh on pressurization.) The rhh–fcc phase line had a very large slope, about 300 K/GPa, between 0.4 and 1.5 GPa and to agree with the usual transition temperature at zero pressure an even larger slope must be assumed below 0.4 GPa. The magnitude of the volume change at the transition decreased with increasing pressure to almost zero near 1.5 GPa, which from the Clausius–Clapeyron relation $dT_c/dp = \Delta V/\Delta S$ implies that either ΔS or the slope of the phase transition line decreases at higher pressure. A decrease in the slope is compatible with the experimental data (Fig. 15.8). At RT, the rhombohedral angle of the lattice decreased with increasing p but saturated near 1 GPa, and it was suggested that at this point a transition into the mcc phase occurred. At both 0.9 and 1.5 GPa an amorphous carbon phase, which, according to Raman studies, contained both sp^2 and sp^3 bonds, was observed above 1100 K. Qualitatively similar results were obtained in a RT XRD study by Christides et al. [62]. An initially fcc sample transformed slowly into the rhh phase above 0.35 GPa with the transformation being completed near 1 GPa. Once formed, the rhh structure remained stable both at zero pressure and if submitted to higher pressures.

Lundin et al. [63] carried out mechanical compression studies over a wide temperature range. No transitions were observed below 236 K, but a very rapid initial drop in sample volume at very low pressures was observed at 296 and 343 K, and it was suggested that this might be due to an orientational ordering of an initial zero-pressure fcc phase to the rhh structure. Thermal conductivity studies versus T under pressure [64] on the same samples showed sharp anomalies near room temperature, similar to those observed at the fcc-to-sc transition in C_{60} [38]. The anomalies, which correspond to the final arrest of

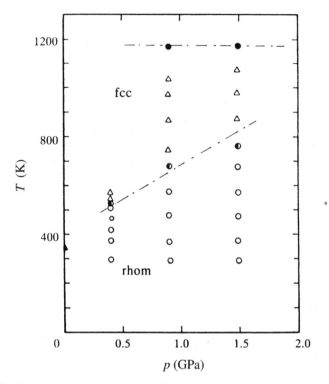

Figure 15.8   Pressure–temperature phase diagram for C_{70}. Solid dots indicate locations where the diffraction peaks from C_{70} disappear. Reprinted from *J. Phys. Chem. Solids,* Vol. 54, H. Kawamura, Y. Akahama, M. Kobayashi, H. Shinohara, H. Sato, Y. Saito, T. Kikegawa, O. Shimomura, and K. Aoki, "Solid C_{70} at high pressure and high temperature", pp. 1675–1678, Copyright © 1993, with permission from Elsevier Science.

molecular rotation, shifted with pressure at a rate $dT/dp = 70$ K/GPa. Samara et al. [65] carried out DTA studies on both fcc and hcp C_{70}, in both He and N_2. As for C_{60}, He was found to intercalate, modifying significantly the phase behavior under pressure. However, the slopes of the phase lines found in these experiments are very much larger than those found by others, reaching >700 K/GPa for the rhh–fcc line. Finally, anomalies have been observed in both IR [66] and Raman [67,68] studies at RT in the range 1–2 GPa, indicating structural phase transitions.

The information above can be used to synthesize the approximate "rotational" phase diagram shown in Figure 15.9, adapted from Refs. 3, 63, and 64. As in C_{60}, molecular rotation and orientation are the basis for the structural evolution with T and p. However, because the intermolecular potential is weak, orientational disorder may remain in C_{70} even at low temperatures, and the high-T fcc phase has in fact been observed to be metastable even at 5 K [62].

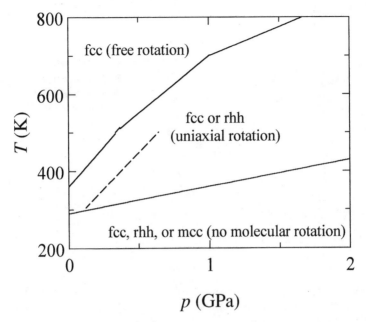

Figure 15.9 Pressure–temperature phase diagram for molecular C_{70} as discussed in text. The phase boundaries shown are based on the molecular rotation rather than the translational structure. Dashed line indicates the mcc–rhh transition line given by Samara et al. [65].

When cooled from the high-T fcc phase into the state with uniaxially rotating molecules, the crystal may either remain fcc, with randomly oriented molecular axes, or order orientationally into a rhh structure with all molecular axes lined up in parallel along the $\langle 111 \rangle$ axis. When uniaxial rotation stops, all three structural phases (fcc, rhh, or mcc) may occur, depending on how well the molecules are able to order. This diagram agrees with that of Kawamura et al. [61] and fits both the thermal conductivity data of Soldatov and Sundqvist [64] and the compression data of Lundin et al. [63] and Christides et al. [62]. An initially fcc sample at RT can, when compressed, order into the rhh phase [61] with its smaller molecular volume without changing its (uniaxial) rotation. XRD studies in a diamond anvil cell usually cannot distinguish between the rhh and mcc structures, but thermal conductivity studies and spectroscopic studies will detect the freezing of molecular rotation above 1 GPa at RT.

15.2.3 Physical Properties of Molecular Fullerenes Under Pressure

15.2.3.1 *Elastic Properties of Fullerenes* High pressure decreases the intermolecular distance in a material. Since most physical properties depend on this distance the relation between pressure and the lattice parameter is a very important one, usually expressed in terms of the compressibility or its

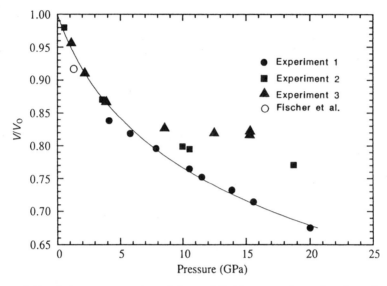

Figure 15.10   Volume compression of C_{60} at RT. The curve shown is a fit to the Vinet equation of state. Reprinted with permission from *Nature* (Ref. 73). Copyright © 1991 Macmillan Magazines Limited.

inverse, the bulk modulus *B*. Experimental data for *B* are very important as benchmarks against which to test theoretical models for the intermolecular potential, but they are also of wide technical and practical interest.

The first high-pressure study on a fullerene was carried out by Fischer et al. [69] to compare the intermolecular interaction in C_{60} with the interplane interaction in graphite. The bulk modulus found was close to one-third that of graphite, making C_{60} a reasonably good three-dimensional analog of graphite. After Ruoff and Ruoff [70] had suggested that the bulk modulus of compressed C_{60} might exceed that of diamond, a number of very-high-pressure studies were carried out. As one example, Duclos et al. [71] measured the compression of C_{60} up to 20 GPa under both hydrostatic and nonhydrostatic conditions, as shown in Figure 15.10. Under hydrostatic conditions ("Experiment 1") the volume decreased smoothly with increasing pressure, and fits to standard equations of state gave an initial zero pressure bulk modulus $B(0) = 18.1$ GPa, with $B' = dB/dp = 5.7$. All experimental data were taken above 0.5 GPa in the sc phase. Under nonhydrostatic conditions ("Experiments 2 and 3"), a much smaller compression was observed above about 5–7 GPa, which, in retrospect, probably indicates that Duclos and co-workers were the first to observe polymerization of C_{60} at RT. Unfortunately, only the very-high-pressure data of Duclos and co-workers have been confirmed by later measurements, which invariably show a larger compression at low pressures. This is clearly shown in Figures 15.11 and 15.12, which display data from a number of studies at RT over the low-pressure range up to 0.6 GPa and to above 20 GPa, respectively.

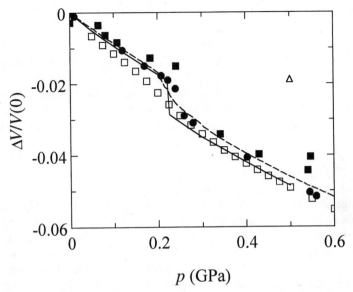

Figure 15.11 Volume compression of C_{60} at RT up to 0.6 GPa. Data from Schirber et al. [6] (dots), Pintschovius et al. [31] (full curve), Lundin and Sundqvist [45,46] (open squares), Duclos et al. [71] (triangle), Soldatov and Simu [72] (dashed curve), and Ludwig et al. [73] (solid squares).

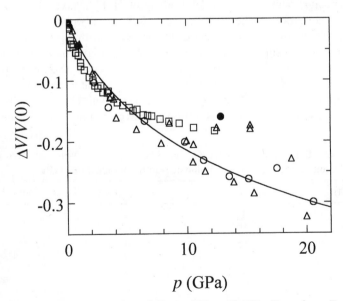

Figure 15.12 Volume compression of C_{60} at RT to 22 GPa. Data from Duclos et al. [71] (triangles), Ludwig et al. [73] (squares), Haines and Léger [81] (circles), Nguyen et al. [82] (dots), and Wang et al. [83] (full curve).

The differences between different data sets is surprisingly large [3], considering that reasonably pure samples have been available for study under well-defined conditions for a long time. Note that the data in Figures 15.11 and 15.12 have been selected for their probable accuracy from the larger set discussed elsewhere [3].

In the low-pressure range below 1 GPa, recent results have converged to give an internally consistent set of data (Fig. 15.11). Surprisingly, data for the fcc phase have been the last to show such a convergence. Early data from Lundin and Sundqvist [45] indicated very low values for $B(0)$, about 6.8 GPa. Higher values are found in more recent studies under hydrostatic conditions [6,18,31,72,73], which agree very well for the relative volume compression, $\Delta V / V(0)$, but give rather different values for $B(0)$ because of the limited pressure range available and the use of different fitting functions. The capacitive technique of Grube [18] gives $B(0) = 8.5$ GPa, while neutron scattering studies by Schirber et al. [6] and Pintschovius et al. [31] give values of about 12 and 9.6 GPa, respectively, all obtained using Ar gas as the pressure-transmitting medium. Data for $\Delta V / V(0)$ from these studies are in excellent agreement. The closely spaced data points of Grube and of Pintschovius and co-workers allow standard equations of state to be fitted, clearly reproducing the curvature in $\Delta V / V(0)$ and giving a pressure-dependent B and a low $B(0)$. Schirber and co-workers find a larger $B(0)$ because of the difficulty to fit their more sparse data. Soldatov and Simu [72] used the same piston-and-cylinder device as Lundin and Sundqvist [45] but immersed the sample in liquid glycerol and obtained data in excellent agrement with those of Ludwig et al. [73], taken by XRD under hydrostatic conditions in liquid Fluorinert. $B(0)$ can also be measured directly at zero pressure by optical or acoustic methods. Although data from such measurements also show a large scatter [3], an adiabatic bulk modulus of 8.4 GPa has been reported from ultrasonic studies [74]. Table 15.2 shows published values for $B(0)$ and B' from selected high-pressure studies together with averaged values B_{av} between 0 and 0.2 GPa. The agreement between the latter data is very good. Considering the resolution and expected accuracy in the various

TABLE 15.2 Bulk Modulus of fcc C_{60} at RT: $B(0)$ Is the Value at Zero Pressure, $B' = dB/dp$, and B_{av} Is the Average Value of B in the Range 0–0.2 GPa

$B(0)$ (GPa)	B'	B_{av} (GPa)	Reference
8.5	32	11.7	18
9.6	19	11.5	31
11.8	5	12.3	6
11.6	6.9	12.3	72
13.4[a]	—	13.4	73
11.9[b]	—	—	75
14[b]	18[b]	15.8[b]	76

[a] Average value (linear approximation).

[b] Theoretical calculation.

studies the "recommended" RT value for $B(0)$ in fcc C_{60} is 9.6 GPa. A number of calculations have been carried out for this relatively "simple" material, most of these using a spherically symmetric intermolecular potential due to Girifalco [75]. Data from two of these [75,76] are shown in Table 15.2 for comparison with experiments, and the agreement is seen to be excellent.

The very low values 6–7 GPa obtained for $B(0)$ in several studies [45,77–80] might be caused by disorder, since all samples had been submitted to strong shear stress, ground, or uniaxially pressed before the experiment, or consisted of disordered films. Remarkably, although Lundin and Sundqvist [45] found very low values for $B(0)$ in fcc C_{60}, the results in sc C_{60} were identical to those found in other recent studies (Fig. 15.11), and it is not understood why the correlation between disorder and B should turn out different in the fcc and sc phases.

The bulk modulus might be expected to be higher in the sc phase than in the fcc phase because the lattice constant is much smaller. However, the effects of the molecular reorientation under pressure are sufficiently strong to offset the effect of the smaller intermolecular distance. An increase in pressure forces a fraction of the molecules to reorient from the P state into the H state, and the resulting decrease in molecular volume is observed as a low apparent bulk modulus. This is well illustrated in Figure 15.13 [6], showing B as a function of

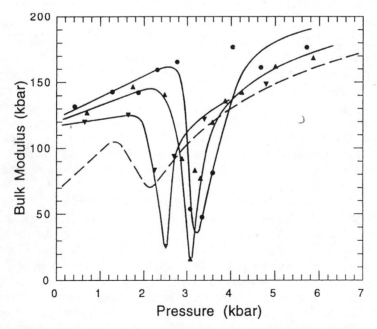

Figure 15.13 Bulk modulus of C_{60} at RT as a function of pressure. The dip near 3 kbar is the fcc–sc transformation, and the full curves have been obtained in, from top down, He, Ne, and Ar. Reprinted with permission from Ref. 6. Copyright © 1995 American Physical Society.

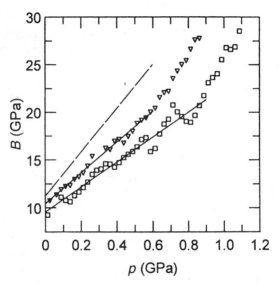

Figure 15.14 Bulk modulus of C_{60} as functions of pressure at 152 K (triangles) and 236 K (squares). The dashed line has been obtained by correcting the data at 152 to constant fractions of P- and H-oriented molecules. Reprinted with permission from Ref. 46. Copyright © 1996 American Physical Society.

pressure at RT. B clearly does not increase at the fcc–sc transformation near 0.25 GPa in spite of a step volume decrease by 1%. The largest effects are observed at low temperatures, where C_{60} is transformed from a reasonably well P-oriented state to an almost fully H-oriented state over a few hundred mega-pascals (a few kilobars) [3,28,31]. Figure 15.14 shows experimental data for B as a function of p at 150 and 236 K [46]. The sharp changes in the slope dB/dp occur at pressures corresponding to about 90% H-oriented molecules (Fig. 15.3), while the dashed line is the result of an approximate calculation to find the "intrinsic" value for B at 150 K for a hypothetical C_{60} solid with a constant fraction of P-oriented molecules.

Figure 15.11 shows there is very good agreement between the results from several groups for the compression behavior of sc C_{60}. Data for $\Delta V/V(0)$ at RT from Fischer et al. [69] (outside range of plot), Lundin and Sundqvist [45,46], Pintschovius et al. [31], Schirber et al. [6], and Soldatov and Simu [72] are practically identical, while data from Ludwig et al. [73] deviate slightly at high pressures. The low-pressure data of Duclos et al. [71] do not agree well with the results from other groups. Again, scatter in the experimental data gives rather wide deviations between reported values for $B(0)$ (extrapolated to zero pressure) from different groups. Table 15.3 presents RT data for $B(0)$, B', and the average value B_{m} over the range 0.3–0.6 GPa (just above the fcc–sc trans-formation) from the groups mentioned above. While the values for $B(0)$ differ widely, the average values B_{m} are very similar, reflecting the fact that scatter in

TABLE 15.3 Reported Values for $B(0)$ and B' in the Orientationally Ordered sc Phase of C_{60}

T (K)	$B(0)$ (GPa)	B'	$B_m{}^a$ (GPa)	Reference
RT	13.4[b]	2.53[b]	14.5	73
298	9.5	11.1	14.5	45,46
299	10.2	9.3	14.4	31
RT	13.2[c]	—	13.2	6
RT	9.6	12.3	15.1	72
262	10.4	10.1	—	31
236	9.6	13.1	—	45,46
180	11.5[c]	—	—	18
170	14.2[b]	—	—	73
152	10.4	16.2	—	45,46
150	12.75[c]	—	—	27
150	12.3[d]	—	—	31
110	12.5	—	—	18
110	13[d]	—	—	31
70	13.2[e]	10[e]	—	31
70	14.7[b]	—	—	73
60	13.75	—	—	18

[a] B_m indicates the average bulk modulus between 0.3 and 0.6 GPa (close above the fcc–sc transformation).

[b] Nonstandard, modified Birch equation of state.

[c] Average value over pressure range indicated (linear approximation).

[d] B nonlinear in p; negative initial dB/dp.

[e] 90% *hexagon* ordered.

the experimental data has a large effect on the curvature of the fitted functions used for extrapolation to zero pressure.

Below 260 K, the sc phase is stable at zero pressure and no extrapolations are necessary to find $B(0)$. In this range, measurements have been carried out by Lundin and Sundqvist [45,46], Ludwig et al. [73], and Pintschovius et al. [31], but also by David and Ibberson [27] and Grube [18]. Again, the agreement between data from these groups is excellent. At 0.28 GPa and 150 K the data for $\Delta V/V(0)$ by Pintschovius and co-workers differ by $\pm 1\%$ from those of Lundin and Sundqvist and those of David and Ibberson, while at 1 GPa the data of Lundin and Sundqvist (at 152 K) and those of Ludwig and co-workers (at 170 K) differ even less. Selected low-T data for $B(0)$ and B' are also given in Table 15.3. A straight line fitted to available data for $B(0)$ as a function of T below 260 K can be extrapolated to give a RT value of 8.9 GPa [3], in good agreement with measured values for $B(0)$ at RT. A reliable composite set of data thus exists for the compression properties of sc C_{60} at all temperatures below 300 K up to about 2 GPa, and a reasonable "recommended" RT value for $B(0)$ is 9.5 ± 0.5 GPa. As pointed out above this is similar to that for fcc C_{60}, in spite of the smaller (extrapolated) lattice parameter, and differs by a factor of two from the early value given by Duclos et al. [71]. Moreover, B increases more rapidly with p in the less dense fcc phase [31], which thus becomes much "harder" than the denser sc phase under pressure.

In the glassy crystal phase below 90 K, only three studies have been carried out. Near 70 K the average measured $B(0)$ for mainly P-oriented C_{60} is 14.2 ± 0.5 GPa, somewhat higher than in the orientationally ordered sc phase, as might be expected when the effect of molecular reorientation with pressure has been frozen out (see also Fig. 15.14). Measurements on samples with different frozen-in ratios of P- and H-oriented molecules have recently shown that B is 10% higher in the P-oriented state than in the H-oriented state [31].

Turning to the high-pressure data in Figure 15.12, the scatter between experiments is rather large, but data from most groups tend to fall in between the extreme curves found by Duclos et al. [71] for hydrostatic and nonhydrostatic conditions, respectively (Fig. 15.10). Under nonhydrostatic conditions there is a very strong curvature in the curves of V versus pressure. Ludwig et al. [73] find that standard equations of state are unable to describe the properties of C_{60} and suggest a modified Birch–Murnaghan equation of state, in which the volume of the molecules is subtracted from that of the crystal. The remaining "compressible fraction" of the lattice then behaves in a more normal way under pressure. However, the actual physical effect behind the rapid increase in B with pressure is probably polymerization, as will be discussed further below. The translational, orientational, and even molecular structures probably differ between experiments because of differences in the pressure media and the rate of compression used. The bulk modulus B_{mol} of the individual C_{60} molecules has been estimated to be at least 670 GPa [71], since the molecular form factor changes very little up to 20 GPa as deduced from the disappearance of the $\langle 200 \rangle$ diffraction peak. Theoretical estimates [70,84,85] for B_{mol} range from 720 to 900 GPa.

The bulk modulus of C_{70} has been measured by three groups [61–63] over a wide range in T and data are given in Table 15.4. At 365 K, nominally in the fcc phase, Lundin et al. [63] report $B(0) = 7.9$ GPa, not very different from their data for fcc C_{60}. At RT, Kawamura et al. [61] find an average bulk modulus of

TABLE 15.4   Experimental Data for the Bulk Modulus B of C_{70}

T (K)	Phase	$B(0)$ (GPa)	B'	Reference
365	fcc	7.9	16	63
343	rhh	8.4	14	63
296	rhh	8.5	14	63
293	rhh	11[a]	—	61
293	rhh (mcc?)	25	10.6	62
236	mcc	10.1	15	63
185	mcc	13.1	14	63

[a] Average value, 0.4–1.5 GPa.

TABLE 15.5   Bulk Moduli of Cubic Alkali Metal Intercalated Fullerenes

Material	T (K)	$B(0)$	B'	Reference
K_3C_{60}	300	28	—	86
Rb_3C_{60}	300	22	—	86
Rb_3C_{60}	300	20.5[a]	4.5[a]	87
Rb_3C_{60}	295	17.4	3.9	88
Rb_3C_{60}	14	18.3	3.9	88
Na_2CsC_{60}	300	21.2	—	89
Na_2CsC_{60}	82	21.2	—	89

[a] From a modified equation of state [73,87].

11 GPa for rhh C_{70} between 0.4 and 1.5 GPa, much lower than would be expected for sc C_{60}. Lundin and co-workers report a very strong initial compression, assumed to be due to an initial ordering of metastable fcc phase to rhh, and a value of 8.5 GPa for $B(0)$ as extrapolated from above 0.2 GPa. For comparison with Kawamura and co-workers, the average B between 0.4 and 1.5 GPa is about 22 GPa. Christides et al. [62] report an initial transformation from fcc to rhh, ending near 1 GPa, and an extrapolated value $B(0) = 25$ GPa for the rhh phase. Considering the phase diagram suggested in Figure 15.9, this value might instead be valid for the mcc phase. Finally, below 250 K Lundin and co-workers find an increase in $B(0)$ to 13.1 GPa at 185 K, very similar to $B(0)$ for C_{60}. Results from different studies thus differ significantly, and both the phase diagram and the elastic properties of C_{70} need further studies.

Intercalation of atoms or molecules into C_{60} (or C_{70}) increases the intermolecular interactions and thus the bulk modulus, even though the intermolecular distance usually increases. Ne and He intercalate easily and have large effects on the compression behavior (Fig. 15.6). For fcc C_{60} intercalation increases $B(0)$ to 12.5 GPa in He and 12 GPa in Ne [6]. A similar effect was observed for sc C_{60} but the available pressure range was too small for a quantitative analysis. Intercalation of alkali metal ions gives even larger changes in B because the ion–ion interaction is much stronger than the van der Waals interaction with the (small) noble gas atoms. Alkali-metal-doped fullerenes are discussed in detail by Buntar (Chapter 16 in this volume), and here it is only noted that data exist only for K_3C_{60}, Rb_3C_{60}, and cubic Na_2CsC_{60} (the linear chain phases of the latter and of the AC_{60} compounds are discussed together with other polymeric phases later). Available data are collected in Table 15.5, and a comparison with Tables 15.2 and 15.3 shows that these compounds are much less compressible than pure C_{60}. Figure 15.15 shows a comparison [88] between available data for the compressibility of Rb_3C_{60}. Although the three sets are in fairly good agreement, the data of Ludwig et al. [87] show a somewhat higher B than the others, as also found by the same group for C_{60} at RT [73].

The elastic moduli of C_{60} and C_{70} have also been measured by various methods, but the results differ strongly between different groups [3]. For fcc C_{60} at RT, Soifer et al. [74] report a Young's modulus E of 9.8 GPa for compacted

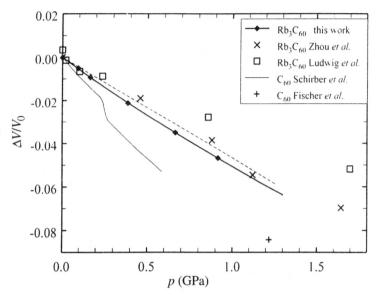

Figure 15.15   Volume compression of Rb_3C_{60} and pure C_{60} versus pressure. Data for Rb_3C_{60} from Ref. 88 (filled diamonds and full curve), Ref. 87 (squares), and Ref. 86 (crosses and dashed line); data for C_{60} from Ref. 69 (plus sign) and Ref. 6 (thin full curve). Reprinted from *J. Phys. Chem. Solids*, Vol. 58, J. Diederichs, J. S. Schilling, K. W. Herwig, and W. B. Yelon, "Dependence of the superconducting transition temperature and lattice parameter on hydrostatic pressure for Rb_3C_{60}", pp. 123–132. Copyright © 1997, with permission from Elsevier Science.

powder, while for single crystals values up to 20 GPa [90] have been reported. Reported data for the various elastic constants also depend strongly on temperature. Calculated values for the elastic constants c_{11}, c_{12}, and c_{44} from various sources have been tabulated by Sundqvist [3].

15.2.3.2 Lattice Properties   In solid fullerenes the intramolecular bonds are very strong and the intermolecular bonds are weak, and properties determined by the intermolecular potential should thus be quite sensitive to pressure while those determined by intramolecular interactions should not. In this section the pressure dependence of the vibrational spectrum of solid fullerenes will be reviewed as observed by various spectroscopies. The pressure dependence of some macroscopic thermophysical properties with magnitudes determined mainly by acoustic vibrations will also be discussed. Because of the large number of atoms in the primitive cell, there are a large number of vibrational modes, ranging from acoustic phonon and libron modes with energies of a few millielectron volts, to intramolecular optical modes with energies up to 200 meV. A very schematic but clear picture of a typical vibrational spectrum [91] is shown in Figure 15.16. This spectrum is drawn for an alkali-metal-doped

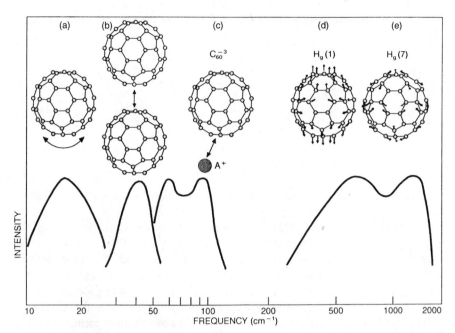

Figure 15.16 Schematic drawing of the vibrational spectrum of solid, alkali metal intercalated C_{60}, showing (a) librational and acoustic lattice modes, (b) optical lattice modes, (c) metal ion modes, and (d) radial and (e) tangential intramolecular modes. Reprinted with permission from A. F. Hebard, *Phys. Today* **1992**, *45*(11), 26–32. Copyright © 1992 American Institute of Physics.

material, but elimination of the intermediate-frequency fullerene–metal ion vibrations leaves a typical fullerene spectrum. For C_{60}, acoustic phonon modes have energies up to almost 7 meV or 55 cm^{-1}, overlapping both libron modes (hindered molecular rotation or rocking) between 2 and 4 meV and optical lattice modes between 4 and 7 meV. There is a wide gap between these intermolecular modes and the 46 allowed high-frequency intramolecular optical modes, which appear above 33 meV or 270 cm^{-1}, and the two groups can usually be considered decoupled and independent. Intramolecular modes above about 700 cm^{-1} (85 meV) are usually associated with radial vibrations and those below with tangential ones. Further details can be found in recent excellent reviews of the vibrational spectra of fullerenes [1,92,93].

The optical modes of fullerenes are usually studied by Raman and IR spectroscopy. C_{60} has only ten Raman-active and four IR-active modes because of the high symmetry of the molecule, but because natural carbon contains about 1% ^{13}C (which breaks the mass symmetry) many other weak lines may often be seen. However, high-pressure studies are usually limited to at most the 14 Raman- and IR-active modes. Also, both C_{60} and C_{70} polymerize when exposed to phonons in the visible or UV range [1,94], which further complicates

TABLE 15.6 Pressure Slopes $d\omega/dp$ (in cm^{-1}/GPa) for the Four IR-Active Modes of C_{60}

$F_{1u}(1)$ (526 cm^{-1})	$F_{1u}(2)$ (576 cm^{-1})	$F_{1u}(3)$ (1183 cm^{-1})	$F_{1u}(4)$ (1429 cm^{-1})	Reference
−1.55	2.87	4.39	6.28	95
−0.87	2.5	3.4	4.4	66
−0.5	2.3	3.8	4.3	96
0.0	1.2	2.4	4.5	84[a]

[a] Theoretical calculation.

the analysis of experimental data since polymerization by the Raman excitation laser changes the vibrational properties of the lattice.

Experimental data from three studies of the IR absorption spectrum of C_{60} under pressure are summarized in Table 15.6. For three of the four absorption lines, a positive slope $d\omega/dp$ is observed, as expected when the lattice stiffens under pressure, but for the lowest frequency $F_{1u}(1)$ mode, a negative slope is found. The data of Huang et al. [66] and Klug et al. [96] are in good agreement, while Aoki et al. [95] find slightly larger pressure dependences for all modes. Martin et al. [97] carried out a very extensive IR absorption experiment on C_{60} in which they identified over 180 vibration modes (including higher-order and combination modes) in the range 100–4000 cm^{-1} and studied their dependence on temperature and pressure. Figure 15.17 shows the pressure dependence for 46 modes between 0.5 and 2.5 GPa in the sc phase, measured while decreasing the pressure from 2.5 GPa. Martin and co-workers used the IR data given by Klug et al. [96] as a pressure calibration and thus only the relative slopes can be compared. Again, most modes show a weak positive slope, but the $F_{1u}(1)$ mode and some modes near 700 and 1500 cm^{-1} show negative slopes. Several additional peaks are apparent in the range 1.5–2 GPa, where the formation of the H-oriented phase has been suggested to occur. A number of these unidentified lines appeared under pressure, and some of them also were present at low temperature, which suggests that they are connected with orientational ordering. A strong line at 611 cm^{-1}, persisting at zero pressure after the experiment, was suggested [97] to result from polymerization under pressure. Although RT polymerization has not been observed by other groups at these pressures, a feature near 608–610 cm^{-1} is indeed often observed in samples intentionally polymerized under pressure [98,99]. For all modes the pressure dependence is weak, with mode Grüneisen parameters $\gamma_i = -d(\ln \omega_i)/d(\ln V) = B\, d(\ln \omega_i)/dp$ well below 0.1, as would be expected for intramolecular modes because the effective bulk modulus B_{mol} of the molecule is very large compared to B (see above). Jishi et al. [84] calculated the pressure dependence of the IR modes using a force-constant model based on bond stretching and angle bending and obtained parameter values by fitting to zero-pressure Raman data. The results, shown together with the experimental data in Table 15.6, are in quite good agreement with the data except that no negative values are predicted.

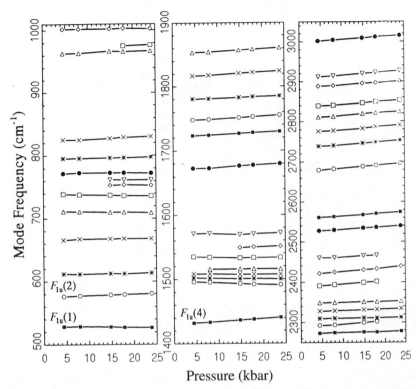

Figure 15.17 The pressure dependence of 46 vibrational modes of C_{60}, studied by IR absorption spectroscopy. Reprinted with permission from Ref. 97. Copyright © 1994 American Physical Society.

Although ten modes are Raman active in C_{60}, only a few of these are usually observed in high-pressure studies. In diamond anvil cells the very strong Raman line from diamond near 1333 cm^{-1} makes it difficult to detect the C_{60} lines at 1100 and 1250 cm^{-1}. Early Raman studies of C_{60} [100–103] were primarily aimed at identifying very-high-pressure phases and the slopes $d\omega/dp$ reported often differ significantly between different groups, even as to sign. Very complete investigations of the Raman spectrum of C_{60} under pressure at RT have been carried out by Meletov et al. [104–106]. They observed three well-defined phases, the fcc phase below 0.4 GPa, the sc phase between 0.4 and 2.4 GPa, and a third phase above 2.4 GPa. The transition at 2.4 GPa was tentatively identified as the glassy crystal transition, but it may also be connected with the formation of a H-oriented structure (see above). Typical data are shown in Figure 15.18 [105]. Contrary to what is observed at zero pressure [107], Meletov and co-workers find a step decrease in most mode frequencies at the transition from the fcc to the sc phase. Above 0.4 GPa, ten new lines were observed because of the lower symmetry of the sc phase and all lines broadened

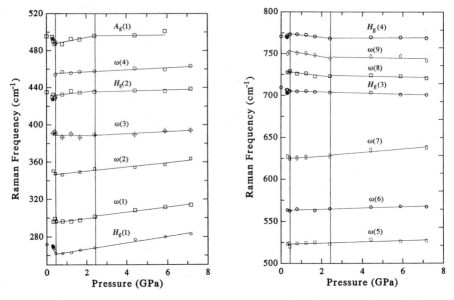

Figure 15.18 Pressure dependence of some Raman modes of C_{60}. Reprinted with permission from Ref. 105. Copyright © 1995 American Physical Society.

with increasing pressure. All changes in the spectrum were reversible on decreasing pressure. Table 15.7 shows selected data for the pressure dependence of the eight Raman modes that can be observed in diamond anvil cells. While most authors give average slopes, the data of Meletov and co-workers are given above and below 2.4 GPa. Most mode frequencies increase with pressure but for the lines at intermediate frequencies a softening can be seen, and again the pressure dependence is weak and the magnitudes of most mode Grüneisen pa-

TABLE 15.7 Pressure Dependence $d\omega/dp$ (in cm^{-1}/GPa) of the Raman-Active Modes of C_{60}

Mode	ω (cm^{-1})	Ref. (101)	Ref. (106) <2.4 GPa	Ref. (106) >2.4 GPa	Ref. (84)[a]
$H_g(1)$	273	1.1	3.2	3.3	0.1
$H_g(2)$	421	2.4	2.4	0.5	0.0
$A_g(1)$	496	0.94	4.2	0.5	1.0
$H_g(3)$	710	−0.55	−0.8	−0.8	0.5
$H_g(4)$	774	−0.50	−2.7	0.1	1.2
$H_g(7)$	1428	2.4	9.8	3.9	4.5
$A_g(2)$	1469	1.7	5.5	5.5	5.0
$H_g(8)$	1575	3.7	4.8	4.8	4.0

[a] Theoretical calculation.

TABLE 15.8 Pressure Dependence $d\omega/dp$
(in cm^{-1}/GPa) of Some IR-Active Modes of C_{70}

ω (cm^{-1})	Ref. (66)	Ref. (108)
1493	—	3.01
1460	1.3	5.38
1415	−7.6	5.44
1289	10.2	—
1260	0.7	—
1134	3.9	3.1
796	−0.6	−0.44
674	−0.5	−0.39
642	0.3	0.28
578	—	−0.2
566	3.4	3.80
535	2.7[a]	2.50
458	2.8	0.54

[a] Splits above 1 GPa.

rameters are below about 0.1 [106]. Data for other modes are given in tabular form by Meletov et al. [106], Snoke et al. [101], and Sundqvist [3].

For C_{70} IR and Raman spectra are much more complicated because of the lower symmetry, with 53 Raman-active and 31 IR-active modes [1,93]. A few spectroscopic studies have been carried out under pressure, often on mixtures of C_{60} and C_{70}, but results from different groups are not always in good agreement. Huang et al. [66] and Yamawaki et al. [108] studied the IR absorption up to 3 and 7 GPa, respectively, and selected data are tabulated in Table 15.8. Huang and co-workers found that several modes disappeared at low pressure and that one mode near 535 cm^{-1} split at 1 GPa, possibly as a result of the fcc-to-rhh transition (see Fig. 15.9). Again, while most mode frequencies increase with pressure, some are found to decrease, but the two groups only agree about two of these, both near 700 cm^{-1}. Raman data are presented by three groups as tabulated in Table 15.9; data from Meletov et al. [68] and Sood et al. [109] agree reasonably well. (Older data from the groups of Meletov and co-workers and Snoke and co-workers have not been included.) Measured data usually show strong anomalies near 1 GPa, or at 4–5 GPa, or both, as shown in Figure 15.19, and the lower pressure agrees reasonably well with the nominal position of the rhh–mcc transition (or arrest of uniaxial rotation) discussed above, especially if local heating by the excitation laser is allowed for. Although Meletov et al. [68] place the lower anomaly near 2 GPa, Figure 15.19 shows no experimental points between 1.2 and 2 GPa. Again, while a few modes at intermediate frequencies have negative slopes, most mode frequencies increase with pressure and the mode Grüneisen parameters are small, as shown in Figure 15.20 [68].

TABLE 15.9 **Pressure Dependence** $d\omega/dp$ **(in cm^{-1}/GPa) of Selected Raman Modes in** C_{70}

ω (cm^{-1})	Ref. (101)	Ref. (68)	Ref. (109)
1567	2.73	4.2/5.9/3.3[a]	5.15/3.9[b]
1513	4.5	3.8/7.8/2.2[a]	4.1/1.65[b]
1470	—	6.0/6.3/5.0[a]	8.07/5.04[b]
1449	—	6.0/7.6/3.6[a]	5.49/1.59[b]
1432	—	5.2/4.5/3.0[a]	—
1370	1.1	6.3/8.2/5.3[a]	—
1261	—	7.8/7.5/0.9[a]	4.54/—[b]
1224	3.2	5.3/6.3/3.1[a]	5.04/5.29[b]
1182	4/−10.3[c]	7.1/9.4/3.5[a]	11.96/—[b]
1065	1.1	3.7/5.4/1.5[a]	4.23/1.61[b]
776	—	0.8	—
740	0.12	0.9	—
708	—	−0.3	—
564	−0.06	−0.3	—
517	0.38	−/1.0/1.0[a]	—
412	—	0.1	—
256	1.65	3.2	3.38/1.43[b]

[a] In the pressure ranges 0–2/2–5/>5 GPa.

[b] Below/above 1.2 GPa.

[c] Below/above 4 GPa.

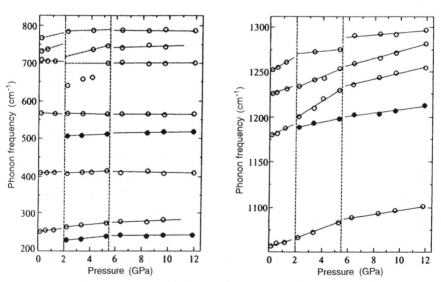

Figure 15.19 Pressure dependence of some Raman modes of C_{70}. Reprinted with permission from K. P. Meletov, A. A. Maksimov, and I. I. Tartakovskii, *JETP* **1997**, *84*, 144–150. Copyright © 1997 American Institute of Physics.

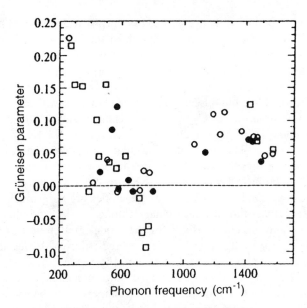

Figure 15.20 Mode Grüneisen parameters of intramolecular phonon modes of C_{60} and C_{70}. Squares show data from Raman studies on C_{60} [106], dots show IR data for C_{70} [108], and open circles show Raman data for C_{70} [68]. Reprinted with permission from K. P. Meletov, A. A. Maksimov, and I. I. Tartakovskii, *JETP* **1997**, *84*, 144–150. Copyright © 1997 American Institute of Physics.

Apart from the calculations by Jishi et al. [84], few attempts have been made to calculate the pressure dependence of intramolecular vibrations in fullerenes. Yu et al. [110] calculated the pressure dependence of the whole phonon spectrum using a tight-binding model for intramolecular modes and van der Waals-type intermolecular interactions and predicated a strong broadening of the spectrum and positive frequency shifts up to about 9 cm^{-1} GPa^{-1} for intramolecular modes, in general agreement with the experimental results, and also splittings of many IR and Raman lines. The low-frequency intermolecular vibrational modes should also increase rapidly with increasing pressure, and calculations show a broadening by about a factor of two up to 5.6 GPa. Prilutski and Shapovalov [111] also calculated the pressure dependence of the acoustic intermolecular modes and the sound velocities and found similar results. While the positive pressure shifts observed are reasonably well reproduced by calculations, there is still no obvious explanation for the negative pressure shifts observed in the intermediate range between 650 and 800 cm^{-1}.

Low-frequency acoustic modes are usually studied by inelastic neutron scattering as reviewed by Pintschovius [92], but no such studies have been carried out on fullerenes under pressure. However, a weighted average of the pressure dependence can be obtained from the thermal Grüneisen parameter $\gamma = 3\alpha B/dc_p = -d(\ln \Theta_D)/d(\ln V)$, where α is the linear thermal expansivity, d the density, and Θ_D the Debye temperature. In contrast to the optical modes, very large mode Grüneisen parameters are found for the acoustic modes because of the weak and strongly anharmonic intermolecular interactions. Values for γ near 8–10 have been found for fcc C_{60} by Sundqvist [112] and for sc C_{60} by White et al. [113], and by theoretical calculations on fcc C_{60} by Girifalco [79] and Zubov et al. [114], but γ seems to decrease rapidly with increasing p [79] and T [114]. The situation for C_{70} is similar. Kniaz et al. [115] have calculated a theoretical value $\gamma = 9.5$ and a Debye temperature of 73 K, very similar to values found for C_{60}. The only intermolecular modes that have been studied under pressure are the very-low-frequency libron modes in C_{60} in the glassy crystal state, which have been studied by Horoyski et al. [116] up to 1.5 GPa using Raman scattering. As expected, these modes are very much more sensitive to pressure than the intramolecular Raman modes and the frequencies increase by 3.7 and 5.2 cm^{-1} GPa^{-1} for the modes at 17.6 and 21.6 cm^{-1}, respectively. These slopes agree well with the predictions of Yu et al. [110] and Prilutski and Shapovalov [111] and correspond to mode Grüneisen parameters $\gamma > 3$, an order of magnitude larger than for intramolecular modes.

The intermolecular phonon spectrum determines a number of thermophysical properties and thermal transport studies can potentially give information on changes in the lattice phonon spectrum. The thermal conductivities of C_{60} and C_{70} have been studied in several experiments, and while the results have been very important for our understanding of the phase diagrams they have not yet improved our understanding of the lattice dynamics. Andersson et al. [38] measured the thermal conductivity λ of polycrystalline C_{60} to above 1 GPa over large ranges in T (50–300 K). The experimental data shown in Figure 15.21 are typical for disordered materials, with no sign of the $\lambda \sim T^{-1}$ depen-

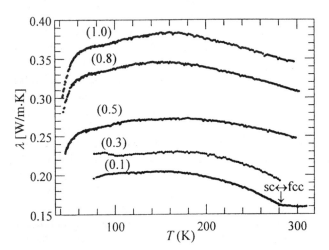

Figure 15.21 Thermal conductivity of C_{60} versus temperature at the pressures (GPa) indicated. Reprinted with permission from Ref. 38. Copyright © 1996 American Physical Society.

dence typical for "good" crystals. In the rotationally disordered fcc phase, λ is practically independent of T because of the very strong phonon scattering and the pressure dependence of λ can be expressed in terms of the Bridgman parameter $g = -d(\ln \lambda)/d(\ln V)$ as $g = 5.5$, a low value typical for this type of structure. In the sc phase there is a significant temperature dependence and a stronger pressure dependence, $g = 9.5$, as expected for a disordered crystalline phase. However, attempts to analyze the temperature dependence of λ in detail were not successful [38] because the break in $d\lambda/dT$ near 150–170 K could not be fit in any reasonable model, and it was shown that a corresponding anomaly also exists in the single crystal data of Yu et al. [117]. Models tested included phonon, impurity, defect, boundary, and orientational disorder scattering with a Debye model for the lattice, as well as an Einstein oscillator model often used for disordered materials. Either some other scattering mechanism, such as libron scattering, dominates above 170 K, or an extra heat transport channel opens between 100 and 170 K, and more theoretical work is needed to understand these data. For C_{70}, a slightly more "normal" temperature dependence was found but no detailed analysis was carried out [64]. At RT, an increase in g from a value near 4.4, typical for a disordered material, at very low pressures, to a value near 8.8, typical for an ordered crystal, at 0.7 GPa verified the effect of pressure in improving the orientational order and changing the structure from fcc to rhh, as discussed above.

15.2.3.3 Electronic Properties

Molecular fullerenes are semiconductors with band gaps between 1.5 and 2 eV [1], and pressure should increase the overlap of the molecular orbitals, which might eventually lead to a closing of the gap and thus a semiconductor-to-metal transformation. The electronic band structures of C_{60} and C_{70} have been studied under pressure both directly, by optical and other methods, and indirectly, by resistivity measurements.

The pressure dependence of the band gap in C_{60} has been determined optically from the photoluminescence [106,118] and the optical absorption edge [106,119–123]. Sood et al. [118] calculated the shift of the band gap from the position of the luminescence peak and found that the band gap energy E decreased with increasing pressure at the rate $dE/dp = -0.14$ eV/GPa. Most band gap studies of C_{60} have measured the optical absorption edge near 1.7 eV, on either films [119–121] or small crystals [106,122,123], and generally under hydrostatic conditions. All studies agree that the absorption edge of C_{60} has a two-step structure at atmospheric pressure with a first onset just above 1.7 eV and an extrapolated second threshold near 2.1 eV, and that the absorption thresholds shift rapidly toward lower energies with increasing pressure as the increasing molecular interactions in the compressed lattice broaden the energy bands, as shown in Figure 15.22. The 1.7 eV feature rapidly disappears under pressure, and most studies concentrate on the high-energy threshold for which early studies found a quasilinear dependence on pressure with rather small average slopes for E, between $dE/dp = -0.055$ eV/GPa [119] and -0.07 eV/GPa [121]. Later, Snoke et al. [120] showed the gap energy to be a nonlinear function

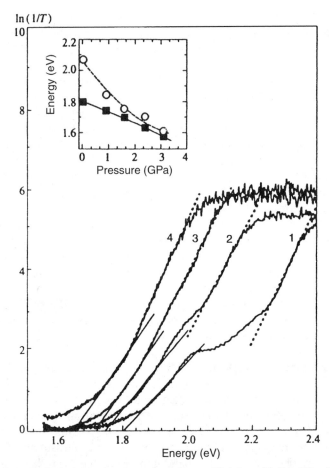

Figure 15.22   Optical absorption spectra of C_{60} at (1) 1 bar, (2) 0.9 GPa, (3) 1.6 GPa, and (4) 2.4 GPa. Inset shows pressure dependence of the two absorption edges indicated. Reprinted with permission from K. P. Meletov and V. K. Dolganov, *JETP* **1998**, *86*, 177–181. Copyright © 1998 American Institute of Physics.

of pressure with an initial zero-pressure slope $dE/dp = -0.14$ eV/GPa, decreasing to about -0.02 eV/GPa above 10 GPa. The initial slope is almost identical to that found by Meletov et al. [122,123], -0.15 eV/GPa, and in excellent agreement with the luminescence data of Sood et al. [118]. Theoretical calculations [124] agree that the absorption threshold energy should decrease with an initial slope near -0.12 eV/GPa and a rather large curvature, and the pressure dependence of the band gap is thus well understood. The strong nonlinearity rules out a semiconductor-to-metal transition below at least 50 GPa, and because polymerization occurs well below this, no metallic phase exists. Xu et al. [124] also calculated the dielectric functions and showed that the static dielectric function ε_o should increase rapidly with pressure as $d(\ln \varepsilon_o)/dp = 0.11$ GPa^{-1}.

Although the general behavior is thus well understood, the low-energy threshold near 1.7 eV is not. This feature shifts slowly with pressure, $dE/dp = -0.071$ eV/GPa [106,122], and the two thresholds merge near 2 GPa (Fig. 15.22). The behavior of the low-energy feature does not correlate with the fcc-to-sc transition, nor with exciton transitions or impurity absorption [122], but since the disappearance of this feature coincides with the final stage of the P-to-H reorientation transition, it might be caused by the orientational structure and/or disorder in C_{60}. Molecular orientation effects significantly modify the band structure [125] of C_{60}, and the P- and H-oriented phases might simply have different band gaps. Because disorder broadens the bands [125], orientational disorder in the sc phase could also create gap states that disappear when the orientational order improves with increasing pressure. Disorder might also break the symmetry and allow forbidden transitions over the smaller band gap near the X point [126]. Pressure also causes significant redistributions of the absorption intensity between wavelengths above the absorption threshold, and these have been discussed by Meletov et al. [123] in terms of crystal-field-induced mixing of electron states.

Meletov et al. [127] have also measured optical absorption spectra in C_{70} under hydrostatic pressures to 10 GPa. The absorption edge at zero pressure is 1.78 eV and shifts toward lower energies under pressure at approximately the same rate as for C_{60}, with an initial slope $dE/dp = -0.1$ eV/GPa, which decreases to about -0.03 eV/GPa at 10 GPa. Again, no metallization should occur below at least 50 GPa. Photoluminescence measurements by Sood et al. [128] up to 2.8 GPa showed a very similar value $dE/dp = -0.09$ eV/GPa.

In the glassy crystal phase of C_{60}, hydrostatic low-temperature (30 K) experiments up to 1.5 GPa yielded pressure coefficients between -0.045 and -0.080 eV/GPa for the energies of various exciton transitions near the gap edge [129]. Although the possible error is large, these values agree well with results from ab initio calculations [130]. In principle, the electronic band structure can also be investigated by positron annihilation, but so far the resolution in the experiments [131] has not been high enough to make this possible.

The band structure can also be investigated by transport measurements, and the first search for a semiconductor-to-metal transformation was carried out by Núñez-Regueiro et al. [132], who measured the electrical resistance over a large range in T up to 20 GPa. The results showed the expected decrease in the band gap under pressure, but also that metallization was prevented by a transition into an insulating state at 15–20 GPa. Only very few investigations have since been carried out on the conduction properties of fullerenes [44,133–138] and there is no detailed agreement between the results, possibly because the resistivity is very sensitive to the presence of impurities.

For a semiconductor the resistivity ρ should follow the standard formula

$$\rho = \rho_o \exp(-E_a/k_B T) \tag{15.2}$$

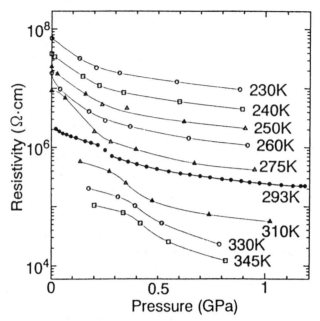

Figure 15.23 Electrical resistivity of C_{60} as a function of pressure and temperature. Reprinted from Ref. 136, p. 440, by courtesy of Marcel Dekker, Inc.

with an activation energy E_a of the order of the band gap. However, for full-erenes, E_a is often only 0.15–0.4 eV, much smaller than the optical gap. This is often blamed on carriers derived from disorder or impurity states in the gap. Also, log ρ is not always a very linear function of $1/T$. Núñez-Regueiro et al. [133] suggested that the resistivity was better described by a 3D variable range hopping model with $\rho = \rho_o \exp[2(T_o/T)^{1/4}]$, since plotting log ρ versus $T^{-1/4}$ was found to give much better linearity at all pressures than Eq. (15.2) for both C_{60} and C_{70} [133]. However, the standard semiconductor model is still used in most studies.

The only studies on single crystal C_{60} are those of Matsuura et al. [136] and Wang et al. [44]. Both were carried out on large crystals in both the fcc and sc phases under hydrostatic (oil) pressures up to 1 GPa. The data in Figure 15.23 show a step decrease in ρ on going from the fcc phase into the sc phase, with zero-pressure values for E_a of 0.32 and 0.27 eV in the fcc and sc phases, respectively. In the sc phase, little or no variation of E_a with pressure was found over the range investigated, but in the fcc phase E_a decreased very rapidly with increasing pressure. Wang et al. [44] measured the resistivity versus pressure at several constant temperatures and found a linear increase in the conductance in the sc phase. However, in the fcc phase a sharp cusp was always observed in the conductivity near 0.2 GPa.

All other studies on C_{60} and C_{70} have been carried out using anvil devices and solid pressure media, and in many cases only at very high pressures. Ramasesha and Singh [137] and Ramasesha et al. [138] studied compacted powders of C_{60} and C_{70}, respectively. The results for C_{60} indicated energy gaps between 0.4 and 0.75 eV with little pressure dependence. Other groups [132–135] carried out studies to very high pressures with little detailed agreement between results. As discussed elsewhere [3], log ρ often decreases linearly with increasing p, and the curves of resistance versus pressure often show minor kinks. Still, no clear anomalies have been observed that can be connected with polymerization or other transitions under pressure. At RT, most data for C_{60} can be described by a function $d(\log_{10} \rho)/dp = 0.36 \pm 0.02$ GPa^{-1} (or $d(\ln \rho)/dp = 0.83$ GPa^{-1}) up to about 15 GPa [3]. Matsuura et al. [136] pointed out that optical studies show a much stronger pressure dependence for the band gap, which indicates that the energies of the assumed disorder or impurity states in the gap do not scale with pressure in the same way as the band structure in general. However, over most of the pressure range studied, samples should be at least partially polymerized into the orthorhombic state, and it is thus uncertain if these results are indeed valid for monomeric C_{60}. While the pressure dependence of the electronic band structures is well understood, the interplay between the band structure and the conduction mechanism in fullerenes is not, and further studies are clearly needed.

Finally, while the properties of the superconducting A_3C_{60} phases are discussed in detail by Buntar (see Chapter 16 in this volume), their normal state conduction properties under pressure will be mentioned here. As reviewed by Zettl [139] single crystal A_3C_{60} usually shows a "metallic" temperature dependence $(d\rho/dT > 0)$, but the magnitude of ρ near RT is much higher than expected for a metal. Also, like pure alkali metals [140], A_3C_{60} has a resistivity that is very nonlinear in T because of thermal expansion effects [141], and early studies even suggested that unusual scattering mechanisms must be invoked to explain this nonlinearity. Later studies on Rb_3C_{60} have shown that, as for the pure alkali metals [140], ρ has a large enough pressure coefficient [139,142,143] that the resistivity at constant *volume* is linear in T [142,143] as predicted by theory. This is important for our understanding of the electron–phonon interaction in the A_3C_{60} phases. From the constant-pressure data [143], a very small electron–phonon interaction factor λ_{tr} is found, while the constant-volume data give $\lambda_{tr} = 0.66$–0.8, indicating quite high transition temperatures in agreement with the measured T_c. Theoretical calculations [144] also show ρ to be linear in temperature at constant volume, and the calculated pressure coefficient at RT is in very good agreement with the experimental data.

15.3 POLYMERIC FULLERENES

Early studies on fullerenes sometimes showed anomalies in spectroscopic and other properties, indicating amorphization or a decrease in symmetry, after storage or during high-pressure treatment [100–103,119]. Rao et al. [94] realized that radiation by visible or UV light polymerized the fullerene molecules into larger clusters and they suggested that a $[2 + 2]$ cycloaddition mechanism, in which a double bond on each molecule breaks up and two single C–C bonds form between each pair of molecules, might be responsible for the formation of intermolecular bonds. The resulting intermolecular four-membered carbon ring would have strongly strained bond angles and a sp^3 type orbital hybridization, in contrast to the normal intramolecular sp^2 bond character. Soon afterward, Iwasa et al. [145] showed that polymerized phases could also be obtained by heat treatment under high pressure. This method made it possible to produce bulk polymer samples on which it was possible to verify the mechanism suggested by detecting the presence of covalent sp^3 intermolecular bonds by NMR in both linear [146] and two-dimensional [147] polymers. Polymerization of C_{60} is actually a thermally forbidden process [148], and, surprisingly, the C_{60} polymer/dimer state has a *lower* energy than the monomer. Still, the energy threshold for polymerization is quite high, more than 4 eV per bond [149], while several polymer/dimer phases break down to monomers on heating to 450–550 K with a characteristic energy of less than 2 eV. Pressure-induced $[2 + 2]$ polymerization has recently been investigated by Ozaki et al. [149] by molecular dynamics simulations, which showed that the reaction paths for polymerization and breakdown are indeed different. The initial stage of pressure polymerization includes a large molecular deformation, which explains the larger energies needed. Under some circumstances single C–C bonds will also form between fullerene molecules.

Polymeric phases of C_{60} and C_{70} have been obtained by several methods such as radiation by light, by electrons, or by ions, doping by alkali metal, direct chemical synthesis, and high-pressure treatment. Photopolymerization is unavoidable whenever fullerenes are studied using sufficiently energetic laser light excitation, and in retrospect, many early spectroscopic studies under pressure show clear signs of photopolymerization (or photoassisted pressure polymerization). This chapter will treat in detail only polymers produced under high pressure, but since many fullerene polymers have similar properties, reference will sometimes be made to materials produced by other methods, especially, of course, if they have been studied under pressure. The polymerized phases of C_{60} have recently been reviewed by a couple of groups. Rao and Eklund [150] concentrate mainly on photopolymerized C_{60}, while Sundqvist [3] gives a very general review of high-pressure polymerized materials and Blank et al. [2] concentrate on very-high-pressure polymerized materials. Blank and co-workers also give a rather complete reaction phase diagram for C_{60}, reproduced in Figure 15.24. This diagram is not a phase diagram in the "equilibrium" sense, but instead shows the phases and structures observed after compressing and heating C_{60} to a particular p–T coordinate. The various phases indicated in this diagram will be discussed below, and the text will refer to this figure whenever the need arises.

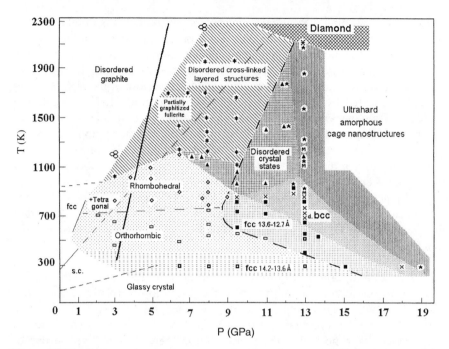

Figure 15.24 Pressure–temperature map showing the structural phases created by treating C_{60} under various conditions. Symbols indicate selected experimental conditions investigated. Reprinted from *Carbon*, Vol. 36, V. D. Blank, S. G. Buga, G. A. Dubitsky, N. R. Serebryanaya, M. Yu. Popov, and B. Sundqvist, "High-pressure polymerized phases of C_{60}", pp. 319–343, Copyright © 1998, with permission from Elsevier Science.

15.3.1 Polymerized Phases of C_{60} and Its Compounds

15.3.1.1 Dimers Three types of C_{60}-based dimers have been produced in reasonably pure form. The pure C_{60} dimer, C_{120}, forms by $[2 + 2]$ cycloaddition and has been produced by chemical [151] and high-pressure [152] synthesis, and also by photopolymerization at elevated temperatures [153]. Alkali metal (A) doped AC_{60} can be transformed into an insulating dimer form with a single intermolecular C–C bond by suitable heat treatment [154], and the azafullerene $C_{59}N$ [155] spontaneously dimerizes to $(C_{59}N)_2$, with a similar single C–C bond between neighbor atoms to the nitrogens. Only $(C_{59}N)_2$, which has a hexagonal lattice, has actually been studied under pressure, using XRD to 22 GPa at RT [156]. A fit of all data for volume versus pressure gave a bulk modulus for $(C_{59}N)_2$ of $B(0) = 21.5$ GPa, with $B' = 4.2$, but a possible transformation into a polymerized phase with identical inter- and intradimer intermolecular distances was indicated by a saturation of the c/a ratio near 6 GPa. Again, an interference between the lattice lines and the molecular form factor indicating a very high molecular bulk modulus is observed. $B(0)$ should be expected to be higher than for C_{60} because the intradimer intermolecular bond should be very incompressible. Regarding C_{120}, at least one such molecule must be formed as a first step whenever C_{60} is polymerized, and information on how to identify this compound is therefore of interest. Wang et al. [151] presented IR spectra of C_{120} and Lebedkin et al. [157] have recently studied its Raman spectrum. In addition to the appearance of a large number of new modes because of the decrease in symmetry on polymerization, transfer of charge from the intramolecular bonds to the intermolecular ones leads to a slight weakening of the former and thus to a small decrease in the frequencies of the intramolecular phonon modes. In particular, the well-known $A_g(2)$ "pentagonal pinch" mode shifts down from 1469 to 1463 cm^{-1} and low-frequency intermolecular modes appear, the most prominent one at 96 cm^{-1}. Note that this frequency shift is opposite in sign to that observed on compression of pristine C_{60} (Table 15.7).

15.3.1.2 Linear Chain Polymers

It is intuitively clear that polymerization under gentle conditions should result in the formation of relatively few intermolecular bonds per molecule. Next to the dimer, the linear polymeric chain has the fewest intermolecular bonds since the typical molecule is connected to at most two other molecules. (The term "linear" here does not necessarily imply *straight* linear chains.) A number of such polymers can be produced by various methods. As mentioned earlier, pure C_{60} can be polymerized either by radiation or by high-pressure treatment, and in both cases linear chains are formed under certain conditions. Doping by alkali metals to form AC_{60} followed by slow cooling to RT also results in the formation of linear chain polymers [158]. The formation of intermolecular bonds decreases the intermolecular distance from 10 Å to about 9.15 Å. Polymerization between nearest neighbors in the $\langle 110 \rangle$ direction should thus in principle distort an initially fcc lattice into a well-ordered orthorhombic (orh) lattice consisting of parallel linear chains, as shown in Figure 15.25.

Under optimum conditions, slow polymerization of C_{60} under pressure may be observed even below 0.8 GPa [159], but the reaction rate increases significantly at higher pressures. Bashkin et al. [78] studied polymerization at low pressures using a volume compression method and deduced the phase diagram shown in Figure 15.26. Polymerization occurs in the area to the upper right, but the polymerization reaction is reversible and the polymeric phase breaks down if heated into the area in the upper left. It is now generally accepted that polymerization below about 650 K results in the formation of basically orthorhombic phases at all pressures up to about 9 GPa (Fig. 15.24). However, at low pressures, including atmospheric, the molecules are rather far apart. In principle, polymerization then occurs only when double bonds on neighboring molecules face each other across the intermolecular gap. In the fcc phase quasi-free rotation may randomly place double bonds in this position, but in the sc phase this is not very likely and below some minimum pressure, neither photo-

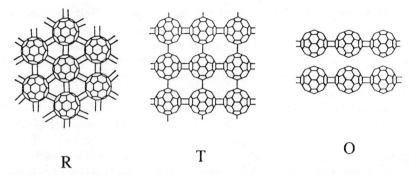

R   T   O

Figure 15.25 Structural arrangements of C_{60} molecules in rhombohedral (*left*) and tetragonal (*center*) two-dimensional polymers and in orthorhombic one-dimensional polymer (*right*). Reprinted with permission from Ref. 99. Copyright © 1997 American Physical Society.

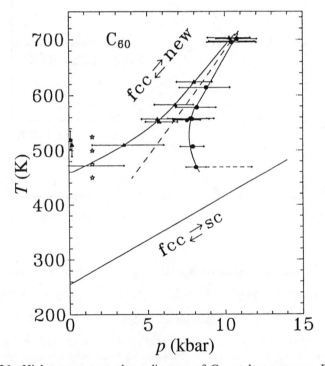

Figure 15.26 High-temperature phase diagram of C_{60} at low pressure. Dashed line shows the equilibrium phase line between molecular C_{60} and the orthorhombic polymer, while symbols and full curves show observed polymerization and depolymerization coordinates. The fcc–sc phase boundary is shown for comparison. Reprinted with permission from Ref. 78. Copyright © 1994 IOP Publishing Ltd.

polymerization nor pressure-induced polymerization are likely to occur in sc C_{60}. The random rotation implies that at least the initial stage of the polymerization occurs in random directions, and it is generally observed [78,98,99,146] that "low"-pressure polymerized C_{60} is very disordered. Early studies [78,146] could not identify the actual structure, and later studies have shown that reasonable fits to measured XRD diagrams can only be found if the presence of orh, rhh, and tetragonal (tet) structures (Fig. 15.25) are assumed. The polymerization reaction is also rather slow, and short-time treatments [146,160] invariably result in XRD results that can still be indexed as fcc, but with broadened lines and changes in relative intensities. Raman spectra [146,160] also show lines and shifts characteristic of dimers, as discussed earlier. The evolution with reaction time from pristine C_{60} over dimers to polymers is very clear from the Raman results shown in Figure 15.27, which were obtained as a function of reaction time at 1.1 GPa and 530–570 K [146]. The figure shows that the high-

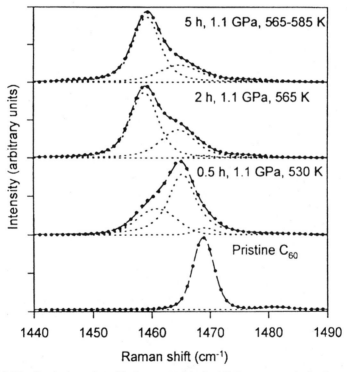

Figure 15.27 Evolution of $A_g(2)$ "pentagonal pinch" Raman mode in C_{60} with reaction time during polymerization. The characteristic frequencies are approximately 1469 cm^{-1} for molecular C_{60}, 1464 cm^{-1} for dimers, and 1459 cm^{-1} for linear chains. Reprinted from *Chem. Phys. Lett.*, Vol. 258, P.-A. Persson, U. Edlund, P. Jacobsson, D. Johnels, A. Soldatov, and B. Sundqvist, "NMR and Raman characterization of pressure-polymerized C_{60}", pp. 540–546, Copyright © 1996, with permission from Elsevier Science.

pressure treatment initially results in the formation of dimers. Since X-ray diffraction diagrams taken on the same sample still show broadened peaks at basically fcc locations, but with relative intensities different from the initial state, the dimer axes probably point in random directions. With time, the dimers coalesce to form polymer chains, which at least locally order into the orh phase. After a reaction time of 5 h, the line characteristic for pristine C_{60} (1469 cm^{-1}) has disappeared, the dimer line (1464 cm^{-1}) has decreased significantly from its peak value after 30 min, and the "linear chain" line at 1459 cm^{-1} dominates. A similar development is found for the low-frequency intermolecular modes, where a dimer peak at 96 cm^{-1} gives way to a linear chain peak near 118 cm^{-1}. The structures of photopolymerized and "low"-pressure polymerized C_{60} are probably very similar, at least on a local scale, because a comparison between the Raman spectra of thin films polymerized by different routes [161] showed them to be practically identical above 150 cm^{-1}.

Although "low"-pressure, nominally orh polymers are usually disordered, well-ordered orh polymers can be produced by heating to above 350 K in the range 3–8 GPa, as observed already in the pioneering study by Núñez-Regueiro et al. [162]. As discussed in detail by Marques et al. [163], at these pressures C_{60} is in a well-ordered H-oriented phase with the double bonds on molecular neighbors in rather close vicinity, and an increase in libration amplitude with T (without molecular rotation) is enough to initiate polymerization. Another factor that might contribute to the production of well-ordered materials is the use of solid pressure media under these conditions, causing some degree of uniaxial stress, which will help to define an initial polymerization direction. Marques and co-workers deduced a structure in which the intermolecular bonds in neighboring chains lay in the same plane and suggested a very simple mechanism for an evolution from the orh phase into two-dimensional rhh and tet phases by simple compression and sliding of the chains [163]. However, recent studies [164–166] have thrown doubts on this simple mechanism. It has recently become possible to polymerize large single crystals under hydrostatic pressures near 1.2 GPa, and careful structural studies on these crystals show that although the basic orh structure is similar to that suggested previously [162,163], neighboring chains are rotated relative to each other [164]. The new structure, which has a symmetry *Pmnn*, is tentatively termed orh$'$ by Agafonov et al. [165], who found very similar results in a study on powder specimens pressurized in the same range. Although experimental studies have not yet been able to pinpoint the exact angles, theoretical studies [166] suggest that it is close to 60°. For a structure with this chain orientation, a transformation from the orh$'$ to the rhh phase should not be energetically possible at reasonable temperatures.

The question of whether the two orthorhombic structures are actually different or not has not yet been resolved, and Agafonov et al. [165] suggest that the well-ordered orh phase found by Marques et al. [163] and others above 3 GPa may actually be rhh. The question of rotated chains also has a bearing on the alkali-metal-doped phases AC_{60}, where CsC_{60}, RbC_{60}, and KC_{60} have surprisingly different band structures and conduction properties [139]. It has been suggested [167] on the basis of NMR studies that the different band structures are caused by different orientations of the chains in different compounds, and a careful recent study [168] indeed shows that the rotational angles of the chains are different in KC_{60} and RbC_{60}.

Many groups have also observed changes in the properties of C_{60} at RT after compression to above 5–6 GPa. Changes have been observed, for example, in the bulk modulus [71] and the Raman [96,169] and IR [108] spectra. These changes are probably also connected with dimerization or a polymerization into a linear chain phase, but no detailed study has yet been carried out and it is not yet clear whether RT polymerization is caused by pressure alone or whether it is assisted by radiation. Some studies [3,71,169] also indicate that at RT polymerization occurs preferably under nonhydrostatic conditions.

Because linear chain polymers are reasonably simple to produce, their properties have been investigated in some detail. As for the dimers, the reduced symmetry after polymerization gives rise to a very large number of new lines in both the IR and the Raman spectra, as illustrated in Figures 15.28 to 15.30. Figures 15.28 and 15.29 are taken from the work of Rao et al. [98,99] and show spectra for a number of different pressure polymerized samples, and the curve denoted O has been measured on nominally orh phase, produced at 4.8 GPa. Figure 15.30 shows the Raman spectrum of the single crystal orh′ sample studied by Moret et al. [164]. For all samples the 14 allowed Raman plus IR lines for pristine C_{60} split on polymerization and their intensities decrease, while many new lines and bands appear. Most of the new lines are previously forbidden lines, which now appear because of the broken symmetry, but some of them are new and characteristic of the new polymeric structure. Although significant shifts are observed for the high-frequency modes like the $A_g(2)$ mode discussed earlier, spectral lines below about 1000 cm^{-1}, correlated with radial vibrations, seem to be affected even more. The intensity of the radial "breathing" $A_g(1)$ mode at 496 cm^{-1} decreases strongly and a large number of new modes appear in the range below 1000 cm^{-1} in both the IR and Raman spectra. As already mentioned, for the linear chain phase the new low-frequency mode at 118 cm^{-1} (Fig. 15.30) is of particular interest because it can be identified as an intermolecular chain mode. In contrast to optical spectroscopy, inelastic neutron scattering can, in principle, study all phonons in the material, but because only polycrystalline samples are available only the density of phonon states can be obtained. Recent neutron scattering studies [170–173] show the appearance of new intermolecular translational and librational modes at low energies in linear chain polymers, mainly between 12 and 20 meV, as shown in Figure 15.31. Calculations [171,172] show that the librational modes dominate above 16 meV. Results for pressure polymerized C_{60} and RbC_{60} [171] were practically identical, in agreement with the results of Raman studies [171,174]. At higher frequencies a splitting of the radial intramolecular modes was also observed [170,173], as also indicated by Raman studies.

Theoretical calculations [175] show that orthorhombic $(C_{60})_n$ should still be a semiconductor with a slightly reduced bang gap compared to the pristine

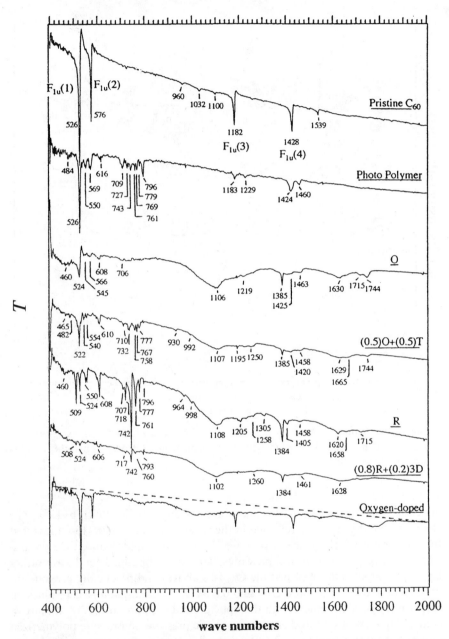

Figure 15.28 IR absorption spectra for several types of C_{60} polymers. Pristine (*top*) and oxygen-doped (*bottom*) C_{60} are both molecular solids, while the other curves are for polymers: from second curve downward photopolymerized, orh (O), mixed orh–tet (T), rhh (R), and mixed rhh and three-dimensionally polymerized materials. Reprinted with permission from Ref. 99. Copyright © 1997 American Physical Society.

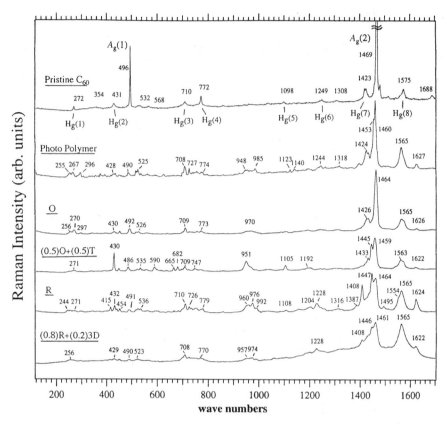

Figure 15.29 Raman spectra for the same materials as in Figure 15.28, except for the oxygen-doped material. Reprinted with permission from Ref. 99. Copyright © 1997 American Physical Society.

material, and this is verified by photoluminescence studies [98,176–178]. The change in band gap is also reflected in the resistivity. Saito et al. [134] measured the temperature dependence of ρ above RT to 700 K and deduced the parameter E_a in Eq. (15.2) at several pressures. The values obtained for E_a are rather large compared to those of pristine C_{60} (see above) and practically pressure independent: 0.8 eV at 3 and 6 GPa and 0.6 eV at 10 GPa. Even larger values, in surprisingly good agrement with the optical band gap, are found by Makarova et al. [179] in resistivity studies at atmospheric pressure on pressure polymerized material. Samples treated at 2 GPa showed a continuous decrease in E_a with increasing treatment temperature, from 1.50 eV at 100 °C (probably only partly polymerized) to 1.02 eV at 650 °C, well into the range of stability of the rhh phase. Surprisingly, polymerization into the orh phase at 100 °C at 6 and 8 GPa gave a much larger band gap near 2.35 eV, larger than for pristine C_{60}. In principle, this agrees with the earlier discussion about two different ortho-

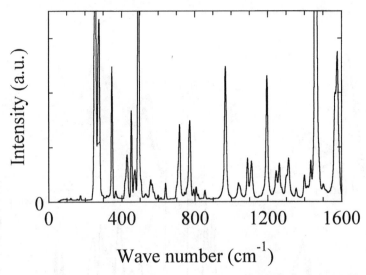

Figure 15.30   Detailed Raman spectrum for a polymerized C_{60} single crystal in the orh phase. (Spectrum taken by Per-Axel Persson, Umeå University, Sweden.)

rhombic phases with different chain orientations and band structures, but the difference is surprisingly large. However, Takahashi et al. [180] also polymerized C_{60} into the orh ("fcc") phase at 5.4 GPa and 673 K, and then measured $\rho(T)$ during a retransformation to molecular C_{60} by heating. Their results verified that the RT resistivity of orh C_{60} was *higher* than that of pristine C_{60}, in agreement with the results of Makarova et al. [179].

Significant changes in the bulk lattice properties are also found on polymerization because of the strong covalent intermolecular bonds formed. Compared to the intermolecular interaction in pristine C_{60}, these covalent bonds are little affected by temperature and practically incompressible. In the direction of these bonds we would therefore expect the properties of the polymer to be similar to those of diamond or to the in-plane properties of graphite. To a first approximation, both the compressibility and the thermal expansivity should thus be about two-thirds those of pristine C_{60}. The measured thermal expansion coefficient [181,182] is in reasonable agreement with this model (Fig. 15.32), while data for $B(0)$ range from 24 to 33 GPa (Table 15.10). The latter values are much more than the expected 50% higher than the value for pristine C_{60}, but the data for $B(0)$ cannot be compared directly to data for pristine C_{60} because the molecules in the polymers are orientationally fixed, which should give a higher bulk modulus (see the dashed line in Fig. 15.14). Alkali-metal-doped chain polymers should have an even higher bulk modulus because of the increased interchain interaction due to the alkali metal ions, in analogy with the case for cubic doped C_{60} (see above), and this is also observed in practice (Table 15.10). However, it is interesting to note that Na_2CsC_{60} is much more

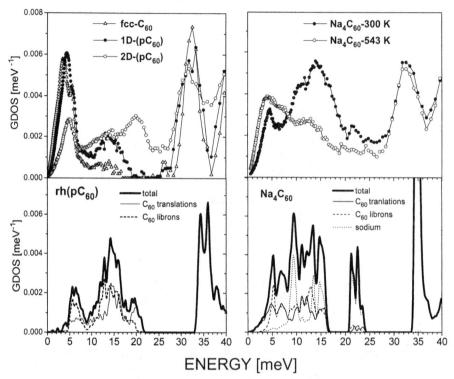

Figure 15.31   Generalized densities of states (GDOS) of lattice modes for C_{60}, one- and two-dimensionally polymerized C_{60}, and the doped two-dimensional polymer Na_4C_{60} as measured by neutron scattering. The lower panels show, for comparison, theoretical calculations for the two-dimensional polymers. Reprinted with permission from B. Renker, H. Schober, R. Heid, and B. Sundqvist, in *Electronic Properties of Novel Materials-Progress in Molecular Nanostructures*, H. Kuzmany, J. Fink, M. Mehring, and S. Roth (eds.), AIP, Woodbury, 1998, pp. 322–326. Copyright © 1998 American Institute of Physics.

compressible than its alkali-metal-doped cousins KC_{60} and RbC_{60}, since the very similar compound Na_2RbC_{60} has very recently been shown to have *single* C–C intermolecular bonds [184]. If we speculate that the two Na_2AC_{60} compounds are similar in this respect, the single C–C bond is much more compressible than the four-membered carbon rings of the AC_{60} compounds.

Thermal expansion studies of the thermal breakdown of dimers and polymers indicate that a thermal energy of 1.7 and 1.9 eV per bond is needed to break the intermolecular bonds in dimers and polymers, respectively [181]. Polymerization changes the acoustic phonon spectrum as discussed earlier and shown in Figure 15.31, resulting in changes in the specific heat capacity. In a simple Born–von Karman type model, two-thirds of the acoustic modes should be left basically intact on polymerization while the modes in the chain direction

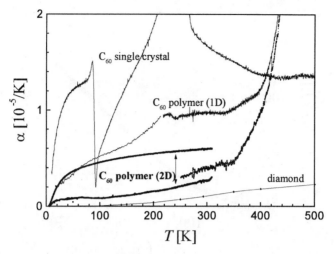

Figure 15.32 Thermal expansion coefficient α for molecular C_{60} and the one- and two-dimensional polymers. Data for the latter have been measured on a macroscopically textured sample, in directions preferentially perpendicular to (upper curve) and parallel with (lower curve) the polymerized planes. Reprinted with permission from P. Nagel, V. Pasler, S. Lebedkin, C. Meingast, B. Sundqvist, T. Tanaka, and K. Komatsu, in *Electronic Properties of Novel Materials-Progress in Molecular Nanostructures*, H. Kuzmany, J. Fink, M. Mehring, and S. Roth (eds.), AIP, Woodbury 1998, pp. 194–197. Copyright © 1998 American Institute of Physics.

move to higher energies with a peak near 15 meV [171,172]. Above about 70 K the specific heat capacity of C_{60} is dominated by Einstein-like intramolecular modes, as shown in Figure 15.33 [182]. Because the new chain modes have Debye temperatures well above 70 K, their contributions to c_p will be masked by those of the Einstein modes, and the low-T Debye "shelf" in the figure decreases in amplitude by about one-third with only a rather small change in effective Debye temperature for the remaining cross-chain modes. The increase in average bond strength should also give a significant increase in the thermal conductivity. Surprisingly, studies as a function of temperature and pressure

TABLE 15.10 Bulk Moduli of Polymerized Fullerenes with Orthorhombic Structures

Material	$B(0)$ (GPa)	Reference
C_{60}	32.9	78
C_{60}	23.8	74
KC_{60}	40	183
KbC_{60}	58	183
Na_2CsC_{60}	28.9	89

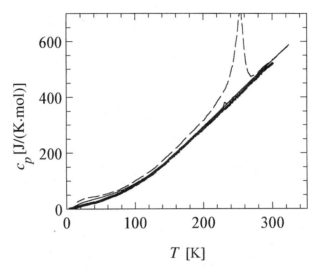

Figure 15.33 Specific heat capacity below RT for one-dimensional polymer (full curve), two-dimensional polymer (dots), and depolymerized C_{60} (dashed curve). Reprinted with permission from Ref. 182. Copyright © 1998 The Electrochemical Society.

show no such increase [159]. Instead, the thermal conductivity actually decreases slightly below RT, as shown in Figure 15.34. However, the linear dependence on temperature indicates a strong, almost glass-like structural disorder, which probably more than offsets the increase in intermolecular interaction.

Finally, the alkali-metal-doped polymers are metallic, and their conduction properties have been investigated under pressure, as reviewed by Zettl [139].

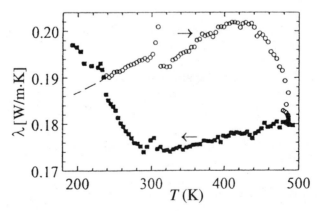

Figure 15.34 Thermal conductivity λ for orh C_{60} at 0.08 GPa. Open symbols show λ in the orh phase during heating. The drop in λ above 430 K indicates depolymerization. Squares show λ for depolymerized C_{60} during cooling. The fcc–sc transformation can clearly be observed. Anomalies near 300 K are experimental artifacts. Reprinted with permission from Ref. 159. Copyright © 1997 Springer-Verlag.

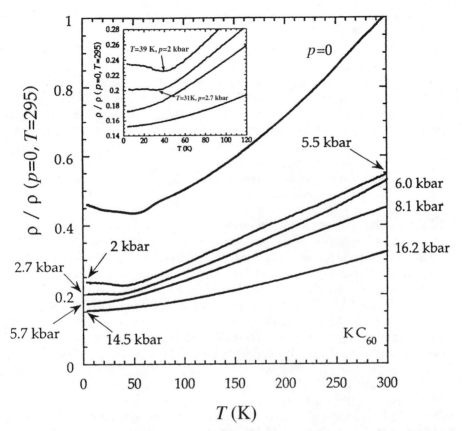

Figure 15.35 Normalized resistivity of KC_{60} under pressure. Inset shows the low-temperature range in more detail. Note that the pressure in each experiment depends on temperature. Reprinted with permission from Ref. 183. Copyright © 1997 Springer-Verlag.

KC_{60} and RbC_{60} have different band structures because of different chain orientations [168], and this is also reflected in their behavior under pressure. ESR studies by Forró et al. [185] show that CsC_{60} has the strongest 1D character, that KC_{60} is the most 3D-like of these materials and that RbC_{60} also shows a charge density wave below 50 K. The resistivity of orthorhombic polymeric KC_{60} is nonlinear but metallic from 50 to 400 K at zero pressure [183,186,187]. As for A_3C_{60}, a correction to constant volume gives a basically linear temperature dependence. Figure 15.35 [184] shows the relative resistance versus T at several pressures. At zero pressure $d\rho/dT$ changes sign near 50 K. Anomalies near 50 K have also been observed in other types of experiments and it has been speculated that they are caused by the formation of spin- or charge-density waves on the chains, opening a Fermi surface gap, but this model does not agree with the linear $\rho(T)$ below 50 K. The initial pressure coefficient

$d(\ln\rho)/dp \approx -1.5$ GPa^{-1} at RT very rapidly decreases to a much lower value above 0.2 GPa [183,186,187], and near 1.5 GPa the resistivity is practically linear in T. The low-T anomaly disappears under pressure, and the transition temperature extrapolates to zero near 0.45 GPa. In contrast, for RbC_{60} the resistivity is "semiconductor-like" at zero pressure with a transition anomaly near 200 K, but a metallic state with a linear constant-volume resistivity appears under pressure [183,186,188]. At zero pressure RbC_{60} has an antiferromagnetic phase with a quasi-1D spin-density wave below 35 K [189] but this is suppressed in the metallic state. Surprisingly, the 200 K anomaly appears at all pressures and divides two insulating phases at zero pressure, an insulator and a metal at intermediate pressures, and two metallic states at high pressures. So far, this behavior is not well understood.

15.3.1.3 Two-Dimensional Polymers

As discussed earlier and shown in Figure 15.24, treatment of C_{60} in the range below 9 GPa and about 725 K always results in the formation of linear chain polymers, probably with either of two slightly different orientational structures. At higher temperatures a larger number of intermolecular bonds are formed and new phases, shown schematically in Figure 15.25, appear. Complete polymerization of the close-packed (111) plane results in a dense rhombohedral (rhh) phase, observed already in the pioneering study of Iwasa et al. [145], while a cross-polymerization of the linear chains along the $\langle 110 \rangle$ directions instead gives a tetragonal (tet) structure. While the latter has only very recently been produced in reasonably pure form, systematic investigations of the structure as a function of treatment conditions [163,190,191] show that the rhh phase is produced as the majority phase above 725 K at almost all pressures from the lower stability limit near 1.5 GPa [192] to about 9 GPa (see Fig. 15.24). Marques et al. [163] showed that a mixed two-dimensional polymer with the tet structure as the majority phase is usually found at pressures below about 4 GPa, and in fact the tet phase seems to form more easily from initially fcc C_{60} [193] while the rhh phase is the preferred structure obtained from orientationally ordered sc material. To produce a pure tetragonal phase it is necessary to pressurize C_{60} at an elevated temperature [194] to avoid an initial nucleation of the orthorhombic phase during heating to suitable reaction conditions (see Figure 15.24). In this way it is even possible to polymerize single crystals into an almost purely tetragonal state. X-ray diffraction studies of such a polymerized crystal show that the symmetry of the tetragonal phase is $P4_2/mmc$ [195], rather than $Immm$ as was assumed earlier [162,163]. The detailed polymerization mechanism for the rhh and tet phases is presently a subject of some debate and is not fully understood, and whether the tetragonal structure is preferred at low pressures because it is less dense than the rhh phase or because it can only form from the rotationally disordered fcc phase is not known. Like the linear chain polymers, the two-dimensional polymers are insoluble and break down to molecular C_{60} on heating at zero pressure. While the dimer breaks down already near 420 K [151] and the linear chains near 500 K [181], the two-dimensional structures are stable to about 550

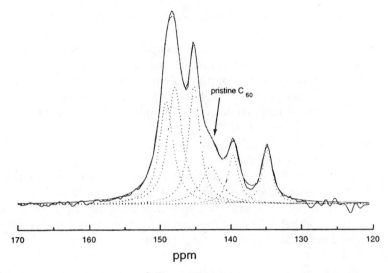

Figure 15.36 Main NMR peak (sp^2 peak) for rhh C$_{60}$, showing that the large number of bonds deforms the molecules and splits the peak into several subpeaks. Reprinted with permission from Ref. 147. Copyright © 1996 American Physical Society.

K [181]. Other two-dimensional fullerene polymers have also been produced. Long-time radiation treatment has recently been claimed to result in the formation of up to six intermolecular bonds per molecule and presumably a two-dimensional polymer [196], while doping with sodium has been shown to give the polymeric compound Na$_4$C$_{60}$, which like the alkali-metal-doped dimer has only a single C–C bond between each C$_{60}$ molecule [197]. Neither of these, however, has been studied or produced under pressure.

The rhh and tet phases, with their stacked fullerene planes, can to some extent be considered as analogs to graphite, and attempts have been made to intercalate alkali metal ions between the polymerized layers [198]. However, as for the linear chain phases [199], all attempts so far have resulted in the breakdown of the polymer and the formation of doped molecular C$_{60}$. Each molecule in the rhh phase is slightly deformed (flattened) by the strain of the six covalent intermolecular bonds connecting it to its neighbors [200], and as a result, the carbon atoms have slightly nonequivalent bond states. NMR studies [147] show that the sp^2 peak near 144 ppm, which, for molecular C$_{60}$, is a very sharp line [1], is broadened and split into five subpeaks as shown in Figure 15.36. The larger number of intermolecular bonds compared to that in the linear polymers also gives a higher intensity of the sp^3 peak at 77 ppm. Other types of spectroscopic studies have also been carried out, and the spectroscopic properties of the two-dimensional polymers are reasonably well known [98,99,194,195]. IR and Raman data for both rhh and mixed orh/tet phases are shown in Figures 15.28 and 15.29. Although, in principle, the symmetry of the rhh phase is higher than for the linear polymers, the former generally have richer spectra, with an even larger number of lines and bands than for the latter. For the rhh phase an especially intense and characteristic IR band appears between 700 and 800 cm^{-1}. A possible reason for the large number of new lines and bands might be that the higher treatment temperatures lead to more lattice disorder and a reduced local symmetry. The larger number of bonds per molecule also gives a larger charge transfer from the molecular "bellies," increasing the bonding-induced shift of the pentagonal-pinch Raman mode toward lower frequencies. The IR absorption spectra of rhh C$_{60}$ and RbC$_{60}$ have been analyzed in some detail by Kamarás et al. [201], who are able to identify a large number of modes from symmetry considerations.

Calculations [202,203] show that the electron band structure of rhh C$_{60}$ should be more three-dimensional than that of the orthorhombic phase. Although rhh C$_{60}$ should still be a semiconductor, the band gap is even smaller than for orh C$_{60}$, about 0.7 eV smaller than for pristine C$_{60}$. This is verified by photoluminiscence studies [176,178,204]. The luminiscence spectrum has also been studied under pressure up to 1.7 GPa by Venkateswaran et al. [204] and the pressure dependence of the band gap is found to be very much smaller than for pristine C$_{60}$, as might be expected for a less compressible material. Above 2 GPa, Venkateswaran et al. [204] found that the spectral intensity decreases strongly and they suggest the reversible formation of interlayer bonds. At zero pressure the rhh phase is found to have a characteristic and strongly temperature-dependent peak at 1.75 eV, possibly connected with lattice defects induced during synthesis by the high temperatures involved [176,178]. Again, the band structure can also be studied by transport measurements. While the resistivity study of Ma and Zou [205] from 300 to 950 K at 5 GPa shows a number of changes probably indicating transitions between the sc, orh, and rhh phases, no data for the excitation energy E_a can be deduced from their results. As mentioned earlier, Makarova et al. [179] find an E_a of 1.02–1.04 eV for this phase.

Again, the increase in the density of intermolecular bonds strongly modifies the lattice dynamics and many macroscopic physical properties. Polymerization in two dimensions leaves only one dimension "soft" and the properties of these phases should not be unlike those of graphite. Using the same simple model as above, we would expect the average bulk modulus to be about three times higher than for pristine C$_{60}$ and the thermal expansion coefficient three times smaller. Recent reports [182,192] give a value $B(0) = 45$ GPa in a random direction of a textured anisotropic sample, in reasonable agreement with the model. Calculations [203] show that the intraplane bulk modulus should be about 110 GPa and the energy needed to break intermolecular bonds about 1.6 eV, in reasonable agreement with the experimental value 1.9 eV [181]. Thermal expansion measurements in two directions on a similar sample [181] shown in Figure 15.32 show that the thermal expansion along the dense, polymerized planes is very similar to that of diamond, while that in a perpendicular direction is indeed about one-third that of pristine C$_{60}$. The specific heat data in Figure 15.33 also follow the trend described above, with a very much reduced "Debye shelf" in comparison with pristine C$_{60}$ but only a small shift in effective

Debye temperature. Although this might be taken to indicate only small changes in the lattice properties, Figure 15.31 shows that there are large changes in the acoustic spectrum compared to both linear chain polymers and pristine C_{60} [171,172,182]. While the formation of the linear chain polymer added "stiff" acoustic (chain) modes between 6 and 20 meV without significantly decreasing the very-low-frequency libration and vibration modes of pristine C_{60} below 7 meV, the formation of the two-dimensional polymer gives much more dramatic changes in the phonon spectrum. A strong decrease is now observed for the modes below 7 meV and the density of higher frequency modes up to 25 meV increases significantly, with a maximum density near 20 meV. These changes are reversible and disappear on depolymerization.

15.3.1.4 Three-Dimensional Polymers

Crystalline Phases Above about 9 GPa the C_{60} lattice has been compressed to the point where even rather mild heating results in the formation of a large number of intermolecular bonds in random directions. Again, several systematic studies [190,206–209] have been carried out in which a series of samples have been heated to high temperatures, then quenched back to RT and brought back to zero pressure for study. In the range 9–15 GPa, the general structural evolution with treatment temperature seems to be basically independent of pressure, although the actual temperature scale is not. According to the discussion above, the initial zero-pressure structure should be either the orientationally ordered sc phase or the linear-chain orh polymeric phase, or most probably a mixture of both. Most studies, however, identify the structure as fcc, probably because of pressure- or polymerization-induced disorder. At the lowest pressures, 9–11 GPa, an initial growth of the orh phase can usually be observed on heating before a polymeric fcc structure appears at high temperatures. At higher pressures the lattice remains basically fcc at all temperatures, at 13 GPa up to about 800 K, as shown in Figure 15.37. At this pressure the lattice parameter decreases sharply with increasing treatment temperature above 500 K, indicating a continuous formation of intermolecular bonds [207]. A similar behavior, with higher threshold temperatures up to 550 K, is observed at all lower pressures down to 9 GPa [202–209]. Also observed in very-high-pressure studies discussed earlier, the decrease in lattice constant is strong enough that the (200) reflection, previously hidden by the molecular form factor, reappears. With increasing reaction temperature, diffraction peaks broaden and lose intensity, but fcc peaks corresponding to lattice constants down to 12.2 Å can still be detected at 820 K, and one study on a sample treated at 13 GPa and 900 K even reports a value of 11.7 Å [210]. At such small lattice parameters the molecules are strongly strained and new features appear in the XRD diagrams and Raman spectra. The (111) line, which at this lattice parameter should be erased by the molecular form factor, instead becomes very prominent, and the relative intensities of the lines observed are quite different for this highly dense high-temperature phase, for pristine fcc C_{60}, and for the three-

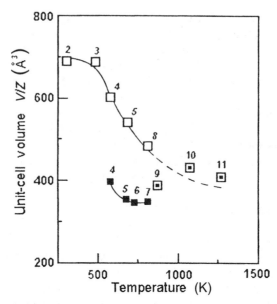

Figure 15.37 Evolution of the unit-cell volume of C_{60} at zero pressure with treatment temperature after treatment at 13 GPa. Symbols denote different observed lattice structures: open squares—fcc; solid squares—distorted bcc; square/dot—monoclinic phases. Reprinted from *Phys. Lett. A*, Vol. 220, V. D. Blank, S. G. Buga, N. R. Serebryanaya, G. A. Dubitsky, S. N. Sulyanov, M. Yu Popov, V. N. Denisov, A. N. Ivlev, and B. N. Mavrin, "Phase transformations in solid C_{60} at high-pressure—high-temperature treatment and the structure of 3D polymerized fullerites", pp. 149–157, Copyright © 1996, with permission from Elsevier Science.

dimensionally polymerized fcc phase with $a = 13.6$ Å obtained at the same pressure near 500 K, that is, at the threshold for lattice collapse in Figure 15.37. Thus, partial molecular collapse might modify the form factor, or small local clusters of carbon atoms may form between the distorted remains of original molecules.

Above 10 GPa, other crystalline phases are indicated in XRD studies on samples treated at very high temperatures, and most samples are two- (or multi-) phase. As shown in Figure 15.37, the presence of body-centered (bc) and monoclinic (mc) structures have been deduced from the diffraction diagrams above about 600 K. Although the basic structure is body-centered, they are always distorted and no longer exactly cubic [207]. Blank et al. [2] report an orthorhombic pseudotetragonal body-centered structure with a density of 3.39 g/cm³ for the sample treated at 720 K in Figure 15.37. Both the orthorhombic distortion and the structural disorder increase with treatment temperature. Because of the large pressure and temperature gradients in the anvil devices used, samples may retain a fcc structure with a lattice constant down to 12.3 Å simultaneously with the orthorhombic phase, making the analysis rather complicated. Above 850 K a monoclinic structure appears, and in some samples, a fraction of the material remains in this phase up to almost 1300 K, where most of the material has become more or less disordered or amorphous. A polymerized bcc structure with a lattice parameter of 9.54 Å, similar to those observed here, was predicted by O'Keeffe [211] assuming polymerization by a hexagon–hexagon (i.e. 6 + 6 atom) interaction, and it is possible that the molecules in the "bc" and mc range are severely distorted and partially destroyed by the formation of a large number of intermolecular bonds. Another mechanism was suggested by Blank et al. [2,212], who proposed that the origin of the body-centered structure in three-dimensionally polymerized C_{60} is new fullerene-like cages, formed in the center of each close-packed, completely polymerized molecular tetrahedron by six hexagons, two pentagons, and five squares on the surfaces of the original molecules.

The crystalline phases observed have been studied by Raman spectroscopy at zero pressure [206,207] with the results shown in Figure 15.38. The initial evolution with treatment temperature is similar to that in Figure 15.29, with an increase in the number of lines, a decrease in their intensity, and a broadening of the lines, reflecting the increasing number of bonds per molecule and the increasing disorder. Raman spectra of fcc and bcc samples are noticeably different, although the low-frequency bands are similar. In the range between 1300 and 1700 cm^{-1} a single band was observed in the spectra of bc samples treated at 670 to 1070 K, instead of the double peak observed for fcc samples. The Raman spectra of the monoclinic phases synthesized at 13 GPa are very similar to those observed for samples treated at very high temperatures, 1270 and 1470 K, samples for which X-ray diffraction shows an amorphous structure (see below), and the shape and position of this Raman band may thus reflect a high degree of sp^3 hybridization in both the bc structures and the disordered states discussed below. On the other hand, distinct lines near 500, 710, and 770 cm^{-1} can be correlated with the $A_g(1)$, $H_g(3)$, and $H_g(4)$ modes of C_{60} and thus show that at least parts of the molecule preserve their basic structure.

Although no change in lattice structure can be observed on polymerization in the fcc phase, large changes are found in the macroscopic properties of the material. A very large increase in hardness can be correlated with the sharp decrease in lattice constant for treatment temperatures near 500–570 K, and this phase has therefore been dubbed "hard fcc." Reported values for the hardness H_v range from 30–34 GPa [208,209] to 70 GPa [2] at a lattice constant of 12.9 Å, and to >90 GPa at 12.4 Å [2]. Materials treated in this range have been reported [2,206,213] to cut cubic BN, the second hardest material known until now. The increase in hardness is described by Brazhkin et al. [208] in terms of a rigidity percolation model, in which sparse, random three-dimensional supramolecular clusters form at low treatment temperatures and gradually grow until a randomly bonded, three-dimensional, still basically fcc polymeric structure is formed when the average number of bonds per molecule exceeds a critical value near 2.4 [214]. Fullerene materials with a large number of intermolecular bonds would be expected to have interesting mechanical and thermal

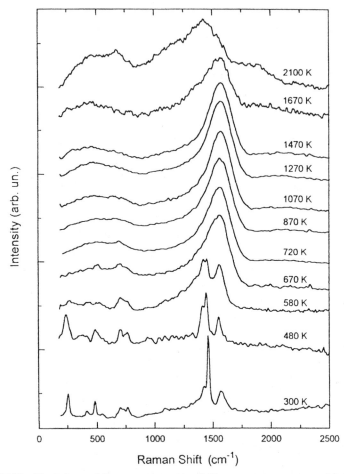

Figure 15.38 Evolution of Raman spectum of C_{60} at zero pressure with treatment temperature after treatment at 13 GPa. Reprinted from *Phys. Lett. A*, Vol. 220, V. D. Blank, S. G. Buga, N. R. Serebryanaya, G. A. Dubitsky, S. N. Sulyanov, M. Yu. Popov, V. N. Denisov, A. N. Ivlev, and B. N. Mavrin, "Phase transformations in solid C_{60} at high-pressure—high-temperature treatment and the structure of 3D polymerized fullerites", pp. 149–157, Copyright © 1996, with permission from Elsevier Science.

properties. Brazhkin et al. [209] report values for the Young's modulus E of 350–400 GPa for fcc C_{60} polymer with $a = 12.27$ Å, while Blank et al. [213] find that the specific heat capacity of fcc polymers is up to 12% smaller than that of pristine C_{60} in the range 400–600 K. A low specific heat indicates a high Debye temperature, and it may be noted that diamond has the lowest value known for the specific heat capacity per mole near RT because of its high Debye temperature. These materials also continue the trend regarding the conduction properties, having a semiconducting behavior with a very small band gap and a measured excitation energy of only 0.015 eV [2,213].

Noncrystalline Phases After treatment at high temperatures, about 1000 K below 7 GPa or 900 K near 10 GPa, or even at RT above 20 GPa, the intensity of X-ray diffraction lines decreases rapidly with increasing treatment temperature and diffuse halos appear, reflecting a strong disorder in the structure, and at sufficiently high temperatures, almost featureless diffraction diagrams are obtained. Below about 10 GPa, studies by several groups [2,163,215] show that high-temperature treatment results in breakdown of the C_{60} molecules and the formation of a very disordered, hard black phase first found by Kozlov et al. [215] after heating C_{60} to near 1000 K at 2.5–3 GPa. Blank et al. [191] denote these phases as "partially graphitized fullerenes" (Fig. 15.24), since the structure probably consists of a layered matrix that is turbostratic and graphite-like on a microscopic scale but strongly cross-linked by sp^3 bonds. Samples treated at high temperatures are not always stable and often develop with time into composites of very hard particles in a softer matrix. Although XRD and Raman studies [191,215] show similarities with turbostratic graphite or glassy carbon, the material is very much harder than these because of a large number of interplanar bonds. Blank et al. [2] compare the structure with various kinds of amorphous and glassy carbons but conclude that this is a truly novel form of amorphous carbon with medium-range order based on cross-linked and partially destroyed C_{60} molecules. As might be expected there is a correlation between the hardness and the interplanar distances such that for the hardest samples, with $H_v = 38$ GPa, XRD indicated an interplanar spacing of 3.32 Å, smaller than $d_{002} = 3.36$ Å for graphite, probably due to the presence of many interplanar bonds. The similarity to amorphous carbon and graphite also extends to the conduction properties, since this material is a semimetal with a practically temperature-independent resistivity of 10^{-3}–10^{-1} $\Omega \cdot$ cm [2,215]. Kozlov et al. [216] also used an excimer laser to ablate this material and produced transparent insulating thin films, which were found to contain crystalline diamond.

A different black, amorphous phase can be obtained by rapidly heating C_{60} to above 900 K at pressures of 8–12 GPa [190,206,207,213]. In this pressure range the material is well above the graphite–diamond boundary in the phase diagram of atomic carbon [55] and there is no longer any tendency for the material to graphitize. This material has a density of 2.4–2.8 g/cm^3 and has been reported to be comparable to crystalline diamond in hardness [206]. XRD studies of samples treated at 9.5 GPa show a (002) halo corresponding to an interplanar distance of 3.14–3.3 Å, decreasing to 3.0 Å for samples treated at 12.5 GPa and very high T. The (100) and (101) reflections at 2.0–2.2 Å are very weak, indicating a low degree of order in the (002) planes. The structure of this state is probably still based mainly on intact C_{60} molecules, covalently linked into a disordered structure with some remains of a crystalline structure [2]. Minor variations between reported diffraction results for samples obtained at different pressures might stem from the fact that they should originally poly-merize into two-dimensional structures at low pressure and three-dimensional at higher pressures. Again, although some properties of this phase are similar to those of amorphous graphite- or diamond-like films, a closer analysis shows that it is probably a new, unique structural phase in the carbon system [2]. Davydov et al. [190] have denoted these materials as "polycondensed," and in an atomic force microscopy (AFM) study [217], they found that the inter-molecular distances were only 6.5–7.7 Å, similar to the diameter of the C_{60} molecule itself. They suggested that the molecules had partially broken down and formed a structure in which a complete hexagon (or pentagon) was re-moved from one molecule and the dangling bonds reconnected to a corre-sponding hexagon (or pentagon) on a neighbor. However, this scheme is diffi-cult to reconcile with the fact that mass spectrometry still showed intact C_{60} units to be the basic building blocks [190], and it has also been pointed out [2,3] that STM pictures of photopolymerized C_{60} sometimes show [218] a periodic modulation along chain with a period of only 4.5 Å because the electron density has two maxima, on the "belly" of the molecule and near the intermolecular sp^3 bonds. Other structural models have also been suggested. A transmission electron microscopy (TEM) study by Blank et al. [2,219] showed a diffuse (002) ring corresponding to a lattice spacing of 3.14–3.3 Å with four intense arcs, interpreted as originating from tetrahedrons formed by (111) planes of a cubic lattice. This structure may be based on tetrahedrons of sp^3-linked molecules, ordered within small domains but slightly out of order at larger distances, thus destroying long-range order. As discussed earlier regarding the bcc phase, a 22-atom polyhedron is created between the molecules in such polyhedra [2,212]. Chernozatonskii et al. [220] recently suggested that the C_{60} molecules them-selves are transformed into "barrelenes," with six pentagons clustering directly around a hexagon at each pole, and that these form the building blocks of high-temperature polymers. Calculated dimensions of such barrelene polymers agree well with those measured by Keita et al. [217].

The existence of intact C_{60} molecules in the structure is confirmed by mass spectrometry [190] and Raman data [206]. The latter technique also confirms the expected existence of intermolecular sp^3 bonds in the structures. Interest-ingly, the characteristic "pentagonal pinch" Raman mode disappears in this structure. Although the most probable reason is that the very large number of intermolecular bonds no longer allows this mode of vibration, it should be noted that this would also be expected in the barrelene structures discussed above [220]. The density of the material produced at 9.5 GPa is lower than that of the densest crystal modification obtained at the same pressure, and an in-crease in treatment temperature from 800 to 1700 K leads to a further decrease in the measured density [2,212], indicating that the low density results from thermally induced disorder in an array of intact molecules. As already men-tioned, the hardness of these materials is also very high [206,207,221]. Figure 15.39 [221] shows the evolution of the hardness H_v with treatment temperature at 9.5 GPa and indicates that $H_v > 100$ GPa for optimally treated samples,

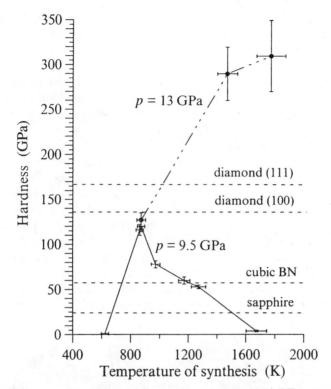

Figure 15.39 Measured hardness of C_{60} at zero pressure and RT versus treatment temperature after treatment at two pressures, 9.5 and 13 GPa. For comparison, RT data for some hard materials are shown as dashed lines. Reprinted from *Diamond Relat. Mater.*, Vol. 7, V. Blank, M. Popov, G. Pivovarov, N. Lvova, K. Gogolinsky, and V. Reshetov, "Ultrahard and superhard phases of fullerite C_{60}: comparison with diamond on hardness and wear", pp. 427–431, Copyright © 1998, with permission from Elsevier Science.

easily cutting cubic BN and almost reaching the hardness of the (100) surface of diamond. In contrast to the semimetallic, partially graphitized samples discussed earlier, these materials are semiconductors, although the measured conductivity is high and the activation energy much lower than for the crystalline phases [179,206,213].

Finally, under extreme conditions, C_{60} is transformed into a disordered material believed to consist of the remnants of collapsed molecules. Many early studies [71,101,102,119,120,169] observed a transition at RT and pressures near 20 GPa into a disordered, transparent phase, hard enough to leave scratch marks on the diamond anvil face [169]. As shown in Figure 15.24, bulk samples of a very hard ("ultrahard") phase with high density of 3.1–3.3 g/cm^3, probably identical to the very-high-pressure RT phase, have been synthesized at temperatures above 1000 K at 12.5–13 GPa [206–209]. The material is reported to be insulating and transparent with a yellowish-brown color, but like other high-temperature phases, it has been very difficult to find homogeneous samples and most studies have been carried out on finely grained mixed samples containing also black, opaque grains. Attempts have been made to deduce the structure of this phase from XRD and Raman studies, but so far with little success. X-ray diffraction studies show [2] strong halos with maxima corresponding to interplane distances of 2.87–2.94 Å and low-intensity bands corresponding to 1.7–1.9 and 5.2 Å. These patterns are similar to those obtained for tetrahedral amorphous carbon (ta-C) films by Gaskell et al. [222], but differences in the Raman spectra and mechanical properties show that the ta-C and amorphous fullerites probably have quite different structures. Zhang et al. [223] simulated the collapse of C_{60} under high pressure and high temperature and found the final structure shown in Figure 15.40, which has a density of 3.35 g/cm^3 and a 21%/79% ratio of sp^2 to sp^3 bonds. The question of bond types is of particular interest since sp^2 bonds are stronger than the diamond-like sp^3 ones. A material based on sp^2 type bonds might therefore be stronger than diamond. Attempts have been made to use Raman scattering as a probe to find the fractions of sp^2 and sp^3 bonds in the real material but results are inconclusive. No lines for truly sp^3-bonded "amorphous diamond" can be observed because of

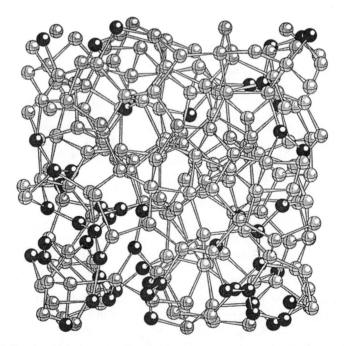

Figure 15.40 Atomic network of carbon atoms in collapsed C_{60} as obtained in a molecular dynamics simulation. Dark dots indicate sp^2 bonded atoms, light dots sp^3 bonded ones. Reprinted with permission from Ref. 223. Copyright © 1994 ECS Sciences.

the much higher intensity of sp^2-related bands near 1560 cm^{-1} [2,207,208] and a peak near 1350 cm^{-1} [207,208], probably arising from the presence of microscopic grains of graphite. Blank et al. [207] found a shift of the Raman bands when excited with lasers of different wavelengths and connected this with the presence of a large range of cluster sizes, indicating that a particular kind of intermediate-range order exists in these materials. Again, the analysis of spectroscopic and structural properties [2] shows that this is another unique, disordered phase of carbon.

As already hinted at, this phase has an extremely high hardness H_v, although data from different sources disagree. Brazhkin et al. [208,209] find values for H_v in the range 70–90 GPa, while Blank et al. [221,225] claim that the hardness may reach 150–300 GPa, as shown in Figure 15.39. Such high values are well outside the range of standard diamond indenter techniques, and a new technique based on scanning force microscopy has been used in these measurements [224]. Using an amorphous fullerite tip the hardness of diamond (100) and (111) faces could be found as 137 and 167 GPa, respectively. The elastic properties of amorphous fullerene phases have also been investigated using acoustic microscopy [2,212,225]. The sound velocities were measured in samples synthesized under various temperatures at 12.5 and 13 GPa, and elastic constants were calculated from these data and measured densities. The longitudinal sound velocities were similar to that of diamond except for material treated above 1700 K, where they were higher, while the transverse wave velocities were lower [2,225]. The adiabatic bulk moduli increased with increasing treatment temperature from 540 to 1360 GPa, all significantly larger than the value 445 GPa for diamond, while the measured shear moduli G and Young's moduli E were about one-half those of diamond at 155–280 GPa and 450–710 GPa, respectively. Poisson's ratio is in the range 0.35–0.45, which is typical for polymers and much higher than the value for diamond, 0.08. A linear relationship was found between H_v and the bulk modulus. The very high values of the elastic moduli arise from very strong bonds, the presence of which are also reflected in the lattice properties. Measurements of the specific heat capacity have shown that the Debye temperature of this phase is about 1450 K, somewhat lower than that of diamond [226]. Finally, measurements of the electrical properties show that these materials are semiconducting with resistivities of the order of 10^5–10^6 $\Omega \cdot$ cm and activation energies of about 0.1 eV.

15.3.2 Polymerization of C_{70} Under Pressure

Although very few studies have been carried out on C_{70} under pressure it is clear that the material can be polymerized not only by radiation treatment [227] but also by high-temperature treatment under high pressure, just like C_{60}. Early Raman studies showed transition anomalies at 4 GPa at RT [101], and Chandrabhas et al. [67,128] reported that the intramolecular Raman modes of C_{70} broadened under pressure but could be observed up to 12 GPa. Above 20 GPa the Raman lines of C_{70} disappeared and a new line at 1650 cm^{-1} characteristic of amorphous carbon appeared, in agreement with the report by Christides et al. [62] that C_{70} became amorphous at 18 GPa. Chandrabhas and co-workers also report that amorphization is reversible, such that normal Raman lines reappear on decreasing the pressure. Since the amorphization observed could be a signature of polymerization, Sunder et al. [60] submitted hcp C_{70} to temperatures up to 1100 K for up to 6 h at 5 and 7.5 GPa, trying to find indications of polymerized structures, but they found no such evidence.

More recent work by Premila et al. [228] and Iwasa et al. [229], however, clearly show that polymerization occurs at elevated pressure and temperature. Iwasa and co-workers treated C_{70} at temperatures up to 800 °C for 1 h at 5 GPa and observed clear signs of polymerization by X-ray diffraction studies, IR spectroscopy, and UV/visible absorption spectroscopy. Identical new, insoluble polymerized structures were obtained on heating both hcp and fcc C_{70} to 2–300 °C. As for C_{60}, IR spectroscopy of the polymerized phases shows a large number of new modes, as shown in Figure 15.41, and the observed IR frequencies were in good agreement with the results of a theoretical calculation for

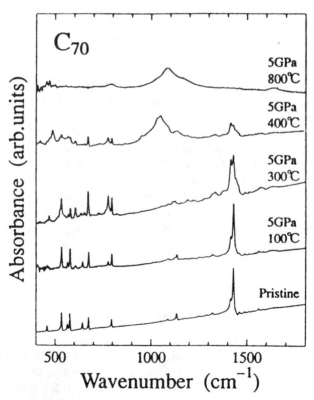

Figure 15.41 Zero-pressure IR absorption spectra for pristine C_{70} and for samples treated under the conditions indicated. Reprinted with permission from Ref. 229. Copyright © 1997 Springer-Verlag.

C_{70} dimers. At and above 400 °C an increasing irreversible transformation into a conducting disordered graphite-like phase was observed, indicating that the C_{70} molecules might collapse under relatively mild conditions under pressure, while heating at 100 °C resulted in partial polymerization only. However, it is possible that partial polymerization may occur in C_{70} also at RT, since Christides et al. [62] observe a reduction in the intensity of many X-ray diffraction lines above 6.5 GPa and Meletov et al. [68] observe an anomaly in their Raman data near 5 GPa. Iwasa et al. [229] report that if heated at zero pressure the C_{70} polymers reverted to pristine C_{70} at 300 °C. The final structure found was always fcc, irrespective of the initial lattice structure, indicating that this is indeed the most stable phase of C_{70} at 300 °C [59]. Similar results were found by Premila et al. [228].

Although these studies indicate that a pressure of several gigapascals is necessary for polymerization, recent studies of the thermal conductivity of C_{70} at temperatures up to 450 K near 1 GPa (A. Soldatov, *private communication*) also show a continuously increasing degree of disorder in the samples over several days. Polymerization of C_{70} may thus be possible also at much lower pressures and temperatures than usually assumed. However, the polymerization rate is always lower for C_{70} than for C_{60} because only double bonds in the "polar" region of the molecule take part in such reactions.

15.4 TRANSFORMATION INTO OTHER FORMS OF GRAPHITE

As discussed earlier, both C_{60} and C_{70} break down to graphite-like materials if heated to above about 1000 K [51] at atmospheric pressure. Figure 15.24 shows that the application of very high temperatures or pressures, or both, always results in molecular breakdown and the formation of graphite-like or diamond-like material. From the phase diagram of carbon [55] we would expect that excessive heating in the range below 3 GPa should result in the formation of graphite, while high-pressure treatment would give diamond or diamond-like materials. This is observed in practice. The high-temperature stability limit of the C_{60} molecule changes little with pressure, and C_{60} breaks down into various kinds of disordered carbon if heated much above 1000 K. Below 2–3 GPa the resulting material is basically "graphitic" but disordered and, if treated at sufficiently high temperatures, mechanically soft [215] in contrast to the very hard disordered phases discussed earlier. The breakdown products seem to tend toward turbostratic graphite after long anneals at high temperatures.

At higher pressures a number of very hard phases are obtained on heating. Since these phases are produced in the pressure range where diamond is more stable than graphite, it might be expected that excessive heating might result in the formation of diamond. However, up to 12 GPa it is observed that at the highest temperatures the very hard phases give way to softer, more graphite-like structures (Fig. 15.39), possibly because most of the structure of the basically graphite-like molecules remains intact.

At very high pressure, however, at least C_{60} seems to collapse into diamond or diamond-like material. Núñez-Regueiro et al. [230] claimed to identify crystalline diamond in C_{60} treated at RT under highly nonhydrostatic conditions at >20 GPa, but this result has never been reproduced. Blank et al. [207] and Ma and Zou [205] report that crystalline diamond is produced only if very high pressures are combined with extreme temperatures, for example, 13 GPa at 2100 K. Fullerenes can also be used as starting material for diamond synthesis using the standard catalyst method [55], and many groups have reported work along these lines. Usually, pressures in the range 5.5–8 GPa and temperatures from 1500 to 1800 K are used, with Ni–Co catalysts [205,207,231,232].

15.5   CONCLUSIONS

Because of the interesting mixture of very weak and very strong bonds in solid fullerenes and the strong effect of pressure on the reactivity of the fullerene molecules, high-pressure studies of fullerenes have proved to be a very rich and fertile field in condensed matter science. Still, in spite of the enormous work already carried out, much remains to be done. The material that has been the most studied under pressure, sc C_{60}, is now reasonably well understood even to the details of the orientational structure. However, to name but one example, the effect of the orientational structure on the electronic structure and on the resistivity is still not understood. Very few studies have been carried out in the glassy state and in the fcc phase. As mentioned earlier, there are tantalizing indications from recent resistivity studies that the orientational potential, which drives the molecular reorientations in the sc phase, might also have interesting effects on local orientational order in fcc C_{60}, and thus also on the electron band structure. In some areas, even the phase boundaries have not yet been finally determined. In general, however, C_{60} is very well understood compared to the higher fullerenes, for which materials very little is known even about the structural phase diagrams.

Likewise, the polymeric high-pressure states of C_{60} have now been reasonably well investigated over much of the range of stability of the molecules. It should be noted, though, that very little is known about the equilibrium phases and structures, or about the reversibility of various phase transformations. The one- and two-dimensional polymers have been investigated both experimentally and theoretically, and most physical properties are already fairly well known. This is not the case for the three-dimensional polymers. Because of the extreme conditions necessary for their production, only very small samples can be made, and most properties are still unknown. This is also true for the polymeric states of higher fullerenes, and in fact only C_{70} has yet been polymerized under pressure.

Considering that not only pure fullerenes but also doped materials and fullerene derivatives can be expected to show very interesting behaviors under high pressures, it is obvious that even though many high-pressure studies have already been carried out, much more still remains to be done, and high-pressure studies will continue to be of large scientific interest for many years to come.

ACKNOWLEDGMENTS

I would like to thank Per-Axel Persson for permission to show the data in Figure 15.30 before publication, and the Swedish Research Councils for Natural Sciences (NFR) and Engineering Sciences (TFR) for their financial support for our studies of fullerenes.

REFERENCES

1. M. S. Dresselhaus, G. Dresselhaus, and P. C. Eklund, *Science of Fullerenes and Carbon Nanotubes*. Academic Press, San Diego, 1996.

2. V. D. Blank, S. G. Buga, G. A. Dubitsky, N. R. Serebryanaya, M. Yu. Popov, and B. Sundqvist, *Carbon* **1998**, *36*, 319–343.

3. B. Sundqvist, *Adv. Phys.* **1999**, *48*, 1–134.

4. W. F. Sherman and A. A. Stadtmuller, *Experimental Techniques in High-Pressure Research*. Wiley, Chichester, 1987. M. Eremets, *High Pressure Experimental Methods*. Oxford University Press, Oxford, 1997. W. B. Holzapfel and N. S. Isaacs, *High-Pressure Techniques in Chemistry and Physics*. Oxford University Press, Oxford, 1997.

5. G. A. Samara, L. V. Hansen, R. A. Assink, B. Morosin, J. E. Schirber, and D. Loy, *Phys. Rev. B* **1993**, *47*, 4756–4764.

6. J. E. Schirber, G. H. Kwei, J. D. Jorgensen, R. L. Hitterman, and B. Morosin, *Phys. Rev. B* **1995**, *51*, 12014–12017.

7. B. Morosin, J. D. Jorgensen, S. Short, G. H. Kwei, and J. E. Schirber, *Phys. Rev. B* **1996**, *53*, 1675–1678.

8. G. H. Kwei, J. D. Jorgensen, J. E. Schirber and B. Morosin, *Fullerene Sci. Technol.* **1997**, *5*, 243–256.

9. G. E. Gadd, M. James, S. Moricca, P. J. Evans, and R. L. Davis, *Fullerene Sci. Technol.* **1996**, *4*, 853–862. G. E. Gadd, S. J. Kennedy, S. Moricca, C. J. Howard, M. M. Elcombe, P. J. Evans, and M. James, *Phys. Rev. B* **1997**, *55*, 14794–14799. G. E. Gadd, S. Moricca, S. Kennedy, M. Elcombe, P. J. Evans, M. Blackford, D. Cassidy, C. J. Howard, P. Prasad, J. V. Hanna, A. Burchwood, and D. Levy, *J. Phys. Chem. Solids* **1997**, *58*, 1823–1826.

10. G. E. Gadd, M. M. Elcombe, J. Dennis, S. Moricca, N. Webb, D. Cassidy, and P. J. Evans, *J. Phys. Chem. Solids* **1998**, *59*, 937–944.

11. R. A. Assink, J. E. Schirber, D. A. Loy, B. Morosin, and G. A. Carlsson, *J. Mater. Res.* **1992**, *7*, 2136–2143.

12. J. E. Schirber, R. A. Assink, G. A. Samara, B. Morosin, and D. Loy, *Phys. Rev. B.* **1995**, *51*, 15552–15554.

13. A. I. Kolesnikov, V. E. Antonov, I. O. Bashkin, G. Grosse, A. P. Moravsky, A. Yu. Muzychka, E. G. Ponyatovsky, and F. E. Wagner, *J. Phys. Condens. Matter* **1997**, *9*, 2831–2838.

14. B. Morosin, R. A. Assink, R. G. Dunn, T. M. Massis, J. E. Schirber, and G. H. Kwei, *Phys. Rev. B.* **1997**, *56*, 13611–13614.

15. I. Holleman, G. von Helden, E. H. T. Olthof, P. J. M. van Bentum, R. Engeln, G. H. Nachtegaal, A. P. M. Kentgens, B. H. Meier, A. van der Avoird, and G. Meijer, *Phys. Rev. Lett.* **1997**, *79*, 1138–1141.

16. I. Holleman, G. von Helden, A. van der Avoird, and G. Meijer, *Phys. Rev. Lett.* **1998**, *80*, 4899–4902.

17. C. Collins, W. Kolodziejski, J. Foulkes, and J. Klinowski, *Chem. Phys. Lett.* **1998**, *289*, 338–340.

18. K. Grube, *Thesis*, Report FZKA 5611, Karlsruhe, 1995 (unpublished).

19. S. Menu, L. Duclaux, J. Conard, P. Lauginie, and V. A. Nalimova, *J. Phys. Chem. Solids* **1996**, *57*, 967–975.

20. V. A. Nalimova, D. E. Sklovsky, G. N. Bondarenko, H. Alvergnat-Gaucher, S. Bonnamy, and F. Béguin, *Synth. Metals* **1997**, *88*, 89–93.

21. J. E. Schirber, L. Hansen, B. Morosin, J. E. Fischer, J. D. Jorgensen, and G. H. Kwei, *Physica C* **1996**, *260*, 173–176.

22. R. Tellgmann, N. Krawez, S.-H. Lin, I. V. Hertel, and E. E. B. Campbell, *Nature* **1996**, *382*, 407.

23. M. Saunders, H. A. Jiménez-Vázquez, R. J. Cross, S. Mroczkowski, M. L. Gross, D. E. Giblin, and R. J. Poreda, *J. Am. Chem. Soc.* **1994**, *116*, 2193–2194.

24. M. Saunders, R. J. Cross, H. A. Jiménez-Vázquez, R. Shimshi, and A. Khong, *Science* **1996**, *271*, 1693–1697.

25. J. Laskin, T. Peres, C. Lifschitz, M. Saunders, R. J. Cross, and A. Khong, *Chem. Phys. Lett.* **1998**, *285*, 7–9.

26. G. E. Gadd, M. Blackford, S. Moricca, N. Webb, P. J. Evans, A. M. Smith, G. Jacobsen, S. Leung, A. Day, and Q. Hua, *Science* **1997**, *277*, 933–936.

27. W. I. F. David and R. M. Ibberson, *J. Phys. Condens. Matter* **1993**, *5*, 7923–7928.

28. B. Sundqvist, O. Andersson, A. Lundin, and A. Soldatov, "Structure, disorder, and phase diagram of C_{60} up to 1 GPa and below 300 K" in *High Pressure in Materials Science and Geoscience*, J. Kamarád, Z. Arnold, and A. Kapicka (eds.). Prometheus Press, Prague, 1994, pp. 109–112. B. Sundqvist, O. Andersson, A. Lundin, and A. Soldatov, *Solid State Commun.* **1995**, *93*, 109–112.

29. J. A. Wolk, P. J. Horoyski, and M. L. W. Thewalt, *Phys. Rev. Lett.* **1995**, *74*, 3483–3486.

30. O. Blaschko, W. Rom, and I. N. Goncharenko, *J. Phys. Condens. Matter* **1996**, *8*, 4235–4243.

31. L. Pintschovius, O. Blaschko, G. Krexner, and N. Pyka *Phys. Rev. B* **1999**, *59*, 11020–11026.

32. S. van Smaalen, R. Dinnebier, I. Holleman, G. von Helden, and G. Meijer, *Phys. Rev. B* **1998**, *57*, 6321–6324.

33. O. Andersson, A. Soldatov, and B. Sundqvist, *Phys. Lett. A* **1995**, *206*, 260–264.

34. P. Launois, S. Ravy, and R. Moret, *Phys. Rev. B* **1997**, *55*, 2651–2665.

35. S. Savin, A. B. Harris, and T. Yildirim, *Phys. Rev. B* **1997**, *55*, 14182–14199.

36. E. Burgos, E. Halac, and H. Bonadeo, *Phys. Rev. B* **1994**, *49*, 15544–15549.

37. C. Meingast, G. Roth, L. Pintschovius, R. H. Michel, C. Stoermer, M. M. Kappes, P. A. Heiney, L. Brard, R. M. Strongin, and A. B. Smith III, *Phys. Rev. B* **1996**, *54*, 124–131.

38. O. Andersson, A. Soldatov, and B. Sundqvist, *Phys. Rev. B* **1996**, *54*, 3093–3100.

39. J. Jäckle, *Rep. Progr. Phys.* **1986**, *49*, 171–231.

40. A. Lundin, A. Soldatov, B. Sundqvist, R. M. Strongin, L. Brard, J. E. Fischer, and A. B. Smith III, *Carbon* **1996**, *34*, 1119–1121.

41. G. Kriza, J.-C. Ameline, D. Jérome, A. Dworkin, H. Szwarc, C. Fabre, D. Schütz, A. Rassat, P. Bernier, and A. Zahab, *J. Phys. I (France)* **1991**, *1*, 1361–1364.

42. G. A. Samara, J. E. Schirber, B. Morosin, L. V. Hansen, D. Loy, and A. P. Sylwester, *Phys. Rev. Lett.* **1991**, *67*, 3136–3139.

43. R. Kerkoud, P. Auban-Senzier, J. Godard, D. Jérome, J.-M. Lambert, and P. Bernier, *Adv. Mater.* **1994**, *6*, 782–785.

44. H. Wang, B. Xie, Y. Li, Q. Wang, J. G. Hou, B. Xu, M. Tan, H. Li, and Y. Xu, *phys. status solidi (b)* **1998**, *207*, 243–248.

45. A. Lundin and B. Sundqvist, *Europhys. Lett.* **1994**, *27*, 463–466.

46. A. Lundin and B. Sundqvist, *Phys. Rev. B* **1996**, *53*, 8329–8336.

47. D. Lamoen and K. H. Michel, *Phys. Rev. B* **1993**, *48*, 807–813.

48. J. E. Fischer, A. R. McGhie, J. K. Estrada, M. Haluska, H. Kuzmany, and H.-U. ter Meer, *Phys. Rev. B* **1996**, *53*, 11418–11424.

49. L. Pintschovius, S. L. Chaplot, G. Roth, and G. Heger, *Phys. Rev. Lett.* **1995**, *75*, 2843–2846.

50. E. Kolodney, B. Tsipinyuk, and A. Budrevich, *J. Chem. Phys.* **1994**, *100*, 8542–8545. T. Sommer, T. Kruse, and P. Roth, *J. Phys. B At. Mol. Opt. Phys.* **1996**, *29*, 4955–4964.

51. C. S. Sundar, A. Bharathi, Y. Hariharan, J. Janaki, V. Sankara Sastry, and T. S. Radhakrishnan, *Solid State Commun.* **1992**, *84*, 823–826. S. D. Leifer, D. G. Goodwin, M. S. Anderson, and J. R. Anderson, *Phys. Rev. B* **1995**, *51*, 9973–9977.

52. M. H. J. Hagen, E. J. Meijer, G. C. A. M. Mooij, D. Frenkel, and H. N. W. Lekkerkerker, *Nature* **1993**, *365*, 425–426.

53. A. Cheng, M. L. Klein, and C. Caccamo, *Phys. Rev. Lett.* **1993**, *71*, 1200–1203. M. C. Abramo and C. Caccamo, *J. Phys. Chem. Solids.* **1996**, *57*, 1751–1755.

54. J. Q. Broughton, J. V. Lill, and J. K. Johnson, *Phys. Rev. B* **1997**, *55*, 2808–2817.

55. F. P. Bundy, W. A. Bassett, M. S. Weathers, R. J. Hemley, H. K. Mao, and A. F. Goncharov, *Carbon* **1996**, *34*, 141–153.

56. P. Nagel, C. Meingast, M. A. Verheijen, and G. Meijer, "Orientational ordering transitions of C_{70} single crystals with fcc and hcp morphologies" in *Fullerenes and Fullerene Nanostructures*, H. Kuzmany, J. Fink, M. Mehring, and S. Roth (eds.). World Scientific Publishing, Singapore, 1996, pp. 68–71.

57. A. R. McGhie, J. E. Fischer, P. A. Heiney, P. W. Stephens, R. L. Cappelletti, D. A. Neumann, W. H. Mueller, H. Mohn, and H.-U. ter Meer, *Phys. Rev. B* **1994**, *49*, 12614–12618.

58. H. Kawamura, Y. Akahama, M. Kobayashi, H. Shinohara, and Y. Saito, *J. Phys. Soc. Jpn.* **1994**, *63*, 2445–2446.

59. R. Komori, T. Nagaosa, T. Hatae, and Y. Miyamoto, *Jpn. J. Appl. Phys.* **1997**, *36*, 5600–5604.

60. C. S. Sundar, P. Ch. Sahu, V. S. Sastry, G. V. N. Rao, V. Sridharan, M. Premila, A. Bharathi, Y. Hariharan, T. S. Radhakrishnan, D. V. S. Muthu, and A. K. Sood, *Phys. Rev. B* **1996**, *53*, 8180–8183.

61. H. Kawamura, Y. Akahama, M. Kobayashi, H. Shinohara, H. Sato, Y. Saito, T. Kikegawa, O. Shimomura, and K. Aoki, *J. Phys. Chem. Solids* **1993**, *54*, 1675–1678.

62. C. Christides, I. M. Thomas, T. J. S. Dennis, and K. Prassides, *Europhys. Lett.* **1993**, *22*, 611–618.

63. A. Lundin, A. Soldatov, and B. Sundqvist, *Europhys, Lett.* **1995**, *30*, 469–474.

64. A. Soldatov and B. Sundqvist, *J. Phys. Chem. Solids* **1996**, *57*, 1371–1375.

65. G. A. Samara, L. V. Hansen, B. Morosin, and J. E. Schirber, *Phys. Rev. B* **1996**, *53*, 5211–5216.

66. Y. Huang, D. F. R. Gilson, and I. S. Butler, *J. Phys. Chem.* **1991**, *95*, 5723–5725.

67. A. K. Sood, N. Chandrabhas, D. V. S. Muthu, Y. Hariharan, A. Bharathi, and C. S. Sundar, *Philos. Mag. B* **1994**, *70*, 347–358.

68. K. P. Meletov, A. A. Maksimov, and I. I. Tartakovskii, *JETP* **1997**, *84*, 144–150.

69. J. E. Fischer, P. A. Heiney, A. R. McGhie, W. J. Romanow, A. M. Denenstein, J. P. McCauley, and A. B. Smith III, *Science* **1991**, *252*, 1288–1290.

70. R. S. Ruoff and A. L. Ruoff, *Nature* **1991**, *350*, 663–664.

71. S. J. Duclos, K. Brister, R. C. Haddon, A. R. Kortan, and F. A. Thiel, *Nature* **1991**, *351*, 380–382.

72. A. Soldatov and J. Simu (to be published).

73. H. A. Ludwig, W. H. Fietz, F. W. Hornung, K. Grube, B. Wagner, and G. J. Burkhart, *Z. Phys. B* **1994**, *96*, 179–183.

74. Ya. M. Soifer, N. P. Kobelev, I. O. Bashkin, A. P. Moravsky, and E. G. Ponyatovsky, *J. Phys. IV Colloque C8* **1996**, *6*, 621–624.

75. L. A. Girifalco, *Phys. Rev. B* **1995**, *52*, 9910–9916.

76. V. I. Zubov, J. F. Sanchez-Ortiz, N. P. Tret'yakov, and A. A. Caparica, *J. Adv. Mater.* **1996**, *3*(3), 163–169.

77. H. Coufal, K. Meyer, R. K. Grygier, M. de Vries, D. Jenrich, and P. Hess, *Appl. Phys. A* **1994**, *59*, 83–86.

78. I. O. Bashkin, V. I. Rashchupkin, A. F. Gurov, A. P. Moravsky, O. G. Rybchenko, N. P. Kobelev, Ya. M. Soifer, and E. G. Ponyatovskii, *J. Phys. Condens. Matter* **1994**, *6*, 7491–7498.

79. N. P. Kobelev, Ya. M. Soifer, I. O. Bashkin, A. F. Gurov, A. P. Moravskii, and O. G. Rybchenko, *phys. status solidi b* **1995**, *190*, 157–162.

80. R. Komori and Y. Miyamoto, *J. Phys. Chem. Solids* **1995**, *56*, 535–537.

81. J. Haines and J. M. Léger, *Solid State Commun.* **1994**, *90*, 361–363.

82. J. H. Nguyen, M. B. Kruger, and R. Jeanloz, *Solid State Commun.* **1993**, *88*, 719–721.

83. J. Wang, L. Wang, L. Chen, H. Chen, R. Wang, Z. Zhang, R. Che, and L. Zhou, *Chinese Phys. Lett.* **1993**, *10*(3), 159–162.

84. R. A. Jishi, R. M. Mirie, and M. S. Dresselhaus, "A model calculation for the vibrational modes in C_{60}" in *Novel Forms of Carbon*, C. L. Renschler, J. J. Pouch, and D. M. Cox (eds.). Materials Research Society, Pittsburgh, 1992, pp. 197–202.

85. S. J. Woo, S. H. Lee, E. Kim, K. H. Lee, Y. H. Lee, S. Y. Hwang, and I. C. Jeon, *Phys. Lett. A* **1992**, *162*, 501–505.

86. O. Zhou, G. B. M. Vaughan, Q. Zhu, J. E. Fischer, P. A. Heiney, N. Coustel, J. P. McCauley, Jr., and A. B. Smith III, *Science* **1992**, *255*, 833–835.

87. H. A. Ludwig, W. H. Fietz, F. W. Hornung, K. Grube, B. Renker, and G. J. Burkhart, *Physica C* **1994**, *234*, 45–48.

88. J. Diederichs, J. S. Schilling, K. W. Herwig, and W. B. Yelon, *J. Phys. Chem. Solids* **1997**, *58*, 123–132.

89. B. Morosin, J. E. Schirber, J. D. Jorgensen, G. H. Kwei, T. Yildirim, and J. E. Fischer, "On the pressure induced phase of Na_2CsC_{60}" in *Fullerenes: Recent Advances in the Chemistry and Physics of Fullerenes and Related materials*, Vol. 3, K. M. Kadish and R. S. Ruoff (eds.). The Electrochemical Society, Pennington, NJ, 1996, pp. 446–456.

90. S. Hoen, N. G. Chopra, X.-D. Xiang, R. Mostovoy, J. Huo, W. A. Vareka, and A. Zettl, *Phys. Rev. B* **1992**, *46*, 12737–12739.

91. A. F. Hebard, *Phys. Today* **1992**, *45*(11), 26–32.

92. L. Pintschovius, *Rep. Prog. Phys.* **1996**, *57*, 473–510.

93. M. S. Dresselhaus, G. Dresselhaus, and P. C. Eklund, *J. Raman Spectrosc.* **1996**, *27*, 351–371.

94. A. M. Rao, P. Zhou, K.-A. Wang, G. T. Hager, J. M. Holden, Y. Wang, W.-T. Lee, X.-X. Bi, P. C. Eklund, D. S. Cornett, M. A. Duncan, and I. J. Amster, *Science* **1993**, *259*, 955–957.

95. K. Aoki, H. Yamawaki, Y. Kakudate, M. Yoshida, S. Usuba, H. Yokoi, S. Fujiwara, Y. Bae, R. Malhotra, and D. Lorents, *J. Phys. Chem.* **1991**, *95*, 9037–9039.

96. D. D. Klug, J. A. Howard, and D. A. Wilkinson, *Chem. Phys. Lett.* **1992**, *188*, 168–170.

97. M. C. Martin, X. Du, J. Kwon, and L. Mihaly, *Phys. Rev. B* **1994**, *50*, 173–183.

98. A. M. Rao, P. C. Eklund, U. D. Venkateswaran, J. Tucker, M. A. Duncan, G. M. Bendele, P. W. Stephens, J.-L. Hodeau, L. Marques, M. Núñez-Regueiro, I. O. Bashkin, E. G. Ponyatovsky, and A. P. Morovsky, *Appl. Phys. A* **1997**, *64*, 231–239.

99. A. M. Rao, P. C. Eklund, J.-L. Hodeau, L. Marques, and M. Núñez-Regueiro, *Phys. Rev. B* **1997**, *55*, 4766–4773.

100. N. Chandrabhas, M. N. Shashikala, D. V. S. Muthu, A. K. Sood, and C. N. R. Rao, *Chem. Phys. Lett.* **1992**, *197*, 319–323. Y. S. Raptis, D. W. Snoke, K. Syassen, S. Roth, P. Bernier, and A. Zahab, *High Pressure Res.* **1992**, *9*, 41–46. S.-J. Jeon, D. Kim, S. K. Kim, and I. C. Jeon, *J. Raman Spectrosc.* **1992**, *23*, 311–313.

101. D. W. Snoke, Y. S. Raptis, and K. Syassen, *Phys. Rev. B* **1992**, *45*, 14419–14422.

102. S. H. Tolbert, A. P. Alivisatos, H. E. Lorenzana, M. B. Kruger, and R. Jeanloz, *Chem. Phys. Lett.* **1992**, *188*, 163–167.

103. C. S. Yoo and W. J. Nellis, *Chem. Phys. Lett.* **1992**, *198*, 379–382.

104. K. P. Meletov, D. Christofilos, G. A. Kourouklis, and S. Ves, *Chem. Phys. Lett.* **1995**, *236*, 265–270. K. P. Meletov, D. Christofilos, S. Ves, and G. A. Kourouklis, *phys. status solidi b* **1996**, *198*, 553–558.

105. K. P. Meletov, D. Christofilos, S. Ves, and G. A. Kourouklis, *Phys. Rev. B* **1995**, *52*, 10090–10096.

106. K. P. Meletov, G. A. Kourouklis, D. Christofilos, and S. Ves, *JETP* **1995**, *81*, 798–805.

107. P. H. M. van Loosdrecht, P. J. M. van Bentum, and G. Meijer, *Phys. Rev. Lett.* **1992**, *68*, 1176–1179.

108. H. Yamawaki, M. Yoshida, Y. Kakudate, S. Usuba, H. Yokoi, S. Fujiwara, K. Aoki, R. Ruoff, R. Malhotra, and D. Lorents, *J. Phys. Chem.* **1993**, *97*, 11161–11163.

109. A. K. Sood, N. Chandrabhas, D. V. S. Muthu, Y. Hariharan, A. Bharathi, and C. S. Sundar, *Philos. Mag. B* **1994**, *70*, 347–358.

110. J. Yu, L. Bi, R. K. Kalia, and P. Vashishta, *Phys. Rev. B* **1994**, *49*, 5008–5019. J. Yu, R. K. Kalia, and P. Vashishta, *Appl. Phys. Lett.* **1993**, *63*, 3152–3154. J. Yu, R. K. Kalia, and P. Vashishta, *J. Chem. Phys.* **1993**, *99*, 10001–10010. P. Vashishta, R. K. Kalia, A. Nakano, and J. Yu, *Int. J. Mod. Phys. C* **1994**, *5*, 281–283.

111. Yu. I. Prilutski and G. G. Shapovalov, *phys. status solidi b* **1997**, *201*, 361–370.

112. B. Sundqvist, *Phys. Rev. B* **1993**, *48*, 14712–14713.

113. M. A. White, C. Meingast, W. I. F. David, and T. Matsuo, *Solid State Commun.* **1995**, *94*, 481–484.

114. V. I. Zubov, J. F. Sanchez, N. P. Tretiakov, A. A. Caparica, and I. V. Zubov, *Carbon* **1997**, *35*, 729–734.

115. K. Kniaz, L. A. Girifalco, and J. E. Fischer, *J. Phys. Chem.* **1995**, *99*, 16804–16806.

116. P. J. Horoyski, J. A. Wolk, and M. L. W. Thewalt, *Solid State Commun.* **1995**, *93*, 575–578.

117. R. C. Yu, N. Tea, M. B. Salamon, D. Lorents, and R. Malhotra, *Phys. Rev. Lett.* **1992**, *68*, 2050–2053.

118. A. K. Sood, N. Chandrabhas, D. V. S. Muthu, A. Jayaraman, N. Kumar, H. R. Krishnamurthy, T. Pradeep, and C. N. R. Rao, *Solid State Commun.* **1992**, *81*, 89–92.

119. F. Moshary, N. H. Chen, I. F. Silvera, C. A. Brown, H. C. Dorn, M. S. de Vries, and D. S. Bethune, *Phys. Rev. Lett.* **1992**, *69*, 466–469.

120. D. W. Snoke, K. Syassen, and A. Mittelbach, *Phys. Rev. B* **1993**, *47*, 4146–4148.

121. B. C. Hess, E. A. Forgy, S. Frolov, D. D. Dick, and Z. V. Vardeny, *Phys. Rev. B* **1994**, *50*, 4871–4874.

122. K. P. Meletov, V. K. Dolganov, O. V. Zharikov, I. N. Kremeenskaya, and Yu. A. Ossipyan, *J. Phys. I (France)* **1992**, *2*, 2097–2105.

123. K. P. Meletov and V. K. Dolganov, *JETP* **1998**, *86*, 177–181.

124. Y.-N. Xu, M.-Z. Huang, and W. Y. Ching, *Phys. Rev. B* **1992**, *46*, 4241–4245.

125. M. P. Gelfand and J. P. Lu, *Phys. Rev. Lett.* **1992**, *68*, 1050–1053. S. G. Louie and E. L. Shirley, *J. Phys. Chem. Solids* **1993**, *54*, 1767–1777. N. Laouini, O. K. Andersen, and O. Gunnarsson, *Phys. Rev. B* **1995**, *51*, 17446–17478.

126. W. Y. Ching, M.-Z. Huang, Y.-N. Xu, and F. Gan, *Mod. Phys. Lett.* **1992**, *6*, 309–321.

127. K. P. Meletov, V. K. Dolganov, and Yu. A. Ossipyan, *Solid State Commun.* **1993**, *87*, 639–641.

128. N. Chandrabhas, A. K. Sood, D. V. S. Muthu, C. S. Sundar, A. Bharathi, Y. Hariharan, and C. N. R. Rao, *Phys. Rev. Lett.* **1994**, *73*, 3411–3414.

129. C. Hartmann, M. Zigone, G. Martinez, E. L. Shirley, L. X. Benedict, S. G. Louie, M. S. Fuhrer, and A. Zettl, *Phys. Rev. B* **1995**, *52*, R5550–5553.

130. E. L. Shirley, L. X. Benedict, and S. G. Louie, *Phys. Rev. B* **1996**, *54*, 10970–10977.

131. Y. C. Jean, X. Lu, Y. Lou, A. Bharathi, C. S. Sundar, Y. Lyu, P. H. Hor, and C. W. Chu, *Phys. Rev. B* **1992**, *45*, 12126–12129. H.-E. Schaefer, M. Forster, R. Würschum, W. Krätschmer, and D. R. Huffman, *Phys. Rev. B* **1992**, *45*, 12164–12166.

132. M. Núñez-Regueiro, P. Monceau, A. Rassat, P. Bernier, and A. Zahab, *Nature* **1991**, *354*, 289–291.

133. M. Núñez-Regueiro, O. Bethoux, J.-M. Mignot, P. Monceau, P. Bernier, C. Fabre, and A. Rassat, *Europhys. Lett.* **1993**, *21*, 49–53.

134. Y. Saito, H. Shinohara, M. Kato, H. Nagashima, M. Ohkohchi, and Y. Ando, *Chem. Phys. Lett.* **1992**, *189*, 236–240.

135. Z.-X. Bao, H.-C. Gu, J.-F. Wang, H. Chen, Y.-L. Li, Y.-X. Yao, and D.-B. Zhu, *Chin. Sci. Bull.* **1993**, *38*, 1079–1081.

136. S. Matsuura, T. Ishiguro, K. Kikuchi, Y. Achiba, and I. Ikemoto, *Fullerene Sci. Technol.* **1995**, *3*, 437–445.

137. S. K. Ramasesha and A. K. Singh, *Solid State Commun.* **1994**, *91*, 25–28.

138. S. K. Ramasesha, A. K. Singh, R. Seshadri, A. K. Sood, and C. N. R. Rao, *Chem. Phys. Lett.* **1994**, *220*, 203–206.

139. A. Zettl, "Transport properties of alkali-fullerides" in *Fullerenes and Fullerene Nanostructures*, H. Kuzmany, J. Fink, M. Mehring, and S. Roth (eds.). World Scientific Publishing, Singapore, 1996, pp. 14–33.

140. J. S. Dugdale and D. Gugan, *Proc. R. Soc. A* **1962**, *270*, 186–211.

141. N. F. Mott and H. Jones, *The Theory of the Properties of Metals and Alloys.* Clarendon Press, Oxford, 1936, p. 268.

142. W. A. Vareka, M. S. Fuhrer, and A. Zettl, *Physica C* **1994**, *235–240*, 2507–2508.

143. W. A. Vareka and A. Zettl, *Phys. Rev. Lett.* **1994**, *72*, 4121–4124.

144. S. C. Erwin and W. E. Pickett, *Phys. Rev. B* **1992**, *46*, 14257–14260.

145. Y. Iwasa, T. Arima, R. M. Fleming, T. Siegrist, O. Zhou, R. C. Haddon, L. J. Rothberg, K. B. Lyons, H. L. Carter, Jr., A. F. Hebard, R. Tycko, G. Dabbagh, J. J. Krajewski, G. A. Thomas, and T. Yagi, *Science* **1994**, *264*, 1570–1573.

146. P.-A. Persson, U. Edlund, P. Jacobsson, D. Johnels, A. Soldatov, and B. Sundqvist, *Chem. Phys. Lett.* **1996**, *258*, 540–546.

147. C. Goze, F. Rachdi, L. Hajji, M. Núñez-Regueiro, L. Marques, J.-L. Hodeau, and M. Mehring, *Phys. Rev. B* **1996**, *54*, R3676–3679.

148. J. Fagerström and S. Stafström, *Phys. Rev. B* **1996**, *53*, 13150–13158. S. Stafström and J. Fagerström, *Appl. Phys. A* **1997**, *64*, 307–314.

149. T. Ozaki, Y. Iwasa, and T. Mitani, *Chem. Phys. Lett.* **1998**, *285*, 289–293.

150. A. M. Rao and P. C. Eklund, *Mater. Sci. Forum* **1996**, *232*, 173–206.

151. G.-W. Wang, K. Komatsu, Y. Murata, and M. Shiro, *Nature* **1997**, *387*, 583–586.

152. Y. Iwasa, K. Tanoue, T. Mitani, A. Izuoka, T. Sugawara, and T. Yagi *Chem. Commun.* (Cambridge) **1998**, 1411–1412.

153. B. Burger, J. Winter, and H. Kuzmany, *Z. Phys. B* **1996**, *101*, 227–233. A. Hassanien, J. Gasperic, J. Demsar, I. Musevic, and D. Mihailovic, *Appl. Phys. Lett.* **1997**, *70*, 417–419.

154. G. Oszlányi, G. Bortel, G. Faigel, L. Gránásy, G. M. Bendele, P. W. Stephens, and L. Forró, *Phys. Rev. B* **1996**, *54*, 11849–11852.

155. J. C. Hummelen, B. Knight, J. Pavlovich, R. González, and F. Wudl, *Science* **1995**, *269*, 1554–1556.

156. C. M. Brown, E. Beer, C. Bellavia, L. Cristofolini, R. González, M. Hanfland, D. Häusermann, M. Keshavar-K., K. Kordatos, K. Prassides, and F. Wudl, *J. Am. Chem. Soc.* **1996**, *118*, 8715–8716.

157. S. Lebedkin, A. Gromov, S. Giesa, R. Gleiter, B. Renker, H. Rietschel, and W. Krätschmer, *Chem. Phys. Lett.* **1998**, *285*, 210–215.

158. P. W. Stephens, G. Bortel, G. Faigel, M. Tegze, A. Jánossy, S. Pekker, G. Oszlányi, and L. Forró, *Nature* **1994**, *370*, 636–639.

159. A. Soldatov and O. Andersson, *Appl. Phys. A* **1997**, *64*, 227–229.

160. V. A. Davydov, L. S. Kashevarova, A. V. Rakhmanina, V. Agafonov, R. Céolin, and H. Szwarc, *Carbon* **1997**, *35*, 735–743.

161. T. Wågberg, P.-A. Persson, B. Sundqvist, and P. Jacobsson, *Appl. Phys. A* **1997**, *64*, 223–226.

162. M. Núñez-Regueiro, L. Marques, J.-L. Hodeau, O. Béthoux, and M. Perroux, *Phys. Rev. Lett.* **1995**, *74*, 278–281.

163. L. Marques, J.-L. Hodeau, M. Núñez-Regueiro, and M. Perroux, *Phys. Rev. B* **1996**, *54*, R12633–12636.

164. R. Moret, P. Launois, P.-A. Persson, and B. Sundqvist, *Europhys. Lett.* **1997**, *40*, 55–60.

165. V. Agafonov, V. A. Davydov, L. S. Kashevarova, A. V. Rakhmanina, A. Kahn-Harari, P. Dubois, R. Céolin, and H. Szwarc, *Chem. Phys. Lett.* **1997**, *267*, 193–198.

166. V. A. Davydov, L. S. Kashevarova, A. V. Rakhmanina, A. V. Dzyabchenko, V. Agafonov, P. Dubois, R. Céolin, and H. Szwarc, *JETP Lett.* **1997**, *66*, 120–125.

167. V. Brouet, H. Alloul, A. Janossy, and L. Forró, "Role of relative orientation of neighboring C_{60} chains in the electronic properties of the A_1C_{60} polymers" in *Molecular Nanostructures*, H. Kuzmany, J. Fink, M. Mehring, and S. Roth (eds.). World Scientific Publishing, Singapore, 1998, pp. 328–332.

168. P. Launois, R. Moret, J. Hone, and A. Zettl, *Phys. Rev. Lett.* **1998**, *81*, 4420–4423.

169. V. Blank, M. Popov, S. Buga, V. Davydov, V. N. Denisov, A. N. Ivlev, B. N. Mavrin, V. Agafonov, R. Ceolin, H. Szwarc, and A. Rassat, *Phys. Lett. A* **1994**, *188*, 281–286.

170. A. I. Kolesnikov, I. O. Bashkin, A. P. Moravsky, M. A. Adams, M. Prager, and E. G. Ponyatovsky, *J. Phys. Condens. Matter* **1996**, *8*, 10939–10949.

171. B. Renker, H. Schober, R. Heid, and P. V. Stein, *Solid State Commun.* **1997**, *104*, 527–530.

172. B. Renker, H. Schober, R. Heid, and B. Sundqvist, "Fingerprints of solid-state chemical reactions in the dynamics of fullerenes" in *Electronic Properties of Novel Materials*—Progress in Molecular Nanostructures, H. Kuzmany, J. Fink, M. Mehring, and S. Roth (eds.) American Inst. of Physics, Woodbury, NY, 1998, pp. 322–326.

173. A. Soldatov, O. Andersson, B. Sundqvist, and K. Prassides, *Mol. Mater.* **1998**, *10*, 271–276. A. Soldatov, K. Prassides, O. Andersson, and B. Sundqvist, "Vibrational and thermal properties of pressure polymerized C_{60}" in *Fullerenes: Recent Advances in the Chemistry and Physics of Fullerenes and Related Materials*, Vol. 6, K. M. Kadish and R. S. Ruoff (eds.). The Electrochemical Society, Pennington, NJ, 1998, pp. 769–779.

174. J. Winter, H. Kuzmany, A. Soldatov, P.-A. Persson, J. Jacobsson, and B. Sundqvist, *Phys. Rev. B* **1996**, *54*, 17486–17492.

175. S. Stafström, M. Boman, and J. Fagerström, *Europhys. Lett.* **1995,** *30*, 295–300. K. Harigaya, *Phys. Rev. B* **1995**, *52*, 7968–7971.

176. M. E. Kozlov, M. Tokumoto, and K. Yakushi, *Appl. Phys. A* **1997**, *64*, 241–245.

177. I. O. Bashkin, A. N. Izotov, A. P. Moravsky, V. D. Negrii, R. K. Nikolaev, Yu. A. Ossipyan, E. G. Ponyatovsky, and E. A. Steinman, *Chem. Phys. Lett.* **1997**, *272*, 32–37.

178. U. D. Venkateswaran, D. Sanzi, A. M. Rao, P. C. Eklund, L. Marques, J.-L. Hodeau, and M. Núñez-Regueiro, *Phys. Rev. B* **1998**, *57*, R3193–3196.

179. T. L. Makarova, N. I. Nemchuk, A. Ya. Vul', V. A. Davydov, L. S. Kashevarova, A. V. Rakhmanina, V. Agafonov, R. Ceolin, and H. Szwarc, *Tech. Phys. Lett.* **1996**, *22*, 985–988.

180. Y. Takahashi, Y. Takada, S. Kotake, A. Matsumuro, and M. Senoo, *J. Jpn. Inst. Metals* **1996**, *60*, 700–707.

181. P. Nagel, V. Pasler, S. Lebedkin, C. Meingast, B. Sundqvist, T. Tanaka, and K. Komatsu, "Intermolecular bond stability of C_{60} dimers and 2D pressure-polymerized C_{60}" in *Electronic Properties of Novel Materials—Progress in Molecular Nanostructures*, H. Kuzmany, J. Fink. M. Mehring and S. Roth (eds.) American Inst. of Physics, Woodbury, NY, 1998, pp. 194–197.

182. B. Sundqvist, Å. Fransson, A. Inaba, C. Meingast, P. Nagel, V. Pasler, B. Renker, and T. Wågberg, "Physical properties of two-dimensionally polymerized C_{60}" in *Fullerenes: Recent Advances in the Chemistry and Physics of Fullerenes and Related Materials*, Vol. 6 K. M. Kadish and R. S. Ruoff (eds.). The Electrochemical Society, Pennington, NJ, 1998, pp. 705–16.

183. K. Khazeni, J. Hone, N. G. Chopra, A. Zettl, J. Nguyen, and R. Jeanloz, *Appl. Phys. A* **1997**, *64*, 263–269.

184. G. M. Bendele, P. W. Stephens, K. Prassides, K. Vavekis, K. Kordatos, and K. Tanigaki, *Phys. Rev. Lett.* **1998**, *80*, 736–739.

185. L. Forró, G. Baumgartner, A. Sienkiewicz, S. Pekker, O. Chauvet, F. Beuneu, and H. Alloul, "Pressure and disorder effect on the magnetic properties of the A_1C_{60} conductors" in *Fullerenes and Fullerene Nanostructures*, H. Kuzmany, J. Fink, M. Mehring, and S. Roth (eds.). World Scientific Publishing, Singapore, 1996, pp. 102–109.

186. J. Hone, M. S. Fuhrer, K. Khazeni, and A. Zettl, *Phys. Rev. B* **1995**, *52*, R8700–8702.

187. K. Khazeni, V. H. Crespi, J. Hone, A. Zettl, and M. L. Cohen, *Phys. Rev. B* **1997**, *56*, 6627–6630.

188. J. Hone, K. Khazeni, and A. Zettl, "Electrical transport and phase transitions in polymerized AC_{60} (A = K, Rb)" in *Fullerenes and Fullerene Nanostructures*, H. Kuzmany, J. Fink, M. Mehring, and S. Roth (eds.). World Scientific Publishing, Singapore, 1996, pp. 115–118.

189. A. Jánossi, N. Nemes, T. Fehér, G. Oszlányi, G. Baumgartner, and L. Forró, *Phys. Rev. Lett.* **1997**, *79*, 2718–2721.

190. V. A. Davydov, L. S. Kashevarova, A. V. Rakhmanina, V. N. Agafonov, R. Céolin, and H. Szwarc, *JETP Lett.* **1996**, *63*, 818–824.

191. V. D. Blank, V. N. Denisov, I. N. Ivlev, B. N. Mavrin, N. R. Serebryanaya, G. A. Dubitsky, S. N. Sulyanov, M. Yu. Popov, N. A. Lvova, S. G. Buga, and G. N. Kremkova, *Carbon* **1998**, *36*, 1263–1267.

192. B. Sundqvist, P. Jacobsson, J. Jun, P. Launois, R. Moret, A. Soldatov, and T. Wågberg, "Polymerization of C_{60} under hydrostatic and non-hydrostatic pressure conditions" in *Fullerenes: Recent Advances in the Chemistry and Physics of Fullerenes and Related Materials*, Vol. 5, K. M. Kadish and R. S. Ruoff (eds.). The Electrochemical Society, Pennington, NJ, 1997, pp. 439–449.

193. B. Sundqvist, *Phys. Rev. B* **1998**, *57*, 3164.

194. V. A. Davydov, L. S. Kashevarova, A. V. Rakhmanina, V. Agafonov, H. Allouchi, R. Ceolin, A. V. Dzyabchenko, V. M. Senyabin, and H. Szwarc, *Phys. Rev. B* **1998**, *58*, 14786–14790.

195. R. Moret, P. Launois, T. Wågberg and B. Sundqvist, *Eur. Phys. J. B* **2000** (in press).

196. J. Onoe and K. Takeuchi, *Phys. Rev. Lett.* **1997**, *79*, 2987–2989.

197. G. Oszlányi, G. Baumgartner, G. Faigel, and L. Forró, *Phys. Rev. Lett.* **1997**, *78*, 4438–4441.

198. C. Kugler, J. Winter, H. Kuzmany, and Y. Iwasa, "Doping of rhombohedral pressure polymerized C_{60} with alkali metals" in *Molecular Nanostructures*, H. Kuzmany, J. Fink, M. Mehring, and S. Roth (eds.). World Scientific Publishing, Singapore, 1998, pp. 369–372.

199. J. Winter, B. Burger, M. Hulman, H. Kuzmany, and A. Soldatov, *Appl. Phys. A* **1997**, *64*, 257–262.

200. G. Oszlanyi and L. Forró, *Solid State Commun.* **1995**, *93*, 265–267.

201. K. Kamarás, Y. Iwasa, and L. Forró, *Phys. Rev. B* **1997**, *55*, 10999–11002.

202. S. Okada and S. Saito, *Phys. Rev. B* **1997**, *55*, 4039–4041.

203. C. H. Xu and G. E. Scuseria, *Phys. Rev. Lett.* **1995**, *74*, 274–277.

204. U. D. Venkateswaran, D. Sanzi, J. Krishnapa, L. Marques, J.-L. Hodeau, M. Núñez-Regueiro, A. M. Rao, and P. C. Eklund, *phys. status solidi b* **1996**, *198*, 545–552.

205. Y. Ma and G. Zou, "Fullerenes transition to diamond under high pressure and high temperature" in *High Pressure Science and Technology*, W. A. Trzeciakowski (ed.). World Scientific Publishing, Singapore, 1996, pp. 702–706.

206. V. D. Blank, S. G. Buga, N. R. Serebryanaya, V. N. Denisov, G. A. Dubitsky, A. N. Ivlev, B. N. Mavrin, and M. Yu. Popov, *Phys. Lett. A* **1995**, *205*, 208–216.

207. V. D. Blank, S. G. Buga, N. R. Serebryanaya, G. A. Dubitsky, S. N. Sulyanov, M. Yu. Popov, V. N. Denisov, A. N. Ivlev, and B. N. Mavrin, *Phys. Lett. A* **1996**, *220*, 149–157.

208. V. V. Brazhkin, A. G. Lyapin, S. V. Popova, R. N. Voloshin, Yu. V. Antonov, S. G. Lyapin, Yu. A. Kluev, A. M. Naletov, and N. N. Mel'nik, *Phys. Rev. B* **1997**, *56*, 11465–11471.

209. V. V. Brazhkin, A. G. Lyapin, S. V. Popova, Yu. A. Klyuev, and A. M. Naletov, *J. Appl. Phys.* **1998**, *84*, 219–226.

210. V. D. Blank, O. M. Zhigalina, B. A. Kulnitsky, and Ye. V. Tatyanin, *Crystallography Reports* **1997**, *42*, 588–591.

211. M. O'Keeffe, *Nature* **1991**, *352*, 674.

212. V. D. Blank, S. G. Buga, G. A. Dubitsky, N. R. Serebryanaya, S. N. Sulyanov, and M. Yu. Popov, "Structure and properties of 3D-polymerized C_{60}: comparison with diamond and C_{70}" in *Molecular Nanostructures*, H. Kuzmany, J. Fink, M. Mehring, and S. Roth (eds.). World Scientific Publishing, Singapore, 1998, pp. 506–510.

213. V. D. Blank, S. G. Buga, N. R. Serebryanaya, G. A. Dubitsky, R. H. Bagramov, M. Yu. Popov, V. M. Prokhorov, and S. A. Sulyanov, *Appl. Phys. A* **1997**, *64*, 247–250.

214. B. R. Djordjevic and M. F. Thorpe, *J. Phys. Condens. Matter* **1997**, *9*, 1983–1994.

215. M. E. Kozlov, M. Hirabayashi, K. Nozaki, M. Tokumoto, and H. Ihara, *Appl. Phys. Lett.* **1995**, *66*, 1199–1201.

216. M. E. Kozlov, P. Fons, H.-A. Durand, K. Nozaki, M. Tokumoto, K. Yase, and N. Minami, *J. Appl. Phys.* **1996**, *80*, 1182–1185. M. E. Kozlov, K. Yase, N. Minami, P. Fons, H.-A. Durand, A. N. Obraztsov, K. Nozaki, and M. Tokumoto, *J. Phys. D Appl. Phys.* **1996**, *29*, 929–933.

217. B. Keita, L. Nadjo, V. A. Davydov, V. Agafonov, R. Céolin, and H. Szwarc, *New J. Chem.* **1995**, *19*, 769–772.

218. P. R. Súrjan, J. G. Ángyán, A. Lázár, K. Németh, and L. P. Bíró, "On the nature of the chemical bond between the C_{60} units of the fullerene polymer" in *Fullerenes and Fullerene Nanostructures*, H. Kuzmany, J. Fink, M. Mehring, and S. Roth (eds.). World Scientific Publishing, Singapore, 1996, pp. 319–322.

219. V. D. Blank, Ye. V. Tatyanin, and B. A. Kulnitskiy, *Phys. Lett A* **1997**, *225*, 121–126.

220. L. A. Chernozatonskii, E. G. Gal'pern, and I. V. Stankevich, *JETP Lett.* **1998**, *67*, 712–719.

221. V. Blank, M. Popov, G. Pivovarov, N. Lvova, K. Gogolinsky, and V. Reshetov, *Diamond Relat. Mater.* **1998**, *7*, 427–431.

222. P. H. Gaskell, A. Saeed, P. Chieux, and D. R. McKenzie, *Phys. Rev. Lett.* **1991**, *67*, 1286–1289.

223. B. L. Zhang, C. Z. Wang, K. M. Ho, and C. T. Chan, *Europhys. Lett.* **1994**, *28*, 219–224.

224. V. Blank, M. Popov, N. Lvova, K. Gogolinsky, and S. Reshetov, *J. Mater. Res.* **1997**, *12*, 3109–3114.

225. V. D. Blank, S. G. Buga, N. R. Serebryanaya, G. A. Dubitsky, V. M. Prokhorov, M. Yu. Popov, N. A. Lvova, V. M. Levin, and S. N. Sulyanov, "Cluster structure and elastic properties of superhard and ultrahard fullerites" in *Electronic Properties of Novel Materials—Progress in Molecular Nanostructures*, H. Kuzmany, J. Fink, M. Mehring, and S. Roth (eds.), American Institute of Physics, Woodbury, NY, 1998, pp. 499–503.

226. V. D. Blank, A. A. Nuzhdin, V. M. Prokhorov, and R. Kh. Bagramov, *Phys. Solid State* **1998**, *40*, 1261–1263.

227. A. M. Rao, M. Menon, K.-A. Wang, P. C. Eklund, K. R. Subbaswamy, D. S. Cornett, M. A. Duncan, and I. J. Amster, *Chem. Phys. Lett.* **1994**, *224*, 106–112.

228. M. Premila, C. S. Sundar, P. Ch. Sahu, A. Bharathi, Y. Hariharan, D. V. S. Muthu, and A. K. Sood, *Solid State Commun.* **1997**, *104*, 237–242.

229. Y. Iwasa, T. Furudate, T. Fukawa, T. Ozaki, T. Mitani, T. Yagi, and T. Arima, *Appl. Phys. A* **1997**, *64*, 251–256.

230. M. Núñez-Regueiro, P. Monceau, and J.-L. Hodeau, *Nature* **1992**, *355*, 237–239.

231. Y. Ma, G. Zou, H. Yang, and J. Meng, *Appl. Phys. Lett.* **1994**, *65*, 822–823.

232. G. Bocquillon, C. Bogicevic, F. Clerc, J. M. Léger, C. Fabre, and A. Rassat, "Synthesis of diamond at high pressure and high temperature from C_{60} fullerene" in *High-Pressure Science and Technology—1993*, S. C. Schmidt, J. W. Shaner, G. A. Samara, and M. Ross (eds.). American Institute of Physics, New York, 1994, pp. 647–649.

CHAPTER 16

SUPERCONDUCTIVITY IN FULLERENES

VICTOR BUNTAR

16.1 INTRODUCTION

There has been a large development in the science and technology of super-conductivity for the last twelve years, and recently discovered fullerene-based superconductivity [1] is one of the most interesting and spectacular subjects in this area.

After many years of attempts to increase the superconducting transition temperature, T_c, in the conventional superconductors, the highest T_c obtained was 23.2 K in Nb_3Ge in 1973, and it was widely believed that this transition temperature could hardly be improved except by one or two degrees. The new era in superconductivity started in 1986 after Bednorz and Müller [2] made a remarkable discovery of superconductivity in a new class of superconducting materials—cuprates (La_2CuO_4) with T_c in excess of 30 K. Very shortly after the discovery, the two scientists were awarded the 1987 Nobel Prize in physics. Soon after, it was followed by discoveries of a large number of cuprate super-conductors with transition temperatures up to 150 K, well above the boiling temperature of liquid nitrogen (77 K), giving strong promise of commercial applications.

Only five years after Bednorz and Müller's publication, superconductivity in alkali-doped fullerenes was discovered [1] with transition temperatures as high as 33 K in $RbCs_2C_{60}$ [3,4] at ambient pressure. The highest $T_c = 40$ K in full-erides so far was observed in Cs_3C_{60} [5] under pressure. Numerous experi-mental investigations of these superconductors show unique properties and distinguish them from other superconducting materials. The high transition temperature and the small coherence length do not allow us to consider them as conventional superconductors. In some respects, fullerene superconductors (FSs) are very similar to the high-T_c superconductors. Both are strong type-II

Fullerenes: Chemistry, Physics, and Technology, Edited by Karl M. Kadish and Rodney S. Ruoff.
0-471-29089-0 Copyright © 2000 John Wiley & Sons, Inc.

superconductors with Ginzburg–Landau parameter larger than 100 and have a very short coherence length. However, in contrast to cuprates, graphite intercalated compounds, borocarbides, and many organic superconductors, the lattice and the electronic structure of FSs are almost isotropic.

The great interest in superconductivity of the alkali-doped fullerenes is, in particular, due to these systems being a completely new class of superconductors, the large T_c, and the question of whether or not such a large value of T_c can be caused by coupling to phonons alone. This led to numerous experimental and theoretical investigations of these materials.

In this chapter we attempt to show a state-of-the-art picture in the science of fullerene superconductivity. Because of the lack of space, we cannot discuss all the aspects in detail showing all experimental and theoretical results. For other general reviews of fullerene superconductivity we refer readers to the literature [6–8]. There are also a number of publications reviewing some special subjects like theoretical approaches [9,10], electronic structure [9,11], nuclear magnetic resonance (NMR) investigations [12], pressure [13,14] and magnetic properties [15], parameters controlling T_c [16,17], and some other properties [18–20] in a long list of reviews of fullerene superconductivity.

This chapter is organised as follows. In Section 16.2 we discuss fullerene-based superconducting compounds and their solid structures. In Section 16.3 we discuss the electronic structures of these materials. Preparation procedures of different FS compounds are given in Section 16.4. In Section 16.5 we summarize results from experimental investigations of the superconducting properties of fullerides, mainly related to the determination of parameters reflecting the coupling mechanism. In Section 16.6, magnetic properties of fullerides in the superconducting state are presented. In this chapter we do not discuss theoretical models in detail but refer interested readers to the excellent reviews by Gelfand [9] and Gunnarsson [10]. Instead, a brief discussion of different theoretical approaches to fullerene superconductivity is given in Section 16.7. Finally, we put conclusions and some speculations in Section 16.8.

16.2 SUPERCONDUCTING COMPOUNDS AND THEIR SOLID STRUCTURES

16.2.1 Subclasses of Fullerene Superconductors

In the first seven years after the discovery of superconductivity in fullerides a large number of superconducting fulleride compounds has been found. These compounds can be divided into five general subclasses according to their structural and valence features. These are summarized in Table 16.1.

The first subclass is the alkali system $A_{3-x}A'_xC_{60}$ (A,A' = K, Rb, and Cs) with valence $n_v = 3$. In these materials exactly three electrons are doped into the lowest unoccupied molecular orbital (LUMO), which is triply degenerate with t_{1u} symmetry. These compounds have face-centered cubic (fcc) lattices with space group $Fm\overline{3}m$. In the second subclass, the sodium system $Na_2A_xC_{60}$ with $x < 1$, and the third subclass, the alkali/alkaline-earth system $A_{3-x}AE_xC_{60}$ with $x > 3$, the filling of the t_{1u} band is less than and larger than 3, respectively. These subclasses have the same solid structure as the first one. In the fourth subclass, the alkaline-earth system, the t_{1u} band is full and the next t_{1g} level, which is the triply degenerate LUMO+1 state, is partly occupied. $K_3Ba_3C_{60}$ is very different from other compounds of the alkali/alkaline-earth system. Simple electron counting for $K_3Ba_3C_{60}$ gives a nominal valence of $n_v = 9$, corresponding to half-filling of the t_{1g} band. This compound has a body-centered cubic (bcc) lattice with space group $Im\overline{3}$. The rare-earth subclass RE_xC_{60}, $Yb_{2.75}C_{60}$ and Sm_3C_{60} has valence $n_v = 5.5$ and $n_v = 6$, respectively, and an orthorhombic solid lattice. Finally, the fifth subclass, the so-called modified system, is obtained by an expansion of the first subclass of superconducting compounds by the introduction of ammonia into their solid lattices. Superconductivity in cubic $K_3Hg_xC_{60}$ ($x = 1, 2, 3$) was also observed at $T_c = 2.7, 2.8,$ and 3.3 K, respectively [21]. Because of a lack of information about the solid state and electronic structure of this compound, it is difficult to attribute it to one of the subclasses.

16.2.2 Solid Structures of Fullerene Superconductors

The alkali-metal-doped system has been studied the most because it was the first superconducting fullerene system discovered and these materials are often easier to prepare. In contrast to cuprate superconductors, which have a wide region of superconducting stoichiometries, in A_xC_{60} only the 3:1 stoichiometry is superconducting. All other structures are insulating, for example, the body-centered tetragonal (bct) A_4C_{60} [22,23] and the bcc A_6C_{60} [22,24], or metallic under certain conditions, for example, K_1C_{60} [25].

Almost all alkali metal $A_{3-x}A'_xC_{60}$ compounds, except for Cs_3C_{60}, crystallize in a fcc structure with space group $Fm\overline{3}m$.[1] (However, LiK_2C_{60}, Li_2KC_{60}, and NaK_2C_{60} are multiphase, and KCs_2C_{60} is unstable [26]). There are three natural interstices for each C_{60}, one octahedrally coordinated, located at the center of six C_{60} molecules at the points of an octahedron, and two tetrahedrally coordinated as shown in Figure 16.1. Therefore, only $0 \leq x \leq 3$ is possible for the fcc structure. The "radius" for the smaller tetrahedral sites is $R_{tet} \approx 0.11$ nm and for the larger octahedral site $R_{oct} \approx 0.21$ nm. For each A_3C_{60} formula unit one alkali atom occupies the octahedral site and the other two atoms occupy tetrahedral sites [27]. This structure is shown in Figure 16.2. If the compound, for example, K_2RbC_{60}, consists of different alkali atoms, then the larger atom occupies the octahedral position. The ionic radii of K, Rb, and

---

[1] Detailed description of the structures of alkali-metal-doped C_{60} can be found in Chapter 14 of this volume or in Ref. 8.

TABLE 16.1 Solid Structure and Superconducting Transition Temperature for Fullerene-Based Superconductors

Compound	Structure	Dopant Site symmetry	Lattice Parameter (nm)	Transition Temperature (K)
		Alkali Subclass		
Li_2CsC_{60}	fcc $Fm\bar{3}m$	Li(T)Li(T)Cs(O)	1.412 [65]	12 [65]
K_3C_{60}	fcc $Fm\bar{3}m$	T and O	1.4253 [98]	19.5
K_2RbC_{60}	fcc $Fm\bar{3}m$	K(T)K(T)Rb(O)	1.4243 [242]	23 [65], 21.8 [3]
K_2CsC_{60}	fcc $Fm\bar{3}m$	K(T)K(T)Cs(O)	1.4292 [65]	24 [65,243]
KRb_2C_{60}	fcc $Fm\bar{3}m$	K(T)Rb(T)Rb(O)	1.4243 [65]	27 [65], 24 [3]
KCs_2C_{60}	Unstable			
$K_{1.5}Rb_{1.5}C_{60}$	fcc $Fm\bar{3}m$	T and O	1.4341 [244]	22.15 [3]
Rb_3C_{60}	fcc $Fm\bar{3}m$	T and O	1.4384 [65]	29.5
Rb_2CsC_{60}	fcc $Fm\bar{3}m$	Rb(T)Rb(T)Cs(O)	1.4431 [65]	31.3 [3], 31 [4]
$RbCs_2C_{60}$	fcc $Fm\bar{3}m$	Rb(T)Cs(T)Cs(O)	1.4555 [65]	33 [65]
$Cs_3C_{60}^{a}$	—	T and O	—	40 (12 kbar) [5]
		Sodium-Containing Subclass		
Na_2KC_{60}	sc $Pa\bar{3}$	Na(T)Na(T)K(O)	1.4025 [65]	2.5 [65]
Na_2RbC_{60}	sc $Pa\bar{3}$	Na(T)Na(T)Rb(O)	1.4097 [66]	2.5 [65], 3.5 [109]
Na_2CsC_{60}	sc $Pa\bar{3}$	Na(T)Na(T)Cs(O)	1.4134 [65]	12 [65], 10.5 [59]
$Na_2Rb_xCs_{1-x}C_{60}$	sc $Pa\bar{3}$	T and O		$3 \div 12$ [66]
$Na_xN_yC_{60}$ $(x = 3{-}4)$	fcc	T and O	1.433 [160]	12–15 [160]
$Na_3(Na_xN_y)_zC_{60}$	fcc	T and O		16 [245]
$(NaH)_xC_{60}$	fcc	T and O	1.4356 [246]	~15 [246]

Alkali/Alkaline-Earth Subclass

Compound	Structure	Sites	Lattice	T_c (K)
$K_{3-x}Ba_xC_{60}$	fcc $Fm\bar{3}m$	T and O		15 (x = 0.25) [30]
$Rb_{3-x}Ba_xC_{60}$	fcc $Fm\bar{3}m$	T and O	1.421 (x = 2) [66]	24 (x = 0.5) [30]
$K_3Ba_3C_{60}$	bcc $Im\bar{3}$		1.124 [75]	5.6 [75]
$Rb_3Ba_3C_{60}$	bcc $Im\bar{3}$		1.1245 [247]	2 [247]

Alkaline-Earth Subclass

Compound	Structure	Sites	Lattice	T_c (K)
Ba_4C_{60}	bco $Immm$		1.0975 [33]	7 [35,170,248]
Sr_6C_{60}	bcc $Im\bar{3}$			4 [33]
Ca_5C_{60}	sc $Pa\bar{3}$	T and O	1.401 [32]	8.4 [32,35,249]

Rare-Earth Subclass

Compound	Structure	Sites	Lattice	T_c (K)
$Yb_{2.75}C_{60}$	Orthorhombic	T and O	$a = 2.817$,	6 [34,39]
Sm_3C_{60}	Orthorhombic	T and O	$b = 2.807$,	8 [40]
			$c = 2.827$ [40]	

Ammonia-Containing Compounds

Compound	Structure	Sites	Lattice	T_c (K)
$Na_2Cs(NH_3)_4$	fcc $Fm\bar{3}m$	T and O	1.4473 [59]	29.6 [59]
$K_3(NH_3)C_{60}$	Orthorhombic	T and O	$a = 1.497$,	— [58]
			$b = 1.4895$,	
			$c = 1.3687$ [58]	

[a] Superconductivity occurs only under pressure and T_c increases with pressure, reaching a maximum of $T_c \sim 40$ K at 12 kbar [5], and the structure is believed to be noncubic.

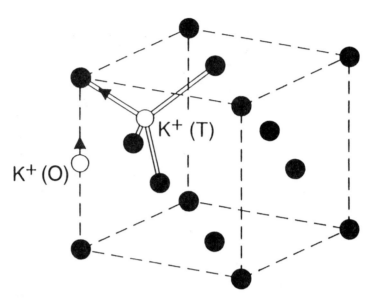

Figure 16.1 The fcc lattice structure of solid C_{60} (solid circles) showing the tetrahedral $(K^+(T))$ and octahedral $(K^+(O))$ alkali sites. The ratio of K(1) to K(2) sites is 2:1. Reprinted with permission from Ref. 7.

Cs are 1.33, 1.48, and 1.67, respectively, all of which are larger than the tetrahedral site radius. Thus, the tetrahedral ions exist in very cramped quarters, and their radius is probably the primary factor controlling the lattice constant [28]. The superconducting transition temperature T_c is strongly dependent on the lattice parameter a in very different ways in different types of FS. For alkali-metal-doped fullerenes, T_c increases with increasing a (this subject will be discussed later). Increasing the ionic radii from K to Rb and Cs increases the fcc lattice constant a from 1.424 to 1.4555 nm [3].

The radius of a Na atom is slightly smaller than the radius of the tetrahedral site and is much smaller than the octahedral one. This means that more than one Na atom can occupy one octahedral position. Structural considerations for Na and Na-containing C_{60} are complex and these materials behave differently from other alkali-doped materials in important ways. This is why they form a separate subclass of FSs.

The relationship between the measured lattice parameter of $A_x C_{60}$ and the average cation size was considered by Rosseinsky et al. [29]. Figure 16.3 shows the unit cell size plotted against the average cation volume, exhibiting a general linear trend for the homologous fcc compounds. Inclusion of Na into the host C_{60} lattice tends to shrink the lattice. This can be understood if the lattice expands when an ion larger than Na goes into the tetrahedral site, whereas it contracts when any alkali goes into the octahedral site. Superconducting and

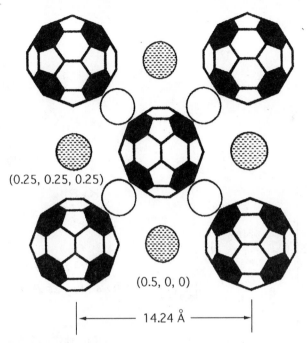

(0.25, 0.25, 0.25)

(0.5, 0, 0)

|← 14.24 Å →|

Figure 16.2 The structure of K_3C_{60}. The open and hatched spheres represent the potassium in the tetrahedral and octahedral sites, respectively. Reprinted with permission from Ref. 27.

structural properties of $Na_2Cs_xC_{60}$ were investigated by Tanigaki et al. [26] and Yildirim et al. [30]. Na ions fully occupy the tetrahedral sites for all x, while a variable fraction of octahedral sites are occupied by Cs, the refined occupancies being in good agreement with the nominal composition. At low temperature, C_{60} molecules are orientationally ordered, space group $Pa\overline{3}$, with preferred angles and defect orientations similar to those found in the Na ternaries and quaternaries. At high temperature, the structure is fcc with space group $Fm\overline{3}m$ and a first-order transition near 300 K for all x. The lattice constant decreases linearly from 1.419 to 1.413 nm as x increases from 0 to 1, confirming solid solution behavior. The lattice contracts upon filling octahedral vacancies with Cs because electrostatic interactions dominate over steric effects. The low-temperature $Pa\overline{3}$ reflections decrease in intensity as x increases. This is attributed [30] to frustrations of the C_{60} molecules between preferred and defect orientations by octahedral Cs.

Intercalation of alkaline-earth and rare-earth metals in the fcc C_{60} host lattice results in the formation of other structure types with transition temperatures much lower than in alkali-metal-doped FSs.

X-ray results for $A_3Ba_3C_{60}$ show single phase bcc patterns with space group

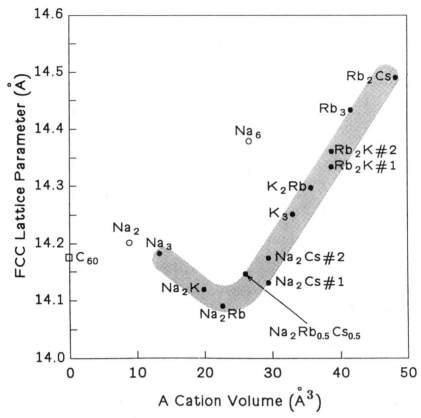

Figure 16.3 The fcc lattice parameter of A_3C_{60} as a function of the total volume of the intercalated cations. Reprinted with permission from Ref. 29.

$Im\bar{3}$ [31]. The crystal structure is the same as for A_6C_{60} and Ba_6C_{60}. Simulation of diffraction patterns indicates that occupation of the interstitial sites by alkali metals and Ba is random, and thus the structure is regarded as a 1:1 solid solution of isostructural A_6C_{60} and Ba_6C_{60}.

The structure of Ca_5C_{60} has the same space group $Pa\bar{3}$ as the C_{60} host lattice, because of the small size of the Ca atoms, as for Na-containing compounds. Ca ions occupy both tetrahedral and octahedral sites and it is believed that up to four Ca^{2+} cations are accommodated in a single octahedral site, despite the Coulomb repulsion between the Ca cations [32]. The lattice constant $a = 1.401$ nm is smaller than that for C_{60} and is associated with attraction between the C_{60} anions and the charged Ca ions.

Sr_xC_{60} and Ba_xC_{60} compounds have fcc and bcc phases, which compete in the same compositional range [33]. Both fcc and bcc-A15 phases have an equilibrium stoichiometry of Sr_3C_{60}. Increasing the strontium composition above $x = 3$ suppresses the fcc and bcc-A15 structures and leads to appearance of the bcc $Im3$ phase with $a = 1.0975$ nm for Sr_6C_{60} and unit occupancy for all sites [34].

Superconductivity in Ba intercalated fullerides was first reported for samples with nominal composition Ba_6C_{60} [35]. However, in order to clarify the stoichiometry of the superconducting phase, this system was subject to further investigations. A series of Ba_xC_{60} samples was prepared with different stoichiometries [36,37]. This clearly showed that orthorhombic Ba_4C_{60} is superconducting [36–38].

The Yb_xC_{60} system was studied [39] in the range $0 < x < 5$ and a single phase, orthorhombic $Yb_{2.75}C_{60}$ was isolated, which became superconducting below $T_c = 6$ K. Long-range ordering gives rise to a new cation-vacancy-ordered superstructure, which appears to be stabilized by a short-range interaction between charge deficient five-membered rings of C_{60} and exclusively divalent cations. Here, ytterbium cations occupy off-centered interstitial sites of the fcc-packed molecules and leave eight tetrahedral sites vacant in an orthorhombic unit cell doubled in each direction. This result suggests that superconductivity is not limited to C_{60} compounds of high structural symmetry. Recently, an identical structure has also been reported in a superconducting Sm_3C_{60} compound [40].

In the beginning of the fullerene superconductivity era it was believed that only materials with close-packed cubic structure and exactly half-doped t_{1u} energy band could be superconducting. This picture was not correct. Table 16.1 shows that non-close-packed cubic with different space group symmetry materials exhibit superconductivity and that crystal structure is not the key parameter regulating superconducting–nonsuperconducting features. An expectation that for some compounds with valence not equal to 3 the transition temperature could be higher [41,42] also does not seem to be the case. T_c goes up with increasing lattice parameter and rapidly decreases when the nominal valence shifts from half-filling of the t_{1u} band [30]. The latter result suggests a failure of both one-electron and strong correlation pictures in A_3C_{60} compounds.

In addition, it is worth mentioning that the presence of dopant atoms in the C_{60} host lattice strongly influences the orientation of the C_{60} molecules in such a way that a hexagonal face points along each of the crystal $\langle 111 \rangle$ directions [27]. Figure 16.2 shows that for each C_{60} molecule there are two equivalent orientations that can accomplish this. Both orientations exist with more or less equal populations at low temperature. This residual low-temperature disorder is known as "merohedral disorder." The orientational state of the C_{60} molecules is controlled mainly by the size of the tetrahedral ion. Large ions (for alkali metals these are K, Rb, and Cs) lock the C_{60} molecules at random into one of the two orientations that minimize the ion–fullerene repulsive overlap [43]. This interaction is negligible for smaller tetrahedral ions (for alkali metals these are Li and Na) in which case there is no frustration between Na^+–C_{60}^{n-} Coulomb attraction and the strictly intermolecular interactions. These compounds have the same orientational ground state as C_{60}, with slightly higher orientational melting temperatures [44].

16.3 ELECTRONIC STRUCTURE

The electronic structure of fullerene superconductors is fundamental to the understanding of the mechanism of superconductivity in these materials. In this section only some basic features of the electronic structure of doped fullerenes will be discussed. A detailed discussion and an analysis of theoretical approaches to calculation of electronic levels in fullerenes, and the band structures in fullerites and fullerides, can be found in the book by Dresselhaus et al. [8] and, for the alkali fullerides, the review by Gelfand [9].

Because the interactions between buckyballs in the solid lattice are relatively weak compared to intramolecular ones, the C_{60} molecules form highly molecular solids with narrow energy bands. This means that the electronic structure of fullerides is determined mainly by the electronic levels of the C_{60} molecule itself. If it is assumed that all vertices of the truncated icosahedron are equivalent, then all carbon atoms in C_{60}, at these sites, can be considered equivalent. Each C atom has four valence electrons. Three of these occupy low-energy σ orbitals lying roughly 3–6 eV below the Fermi level E_F. The fourth electron occupies a π orbital close to the Fermi level, which is significant for electron transport. Occupancy of the molecular energy levels, based on a one-electron Hükel calculation, are shown in Figure 16.4 [18]. Electrons fill all the levels up to h_u, which is the highest occupied molecular orbital (HOMO). This level is completely filled. The next level t_{1u} is the lowest unoccupied molecular orbital (LUMO) and is completely empty. (Here h (t) refers to 5 (3)-fold degeneracy, an u (g) refers to odd (even) parity.) The empty t_{1u} band causes the insulating properties of fullerenes. Figure 16.4 shows a representative t_{1u} orbital with the filled states indicated by positive sign and dark shading and the empty states indicated by negative sign and light shading. The sign of the wave function changes very rapidly with position within the molecule. This variation should reduce overlap with the atomic wave functions of neighboring alkali atoms, thereby reducing the hyperfine coupling to the alkali nuclei [12].

Since the shortest distance between carbon atoms in the molecule is about 0.14 nm and the shortest distance between two carbon atoms in different molecules is about 0.3 nm, the intramolecular interactions dominate over the intermolecular ones. Thus, the C_{60} molecules condense to a solid in which many molecular properties still survive. In this molecular solid, the bands, which are formed when the molecular orbitals of the neighboring molecules mix, are only weakly broadened and result in a set of rather narrow nonoverlapping bands. From calculations based on the local density approximation (LDA), important features of the electronic structure such as the precise bandwidths and bandgaps and the value of the density of states at the Fermi level $N(E_F)$ can be determined. (Reviews of the LDA approach can be found in Refs. 45 and 46.) Figure 16.5 [47] shows the result of a LDA calculation of the subbands around the Fermi energy for solid C_{60}. The t_{1u} band has a width of about 0.5 eV and can take six electrons. The distance between HOMO and LUMO bands is about

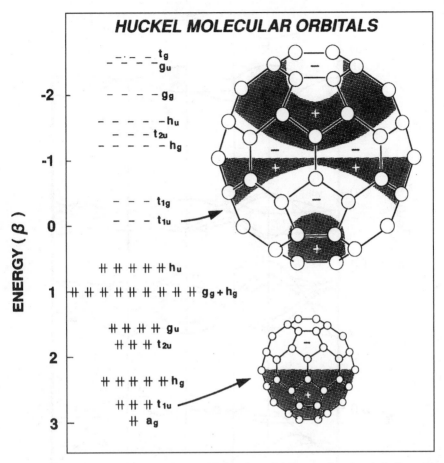

Figure 16.4 A representative t_{1u} orbital. The sign of the wave function is indicated by the dark or light shading. Reprinted with permission from Ref. 9.

1.5 eV. In the one-electron band calculation approach, the h_u level in the solid forms the valence band and the lowest unoccupied state forms the conduction band. Doping of the C_{60} solid by alkali metal leads to a full charge transfer of one valence electron per atom. These electrons occupy the t_{1u} band and for A_3C_{60} this band is half-filled. Thus, these materials are metals.

As shown in Figure 16.6, the bandwidth, electronic structure, and density of states profiles are very sensitive to structural symmetries and disorder [48]. As for doped fullerenes, the electronic structure is also influenced by distortions of the icosahedral shell and by perturbations of bond lengths and structural geometry due to charge-transfer effects. Distortion of the molecules' icosahedral symmetry is caused by an expansion of the molecule when electrons are added,

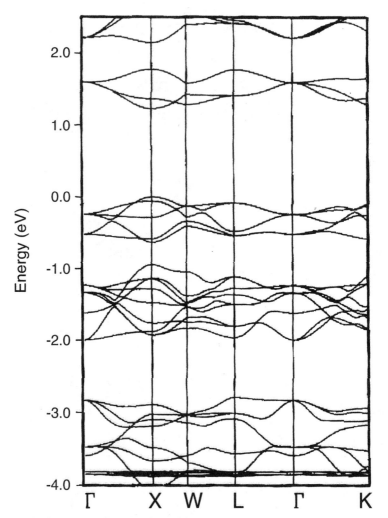

Figure 16.5   Some of the subbands around the Fermi energy for solid C_{60} in the $Fm\bar{3}$ structure. The bands at about -0.5 eV are the h_u bands, which are occupied in solid C_{60}, and the bands at about 1.5 eV are the t_{1u} bands, which become populated in A_nC_{60}. The bands around -1.5 eV results from the overlapping h_g and g_g bands. Reprinted with permission from Ref. 256.

as should be expected because the LUMOs are antibonding. Changes in molecular geometry would be associated with shifts in the energies of molecular orbitals, the positions of band centers, and the values of bandgaps [9].

The density of states, calculated within the LDA approximation, has a two-hump structure with roughly two-thirds of the states in the lower-energy hump, as shown in Figure 16.7 [48]. Adding three electrons to one C_{60} molecule puts

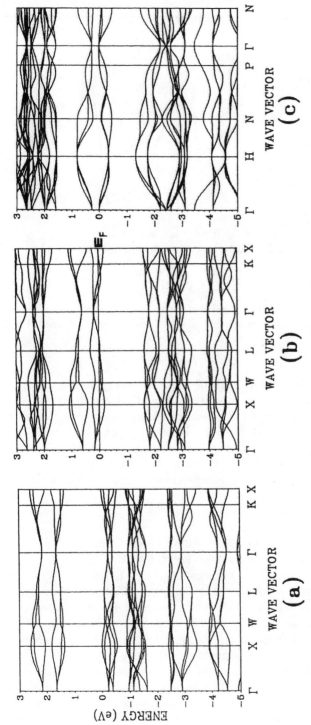

Figure 16.6 · Band structure of (a) fcc C_{60}, (b) fcc K_3C_{60}, and (c) bcc K_6C_{60}. The zero energy is either at the top of the valence band or at the Fermi level. Reprinted with permission from Ref. 51.

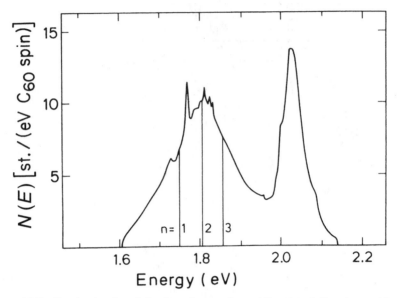

Figure 16.7   Conduction band density of states for unidirectional C_{60}; the positions of the Fermi levels for A_nC_{60}, with $n = 1$, 2, and 3, are indicated. Reprinted with permission from Ref. 48.

the Fermi level on the falling side of the lower-energy hump. Since dN/dE near the Fermi level is quite large, results of different calculations of the precise value of $N(E_F)$ could be expected to be sensitive to the details of the calculations and give a fairly large spread in values. Results of such calculations are given in Table 16.2 and the density of states lies between 6.6 and 9.8 states/eV/C_{60}/spin. From results in which both $N(E_F)$ and the bandwidth W_b can be found [48–50], a better approximation for unidirectionally ordered A_3C_{60} is [9]

$$N(E_F) = (4.0 \pm 0.1)/W_b \tag{16.1}$$

From the values of the density of states listed in Table 16.2, one can estimate the bandwidth to range between 400 and 600 meV.

Calculations of the Fermi surface in K_3C_{60} by Erwin and Pickett [49] and by Xu et al. [51] showed that it is anisotropic and consists of two inequivalent sheets. One sheet is a closed spheroidal surface with protrusions in the $\langle 111 \rangle$ directions, reflecting contact of the Fermi surface with the Brillouin zone boundary with two (rather then four) necks along the $\langle 100 \rangle$ directions, indicative of the T_h^6 ($Pa3$) lower-symmetry space group.

Another feature to be mentioned in this section is that merohedral disorder strongly influences the electronic structure. Below the phase transition temperature $T_{01} = 261$ K (see Chapter 14 in this volume) each molecule tries to align itself with respect to its 12 nearest neighbors, so that its 12 electron-poor pen-

TABLE 16.2 Results of LDA Calculations for the
Density of States at the Fermi Energy $N(E_F)$ for K_3C_{60}
in states/eV/C_{60}/spin, and the Ratio $N(E_F)$ for Rb_3C_{60} to
that in K_3C_{60}[a]

$N(E_F)$ (K_3C_{60})	$N(E_F)$ (Rb_3C_{60})/ $N(E_F)$(K_3C_{60})	Source
6.6	1.26	49, 250
8.9	1.18	50
8.6	1.21	48[b]
9.0	1.14	259[b]
9.8	1.27	251[c]
12.5	1.25	52[d]

Source: From Ref. 9.

[a] To make comparison possible, in every case the results correspond to lattice constant $a = 1.424$ nm for K_3C_{60} and 1.444 nm for Rb_3C_{60}. Interpolated values are subject to uncertainties at the level of ± 0.1 in the first column and ± 0.03 in the second column.

[b] Interpolated results.

[c] Only W_b was reported explicitly, so Eq. (16.1) was used [9] to obtain the values presented; the lattice constant dependence was given as $dW/d\ln(a^3) = -2.03$ eV.

[d] Probably less reliable than others; see Ref. 9.

tagons are located opposite an electron-rich double bond on the nearest-neighbor C_{60}. However, substantial orientational disorder is present even at low temperatures. The effect of orientational disorder on the band structure has been considered by several groups [52,53] showing that details in the band structure and density of states are sensitive to this [54].

Gelfand and Lu [53] have shown that merohedral disorder greatly influences electronic structure and through this the superconducting properties of FSs. There have been frequent suggestions that correlation effects and orientational disorder are responsible for the complications encountered by band theory [7], but the ways in which these effects operate are still unknown. Ceulemans et al. [55] developed a model calculation on the $Fm\bar{3}m$ family of A_3C_{60} compounds, which provided evidence that the unusual band structure of the fullerides can be explained in a simple way by charge disproportionation of conduction electron density.

16.4 PREPARATION OF FULLERENE-BASED SUPERCONDUCTORS

Superconducting fullerene compounds, as well as practically all dopants, are extremely air sensitive. The superconductivity can be destroyed by simply exposing the superconducting material to air for a fraction of a second. Preparation must take place in the total absence of oxygen and moisture, usually in argon glove boxes with $O_2 < 1$ ppm and $H_2O < 1$ ppm. Before the doping procedure, the C_{60} solid has to be dried and degassed under heating in vacuum. An appropriate amount of dopant is then introduced to the C_{60} (film, powder, crystal). The mixture is then sealed in a quartz or glass tube and heated at a certain temperature to complete the reaction. The annealing temperature and time of the treatment depend on the type of dopant and size of the C_{60} particles.

16.4.1 Alkali Intercalated Fullerenes

16.4.1.1 *Vapor Phase Intercalation*   The intercalation of C_{60} with alkali metals is, in principle, very simple thanks to the low vapor pressure of most of the alkali metals (with the exception of Li) and the large interstices in the C_{60} fcc lattice, which allow a large diffusivity of the alkali atoms and a homogeneous doping after annealing. This method has proved to be a popular and successful technique. The direct reaction of a stoichiometric amount of alkali metal vapor with C_{60} powder, followed by an appropriate annealing, was the first method used for the preparation of fullerides [1]. The preparation procedure is as follows. C_{60} solid is placed, in a glove box with an inert atmosphere, into a Pyrex or quartz tube together with a stoichiometric amount of alkali metal. The reaction tube containing C_{60} and alkali metal is sealed with a high vacuum valve and removed from the glove box. Then the inert atmosphere is evacuated and the tube is finally sealed under high vacuum. Next, the reaction tube is placed in an oven and the A/C_{60} mixture is annealed at $T = 200–400\,^{\circ}C$ for the diffusion of the alkali metal in the C_{60} lattice. The annealing time depends on the size of the C_{60} particles and is usually between several hours for powders and weeks or months for crystals. In thin films, the time required to achieve homogeneous doping appears to be much shorter—several minutes.

This method, however, presents some experimental difficulties with regard to weighing the alkali metals, since very precise balances do not work properly in glove boxes. If the alkali metal is loaded into a capillary with precisely known inner diameter, the appropriate amount of alkali metal can be measured from the length of the capillary. This method has an advantage because the "open" surface of the alkali metal filled into the capillary, which can react with a small amount of oxygen probably present in the glove box, is very small. Another procedure sometimes adopted for avoiding the difficulty of measuring the precise amount of A is overdoping the C_{60} with an excess of alkali metal in order to obtain a compound with stoichiometry A_6C_{60}. Because no more than six alkali ions (K, Rb, and Cs, but not Li and Na) can be intercalated into the C_{60} lattice, it is possible to synthesize A_6C_{60} as a well-defined homogeneous compound and then to anneal a proper mixture of A_6C_{60} and C_{60} to produce the desired final stoichiometry.

16.4.1.2 *Intercalation with Azides*   An alternative doping procedure that does not require the use of a glove box while weighing is based on the thermal decomposition of alkali metal azides [56]. These compounds are not air sensitive and, when heated above their decomposition temperature, produce pure alkali metals and nitrogen:

$$2AN_3 \rightarrow 2A + 3N_2 \quad (A = Li, Na, K, Rb, Cs)$$

Good-quality superconducting fullerides can be prepared following this procedure. The decomposition temperature (T_d) ranges from 360 to 520 °C (depending on A). The mixture C_{60}/AN_3 is slowly heated from room temperature to T_d in dynamic vacuum while monitoring the pressure and the heating process is stopped just after decomposition has been detected as a (narrow) pressure peak. A further annealing is sometimes required to enhance crystallinity and the SC fraction.

16.4.1.3 Intercalation with Ammonia A third method based on doping C_{60} in a solution of liquid ammonia gives good-quality homogeneously doped samples and does not require the use of high-temperature treatments [57]. It takes advantage of the good solubility of alkali metals and C_{60} anion in liquid ammonia, which can be evaporated slowly after the reaction has taken place. Some compounds tend to trap ammonia in their lattice and in these cases very long heat treatments are required for completely deintercalating ammonia. On the other hand, ammonia intercalated fullerides are quite interesting systems [58,59], and this is a possible preparation route for these compounds as an alternative to the simple exposure of the bare fulleride to ammonia vapors. However, a possible disadvantage of this method is that ammonia may be incorporated as an impurity in the A_3C_{60} lattice.

16.4.1.4 Intercalation with Metal Alloys One more approach to doping is to use metal alloys [60,61]. The procedure is similar to those described earlier, only instead of azides or ammonia, metal alloys such as AHg are used. The advantage of this method, as for intercalation with azides or ammonia, is that the molar mass is large compared to pure alkali metal. This allows the preparation of a more precise stoichiometric mixture. At the same time, these alloys are hard solids and can be ground and mixed directly with C_{60} powder. A very important feature is that AHg is less sensitive to oxygen and moisture.

16.4.1.5 Intercalation in an Organic Solvent Wang et al. [62,63] reported that alkali-doped C_{60} could be prepared in an organic solvent. The procedure was the following. A toluene solution containing pure C_{60} was freeze–thaw–degassed three times. Small potassium chips were then added while the $C_{60}/$ toluene solution was kept frozen. The reaction flask was immediately evacuated and back-filled with Ar three times. The mixture was warmed to room temperature and then refluxed for two hours with vigorous magnetic bar stirring. The color of the solution turned from purple (pure C_{60}) to burgundy and finally to black with a large amount of black precipitate being formed. It was filtered to remove the almost colorless toluene solution and vacuum-dried at room temperature [62]. However, the superconducting fraction in the obtained material was very small, <10%. Schlueter et al. [64] demonstrated that AC_{60} is the dominant phase obtained from reduction in toluene.

16.4.1.6 Sodium Intercalated Compounds Preparation of Na intercalated fullerides is a more laborious procedure, compared to other alkali dopants, because Na is very reactive with glass. For synthesis of these compounds, Rosseinsky et al. [29] used the NaH and Na_5Hg_2 reagents annealing procedure for a week or longer. However, Tanigaki et al. [65] reported that, in order to prepare Na_2CsC_{60} and Li_2CsC_{60}, a direct reaction of sodium and lithium with a C_{60} host previously doped with Cs can also be successful. This method was later used by Yildirim et al. [30,66], where OFHC cooper tube was used as a container for the direct reaction of alkali metals and C_{60}.

16.4.2 Preparation of A_3C_{60} Single Crystals

Powder samples are inappropriate for some experiments. For instance, this is the case for any magnetic measurements because the penetration depth in FSs is of the order of 1 μm [67]. Therefore, powder with average grain size ~1–3 μm (this is the most common grain size in fullerene powder) is completely penetrated by a magnetic field. Thus, crystalline samples with sizes of at least one-tenth of a millimeter are required for these measurements.

First sizeable K_3C_{60} crystals were prepared by Xiang et al. [78] for transport measurements. Later on, bulk crystals (~1 mm^3) were prepared by several groups [68–73], but none of these crystals (except two in [73]) had 100% superconducting fraction and, therefore, they had a very inhomogeneous and defect-filled structure. The problems in producing homogeneous A_3C_{60} bulk crystals arise from all the difficulties discussed previously plus a very slow diffusion of alkali metals into a large sample. The final quality of the crystals depends on many parameters, such as the amount of reacting K and C_{60}, quality of C_{60} crystals and type and concentration of defects inside them, doping temperature, and annealing time.

Recently, Haluska et al. [74] reported an investigation into preparation of high-quality superconducting crystals, that is, crystals with exact A_3C_{60} stoichiometry, with homogeneous dopent distribution and minimum defects.

For doping experiments, crystals with smooth and shiny faces were selected. One part of them was twinned or multiply twinned and another part had a morphology of fcc single crystals. The crystals were placed into reaction glass tubes together with a glass micropipette containing a precise weight of pure (99.6%) potassium. Four batches were prepared for doping with different amounts of potassium, corresponding to 100%, 110%, 112%, and 125% of the stoichiometric amount of K_3C_{60}. The reaction tubes were sealed and separately accommodated in a three-zone horizontal furnace with windows. The crystals and the micropipette with potassium were located in the central part of the tube. This part of the tube was visible through the window. Temperatures at both ends of the tube, T_{ends}, and in the central part, T_{cnt}, were controlled independently. During the whole doping experiment the condition $T_{ends} > T_{cnt}$ was satisfied.

At first the doping temperature was adjusted. During this stage the central part of the reaction tube with the crystals was at $T_{dop} = 205$ or $230\,^{\circ}C$. The evaporation rate at 205 and $230\,^{\circ}C$ was estimated to be higher than 0.27 mg/h and lower than 0.54 mg/h, respectively. The corresponding "doping" rate calculated from the evaporation time of potassium and inserted amount of pure C_{60} was lower than 1.1 and higher than 2.4 mg of created K_3C_{60} per hour for 205 and $230\,^{\circ}C$, respectively. After all the potassium had evaporated from the pipette the temperature was increased to an annealing temperature $T_{cnt} = T_{ann}$. Crystals were kept at $T_{ann} = 385\text{--}405\,^{\circ}C$ for one or two weeks. Preliminary information about the doped C_{60} crystals was obtained by optical microscopy. This was obtained on crystals still sealed in the reaction tube. A larger number of crystals was observed in the tube compared to the number originally inserted. Macroscopic cracks and splitting along $\langle 111 \rangle$ planes were clearly recognized in some crystals. The observation of a larger number of crystals than inserted originally can be explained by the fact that one of the twinned crystals was cleaved along the $\langle 111 \rangle$ twinning plane during the doping and/or the heat treatment.

Prepared crystals were characterized by dc magnetization and ac susceptibility measurements. The dc magnetization experiments showed that the molar ratio K/C_{60} has to be slightly larger than 3 in order to obtain a fully doped crystal with 100% superconducting (K_3C_{60}) fraction. This means that a certain portion of the alkali metal either reacts with a small amount of oxygen, which can be present in the glove box, or than not all of the potassium reacts with the fullerenes. The ac susceptibility measurements showed that the quality of the superconducting material of the best samples obtained at lower doping rate was higher than the quality of the best samples obtained at higher doping rate for both shorter and longer annealing times. This is demonstrated in Figure 16.8, which depicts ac susceptibility measurements on samples K17 and K52 prepared at higher doping rate and K22 and K27 prepared at lower doping rate. However, some of the crystals prepared at lower doping rate show low superconducting quality. This means that the doping temperature is not the only quality-controlling factor. A higher superconducting quality of crystals prepared at the same doping rate was found for samples that were annealed longer. Crystals K27 and K52 were annealed for two weeks while crystals K18 and K22 were annealed for only one week. Crystals, annealed for less than two weeks, showed either incomplete shielding (K17) or two steps in the dc $\chi(T)$ dependence or a very strong and wide intergrain peak in the imaginary part of the ac susceptibility (K22). The superconducting transitions are rather broad. In K17, for instance, the width of the transition between 10% and 90% of the diamagnetic signal is $\Delta T = 2.15$ K and in K22 $\Delta T = 3.3$ K. Moreover, the transition temperature $T_c = 18$ K in K22 is much lower than that in K17, where $T_c = 19.4$ K. The transition temperature is low because of a large number of

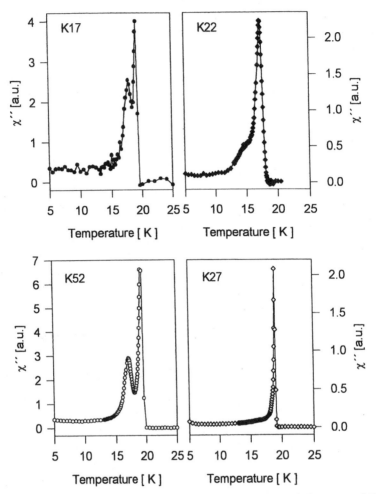

Figure 16.8 Temperature dependence of the imaginary part of the susceptibility at dc field $H = 0$, ac field $\mu_0 h_{ac} = 10^{-4}$ T, and ac frequency $f = 31$ Hz. *Left panels*: $T_{dop} = 503$ K. *Right panels*: $T_{dop} = 477$ K. *Top panels*: $t_{ann} = 1$ week. *Bottom panels*: $t_{ann} = 2$ weeks. Reprinted with permission from Ref. 74.

structural defects in the sample. If the doping rate was low, the crystals were annealed after doping for two weeks or longer, and the molar ratio was slightly larger than 3, then all the samples showed 100% superconducting fraction, perfect shielding, and no intergrain peak in $\chi''(T)$. An example (K27) is shown in Figure 16.8. The peak in $\chi''(T)$ is very close to the transition temperature and certainly corresponds to dissipation in the bulk material. According to ac measurements, the current in this sample is not impeded by granularity and from this point of view the sample can be considered as a single crystal. Around

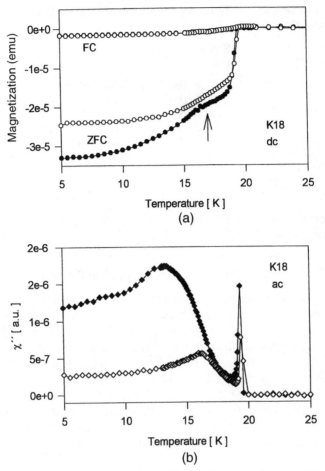

Figure 16.9 The (a) dc and (b) ac magnetization obtained for K18. Black symbols: 6 days annealing, while symbols: 6 + 12 days annealing. Reprinted with permission from Ref. 74.

50% of crystals from the same batch as K27 showed high superconducting quality.

The influence of annealing time on crystal K18, which had large intergranular dissipation compared to the intragranular one (Fig. 16.9b), was demonstrated by further annealing. After the first series of magnetic investigations (these results are given by solid symbols in Fig. 16.9), K18 was annealed for an additional 12 days. The annealing caused an increase in the ratio of intra- to intergranular dissipation. The second step of the transition on the zero-field-cooled (ZFC) curve in dc magnetization obtained for the crystal before annealing (indicated by arrow in Fig. 16.9a) disappeared after annealing. In Table

TABLE 16.3   Parameters of the Doping Procedure and Superconducting Parameters of the Investigated Crystals

Sample	K_3/C_{60}	t_{anneal} (days)	Superconducting Fraction, x_{sc} (%)	x_{sc} in the Batch (%)	T_c (K)	ΔT_c (K) ($m = 10-90\%$)
K17	1.0	5	≤ 100	$\sim 50-100$	19.4	2.15
K22	1.10	7	~ 100	~ 100	18	3.3
K25	1.10	7	~ 100	~ 100	19.6	—
K27	1.12	15	~ 100	~ 100	19.2	0.75
K52	1.25	15	≤ 100	≤ 100	19.2	Two steps

Source: From Ref. 189.

16.3 superconducting properties of some of the crystals prepared under different conditions are shown.

16.4.3   Alkaline-Earth (AE) and Rare-Earth (RE) Compounds

The alkaline-earth metals are very reactive with glass, which gives rise to the same problems as for Na, discussed in Section 16.4.1.6. Synthesizing Ca_5C_{60}, Kortan et al. [32] in order to perform direct reaction between AE and C_{60} pressed a stoichiometric mixture of Ca/C_{60} into pellets within tantalum cells. After that, the cells were sealed in quartz tubes and heated to 500–600 °C for 20 hours. A similar procedure was used for preparation of Ba-doped [35] and Sr-doped [33] C_{60}. The Sr/C_{60} mixture was annealed at 450–600 °C, while Ba/C_{60} was annealed at 800 °C. Synthesis of Ba_4C_{60} was successfully performed by Sparn et al. [38]. The temperature during the intercalation process was kept below 450 °C in order to minimize reaction of Ba with the wall of the quartz tube.

To synthesize $K_3Ba_3C_{60}$, the starting Ba_3C_{60} powders were first obtained from a direct reaction of stoichiometric amounts of Ba and C_{60} powders [75] and then doped with potassium. The sealed mixture was heated at 600 °C for three or four days. This reaction reproducibly yielded a single phase of Ba_3C_{60}. The reaction of potassium and Ba_3C_{60} was made in a way similar to that of potassium doping into pure C_{60}. The reaction was carried out at 260 °C for three days. Reaction at 400 °C, or longer annealing time at 260 °C, tends to result in phase separation into Ba_6C_{60} and K_6C_{60}.

Rare-earth fullerides $Yb_{2.75}C_{60}$ [39] and Sm_3C_{60} [40] were prepared in a way similar to the AE-C_{60} doping. Powders were mixed and loaded into tungsten [39] and tantalum [40] cells and pressed into pellets. Yb compounds were then heated at 630 °C for 6 hours (with a few intermediate grindings) and cooled to room temperature at a rate of 5 °C/min. Bulk synthesis of $Yb_{2.75}C_{60}$ could only be achieved in a narrow temperature range: Above 670 °C the structure rapidly decomposed to an amorphous phase, and below 630 °C Yb diffusion limited the reaction.

The Sm compounds were heated between 550 and 600 °C with a duration of several hours. It was found [40] that the optimum reaction temperatures are different for different concentrations of samarium. Higher concentration of samarium required lower temperatures and shorter reaction time. For high temperatures and long duration, the products are almost completely amorphous.

16.5 SUPERCONDUCTING PROPERTIES (EXPERIMENT)

16.5.1 Normal State Properties

16.5.1.1 *Normal State Resistivity*   It is not accidental that we start with a discussion of the normal state properties. Investigation of the normal state is one of the basic avenues of research in order to understand the superconducting state because superconductivity is just a low-temperature consequence of the normal state features. Measurements of the temperature-dependent resistivity provide information about the electron scattering mechanism, mean free path, electron–phonon coupling constant, and so on.

Transport experiments on fullerene films [76] and crystals [77–79] show that the resistivity, ρ, monotonically decreases with decreasing temperature and $\rho(T)$ has clear "metallic-like" behavior. Different experimental methods to determine the absolute value of the resistivity give slightly different $\rho(0)$ but all of them lie in the range 0.12–0.5 m$\Omega \cdot$ cm. Values of $\rho(0)$ obtained from transport measurements may not be intrinsic because of possible granularity of crystals. So, indirect measurements are useful. Klein et al. [80] found $\rho(0) = 0.41$ m$\Omega \cdot$ cm in K_3C_{60} pressed pellets using a contactless microwave surface impedance technique. Indirect determination [79] from $\rho(T)$ near the transition temperature using an analysis of fluctuation (or para-) conductivity gave $\rho(0) = 0.12$ and 0.22 m$\Omega \cdot$ cm for K_3C_{60} and Rb_3C_{60}, respectively. One more indirect method to evaluate ρ from values of the upper critical field [81,82] led to $\rho(0) = 0.18$ and 0.57 m$\Omega \cdot$ cm for K_3C_{60} and Rb_3C_{60} crystals. However, the difference between these two values seems to be large. Anyhow, all obtained $\rho(0)$ allow us to establish the fact that these materials are true metals with conductivity higher than Mott's minimum metallic conductivity.

The $\rho(T)$ dependence is unusual in that (1) the residual resistivity ratio is rather high $\rho(300)/\rho(0) \approx 2$ for both K_3C_{60} and Rb_3C_{60} [78]; (2) $\rho(T)$ has quadratic behavior [76–78,80]; and (3) no saturation was observed (at least on films) at high temperatures.

In order to explain the quadratic behavior several models have been suggested, such as electron–electron scattering [83] and electron–phonon scattering with only intramolecular phonons involved, which could give $\rho \sim T^2$ in certain situations [77,84–86] and including intermolecular phonons [77]. However, most probably there is no special mechanism behind this quadratic dependence. Since the thermal expansivity is large in these materials, the density of states is expected to be highly temperature dependent. A significant difference between

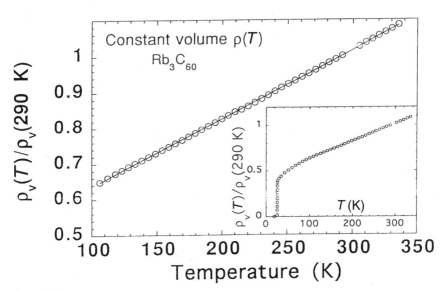

Figure 16.10 Constant-sample-volume resistivity versus temperature for Rb_3C_{60}. A linear temperature dependence is apparent. The inset shows the constant-volume resistivity over an extended range, but data at lower temperatures become less reliable due to increasing uncertainties in the data corrections applied. Reprinted with permission from Ref. 89.

the resistivities at constant pressure (which is usually measured experimentally) and at constant volume was predicted by Sundqvist and Nilsson [87]. Erwin and Pickett in their analysis found that ρ should be a linear function of temperature. All these predictions were confirmed experimentally [88,89]. In the temperature range from 100 to 350 K a linear temperature dependence was observed (Fig. 16.10), as expected for a metallic material with a Debye temperature Θ_D well below 100 K. From the high-temperature slope $\delta\rho/\delta T$, the transport electron–phonon coupling constant λ_{tr} was estimated using [90]

$$\rho(T) = \frac{8\pi^2 \lambda_{tr} k_B T}{\omega_p^2 \hbar} \tag{16.2}$$

(where ω_p is the plasma frequency) and was found to be $\lambda_{tr} = 0.65$–0.80 [89].

Using recent Raman data [91] Huang and Tanaka [92] theoretically obtained the linear $\rho(T)$ from the Zimans equation [93]. They have shown that intermolecular vibration modes play the main role in determination of the temperature-dependent resistivity at constant volume. This result, giving an explanation to some experimental data, is not in agreement with the picture that the intramolecular high-frequency modes drive superconductivity in fullerenes (see discussion in Section 16.7).

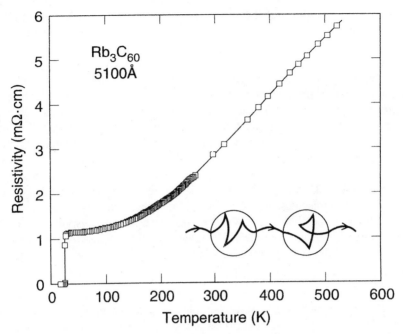

Figure 16.11 Plot of resistivity as a function of temperature for a 510 nm thick crystalline Rb_3C_{60} film. The inset depicts schematically electronic transport where the charge carriers are confined predominantly on individual molecules. Reprinted with permission from Ref. 76.

16.5.1.2 *Mean Free Path*

At high temperatures where the electronic mean free path, l_{tr}, approaches the intrinsic interatomic distance, the resistivity is expected to saturate. Hebard et al. [76] measured the temperature dependence of ρ in crystalline Rb_3C_{60} film up to 520 K and found a linear dependence above room temperature and no saturation, as shown in Figure 16.11. However, on close examination of the obtained dependence, one can find for $\rho(T)$ a tendency to saturate at $T \approx 450$–470 K. Most probably, the linear dependence is an accidental result of an interplay between the square dependence dominating at low temperatures and the saturation that is expected at high temperatures.

Measurements on crystalline Rb_3C_{60} up to 700 K [94] did show saturation at $T \geq 450$–500 K. But no saturation was reported in K_3C_{60} crystals [94] up to $T = 800$ K. This result remains unclear because the characteristic intrinsic interatomic distances in K_3C_{60} and Rb_3C_{60} are expected to be very similar and thus the saturation could also be expected to be in a similar temperature range.

From the conversion $\tau/m_b = l_{tr}/\hbar k_F$ (where τ is a scattering time, m_b is the band mass, and k_F is the Fermi wave factor) and using a spherical Fermi surface approximation, Hebard et al. [76] obtained a very short transport mean free path $l_{tr} = 0.063$ nm. The Fermi surface in alkali metal fullerides has a complicated shape [49] and the spherical Fermi surface approximation could certainly affect the resulting value of l_{tr}. However, Hou et al. [94] fitting their data to a Ziman resistivity model [90] also obtained short $l_{tr} = 0.1$ and 0.15 nm in Rb_3C_{60} and K_3C_{60}, respectively. This length is significantly shorter than both the nearest-neighbor center-to-center distance between C_{60}'s (~ 1 nm) and the average separation between conduction electrons (~ 0.6 nm) and is close to the distance between C atoms on each C_{60} molecule (~ 0.14 nm). It was suggested [76] that an electron is for the most part confined to the surface of a given molecule and is able to interact strongly with the intramolecular phonons before hopping to the next molecule. This behavior is represented schematically in the inset in Figure 16.11. Huang and Tanaka [92] estimated the mean free path to be about 0.66 nm at 350 K. Considering the closest distance between the surfaces of molecules, which is ~ 0.3 nm, rather than the center-to-center distance between neighboring C_{60}'s, the mean free path corresponds to a scattering probability after two molecules. However, the reason why the mean free path is shorter than the molecular separation is not clear and the question remains open.

16.5.1.3 Hall Coefficient

In the simple free electron model, the Hall coefficient is given by $R_H = -1/nec$, with n the carrier concentration. Assuming that there are three electrons donated by alkali metals to one C_{60}, one obtains $R_H = -1.4 \times 10^{-9}$ m^3/C. $R_H(T)$ determined for K_3C_{60} and Rb_3C_{60} crystals was found [95] to be -3×10^{-9} to -1×10^{-9} m^3/C and -0.5×10^{-9} to 1×10^{-9} m^3/C, respectively, in the temperature range 50–300 K. The difference in R_H between K_3C_{60} and Rb_3C_{60} at a given temperature can be ascribed to the difference in the lattice constant of these compounds. Figure 16.12 represents R_H versus the lattice constant [95] and shows that R_H is a linear function of a. Erwin and Pickett [96] calculated the pressure dependence of the Hall coefficient as $\delta \ln R_H / \delta P = -0.03$. There is no experimental data yet on R_H versus P but from the $R_H(a)$ dependence in Figure 16.12 one can expect that the true pressure dependence of R_H may be larger [14] than is suggested by Erwin and Pickett [96].

16.5.2   Transition Temperature Versus Lattice Parameter

16.5.2.1 Universal Behavior

Investigation of the influence on T_c of different parameters, such as density of states at the Fermi level, lattice parameter, and valence, has attracted a lot of attention since the discovery of superconductivity in fullerides because it can give important information about the mechanism of superconductivity. Assuming that superconductivity is driven by electron–phonon weak coupling (different models of the mechanism of superconductivity and their applicability will be discussed later in Section 16.7), a simple estimation for T_c based on the Bardeen–Cooper–Schrieffer (BCS) theory gives (e.g., see Ref. 97)

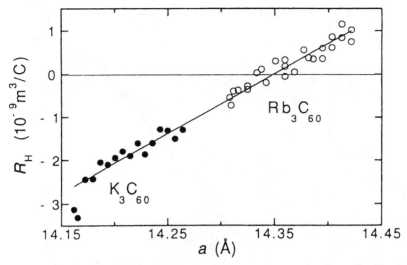

Figure 16.12 Hall coefficient as a function of lattice constant for the optimally doped K_3C_{60} and Rb_3C_{60} samples. Reprinted with permission from Ref. 95.

$$k_B T_c = 1.13\bar{\omega}_{ph} \exp\left(-\frac{1}{N(E_F)v_i}\right) \tag{16.3}$$

where $\bar{\omega}_{ph}$ is an average phonon frequency mediating the electron pairing and v_i is the superconducting pairing interaction. The density of electron states at the Fermi level and the pairing interaction give the electron–phonon coupling constant $\lambda_{ep} = N(E_F)v_i$. According to Eq. (16.3), increasing $N(E_F)$ should increase the transition temperature. Variation of the density of states at the Fermi level in FSs can be achieved by varying the lattice constant. The larger the distance between fullerene molecules, the smaller the intermolecular C_{60}–C_{60} coupling. This effect narrows the conducting t_{1u} LUMO band and thereby increases the density of states. Experimentally this can be achieved either by chemical substitution of smaller alkali atoms by larger ones, or the lattice constant can be decreased by application of external pressure. Many studies have been done in both directions, especially on alkali binary A_3C_{60} and ternary $A_{3-x}A'_xC_{60}$ compounds. This system is rather "clean" for investigation of the $T_c(a)$ dependence because all alkali compounds have fcc structure with space group $Fm\bar{3}m$ (see Table 16.1), and they all have merohedral disorder in which C_{60} molecules are locked at random into one of two energetically degenerate "standard orientations" [27,43]. The only parameter that can be changed is the distance between buckyballs.

The first work on this subject [3], made on $K_{3-x}Rb_xC_{60}$ compounds, established a universal $T_c(a)$ behavior where T_c monotonically increases with increasing lattice parameter. A point for $RbCs_2C_{60}$, which was obtained later [98], continued this almost linear dependence till $a \cong 1.45$ nm. Figure 16.13

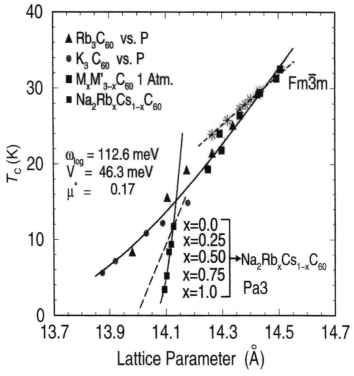

Figure 16.13 T_c versus a for $Fm\bar{3}m$ and $Pa\bar{3}$-ordered superconductors. Solid lines are fits to McMillan's formula. The slope of the $Pa\bar{3}$ curve is much larger than the $Fm\bar{3}m$ one, indicating that a small increase in a without destroying $Pa\bar{3}$ ordering should cause T_c to increase very rapidly. Dashed lines represent the recent hydrostatic pressure studies indicating that the T_c–a curve obtained from high pressure and from chemical substitution gives rise to different results. Reprinted with permission from Ref. 17.

summarizes experimental data for fcc alkali systems. Substitution of K atoms by larger cations raises T_c from ~19 K (K_3C_{60}) up to ~33 K ($RbCs_2C_{60}$). These chemical substitution data are shown by big solid squares in Figure 16.13. $T_c(a)$ data obtained by decreasing the lattice parameter in K_3C_{60} [99] and Rb_3C_{60} [100,101] by applying external pressure are shown by circles and up-triangles in Figure 16.13. These data fall on the same curve as the previous results. Some of the high-pressure results are summarized in Table 16.4. However, T_c cannot grow to "infinity" with increase of the lattice parameter. The overlap of the electron wave function between two adjusted C_{60} molecules decreases and this effect tends to decrease T_c. Moreover, there is a limit for the maximum a above which the fcc structure is no longer stable. For instance, this is the case for Cs_sC_{60}, which exhibits superconductivity only under pressure and its structure is believed to be noncubic.

TABLE 16.4   Results of High-Pressure Studies on Fullerene Superconductors

System	$T_c(0)^a$	dT_c/dP^b	Medium^c	Reference	Lattice	Band^d
K_3C_{60}	19.3 K	−7.8	Fluorinert	99	fcc	t_{1u}
Rb_3C_{60}	29.6 K	−9.7	Fluorinert	100	fcc	t_{1u}
Rb_3C_{60}	29.5 K	−8.73	Fluorinert	101	fcc	t_{1u}
Na_2CsC_{60}	12.0 K	−12.5	Fluorinert	112	sc	t_{1u}
Na_2CsC_{60}	10.6 K	−2 to −13	Solid Ne	115	sc	t_{1u}
Cs_3C_{60}	18 K	∼ +15	Oil	5	bcc	t_{1u}
$Yb_{2.75}C_{60}$	6.25 K	+0.3	Solid He	116	Orthorhombic	t_{1u}
Ca_5C_{60}	8.4 K	+1.1	Solid He	116,117	sc	t_{1g}
Ba_4C_{60}	6.8 K	−1.9	Fluorinert	38	Orthorhombic	t_{1g}

Source: From Ref. 13.

^{a} Gives value of transition temperature at ambient pressure.

^{b} In units of K/GPa.

^{c} Gives pressure medium used.

^{d} Names partially filled band.

In order to experimentally examine the validity of the proposed mechanism for the rapid decrease of T_c under pressure, that is, that this is a rapid decrease in the density of states, Diederichs et al. [102] measured the pressure dependence of both T_c and $N(E_F)$ on the same sample in the same experiment. The obtained dependence of the transition temperature on pressure is shown in Figure 16.14. In Figure 16.15 the total measured susceptibility of both C_{60} and Rb_3C_{60} as a function of pressure is shown. The measurement on pure insulating C_{60} serves as a test measurement since its diamagnetic susceptibility would be expected to have a negligible pressure dependence. This expectation is due to both the extreme hardness of the C_{60} molecule and the absence of any nonmolecular contributions to the susceptibility. At 300 K, as well as at 50 K, the electronic density of states in Rb_3C_{60} decreases rapidly under pressure with the rate $d \ln N(E_F)/dP \approx -14.5\%/GPa$, if a Stoner enhancement factor is assumed of $S = 1.3$. The experiment [102] clearly confirms the mechanism of the $T_c(P)$ and $T_c(a)$ dependence.

However, not everything is so clear-cut in the issue of transition temperature dependence on the lattice parameter and pressure as it appears at first sight. Let us discuss some examples of nonuniversal behavior.

16.5.2.2 *High-Pressure Effects*   Schilling et al. [13] showed that a diamond-anvil cell used in early work on Rb_3C_{60} [103,104] tends to underestimate the initial compressibility. This technique is unparalleled for the lattice parameter studies in the extremely high-pressure region. Indeed, for Rb_3C_{60}, the results of two groups [103,104] differ considerably (see discussion in Ref. 101). Using a neutron diffractometer and a self-constructed pressure cell, Diederichs et al. [101] found that the compressibility of Rb_3C_{60} was noticeably higher ($\geq 26\%$) than that from previous studies. From these data, together with the

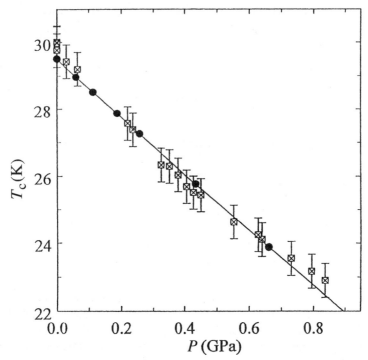

Figure 16.14   Measured dependence of T_c on hydrostatic pressure for Rb_3C_{60} given by solid line through data points ($\bullet$), from Refs. 101 and 102. Calculated T_c values from the McMillan equation using $SN(E_F)$ data from Figure 16.15 are represented by open squares for $S = 1$ and crosses for $S = 2$. Reprinted with permission from Ref. 13.

determined $T_c(P)$ (Fig. 16.14), T_c as a function of the lattice constant was evaluated. These data (dashed line in Fig. 16.13) clearly lie above the data from the mixed alkali-doped fullerenes at ambient pressure for a similar lattice constant. When the lattice of Rb_3C_{60} is reduced to that of K_3C_{60} at ambient pressure, T_c in Rb_3C_{60} is higher by ~4–5 K. As discussed in Ref. 17, some of the earlier data [103,105] upon close examination are also consistent with a T_c versus a correlation that depends weakly on the specific alkali metal dopant. The misfitting between results of the substitution of alkali metals and high-pressure experiments suggests that the superconducting transition temperature in the merohedrally disordered fcc compounds is not a universal function of the lattice parameter. Some suggested explanations [17], namely, that either coupling strength or phonon frequency are pressure dependent, or that there is a contribution to the coupling from external modes involving alkali motion, are in contradiction with the accepted picture. Another scenario [13] is that the merohedral disorder is suppressed, which leads to either pentagon or hexagon ordering. Reduction of the merohedral disorder would reduce the smearing of the density of states, thus allowing a possible enhancement in $N(E_F)$.

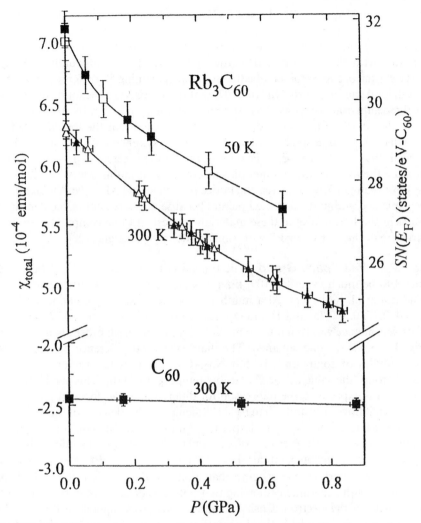

Figure 16.15 Measured magnetic susceptibility per mole of Rb_3C_{60} versus hydrostatic pressure at 50 K and 300 K for increasing/decreasing (closed/open symbols) pressure. The scale on the right gives $SN(E_F)$. Date for C_{60} at 300 K are also shown. Solid lines are guides to eye. Reprinted with permission from Ref. 102.

16.5.2.3 Ammoniated Compounds

Intercalation of neutral molecules into A_3C_{60} lattice expands the unit cell and therefore T_c in these compounds is expected to be larger. Ammoniated $Na_xA_{3-x}C_{60}$ compounds retain a fcc cubic structure with the same space group as before the ammoniation. Ammoniation of Na_2CsC_{60} to $(NH_3)_4Na_2CsC_{60}$ increases the transition temperature from 10.5 to 29.6 K [59]. The point for $(NH_3)_4Na_2CsC_{60}$ falls on a uniform $T_c(a)$

dependence. However, T_c values of $(NH_3)_xNaK_2C_{60}$ and $(NH_3)_xNaRb_2C_{60}$ are dramatically lower than would be expected from uniform behavior and the transition temperature decreases with increasing lattice parameter in sharp contrast with the conventional relation [106]. In these compounds, the Na–NH$_3$ cluster occupies the octahedral site and remaining K or Rb occupies the tetrahedral site. It was shown [106,107] that there exists significant displacement of the octahedral ion away from the high-symmetry position by 0.04–0.06 nm from the center of the octahedral site. It was argued that the low T_c could be explained by this displacement and that high symmetry of the molecule is crucial. In $(NH_3)_4Na_2CsC_{60}$, the octahedral Na ion is centered at (0.5, 0.5, 0.5) because the four solvating ammonias nicely center it in this interstice.

Decreasing of T_c with increasing lattice parameter in $(NH_3)_xNaA_2C_{60}$ cannot be explained within the accepted picture because measurement of the density of states at the Fermi level in these materials suggests it to be comparable or even larger than that for other FS with similar lattice parameters [108].

16.5.2.4 Pa$\bar{3}$ Space Group

The transition temperature of Na$_2$RbC$_{60}$ was found to be much lower [65,109] than was expected from its lattice parameter. Yildirim et al. [66] prepared a number of Na$_2$Rb$_x$Cs$_{1-x}$C$_{60}$ compounds (with $x = 0$, 0.25, 0.5, 0.75, and 1) and reported a constant correlation of T_c versus a. This dependence was found to be much steeper than for all other alkali FSs (see Fig. 16.13, small solid squares). The main structural difference between these two families of compounds is that Na$_2$AC$_{60}$ has a cubic structure with Pa$\bar{3}$ space group, the same as for C$_{60}$ at low temperature [110]. This symmetry difference was expected to be the reason for different dT_c/da slopes in these families of fullerides. Lu and Zhang [111] calculated the density of states at the Fermi level as a function of the lattice constant for both Fm$\bar{3}$ and Pa$\bar{3}$ ordering, taking into account the effects of molecular orientational disorder using the tight-binding recursion method. They found that $N(E_F)$ for Pa$\bar{3}$ rises much steeper with interfullerene spacing than that for Fm$\bar{3}$. This result is in good agreement with experiment, showing that both sets of data could be universally dependent on the electron density of states. This gives support to the assumption that the BCS model is the dominating mechanism of superconductivity in FSs.

Miziki et al. [112] found that T_c versus pressure-reduced a in Na$_2$CsC$_{60}$ decreased less than it should when compared with data obtained from chemical substitution (dashed line in Fig. 16.13). This effect is similar to that discussed earlier for Rb$_3$C$_{60}$. Zhu [113] and Morosin et al. [114] reported a structural transition in Na$_2$CsC$_{60}$ to a polymeric phase at pressures above 0.3–0.45 GPa. Structural transitions strongly affect high-pressure data (e.g., see Ref. 14) because they often lead to a sharp change in the lattice parameter, molecule orientation, and so on. Thus, the transition to the polymeric phase could be responsible for the anomalous dT_c/dP dependence observed by Schirber et al. [115] and for the strong differences in the $T_c(a)$ between high-pressure and chemical substitution experiments.

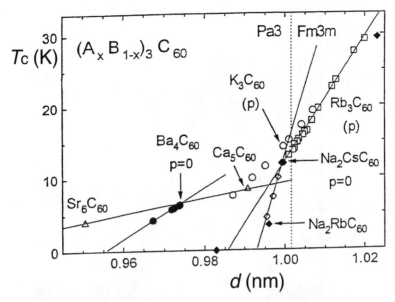

Figure 16.16 T_c relations obtained from pressure experiments on intercalated fullerenes (d is the intermolecular distance of the C_{60} molecules). Reprinted with permission from Ref. 258.

16.5.2.5 Alkaline-Earth and Rare-Earth Compounds

In contrast to the alkali FSs, no uniform behavior of the transition temperature with lattice parameter is found within the group of alkaline-earth and rare-earth compounds. Figure 16.16 shows a comparison of the experimental data obtained on these materials compared to those in alkali fullerides. First, they strongly deviate from the universal (for alkali FS) line and the values of dT_c/P differ strongly from those in the alkali fullerides. Second, some of these compounds exhibit positive dT_c/P slope. This effect cannot be explained by solid or electronic structural features because a positive sign of dT_c/P is observed, for instance, in $Yb_{2.75}C_{60}$ [116] with orthorhombic structure and half-filled t_{1u} band, and in Ca_5C_{60} [117] with simple cubic structure and partly filled t_{1g} band. At the same time, Ba_4C_{60} [38] with orthorhombic structure and half-filled t_{1g} band exhibits negative dT_c/P dependence.

16.5.3 Valence-Dependent Transition Temperature

Investigation of valence dependence of the transition temperature could give information about correlation effects in FSs. Some theoretical models suggest that the A_3C_{60} phases with n_v equal exactly to three are not stoichiometric [118,119], otherwise they would be insulators. Careful structure refinements showed significant vacancy concentration [120–122]. However, thermopower

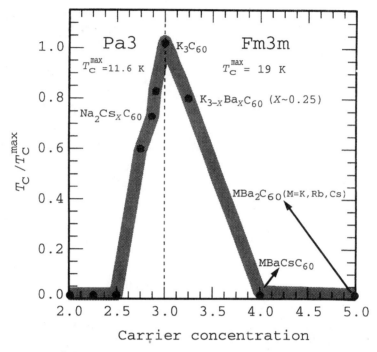

Figure 16.17 T_c/T_c^{max} versus conducting electron concentration n in fulleride super-conductors. Here T_c versus n is scaled by the T_c values of the end members Na_2CsC_{60} and K_3C_{60}. The heavy line connecting the points is a guide for the eye and indicates upper bounds on the uncertainties in T_c and n. Reprinted with permission from Ref. 30.

measurements [123] indicated n-type doping instead of the p-type doping, which would be expected from nonstoichiometric phases [119].

Direct measurements [30] of the transition temperature in compounds with $n_v \leq 3$ and $n_v > 3$ were performed. The first group of investigated compounds was $Na_2Cs_xC_{60}$ with $x \leq 1$. All these phases make solid solutions, space group $Pa\bar{3}$, with $n_v < 3$ if $x < 1$. With decreasing n_v, the transition temperature drops sharply down, $T_c = 11.6$, 9.6, 8.3, and 7.0 K for $x = 1$, 0.93, 0.88, and 0.75, respectively. Compounds with $n_v = 2.0$, 2.25, and 2.5 were not superconducting down to 0.5 K. These results are plotted in Figure 16.17, left side. However, from the spin susceptibility measurements all these compounds, including non-superconducting ones, were found to be metallic down to at least $n_v = 2.25$.

The second group of compounds, with valence larger than 3, is $A_{3-y}Ba_yC_{60}$, which can be obtained by substitution of alkali metal ions by Ba. Ba provides two valence electrons to the C_{60} molecule instead of one electron from alkali metals. These lead to the increase in n_v; that is, the t_{1u} band is more than half full. The valences of $ABaCsC_{60}$ and of $RbBa_2C_{60}$ are expected to be -4 and -5, respectively [30]. These compounds were found to be nonsuperconducting above $T = 0.5$ K; they showed a very small Pauli susceptibility and it was tentatively concluded that they are weakly metallic [30]. The $K_{3-x}Ba_xC_{60}$ compound with $x \approx 0.25$ had the superconducting transition temperature $T_c = 15$ K. All these data are plotted in Figure 16.17 (right side).

The reduction of T_c for $n_v > 3$, but not $n_v < 3$ (see discussion in Ref. 10), may be explained by reduction of the density of states with increasing filling, taking the merohedral disorder into account [66]. The drop of T_c for $n_v < 3$ seems not to be understood. However, these results are inconsistent with some theoretical predictions [41,42,119], which conclude that the stoichiometric A_xC_{60} is an insulator.

16.5.4 Superconductivity Gap

The value of the superconductivity energy gap, Δ, provides information about the energy of the coupling effects in superconductors. For the reduced energy gap, weak coupling BCS theory gives

$$2\Delta/kT_c = 3.53$$

If the reduced gap is larger, this indicates strong coupling. Experimentally, the superconductivity gap can be obtained from different types of measurements and, particularly for A_3C_{60}, it has been obtained from tunneling microscopy [124–127], NMR [128–131], infrared [132,133], optical reflectivity [134,135], photoemission [136], and muon spin relaxation (μSR) [137] measurements. The first measurements of 2Δ for Rb_3C_{60} using a tunneling spectroscopy technique [124,125] and later scanning tunneling microscopy (STM) measurements [126,127] gave $2\Delta/kT_c = 5.3$ [124], 5.4 [126], and between 2 and 4 [127]. These results suggested that electrons could couple strongly to phonons. Infrared experiments [132,133] also showed a rather large upper limit for the reduced gap: $2\Delta/kT_c = 3–5$ [132] and $2\Delta/kT_c = 2–5$ [133]. From these results no conclusion about the coupling strength could be made because they range from weak to strong coupling. However, later optical measurements [134] gave $2\Delta/kT_c = 3.6$ in K_3C_{60}, 2.98 in Rb_3C_{60}, 3.44 in K_3C_{60} [138], and 3.45 in Rb_3C_{60} [135], in agreement with weak coupling theory. NMR measurements also give values of the reduced gap close to the BCS theory. Tycko et al. [128] obtained $2\Delta/kT_c = 3.0$ and 4.1 (for K_3C_{60} and Rb_3C_{60} respectively). Sasaki et al. [130] obtained 4.3, and Auban-Senzier et al. [129] obtained 3.4. A similar result, $2\Delta/kT_c = 3.6$ [137] and 3.5–4.0 was evaluated from μSR measurements [139] while photoemission experiments [136] gave 4.1 for the reduced gap in Rb_3C_{60}, which is slightly larger than the BCS value.

Summarizing all these data, one can see that there is a rather wide distribution for the value of the reduced superconductivity gap. However, as concluded in Ref. 10, most of the recent results tend to scatter around the weak coupling limit and are consistent with this limit, if a reasonable uncertainty is assumed for these experiments.

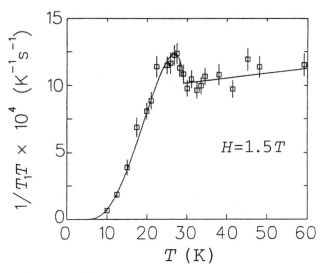

Figure 16.18   Muon spin relaxation measurements of $1/T_1 T$ in $Rb_3 C_{60}$ at a magnetic field of 1.5 T. The solid curve is a fit to the theory of Hebel and Slichter with a broadened density of states. Reprinted with permission from Ref. 137.

16.5.5   Hebel–Slichter Coherence Peak

Nuclear magnetic resonance experiments have been used extensively for studying superconductivity in fullerenes [12] and, particularly, for investigation of the Hebel–Slichter coherence peak. This peak appears in the temperature dependence of $1/T_1 T$ (where T_1 is the nuclear spin–lattice relaxation time) for s-wave superconductors as the temperature is lowered below T_c. For strong coupling, however, there is no increase in $(T_1 T)^{-1}$ below the superconductivity transition; that is, the Hebel–Slichter peak is absent.

The very first NMR experiment performed by Tycko et al. [128] showed, at first, inconsistent results. The reduced energy gap obtained from this experiment was found to be in agreement with weak coupling BCS theory. At the same time, the authors did not observe the Hebel–Slichter peak but a monotonic dropoff of $1/T_1 T$, which suggested the strong coupling limit. This NMR measurement was performed at a value of the applied magnetic field $\mu_0 H = 9.4$ T. Later on, Kiefl et al. [137] from µSR measurements showed that there is a well-pronounced coherence peak at $\mu_0 H = 1.5$ T. This result is shown in Figure 16.18. Stenger et al. [131,140] from their NMR measurements on $Rb_2 CsC_{60}$ showed that the strength of the applied magnetic field was a very important parameter. The Hebel–Slichter peak is strongly suppressed by magnetic fields larger than a few tesla while, as shown in Figure 16.19, at $\mu_0 H < 3$ T this peak is well pronounced. This peak in FSs was also observed in later NMR ^{13}C experiments [141,142], ^{1}H relaxation in $(NH_3)_x NaRb_2 C_{60}$ [143], µSR [139], and some others.

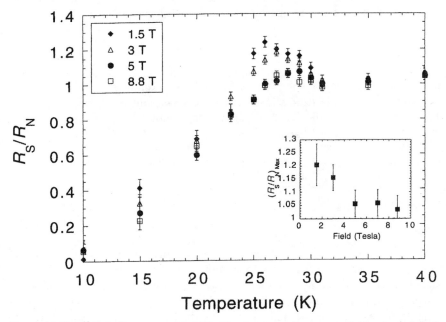

Figure 16.19 ^{13}C R_s/R_n versus T of Rb$_3$C$_{60}$ for several magnetic fields. *Inset*: Maximum of R_s/R_n versus applied fields. Reprinted with permission from Ref. 131.

The mechanism whereby the peak is suppressed, however, was not clearly understood. The effect might be related to (1) the gap anisotropy, which broadens the singularity in the BCS density of states and thus reduces the Hebel–Slichter peak, (2) time reversal symmetry breaking, such as magnetic impurities or applied magnetic field, or (3) damping effects in the superconducting state. All these mechanisms were discussed in detail by Pennington and Stenger [12] and by Choi [144]. Choi [144] showed that the last mechanism is the most important. The author concluded that frequency-dependent Coulomb interactions between conduction electrons induced substantial damping in the superconducting state and, consequently, suppressed the NMR coherence peak in FSs.

16.5.6 Isotope Effect

Substitution of atoms by their isotopes provides direct information about the role, if any, played by phonons in the pairing mechanism for superconductivity. The transition temperature is expected to be sensitive to the isotope substitution and in BCS theory it is proportional to the mass ratio of the isotopes as $T_c \propto (m/m')^\alpha$ with $\alpha = 0.5$. In the A$_3$C$_{60}$ compounds isotopes can be substituted either in the alkali metal subsystem or C atoms in fullerene molecules. Because the intramolecular phonons are expected to be dominant in the pairing

TABLE 16.5 **Experimental Values of the Isotope Shift Exponent from Various Experiments**

α	Comments and References
0.30 ± 0.06	For K_3C_{60} [145] and Rb_3C_{60} [146], full substitution (99%)
0.8	For Rb_3C_{60} [146], partial molecular substitution (50%)
1.4 ± 0.5	For Rb_3C_{60} [252], partial atomic substitution (33%)
1.2 ± 0.2	For K_3C_{60} [253], partial atomic substitution (60%)
2 ± 0.25	For Rb_3C_{60} [253], partial atomic substitution (60%)
1.45 ± 0.3	For Rb_3C_{60} [129], partial atomic substitution (82%)
0.37 ± 0.05	For Rb_3C_{60} [254], partial atomic substitution (10%) and molecular substitution (15%)

Source: From Ref. 17.

mechanism, the isotope $^{12}C/^{13}C$ effect could be more important. Substitution of the carbon atom isotopes can be done in three ways. The first one is a complete substitution so that all molecules are $^{13}C_{60}$. Two other types of substitution are partial substitution, which can be realized either when each C_{60} molecule consists of different isotopes (single mass peak distribution), which is atomic substitution, or when there are two "different" fullerenes $^{12}C_{60}$ and $^{13}C_{60}$ present (two mass peak distribution), which is molecular substitution.

For the case of complete (99%) substitution, α was reported to be $\alpha = 0.3$ for both K_3C_{60} [145] and Rb_3C_{60} [146]. This value is smaller than that predicted by BCS theory. Schlüter et al. [147] theoretically calculated α assuming that electrons couple to intramolecular phonons and obtained $\alpha \approx 0.3$. Gunnarsson et al. [148] within Eliashberg theory obtained $\alpha = 0.32$ and 0.37 for K_3C_{60} and Rb_3C_{60}, respectively. Therefore, the isotope effect for the complete substitution is well understood and can be described by a weak coupling mechanism.

The situation with partial substitution is much more complicated. The experimental results are summarized in Table 16.5. For partial atomic substitution, in $Rb_3(^{13}C_{0.56}{}^{12}C_{0.45})_{60}$ [145], α was found to be consistent with the results shown above in $Rb_3{}^{13}C_{60}$ (99%). However, for partial molecular substitution, $\alpha \approx 0.8$ was reported [146]. Such a large value of α is very unusual. However, other groups found α values even up to 2 (see Table 16.5) and different experiments give a wide distribution of results. Thus, the isotope effect in partially substituted samples remains very unclear and if large α is confirmed by further experiments, new theoretical models will have to be invoked in order to explain this phenomenon.

Substitution of the alkali metal isotopes does not lead to a shift of the superconducting transition temperature [149–151]. This suggests that alkali metal phonons are weakly responsible, or not responsible at all, for the coupling. Experimental results on the isotope effect show that the main mechanism for superconductivity in FS is electron–phonon coupling with a relatively large electron–phonon coupling constant. The observation of a ^{13}C isotope effect on

T_c indicates that intramolecular vibration modes are involved in the pairing mechanism.

16.6 MAGNETIC PROPERTIES OF FS

The first experiments on K_3C_{60} [152] and Rb_3C_{60} [100,153] established that alkali-doped fullerenes were "strong" type-II superconductors and that their main superconducting parameters, the Ginzburg–Landau parameter κ, the penetration depth λ, the coherence length ξ, the thermodynamic field H_c, and the critical fields H_{c1} and H_{c2}, were very similar to those of the high-T_c oxides.

If a type-II superconductor is subjected to a small magnetic field $H < H_{c1}$, the field is completely expelled. Shielding currents, which flow in the surface of the superconductor, prevent any penetration of flux. The superconductor is in the Meissner state and behaves like a type-I superconductor. When the external magnetic field increases beyond the lower critical field H_{c1}, flux penetrates the sample in the form of quantized flux lines (vortices). This happens when the vortex energy is smaller than the magnetic energy associated with the current flow shielding the superconductor. At $H > H_{c2}$, the induction B inside the sample is equal to the external field H and the superconductivity is completely destroyed. Type-II superconductors in the mixed state ($H_{c1} < H < H_{c2}$) are no longer ideal diamagnets.

The temperature dependence of the magnetization illustrates these features; that is, the zero-field-cooled (ZFC) curve represents a shielding of the magnetic field and the field-cooled (FC) curve represents the Meissner effect. A typical $M(T)$ dependence for a K_3C_{60} crystal and for the same crystal crushed into powder is shown in Figure 16.20. A large difference between the ZFC and FC curves and a strong hysteresis in the magnetic field dependence of the magnetization at fixed temperature (Fig. 16.21) indicate pinning of the magnetic vortices. The fact that the FC signal of the K_3C_{60} crystal (Fig. 16.20a) is very small and lies close to the zero-magnetization line shows that pinning in the sample is extremely strong and that there is almost no expulsion of the magnetic field. It is interesting to note that the FC curve for the same sample, after it was crushed (Fig. 16.20b) has smooth behavior with a stronger diamagnetic signal. This is most probably because the average grain size in the powder (in this case it was ~6 μm) is of the order of the penetration depth and, therefore, pinning in these small grains appears to be much weaker.

16.6.1 Critical Fields and Characteristic Lengths

Two fields, the lower critical field, H_{c1}, and the upper critical field, H_{c2}, are important for the characterization of a type-II superconductor. Both fields can be determined from magnetization measurements. In order to evaluate parameters commonly used to characterize the mixed state of type-II superconductors, the Ginzburg–Landau relations [97] can be applied:

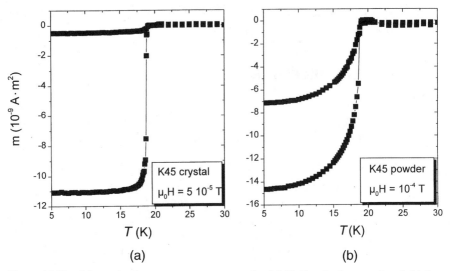

Figure 16.20 Magnetization versus temperature for (a) K_3C_{60} single crystal and (b) for the same crystal crushed into powder.

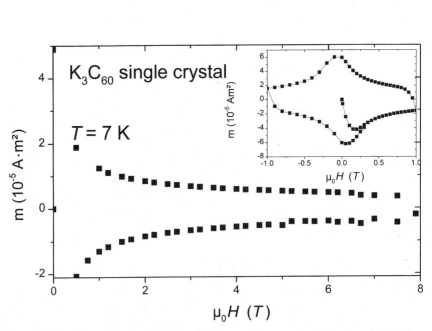

Figure 16.21 Magnetization in K_3C_{60} single crystal versus applied magnetic field up to 8 T at $T = 7$ K. Inset shows "full" hysteresis curve up to 1 T.

$$\mu_0 H_{c2} = \frac{\Phi_0}{2\pi\xi^2}; \quad \mu_0 H_{c1} = \frac{\Phi_0}{4\pi\lambda^2} \ln\kappa; \quad \kappa = \frac{\lambda}{\xi} \tag{16.4}$$

with $\Phi_0 = h/2e = 2 \times 10^{-15}$ Wb, where Φ_0 is the flux quantum, h is Planck's constant, and e is the electron charge. This approach is valid for all temperatures in the dirty limit.

16.6.1.1 Upper Critical Field Many experiments were made to determine the upper critical field for crystalline [78,81,82,136,154,155] and powdered [100,152,153,156–165] fullerenes as well as for thin films [166] using different techniques such as magnetization [100,152,153,159,165], ac susceptibility [157,162], transport [166,164], and rf absorption [163]. Almost all measurements were done on K_3C_{60} and Rb_3C_{60} and only a few results are available for $RbCs_2C_{60}$ [159], K_2CsC_{60}, and Rb_2CsC_{60} [170] and Ba_4C_{60} [167].

Usually, $H_{c2}(T)$ obtained in FS is linear at temperatures not far below T_c except for small fields where a positive upturn is observed. The nature of this upturn will be discussed later. Due to the rather large values of H_{c2} and the experimental limitations of the magnetic field window (5–8 T in superconducting quantum interference device (SQUID) magnetometers), measurements of the upper critical field are usually performed at temperatures close to the transition temperature. The extrapolation of $H_{c2}(T)$ to zero temperature is subject to a large uncertainty and depends on the fitting scheme. The standard theory by Werthamer–Helfand–Hohenberg (WHH) [168] is usually employed. This theory predicts an $H_{c2}(T)$ dependence, which roughly follows a power law $h = 0.6(1 - t^{\alpha})$, where $t_1 = T/T_c h = H/H_{c2}$ and $\alpha \approx 0.75$. $H_{c2}(0)$ can be evaluated from the slope, $\mu_0 H'_{c2} = \mu_0 \delta H_{c2}/\delta T$, of the linear dependence near T_c by the relation

$$\mu_0 H_{c2}(0) = 0.69 T_c \left. \frac{|\mu_0 \delta H_{c2}|}{\delta T} \right|_{T_c} \tag{16.5}$$

In order to verify the applicability of this relation to C_{60}-based materials, H_{c2} was measured at high magnetic fields in several experiments [161–163]. Good agreement of the experimental data with the WHH prediction was obtained [161,163], demonstrating that this theory is successful in describing fullerene superconductors (Fig. 16.22). However, Boebinger et al. [162], who performed measurements on K_3C_{60} powder, found an enhancement of $H_{c2}(T)$ compared to theory. The authors proposed that the deviations could be attributed to flux motion in the superconducting powder, although other intrinsic mechanisms might also play a role. Additional enhancements of H_{c2} at low temperatures could result from strong electron–phonon coupling, which can lead to a relatively large increase of $H_{c2}(0)$, such that the temperature dependence of H_{c2} becomes roughly linear [169]. Fermi-surface anisotropy [49] can also result in low-temperature enhancements of the upper critical field. Another

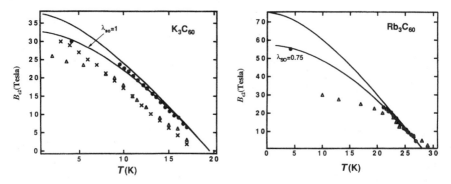

Figure 16.22   Upper critical field against temperature for K_3C_{60} powder and Rb_3C_{60} powder. The low-field data are taken in dc fields and the data points at 4.2 K are taken in pulsed fields. The lower solid curves are fits to the data ($\bullet$) including Pauli paramagnetic limiting and the upper curves assume no paramagnetic limiting. Both experimental dependencies follow roughly the WHH prediction [168]. After Ref. 163.

mechanism proposed by Boebinger et al. [162], which we consider to be more likely, is the dirty-limit effect. In the dirty limit, the Ginzburg–Landau coherence length ξ_{GL} is expected to be limited by the mean free path l, $\xi_{GL} \approx (\xi_0 l)^{1/2}$, where ξ_0 is the BCS coherence length. Thus, in C_{60}-based superconductors, high upper critical fields might partly result from a reduction of l [162] by different types of defects. One of these could be orientational disorder between adjacent C_{60} molecules [53].

This limitation from the mean free path could be the reason for the strong scatter of experimental data for $H_{c2}(0)$ and H'_{c2}. For example, for K_3C_{60} H'_{c2} varies from -2 to -5.5 T/K and for Rb_3C_{60} from -2 to -3.9 T/K (see Table I in Ref. 15). Buntar et al. [155] performed detailed magnetic investigations on K_3C_{60} single crystals with volumes between 1 and 6 mm^3; that is, their dimensions were much larger than their penetration depth. Experiments were done on samples of different quality with various degrees of nonsuperconducting imperfections in order to determine the influence of these factors on the superconducting properties. The upper critical field obtained was $\mu_0 H_{c2}(0) = 28$ T (compared to 17 T $< \mu_0 H_{c2}(0) < 49$ T, see Table 1 in Ref. 15) and no influence of sample quality on $H_{c2}(T)$ was found. Relation (16.4) leads to $\xi(T = 0) = 3.4$ nm (compared to 2 nm $< \xi < 4.5$ nm, Table 1 in Ref. 15). These values are very close to those obtained by Johnson et al. [161] on powder samples. The slope of -2.1 T/K is close to that obtained by Boebinger et al. [162].

One further interesting effect, that we would like to discuss briefly here, is the upturn of $H_{c2}(T)$ at temperatures very close to T_c. This upturn was observed in almost all experiments on all superconducting compounds [81,82,152,153,157, 158,160,162,170,171]. Different authors suggested different explanations for this effect, such as a slight variation in the local T_c [152], a crossover from three- to two- dimensionality [157], and sample imperfections [166]. In our opinion, the

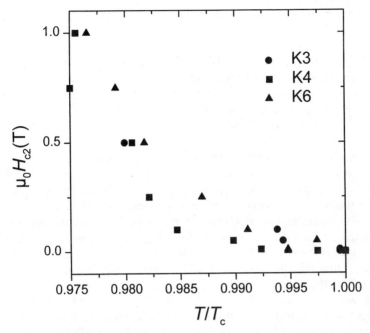

Figure 16.23 Temperature dependence of the upper critical field close to the transition temperature for K_3C_{60} crystals with different shielding fractions. Symbols ● and ▲ are for samples with 100% shielding; ■ symbol is for 65% shielding. Reprinted with permission from Ref. 155.

last explanation is not very likely because the effect has been seen in samples with superconducting fractions between 1% and 75%, that is, with strongly varying degrees of imperfections. Buntar et al. [155] performed detailed mea surements of the upper critical field close to T_c in K_3C_{60} crystals of different quality with shielding fractions ranging from 25% to 100%. The effect was found in all samples irrespective of the degree of imperfections (Fig. 16.23). It was proposed that the upturn is a consequence of the anisotropy of the Fermi surface for fullerene superconductors [49] and a preliminary experiment [72] showed an influence of the anisotropy on critical fields. Strong effects of anisotropy on magnetic properties of conventional superconductors, specifically on $H_{c2}(T)$ dependence, are well known [172]. For example, anisotropy effects on the magnetic properties of superconducting niobium were analyzed [173] and very good agreement between theory and experiment was found.

16.6.1.2 Lower Critical Field More than 10 different methods are available for measuring H_{c1} (see discussion in Ref. 15). Each one of these methods requires a model. Hence, the reliability of the results depends substantially on how well the relationship between H_{c1} and the measured quantities can be established.

The first results on H_{c1} for K_3C_{60} [152], Rb_3C_{60} [100,153], $RbCs_2C_{60}$ [159], and Ba_4C_{60} [167] were obtained by a dc-magnetization technique. $H_{c1}(\text{zero})$ evaluated from these measurements lies in the range $\mu_0 H_{c1}(0) = 10\text{--}16$ mT. From $H_{c1}(0)$, the penetration depth λ is evaluated using the value of the coherence length known from independent measurements and Eq. (16.4). The magnitude of λ obtained in this way is of the order of 200–250 nm.

In these experiments the lower critical field was defined as the field at which a deviation from linearity in $M(H)$ first appeared. Indeed, an ideal superconductor exhibits linear $M(H)$ behavior up to H_{c1}, where a sharp cusp occurs. However, it is extremely difficult to obtain the point of first deviation from such a curve, since the deviations themselves are very small. In order to make this procedure more quantitative, it is tempting to apply Bean's critical state model [174] for the entry of vortices into hysteretic superconductors. According to this theory, the magnetization is related to the critical current density J_c (which is assumed to be field independent for simplicity), at fields above H_{c1} by $(M + H) \approx (H_a^2 - H_{c1}^2)/J_c D$, where D is a characteristic length for the sample geometry studied. This relation holds in the range $H_{c1} < H < H^*$, where $H^* \approx J_c D$ is the field at which flux completely penetrates the sample. This analysis was performed for $Rb_3 C_{60}$ [175] and $RbCs_2 C_{60}$ [159]. In Buntar et al. [175], the data obtained from Bean's analysis showed much smaller values of H_{c1} and the authors related their data to the field at which breaking of intergranular Josephson junctions occurs.

Politis et al. [158] and Buntar et al. [159] used a third method, based on measurements of the reversible magnetization at high external fields, to calculate H_{c1} from the well-known Ginzburg–Landau relations [176]:

$$-M = \frac{H_{c2}(T) - H}{(2\kappa^2 - 1)\beta_a}; \quad -M = \frac{\alpha \Phi_0}{8\pi \mu_0 \lambda^2(T)} \ln\left(\frac{\beta H_{c2}(T)}{H}\right) \quad (16.6)$$

for high and intermediate magnetic fields, respectively. (β and β_a are coefficients and can be found in Ref. 260.) This analysis led to lower critical fields, which were slightly smaller but comparable to the data obtained by the first two methods. The last method, being based on direct measurements of the reversible magnetization, is more accurate for the determination of the lower critical field. However, the resulting values of the Ginzburg–Landau parameter κ and, hence, of H_{c1} depend strongly on the value of the superconducting fraction assumed (see Ref. 159, Eq. (4)). This leads to a large uncertainty in quantitative calculations, especially in the case of powdered samples with a large distribution of grain sizes.

In order to avoid the difficulties associated with the three previous methods, another way to evaluate H_{c1} must be found. In Buntar et al. [67], H_{c1} is determined through measurements of the trapped magnetization. This method is far more accurate than measurements of δM because of the cancellation of a large linear contribution [177]. It is based on the fact that trapped magnetic flux, M_t, can be built up in a sample only when the field has increased beyond H_{c1}. The

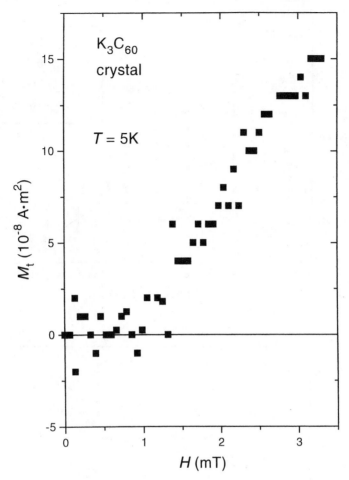

Figure 16.24 Magnetic field dependence of the trapped magnetization M_t at $T = 5$ K for a K_3C_{60} crystal. Reprinted with permission from Ref. 67.

advantage of this method for type-II superconductors with strong pinning was illustrated in Ref. 177, where the $M(H)$ behavior was shown to be quite linear in the vicinity of H_{c1}, whereas $M_t^{1/2}$ against H showed a well-resolved kink at the field corresponding to H_{c1}.

Such measurements were performed [67] on Rb_3C_{60} and K_3C_{60} crystals and $RbCs_2C_{60}$ powder. The magnetic field dependence of M_t at $T = 5$ K for a K_3C_{60} single crystal with a shielding fraction of 100% is shown in Figure 16.24. The ac measurements indicate that there is no granularity for current flow in the K_3C_{60} single crystal; that is, there should be no influence from weak links or intergranular boundaries. The $H_{c1}(T)$ dependence obtained in these experiments for all three compounds is shown in Figure 16.25. The penetration depth

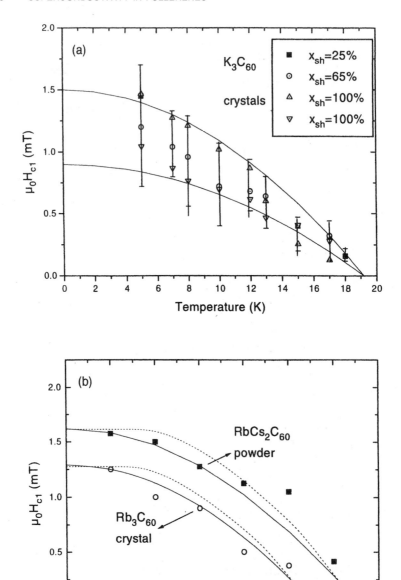

Figure 16.25 Temperature dependence of the lower critical field of (a) K_3C_{60} crystals and (b) Rb_3C_{60} crystal (open circles) and $RbCs_2C_{60}$ powder (solid squares). Solid lines correspond to the parabolic law $H_{c1}(T) = H_{c1}(0)[1 - (T/T_c)^2]$ and dashed lines correspond to the weak coupling BCS theory. Reprinted with permission from Ref. 67.

λ obtained from the measured H_{c1} using Eq. (16.4) was found to be 890, 850, and 720 nm for K_3C_{60}, Rb_3C_{60}, and $RbCs_2C_{60}$, respectively.

These values of H_t are very small (no higher than 1 mT at zero temperature) in comparison with those values of $H_{c1} \approx 10\text{--}16$ mT obtained previously from δM measurements. Such small values of H_t are observed for powders and crystals of different quality. Therefore, it is very unlikely that the trapped magnetization appears at small fields because of granularity or imperfections of the samples. It is clear that the lower critical fields of FSs are no higher than the values of H_t observed in these experiments, because magnetic field has clearly penetrated the sample (Fig. 16.24). The smallness of H_{c1} can be explained by the fact that the electron wavefunctions between adjacent C_{60}^{3-} ions overlap relatively weakly (see Section 16.2.2). This weak overlap can easily be destroyed and magnetic field starts to penetrate the sample between the C_{60} molecules.

Small values of the field at which the trapped magnetization appeared, similar to the above results, were also observed by Kraus et al. [178], who measured the irreversible $M(T)$ curves. In addition, in several publications small values of the penetration field were obtained [68,175,179] but were attributed to breaking of intergranular coupling. Moreover, in μSR measurements of λ [180,181] the penetration depth was found to be 480 nm for K_3C_{60} and 420 nm for Rb_3C_{60}, which leads to $H_{c1}(0) = 4.0$ and 4.9 mT, respectively. The results obtained by Buntar et al. [67] are in very good agreement with optical [134] and with NSR measurements [182]. Considering the indirect evaluation procedure employed to obtain λ from magnetization [67] and μSR [180,181] measurements, the difference between these results is not very large. This difference could be related to the fact that the μSR experiments were done on powder samples with an average grain radius r of the order of the penetration depth. The distribution of the magnetic field (which is actually measured by μSR) in grains with $r \approx \lambda$ is effectively "cut off" at least at half the amplitude. Therefore, the effective "μSR penetration depth" should be shorter than the real one. On the other hand, the distribution of the magnetic field in a *system* of vortices with an intervortex distance much shorter than the penetration depth could be calculated more carefully. However, models have to be used for such a calculation, and this can lead to a discrepancy between results with a factor of 2. From very recent μSR experiments [139] λ was estimated to be in the range 300--700 nm, with the upper limit close to the values obtained from optical, magnetic, and NMR measurements.

16.6.1.3 Coherence Length and Penetration Depth

Because the average grain size r of powder samples is comparable to the penetration depth λ, grains are penetrated by magnetic flux at all fields. Therefore, a correction $r/\lambda(0)$ has to be included when determining the superconducting volume. This was done in Ref. 183 for Rb_3C_{60} powder. The authors fitted theoretical curves to the experimental data of Ref. 158 using three different theories for $\lambda(T)$—the Ginzburg–Landautheory, the BCS theory, and the two-fluid model:

$$\lambda(T) = \lambda_{\mathrm{GL}}(0)\left[1 - \left(\frac{T}{T_c}\right)^n\right]^{-1/2} \tag{16.7}$$

with $n = 1$, 2, and 3, respectively. In the temperature range $5\,\mathrm{K} = 0.17$ $T_c \leq T \leq 0.71\,T_c = 20\,\mathrm{K}$, the two-fluid model was found to be completely inappropriate, and the best fit was achieved with the Ginzburg–Landau temperature dependence, which gave an increase of the superconducting fraction compared to the experimental data by a factor of $2(r/\lambda(0) = 4.8)$. The BCS dependence resulted in a much higher increase of the superconducting fraction, by a factor of 7. The authors, assuming the penetration depth to be of the order of 200–300 nm, found this result unacceptable. However, in light of the data for λ [67], $r(\sim 1\,\mu\mathrm{m})/\lambda(\sim 0.87\,\mu\mathrm{m}) \approx 1$, the BCS result seems to be much better.

$\xi(T)$ and $\lambda(T)$ were evaluated from the experimentally obtained temperature dependence of the upper and lower critical fields [155], and the results are shown in Figure 16.26. The authors [155] also fitted the experimental data to the Ginzburg–Landau theory, to the BCS theory, and to the two-fluid model using $\xi(0)$ and $\lambda_{\mathrm{GL}}(0)$ as fit parameters. All three theories describe both $\xi(T)$ and $\lambda(T)$ well. However, this is only true in the high-temperature range shown in Figure 16.26, since the fit parameters $\xi_{\mathrm{fit}}(0)$ and $\lambda_{\mathrm{fit}}(0)$ are quite different. Both the coherence length and the penetration depth at zero temperature, obtained from the BCS fit, which is shown by dashed line in Figure 16.26 ($\xi_{\mathrm{fit}}(0) = 3.8$ nm and $\lambda_{\mathrm{fit}}(0) = 887$ nm), agree well with those obtained from $H_{c2}(0)$ and $H_{c1}(0)$, $\xi(0) = 3.4$ nm, and $\lambda_{\mathrm{GL}}(0) = 870$ nm. The Ginzburg–Landau fit is also close. The fit parameters, obtained from the two-fluid model, do not agree at all with these results.

It can be concluded, therefore, that the best fit for $\xi(T)$ and $\lambda(T)$ is the BCS one and that the Ginzburg–Landau theory also describes the data reasonably well. The two-fluid model, in agreement with the results in Ref. 183, is found to be completely inappropriate.

16.6.2 Flux Pinning and Related Topics

Vortex lines in an ideal type-II superconductor are not fixed in the crystal lattice and their motion starts at any nonzero value of the transport current J. The movement of vortices under a Lorentz force, F_L, leads to dissipation of energy [184]. In nonideal type-II superconductors, the vortex lines are fixed by crystal structure defects. This phenomenon is called pinning. The movement of vortices in these superconductors starts at the critical current density J_c. From a thermodynamic point of view, pinning in a type-II superconductor means that the free energy of the vortices depends on their position in the sample. The vortex lines can be pinned by different structural defects, which are called pinning centers. They can be dislocations, grain boundaries, inhomogeneities in the sample, twin boundaries, and so on [185]. The pinning force in a type-II superconductor is the main parameter affecting all important superconducting properties for applications, such as critical current density and irreversibility line.

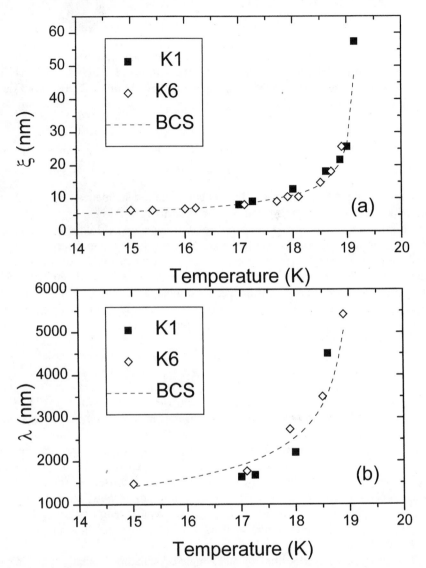

Figure 16.26 (a) Coherence length and (b) penetration depth versus temperature. Open and closed symbols correspond to two different K_3C_{60} crystals. The dashed line shows the BCS dependence with fit parameters $\xi_{fit}(0) = 3.8$ nm and $\lambda_{fit} = 887$ nm. Reprinted with permission from Ref. 155.

16.6.2.1 *Critical Current Density*

In a nonideal type-II superconductor the vortex line begins to move when the Lorentz force F_L is strong enough to overcome the pinning force F_p. Therefore, energy dissipation occurs when $F_L > F_p$. The pinning force can be written as

$$\vec{F}_p = \vec{B} \times \vec{J}_c \qquad (16.8)$$

where J_c is the critical current density. The critical current density decreases in general monotonically with increasing temperature and magnetic field. For typical oxide superconductors, for instance, $J_c(T)$ exhibits a rather rapid decrease with increasing T:

$$J_c = J_c(0) \left[1 - \left(\frac{T}{T_c} \right)^2 \right]^n \qquad (16.9)$$

with an exponent $n = 3$–4 [186]. The magnetic field dependence $(H \ll H_{c2})$ of the critical current density is often proportional to $1/B$.

All information about the temperature and magnetic field dependence of J_c in the fullerene superconductors has been obtained from magnetization, which avoids problems with direct transport measurements on bulk samples. Measurements of the magnetization at various magnetic fields and temperatures were performed to obtain J_c using Bean's critical state model [174]. From an experimental point of view, J_c can be determined simply from the dc magnetization curve (Fig. 16.21) by the following equation [174]:

$$J_c = A \frac{M_+ - M_-}{R} \qquad (16.10)$$

In Eq. (16.10) A is a coefficient dependent on the sample geometry [174,187], M_+ and M_- denote the magnetization measured in increasing and decreasing fields at a certain magnetic field, and R is the region screened by supercurrent. Lee at al. [188] and Buntar [189] made ac and dc magnetic measurements to evaluate J_c and showed that the dc and ac results matched reasonably well. The authors argue that the Bean critical state model can be used to describe fullerene superconductors. Some values of J_c for fullerene powders are crystals obtained in this way are shown in Table 16.6. Typical temperature and magnetic field dependences obtained in K_3C_{60} bulk crystal are shown in Figures 16.27 and 16.28.

The critical current densities of bulk single crystals are typically two orders of magnitude smaller than those of powders (see Table 16.6). If we compare them with those of $YBa_2Cu_3O_{7-\delta}$ single crystals, the material with the lowest anisotropy of all high-T_c superconductors, we find again that the critical current density J_c of the FSs is significantly smaller (by about a factor of 100, see, e.g., Ref. 190). This represents a serious problem, because flux pinning depends only on a few intrinsic parameters, such as the thermodynamic critical field, H_c, and the coherence length, both of which are comparable in $YBa_2Cu_3O_{7-\delta}$, and on the size and the density of the defects. The defect size can hardly be much smaller than the lattice parameter and, therefore, has to be comparable to ξ in the fullerenes. This would leave only a small density of defects as an explana-

TABLE 16.6 Values of the Critical Current Density Obtained from Magnetic Measurements on Powders and Crystals

Compound	R (mm)	J_c (A/m^2)	ΔM (10^3 A/m) 5 K, 0.5 T	ΔM (10^3 A/m) 5 K, 0.1 T	Reference	m^a (Eq. (16.11)) and Reference
K$_3$C$_{60}$	10^{-3}	10^9	0.7	1.97	157	1.47 [205]
	10^{-3}	1.2 × 10^9			152	1.5 [73]
	10^{-3}	~10^9	~0.7		179	
	10^{-3} to 3 × 10^{-4}		3.4	7.63	191	
	1	~10^7	14.6	25.33	68	
	1	6 × 10^7	9.3	25	155	
	1	4 × 10^7	7.1	18.5	155	
Rb$_3$C$_{60}$	10^{-3}	4 × 10^{10}	0.68	0.97	153	1.59 [205]
	10^{-3}	1.5 × 10^{10}	0.23	0.32	188	1.8–2
	10^{-3}	2 × 10^{10}			165	[255]
	10^{-3}	~10^{10}			179	
	10^{-3} to 3 × 10^{-4}		1.72	3.71	191	
	1	~10^7	19.17		68	
	1	6 × 10^8	21.9		71	

^{a}This is the experimentally obtained irreversibility line exponent.

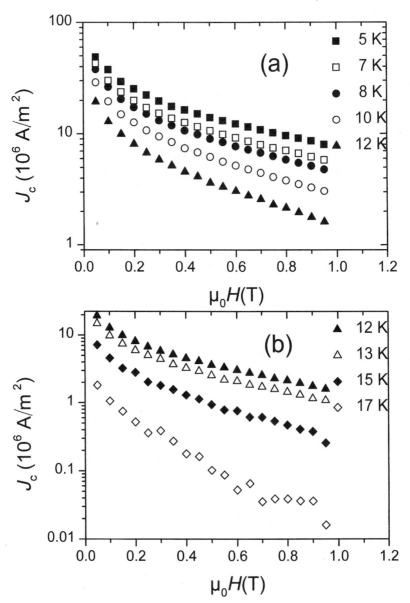

Figure 16.27 Magnetic field dependence of the critical current density at fixed temperatures: (a) 5, 7, 8, 10, and 12K; and (b) 12, 13, 15, and 17 K. Reprinted with permission from Ref. 155.

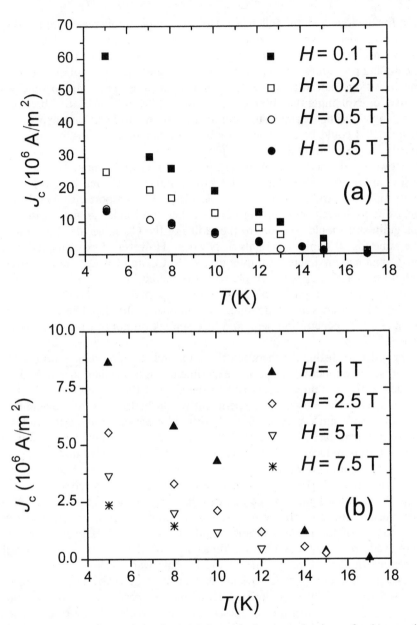

Figure 16.28 Temperature dependence of the critical current density at fixed magnetic fields: (a) 0.1, 0.2, and 0.5 T; and (b) 1, 2.5, 5, and 7.5 T. Reprinted with permission from Ref. 155.

tion for the low J_c in the fullerenes. However, typical fullerene samples are far from being perfect single crystals and are expected to have many defects, especially since the C_{60} lattice is damaged by diffusion of alkali metal atoms during the doping procedure. Therefore, pinning in single crystals is not expected to be much weaker than in powders or in high-T_c crystals. The only possible alternative for explaining the observed low J_c would be an "automatic" breakup of the sample into many small superconducting grains. In fact, some experimental results [191] could be explained *only* by the assumption of an "intrinsic granularity" in FSs with grain sizes ≤ 1 µm.

However, this assumption of "intrinsic granularity" cannot be understood easily, unless we assume "molecular superconductivity," that is, superconductivity that is restricted to the C_{60} molecules, and transport currents that one only due to Josephson coupling between these molecular grains. In this case, the grain size should be equal to the radius of the C_{60} molecule and should be the same for all samples, crystals or powders. However, if experimental results on J_c obtained on different samples of both K_3C_{60} and Rb_3C_{60} are compared, a certain increase of J_c with sample size is observed (Table 16.6), although not exactly inversely proportional, as would be expected from Bean's critical state model. Moreover, some experimentally obtained properties like nonzero H_{c1}, large (~ 0.1 T) penetration field, and flux pinning do not fit to the "molecular" model.

In order to distinguish between the bulk and the molecular nature of fullerene superconductivity, special experiments and evaluation methods are needed, which are directly sensitive to the sizes of the regions screened by the supercurrent. The most direct experiment would be to measure the magnetization of a bulk single crystal ($R \gg \lambda$) without structural granularity and weak links, and then to crush the crystal. If the crystal is a true bulk superconductor, then this only reduces the "grain size." From a comparison of the calculated J_c values before and after crushing, any intrinsic granularity could be detected.

In Buntar et al. [192] single crystals of the fullerene superconductor K_3C_{60} were investigated by dc and ac magnetic techniques. Several evaluation methods (such as the slope of the hysteresis loop upon reversing the field, the field dependence of the trapped moment and of the trapped magnetization) were employed to obtain the relevant dimension of the crystals for unimpeded supercurrent flow. Finally, one of the crystals was crushed into powder. A consistent description of the results can only be achieved by invoking *bulk* superconductivity in the fullerenes. Discrepancies pertaining to the seemingly low critical current densities in these materials are also removed. The critical current density of K_3C_{60} single crystal with well-defined R (from Eq. (16.10)) was found to be $J_c = 1.26 \times 10^9$ A/m^2. The data presented in Buntar et al. [192] are, therefore, suited to rule out "exotic" forms of superconductivity in the fullerenes and to establish its bulk character in an unambiguous way.

16.6.2.2 Irreversibility Line The irreversibility line $H_{irr}(T)$, which separates the regions in the H–T diagram, where the magnetization is reversible from the region where irreversible effects are strong, has been the subject of detailed investigations. $H_{irr}(T)$ must be known for applications, because the irreversibility line indicates the values of the temperature and the magnetic field, above which the critical current density is zero and the material cannot be used for current transport. From the point of view of fundamental research, $H_{irr}(T)$ provides us with important information about pinning in the sample and about structural features of the vortex system. Many experimental results show that the irreversibility line can be well described by

$$H_{irr} = H_0 \left(1 - \frac{T}{T_c} \right)^m \qquad (16.11)$$

where H_0 is the value of H_{irr} at zero temperature. For the model of thermally activated flux motion with collective pinning [193–195], m in Eq. (16.11) is equal to 1.5, while for vortex–glass models [196–198] $m = \frac{4}{3}$ and for melting theories [199–204] $m = 2$.

Experimentally, the irreversibility line is usually determined from $M(T)$ measurements as the points where the ZFC and the FC curves merge (as used for K_3C_{60} and Rb_3C_{60} powders in Ref. 205), or as the points at which the current densities drop below a certain value. For fullerenes, the irreversibility lines obtained by these methods were found to be well described by Eq. (16.11) with values of m between 1.47 and 2.15 (see Table 16.6). However, the determination of the irreversibility line from the merging point of either $M(T)$ or $M(H)$ curves is as complicated as the determination of the lower critical field from the first deviation from linearity (see the discussion in Section 16.6.1.2). Another method was used in Ref. 155, where $H_{irr}(T)$ was obtained as the field at which the critical current density dropped sharply to zero. In this case, a strong kink in the linear $1/J_c(H)$ dependence appeared, as shown in Figure 16.29. The irreversibility line obtained in this work is shown in the inset of Figure 16.29 and m for the K_3C_{60} single crystal was evaluated to be 1.5.

In summary, because of the small number of experimental data, it is too early to give any final conclusions about the nature of the irreversibility line in fullerene superconductors.

16.6.2.3 Neutron Irradiation of Single K_3C_{60} Crystals

Irradiation of superconductors could lead to a large increase in values of the critical current density [206]. Very recently, measurements of the magnetic properties of bulk superconducting K_3C_{60} crystals, as a function of defect structure after irradiation with neutrons, were performed [207] in order to investigate the maximum possible critical current density and its temperature and magnetic field dependence. Two crystals with sizes ~1 mm^3 were irradiated with neutrons. One of the crystals was of "bad" quality with only 66% shielding fraction. The other had 100% superconducting fraction and no intergranular dissipation peak in $\chi''(T)$. While no influence of irradiation was observed for the "bad" crystal, irradiation of the "good" one with neutron fluxes 0.5×10^{21} m^{-2}, 1×10^{21} m^{-2},

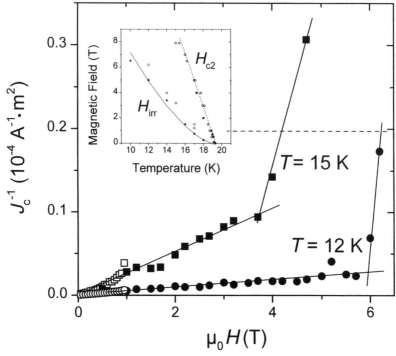

Figure 16.29 Magnetic field dependence of the reciprocal critical current density at $T = 15$ K (squares) and $T = 12$ K (circles). The dashed line shows the resolution limit of the experimental device. The inset shows the irreversibility line, evaluated from $1/J_c \sim H$, for two different K_3C_{60} crystals. The irreversibility line is compared to the H_{c2} versus T dependence. Reprinted with permission from Ref. 155.

and 2×10^{21} m^{-2} ($E > 0.1$ MeV) led to an increase in the critical current density by 1.45, 1.66, and 1.98 (at $\mu_0 H = 4$ T), respectively (Fig. 16.30). Irradiation at the lowest flux did not produce any influence on the irreversibility line $H_{irr}(T)$. However, irradiation at higher fluxes *decreased* the irreversibility line (Fig. 16.31). This effect is in contrast to results on cuprates.

These results are very preliminary and in order to give an explanation of the "negative" influence of the irradiation on the irreversibility line, despite an increase in J_c at the same flux, further investigations of the structural defects in fullerene superconductors, before and after irradiation, are required.

16.6.2.4 Magnetic Relaxation One of the most remarkable features of some hysteretic type-II superconductors, reflecting the dynamics of flux motion, is the relaxation of the magnetization $M(t)$ at fixed temperature and magnetic field. This relaxation has been the subject of intensive study because it heavily affects the current-carrying capability of superconductors. Usually, the relaxation is described in terms of the flux-creep activation energy U_0. The

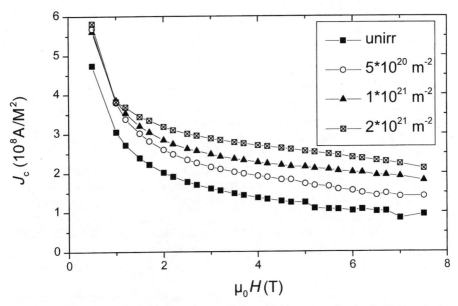

Figure 16.30 Magnetic field dependence of the critical current density in K_3C_{60} single crystal at $T = 7$ K, unirradiated (■) and irradiated with neutron fluences 0.5×10^{20} (○), 1×10^{21} (▲), and 2×10^{21} m^{-2} (⊠). Reprinted with permission from Ref. 207.

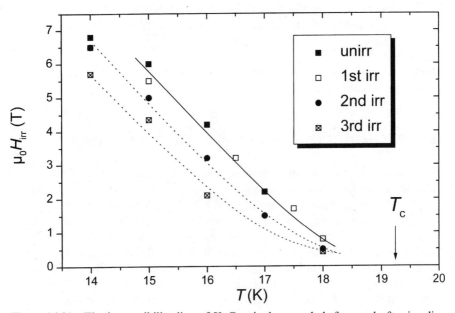

Figure 16.31 The irreversibility line of K_3C_{60} single crystals before and after irradiation. Lines are guides for the eye. Reprinted with permission from Ref. 207.

Anderson–Kim model [208] assumes a uniform barrier U_0 for depinning of vortex bundles, resulting in a logarithmic time dependence of the magnetization:

$$M(t) = M_0 \left(1 - \frac{kT}{U_0} \ln \frac{t}{t_0} \right) \tag{16.12}$$

where M_0 is the unrelaxed value of M and t_0 is a time constant.

Investigation of magnetic relaxation has also been carried out in fullerene superconductors [153,209–211]. In Refs. 153, 209, and 210 the flux-creep activation energy was estimated to be of the order of 10^{-2} eV. However, it should be pointed out that all of these measurements were made on powder samples and that the magnetic relaxation did not follow a logarithmic dependence [209]. Some peaks were even observd in $M(t)$ curves during short-term relaxation [211]. This behavior can be connected to intergranular coupling between grains in powder samples as well as to weak links, which may exist in samples of poor quality. Very recently [73], magnetization relaxation was studied in K_3C_{60} crystals of different quality. It was shown [73] that inhomogeneities in the superconductor strongly affect the relaxation process, possibly masking the logarithmic $M(t)$ dependence, while for samples of good quality the magnetic relaxation followed the $M \sim \ln(t)$ behavior, as shown in Figure 16.32. It was also observed that the flux-creep rate increased progressively with temperature up to, at least, $T = 17$ K $\approx 0.88 T_c$, and that the temperature dependence of the flux-creep activation energy showed a peak at some characteristic temperature T_p (Fig. 16.33), which had a roughly linear field dependence $T_p(H)$. The values of the flux-creep activation energy varied from 10 to 80 meV and are similar to those obtained in oxide superconductors.

Additional investigations of magnetic relaxation will have to be done on samples of good quality and on different compounds, in order to establish the underlying mechanisms.

16.7 MECHANISM OF SUPERCONDUCTIVITY IN FULLERIDES

The most crucial point to understanding the mechanism of superconductivity in fullerides is the Coulomb interaction. Because of the large Coulomb repulsion interaction, A_3C_{60} could be expected to be a Mott–Hubbard insulator [118,212] and the superconducting compounds, therefore, are $A_{3-\delta}C_{60}$. Another model suggests superconductivity being driven by the effective electron–electron attraction [41,213–216]. On the other hand, in case of the electron–phonon pairing mechanism, there should be very strong retardation or other effects to decrease the large Coulomb repulsion between electrons on the C_{60} cage.

Estimates of the Coulomb interaction, U_{mol}, between electrons that are added to the C_{60} molecule give $U_{mol} = 2.7$–3.3 eV (see Ref. 10 and references therein). In the solid, this interaction is reduced due to polarization of the

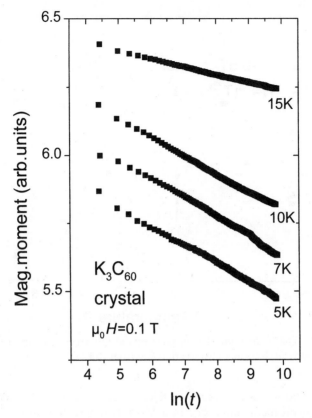

Figure 16.32   Decay of the remanent magnetization m with $\ln(t)$ at different temperatures. Reprinted with permission from Ref. 73.

neighboring molecules and is calculated to be 0.8–1.3 eV [217] or, from Auger spectroscopy experiments, estimated to be $\sim$1.5 eV [118,218].

16.7.1   Metal–Insulator

Such a strong interelectron repulsion could lead to insulating properties of the material. In narrow-band systems in the limit where the ratio of U/W is larger than unity (where W is the energy of the intermolecular interactions), the t_{1u} band is split into two bands and, as a result, Mott–Hubbard insulating behavior is expected. If this is the case for fullerides, where $U/W \sim 2$, the exact A_3C_{60} compounds are insulators and the superconducting compounds are $A_{3-\delta}C_{60}$; where δ could be very small of the order of $\delta \sim 10^{-3}$. Gunnarsson et al. [219–221] theoretically studied the effect of the doping by alkali atoms on the t_{1u} band in A_nC_{60}. From Monte Carlo calculations for a Hubbard-like model, the authors show that frustrations increase the critical ratio of U/W,

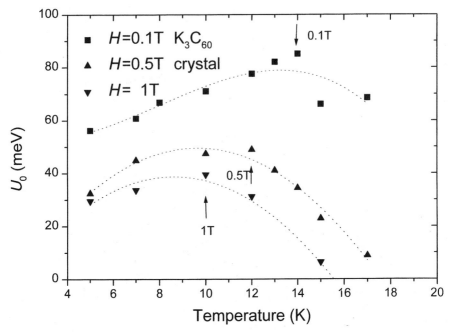

Figure 16.33 The temperature dependence of the flux-creep activation energy for a K_3C_{60} single crystal. Reprinted with permission from Ref. 73.

where the metal-insulating transition takes place, to $U/W \sim 2.5$. This suggests that A_3C_{60} is on the metallic side of a Mott–Hubbard transition. Since A_3C_{60} has a frustrated fcc lattice but A_4C_{60} has a nonfrustrated bcc lattice, this tends to put A_4C_{60} on the insulating side of the transition. This could explain why A_4C_{60} with partly filled t_{1u} band is not a superconductor.

In addition, as discussed in Section 16.5.3, experiments show that the transition temperature drops sharply down as the valence is changed above or below 3; that is, A_3C_{60} is the superconducting stoichiometry. Still, from the asymmetric shape of the peak in Figure 16.17, one could expect that the maximum of T_c appears slightly to one side of $n = 3$. But if A_3C_{60} is an insulator, then T_c should be increasing at n close to, but not equal to, 3, while experimentally it is not so. Both theory and experiment strongly suggest that fullerenes with exactly half-filled t_{1u} band are not Mott–Hubbard insulators but metals.

16.7.2 Electron–Electron Attraction

Chakravarty et al. [41,216,222] suggested a mechanism of superconductivity based on attraction between two electrons localized on a single C_{60} anion. White et al. [216] calculated ground state energies for small Hubbard-model molecules and showed that two electrons added to the molecule can experience an effective attraction despite their bare repulsive interaction. They pointed out that this phenomenon takes place only for intermediate length scales. In this model, the pair-binding energy, E'_{pair}, needs to be positive [216]. In this case, one of the two adjacent molecules has $i + 1$ additional electrons, while the other has $i - 1$ electrons. The mechanism of superconductivity was proposed to be based on this transfer of electrons. However, calculations of the total energy of the C_{60}^{n-} ion [223] show that the pair-binding energy E'_{pair} is always negative, which contradicts the effective attractive interaction between two electrons in a molecule.

The effective attraction was also studied as a function of the range of the interaction [224,225]. It was shown that as the range of the interaction increases toward physically reasonable estimates, the direct interaction rapidly increases, leading to a repulsive effective interaction. The authors found that, in contrast, the proposed pair-binding model needs extended-range interaction to explain the high critical transition temperature. Discussion about the electronic mechanism of superconductivity in fullerides is still open and there are no decisive results proving or disproving an electron–electron pairing mechanism.

16.7.3 Electron–Phonon Coupling

As discussed in Section 16.5, experimental results tend to support the electron–phonon coupling mechanism. Because of the strong Coulomb repulsion interaction, one could expect that there should be strong coupling to keep electron pairs. However, it is widely believed that fullerides are s-wave, weak-coupling BCS superconductors with superconductivity driven by high-energy intramolecular phonons.

For the electron coupling mechanism, a simple estimation of the transition temperature can be obtained from the MacMillan equation [226]:

$$\kappa_B T_c = \frac{\hbar \omega_{ln}}{1.2} \exp\left(-\frac{1.04(1+\lambda)}{\lambda - \mu^*(1+0.62\lambda)} \right) \quad (16.13)$$

where ω_{ln} is a logarithmic averaged phonon frequency, λ is the electron–phonon coupling parameter, and μ^* is the Coulomb pseudopotential, which describes the Coulomb repulsion effects. Within the Eliashberg equation [227], the electron–phonon coupling can be determined as

$$\lambda_{ep} = 2 \int_0^\infty d\omega \frac{\alpha_e^2 F(\omega)}{\omega} \quad (16.14)$$

($\alpha_l^2 F(\omega)$ is a spectral function), and the average phonon frequency is given as

$$\bar{\omega}_{ph} = \exp\left(\frac{2}{\lambda_{ep}} \int_0^\infty \frac{d\omega}{\omega} \alpha_e^2 F(\omega) \ln \omega \right) \quad (16.15)$$

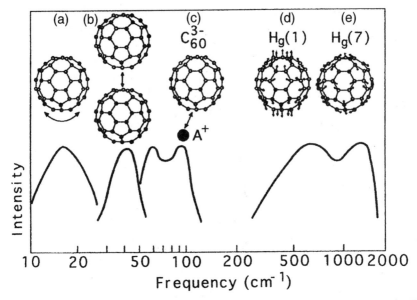

Figure 16.34   Schematic representation of various phonons in A_3C_{60} compounds. The figure shows (a) librations, (b) optic C_{60}–C_{60} phonons (c) A–C_{60} phonons, and (d,e) intramolecular H_g modes. The figure indicates the radial and tangential character of the low-lying and the high-lying H_g modes, respectively. Reprinted with permission from Ref. 18.

From Eqs. (16.14) and (16.15) it can clearly be seen that the higher the phonon frequency, the weaker the coupling and, at the same time, the higher the transition temperature.

Let us briefly consider the phonon spectra of A_3C_{60} fullerides. Phonons in the fulleride solid can be divided into several groups and their frequencies range from 10 to 2000 cm^{-1}. These groups represent different vibration modes and are schematically presented in Figure 16.34. The lowest energy group corresponds to librations with energy 10–25 cm^{-1} (4–5 meV) [228,229]. The intermolecular spectrum has two main peaks that correspond to C_{60}–C_{60} and A–C_{60} vibrations with corresponding frequencies of 35–50 cm^{-1} and 60–120 cm^{-1}. The highest energy modes are intramolecular phonons, both radial and tangential vibrations, with energies between 250 and 1700 cm^{-1}.

In order to overcome the strong Coulomb repulsion between electrons, one could expect the coupling to be strong, that is, according to Eqs. (16.14) and (16.15), coupling to the low-frequency phonons. This would suggest coupling to librations or intermolecular modes.

16.7.3.1   Librations   Antropov et al. [85] within a tight-binding formalism found that the coupling to librations had to be very weak. Experimentally, librations on both sides of the superconducting transition were studied by

Christides et al. [228]. It was expected that at small temperatures broadening of the librations due to decay in electron–hole pairs should go to zero. However, no effect on the width was found at the transition [228] and it was concluded that the parameter for coupling to librations is very small, $\lambda_l < 0.08$.

16.7.3.2 Intermolecular Phonons

Zhang et al. [230] suggested that alkali metal–C_{60} vibrations should lead to a strong electron–phonon coupling. The authors estimated that the alkali phonons could contribute an attractive interaction of the order of 0.9 eV, which corresponds to $\lambda \sim 2$. Isotope effect experiments showed that there was no effect on T_c with the substitution of alkali metal isotopes. This means that alkali modes do not couple or contribute very little to the coupling. This was rationalized in terms of an efficient metallic screening [231], showing that the coupling to the alkali modes is reduced by almost two orders of magnitude.

16.7.3.3 Intramolecular Phonons

As discussed in Section 16.5, experiments showed that, most probably, the coupling is weak and the high-energy intramolecular phonons play the main role in this process. The open question is which modes are dominating in the coupling and how big is the coupling parameter λ for these phonons. Symmetry arguments show [232,233] that only the two phonons with A_g and the eight phonons with H_g symmetry can couple. The contribution from A_g modes is believed to be screened out [147] and estimates within RPA suggest that the coupling to these modes is reduced by one or two orders of magnitude [85]. This conclusion is supported by Raman experiments [234,235], which show no extra broadening of $A_g(2)$ in the metallic phase. Theoretical and experimental estimations of the coupling parameter to different intramolecular H_g modes are given in Table 16.7. There is a rather strong discrepancy between theory and experiments for λ_ν (where λ_ν is the coupling to the ν mode). Different theoretical calculations conclude that the coupling to the two highest modes should be stronger than for the six lower vibrational modes. At the same time, experiments suggest a strong coupling to $H_g(2)$ [236,237] and/or to $H_g(1)$ [91,237] and the total coupling to the lowest two modes is several times larger than theoretical estimations.

The partial electron–phonon coupling constant in the Table 16.7 data is given as $\lambda_\nu/N(0)$. In order to obtain λ, the density of states at the Fermi level should be estimated. There is no complete agreement between $N(0)$ results from different experiments and theoretical calculations and values of $N(0)$ lie usually between 7 and 12 eV^{-1}. Using an average value of $N(0)$, the electron–phonon coupling λ is about 0.5–0.9 and the attractive interaction between electrons may be of the order of 0.05–0.1 eV.

Because the coupling energy is much smaller than the Coulomb repulsion interaction (1–1.5 eV), the latter is expected to be renormalized by retardation or screening by conducting electrons, similar to that in conventional superconductors. In fullerenes, the retardation is supposed to be rather ineffective with the Coulomb pseudopotential $\mu^* > 0.3$ due to scattering of the electrons

TABLE 16.7   The Partial Electron–Phonon Coupling Constants $\lambda_v/N(0)$ (in eV) According to Different Theoretical Calculations and Experimental (last two columns) Estimations

Mode	Energy[a] (cm^{-1})	$\lambda_v/N(0)$					
		Varma [232]	Schluter [147]	Antropov [85]	Faulhaber [256]	Neutron [236]	Raman [91]
$H_g(8)$	1575	0.011	0.009	0.022	0.009		0.002
$H_g(7)$	1428	0.034	0.013	0.020	0.015		0.003
$H_g(6)$	1250	0.000	0.003	0.008	0.002		0.001
$H_g(5)$	1099	0.006	0.001	0.003	0.002		0.001
$H_g(4)$	774	0.000	0.007	0.003	0.010	0.005	0.002
$H_g(3)$	710	0.001	0.004	0.003	0.001	0.001	0.002
$H_g(2)$	437	0.001	0.007	0.006	0.010	0.023	0.014
$H_g(1)$	273	0.003	0.008	0.003	0.001	0.014	0.034
$\sum H_g$		0.056	0.052	0.068	0.049		0.059

Source: After Ref. 10.

[a] For modes in undoped system.

into subbands. Studying the screening effects, Gunnarsson et al. [238,239], taking into account the detailed electronic structure of fullerides by a tight-binding model with four orbitals per carbon atom, found μ^* to be of the order of 0.4. This Coulomb repulsion is rather large, while authors suggest that even with this μ^* superconductivity in fullerenes can still be driven by an electron–phonon mechanism.

16.7.3.4 Polaronic Superconductivity

Alexandrov and Kabanov [240] developed the nonadiabatic polaron theory of superconductivity in A_3C_{60}, taking into account polaron band narrowing and electron–phonon and Coulomb interactions. Presence of nonadiabaticity should lead to a big difference between polaronic superconductors and BCS superconductors. It is expected [241] that in FSs the nonadiabaticity of electrons, measured by the ratio of characteristic phonon frequency ω to the Fermi energy E_F, is not small. Alexandrov and Kabanov [240] argue that at the BCS coupling constant $\lambda \sim 1$, practically independent of the adiabatic ratio where the Migdal theorem breaks down, a crossover from BCS weak-coupling superconductivity to strong-coupling polaronic superconductivity takes place. It is found that the strongest coupling is with the $A_g(2)$ mode with the Frenkel exciton. At the same time, for low-frequency phonons the Eliashberg-type equation is solved and results for the absolute value of T_c and isotope effect are in agreement with experiment.

However, it remains unclear how the strong coupling to the $A_g(2)$ mode can take place if the Raman experiments [234,235] do not show an extra broadening of this mode in the metallic phase.

16.8 CONCLUSIONS

Almost all superconducting properties of FSs, measured in early experiments, exhibited many unusual features. Some of them, for instance, a large value of the isotope coefficient α, were carefully remeasured and expected values were obtained. Some of the features, like $\rho \sim T^2$, nonmetallic A_4C_{60}, and absence of the Hebel–Slichter peak, were explained. And some results, particularly a drop in T_c with valence $x < 3$ and the isotope effect for particular substitutions, still remain unclear. From the time of their discovery up until now the FSs have changed from being very exotic superconductors to more clear and usual, albeit unique, systems.

While it is widely believed that FSs are s-wave superconductors with weak intramolecular phonon-mediated coupling, the electronic mechanism or the strong coupling polaronic mechanism is not yet ruled out and cannot be excluded from consideration.

If weak electron–phonon coupling is believed to drive superconductivity in FSs, the detailed mechanism of the reduction of the Coulomb repulsion is still unclear and a subject for future investigation.

Rather high transition temperatures in FSs and their magnetic properties, which are very similar to those in high-T_c superconductors, give some promise for practical applications of these materials. However, they are extremely sensitive to air and this problem has to be overcome.

Only C_{60} fullerene doped with alkali or other metals exhibits superconductivity. No other fullerenes have been observed to be superconducting. Recently discovered C_{36} could form the basis for a new family of superconducting fullerenes. If so, the transition temperature in these materials is expected to higher than in A_xC_{60} because of the bigger curvature of the C_{36} molecule. Some of the single-wall nanotubes are metallic. These doped or undoped materials could provide one more direction in which to search for superconductivity.

ACKNOWLEDGMENTS

First, I would like to thank Harald Weber and Franz Sauerzopf (Atominstitut, Vienna) for the huge amount of support, help, and countless number of discussions that enabled me to finish this work and most of my other projects. I appreciate very much the help of Jenny Henderson during writing and preparation of this chapter. I would like to thank Mauro Ricco (Parma University), Jack Fischer (University of Pennsylvania), Hans Kuzmany and Miro Haluska (Vienna University), Olle Gunnarsson (Max Planck Institut, Stuttgart), Alex Alexandrov (Loughborou University), Bertil Sundqvist (Umeå University), Alex Gurevich (University of Wisconsin–Madison), my colleagues from the Low Temperature Laboratory, Atominstitut, and many other persons with whom I have discussed superconducting properties of fullerides.

This work was supported in part by the Austrian Science Foundation (FWF) under Grant No. P12098-PHY.

REFERENCES

1. A. F. Hebard, M. J. Rosseinsky, R. C. Haddon, D. W. Murphy, S. H. Glarum, T. T. M. Palstra, A. P. Ramirez, and A. R. Kortan, *Nature* **1991**, *350*, 600.
2. J. G. Bednorz and K. A. Müller, *Z. Phys. B* **1986**, *46*, 189.
3. R. M. Fleming, A. P. Ramirez, M. J. Rosseinsky, D. W. Murphy, R. C. Haddon, S. M. Zahurak, and V. Makhija, *Nature* **1991**, *352*, 787.
4. K. Tanigaki, I. Hirosawa, T. W. Ebbesen, J.-I. Mizuki, and J.-S. Tsai, *J. Phys. Chem. Solids* **1994**, *54*, 1645.
5. T. T. M. Palstra, O. Zhou, Y. Iwasa, P. E. Sulevski, R. M. Flemming, and B. R. Zegarski, *Solid State Commun.* **1995**, *93*, 327.
6. C. M. Lieber and Z. Zhang, "Physical Properties of Metal-Doped Fullerene Superconductors" in *Solid State Properties of Fullerenes and Fullerene-Based Materials*, Solid State Physics Vol. 48, H. Ehrenreich and Frans Spaepen (eds.). Academic Press, Boston, 1994, pp. 349–384.
7. A. P. Ramirez, *Supercond. Rev.* **1994**, *1*, 1.
8. M. S. Dresselhaus, G. Dresselhaus, and P. C. Eklund, *Science of Fullerenes and Carbon Nanotubes*. Academic Press, Boston, 1996.
9. M. P. Gelfand, *Supercond. Rev.* **1994**, *1*, 103.
10. O. Gunnarsson, *Rev. Mod. Phys.* **1997**, *69*, 575.
11. W. E. Pickett, "Electrons and Phonons in C_{60}-Based Materials" in *Solid State Properties of Fullerenes and Fullerene-Based Materials*, Solid State Physics, Vol. 48, H. Ehrenreich and Frans Spaepen (eds.). Academic Press, Boston, pp. 226–347.
12. C. H. Pennington and V. A. Stenger, *Rev. Mod. Phys.* **1996**, *68*, 855.
13. J. S. Schilling, J. Diederichs, and A. K. Gangopadhyay, in *Recent Advances in the Chemistry and Physics of Fullerenes and Related Materials*, Vol. 5, K. M. Kadish and R. S. Ruoff (Eds.). The Electrochemical Society, Pennington, NJ, 1997, p. 981.
14. B. Sundqvist, *Adv. Phys.* **1998**, in press.
15. V. Buntar and H. W. Weber, *Supercond. Sci. Technol.* **1996**, *9*, 599.
16. T. Yildirim, in *Recent Advances in the Chemistry and Physics of Fullerenes and Related Materials*, Vol. 3, K. M. Kadish and R. S. Ruoff (eds.). The Electrochemical Society, Pennington, NJ, 1996, p. 1155.
17. T. Yildirim, in *Recent Advances in the Chemistry and Physics of Fullerenes and Related Materials*, Vol. 6, K. M. Kadish and R. S. Ruoff (eds.). The Electrochemical Society, Pennington, NJ, 1998.
18. A. F. Hebard, *Phys. Today* **1992**, *45*, 26.
19. J. D. Axe, S. C. Moss, and D. A. Neumann, "Structure and Dynamics of Crystalline C_{60}" in *Solid State Properties of Fullerenes and Fullerene-Based Materials*, Solid State Physics, Vol. 48, H. Ehrenreich and Frans Spaepen (eds.). Academic Press, Boston, 1994, pp. 149–224.
20. C. M. Lieber and C.-C. Chen, "Preparation of Fullerenes and Fullerene based Materials" in *Solid State Properties of Fullerenes and Fullerene-Based Materials*, Solid State Physics, Vol. 48, H. Ehrenreich and Frans Spaepen (eds.). Academic Press, Boston, 1994, pp. 109–148.

21. N. Satoh, S. Tamuma, H. Takenaka, T. Nishizaki, and N. Kobayashi, *Jpn. J. Appl. Phys.* **1996**, *35*, 3392.

22. J. E. Fischer, P. A. Heiney, and A. B. Smith, *Accts. Chem. Res.* **1992**, *25*, 115.

23. R. M. Fleming, M. J. Rosseinsky, A. P. Ramirez, D. W. Murphy, J. C. Tully, R. C. Haddon, T. Siegrist, R. Tycko, S. H. Glarum, P. Marsh, G. Dabbagh, S. M. Zahurak, A. V. Makhija, and C. Hampton, *Nature* **1991**, *352*, 701.

24. O. Zhou, J. E. Fischer, N. Coustel, S. Kycia, Q. Zhu, A. R. McGhie, W. J. Romanow, J. P. McCauley, Jr., A. B. Smith III, and D. E. Cox, *Nature* **1991**, *351*, 462.

25. See, for instance, J. Robert, P. Petit, J.-J. Andre, and J. E. Fischer, *Solid State Commun.* **1995**, *96*, 143, and references therein.

26. K. Tanigaki, I. Hirosawa, T. W. Ebbesen, J. Mizuki, and S. Kuroshima, *Chem. Phys. Lett.* **1993**, *203*, 33.

27. P. W. Stephens, L. Michaly, P. L. Lee, R. L. Whetten, S.-M. Huang, R. Kaner, F. Diederichs, and K. Holczer, *Nature* **1991**, *351*, 632.

28. R. Ziebarth, S.-M. Lee, V. A. Stenger, and C. Pennington, in *Recent Advances in the Chemistry and Physics of Fullerenes and Related Materials*, Vol. 2, K. M. Kadish and R. S. Ruoff (eds.). The Electrochemical Society, Pennington, NJ, 1995.

29. M. J. Rosseinsky, D. W. Murphy, R. M. Fleming, R. Tycko, A. P. Ramirez, T. Siegrist, G. Dabbagh, and S. E. Barrett, *Nature* **1992**, *356*, 416.

30. T. Yildirim, L. Barbedette, J. E. Fischer, C. L. Lin, J. Robert, P. Patit, and T. T. M. Palstra, *Phys. Rev. Lett.* **1996**, *77*, 167.

31. Y. Iwasa, M. Kawaguchi, H. Iwasaki, T. Mitani, N. Wada, and T. Hasegawa, *Phys. Rev. B* **1998**, *57*, 13395.

32. A. R. Kortan, N. Kopylov, S. Glarum, E. M. Gyorgy, A. P. Ramirez, R. M. Fleming, F. A. Thiel, and R. C. Haddon, *Nature* **1992**, *355*, 529.

33. A. R. Kortan, N. Kopylov, E. Özdas, A. P. Ramirez, R. M. Fleming, and R. C. Haddon, *Chem. Phys. Lett.* **1994**, *223*, 501.

34. A. R. Kortan, N. Kopylov, and E. Özdas, in *Recent Advances in the Chemistry and Physics of Fullerenes and Related Materials*, Vol. 3, K. M. Kadish and R. S. Ruoff (eds.). The Electrochemical Society, Pennington, NJ, 1996, p. 423.

35. A. R. Kortan, N. Kopylov, S. Glarum, E. M. Gyorgy, A. P. Ramirez, R. M. Fleming, O. Zhou, F. A. Thiel, P. L. Trevor, and R. C. Haddon, *Nature* **1992**, *360*, 566.

36. M. Baenitz, M. Heinze, K. Lüders, H. Werner, R. Schlögl, M. Weiden, G. Sparn, and F. Steglich, *Solid State Commun.* **1995**, *96*, 539.

37. Y. Chen, F. Stepniak, J. H. Weaver, L. P. F. Chibante, and R. E. Smalley, *Phys. Rev. B* **1992**, *45*, 8845.

38. G. Sparn, F. Laube, A. Link, F. Steglich, M. Baenitz, K. Lüders, H. Werner, and R. Schlögl, *J. Low. Temp. Phys.* **1996**, *105*, 1703.

39. E. Özdas, A. R. Kortan, N. Kopylov, A. P. Ramirez, T. Siegrist, K. M. Rabe, H. E. Bair, S. Schuppler, and P. H. Citrin, *Nature* **1995**, *375*, 126.

40. X. H. Chen and G. Roth, *Phys. Rev. B* **1995**, *52*, 15534.

41. S. Chakravarty, M. P. Gelfand, and S. Kievelson, *Science* **1991**, *254*, 970.

42. S. K. Sarker, *Phys. Rev. B* **1994**, *49*, 12047.

43. T. Yildirim, S. Hong, A. B. Harris, and E. J. Mele, *Phys. Rev. B* **1993**, *48*, 12262.

44. T. Yildirim, J. E. Fischer, A. B. Harris, P. W. Stephens, D. Liu, L. Brard, R. M. Strongin, and A. B. Smith III, *Phys. Rev. Lett.* **1993**, *71*, 1383.

45. R. O. Jones and O. Gunnarsson, *Rev. Mod. Phys.* **1991**, *61*, 689.

46. S. Lundqvist and N. H. March (eds.), *Theory of the Inhomogeneous Electron Gas.* Plenum Press, New York, 1983.

47. S. C. Erwin, in *Buckminsterfullerenes*, W. E. Billups and M. A. Ciufolini (eds.). VCH Publishers, New York, 1993, p. 217.

48. S. Satpathy, V. P. Antropov, O. K. Andersen, O. Jepson, O. Gunnarsson, and A. I. Liechtenstein, *Phys. Rev. B* **1992**, *46*, 1773.

49. S. C. Erwin and W. E. Pickett, *Science* **1991**, *254*, 842.

50. M.-Z. Huang, Y.-N. Xu, and W. Y. Ching, *Phys. Rev. B* **1992**, *46*, 6572.

51. Y.-N. Xu, M.-Z. Huang, and W. Y. Ching, *Phys. Rev. B* **1992**, *44*, 13171.

52. A. Oshiyama, S. Saito, N. Hamada, and Y. Miyamoto, *J. Phys. Chem. Solids* **1992**, *53*, 1457.

53. M. P. Gelfand and J. P. Lu, *Phys. Rev. Lett.* **1992**, *68*, 1050.

54. B. L. Gu, Y. Maruyama, J. Z. Yu, K. Ohno, and Y. Kawazoe, *Phys. Rev. B* **1994**, *49*, 16202.

55. A. Ceulemans, L. F. Chibotaru, and F. Cimpoesu, *Phys. Rev. Lett.* **1997**, *78*, 3725.

56. F. Bensebaa, B. Xiang, and L. Kevan, *J. Phys. Chem.* **1992**, *96*, 6118.

57. D. R. Buffinger, R. P. Ziebarth, V. A. Stenger, C. Recchia, and C. H. Pennington, *J. Am. Chem. Soc.* **1993**, *115*, 9267.

58. M. J. Rosseinsky, D. W. Murphy, R. M. Fleming, and O. Zhou, *Nature* **1993**, *364*, 425.

59. O. Zhou, R. M. Fleming, D. W. Murphy, M. J. Rosseinsky, A. P. Ramirez, R. B. van Dover, and R. C. Haddon, *Nature* **1993**, *362*, 433.

60. S. P. Kelty, C.-C. Chen, and C. M. Lieber, *Nature* **1991**, *352*, 223.

61. C.-C. Chen, S. P. Kelty, and C. M. Lieber, *Science* **1991**, *253*, 886.

62. H. H. Wang, A. M. Kini, B. M. Savall, K. D. Carlson, J. M. Williams, K. R. Lykke, P. Wurz, D. H. Paker, M. J. Pellin, D. M. Gruen, U. Welp, W.-K. Kwok, S. Fleshler, and G. W. Crabtree, *Inorg. Chem.* **1991**, *30*, 2838.

63. H. H. Wang, A. M. Kini, B. M. Savall, K. D. Carlosn, J. M. Williams, M. W. Lathrop, K. R. Lykke, D. H. Parker, P. Wurz, M. J. Pellin, D. M. Gruen, U. Welp, W.-K. Kwok, S. Fleshler, G. W. Crabtree, J. E. Schirber, and D. L. Overmyer, *Inorg. Chem.* **1991**, *30*, 2962.

64. J. A. Schlueter, H. H. Wang, M. W. Lathrop, U. Geiser, K. D. Carlson, J. D. Dudek, J. Yaconi, and J. M. Williams, *Chem. Mat.* **1993**, *5*, 720.

65. K. Tanigaki, I. Hirosawa, T. W. Ebbesen, J. Mizuki, Y. Shimakawa, Y. Kubo, J. S. Tsai, and S. Kuroshima, *Nature* **1992**, *356*, 419.

66. T. Yildirim, J. E. Fischer, R. Dinnebier, P. W. Stephens, and C. L. Lin, *Solid State Commun.* **1995**, *93*, 269.

67. V. Buntar, F. M. Sauerzopf, and H. W. Weber, *Phys. Rev. B* **1996**, *54*, R9651.

68. S. H. Irons, J. Z. Liu, P. Klavins, and R. N. Shelton, *Phys. Rev. B* **1995**, *52*, 15517.

69. M.-W. Lee, M. F. Tai, and J.-B. Shi, *Physica C* **1996**, *272*, 137.

70. M.-W. Lee, J.-B. Shi, and C.-S. Chen, *Jpn. J. Appl. Phys.* **1997**, *36*, 56.

71. S. Chu and M. E. McHenry, *Phys. Rev. B* **1997**, *55*, 11722.

72. S. Chu and M. E. McHenry, in *Recent Advances in the Chemistry and Physics of Fullerenes and Related Materials*, Vol. 5, K. M. Kadish and R. S. Ruoff (eds.). The Electrochemical Society, Pennington, NJ, 1997, p. 1005.

73. V. Buntar, F. M. Sauerzopf, H. W. Weber, J. E. Fischer, H. Kuzmany, M. Haluska, and C. L. Lin, *Phys. Rev. B* **1996**, *54*, 14952.

74. M. Haluska, V. Buntar, C. Krutzler, and H. Kuzmany, in *Recent Advances in the Chemistry and Physics of Fullerenes and Related Materials*, Vol. 7, K. M. Kadish and R. S. Ruoff (eds.). The Electrochemical Society, Pennington, NJ, 1998, p. 436.

75. Y. Iwasa, H. Hayashi, T. Furudate, and T. Mitani, *Phys. Rev. B* **1996**, *54*, 14960.

76. A. F. Hebard, T. T. M. Palstra, R. C. Haddon, and R. M. Fleming, *Phys. Rev. B* **1993**, *48*, 9945.

77. V. H. Crespi, J. G. Hou, X.-D. Xiang, M. L. Cohen, and A. Zettl, *Phys. Rev. B* **1992**, *46*, 12064.

78. X.-D. Xiang, J. G. Hou, G. Briceno, W. A. Vareka, R. Mostovoy, A. Zettl, V. H. Crespi, and M. L. Cohen, *Science* **1992**, *256*, 1190.

79. X.-D. Xiang, J. G. Hou, V. H. Crespi, A. Zettl, and M. L. Cohen, *Nature* **1993**, *361*, 54.

80. O. Klein, G. Grüner, S.-M. Huang, J. B. Wiley, and R. B. Kaner, *Phys. Rev. B* **1992**, *46*, 11247.

81. J. G. Hou, V. H. Crespi, X.-D. Xiang, W. A. Vareka, G. Briceno, A. Zettl, and M. L. Cohen, *Solid State Commun.* **1993**, *86*, 643.

82. J. G. Hou, X.-D. Xiang, V. H. Crespi, M. L. Cohen, and A. Zettl, *Physica C* **1994**, *228*, 175.

83. T. T. M. Palstra, A. F. Hebard, R. C. Haddon, and P. B. Littlewood, *Phys. Rev. B* **1994**, *50*, 3462.

84. S. C. Erwin and M. R. Pederson, *Phys. Rev. B* **1993**, *47*, 14657.

85. V. P. Antropov, O. Gunnarsson, and A. I. Liechtenstein, *Phys. Rev. B* **1993**, *48*, 7651.

86. J. S. Lannin and M. G. Mitch, *Phys. Rev. B* **1994**, *50*, 6497.

87. B. Sundqvist and E. M. C. Nilsson, *Physica C* **1994**, *235–240*, 1407.

88. W. A. Vareka, M. S. Fuhrer, and A. Zettl, *Physica C* **1994**, *235–240*, 2507.

89. W. A. Vareka and A. Zettl, *Phys. Rev. Lett.* **1994**, *72*, 4121.

90. G. Grimvall, *The Electron–Phonon Interaction in Metals*. North-Holland, Amsterdam, 1981.

91. J. Winter and H. Kuzmany, *Phys. Rev. B* **1996**, *53*, 655.

92. Y. Huang and K. Tanaka, *Phys. Rev. B* **1998**, *57*, 7462.

93. J. M. Ziman, *Adv. Phys.* **1967**, *13*, 578.

94. J. G. Hou, L. Lu, V. H. Crespi, X.-D. Xiang, A. Zettl, and M. L. Cohen, *Solid State Commun.* **1995**, *93*, 973.

95. L. Lu, V. H. Crespi, M. S. Fuhler, A. Zettl, and M. L. Cohen, *Phys. Rev. Lett.* **1995**, *74*, 1637.

96. S. C. Erwin and W. E. Pickett, *Phys. Rev. B* **1992**, *46*, 14257.

97. M. Tinkham, *Introduction to Superconductivity*. Robert E. Krieger Publishing, Malabar, FL, 1980.

98. K. Tanigaki, T. W. Ebbesen, S. Saito, J. Mizuki, J. S. Tsai, Y. Kubo, and S. Kuroshima, *Nature* **1991**, *352*, 222.

99. G. Sparn, J. D. Thompson, S.-M. Huang, R. B. Kaner, F. Diederich, R. L. Whetten, G. Grüner, and K. Holczer, *Science* **1991**, *252*, 1829.

100. G. Sparn, J. D. Thompson, R. L. Whetten, S.-M. Huang, R. B. Kaner, F. Diederich, G. Grüner, and K. Holczer, *Phys. Rev. Lett.* **1992**, *68*, 1228.

101. J. Diederichs, J. S. Schilling, K. W. Herwig, and W. B. Yelon, *J. Phys. Chem. Solids* **1997**, *58*, 123.

102. J. Diederichs, A. K. Gangopadhyay, and J. S. Schilling, *Phys. Rev. B* **1996**, *54*, R9662.

103. O. Zhou, G. B. M. Vaughan, Q. Zhu, J. E. Fischer, P. A. Heiney, N. Coustel, J. P. McCauley, Jr, A. B. Smith III, *Science* **1992**, *255*, 833.

104. H. A. Ludwig, W. H. Fietz, F. W. Hornung, K. Grube, B. Renker, and G. J. Burkhart, *Physica C* **1994**, *234*, 45.

105. J. E. Schirber, D. L. Overmeyer, W. R. Bayless, M. J. Rosseinsky. O. Zhou, D. W. Murphy, Q. Zhu, K. Kniaz, and J. E. Fischer, *J. Phys. Chem. Solids* **1993**, *54*, 1427.

106. H. Shimoda, Y. Iwasa, Y. Miyamoto, Y. Maniwa, and T. Mitani, *Phys. Rev. B* **1996**, *54*, R15653.

107. H. Shimoda, Y. Iwasa, and T. Mitani, *Synthetic Metals* **1997**, *85*, 1593.

108. Y. Maniwa, D. Sugiura, K. Kume, K. Kikuchi, S. Suzuki, Y. Achiba, I. Hirosawa, K. Tanigaki, H. Shimoda, I. Iwasa, *Phys. Rev. B* **1996**, *54*, R6861.

109. K. Tanigaki, T. W. Ebbesen, J. S. Tsai, I. Hirosawa, and J. Mizuki, *Europhys. Lett.* **1993**, *23*, 57.

110. K. Kniaz, J. E. Fischer, Q. Zhu, M. J. Rosseinsky, O. Zhou, and D. W. Murphy, *Solid State Commun.* **1993**, *88*, 47.

111. J. Lu and L. Zhang, *Solid State Commun.* **1998**, *105*, 99.

112. J. Mizuki, M. Takai, N. Mori, K. Tanigaki, I. Hirosawa, and K. Prassides, *Phys. Rev. B* **1994**, *50*, 3466.

113. Q. Zhu, *Phys. Rev. B* **1995**, *52*, R723.

114. B. Morosin, J. E. Schirber, J. D. Jorgensen, G. H. Kwei, T. Yildirim, and J. E. Fischer, in *Recent Advances in the Chemistry and Physics of Fullerenes and Related Materials*, Vol. 3, K. M. Kadish and R. S. Ruoff (eds.). The Electrochemical Society, Pennington, NJ, 1996, p. 446.

115. J. E. Schirber, L. Hansen, B. Morosin, J. E. Fischer, J. D. Jorgensen, and G. H. Kwei, *Physica C* **1996**, *260*, 173.

116. J. E. Schirber, W. R. Bayless, A. R. Kortan, M. J. Rosseinsky, E. Özdas, O. Zhou, R. M. Fleming, and D. Murphy, in *Recent Advances in the Chemistry and Physics of Fullerenes and Related Materials*, Vol. 3, K. M. Kadish and R. S. Ruoff (eds.). The Electrochemical Society, Pennington, NJ, 1996, p. 556.

117. J. E. Schirber, W. R. Bayless, A. R. Kortan, and N. Kopylov, *Physica C* **1993**, *213*, 190.

118. R. W. Lof, M. A. van Veenendaal, B. Koopmans, H. T. Jonkman, and G. A. Sawatsky, *Phys. Rev. Lett.* **1992**, *68*, 3924.

119. G. A. Sawatsky, in *Physics and Chemistry of Fullerene and Derivatives*, H. Kuzmany, J. Fink, M. Mehring, and S. Roth (eds.). World Scientific, Singapore, 1995, p. 373.

120. J. E. Fischer et al., *J. Phys. Chem. Solids* **1995**, *56*, 1445.

121. Q. Zhu, O. Zhou, N. Coustel, G. B. M. Vaughan, J. P. McCauley, Jr, W. M. Romanow, J. E. Fischer, and A. B. Smith III, *Science* **1991**, *254*, 545.

122. Q. Zhu, J. E. Fischer, and D. E. Cox, *Springer Ser. Solid State Sci.* **1994**, *117*, 168.

123. K. Siguhara, T. Inabe, Y. Maruyama, and Y. Achiba, *J. Soc. Phys. Jpn.* **1993**, *62*, 2757.

124. Z. Zhang, C.-C. Chen, and C. M. Lieber, *Science* **1991**, *254*, 1619.

125. Z. Zhang and C. M. Lieber, *Mod. Phys. Lett. B* **1991**, *5*, 1905.

126. P. Jess, S. Behler, M. Bernasconi, V. Thommen-Geiser, H. P. Lang, M. Baenitz, K. Lüders, and H.-J. Güntherod, *Physica C* **1994**, *235*, 2499.

127. P. Jess et al., *J. Phys. Chem. Solids* **1997**, *58*, 1803.

128. R. Tycko, G. Dabbach, M. J. Rosseinsky, D. W. Murphy, A. P. Ramirez, and R. M. Fleming, *Phys. Rev. Lett.* **1992**, *68*, 1912.

129. P. Auban-Senzier, G. Quirion, D. Jerome, P. Bernier, S. Della-Hegra, C. Farbe, and A. Rassat, *Synthetic Metals* **1993**, *55–57*, 3027.

130. S. Sasaki, A. Matsuda, and C. W. Chu, *J. Soc. Phys. Jpn.* **1994**, *63*, 1670.

131. V. A. Stenger, C. H. Pennington, D. R. Bufinger, and R. P. Ziebarth, *Phys. Rev. Lett.* **1995**, *74*, 1649.

132. L. D. Rotter, Z. Schlesinger, J. P. McCauley, Jr., N. Coustel, J. E. Fischer, and A. B. Smith III, *Nature* **1992**, *355*, 532.

133. S. A. Fitzgerald, S. G. Kaplan, A. Rosenberg, A. J. Sievers, and R. A. McMordie, *Phys. Rev. B* **1992**, *45*, 10165.

134. L. Degiorgi, P. Wachter, G. Grüner, S.-M. Huang, J. Willey, and R. B. Kaner, *Phys. Rev. Lett.* **1992**, *69*, 2987.

135. L. Degiorgi, *Mod. Phys. Lett. B* **1995**, *9*, 445.

136. C. Gu, B. W. Veal, R. Liu, A. P. Paulikas, P. Kostic, H. Ding, K. Gofron, J. C. Campuzano, J. A. Schlueter, H. H. Wang, U. Geiser, and J. M. Williams, *Phys. Rev. B* **1994**, *50*, 16566.

137. R. F. Kiefl, W. A. MacFarlane, K. H. Chow, S. Dunsiger, T. L. Duty, T. M. S. Johnston, J. W. Schneider, J. Sonier, L. Brard, R. M. Strongin, J. E. Fischer, and A. B. Smith III, *Phys. Rev. Lett.* **1993**, *70*, 3987.

138. L. Degiorgi, G. Briceno, M. S. Fuhrer, A. Zettl, and P. Wachter, *Nature* **1994**, *369*, 541.

139. W. A. MacFarlane, R. F. Kiefl, S. Dunsiger, J. E. Sonier, J. Chakhalian, J. E. Fischer, T. Yildirim, and K. H. Chow, *Phys. Rev. B* **1998**, *58*, 1004.

140. V. A. Stenger, C. Recchia, C. H. Pennington, D. R. Buffinger, and R. P. Ziebarth, *J. Superconductivity* **1994**, *7*, 931.

141. S. Sasaki, A. Matsuda, and C. W. Chu, *J. Phys. Soc. Jpn.* **1994**, *63*, 1670.

142. S. Sasaki, A. Matsuda, and C. W. Chu, *Physica C* **1997**, *278*, 238.

143. M. Ricco, L. Menozzi, R. De Renzi, and F. Bolzoni, *Physica C* **1998**, *306*, 136.

144. H.-Y. Choi, *Phys. Rev. Lett.* **1998**, *81*, 441.

145. C.-C. Chen and C. M. Lieber, *J. Am. Chem. Soc.* **1992**, *114*, 3141.

146. C.-C. Chen and C. M. Lieber, *Science* **1993**, *259*, 655.

147. M. Schlüter, M. Lannoo, M. Needels, G. A. Baraff, and D. Tomanek, *J. Phys. Chem. Solids* **1992**, *53*, 1473.

148. O. Gunnarsson, H. Handschuh, P. S. Behthold, B. Kessler, G. Ganteför, and W. Eberhardt, *Phys. Rev. Lett.* **1995**, *74*, 1875.

149. T. W. Ebbesen, J. S. Tsai, K. Tanigaki, H. Hiura, Y. Shimakawa, Y. Kubo, I. Hirosawa, and J. Mizuki, *Physica C* **1992**, *203*, 163.

150. B. Burk, V. H. Crespi, M. S. Fuhrer, A. Zettl, and M. L. Cohen, *Physica C* **1994**, *235–240*, 2493.

151. B. Burk, V. H. Crespi, A. Zettl, and M. L. Cohen, *Phys. Rev. Lett.* **1994**, *72*, 3706.

152. K. Holczer, O. Klein, G. Grüner, J. D. Thompson, F. Diederich, and R. L. Whetten, *Phys. Rev. Lett.* **1991**, *67*, 271.

153. C. Politis, V. Buntar, W. Krauss, and A. Gurevich, *Europhys. Lett.* **1992**, *17*, 175.

154. H. Ogata, T. Inabe, H. Hishi, Y. Maruyama, Y. Achiba, K. Kikuchi, and I. Ikemoto, *Jpn. J. Appl. Phys.* **1992**, *31*, L166.

155. V. Buntar, S. M. Sauerzopf, H. W. Weber, J. E. Fischer, H. Kuzmany, and M. Haluska, *Phys. Rev. B* **1997**, *56*, 14128.

156. R. Tycko, G. Dabbagh, M. J. Rosseinsky, D. W. Murphy, R. M. Fleming, A. P. Ramirez, and J. C. Tully, *Science* **1991**, *253*, 884.

157. M. Baenitz, M. Heinze, E. Straube, H. Werner, R. Schlögl, V. Thommen, H.-J. Güntherod, and K. Lüders, *Physica C* **1994**, *228*, 181.

158. C. Politis, A. I. Sokolov, and V. Buntar, *Mod. Phys. Lett. B* **1992**, *6*, 351.

159. V. Buntar, M. Ricco, L. Cristofolini, H. W. Weber, and F. Bolzoni, *Phys. Rev. B* **1995**, *52*, 4432.

160. I. I. Khairullin, K. Imaeda, K. Yakushi, and H. Inokuchi, *Physica C* **1994**, *231*, 26.

161. C. E. Johnson, H. W. Jiang, K. Holtzer, R. B. Kaner, R. L. Whetten, and F. Diederich, *Phys. Rev. B* **1992**, *46*, 5880.

162. G. S. Boebinger, T. T. M. Palstra, A. Passner, M. J. Rosseinsky, D. W. Murphy, and I. I. Mazin, *Phys. Rev. B* **1992**, *46*, 5876.

163. S. Foner, E. J. McNiff, Jr., D. Heiman, S.-M. Huang, and R. B. Kaner, *Phys. Rev. B* **1991**, *46*, 14936.

164. J. G. Hou, X.-D. Xiang, M. L. Cohen, and A. Zettl, *Physica C* **1994**, *232*, 22.

165. C. Polits, V. Buntar, and V. P. Seminozhenko, *Int. J. Mod. Phys. B* **1993**, *7*, 2163.

166. T. T. M. Palstra, R. C. Haddon, A. F. Hebard, and J. Zaanen, *Phys. Rev. Lett.* **1992**, *68*, 1054.

167. V. Korenivski, K. V. Rao, and Z. Iqbal, *Phys. Rev. B* **1994**, *50*, 13890.

168. N. R. Werthamer, E. Helfand, and P. C. Hohenberg, *Phys. Rev.* **1966**, *147*, 295.

169. J. P. Carbotte, *Rev. Mod. Phys.* **1990**, *62*, 1027.

170. M. Baenitz, M. Heinze, K. Lüders, H. Werner, and R. Schlögl, in *Physics and Chemistry of Fullerenes and Derivatives*, H. Kuzmany, J. Fink, M. Mehring, and S. Roth (eds.). World Scientific, Singapore, 1995., p. 647.

171. M. Baenitz, M. Heinze, K. Lüders, H. Werner, and R. Schlögl, *Solid State Commun.* **1994**, *91*, 337.

172. H. W. Weber (ed.), *Anisotropy Effects in Superconductors*. Plenum Press, New York, 1997, p. 316.

173. H. W. Weber, E. Seidl, C. Laa, E. Schachinger, M. Prohammer, A. Junod, and D. Eckert, *Phys. Rev. B* **1991**, *44*, 7585.

174. C. P. Bean, *Phys. Rev. Lett.* **1992**, *8*, 250.

175. V. Buntar, U. Eckern, and C. Politis, *Mod. Phys. Lett. B* **1992**, *6*, 1037.

176. A. A. Abrikosov, *Zh. Eksp. Teor. Phys.* **1957**, *32*, 1442.

177. V. Moshchalkov, J. V. Henry, C. Marin, J. Rossat-Mignod, and J. F. Jacquot, *Physica C* **1991**, *175*, 407.

178. M. Kraus, H. Sindlinger, H. Werner, R. Schlögl, V. Thommen, H. P. Lang, H.-J. Güntherod, and K. Lüders, *J. Phys. Chem. Solids* **1996**, *57*, 999.

179. J. D. Thompson, G. Sparn, K. Holczer, O. Klein, G. Grüner, R. B. Kaner, F. Diederich, and R. L. Whetten, in *Physical and Material Properties of High Temperature Superconductors*, S. K. Malic and S. S. Shah (eds.). Nova Science Publishers, Commack, NJ, 1994, p. 139.

180. Y. J. Uemura, A. Keren, L. P. Le, G. M. Luke, B. J. Sternlieb, W. D. Wu, J. H. Brewer, R. L. Whetten, S. M. Huang, S. Lin, R. B. Kaner, F. Diederich, S. Donovan, G. Grüner, and K. Holtzer, *Nature* **1991**, *352*, 606.

181. Y. J. Uemura, A. Keren, L. P. Le, G. M. Luke, W. D. Wu, J. S. Tsai, K. Tanigaki, K. Holtzer, S. Donovan, and R. L. Whetten, *Physica C* **1994**, *235–240*, 2501.

182. S. Sasaki, in *Recent Advances in the Chemistry and Physics of Fullerenes and Related Materials*, Vol. 6, K. M. Kadish and R. S. Ruoff (eds.). The Electrochemical Society, Pennington, NJ, 1998, p. 636.

183. A. I. Sokolov, Yu. A. Kufaev, and E. B. Sonin, *Physica C* **1993**, *212*, 19.

184. A. V. Gurevich, R. G. Minz, and A. L. Rakhmanov, *Physics of Composed Superconductors*. Nauka, Moskow, 1987.

185. A. M. Campbell and J. E. Evetts, *Critical Currents in Superconductors*. Taylor and Francis, London, 1972.

186. T. Matsushita, E. S. Otabe, B. Ni et al., *Jpn. J. Appl. Phys.* **1991**, *30*, L342.

187. W. A. Fietz and W. W. Webb, *Phys. Rev.* **1969**, *178*, 657.

188. M. W. Lee, M. F. Tai, S. C. Luo, and J. B. Shi, *Physica C* **1995**, *245*, 6.

189. V. Buntar, *Physica C* **1998**, *309*, 98.

190. F. M. Sauerzopf, *Phys. Rev. B* **1998**, *57*, 10959.

191. R. D. Boss, J. C. Broggs, E. W. Jacobs, T. E. Jones, and P. A. Mosier-Boss, *Physica C* **1995**, *243*, 29.

192. V. Buntar, F. M. Sauerzopf, C. Krutzler, and H. W. Weber, *Phys. Rev. Lett.* **1998**, *81*, 3749.

193. Y. Yeshurun and A. P. Malosemoff, *Phys. Rev. Lett.* **1988**, *60*, 2202.

194. R. X. Tinkham, *Phys. Rev. Lett.* **1988**, *61*, 1658.

195. C. W. Hagen and R. Griessen, *Phys. Rev. Lett.* **1989**, *62*, 2857.

196. D. S. Fisher, M. P. A. Fisher, and D. A. Huse, *Phys. Rev. B* **1991**, *43*, 130.

197. P. L. Gammel, L. F. Schneemeyer, and D. J. Bishop, *Phys. Rev. Lett.* **1991**, *66*, 953.

198. R. H. Koch, V. Foglietti, W. J. Gallagher, G. Koren, A. Gupta, and M. P. A. Fisher, *Phys. Rev. Lett.* **1989**, *63*, 1511.

199. A. Houghton, R. A. Pelcovits, and A. Sudbø, *Phys. Rev. B* **1989**, *40*, 6763.

200. E. H. Brandt, *Phys. Rev. Lett.* **1989**, *63*, 1106.

201. D. E. Farrel, J. P. Rice, and D. M. Ginsberg, *Phys. Rev. Lett.* **1991**, *67*, 1165.

202. R. G. Beck, D. E. Farell, J. P. Rice, D. M. Ginzberg, and V. G. Kogan, *Phys. Rev. Lett.* **1992**, *68*, 1594.

203. L. I. Glazman and A. E. Koshelev, *Phys. Rev. B* **1991**, *43*, 2835.

204. M. F. Schmidt, N. E. Israeloff, and A. M. Goldman, *Phys. Rev. Lett.* **1993**, *70*, 2162.

205. C. L. Lin, T. Mihalisin, N. Bykovetz, Q. Zhu, and J. E. Fischer, *Phys. Rev. B* **1994**, *49*, 4285.

206. H. W. Weber, in *Proceedings 10th Anniversary HTS Workshop*, March 1996.

207. V. Buntar, F. M. Sauerzopf, C. Krutzler, H. W. Weber, M. Haluska, and H. Kuzmany, in *Proceedings of 3rd Canadian Applied Superconductivity Workshop, CASW '98*, Vancouver, Canada, 1998, p. 38.

208. P. W. Anderson and Y. B. Kim, *Rev. Mod. Phys.* **1964**, *36*, 39.

209. C. L. Lin, T. Mihalisin, M. M. Labes, N. Bykovetz, Q. Zhu, and J. E. Fischer, *Solid State Commun.* **1994**, *90*, 62.

210. M. W. Lee, M. F. Tai, and S. C. Luo, *Jpn. J. Appl. Phys.* **1995**, *34*, 4774.

211. V. Buntar et al., to be published.

212. R. W. Lof, M. A. van Veenendaal, B. Koopmans, A. Heessels, H. T. Jonkman, and G. A. Sawatsky, *Int. J. Mod. Phys. B* **1992**, *6*, 3915.

213. S. Chakravarty and S. Kivelson, *Europhys. Lett.* **1991**, *16*, 751.

214. G. Baskaran and E. Tosatti, *Curr. Sci.* **1991**, *33*, 61.

215. R. Friedberg, T. D. Lee, and H. C. Ren, *Phys. Rev. B* **1992**, *46*, 14150.

216. S. R. White, S. Chakravarty, M. P. Gelfand, and S. A. Kivelson, *Phys. Rev. B* **1992**, *45*, 5062.

217. V. P. Antropov, O. Gunnarsson, and O. Jepsen, *Phys. Rev. B* **1992**, *46*, 13647.

218. P. A. Brühwiler, A. J. Maxwell, A. Nilson, N. Martensson, and O. Gunnarsson, *Phys. Rev. B* **1992**, *48*, 18296.

219. O. Gunnarsson, V. Eyert, M. Knupfer, J. Fink, and J. F. Armbruster, *J. Phys. Condens. Matter* **1996**, *8*, 2557.

220. O. Gunnarsson, E. Koch, R. M. Martin, and F. Aryasetiawan, in *Molecular Nanostructures*, H. Kuzmany, J. Fink, M. Mehring, and S. Roth (eds.). World Scientific, Singapore, 1997, p. 229.

221. O. Gunnarsson, in *Recent Advances in the Chemistry and Physics of Fullerenes and Related Materials*, Vol. 6, K. M. Kadish and R. S. Ruoff (eds.). The Electrochemical Society, Pennington, NJ, 1998.

222. S. Chakravarty, L. Chayes, and S. A. Kivelson, *Lett. Math. Phys.* **1992**, *23*, 265.

223. R. Saito, G. Dresselhaus, and M. S. Dresselhaus, *Chem. Phys. Lett.* **1993**, *210*, 159.

224. W. E. Goff and P. Phillips, *Phys. Rev. B* **1992**, *46*, 603.

225. W. E. Goff and P. Phillips, *Phys. Rev. B* **1993**, *48*, 3491.

226. W. L. McMillan, *Phys. Rev.* **1968**, *167*, 331.

227. G. M. Eliashberg, *J. Eksp. Teor. Phys.* **1960**, *38*, 966.

228. C. Christides, D. A. Neunman, K. Prassides, J. R. D. Copley, J. J. Rush, M. J. Rosseinsky, D. W. Murphy, and R. C. Haddon, *Phys. Rev. B* **1992**, *46*, 12088.

229. H. Schober, B. Renker, F. Gompf, and P. Adelmann, *Physica C* **1994**, *235–240*, 2487.

230. F. C. Zhang, M. Ogata, and T. M. Rice, *Phys. Rev. Lett.* **1991**, *67*, 3452.

231. O. Gunnarsson and G. Zwicknagl, *Phys. Rev. Lett.* **1992**, *69*, 957.

232. C. M. Varma, J. Zaanen, and K. Raghavachari, *Science* **1991**, *254*, 989.

233. M. Lanoo, G. A. Baraff, M. Schluter, and D. Tomanek, *Phys. Rev. B* **1991**, *44*, 12106.

234. S. J. Duclos, R. C. Haddon, S. Glarum, A. F. Hebard, and K. B. Lyons, *Science* **1991**, *254*, 1625.

235. T. Pichler, M. Matus, J. Kürti, and H. Kuzmany, *Phys. Rev. B* **1992**, *45*, 13841.

236. K. Prassides, C. Christides, M. J. Rosseinsky, J. Tomkinson, D. W. Murphy, and R. C. Haddon, *Europhys. Lett.* **1992**, *19*, 629.

237. O. Gunnarsson, *Phys. Rev. B* **1995**, *51*, 3493.

238. O. Gunnarsson and G. Zwicknagl, *Phys. Rev. Lett.* **1992**, *69*, 957.

239. O. Gunnarsson, D. Rainer, and G. Zwicknagl, *Int. J. Mod. Phys. B* **1992**, *6*, 3993.

240. A. S. Alexandrov and V. V. Kabanov, *Phys. Rev. B* **1996**, *54*, 3655.

241. V. L. Aksenov and V. V. Kabanov, *Phys. Rev. B* **1998**, *57*, 608.

242. I. Hirosawa, J. Mizuki, K. Tanigaki, and H. Kimura, *Solid State Commun.* **1994**, *89*, 55.

243. S. E. Barrett and R. Tycko, *Phys. Rev. Lett.* **1992**, *69*, 3754.

244. R. C. Haddon, A. F. Hebard, M. J. Rosseinsky, D. W. Murphy, S. J. Duclos, K. B. Lyond, B. Miller, J. M. Rosamilla, R. M. Fleming, A. R. Kortan, S. H. Glarum, A. V. Makhija, A. J. Müller, R. H. Elck, S. M. Zahurak, R. Tycko, G. Dabbagh, and F. A. Thiel, *Nature* **1991**, *350*, 320.

245. N. Yamasaki, H. Araki, A. A. Zakhidov, and K. Yoshino, *Solid State Commun.* **1994**, *92*, 547.

246. K. Imaeda, J. Kröber, H. Inokuchi, Y. Yonehara, and K. Ichimura, *Solid State Commun.* **1996**, *99*, 479.

247. Y. Iwasa, M. Kawaguchi, S. Taga, T. Takenobu, T. Mitani, K. Prassides, and C. M. Brown, in *Recent Advances in the Chemistry and Physics of Fullerenes and Related Materials*, Vol. 6, K. M. Kadish and R. S. Ruoff (eds.). The Electrochemical Society, Pennington, NJ, 1998, p. 428.

248. M. Tokumoto, Y. Tanaka, N. Kinoshita, T. Kinoshita, S. Ishibashi, and H. Ihara, *J. Chem. Phys. Solids* **1993**, *54*, 1667.

249. H. Araki, N. Yamasaki, A. A. Zakhidov, and K. Yoshino, *Physica C* **1994**, *233*, 242.

250. S. E. Erwin and M. R. Pederson, *Phys. Rev. Lett.* **1992**, *67*, 1610.

251. N. Troullier and J. L. Martins, *Phys. Rev. B* **1992**, *46*, 1766.

252. T. W. Ebbesen, J. S. Tsai, K. Tanigaki, J. Tabuchi, Y. Shimakawa, Y. Kubo, I. Hirosawa, and J. Mizuki, *Nature* **1992**, *355*, 620.

253. A. A. Zakhidov, K. Imaeda, D. M. Patty, K. Yakushi, H. Inokuchi, K. Kikuchi, I. Ikemoto, S. Suzuki, and Y. Achiba, *Phys. Lett. A* **1992**, *164*, 355.

254. A. P. Ramirez, A. R. Kortan, M. J. Rosseinsky, S. J. Duclos, A. M. Mujsce, R. C. Haddon, D. W. Murphy, A. V. Makhija, S. M. Zahurak, and K. B. Lyons, *Phys. Rev. Lett.* **1992**, *68*, 1058.

255. J. W. White, G. Lindsell, L. Pang, A. Palmisano, D. S. Sivia, and J. Tomkinson, *Chem. Phys. Lett.* **1991**, *84*, p. 1671.

256. J. C. R. Faulhaber, D. Y. K. Ko, and R. P. Briddon, *Phys. Rev. B* **1993**, *48*, 661.

257. S. C. Erwin, in *Buckminsterfullerenes*, W. E. Billups and M. A. Ciufolini (eds.). VCH Publishers, New York, 1993. p. 217.

258. G. Sparn, Habilitation thesis, Johann Wolfgang Goethe University, Frankfurt am Main, Germany, 1996.

259. D. L. Novikov, V. A. Gubanov, and A. J. Freeman, *Physica C* **1992**, *191*, 399.

260. Z. Hao and J. R. Clem, *Phys. Rev. Lett.* **1991**, *67*, 2371.

CHAPTER 17

BORON-NITRIDE-CONTAINING NANOTUBES

NASREEN G. CHOPRA and ALEX ZETTL

17.1 INTRODUCTION

Graphite is a layered material in which each sheet is a hexagonal grid of carbon atoms covalently bonded to each other in the plane with van der Waals bonds between layers. The covalent bond is one of the strongest in nature while the van der Waals bond is weak; thus, planes may easily slide relative to each other. A nanotube can be described mathematically as a long thin strip, cut out of a single atomic plane of material, rolled to form a cylinder with a diameter of nanometer scale and a length on the order of micrometers. Carbon nanotubes were first discovered by Iijima in 1991 [1] while performing transmission electron microscopy (TEM) on a fullerene sample taken from the chamber where C_{60} [2] is produced. The discovery has led to extensive interest in many fields due to the versatility and potential applications of these nanostructures. Experimental observations of carbon nanotubes demonstrate that tubes may be single-walled or multiwalled structures, sometimes with over 50 walls. Multiwalled tubes are hollow, seamless cylinders that are concentrically organized such that the spacing between walls approximately equals the graphitic interplanar distance. Perhaps the most unique feature of these nanostructures is their high aspect ratio. Inner diameters of nanotubes range from 7 Å to about 4 nm, while their lengths are typically several micrometers, even up to several hundred micrometers.

Nanotubes have been a rich source of inspiration for theorists. Indeed, much of the excitement about nanotubes stems from the prediction of their interesting electrical and mechanical properties. Calculations show that the electrical behavior of nanotubes is integrally related to the specific geometry of the tube structure [3]; that is, a carbon nanotube can be semiconducting or metallic, depending on its radius and chirality. Empirical formulation of the rigidity of

Fullerenes: Chemistry, Physics, and Technology, Edited by Karl M. Kadish and Rodney S. Ruoff.
0-471-29089-0 Copyright © 2000 John Wiley & Sons, Inc.

nanotubes finds nanotubes to be extremely strong in the axial direction with predicted elastic moduli exceeding 1 TPa [4] for these structures.

Theorists have also predicted the existence of tubes made from other layered materials such as boron nitride [5], BC_2N [6], and BC_3 [7]. All of these materials are hexagonal networks of atoms, which imitate the planar sp^2 bonding in graphite. Band structure calculations predict unique, and equally interesting, electrical properties for each of these novel tubes. The mechanical properties of $B_xC_yN_z$ nanotubes are again expected to be exceptional.

This chapter summarizes theoretical and experimental investigations of boron-nitride-containing nanotubes, including pure BN nanotubes and nanotubes of stoichiometries BC_2N and BC_3. Section 17.2 compares the crystal structures of graphite, BN, BC_2N, and BC_3. In the case of BC_2N and BC_3, only the single-sheet structure is considered. Section 17.3 investigates the theoretical band structure of graphite and pure BN, both in bulk and nanotube form. Further theoretical predictions for the electronic properties of boron-nitride-containing nanotubes are presented in Section 17.4. Section 17.5 introduces experimental synthesis and characterization of $B_xC_yN_z$ nanotubes and proposes possible nanotube growth mechanisms. The mechanical (elastic) properties of pure BN nanotubes are described in Section 17.6. It is found that a BN nanotube is the stiffest insulating fiber known, most likely due to the inherent strength of the BN bond and the high degree of crystallinity of BN nanotubes. The last section summarizes BN, BC_2N, and BC_3 nanotube properties and comments on additional $B_xC_yN_z$ nanostructures, including carbon/BN nano "coaxial cables."

17.2 COMPARISON OF CRYSTAL STRUCTURES OF GRAPHITE, BN, BC_2N, AND BC_3

The crystal structures of graphite, BN, BC_2N, and BC_3 are quite similar. They are all hexagonal layered structures, with ABAB packing being the most common arrangement of the layers. The crystal structure is best described by schematics of the ideal structures.

17.2.1 Graphite

See Figure 17.1.

17.2.2 BN

See Figure 17.2.

17.2.3 BC_2N

See Figures 17.3 (Type I) and 17.4 (Type II).

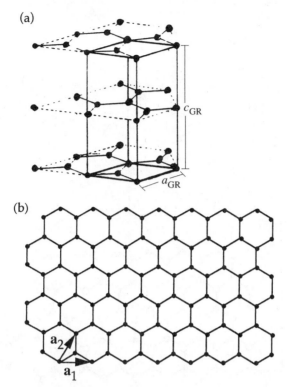

Figure 17.1 Crystal structure of graphite. (a) Unit cell has lattice constant of $a_{GR} = 2.46\,\text{Å}$ [8], which corresponds to the second-nearest-neighbor distance, and the interplanar spacing is $c_{GR}/2 = 3.36\,\text{Å}$. (b) Graphite is a hexagonal sheet with carbon atoms at each vertex. The bond length is $d_{C-C} = 1.42\,\text{Å}$. Lattice translation vectors are shown and have magnitudes equal to $|\vec{a_1}| = |\vec{a_2}| = a_{GR}$. Due to the homogeneity of the material, this sheet structure has inversion symmetry.

17.2.4 BC_3

See Figure 17.5.

17.3 THEORETICAL BAND STRUCTURE OF CARBON AND BN NANOTUBES

The band structure for carbon nanotubes is derived primarily from the band structure of graphite. Thus, a brief discussion of the unique features of the graphite band structure precedes that of nanotubes. Emphasis is placed on a pictorial description rather than involved calculations.

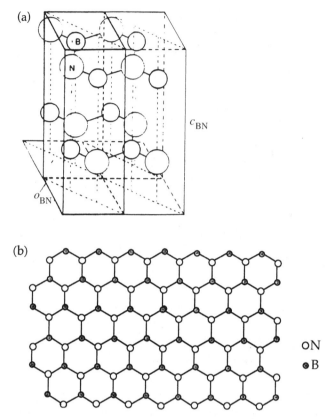

Figure 17.2 Crystal structure of BN. (a) Unit cell has lattice constant of $a_{BN} = 2.50\,\text{Å}$, which corresponds to the second-nearest-neighbor distance, and the interplanar spacing is $c_{BN}/2 = 3.33\,\text{Å}$ [9]. (b) BN is a hexagonal sheet with boron atoms surrounded by nitrogen atoms and vice versa. The bond length is $d_{B-N} = 1.44\,\text{Å}$. Lattice translation vectors are defined similar to the graphite sheet with $|\vec{a_1}| = |\vec{a_2}| = a_{BN}$. Because of the dissimilar B and N atoms, this sheet arrangement does not have inversion symmetry.

17.3.1   Band Structure of Graphite

There are two common orientations of the two-dimensional hexagonal lattice of graphite. The lattice (grid) may be oriented such that a side of the hexagonal unit is parallel to either (1) the x axis or (2) the y axis. If the repeat unit in real space is the hexagon oriented as shown in Figure 17.6a (i.e., parallel to the y axis), the Brillouin zone (BZ) of the sheet consists of hexagons rotated[1] by 90° as displayed in Figure 17.6b (parallel to the x axis). The central point ($\vec{k} = 0$) is

---

[1] This rotation is a consequence of the mathematical formalism used when transforming from real space to reciprocal space (see Chapter 2 in *Introduction to Solid State Physics* by Kittel).

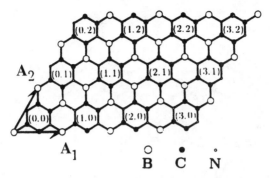

Figure 17.3 Structure of BC$_2$N (Type I) sheet. The unit cell is about twice that of either graphite or BN as seen from the lattice translation vectors, $\vec{A}_1$ and $\vec{A}_2$. Bond lengths determined from theoretical work are the following: $d_{C-C} = 1.42$ Å, $d_{B-C} = 1.55$ Å, $d_{B-N} = 1.45$ Å, and $d_{C-N} = 1.32$ Å [6]. The distance between layers is determined from experiment to be 3.35 Å for this material [10]. Note that for this particular arrangement of B, C, and N atoms, the sheet has inversion symmetry.

Γ while the corner point is labeled **K** (all the corner points are equivalent). The unique feature of graphite is that it has a Fermi *point*, instead of a surface, because the valence and conduction bands meet exactly at the **K** points as is apparent in the graphite band structure plot of Figure 17.6c.

17.3.2 Band Structure of Carbon Tubes

Since a tube may be formed mathematically by cutting a strip from a graphite sheet and rolling it into a cylinder, the two-dimensional sheet becomes a one-

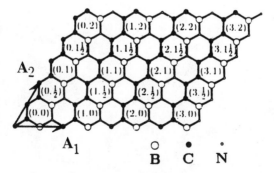

Figure 17.4 Structure of BC$_2$N (Type II) sheet. Lattice translation vectors, $\vec{A}_1$ and $\vec{A}_2$, are identical to those shown for the Type I material (Fig. 17.3). Bond lengths determined from theoretical work are the same as for Type I [6]. The distance between layers is determined from experiment to be 3.35 Å [10]. Note that for this particular arrangement of B, C, and N atoms, the sheet does not have inversion symmetry.

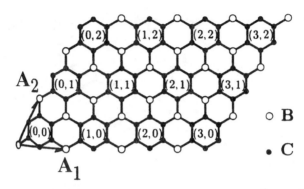

Figure 17.5   Structure of BC$_3$ sheet. Lattice translation vectors, $\vec{A}_1$ and $\vec{A}_2$, are similar to those shown for BC$_2$N material. Bond lengths determined from theoretical work are the following: $d_{C-C} = 1.42\,\text{Å}$ and $d_{C-B} = 1.55\,\text{Å}$ [7]. The distance between layers is determined from experiment to be $3.35\,\text{Å}$ [10]. The unique arrangement of B and C atoms causes this heterogeneous sheet to have inversion symmetry.

dimensional tube structure. The curvature induces σ-π hybridization, which introduces some sp^3 character into the planar sp^2 nature of the carbon bonds. This hybridization, however, only significantly affects small tubes; thus, for the most part, the energy bands of nanotubes may be derived explicitly from the graphite band structure. Due to the high aspect ratio of nanotubes, the axial dimension is still considered infinite but the periodic boundary conditions around the tube circumference now constrain the allowed k values. The *sheet* wave vector $\vec{k}$ will be included if it satisfies the condition of single-valuedness of

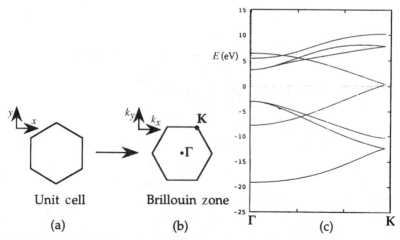

Figure 17.6   Band structure of graphite. (a) Unit cell of graphite. (b) BZ of graphite. (c) Band structure of graphite.

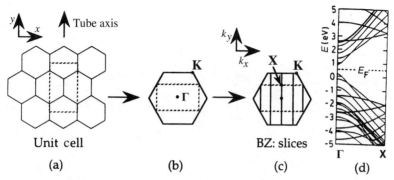

Figure 17.7 Band structure of zigzag carbon nanotube. (a) Tubular unit cell. (b) BZ of zigzag tube relative to graphite BZ. (c) Tubes have BZ lines as opposed to areas. These give the allowed k values for the (4, 0) tube. (d) Band structure for (13, 0) tube [3].

the *tube* wave function:

$$\vec{C} \cdot \vec{k} = 2\pi \mathbf{J}, \quad \mathbf{J} = 0, 1, 2, 3, \ldots \tag{17.1}$$

where $\vec{C}$ is the circumference vector of the tube. This constraint manifests itself as allowed *lines* in the reciprocal lattice instead of an area. The construction of the band structure of nanotubes is best illustrated by example.

17.3.2.1 Zigzag Tubes
Figure 17.7 describes the formation of the band structure for an $(n, 0)$ tube, which is referred to as a zigzag tube. The repeat unit for this tube along the tube axis is shown in Figure 17.7a. Because this unit cell is bigger than the hexagonal unit cell of graphite, the BZ is smaller. Figure 17.7b shows the BZ of a zigzag tube relative to that of graphite. k_y, corresponding to the direction along the tube axis, can take on a continuum of values but k_x is discretized. Now, the allowed k values depend on the particular circumference vector $\vec{C}$. Let us take the (4, 0) tube specifically. The allowed k values then lie along the lines drawn in Figure 17.7c and the tube bands are given by the corresponding graphite bands for those k values. Since the allowed points in this particular case do not include the **K** point, the tube is semi-conducting. From symmetry, the bands can be folded onto each other so that complete band information may be conveyed between the Γ and X points. As labeled in Figure 17.7c, the X refers to the edge of the BZ in the k_y direction; thus, the coordinates are determined directly from the size of the tubular unit cell. In this case $\mathbf{X} = \pi/\sqrt{3}a_{GR}$ for all $(n, 0)$ tubes. The $(n, 0)$ carbon nanotubes, which are semiconducting, have direct gaps because the tube band may be folded onto the $k_x = 0$ line from symmetry arguments. Continuing this kind of construction for other values of n gives the following results: Tubes that have $n = 3J$, where $J = 1, 2, 3 \ldots$, are narrow-gap semiconductors and the rest are semiconductors with gaps ranging up to 1.25 eV. Figure 17.7d shows the band structure of the (13, 0) tube.

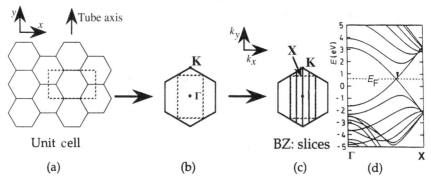

Figure 17.8 Band structure of armchair carbon nanotube. (a) Tubular unit cell. (b) BZ of armchair tube relative to graphite BZ. (c) BZ of the (2, 2) tube. (d) Band structure for (6, 6) tube [3].

17.3.2.2 Armchair Tubes As another example, let us take the case of an (n, n) armchair tube. Figure 17.8 is analogous to Figure 17.7. Note the repeat unit has now rotated by 90° from the previous case. This time, $\mathbf{X} = \pi/a_{GR}$ in accordance with the armchair tubular unit cell, and the $\mathbf{K}$ point is always included because $k = 0$ is always allowed. Thus, all armchair chirality tubes are metallic. In fact, these have the distinction of being the only tubes calculated to be metallic. The band structure of the (6, 6) tube is shown in Figure 17.8c.

Simular analyses may be done for the other (n_1, n_2) carbon nanotubes. The BZ in each case will be oriented slightly differently with the angle of rotation reflecting the helicity designated by the particular $\check{\mathbf{C}}$.

17.3.3 BN Band Structure

Planar hexagonal BN has a lattice almost identical to that of graphite. Thus, the BZ of the BN sheets and tubes may be constructed in a similar fashion. For comparison, the band structures of a BN sheet and the (4, 4) BN nanotube are shown in Figure 17.9. Note that the conduction and valence bands are separated by a gap, which arises from the asymmetry in the crystal potential due to the heterogeneous lattice of B and N atoms.

17.4 PREDICTED PROPERTIES OF BN, BC$_2$N, AND BC$_3$ NANOTUBES

17.4.1 Theoretical Predictions of BN Nanotubes

17.4.1.1 Nanotubular Structure As discussed earlier, hexagonal boron nitride (BN) is a layered material with approximately the same lattice param-

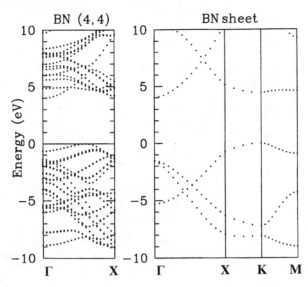

Figure 17.9 Band structure of (a) BN sheet and (b) a BN (4, 4) nanotube [11].

eters as graphite. The main difference, of course, is that while in graphite each vertex of the hexagon is occupied by a carbon atom, in BN a boron atom is surrounded by three nitrogen atoms or vice versa. The bonding between the boron and nitrogen atoms is also sp^2-like in nature, but the asymmetry in the crystal potential, arising from the dissimilar atoms, causes bulk hexagonal BN to be an insulator with a gap of 5.8 eV [12]. Authors of a tight-binding calculation originally proposed that nanotubes might also be formed from hexagonal BN [5]. Theoretical formulation of a BN tube is identical to the carbon case described earlier, where a strip is cut out of a hexagonal sheet and rolled to form a tube. Thus, BN tubes may also be formed with a variety of diameters and chiralities depending on the circumference vector $\vec{C}$.

Total energy calculation results of the strain energy needed to form a tube of a given diameter are shown in Figure 17.10 [11]. The closed circles represent the energy needed to form a BN tube relative to a sheet of hexagonal BN, while the open circles indicate the energy of a carbon nanotube relative to graphite. Clearly, the tubes are higher energy structures, but compared to their respective sheet material, BN nanotubes are energetically even more favorable than carbon ones and therefore likely to form.

17.4.1.2 Electrical and Mechanical Properties
Local density approximation (LDA) and quasi-particle band structure calculations predict BN tubes to be semiconducting with a gap of roughly 4–5.5 eV *independent* of tube diameter, chirality, and number of tube walls [11]. This uniformity in the calculated electronic properties of BN nanotubes contrasts sharply with the heter-

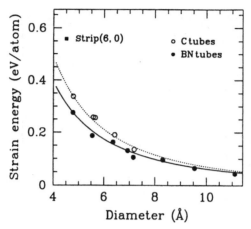

Figure 17.10   Strain energy versus diameter for the formation of BN and carbon nano-tubes relative to their sheet structures. Closed and open circles indicate the energy for BN and carbon nanotubes, respectively. (Courtesy of X. Blase.)

ogeneity of carbon tubes and suggests that BN tubes may present significant advantages from an applications point of view. Details of the tube band structure show that the lowest lying state is a nearly free-electron-like state, which has a maximum charge density about 2 Å interior to the tube wall. Thus, if the BN tubes were doped with, say, carbon, the resulting metallic tube would carry a cylinder of charge internally along its length.

Theoretical calculations of the elastic properties predict that BN nanotubes will be slightly softer than carbon tubes. The predicted Young's modulus of BN tubes is 0.95 times the elastic modulus of carbon tubes, or of order 1 TPa.

17.4.2   Theoretical Predictions of BC_2N and BC_3

Due to the greater complexity of BC_2N, the unit cell is double that of graphite, and there are two possible arrangements of the B, C, and N atoms in the sheet, resulting in two different tube structures. Figure 17.11 shows the structure of tubes (Types I and II) of the same diameter but made from different isomers of the sheet material [6]. The Type I sheet has inversion symmetry (as does graphite) while the Type II sheet does not (similar to BN). Thus, it is not surprising that band structure calculations on the electrical properties of the Type I and Type II tubes parallel the properties of carbon and BN nanotubes, respectively. Type I tubes range from semiconducting to metallic depending on diameter and chirality, while the Type II tubes are predicted to be semiconducting independent of tube parameters. The most unique feature of tubes made from this material is that the arrangement of atoms (chain of conducting carbon alternating with a string of insulating BN) in the Type II tubule resembles a solenoid. Doping this semiconducting Type II tubule to metallicity

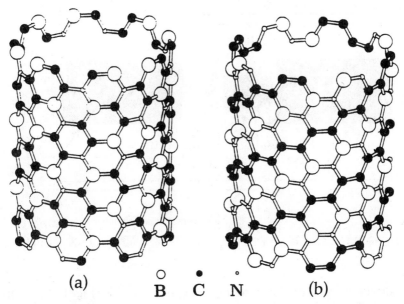

(a)   ○   ●   ○   (b)
      B   C   N

Figure 17.11   Theoretically determined tubules of isomers of BC_2N. These are the (4, 4) tubes in the indexing numerology of tubes. (Courtesy of Yoshiyuki Miyamoto.)

would cause the electrons to flow in a helical pattern along the chain of carbon atoms, becoming a *nanocoil*!

The electrical behavior of BC_3 tubes is rather complex, but the most significant result from the theoretical calculations predicts concentric tubes of BC_3 to be metallic [7]. Thus, all multiwalled structures made from this material are likely to be good conductors. On the other hand, single-wall, isolated BC_3 tubes are predicted to be semiconducting. Interestingly, when a number of such semiconducting tubes are aligned and brought into contact, the resulting bundle of tubes constitutes a metallic system.

17.5   SYNTHESIS OF BORON-NITRIDE-CONTAINING NANOTUBES

17.5.1   Synthesis of BN Nanotubes

The similarity between graphite and hexagonal BN suggests that some of the successful synthesis methods used for carbon nanotube production might be adapted to $B_xC_yN_z$ nanotube growth. This is indeed the case. A nonequilibrium plasma arc technique has been used to produce pure BN nanotubes [13]. To avoid the possibility of carbon contamination, no graphite components are used in this synthesis. The insulating nature of bulk BN prevents the use of a pure BN electrode. Instead, a pressed rod of hexagonal BN (white in color) is inserted into a hollow tungsten (W) electrode, forming a compound anode. The

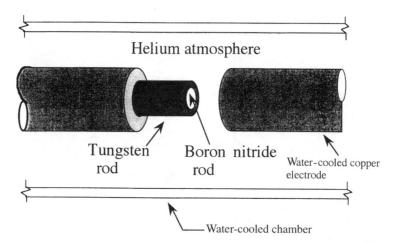

Figure 17.12 Electrode configuration used for the synthesis of BN nanotubes.

cathode consists of a rapidly cooled pure copper electrode (Fig. 17.12). During discharge the environmental helium gas is maintained at 650 torr and a dc current between 50 and 140 A is applied to maintain a constant potential drop of 30 volts between the electrodes while arcing.

Arcing the BN/W compound electrode results in a dark gray soot deposit on the copper cathode, in contrast to the cohesive cylindrical boule, which typically grows on the cathode upon graphite arcing. Due to instabilities, however, the BN/W arc burns only for a short time, thus yielding a limited quantity of soot. Pieces of solidified tungsten are often found spattered inside the chamber, indicating that the temperature at the anode during synthesis exceeds 3700 K, the melting point of tungsten. Other synthesis methods, using different catalysts, are also possible (see Section 17.7).

17.5.2 Characterization

17.5.2.1 High-Resolution TEM
The preferred analytical tool for nanotube characterization is transmission electron microscopy (TEM).[2] Figure 17.13 shows a typical TEM image of the dark gray cathodic deposit produced in the arc-discharge chamber of a BN/W arc run. There are apparent numerous structures of distinct and contrasting morphologies. The large amorphous band covering nearly the entire lower half of the image is a portion of the support grid. The dark clusters scattered throughout the upper half of the image are tentatively identified as tungsten. Most importantly, Figure 17.13 clearly shows structures that appear to be multiwalled nanotubes, with inner (outer) diame-

---

[2] Cathodic deposit was characterized using a JEOL JEM 200CX TEM with 200 keV accelerating voltage.

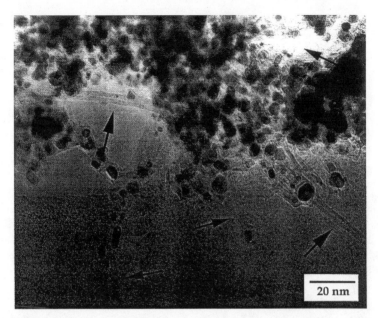

Figure 17.13 TEM image of soot produced in BN nanotube synthesis. Black arrows point out to sections of tubes. Big black arrows point to one long tube.

ters on the order of 1–3 nm (6–8 nm) and lengths exceeding 200 nm. The two dark arrows in Figure 17.13 identify one such tube, which extends beyond the left and right borders of the image. The light arrow in Figure 17.13 identifies another multiwalled tube of apparently shorter length.

Figure 17.14 shows a high-resolution TEM image of a portion of an observed nanotube. This and other similar images exhibit sharp lattice fringes, indicating that the walls of the tubes are well ordered with an interlayer distance of ∼3.3 Å consistent with the interplanar distance of 3.33 Å in bulk hexagonal boron nitride [9]. The particular tube shown in Figure 17.14 has eight walls; similar tubes with wall numbers ranging from two to nine have been observed.

17.5.2.2 Electron Energy Loss Spectroscopy

Although no graphite is used in the particular synthesis process described in detail above, confirmation of the chemical makeup and stoichiometry is crucial for conclusive evidence of BN nanotube discovery. Determination of the chemistry and stoichiometry of individual tubes is possible using electron energy loss spectroscopy (EELS) inside the TEM. High spatial resolution EELS studies have been performed [13] on portions of tubes suspended over holes in the carbon support grid as characterized in Figure 17.15. Figure 17.16 shows a characteristic tube energy loss spectrum, collected by probing a 10 nm region of the tube. Two distinct absorption features are revealed, one beginning at 188 eV and another at 401 eV. These correspond to the known K-edge onsets for boron and nitrogen, respec-

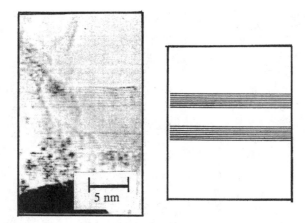

Figure 17.14 High-resolution TEM image of a multiwalled BN nanotube clearly showing the inner diameter and the equal number of lattice fringes on either side representing the number of tube walls. This tube has eight walls.

tively. The fine structure in the spectrum reveals the sp^2 bonding between boron and nitrogen [14]. Noteworthy is the absence of any feature at 284 eV, the K-edge absorption for carbon. Quantification of the tube EELS spectrum gives a B/N ratio of 1.14, consistent with a stoichiometry of BN (due to uncertainties in baseline corrections, the given B/N ratio has an estimated error of 20%).

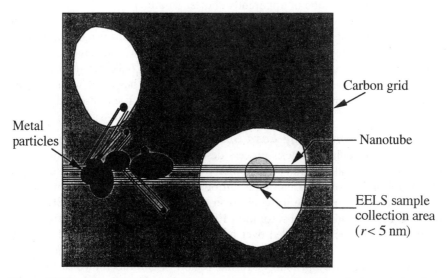

Figure 17.15 Schematic of configuration used in collecting the electron energy loss spectra on a nanotube.

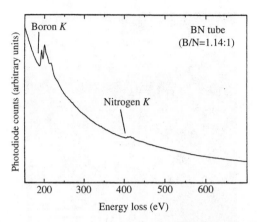

Figure 17.16 Electron energy loss spectrum of a BN nanotube confirming the sp^2 bonding of the boron and nitrogen and the absence of carbon in the tube structure. Quantitative analysis gives a B/N ratio of 1:1.14.

17.5.3 Possible Growth Scenario

A careful study of the ends of BN nanotubes synthesized using the tungsten electrode method reveals an interesting feature. As seen in Figure 17.17, the observed end contains a metal particle, most likely tungsten or a tungsten compound with boron and nitrogen. In contrast to the carbon nanotube, where the capping is fullerene-like or involves pentagons and heptagons, the BN tube closure by pentagon formation is suppressed due to the necessity of unfavorable B–B or N–N bonds. Nature seems to solve this problem by using, if available, a small metal cluster. The presence of many metal particles wrapped in layers of planar boron nitride, as is evident in Figure 17.13 and seen at higher magnifi-

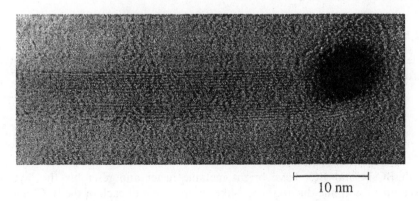

10 nm

Figure 17.17 Micrograph of the end of a BN nanotube showing termination by a metal particle.

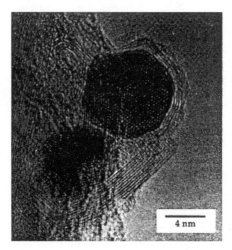

Figure 17.18 Micrograph of a metal particle covered with crystalline BN.

cation in Figure 17.18, suggests that in a high-curvature region of a covered particle, the outer layer of the BN coating may pull from the innermost layer (since the layer–layer interaction in BN is rather weak) and grow outward to form a nanotube. Given this scenario, the tube growth is likely to terminate when a metal particle collides with the open end of a growing tube and attaches to the dangling bonds, particularly if the metal forms stable nitrides and borides, as tungsten does.

Other experiments, using a slightly different arc-discharge synthesis configuration, have successfully produced single-walled BN nanotubes [15]. Tubes made by Loiseau et al. [15], however, do not have metal particles at the end, but terminate with flat tops, suggesting a square B–N arrangement at the ends compared to the hexagons in the wall. (Any even number polygon accommodates the preferred B–N bonding.)

17.5.4 Synthesis of BC$_2$N and BC$_3$

The synthesis of bulk (i.e., layered sheets of) BC$_2$N and BC$_3$ is achieved through the following chemical reactions:

$$CH_3CN + BCl_3 \xrightarrow{>800\,°C} BC_2N + 3HCl$$

$$2BCl_3 + C_6H_6 \xrightarrow{\sim800\,°C} 2BC_3 + 6HCl$$

Both BC$_2$N and BC$_3$ have bright metallic luster and resemble the layered structure of graphite due to the sp^2-like environment of each of the B, C, and N atoms in the bulk materials, signifying two-dimensionality. Resistivity measurements indicate that BC$_2$N is semiconducting with a gap of about 0.03 eV

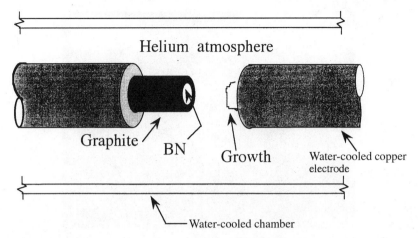

Figure 17.19 Schematic of the electrode configuration used to produce multiwalled BC_2N and BC_3 nanotubes in the arc-discharge chamber.

and BC_3 is semimetallic. The synthesis and characterization of both of these novel materials were done by Kouvetakis et al. in 1989 [16].

Along with the synthesis of carbon and BN nanotubes, the arc discharge allows for the successful production of tubes with other novel stoichiometries of boron, carbon, and nitrogen as well [17]. The arc configuration used in one experiment is described schematically in Figure 17.19. An insulating BN piece is inserted into a hollowed graphite rod resulting in a compound electrode. The cathode is the water-cooled copper piece as in the other experiments. The chamber parameters are similar to those used in multiwalled carbon nanotubes. The 450 torr helium pressure is bled into the chamber and the current is set to 55 A. Reminiscent of the carbon scenario, the compound BN/C electrode erodes and a growth occurs on the cooled copper surface. Detailed examination of the boule after arcing finds the inner core to be harder than the surrounding layers, which is in direct contrast to observations of pure carbon boule samples. However, in the case of a compound BN/C electrode, the central region contains hexagonal BN, a ceramic; therefore, from simple geometric arguments, a harder inner core is reasonable. The color of the boule, too, is gray instead of the typical carbon black, suggesting that the originally white BN has indeed been consumed in the arcing process and mixed with the black graphitic material. Figure 17.20 is a representative scanning electron micrograph (SEM) of the inner core boule material from this experiment. Tubes are clearly visible in the image along with significant quantities of bulk material. As is evident from Figure 17.20, the yield of nanotube structures is low in this experiment. TEM and EELS studies [17] confirm the crystallinity of the tube structure and individual tube stoichiometries of BC_2N and BC_3. In fact, EELS analysis also revealed that the sample contains nanotubes of pure carbon. Thus, using the configuration shown in Figure 17.19, the nonequilibrium arc-discharge tech-

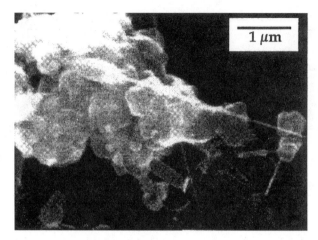

Figure 17.20 SEM image of boule sample from a combined boron, carbon, and nitrogen experiment.

nique simultaneously produces multiwalled nanotubes of BC_2N, BC_3, and pure carbon.

Other studies using boron, carbon, and nitrogen have bound similar results [18–20].

17.6 ELASTIC PROPERTIES OF BN NANOTUBES

The elastic properties of an individual boron nitride (BN) nanotube have been experimentally determined through in situ studies in the transmission electron microscope [21]. These experiments are motivated by similar experiments on carbon nanotubes, first performed by Treacy et al. [22]. Analysis using the thermal vibration amplitude of a cantilevered BN nanotube yields a Young's modulus of 1.22 TPa. Because elastic measurements probe the microstructure of a material, the high value of the elastic constant suggests the BN nanotubes are indeed crystalline with few defects as observed in high-resolution micrographs.

Comparison with other materials finds BN nanotubes to be the stiffest insulating fiber known. Hexagonal BN is well known for its high-temperature resistance, and combined with these high-strength properties, BN nanotubes have the potential for unique applications in many different areas.

17.6.1 Vibrating BN Nanotube

Figure 17.21 is a TEM image of a BN nanotube specimen at 300 K. Two individual BN nanotubes are clearly visible, both cantilevered over a hole in the support grid. A short BN tube in the lower right region of the image (identified

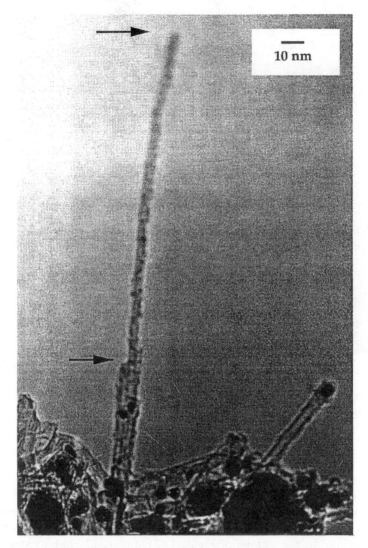

Figure 17.21 Vibrating BN nanotube at $T = 300$ K.

by a white arrow) has a clearly visible metal particle at its tip; the entirety of the tube is in clear focus, indicating a small vibration amplitude. The central region of Figure 17.21 shows a single long cantilevered BN nanotube. The base (lower black arrow) of this nanotube is in clear focus, while closer to the tip region (upper black arrow) the image becomes successively more blurred. Rotation studies verified that the blurring nature of the tip was not due to sample tilt, and variations in electron flux had no effect on the image, confirming that the tube vibration was due to thermal effects.

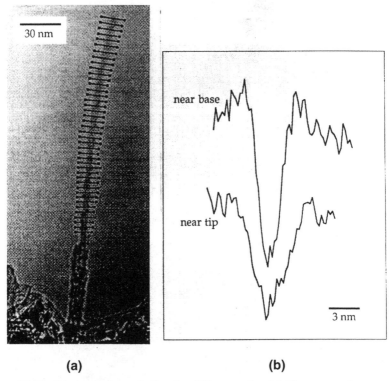

(a)   **(b)**

Figure 17.22   Line scan data of vibrating BN nanotube. (a) Placement of scans on the tube image. (b) Line scans near base and tip.

17.6.2   Data Analysis

To quantify the amplitudes of the vibration modes of the nanotube in the image plane, a series of intensity line scans perpendicular to the nanotube axis were performed on the micrograph of Figure 17.21. Figure 17.22a indicates the positions of the scans (short horizontal bars) while Figure 17.22b shows two representative intensity scans, one near the supported base and one near the nanotube tip. Because the tube extends out over unsupported area, the sudden drop from the background signifies the high contrast due to the tube. Near the supported base, the scan shows a sharper drop in contrast as compared to the line scan near the tip. This is consistent with the condition of focus at the base versus blurring at the end, which is depicted visually in Figure 17.21. From the width of the high-contrast region, the apparent tube diameter at points along the length of the structure is determined. As expected, the apparent tube width at the tip is greater than that at the base. The width of the nanotube at the base, where the vibration amplitude falls to zero, is 3.5 nm.

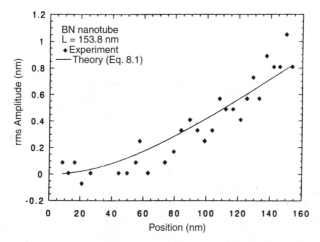

Figure 17.23 Plot of rms amplitude versus position for oscillating BN nanotube. Note general trend looks like the fundamental mode of a cantilever.

Figure 17.23 shows (as diamonds) the rms amplitude of the BN nanotube as a function of distance from the nanotube base, determined from a deconvolution of the intensity line scans from the baseline scan. As expected, the vibration amplitude increases with increasing distance from the clamped base of the nanotube.

17.6.3 Derivation of Y for a Thermally Excited Cantilever

The unique size of these nanotubes leads to interesting questions regarding the calculation of their mechanical properties. Can the tubes be treated as continuous hollow cylindrical structures or does their nanometer size call for a more discrete treatment? Doublet mechanics, a recently developed analytic approach to mechanics that incorporates the discrete nature of matter in the calculation of bulk behavior, has the potential to answer such a question [23]. Meanwhile, empirical potentials and first-principles total-energy calculations indicate that relationships derivable from continuum elasticity theory are applicable even for tubes with diameters as small as a C_{60} molecule (7 Å) [24].

The nanotube is approximated as a cantilever of length L, rigidly clamped at one end, freely vibrating at the other with a uniform circular cross section of outer diameter a and inner diameter b and mass per unit length μ. A schematic of the mechanical system is shown in Figure 17.24. Ideally a multiwalled tube is not a Bernoulli–Euler beam because the elastic property of the tube walls is, in fact, different from that of the area in-between, where the van der Waals force acts. However, the region that distinguishes the wall from the area in-between is so small that the error in assuming a multiwalled nanotube to be a uniform rigid beam is negligible. The displacement, $u(x, t)$, of the vibrating nanotube is

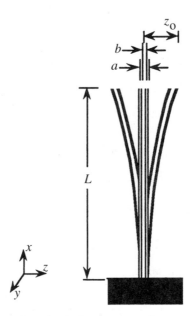

Figure 17.24 Schematic of mechanical system used to approximate a cantilevered nanotube.

a function of distance x and time t and can be described as a superposition of normal modes,

$$u(x, t) = \sum_{n=1}^{\infty} u_n(x, t) = z_0 \sum_{n=1}^{\infty} \alpha_n \phi_n(x) \sin \omega_n t \qquad (17.2)$$

where z_0 is the maximum amplitude and α_n is the relative amplitude of the normal mode $\phi_n(x)$ at frequency ω_n. The complete set of normalized normal modes of a cantilever beam are

$$\phi_n(x) = \frac{1}{2}\left(\cosh \beta_n x - \cos \beta_n x - \frac{\cosh \beta_n L + \cos \beta_n L}{\sinh \beta_n L + \sin \beta_n L}(\sinh \beta_n x - \sin \beta_n x)\right) \qquad (17.3)$$

with $\beta_n L = 1.8751, 4.6941, 7.8548,$ and 10.996 for $n = 1, 2, 3,$ and 4, respectively, and approximately $((2n - 1)\pi)/2$ for $n > 4$. Young's modulus, Y, is embedded in the associated frequency expression,

$$\omega_n = (\beta_n L)^2 \sqrt{\frac{YI}{\mu L^4}} = (\beta_n L)^2 \sqrt{\frac{Y\pi(a^4 - b^4)}{64 \mu L^4}} \qquad (17.4)$$

where the second moment of area $I = (\pi(a^4 - b^4))/64$ is for a hollow circular cross section.

In order to extract a value for Y from the experimental data, let us consider the energy of the system. The average kinetic energy of a given mode is

$$\langle E_n^{\text{kinetic}} \rangle = \int_0^L \tfrac{1}{2} \mu \langle [\dot{u}_n(x, t)^2] \rangle \, dx = \tfrac{1}{2} \mu \omega_n^2 z_0^2 \alpha_n^2 \int_0^L [\phi_n(x)]^2 \, dx \langle \cos^2 \omega_n t \rangle \quad (17.5)$$

From the equipartition theorem, the average kinetic energy in each mode is $k_B T/2$, where k_B is Boltzmann's constant and T is the temperature. Equation Eq. (17.5) to $k_B T/2$ and using Eq. (17.4) yields

$$\alpha_n = \sqrt{\frac{8 k_B T}{\mu \omega_n^2 z_0^2 L}} = \frac{1}{(\beta_n L)^2} \sqrt{\frac{512 L^3 k_B T}{Y \pi (a^4 - b^4) z_0^2}} \quad (17.6)$$

Hence, Eq. (17.2) becomes

$$u(x, t) = \sqrt{\frac{512 L^3 k_B T}{Y \pi (a^4 - b^4)}} \sum_{n=1}^{n_{\max}} \frac{1}{(\beta_n L)^2} \phi_n(x) \sin \omega_n t \quad (17.7)$$

Thus, the equipartition theorem fixes the relative amplitude of each mode. The amplitude contribution of higher modes falls off as $\sim(1/n^2)$; therefore, the vibration amplitude profile is dominated by the first few modes. The sum is ultimately limited at room temperature by $n_{\max} \approx 85$, where $\hbar \omega_{n_{\max}} \approx k_B T$.

The present formulation is an exact solution to the problem of a cantilever oscillating due to thermal effects. However, as stated earlier, the nanotube walls are treated as a single uniform material instead of discrete layers, a clear approximation.

17.6.4 Calculating Y_{BN}

The BN nanotube is approximated as a cantilevered Bernoulli–Euler beam of length L_{BN}, rigidly clamped at one end, freely vibrating at the other with a uniform circular cross section of outer diameter a_{BN} and inner diameter b_{BN}. Rewriting Eq. (17.7) for the displacement of a cantilevered tube oscillating at temperature T in terms of BN nanotube parameters gives

$$u_{BN}(x, t) = \sqrt{\frac{512 L_{BN}^3 k_B T}{Y_{BN} \pi (a_{BN}^4 - b_{BN}^4)}} \left(\sum_{n=1}^{n_{\max}} \frac{1}{(\beta_n L_{BN})^2} \phi_n(x) \sin \omega_n t \right) \quad (17.8)$$

The markers in Figure 17.23 are the experimental data and the solid line is the fit using Eq. (17.8). From the fit, the maximum rms amplitude is found to be

0.8 nm. This fit together with the measured dimensions of the nanotube, $a_{BN} = 3.5$ nm, $b_{BN} = 2.2$ nm, and $L_{BN} = 153.8$ nm, yields an elastic modulus of $Y_{BN} = 1.22$ TPa for the BN nanotube at 300 K. Incorporating in the analysis the metal particle at the nanotube tip does not alter the result.

17.6.5 Discussion of Experimental Parameters for Y_{BN} Measurement

In the above calculation, the base of the cantilever has been taken to be where the single tube extends over the hole, as pointed to by the lower black arrow in Figure 17.21. (The focus condition supports this assumption since the region below the lower black arrow in Fig. 17.21 is entirely in focus.) If, instead, the point of rigid clamping is assumed to be directly at the clump, the mechanical system is then described by a beam with a varying second moment of area, $I = I(x)$. However, the second moment of area of the region where several tubes extend from the clump a short distance is much large than that of the single tube, and thus the combined system dynamically behaves as if the canti-levered tube was rigidly clamped at the position pointed out by the lower black arrow. Thus, the calculations for the combined system are identical to the mechanical system assumed in deriving Eq. (17.8), and the result for the elastic modulus of the BN nanotube is the same.

As mentioned previously, BN nanotubes synthesized by the tungsten-arc technique often terminate with a metal particle. If we assume there is a tungsten particle with a diameter on the order of the outer tube diameter and repeat the calculation, we find that the fundamental frequency goes down by 17% but the BN tube elastic modulus does not change. This is consistent with a simple mass–spring system, where the frequency depends on the mass and elastic con-stant of the spring, but the spring constant is independent of the mass as long as it is under its elastic limit.

17.6.6 Comparison of Elastic Moduli

Theoretically, the elastic modulus of a BN nanotube should be slightly smaller than that of a carbon nanotube. Although the structures are similar, the pho-non frequencies [25] in graphite are calculated to be higher than those in planar BN, suggesting that the carbon–carbon bond is stronger than the boron–nitrogen bond. The experimental result for BN reflects this difference when compared to measured values of multiwalled carbon nanotubes [22].

The BN nanotube Young's modulus is 14 times greater than the measured in-plane modulus of bulk hexagonal BN material [33]. This difference is possibly due to the tube being a defect-free single crystalline piece, while the bulk hex-agonal material is a composite of defected layers. Also, it is conceivable that the curvature of the tube strengthens the sp^2 bonding between the boron and nitrogen atoms, resulting in a material with improved mechanical properties. Thus, this experimental measurement shows the impressive change in elastic properties of a material due to nanometer scale sample geometry.

TABLE 17.1   Table of Elastic Moduli of a Variety of Fiber Materials

Material	Y (TPa)	Reference
Carbon fiber	0.2–0.8	26
Carbon nanotube	0.4–4	22
BN nanotube	1.22	21
Kevlar 49	0.112	27, p. II-2
E-glass fiber	0.074	28
SiC fiber	0.2	29
Steel wire	0.2	30
Copper wire	0.110	31
$Bi_2Sr_2Ca_1Cu_2O_x$ whisker	0.02	32
$Bi_2Sr_2Ca_2Cu_3O_x$ whisker	0.03	32

Indeed, fibers often have improved elastic properties in comparison to the bulk material [34]. Table 17.1 shows the Young's moduli of a range of fiber materials. The BN nanotube is an order of magnitude stiffer than any other insulating fiber and the second stiffest material after carbon nanotubes.

17.7   SUMMARY OF BN, BC$_2$N AND BC$_3$ NANOTUBE PROPERTIES AND OTHER B$_x$C$_y$N$_z$ STRUCTURES

The pure BN and carbon-containing sp^2-bonded host materials discussed in this chapter form the basis for a number of interesting structures. The synthesis and characterization of such B$_x$C$_y$N$_z$ materials represents an area of active experimental and theoretical research. Table 17.2 summarizes much of the theoretical and experimental work that has been presented here on nanotubes synthesized from boron, carbon, and nitrogen. From Table 17.2 it is evident that,

TABLE 17.2   Summary of Predicted and Measured Properties of B$_x$C$_y$N$_z$ Nanotubes

Type of Nanotube	Predicted Properties			Experimentally Determined	
	Electrical	E_{gap} (eV)	Y (TPa)	Y (TPa)	Single-Walled Nanotube Found?
Carbon	Semiconducting or metallic	0–1.5	1–7	1.35	Yes
BN	Semiconducting	4–5.5	0.95–6.65	1.18	Yes
BC$_2$N (I)	Semiconducting or metallic	—	—	—	No
BC$_2$N (II)	Semiconducting	1.28	—	—	No
BC$_3$	Metallic	—	—	—	No

so far, the largest amount of work, both theoretcial and experimental, has been done on carbon and BN nanotubes. Clearly, minimal experimental work has been done on BC_2N and BC_3. Carbon nanotubes have existed for the longest period of time and large quantities of tube material are readily available due to the stable arcing of graphite, which produces the plentiful tube-containing boule. BN nanotubes, due to their potentially interesting electrical application, have also been studied, although techniques for large-scale production need to be explored.

Loiseau et al. [15] have used an arc-discharge method employing HfB_2 electrodes in a nitrogen atmosphere to synthesize BN nanotubes with wall number ranging from many to one. In this synthesis method, the Hf is apparently not incorporated into the tube itself, but rather acts as a catalyzing agent. The source of nitrogen for tube growth is from the N_2 environmental gas. The ends of the BN nanotubes so produced have flat layers perpendicular to the tube axis and are representative of the bond frustration that occurs upon tube closure (the simple six pentagon addition, which so beautifully closes a pure carbon nanotube, is not realized in BN nanotubes because the B–B bond is not favored).

Tantalum has also been used as the catalyzing agent in the synthesis of various nanoscale BN structures using arc-vaporization methods [19]. Pure BN nanotubes are produced, along with other nanoparticles including onion-like spheres similar to those produced by Banhart et al. [35] using high-intensity electron irradiation. In the study of Terrones et al. [19], circumstantial evidence is found for the presence of B_2N_2 squares at the BN nanotube tips, as well as B_3N_3 hexagons in the main fabric of the nanotubes.

Perhaps one of the most intriguing new developments in BN tube synthesis is the realization of nanotubes with segregated tube-wall stoichiometry. For example, Suenaga et al. [36] have produced multiwalled nanotubes containing pure carbon walls adjacent to pure BN walls, forming a sort of nanotube co-axial cable. In one specific tube studied carefully by electron energy loss spectroscopy, the innermost three walls of the tube contained only carbon, the next six walls were comprised of BN, and the last five outermost walls were again pure carbon. The entire 14-walled composite nanotube was 12 nm in diameter. Similar layer segregation is obtained for onion-like coverings over nanoparticles (quite often the core nanoparticle is composed of the catalyst material). The results of Suenaga and co-workers are consistent with the earlier findings of Redlich et al. [37], who observed B–C–N nanotubes consisting of concentric cylinders of BC_2N and pure carbon.

More recently, Zhang et al. [38] have used a reactive laser ablation method to synthesize multi-element nanotubes containing BN. The nanotubes contain a silicon carbide core followed by an amorphous silicon oxide intermediate layer; this composite nanorod is then sheathed with BN and carbon nanotube layers, segregated in the radial direction. It has been speculated that merging BN and carbon nanotube structures may be the basis for novel electronic device architectures.

ACKNOWLEDGMENTS

We thank the following people for helpful interactions: L. Benedict, K. Cherrey, M. L. Cohen, P. Collins, V. Crespi, R. Gronsky, S. G. Louie, R. Luyken, Y. Miyamoto, M. O'Keefe, A. Rubio, and Z. Weng-Sieh. This work was supported in part by the sp^2 Materials Initiative, Office of Energy Research, Office of Basic Energy Sciences, Materials Sciences Division of the U.S. Department of Energy under Contract No. DE-AC03-76SF00098.

REFERENCES

1. S. Iijima, *Nature* **1991**, *354*, 56–58.

2. H. W. Kroto, J. R. Heath, S. C. O'Brien, R. F. Curl, and R. E. Smalley, *Nature* **1985**, *318*, 162–163.

3. N. Hamada, S. Sawada, and A. Oshiyama, *Phys. Rev. Lett.* **1992**, *68*, 1579–1581.

4. G. Overney, W. Zhong, and D. Tománek, *Z. Phys. D* **1993**, *27*, 93–96.

5. A. Rubio, J. L. Corkill, and M. L. Cohen, *Phys. Rev. B* **1994**, *49*, 5081–5084.

6. Y. Miyamoto, A. Rubio, M. L. Cohen, and S. G. Louie, *Phys. Rev. B* **1994**, *50*, 4976–4979.

7. Y. Miyamoto, A. Rubio, M. L. Cohen, and S. G. Louie, *Phys. Rev. B* **1994**, *50*, 18360–18364.

8. M. S. Dresselhaus, G. Dresselhaus, and R. Saito, *Carbon* **1995**, *33*, 883–891.

9. R. S. Pease, *Acta Crystallogr.* **1952**, *5*, 356.

10. J. Kouvetakis, R. B. Kaner, M. L. Sattler, and N. Bartlett, *J. Chem. Soc. Chem. Commun.* **1986**, 1758.

11. X. Blase, A. Rubio, S. G. Louie, and M. L. Cohen, *Europhys. Lett.* **1994**, *28*, 335–340.

12. A. Zunger, A. Katzir, and A. Halperin, *Phys. Rev. B* **1976**, *13*, 5560–5573.

13. N. G. Chopra, R. J. Luyken, K. Cherrey, V. H. Crespi, M. L. Cohen, S. G. Louie, and A. Zettl, *Science* **1995**, *269*, 966–967.

14. L. Reimer, *Transmission Electron Microscopy*, P. W. Hawkes (ed.), Springer Series in Optical Sciences, Vol. 36. Springer-Verlag, Berlin, 1993.

15. A. Loiseau, F. Williame, N. Demoncy, G. Hug, and H. Pascard, *Phys. Rev. Lett.* **1996**, *76*, 4737–4740.

16. J. Kouvetakis, T. Sasaki, C. Chen, R. Hagiwara, M. Lerner, K. M. Krishnan, and N. Bartlett, *Synthetic Materials* **1989**, *34*, 1–7.

17. Z. Weng-Sieh, K. Cherrey, N. G. Chopra, X. Blase, Y. Miyamoto, A. Rubio, M. L. Cohen, S. G. Louie, A. Zettl, and R. Gronsky, *Phys. Rev. B* **1995**, *51*, 11229–11232.

18. O. Stephan, P. M. Ajayan, C. Colliex, Ph. Redlich, J. M. Lambert, P. Bernier, and P. Lefin, *Science* **1994**, *266*, 1683–1685.

19. M. Terrones, A. M. Benito, C. Manteca-Diego, W. K. Hsu, O. I. Osman, J. P. Hare, D. G. Reid, H. Terrones, A. K. Cheetham, K. Prassides, H. W. Kroto, and D. R. M. Walton, *Chem. Phys. Lett.* **1996**, *257*, 576–582.

20. Ph. Redlich, J. Loeffler, P. M. Ajayan, J. Bill, F. Aldinger, and M. Rühle, *Chem. Phys. Lett.* **1996**, *260*, 465–470.

21. N. G. Chopra and A. Zettl, *Solid State Commun.* **1998**, *105*, 297–300.

22. M. M. J. Treacy, T. W. Ebbesen, and J. M. Gibson, *Nature* **1996**, *381*, 678–680.

23. M. Ferrari, V. T. Granik, A. Imam, and J. C. Nadeau (eds.), *Advances in Doublet Mechanics*, Lecture Notes in Physics, Monogram Vol. 45. Springer-Verlag, Berlin, 1997.

24. D. H. Robertson, D. H. Brenner, and J. W. Mintmire, *Phys. Rev. B* **1992**, *45*, 12592–12595.

25. Y. Miyamoto, M. L. Cohen, and S. G. Louie, *Phys. Rev. B* **1995**, *52*, 14971–14975.

26. R. L. Jacobsen, T. M. Tritt, J. R. Guth, A. C. Ehrlich, and D. J. Gillespie, *Carbon* **1995**, *33*, 1217–1221.

27. DuPont, *Kevlar Aramid Fiber.* DuPont Advanced Fibers Systems, Wilmington, DE, 1992.

28. C. Matotzke, *Composites Sci. Technol.* **1994**, *50*, 393–405.

29. M. W. Barsoum, P. Kangutkar, and A. S. D. Wang, *Composites Sci. Technol.* **1992**, *44*, 257–269.

30. E. P. Popov, *Engineering Mechanics of Solids.* Prentice Hall, New York, 1990.

31. C.-L. Tsai and I. M. Daniel, *Composites Sci. Technol.* **1994**, *50*, 7–12.

32. T. M. Tritt, M. Marone, A. C. Ehrlich, M. J. Skove, D. J. Gillespie, R. L. Jacobsen, G. X. Tessema, J. P. Franck, and J. Jung, *Phys. Rev. Lett.* **1992**, *68*, 2531–2534.

33. Data supplied by Carborundum Corporation, Latrobe, PA.

34. X. Peters, *Composites* **1995**, *26*, 108–114.

35. F. Banhart, T. Fuller, P. Redlich, P. M. Ajayan et al., *Chem. Phys. Lett.* **1997**, *269*, 349–355.

36. K. Suenaga, C. Carbon, N. Demoncy, A. Loiseau et al., *Science* **1997**, *278*, 653–655.

37. Ph. Redlich, J. Loeffler, P. M. Ajayan et al., *Chem. Phys. Lett.* **1996**, *260*, 465–470.

38. Y. Zhang, K. Suenaga, and S. Iijima, *Science* **1998**, *281*, 973–975.

CHAPTER 18

SYNTHESIS AND CHARACTERIZATION OF MATERIALS INCORPORATED WITHIN CARBON NANOTUBES

JEREMY SLOAN and MALCOLM L. H. GREEN

18.1 Introduction

The discovery by Iijima in 1991 [1] that sp^2 graphene sheets can be made to curve in a helical fashion about an axis, thereby forming hollow concentric tubules 1–20 nm in internal diameter and from about 100 nm up to several micrometers in length, was followed, shortly thereafter, by the observation that their internal capillaries could be filled with molten lead [2]. Subsequently, both multiple-walled carbon nanotubes (MWNTs) and, more recently, single-walled carbon nanotubes (SWNTs) have been filled with a wide variety of different materials, including metal oxides [2–9], base metals [2,5,7,10–19], mixed metal oxides [8], metal salts [3,6–9,15,20–22], metal sulfides [8], metal carbides [10,19,23–33], main group elements [5,33], gases [34–36], proteins [37–39], and molecules of C_{60} [40]. The ability of carbon nanotubes to incorporate many of these materials arises from the fact that they terminate in fullerenic caps that are relatively more prone to oxidation than their cylindrical graphene bodies [3,12,16,41–44], therefore making their capillaries accessible to filling via solution-deposition or capillarity. An alternative filling strategy involves a process whereby the encapsulate effectively catalyzes the formation of its own carbon shell, which originates from an external carbon source. This has been achieved by coevaporation of the encapsulate with carbon in an electric arc [10,11,17,19,23–33]; by gas–solid deposition of carbon onto the encapsulate [45–47]; by chemical vapor deposition (CVD) onto metal deposited into an alumina membrane [48]; or, in the liquid phase, by the electrolysis of carbon rods in a mixture of molten alkali halide together with the halide of the material to be encapsulated [49,50].

Fullerenes: Chemistry, Physics, and Technology, Edited by Karl M. Kadish and Rodney S. Ruoff.
0-471-29089-0 Copyright © 2000 John Wiley & Sons, Inc.

While these studies are clearly interesting, the question must at some point be asked: what are the scientific and practical benefits of studying encapsulation phenomena within carbon nanotubes? The answer to this lies partly in terms of the physical properties of the nanotubes themselves and also in terms of the relative *scale* upon which the encapsulates are formed. SWNTs, in particular, are truly molecular capillaries consisting of just a single capped graphene cylinder with internal diameters spanning approximately 1–6 nm in magnitude, depending on the method of formation [51–54]. MWNTs, which according to the "Russian doll" structural model consist of groups of 2–20+ concentric capped graphene cylinders, have internal diameters spanning a somewhat larger range, typically 1–20 nm [1,55]. Even with respect to the densest of materials—for example, metals or metal carbides—these dimensions correspond to only a few atomic layers. On this scale, the surface, quantum, and nonlinear properties of many materials, as appropriate, will tend to predominate over their bulk properties [56–59]. Such materials could find uses, for example, as spectroscopic enhancers or chemical sensors in the visible range [60]. Carbon nanotubes themselves are excellent field emitters and several devices have already made use of this property [61–66]. Materials encapsulation within nanotubes may permit tuning or modification of some or all of these novel physical properties. In more general terms, a knowledge of how materials form inside MWNTs and SWNTs will contribute to a deeper understanding of their physical properties and also how they can crystallize in other aligned porous materials such as zeolites and phases such as MCM-41, thereby leading to the design of better catalysts for use in industrial applications.

In this chapter we review general approaches toward the encapsulation of materials within carbon nanotubes. The relative merits of the techniques used in the characterization of the resulting composite materials in terms of both their structural and physical properties are briefly assessed and the obtained crystallization behavior within MWNTs and SWNTs, as a function of the filling methodology employed, is described and discussed.

18.2 METHODOLOGIES FOR FILLING MWNTs AND SWNTs

18.2.1 Synthesis, Opening, and Purification of MWNTs and SWNTs

MWNTs can be synthesized efficiently in bulk by the modified Krätschmer–Huffman procedure [67,68], although they have also been produced by a variety of other methods including their synthesis in benzene and other hydrocarbon flames [69–71], by laser ablation of graphite [72–74], by stable glow discharge [75], by template synthesis in porous solids [48,76–79], by catalytic growth on etched substrates [80], by pyrolysis [81,82], and by Ar^+ ion sputtering of graphite targets [83]. In the case of synthesis via arc discharge, laser ablation, or pyrolysis, both the microstructure and bulk morphology of the resulting nanotubes and their by-products can be altered by the introduction of catalytic

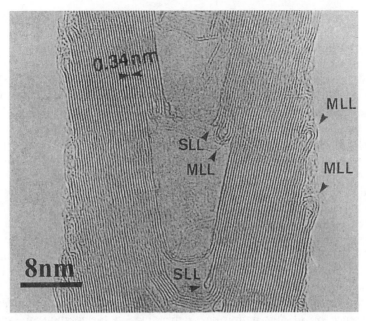

Figure 18.1 Heat-treated (at 2800 °C) carbon "bamboo"-shaped nanotube produced by pyrolysis of polyethylene over nickel. The walls of the nanotube contain a variety of carbon loops as a result of the heat treatment. (MLL, multilayer loop; SLL, single-layer loop). Adapted from Ref. 47.

species either into the carbon feedstock or onto the target substrate. This modification is also the principal route to the synthesis of SWNTs, which can be produced in 20–70% yield, either by arc evaporation of metals such as Co, Cu, Fe, Ni, or Ni/Y in doped carbon rods [52–54,84–87], or by laser ablation of graphite targets doped with similar catalytic species [88–91] and also laser-induced disproportionation of carbon dioxide [92]. When other catalytic particles and carbon deposition conditions are employed, a variety of different nanotube morphologies can result, which we describe here as "defect" nanotubes because of their deviation from the idealized "Russian doll" MWNT model; cone-shaped, "bamboo"-shaped (e.g., Fig. 18.1), "herring-bone," supra-helical (i.e., "spring-shaped"), or toroidal MWNTs can be produced, depending on the conditions of formation [93–97]. From the point of view of filling experiments, these types of nanotube are of poor utility as they contain a very high density of internal caps and wall defects. Attempts to selectively oxidize these types of nanotubes (vide infra) result in their destruction, as both the walls and internal cap regions are attacked simultaneously [47]. Most recently, attention has focused on the preparation of aligned nanotubes with a view to synthesizing field-emitting arrays of nanotubes to be used, for example, in flat panel devices [98]. The development of nanotube array synthesis will be described separately in Section 18.2.5.

With respect to the purification of MWNTs, many of the earlier developed techniques made use of the fact that the extraneous (i.e., non-nanotube) carbons contain a higher density of non-six-membered rings than MWNTs, making the former more prone to chemical attack. Ebbesen et al. [99] reported a low yield (~1–2%) technique that produces complete purification of MWNTs by oxidation in air at 750 °C. The same group also described a liquid-phase method that used $KMnO_4$ and H_2SO_4 mixtures as the oxidizing medium [100]. However, while the carbon residues produced by this method contained purified MWNTs, these were found to be severely damaged. Ikazaki et al. [101] have shown that the small graphitic nanoparticles can be removed by their intercalation with $CuCl_2$ followed by heating in air at 550 °C, resulting in the preferential destruction of the intercalated carbon. The removal of the residual copper by acid washing then gives purified MWNTs. Chen et al. [102] used liquid bromine to selectively intercalate the nanoparticles and achieved their near complete exfoliation by subsequent air oxidation. The graphene bodies of MWNTs purified by this method were only slightly damaged and the majority of the tubules were open at both ends. The gross yield was found to be 10–20 wt%.

As with MWNTs, the production of SWNTs is invariably accompanied by the presence of 30–80% impurities, depending on the method of synthesis. These impurities consist of microporous carbons and carbon-encapsulated catalytic particles. The techniques developed for the purification of SWNTs have also tended to rely heavily on the selective oxidation of the extraneous carbons and subsequent removal of the catalyst species by a combination of centrifugation and size-selective filtration techniques (e.g., Fig. 18.2). Bandow et al. [103] have separated SWNTs from the other materials by use of a cationic surfactant in aqueous solution, which renders the latter into suspension, while the SWNTs are trapped on a porous membrane following microfiltration. Amorphous and crystalline carbon impurities can also be removed from SWNT samples by ultrasonically assisted microfiltration [104,105]. This method "cuts up" the normally long, high aspect ratio SWNTs into smaller lengths and also encourages their propensity to form aligned bundles. Another reported method involves the hydrothermal treatment of the impure SWNT starting materials followed by extraction of the SWNTs and oxidation and dissolution of the impurity metal particles [106]. These procedures were recently refined by Rinzler et al. [107] and bulk purification of SWNTs is now possible by a multistep procedure in which the nanotubes are refluxed in ~3 M HNO_3, centrifuged, pH neutralized, resuspended in deionized water and then basic media (i.e., NaOH), and separated in a cross-flow filtration (CFF) apparatus. The gross yield of this technique is quite low (~10–20%) but the technique showed that the bulk separation of purified SWNTs is both feasible and can be performed on gram-scale quantities. An alternative strategy to those described above has been developed in which SWNTs (also MWNTs) are separated in a size-selective fashion by controlled pore glass (CPG) column chromatography, as described by Duesberg et al. [108,109]. This method is also scalable although

Figure 18.2 Typical purification of SWNTs. (a) TEM micrograph showing "web"-like raw soot product from a catalytic arc synthesis preparation of SWNTs. (b) Same material following bromination and oxidation. (c) Purified product following oxidation, centrifugation, and sonically assisted filtration. From Oxford Inorganic Chemistry Laboratory, unpublished results.

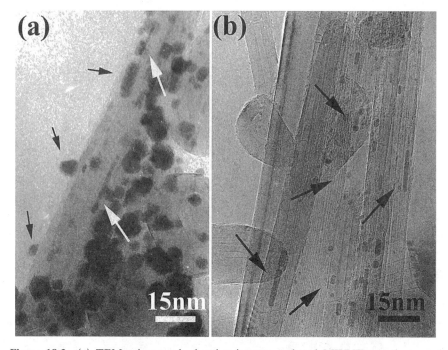

Figure 18.3 (a) TEM micrograph showing impure, reduced MWNT sample treated with RhCl₃. The extraneous filling material, visible here as the reduced metal, is clearly visible (small black arrows). Some MWNT filling is also visible (white arrows). (b) TEM micrograph obtained from a similar region from the reduced DAP treated RhCl₃/MWNT sample. In this case, only the filling material (black arrows) is visible. Adapted from Ref. 110.

purified nanotubes have so far been produced by this method only on the milligram scale.

Relatively few techniques have been developed with respect to the purifying of filled nanotubes from excess filling materials. Cook et al. [110] showed how this could be done for MWNTs filled from solution with RhCl₃. This purification proceeds by washing the nanotube-encapsulated sample with a reversed micelle mixture consisting of dodecylammonium propionate (DAP) solubilized in water and dispersed in benzene. This method effectively removed all of the excess RhCl₃ from the exterior of the MWNTs (Fig. 18.3). By subsequent heat treatment of the MWNT/RhCl₃ composite, a sample containing nanotubes filled with discrete crystals of rhodium was obtained. So far other techniques for the systematic purification of filled nanotube samples have yet to be published.

18.2.2 Chemical Methods for Filling Carbon Nanotubes

All of the chemical methods employed to fill MWNTs and SWNTs rely jointly on the principles of solution-deposition and capillarity (vide infra). The nanotubes are opened by selective oxidation of the nanotube tips and a material is introduced into the nanotube cavity from solution that enters the nanotube by capillary wetting. This process can be achieved in a single step by treating the closed nanotubes with refluxing nitric acid containing a soluble metal salt such as a metal nitrate for 4–24 h, as shown by Tsang et al. [3]. The resulting composite materials are dried in air and then calcined in the presence of a non-oxidizing gas such as nitrogen or argon for several hours. At this temperature, salts such as metal nitrates are decomposed to the oxides. In many cases, it is possible to unambiguously identify the encapsulated materials by direct observation of the characteristic lattice fringes of encapsulated single crystals (see Table 18.1) [8]. Figure 18.4 shows high-resolution transmission electron mi-

TABLE 18.1 Observed and Literature d-Spacings of Materials Incorporated Within MWNTs Via Solution-Deposition

Encapsulated Metals and Metal Oxides	Starting Materials	Observed Lattice Spacing (nm)	{hkl} Planes	Literature Spacing (nm)
La_2O_3	$La(NO_3)_3 \cdot 6H_2O$	0.28, 0.35	{−303}, {202}	0.28, 0.36
Pr_2O_3	$Pr(NO_3)_3 \cdot 6H_2O$	0.28, 0.31, 0.35, 0.41	{012}, {302}, {400}, {102}	0.29, 0.31 0.36, 0.42
CeO_2	$Ce(NO_3)_3 \cdot 6H_2O$	0.31	{111}	0.31
Y_2O_3	$Y(NO_3)_3 \cdot 5H_2O$	0.26	{400}	0.27
Nd_2O_3	$Nd(NO_3)_3 \cdot xH_2O$	0.30	{002}	0.30
Sm_2O_3	$Sm(NO_3)_3 \cdot 6H_2O$	0.32	{222}	0.32
$FeBiO_3$	$Fe(NO_3)_3 \cdot 9H_2O$	0.28	{110}	0.28
	$Bi(NO_3)_3 \cdot 5H_2O$	0.39	{100}	0.39
UO_{2-x}	$UO_2(NO_3)_2 \cdot 6HO$	0.32	{111}	Stoichiometry varies
NiO	$Ni(NO_3)_2 \cdot 6H_2O$	0.24	{111}	0.24
MoO_3	MoO_3	0.39	{110}	0.38
MoO_2	MoO_3	0.34	{−111}	0.34
		0.25	{111}	0.24
ZrO_2	$ZrCl_4$	0.32	{−111}	0.32
		0.29	{1−11}	0.28
ZrO_2	$ZrO(NO_3)_2 \cdot xH_2O$	2.98	{111}	2.96
Re metal	$KReO_4$	0.23	{002}	0.23
Pd metal	$Pd(NO_3)_2$	0.23	{111}	0.25
Ag metal	$Ag(NO_3)$	0.24	{111}	0.24
AuCl	$AuCl_3$	0.53	{101}	0.53
Au metal	$AuCl_3$	0.23	{111}	0.24
CdO	$Cd(NO_3)_2 \cdot 4H_2O$	0.19	{220}	0.19
		0.26	{200}	0.27
CdS	$CdO + H_2S$	0.25, 0.18	{102}, {200}	0.25, 0.18
		0.17	{004}	0.17

Source: Adapted from Ref. 8.

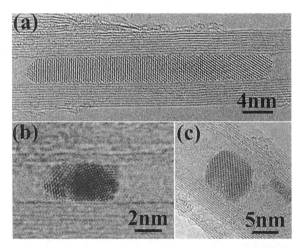

Figure 18.4   (a) HRTEM micrograph showing the filling of a MWNT capillary with an elongated single crystal of Sm_2O_3. The diagonal lattice fringes (on the right) correspond to {222} Sm_2O_3 lattice planes while alternating vertical fringes correspond to (200) planes. (b) HRTEM micrograph showing a reduced Ni crystal originating from NiO. This material was produced by in situ reduction of the oxide. Faceted crystals were also observed within a CETEM [14]. (c) HRTEM micrograph showing a single crystal of Nd_2O_3 within a wide-bore MWNT capillary. The observed lattice fringes correspond to {020} planes of Nd_2O_3. As two planes are viewed orthogonally, the Nd dots image as atoms rather than atomic planes. Adapted from Ref. 8.

croscopy (HRTEM) images of encapsulated crystallites of Sm_2O_3, Ni metal (reduced from the oxide, see below), and Nd_2O_3, all of which have been encapsulated by the one-step method. MWNTs containing Eu, La, Ce, Y, and Cd oxides have also been prepared by this approach (Table 18.1). Mixed metal oxides may be prepared in a similar manner. For example, refluxing closed MWNTs in nitric acid solution containing equimolar amounts of $Fe(NO_3)_3 \cdot 9H_2O$ and $Bi(NO_3)_3 \cdot 5H_2O$ gives crystals of the mixed oxide $FeBiO_3$ [8].

In a "two step" modification, the closed nanotubes are first opened with nitric acid and dried at $160\,°C$. These tubes frequently contain acid functionalities, such as —COOH or —OH, which can be removed by decarboxylation, as a separate step, that involves heating the nanotubes in vacuo or in an inert atmosphere to approximately $600\,°C$ [6]. In some cases, these functional groups can interfere with the filling process and their removal is often considered to be desirable. The empty, opened, and decarboxylated MWNTs can then be filled by stirring with a concentrated aqueous solution of a metal salt. For example, stirring nanotubes with a concentrated solution of $H_4SiW_{12}O_{40}$ in deionized water for 16 hours gives tubes filled with the same material [8]. When this method is applied to solutions of $AuCl_3$ or $AgNO_3$, a high percentage of opened nanotubes are filled with gold chloride or silver, respectively [15]. Calcination of the $AuCl_3$ containing specimen gives spherical gold crystals 1–5 nm

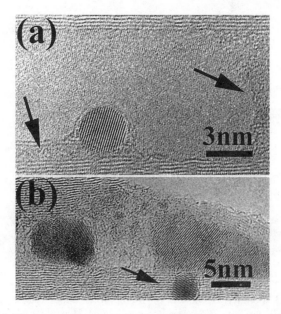

Figure 18.5 (a) Single crystal of Au observed on the inner wall of a MWNT. EDX scans obtained from amorphous regions (arrows) showed them to correspond to $AuCl_x$. (b) Two crystallites of Ag metal formed inside the open tip of a MWNT. A small Ag crystal formed on the exterior of the nanotube (arrow) is also visible. Adapted from Ref. 15.

in diameter (Figure 18.5(a)) while calcination of the $AgNO_3$ containing sample gives crystals of silver metal 3.5–8.5 nm in diameter (Fig. 18.5b). In the case of the gold filling, an amorphous or lower contrast crystalline filling was also observed, the presence of which was demonstrated by energy-dispersive X-ray (EDX) microanalysis (vide infra) and inspection of the lattice d-spacings (i.e. in the case of the crystalline filling) to be AuCl. Layers of this material were also observed on the outside surfaces of the MWNTs. The AuCl was completely converted to Au after treatment with H_2 at 300 °C.

In a modification to the two-step procedure [9], polycrystalline SnO has been encapsulated by mixing opened MWNTs with $SnCl_2 \cdot 2H_2O$ in hot concentrated HCl followed by dropwise addition of an aqueous Na_2CO_3 solution. At a pH of 10.2, precipitation occurs and, following refluxing (3 h) and drying (160 °C for 12 h), the MWNTs are filled with polycrystalline SnO. HRTEM studies indicate that $\sim$80% of the nanotubes contain SnO filling as shown in Figure 18.6. The lattice fringes of the crystallites denoted **I**, **II**, and **III** can be attributed to the (101), (110), and (002) planes of SnO, respectively. The packing of this material inside the MWNT capillaries has proved to be a useful model system for the investigation of crystallization inside nanotubes and this phenomenon is described in more detail later.

The solution method has been applied successfully to the filling of SWNTs,

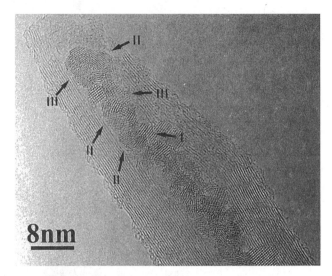

Figure 18.6 HRTEM micrograph showing packing of polycrystalline SnO into the bore of a MWNT capillary. As viewed, the filling forms randomly orientated crystallites within the MWNT capillary. The crystallites denoted **I**, **II**, and **III** are (101), (110), and (002) lattice fringes parallel to the incident electron beam. Adapted from Ref. 9.

although to date this has been achieved in extremely low yield. The formation of either small crystals or elongated and preferentially aligned crystals of Ru metal (Fig. 18.7) was achieved from solution, starting from $RuCl_3$ [18]. The synthesis included an opening step that involved refluxing in concentrated HCl. However, as subsequent capillary filling experiments have shown, this step may have been unnecessary (vide infra). Filling of nanotubes from solution has so far not been achieved in greater than 2% yield.

A final point is that, for both SWNTs and MWNTs, filling is not always restricted to their internal cavities. Multilayer MWNTs have the ability to incorporate materials into their graphene walls, therefore reducing the thickness of the incorporated species down to 1–3 atomic layers [4,5]. SWNTs, which consist of single cylinders, do not have this ability but, as they readily cluster together by van der Waals interactions to form ropes, species such as elemental potassium, rubidium, molecular bromine, molecular iodine, and charged iodine chains can be incorporated into the remanent interstitial spaces defined by the hexagonal or square packing of the SWNT cylinders [111–113].

18.2.3 Direct Capillary Filling of Carbon Nanotubes

Opened nanotubes can be filled directly by capillary action using molten media [2,4,5,20–22]. Typically, the tubes are stirred in either a crucible or a sealed silica quartz tube containing an approximately 1:1 mixture of MWNTs together with a melt of the filling material. The surface tension of the melt should

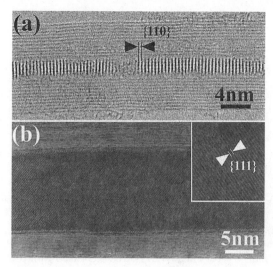

Figure 18.7 (a) HRTEM micrograph showing continuous filling of a MWNT with MoO$_3$. A typical feature of this type of filling is the aligned nature of the filling. In this case, the {110} lattice fringes of the MoO$_3$ filling are aligned at 90° with respect to the axis of the nanotube. (b) HRTEM micrograph showing continuous MoO$_2$ filling obtained by reducing MoO$_3$ in situ. In this case, the {111} lattice fringes of the encapsulated product are aligned at about 55° to the nanotube axis. Reprinted with permission from Ref. 114.

be less than ~100–200 mN/m, as established by Ebbesen [5]. In general, this method gives long, continuous crystals that occupy the entire internal volume of the nanotubes, as has been observed for the oxide phase of lead, as described in Section 18.1 [2] and also V$_2$O$_5$, as described by Ajayan et al. [4]. In a typical procedure, MoO$_3$ is mixed with opened tubes and heated to 800 °C for 3 h [114]. Approximately 50% of the nanotubes are filled in this way with long single crystals of MoO$_3$, which were typically several hundred nanometers long, as shown in Figure 18.7a. Attempts to fill empty tubes with molten metals directly were unsuccessful. It has been suggested that the surface tensions and melting temperatures of most metals are too high to wet and fill the tubes [5]. Currently, the best methods for filling MWNTs and SWNTs with elemental metals are either the catalytic method (MWNTs only) or the solution techniques, as described earlier.

A modification to the liquid-phase method developed recently uses low-melting mixtures of phases in binary or higher eutectic systems such that the resulting composition (1) has an overall surface tension lower than the determined threshold value (i.e., 100–200 mN/m; see above); (2) has a sufficiently low melting temperature (<900 °C) such that the nanotubes are not damaged thermally; and (3) does not attack the MWNTs chemically. We recently described a study of molten media filling of MWNTs achieved using various

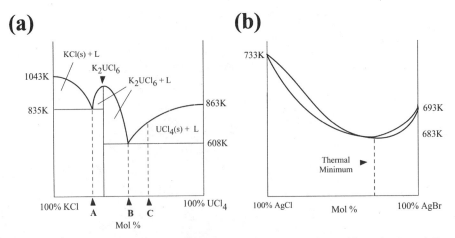

Figure 18.8   (a) Pseudo-binary KCl–UCl₄ phase diagram. Compositions A, B, and D were used to fill both MWNTs and SWNTs [20]. Reprinted with permission from Ref. 20. (b) AgCl–AgBr pseudo-binary phase diagram. Both end member compositions and the thermal minimum, corresponding to the solid solution $AgBr_{0.2}Cl_{0.8}$, were used to fill SWNTs. Reprinted with permission from Ref. 21.

eutectic and noneutectic mixtures of KCl with UCl₄ [20]. The technique makes direct use of the phase relations of the KCl–UCl₄ system, represented here by the pseudo-binary phase diagram depicted in Figure 18.8a [115]. The study also made use of surface tension/composition data recorded by Janz et al. [116] for the KCl–UCl₄ system, collectively defined in terms of Eq. (18.1):

$$s = a - bT \tag{18.1}$$

where s is the surface tension in mN/m, T is the temperature in K, and a and b are coefficients determined for a range of compositions and temperatures (Table 18.2). For example, the surface tension of a noneutectic 39.33 mol % KCl/60.67 mol % UCl₄ mixture (composition A, Fig. 18.8a) is 46.94 mN/m at 750 K, which is below the threshold value for which filling of MWNTs with molten media is predicted and is also too low to cause them significant thermal damage. Similarly, the surface tension of pure UCl₄ at 900 K is 38.00 mN/m, which is also below the threshold value for which filling of MWNTs is predicted. Filling experiments were attempted by heating each of the mixtures indicated in Figure 18.8a to ~150 K above the liquidus of the respective composition together with a 1:1 mass ratio of opened, decarboxylated MWNTs. The mixtures were then allowed to furnace cool to room temperature.

The modified molten media method was successful in forming a variety of fillings inside MWNT capillaries, which individually depended on the KCl/UCl₄ molar ratio employed during a given experiment. An important factor also, however, was the melting properties of the particular composition used to

TABLE 18.2 Compositions, *a* and *b* Coefficients, and Temperature Ranges for the KCl–UCl$_4$ Pseudo-binary Eutectic Melting System [116]

KCl/UCl$_4$ (mol %)	*a*	*b*	Temperature Range (K)
0:100	204.95	0.185	880–960
39.33:60.67	102.63	0.0621	890–980
56.04:43.96	93.04	0.04808	880–990
64.31:35.69	59.73	0.00865	930–1010
73.68:26.32	77.59	0.01349	870–1050
96.00:4.00	136.65	0.06732	1020–1150
97.33:2.67	164.82	0.08871	1050–1090
100:0	182.51	0.0782	1080–1170

fill the MWNT capillaries, as summarized in Table 18.3. When either pure UCl$_4$ or KCl was used, no significant crystalline filling was observed. However, when UCl$_4$-enriched composition A was used (Fig. 18.8a), for example, preferential filling with predominantly UCl$_4$ only was obtained (Fig. 18.9a). When filling was attempted with either composition B or composition D from Figure 18.8a, continuous filling with the glassy eutectic composition was invariably obtained, except when filling is obtained inside thin capillaries with diameters approaching 1 nm. The filling obtained with composition B in thick and thin MWNT capillaries is shown in Figures 18.9b and 18.9c, respectively. The crystallization behavior of this material will be discussed more fully in Section 18.4.2.

The molten media filling method has recently been extended to SWNTs [21,22]. As with MWNTs, similar limitations apply toward the filling of SWNTs with respect to the melting temperature and surface tension required of the molten media used to fill them. In the case of SWNTs, these conditions are made more stringent by the fact that they only consist of one graphene shell per nanotube; even partial thermal or chemical attack to the body of the nanotubes will be catastrophic. This consideration notwithstanding, SWNTs can now be

TABLE 18.3 KCl–UCl$_4$ Compositions Used to Fill MWNTs (Corrected from Ref. 20)

Composition (see Fig. 18.8a)	KCl/UCl$_4$ (mol %)	Liquidus Temperature (K)	Calculated Surface Tension (mN/m)
UCl$_4$	0:100	863	38.00 at 900 K
A	39.33:60.67	742	46.94 at 900 K
B	50:50 (1:1)	608	56.98 at 750 K
C	66.7:33.3 (2:1)	~893	51.51 at 950 K
D	73.2:26.8	835	64.37 at 980 K
KCl	100:0	1043	94.93 at 1120 K

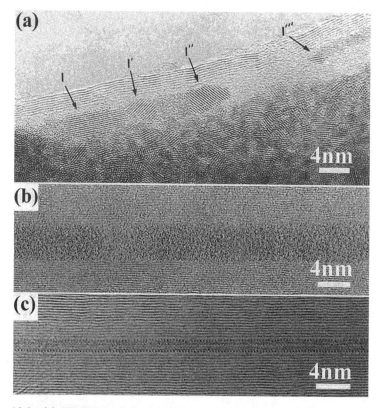

Figure 18.9 (a) HRTEM micrograph showing MWNT (embedded in a mixture of extraneous filling material) filled with composition A (see Fig. 18.8a). Note polycrystalline nature of the product. All of the observed lattice fringes correspond to the {211} planes of UCl₄. (b) HRTEM micrograph showing filling of a wide bore (∼8 nm I.D.) filled with the eutectic composition B (see Fig. 18.8a). Note the "glassy" nature of the filling. (c) Thin bore (∼1 nm I.D.) MWNT continuously filled with composition B. Note crystalline nature of filling compared to (b) Adapted from Ref. 20.

filled with a variety of metal halides by the molten media technique, including alkali halides, mixtures of alkali metal halides with actinide halides, such as UCl₄, and mixtures and pure phases of noble metal halides, such as those of Ag. Fillings obtained with other metal halides will be described later.

In the case of the KCl–UCl₄ system (Fig. 18.8a), the same eutectic and noneutectic compositions were used to fill SWNTs (see Fig. 18.10a,b). In the case of the AgCl–AgBr system, either compositions along the AgCl–AgBr solid solution system (Fig. 18.8b) or the end member components can be used to fill SWNTs (Fig. 18.10c). As is the case with MWNTs, the filling yield is higher and more continuous than that obtained by solution deposition, approaching 50% in the case of the lowest surface tension Ag halides (i.e., either the thermal

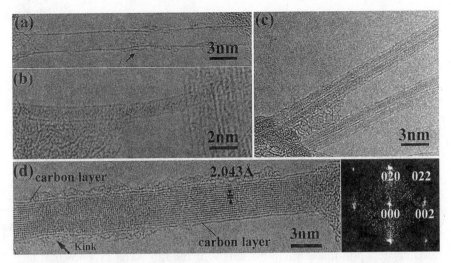

Figure 18.10 (a) HRTEM micrograph showing an unfilled, wide-bore SWNT. Note "kink" in the lower wall as viewed. This has probably formed in response to defects (i.e., non-six-membered rings) in the SWNT wall. (b) SWNT continuously filled with composition B (see Fig. 18.8a). In this case, the obtained filling is amorphous, although this type of filling is rare within SWNTs. (c) SWNT continuously filled with the AgBr end member composition from Figure 18.8b. In this case, the filling is both continuous and crystalline. $AgCl_{0.8}Br_{0.2}$ forms a similar product within SWNTs. (d) Ag "nanowire" formed by photolytic decomposition of AgCl within SWNTs. This type of product has so far only been observed in wide-bore SWNTs. The inset FFT, obtained from a region of the filling material, unambiguously identifies the filling as Ag metal. Adapted from Ref. 21.

minimum, $AgCl_{0.2}Br_{0.8}$, or pure AgBr) (Table 18.4). A further feature of this process is that the silver halides can readily be reduced to their metal by photolytic reduction. This has resulted in the formation of a small quantity of Ag wires although so far this kind of reduction seems readily to occur only within

TABLE 18.4   AgCl–AgBr Compositions Used to Fill SWNTs [21]

Composition	Filling Temperature	Calculated Surface Tension (mN/m)
AgCl	833 K	173.08
AgBr	800 K	151.3
$AgBr_{0.8}Cl_{0.2}$	783 K	154.4

wide bore (i.e., >2 nm) SWNTs, as shown by the example in Figure 18.10d. Apparently reduction does not occur rapidly for Ag halides incorporated within smaller diameter SWNTs (i.e., >2 nm) and both the metal and corresponding halide are still identifiable via EDX [21].

18.2.4 Catalytic Methods

Closed nanotubes and the closely related carbon nanoparticles filled with metals, or their carbides, can been formed by in situ arc-evaporation of composite carbon electrodes made by blending graphite powder with the appropriate metal or metal oxide [10,11,17,19,23–33]. The resulting products normally consist of nanoparticles with closed carbon shells with the encapsulated materials being either the pure metals or metal carbides. In a limited number of cases, this method can also give MWNTs filled with crystalline materials, which vary from continuous crystals that fill the entire hollow cavity to discrete separated crystals positioned along the length of the tube. The experimental technique consists of packing a hollow graphite anode with the element to be encapsulated and then proceeding with the Krätschmer–Huffman method as described earlier. During the arcing process, the tip of the anode and its contents are rendered into the vapor phase and the carbon shell then grows catalytically from the condensing species. The metals Cr, Fe, Co, Ni, Pd, Gd, Dy, and Yb all form nanowires inside MWNT-like nanotubes [24,33]. All other metals tend to form single crystals of either the metals or the respective metal carbide inside carbon nanoparticles. A mixture of both types of filling is observed, for example, in the case of the arc-encapsulated metals or carbides of Y [23,27] and Mn [30].

Defect MWNTs filled with either metal carbides or catalytic metal particles can also be produced by direct gas–solid carbon deposition onto catalytic metal. The carbon carrier can vary from gaseous carbon oxides [117–119], hydrocarbon gases such as methane [117,120], and acetylene [121,122], to aromatics, including benzene [81,123], to polymers, such as polyethylene [47] and even more complex organic species [17,45]. Certain metals efficiently promote carbon growth from the gas phase and these include Co, Fe, Ni, Pt, Cu, and Ge [17,45,52,81,120,121,124]. In nearly all cases, the goal has been to produce either pure MWNTs, SWNTs, or defect nanotubes, rather than the encapsulated species, which, in these cases, are considered to be an unwanted byproduct. However, Setlur et al. [17] have deliberately prepared encapsulated nanowires containing copper and germanium by pyrolysis of polycyclic aromatic hydrocarbons over finely divided copper and germanium catalytic particles, respectively. More recently, Grobert et al. [125] have achieved Ni nanowire encapsulation by the thermolysis at 950 °C of alternating thin films of C_{60} with Ni deposited on silica. The product obtained by this process consisted of tapered needle-like nanotubes with graphitic walls completely filled with Ni metal.

18.2.5 Methods for Preparing, Decorating, and Filling Two-Dimensional Arrays of Carbon Nanotubes

A rapidly developing area of research with respect to nanotube synthesis concerns their formation into aligned two-dimensional (i.e., 2D) arrays. One of the first methods for preparing 2D arrays involved cutting a MWNT-embedded polymer resin composite into 50–500 nm thick slices [126]. Alignment of the nanotubes in the presolidified resin was produced by flow-induced anisotropy. This technique has subsequently proved to be valuable for determining the mechanical, fracture, and wear properties of polymeric nanocomposites containing nanotube arrays [127,128]. A more popular strategy for the formation of "free" aligned nanotubes involves their formation on different substrates by a variety of carbon deposition methodologies. Li et al. [129] have shown that this can be done by carbon chemical vapor deposition (CVD) onto iron particles contained within the open pores of mesoporous silica, a method referred to as "template synthesis." Scanning electron microscopy (SEM) images obtained from the surface of the substrate showed that well graphitized (i.e., multilayer) nanotubes formed in this way grow approximately perpendicular to the silica surface. In this case the tubes are regularly spaced at 100 nm intervals on the silica surface. A similar approach, developed by Hulteen et al. [130], achieved the same result but in the absence of catalyst. In this case the carbon was formed by the polymerization of acrylonitrile within the pores of aluminum oxide membranes. Heat treatment of the composite membrane results in the formation of partially graphitized and well-separated nanotube structures with diameters of the order of 200 nm and lengths of up to 60 μm.

Greater control over the length, uniformity, and physical dispersion of nanotubes on substrates has been obtained by their formation onto flat or patterned surfaces. Ren et al. [131] have formed free-standing, well-aligned nanotubes onto nickel-coated glass by using acetylene as a carbon source together with ammonia present both as a diluent and also as a catalyst gas. This method produces multilayered nanotubes with intrinsically controllable diameters from 20 to 400 nm and lengths from 0.1 to 50 μm. Terrones et al. [132] have taken this process a step further by forming the nanotubes into laser-etched cobalt thin films prepared on an inverted silica substrate. In this case, the carbon source was pyrolyzed 2-amino-4,6-dichlorotriazine and the nanotubes were produced as aligned bundles that traversed the grooves etched into the substrate. The nanotubes produced by this method were both highly uniform, being mainly straight with diameters of 30–50 nm and lengths of ∼100 μm. The nanotubes were also shown to exhibit a preferred conformation (e.g., armchair 25–30%).

Filling or decorating of arrays of nanotubes has been achieved by a combined synthesis procedure whereby the filling medium is introduced into the nanotubes by deposition following synthesis, as suggested by Hulteen and Martin [133]. This has been achieved, for example, by Li et al. [134] who deposited nickel and cobalt particles into aligned MWNTs by electroless deposition. The aligned nanotubes were fabricated by pyrolyzing acetylene within the pores of an anodic aluminum oxide template. The deposited materials, which in both cases contained impurities of phosphorus or boron, consisted of discontinuous, polycrystalline metal filling. An alternative strategy was employed by Pradhan et al. [135] for the attempted synthesis of iron nanoparticles in aligned carbon nanotubes, which has also been prepared in an aluminum oxide membrane template. Ferrocene vapor was used as an iron precursor but, after liberation of the iron/nanotube composites from the template, the nanotubes were instead found to contain crystalline nanoparticles of Fe_3O_4.

18.3   TECHNIQUES FOR CHARACTERIZING CARBON NANOTUBE COMPOSITES

18.3.1   Nanoscopic Methods

The most widely and successfully applied methods with respect to the characterization of both unfilled and filled carbon nanotubes have been nanoscopic techniques capable of imaging materials with a resolution in the range <0.1 to 10 nm. These include the techniques of transmission electron microscopy (TEM) or its high-resolution analog (HRTEM) and the closely related techniques of scanning transmission electron microscopy (STEM) and scanning electron microscopy (SEM); scanning probe techniques, such as atomic force microscopy (AFM), scanning tunneling microscopy (STM), and scanning probe microscopy (SPM). While many of these techniques are valued intrinsically for their imaging properties, most have the ability to combine additional analytical features in the same instrument. TEM and STEM have been particularly successful in this regard and, in addition to electron diffraction (ED) and direct lattice imaging, techniques such as energy-dispersive X-ray spectroscopy (EDX) or wavelength-dispersive spectroscopy (WDS), parallel electron energy loss spectroscopy (PEELS), and Z-contrast or energy filtered imaging (EFI) can now be combined in the same instrument.

Many of these techniques have been boosted by the advent of TEMs and STEMs incorporating field emission gun (i.e., FEG) electron sources. The advantage of this type of electron source is that the beam can be focused to a much smaller probe diameter (down to ~0.5 nm for a FEGTEM versus ~3 nm for a conventional TEM) with a far higher electron density (i.e., 100 × compared to a conventional LaB_6 filament), therefore considerably enhancing the signal-to-noise ratio with respect to the spectroscopic method being employed, in particular with respect to EDX, WDS, and PEELS. Furthermore, for a FEG electron source electron diffraction can be obtained from regions down to 1 nm in diameter [136]. The disadvantage of such high-density electron probes is that they can often be very destructive to the specimen under examination—to carbon in particular—and care must therefore be used in the interpretation

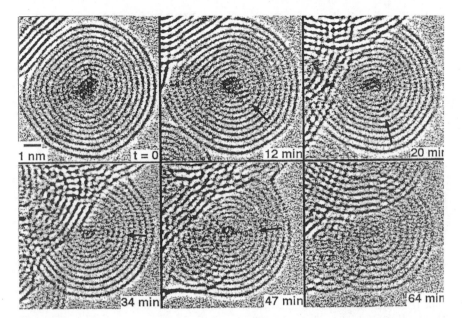

Figure 18.11 A sequence of six images showing the migration of a small gold crystal from within a carbon "onion" at 1050 K induced by the influence of an electron beam in a HRTEM. Clusters of a few atoms, seen here as black dots (arrows), split from the core and move through the shells. The gold atoms leave the onion and the black dots vanish. Reprinted with permission from Ref. 138.

of results obtained from them. However, FEGTEMs can be appropriately modified to incorporate cooling (i.e., using liquid N_2 or He as a coolant) so these problems can be obviated. With respect to the problem of damage caused to carbon in an electron beam, Banhart et al. [137] have shown that carbon destruction within carbon onions can be prevented by *heating* not cooling the specimen in situ as the annealing process supplies carbon atoms that "pop out" of the carbon superstructure with enough energy to recombine with the defected onion with the consequence that the more thermodynamically stable structure is retained. They also showed that, under conditions of high temperature, metals such as gold or cobalt [138] can diffuse out of the carbon shells (Fig. 18.11).

HRTEM was the first method applied toward the characterization of MWNTs [1]. In his initial study, Iijima was able not only to characterize MWNTs in terms of their concentric tubular nature but also, by careful analysis of the $\{hki0\}$ spot splitting of ED patterns obtained from individual tubules, to determine their helical nature as well. The same author was also responsible for the first imaging of discrete SWNTs [51], although these had also been identified from CO_2 thinned MWNTs [41]. Subsequently, both HRTEM and ED have been used extensively to probe both the local and bulk structures of both MWNTs and SWNTs, particularly with regard to their defect structure and their respective helicities. With respect to MWNTs, particular attention has been paid to the helicity values of concentric tubes *within* a single nanotubule, as these vary considerably [139,140]. For SWNTs, the study of helicity ought to be relatively straightforward but, because they are only a single graphene cylinder thick, they diffract extremely weakly and, in general, it has only been possible to determine average helicities from bundles of SWNTs by ED and dark field imaging [141,142]. Nanobeam diffraction in STEM has also been used to probe the local defect populations within aligned bundles of SWNTs [143].

With respect to the characterization of materials incorporated within carbon nanotubes, HRTEM has been to date the most widely applied technique. The full characterization of materials within structures as small as MWNTs and SWNTs (and even smaller encapsulates such as carbon nanoparticles) represents a considerable technical challenge, however, particularly where the chemical identity of the incorporated material is uncertain. In general, a HRTEM lattice image of a crystalline material contains lattice spacing information rather than direct structural information and can only be used to interpret the structure of the incorporated materials with care. An additional limitation is imposed by the fact that heavier atoms image much more strongly than lighter atoms, due to their relatively higher scattering power, and information concerning the lighter atoms can be lost. The HRTEM image in Figure 18.4a shows a well-resolved image of a crystal of Sm_2O_3 incorporated within an MWNT. In this image, the dark spots represent the positions of the Sm^{3+} atoms, whereas the positions of the oxygen atoms are effectively invisible. Fortunately, more information can be gained from electron diffraction (ED) or, in the case of Figure 18.10d, calculated fast Fourier transforms (FFTs) from suitably aligned lattice images. With care, these techniques can be used to identify the zone axis and phase of nanotube incorporated crystals. For example, using a combination of ED and lattice imaging, Liu and Cowley [30] were able to identify no fewer than four different manganese carbides incorporated into both MWNTs and carbon nanoparticles. In the case of Ag-incorporated SWNTs, the FFTs were calculated from the direct lattice image to confirm that the imaged material in Figure 18.10d is a [100] projection of Ag metal rather than the parent AgCl material [21].

As stated above, additional information can be gained from a specimen visualized with TEM or STEM by the use of associated analytical techniques. Ajayan et al. [4] used PEELS incorporated within STEM to probe the interfacial properties of V_2O_5 formed on the surfaces of MWNTs and showed, from the fine structure of the obtained oxygen L-edge, that it was oxygen deficient relative to the bulk material. This information, combined with HRTEM imaging data, was used to predict that surface monolayers of V_2O_5 were reduced pyramidal layers [4]. PEELS has also been used to probe the intercalation and deintercalation of Cs and K into SWNT bundles by Suzuki et al. [144] The authors showed that the elements could be reversibly inserted into the interstitial spaces within the nanotubes by vacuum deposition. EDX, in particular, can be used to study local compositional variations within nanotube incorporated materials. Cook et al. [145] showed, for example, that partially decomposed AuCl incorporated into MWNTs from solution could be resolved into mainly amorphous (i.e., AuCl) and crystalline (i.e., Au) components using a 0.5 nm EDX probe. The Au crystallites were further identified via nanobeam electron diffraction [145]. Small amounts of surface and capillary incorporated crystalline AuCl were also identified via direct lattice imaging. With respect to SWNTs, Z-contrast imaging was used by Grigorian et al. [113] to establish the presence of iodine chains in interstitial spaces in SWNT bundles.

Scanning or "atomic" probe methods (SPMs), such as AFM and STM, can produce images at even higher magnification to those produced by HRTEM with resolutions of the order of ~0.1 nm. These techniques have been effective in terms of directly imaging the external atomic structure of carbon nanotubes. For example, the helical character of MWNTs and SWNTs has been unambiguously imaged by AFM and STM. In the case of MWNTs, the helical nature of the outermost concentric tubules can be imaged directly [146], whereas for SWNTs, the helical nature of the entire nanotube may be determined by direct imaging [147, 148].

In addition to probing the local atomic order of SWNTs, atomic probe methods can also detect defect behavior and, additionally, modifications produced on nanotube surfaces. For example, Tsang et al. [149] used HRTEM combined with AFM to probe both the external surface morphology of MWNTs and their local structure. By both techniques, it was possible to characterize regions of local positive curvature, caused by the presence of five (or smaller) membered rings, or negative curvature, caused by the presence of seven (or larger) membered rings. Biro et al. [150] further showed that local damage of carbon nanotubes caused by high-energy focused ion beam (FIB) irradiation could be studied effectively by both AFM and STM. Using STM, Clauss et al. [151] have also shown that SWNTs arranged in bundles frequently display quenched twist distortions, which they postulate may lead to the observed insulator-like behavior of SWNTs at low temperatures.

Thus far, atom probe techniques have yet to be applied to the study of filled nanotubes. However, STM has been used to probe the local conductivity and resistivity of both MWNTs and SWNTs by Collins et al. [152]. Ohnishi et al. [153] have shown that it is possible to measure quantized conductance of one- and two-dimensional "nanowires" of gold atoms suspended on the tips of ultra-high-vacuum STM tips, and, as similar experiments have already been performed with respect to the mounting of unfilled SWNTs and MWNTs on STM tips [154], it is interesting to speculate that similar experiments ought to be possible with nanotube composites. Furthermore, a number of experiments have been conducted with respect to the direct measurement of the physical properties of individual MWNTs, SWNTs, and SWNT "ropes." In 1996, Kasumov et al. [155] measured the resistivity, transverse magnetoresistance, and temperature dependence of resistance of a single MWNT, although low-resistance contacts were not employed. Subsequently, Bachtold et al. [156] have devised a method for contacting low-ohmic contacts onto MWNTs via a four-terminal (i.e., four-probe) contact configuration. Using this technique, they established values for the resistances of all measured MWNTs in the range 0.35–2.6 kΩ at room temperature. Grigorian et al. [157] have perfomed similar measurements on Cs- and Li-doped SWNT "mats," but this time measuring their resistance as a function of temperature and the amount of doping. With increasing Cs dopant concentration, the mat resistance was found to first decrease and then increase, exhibiting a minimum for optimum doping. For Li doping, the mat resistance decreased monotonically and then saturated. The optimally Cs-doped sample exhibited a positive dR/dT over the entire range of measurement ($80 < T < 300$ K). Tans et al. [158] have subsequently shown how electrical devices such as field-effect transistors can be constructed from SWNTs contacted to a three-terminal switching device. By applying a voltage to a gate electrode, the nanotube can be switched from a conducting to an insulating state.

18.3.2   In Situ Studies Inside Carbon Nanotubes

In favorable circumstances, chemical reactions can be identified in situ within MWNTs. This may be achieved most efficiently by the consecutive character-ization of nanotube samples pre- and postreaction or even in real time (see below). Characterization of the specimens can be accomplished either by direct HRTEM imaging, when the chemical transformations are gross enough, or by electron probe techniques, such as EDX, WDS, or PEELS, as described in Section 18.3.1. The example of transformation of $AuCl_3$ to either $AuCl$ or the base metal, as described earlier, is an example of the observation of reduction where the starting, intermediate, and end compositions can readily be dis-tinguished in the HRTEM either by relative differences in image contrast (arising as a result of varying electron density) and/or where the products can be distinguished due to their respective $\{hkl\}$ d-spacings [15]. The observation of reduction of NiO to nickel metal in real time has also been performed by controlled environment transmission electron microscopy (CETEM) [14]. A schematic representation of a typical modified pole-piece arrangement for a CETEM is give in Figure 18.12. This arrangement allows for the introduction of a small partial pressure (up to 5 mbar) of a reducing or oxidizing gas into a conventional TEM column under dynamic pumping conditions, the gas being removed through pumped apertures. In the case of NiO, the reducing medium was 5 mbar H_2 and the direct observation of the formation of faceted nickel crystals, as shown in Figure 18.4b, was observed inside MWNT capillaries. The prominent lattice fringes of 0.205 nm in the faceted crystallite correspond to the spacing between the $\{111\}$ planes of Ni metal. Similarly, encapsulated metal crystals of pure Pd and Ag [6,15] have all been prepared inside MWNTs from oxides originating from the corresponding metal nitrates. Reduction of $KReO_4$, inserted into MWNTs from solution, gives crystallites of Re metal (potassium

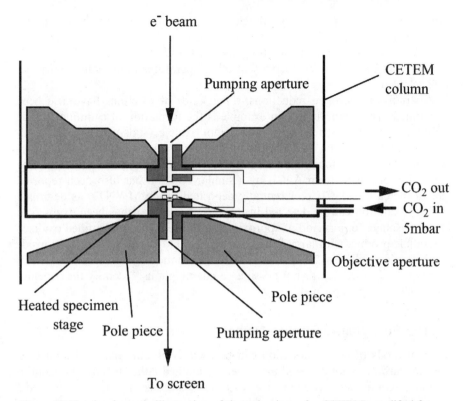

Figure 18.12 A schematic illustration of the pole piece of a CETEM, modified from a conventional TEM, to introduce a small partial pressure (~5 mbar) of a gas into the microscope column. Gases including H_2, O_2, and CO_2 can be introduced so that chemical reactions may be observed directly with a TEM. The gas is removed from the TEM column by differential pumping.

oxide was not observed) with all of the imaged and measured $\{hkl\}$ lattice spacings corresponding to metallic Re [8].

Reductions of continuous MWNT fillings have also been followed by consecutive specimen examination using HRTEM both prior to and following reduction [114]. Treatment of the nanotubes filled with continuous crystals of MoO_3 (Fig.18.7a) with hydrogen at 450–500 °C causes reduction to pure elongated MoO_2 filling (see Fig. 18.7b). The MoO_2 crystals show a strong preferred orientation with the $\{111\}$ planes lying at an angle of 55° to the same axis. It is perhaps surprising that the reduction of long MoO_3 crystals to similarly long MoO_2 crystals proceeds without apparent degradation of the crystallinity or morphology of the latter. Hydrogen must penetrate and react along the entire length of the MWNT capillary and water vapor is then subsequently expelled. The loss of the water must then be accompanied by shrinkage of the original dimensions of the MoO_3 crystal with a longitudinal rather than a cross-sectional contraction. The properties of MoO_2-filled MWNTs are potentially interesting as single crystals of MoO_2 have a resistivity of $\sim 2.7 \times 10^{-4}$ W/cm at 400 K and a higher conductivity at all temperatures than many of the metallic elements.

Apart from hydrogen reductions, other kinds of reactions have also been performed inside nanotubes. An example is the formation of cadmium sulfide crystals by the in situ treatment of cadmium oxide crystals with hydrogen sulfide at 400 °C. Electron diffraction patterns, obtained from these crystals, correspond to the thermodynamically stable form of hexagonal CdS [8]. The formation of Au_2S_3 from $AuCl_3$ by a similar reaction has also been reported [8]. In the case of UCl_4 preferentially deposited inside MWNTs, as described earlier, oxidation can be detected in the specimen after it has been exposed to the atmosphere for a period of approximately 2 weeks. An unidentified oxidized crystalline product with large fringe d-spacings (0.512 nm) was observed inside the MWNT capillaries [20]. Unfortunately, due to the large number of $U_xCl_yO_z$ compounds formed, it was not possible to unambiguously identify the oxychloride phase that was formed.

18.3.3 Bulk Characterization Methods

Bulk methods of characterization can be used to obtain analytical data concerning carbon nanotubes and are generally applied either toward the identification of components present in a nanotube synthesis or with respect to the determination of gross (i.e., bulk) physical properties. With respect to the former, the process of identification can also be tuned toward the determination of overall sample purity. In this respect, bulk sample characterization is most effective when backed up either by other bulk techniques or by nanoscopic characterization techniques such as those described in Section 18.3.2. For example, with respect to the bulk purification of SWNTs, Rinzler et al. [107] applied a battery of bulk characterization methods to verify as quantitatively as possible the efficacy of each purification step. At each stage, their samples were successively characterized by SEM, TEM, Raman scattering, and resistivity measurements. Such measurements also provide a useful standard against which samples purified by other techniques may be assessed.

With respect to bulk sample component identification via diffraction, Sorokin et al. [159] have used a combination of XRD and HRTEM to identify the various components in fullerene soot and to evaluate the contributions of the nanoparticle, nanotube, and C_{60} fractions to the bulk diffraction pattern. Neutron diffraction studies have also been used by Burian et al. [160] to differentiate between the bulk diffraction behavior of graphite and turbostratic and nanotube carbon and to quantify the differences between the short-range and medium-range order of these different types of carbon.

Unfortunately, the application of powder diffraction techniques toward the characterization of nanotube encapsulates is severely limited by a number of factors, not the least of which are problems associated with particle size and sample purity. The extent to which the former will be significant is given by the Scherrer equation:

$$t = \frac{0.9\lambda}{B \cos \theta_B} \qquad (18.2)$$

in which t is the crystal thickness (specified in this case by the nanotube diameter), λ is the incident radiation wavelength, θ_B is the Bragg angle, and B is the width of the diffracted intensity measured at half the maximum intensity, expressed in radians [161]. With respect to nanotube encapsulation, the magnitude of t will effectively be determined by the nanotube diameter, which is in the range of 1–20 nm for MWNTs and 1–6 nm for SWNTs. With respect to sample purity, two considerations will be paramount: first, that the amount of filling must be such that the encapsulated material can make a significant contribution to an overall diffraction pattern and, second, that any extraneous material formed during a filling experiment should be removed so that its contribution to the diffraction pattern can be ignored. The first has so far only been demonstrated for a very few cases, such as the identification of GdC and HfC_2 within carbon nanotubes as described by Ata et al. [162]. A method for the purification of MWNTs from excess filling material was already described in Section 18.2.1; however, this specimen was not examined by XRD.

One of the most frequently applied bulk techniques used with respect to the characterization of both MWNTs and SWNTs other than XRD has been Raman spectroscopy. In addition to its use with respect to the determination of sample purity, as described earlier, Raman scattering has also been employed by Pimenta et al. [163,164] to probe the 1D density of states, particularly of metallic and semiconducting SWNTs. Raman scattering has been obtained by Rao et al. [165] from ropes of SWNTs formed by laser ablation, which were separately shown by HRTEM to be close-packed arrays. The observed peaks were correlated with vibrational modes of SWNTs of mainly armchair symmetry ((n,n) SWNTs). The authors observed additional Raman resonances, which they attributed to 1D quantum confinement of the electrons in the nanotube. Sun et al. [166] have recently reported polarized Raman spectra obtained from SWNTs incorporated within the 1D channels of $AlPO_4$ single crystals. Finally, Raman intensities for both chiral and nonchiral nanotubes have been calculated ab initio from nonresonant bond-polarization theory by Saito et al. [167]. The authors showed that the polarization and sample orientation dependence of the calculated Raman intensities can be obtained by varying the direction of the nanotube axis and by keeping the polarization vectors of the incident light fixed.

Other spectroscopic techniques described for the characterization of bulk nanotube specimens have included X-ray absorption near-edge structure (XANES) spectroscopy, which was used to probe the differences between the band structures of highly oriented pyrolytic graphite (HOPG) and MWNTs;

electron spin resonance (ESR), which has been used to probe the metallic behavior of SWNTs produced by laser ablation [168] and also the conductivity and magnetic susceptibility of potassium-doped aligned carbon nanotubes [169]; and, similarly, electron paramagnetic resonance techniques, which have been used to detect the presence of "solubilized" SWNTs in aqueous or organic solvents [170]. Low-dose X-ray emission spectroscopy has also been used to identify both fullerene and nanotube components in bulk solid materials and also the fine structure behavior associated with the density of states [171].

18.4 CRYSTALLIZATION BEHAVIOR OF MATERIALS INSIDE CARBON NANOTUBES

The behavior of crystalline materials inside MWNTs and SWNTs—whether formed by precipitation from solution, by direct capillary action, or during catalytic growth—allows for the study of crystallization on the most intimate scale. In the case of solution-precipitation, crystallization will proceed according to conventional nucleation and growth mechanism, but once the size of the encapsulated crystallite approaches that of the internal diameter of the nanotube cavity, then the nanotube capillaries must exert some form of morphological control over any future growth. Crystallization by direct capillarity of liquid-phase materials occurs in a different fashion as solidification of all of the encapsulated material occurs virtually simultaneously. The exception to this is when materials crystallize from a noneutectic mixture of salts, in which case the situation will more closely resemble that of solution-deposition. In the instance of materials encapsulated catalytically, the situation is made more complex due to the much closer (i.e., chemical) interaction between the encapsulating carbon and the bound encapsulate. Here we attempt to elucidate some of the crystallization phenomena observed within carbon nanotubes.

18.4.1 Nucleation and Growth of Materials Formed Within Carbon Nanotubes from Solution

SnO precipitated from solution forms a polycrystalline filling within MWNTs, resulting in the encapsulation of either spherical or ellipsoidal encapsulates with external diameters ranging from 2 to 6 nm [9] inside MWNT capillaries with internal diameters that vary between 2 and 9 nm. The packing of this material inside MWNTs therefore provides a model system for the investigation of thin capillary control over crystallization [172]. Figure 18.13a shows a HRTEM micrograph of a large agglomeration of randomly oriented SnO crystallites inside a nanotube with a large internal diameter ($\sim$9 nm). In this example, the nanotube exerts morphological control over the agglomeration but none over individual crystallites. In Figure 18.13b, a second nanotube with a smaller diameter ($\sim$3.5 nm) can be seen in which three crystallites (indicated **I**, **II**, and **III**) reside in the central cavity. Whereas crystallite **I** has its growth axis (arrow)

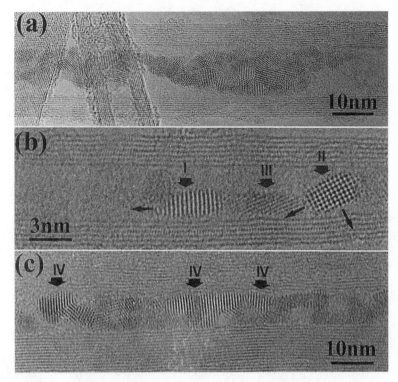

Figure 18.13   (a) HRTEM micrograph showing an agglomeration of SnO crystallites confined within a MWNT. The capillary exerts no influence over the growth behavior of individual crystallites but controls the formation of the agglomerate. (b) Three SnO crystallites showing different orientations within a thin MWNT capillary. Note how the crystal growth of **II** is constrained by the capillary. (c) Micrograph showing densely packed SnO crystallites within a thin MWNT capillary. Several of the indicated crystallites apparently exhibit preferred orientations. Reprinted with permission from Ref. 172.

aligned parallel to the nanotube axis, crystallite **II** is tilted at an angle of ~25° to the tube axis. Regardless of where the nucleation of crystallite **II** started, crystal growth in either of the directions indicated in Figure 18.13b will be impeded by the walls of the MWNT capillary. By contrast, crystallite **I** can continue crystal growth along its current axis until either it meets an obstruction or the crystallization terminates. In Figure 18.13c, the micrograph shows an MWNT densely packed with SnO crystallites in which the smaller crystallites (denoted **IV**) have apparently random orientations while the larger crystallites (denoted **V**) have their (101) lattice fringes oriented at 90° to the nanotube axis as in a preferred orientation. In Figure 18.14, we see examples of two more polycrystalline materials, SnO_2 and ZrO_2, both of which are also formed from solution [172]. Whereas polycrystalline SnO_2 crystallites form small crys-

Figure 18.14   (a) HRTEM showing packing of SnO_2 crystallites within a MWNT capillary. As the external diameters of each of the crystallites are, on average, smaller than that of the I.D. of the MWNT capillary, the crystallites are apparently randomly orientated. (b) Packing of ZrO_2 crystallites within MWNT capillary. In this case, the crystallites are elongated (i.e., compared to those of SnO_2) and pack with a pronounced preferred orientation along the capillary. Reprinted with permission from Ref. 172.

tallites with diameters approximately equal to the internal diameter of the MWNT and apparently randomly orientate with respect to the capillary (Fig. 18.14a), polycrystalline ZrO_2 crystallites are elongated and several crystallites (Fig. 18.14b) can be seen to align with their lattice fringes arranged parallel to the nanotube axis.

Occasionally, other size-related crystallization effects are observed. Where nucleation conditions permit, we see the formation of chains of SnO crystals that are arranged into "zigzags" or spirals inside the capillaries. Figure 18.15

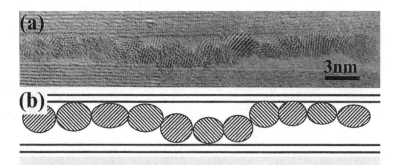

Figure 18.15   (a) HRTEM micrograph showing "zigzag" or spiraling SnO crystal growth within a MWNT capillary. The most likely explanation for this sort of behavior is that the crystallites buckle due to compressive forces that occur between adjoining crystallites during nucleation and growth. (b) Schematic representation of (a). Adapted from Ref. 9.

shows a micrograph and a schematic representation of crystal spiral formation from a chain of single SnO crystallites of approximately equal size observed inside a MWNT capillary. In order to produce such a chain, the nucleation of crystallites from solution must occur simultaneously along a row of evenly spaced sites along the capillary wall, thus producing a chain of crystallites of approximately equal size. Then spiraling can occur according to one of two possible mechanisms. First, the chains could spiral as a result of a van der Waals type interaction between the helically arranged MWNT walls and the crystallite chains. A second, more likely explanation is that chains of crystals buckle to form spirals as a result of localized compressive forces arising as the crystallites nucleate, grow, and then press together within the capillary. Such an effect can be modeled by observing the packing of hard spheres (e.g., ball bearings) of diameter x inside a smooth vertical glass capillary tube of diameter $> 0.5x$ but less than $2x$ under the influence of gravity. Under these conditions, the chains of ball bearings always form a spiral as the forces due to gravitation on them are minimized. This type of spiraling crystal growth has also been noted for UCl_4 crystallized from a noneutectic mixture [20].

With respect to larger elongated crystallites incorporated within nanotubes from solution, we can gain information with respect to the crystallographic relationships between the incorporated crystal and the encapsulating nanotube from well-resolved HRTEM lattice images [172]. Figure 18.4a shows an elongated Sm_2O_3 crystallite that completely fills the internal volume of a MWNT for a distance of ~ 60 nm. Figures 18.16a and 18.16b show two enlargements obtained at intervals of 20 nm along the capillary in which the arrangements of the Sm^{3+} cations (i.e., the dark dots, see above) can be discerned at the MWNT wall/encapsulate interface. These cations are arranged in a triangular motif that extends along the wall of the carbon nanotube and that represents the point at which parallel Sm_2O_3 [222] lattice planes, arranged at an angle of $\sim 30°$ to the MWNT wall, terminate at precise intervals of 0.546 nm along the wall. Although these lattice terminations represent a very precise stacking arrangement with respect to Sm_2O_3 within the tubule, the lattice image gives no indication as to the corresponding arrangement of the carbon atoms in the surrounding wall, and we also do not know what is the precise (n, m) conformation of the innermost encapsulating tubule. If we assume that MWNTs consist of a "Russian doll" arrangement of concentric SWNTs, then for a "zigzag" or $(m, 0)$ conformation tubule (i.e., according to the $\mathbf{C}_h = m\mathbf{a}_1 + m\mathbf{a}_2$ "roll-up" vector notation used to described SWNTs [173]), the periodicity of the innermost graphene cylinder will be 0.21 nm along the nanotube axis and, for an "armchair" tubule, the graphene cylinder will observe a 0.12 nm periodicity (Fig. 18.16c). For all other tubules, which will consist of a chiral or helical net wrapped with $n \neq m$, an irregular periodicity along the tubule wall will be observed as a result of the "twisting" of the graphene cylinder to form a helix. By inspection, we see that none of these possible conformations are commensurable with respect to the stacking arrangement of the incorporated crystal. It therefore appears, in this example, that a direct crystallographic relationship between the wall and the

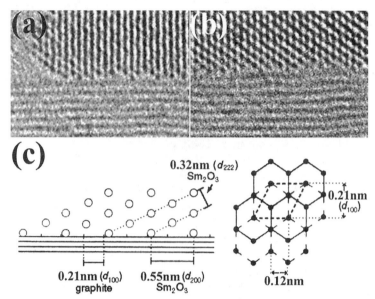

Figure 18.16 (a,b) Two enlargements obtained from 20 nm intervals from the HRTEM lattice in Figure 18.4a showing the very precise stacking of Sm_2O_3 in a MWNT capillary. In these micrographs, the dark spots represent the positions of the Sm^{3+} cations, the O^{2-} anions being effectively invisible. (c) A schematic representation of the stacking arrangement along the Sm_2O_3–MWNT interface. On the right is a depiction of two layers of the carbon lattice, which will actually be staggered within the capillary and not *ab* stacked as it appears here. The two repeats (measured relative to the parent d_{100} graphite structure) for the "zigzag" (i.e., $(n,0)$; 0.21 nm) and for the "armchair" (i.e., (n,m); 0.12 nm) are indicated. Adapted from Ref. 172.

encapsulating tubule is *not* obtained, which indicates that such a relationship is not a necessary prerequisite to either encapsulation or preferred orientation within MWNTs. This result also suggests that the way a crystal grows within a medium-sized MWNT capillary may be more important than any direct crystallographic relationship with the nanotube wall.

In the case of SWNTs, only one successful filling has been reported from solution in which Ru metal was formed via reduction of $RuCl_3$, as described earlier. In that case, filling with either small discrete metal crystals (Fig. 18.17a) or elongated metal crystallites formed with their [101] planes aligned parallel to the SWNT axes (Fig. 18.17b) were observed. Although this filling was formed via reduction from the halide, the formation of continuously aligned filling is wholly consistent with what was described earlier for other reduced aligned fillings. This result and results obtained for SnO, SnO_2, and ZrO_2 all indicate that larger elongated crystallites exhibit a definite tendency to align preferentially within nanotube capillaries. This trend is not observed for smaller crystallites, with longitudinal dimensions smaller than the MWNT internal diameters, which appear to randomly orientate within MWNT capillaries.

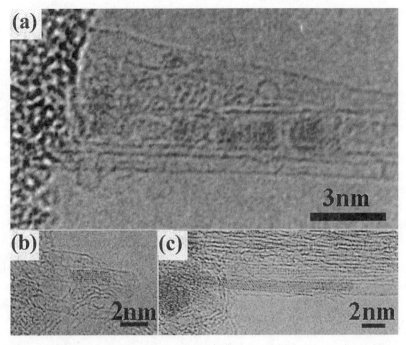

Figure 18.17 (a) HRTEM micrograph showing small discrete Ru crystallites incorporated into SWNTs by solution deposition followed by reduction. (b,c) Elongated Ru crystallites formed within SWNT capillaries. In both cases, the crystallites form with their (101) planes aligned parallel to the SWNT axes. Reprinted with permission from Ref. 18.

18.4.2 Crystallization of Materials Within MWNTs and SWNTs Via Direct Capillary Action

When materials are inserted into MWNTs by capillary action, orientated crystal growth is almost invariably obtained as Ajayan et al. [4] have noted for V_2O_5 and Chen et al. [114] have noted for MoO_3 (e.g., see Fig. 18.5a). In the case of orientated V_2O_5, the [010] zone axis corresponding to the parent orthorhombic unit cell aligns parallel to the axis of the MWNT capillary with imaged b repeat (equivalent to 0.356 nm) forming along the same direction. The behavior of MoO_3 is very similar (Fig. 18.5a) with continuously orientated filling in which the 0.37 nm c lattice repeat is aligned with the axis of the MWNT capillary. When this filling is reduced, it forms the denser filling within the MWNT capillaries, as shown in Figure 18.7b.

A more complex situation occurs when capillary filling is achieved using partially miscible phases in a eutectic melting system, as described earlier for the $KCl-UCl_4$ system. When a noneutectic composition is used to fill MWNTs, the obtained filling is polycrystalline. For example, Figure 18.9a shows a region of bulk polycrystalline UCl$_4$ with a partially filled MWNT on the surface. All of the encapsulated crystallites denoted **I** have lattice fringes conforming to the [211] planes of UCl$_4$. Several crystallites denoted **II** in the material on the exterior of the nanotube exhibit the same d-spacing. Two of the encapsulated crystallites denoted **I** can be seen with their [211] lattice fringes orientated parallel to the MWNT axis. Two further [211] crystallites, denoted **I′** and **I″**, are visible in which the respective orientations are at 32° and 44° to the MWNT axis. A crystallite with a diameter smaller than the nanotube capillary, **I‴**, can be seen with an orientation of 10° to the MWNT axis. When a eutectic composition is used to fill MWNTs—in this case, a 1:1 molar ratio for KCl/UCl$_4$— a glassy, amorphous filling is obtained instead, within thick capillaries of >1 nm in internal diameter (I.D.), as shown in Figure 18.9b. This would be the expected cystallinity for this type of material were it to be formed on the exterior of the nanotube. The fact that it is seen on the inside of a wide bore ($\sim$8 nm I.D.) should therefore be no surprise. However, when the same composition is observed within *thin* MWNT capillaries (i.e. <1–2 nm I.D.), a quite different crystallinity is obtained. The example in Figure 18.9c shows the same filling (i.e., from the same specimen) in a 1 nm I.D. MWNT. The crystallinity is quite pronounced and the walls of the MWNT are lined with a series of dark dots, which possibly correspond to the heavy metal centers within UCl$_x$ polyhedra. The form of the filling suggests that either the filling is wetting the inside of the MWNT or that a crystalline "cylinder" formed on the inside of the MWNT surrounds an amorphous filling in the center cavity.

The crystallization of materials within SWNTs, as formed by capillarity, presents possibly the most exciting prospect for the study of fine-scale crystallization. As the diameters of SWNTs exist in a lower range than for MWNTs ($\sim$1–6 nm), the interactions between the capillaries and the incorporated materials will be even more profound than for MWNTs and for any other conceivable aligned porous material. The cases of CsCl and CdCl$_2$ formed inside SWNT capillaries help to make this point. The former can conventionally form two different, though closely related, structures in the bulk, either body-centered cubic (bcc) or the classical *Pm3m* "CsCl" type structure, or, alternatively, it can adopt the *Fm3m* sodium chloride or β-CsCl face-centered cubic (fcc) structure, which it generally forms upon heating to above 469 °C [174]. When this material is inserted into SWNTs from the melt [22], it adopts the latter structure within narrow (i.e., 1–2 nm diameter) SWNTs and forms a very precise atomic arrangement within the SWNT capillary (Fig. 18.18).

The example of CdCl$_2$ formed within SWNTs is even more intriguing than that of CsCl, which is a relatively simple structure. Conventionally, CdCl$_2$ forms a structure in the bulk that consists of stacked layers of edge-sharing CdCl$_6$ octahedra as depicted in Figure 18.19a. This layered structure is too bulky to fit within SWNTs and a HRTEM micrograph (Fig. 18.19b) shows that the CdCl$_2$ that forms within SWNTs is extremely distorted and, in the micrograph, a "twisted" crystal of CdCl$_2$ is visible. In actual fact, the SWNT capillary can only accommodate segments of polyhedral chains isolated from

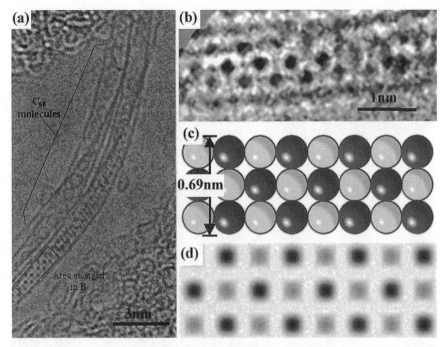

Figure 18.18 (a) HRTEM micrograph showing CsCl incorporated into a SWNT on the lower right. Note also the incoproration of C_{60} molecules; these have previously been observed within SWNTs formed by laser ablation (40) (b) Enlargement of region at lower left of (a). (c,d) A structural model and corresponding calculated lattice image corresponding to the rock salt (i.e., $Fm3m$) form of CsCl. Reprinted with permission from Ref. 21.

the $CdCl_2$ layered structure and it is this structure that is formed within the capillary, as represented schematically in Figure 18.19c. A van der Waals surface model (Fig. 18.19d) shows clearly how the SWNT capillary can accommodate the constrained $CdCl_2$ crystal. The presence of twisting in the SWNT is intriguing and suggests a much stronger interaction between the SWNT capillary with the incorporated crystal than for the case of SnO and Sm_2O_3 as described earlier (see Fig. 18.15 and 18.16). Two explanations can be put forward to account for this phenomenon—first, the crystal can distort in response to a helically arranged SWNT, and second, the crystal can distort in response to defects incorporated in the walls of the SWNTs similar to that shown for an empty SWNT in Figure 18.10a. It is noteworthy that the SWNT itself appears to distort in response to the presence of the incorporated crystal (although the distortions could be the result of defects). This could be a result if the van der Waals surface of the particular crystal defining an inner capillary is too narrow to accommodate the crystal. As the $CdCl_2$ crystal "twists" in response to the helicity of the capillary, the SWNT "squashes" in order to accommodate it.

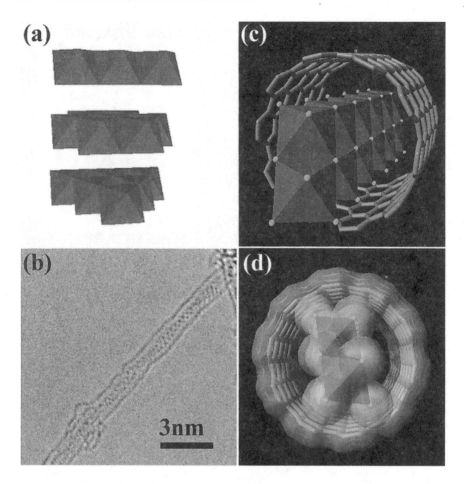

Figure 18.19 (a) Schematic depiction of the stacked, layered $CdCl_2$ bulk structure. This structure is too bulky to fit within a SWNT capillary. (b) HRTEM micrograph showing the incorporation of $CdCl_2$ into a SWNT capillary. Note the distorted ("twisted") nature of the incorporated crystal. (c,d) Schematic structural representation and van der Waals surface model of a polyhedral chain of $CdCl_2$ incorporated into a SWNT capillary. This is an idealized structural representation. The coordination of the $CdCl_2$ must be reduced (i.e. by removing some Cl atoms) to change balance the structure. Reprinted with permission from Ref. 21.

18.4.3 Crystallization Observed in Catalytically Formed MWNTs

There is still much uncertainty concerning the mechanism of formation of materials encapsulated by the arc method. Guerret-Piécourt et al. [24] have claimed that the propensity for the formation of long metallic "nanowires" within nanotubes rather than small encapsulates inside nanocapsules is correlated to the existence of an incomplete shell in the most stable ionic state of the element. They investigated encapsulation behavior for the elements Ti, Cr, Fe,

Co, Ni, Cu, Zn, Mo, Pd, Sn, Ta, W, Gd, Dy, and Yb. Subsequently, the same authors have recognized that sulfur plays a role with respect to the formation of some of these encapsulates, as it was identified in many of their products as an impurity [175]. Saito [176] has indicated that there is an additional correlation for rare earths between their volatility and their ability to form encapsulates. Seraphin et al. [177] have indicated that neither of the proposed models are wholly without exceptions and have advanced their own model defined in terms of the interfacial compatibility of the carbide with the encapsulating graphitic network.

The phenomenon of arc encapsulation is further complicated by the fact that, in some cases, mixed and often metastable products are often obtained. In the case of encapsulated manganese carbides, for example, Liu and Cowley [30] have observed no fewer than four different encapsulated carbides—Mn_3C, Mn_5C_2, Mn_7C_3, and $Mn_{23}C_6$—some of which are incommensurate and presumably metastable. Guerret-Piécourt et al [24] have also observed the formation of both microcrystalline ytterbium and "spiral" dysprosium products in their nanowires, both of which indicate that their respective carbon solution–dissolution phenomena during condensation are complex. In the case of the latter, the authors suggest that the growths of the spiral dysprosium product and helical encapsulating MWNT are coupled, which is interesting in view of the situation described earlier for SnO. However, it seems unlikely that the phenomena of spiraling crystal growth observed occurring as a result of crystallization from solution (or from a melt) and by condensation from an electric arc can readily be related, given the differences between the respective conditions of formation. It is also worth pointing out again that the spiraling crystal growth observed within MWNTs may not be caused by interactions with helicity, but rather by localized compressive forces arising as a result of crystallization.

The nature of the encapsulated products obtained in the case of gas-phase catalytic deposition onto catalytic particles are relatively simpler to interpret than those formed by arc codeposition. In general, encapsulation has been proposed to proceed via a solution–precipitation mechanism [178,179], which involves absorption of carbon onto the surface of the catalytic particle resulting in the formation of a small amount of interfacial carbide, thus leaving the remainder of the encapsulate in its native elemental state. Carbon formation then occurs as a result of carbon dissolution from the catalytic species. Frequently, it has been possible to study the interfacial behavior at the carbon–encapsulate interface. The extent to which carbon diffuses into the catalyst is a function of the operating conditions of the process. Carbon formation on nickel in the temperature range 500–800 °C, for example, results in encapsulates that are rich in carbon at the surface, normally forming a carbide at the carbon–metal interface [47]. In the example given in Figure 18.20, a catalyst particle is shown, which has clearly been involved in carbon formation but which has been "released" by degradation of the encapsulating nanotube. In the center of the particle, Ni metal, with a very fine d-spacing, is visible. Closer to the surface of the crystal, regions of Ni_3C and NiO are clearly visible. The latter formed presumably as a result of partial oxidation of the nickel catalyst during condensation.

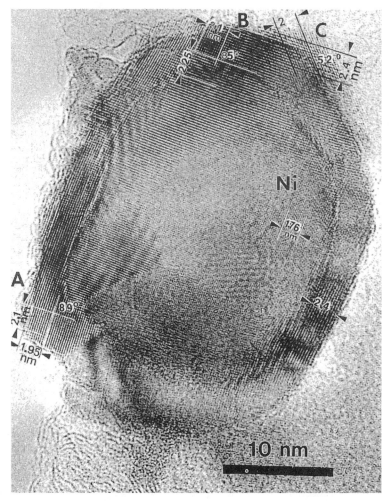

Figure 18.20 HRTEM micrograph showing a Ni particle released from the tip of an incorporating conical layered MWNT. Regions A and B of the outer polycrystalline layer most probably correspond to Ni_3C, while region C corresponds to the formation of NiO. The central region of the Ni crystal corresponds to Ni metal. Reprinted with permission from Ref. 47.

18.5 CONCLUDING REMARKS

The dimensions of the internal capillaries of both multiple- and single-walled carbon nanotubes occur at the limits to which nucleation and crystal growth can be observed for many materials. In such an environment, the physical and structural properties of the incorporated materials frequently undergo considerable modification in comparison with their properties exhibited in the bulk.

Studies with respect to the chemistry and physics of incorporation within nanotubes can therefore lead to both the development of new kinds of hybrid materials and also to a broader understanding of how incorporated materials behave in an atomistically constrained, low-dimensional environment.

REFERENCES

1. S. Iijima, *Nature* **1991**, *56*, 354.

2. P. M. Ajayan and S. Iijima, *Nature* **1991**, *361*, 6410.

3. S. C. Tsang, Y. K. Chen, P. J. F. Harris, and M. L. H. Green, *Nature* **1994**, *372*, 159.

4. P. M. Ajayan, O. Stephan, P. Redlich, and C. Colliex, *Nature* **1995**, *375*, 564.

5. T. W. Ebbesen, *J. Phys. Chem. Solids* **1996**, *57*, 951.

6. R. M. Lago, S. C. Tsang, K. L. Lu, Y. K. Chen, and M. L. H. Green, *J. Chem. Soc. Chem. Commun.*, **1995**, 1355.

7. D. Ugarte, A. Chatelain, and W. A. Deheer, *Science* **1996**, *274*, 1897.

8. Y. K. Chen, A. Chu, J. Cook, M. L. H. Green, P. J. F. Harris, R. J. R. Heesom, M. Humphries, J. Sloan, S. C. Tsang, and J. F. C. Turner, *J. Mater. Chem.* **1997**, *7*, 545.

9. J. Sloan, J. Cook, R. J. R. Heesom, M. L. H. Green, and J. L. Hutchison, *J. Cryst. Growth* **1997**, *173*, 81.

10. Y. Saito, T. Yoshikawa, M. Okuda, N. Fujimoto, K. Sumiyama, K. Suzuki, A. Kasuya, and Y. Nishina, *J. Phys. Chem. Solids* **1993**, *54*, 1849.

11. P. M. Ajayan, C. Colliex, J. M. Lambert, P. Bernier, L. Barbedette, M. Tence, and O. Stephan, *Phys. Rev. Lett.* **1994**, *72*, 1722.

12. R. Seshadri, A. Govindaraj, H. N. Aiyer, R. Sen, G. N. Subbanna, and C. N. R. Rao, *Curr. Sci.* **1994**, *66*, 839.

13. R. M. Lago, S. C. Tsang, K. L. Lu, and M. L. H. Green, *J. Chem. Soc. Chem. Commun.* **1995**, 1355.

14. E. G. Bithell, A. Rawcliffe, S. C. Tsang, M. J. Goringe, and M. L. H. Green, *Inst. Phys. Conf. Ser.* **1995**, *147*, 361.

15. A. Chu, J. Cook, R. J. R. Heesom, J. L. Hutchison, M. L. H. Green, and J. Sloan, *Chem. Mater.* **1996**, *8*, 2751.

16. B. C. Satishkumar, A. Govindaraj, J. Mofokeng, G. N. Subbanna and C. N. R. Rao, *J. Phys. B* **1996**, *29*, 4925.

17. A. A. Setlur, J. M. Lauerhaas, J. Y. Dai, and R. P. H. Chang, *Appl. Phys. Lett.* **1996**, *69*, 345.

18. J. Sloan, J. Hammer, M. Zweifka-Sibley, and M. L. H. Green, *J. Chem. Soc. Chem. Commun.* **1998**, 348.

19. M. Terrones, W. K. Hsu, A. Schilder, H. Terrones, N. Grobert, J. P. Hare, Y. Q. Zhu, M. Schwoerer, K. Prassides, H. W. Kroto, and D. R. M. Walton, *Appl. Phys. A.* **1998**, *66*, 307.

20. J. Sloan, J. Cook, A. Chu, M. Zweifka-Sibley, M. L. H. Green, and J. L. Hutchison, *J. Solid State Chem.* **1998**, *140*, 83.

21. J. Sloan, D. M. Wright, H. G. Woo, S. Bailey, G. Brown, A. P. E. York, K. S. Coleman, J. L. Hutchison, and M. L. H. Green, *J. Chem. Soc. Chem. Commun.* **1999**, 699.

22. J. Sloan, S. Bailey, G. Brown, K. S. Coleman, R. E. Dunin-Borkowski, D. M. Wright, H. G. Woo, J. B. Claridge, A. P. E. York, C. Xu, V. C. Williams, S. Friedrichs, R. L. Callender, J. L. Hutchison, and M. L. H. Green, recently resubmitted.

23. S. Seraphim, D. Zhou, J. Jiao, J. C. Withers, and R. Loufty, *Nature* **1993**, *362*, 503.

24. G. C. Guerret-Piécourt, Y. Le Bouar, A. Loiseau, and H. Pascard, *Nature* **1994**, *372*, 761.

25. S. Subramoney, R. S. Ruoff, D. C. Lorents, B. Chan, R. Malhotra, and R. Dyer, *Carbon* **1994**, *32*, 507.

26. M. Ata, Y. Kijima, A. J. Hudson, H. Imoto, N. Matsuzawa, and N. Takahashi, *Adv. Mater.* **1994**, *6*, 590.

27. J. M. Cowley and M. Q. Liu, *Micron* **1994**, *53*, 25.

28. M. Ata, A. J. Hudson, K. Yamaura, and K. Kurihara, *Jpn. J. Appl. Phys. I* **1995**, *34*, 4207.

29. M. Q. Liu and J. M. Cowley, *Carbon* **1995**, *33*, 225.

30. M. Q. Liu and J. M. Cowley, *Carbon* **1995**, *33*, 749.

31. W. K. Maser, P. L. Bernier, J. M. Lambert, O. Stephan, P. M. Ajayan, C. Colliex, V. Brotons, J. M. Planeix, B. Coq, P. Molinie, and P. Lefrant, *Synth. Met.* **1996**, *81*, 243.

32. Y. Yosida, *Physica B* **1997**, *229*, 301.

33. A. Loiseau and H. Pascard, *Chem. Phys. Lett.* **1996**, *256*, 246.

34. A. C. Dillon, K. M. Jones, T. A. Bekkedahl, C. H. Kiang, D. S. Bethune, and M. J. Heben, *Nature* **1997**, *386*, 377.

35. G. E. Gadd, M. Blackford, S. Moricca, N. Webb, P. J. Evans, A. N. Smith, G. Jacobsen, S. Leung, A. Day, and Q. Hua, *Science* **1997**, *277*, 933.

36. G. Stan and M. W. Cole, *Surf. Sci.* **1998**, *395*, 280.

37. S. C. Tsang, J. J. Davis, M. L. H. Green, H. A. O. Hill, Y. C. Leung, and P. J. Sadler, *J. Chem. Soc. Chem. Commun.* **1995**, 1803.

38. J. J. Davis, R. J. Coles, and H. A. O. Hill, *J. Electroanal. Chem.* **1997**, *279*, 440.

39. J. J. Davis, M. L. H. Green, H. A. O. Hill, Y. C. Leung, P. J. Sadler, J. Sloan, A. V. Xavier, and S. C. Tsang, *Inorg. Chim. Acta* **1998**, *272*, 261.

40. B. W. Smith, M. Monthioux, and D. E. Luzzi, *Nature* **1998**, *396*, 323.

41. S. C. Tsang, P. J. F. Harris, and M. L. H. Green, *Nature* **1993**, *362*, 520.

42. P. M. Ajayan, T. W. Ebbesen, T. Ichihashi, S. Iijima, K. Tanigaki, and H. Hiura, *Nature* **1993**, *362*, 522.

43. H. Hiura, T. W. Ebbesen, and K. Tanigaki, *Adv. Mater.* **1995**, *7*, 275.

44. Y. K. Chen, M. L. H. Green, J. L. Griffin, J. Hammer, R. M. Lago, and S. C. Tsang, *Adv. Mater.* **1996**, *8*, 1012.

45. J. Y. Dai, J. M. Lauerhaas, A. A. Setlur, and R. P. H. Chang, *Chem. Phys. Lett.* **1996**, *258*, 547.

46. T. E. Muller, D. G. Reid, W. K. Hsu, J. P. Hare, H. W. Kroto, and D. R. M. Walton, *Carbon* **1997**, *35*, 951.

47. A. K. Kiselev, J. Sloan, D. N. Zakharov, E. F. Kukovitskii, J. L. Hutchison, J. Hammer, and A. S. Kotosonov, *Carbon* **1998**, *36*, 1149.

48. G. Che, B. B. Lakshmi, C. R. Martin, E. R. Fisher, and R. S. Ruoff, *Chem. Mater.* **1998**, *10*, 260.

49. W. K. Hsu, M. Terrones, J. P. Hare, H. Terrones, H. W. Kroto, and D. R. M. Walton, *Chem. Phys. Lett.* **1996**, *262*, 161.

50. W. K. Hsu, M. Terrones, H. Terrones, N. Grobert, A. I. Kirkland, J. P. Hare, K. Prassides, P. D. Townsend, H. W. Kroto, and D. R. M. Walton, *Chem. Phys. Lett.* **1998**, *177*, 284.

51. S. Iijima and T. Ichihashi, *Nature* **1993**, *363*, 603.

52. D. S. Bethune, C. H. Kiang, M. S. Devries, G. Gorman, R. Savoy, J. Vazquez, and R. Beyers, *Nature* **1993**, *363*, 605.

53. C. H. Kiang, W. A. Goddard, R. Beyers, J. R. Salem, and D. S. Bethune, *J. Phys. Chem.* **1994**, *98*, 6612.

54. P. Nikolaev, A. Thess, A. G. Rinzler, D. T. Colbert, and R. E. Smalley, *Chem. Phys. Lett.* **1997**, *266*, 422.

55. M. S. Dresselhaus, G. Dresselhaus, and R. Saito, *Carbon* **1995**, *33*, 883.

56. V. V. Poborchii, *Jpn. J. Appl. Phys. Lett.* **1994**, *34*, 271.

57. M. S. Ivanova, Y. A. Kumzerov, V. V. Poborchii, Y. V. Ulashkevich, and V. V. Zhuravlev, *Microp. Mater.* **1995**, *4*, 319.

58. V. Dneprovskii, N. Gushina, O. Pavlov, V. Poborchii, I. Salamatina, and E. Zhukov, *Phys. Lett. A* **1995**, *59*, 204.

59. V. V. Poborchii, V. I. Alperovich, Y. Nozue, N. Ohnishi, A. Kasuya, and O. Terasaki, *J. Phys. Cond. Mater.* **1997**, *9*, 5687.

60. F. J. Garciavidal, J. M. Pitarke, and J. B. Pendry, *Phys. Rev. B.* **1998**, *58*, 6783.

61. W. A. Deheer, A. Chatelain, and D. Ugarte, *Science* **1995**, *270*, 1179.

62. P. G. Collins and A. Zettle, *Appl. Phys. Lett.* **1996**, *69*, 1969.

63. Y. Saito, K. Hamaguchi, T. Nishino, K. Hata, K. Tohji, A. Kasuya, and Y. Nishina, *Jpn. J. Appl. Phys.* **1997**, *36*, L1340.

64. J. M. Bonard, J. P. Salvetat, T. Stockli, W. A. Deheer, L. Forro, and A. Chatelain, *Appl. Phys. Lett.* **1998**, *73*, 918.

65. Q. H. Wang, A. A. Setlur, J. M. Lauerhaas, J. Y. Dai, E. W. Seelig, and R. P. H. Chang, *Appl. Phys. Lett.* **1998**, *72*, 2912.

66. Y. Saito, K. Hamaguchi, S. Uemura, K. Uchida, Y. Tasaka, F. Ikazaki, M. Yumura, A. Kasuya, and Y. Nishina, *Appl. Phys. A* **1998**, *67*, 95.

67. W. Krätschmer, L. D. Lamb, K. Fostiropoulos, and D. R. Huffman, *Nature* **1990**, *347*, 354.

68. T. W. Ebbesen and P. M. Ajayan, *Nature* **1992**, *358*, 220.

69. H. M. Duan and J. T. McKinnon, *J. Phys. Chem.* **1994**, *98*, 12815.

70. H. Richter, K. Hernadi, R. Caudano, A. Fonseca, H. N. Migeon, J. B. Nagy, S. Schneider, J. Vandooren, and P. J. Vantiggelen, *Carbon* **1996**, *34*, 427.

71. K. Daschowdhury, J. B. Howard, and J. B. Vandersande, *J. Mater. Res.* **1996**, *11*, 341.

72. D. D. Cheng, R. Q. Yu, Z. Y. Liu, Q. Zhang, Y. H. Wang, R. B. Huang, M. X. Zhan, and L. S. Zheng, *Chem. J. Chin. Univ.* **1995**, *16*, 948.

73. L. C. Qin and S. Iijima, *Chem. Phys. Lett.* **1997**, *269*, 65.

74. M. Yudasaka, T. Komatsu, T. Ichihashi, and S. Iijima, *Chem. Phys. Lett.* **1997**, *278*, 102.

75. X. K. Wang, X. W. Lin, V. P. Dravid, J. B. Ketterson, and R. P. H. Chang, *Appl. Phys. Lett.* **1995**, *66*, 427.

76. R. V. Parthasarathy, K. L. N. Phani, and C. L. R. Martin, *Adv. Mater.* **1995**, *7*, 896.

77. W. Z. Li, S. S. Xie, L. X. Qian, B. Chang, B. S. Zou, W. Y. Zhou, R. A. Zhao, and G. Wang, *Science* **1996**, *274*, 1701.

78. T. Kyotani, L. F. Tsai, and A. Tomita, *Chem. Mater.* **1996**, *8*, 2109.

79. K. Hernadi, A. Fonseca, J. B. Nagy, D. Bernaerts, A. Fudala, and A. A. Lucas, *Zeolites* **1996**, *17*, 416.

80. M. Terrones, N. Grobert, J. P. Zhang, H. Terrones, J. Olivares, W. K. Hsu, J. P. Hare, A. K. Cheetham, H. W. Kroto, and D. R. M. Walton, *Chem. Phys. Lett.* **1998**, *285*, 299.

81. M. Endo, K. Takeuchi, S. Igarashi, K. Kobori, M. Shiraishi, and H. W. Kroto, *J. Phys. Chem. Solids* **1993**, *54*, 1841.

82. M. Endo, K. Takeuchi, K. Kobori, K. Takahashi, H. W. Kroto, and A. Sarkar, *Carbon* **1995**, *33*, 873.

83. Z. X. Wang, F. Y. Zhu, W. M. Wang, and M. L. Ruan, *Phys. Lett. A* **1998**, *242*, 261.

84. X. Lin, X. K. Wang, V. P. Dravid, R. P. H. Chang, and J. B. Ketterson, *Appl. Phys. Lett.* **1994**, *64*, 181.

85. S. Seraphin and D. Zhou, *Appl. Phys. Lett.* **1994**, *64*, 2087.

86. J. M. Lambert, P. M. Ajayan, and P. Bernier, *Synth. Met.* **1995**, *70*, 1475.

87. C. Journet, W. K. Maser, P. Bernier, A. Loiseau, M. L. de la Chapelle, S. Lefrant, P. Deniard, R. Lee, and J. E. Fischer, *Nature* **1997**, *388*, 6644.

88. T. Guo, P. Nikolaev, A. Thess, D. T. Colbert, and R. E. Smalley, *Chem. Phys. Lett.* **1995**, *243*, 49.

89. M. Yudasaka, T. Komatsu, T. Ichihashi, and S. Iijima, *Chem. Phys. Lett.* **1997**, *278*, 103.

90. L. C. Qin and S. Iijima, *Chem. Phys. Lett.* **1997**, *269*, 65.

91. W. K. Maser, E. Munoz, A. M. Benito, M. T. Martinez, G. F. de la Fuente, Y. Maniette, E. Anglaret, and J. L. Sauvajol, *Chem. Phys. Lett.* **1998**, *292*, 587.

92. H. J. Dal, A. G. Rinzler, P. Nikolaev, A. Thess, D. T. Colbert, and R. E. Smalley, *Chem. Phys. Lett.* **1996**, *471*, 260.

93. M. H. Ge and K. Sattler, *Chem. Phys. Lett.* **1994**, *220*, 192.

94. M. Endo, K. Takeuchi, K. Kobori, K. Takahashi, H. W. Kroto, and A. Sarkar, *Carbon* **1995**, *33*, 873.

95. J. Jiao and S. Seraphin, *Chem. Phys. Lett.* **1996**, *249*, 93.

96. M. Terrones, W. K. Hsu, J. P. Hare, H. W. Kroto, H. Terrones, and D. R. M. Walton, *Philos. Trans. R. Soc. London* **1996**, *354*, 2025.

97. A. Krishnan, E. Dujardin, M. M. J. Treacy, J. Hugdahl, S. Lynum, and T. W. Ebbesen, *Nature* **1997**, *388*, 451.

98. P. G. Collins and A. Zettl, *Phys. Rev. B* **1997**, *55*, 9391.

99. T. W. Ebbesen, P. M. Ajayan, H. Hiura, and K. Tanigaki, *Nature* **1994**, *367*, 519.

100. H. Hiura, T. W. Ebbesen, and K. Tanigaki, *Adv. Mater.* **1995**, *7*, 275.

101. F. Ikazaki, S. Ohshima, K. Uchida, Y. Kuriki, H. Hayakawa, M. Yumura, K. Takahashi, and K. Tojima, *Carbon* **1994**, *32*, 1539.

102. Y. K. Chen, M. L. H. Green J. L. Griffin, J. Hammer, R. M. Lago, and S. C. Tsang, *Adv. Mater.* **1996**, *8*, 1012.

103. S. Bandow, A. M. Rao, K. A. Williams, A. Thess, R. E. Smalley, and P. C. Eklund, *J. Phys. Chem. B.* **1997**, *101*, 8839.

104. K. B. Shelimov, R. O. Esenaliev, A. G. Rinzler, C. B. Huffman, and R. E. Smalley, *Chem. Phys. Lett.* **1998**, *282*, 429.

105. P. J. Boul, A. Lu, T. Iverson, K. Shelimov, C. B. Huffman, F. Rodriguez-Macias, Y. S. Shon, T. R. Lee, D. T. Colbert, and R. E. Smalley, *Science* **1998**, *280*, 1253.

106. K. Tohji, H. Takahashi, Y. Shinoda, N. Shimizu, B. Jeyadevan, I. Matsuoka, Y. Saito, A. Kasuya, S. Ito, and Y. Nishina, *J. Phys. Chem. B.* **1997**, *101*, 1974.

107. A. G. Rinzler, J. Liu, H. Dai, P. Nikolaev, C. B. Huffman, F. J. Rodriguez-Macias, P. J. Boul, A. H. Lu, D. Heymann, D. T. Colbert, R. S. Lee, J. E. Fischer, A. M. Rao, P. C. Eklund, and R. E. Smalley, *Appl. Phys. A.* **1998**, *67*, 29.

108. G. S. Duesberg, M. Burghard, J. Muster, G. Philipp, and S. Roth, *J. Chem. Soc. Chem. Commun.* **1998**, 435.

109. G. S. Duesberg, J. Muster, V. Krstic, M. Burghard, and S. Roth, *Appl. Phys. A.* **1998**, *67*, 117.

110. J. Cook, J. Sloan, R. J. R. Heesom, J. Hammer, and M. L. H. Green, *J. Chem. Soc. Chem. Commun.* **1996**, 2673.

111. R. S. Lee, H. J. Kim, J. E. Fischer, A. Thess, and R. E. Smalley, *Nature* **1997**, *388*, 255.

112. A. M. Rao, P. C. Eklund, S. Bandow, A. Thess, and R. E. Smalley, *Nature* **1997**, *388*, 257.

113. L. Grigorian, K. A. Williams, S. Fang, G. U. Sumanasekera, A. L. Loper, E. C. Dickey, S. J. Pennycook, and P. C. Eklund, *Phys. Rev. Lett.* **1998**, *80*, 5560.

114. Y. K. Chen, M. L. H. Green, and S. C. Tsang, *J. Chem. Soc. Chem. Commun.* **1996**, 2489.

115. V. N. Desyatnik and S. P. Raspopin, *Russ. J. Inorg. Chem.* **1975**, *20*, 780.

116. G. J. Janz, *J. Phys. Chem. Ref. Data* **1998**, *17*, 129.

117. J. K. Wang, Y. H. Wang, W. Z. Weng, L. S. Zheng, Y. H. Hu, and H. L. Wan, *Acta Chim. Sin.* **1997**, *55*, 271.

118. J. Jiao and S. Seraphim, *J. Appl. Phys.* **1998**, *83*, 2442.

119. P. E. Nolan, D. C. Lynch, and A. H. Cutler, *J. Phys. Chem. B* **1998**, *102*, 4195.

120. Y. Chen, Z. L. Wang, J. S. Yin, D. J. Johnson, and R. H. Prince, *Chem. Phys. Lett.* **1997**, *272*, 178.

121. V. Ivanov, J. B. Nagy, P. Lambin, A. Lucas, X. B. Zhang, D. Bernaerts, G. van Tendeloo, S. Amelinckx, and J. van Landuyt, *Chem. Phys. Lett.* **1994**, *223*, 329.

122. K. Hernadi, A. Fonseca, P. Piedigrosso, M. Delvaux, J. B. Nagy, D. Bernaerts, and J. Riga, *Catal. Lett.* **1997**, *48*, 229.

123. K. Hernadi, A. Fonseca, J. B. Nagy, D. Bernaerts, and A. A. Lucas, *Carbon* **1996**, *34*, 1249.

124. J. H. Hafner, M. J. Bronikowski, B. R. Azamian, P. Nikolaev, A. G. Rinzler, D. T. Colbert, K. A. Smith, and R. E. Smalley, *Chem. Phys. Lett.* **1998**, *296*, 195.

125. N. Grobert, M. Terrones, A. J. Osborne, W. K. Hsu, S. Trasobares, Y. Q. Zhu, J. P. Hare, H. W. Kroto, and D. R. M. Walton, *Appl. Phys. A.* **1998**, *67*, 595.

126. P. M. Ajayan, O. Stephan, C. Colliex, and D. Trauth, *Science* **1994**, *265*, 1212.

127. O. Lourie, D. M. Cox, and H. D. Wagner, *Phys. Rev. Lett.* **1998**, *81*, 1638.

128. H. D. Wagner, O. Lourie, and X. F. Zhou, *Compos. A Appl. Sci. Math.* **1999**, *30*, 59.

129. W. Z. Li, S. S. Xie, L. X. Qian, B. H. Chang, B. S. Zou, W. Y. Zhou, R. A. Zhao, and G. Wang, *Science* **1996**, *274*, 1701.

130. J. C. Hulteen, H. X. Chen, C. K. Chambliss, and C. R. Martin, *Nanostr. Mater.* **1997**, *9*, 133.

131. Z. F. Ren, Z. P. Huang, J. W. Xu, J. H. Wang, P. Bush, M. P. Siegal, and P. N. Provencio, *Science* **1998**, *282*, 1105.

132. M. Terrones, N. Grobert, J. P. Zhang, H. Terrones, J. Olivares, H. K. Hsu, J. P. Hare, A. K. Cheetham, H. W. Kroto, and D. R. M. Walton, *Chem. Phys. Lett.* **1998**, *285*, 299.

133. J. C. Hulteen and C. R. Martin, *J. Mater. Chem.* **1997**, *7*, 1075.

134. J. Li, M. Moskovits, and T. L. Haslett, *Chem. Mater.* **1998**, *10*, 1963.

135. B. K. Pradhan, T. Toba, T. Kyotani, and A. Tomita, *Chem. Mater.* **1998**, *10*, 2510.

136. J. M. Cowley, *J. Electron Microsc.* **1996**, *45*, 3.

137. F. Banhart, T. Fuller, P. Redlich, and P. M. Ajayan, *Chem. Phys. Lett.* **1997**, *269*, 349.

138. F. Banhart, P. Redlich, and P. M. Ajayan, *Chem. Phys. Lett.* **1998**, *192*, 554.

139. C. T. White, D. H. Robertson, and J. W. Mintmire, *Phys. Rev. B.* **1993**, *47*, 5485.

140. M. Q. Liu and J. M. Cowley, *Ultramicroscopy*, **1995**, *53*, 333.

141. J. M. Cowley, P. Nikolaev, A. Thess, and R. E. Smalley, *Chem. Phys. Lett.* **1997**, *265*, 379.

142. D. Bernaerts, A. Zettl, N. G. Chopra, A. Thess, and R. E. Smalley, *Solid State Commun.* **1998**, *105*, 145.

143. J. M. Cowley and F. A. Sundell, *Ultramicroscopy* **1997**, *68*, 1.

144. S. Suzuki, C. Bower, and O. Zhou, *Chem. Phys. Lett.* **1998**, *285*, 230.

145. J. Cook, J. Sloan, A. Chu, R. Heesom, M. L. H. Green, J. L. Hutchison, M. L. H. Green, and M. Kawasaki, *JEOL News* **1996**, *32E*, 2–6.

146. K. Sattler, *Carbon* **1995**, *33*, 915.

147. L. C. Venema, J. W. G. Wildoer, C. Dekker, G. A. Rinzler, and R. E. Smalley, *Appl. Phys. A* **1998**, *66*, S153.

148. A. Hassanien, M. Tokumoto, Y. Kumazawa, H. Kataura, S. Susuki and Y. Achiba, *Appl. Phys. Lett.* **1998**, *73*, 3839.

149. S. C. Tsang, P. de Oliveira, J. J. Davis, M. L. H. Green, and H. A. O. Hill, *Chem. Phys. Lett.* **1996**, *249*, 413.

150. L. P. Biro, J. Gyulai, P. Lambin, J. B. Nagy, S. Lazarescu, G. I. Mark, A. Fonseca, P. R. Surjan, Z. Szekeres, P. A. Thiry, and A. A. Lucas, *Carbon* **1998**, *36*, 689.

151. W. Clauss, D. J. Bergeron, and A. T. Johnson, *Phys. Rev. B* **1998**, *58*, R4266.

152. P. G. Collins, A. Zettl, H. Bando, A. Thess, and R. E. Smalley, *Science* **1997**, *278*, 100.

153. H. Ohnishi, Y. Kondo, and K. Takayanagi, *Nature*, **1998**, *395*, 780.

154. H. J. Dai, J. H. Hafner, A. G. Rinzler, D. T. Colbert, and R. E. Smalley, *Nature* **1996**, *384*, 147.

155. A. Y. Kasumov, I. L. Khodos, and P. M. Ajayan, *Europhys. Lett.* **1996**, *34*, 429.

156. A. Bachtold, M. Henny, C. Tarrier, C. Strunk, C. Schonenberger, J. P. Salvetat, J. M. Bonard, and L. Forro, *Appl. Phys. Lett.* **1998**, *73*, 274.

157. L. Grigorian, G. U. Sumanasekera, A. L. Loper, S. Fang, J. L. Allen, and P. C. Eklund, *Phys. Rev. B.* **1998**, *58*, R4185.

158. S. J. Tans, A. R. M. Verschueren, and C. Dekker, *Nature* **1998**, *393*, 6690.

159. L. M. Sorokin, V. V. Ratnikov, G. N. Mosina, G. A. Dyuzhev, A. A. Bogdanov, and J. L. Hutchison, *Molec. Cryst. Liq. Cryst. Sci. Tech. C* **1996**, *7*, 111.

160. A. Burian, J. C. Dore, H. E. Fischer, and J. Sloan, *Phys. Rev. B* **1999**, *59*, 1665.

161. B. D. Cullity, in *Elements of X-ray Diffraction*, 2nd ed. Addison-Wesley, London, 1978, pp. 99–111.

162. M. Ata, A. J. Hudson, K. Yamaura, and K. Kurihara, *Jpn. J. Appl. Phys. 1* **1995**, *34*, 4207.

163. M. A. Pimenta, A. Marucci, S. A. Empedocles, M. G. Bawendi, E. B. Hanlon, A. M. Rao, P. C. Eklund, R. E. Smalley, G. Dresselhaus, and M. S. Dresselhaus, *Phys. Rev. B* **1998**, *58*, 16016.

164. M. A. Pimenta, A. Marucci, S. D. M. Brown, M. J. Matthews, A. M. Rao, P. C. Eklund, R. E. Smalley, G. Dresselhaus, and M. S. Dresselhaus, *J. Mater. Res.* **1998**, *13*, 2396.

165. A. M. Rao, E. Richter, S. Bandow, B. Chase, P. C. Eklund, K. A. Williams, S. Fang, K. R. Subbaswamy, M. Menon, A. Thess, R. E. Smalley, G. Dresselhaus, and M. S. Dresselhaus, *Science* **1997**, *275*, 187.

166. H. D. Sun, Z. K. Tang, J. Chen, and G. Li, *Solid State Commun.* **1999**, *109*, 365.

167. R. Saito, T. Takeya, T. Kimura, G. Dresselhaus, and M. S. Dresselhaus, *Phys. Rev. B* **1998**, *57*, 4145.

168. E. Jounguelet, P. Petit, J. E. Fischer, A. Thess, and R. E. Smalley, *J. Chim. Phys. Phys. Chim. Biol.* **1998**, *95*, 337.

169. O. Chauvet, G. Baumgartner, M. Carrard, W. Bacsa, D. Ugarte, W. A. Deheer, and L. Forro, *Appl. Phys. B* **1996**, *53*, 13996.

170. Y. Chen, J. Chen, H. Hu, M. A. Hamon, M. E. Itkis, and R. C. Haddon, *Chem. Phys. Lett.* **1999**, *299*, 532.

171. V. L. Korotkikh, A. B. Ormont, and E. F. Kukovitskii, *Molec. Cryst. Liq. Cryst. Sci. Tech. C.* **1998**, *10*, 159.

172. J. Sloan, J. Cook, M. L. H. Green, J. L. Hutchison, and R. Tenne, *J. Mater. Chem.* **1997**, *7*, 1089.

173. M. S. Dresselhaus, G. Dresselhaus, and R. Saito, *Carbon* **1995**, *33*, 883.

174. A. F. Wells, in *Structural Inorganic Chemistry*, 5th ed. Clarendon Press, London, **1984**, pp. 384–444.

175. N. Demoncy, O. Stephan, N. Brun, C. Colliex, A. Loiseau, and H. Pascard, *Eur. Phys. J. B* **1998**, *4*, 147.

176. Y. Saito, *Carbon* **1995**, *33*, 979.

177. S. Seraphin, D. Zhou, and J. Jiao, *J. Appl. Phys.* **1996**, *80*, 2097.

178. R. Lamber, N. Jaeger, and G. Schulz-Ekloff, *Surf. Sci.* **1988**, *197*, 402.

179. P. E. Nolan, D. C. Lynch, and A. H. Cutler, *Carbon* **1996**, *34*, 817.

CHAPTER 19

SYNTHESIS, STRUCTURE, AND PROPERTIES OF CARBON ENCAPSULATED METAL NANOPARTICLES

MICHAEL E. MCHENRY and SHEKHAR SUBRAMONEY

19.1 INTRODUCTION

Submicron particles find usage in a number of common applications of every-day life. For example, titania white pigments, which have typical diameters ranging from 200 to 300 nm, are used extensively in paints, papers, and plastics that we use almost every day. Magnetic pigments such as ferric oxide and chromia are used in magnetic recording media. Pigments such as zinc oxide find use in applications such as personal care. Particulate carbon also is used extensively as fillers in tires, paints, and inks. One of the key characteristics of any material in particulate form is its large surface area to volume ratio, and this particular feature leads to unique properties of the particles that are quite different from their bulk counterparts. This is particularly true of particles whose average diameter is 100 nm or less (in this size realm these materials are referred to as nanoparticles). It is believed that the properties of nanoparticles should enable the development of novel materials with unique properties useful in a host of applications including nonlinear optics, magnetism, mechanical durability, chemical catalysis, and nanoscale probes and fabrication. Typically, pure metal particles become significantly more reactive and their magnetic properties change once they get into the nanoparticulate regime. In fact, several metal nanoparticles are pyrophoric and spontaneously ignite when exposed to air at room temperature. The enhanced reactivity and changes in magnetic properties would be of particular concern to several applications. It is with respect to some of these concerns that the field of encapsulated nanoparticles takes on added significance. Encapsulation of metal nanoparticles in shells of

Fullerenes: Chemistry, Physics, and Technology, Edited by Karl M. Kadish and Rodney S. Ruoff.
0-471-29089-0 Copyright © 2000 John Wiley & Sons, Inc.

graphitic layers would not only protect the metal from the harmful effects of the environment but it would also allow the utilization of the specific properties of the encapsulating medium as well as the encapsulant.

The concept of the encapsulation of metal nanoparticles into carbon layers is not a novel one. In the chemical (particularly the petrochemical) industry, the deleterious effects of coking on the surface of supported metal catalysts and the resultant drop in the chemical activity of the catalyst, blockage of the reactors, and reduction of the heat-transfer performance within the reactors have been known for decades [1–4]. In several related studies [5–7], transmission electron microscopy (TEM) was used effectively to demonstrate that the metal nanoparticles responsible for the catalysis had been encapsulated in graphitic layers of carbon, resulting in a loss of performance of the catalyst.

Significant interest in the research and potential application of nanoscale materials was spurred by the 1985 discovery [8] of the third form of ordered carbon (after graphite and diamond) that is commonly known as fullerenes. The existence of fullerenes was discovered as a result of laser pyrolysis (ablation) experiments designed to mimic temperature and pressure conditions for the formation of cluster species in interstellar dust. The efficiency of fullerene synthesis was significantly enhanced by the use of the carbon-arc process, which is by convention referred to as the Krätschmer–Huffman process [9]. The Krätschmer–Huffman process, in turn, was a source of considerable research interest as any or all of a number of the experimental parameters could be varied to produce a wide range of novel and exciting nanostructures, as illustrated schematically in Figure 19.1. In chronological order, the nanoparticles discovered using the Krätschmer–Huffman process include the *multiwalled carbon nanotubes and polyhedra* [10], *carbon encapsulated metal nanoparticles* (*CEMNs*) [11,12], and *single-walled tubes* [13,14]. CEMNs were obtained for the first time in small quantities in the Krätschmer–Huffman process by the incorporation of rare-earth metal oxides into the graphitic anode of the arc-discharge experiment. This chapter primarily deals with the synthesis, structure, properties, and potential applications of CEMNs.

19.2   INTERACTION BETWEEN METALS AND CARBON

The interaction between metals and carbon has been a subject of much scrutiny for over 50 years, since the discovery of carbon filaments, most often by accident, inside furnaces containing hydrocarbon gases or carbon monoxide [15,16]. These filaments, which are believed to have been catalyzed by small metal particles as well as metallic inhomogeneities on the reactor walls of the furnaces, are typically several nanometers in diameter and several micrometers long. Their tubular microstructure led them to being conventionally referred to as *vapor-grown carbon filaments* (*VGCFs*). The formation of VGCFs and the degradation of metal catalysts through coking have been the subject of a number of studies addressing the relationship between metals and carbons, with

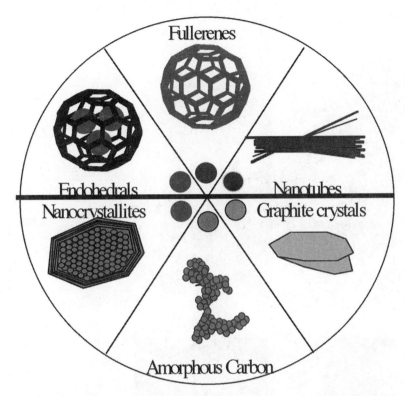

Figure 19.1 Schematic showing the more common reaction products observed in the C-arc (courtesy of S. A. Majetich)

specific emphasis on the ability of these metals to convert disordered carbon to more ordered forms such as graphite or turbostratic carbon.

When disordered carbon is converted to more ordered forms by interaction with metal catalysts, several factors contribute to the development of the microstructure of the graphitized material. The disorderd carbon could be one or a combination of several species such as hydrocarbons, carbon monoxide, or solid amorphous carbon. Typically, carbon nanofibers are formed by the catalytic process when a small crystallite is broken away, either from the support or from the bulk metal, and deposits a tail of ordered carbon behind it as it reacts with the disordered carbon species. The size and orientation of the catalyst particles, in most cases transition metals such as nickel, iron, or cobalt, are, for example, important parameters that determine the morphology of the filaments formed. Figure 19.2 clearly illustrates how the microstructure of catalytically grown nanofilaments can be altered as a function of catalyst orientation (in this case the catalyst was an iron-based particle) [17].

Currently, there are two main trains of thought regarding ordered carbon formation through the catalytic process:

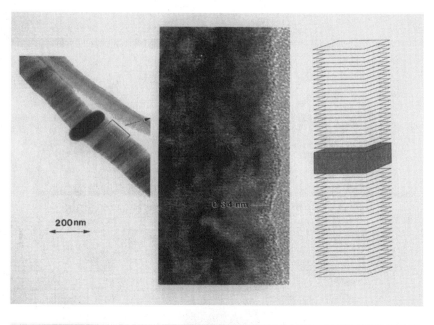

a

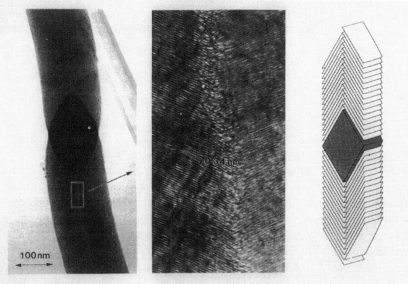

b

Figure 19.2 Manipulation of carbon nanofiber microstructure by controlling the orientation of catalyst particle: (a) graphene basal planes stacked perpendicular to the filament axis, (b) "herringbone" stacking of graphene basal planes (angular stacking with respect to the filament axis), and (c) graphene basal planes stacked parallel to the filament axis. Reprinted from *Langmuir* (1995), **11**, N. M. Rodriguez, A. Chambers, R. T. K. Baker, *Catalytic Engineering of Carbon Nanostructures*, pp. 3862–3866, with kind permission from the American Chemical Society.

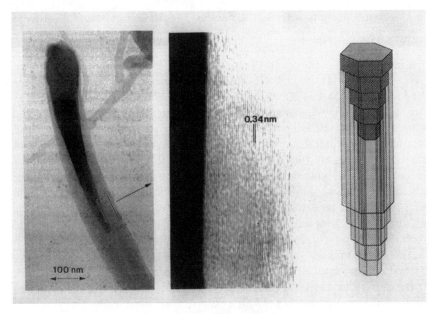

Figure 19.2 (continued)

$$\text{metal} + \text{disordered carbon} \rightarrow \text{metal carbide}$$

$$\rightarrow \text{decompose to metal} + \text{ordered carbon}$$

$$\text{carbon dissolves in metal} \rightarrow \text{reprecipitates as ordered carbon}$$

Work by several investigators in the recent past appear to support the latter hypothesis, namely, the adsorption and decomposition of the carbon-containing gas on one side of the catalytic particle followed by the subsequent diffusion of the resultant carbon atoms through the bulk of the catalyst and redeposition on the opposite side in a more ordered structure [18–22]. Graphite is considered to be energetically the most stable form of carbon and the conversion of any of the disordered forms of carbon to an ordered structure would be a natural route for the reduction of its free energy.

A detailed analysis of the carbon ordering process indicates that it could be separated into three distinct reactions—at the carbon source/metal interface, the bulk of the metal, and the metal/ordered carbon interface. At the carbon source/metal interface, the solubility of the metal for carbon as well as the crystallographic orientation of the metal particle have long been recognized as extremely important factors in controlling the adsorption and decomposition processes. The nickel$_{(111)}$ set of crystallographic planes is extremely active in this reaction. On the other hand, the copper$_{(111)}$ set of crystallographic planes, though identical in appearance to their nickel counterparts, do not cause the rupture of the carbon–carbon bonds. This is one of the key reasons as to why nickel is a significantly more powerful catalyst for the conversion of disordered carbon to ordered structures compared to copper.

In the bulk of the metal catalyst, although the carbon source is considered to be of no importance in the ordering process, the diffusion of carbon through the catalyst is regarded as the rate-determining step controlling the growth of the ordered structures. Metals that do not have a high solubility for carbon, such as copper, silver, and gold, however, do not catalyze the formation of ordered structures from any carbon source. The low solubility leads to the diffusional flux for carbon through these metals to be extremely low. Although the role of bulk metal carbides as the primary catalysts in this reaction has been discounted by experimental data [19], there is a possibility that some of these metals could form a very thin surface carbide layer, which is defined as carbidic carbon, that might be extremely active in promoting the grapitization of the disordered carbon. In a recent analysis of the diffusion process involved in the formation of hollow-cored nanofilaments by transition metal catalysts [21], it was summarized that the rate of carbon deposition is controlled largely by the isothermal and not by the Soret diffusion of carbon through the metal particle.

At the catalyst/precipitated carbon interface, the size of the catalyzed carbon structures has been observed to the time–temperature dependent. The graphene layer planes that precipitate out of the catalyst usually are arranged epitaxially on preferred crystallographic planes in the catalyst particle. It has also been shown that the geometry of the catalyst particle interacting with the precipitated ordered carbon phase is governed to a large extent by the strength of the interaction between the metal and carbon atoms at that interface [23], such as whether the metal wets the carbon or not. Although most of the literature surveyed discuss the formation of graphitic nanofilaments by the catalytic route, the formation of encapsulating shells of graphitic carbon over the metal particles has been observed at temperatures higher than 500 °C when hydrogen was present in the reaction gas [6], or at lower temperatures when no hydrogen was present [7]. When the catalytic particles are encapsulated in shells of graphitic carbon, their microstructure is analogous to the CEMNs that are discussed in detail in this chapter. Figure 19.3 is a transmission electron micrograph of a portion of a cobalt particle in a cobalt/alumina catalyst following encapsulation in a shell consisting of roughly 20 graphene layers [24].

19.3 SYNTHESIS OF NANOENCAPSULATED MATERIALS

19.3.1 Krätschmer–Huffman Process

One of the interesting nanostructures that was discovered as a by-product of multiwalled carbon nanotubes was the multiwalled polyhedron. Figure 19.4 is a high-resolution TEM image depicting typical polyhedral nanoparticles consisting of a regular faceted shape enclosing a void in the middle. It is apparent that each polyhedron consists of 12 pentagons (as dictated by Euler's theorem), each

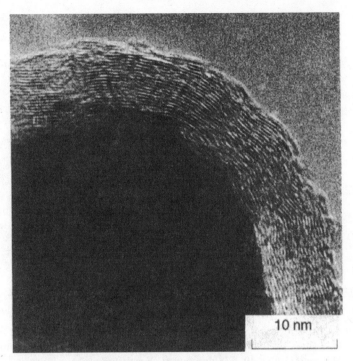

Figure 19.3 High resolution TEM image of a portion of a cobalt particle in a Co/alumina catalyst following encapsulation in about 20 graphene layers following disproportionation of CO. Reprinted from *Carbon* (1994), **32**, P. E. Nolan, D. C. Lynch, A. H. Cutler, *Catalytic Disproportionation of CO in the absence of Hydrogen: Encapsulating Shell Carbon Formation*, pp. 477–483, with kind permission from Elsevier Science Ltd, The Boulevard, Langford Lane, Kidlington, OX5 1GB, UK.

arranged at the intersection of the faceted planes. Analogous to multiwalled nanotubes, the nanopolyhedra are also found in the growth that occurs on the face of the cathode. As with carbon nanotubes, one of the most exciting aspects of the nanopolyhedra was the empty cavity in the middle, offering the potential to be filled with nanoparticles with exciting properties, while at the same time protecting these nanoparticles from the influences of the atmosphere. This was contemplated to be analogous to endohedral doping of fullerenes such as C_{60} [26]—which involves the addition of an atom or an ion into the core of the fullerene molecule to form an endohedrally doped molecular unit. In fullerene parlance these molecules are referred to as metallofullerenes or endofullerenes. Mass spectroscopy of the material produced following laser evaporation of a graphitic target doped with lanthanum chloride in water suggested that the C_{60} molecules were endohedrally doped with single ions of La^+. So it was but natural for the initial experiments to be run with the lanthanide or rare-earth (RE)

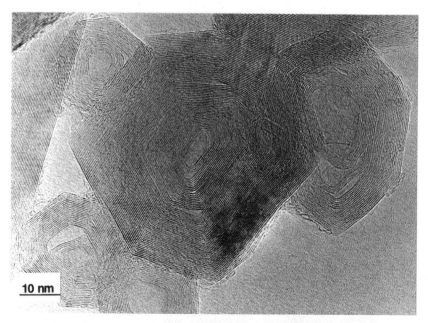

Figure 19.4   High resolution TEM image of multi-walled nanopolyhedral particles, which are typically a by-product of the synthesis of multi-walled tubes. Reprinted from *Advanced Materials* (1998), **10**, S. Subramoney, *Novel Nanocarbons—Structure, Properties, and Potential Applications*, in press, with kind permission from Wiley-VCH.

metal stuffed anodes but at conditions necessary for the formation of nanotubes and nanopolyhedra [27].

Transmission electron microscopy of the material obtained from the core of the growth on the cathode following our experiments with graphitic anodes incorporating a small percentage of lanthanum oxide [11] revealed that a nominal fraction of the product consisted of CEMNs. Figure 19.5 is a high-resolution TEM image of such a CEMN, with the metal in this particular micrograph being gadolinium. In reality, the encapsulates were mostly MC_2, where M is a RE metal, as determined by lattice spacings and electron diffraction. Most of the encapsulates were single crystals as well, as evidenced by the lack of grain boundaries in the high-resolution images, and the crystal boundaries conformed to the interior carbon shell walls, suggesting that the particle was completely enclosed within the carbon shell. This was also confirmed by the fact that the RE carbides did not degrade after exposure to air for considerable lengths of time; RE carbides are hydroscopic and are known to readily hydrolyze in air. Although the first metal to be encapsulated in the form of a CEMN was lanthanum [11,12], it was subsequently demonstrated that other RE metals also exhibited the same behavior [28,29].

Nanoencapsulates formed in the gas phase by a variety of methods including

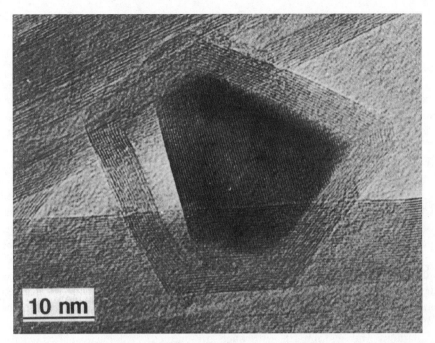

Figure 19.5 High resolution TEM image of carbon encapsulated single crystals of dicarbide of rare earth metal (Gd). Reprinted from *Carbon* (1994), **32**, S. Subramoney, R. S. Ruoff, D. C. Lorents, B. Chan, R. Malhotra, M. J. Dyer, K. Parvin, *Magnetic Separation of GdC$_2$ Encapsulated in Carbon Nanoparticles*, pp. 507–513, with kind permission from Elsevier Science Ltd, The Boulevard, Langford Lane, Kidlington, OX5 1GB, UK.

laser pyrolysis, thermal evaporation, carbon arc production, and thermal plasma share a common growth mechanism involving formation of particles from a supersaturated vapor [30,31]. A vapor becomes supersaturated with respect to a given constituent when its supersaturation ratio,

$$S = \frac{p_{\text{vap}}}{p_{\text{sat}}(T)} \tag{19.1}$$

becomes larger than unity. The quantity S is defined in terms of the temperature-dependent equilibrium vapor pressure of a species, p_{sat}, and the actual partial vapor pressure of the species in the plasma, p_{vap}. As hot vapor migrates from the source into a cooler gas region by a combination of convection and diffusion, quench rates of 104–106 K/s can be realized, especially in gases with high thermal conductivities such as hydrogen or helium. This rapid drop in temperature results in a drop in the equilibrium vapor pressure, causing the super-saturation ratio to rise quickly. This occurs despite a decrease in p_{vap} due to dilution of the vapor in the cooler gas. As a general rule, when S rises to between 6 and 10, particles begin to spontaneously nucleate from the vapor. Although the precise value of S required to induce nucleation depends on many factors and can vary locally within the reactor, the final property distributions of the resulting particle dispersion are insensitive to the initial value of S.

As with any discovery, several theories were empirically proposed to explain the encapsulation of the RE carbides into the graphitic shells by the Krätschmer–Huffman process. These include the ion bombardment encapsulation mechanism [32], which proposed that the cathode surface of the arc-discharge setup is bombarded with ions of M^+ and C^+. When this molten ion/metal/carbon mix begins to cool, graphitic layers begin to crystallize on the outer surface of the metal/carbon core, which also shrinks as the carbon segregates to form graphitic layers, accounting for the void present in most of these CEMNs as depicted in Figure 19.5. Another reason proposed for the presence of the void in the CEMNs was the placement of the 12 pentagons to close the polyhedral shells being simply incompatible with the shape of the encaged MC_2 crystal [29]. The ionization theory [33] suggested that the anode materials (metal and carbon) are ionized and that the encapsulates form on the cathode face from the ionized material. The propensity for a metal to get encapsulated in a carbon shell was explained on the basis of an unfilled electron shell in the most stable oxidation state of the metal under consideration; metals such as palladium and gadolinium with such an unfilled shell encapsulated easily, whereas metals that did not have an unfilled shell in their most stable oxidation state, such as copper, tungsten, and molybdenum, did not encapsulate. Some of the other mechanisms proposed to explain the formation of CEMNs are similar to the more conventional catalytic methods, such as the segregation of graphitic carbon from a solidifying liquid metal/carbon solution [34,35], as well as due to catalytic reactions on the metal surface, especially in the case of transition metals (TMs) [36,37].

The demonstration that metal nanoparticles could indeed be encapsulated into protective carbon shells immediately raised several possibilities for potential applications—ranging from magnetic nanoparticles for a variety of applications such as magnetic data storage, magnetic toners in xerography, magnetic inks, ferrofluids, and as contrast agents for magnetic resonance imaging, to confinement of radioactive waste particles, and to the easy handling of air-sensitive material for the research and evaluation of specific material properties at the nanoscale. However, the conventional arc-discharge technique was clearly successful only for the encapsulation of REs in any significant quantities. Attempts at the encapsulation of TMs such as iron, nickel, and cobalt by the conventional Krätschmer–Huffman process met with very limited success and this was disappointing considering the potential uses of ferromagnetic nano-crystals for a number of applications where fine particle magnetism is important and where adherent, protective carbon coatings would be important for aspects such as the oxidation resistance of these particles.

Some rather unique applications developed from the combination of the attributes of carbon encapsulated magnetic metal nanoparticles. For example, the superparamagnetic response of carbon encapsulated gadolinium dicarbide nanocrystals, their nanometer scale dimensions (typically 20–50 nm), and the highly adsorbent carbon coating made these CEMNs ideal candidates for the delivery of biological substances into plant or animal cells by the Biolistics® process [38]. The magnetic nature of the particles enabled the preferential separation of the mutated cells (containing the DNA-coated CEMNs) from the nonmutated cells, which in turn enables the successful introduction of the biological substance into a biological target. These experiments would have benefited immensely from stronger magnetic materials such as cobalt or iron.

While RE metals are typically encapsulated as MC_2 crystals within graphitic shells by the Krätschmer–Huffman process, carbon encapsulated TMs free of the carbide phases have been synthesized by several groups [39–43]. These are particularly interesting because of their magnetic properties and the ability to form monodomain magnetic particles. Encapsulation of iron, which is a good carbide former, has been investigated with stable carbide nanoencapsulates typically being formed. In magnetic transition metal systems, carbide formation is deleterious to the magnetic induction and therefore usually undesirable. In RE/TM systems interstitial carbon can be advantageous in increasing induction. Attempts at synthesizing a variety of alloy nanoencapsulates have resulted in a modicum of success. Among the systems that have been investigated are the following.

1. *Late Transition Metal/Late Transition Metal Alloys.* McHenry et al. [44,45] and Gallagher et al. [46] have employed a Krätschmer–Huffman carbon arc method to synthesize carbon-coated transition metal alloy particles with nanocrystalline dimensions for the $Fe_{1-x}Co_x[C]$ system (Fig. 19.6). The purpose of this work was to produce materials with larger inductions in alloys than in elemental iron or cobalt nanoparticles. Structural characterization by X-ray diffraction, TEM, and high-resolution TEM (HRTEM) has been used to illus-

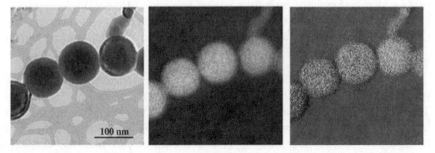

Figure 19.6 Bright field image of FeCo nanoparticles and EFTEM map of Fe and Co in the same (picture courtesy of J. H. Scott, reproduced from K. Gallagher, F. Johnson, E. Kirkpatrick, J. H. Scott, S. Majetich, and M. E. McHenry, *IEEE Trans. Mag.* **32**, 4842 (1996); reproduced with the permission of M. E. McHenry)).

trate the phases present, their morphology, and the particle size distribution. In particular, they have observed the face-centered cubic (fcc) and body-centered cubic (bcc) structures in the alloys, with no observation of carbides or hexagonal close-packed (hcp) cobalt phase. Cobalt nanoparticle containing soots have been produced with saturation magnetization (per gram of soot) in excess of 100 emu/g, and iron and cobalt containing soots with over 200 emu/g and this does not appear to be limiting. Note that pure iron has a 220 emu/g saturation magnetization. Iron and cobalt have similar sizes, which have led to the notable lack of chemical segregation in carbon arc produced iron/cobalt alloy nanoparticles [46]. Recently, Delaunay et al. [47] have produced nanogranular cobalt/platinum–carbon thin films demonstrating the encapsulation of the hard magnetic cobalt/platinum phase.

2. *Late Transition Metal/Simple Metal Alloys.* Mn/Al[C] alloys nanocrystals have been synthesized by the carbon arc technique. The τ-phase in equiatomic Mn/Al[C] alloys is an interesting permanent magnet material and the only reported ferromagnetic material in the manganese/aluminum–carbon system. Observation of ferromagnetic response in Mn–Al–C nanoparticles produced by the carbon arc processing was taken as preliminary evidence of formation of the τ-phase, although this was a minority phase in the nanocrystalline materials.

3. *Rare Earth/Transition Metal Alloys.* Nanoparticles are of interest as precursors in synthesizing oriented, pressed, and sintered permanent magnets. In producing interesting nanocrystalline RE permanent magnet precursors, the extreme reactivity of rare-earth elements with oxygen is a crucial issue and one where carbon arc research could provide potentially important new materials. Kirkpatrick [48] has studied the possibility of synthesizing neodymium/iron/boron and samarium/cobalt RE/TM hard magnetic materials using the carbon arc. While moderately hard magnetic materials did result from this synthesis, TEM chemical mapping revealed that invariably these materials contained significant segregation of the RE and TM species. State of the art analytical TEM facilities (filtered imaging, GIF [49]; scanning transmission electrom microscopy, STEM; parallel electron energy loss spectroscopy, PEELS) have allowed probes of the chemistry on a nanometer size scale. Figure 19.7 shows GIF images of some of our initial samarium/cobalt–carbon nanoparticles in which chemical segregation is evident. Similarly produced neodymium/iron/boron–carbon nanoparticles were notable in less segregation.

Early developments in carbon arc generators included efforts to improve speed, efficiency, and yield. Glass reactors have been designed by Scrivens and Tour [50] with the advantage of being able to observe the carbon arc. A stainless steel horizontal reactor was commercialized by Terrasimco with copper bath reactor with the capability of consuming up to $\frac{1}{2}$ in. diameter carbon or composite electrodes. Experiments [51] indicate that fullerene yields decrease with larger electrode size but this has not been observed for nanoencapsulate production. Evolution of carbon arc reactors is illustrated in Figure 19.8, which shows a water-cooled bath reactor that has been employed by the group at

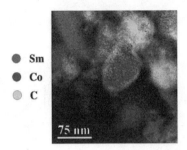

Figure 19.7   GIF chemical map of Sm-Co-C nanoparticles illustrating chemical segregation (courtesy of J. H. Scott).

Carnegie Mellon University to synthesize nanoencapsulates of a variety of materials. Another goal of early studies was to understand reactions in the carbon arc in greater detail. A discussion of growth mechanisms of nano-encapsulates in the arc is summarized above. While the products are similar, there are variations in the carbon arc preparation procedure (Fig. 19.8). In all cases, a composite anode is vaporized and transformed into products while a graphitic cathode is not consumed. The cathode diameter is usually the same as the anode, or larger. The cathode deposit depends on the size and relative position of the anode. In some cases the anode is a graphite rod, typically $\frac{1}{4}$ in. in diameter, drilled out and filled with a mixture of metal or metal oxide powder and graphite powder. A few groups use molded composite anodes prepared from powders with a pitch or dextrin binder, carbonized at elevated temperatures. With more complicated mixtures of elements, homogeneity is important; alloy or alloy oxide powders are preferable to mixtures of powders to obtain uniform products. The metal fraction can range from ≪1 to >70 wt %, with bound composites required to reach the higher values [48].

Both alternating current (ac) and direct current (dc) carbon arcs produce nanoparticles, but the dc method generates them at a higher rate. It also enables a cathode deposit to grow. Since the cathode deposit contains the largest fraction of multiwalled nanotubes and faceted nanoparticles, the dc arc is standard for the field. The exact conditions vary, with voltages ranging from 25 to 45 V, and currents from 70 to 200 A. Although $\frac{1}{2}$ in. diameter rods have been used to prepare nanoparticles, $\frac{1}{4}$ in. is by far the most common anode diameter. The arc gap is generally about 1 mm, but gaps ranging from 0.5 mm to several millimeters have been used successfully. The He pressure has been varied between 50 and 900 torr, but typical conditions are either standard fullerene (100–125 torr) or carbon nanotube (500–600 torr) pressures. The He flow rate can affect the average nanoparticle size, but values are seldom reported in the literature. Vaporizing the composite anode leads to a hard cathode deposit plus powdery soot on the reactor walls. The cathode deposit has an intricate structure consisting of a hard pencil-like core (1 mm to 4 cm long) and a "pancake" at its base. The core has a hard outer shell and a soft black inner region. The inner

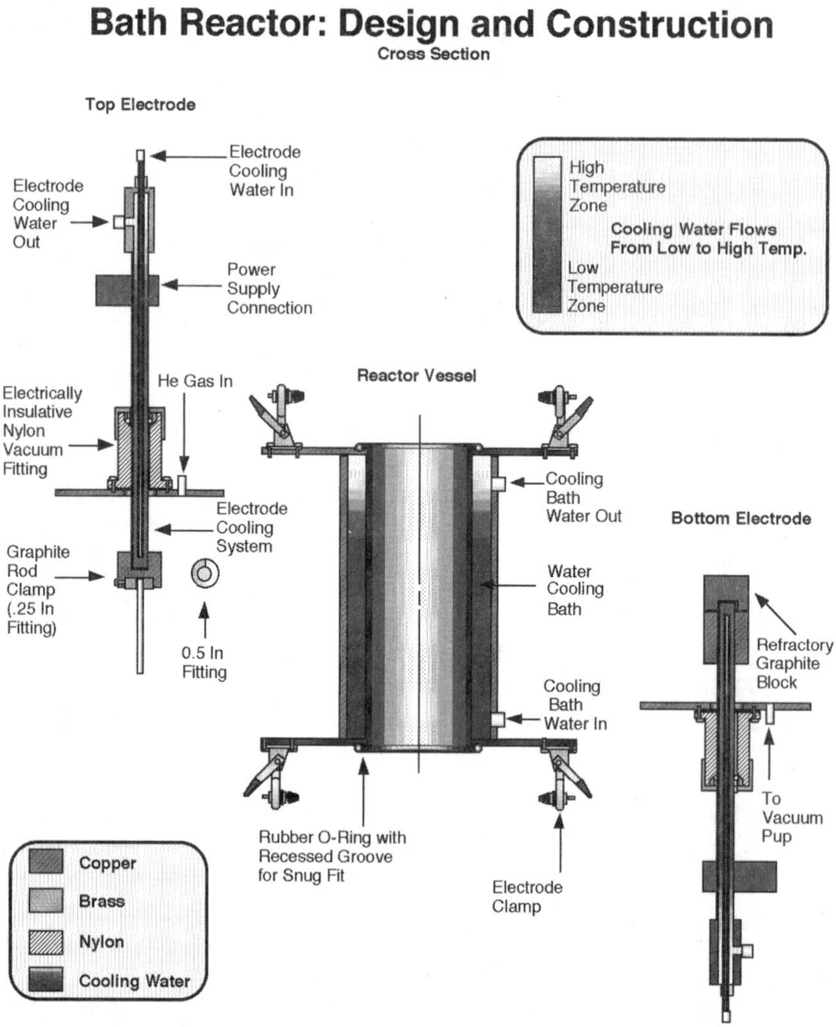

Bath Reactor: Design and Construction
Cross Section

Figure 19.8 Typical carbon arc reactor schematic. A cylindrical chamber is evacuated and then a He flow is used. In the center, the two electrodes are placed with a small gap between them. A DC arc is struck, and the anode position is adjusted to maintain a constant gap as it is consumed. Both the electrodes and the reactor walls are water-cooled (picture courtesy of F. Johnson and S. Curtin; reproduced from M. E. McHenry, Y. Nakamura, S. Kirkpatrick, F. Johnson, S. A.Majetich, and E. M. Brunsman, in Fullerenes: Physics, Chemistry, and New Directions VI, eds. R. S. Ruoff and K. M. Kadish, The Electrochemical Society, Pennington, NJ, 1463 (1994b); reproduced with the permission of M. E. McHenry).

core has a laminar structure, with 40 mm diameter columns aligned to form layers 100–300 mm thick. Carbon nanotubes are sometimes found in bundles here, but their alignment is random, and not along the long axis of the cathode deposit.

19.3.2 Tungsten Arc Metal Pool Synthesis

An innovative variant of carbon arc synthesis produces graphite encapsulated metal nanocrystals using a tungsten arc over a carbon saturated metal pool [52,53]. In this technique a tungsten electric arc is used to coevaporate metal and carbon from a liquid metal pool. In the technique described by Dravid and co-workers, an arc is struck between a 6.5 mm diameter tungsten electrode and a 40 mm diameter metal anode (supported by a graphite crucible) in a helium atmosphere. The arc between the liquid metal/carbon pool can be maintained for periods of hours with helium provided in a small flow near the liquid metal surface. A mixture of uncoated and nanoencapsulated metal particles are collected on the walls of a reactor. This technique offers the advantage of limiting the amount of carbonaceous debris that forms with the nanoencapsulates. Unencapsulated nanocrystals are easily removed using an acid bath, yielding a final sample composed primarily of graphite encapsulated nanocrystals. Jiao et al. [54] have applied this technique to the synthesis of transition metal nanoencapsulates.

Host et al. [35] have studied graphite encapsulated nickel and cobalt nanocrystals produced by the tungsten arc method. They have done a systematic investigation of the particle size and morphology as well as postsynthesis annealing studies on the same. As in earlier carbon arc produced cobalt, a heavily faulted fcc cobalt phase was observed in the nanoparticles. For nickel nanoencapsulates the nanoparticles were also fcc. Log-normal particle size distributions,

$$f(x) = \frac{1}{\sqrt{2\pi}\ln\sigma} \exp\left(\frac{-(\ln x - \ln\langle x\rangle)^2}{2\ln^2\sigma}\right) \tag{19.2}$$

have been observed for these particles based on TEM observations. After annealing the nanoparticles, the particle size was observed to increase, as detailed in Table 19.1. This size increase was attributed to diffusion of metal atoms from smaller particles to larger particles in a manner analogous to Ostwald ripening of precipitates, Host et al. [35] determined parameters for the diffusional growth of the nanopaticles considering metal diffusion through the encapsulation layers. The diffusion distance D_d was estimated from the random walk equation:

$$D_d = [tD_0 \exp(-Q/RT)]^{1/2} \tag{19.3}$$

to determine $D_0 = 2.2$ cm^2/s and the activation energy for diffusion, $Q = 53.3$ kcal/mol.

TABLE 19.1   Best-Fit Parameter to the Mean, $\langle x \rangle$, and Standard Deviation, σ, of the Particle Size Distribution for N, Co, and Ni Nanoparticles Before and After Annealing

Parameter	Ni (Prior to Anneal)	Ni (After 650°C Anneal)	Co (Prior to Anneal)	Ni (After 900°C Anneal)
N	363	351	692	411
$\langle x \rangle$	11.4 ± 0.2	13.6 ± 0.3	10.7 ± 0.2	15.7 ± 0.4
σ	1.96 ± 0.03	1.65 ± 0.03	1.65 ± 0.03	1.82 ± 0.04

The successful encapsulation of transition metals by the tungsten arc metal pool method with its relatively higher concentration of CEMNs (compared to the conventional carbon arc process) led to the examination of parameters such as the carbon/metal ratio as being critical factors controlling the success of the process [35,55]. These studies clearly illustrated that the major drawback of the conventional arc process was the significantly higher carbon/metal ratio, which resulted in a substantial amount of unwanted carbonaceous debris in the product. These studies, based on aerosol and gas phase chemistries, proposed that several competing mechanisms are in operation during the formation of CEMNs, and, to a large extent, the dominant mechanisms is determined by either the local or global carbon/metal ratio. While the local carbon/metal ratio is believed to be the driving mechanism controlling the formation of the core nanoscale products, the global carbon/metal ratio is believed to be responsible for determining the morphology of the final collected product such as the encapsulating carbon shell structure and whether the encapsulant is metal or carbide. Although none of the models proposed to date to describe the CEMN formation is all-encompassing, it is fairly apparent that a number of processing parameters such as arc current, electrode spacing, gas environment and pressure, gas velocity in the case of blown arcs, target composition, electrode design and configuration, and possibly as yet unknown parameters could potentially be controlled to make CEMN processing a commercial reality.

19.3.3   Plasma Torch Synthesis

Plasma synthesis is a recent development in synthetic routes for the production of magnetic nanoparticles. This plasma technology is based around a radio frequency (rf) plasma torch; thermal plasma synthesis has already proved its scalability to production of industrial quantities. A wide variety of starting materials and reactant gases can be used as feedstocks and many classes of materials, including metals, alloys, carbides, and oxides, can be synthesized. It offers the significant advantage of providing a continuous method for the production of fullerenes or other carbon coated nanoparticles. Typical plasma torch techniques are used to deposit low to intermediate melting temperature metals. In order to vaporize carbon to produce fullerenes, nanotubes, carbon

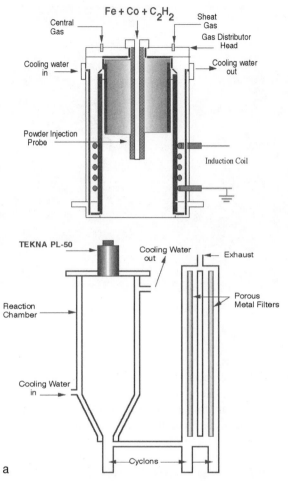

Figure 19.9 (a) Schematic of an RF plasma torch reactor showing continuous (gas) feed injection probe, RF induction coil and plasma reaction and collection chambers. (Z. Turgut, J. H. Scott, M.-Q. Huang, S. A.Majetich, and M. E. McHenry. J. *Appl. Phys.* **83**, 6468 (1998), reproduced with the permission of M. E. McHenry), and (b) TEM image of typical plasma torch synthesized FeCo nanoparticles (courtesy N. T. Nuhfer).

coated nanoparticles, and so on, a much higher power density needs to be present in the plasma. For this reason standard plasma torch methods are not sufficient to produce these materials. Instead, more sophisticated hybrid plasma and rf plasma techniques have been used.

Radio-frequency plasma torches (Fig. 19.9) are electrodeless discharges, which use magnetic induction to create eddy currents in a flowing gas stream, coupling radio-frequency electrical energy in a copper coil into the plasma gas. Unlike dc plasma torches, there are no electrodes to degrade over time and

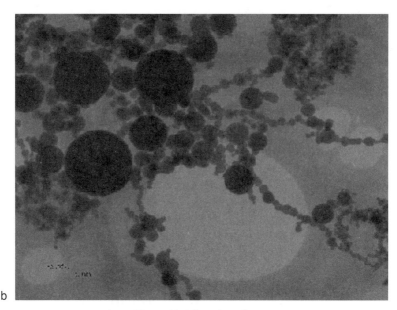

b

Figure 19.9 (continued)

contaminate the products. Thermal plasma synthesis is not a new technology and its adaptation to the production of nanoencapsulates benefits from a large body of previous work. The operator has control over many variables during thermal plasma synthesis, and the choice of operating conditions has a pronounced effect on the quality of the product. Fortunately, the temperature, flow, and concentration fields in the induction plasma have been well characterized through extensive mathematical modeling and measurements in the group of Maher Boulos of Sherbrooke University. This knowledge is important in ensuring the stability of the plasma and in optimizing the efficiency of heating of materials injected into the flow.

The optimum rf frequency, f, and torch radius, r, are related to the coupling efficiency and torch stability. The skin depth, d, of the plasma is given by $d = (\pi \mu \sigma f)^{1/2}$, where μ is the magnetic permeability of the plasma (approximately equal to μ_0), and σ is the electrical conductivity of the plasma [56,57]. For an argon plasma at one atmosphere and an average temperature of 8000 K, $s = 10^3$ W^{-1} · m^{-1} [56,57]. With a frequency of 3 MHz the skin depth is ~ 9 mm. The optimum coupling of electrical energy into the plasma occurs when the ratio of the plasma radius to the skin depth, r/d, is between 1.5 and 2.5 [56,57]. Most of the energy is dissipated in the cylindrical outer shell of the plasma defined by the skin depth.

Turgut et al. [58–60] have employed plasma torch synthesis techniques to produce a variety of encapsulated and unencapsulated magnetic nanoparticles. Carbon coated Fe_xCo_{100-x} ($x = 50, 45, 40, 35, 30, 25$ at %) nanoparticles were produced using a rf plasma torch, where the only carbon source was acetylene used as a carrier gas. Structural determination X-ray diffraction (XRD) indicated a single disordered bcc α-iron/cobalt phase along with graphitic carbon for all compositions. After annealing, X-ray diffraction indicated some carbide formation. A Scherrer analysis of the peak widths revealed particles to have an average diameter of 50 nm. TEM indicated a log-normal size distribution. Energy-dispersive X-ray spectrometry (EDXS) indicated compositional fluctuations of a few atomic % for individual particles of a nominal composition attributed to starting with elemental rather than alloy precursors.

Magnetic properties of nanoparticles with alloy compositions $Fe_{50}Co_{50}$, $Fe_{55}Co_{45}$, $Fe_{60}Co_{40}$, $Fe_{65}Co_{35}$, $Fe_{70}Co_{30}$, and $Fe_{75}Co_{25}$ have been measured within the temperature range 4–300 K. Temperature-dependent magnetization measurements have revealed the effects of atomic ordering in the nanoparticles. The variation of the saturation magnetization discussed below with temperature showed a discontinuity near the bulk order–disorder $(\alpha \rightarrow \alpha')$ transformation temperature, as well as loss of magnetization at the $\alpha \rightarrow \gamma$ structural phase transition temperature. Other features of $M(T)$ near 500–550 °C are consistent with prior observations of a "550 °C structural anomaly," which has been observed in bulk alloys with less than perfect order.

Scott et al. [61–63] have examined the issue of chemical homogeneity in plasma torch synthesized iron/cobalt alloy powders. To distinguish between inherent fluctuations and inhomogeneity introduced by poorly mixed starting material, $Fe_{50}Co_{50}$ nanoparticles were prepared using mixtures of elemental powders and a prealloyed $Fe_{50}Co_{50}$ starting material produced by ultrasonic gas atomization. The resulting thermal plasma product was examined with energy-dispersive X-ray spectrometry. The nanoparticles produced from alloy precursor showed a smaller spread in iron/cobalt ratio than nanoparticles produced from mixed elemental powders. Figure 19.10 compares the compositional spread of the two runs in a scatter plot of cobalt abundance (as estimated from the cobalt X-ray fraction) for several particles sampled from each powder. The shaded boxes are intended only to highlight the compositional spread for each set of particles. The increased iron/cobalt dispersion for the mixed element powders suggests that precursor homogeneity can persist through the plasma processing to impact the quality of the final nanoparticle powder.

19.3.4   Ion Beam Deposition for Thin Film Growth

Hayashi et al. [64] have reported on the growth of films of carbon-coated cobalt nanocrystals grown on a glass and cleaved sodium chloride crystals with carbon underlayers. These films contained nanoencapsulated hcp cobalt and a resultingly higher coercive field (370 Oe after annealing). These small isolated particles offer the advantage of relatively high coercivities in particles, which are not exchange coupled because of the intervening carbon encapsulation. These materials have thus been suggested as potential high-density magnetic recording media. These cobalt–carbon films were prepared by co-sputtering in an ion

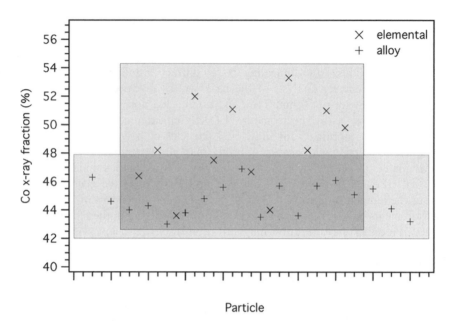

Particle

Figure 19.10 Scatter plot showing compositional variations in nanoparticles produced from mixed elemental and pre-alloyed precursors. Some increased inhomogeneity of mixed elemental precursors has survived to manifest as a greater range of Fe/Co ratio in the final product. Shaded boxes roughly delineate the compositional range of each particle population (courtesy J. H. Scott, reproduced from J. H. Scott, Z. Turgut, K. Chowdary, M. E. McHenry and S.A. Majetich, *Mater. Res. Soc. Proc.* 501, (1998); reproduced with the permission of M. E. McHenry).

beam deposition apparatus as illustrated in Figure 19.11. Here the carbon concentration of the deposited films can be varied by varying the position of the ion beam center with respect to be boundary between the cobalt and carbon sources. Hayashi et al., [64] and Delaunay et al. [65] have reported carbon concentrations in the films that vary from 27 to 57 at %. In these studies a Kaufman-type ion beam source was employed with 1.5 keV Ar ions and a beam diameter of 30 mm. Furthermore, they were able to vary substrate temperature between 100 and 300 °C. While the higher coercivities that apparently result from the formation of hcp cobalt nanoparticles with uniaxial anisotropy are desirable for magnetic recording applications, they are still not competitive with state of the art (∼several 1000 Oe). For this reason encapsulation of alloys with larger magnetocrystalline anisotropies has been explored. Delaunay et al. [47] have investigated cobalt/platinum–carbon nanogranular films. These experiments sought to take advantage of the large magnetocrystalline anisotropy hcp cobalt/platinum phases. These films had a nanogranular morphology with 5–15 nm grains. In-plane coercivities of 1500 Oe were achieved, reflecting the larger anisotropy. Experiments investigating encapsulation of the ordered $L1_0$

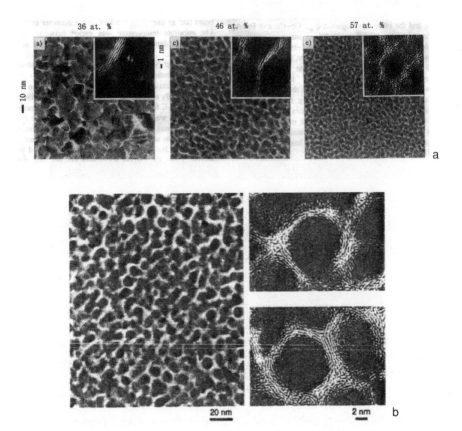

Figure 19.11 (a) Schematic diagram of an ion beam deposition system (courtesy of Commonwealth Scientific) and (b) example of ion beam deposited Co-C nanogranular films with 36, 46, and 57 at % carbon, respectively [J. J. Delaunay, T. Hayashi, M. Tomita, and S. Hirono; J. *Appl. Phys.* 82, 2200 (1997), reproduced with the permission of T. Hayashi and J. J. Delaunay].

equiatomic cobalt/platinum phase have not been performed. The large magnetocrystalline anisotropy of the ordered $L1_0$ phase suggests that even larger coercivities may be possible in nanogranular media.

19.3.5 Postprocessing Techniques: Filled Nanotube and Nanorod Growth

An interesting postprocessing method to synthesize nanoencapsulates uses carbon nanotubes in the production of carbon nanorods. Although in situ growth [32,66] of filled nanotubes has been reported, more promising techniques (e.g., in terms of yield and morphology) have been demonstrated in postprocessing techniques. Of these, a liquid reagent route has been used as an additional processing step to fill open carbon nanotubes in the pursuit of synthesis of so-

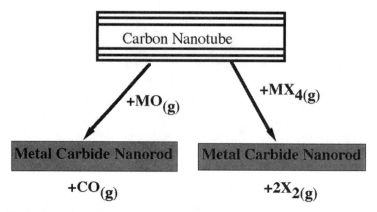

Figure 19.12 Reaction scheme used to prepare metal carbide nanorods (adapted after Dai et. al., 1995) MO is a volatile metal or non-metal oxide and MX_4 is a volatile metal or non-metal halide.

called quantum wires. Another interesting approach has been in the synthesis of carbon nanorods, where carbon nanotubes are converted to carbide nanorods through reaction with volatile oxide or halide species (Fig. 19.12). TaC nanorods and nanoparticles have been synthesized using a vapor–solid reaction path starting with CVD grown carbon nanotube precursors by Fukanaga et al. [68,69]. Their structures were studied using XRD, TEM, and HRTEM. The modification of carbon nanotubes by filling with a metal species opens up several possibilities for potential uses. Because the properties of modified fibers could be very different from the original carbon material, encapsulation might have interesting effects on the properties (such as superconductivity or magnetism) of the interior phase. Superconducting properties were characterized using superconducting quantum interference device (SQUID) magnetometry. For reactions at lower temperatures, carbide nanorods, which replicate the ~14 nm diameter of the precursor carbon nanotubes, are observed. For higher temperature reactions, coarsened carbide nanoparticles (100–200 nm) are observed, which have spherical or cubic-faceted morphologies. A morphological Rayleigh instability [70,71] was postulated as governing the transition from nanorod to nanoparticle morphologies. Stoichiometric bulk TaC crytallizes in the rock salt structure and has a superconducting transition temperature of 9.7 K. In TaC nanorods and nanoparticles, the superconducting properties correlate with the lattice parameter. Nanoparticles with a little higher lattice parameter than the ideal one show higher T_c and superconductivity disappears at higher fields.

19.3.6 Miscellaneous Synthesis Techniques

While most of the research on CEMNs was focused on materials processed by techniques such as arc discharge or plasma torch, encapsulation of metallic particles was also accomplished by other techniques, most often by serendipity. For example, well-encapsulated single crystals of CaS were detected on the surface of pitch-based carbon fibers when they were subjected to high-temperature heat treatments in inert atmospheres [72]. It was speculated that calcium and sulfur, common impurities present in pitch, reacted with each other during the heat treatment process and were subsequently encapsulated by carbon atoms provided by the fibers, with graphitization of the fibers initiating at about 800 °C and highly ordered graphitic shells forming at about 2000 °C. The CEMNs formed on the carbon fiber surfaces were found to be easily detachable by conventional ultrasonic methods. Recently, sharp needle-like structures consisting of highly graphitized nickel-filled carbon nanotubes were produced by the heat treatment of a sample consisting of sublimed alternate layers of C_{60} and nickel thin films [73].

As reviewed in this chapter, a significant number of scientific papers have been published over the course of the last five years, outlining a number of fascinating discoveries of nanostructured materials, such as the CEMNs, processed by a variety of techniques. With specific reference to the CEMNs, a large number of potential applications for magnetic CEMNs are discussed later in this chapter. Although the blown-arc method [37] was significantly more successful in producing higher concentrations of uniform and well-encapsulated CEMNs than the conventional arc discharge process, it still would not be a commercially viable process for applications requiring large quantities of CEMNs. Methods such as the use of rf plasma torches to synthesize large quantities of CEMNs are being investigated [58–60], and it is fairly certain that such methods, once fine tuned, will significantly enhance the level of commercial interest in nanocarbons.

19.4 MAGNETIC PROPERTIES OF NANOENCAPSULATED MATERIALS

A variety of magnetic properties have now been observed in various nanoencapsulated products of the carbon arc and other similar processes. Among them have been *paramagnetism* observed for magnetic ions trapped endohedrally in fullerenes, paramagnetism in carbon-coated rare-earth carbide (RE-C) nanocrystals, *ferromagnetism* in elemental transition metal (TM), transition metal carbide (TM-C), TM alloys, and rare-earth (RE) transition metal alloys. Also of interest have been the observation of *fine-particle magnetism* and the related phenomenon of *superparamagnetism*. Nanoencapsulated magnets have potential interest for applications especially when the particle size is such that particles remain monodomain. Carbon-coated magnetic nanoparticles have the additional advantage of being corrosion and oxidation resistant. Magnetic nanoparticles can be considered as particulates for recording media or as magnetic particles for inks, toners, or ferrofluidic applications, among others. Magnetic nanoparticles with sizes much less than the width of a magnetic domain wall can be considered for soft magnetic applications if after compaction grains remain nanocrystalline. Magnetic nanoparticles with sizes tailored to maximize the magnetic coercivity could prove to the excellent precursor materials for producing high-energy permanent magnets.

19.4.1 Paramagnetic Nanoencapsulates

Paramagnetism results from permanent atomic magnetic dipole moments. This atomic dipole moment arises from the incomplete cancellation of the electrons' angular momentum vector in an open shell configuration. In the absence of a field, the local atomic moments are uncoupled and will be aligned in random directions so that the average of the vector dipole moments, $\langle \mu_m \rangle$, is zero and therefore the magnetization $\mathbf{M} = 0$. To further describe the response of the quantum mechanical paramagnet we again consider the behavior of the dipole moments in an applied field. The magnetic induction, B, will serve to align the atomic dipole moments. The potential energy, U_p, of a dipole oriented at an angle θ with respect to the magnetic induction is given by

$$U_p = -\mathbf{M} \cdot \mathbf{B} = -\mu_B \cos\theta = g\mu_B \mathbf{J} \cdot \mathbf{B} \tag{19.4}$$

where g is the Lande g-factor and $\mathbf{J}$ is the total angular momentum vector. A quantum mechanical description of angular momentum tells us that $|\mathbf{J}_z|$, the projection of the total angular momentum on the field axis, must be quantized, so that $U = (g\mu_B)m_j B$, where $m_J = J, J-1, \ldots, -J$ and $J = |\mathbf{J}| = |\mathbf{L} + \mathbf{S}|$ (where $\mathbf{L}$ and $\mathbf{S}$ are the orbital and spin angular momenta, respectively), which may take on integral or half-integral values. In the case of spin only, $m_J = m_S = \pm\frac{1}{2}$. Quantization of J_Z requires that only certain angles q are possible for the orientation of μ with respect to $\mathbf{B}$ and the ground state corresponds to $\mu \| \mathbf{B}$. With increasing thermal energy it is possible to misalign $\mathbf{m}$ so as to occupy excited angular momentum states. If we consider the simple system with spin only we can consider the Zeeman splitting, where eigenstates are split by $\mu_B B$. The lower lying state corresponds to $m_J = m_S = -\frac{1}{2}$ with the spin moment parallel to the field. The higher energy state corresponds to $m_J = m_S = \frac{1}{2}$ and an antiparallel spin moment. Boltzmann statistics describes the population of these two states. If we have N isolated atoms per unit volume, in a field, we can define

$$N_1 = N\uparrow = A\exp\left(\frac{\mu_B B}{kT}\right) \quad N_2 = N\downarrow = A\exp\left(\frac{-\mu_B B}{kT}\right) \tag{19.5}$$

Recognizing that $N = N_1 + N_2$, the net magnetization (dipole moment/volume) is $M = (N_1 - N_2)\mu_B$ or

$$M = \frac{N}{A(e^x + e^{-x})} \cdot A(e^x - e^{-x})\mu = Nm\frac{e^x - e^{-x}}{e^x + e^x} = N\mu\tanh(x) \tag{19.6}$$

where $x = \dfrac{\mu_B B}{kT}$. Note that for small values of x, $\tanh(x) \sim x$ and we can approximate m:

$$M = N\frac{\mu^2 B}{kT} \quad \text{and} \quad \chi = \frac{M}{B} = \frac{N\mu^2}{kT} \qquad (19.7)$$

This expression relating χ to $1/T$ is called the *paramagnetic Curie law*.

For the case where we have both spin and orbital angular momentum, we are interested in the J quantum number and the $2J + 1$ possible values of m_J, each giving a different projected value (J_Z) of $\mathbf{J}$ along the field (z) axis. In this case we no longer have a two-level system but instead a $(2J + 1)$-level system. The $2J + 1$ projections are equally spaced in energy. Again considering a Boltzmann distribution to describe the thermal occupation of the excited states, we find that

$$\boldsymbol{\mu} \cdot \mathbf{B} = \frac{J_Z}{J}\mu_B B, \quad \boldsymbol{\mu} = S\sum_J \frac{J_Z}{J}\mu_B \exp\left(\frac{J_Z}{J}\frac{\mu_B B}{kT}\right), \quad \text{and}$$

$$N = S\sum_J \exp\left(\frac{J_Z}{J}\frac{\mu_B B}{kT}\right)$$

so that finally

$$M = N\mu = NgJ\mu_B B_J(x) \qquad (19.8)$$

where $x = gJ\mu_B B/kT$ and $B_J(x)$ is called the *Brillouin function* and is expressed as

$$B_J(x) = \frac{2J + 1}{J}\coth\left(\frac{(2J + 1)x}{2J}\right) - \frac{1}{2J}\coth\left(\frac{x}{2J}\right) \qquad (19.9)$$

Note that for $J = \frac{1}{2}$, $B_J(x) = \tanh(x)$ as for the spin only case before. For small x we then see that

$$\chi = \frac{M}{B} = \frac{Ng^2\mu_B^2 J(J + 1)}{3kT} = \frac{np_{\text{eff}}^2}{3kT}$$

where $p_{\text{eff}} = g[J(J + 1)]^{1/2}\mu_B$ is called the effective local moment. This expression is also a *Curie law* with $\chi = C/T$ and $C = Np_{\text{eff}}^2/3k$. Experimentally derived magnetic susceptibility versus T data can be plotted as $1/\chi$ versus T to determine C (from the slope) and therefore p_{eff} (if the concentration of paramagnetic ions is known).

19.4.1.1 *Paramagnetic Properties of Endohedral Fullerenes*

Mass spectroscopy was performed for a collection of endohedrally doped $Ho@C_{2n}$ endohedral fullerenes. Of particular interest is the determination of the ground state configuration of endohedrally trapped species in fullerenes. While mass spectra clearly showed that it is possible to trap magnetic ions in fullerenes, unfortunately it has proved nearly impossible to separate out enough of a single endohedral species so as to determine the ground state configuration unambiguously. On the other hand, the absence of pure holmium peaks in the mass spectra suggests that holmium atoms are indeed trapped endohedrally and not exohedrally in several fullerene species. Peaks in the mass spectra suggest $Ho@C_{78}$, $Ho@C_{80}$, and $Ho@C_{82}$ to be the most abundant endohedral species.

Magnetization data for a collection of these endohedral fullerenes reveals paramagnetic response. Figure 19.13 shows the behavior of scaled magnetization data versus H/T for a collection of $Ho@C_{2n}$ endohedral fullerenes. The H/T scaling is roughly satisfied, with slight deviation from the scaling law perhaps explained by several different holmium environments as indicated by the mass spectra. The magnetic data in Figure 19.13 can be fit by a Brillouin function $B_J(x)$, with an average moment $m_{eff} = 6.4\,\mu_B$ inferred. While several holmium environments are indicated, potentially with different moments, the fitted moment (viewed as an average) deviates significantly from the 10.6 μ_B moment, which would exist for an ionic ground state. This suggests that a Ho^{3+} ionic ground state is not observed and that crystal field effects or holmium–holmium interactions may be important in describing the magnetic configuration of holmium trapped in fullerenes. A more quantitative description of the

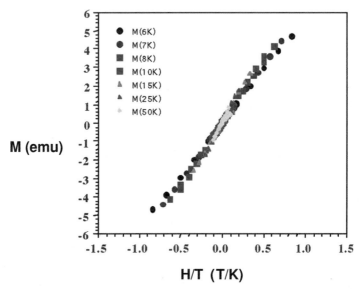

Figure 19.13 Magnetic response of Ho ions in fullerenes exhibiting rough paramagnetic H/T scaling [E. Kirkpatrick, S. A. Majetich and M. E. McHenry, unpublished data].

ground state of endohedrally trapped rare-earth species awaits development of techniques to efficiently extract various endohedral species from one another. As noted in the next section, the paramagnetic response of the holmium ions is roughly analogous to that observed in the Ho_2C_3 (sesquicarbide) phase in which a crystal field with C_3 symmetry is postulated as being important in determining the magnetic ground state.

19.4.1.2 Paramagnetic Properties of RE Carbide Nanocrystals

Majetich et al. [49] have analyzed magnetization data for powder samples of Gd_2C_3 nanocrystallites. $M(H, T)$ has been determined in fields between $\pm 5T$ and for temperatures from 4 to 300 K. Figure 19.14a shows scaled magnetization data as a function of H/T for data sets collected at 4, 5, 6, 7, 10, and 100 K. A universal curve described by a $J = \frac{7}{2}$ Brillouin function, $B_J(x)$, where $x = gJ\mu_B H/k_B T$ and $g = 2$ fits the data. This is consistent with the Gd^{3+} ionic ground state observed in several GdC_x phases. Curie–Weiss law fits, $\chi - \chi_0 = C/(T + \theta)$, yield a Curie temperature $\theta = 10.4$ K, and a Curie constant $C = 4.56 \times 10^{-3}(\text{emu} \cdot \text{K})/(\text{g} \cdot \text{Oe})$ (per gram of sample).

Room temperature electron paramagnetic resonance spectra of the powder at 9.104 GHz showed a single broad derivative centered at 3.13 kG, corresponding to a g-value of 2.08. This is nearly consistent with the $g = 2$ predicted for a $J = S = \frac{7}{2}$ ground state for a Gd^{3+} ion. It was concluded that the magnetic data for Gd_2C_3 nanocrystals was consistent with a Gd^{3+} ground state. Since $L = 0$ for the Gd^{3+} ion, crystal field effects are unlikely to be important.

Figure 19.14b shows similar data for Ho_2C_3 nanocrystals. At low temperature $M(x) \sim x/3$. $M(H)$ is well described by a Brillouin function with a characteristic H/T scaling bringing curves from different temperatures into coincidence. Brillouin function fits of this data have been used to infer an effective moment, $m_{\text{eff}} = 7.4\ \mu_B$. This is much smaller than the 10.6 μ_B effective moment predicted for a Ho^{3+} simple ionic ground state. This suggests that crystal field effects need to be considered in determining the magnetic ground state. Ho ions in α-Ho_2C_3 (having an $I43d$ space group) sit in sites of C_3 symmetry. A crystal field analysis [49] indicates that the $J = 8$ ionic ground state configuration for Ho^{3+} splits into a configuration with five singlets (A) and six doublets (E). Thermal population of these states has been shown to yield a temperature-dependent susceptibility that is consistent with the experimentally determined magnetization data.

19.4.2 Ferromagnetic Nanoencapsulates

A description of ferromagnetism also starts with the permanent local atomic dipole moments. Ferromagnetic response is distinct in that the local atomic moments remain coupled even in the absence of an applied field. As a result, a ferromagnetic material possesses a nonzero magnetization (even for $H = 0$) over a macroscopic volume, called a domain. Ferromagnetism is a collective phenomenon since individual atomic moments interact so as to promote align-

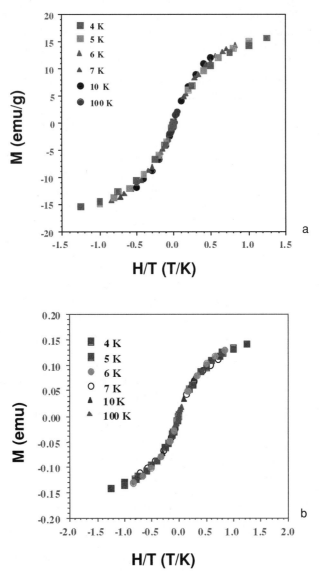

Figure 19.14 (a) Magnetic response of Gd^{3+} ions in Gd_2C_3 nanocrystals exhibiting paramagnetic H/T scaling. (B. Diggs, A. Zhou, C. Silva, S. Kirkpatrick, N. T. Nuhfer, M. E. McHenry, D. Petasis, S. A. Majetich, B. Brunett, J. O. Artman, and S. W. Staley, *J. Appl. Phys.* **75**, 5879 (1994).) and (b) magnetic response of Ho^{3+} ions in Ho_2C_3 nanocrystals exhibiting H/T scaling.

ment with one another. Two models have explained the interaction between atomic moments. *Mean field theory* considers the existence of a nonlocal internal magnetic field, called the Weiss field, which acts to align magnetic dipole moments even in the absence of an applied field, $\mathbf{H_a}$. *Heisenberg exchange theory* considers a local (usually nearest-neighbor) interaction between atomic moments (spins), which acts to align adjacent moments even in the absence of a field. Both theories will be considered. The two theories do lead to somewhat different pictures of certain aspects of the collective phenomena and the ferromagnetic phase transformation. The Heisenberg theory also lends itself to quite convenient representations of other collective magnetic phenomena such as antiferromagnetism, ferrimagnetism, and helimagnetism.

Weiss postulated the existence of an *internal magnetic field* (the Weiss field), $\mathbf{H_{INT}}$, which acts to align the atomic moments even in the absence of an external applied field, $\mathbf{H_a}$. The basic assumption of the mean field theory is that this internal field is nonlocal and is directly proportional to the sample magnetization, $\mathbf{H_{INT}} = \lambda\mathbf{M}$, where the constant of proportionality, λ, is called the *Weiss molecular field constant*. To consider ferromagnetic response in an applied field, $\mathbf{H_a}$, as well as the randomizing effects of temperature, we treat this problem as we did for a paramagnet except that we now consider the superposition of the applied and internal magnetic fields. By analogy it can be concluded that $\langle\mu_m\rangle = \mu_m^{atom}B_J(a')$, where $a' = (\mu_0\mu_m^{atom}/KT)[H + \lambda M]$ for a collection of classical dipole moments. Similarly, $M_S = N_m\langle\mu_m^{atom}\rangle$ and

$$\frac{M}{N_m m_m^{atom}} = \frac{M}{M_S} = B_J\left(\frac{\mu_m^{atom}m_0}{KT}[H + \lambda M]\right) \qquad (19.10)$$

where this rather simple expression represents a formidable transcendental equation to solve. Under appropriate conditions, this leads to solutions for which there is a nonzero magnetization (*spontaneous magnetization*) even in the absence of an applied field. For $T > \Theta$, the ferromagnetic Curie temperature, the only solution to Eq. (19.10) is $M = 0$, that is, no spontaneous magnetization and paramagnetic response. For $T < \Theta$, we obtain solutions with a nonzero, spontaneous magnetization, the defining feature of a ferromagnet.

The *Heisenberg model* considers ferromagnetism and the defining spontaneous magnetization to result from nearest-neighbor exchange interactions, which act to align spins in a parallel configuration, instead of a nonlocal, mean field. The Heisenberg model can be further generalized to account for atomic moments of different magnitude, that is, in alloys, and for exchange interactions, which act to align nearest-neighbor moments in an antiparallel fashion, or in a noncollinear relationship. Let us consider first the Heisenberg ferromagnet. Here we assume that the atomic moments on nearest-neighbor sites are coupled by a nearest-neighbor exchange interaction giving rise to a potential energy:

$$E_p = J_{ex}\mathbf{S}_i\cdot\mathbf{S}_{i+1} \qquad (19.11)$$

which for $J_{ex} > 0$ favors parallel alignment of the spins.

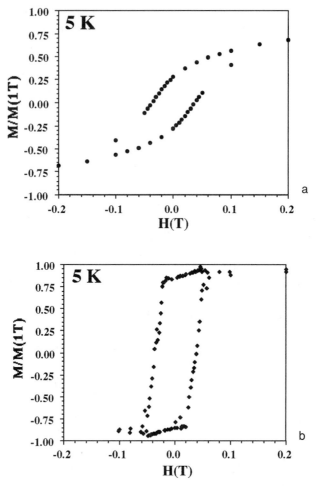

Figure 19.15 Low field magnetization curves at 5 and 50 K for Co/C particles (a) randomly oriented and (b) magnetically aligned. (M. E. McHenry, S. A. Majetich, J. O. Artman, M. DeGraef and S. W. Staley, Phys. Rev. B **49**, 11358 (1994a), reprinted with the permission of M. E. McHenry).

19.4.2.1 Magnetization and Ordering Temperatures in Nanoencapsulates

Figure 19.15a shows magnetic hysteresis for a randomly aligned nanocrystalline cobalt sample viewed on a small field scale [39–41]. The sheared and significantly rounded nature of the $M(H)$ loops at 5 K can be associated with demagnetization in nominally spherical particles (shearing) and rotational barriers to saturation (rounding). Note that at the higher temperature of 50 K, the hysteresis is significantly reduced and the rounding is less significant, but the shearing is the same. These observations are consistent with thermally activated rotational processes. Figure 19.15b shows the hysteretic response for cobalt nanocrystallites magnetically aligned with easy axes parallel to the field. Again some demagnetization-related shearing is observed in these loops. $M(H)$ for the aligned material, however, exhibits significantly reduced rounding, consistent with the reduction of the rotational barriers to saturation. The loops reflect a simple 180° switching of the magnetization vector in the monodomains (i.e., nearly square loops). The coercivity, H_c, is the same for the aligned and unaligned samples at similar temperatures.

Specific Magnetizations One of the goals of early studies of the magnetic properties of nanoencapsulates was the determination of the extent to which the specific magnetization could be increased as a result of control of the M/C ratios in carbon-coated ferromagnetic nanoparticles and control of the alloy chemistry in carbon-coated alloy ferromagnetic nanoparticles. McHenry et al. [44,45] and Kirkpatrick et al. [75] have demonstrated that they could encapsulate up to 66 wt % cobalt in nanoparticles. Figure 19.16 illustrates the variation of the specific magnetization of cobalt[carbon] nanocrystals with composition. The lower compositions represent nominal concentrations based on a mass balance in the precursor rods. The 66 wt % point represents the results of chemical analysis of the resulting soot. The line drawn is that expected for soots of a given composition assuming the nanoparticles to have the same specific magnetization as pure elemental cobalt. The largest specific magnetization achieved in this system was 110 emu/g.

Iron/cobalt alloys (e.g., *Carpenter Steel's* Hiperco50) are high-induction soft magnetic materials that are currently used in applications such as rotor materials in aircraft engines. Iron/cobalt alloys exhibit the largest magnetic induction of any material, as illustrated in the famous Slater–Pauling curve. Alloys near the equiatomic composition are particularly soft and exhibit large permeabilities. Efforts to produce iron/cobalt alloy nanoparticles have aimed at further increasing the magnetization as well as taking advantage of attractive technical magnetic properties in these materials. While cobalt nanoparticle-containing soots have been produced with saturation magnetization (per gram of soot) in excess of 100 emu/g, iron/cobalt-containing soots, reported by Gallagher et al. [46], have over 200 emu/g magnetizations. Note that pure iron has a $\sim$220 emu/g saturation magnetization. The 5 K magnetization for these materials was observed to be nonsaturating, consistent with paramagnetic or superparamagnetic response of some of the particles. Figure 19.16b shows the $H = 1$ T magnetization to be reminiscent of the Slater–Pauling curve for bulk alloys. The curve, however does peak at alloys that are more cobalt rich than bulk and falls far below saturation values for bulk iron-rich alloys. This could be explained by the greater potential for carbide formation in the iron-rich alloys. Carbon is well known to reduce the moment of bcc iron, though no evidence for carbide formation was observed.

Curie Temperatures In general, the mean field theory describing Curie temperatures for nanoparticles is not thought to deviate strongly from that of the

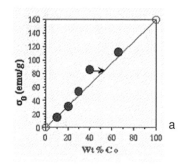

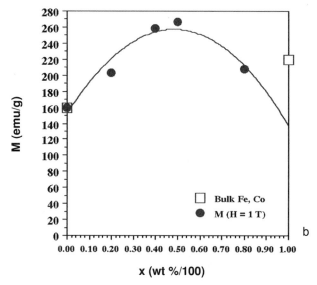

Figure 19.16 Specific magnetization of Co[C] nanocrystals [M. E. McHenry, E. M. Brunsman, and S. A. Majetich, IEEE Trans. Mag. 31, 3787, (1995); reproduced with the permission of M.E. McHenry)] and (b) M(H = 1 T) vs. x for FeCo nanoparticles [K. Gallagher, F. Johnson, E. Kirkpatrick, J. H. Scott, S. Majetich, and M. E. McHenry, *IEEE Trans. Mag.* **32**, 4842 (1996); reproduced with the permission of M.E. McHenry)] and data for bulk Fe and Co for comparison. Parabola is a guide to the eye.

bulk materials. This may not be the case in very small nanoparticles where the fraction of surface atomic sites is large. Here the reduced coordination of surface atoms should in fact alter the Curie temperature. For most observations in the literature, however, the T_c does not deviate strongly from that of the bulk materials. An example of $M(T)$ for carbon arc produced nanocrystalline nickel is shown in Figure 19.17. Figure 19.17 shows which reveals a T_c of ~360 °C (630 K), for nickel[carbon] nanocrystals, in excellent agreement with that of bulk nickel [41]. $M(T)$ for iron nanoencapsulates has also been reported by

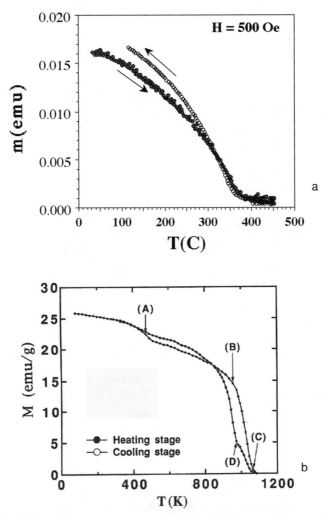

Figure 19.17 Magnetic moment vs. temperature for Ni[C] nanocrystals produced by C-arc synthesis [M. E. McHenry, Y. Nakamura, S. Kirkpatrick, F. Johnson, S. A. Majetich, and E. M. Brunsman, in Fullerenes: Physics, Chemistry, and New Directions VI, eds. R.S. Ruoff and K.M. Kadish, The Electrochemical Society, Pennington, NJ, 1463 (1994b)], and for Fe-containing nanoencapsulates [T. Hihara, H. Onodera, K. Sumiyama, K. Suzuki, A. Kasuta, Y. Nishina, Y. Saito, T. Yoshikawa, and M. Okuda; *Jpn. J. Appl. Phys.* **33**, L24 (1994).] showing Tc of Fe_3C (A) and a-Fe (C) (B is $\alpha \sim \gamma$ and C is $\gamma \sim \alpha$ phase transition).

Hihara et al. [76]. Here, $M(T)$ is complicated by the existence of iron carbides (notably cementite, Fe_3C) and small amounts of γ-iron. It is typical for α-iron to be the prominent phase, along with second-phase carbides and typically a very small amount of the γ-phase. The γ-phase is nonmagnetic and the α-phase and Fe_3C phases have Curie temperatures of 1040 K and ~480 K, respectively. Host et al. [35,77] have inferred a T_c of 1366 K for cobalt[carbon] nanoparticles, from fits of $M(T)$ data, in good agreement with the 1388 K value for bulk cobalt.

19.4.2.2 Magnetic Observations of Chemical Ordering Phenomena in Nanoencapsulates

Iron/cobalt alloys undergo an order–disorder transformation at a maximum temperature of 725 °C at the composition $Fe_{50}Co_{50}$, with a change in structure from the disordered α-bcc(A1) to the ordered β-CsCl(B2)-type structure. This ordering is important to the mechanical, electrical, and magnetic properties of this alloy. Carbon-coated Fe_xCo_{1-x} (x = 0.50, 0.45,0.40, 0.35, 0.30, 0.25) nanoparticles have been produced using a rf plasma torch by Turgut et al. [58]. The only carbon source was acetylene used as a carrier gas. Structural determination by X-ray diffraction indicated a single disordered bcc α-iron/cobalt phase along with graphitic carbon for all compositions. EDX analysis [61] indicated compositional fluctuations of a few atomic (%) for individual particles of one nominal composition, which is attributed to starting with elemental rather than alloy precursors. Magnetic hysteresis loops have been measured to $T > 1050$ K and revealed relatively high room temperature coercivities (200–400 Oe), with a strong compositional variation similar to that observed in bulk alloys. Larger coercivities are consistent with particles near the monodomain size for these alloys. The temperature dependence of the magnetization revealed the effects of atomic ordering.

Figure 19.18a illustrates $M(T)$, at $H = 500$ Oe, for equiatomic iron/cobalt nanocrystalline powders measured on a second heating cycle after initial heating to 920 °C. The influence of ordering is observed prominently in features that are quite similar to thermomagnetic observations for bulk $Fe_{49}Co_{49}V_2$ alloys. These features include a discontinuity in $M(T)$ at ~600 °C (due to chemical ordering) and a return to the extrapolated low-temperature branch of the curve at the disordering temperature of ~730 °C. Also of note is that at ~950 °C we observe an abrupt drop in the magnetization that corresponds to the α → γ structural phase transformation. A prominent Curie tail is observed above this transformation temperature, indicating paramagnetic response. It can be concluded from the abruptness in the drop of $M(T)$ that the $Fe_{50}Co_{50}$ alloy has a Curie temperature that exceeds the α → γ phase transformation temperature. Thus, a magnetic phase transition is not observed; instead it is concluded that (ferro)magnetic α-$Fe_{50}Co_{50}$ transforms to non(para)magnetic γ-$Fe_{50}Co_{50}$. This is corroborated in the differential thermal analysis (DTA) data, where the order–disorder and α → γ phase transformations have also been clearly observed.

Figure 19.19a shows thermomagnetic data (4 °C/min) at $H = 500$ Oe for various $Fe_xCo_{1-x}[C]$ alloy nanoparticles. DTA results in the work of Turgut et al. [58] demonstrated that the peak in the heating branch corresponds to reordering. The contribution of reordering to an increase in magnetization was inferred. Figure 19.19b shows high-field thermomagnetic data for $Fe_xCo_{1-x}[C]$ alloy nanocrystals ($H = 10$ kOe, 2 °C/min). No compositional dependence of

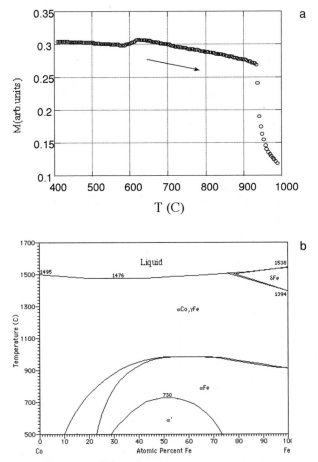

Figure 19.18   (a) M(T) data for C-arc synthesized equiatomic Fe-Co nanoparticles measured using a Lakeshore model 7300 VSM and oven assembly at 500 Oe measured on a second heating cycle to 995 C after initial heating to 920 °C (Z. Turgut, M.-Q. Huang, K. Gallagher, S. A. Majetich and M. E. McHenry, J. Appl. Phys. 81, 4039, (1997) reproduced with the permission of M.E. McHenry); (b) Fe-Co phase diagram (produced using TAPP @TM software, ES Microware).

the reordering temperature was observed for the saturation magnetization. A "550 °C anomaly" was observed, which occurs as a decrease in saturation magnetization with increasing temperatures up to 510 °C. The $Fe_{70}Co_{30}$ composition exhibits only a very small anomaly in agreement with the data of Velisek [78], who showed that while the $Fe_{50}Co_{50}$ composition shows 550 °C anomaly accompanied with a change in specific heat, an Fe_3Co composition showed only a small anomaly. In nanocrystalline materials the anomaly occurs at temperatures well below 550 °C. This has been attributed to the high diffusivity in these particles, which aids reordering.

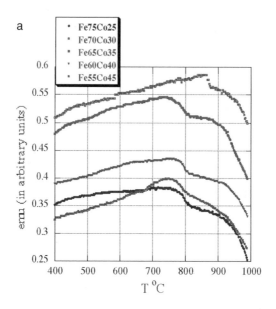

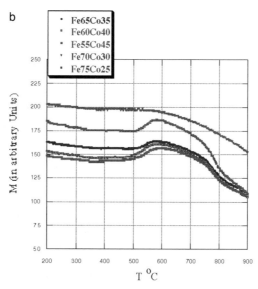

Figure 19.19 (a) Magnetization versus temperature data, showing ordering anomalies, taken at 500 Oe and 4 °C/min. heating rate for various FexCo1-x[C] nanoparticles. (b) Same response taken at 10 kOe. and 2 °C/min. heating rate (Z. Turgut, J. H. Scott, M.-Q. Huang, S. A. Majetich, and M. E. McHenry. J. *Appl. Phys.* **83**, 6468 (1998), reproduced with the permission of M. E. McHenry).

19.4.2.3 Superparamagnetism in Nanoencapsulates

For fine particle magnets, the possibility of thermally activated switching, and reduction of the coercivity as a function of temperature, must be considered as a consequence of a superparamagnetic response. Superparamagnetic response refers to the probability of thermally activated switching of magnetic fine particle moments. This thermally activated switching can be described by an Arrhenius law for which the activation energy barrier is $K_u \langle V \rangle$ and $\langle V \rangle$ is the particle volume. The switching frequency becomes larger for smaller particle size, smaller anisotropy energy density, and at higher temperatures. Above a *blocking temperature*, T_B, the switching time is less than the experimental time and the hysteresis loop is observed to collapse; that is, the coercive force becomes zero. Above T_B, the magnetization scales with field and temperature in the same manner as does a classical paramagnetic material. However, the inferred dipole moment is a particle moment and not an atomic moment. The magnetization, $M(H, T)$ can be fit to a Langevin function, L, using the relation

$$\frac{M}{M_0} = L(a) = \coth(a) - \frac{1}{a}$$

where M_0 is the 0 K saturation magnetization and $a = mH/k_B T$. The effective moment m is given by the product $M_S \langle V \rangle$, where M_S is the saturation magnetization and $\langle V \rangle$ is the particle volume. Below the blocking temperature, T_B, hysteretic magnetic response is observed. The coercivity has the temperature dependence $H_c = H_{c0}[1 - (T/T_B)^{1/2}]$. In the theory of superparamagnetism, the blocking temperature represents the temperature at which the metastable hysteretic response is lost in a particular experimental time frame. In other words, below the blocking temperature hysteretic response is observed since thermal activation is not sufficient to allow the immediate alignment of particle moments with the applied field. For spherical particles with a uniaxial anisotropy axis, the rotational energy barrier to alignment is $(K_u V)$. For hysteresis loops taken over $\sim$1 hour, the blocking temperature, T_B, should roughly satisfy the relationship

$$T_B = \frac{K_u \langle V \rangle}{30 k_B}$$

The factor of 30 represents $\ln(\omega_0/\omega)$, where ω is the inverse of the experimental time constant $(\sim 10^{-4}$ Hz$)$ and ω_0 is an attempt frequency for switching $(\sim 1$ GHz$)$.

In early observations of the temperature-dependent magnetic response of carbon arc produced nanoparticles, collections of smaller nanoparticles were observed to exhibit superparamagnetic response. Signatures of this response included the collapsing of the hysteretic response above the blocking temperature, a Langevin H/T scaling of the magnetization above the blocking temperature, and time-dependent magnetic response below the blocking temperature. Figure 19.20a shows a Langevin function fit to $M(T)$ for iron[carbon] nanoparticles having a blocking temperature of $\sim$80 K. Figure 19.20b shows

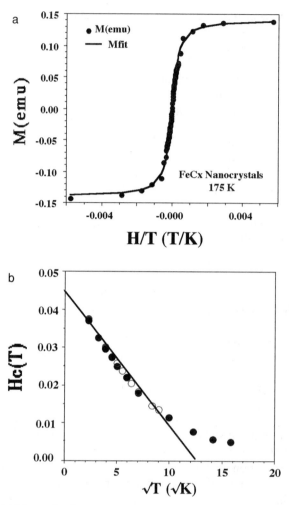

Figure 19.20 (a) Magnetization at temperatures above the blocking temperature (80 K) for Fe[C] particles plotted as a function of H/T and (b) temperature dependence of the coercivity, Hc, for Co[C] particles showing Hc(T) to obey a $\sim$ T dependence with a blocking temperature, T_B, of $\sim$160 K (M. E. McHenry, S.A. Majetich, J. O. Artman, M. DeGraef and S. W. Staley, *Phys. Rev. B* **49**, 11358 (1994a), reprinted with the permission of M. E. McHenry).

the temperature dependence of the coercivity for Co[C] nanoparticles produced by a carbon arc route. A low-temperature coercivity of 450 Oe is observed, which disappears with roughly the $[1 - (T/T_B)^{1/2}]$ temperature dependence predicted in the theory of superparamagnetism. The deviation from the theory has been attributed to having a distribution of sizes as opposed to a monodisperse set of magnetic nanoparticles. A blocking temperature for these particles is $\sim$160 K.

Superparamagnetic response is a phenomenon to be avoided in such applications as magnetic recording media since written information is lost due to thermally activated switching of the magnetization. On the other hand, a *magnetocaloric effect*, due to the strong temperature dependence of the magnetization observed in superparamagnetic response, can be exploited in magnetic refrigeration. The magnetic specific heat, C_M, can be shown to be proportional to the temperature derivative of the magnetization, dM/dT, and therefore is large near a magnetic phase transition or at the blocking temperature for superparamagnetic particles. This then has been suggested by McMichael et al. [79], as a method for magnetic refrigeration at cryogenic temperatures.

19.4.2.4 Technical Magnetic Properties of Nanoencapsulates The

technical magnetic properties of a ferromagnetic material are described by the metastable magnetic response depicted in a magnetic hysteresis curve. Two of the important technical magnetic properties are the so-called *remnant magnetization*, M_r, and the *coercive field*, H_c. For large magnetic fields H, all of the atomic magnetic dipole moments are aligned with the field and the ferromagnet is said to be saturated, having a *saturation magnetization* $M_s = N_m m$, where N_m is the number of magnetic dipoles per unit volume and m is the atomic dipole moment. The saturation magnetization is an intrinsic magnetic property of the material. Upon reduction of the applied field to zero, the magnetization is reduced due to rotational processes. The remaining or remnant magnetic moment at zero field results from energy barriers to the reversal of the magnetization. It is not until a reverse field equal to the coercivity, H_c, is applied that the magnetization is returned to zero. For applications in which the remnant magnetization is to be exploited, it is important for H_c to be large enough to prevent switching of the magnetization due to fluctuations in the field. For applications in magnetic recording, for example, a moderate coercivity is desirable to prevent fluctuation-induced switching; however, too large a coercivity proves problematic for writing information (i.e., reversal in an applied field).

An upper bound for the coercivity, H_c, is the so-called *anisotropy field*, $H_K = 2K_u/M_S$, where K_u is a *uniaxial energy density* for a magnet having its magnetization vector aligned parallel or antiparallel to an *easy axis of magnetization*. The common origins of magnetic anisotropy include *magnetocrystalline anisotropy* (*bulk and surface*), *magnetic shape anisotropy*, and *magnetomechanical* (*magnetostrictive*) *anisotropy*. Typically, the coercivity takes on values much less than H_K. In large volumes the reversal field is reduced due to the nucleation and growth (via domain wall motion) of reverse domains.

The interest in fine particle magnets stems mainly from the strong variation of their technical magnetic properties, notably H_C, with respect to particle size. The coercivities for a variety of elemental alloy and oxide ferromagnets were observed to have maxima, H_c^{max}, as a function of particle size (several 100 nm) corresponding to the limiting size for monodomain ferromagnets. Above this maximum size multidomain configurations are possible and domain wall nucleation and motion must be considered to understand the reversal process.

Below the size corresponding to H_c^{max}, the particles are typically monodomain and the reversal of the magnetization can be ascribed to coherent rotation of the particle moments, if the particles do not interact. To ascertain whether a particle will be mono- or multidomain, one needs to compare the energy of a particle with and without a domain wall. In the latter case the magnetostatic self-energy (demagnetization) results from free poles on the particle surface. In the former case the self-energy is reduced through domain formation but at the expense of domain wall energy. If the reduction in self-energy exceeds the domain wall energy, then a multidomain configuration will be stable. The self-energy is of the form $N_d(4_\pi M_s^2)$, where N_d is the demagnetization factor $(0 < N_d < 1)$ and the domain wall energy is calculated by minimizing the sum of the exchange (JV) and anisotropy (KV) energies in the volume V, over which the magnetization rotates from one direction to another (the domain wall). For a 180° Bloch wall in a crystal with lattice constant a_0, the domain wall thickness δ, which minimizes this total energy, is given by $\delta = Na = \pi S(J/Ka_0)^{1/2}$, where S is the spin moment per atom and N is the number of atoms over which the magnetization rotates in the wall. This yields a wall energy (per unit area) of $\gamma = 2\pi S(JK/a_0)^{1/2}$. For bcc iron $\gamma = \sim 1.1$ ergs/cm^2 and $\delta = \sim 45$ nm. A comparison of the magnetostatic self-energy and the domain wall energy is necessary to predict the size of spherical particles that will be stable as monodomain ferromagnets; this should exceed the domain wall thickness, δ.

In general, for magnetic nanoparticles with slightly larger sizes than the cobalt particles, described in Section 19.4.2.3, or with materials possessing larger magnetic anisotropies, the particles are not superparamagnetic and exhibit hysteretic magnetic response at room temperature. The technical magnetic properties that are important to a variety of applications include the remnant induction and the coercivity (in addition to the specific magnetization, saturation magnetization, and Curie temperatures, which are intrinsic magnetic properties that are described above). These properties have now been observed for a variety of elemental and alloy ferromagnetic nanoparticles that have been produced by a variety of synthesis routes. We summarize some of the more important observations and trends in the literature here. In early observations [39] small cubic iron, cobalt, and nickel nanoparticles were observed to be superparamagnetic with low-temperature coercivities of several tens to several hundreds of oersteds. With blocking temperatures well below room temperature, little or no hysteresis was observed at room temperature for these particles.

Efforts to stabilize magnetic nanoparticles against the thermally activated switching characteristic of superparamagnetic response led to efforts to increase particle size and to investigate alloys with larger magnetocrystalline anisotropies. Observations of room temperature coercivities of 100–200 Oe have been reported for a variety of cubic elemental iron, cobalt, and nickel nanoparticles with sizes ranging from 10 to 50 nm [35,77]. Even larger coercivities (370 Oe) have been observed for 8 nm hcp cobalt nanoparticles in ion beam deposited cobalt[carbon] thin films by Hayashi et al. [64]. This larger coercivity in smaller particles can be attributed to the larger magnetocrystalline anisotropy in hcp as compared with fcc cobalt. Coercivities as large as 300 Oe have also been observed in small iron/cobalt nanoparticles (without carbon coatings), where a contribution from surface anisotropy, due to a thin protective oxide coating, has been suggested as a coercivity mechanism [59,60].

Attempts to take advantage of the large magnetocrystalline anisotropies in rare-earth/transition metal alloy systems include $Fe_{1-x-y-z}Nd_xB_yC_z$ and $Sm_{1-x-y}Co_xC_y$ nanocrystals produced by a carbon arc. Coercivities of 500–600 Oe have been reported by McHenry et al. [44,45] and Kirkpatrick [48] in $Sm_{1-x-y}Co_xC_y$ nanocrystals produced using a Sm_2Co_7 precursor (which typically contained multiple phases due to chemical segregation of the rare-earth species). A 5 K coercivity of 900 Oe has been observed in similar multiphase $Fe_{1-x-y-z}Nd_xB_yC_z$ nanocrystals by McHenry et al. [44,45]. Perhaps the most exciting development in producing large coercivities in nanoencapsulates is the observation of a 1500 Oe coercivity in nanogranular hcp $Co_{50}Pt_{15}C_{35}$ thin films produced by ion beam deposition [47,65,80,81]. These films had 5–15 nm carbon encapsulated grains of an hcp cobalt/platinum phase. Platinum substitution is well known for its effect in increasing the magnetocrystalline anisotropy of hcp cobalt.

19.4.3 Applications of Magnetic Nanoparticles

19.4.3.1 Particulates A variety of applications for fine particle magnets have been suggested or already exist. We distinguish between two types of applications. These applications are (1) those simply requiring large particle moments and therefore large magnetic susceptibility (permeability), χ ($\chi = M/H$), and (2) those requiring the moment to be stable with respect to thermal fluctuations and fluctuations in field. Among these interesting applications are the following.

> *Ferrofluids.* Ferrofluids are colloidal suspensions of magnetic particles. These particles can be used to decorate the surface of a ferromagnetic material where they are oriented and aligned by the magnetic poles on the surface of the ferromagnet. These decoration techniques (called Bitter pattern techniques) allow, for example, the imaging of magnetic domains. Similar techniques involving "smokes" of fine particle magnets have been used to produce Bitter patterns that decorate the cores of magnetic flux lines in Type II superconductors. Colloidal suspensions of fine magnetic particles have found recent applications in magnetic seals and bearings. These applications of suspended fine particles do not necessarily require the moment to be stable with respect to thermal fluctuations and therefore superparamagnetic particles can and are typically used in such applications.

Magnetic Toners. Carbon-coated ferromagnetic particles have potential applications in magnetic toners for use in xerography. Magnetic brushes can be used to move magnetic toner particles in a xerographic process.

Magnetic Inks. Magnetic inks are currently used, for example, in encryption techniques and as a means of circumventing counterfeiting of documents or currency. We have demonstrated suspension of fine particle magnets in water and methanol. One could envision delivering these particles using droplet atomization as is now commonly used in ink-jet printers. The use of magnetic inks allows for magnetic "reading" of documents.

MRI Contrast Agents. Fine particle, biocompatible, ferromagnetic or paramagnetic materials with large permeabilities have possible applications as image enhancement agents in magnetic resonance imaging (MRI) techniques. Current imaging enhancement agents often employ molecular magnets (paramagnets) containing large moment Gd^{3+} and biocompatible ligands. Spin–spin coupling of hydrogen, in water molecules of tissue, to the contrast agent aids in accelerating the proton (T_2) relaxation rate.

Particulate Magnetic Storage Media. Fine ferromagnetic particles are commonly used in particulate recording media, for example, in magnetic recording tapes. Common particulate materials include Fe_2O_3, Cr_2O_3, and $BaO \cdot 6Fe_2O_3$. These applications of fine particle magnets have stringent requirements as to the stability of the moment with respect to thermal fluctuations and fluctuations in the field. In such applications two antiparallel orientations of the particle moment are used to record binary information that can then be read with a recording head. Errors in data storage are directly related to magnetization reversal due to thermal or field fluctuations.

Magnetic Refrigerants. The magnetic heat capacity of superparamagnetic particles can be exploited in a variety of magnetic refrigeration techniques, the principal of which is discussed above.

19.4.3.2 Precursors for Exchange Coupled Soft Magnetic Materials

Magnetic hysteresis is a useful attribute of a permanent magnet material, where we wish to permanently preserve a large remnant magnetic moment, that is, store a large metastable magnetization. For such applications a large stored energy in the hysteresis cycle is desired. On the other hand, a large class of applications requires small hysteresis losses per cycle. These include applications as inductors, low- and high-frequency transformers, alternating current machines, motors, generators, and magnetic amplifiers. The desired technical properties of interest for *soft magnetic materials* include:

1. *High Permeability.* Permeability, $m = B/H = (1 + \chi)$, is the materials parameter that describes the flux density, B, produced by a given applied field, H. In high-permeability materials we can produce very large changes in magnetic flux density in very small fields.

2. *Low Loss.* Hysteresis loss is the energy consumed in cycling a material between a field H and $-H$ and then back again. The energy consumed in one cycle is $W_h = \oint M \, dB$ or the area inside the hysteresis loop.

3. *Large Inductions.* A large saturation magnetization is usually desirable in most applications of soft magnetic materials.

Conventional physical metallurgy approaches to improving soft ferromagnetic properties involve tailoring chemistry and optimizing microstructure. Significant in the tailoring of microstructure is recognition of the fact that a measure of the magnetic hardness (the coercivity, H_c) is roughly inversely proportional to the grain size (D_g) for grain sizes exceeding ~ 0.1–1 mm (where the grain size exceeds the Bloch wall thickness, d). In such cases grain boundaries act as impediments to domain wall motion, and thus fine-grained materials are usually harder than large-grained materials. Significant recent developments in the understanding of magnetic coercivity mechanisms have lead to the realization that for very small grain sizes $D_g < \sim 100$ nm [82–85], H_c decreases quickly with decreasing grain size. This can be rationalized by the fact that the domain wall, whose thickness, d, exceeds the grain size, now samples several or many grains so that fluctuations in magnetic anisotropy on the crystal length scale are irrelevant to domain wall pinning. This important concept suggests nanocrystalline and amorphous alloys as having significant potential as soft magnetic materials. Soft magnetic properties require that nanocrystalline grains be exchange coupled and therefore any of the processing routes described above must be coupled with a compaction method in which the magnetic nanoparticles end up exchange coupled. Turgut et al. [58] have described initial efforts aimed at producing iron/cobalt nanoparticles for use as precursors for such novel soft magnetic materials.

ACKNOWLEDGMENTS

M.E.M. thanks the National Science Foundation for support through NYI award DMR-9258450. The material reviewed here is also based in part on work supported by the National Science Foundation under Grant No. ECD-8907068. Efforts reviewed here have also been sponsored by the Air Force Office of Scientific Research, Air Force Materiel Command, under grant number F49620-96-1-0454. The assistance of the CMU SURG program and the participation of CMU Buckyball Project members have been invaluable.

REFERENCES

1. T. J. Hirt and H. B. Palmer, *Carbon* **1963**, *1*, 65.

2. D. L. Trimm, *Catal. Rev. Sci. Eng.* **1977**, *16*(2), 155.

3. R. T. K. Baker and P. S. Harris, in *Chemistry and Physics of Carbon*, Vol. 14, P. L. Walker, Jr. and P. A. Thrower (eds.). Marcel Dekker, New York, 1978, p. 83.

4. McCarty, P. Y. Hou, D. Sheridan, and H. Wise, in *ACS Symposium Series 202*, L. F. Albright and R. T. K. Baker, eds. American Chemical Society, Washington, DC, 1982, p. 253.

5. T. Baird, J. R. Fryer, and B. Grant, *Nature* **1961**, *233*, 329.

6. M. Audier, J. Guinot, M. Coulon, and L. Bonnetain, *Carbon* **1981**, *19*, 99.

7. G. A. Jablonski, F. W. Guerts, A. Sacco, Jr., and R. R. Biederman, *Carbon* **1992**, *30*, 87.

8. H. W. Kroto, J. R. Heath, S. C. O'Brien, R. F. Curl, and R. E. Smalley, *Nature* **1985**, *318*, 162.

9. W. Krätschmer, L. D. Lamb, K. Fostiropoulos, and D. R. Huffman, *Nature* **1990**, *347*, 354.

10. S. Iijima, *Nature* **1991**, *354*, 56.

11. R. S. Ruoff, D. C. Lorents, B. Chan, R. Malhotra, and S. Subramoney, *Science* **1993**, *259*, 336.

12. M. Tomita, Y. Saito, and T. Hayashi, *Jpn. J. Appl. Phys.* **1993**, *32*, L280.

13. S. Iijima and T. Ichihashi, *Nature* **1993**, *363*, 603.

14. D. S. Bethune, C. H. Kiang, M. S. de Vries, G. Gorman, R. Savoy, J. Vazquez, and R. Beyers, *Nature* **1993**, *363*, 605 (1993); D. S. Bethune, R. D. Johnson, J. R. Salem, M. S. de Vries, and C. S. Yannoni, *Nature* **1993**, *366*, 123.

15. W. R. Davis, R. J. Slawson, and G. R. Rigby, *Nature* **1953**, *171*, 756.

16. L. J. E. Hofer, E. Sterling, and J. T. MacCartney, *J. Phys. Chem.* **1955**, *59*, 1153.

17. N. M. Rodriguez, A. Chambers, and R. T. K. Baker, *Langmuir* **1995**, *11*, 3862.

18. S. B. Austerman, S. M. Myron, and J. W. Wagner, *Carbon* **1967**, *5*, 549.

19. F. J. Derbyshire, A. E. B. Presland, and D. L. Trimm, *Carbon* **1975**, *13*, 111.

20. N. M. Rodriguez, *J. Mater. Res.* **1993**, *8*, 3233.

21. W. L. Holstein, *J. Catal.* **1995**, *152*, 42.

22. T. J. Konno and R. Sinclair, *Acta Metall. Mater.* **1995**, *43*(2), 471.

23. R. M. Pilliar and J. Nutting, *Philos. Mag.* **1967**, *6*, 181.

24. P. E. Nolan, D. C. Lynch, and A. H. Cutler, *Carbon* **1994**, *32*, 477.

25. S. Subramoney, *Adv. Mater.* **1998**, *10*, 1157.

26. J. R. Heath, S. C. O'Brien, Q. Zhang, Y. Liu, R. F. Curl, H. W. Kroto, F. K. Titell, and R. E. Smalley, *J. Am. Chem. Soc.* **1985**, *107*, 7779.

27. T. W. Ebbesen and P. M. Ajayan, *Nature* **1992**, *358*, 220.

28. Y. Saito, T. Yoshikawa, M. Okuda, N. Fujimoto, K. Sumiyama, K. Suzuki, A. Kasuya, and Y. Nishina, *J. Phys. Chem. Solids* **1993**, *54*, 1849.

29. S. Subramoney, R. S. Ruoff, D. C. Lorents, B. Chan, R. Malhotra, M. J. Dyer, and K. Parvin, *Carbon* **1994**, *32*, 507.

30. J. H. Scott and S. A. Majetich, *Phys. Rev. B* **1995**, *52*, 12564.

31. J. H. Scott, Ph.D. thesis, Carnegie Mellon University, Department of Physics, 1996.

32. Y. Saito, T. Yoshikawa, M. Okuda, M. Ohkohchi, Y. Ando, A. Kasuya, and Y. Nishina, *Chem. Phys. Lett.* **1993**, *209*, 72.

33. C. Guerret-Piecourt, Y. Le Bouar, A. Loiseau, and H. Pascard, *Nature* **1994**, *372*, 761.

34. Y. Saito, *Carbon* **1995**, *33*, 989.

35. J. J. Host, M. H. Teng, B. R. Elliott, J.-H. Hwang, T. O. Mason, D. L. Johnson, and V. P. Dravid, *J. Mater. Res.* **1997**, *12*, 1268.

36. S. Seraphin and D. Zhou, *Appl. Phys. Lett.* **1994**, *64*, 2087.

37. V. P. Dravid, J. J. Host, M. H. Teng, B. R. Elliott, J.-H. Hwang, D. L. Johnson, T. O. Mason, and J. R. Weertman, *Nature* **1995**, *374*, 602.

38. U.S. Patent 5,466,587 (Nov. 14, 1995), S. G. Fitzpatrick-McElligott, J. G. Lavin, G. F. Rivard, and S. Subramoney (to DuPont Company).

39. M. E. McHenry, S. A. Majetich, J. O. Artman, M. DeGraef, and S. W. Staley, *Phys. Rev. B* **1994**, *49*, 11358.

40. M. E. McHenry, S. A. Majetich, M. De Graef, J. O. Artman, and S. W. Staley, *Phys. Rev. B* **1994**, *49*, 11358.

41. M. E. McHenry, Y. Nakamura, S. Kirkpatrick, F. Johnson, S. A. Majetich, and E. M. Brunsman, in *Fullerenes: Physics, Chemistry, and New Directions VI*, R. S. Ruoff and K. M. Kadish (eds.). The Electrochemical Society, Pennington, NJ, 1994, p. 1463.

42. Y. Saito, M. Okuda, T. Yoshikawa, A. Kasuya, and Y. Nishina, *J. Phys. Chem.* **1994**, *98*, 6696.

43. Y. Saito, in *Recent Advances in the Chemistry and Physics of Fullerenes and Related Materials*, K. M. Kadish and R. S. Ruoff (eds.). The Electrochemical Society, Pennington, NJ, 1994, p. 1419.

44. M. E. McHenry, E. M. Brunsman, and S. A. Majetich, *IEEE Trans. Magn.* **1995**, *31*, 3787.

45. M. E. McHenry, S. A. Majetich, and E. M. Brunsman, *Mater. Sci. Eng.* **1995**, *A204*, 19.

46. K. Gallagher, F. Johnson, E. Kirkpatrick, J. H. Scott, S. Majetich, and M. E. McHenry, *IEEE Trans. Mag.* **1996**, *32*, 4842.

47. J.-J. Delaunay, T. Hayashi, M. Tomita, S. Hirono, and S. Umemura, *Appl. Phys. Lett.* **1997**, *71*, 3427.

48. E. Kirkpatrick, Ph.D. thesis, Carnegie Mellon University, Department of Physics, 1997.

49. S. A. Majetich, J. H. Scott, E. M. Brunsman, M. E. McHenry, and N. T. Nuhfer, in *Fullerenes: Physics, Chemistry, and New Directions VI*, R. S. Ruoff and K. M. Kadish (eds.). The Electrochemical Society, Pennington, NJ, 1994, p. 1448.

50. W. A. Scrivens and J. M. Tour, *J. Org. Chem.* **1993**, *57*, 6932.

51. H. Schwarz, *Chem. Int. Ed. Engl.* **1993**, *32*, 1412.

52. U.S. Patent No. 5,472,749 (1995), V. P. Dravid, M. H. Teng, J. J. Host, B. R. Elliot, D. L. Johnson, T. O. Mason, J. R. Weertman, and J.-H. Hwang (to Northwestern University).

53. M. H. Teng, J. J. Host, J.-H. Hwang, B. R. Elliot, J. R. Weertman, T. O. Mason, V. P. Dravid, and D. L. Johnson, *J. Mater. Res.* **1995**, *10*(2), 233.

54. J. Jiao, S. Seraphin, X. Wang, and J. C. Withers, *J. Appl. Phys.* **1996**, *80*, 103.

55. B. R. Elliott, J. J. Host, V. P. Dravid, M. H. Teng, and J.-H. Hwang, *J. Mater. Res.* **1997**, *12*, 3328.

56. M. I. Boulos, *Pure Appl. Chem.* **1985**, *57*, 1321.

57. M. I. Boulos, *J. High Temp. Mat. Process.* **1997**, *1*, 17.

58. Z. Turgut, M.-Q. Huang, K. Gallagher, S. A. Majetich, and M. E. McHenry, *J. Appl. Phys.* **1997**, *81*(8), 4039.

59. Z. Turgut, J. H. Scott, M.-Q. Huang, S. A. Majetch, and M. E. McHenry *J. Appl. Phys.* **1998**, *83*, 6468.

60. Z. Turgut, N. T. Nuhfer, H. R. Piehler, and M. E. McHenry, *J. Appl. Phys.* **1999**, *85*, 4406.

61. J. H. Scott, S. A. Majetich, Z. Turgut, M. E. McHenry, and M. Boulos, "Carbon coated nanoparticle composites synthesized in an RF plasma torch," in *Nanostructured Materials*, J. C. Parker (ed.), MRS Symposium Proceedings. Materials Research Society, Pittsburgh, PA, 1997.

62. J. H. Scott, Z. Turgut, M. E. McHenry, and S. A. Majetich, *Mater, Res. Soc. Proc.* **1998**, 501.

63. J. H. J. Scott, K. Chowdary, Z. Turgut, S. A. Majetich, and M. E. McHenry, *J. Appl. Phys.* **1999**, *85*, 4409.

64. T. Hayashi, S. Hirono, M. Tomita, and S. Umemura, *Nature* **1996**, *381*, 772.

65. J.-J. Delaunay, T. Hayashi, M. Tomita, and S. Hirono, *J. Appl. Phys.* **1997**, *82*, 2200.

66. S. Seraphin, D. Zhou, J. Jiao, J. C. Withers, and R. Loutfy, *Nature* **1993**, *362*, 503.

67. H. Dai, E. W. Wong, Y. Z. Lu, S. Fan, and C. M. Lieber **1995**, *375*, 769.

68. A. Fukanaga, S.-Y. Chu, and M. E. McHenry, Low temperature electronics and high temperature superconductivity, *ECS Proc.* **1997**, *97-2*, 9.

69. A. Fukunaga, S.-Y. Chu and M. E. McHenry, J. Mat. Res., 13(9), 2465 (1998).

70. F. A. Nichols and W. W. Mullins, *Trans. Met. Soc. AIME* **1965**, *233*, 1840.

71. F. A. Nichols, *J. Mater. Sci.* **1976**, *11*, 1077.

72. M. Kusunoki and Y. Ikuhara, in *Fullerenes: Physics, Chemistry, and New Directions VI*, R. S. Ruoff and K. M. Kadish (eds.). The Electrochemical Society, Pennington, NJ, 1996, p. 625.

73. N. Grobert, M. Terrones, A. J. Osborne, H. Terrones, W. K. Hsu, S. Trasobares, Y. Q. Zhu, J. P. Hare, H. W. Kroto, and D. R. M. Walton in *Kirchberg Proceedings in Molecular Nanostructures*, H. Kuzmany, J. Fink, M. Mehring, and S. Roth (eds.) **1998**.

74. B. Diggs, C. Silva, B. Brunett, S. Kirkpatrick, A. Zhou, D. Petasis, N. T. Nuhfer, S. A. Majetich, M. E. McHenry, J. O. Artman, and S. W. Staley, *J. Appl. Phys.* **1994**, *75*, 5879.

75. E. Kirkpatrick, S. A. Majetich, and M. E. McHenry, *IEEE Trans. Magn.* **1995**, *31*, 3787.

76. T. Hihara, H. Onodera, K. Sumiyama, K. Suzuki, A. Kasuta, Y. Nishina, Y. Saito, T. Yoshikawa, and M. Okuda, *Jpn, J. Appl. Phys.* **1994**, *33*, L24.

77. J. J. Host, J. A. Block, K. Parvin, V. P. Dravid, J. L. Alpers, T. Sezen, and R. LaDuca, *J. Appl. Phys.* **1998**, *83*, 793.

78. J. Velisek, *Czech, J. Phys.* **1970**, *B20*, 250.

79. R. D. McMichael, R. D. Shull, L. J. Swartzendruber, L. H. Bennett, and R. E. Watson, *J. Magn. Magn. Mater.* **1992**, *111*, 29 (1992).

80. J.-J. Delaunay, T. Hayashi, M. Tomita, and S. Hirono, *Jpn. J. Appl. Phys.* **1997**, *36*, 7801.

81. J.-J. Delaunay, T. Hayashi, M. Tomita, and S. Hirono, *IEEE Trans. Magn.* **1998**, *34*, 1627.

82. G. Herzer and H. R. Hilzinger, *J. Magn. Magn. Mater.* **1986**, *62*, 143.

83. G. Herzer and H. R. Hilzinger, *Phys. Scr.* **1989**, *39*, 639.

84. G. Herzer, *IEEE Trans. Magn.* **1990**, *26*, 1397.

85. G. Herzer, *J. Magn. Magn. Mater.* **1992**, *112*, 258.

CHAPTER 20

MOLECULAR AND SOLID C$_{36}$

JEFFREY C. GROSSMAN, CHARLES PISKOTI, STEVEN G. LOUIE,
MARVIN L. COHEN, and ALEX ZETTL

20.1 INTRODUCTION

The synthesis and identification of the pure carbon-cage molecule C$_{60}$ in 1985 [1] set the stage for a flurry of experimental and theoretical activity [2] catalyzed by the discovery of a bulk synthesis method [3]. In its molecular form, C$_{60}$ is a highly spherical hollow-cage molecule with the carbon atoms in the shell arranged in a network of hexagons and pentagons. In the experimentally observed and energetically most favorable "soccer ball" configuration, all 12 pentagons are "isolated." The fact that C$_{60}$ is the smallest fullerene obeying the isolated pentagon rule [4] has prompted some to suggest that it is in fact the smallest possible stable fullerene: smaller pure-carbon-cage molecules formed from hexagons and pentagons must necessarily contain adjacent pentagons with a resultant large strain energy. While the exceptional stability of C$_{60}$ is undisputed, the impossibility of stable "lower-order" fullerene-like molecules and molecular solids formed from such carbon networks is by no means a foregone conclusion. Indeed, such structures, precisely because of their enhanced strain energy and higher chemical reactivity, might be expected to display chemical, electronic, magnetic, and mechanical properties significantly different from (and possibly technologically more important than) those of the more conventional fullerenes.

Recent theoretical and experimental work [5–10] has shown that C$_{36}$ is a stable carbon-cage molecule and is the basis of novel pure-carbon solids. Both the molecule and the solid display a rich spectrum of physical properties. We here summarize recent research progress on this interesting new material.

Pseudopotential density functional calculations show that the structure with D_{6h} symmetry is one of two most energetically favorable. Based on this result and the fact that D_{6h} is conducive to forming a periodic system, a new solid

Fullerenes: Chemistry, Physics, and Technology, Edited by Karl M. Kadish and Rodney S. Ruoff.
0-471-29089-0   Copyright © 2000 John Wiley & Sons, Inc.

phase of carbon using C$_{36}$ fullerenes as a basis is proposed. The lowest energy crystal is a highly bonded network of hexagonal planes of C$_{36}$ units with AB stacking. The electron–phonon interaction potential for C$_{36}$ is substantially enhanced compared to C$_{60}$, leading to the possibility of larger superconducting transition temperatures than in alkali-doped C$_{60}$ solids. The reaction pathway to form a neutral C$_{36}$ dimer is predicted to be barrierless, while negatively charged C$_{36}$ molecules are less likely to bond due to a substantial barrier of formation. Calculations on doping demonstrate that substitutional doping with nitrogen can lead to a 10% decrease in the C–C bond lengths of C$_{36}$. Calculated endohedral binding energies show that C$_{36}$ is perhaps the smallest fullerene size that can easily trap a range of atoms. For the lowest-energy solid, it is predicted that Na is the largest alkali atom that can be intercalated into the new crystal structure without causing severe structural distortion. Further molecular properties, such as nuclear magnetic resonance (NMR) chemical shifts and infrared (IR) absorption spectra, are evaluated for the two lowest-energy fullerene isomers. It is shown that these isomers are sufficiently different chemically to distinguish by these methods.

Although evidence for isolated C$_{36}$ molecules can be inferred from early gas-phase carbon studies [11–13], it is only recently that bulk quantities of C$_{36}$ have been obtained using variations on the original Krätschmer–Huffman arc-plasma technique [3]. The molecular form of C$_{36}$ has been investigated via mass spectroscopy, and bulk solids have been studied experimentally via electron diffraction, transport measurements, and scanning tunneling spectroscopy. Electron diffraction suggests that C$_{36}$ forms a closed-packed solid with a lattice constant significantly smaller than that of the C$_{60}$ counterpart. In contrast to solids formed from C$_{60}$ and higher-order fullerenes, the C$_{36}$ solid is not a purely van der Waals solid but has covalent-like bonding, leading to a solid with enhanced structural rigidity. Tunneling spectroscopy gives evidence for clustering of C$_{36}$ molecules, where the clusters retain energy level signatures of the original molecular orbitals. Alkali doping increases the conductivity of C$_{36}$ and helps dissociate the solid into molecular units under laser irradiation.

20.2 METHODS

Calculations on solids reported in this chapter have been carried out using a plane wave pseudopotential total-energy scheme [14]. We employ the local density approximation [15,16] and use the Ceperley–Alder interpolation formula [17] for the exchange-correlation energy. Ab initio pseudopotentials are generated using the method of Martins et al. [18] and include semirelativistic corrections. Irreducible k-points are generated according to the Monkhorst–Pack scheme [19], and convergence is tested for each solid and is achieved with a range of 6–27 k-points in the irreducible part of the Brillouin zone, depending on the crystal structure. As energy cutoff of 60 Ry is used for the carbon atoms to ensure convergence in the total energy to less than 1 mRy per atom. For all

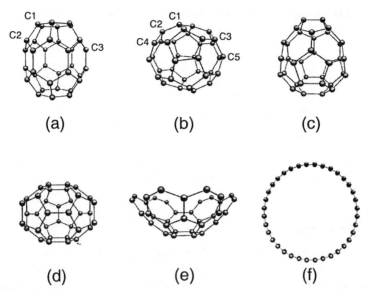

Figure 20.1 Six structural isomers of C_{36}: (a) the D_{6h}, (b) D_{2d}, (c) C_{2v}, and (d) D_{3h} fullerenes, (e) the C_{3v} bowl, and (f) the D_{18d} ring. Symmetry unique atoms are labeled for the first two structures.

molecular calculations, electronic and structural properties are calculated with DMol version 960 [20] using the frozen-core approximation, a fine integration grid mesh, and double numerical plus polarization basis sets. Both the local density [21] and generalized gradient [22,23] approximations (LDA and GGA) are used.

20.3 MOLECULAR PROPERTIES

20.3.1 Structures

The 36-atom carbon clusters lie in an interesting size regime, intermediate between the fullerene ($N = 40$–90) and ring ($N = 10$–28) dominated sizes. Five distinct peaks are observed in mobility measurements [24] of pure C_{36} isomers, corresponding to fullerenes, rings, and planar graphite-like structures. Theoretical mobility studies [25] matched several of these peaks very well, although it is not possible to distinguish among the fullerenes using mobility measurements alone. Here we consider six distinct equilibrium structures (see Fig. 20.1): the D_{6h}, D_{2d}, C_{2v}, and D_{3h} fullerenes, a corranulene-like bowl (C_{3v}), and a monocyclic ring (D_{18d}). Eleven other classical fullerene structures exist for 36 atoms [26]; however, most of these are less likely to form due to the considerable strain caused by clustering too many pentagons next to one another.

TABLE 20.1 Energy Differences (eV) Within the LDA and GGA (B-PW91) Methods for Five C$_{36}$ Isomers, Relative to the D_{6h} Fullerene^a

	Fullerenes				Bowl	Ring
	D_{6h}	D_{2d}	C_{2v}	D_{3h}	C_{3v}	D_{18d}
LDA	0.0	0.0	0.5	1.4	9.5	11.6
BPW91	0.0	0.0	0.5	1.8	7.1	20.3
Δ	0.5	0.4	0.4	0.4	2.0	0.5
EA	−3.3	−3.0	−3.3	−3.4	—	—

^{a} Each structure has been fully relaxed within the given symmetry by both methods. LDA HOMO–LUMO gaps (Δ) and electron affinities (EA) are also given.

20.3.2 Energetics

For the six geometries of Figure 20.1, full structural relaxations within the given symmetries have been carried out using both LDA and GGA. Total energy differences for these structures are listed in Table 20.1. We find the fullerenes to be substantially more stable than the bowl and the ring, in agreement with previous theoretical studies [27]. Our calculations predict the D_{6h} and D_{2d} fullerenes to be the most energetically favorable structures and to be essentially isoenergetic.

As seen in Table 20.1, there is a ~0.25 eV/atom difference between LDA and GGA in predicting the ring–cage energy differences. This is consistent with previous results, which indicated that LDA is a poor predictor when comparing carbon structures with dramatically different bonding character [28,29]. Indeed, the energy differences between the four fullerene structures, which have essentially identical bonding character, are very similar for LDA and GGA. In contrast to the calculations for C$_{20}$ isomers, this case does not appear to require further, more accurate calculations by, for example, quantum Monte Carlo approaches, since for C$_{36}$ discrepancies between the LDA and GGA do not qualitatively change the energetic ordering of the structures.

In addition to total energies, we compute LDA gaps between highest occupied and lowest unoccupied molecular orbitals (HOMO–LUMO) as well as electron affinities (see Table 20.1). The four fullerenes all have roughly the same HOMO–LUMO gap and similar electron affinities. Note that for each fullerene, an additional electron is well bonded to the molecule, as is the case for many of the intermediate-size carbon fullerenes. The remainder of this chapter focuses on the two lowest-energy structures as they are likely to be more abundant and easier to synthesize experimentally.

Of the two lowest-energy structures, the higher symmetry (D_{6h}) structure can form simple solids and will therefore be the main focus of the rest of this chapter. The C$_{36}$ (D_{6h}) is a good prototypical system for our theoretical study since it is small yet highly symmetric. Furthermore, the D_{6h} structure has the advantage that higher degeneracy in the electronic levels, because of its higher symmetry, can lead to a potentially large density of states. Later in this chapter, we will discuss possible means for distinguishing it from the D_{2d} structure.

20.3.3 D_{6h} C$_{36}$

The relaxed D_{6h} C$_{36}$ structure has a height of 5.2 Å and a width of 4.9 Å as measured from the atomic positions. Any closed structure consisting of three-fold coordinated atoms that form only pentagons and hexagons must have 12 pentagons according to Euler's relations. In the D_{6h} structure, the pentagons form two belts around the top and bottom hexagons and there is a row of hexagons separating these pentagon belts (Fig. 20.1a).

The calculated LDA binding energy is 8.14 eV/atom for this structure, compared with C$_{60}$, which has a binding energy of 8.42 eV/atom within the same method. This result shows that, as expected due to its smaller size, there is a higher energy cost per atom in forming the C$_{36}$ cage than C$_{60}$. In fact, most of the energy cost in forming C$_{60}$ can be attributed to the strain energy involved in curving the structure. It is possible to estimate this strain energy by considering the deviation of the π-orbital axis vector (POAV) from the planar configuration [30–32]. In its simplest form, the POAV is defined as the vector that makes equal angles ($\theta_{\sigma\pi}$) to the three σ bonds at a conjugated carbon atom. A simple formula for estimating the strain energy is obtained using a fit to a set of carbon molecules [32]:

$$E_{\text{strain}}(\text{eV/atom}) \approx 8.7\frac{1}{N}\sum_{}^{N}(\theta_{\sigma\pi} - \pi/2)^2$$

where the angle is expressed in radians. The quantity $(\theta_{\sigma\pi} - \pi/2)$ can be evaluated from the structure and for C$_{60}$ it is found to be 11.64°. Using the above formula, C$_{60}$ is estimated to have a strain energy of 0.36 eV/atom, which accounts for most of the energy difference between its binding energy and that of graphite (8.86 eV/atom within this method). The D_{6h} C$_{36}$ structure has three nonequivalent atoms, labeled in Figure 20.1. For the C1, C2, and C3 sites, the quantities $(\theta_{\sigma\pi} - \pi/2)$ are found to be 17.10°, 16.01°, and 12.82°, respectively. Using the above equation, we obtain an estimate for the strain energy of 0.62 eV/atom, which is in very good agreement with our LDA calculated energy difference relative to graphite. Thus, as in C$_{60}$, the energy required to form C$_{36}$ is due to the hybridization caused by curvature. Since clustering of pentagons creates severely strained atomic sites, it is preferable for the pentagons to be spread over the whole structure as in the case of the D_{6h} and the D_{2d} fullerenes. The values of the $\theta_{\sigma\pi}$ angles also suggest that the C1 and C2 atoms should be the most reactive sites in this structure.

In Figure 20.2 we show the energy levels near the Fermi gap. The calculated energy gap is found to be 0.5 eV between singlet states of representation B_{1u} and B_{2g} for the occupied and unoccupied state, respectively. One should keep in

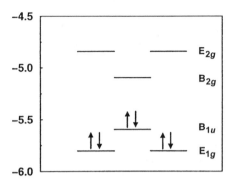

Figure 20.2 Energy levels of the D_{6h} structure near the Fermi energy. We show here the last two sets of occupied states and the first two sets of unoccupied states.

mind that the LDA has the tendency to underestimate the real gap of a structure. The second to last occupied states and next unoccupied states above the singlet are both part of doublets with symmetry E_{1g} and E_{2g}, respectively.

20.4 DIMER FORMATION

As a first step toward building a C$_{36}$ solid, we have investigated the possibility of intermolecular bonding through a series of calculations for a variety of C$_{36}$ dimers and trimers [9]. All geometries are fully relaxed within the given symmetries. Dimerization was tested at each of the three nonequivalent sites of the C$_{36}$ molecule, with either one or two single bonds between two C$_{36}$ molecules. Stable dimer formation occurred between C atoms located on the top hexagon or on the pentagon ring (C1 and C2 of Fig. 20.1a), but not for the third nonequivalent site on the middle hexagon belt (C3). In fact, completely unbonded dimers resulted from the latter case.

Figure 20.3 depicts three of the different dimers—(C$_{36}$)$_2$-A, (C$_{36}$)$_2$-B, and (C$_{36}$)$_2$-C (hereafter referred to as dimers A, B, and C)—and one of the trimers, (C$_{36}$)$_3$, considered in this chapter. Below each unit is the relevant symmetry and calculated binding energy. In all cases, the intermolecular bond distances can be considered as single C–C covalent bonds with an average length of 1.56 Å. Bonding that involves sites on the pentagon ring appear more favorable than bonding involving atoms on the top hexagon sites, but all of the configurations shown are energetically bound with respect to isolated molecules. Note that dimer C, which is the most stable configuration, can be obtained from dimer B simply by rotating one of the C$_{36}$ units with respect to one another without breaking an existing bond.

Given the potential for C$_{36}$ molecules to bond to one another, we investigate the reaction pathway for the formation of a C$_{36}$ dimer [10]. We choose dimer C,

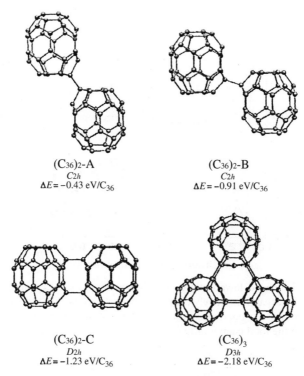

(C₃₆)₂-A
C2h
ΔE = −0.43 eV/C₃₆

(C₃₆)₂-B
C2h
ΔE = −0.91 eV/C₃₆

(C₃₆)₂-C
D2h
ΔE = −1.23 eV/C₃₆

(C₃₆)₃
D3h
ΔE = −2.18 eV/C₃₆

Figure 20.3 Three different configurations of C_{36} dimers, and a C_{36} trimer, with their corresponding symmetries and binding energies with respect to isolated C_{36} molecules. To clarify the structure of the trimer, the z axis has been rotated.

which is also the lowest energy dimer according to our calculations. The reaction coordinate is along the C_{36}–C_{36} bond axis.

In Figure 20.4 we show the results of our LDA total energy calculations for varying intermolecular separations. At each point, a full structural optimization is carried out with the constraint of fixed intermolecular bond distance. As the molecules are brought closer together, the onset of an attractive, van der Waals type of interaction appears at around 4.5 Å and possesses a minimum at about 3.0 Å. At about a 2.6 Å separation, a small barrier of ∼0.1 eV is encountered before the dimer begins to covalently bond (reaching a stable minimum at 1.6 Å). Thus, the LDA calculations imply an effectively barrierless dimerization reaction since the peak of the barrier encountered remains below the initial reactant energies.

The appearance of a kink in the reaction path may imply that the reaction induces a change in the molecular state symmetry. The LDA approach is a single-determinantal method and therefore cannot represent such a symmetry crossing accurately. However, qualitatively the results would not change since

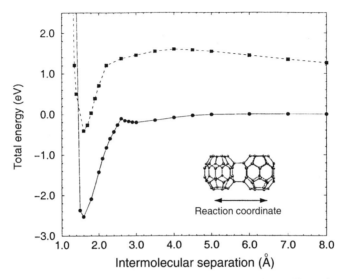

Figure 20.4 Calculated lowest energy reaction pathways of C$_{36}$ dimer formation for both neutral (circles) and negatively charged (squares) species. The reaction coordinate is along the intermolecular bond axis as shown. The neutral and charged curves have been shifted by twice the total energy of the isolated C$_{36}$ and C$_{36}^{-}$ molecules, respectively.

the barrier occurs below the reactant energy and the only difference would be to smooth out the kink in question.

Previous ab initio calculations for C$_{60}$ dimer formation did not include structural relaxation at each point on the reaction path and are therefore difficult to compare with the present calculations. In the work of Adams et al. [33], LDA total energy calculations were carried out on interpolated C$_{60}$ dimer coordinates and a barrier of 2.2 eV was found. Experimentally, it is well known that undoped C$_{60}$ has to be photoexcited or treated at high pressure to polymerize, whereas alkali-doped C$_{60}$ tends to polymerize spontaneously. Our calculations suggest that spontaneous polymerization for C$_{36}$ is highly likely *without* doping. The strong dimer binding of 2.6 eV for C$_{36}$ (compared with 0.6 eV for our calculations of the C$_{60}$ dimer binding) lends further credence to the idea that C$_{36}$ naturally prefers to polymerize.

As an additional comparison with C$_{60}$, we have evaluated total energies along the reaction path for C$_{36}^{-}$ molecules (such that the dimer has a charge of -2). These results, shown as squares in Figure 20.4, indicate that the presence of charge may substantially inhibit dimer formation. The two negatively charged molecules experience an electrostatic $1/r$ repulsion as they approach one another, which accounts for the fact that at 8 Å separation the dimer is more than 1 eV unbounded. At the equilibrium distance of 1.6 Å the charged system is energetically bound by roughly 0.4 eV; however, to reach this minimum the system must pass through an unfavorable barrier, which is 0.35 eV above the energy at 8 Å separation and 1.5 eV above the energy of two isolated C_{36}^- molecules. Therefore, we propose that it may be necessary and desirable to provide a source of negative ions in order to prevent C_{36} molecules from bonding to one another. The opposite is true in C_{60}, for which it has been shown both theoretically and experimentally that negative ions lower the barrier to dimerization and significantly enhance polymerization [33–35].

20.5 SOLID PROPERTIES

The originally proposed C_{36} solid [6] was formed using a rhombohedral crystal lattice because it involved only one C_{36} molecule per unit cell and was therefore computationally less demanding. Subsequently, we found that a very similar crystal, but with hexagonal symmetry imposed, was substantially lower in energy and had completely different electronic properties [7]. In particular, our calculations of the density of states (DOSs) revealed that the rhombohedral crystal is metallic with a large peak at the Fermi energy (E_F), whereas the hexagonal crystal is insulating with a large gap of ~2 eV. Furthermore, the hexagonal structure was found to have a significantly larger binding energy per molecule than the rhombohedral crystal.

The difference between these crystals is perhaps best understood by considering that each is simply a stacking of hexagonally symmetric planes of well-separated C_{36} units. Figure 20.5 shows the two kinds of stacking sequences—AB and ABC—corresponding to the hexagonal and rhombohedral crystals, respectively. In each case, both a top view of the repeating planes and a side view showing relaxed interlayer bonding character are shown. The fundamental stacking unit is a plane of unbonded C_{36} molecules, which we refer to as sheet 1 (S1), and the two crystals are labeled S1-AB and S1-ABC for the two kinds of stacking. The AA stacking sequence (i.e., S1-AA) is not considered here because it does not form a metastable structure, as will be discussed later.

In Table 20.2 we list the binding energy, density, gap, and k-point sampling for each of these crystals. The same information for a single S1 sheet is also given, although these data simply correspond to an isolated C_{36} molecule since the C_{36} units are essentially noninteracting within an S1 sheet. Note the dramatic differences in electronic and structural properties, despite the fact that the densities are roughly the same. The S1-ABC crystal has a binding energy of 3.53 eV/C_{36}, which is larger than our previous result of 2.2 eV/C_{36} [6] for this crystal. The difference is due to k-point convergence: In the present study we employ a $5 \times 5 \times 5$ shifted grid (23 k-points in the irreducible part of the Brillouin zone) whereas in the previous calculation we used a $3 \times 3 \times 3$ shifted grid (6 k-points) for structural relaxation. Here, structural and energetic properties are fully converged with respect to k-points separately for each crystal.

Interestingly, the significant differences in electronic properties between S1-ABC and S1-AB can be reproduced in isolated $C_{36}H_6$ molecules, where the H atoms are attached to the same six sites of adjacent interlayer bonds in the

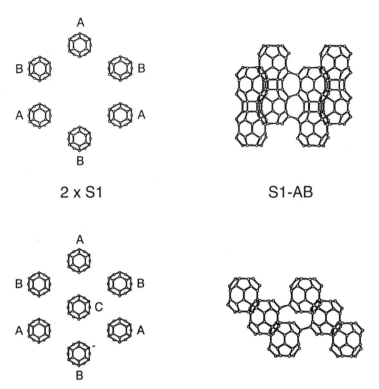

2 x S1

S1-AB

Figure 20.5 Fully relaxed LDA crystal structures formed by stacking planes of unbonded C$_{36}$ units. For the two different stacking types a top view (*left*) shows the repeating planes and a side view (*right*) shows interlayer bonding.

TABLE 20.2 Binding Energy (eV/C$_{36}$), Density (C$_{36}$/au^3), and Gap (eV) for Each Crystal Considered Here[a]

Structure	Binding	Density	Gap	k-Point Grid
S1	0.00	—	0.50	$1 \times 1 \times 1$
S1-AB	6.92	0.49	1.95	$5 \times 5 \times 5$
S1-ABC	3.53	0.51	metal	$5 \times 5 \times 5$
S2	1.67	—	1.09	$6 \times 6 \times 1$
S2-AA	3.98	0.72	0.86	$5 \times 5 \times 5$
S2-AB	7.71	0.76	0.61	$4 \times 4 \times 4$
S2-ABC	1.84	0.63	0.15	$4 \times 4 \times 4$

[a] S1/S2 correspond to sheets without/with intralayer bonding. k-Point grids used for each calculation are also given. For reference, our calculated binding energy of isolated C$_{36}$ is 8.14 eV/atom.

crystals. Upon structural relaxation, we find that D_{3d} $C_{36}H_6$ has a large gap of 2.2 eV and is energetically favored by 3.1 eV over D_{3h} $C_{36}H_6$, which has a much smaller gap of 0.4 eV.

The D_{3d} symmetry form disrupts all six aromatic rings in the belt of six fused benzenes and the D_{3h} form disrupts only alternate benzene rings. However, our calculations for various dihydrogenated forms of C_{36} show that simple resonance concepts that accurately predict the hydrogenation pattern in small polyacenes do not explain the results seen for C_{36}. This could be the case either because the action of resonance stabilization is not valid in such fused ring systems, or because other effects such as ring strain are counteracting the expected resonance pattern. A clear understanding of the difference between these two simple hydrogenated C_{36} molecules should also explain the difference between S1-ABC rhombohedral and S1-AB hexagonal C_{36} crystals. A thorough study of hydrogenation patterns in C_{36} and its component fragments, including the determination of overall strain and resonance energies by considering homodesmotic reactions, is discussed by Colvin et al. [36].

As we showed in the preceding section, the D_{6h} C_{36} molecules form stable dimers and trimers with two intermolecular bonds for the dimer and six intermolecular bonds for the trimer (i.e., dimer C and the trimer in Fig. 20.3). Furthermore, C_{36} molecules may be added in the same manner to form an infinite sheet of bonded units [10]. We refer to this layer as sheet 2 (S2) and note that our calculations give a binding energy for such a sheet of 1.67 eV/C_{36} (see Table 20.2). In Figure 20.6 we show a top view of S2 as well as side views of three different relaxed crystal structures, which form by stacking S2 layers according to AA, AB, or ABC sequences. All three stacking types lead to energetically bound layers.

For S2-AA, there is bonding between all six C atoms of the hexagon rings on top and bottom, as in the case of S1-AB. Such a stacking scheme therefore leads to intermolecular bonds for 24 of the 36 C atoms in each molecule. As shown in Table 20.2, S2-AA has a much higher density than the structures based on S1. Nonetheless, its binding energy is only slightly larger than that of S1-ABC and still 3 eV/C_{36} less than the binding of S1-AB.

Stacking S2 sheets in the AB sequence results in a slightly larger density and a substantially larger binding energy than S2-AA. In fact, S2-AB is the lowest energy crystal structure we have studied to date. It has the same number of intermolecular bonds as S2-AA and a d-spacing of 6.55 Å.

The S2-ABC crystal structure contains no interlayer bonding and has a much smaller total binding energy (1.8 eV/C_{36}) than the other solids. Since the interlayer binding is <0.2 eV/molecule, it should be rather easy for sheets stacked in this manner to "slip" over one another into the much more energetically favorable AB stacking scheme. Thus, we believe that the S2-ABC solid is unlikely to remain stable.

In Figure 20.7 we show the electronic density of states for the lowest energy (S2-AB) structure. Note that the crystal has characteristics typical of a molecular solid, with large peaks in the range of 10–20 states and narrow widths of

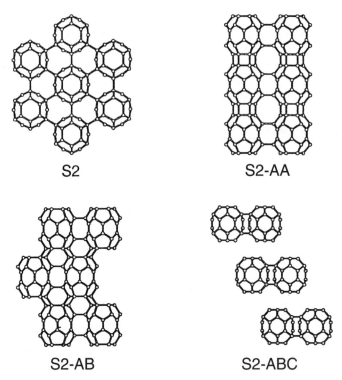

Figure 20.6   Fully relaxed LDA crystal structures formed by stacking planes of bonded C$_{36}$ units. A representative S2 sheet (top view) and side views of three stacking types shows intra- and interlayer bonding, respectively.

less than 0.5 eV. The large, isolated peaks just above and below E_F account for roughly two electrons, which implies that half-filling of a band can be achieved through electron/hole doping with 1 electron donor/acceptor per C$_{36}$.

20.6   SUPERCONDUCTIVITY

It has been argued that the curvature of the C$_{60}$ fullerene is thought to be responsible [37,38] for the substantial increase in T_c in its solid phase compared to intercalated graphite. Since C$_{36}$ is even more curved than C$_{60}$, this argument is suggestive of even higher transition temperatures in C$_{36}$-based solids. To explore this possibility, we have carried out state of the art calculations of the electron–phonon interaction potential, V_{ep}, which is used in the definition of the coupling parameter $\lambda = N(0)V_{ep}$, where $N(0)$ is the DOSs at E_F.

One can extract V_{ep} from the electron–phonon spectral function [39], which is a double average over the Fermi surface connecting states due to the change in the potential caused by a phonon. For fullerene crystals, the computation is

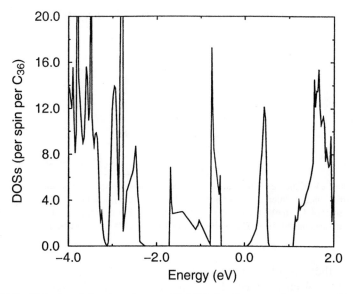

Figure 20.7 Calculated density of states (DOSs) for C_{36} crystal structure S2-AB. The Fermi energy is at 0 eV.

greatly simplified because of the small dispersion in both the electronic and phononic spectra. Therefore, one can use the electronic states and vibrational modes of the isolated molecule to approximate those found in the solid [38,40,41]; this implies that only the intramolecular modes are considered in the evaluation of V_{ep}. In the case of solid C_{36}, because of the intermolecular covalent bonds, one should carry out the full average over the Fermi surface. While such a calculation is prohibitive in practice, the C_{36} crystal appears to have molecular-like features, and we therefore use the procedure described above for the calculation of V_{ep}. However, the presence of covalent bonds in the crystal can cause the states of the isolated C_{36} molecule to rearrange and possibly mix in the solid. In order to keep our results as general as possible, we evaluate V_{ep} for the four molecular states that are nearest in energy to E_F.

The dynamical matrix is constructed using the forces on the atoms where only nonequivalent atoms are moved along the three Cartesian directions, and the rest of the information is obtained from the symmetry properties of this matrix. Diagonalizing the dynamical matrix yields the $3N$-dimensional polarization vectors, ε_α, with normalization $\varepsilon_\alpha \cdot \varepsilon_\beta = \delta_{\alpha\beta}$, which means that we have to use the DOSs per unit cell in the expression for λ. Using the approach outlined in the preceding paragraph, one can write

$$V_{ep} = \sum_\alpha \frac{1}{M\omega_\alpha^2} \frac{1}{g^2} \sum_{i,j=1}^{g} |\langle i|\varepsilon_\alpha \cdot \nabla V|j\rangle|^2 \qquad (20.1)$$

TABLE 20.3   Electron–Phonon Coupling for the D_{6h} C$_{36}$ Molecule^a

Electronic States	Phonons Coupled	V_{ep} (meV)
E$_{2g}$	A$_{1g}$	32
	E$_{2g}$	154
B$_{2g}$	A$_{1g}$	136
B$_{1u}$	A$_{1g}$	126
E$_{1g}$	A$_{1g}$	39
	E$_{2g}$	120
T$_{1u}$(C$_{60}$)	A$_g$	8
	H$_g$	63 (52[38], 56[40], 68[44], 49[45])

^{a} States within 1 eV of the Fermi level are considered and the corresponding phonons that contribute to the electron–phonon interaction potential V_{ep} (see Eq. (20.1)) are listed. For each state, we list calculated values of V_{ep} (meV) for the given phonon modes. For comparison, the last row shows the same quantities calculated for A$_3$C$_{60}$ with previous LDA results listed in parentheses.

where M is the mass of a carbon atom, i and j are degenerate electronic states in the isolated molecule for which we are evaluating the coupling, and g is the degeneracy of these states. The quantity $\langle i|\varepsilon_\alpha \cdot \nabla V|j\rangle$ is obtained by means of a finite difference approach. The application of selection rules to the matrix elements can be used as a consistency check to determine which vibrational modes should produce nonzero contributions. Within the LDA framework, we make no approximations or fits in our evaluation of V_{ep}.

Table 20.3 lists our calculated electron–phonon coupling values for several electronic states near E_F. The couplings due to the different phonon modes are listed separately since the A$_g$ contribution to V_{ep} is expected to be screened out in A$_3$C$_{60}$. This effect has been observed by Raman scattering experiments [42,43] as well as demonstrated theoretically within the random phase approximation for static screening [44].

Our results for C$_{60}$ are in good agreement with previous LDA calculations [38,40,44,45]. The results show that the coupling in C$_{36}$ is substantially larger than in C$_{60}$ for the E$_{1g}$ and E$_{2g}$ degenerate states. The increase in V_{ep} for C$_{36}$ supports arguments based on curvature. Note that, although the A$_g$ phonon modes comprise a slightly larger percent contribution of V_{ep} for C$_{36}$ than for C$_{60}$, even without the inclusion of these modes the coupling strength of the E$_{2g}$ electronic state is still enhanced by more than a factor of two compared with C$_{60}$. Of course, for the singly degenerate B$_{2g}$ and B$_{1u}$ states, removal of the contributions from A$_g$ phonons would reduce the coupling to zero. The strongest coupling, both with and without the A$_g$ contributions, is due to the doublet electronic states.

The strength of the electron–phonon interaction potential plays a crucial role in determining the superconducting transition temperature. Thus, our results imply that T_c in solid C$_{36}$ can be significantly different than in solid C$_{60}$. The evaluation of T_c also requires knowledge of $N(0)$ and μ^*, which describes the Coulomb electron–electron repulsion. As we have shown, $N(0)$ for the solid

C_{36} considered here is expected to be comparable to that of doped C_{60}. Very recent experimental evidence suggests that μ^* is about 0.25 for A_3C_{60} compounds [46]. We may expect μ^* for C_{36} to be close to this value since the range and width of narrow subbands near the Fermi level and typical phonon energies are similar. However, due to the sensitivity of T_c to μ^* and the fact that μ^* is not known for C_{36}, only a very qualitative comparison of T_c can be made here. As an example, if we choose the same $N(0)$ and μ^* for C_{60} and C_{36}, such that $T_c = 18$ K for C_{60}, then a solution of the Eliashburg equations yields $T_c(C_{36}) \approx 6T_c(C_{60})$.

20.7 EFFECTS OF DOPING

20.7.1 Substitutional Doping

Substitutional doping of atoms on the fullerene cages by electron donors and acceptors such as N and B is of considerable interest because of induced changes in the electronic and structural properties of these molecules. Experimental evidence for the existence of N-doped C_{60} and C_{70} fullerenes has been verified by several groups [47,48], and theoretical studies have shown that both N and B can be substituted to form stable structures [49,50].

We have explored several N substitutionally doped C_{36} D_{6h} cages: $C_{34}N_2$, $C_{28}N_8$, and $C_{24}N_{12}$. In all cases no two dopant atoms are placed next to one another due to the strong nitrogen dimer bond, which weakens the overall structural integrity of the ball. This effect was tested for $C_{34}N_2$, where it was found that placing the N atoms next to one another decreases the total binding energy by ~3 eV compared with the cases where the two N atoms are apart. As the fullerene is further substituted with N, the positions of the nitrogen atoms become more important. Our calculations indicate that the most symmetric configurations are also the most energetically favorable. The highest symmetry case is $C_{24}N_{12}$, which can retain the full D_{6h} symmetry of the undoped molecule.

Upon structural relaxation, we find that the C–N bonds shorten by 2–3% with respect to the C–C bond lengths, in agreement with previous calculations on $C_{59}N$ [49]. A comparison of the four unique bond lengths of the D_{6h} $C_{24}N_{12}$ structure is given in Table 20.4. Interestingly, for the highest symmetry D_{6h}

TABLE 20.4 Comparison of the Four Unique Bond Distances (Å) in D_{6h} C_{36} and $C_{24}N_{12}$[a]

	d_1	d_2	d_3	d_4
C_{36}	1.41	1.48	1.43	1.43
$C_{24}N_{12}$	1.39	1.44	1.45	1.33

[a] All values are for fully relaxed LDA structures. The d_4 bond corresponds to the C–C bond across the hexagon belt (see Fig. 20.9).

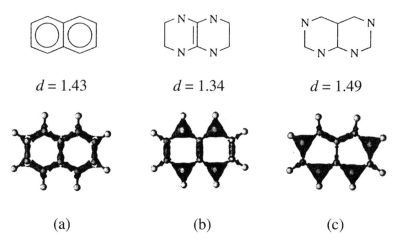

$d = 1.43$ $d = 1.34$ $d = 1.49$

(a)   (b)   (c)

Figure 20.8 Planar C$_{10}$H$_8$ and C$_6$N$_4$H$_8$ molecules, shown with the total charge density and hydrogen atoms on bottom. The top figures illustrate the positions of the nitrogen atoms. The C–C distance of the center vertical bond is given for each case.

case, substitution with nitrogen results in a 10% shortening of the carbon–carbon bonds across the hexagon belt.

In order to understand better this effect, which causes the fullerene to possess six C–C double bonds, we consider the smaller, planar molecule C$_{10}$H$_8$. Figure 20.8 shows this test molecule for the case of (a) pure carbon and (b,c) two different configurations substituted with four nitrogen atoms. Total electronic charge densities are overlayed on each molecule in order to see more clearly the bonding character. The middle C–C bond (without any hydrogens and whose value is listed above each figure) represents the d_4 bond in the D_{6h} C$_{36}$ and C$_{24}$N$_{12}$ fullerenes. In the pure carbon case this bond is 1.43 Å, which is exactly what it is in pure C$_{36}$. Upon introducing the nitrogen (Fig. 20.8b), the middle bond shortens to 1.34 Å, which is similar to the reduction in d_4 seen in the N-substituted fullerene. Note, however, that if the nitrogen is placed slightly differently (Fig. 20.8c), the effect is lost, and in fact the middle carbon bond lengthens rather than shortens.

The results of this simple test molecule show that the C–C bond is significantly strengthened when nitrogen is substituted in the appropriate manner. The effect of shortening the sp^2 C–C bond occurs only when it is "trapped" between four nitrogen atoms. It may appear at first that this second-neighbor effect is due to the excess charge on the nitrogen atoms. However, by looking closely at the charge densities, we observe that all of the excess nitrogen charge remains localized on the N atoms. Instead, the effect of the nitrogen atoms is to push more of the *carbon* charge away from the C–N bond and over to the C–C "trapped" bond. The reason for such a transfer of charge is that the carbon π bonds cannot form in the usual sp^2 network with the nitrogen atoms, resulting in extra bonding with its carbon neighbor.

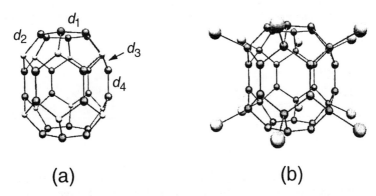

Figure 20.9 The (a) D_{6h} $C_{24}N_{12}$ fullerene and the (b) D_{6h} $C_{36}Cl_{12}$ fullerene. The four unique bonds for these structures are shown in (a).

As a further check, we have disrupted the π bonding of the 12 carbon atoms in C_{36}, which were previously substituted with nitrogen atoms by attaching additional atoms that bond to the carbons at these sites. We choose the highly electronegative Cl atoms, which pull an electron away from the carbon atoms, leaving them with a truly sp^3 bonding character. Upon structural relaxation (see Fig. 20.9b), the exact same shortening of the C–C bonds along the hexagon belt is observed. In fact, this effect has been observed experimentally for the $C_{60}Br_{24}$ compound [51], which had Br atoms situated outside the C_{60} in a tetrahedral configuration such that every C–C bond was surrounded by four C atoms attached to Br, leading to a molecule with 18 trapped C–C bonds. Thus, both substitution with nitrogen and addition of atoms, which bind to the outside of the fullerene, have the same effect on the fullerene's C–C bonds. To test this effect even further, we replaced the C atoms in C_{60} with N at exactly the positions where the Br atoms were attached in the experiment of Tebbe et al. [51]. Our calculations of the resulting $C_{36}N_{24}$ fullerene show almost identical bond lengths to those of the experimental brominated C_{60}.

It is interesting to note that the chlorination of these D_{6h} C_{36} fullerenes may be significantly easier than for C_{60} because the smaller fullerenes are more reactive. Our calculations of $C_{36}Cl_{12}$ show an energetically stable structure with a large LDA gap of 2.5 eV. In C_{60}, the addition of bromine atoms to the surface creates structural distortions that cause considerable strain on the molecule. In contrast, pure C_{36} already has some of these distortions (referred to as the "boat" conformation by Tebbe et al. [51]), so little additional geometric strain is introduced.

20.7.2 Endohedral Doping

Next, we consider endohedral doping of the molecules as an additional means of modifying their electronic properties. The detection of endohedral fullerenes

TABLE 20.5 Binding Energies, ΔE, of Endohedrally Doped Atoms in C$_{36}$ (D_{6h} and D_{2d}) and C$_{28}$ (T_d): $\Delta E = E(M) + E(C_{36}) - E(M@C_{36})$, Where M Is the Dopant Atom

	Mg	Ca	Ge	Si	Zr	Sr
ΔE (C$_{36}$ D_{6h})	+0.9	+4.7	+1.2	+1.4	+6.7	+4.2
ΔE (C$_{36}$ D_{2d})	+0.3	+4.2	+0.7	+0.9	+6.1	+4.9
ΔE (C$_{28}$)[a]	−1.6	−4.5	—	−7.8	+2.8	—

[a] From Ref. 55.

such as La@C$_{60}$ [52] shortly after the discovery of C$_{60}$ led to the suggestion of encapsulating atoms inside carbon cages as a technique for creating new materials [53–55]. When an alkaline-earth atom M is placed in the center of pure C$_{36}$, the outer valence electrons from M jump onto the ball and fill the first LUMO, in accord with a rigid-band type model. Thus, very little changes in the electronic and structural properties of these endohedrally doped fullerenes, except perhaps the degeneracy of the LUMO or spin multiplicity.

We have evaluated the binding energies, ΔE, of various endohedral atoms as the difference of total energies, $\Delta E = E(M) + E(C_{36}) - E(M@C_{36})$, where M is the dopant atom. Results for doping with the alkaline-earth atoms Mg, Ca, and Sr for the D_{6h} and D_{2d} fullerenes are given in Table 20.5. In addition, we calculate the binding energies of Si, Ge, and Zr in order to compare with the same quantities calculated by Guo et al. [55] for C$_{28}$. Note that, while C$_{28}$ binds only one of the atoms (Zr), all of the atoms that we tried are well-bounded by the C$_{36}$ cages. Thus, the C$_{36}$ fullerenes are perhaps the smallest size carbon cages that easily trap additional atoms.

20.7.3 Intercalation in the Solid

Upon closer inspection of Figure 20.6, we observe that in the S2-AB crystal there are two cavities per C$_{36}$ (directly above and below each molecule), which have a radius of roughly 1.4 Å. These are the largest empty spaces and would therefore be appropriate for intercalation with alkali metal atoms. The ionic radii of Na and K are 0.95 and 1.33 Å, respectively, indicating that Na should fit well but that perhaps K is too large.

In order to test the effects of these dopants, we have carried out calculations of the S2-AB crystal with both Na and K intercalated at the above-mentioned sites (two metal atoms per C$_{36}$). Relaxed structures of both Na- and K-doped crystals are shown in Figure 20.10. Note that Na$_2$C$_{36}$ maintains the original crystal structure and little relaxation occurs upon introducing the intercalant. In contrast, K$_2$C$_{36}$ is highly strained in the original crystal and therefore undergoes a structural transition. In this case, the top and bottom hexagons of each C$_{36}$ molecule are "opened up" in order to relieve the strain of accommodating the large K atom, and intermolecular bonds within the planes are broken as the C$_{36}$ molecules push away from one another.

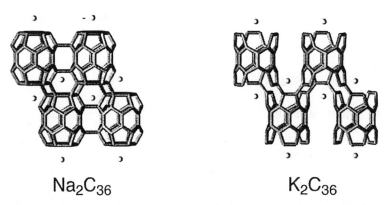

Na$_2$C$_{36}$ K$_2$C$_{36}$

Figure 20.10 Na$_2$C$_{36}$ and K$_2$C$_{36}$ crystal structures. Before relaxation, both crystals appear the same; upon relaxation the K-doped crystal undergoes a structural transition.

Our results for the density of states of the Na$_2$C$_{36}$ crystal (Fig. 20.11) show that the two electrons per C$_{36}$ donated by the Na atoms fill the empty band above E_F of the undoped crystal in a rigid-band manner. There is some broadening of this band, although a small gap of ~0.2 eV remains in the crystal. Further doping of the S2-AB crystal by one or even more alkali atoms should result in a partially filled peak at E_F. While K$_2$C$_{36}$ forms a very interesting crystal structure, our results demonstrate that fabrication of C$_{36}$-based

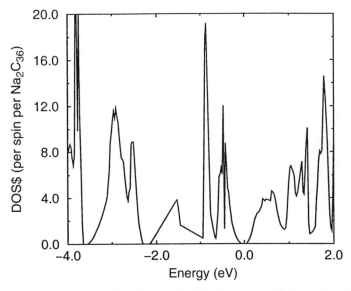

Figure 20.11 Calculated density of states (DOSs) for the alkali-doped Na$_2$C$_{36}$ in the S2-AB crystal structure. The Fermi energy is at 0 eV.

superconductors with K would require the C_{36} molecules to be well separated at the time of intercalation.

20.8 FURTHER MOLECULAR PROPERTIES

20.8.1 Chemical Shifts

An experimental determination of the fullerene structures is not feasible with mobility measurements since the fullerene drift times cannot be separated out. On the other hand, NMR chemical shifts distinguish atoms in different chemical environments and could be used if the bonding configurations of the nonequivalent atoms of the C_{36} fullerenes differ enough. We have performed NMR calculations for the D_{6h} and D_{2d} structures using the gauge-independent atomic orbital (GIAO) method [56,57] with cc-pVTZ basis sets [58].

The chemical shifts are computed with respect to methane at 0 K $(r_e = 1.087)$ and it was assumed that δ_{TMS} $(CH_4) = -14.3$ ppm [59]. Our results are shown in Table 20.6, which lists the isotropic chemical shift with respect to TMS for the two fullerenes. For the D_{6h} structure two of the three nonequivalent atoms have nearly identical shifts, which should cause a large peak at around 159–160 ppm. For the D_{2d} structure, the five peaks are clearly identifiable, and one can therefore distinguish between these two fullerenes due to the different numbers of identifiable peaks.

20.8.2 Infrared Spectra

To further test the possibility of distinguishing between the D_{6h} and D_{2d} molecules, we have performed calculations of IR modes using the Gaussian94 package [60] with the S-VWN5 option (Slater exchange and Vosko–Wilks–Nusair parameterization of the correlation functional) and 6-311G* basis set.

TABLE 20.6 Calculated NMR Chemical Shifts (ppm) Relative to Tetramethylsilane (TMS) for C_{36} Fullerenes with D_{6h} and D_{2d} Symmetries[a]

Molecule	Atom	δ_{TMS}	Intensity
C_{36} (D_{6h})	C1	137.5	1/3
	C2	159.4	1/3
	C3	160.0	1/3
C_{36} (D_{2d})	C1	151.8	1/9
	C2	160.5	2/9
	C3	150.3	2/9
	C4	139.3	2/9
	C5	135.7	2/9

[a] Symmetry unique atoms (see Fig. 20.1) and intensities are also listed.

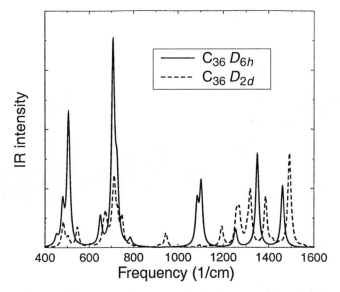

Figure 20.12   LDA calculated IR spectra for the D_{6h} and D_{2d} molecules.

The relative absorption intensities of the different modes are also computed by evaluating the effective-charge tensor, which permits the calculation of the induced dipole due to the motion of the atoms. We perform such calculations for both the D_{6h} and D_{2d} isomers at fully relaxed LDA geometries. The results for the fundamental infrared absorption, represented by upward peaks, are shown in Figure 20.12.

The calculated intensities show that we expect somewhat stronger absorption for the D_{6h} structure than for the D_{2d} structure. This result implies that if the two structures are present in the sample, the D_{2d} structure might not be able to be resolved. From the symmetry of the different molecules we can deduce that the D_{6h} fullerene has 13 IR active modes and the D_{2d} has 37 IR active modes, but only a few of those have significant absorption intensities. Note that only the D_{6h} structure has a significant peak near 1100 cm^{-1}. This observation provides additional support to the idea that experimental characterization can allow one to distinguish between the two different fullerenes.

20.9 EXPERIMENTS

20.9.1 Synthesis

A method of synthesizing C_{36} has been developed using a carbon arc discharge method and preliminary results suggest that it is a closed-cage, fullerene-like structure [8]. Six millimeter (dc) POCO graphite rods were arced together using a 100 ampere direct current source in a 400 torr helium atmosphere. The arc

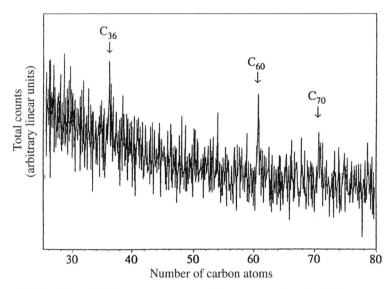

Figure 20.13 Time-of-flight mass spectrum of a sample of crude fullerene soot produced at a helium pressure of 400 torr. The data show the presence of a peak at the mass of C$_{36}$ (432 amu) with approximately the same intensity as the peak due to C$_{60}$ (720 amu).

chamber apparatus was similar to that described by Krätschmer et al. [3]. Films produced by the graphite sublimation were grown directly onto a probe head for use in a micromass time-of-flight mass spectrometer using a 366 nm nitrogen laser for desorption and ionization of the sample. All spectra were taken in negative ion mode. Figure 20.13 shows the mass spectrum of one of these films grown at 400 torr helium. The dominant peaks in the spectrum are at 720 and 432 amu. The former peak corresponds to 60 carbon atoms, while the latter corresponds to 36 carbon atoms. The C$_{60}$ and C$_{36}$ peaks are of comparable magnitude and these molecular species appear to be the most prominent in the sample. Lesser peaks are also observed, for example, at 840 amu (C$_{70}$). The synthesis of C$_{36}$ is very sensitive to experimental conditions, notably the helium pressure: Runs in this series at pressures significantly different from 400 torr failed to produce prominent peaks below 720 amu in the mass spectrum.

To produce bulk amounts of C$_{36}$ suitable for purification, arcing runs at 400 torr helium were repeated and the resulting "soot" was collected from the synthesis chamber walls. To avoid possible reactions with atmospheric impurities, the samples were removed from the arc chamber and handled inside an argon atmosphere glovebox. The soot was initially washed with toluene in a standard Soxhlet extractor, which removed C$_{60}$, C$_{70}$, and trace amounts of even higher-order fullerenes. Mass spectrometry on the toluene-soluble extract showed the expected fullerene peaks but no peak at 432 amu, indicating that C$_{36}$ is not soluble in toluene. Next, the sublimation product was extracted with pyridine.

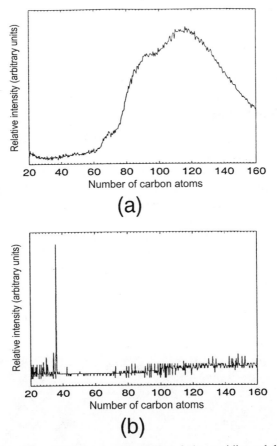

Figure 20.14 (a) Time-of-flight mass spectrum of the pyridine-soluble extract from higher fullerene depleted soot. (b) Time-of-flight mass spectrum of the same material after reduction with potassium metal in liquid ammonia.

This pyridine-soluble fraction formed a dark yellow solution as opposed to the toluene-soluble fraction, which was dark red.

Mass spectrometry of this fraction produced the spectrum shown in Figure 20.14a. Here, a very broad mass distribution is observed, which peaks at about 1500 amu. A similar spectrum has been published previously and attributed to the presence of higher fullerenes [61]. However, Figure 20.14b shows a mass spectrum of the same material after a subsequent reduction with one molar equivalent of potassium stirred for 3 hours in liquid ammonia and then dried. This spectrum shows only one major peak at C_{36} and almost no signal in the higher mass region around 1500 amu [62]. One possible explanation of this difference in spectra is that the pyridine extract is composed of covalently bonded C_{36} solid particles, which fragment upon laser desorption, producing species composed of 100–140 carbon atoms. However, potassium reacts with the material by transferring electrons to the molecules, thereby reducing the intermolecular binding energy and allowing individual molecules to desorb from the surface. This material was also found to dissolve in carbon disulfide (CS$_2$) and liquid ammonia.

20.9.2 Electron Diffraction

Electron diffraction studies were performed on the solid C$_{36}$ obtained from the pyridine extraction. A small amount of the material was ground up, dispersed on a holey carbon grid, and inserted into a JEOL 200CX transmission electron microscope (TEM). The material was observed to be polycrystalline with a crystallite grain size of about 100 nm. Using a field-limiting aperture, the diffraction pattern of selected crystallites was recorded.

Figure 20.15 shows a TEM diffraction pattern for a C$_{36}$ crystallite. The hexagonal diffraction pattern suggests a close-packing arrangement perpendicular to the zone axis. The pattern is reminiscent of diffraction patterns observed for C$_{60}$ and C$_{70}$. However, the d-spacing measured from this pattern for the first-order diffracted spots is 6.68 Å, significantly less than the (100) d-spacing of 8.7 Å reported for C$_{60}$. This d-spacing is consistent with the predicted d-

Figure 20.15 Electron diffraction pattern of C$_{36}$ crystallite. The pattern is hexagonal with a calculated d-spacing of 6.68 Å.

spacing of 6.55 Å for the lowest energy S2-AB C_{36} crystal structure (see Section 20.5). Unfortunately, because the C_{36} crystallites are platelets with high aspect ratios, this was the only zone axis along which the crystallites were thin enough to allow useful TEM imaging, and thus determination of the detailed C_{36} crystal structure was not possible.

Like C_{60}, C_{36} appears to suffer some TEM-induced damage at an electron beam energy of 200 keV. Under continuous TEM observation, the sharp crystalline diffraction patterns were found to deteriorate for extended irradiation times. Additional studies were performed to investigate the long-term stability of solid C_{36} subject (only) to high ambient temperatures. C_{36} powder was heated in vacuum to 1350 °C for 48 hours and then characterized by TEM imaging. Over 50% of the diffraction patterns obtained for the heat-treated material were graphitic, indicating that a large amount of the material had converted to the energetically more favorable graphite.

20.9.3 Scanning Tunneling Spectroscopy

The molecular energy levels of C_{36} thin films have also been studied by scanning tunneling spectroscopy (STS) [9]. The measured local density of electronic states $N(E)$ has been compared to a predicted density of states for the C_{36} monomer as well as several C_{36} dimer and trimer configurations. C_{36} films were grown using two separate methods—thermal evaporation and solution deposition. For thermal evaporation, the pyridine-soluble extract described in Section 20.9.1 was dried and thermally evaporated from a tungsten filament at approximately 10^{-6} torr onto a substrate, which was suspended above the filament. In order to prevent thermal decomposition of the material, the filament was heated very quickly and the films were grown in a matter of seconds. A shutter was used between the source and the substrate to control the film thickness. Films with thicknesses on the order of 0.1 monolayer to 1 monolayer were grown. Alternatively, films were grown by simply depositing a dilute solution of the pyridine extract onto the substrate and allowing the solvent (carbon disulfide) to evaporate. Two substrates were used in this study—an atomically flat gold (111) film grown on mica and the (001) surface of a highly oriented pyrolytic graphite (HOPG) crystal.

Each of these films were immediately loaded into a room temperature scanning tunneling microscope (STM) using a Pt-Ir tip and Oxford instruments TOPS3 controller. For all samples it was found that the C_{36} film tended to aggregate into islands approximately 10–50 nm in diameter and 10–20 Å high. This clustering is consistent with the predicted tendency of these molecules to covalently bind together. Figure 20.16 shows the STS density of states $((V/I)dI/dV)$ plot versis tip bias voltage obtained by placing the tip at a fixed height above a C_{36} island for both substrates. A similar measurement performed above the clean substrates produced the flat line at zero on this plot. Here, several sharp peaks can be seen, which indicate resonant tunneling through discrete molecular states. No differences between the thermally evapo-

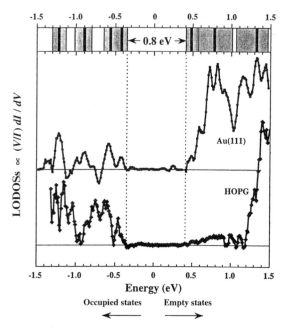

Figure 20.16 $N(E)$ for C$_{36}$ as measured on Au (111) and HOPG substrates. The most reproducible states on both substrates are indicated at the top by black lines with the measured broadening shown in gray.

rated films and the solution grown films could be found. Both spectra indicate the presence of a 0.8 eV gap. There appears to be a slight discrepancy in the features of these graphs, which may be explained by the possibility that the surface is influencing the electronic structure of the clusters. Nevertheless, the features that are most reproducible on both substrates have been indicated by black strips at the top of the figure with the peak widths shown in gray.

Our theoretical LDA calculations for isolated D_{6h} symmetry C$_{36}$ molecules compare favorably with these results and are reproduced at the top of Figure 20.17. After accounting for the standard LDA underestimation of the energy gap, the experimental and theoretical spectra match closely near the Fermi level. The theoretical energy spacing between the two lowest unoccupied and two highest occupied molecular levels, as well as the relative spectral weights of these levels, are all reproduced by the experimental data.

However, there is serious disagreement between experiment and theory further from the Fermi level: In the LDA, the isolated molecule has no eigenvalues in the +1.0 to +2.0 eV and the −1.0 to −2.0 eV energy ranges, while a number of states appear to be present experimentally. There are a number of effects that may explain such a discrepancy, including (1) bonding between C$_{36}$ units, (2) impurities in the sample, and (3) passivation of the C$_{36}$ molecules by other species. As we have already discussed, C$_{36}$ molecules are predicted to be more

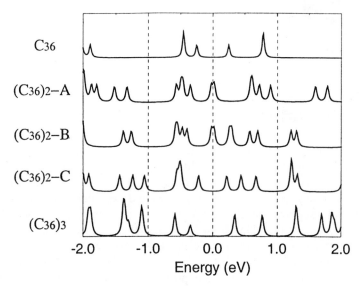

Figure 20.17 Theoretical electronic spectra for five different configurations of C_{36} molecules illustrated in Figure 20.3. $(C_{36})_2$-C and $(C_{36})_3$ provide the best match for the experimental spectrum.

reactive than larger fullerenes like C_{60}, even to the extent that stable covalently bonded crystals may be formed. Although impurities and passivation cannot absolutely be ruled out, the vacuum environment in which the samples were produced would favor intermolecular bonding as the most likely cause of deviation from isolated-molecule behavior, particularly considering that the molecules are mobile enough to cluster into islands.

To investigate the possible effects of intermolecular bonding, we have also calculated the energy spectra for the dimers and the trimer we considered earlier (see Section 20.4). The resulting molecular orbital energy spectra shown in Figure 20.17 indicate how sensitive the eigenvalue spectrum is to the nature of the intermolecular bonding. To facilitate comparison with the experimental results, a small, unweighted Lorentzian broadening is employed and each spectrum has been shifted so that the middle of the HOMO–LUMO gap lies at 0 eV. Doubly degenerate levels are counted twice so that relative peak heights are meaningful. In dimers A and B, a small gap (0.01 eV) results in a metallic eigenvalue spectrum. Dimer C, on the other hand, has a gap that is roughly the same as in the isolated molecule, shown at the top of the figure.

The experimental data in Figure 20.16 may now be compared directly to the various theoretical spectra of Figure 20.17. Neither dimers A nor B have the appropriate gap, as found for the isolated C_{36} molecule or dimer C. Although the isolated molecule fits the experimental data reasonably well, as described earlier, dimer C possesses a more even distribution of states in the range of 1–2 eV above and below the Fermi level. This particular dimer, therefore, resolves the primary discrepancy between the experimental data and the isolated-molecule energy spectrum. Since it is also considerably well bound energetically, it strongly suggests that the C$_{36}$ molecules are dimerized rather than isolated in this study.

ACKNOWLEDGMENTS

We thank M. Côté for significant contributions to the theoretical portion of this work, and K. Bradley, J. Burward-Hoy, P. G. Collins, M. Ishigami, T. Wagberg, M. C. Martin, Lafe Spietz, J. Kriesel, T. D. Tilley, J. Yarger, M. Tomaselli, and A. Pines for aiding in the synthesis and experimental investigation of this material. Support for this work was provided by the National Science Foundation, Grant No. DMR-9404755, and the Director, Office of Energy Research, Office of Basic Energy Sciences, Material Sciences Division of the U.S. Department of Energy under Contract Number DE-AC-03-76SF00098. Computational resources have been provided by NCSA, by SDSC, and by NERSC.

REFERENCES

1. H. W. Kroto, J. R. Heath, S. C. O'Brien, R. F. Curl, and R. E. Smalley, *Nature* **1985**, *318*, 162.

2. H. Ehrenreich and F. Spaepen (eds), *Solid State Physics: Advances in Research and Applications*. Academic Press, San Diego, **1994**.

3. W. Krätschmer, L. D. Lamb, K. Fostiropoulos, and D. R. Huffman, *Nature* **1990**, *347*, 354–358.

4. H. W. Kroto, *Nature* **1987**, *329*, 529.

5. J. C. Grossman, M. Côté, S. G. Louie, and M. L. Cohen, *Chem. Phys. Lett.* **1998**, *284*, 344.

6. M. Côté, J. C. Grossman, S. G. Louie, and M. L. Cohe, *Phys. Rev. Lett.* **1998**, *81*, 697.

7. M. Côté, J. C. Grossman, M. L. Cohen, and S. G. Louie, in *Proceedings 193rd Meeting of the Electrochemical Society*, San Diego, CA, 1998.

8. C. Piskoti, J. Yarger, and A. Zettl, *Nature* **1998**, *393*, 771.

9. P. G. Collins, J. C. Grossman, M. Côté, M. Ishigami, C. Piskoti, S. G. Louie, M. L. Cohen, and A. Zettl, *Phys. Rev. Lett.* **1999**, *82*, 165.

10. J. C. Grossman, S. G. Louie, and M. L. Cohen, *Phys. Rev. B.*, vol. 60, num. 10, R6941 (**1999**).

11. See, for example, W. E. Billups and M. A. Ciufolini (eds.), *Buckminsterfullerenes*. VCH Publishers, New York, 1993.

12. E. A. Rohlfing, D. M. Cox, and A. Kaldor, *J. Chem. Phys.* **1984**, *81*, 3322.

13. S. C. O'Brien, J. R. Heath, R. F. Curl, and R. E. Smalley, *J. Chem. Phys.* **1988**, *88*, 220.

14. M. L. Cohen, *Phys. Scr.* **1982**, *T1*, 5.

15. P. Hohenberg and W. Kohn, *Phys. Rev. B* **1964**, *136*, 864.

16. W. Kohn and L. J. Sham, *Phys. Rev. A* **1965**, *140*, 1133.

17. D. M. Ceperley and B. J. Alder, *Phys. Rev. Lett.* **1980**, *45*, 566.

18. J. L. Martins, N. Troullier, and S.-H. Wei, *Phys. Rev. B* **1991**, *43*, 2213.

19. H. J. Monkhorst and J. D. Pack, *Phys. Rev. B* **1976**, *13*, 5188.

20. B. Delley, *J. Chem. Phys.* **1990**, *92*, 508.

21. S. H. Vosko, L. Wilk, and M. Nusair, *Can. J. Phys.* **1980**, *58*, 1200.

22. J. P. Perdew, in *Electronic Structure of Solids '91*, P. Ziesche and H. Eschrig (eds.). Akademie Verlag, Berlin, 1991, p. 11.

23. A. D. Becke, *J. Chem. Phys.* **1993**, *98*, 5648.

24. G. von Helden, M. T. Hsu, N. G. Gotts, and M. T. Bowers, *J. Phys. Chem.* **1993**, *97*, 8182.

25. L. D. Book, C. Xu, and G. Scuseria, *Chem. Phys. Lett.* **1994**, *222*, 281.

26. P. W. Fowler and D. E. Manolopoulos, *An Atlas of Fullerenes*. Clarendon Press, Oxford, 1995.

27. M. Feyereisen, M. Gutowski, J. Simons, and J. Almlof, *J. Chem. Phys.* **1992**, *96*, 2926.

28. K. Raghavachari, D. L. Strout, G. K. Odom, G. E. Scuseria, J. A. Pople, B. G. Johnson, and P. M. W. Gill, *Chem. Phys. Lett.* **1993**, *214*, 357.

29. J. C. Grossman, L. Mitas, and K. Raghavachari, *Phys. Rev. Lett.* **1995**, *75*, 3870.

30. D. Bakowies and W. Thiel, *J. Am. Chem. Soc.* **1991**, *113*, 3704.

31. D. Bakowies, A. Gelessus, and W. Thiel, *Chem. Phys. Lett.* **1992**, *197*, 325.

32. R. C. Haddon, *Science* **1993**, *261*, 1545.

33. G. B. Adams, J. B. Page, O. F. Sankey, and M. O'Keeffe, *Phys. Rev. B* **1994**, *50*, 17471.

34. J. Kürti and K. Németh, *Chem. Phys. Lett.* **1996**, *256*, 119.

35. J. Fagerström and S. Stafström, *Phys. Rev. B* **1996**, *53*, 13150.

36. M. E. Colvin, N. L. Trian, L. L. Bui, J. C. Grossman, S. G. Louie, M. L. Cohen, and C. Janssen, *Chem. Phys. Lett.*, submitted.

37. J. L. Martins, *Europhys. News* **1992**, *23*, 31.

38. M. Schluter, M. Lannoo, M. Needels, and G. A. Baraff, *Phys. Rev. Lett.* **1992**, *68*, 526.

39. P. B. Allen and B. Mitrović, *Solid State Phys.* **1982**, *37*, 2–91.

40. C. M. Varma, J. Zaanen, and K. Raghavachari, *Science* **1991**, *254*, 989.

41. O. Gunnarsson, H. Handschuh, P. S. Bechthold, B. Kessler, G. Ganteför, and W. Eberhardt, *Phys. Rev. Lett.* **1995**, *74*, 1875.

42. S. J. Duclos, R. C. Haddon, S. Glarun, A. F. Hebard, and K. B. Lyons, *Science* **1991**, *254*, 1625.

43. T. Pichler, M. Matus, J. Kürti, and H. Kuzmany, *Phys. Rev. B* **1992**, *45*, 13841.

44. V. P. Antropov, O. Gunnarsson, and A. I. Liechtenstein, *Phys. Rev. B* **1993**, *48*, 7651.

45. J. C. R. Faulhaber, D. Y. K. Ko, and P. R. Briddon, *Phys. Rev. B* **1993**, *48*, 661.

46. M. S. Fuhrer, K. Cherrey, V. H. Crespi, A. Zettl, and M. L. Cohen, Physical Review Letters. 1999, Vol. 83 p. 404–407.

47. T. Guo, C. Jin, and R. E. Smalley, *J. Phys. Chem.* **1991**, *95*, 4948.

48. S. Glenis, S. Cooke, X. Chen, and M. M. Labes, *Chem. Mater.* **1994**, *6*, 1850.

49. W. Andreoni, F. Gygi, and M. Parrinello, *Chem. Phys. Lett.* **1991**, *190*, 159.

50. X. Xia, D. A. Jelski, J. R. Bowser, and T. F. George, *J. Am. Chem. Soc.* **1992**, *114*, 6493.

51. F. N. Tebbe, R. L. Harlow, D. B. Chase, D. L. Thorn, G. C. Campbell, Jr., J. C. Calabrese, N. Herron, R. J. Young, and E. Wasserman, *Nature* **1992**, *256*, 822.

52. J. R. Heath, S. C. O'Brien, Q. Zhang, Y. Liu, R. F. Curl, H. W. Kroto, F. K. Tittel, and R. E. Smalley, *J. Am. Chem. Soc.* **1985**, *107*, 7779.

53. S. D. Bethune, R. D. Johnson, J. R. Salem, M. S. de Vries, and C. S. Yannoni, *Nature* **1993**, *366*, 123.

54. T. Guo, M. D. Diener, Y. Chai, M. J. Alford, R. E. Haufler, S. M. McClure, T. Ohno, J. H. Weaver, G. E. Scuseria, and R. E. Smalley, *Science* **1992**, *257*, 1661.

55. T. Guo, R. E. Smalley, and G. E. Scuseria, *J. Chem. Phys.* **1993**, *99*, 352.

56. K. Wolinski, J. F. Hilton, and P. Pulay, *J. Am. Chem. Soc.* **1990**, *112*, 8251.

57. J. L. Dodds, R. McWeeny, and A. J. Sadlej, *Mol. Phys.* **1980**, *41*, 1419.

58. T. H. Dunning, Jr., *J. Chem. Phys.* **1989**, *90*, 1007.

59. A. K. Jameson and C. J. Jameson, *Chem. Phys. Lett.* **1987**, *134*, 461.

60. M. J. Frisch et al., *GAUSSIAN94, Revision B.1*. Gaussian, Inc., Pittsburgh, PA, 1995.

61. K. R. Lykke, D. H. Parker, and P. Wurz, *Int. J. Mass Spectrometry Ion Processes* **1994**, *138*, 149–157.

62. M. L. Cohen, J. C. Grossman, S. G. Louie, C. Piskoti, A. Zettl, *Electronic Properties of Novel Materials – Science and Technology of Molecular Nanostructures*, edited by H. Kuzmany, J. Fink, M. Mehring, & S. Roth © 1999 American Institute of Physics p. 183–186. "Effect of Alkali Doping on the Structural Stability of Solid C_{36}".

INDEX